Methods in Enzymology

Volume 403
GTPASES REGULATING MEMBRANE TARGETING
AND FUSION

METHODS IN ENZYMOLOGY

EDITORS-IN-CHIEF

John N. Abelson   Melvin I. Simon

DIVISION OF BIOLOGY
CALIFORNIA INSTITUTE OF TECHNOLOGY
PASADENA, CALIFORNIA

FOUNDING EDITORS

Sidney P. Colowick and Nathan O. Kaplan

Methods in Enzymology

Volume 403

GTPases Regulating Membrane Targeting and Fusion

EDITED BY

William E. Balch

DEPARTMENTS OF CELL AND MOLECULAR BIOLOGY
THE SCRIPPS RESEARCH INSTITUTE
LA JOLLA, CALIFORNIA

Channing J. Der

DEPARTMENT OF PHARMACOLOGY
THE UNIVERSITY OF NORTH CAROLINA AT CHAPEL HILL
CHAPEL HILL, NORTH CAROLINA

Alan Hall

CRC ONCOGENE AND SIGNAL TRANSDUCTION GROUP
MRC LABORATORY FOR MOLECULAR CELL BIOLOGY
UNIVERSITY COLLEGE LONDON
LONDON, ENGLAND

AMSTERDAM • BOSTON • HEIDELBERG • LONDON
NEW YORK • OXFORD • PARIS • SAN DIEGO
SAN FRANCISCO • SINGAPORE • SYDNEY • TOKYO
Academic Press is an imprint of Elsevier

Elsevier Academic Press
525 B Street, Suite 1900, San Diego, California 92101-4495, USA
84 Theobald's Road, London WC1X 8RR, UK

This book is printed on acid-free paper. ∞

For all information on all Elsevier Academic Press publications
visit our Web site at www.books.elsevier.com

ISBN-13: 978-0-12-182808-0
ISBN-10: 0-12-182808-5

PRINTED IN THE UNITED STATES OF AMERICA
05 06 07 08 09 9 8 7 6 5 4 3 2 1

Working together to grow
libraries in developing countries

www.elsevier.com | www.bookaid.org | www.sabre.org

ELSEVIER   BOOK AID
            International   Sabre Foundation

Table of Contents

CONTRIBUTORS TO VOLUME 403   xi

PREFACE   xix

VOLUMES IN SERIES   xxi

1. Exploring Trafficking GTPase Function by mRNA Expression Profiling: Use of the SymAtlas Web-Application and the Membrome Datasets   CEMAL GURKAN, HILMAR LAPP, JOHN B. HOGENESCH, AND WILLIAM E. BALCH   1

2. Use of Search Algorithms to Define Specificity in Rab GTPase Domain Function   MARIA NUSSBAUM AND RUTH N. COLLINS   10

3. Application of Phylogenetic Algorithms to Assess Rab Functional Relationships   RUTH N. COLLINS   19

4. Application of Protein Semisynthesis for the Construction of Functionalized Posttranslationally Modified Rab GTPases   ROGER S. GOODY, THOMAS DUREK, HERBERT WALDMANN, LUCAS BRUNSVELD, AND KIRILL ALEXANDROV   29

5. Assay and Functional Properties of PrBP(PDEδ), a Prenyl-Binding Protein Interacting with Multiple Partners   HOUBIN ZHANG, SUZANNE HOSIER, JENNIFER M. TEREW, KAI ZHANG, RICK H. COTE, AND WOLFGANG BAEHR   42

6. Functional Assays for the Investigation of the Role of Rab GTPase Effectors in Dense Core Granule Release   SÉVERINE CHEVIET, THIERRY COPPOLA, AND ROMANO REGAZZI   57

7. Analysis of Rab1 Recruitment to Vacuoles Containing *Legionella pneumophila*   JONATHAN C. KAGAN, TAKAHIRO MURATA, AND CRAIG R. ROY   71

8. Reconstitution of Rab4-Dependent Vesicle Formation *In Vitro*   ADRIANA PAGANO AND MARTIN SPIESS   81

9. Assay of Rab4-Dependent Trafficking on Microtubules   JOHN W. MURRAY AND ALLAN W. WOLKOFF   92

10. CD2AP, Rabip4, and Rabip4′: Analysis of   PASCALE MONZO,
    Interaction with Rab4a and Regulation of   MURIEL MARI,
    Endosomes Morphology   VINCENT KADDAI,
                                                    TERESA GONZALEZ,
                                                    YANNICK LE
                                                    MARCHAND-BRUSTEL, AND
                                                    MIREILLE CORMONT   107

11. Visualization of Rab5 Activity in Living Cells EMILIA GALPERIN AND
    Using FRET Microscopy   ALEXANDER SORKIN   119

12. Selection and Application of Recombinant   CLÉMENT NIZAK,
    Antibodies as Sensors of Rab   SANDRINE MOUTEL,
    Protein Conformation   BRUNO GOUD, AND
                                                    FRANCK PEREZ   135

13. Analysis of Rab Protein Function in   NASHAAT Z. GERGES,
    Neurotransmitter Receptor Trafficking at   TYLER C. BROWN,
    Hippocampal Synapses   SUSANA S. CORREIA, AND
                                                    JOSÉ A. ESTEBAN   153

14. Use of Rab GTPases to Study Lipid   AMIT CHOUDHURY,
    Trafficking in Normal and Sphingolipid   DAVID L. MARKS,
    Storage Disease Fibroblasts   AND RICHARD E. PAGANO   166

15. Assay of Rab13 in Regulating Epithelial Tight   ANNE-MARIE MARZESCO AND
    Junction Assembly   AHMED ZAHRAOUI   182

16. Tyrosine Phosphorylation of Rab Proteins   JEAN H. OVERMEYER AND
                                                    WILLIAM A. MALTESE   194

17. Assay of Rab25 Function in Ovarian and   KWAI WA CHENG, YILING LU,
    Breast Cancers   AND GORDON B. MILLS   202

18. Functional Analysis of Rab27a Effector   TETSURO IZUMI,
    Granuphilin in Insulin Exocytosis   HIROSHI GOMI, AND
                                                    SEIJI TORII   216

19. Assays for Functional Properties of Rab34 in   PENG SUN AND TAKESHI ENDO 229
    Macropinosome Formation

20. Fluorescent Microscopy-Based Assays to   ROBERTO WEIGERT AND
    Study the Role of Rab22a in   JULIE G. DONALDSON   243
    Clathrin-Independent Endocytosis

21. Purification and Properties of Rab3 GEP   TOSHIAKI SAKISAKA AND
    (DENN/MADD)   YOSHIMI TAKAI   254

22. Biochemical Characterization of Alsin, a   JUSTIN D. TOPP,
    Rab5 and Rac1 Guanine Nucleotide   DARREN S. CARNEY, AND
    Exchange Factor   BRUCE F. HORAZDOVSKY   261

23. Purification and Analysis of RIN Family-Novel Rab5 GEFs — KOTA SAITO, HIROAKI KAJIHO, YASUHIRO ARAKI, HIROSHI KUROSU, KENJI KONTANI, HIROSHI NISHINA, AND TOSHIAKI KATADA — 276

24. Purification and Functional Properties of a Rab8-Specific GEF (Rabin3) in Action Remodeling and Polarized Transport — KATARINA HATTULA AND JOHAN PERÄNEN — 284

25. Assay and Functional Properties of SopE in the Recruitment of Rab5 on Salmonella-Containing Phagosomes — SEETHARAMAN PARASHURAMAN AND AMITABHA MUKHOPADHYAY — 295

26. Purification and Functional Analyses of ALS2 and its Homologue — SHINJI HADANO AND JOH-E IKEDA — 310

27. Polycistronic Expression and Purification of the ESCRT-II Endosomal Trafficking Complex — AITOR HIERRO, JAEWON KIM, AND JAMES H. HURLEY — 322

28. Analysis and Properties of the Yeast YIP1 Family of Ypt-Interacting Proteins — CATHERINE Z. CHEN AND RUTH N. COLLINS — 333

29. Use of Hsp90 Inhibitors to Disrupt GDI-Dependent Rab Recycling — CHRISTINE Y. CHEN, TOSHIAKI SAKISAKA, AND WILLIAM E. BALCH — 339

30. Purification and Properties of Yip3/PRA1 as a Rab GDI Displacement Factor — ULF SIVARS, DIKRAN AIVAZIAN, AND SUZANNE PFEFFER — 348

31. Purification and Analysis of TIP47 Function in Rab9-Dependent Mannose 6-Phosphate Receptor Trafficking — ALONDRA SCHWEIZER BURGUETE, ULF SIVARS, AND SUZANNE PFEFFER — 357

32. Capture of the Small GTPase Rab5 by GDI: Regulation by p38 MAP Kinase — MICHELA FELBERBAUM-CORTI, VALERIA CAVALLI, AND JEAN GRUENBERG — 367

33. Rab2 Purification and Interaction with Protein Kinase C ι/λ and Glyceraldehyde-3-Phosphate Dehydrogenase — ELLEN J. TISDALE — 381

34. Purification and Functional Interactions of GRASP55 with Rab2 — FRANCIS A. BARR — 391

35. Purification and Properties of Rabconnectin-3 — TOSHIAKI SAKISAKA AND YOSHIMI TAKAI — 401

36. Physical and Functional Interaction of   TADAO SHIBASAKI AND
    Noc2/Rab3 in Exocytosis   SUSUMU SEINO   408

37. Functional Analysis of Slac2-a/Melanophilin as TARUHO S. KURODA,
    a Linker Protein between Rab27A and   TAKASHI ITOH, AND
    Myosin Va in Melanosome Transport   MITSUNORI FUKUDA   419

38. Identification and Biochemical Analysis of   TARUHO S. KURODA AND
    Slac2-c/MyRIP as a Rab27A-, Myosin   MITSUNORI FUKUDA   431
    Va/VIIa-, and Actin-Binding Protein

39. Analysis of the Role of Rab27 Effector   MITSUNORI FUKUDA AND
    Slp4-a/Granuphilin-a in Dense-Core   EIKO KANNO   445
    Vesicle Exocytosis

40. Assay and Functional Interactions of Rim2   MITSUNORI FUKUDA   457
    with Rab3

41. Assay of the Rab-Binding Specificity of   MITSUNORI FUKUDA AND
    Rabphilin and Noc2: Target Molecules   AKITSUGU YAMAMOTO   469
    for Rab27

42. Functional Properties of the Rab-Binding   ANDREW J. LINDSAY,
    Domain of Rab-Coupling Protein   NICOLAS MARIE, AND
                                                    MARY W. MCCAFFREY   481

43. Purification and Functional Properties of   ANDREW J. LINDSAY AND
    Rab11-FIP2   MARY W. MCCAFFREY   491

44. Purification and Functional Properties of   CONOR P. HORGAN,
    Rab11-FIP3   TOMAS H. ZURAWSKI, AND
                                                    MARY W. MCCAFFREY   499

45. Class I FIPs, Rab11-Binding Proteins That   ELIZABETH TARBUTTON,
    Regulate Endocytic Sorting and Recycling   ANDREW A. PEDEN,
                                                    JAGATH R. JUNUTULA, AND
                                                    RYTIS PREKERIS   512

46. Expression and Properties of the Rab4,   IOANA POPA, MAGDA DENEKA,
    Rabaptin-5α, AP-1 Complex in   AND PETER VAN DER SLUIJS   526
    Endosomal Recycling

47. Measurement of the Interaction of the   M. DEAN CHAMBERLAIN AND
    p85α Subunit of Phosphatidylinositol   DEBORAH H. ANDERSON   541
    3-Kinase with Rab5

48. Assay and Stimulation of the Rab5 GTPase   DEBORAH H. ANDERSON AND
    by the p85α Subunit of Phosphatidylinositol   M. DEAN CHAMBERLAIN   552
    3-Kinase

49. Ubiquitin Regulation of the Rab5 Family GEF Vps9p   BRIAN A. DAVIES, DARREN S. CARNEY, AND BRUCE F. HORAZDOVSKY   561

50. Analysis of the Interaction between GGA1 GAT Domain and Rabaptin-5   GUANGYU ZHU, PENG ZHAI, NANCY WAKEHAM, XIANGYUAN HE, AND XUEJUN C. ZHANG   583

51. Purification and Properties of Rab6 Interacting Proteins   SOLANGE MONIER AND BRUNO GOUD   593

52. Affinity Purification of Ypt6 Effectors and Identification of TMF/ARA160 as a Rab6 Interactor   SYMEON SINIOSSOGLOU   599

53. Assay and Properties of Rab6 Interaction with Dynein–Dynactin Complexes   EVELYN FUCHS, BENJAMIN SHORT, AND FRANCIS A. BARR   607

54. Assay and Functional Properties of Rabkinesin-6/Rab6-KIFL/MKlp2 in Cytokinesis   RÜDIGER NEEF, ULRIKE GRÜNEBERG, AND FRANCIS A. BARR   618

55. Interaction and Functional Analyses of Human VPS34/p150 Phosphatidylinositol 3-Kinase Complex with Rab7   MARY-PAT STEIN, CANHONG CAO, MATHEWOS TESSEMA, YAN FENG, ELSA ROMERO, ANGELA WELFORD, AND ANGELA WANDINGER-NESS   628

56. Functional Analyses and Interaction of the XAPC7 Proteasome Subunit with Rab7   SANCHITA MUKHERJEE, JIANBO DONG, CARRIE HEINCELMAN, MELANIE LENHART, ANGELA WELFORD, AND ANGELA WANDINGER-NESS   650

57. Expression, Assay, and Functional Properties of RILP   ANNA MARIA ROSARIA COLUCCI, MARIA RITA SPINOSA, AND CECILIA BUCCI   664

58. Assay and Functional Properties of Rab34 Interaction with RILP in Lysosome Morphogenesis   TUANLAO WANG AND WANJIN HONG   675

59. Rabring7: A Target Protein for Rab7 Small G Protein   KOUICHI MIZUNO, AYUKO SAKANE, AND TAKUYA SASAKI   687

60. Analysis of Potential Binding of the   DIEGO SBRISSA,
    Recombinant Rab9 Effector p40   OGNIAN C. IKONOMOV, AND
    to Phosphoinositide-Enriched   ASSIA SHISHEVA   696
    Synthetic Liposomes

61. Assessment of Rab11-FIP2 Interacting   NICOLE A. DUCHARME,
    Proteins *In Vitro*   MIN JIN,
                                                  LYNNE A. LAPIERRE, AND
                                                  JAMES R. GOLDENRING   706

62. Interactions of Myosin Vb with Rab11 Family   LYNNE A. LAPIERRE AND
    Members and Cargoes Traversing the   JAMES R. GOLDENRING   715
    Plasma Membrane Recycling System

63. Properties of Rab13 Interaction with   AHMED ZAHRAOUI   723
    Protein Kinase A

64. Functional Properties of Rab15 Effector   LISA A. ELFERINK AND
    Protein in Endocytic Recycling   DAVID J. STRICK   732

65. Assays for Interaction between Rab7 and   MARIE JOHANSSON AND
    Oxysterol Binding Protein Related Protein   VESA M. OLKKONEN   743
    1L (ORP1L)

66. Characterization of Rab23, a Negative   TIMOTHY M. EVANS,
    Regulator of Sonic Hedgehog Signaling   FIONA SIMPSON,
                                                  ROBERT G. PARTON, AND
                                                  CAROL WICKING   759

67. Purification and Functional Analysis of a   RYUTARO SHIRAKAWA,
    Rab27 Effector Munc13-4 Using a   TOMOHITO HIGASHI,
    Semiintact Platelet Dense-Granule   HIROKAZU KONDO,
    Secretion Assay   AKIRA YOSHIOKA,
                                                  TORU KITA, AND
                                                  HISANORI HORIUCHI   778

68. Analysis of hVps34/hVps15 Interactions with   JAMES T. MURRAY AND
    Rab5 *In Vivo* and *In Vitro*   JONATHAN M. BACKER   789

69. Purification and Functional Properties of   PIERRE-YVES GOUGEON AND
    Prenylated Rab Acceptor 2   JOHNNY K. NGSEE   799

AUTHOR INDEX   809

SUBJECT INDEX   843

Contributors to Volume 403

Article numbers are in parentheses following the name of Contributors.
Affiliations listed are current.

DIKRAN AIVAZIAN (30), *Department of Biochemistry, Stanford University School of Medicine, Stanford, California*

KIRILL ALEXANDROV (4), *Department of Physical Biochemistry, Max-Planck-Institute for Molecular Physiology, Dortmund, Germany*

DEBORAH H. ANDERSON (47, 48) *Cancer Research Unit, Health Research Division, Saskatchewan Cancer Agency, Saskatoon, Canada*

YASUHIRO ARAKI (23), *Department of Physiological Chemistry, Graduate School of Pharmaceutical Sciences, University of Tokyo, Tokyo, Japan*

JONATHAN M. BACKER (68), *Department of Molecular Pharmacology, Albert Einstein College of Medicine, Bronx, New York*

WOLFGANG BAEHR (5), *Department of Ophthalmology, University of Utah, Salt Lake City, Utah*

WILLIAM E. BALCH (1, 29), *Departments of Cell and Molecular Biology, The Scripps Research Institute, La Jolla, California*

FRANCIS A. BARR (34, 53, 54), *Department of Cell and Molecular Biology, Max-Planck-Institute of Biochemistry, Martinsried, Germany*

TYLER C. BROWN (13), *Department of Pharmacology, University of Michigan Medical School, Ann Arbor, Michigan*

LUCAS BRUNSVELD (4), *Department of Chemical Biology, Max-Planck-Institute for Molecular Physiology, Dortmund, Germany*

CECILIA BUCCI (57), *Dipartimento di Scienze e Tecnologie Biologiche ed Ambientali, Universitá di Lecce, Lecce, Italy*

ALONDRA SCHWEIZER BURGUETE (31), *Department of Biochemistry, Stanford University School of Medicine, Stanford, California*

CANHONG CAO (55), *Department of Pathology, University of New Mexico Health Sciences Center, Albuquerque, New Mexico*

DARREN S. CARNEY (22, 49), *Department of Biochemistry and Molecular Biology, The Mayo Clinic Cancer Center, Rochester, Minnesota*

VALERIA CAVALLI (32), *Department of Cellular and Molecular Medicine, Howard Hughes Medical Institution, University of California San Diego, La Jolla, California*

M. DEAN CHAMBERLAIN (47, 48) *Department of Oncology, University of Saskatchewan, Saskatoon, Canada*

CATHERINE Z. CHEN (28), *Department of Molecular Medicine, College of Veterinary Medicine, Cornell University, Ithaca, New York*

CHRISTINE Y. CHEN (29), *Department of Cell Biology, The Scripps Research Institute, La Jolla, California*

KWAI WA CHENG (17), *Department of Molecular Therapeutics, University of Texas—MD Anderson Cancer Center, Houston, Texas*

SÉVERINE CHEVIET (6), *Department of Cell Biology and Morphology, University of Lausanne, Lausanne, Switzerland*

AMIT CHOUDHURY (14), *Department of Biochemistry and Molecular Biology, Thoracic Diseases Research Unit, Mayo Clinic College of Medicine, Rochester, Minnesota*

RUTH N. COLLINS (2, 3, 28) *Department of Molecular Medicine, College of Veterinary Medicine, Cornell University, Ithaca, New York*

ANNA MARIA ROSARIA COLUCCI (57), *Dipartimento di Scienze e Tecnologie Biologiche ed Ambientali, Universitá di Lecce, Lecce, Italy*

THIERRY COPPOLA (6), *IPMC UMR 6097, Valbonne, France*

MIREILLE CORMONT (10), *Inserm U568, Signalisation Moleculaire et Obesite, Faculte de Medecine, Nice, France*

SUSANA S. CORREIA (13), *Department of Pharmacology, University of Michigan Medical School, Ann Arbor, Michigan*

RICK H. COTE (5), *Department of Biochemistry and Molecular Biology, University of New Hampshire, Durham, New Hampshire*

BRIAN A. DAVIES (49), *Department of Biochemistry and Molecular Biology, The Mayo Clinic Cancer Center, Rochester, Minnesota*

MAGDA DENEKA (46), *Department of Cell Biology, University Medical Center Utrecht, Utrecht, The Netherlands*

JULIE G. DONALDSON (20), *Laboratory of Cell Biology, National Heart, Lung, and Blood Institute, National Institutes of Health, Bethesda, Maryland*

JIANBO DONG (56), *Division of Chemical Biology, Stanford University, Stanford, California*

NICOLE A. DUCHARME (61), *Department of Cell and Developmental Biology, Vanderbilt University School of Medicine, Nashville, Tennessee*

THOMAS DUREK (4), *Department of Physical Biochemistry, Max-Planck-Institute for Molecular Physiology, Dortmund, Germany*

LISA A. ELFERINK (64), *Department of Neuroscience and Cell Biology, University of Texas Medical Branch, Galveston, Texas*

TAKESHI ENDO (19), *Department of Biology, Faculty of Science, Graduate School of Science and Technology, Chiba University, Chiba, Japan*

JOSÉ A. ESTEBAN (13), *Department of Pharmacology, University of Michigan Medical School, Ann Arbor, Michigan*

TIMOTHY M. EVANS (66), *Department of Biochemistry, Institute for Molecular Bioscience, The University of Queensland, Brisbane, Queensland, Australia*

MICHELA FELBERBAUM-CORTI (32), *Department of Biochemistry, University of Geneva, Geneva, Switzerland*

YAN FENG (55), *Functional Genomics, Novartis Institutes for Biomedical Research, Cambridge, Massachusetts*

EVELYN FUCHS (53), *Department of Cell Biology, Max-Planck-Institute of Biochemistry, Martinsried, Germany*

MITSUNORI FUKUDA (37, 38, 39, 40, 41), *Fukuda Initiative Research Unit, RIKEN (The Institute of Physical and Chemical Research), Saitama, Japan*

EMILIA GALPERIN (11), *Department of Pharmacology, University of Colorado Health Sciences Center, Aurora, Colorado*

NASHAAT Z. GERGES (13), *Department of Pharmacology, University of Michigan Medical School, Ann Arbor, Michigan*

JAMES R. GOLDENRING (61, 62), *Department of Cell and Developmental Biology, Section of Surgical Sciences, Vanderbilt University School of Medicine, Nashville, Tennessee*

HIROSHI GOMI (18), *Department of Molecular Medicine, Institute for Molecular and Cellular Regulation, Gunma University, Maebashi, Gunma, Japan*

TERESA GONZALEZ (10), *Inserm U568, Signalisation Moleculaire et Obesite, Faculte de Medecine, Nice, France*

ROGER S. GOODY (4), *Department of Physical Biochemistry, Max-Planck-Institute for Molecular Physiology, Dortmund, Germany*

BRUNO GOUD (12, 51), *CNRS UMR144, Institut Curie, Paris, France*

PIERR-YVES GOUGEON (69), *Ottawa Health Research Institute, Cellular and Molecular Medicine, University of Ottawa, Ottawa, Canada*

JEAN GRÜENBERG (32), *Department of Biochemistry, University of Geneva, Geneva, Switzerland*

ULRIKE GRÜNEBERG (54), *Department of Cell Biology, Max-Planck-Institute of Biochemistry, Martinsried, Germany*

CEMAL GURKAN (1), *Department of Cell Biology, The Scripps Research Institute, La Jolla, California*

SHINJI HADANO (26), *Department of Molecular Neuroscience, The Institute of Medical Sciences, Tokai University, Isehara, Japan*

KATARINA HATTULA (24), *Institute of Biotechnology, University of Helsinki, Helsinki, Finland*

XIANGYUA HE (50), *Protein Studies Program, Oklahoma Medical Research Foundation, Oklahoma City, Oklahoma*

CARRIE HEINCELMAN (56), *Department of Pathology, University of New Mexico Health Sciences Center, Albuquerque, New Mexico*

AITOR HIERRO (27), *Laboratory of Molecular Biology, National Institute of Diabetes and Digestive and Kidney Diseases, National Institutes of Health, Bethesda, Maryland*

TOMOHITO HIGASHI (67), *Department of Geriatric Medicine, Graduate School of Medicine, Kyoto University, Kyoto, Japan*

JOHN B. HOGENESCH (1), *Department of Cell Biology, The Scripps Research Institute, La Jolla, California*

WANJIN HONG (58), *Membrane Biology Laboratory, Institute of Molecular and Cell Biology, Singapore*

BRUCE F. HORAZDOVSKY (22, 49), *Department of Biochemistry and Molecular Biology, The Mayo Clinic Cancer Center, Rochester, Minnesota*

CONOR P. HORGAN (44), *Department of Biochemistry, Molecular Cell Biology Laboratory, Biosciences Institute, University College Cork, Cork, Ireland*

HISANORI HORIUCHI (67), *Department of Geriatric Medicine, Graduate School of Medicine, Kyoto University, Kyoto, Japan*

SUZANNE HOSIER (5), *Department of Biochemistry and Molecular Biology, University of New Hampshire, Durham, New Hampshire*

JAMES H. HURLEY (27), *Laboratory of Molecular Biology, National Institute of Diabetes and Digestive and Kidney Diseases, National Institutes of Health, Bethesda, Maryland*

JOH-E IKEDA (26), *Department of Molecular Neuroscience, The Institute of Medical Sciences, Tokai University, Isehara, Japan*

OGNIAN C. IKONOMOV (60), *Department of Physiology and the Center for Molecular Medicine and Genetics, Wayne State University School of Medicine, Detroit, Michigan*

TAKASHI ITOH (37), *Fukuda Initiative Research Unit, RIKEN (The Institute of Physical and Chemical Research), Saitama, Japan*

TETSURO IZUMI (18), *Department of Molecular Medicine, Institute for Molecular and Cellular Regulation, Gunma University, Maebashi, Gunma, Japan*

MIN JIN (61), *Section of Surgical Sciences, Vanderbilt University School of Medicine, Nashville, Tennessee*

MARIE JOHANSSON (65), *Department of Molecular Medicine, National Public Health Institute (KTL), Biomedicum, Helsinki, Finland*

JAGATH R. JUNUTULA (45), *Genentech Inc., San Francisco, California*

VINCENT KADDAI (10), *Inserm U568, Signalisation Moleculaire et Obesite, Faculte de Medecine, Nice, France*

JONATHAN C. KAGAN (7), *Section of Microbial Pathogenesis, Boyer Center for Molecular Medicine, Yale University School of Medicine, New Haven, Connecticut*

HIROAKI KAJIHO (23), *Department of Physiological Chemistry, Graduate School of Pharmaceutical Sciences, University of Tokyo, Tokyo, Japan*

EIKO KANNO (39), *Fukuda Initiative Research Unit, RIKEN (The Institute of Physical and Chemical Research), Saitama, Japan*

TOSHIAKI KATADA (23), *Department of Physiological Chemistry, Graduate School of Pharmaceutical Sciences, University of Tokyo, Tokyo, Japan*

JAEWON KIM (27), *Laboratory of Molecular Biology, National Institute of Diabetes and Digestive and Kidney Diseases, National Institutes of Health, Bethesda, Maryland*

TORU KITA (67), *Department of Geriatric Medicine, Graduate School of Medicine, Kyoto University, Kyoto, Japan*

HIROKAZU KONDO (67), *Department of Geriatric Medicine, Graduate School of Medicine, Kyoto University, Kyoto, Japan*

KENJI KONTANI (23), *Department of Physiological Chemistry, Graduate School of Pharmaceutical Sciences, University of Tokyo, Tokyo, Japan*

TARUHO S. KURODA (37, 38), *Fukuda Initiative Research Unit, RIKEN (The Institute of Physical and Chemical Research), Saitama, Japan*

HIROSHI KUROSU (23), *Department of Physiological Chemistry, Graduate School of Pharmaceutical Sciences, University of Tokyo, Tokyo, Japan*

LYNNE A. LAPIERRE (61, 62), *Section of Surgical Sciences, Vanderbilt University School of Medicine, Nashville, Tennessee*

HILMAR LAPP (1), *Department of Cell Biology, The Scripps Research Institute, La Jolla, California*

YANNICK LE MARCHAND-BRUSTEL (10), *Inserm U568, Signalisation Moleculaire et Obesite, Faculte de Medecine, Nice, France*

MELANIE LENHART (56), *Department of Pathology, University of New Mexico Health Sciences Center, Albuquerque, New Mexico*

ANDREW J. LINDSEY (42, 43), *Department of Biochemistry, Molecular Cell Biology Laboratory, Biosciences Institute, University College Cork, Cork, Ireland*

YILING LU (17), *Department of Molecular Therapeutics, University of Texas—MD Anderson Cancer Center, Houston, Texas*

WILLIAM A. MALTESE (16), *Department of Biochemistry and Cancer Biology, Medical University of Ohio, Toledo, Ohio*

MURIEL MARI (10), *Department of Cell Biology, University Medical Center Utrecht, Utrecht, The Netherlands*

NICOLAS MARIE (42), *Department of Biochemistry, Molecular Cell Biology Laboratory, Biosciences Institute, University College Cork, Cork, Ireland*

DAVID L. MARKS (14), *Department of Biochemistry and Molecular Biology, Thoracic Diseases Research Unit, Mayo Clinic College of Medicine, Rochester, Minnesota*

ANNE-MARIE MARZESCO (15), *Max-Planck-Institute of Molecular Cell Biology and Genetics, Dresden, Germany*

MARY W. McCAFFREY (42, 43, 44), *Department of Biochemistry, Molecular Cell Biology Laboratory, Biosciences Institute, University College Cork, Cork, Ireland*

GORDON B. MILLS (17), *Department of Molecular Therapeutics, University of Texas—MD Anderson Cancer Center, Houston, Texas*

KOUICHI MIZUNO (59), *Department of Biochemistry, Institute of Health Biosciences, The University of Tokushima, Graduate School of Medicine, Tokushima, Japan*

SOLANGE MONIER (51), *CNRS UMR144, Institut Curie, Paris, France*

PASCALE MONZO (10), *Inserm U568, Signalisation Moleculaire et Obesite, Faculte de Medecine, Nice, France*

SANDRINE MOUTEL (12), *CNRS UMR144, Institut Curie, Paris, France*

SANCHITA MUKHERJEE (56), *Department of Pathology, University of New Mexico Health Sciences Center, Albuquerque, New Mexico*

AMITABHA MUKHOPADHYAY (25), *Cell Biology Lab, National Institute of Immunology, New Delhi, India*

TAKAHIRO MURATA (7), *Section of Microbial Pathogenesis, Boyer Center for Molecular Medicine, Yale University School of Medicine, New Haven, Connecticut*

JAMES T. MURRAY (68), *Department of Molecular Pharmacology, Albert Einstein College of Medicine, Bronx, New York*

JOHN W. MURRAY (9), *Marion Bessin Liver Research Center and Department of Anatomy and Structural Biology, Albert Einstein College of Medicine, Bronx, New York*

RUDIGER NEEF (54), *Department of Cell Biology, Max-Planck-Institute of Biochemistry, Martinsried, Germany*

JOHNNY K. NGSEE (69), *Ottawa Health Research Institute, Cellular and Molecular Medicine, University of Ottawa, Ottawa, Canada*

HIROSHI NISHINA (23), *Department of Physiological Chemistry, Graduate School of Pharmaceutical Sciences, University of Tokyo, Tokyo, Japan*

CLÉMENT NIZAK (12), *CNRS UMR144, Institut Curie, Paris, France*

MARIA NUSSBAUM (2), *Department of Molecular Medicine, College of Veterinary Medicine, Cornell University, Ithaca, New York*

VESA M. OLKKONEN (65), *Department of Molecular Medicine, National Public Health Institute (KTL), Biomedicum, Helsinki, Finland*

JEAN H. OVERMEYER (16), *Department of Biochemistry and Cancer Biology, Thoracic Diseases Research Unit, Medical University of Ohio, Toledo, Ohio*

RICHARD E. PAGANO (14), *Department of Biochemistry and Molecular Biology, Thoracic Diseases Research Unit, Mayo Clinic College of Medicine, Rochester, Minnesota*

ADRIANO PAGANO (8), *Biozentrum, University of Basel, Basel, Switzerland*

SEETHARAMAN PARASHURAMAN (25), *Cell Biology Lab, National Institute of Immunology, New Delhi, India*

ROBERT G. PARTON (66), *Department of Biochemistry, Institute for Molecular Bioscience, The University of Queensland, Brisbane, Queensland, Australia*

ANDREW A. PEDEN (45), *Genentech Inc., San Francisco, California*

JOHAN PERÄNEN (24), *Institute of Biotechnology, University of Helsinki, Helsinki, Finland*

FRANCK PEREZ (12), *CNRS UMR144, Institut Curie, Paris, France*

SUZANNE PFEFFER (30, 31), *Department of Biochemistry, Stanford University School of Medicine, Stanford, California*

IOANA POPA (46), *Department of Cell Biology, University Medical Center Utrecht, Utrecht, The Netherlands*

RYTIS PREKERIS (45), *Department of Cellular and Developmental Biology, School of Medicine, University of Colorado Health Sciences Center, Aurora, Colorado*

ROMANO REGAZZI (6), *Department of Cell Biology and Morphology, University of Lausanne, Lausanne, Switzerland*

ELSA ROMERO (55), *Department of Pathology, University of New Mexico Health Sciences Center, Albuquerque, New Mexico*

CRAIG R. ROY (7), *Section of Microbial Pathogenesis, Boyer Center for Molecular Medicine, Yale University School of Medicine, New Haven, Connecticut*

KOTA SAITO (23), *Department of Physiological Chemistry, Graduate School of Pharmaceutical Sciences, University of Tokyo, Tokyo, Japan*

AYUKO SAKANE (59), *Department of Biochemistry, Institute of Health Biosciences, The University of Tokushima, Graduate School of Medicine, Tokushima, Japan*

TOSHIAKI SAKISAKA (21, 29, 35), *Department of Cell Biology, The Scripps Research Institute, La Jolla, California*

TAKUYA SASAKI (59), *Department of Biochemistry, Institute of Health Biosciences, The University of Tokushima, Graduate School of Medicine, Tokushima, Japan*

DIEGO SBRISSA (60), *Department of Physiology and the Center for Molecular Medicine and Genetics, Wayne State University School of Medicine, Detroit, Michigan*

SUSUMU SEINO (36), *Division of Cellular and Molecular Medicine, Kobe University Graduate School of Medicine, Kusunoki-cho, Chuo-ku, Kobe, Japan*

TADAO SHIBASAKI (36), *Division of Cellular and Molecular Medicine, Kobe University Graduate School of Medicine, Kusunoki-cho, Chuo-ku, Kobe, Japan*

RYUTARO SHIRAKAWA (67), *Department of Geriatric Medicine, Graduate School of Medicine, Kyoto University, Kyoto, Japan*

ASSIA SHISHEVA (60), *Department of Physiology and the Center for Molecular Medicine and Genetics, Wayne State University School of Medicine, Detroit, Michigan*

BENJAMIN SHORT (53), *Department of Cell Biology, Max-Planck-Institute of Biochemistry, Martinsried, Germany*

FIONA SIMPSON (66), *Department of Biochemistry, Institute for Molecular Bioscience, The University of Queensland, Brisbane, Queensland, Australia*

SYMEON SINIOSSOGLOU (52), *Cambridge Institute for Medical Research, Cambridge, United Kingdom*

ULF SIVARS (30, 31), *Department of Biochemistry, Stanford University School of Medicine, Stanford, California*

ALEXANDER SORKIN (11), *Department of Pharmacology, University of Colorado Health Sciences Center, Aurora, Colorado*

MARTIN SPIESS (8), *Biozentrum, University of Basel, Basel, Switzerland*

MARIA RITA SPINOSA (57), *Dipartimento di Scienze e Tecnologie Biologiche ed Ambientali, Universitá di Lecce, Lecce, Italy*

MARY-PAT STEIN (55), *Section Microbial Pathogenesis, Yale University, New Haven, Connecticut*

DAVID J. STRICK (64), *Department of Neuroscience and Cell Biology, University of Texas Medical Branch, Galveston, Texas*

PENG SUN (19), *Department of Biology, Faculty of Science, Graduate School of Science and Technology, Chiba University, Chiba, Japan*

YOSHIMI TAKAI (21, 35), *Department of Molecular Biology and Biochemistry, Osaka University Graduate School of Medicine, Suita, Japan*

JUSTIN D. TOPP (22), *Department of Biochemistry and Molecular Biology, The Mayo Clinic Cancer Center, Mayo Clinic, Rochester, Minnesota*

ELIZABETH TARBUTTON (45), *Department of Cellular and Developmental Biology, School of Medicine, University of Colorado Health Sciences Center, Aurora, Colorado*

JENNIFER M. TEREW (5), *Department of Biochemistry and Molecular Biology, University of New Hampshire, Durham, New Hampshire*

MATHEWOS TESSEMA (55), *Department of Pathology, University of New Mexico Health Sciences Center, Albuquerque, New Mexico*

ELLEN J. TISDALE (33), *Department of Pharmacology, School of Medicine, Wayne State University, Detroit, Michigan*

SEIJI TORII (18), *Department of Molecular Medicine, Institute for Molecular and Cellular Regulation, Gunma University, Maebashi, Gunma, Japan*

PETER VAN DER SLUIJS (46), *Department of Cell Biology, University Medical Center Utrecht, Utrecht, The Netherlands*

NANCY WAKEHAM (50), *Crystallography Research Program, Oklahoma Medical Research Foundation, Oklahoma City, Oklahoma*

HERBERT WALDMANN (4), *Department of Physical Biochemistry, Max-Planck-Institute for Molecular Physiology, Dortmund, Germany*

ANGELA WANDINGER-NESS (55, 56), *Department of Pathology, University of New Mexico Health Sciences Center, Albuquerque, New Mexico*

TUANLAO WANG (58), *Membrane Biology Laboratory, Institute of Molecular and Cell Biology, Singapore*

ROBERTO WEIGERT (20), *Laboratory of Cell Biology, National Heart, Lung, and Blood Institute, National Institutes of Health, Bethesda, Maryland*

ANGELA WELFORD (55, 56), *Department of Pathology, University of New Mexico Health Sciences Center, Albuquerque, New Mexico*

CAROL WICKING (66), *Department of Biochemistry, Institute for Molecular Bioscience, The University of Queensland, Brisbane, Queensland, Australia*

ALLAN W. WOLKOFF (9), *Marion Bessin Liver Research Center and Department of Anatomy and Structural Biology, Albert Einstein College of Medicine, Bronx, New York*

AKITSUGU YAMAMOTO (41), *Nagahama Institute of Bio-Science and Technology, Shiga, Japan*

AKIRA YOSHIOKA (67), *Department of Geriatric Medicine, Graduate School of Medicine, Kyoto University, Kyoto, Japan*

AHMED ZAHRAOUI (15, 63), *Laboratory of Morphogenesis and Cell Signaling, Centre National de la Recherche Scientifique, Paris, France*

PENG ZHAI (50), *Crystallography Research Program, Oklahoma Medical Research Foundation, Oklahoma City, Oklahoma*

HOUBIN ZHANG (5), *Department of Ophthalmology, University of Utah, Salt Lake City, Utah*

KAI ZHANG (5), *Department of Ophthalmology, University of Utah, Salt Lake City, Utah*

XUEJUN C. ZHANG (50), *Crystallography Research Program, Oklahoma Medical Research Foundation, Oklahoma City, Oklahoma*

GUANGYU ZHU (50), *Crystallography Research Program, Oklahoma Medical Research Foundation, Oklahoma City, Oklahoma*

TOMAS H. ZURAWSKI (44), *International Centre for Neurotherapeutics, Dublin City University, Dublin, Ireland*

Preface

The Ras superfamily now encompasses Ras-related GTPases involved in cell proliferation, Rho-related GTPases involved in regulation of the cytoskeleton, Rab GTPases involved in membrane targeting and fusion, and a large group of GTPases including Sar1, Arf, Arl, and dynamin involved in vesicle budding as well as control of membrane fission. Because GTPases are fundamentally switches, they have a large number of molecules that regulate switch function to control of GDP loading (guanine nucleotide exchange factors [GEFs]) and GTP hydrolysis (guanine nucleotide activating proteins [GAPs]) as well as an untold number of downstream effectors whose recruitment and discharge from functional protein complexes affect a myriad of cell biological events controlled by the GTPase cycle.

In this new series of Methods of Enzymology, we have striven to bring together the latest thinking, approaches, and techniques that contribute to our understanding of these different GTPase family members. Two volumes (403 and 404) focus on membrane regulating GTPases, with the first volume focused on those that control membrane targeting and fusion (Rabs) and the second dedicated to the GTPases regulating budding and fission (Sar1, Arf, Arl, and dynamin). Volumes 406 and 407 focus on Rho family and Ras-family regulators and effectors, respectively. It is important to emphasize that while each of these volumes focus on different GTPase families, they contain a wealth of methodologies that cross family borders. As such, methodologies pioneered with respect to one class of Ras superfamily GTPases are likely to be equally applicable to other classes for analysis of unanticipated functions. Furthermore, the functional distinctions that have been classically associated with the distinct branches of the superfamily are beginning to blur. There is now considerable evidence for biological and biochemical interplay and crosstalk among seemingly divergent family members. The compilation of a database of regulators and effectors of the whole superfamily by Bernards (volume 407) reflects some of these complex interrelationships. In addition to fostering cross-talk among investigators who study different GTPases, these volumes will also aid the entry of new investigators of into the field.

In volume 403, focused on Rab GTPases that regulate membrane targeting and fusion, it is apparent that the depth and breadth of the Rab GTPase family continues to expand, highlighting their central importance in eukaryotic membrane architecture. With over 70 Rab GTPases identified to date, recent phylogenetic approaches as highlighted in a chapter by Collins and colleagues, provide systematic approaches that allows the investigator to mine functional regions pertinent to activity. From a bioinformatics and systems biology perspective, it is now apparent that Rab-regulated hubs provide the foundation to generate the specialized membrane architectures that dictate organization of exocytic and endocytic pathways across the entire eukaryotic kingdom. This architectural foundation, referred to as the *membrome*, as described by Gurkan and colleagues in the first chapter, provides an integrated coding system that can serve as a starting point to generate a 'top-down' view of Rab GTPase function in eukaryotic cells. By defining the building blocks for membrane trafficking, such a Rab-centric view illustrates a number of facets that directs the overall organization of subcellular compartments of cells and tissues through the activity of dynamic protein interaction networks. Remarkably, it points towards the function of Rab GTPase hubs as tethers to control movement of vesicles and organelles along the actin and microtubule cytoskeletons, and to control events leading to membrane recognition (docking) and fusion by the SNARE machinery. An interactive website for exploring datasets comprising the many components of Rab-regulated hubs that define the *membrome* of different cell and organ systems in both human and mouse is now available at **http://www.membrome.org/.** In addition to these overview chapters on phylogeny and systems biology, a general approach for prenylating Rab GTPases, a chapter by Alexandrov and colleagues focus on important technical issues required to work with Rab GTPases *in vitro*.

Of course, the heart of this series belongs to the many important contributions by experts in the field that highlight the methodologies and proteins involved in the analysis of individual Rab GTPase function. The knowledge base presented in these articles should provide an important foundation for further rapid growth of our understanding of Rab GTPase activities and their role in controlling eukaryotic membrane architecture. We are extremely grateful to the many investigators who have generously contributed their time and expertise to bring this wealth of technical expertise into this and other volumes comprising the Ras-superfamily series.

WILLIAM E. BALCH
CHANNING J. DER
ALAN HALL

METHODS IN ENZYMOLOGY

VOLUME I. Preparation and Assay of Enzymes
Edited by SIDNEY P. COLOWICK AND NATHAN O. KAPLAN

VOLUME II. Preparation and Assay of Enzymes
Edited by SIDNEY P. COLOWICK AND NATHAN O. KAPLAN

VOLUME III. Preparation and Assay of Substrates
Edited by SIDNEY P. COLOWICK AND NATHAN O. KAPLAN

VOLUME IV. Special Techniques for the Enzymologist
Edited by SIDNEY P. COLOWICK AND NATHAN O. KAPLAN

VOLUME V. Preparation and Assay of Enzymes
Edited by SIDNEY P. COLOWICK AND NATHAN O. KAPLAN

VOLUME VI. Preparation and Assay of Enzymes *(Continued)*
Preparation and Assay of Substrates
Special Techniques
Edited by SIDNEY P. COLOWICK AND NATHAN O. KAPLAN

VOLUME VII. Cumulative Subject Index
Edited by SIDNEY P. COLOWICK AND NATHAN O. KAPLAN

VOLUME VIII. Complex Carbohydrates
Edited by ELIZABETH F. NEUFELD AND VICTOR GINSBURG

VOLUME IX. Carbohydrate Metabolism
Edited by WILLIS A. WOOD

VOLUME X. Oxidation and Phosphorylation
Edited by RONALD W. ESTABROOK AND MAYNARD E. PULLMAN

VOLUME XI. Enzyme Structure
Edited by C. H. W. HIRS

VOLUME XII. Nucleic Acids (Parts A and B)
Edited by LAWRENCE GROSSMAN AND KIVIE MOLDAVE

VOLUME XIII. Citric Acid Cycle
Edited by J. M. LOWENSTEIN

VOLUME XIV. Lipids
Edited by J. M. LOWENSTEIN

VOLUME XV. Steroids and Terpenoids
Edited by RAYMOND B. CLAYTON

VOLUME XVI. Fast Reactions
Edited by KENNETH KUSTIN

VOLUME XVII. Metabolism of Amino Acids and Amines
(Parts A and B)
Edited by HERBERT TABOR AND CELIA WHITE TABOR

VOLUME XVIII. Vitamins and Coenzymes (Parts A, B, and C)
Edited by DONALD B. MCCORMICK AND LEMUEL D. WRIGHT

VOLUME XIX. Proteolytic Enzymes
Edited by GERTRUDE E. PERLMANN AND LASZLO LORAND

VOLUME XX. Nucleic Acids and Protein Synthesis (Part C)
Edited by KIVIE MOLDAVE AND LAWRENCE GROSSMAN

VOLUME XXI. Nucleic Acids (Part D)
Edited by LAWRENCE GROSSMAN AND KIVIE MOLDAVE

VOLUME XXII. Enzyme Purification and Related Techniques
Edited by WILLIAM B. JAKOBY

VOLUME XXIII. Photosynthesis (Part A)
Edited by ANTHONY SAN PIETRO

VOLUME XXIV. Photosynthesis and Nitrogen Fixation (Part B)
Edited by ANTHONY SAN PIETRO

VOLUME XXV. Enzyme Structure (Part B)
Edited by C. H. W. HIRS AND SERGE N. TIMASHEFF

VOLUME XXVI. Enzyme Structure (Part C)
Edited by C. H. W. HIRS AND SERGE N. TIMASHEFF

VOLUME XXVII. Enzyme Structure (Part D)
Edited by C. H. W. HIRS AND SERGE N. TIMASHEFF

VOLUME XXVIII. Complex Carbohydrates (Part B)
Edited by VICTOR GINSBURG

VOLUME XXIX. Nucleic Acids and Protein Synthesis (Part E)
Edited by LAWRENCE GROSSMAN AND KIVIE MOLDAVE

VOLUME XXX. Nucleic Acids and Protein Synthesis (Part F)
Edited by KIVIE MOLDAVE AND LAWRENCE GROSSMAN

VOLUME XXXI. Biomembranes (Part A)
Edited by SIDNEY FLEISCHER AND LESTER PACKER

VOLUME XXXII. Biomembranes (Part B)
Edited by SIDNEY FLEISCHER AND LESTER PACKER

VOLUME XXXIII. Cumulative Subject Index Volumes I-XXX
Edited by MARTHA G. DENNIS AND EDWARD A. DENNIS

VOLUME XXXIV. Affinity Techniques (Enzyme Purification: Part B)
Edited by WILLIAM B. JAKOBY AND MEIR WILCHEK

VOLUME XXXV. Lipids (Part B)
Edited by JOHN M. LOWENSTEIN

VOLUME XXXVI. Hormone Action (Part A: Steroid Hormones)
Edited by BERT W. O'MALLEY AND JOEL G. HARDMAN

VOLUME XXXVII. Hormone Action (Part B: Peptide Hormones)
Edited by BERT W. O'MALLEY AND JOEL G. HARDMAN

VOLUME XXXVIII. Hormone Action (Part C: Cyclic Nucleotides)
Edited by JOEL G. HARDMAN AND BERT W. O'MALLEY

VOLUME XXXIX. Hormone Action (Part D: Isolated Cells, Tissues,
and Organ Systems)
Edited by JOEL G. HARDMAN AND BERT W. O'MALLEY

VOLUME XL. Hormone Action (Part E: Nuclear Structure and Function)
Edited by BERT W. O'MALLEY AND JOEL G. HARDMAN

VOLUME XLI. Carbohydrate Metabolism (Part B)
Edited by W. A. WOOD

VOLUME XLII. Carbohydrate Metabolism (Part C)
Edited by W. A. WOOD

VOLUME XLIII. Antibiotics
Edited by JOHN H. HASH

VOLUME XLIV. Immobilized Enzymes
Edited by KLAUS MOSBACH

VOLUME XLV. Proteolytic Enzymes (Part B)
Edited by LASZLO LORAND

VOLUME XLVI. Affinity Labeling
Edited by WILLIAM B. JAKOBY AND MEIR WILCHEK

VOLUME XLVII. Enzyme Structure (Part E)
Edited by C. H. W. HIRS AND SERGE N. TIMASHEFF

VOLUME XLVIII. Enzyme Structure (Part F)
Edited by C. H. W. HIRS AND SERGE N. TIMASHEFF

VOLUME XLIX. Enzyme Structure (Part G)
Edited by C. H. W. HIRS AND SERGE N. TIMASHEFF

VOLUME L. Complex Carbohydrates (Part C)
Edited by VICTOR GINSBURG

VOLUME LI. Purine and Pyrimidine Nucleotide Metabolism
Edited by PATRICIA A. HOFFEE AND MARY ELLEN JONES

VOLUME LII. Biomembranes (Part C: Biological Oxidations)
Edited by SIDNEY FLEISCHER AND LESTER PACKER

VOLUME LIII. Biomembranes (Part D: Biological Oxidations)
Edited by SIDNEY FLEISCHER AND LESTER PACKER

VOLUME LIV. Biomembranes (Part E: Biological Oxidations)
Edited by SIDNEY FLEISCHER AND LESTER PACKER

VOLUME LV. Biomembranes (Part F: Bioenergetics)
Edited by SIDNEY FLEISCHER AND LESTER PACKER

VOLUME LVI. Biomembranes (Part G: Bioenergetics)
Edited by SIDNEY FLEISCHER AND LESTER PACKER

VOLUME LVII. Bioluminescence and Chemiluminescence
Edited by MARLENE A. DELUCA

VOLUME LVIII. Cell Culture
Edited by WILLIAM B. JAKOBY AND IRA PASTAN

VOLUME LIX. Nucleic Acids and Protein Synthesis (Part G)
Edited by KIVIE MOLDAVE AND LAWRENCE GROSSMAN

VOLUME LX. Nucleic Acids and Protein Synthesis (Part H)
Edited by KIVIE MOLDAVE AND LAWRENCE GROSSMAN

VOLUME 61. Enzyme Structure (Part H)
Edited by C. H. W. HIRS AND SERGE N. TIMASHEFF

VOLUME 62. Vitamins and Coenzymes (Part D)
Edited by DONALD B. MCCORMICK AND LEMUEL D. WRIGHT

VOLUME 63. Enzyme Kinetics and Mechanism (Part A: Initial Rate and
Inhibitor Methods)
Edited by DANIEL L. PURICH

VOLUME 64. Enzyme Kinetics and Mechanism
(Part B: Isotopic Probes and Complex Enzyme Systems)
Edited by DANIEL L. PURICH

VOLUME 65. Nucleic Acids (Part I)
Edited by LAWRENCE GROSSMAN AND KIVIE MOLDAVE

VOLUME 66. Vitamins and Coenzymes (Part E)
Edited by DONALD B. MCCORMICK AND LEMUEL D. WRIGHT

VOLUME 67. Vitamins and Coenzymes (Part F)
Edited by DONALD B. MCCORMICK AND LEMUEL D. WRIGHT

VOLUME 68. Recombinant DNA
Edited by RAY WU

VOLUME 69. Photosynthesis and Nitrogen Fixation (Part C)
Edited by ANTHONY SAN PIETRO

VOLUME 70. Immunochemical Techniques (Part A)
Edited by HELEN VAN VUNAKIS AND JOHN J. LANGONE

VOLUME 71. Lipids (Part C)
Edited by JOHN M. LOWENSTEIN

VOLUME 72. Lipids (Part D)
Edited by JOHN M. LOWENSTEIN

VOLUME 73. Immunochemical Techniques (Part B)
Edited by JOHN J. LANGONE AND HELEN VAN VUNAKIS

VOLUME 74. Immunochemical Techniques (Part C)
Edited by JOHN J. LANGONE AND HELEN VAN VUNAKIS

VOLUME 75. Cumulative Subject Index Volumes XXXI, XXXII, XXXIV–LX
Edited by EDWARD A. DENNIS AND MARTHA G. DENNIS

VOLUME 76. Hemoglobins
Edited by ERALDO ANTONINI, LUIGI ROSSI-BERNARDI, AND EMILIA CHIANCONE

VOLUME 77. Detoxication and Drug Metabolism
Edited by WILLIAM B. JAKOBY

VOLUME 78. Interferons (Part A)
Edited by SIDNEY PESTKA

VOLUME 79. Interferons (Part B)
Edited by SIDNEY PESTKA

VOLUME 80. Proteolytic Enzymes (Part C)
Edited by LASZLO LORAND

VOLUME 81. Biomembranes (Part H: Visual Pigments and Purple Membranes, I)
Edited by LESTER PACKER

VOLUME 82. Structural and Contractile Proteins (Part A: Extracellular Matrix)
Edited by LEON W. CUNNINGHAM AND DIXIE W. FREDERIKSEN

VOLUME 83. Complex Carbohydrates (Part D)
Edited by VICTOR GINSBURG

VOLUME 84. Immunochemical Techniques (Part D: Selected Immunoassays)
Edited by JOHN J. LANGONE AND HELEN VAN VUNAKIS

VOLUME 85. Structural and Contractile Proteins (Part B: The Contractile Apparatus and the Cytoskeleton)
Edited by DIXIE W. FREDERIKSEN AND LEON W. CUNNINGHAM

VOLUME 86. Prostaglandins and Arachidonate Metabolites
Edited by WILLIAM E. M. LANDS AND WILLIAM L. SMITH

VOLUME 87. Enzyme Kinetics and Mechanism (Part C: Intermediates, Stereo-chemistry, and Rate Studies)
Edited by DANIEL L. PURICH

VOLUME 88. Biomembranes (Part I: Visual Pigments and Purple Membranes, II)
Edited by LESTER PACKER

VOLUME 89. Carbohydrate Metabolism (Part D)
Edited by WILLIS A. WOOD

VOLUME 90. Carbohydrate Metabolism (Part E)
Edited by WILLIS A. WOOD

VOLUME 91. Enzyme Structure (Part I)
Edited by C. H. W. HIRS AND SERGE N. TIMASHEFF

VOLUME 92. Immunochemical Techniques (Part E: Monoclonal Antibodies and General Immunoassay Methods)
Edited by JOHN J. LANGONE AND HELEN VAN VUNAKIS

VOLUME 93. Immunochemical Techniques (Part F: Conventional Antibodies, Fc Receptors, and Cytotoxicity)
Edited by JOHN J. LANGONE AND HELEN VAN VUNAKIS

VOLUME 94. Polyamines
Edited by HERBERT TABOR AND CELIA WHITE TABOR

VOLUME 95. Cumulative Subject Index Volumes 61–74, 76–80
Edited by EDWARD A. DENNIS AND MARTHA G. DENNIS

VOLUME 96. Biomembranes [Part J: Membrane Biogenesis: Assembly and Targeting (General Methods; Eukaryotes)]
Edited by SIDNEY FLEISCHER AND BECCA FLEISCHER

VOLUME 97. Biomembranes [Part K: Membrane Biogenesis: Assembly and Targeting (Prokaryotes, Mitochondria, and Chloroplasts)]
Edited by SIDNEY FLEISCHER AND BECCA FLEISCHER

VOLUME 98. Biomembranes (Part L: Membrane Biogenesis: Processing and Recycling)
Edited by SIDNEY FLEISCHER AND BECCA FLEISCHER

VOLUME 99. Hormone Action (Part F: Protein Kinases)
Edited by JACKIE D. CORBIN AND JOEL G. HARDMAN

VOLUME 100. Recombinant DNA (Part B)
Edited by RAY WU, LAWRENCE GROSSMAN, AND KIVIE MOLDAVE

VOLUME 101. Recombinant DNA (Part C)
Edited by RAY WU, LAWRENCE GROSSMAN, AND KIVIE MOLDAVE

VOLUME 102. Hormone Action (Part G: Calmodulin and Calcium-Binding Proteins)
Edited by ANTHONY R. MEANS AND BERT W. O'MALLEY

VOLUME 103. Hormone Action (Part H: Neuroendocrine Peptides)
Edited by P. MICHAEL CONN

VOLUME 104. Enzyme Purification and Related Techniques (Part C)
Edited by WILLIAM B. JAKOBY

VOLUME 105. Oxygen Radicals in Biological Systems
Edited by LESTER PACKER

VOLUME 106. Posttranslational Modifications (Part A)
Edited by FINN WOLD AND KIVIE MOLDAVE

VOLUME 107. Posttranslational Modifications (Part B)
Edited by FINN WOLD AND KIVIE MOLDAVE

VOLUME 108. Immunochemical Techniques (Part G: Separation and Characterization of Lymphoid Cells)
Edited by GIOVANNI DI SABATO, JOHN J. LANGONE, AND HELEN VAN VUNAKIS

VOLUME 109. Hormone Action (Part I: Peptide Hormones)
Edited by LUTZ BIRNBAUMER AND BERT W. O'MALLEY

VOLUME 110. Steroids and Isoprenoids (Part A)
Edited by JOHN H. LAW AND HANS C. RILLING

VOLUME 111. Steroids and Isoprenoids (Part B)
Edited by JOHN H. LAW AND HANS C. RILLING

VOLUME 112. Drug and Enzyme Targeting (Part A)
Edited by KENNETH J. WIDDER AND RALPH GREEN

VOLUME 113. Glutamate, Glutamine, Glutathione, and Related Compounds
Edited by ALTON MEISTER

VOLUME 114. Diffraction Methods for Biological Macromolecules (Part A)
Edited by HAROLD W. WYCKOFF, C. H. W. HIRS, AND SERGE N. TIMASHEFF

VOLUME 115. Diffraction Methods for Biological Macromolecules (Part B)
Edited by HAROLD W. WYCKOFF, C. H. W. HIRS, AND SERGE N. TIMASHEFF

VOLUME 116. Immunochemical Techniques (Part H: Effectors and Mediators of Lymphoid Cell Functions)
Edited by GIOVANNI DI SABATO, JOHN J. LANGONE, AND HELEN VAN VUNAKIS

VOLUME 117. Enzyme Structure (Part J)
Edited by C. H. W. HIRS AND SERGE N. TIMASHEFF

VOLUME 118. Plant Molecular Biology
Edited by ARTHUR WEISSBACH AND HERBERT WEISSBACH

VOLUME 119. Interferons (Part C)
Edited by SIDNEY PESTKA

VOLUME 120. Cumulative Subject Index Volumes 81–94, 96–101

VOLUME 121. Immunochemical Techniques (Part I: Hybridoma Technology and Monoclonal Antibodies)
Edited by JOHN J. LANGONE AND HELEN VAN VUNAKIS

VOLUME 122. Vitamins and Coenzymes (Part G)
Edited by FRANK CHYTIL AND DONALD B. MCCORMICK

VOLUME 123. Vitamins and Coenzymes (Part H)
Edited by FRANK CHYTIL AND DONALD B. MCCORMICK

VOLUME 124. Hormone Action (Part J: Neuroendocrine Peptides)
Edited by P. MICHAEL CONN

VOLUME 125. Biomembranes (Part M: Transport in Bacteria, Mitochondria, and Chloroplasts: General Approaches and Transport Systems)
Edited by SIDNEY FLEISCHER AND BECCA FLEISCHER

VOLUME 126. Biomembranes (Part N: Transport in Bacteria, Mitochondria, and Chloroplasts: Protonmotive Force)
Edited by SIDNEY FLEISCHER AND BECCA FLEISCHER

VOLUME 127. Biomembranes (Part O: Protons and Water: Structure and Translocation)
Edited by LESTER PACKER

VOLUME 128. Plasma Lipoproteins (Part A: Preparation, Structure, and Molecular Biology)
Edited by JERE P. SEGREST AND JOHN J. ALBERS

VOLUME 129. Plasma Lipoproteins (Part B: Characterization, Cell Biology, and Metabolism)
Edited by JOHN J. ALBERS AND JERE P. SEGREST

VOLUME 130. Enzyme Structure (Part K)
Edited by C. H. W. HIRS AND SERGE N. TIMASHEFF

VOLUME 131. Enzyme Structure (Part L)
Edited by C. H. W. HIRS AND SERGE N. TIMASHEFF

VOLUME 132. Immunochemical Techniques (Part J: Phagocytosis and Cell-Mediated Cytotoxicity)
Edited by GIOVANNI DI SABATO AND JOHANNES EVERSE

VOLUME 133. Bioluminescence and Chemiluminescence (Part B)
Edited by MARLENE DELUCA AND WILLIAM D. MCELROY

VOLUME 134. Structural and Contractile Proteins (Part C: The Contractile Apparatus and the Cytoskeleton)
Edited by RICHARD B. VALLEE

VOLUME 135. Immobilized Enzymes and Cells (Part B)
Edited by KLAUS MOSBACH

VOLUME 136. Immobilized Enzymes and Cells (Part C)
Edited by KLAUS MOSBACH

VOLUME 137. Immobilized Enzymes and Cells (Part D)
Edited by KLAUS MOSBACH

VOLUME 138. Complex Carbohydrates (Part E)
Edited by VICTOR GINSBURG

VOLUME 139. Cellular Regulators (Part A: Calcium- and Calmodulin-Binding Proteins)
Edited by ANTHONY R. MEANS AND P. MICHAEL CONN

VOLUME 140. Cumulative Subject Index Volumes 102–119, 121–134

VOLUME 141. Cellular Regulators (Part B: Calcium and Lipids)
Edited by P. MICHAEL CONN AND ANTHONY R. MEANS

VOLUME 142. Metabolism of Aromatic Amino Acids and Amines
Edited by SEYMOUR KAUFMAN

VOLUME 143. Sulfur and Sulfur Amino Acids
Edited by WILLIAM B. JAKOBY AND OWEN GRIFFITH

VOLUME 144. Structural and Contractile Proteins (Part D: Extracellular Matrix)
Edited by LEON W. CUNNINGHAM

VOLUME 145. Structural and Contractile Proteins (Part E: Extracellular Matrix)
Edited by LEON W. CUNNINGHAM

VOLUME 146. Peptide Growth Factors (Part A)
Edited by DAVID BARNES AND DAVID A. SIRBASKU

VOLUME 147. Peptide Growth Factors (Part B)
Edited by DAVID BARNES AND DAVID A. SIRBASKU

VOLUME 148. Plant Cell Membranes
Edited by LESTER PACKER AND ROLAND DOUCE

VOLUME 149. Drug and Enzyme Targeting (Part B)
Edited by RALPH GREEN AND KENNETH J. WIDDER

VOLUME 150. Immunochemical Techniques (Part K: *In Vitro* Models of B and T Cell Functions and Lymphoid Cell Receptors)
Edited by GIOVANNI DI SABATO

VOLUME 151. Molecular Genetics of Mammalian Cells
Edited by MICHAEL M. GOTTESMAN

VOLUME 152. Guide to Molecular Cloning Techniques
Edited by SHELBY L. BERGER AND ALAN R. KIMMEL

VOLUME 153. Recombinant DNA (Part D)
Edited by RAY WU AND LAWRENCE GROSSMAN

VOLUME 154. Recombinant DNA (Part E)
Edited by RAY WU AND LAWRENCE GROSSMAN

VOLUME 155. Recombinant DNA (Part F)
Edited by RAY WU

VOLUME 156. Biomembranes (Part P: ATP-Driven Pumps and Related Transport: The Na, K-Pump)
Edited by SIDNEY FLEISCHER AND BECCA FLEISCHER

VOLUME 157. Biomembranes (Part Q: ATP-Driven Pumps and Related Transport: Calcium, Proton, and Potassium Pumps)
Edited by SIDNEY FLEISCHER AND BECCA FLEISCHER

VOLUME 158. Metalloproteins (Part A)
Edited by JAMES F. RIORDAN AND BERT L. VALLEE

VOLUME 159. Initiation and Termination of Cyclic Nucleotide Action
Edited by JACKIE D. CORBIN AND ROGER A. JOHNSON

VOLUME 160. Biomass (Part A: Cellulose and Hemicellulose)
Edited by WILLIS A. WOOD AND SCOTT T. KELLOGG

VOLUME 161. Biomass (Part B: Lignin, Pectin, and Chitin)
Edited by WILLIS A. WOOD AND SCOTT T. KELLOGG

VOLUME 162. Immunochemical Techniques (Part L: Chemotaxis
and Inflammation)
Edited by GIOVANNI DI SABATO

VOLUME 163. Immunochemical Techniques (Part M: Chemotaxis
and Inflammation)
Edited by GIOVANNI DI SABATO

VOLUME 164. Ribosomes
Edited by HARRY F. NOLLER, JR., AND KIVIE MOLDAVE

VOLUME 165. Microbial Toxins: Tools for Enzymology
Edited by SIDNEY HARSHMAN

VOLUME 166. Branched-Chain Amino Acids
Edited by ROBERT HARRIS AND JOHN R. SOKATCH

VOLUME 167. Cyanobacteria
Edited by LESTER PACKER AND ALEXANDER N. GLAZER

VOLUME 168. Hormone Action (Part K: Neuroendocrine Peptides)
Edited by P. MICHAEL CONN

VOLUME 169. Platelets: Receptors, Adhesion,
Secretion (Part A)
Edited by JACEK HAWIGER

VOLUME 170. Nucleosomes
Edited by PAUL M. WASSARMAN AND ROGER D. KORNBERG

VOLUME 171. Biomembranes (Part R: Transport Theory: Cells and Model
Membranes)
Edited by SIDNEY FLEISCHER AND BECCA FLEISCHER

VOLUME 172. Biomembranes (Part S: Transport: Membrane Isolation and
Characterization)
Edited by SIDNEY FLEISCHER AND BECCA FLEISCHER

VOLUME 173. Biomembranes [Part T: Cellular and Subcellular Transport:
Eukaryotic (Nonepithelial) Cells]
Edited by SIDNEY FLEISCHER AND BECCA FLEISCHER

VOLUME 174. Biomembranes [Part U: Cellular and Subcellular Transport:
Eukaryotic (Nonepithelial) Cells]
Edited by SIDNEY FLEISCHER AND BECCA FLEISCHER

VOLUME 175. Cumulative Subject Index Volumes 135–139, 141–167

VOLUME 176. Nuclear Magnetic Resonance (Part A: Spectral Techniques and Dynamics)
Edited by NORMAN J. OPPENHEIMER AND THOMAS L. JAMES

VOLUME 177. Nuclear Magnetic Resonance (Part B: Structure and Mechanism)
Edited by NORMAN J. OPPENHEIMER AND THOMAS L. JAMES

VOLUME 178. Antibodies, Antigens, and Molecular Mimicry
Edited by JOHN J. LANGONE

VOLUME 179. Complex Carbohydrates (Part F)
Edited by VICTOR GINSBURG

VOLUME 180. RNA Processing (Part A: General Methods)
Edited by JAMES E. DAHLBERG AND JOHN N. ABELSON

VOLUME 181. RNA Processing (Part B: Specific Methods)
Edited by JAMES E. DAHLBERG AND JOHN N. ABELSON

VOLUME 182. Guide to Protein Purification
Edited by MURRAY P. DEUTSCHER

VOLUME 183. Molecular Evolution: Computer Analysis of Protein and Nucleic Acid Sequences
Edited by RUSSELL F. DOOLITTLE

VOLUME 184. Avidin-Biotin Technology
Edited by MEIR WILCHEK AND EDWARD A. BAYER

VOLUME 185. Gene Expression Technology
Edited by DAVID V. GOEDDEL

VOLUME 186. Oxygen Radicals in Biological Systems (Part B: Oxygen Radicals and Antioxidants)
Edited by LESTER PACKER AND ALEXANDER N. GLAZER

VOLUME 187. Arachidonate Related Lipid Mediators
Edited by ROBERT C. MURPHY AND FRANK A. FITZPATRICK

VOLUME 188. Hydrocarbons and Methylotrophy
Edited by MARY E. LIDSTROM

VOLUME 189. Retinoids (Part A: Molecular and Metabolic Aspects)
Edited by LESTER PACKER

VOLUME 190. Retinoids (Part B: Cell Differentiation and Clinical Applications)
Edited by LESTER PACKER

VOLUME 191. Biomembranes (Part V: Cellular and Subcellular Transport: Epithelial Cells)
Edited by SIDNEY FLEISCHER AND BECCA FLEISCHER

VOLUME 192. Biomembranes (Part W: Cellular and Subcellular Transport: Epithelial Cells)
Edited by SIDNEY FLEISCHER AND BECCA FLEISCHER

VOLUME 193. Mass Spectrometry
Edited by JAMES A. MCCLOSKEY

VOLUME 194. Guide to Yeast Genetics and Molecular Biology
Edited by CHRISTINE GUTHRIE AND GERALD R. FINK

VOLUME 195. Adenylyl Cyclase, G Proteins, and Guanylyl Cyclase
Edited by ROGER A. JOHNSON AND JACKIE D. CORBIN

VOLUME 196. Molecular Motors and the Cytoskeleton
Edited by RICHARD B. VALLEE

VOLUME 197. Phospholipases
Edited by EDWARD A. DENNIS

VOLUME 198. Peptide Growth Factors (Part C)
Edited by DAVID BARNES, J. P. MATHER, AND GORDON H. SATO

VOLUME 199. Cumulative Subject Index Volumes 168–174, 176–194

VOLUME 200. Protein Phosphorylation (Part A: Protein Kinases: Assays,
Purification, Antibodies, Functional Analysis, Cloning, and Expression)
Edited by TONY HUNTER AND BARTHOLOMEW M. SEFTON

VOLUME 201. Protein Phosphorylation (Part B: Analysis of Protein
Phosphorylation, Protein Kinase Inhibitors, and Protein Phosphatases)
Edited by TONY HUNTER AND BARTHOLOMEW M. SEFTON

VOLUME 202. Molecular Design and Modeling: Concepts and Applications
(Part A: Proteins, Peptides, and Enzymes)
Edited by JOHN J. LANGONE

VOLUME 203. Molecular Design and Modeling:
Concepts and Applications (Part B: Antibodies and Antigens, Nucleic Acids,
Polysaccharides,
and Drugs)
Edited by JOHN J. LANGONE

VOLUME 204. Bacterial Genetic Systems
Edited by JEFFREY H. MILLER

VOLUME 205. Metallobiochemistry (Part B: Metallothionein and
Related Molecules)
Edited by JAMES F. RIORDAN AND BERT L. VALLEE

VOLUME 206. Cytochrome P450
Edited by MICHAEL R. WATERMAN AND ERIC F. JOHNSON

VOLUME 207. Ion Channels
Edited by BERNARDO RUDY AND LINDA E. IVERSON

VOLUME 208. Protein–DNA Interactions
Edited by ROBERT T. SAUER

VOLUME 209. Phospholipid Biosynthesis
Edited by EDWARD A. DENNIS AND DENNIS E. VANCE

VOLUME 210. Numerical Computer Methods
Edited by LUDWIG BRAND AND MICHAEL L. JOHNSON

VOLUME 211. DNA Structures (Part A: Synthesis and Physical Analysis of DNA)
Edited by DAVID M. J. LILLEY AND JAMES E. DAHLBERG

VOLUME 212. DNA Structures (Part B: Chemical and Electrophoretic Analysis of DNA)
Edited by DAVID M. J. LILLEY AND JAMES E. DAHLBERG

VOLUME 213. Carotenoids (Part A: Chemistry, Separation, Quantitation, and Antioxidation)
Edited by LESTER PACKER

VOLUME 214. Carotenoids (Part B: Metabolism, Genetics, and Biosynthesis)
Edited by LESTER PACKER

VOLUME 215. Platelets: Receptors, Adhesion, Secretion (Part B)
Edited by JACEK J. HAWIGER

VOLUME 216. Recombinant DNA (Part G)
Edited by RAY WU

VOLUME 217. Recombinant DNA (Part H)
Edited by RAY WU

VOLUME 218. Recombinant DNA (Part I)
Edited by RAY WU

VOLUME 219. Reconstitution of Intracellular Transport
Edited by JAMES E. ROTHMAN

VOLUME 220. Membrane Fusion Techniques (Part A)
Edited by NEJAT DÜZGÜNEŞ

VOLUME 221. Membrane Fusion Techniques (Part B)
Edited by NEJAT DÜZGÜNEŞ

VOLUME 222. Proteolytic Enzymes in Coagulation, Fibrinolysis, and Complement Activation (Part A: Mammalian Blood Coagulation Factors and Inhibitors)
Edited by LASZLO LORAND AND KENNETH G. MANN

VOLUME 223. Proteolytic Enzymes in Coagulation, Fibrinolysis, and Complement Activation (Part B: Complement Activation, Fibrinolysis, and Nonmammalian Blood Coagulation Factors)
Edited by LASZLO LORAND AND KENNETH G. MANN

VOLUME 224. Molecular Evolution: Producing the Biochemical Data
Edited by ELIZABETH ANNE ZIMMER, THOMAS J. WHITE, REBECCA L. CANN, AND ALLAN C. WILSON

VOLUME 225. Guide to Techniques in Mouse Development
Edited by PAUL M. WASSARMAN AND MELVIN L. DEPAMPHILIS

VOLUME 226. Metallobiochemistry (Part C: Spectroscopic and Physical Methods for Probing Metal Ion Environments in Metalloenzymes and Metalloproteins)
Edited by JAMES F. RIORDAN AND BERT L. VALLEE

VOLUME 227. Metallobiochemistry (Part D: Physical and Spectroscopic Methods for Probing Metal Ion Environments in Metalloproteins)
Edited by JAMES F. RIORDAN AND BERT L. VALLEE

VOLUME 228. Aqueous Two-Phase Systems
Edited by HARRY WALTER AND GÖTE JOHANSSON

VOLUME 229. Cumulative Subject Index Volumes 195–198, 200–227

VOLUME 230. Guide to Techniques in Glycobiology
Edited by WILLIAM J. LENNARZ AND GERALD W. HART

VOLUME 231. Hemoglobins (Part B: Biochemical and Analytical Methods)
Edited by JOHANNES EVERSE, KIM D. VANDEGRIFF, AND ROBERT M. WINSLOW

VOLUME 232. Hemoglobins (Part C: Biophysical Methods)
Edited by JOHANNES EVERSE, KIM D. VANDEGRIFF, AND ROBERT M. WINSLOW

VOLUME 233. Oxygen Radicals in Biological Systems (Part C)
Edited by LESTER PACKER

VOLUME 234. Oxygen Radicals in Biological Systems (Part D)
Edited by LESTER PACKER

VOLUME 235. Bacterial Pathogenesis (Part A: Identification and Regulation of Virulence Factors)
Edited by VIRGINIA L. CLARK AND PATRIK M. BAVOIL

VOLUME 236. Bacterial Pathogenesis (Part B: Integration of Pathogenic Bacteria with Host Cells)
Edited by VIRGINIA L. CLARK AND PATRIK M. BAVOIL

VOLUME 237. Heterotrimeric G Proteins
Edited by RAVI IYENGAR

VOLUME 238. Heterotrimeric G-Protein Effectors
Edited by RAVI IYENGAR

VOLUME 239. Nuclear Magnetic Resonance (Part C)
Edited by THOMAS L. JAMES AND NORMAN J. OPPENHEIMER

VOLUME 240. Numerical Computer Methods (Part B)
Edited by MICHAEL L. JOHNSON AND LUDWIG BRAND

VOLUME 241. Retroviral Proteases
Edited by LAWRENCE C. KUO AND JULES A. SHAFER

VOLUME 242. Neoglycoconjugates (Part A)
Edited by Y. C. LEE AND REIKO T. LEE

VOLUME 243. Inorganic Microbial Sulfur Metabolism
Edited by HARRY D. PECK, JR., AND JEAN LEGALL

VOLUME 244. Proteolytic Enzymes: Serine and Cysteine Peptidases
Edited by ALAN J. BARRETT

VOLUME 245. Extracellular Matrix Components
Edited by E. RUOSLAHTI AND E. ENGVALL

VOLUME 246. Biochemical Spectroscopy
Edited by KENNETH SAUER

VOLUME 247. Neoglycoconjugates (Part B: Biomedical Applications)
Edited by Y. C. LEE AND REIKO T. LEE

VOLUME 248. Proteolytic Enzymes: Aspartic and Metallo Peptidases
Edited by ALAN J. BARRETT

VOLUME 249. Enzyme Kinetics and Mechanism (Part D: Developments in Enzyme Dynamics)
Edited by DANIEL L. PURICH

VOLUME 250. Lipid Modifications of Proteins
Edited by PATRICK J. CASEY AND JANICE E. BUSS

VOLUME 251. Biothiols (Part A: Monothiols and Dithiols, Protein Thiols, and Thiyl Radicals)
Edited by LESTER PACKER

VOLUME 252. Biothiols (Part B: Glutathione and Thioredoxin; Thiols in Signal Transduction and Gene Regulation)
Edited by LESTER PACKER

VOLUME 253. Adhesion of Microbial Pathogens
Edited by RON J. DOYLE AND ITZHAK OFEK

VOLUME 254. Oncogene Techniques
Edited by PETER K. VOGT AND INDER M. VERMA

VOLUME 255. Small GTPases and Their Regulators (Part A: Ras Family)
Edited by W. E. BALCH, CHANNING J. DER, AND ALAN HALL

VOLUME 256. Small GTPases and Their Regulators (Part B: Rho Family)
Edited by W. E. BALCH, CHANNING J. DER, AND ALAN HALL

VOLUME 257. Small GTPases and Their Regulators (Part C: Proteins Involved in Transport)
Edited by W. E. BALCH, CHANNING J. DER, AND ALAN HALL

VOLUME 258. Redox-Active Amino Acids in Biology
Edited by JUDITH P. KLINMAN

VOLUME 259. Energetics of Biological Macromolecules
Edited by MICHAEL L. JOHNSON AND GARY K. ACKERS

VOLUME 260. Mitochondrial Biogenesis and Genetics (Part A)
Edited by GIUSEPPE M. ATTARDI AND ANNE CHOMYN

VOLUME 261. Nuclear Magnetic Resonance and Nucleic Acids
Edited by THOMAS L. JAMES

VOLUME 262. DNA Replication
Edited by JUDITH L. CAMPBELL

VOLUME 263. Plasma Lipoproteins (Part C: Quantitation)
Edited by WILLIAM A. BRADLEY, SANDRA H. GIANTURCO, AND JERE P. SEGREST

VOLUME 264. Mitochondrial Biogenesis and Genetics (Part B)
Edited by GIUSEPPE M. ATTARDI AND ANNE CHOMYN

VOLUME 265. Cumulative Subject Index Volumes 228, 230–262

VOLUME 266. Computer Methods for Macromolecular Sequence Analysis
Edited by RUSSELL F. DOOLITTLE

VOLUME 267. Combinatorial Chemistry
Edited by JOHN N. ABELSON

VOLUME 268. Nitric Oxide (Part A: Sources and Detection of NO; NO Synthase)
Edited by LESTER PACKER

VOLUME 269. Nitric Oxide (Part B: Physiological and Pathological Processes)
Edited by LESTER PACKER

VOLUME 270. High Resolution Separation and Analysis of Biological Macromolecules (Part A: Fundamentals)
Edited by BARRY L. KARGER AND WILLIAM S. HANCOCK

VOLUME 271. High Resolution Separation and Analysis of Biological Macromolecules (Part B: Applications)
Edited by BARRY L. KARGER AND WILLIAM S. HANCOCK

VOLUME 272. Cytochrome P450 (Part B)
Edited by ERIC F. JOHNSON AND MICHAEL R. WATERMAN

VOLUME 273. RNA Polymerase and Associated Factors (Part A)
Edited by SANKAR ADHYA

VOLUME 274. RNA Polymerase and Associated Factors (Part B)
Edited by SANKAR ADHYA

VOLUME 275. Viral Polymerases and Related Proteins
Edited by LAWRENCE C. KUO, DAVID B. OLSEN, AND STEVEN S. CARROLL

VOLUME 276. Macromolecular Crystallography (Part A)
Edited by CHARLES W. CARTER, JR., AND ROBERT M. SWEET

VOLUME 277. Macromolecular Crystallography (Part B)
Edited by CHARLES W. CARTER, JR., AND ROBERT M. SWEET

VOLUME 278. Fluorescence Spectroscopy
Edited by LUDWIG BRAND AND MICHAEL L. JOHNSON

VOLUME 279. Vitamins and Coenzymes (Part I)
Edited by DONALD B. MCCORMICK, JOHN W. SUTTIE, AND CONRAD WAGNER

VOLUME 280. Vitamins and Coenzymes (Part J)
Edited by DONALD B. MCCORMICK, JOHN W. SUTTIE, AND CONRAD WAGNER

VOLUME 281. Vitamins and Coenzymes (Part K)
Edited by DONALD B. MCCORMICK, JOHN W. SUTTIE, AND CONRAD WAGNER

VOLUME 282. Vitamins and Coenzymes (Part L)
Edited by DONALD B. MCCORMICK, JOHN W. SUTTIE, AND CONRAD WAGNER

VOLUME 283. Cell Cycle Control
Edited by WILLIAM G. DUNPHY

VOLUME 284. Lipases (Part A: Biotechnology)
Edited by BYRON RUBIN AND EDWARD A. DENNIS

VOLUME 285. Cumulative Subject Index Volumes 263, 264, 266–284, 286–289

VOLUME 286. Lipases (Part B: Enzyme Characterization and Utilization)
Edited by BYRON RUBIN AND EDWARD A. DENNIS

VOLUME 287. Chemokines
Edited by RICHARD HORUK

VOLUME 288. Chemokine Receptors
Edited by RICHARD HORUK

VOLUME 289. Solid Phase Peptide Synthesis
Edited by GREGG B. FIELDS

VOLUME 290. Molecular Chaperones
Edited by GEORGE H. LORIMER AND THOMAS BALDWIN

VOLUME 291. Caged Compounds
Edited by GERARD MARRIOTT

VOLUME 292. ABC Transporters: Biochemical, Cellular, and Molecular Aspects
Edited by SURESH V. AMBUDKAR AND MICHAEL M. GOTTESMAN

VOLUME 293. Ion Channels (Part B)
Edited by P. MICHAEL CONN

VOLUME 294. Ion Channels (Part C)
Edited by P. MICHAEL CONN

VOLUME 295. Energetics of Biological Macromolecules (Part B)
Edited by GARY K. ACKERS AND MICHAEL L. JOHNSON

VOLUME 296. Neurotransmitter Transporters
Edited by SUSAN G. AMARA

VOLUME 297. Photosynthesis: Molecular Biology of Energy Capture
Edited by LEE MCINTOSH

VOLUME 298. Molecular Motors and the Cytoskeleton (Part B)
Edited by RICHARD B. VALLEE

VOLUME 299. Oxidants and Antioxidants (Part A)
Edited by LESTER PACKER

VOLUME 300. Oxidants and Antioxidants (Part B)
Edited by LESTER PACKER

VOLUME 301. Nitric Oxide: Biological and Antioxidant Activities (Part C)
Edited by LESTER PACKER

VOLUME 302. Green Fluorescent Protein
Edited by P. MICHAEL CONN

VOLUME 303. cDNA Preparation and Display
Edited by SHERMAN M. WEISSMAN

VOLUME 304. Chromatin
Edited by PAUL M. WASSARMAN AND ALAN P. WOLFFE

VOLUME 305. Bioluminescence and Chemiluminescence (Part C)
Edited by THOMAS O. BALDWIN AND MIRIAM M. ZIEGLER

VOLUME 306. Expression of Recombinant Genes in
Eukaryotic Systems
Edited by JOSEPH C. GLORIOSO AND MARTIN C. SCHMIDT

VOLUME 307. Confocal Microscopy
Edited by P. MICHAEL CONN

VOLUME 308. Enzyme Kinetics and Mechanism (Part E: Energetics of
Enzyme Catalysis)
Edited by DANIEL L. PURICH AND VERN L. SCHRAMM

VOLUME 309. Amyloid, Prions, and Other Protein Aggregates
Edited by RONALD WETZEL

VOLUME 310. Biofilms
Edited by RON J. DOYLE

VOLUME 311. Sphingolipid Metabolism and Cell Signaling (Part A)
Edited by ALFRED H. MERRILL, JR., AND YUSUF A. HANNUN

VOLUME 312. Sphingolipid Metabolism and Cell Signaling (Part B)
Edited by ALFRED H. MERRILL, JR., AND YUSUF A. HANNUN

VOLUME 313. Antisense Technology (Part A: General Methods, Methods of
Delivery, and RNA Studies)
Edited by M. IAN PHILLIPS

VOLUME 314. Antisense Technology (Part B: Applications)
Edited by M. IAN PHILLIPS

VOLUME 315. Vertebrate Phototransduction and the Visual Cycle (Part A)
Edited by KRZYSZTOF PALCZEWSKI

VOLUME 316. Vertebrate Phototransduction and the Visual Cycle (Part B)
Edited by KRZYSZTOF PALCZEWSKI

VOLUME 317. RNA–Ligand Interactions (Part A: Structural Biology Methods)
Edited by DANIEL W. CELANDER AND JOHN N. ABELSON

VOLUME 318. RNA–Ligand Interactions (Part B: Molecular Biology Methods)
Edited by DANIEL W. CELANDER AND JOHN N. ABELSON

VOLUME 319. Singlet Oxygen, UV-A, and Ozone
Edited by LESTER PACKER AND HELMUT SIES

VOLUME 320. Cumulative Subject Index Volumes 290–319

VOLUME 321. Numerical Computer Methods (Part C)
Edited by MICHAEL L. JOHNSON AND LUDWIG BRAND

VOLUME 322. Apoptosis
Edited by JOHN C. REED

VOLUME 323. Energetics of Biological Macromolecules (Part C)
Edited by MICHAEL L. JOHNSON AND GARY K. ACKERS

VOLUME 324. Branched-Chain Amino Acids (Part B)
Edited by ROBERT A. HARRIS AND JOHN R. SOKATCH

VOLUME 325. Regulators and Effectors of Small GTPases (Part D: Rho Family)
Edited by W. E. BALCH, CHANNING J. DER, AND ALAN HALL

VOLUME 326. Applications of Chimeric Genes and Hybrid Proteins (Part A: Gene Expression and Protein Purification)
Edited by JEREMY THORNER, SCOTT D. EMR, AND JOHN N. ABELSON

VOLUME 327. Applications of Chimeric Genes and Hybrid Proteins (Part B: Cell Biology and Physiology)
Edited by JEREMY THORNER, SCOTT D. EMR, AND JOHN N. ABELSON

VOLUME 328. Applications of Chimeric Genes and Hybrid Proteins (Part C: Protein–Protein Interactions and Genomics)
Edited by JEREMY THORNER, SCOTT D. EMR, AND JOHN N. ABELSON

VOLUME 329. Regulators and Effectors of Small GTPases (Part E: GTPases Involved in Vesicular Traffic)
Edited by W. E. BALCH, CHANNING J. DER, AND ALAN HALL

VOLUME 330. Hyperthermophilic Enzymes (Part A)
Edited by MICHAEL W. W. ADAMS AND ROBERT M. KELLY

VOLUME 331. Hyperthermophilic Enzymes (Part B)
Edited by MICHAEL W. W. ADAMS AND ROBERT M. KELLY

VOLUME 332. Regulators and Effectors of Small GTPases (Part F: Ras Family I)
Edited by W. E. BALCH, CHANNING J. DER, AND ALAN HALL

VOLUME 333. Regulators and Effectors of Small GTPases (Part G: Ras Family II)
Edited by W. E. BALCH, CHANNING J. DER, AND ALAN HALL

VOLUME 334. Hyperthermophilic Enzymes (Part C)
Edited by MICHAEL W. W. ADAMS AND ROBERT M. KELLY

VOLUME 335. Flavonoids and Other Polyphenols
Edited by LESTER PACKER

VOLUME 336. Microbial Growth in Biofilms (Part A: Developmental and Molecular Biological Aspects)
Edited by RON J. DOYLE

VOLUME 337. Microbial Growth in Biofilms (Part B: Special Environments and Physicochemical Aspects)
Edited by RON J. DOYLE

VOLUME 338. Nuclear Magnetic Resonance of Biological Macromolecules (Part A)
Edited by THOMAS L. JAMES, VOLKER DÖTSCH, AND ULI SCHMITZ

VOLUME 339. Nuclear Magnetic Resonance of Biological Macromolecules (Part B)
Edited by THOMAS L. JAMES, VOLKER DÖTSCH, AND ULI SCHMITZ

VOLUME 340. Drug–Nucleic Acid Interactions
Edited by JONATHAN B. CHAIRES AND MICHAEL J. WARING

VOLUME 341. Ribonucleases (Part A)
Edited by ALLEN W. NICHOLSON

VOLUME 342. Ribonucleases (Part B)
Edited by ALLEN W. NICHOLSON

VOLUME 343. G Protein Pathways (Part A: Receptors)
Edited by RAVI IYENGAR AND JOHN D. HILDEBRANDT

VOLUME 344. G Protein Pathways (Part B: G Proteins and Their Regulators)
Edited by RAVI IYENGAR AND JOHN D. HILDEBRANDT

VOLUME 345. G Protein Pathways (Part C: Effector Mechanisms)
Edited by RAVI IYENGAR AND JOHN D. HILDEBRANDT

VOLUME 346. Gene Therapy Methods
Edited by M. IAN PHILLIPS

VOLUME 347. Protein Sensors and Reactive Oxygen Species (Part A: Selenoproteins and Thioredoxin)
Edited by HELMUT SIES AND LESTER PACKER

VOLUME 348. Protein Sensors and Reactive Oxygen Species (Part B: Thiol Enzymes and Proteins)
Edited by HELMUT SIES AND LESTER PACKER

VOLUME 349. Superoxide Dismutase
Edited by LESTER PACKER

VOLUME 350. Guide to Yeast Genetics and Molecular and Cell Biology (Part B)
Edited by CHRISTINE GUTHRIE AND GERALD R. FINK

VOLUME 351. Guide to Yeast Genetics and Molecular and Cell Biology (Part C)
Edited by CHRISTINE GUTHRIE AND GERALD R. FINK

VOLUME 352. Redox Cell Biology and Genetics (Part A)
Edited by CHANDAN K. SEN AND LESTER PACKER

VOLUME 353. Redox Cell Biology and Genetics (Part B)
Edited by CHANDAN K. SEN AND LESTER PACKER

VOLUME 354. Enzyme Kinetics and Mechanisms (Part F: Detection and Characterization of Enzyme Reaction Intermediates)
Edited by DANIEL L. PURICH

VOLUME 355. Cumulative Subject Index Volumes 321–354

VOLUME 356. Laser Capture Microscopy and Microdissection
Edited by P. MICHAEL CONN

VOLUME 357. Cytochrome P450, Part C
Edited by ERIC F. JOHNSON AND MICHAEL R. WATERMAN

VOLUME 358. Bacterial Pathogenesis (Part C: Identification, Regulation, and Function of Virulence Factors)
Edited by VIRGINIA L. CLARK AND PATRIK M. BAVOIL

VOLUME 359. Nitric Oxide (Part D)
Edited by ENRIQUE CADENAS AND LESTER PACKER

VOLUME 360. Biophotonics (Part A)
Edited by GERARD MARRIOTT AND IAN PARKER

VOLUME 361. Biophotonics (Part B)
Edited by GERARD MARRIOTT AND IAN PARKER

VOLUME 362. Recognition of Carbohydrates in Biological Systems (Part A)
Edited by YUAN C. LEE AND REIKO T. LEE

VOLUME 363. Recognition of Carbohydrates in Biological Systems (Part B)
Edited by YUAN C. LEE AND REIKO T. LEE

VOLUME 364. Nuclear Receptors
Edited by DAVID W. RUSSELL AND DAVID J. MANGELSDORF

VOLUME 365. Differentiation of Embryonic Stem Cells
Edited by PAUL M. WASSAUMAN AND GORDON M. KELLER

VOLUME 366. Protein Phosphatases
Edited by SUSANNE KLUMPP AND JOSEF KRIEGLSTEIN

VOLUME 367. Liposomes (Part A)
Edited by NEJAT DÜZGÜNEŞ

VOLUME 368. Macromolecular Crystallography (Part C)
Edited by CHARLES W. CARTER, JR., AND ROBERT M. SWEET

VOLUME 369. Combinational Chemistry (Part B)
Edited by GUILLERMO A. MORALES AND BARRY A. BUNIN

VOLUME 370. RNA Polymerases and Associated Factors (Part C)
Edited by SANKAR L. ADHYA AND SUSAN GARGES

VOLUME 371. RNA Polymerases and Associated Factors (Part D)
Edited by SANKAR L. ADHYA AND SUSAN GARGES

VOLUME 372. Liposomes (Part B)
Edited by NEJAT DÜZGÜNEŞ

VOLUME 373. Liposomes (Part C)
Edited by NEJAT DÜZGÜNEŞ

VOLUME 374. Macromolecular Crystallography (Part D)
Edited by CHARLES W. CARTER, JR., AND ROBERT W. SWEET

VOLUME 375. Chromatin and Chromatin Remodeling Enzymes (Part A)
Edited by C. DAVID ALLIS AND CARL WU

VOLUME 376. Chromatin and Chromatin Remodeling Enzymes (Part B)
Edited by C. DAVID ALLIS AND CARL WU

VOLUME 377. Chromatin and Chromatin Remodeling Enzymes (Part C)
Edited by C. DAVID ALLIS AND CARL WU

VOLUME 378. Quinones and Quinone Enzymes (Part A)
Edited by HELMUT SIES AND LESTER PACKER

VOLUME 379. Energetics of Biological Macromolecules (Part D)
Edited by JO M. HOLT, MICHAEL L. JOHNSON, AND GARY K. ACKERS

VOLUME 380. Energetics of Biological Macromolecules (Part E)
Edited by JO M. HOLT, MICHAEL L. JOHNSON, AND GARY K. ACKERS

VOLUME 381. Oxygen Sensing
Edited by CHANDAN K. SEN AND GREGG L. SEMENZA

VOLUME 382. Quinones and Quinone Enzymes (Part B)
Edited by HELMUT SIES AND LESTER PACKER

VOLUME 383. Numerical Computer Methods (Part D)
Edited by LUDWIG BRAND AND MICHAEL L. JOHNSON

VOLUME 384. Numerical Computer Methods (Part E)
Edited by LUDWIG BRAND AND MICHAEL L. JOHNSON

VOLUME 385. Imaging in Biological Research (Part A)
Edited by P. MICHAEL CONN

VOLUME 386. Imaging in Biological Research (Part B)
Edited by P. MICHAEL CONN

VOLUME 387. Liposomes (Part D)
Edited by NEJAT DÜZGÜNEŞ

VOLUME 388. Protein Engineering
Edited by DAN E. ROBERTSON AND JOSEPH P. NOEL

VOLUME 389. Regulators of G-Protein Signaling (Part A)
Edited by DAVID P. SIDEROVSKI

VOLUME 390. Regulators of G-protein Sgnalling (Part B)
Edited by DAVID P. SIDEROVSKI

VOLUME 391. Liposomes (Part E)
Edited by NEJAT DÜZGÜNEŞ

VOLUME 392. RNA Interference
Edited by ENGELKE ROSSI

VOLUME 393. Circadian Rhythms
Edited by MICHAEL W. YOUNG

VOLUME 394. Nuclear Magnetic Resonance of Biological Macromolecules
(Part C)
Edited by THOMAS L. JAMES

VOLUME 395. Producing the Biochemical Data (Part B)
Edited by ELIZABETH A. ZIMMER AND ERIC H. ROALSON

VOLUME 396. Nitric Oxide (Part E)
Edited by LESTER PACKER AND ENRIQUE CADENAS

VOLUME 397. Environmental Microbiology
Edited by JARED R. LEADBETTER

VOLUME 398. Ubiquitin and Protein Degradation (Part A)
Edited by RAYMOND J. DESHAIES

VOLUME 399. Ubiquitin and Protein Degradation (Part B)
Edited by RAYMOND J. DESHAIES

VOLUME 400. Phase II Conjugation Enzymes and Transport Systems
(in preparation)
Edited by HELMUT SIES AND LESTER PACKER

VOLUME 401. Glutathione Transferases and Gamma Glutamyl Transpeptidases
(in preparation)
Edited by HELMUT SIES AND LESTER PACKER

VOLUME 402. Biological Mass spectrometry (in preparation)
Edited by A. L. BURLINGAME

VOLUME 403. GTPases Regulating Membrane Targeting and Fusion
Edited by WILLIAM E. BALCH, CHANNING J. DER, AND ALAN HALL

VOLUME 404. GTPases Regulating Membrane Dynamics (in preparation)
Edited by WILLIAM E. BALCH, CHANNING J. DER, AND ALAN HALL

VOLUME 405. Mass Spectrometry: Modified Proteins and Glycoconjugates
(in preparation)
Edited by A. L. BURLINGAME

VOLUME 406. Regulators and Effectors of Small GTPases: Rho Family
(in preparation)
Edited by WILLIAM E. BALCH, CHANNING J. DER, AND ALAN HALL

[1] Exploring Trafficking GTPase Function by mRNA Expression Profiling: Use of the SymAtlas Web-Application and the Membrome Datasets

By CEMAL GURKAN, HILMAR LAPP,
JOHN B. HOGENESCH, and WILLIAM E. BALCH

Abstract

Despite complete sequencing of the human and mouse genomes, functional annotation of novel gene function still remains a major challenge in mammalian biology. Emerging strategies to help elucidate unknown gene function include the analysis of tissue-specific patterns of mRNA expression. A recent study investigated the steady-state mRNA expression profiling of the vast majority of protein-encoding human and mouse genes across a panel of 79 human and 61 mouse nonredundant tissues. The microarray data from this study constitutes the Genomics Institute of Novartis Foundation (GNF) Human and Mouse Gene Atlases and is publicly available for exploration through the SymAtlas web-application (http://symatlas.gnf.org). We have recently reported the use of these data and hierarchical clustering algorithms to generate a global overview of the distribution of Rabs, SNAREs, and coat machinery components, as well as their respective adaptors, effectors, and regulators. This systems biology approach led us to propose Rab-centric protein activity hubs as a framework for an integrated coding system, the *membrome* network, which orchestrates the dynamics of specialized membrane architecture of differentiated cells. Here, we describe the use of the SymAtlas web-application and the Membrome datasets to help explore trafficking GTPase function. The human and mouse *membrome* datasets are available through the Membrome homepage (http://www.membrome.org/) and correspond to subsets of the SymAtlas content restricted to known membrane trafficking components. Considering the fragmentary nature of the current reductionist approaches in elucidating trafficking component functions, the *membrome* datasets provide a more focused systems biology perspective that not only complements our current understanding of transport in complex tissues but also provides an integrated perspective of Rab activity in controlling membrane architecture.

METHODS IN ENZYMOLOGY, VOL. 403   0076-6879/05 $35.00
Copyright 2005, Elsevier Inc. All rights reserved.   DOI: 10.1016/S0076-6879(05)03001-6

Introduction

Traffic along the eukaryotic secretory and endocytic pathways is characterized by the formation and maintenance of numerous subcellular compartments defined by the encapsulating lipid bilayer with varying phospholipid composition and unique sets of integral and peripheral membrane proteins. These subcellular compartments are dynamic structures that are in continuous and specific communication through carrier vesicles and tubules that mobilize cargo to specific destinations (Bonifacino and Glick, 2004; Kirchhausen, 2000; Pfeffer, 2003). By harnessing and regulating the fundamental processes of membrane fission and fusion through the action of protein complexes, the lipid bilayer can be exploited to produce a variety of distinct subcellular compartments that have unique chemical environments and play essential roles in cell and organ function. While it is clearly evident that the key to these dynamic processes is the systematic and reversible regulation of protein interactions, our understanding of the molecular basis for the global organization of the exocytic and endocytic trafficking systems still remains fragmentary.

Traditionally, phylogenetic analysis of proteins in a gene family is commonly used to identify potential functional relationships to other family members, such as that of the Rab GTPases and the SNARE family of docking/fusion proteins directing eukaryotic membrane traffic (Bock *et al.*, 2001; Chen and Scheller, 2001; Pereira-Leal and Seabra, 2000, 2001; Ungar and Hughson, 2003). Computational approaches applying hierarchical clustering algorithms to systematic tissue profiling can complement this annotation by providing insights into the physiological activity of close and distant family members, and to different gene families in different cell types (Panda *et al.*, 2002; Su *et al.*, 2004; Walker *et al.*, 2004). We have also recently described the use of steady-state mRNA expression profiling and hierarchical clustering algorithms to generate a global overview of the distribution of Rabs, SNAREs, and coat machinery components, as well as their respective adaptors, effectors, and regulators in 79 human and 61 mouse nonredundant tissues (Gurkan *et al.*, 2005). Our systems biology approach had led us to propose that membrane trafficking events are largely orchestrated by Rab-regulated protein hubs that can be linked to biochemically characterized components of the coat, and tethering, targeting, and fusion machineries (Gurkan *et al.*, 2005). We refer to this collection of interacting components that define the specific membrane architectures of a given cell type as the *membrome* network. Here, we describe the use of the SymAtlas web-application to help explore trafficking GTPase function by browsing through the Genomics Institute of the Novartis Research Foundation (GNF) Human and Mouse Gene Atlases, as well as their subsets, the human and mouse membrome datasets that we have compiled from the literature (Gurkan *et al.*, 2005).

Materials

Microarray Data

Microarray data mentioned in this chapter are from the GNF Human and Mouse Gene Atlases (Version 2) (Su *et al.*, 2004).

SymAtlas Web-Application

SymAtlas is a web-application (http://symatlas.gnf.org) for publishing experimental gene functionalization datasets (e.g., the GNF Human and Mouse Gene Atlases) integrated with a flexibly searchable gene-centric database of public and proprietary annotations. It is publicly available for searching and visualization by keyword, accession number, gene symbol, genome interval, sequence, expression pattern, and coregulation. By default, visualization of the mRNA expression profile for each gene is provided in the form of a bar chart, following condensation of the raw data using the Microarray Analysis Suite 5.0 (MAS 5.0) Software (Affymetrix, Inc., Santa Clara, CA) or GeneChip RMA (GCRMA) Algorithm (Wu and Irizarry, 2004).

Human and Mouse Membrome Datasets

To help explore trafficking GTPase function, we have compiled human and mouse *membrome* datasets that currently comprise ~450 human/mouse proteins corresponding to known trafficking components within the cell (Gurkan *et al.*, 2005). These datasets can be accessed through the Membrome homepage (http://www.membrome.org) for direct searching and visualization using the SymAtlas web-application as described above. They may also be accessed by selecting either the human or mouse "membrome" option from the dataset selection box/pull-down menu available on any given SymAtlas gene annotation or mRNA expression profile page.

Exploring Trafficking GTPase Function

Use of the SymAtlas Web-Application

SymAtlas web-application (http://symatlas.gnf.org) can be accessed using an internet browser such as the Microsoft Internet Explorer, Apple Safari, Mozilla Firefox, etc. Once at the SymAtlas homepage, the mRNA expression profile of a gene or protein of interest (e.g., Rab3A) can be searched using either of the two query forms (or search fields) available. The smaller query form at the top of the homepage (*and the larger query form immediately below it when using the default parameters provided*) treats the user input as gene symbols, aliases, accession numbers, or identifiers from all species available. Through the use of pull-down menus provided, the larger query form also allows selection of additional search parameters such as that by keywords in order to locate all the related genes (e.g., Rab3A, Rab3A interacting protein Rab3IP, etc.) in SymAtlas. Further selectable search parameters at the SymAtlas homepage include user input of the reference author or title, sequence, protein domain and family, and genome interval, as well as expression pattern and coregulation.

Figure 1 is a snapshot image of a typical search results page available through SymAtlas. In this particular case, the search results page was reached following user input of "Rab3A" in the query form available at the SymAtlas homepage and then implementing a search using the default parameters provided (*for direct access to this page, click **here***). The screen layout of a typical search results page at SymAtlas is organized as follows.

A. A navigation panel on the left-hand side for exploring the search results set (Fig. 1A). The result set shown in this panel may contain a single gene or multiple genes from one or more species, depending on the criteria chosen on the starting search page. All genes in the result set are *hyperlinked* to their corresponding "bioentry" pages at SymAtlas in two ways. The bar chart icon to the left of each gene name is clickable and provides a link to the corresponding mRNA expression profile (*profile view*), whereas the gene name (*shown in blue and underlined*) itself provides a link to the corresponding full gene annotation page (*annotation view*). If there is no bar chart icon present to the left of a gene name, then there is no mRNA expression profile data available for this gene within SymAtlas.

B. A main panel in the middle displaying the mRNA expression profile of the gene of interest (Fig. 1B). Alternatively, when selected from the navigation panel, the main panel may show the full annotation view for a given gene, or an annotation table for multiple genes. While by default the mRNA expression profile is presented in the form of a horizontal bar chart, alternative forms of data presentation are also available for selection through the "render" pull-down menu available above the navigation panel. In the bar chart view, the microarray data for each tissue are presented along with three additional lines for the median, three times the

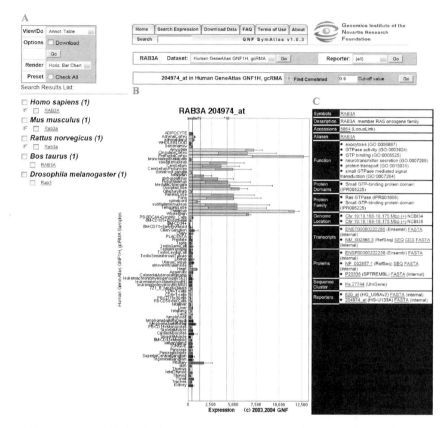

Fig. 1. A snapshot of a typical search results page at SymAtlas. In this case, the results page corresponds to a simple search carried out at the SymAtlas homepage (http://symatlas. gnf.org) as described in the text using the search term "Rab3A" and the default search parameters provided. The screen layout is divided into three sections, namely a navigation panel on the left-hand side for exploring the results list, a main panel in the middle for the graphic display of the selected gene's mRNA expression profile, and finally an annotation panel on the right-hand side corresponding to that of the selected gene. (See color inserts labeled in red fonts with A, B, and C, respectively.)

median, and 10 times the median (*shown as black, blue, and red* in Fig. 1B). This allows the user to quickly assess how the peak expression values relate to the median expression for a gene of interest, and also how tissue specific a gene is being expressed. The median expression of each gene is calculated across all of its replicate-averaged expression levels. We find that the median value differs considerably between genes depending on whether it is a housekeeping gene or a tissue specialist, such as in the case of the Rab GTPases (Gurkan *et al.*, 2005). Expression values below 250 roughly correspond to 1–2 copies/cell and define the lower limits of confidence. Finally, the error bars provided with the majority of samples represent the standard deviation between two or more replicates that have been averaged.

C. An annotation panel on the right-hand side for partial annotation of the gene of interest (Fig. 1C). To reach the full annotation page (*annotation view*), the gene name of interest in the navigation panel needs to be clicked. In both cases, annotation provided includes hyperlinks to relevant entries in various external databases.

Following an initial analysis as described above, the search results obtained can be further extended to locate other genes/proteins that are coclustering or exhibiting similar mRNA expression profiles with the gene of interest. The first assumption is that the level of the mRNA signal reflects the corresponding protein activity. The second assumption is that coclustering of two or more evolutionarily divergent components based on the similarities of their mRNA expression profiles may indicate a potentially direct or indirect interaction between these species and their contribution to a common cellular pathway (Gurkan *et al.*, 2005). To locate such potentially interacting partners, the SymAtlas web-application features a "profile neighborhood" search function to identify genes whose expression profile is correlated with that of a query profile as defined by the given Pearson correlation coefficient. To execute this search, the profile view of the gene of interest needs to be selected by clicking on the corresponding bar chart icon in the navigation panel (Fig. 1A). Next, the user needs to specify the cut-off value for the Pearson correlation coefficient between the selected profile and other profiles in the same dataset using the query form located above each profile (Fig. 1, above B and C). Genes with one or more profiles that are correlated by the threshold value will be returned in a new search results list appearing in the navigation panel. It should be noted that the lower the Pearson correlation coefficient cut-off value specified, the longer it would take for the server to complete the profile neighborhood search, and the more "hits" are likely to be received.

There is also an additional query form available (through the "search expression" tab at the top of each SymAtlas page) for searching genes by fold-over-median expression. Here, the user selects a dataset and the threshold value for the tissue of interest. Executing the search will return all genes with one or more profiles that have an expression level of at least the threshold times the gene's median expression value in the chosen tissue.

Use of the Human and Mouse Membrome Datasets

We have compiled human and mouse membrome datasets based on the current literature and made them available online for direct searching and visualization using the SymAtlas web-application as described above. These datasets can be easily accessed through the Membrome homepage (http://www.membrome.org), which also features direct links to relevant literature and supplementary material. They may also be accessed by selecting either the human or mouse "membrome" option from the dataset selection box available on the SymAtlas gene annotation or mRNA expression profile pages (e.g., above B in Fig. 1). While necessarily restricted to only currently known components of membrane trafficking, membrome datasets may be better suited for the identification of potentially interacting partners by eliminating genes populating different processes (i.e., intermediary metabolism, mitochondrial function, etc.) that show similar expression patterns (Gurkan et al., 2005).

Further Notes on the Use of Tissue mRNA Profiling Data

Use of expression profiling as a systems biology tool for understanding membrane architecture and trafficking GTPase function requires several considerations.

1. Annotation is affected by the presence or absence of the probe set (s) (reporter) corresponding to the gene of interest on the microarray chip and the quality of the target or labeled DNA prepared from each tissue (Su et al., 2004).

2. Even though the SymAtlas web-application indexes as many gene symbols and identifiers as annotated in several source databases, it should be noted that differences in nomenclature in the databases and the literature may still pose a problem in identifying a gene target of interest.

3. The functional significance of the microarray data used for computational clustering (i.e., profile neighborhood search at the SymAtlas web-application) is based on the assumption that the level of mRNA reflects protein activity. This in fact could vary due to either the differential half-life of a given protein, the presence or absence of regulatory posttranslational modifications, and/or the presence of splice variants not discriminated by this technique. However, elevated expression is expected to be diagnostic of the importance of a protein in the trafficking pathways of a given tissue.

4. The complexity of the samples being examined contributes significantly to the profile. In the case of tissue samples, the profiling represents the aggregate of total mRNA message levels for all cell types in

the tissue and the abundance of a particular cell type in the given tissue. Thus, profiling may highlight dominant, cell-specific pathways. More information on the exact tissue samples and cell types analyzed (Su et al., 2004) can be found by following the relevant links provided through the "download data" tab at the top of all SymAtlas pages.

5. It should be noted that for some genes (e.g., human **Rab1A**), more than one mRNA expression profile may be available at SymAtlas. This is due to the fact that the GNF Human and Mouse Gene Atlases in part use commercially available gene expression arrays (Su et al., 2004), wherein multiple probe sets were designed to one gene at times to ensure that different splice variants are all interrogated, or when a single probe set that is specific to the target transcript and meets melting temperature and other sequence composition parameters could not be found. Alternatively, the UniGene (NCBI) clusters used at the time of the microarray design may have been subsequently merged during the reannotation process. Regardless, any significant differences in the expression patterns between multiple probe sets that are annotated as targeting the same gene can be due to a variety of technical and scientific reasons. These include probe sets that despite the initial in silico prediction, show poor hybridization efficiency, cross-hybridization, or fail to recognize their target due to target site obstruction by secondary structure formation in the mRNA or the probe sequence. Since probes need to target the 3' untranslated region (UTR) of a transcript to optimize the likelihood of detection, the design process relies heavily on the correctness of the current 3'UTR annotation of transcripts in public databases. If for less well characterized genes the 3'UTR was overpredicted, a number of probes in the probe set may fail to detect the transcript. Conversely, if the 3'UTR was substantially under-predicted, the reverse transcription step in the hybridization protocol may not yield cDNA fragments long enough to be detectable by a sufficient number of probes in the probe set.

6. Hierarchical clustering methods described in our recent manuscript (Gurkan et al., 2005) and the profile neighborhood search function provided by the SymAtlas web-application simultaneously examine all pathways within a given cell or tissue. A systems biology approach highlights both constitutive and cell-specific pathways that may be linked to accomplish a particular exocytic and/or endocytic activity. Given this limitation, not all relationships established using biochemical approaches will be necessarily highlighted in the clustering profile, for instance, reflecting the relative level of activity of a particular Rab-regulated hub in a given cell type (Gurkan et al., 2005). Conversely, mRNA expression profiling provides a relative measure of the possible relationships and can serve as a *guide* to identify potential protein interactions and/or cell

systems that define particular trafficking pathways. However, the physiological role of a particular *membrome* component will still be best achieved using reductionist approaches that involve the tissue or cell type in which the protein is normally expressed as a component of the appropriate Rab-regulated hub. This does not necessarily negate current studies in heterologous expression systems, but suggests that components of a particular Rab hub that may have an important impact on understanding the mechanism of trafficking by a particular Rab may be missing in these artificial systems where the hallmark of true function is the specialized membrane architecture.

Acknowledgments

These studies are supported by grants from the National Institutes of Health (GM33301, GM42336, and EY11606) to W.E.B. C.G. is a Cystic Fibrosis Foundation Postdoctoral Research Fellowship recipient. This is TSRI Manuscript No. 17372-CB.

References

Bock, J. B., Matern, H. T., Peden, A. A., and Scheller, R. H. (2001). A genomic perspective on membrane compartment organization. *Nature* **409,** 839–841.
Bonifacino, J. S., and Glick, B. S. (2004). The mechanisms of vesicle budding and fusion. *Cell* **116,** 153–166.
Chen, Y. A., and Scheller, R. H. (2001). SNARE-mediated membrane fusion. *Nat. Rev. Mol. Cell. Biol.* **2,** 98–106.
Gurkan, C., Lapp, H., Alory, C., Su, A. I., Hogenesch, J. B., and Balch, W. E. (2005). Large scale profiling of Rab GTPase trafficking networks: The membrome. *Mol. Biol. Cell* **16,** 3847–3864.
Kirchhausen, T. (2000). Three ways to make a vesicle. *Nat. Rev. Mol. Cell. Biol.* **1,** 187–198.
Panda, S., Antoch, M. P., Miller, B. H., Su, A. I., Schook, A. B., Straume, M., Schultz, P. G., Kay, S. A., Takahashi, J. S., and Hogenesch, J. B. (2002). Coordinated transcription of key pathways in the mouse by the circadian clock. *Cell* **109,** 307–320.
Pereira-Leal, J. B., and Seabra, M. C. (2000). The mammalian Rab family of small GTPases: Definition of family and subfamily sequence motifs suggests a mechanism for functional specificity in the Ras superfamily. *J. Mol. Biol.* **301,** 1077–1087.
Pereira-Leal, J. B., and Seabra, M. C. (2001). Evolution of the Rab family of small GTP-binding proteins. *J. Mol. Biol.* **313,** 889–901.
Pfeffer, S. (2003). Membrane domains in the secretory and endocytic pathways. *Cell* **112,** 507–517.
Su, A. I., Wiltshire, T., Batalov, S., Lapp, H., Ching, K. A., Block, D., Zhang, J., Soden, R., Hayakawa, M., Kreiman, G., Cooke, M. P., Walker, J. R., and Hogenesch, J. B. (2004). A gene atlas of the mouse and human protein-encoding transcriptomes. *Proc. Natl. Acad. Sci. USA* **101,** 6062–6067.
Ungar, D., and Hughson, F. M. (2003). SNARE protein structure and function. *Annu. Rev. Cell. Dev. Biol.* **19,** 493–517.

Walker, J. R., Su, A. I., Self, D. W., Hogenesch, J. B., Lapp, H., Maier, R., Hoyer, D., and Bilbe, G. (2004). Applications of a rat multiple tissue gene expression data set. *Genome Res.* **14,** 742–749.

Wu, Z., and Irizarry, R. A. (2004). Preprocessing of oligonucleotide array data. *Nat. Biotechnol.* **22,** 656–658.

[2] Use of Search Algorithms to Define Specificity in Rab GTPase Domain Function

By MARIA NUSSBAUM and RUTH N. COLLINS

Abstract

The continuing explosion of sequencing data has inspired a corresponding effort in the annotation and classification of protein families. Within a particular protein family, however, individual members may have distinct functions, although they share a common fold and broadly defined physiological role. Rab GTPases are the largest subfamily of the Ras superfamily, yet from early in their discovery, it was apparent that each Rab protein has a unique subcellular localization and regulates a particular stage(s) membrane traffic. To gain insight into the contribution of individual residues to unique protein functions a general strategy is outlined. This method should allow the cell and molecular biologist with no specialist expertise to implement an algorithm that makes use of a combination of experimental and phylogenetic data. The algorithm is applicable to the analysis of any protein domain and here is illustrated with the analysis of residues contributing to the individual functions of a pair of Rab GTPases.

Introduction

Rab proteins comprise the largest subfamily of the Ras superfamily. Although individual Rab proteins have the same mechanism of GTP binding and hydrolysis and overall structural homology, they have distinct cellular functions and localizations and there is much interest in understanding this "speciation" of function at the molecular level. Several computational methods have been developed to analyze individual protein function and each offers distinct advantages (del Sol Mesa *et al.*, 2003). From the viewpoint of the cell and molecular biologist, it is desirable to have an algorithm whose output results in a limited number of testable

METHODS IN ENZYMOLOGY, VOL. 403   0076-6879/05 $35.00
DOI: 10.1016/S0076-6879(05)03002-8

predictions that are tractable to experimental testing, as the ability to make predictions far outstrips the ability to address the outcomes experimentally. To facilitate implementation, the construction of the algorithm should be as user-controlled as possible so that the parameters and assumptions built into the algorithm can be critically evaluated prior to "wet-lab" work. Below, we describe the development of a predictive algorithm that attempts to identify, in a quantitative manner, those residues that are relevant for the individual functions of Rab proteins. The method makes use of the power of comparative genomics to provide a systematic comparison of Rab orthologs, and is combined with experimental data to generate quantitative predictions.

Algorithm Method

1. Rab proteins are structurally similar, so their functional specificity and unique properties must be determined by residues that vary among individual Rab proteins. Such residues can be qualitatively identified through global homology alignments as illustrated in Chapter 1 (this volume); however, homology alignments are very inclusive, namely, they can be used to eliminate residues from consideration, but this leaves too large a number of potential specificity-determining residues to be experimentally tractable. The problem is illustrated for the pair of Rab proteins, Ypt1p and Sec4p, that regulates the activity of the early and late secretory pathway in *Saccharomyces cerevisiae*. Ypt1p is the closest paralog to Sec4p and vice versa. However, Ypt1p and Sec4p are each encoded by essential genes, and even when expressed at high levels cannot complement function in the absence of the other. A pairwise alignment of the Rab GTPases Sec4p and Ypt1p reveals 99 identical positions and 119 positions containing divergent amino acids (Fig. 1), among which presumably lie the residues that provide unique Sec4p and Ypt1p function.

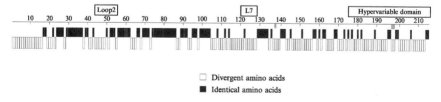

FIG. 1. Amino acid sequence comparison of Sec4p and Ypt1p. Loop2, Loop7, and hypervariable domains are indicated. The numbering is for Sec4p (gray box indicates gap in the alignment). Identical amino acids and divergent amino acids are indicated with black and white marks on the top and bottom rows beneath the ruler, respectively.

The algorithm we have developed is based on the method of Casari *et al.* (1995) who take a theoretical approach to classify residues according to their potential to provide specific function ("tree-determining" residues), without making any prior assumptions as to their positional location. This output of this method generates a large number of potentially important functional residues and can be combined with other considerations to create a smaller list of possibilities that can be experimentally addressed. One adaptation is to incorporate structural considerations, such as solvent accessibility and positions responsive to the nucleotide status of the GTPase (Bauer *et al.*, 1999). We have made use of the adaptation of Heo and Meyer (2003), who incorporated experimental data to provide functional classifications. The main assumption of this algorithm, illustrated in Fig. 2, is that homologous proteins in two different functional classes have diverged from an ancestral gene. Each amino acid position

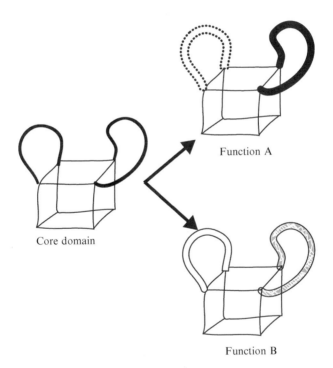

Function A

Core domain

Function B

Fig. 2. Schematic illustrating evolutionary assumption of algorithm. The algorithm assumes that homologous proteins that share a common catalytic mechanism and overall fold but that have different cellular functions have diverged from a common ancestor. Residues, or loops that do not participate in the catalytic mechanism (black loops of the core domain), have evolved to provide unique functions A or B.

relevant for the divergent function of two classes has been under selective pressure to remain conserved within one functional class and to become divergent in the other functional class. The algorithm then calculates a "conservation distance" and a "divergence distance" for each amino acid position.

2. Each of the two functional classes is named according to convenience; in this example we use "Ypt1p group" and "Sec4p group." Each group contains sequences that have been experimentally determined to provide the function of either class. For Ypt1p and Sec4p, we define function as the ability to complement a deletion of the gene, or to restore functionality in cells containing a thermosensitive allele of the gene. Such sequences have been acquired through a meta-analysis, with a literature search of genes that have been documented to functionally substitute for either *YPT1* or *SEC4* (Clement *et al.*, 1998; Dietmaier *et al.*, 1995; Dumas *et al.*, 2001; Fabry *et al.*, 1993; Haubruck *et al.*, 1989; Pertuiset *et al.*, 1995; Saloheimo *et al.*, 2004). Accession numbers for the "Ypt1p group" are AAF33844, CRU13168, L08128, and TRE277108; and for the "Sec4p group" are AF015306, CLI272025, RYL1_YARLI, and VVCYPTV1. A global alignment file is then created using Clustal X (Thompson *et al.*, 1997) with the core groups of amino acid sequences and sequences related to both Ypt1p and Sec4p that were identified through BLAST searches (total number of sequences = 114).

3. The global alignment file is then perused for potentially relevant residues. We include amino acid positions as potentially relevant for consideration only if they are conserved for all members in one core group and divergent from all members in the other core group (and vice versa). The conservation criteria used are amino acid identity, conservation of charge, and conservation of aromatic amino acids. An example demonstrating the residues considered potentially relevant is shown in Fig. 3A where the alignment of Loop2 between Ypt1p and Sec4p is shown. Out of a total of 12 amino acids, 7 are divergent between the two sequences and thus are potentially relevant amino acids. As can be readily appreciated, this is an impractical number of combinations to check experimentally. However, with the inclusion of the core groups, shown in Fig. 3B, it is immediately apparent, for instance, that residue S48 of Sec4p, which might otherwise attract attention because it encodes a neutral residue whereas the corresponding residue of Ypt1p is charged, is most likely insignificant. The reason for this is that in the other sequences that encode Ypt1p function (Haubruck *et al.*, 1990), the corresponding residue is identical to that of Sec4p. Similar considerations would suggest that Sec4p N46 is most probably not functionally relevant and also perhaps K44 because the corresponding residue is not conserved among members of each group.

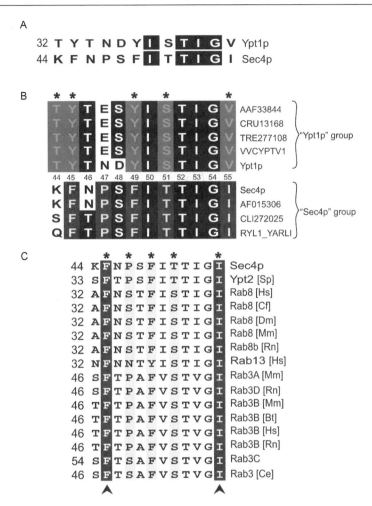

FIG. 3. (A) Alignment of the Loop2 region of Sec4p and Ypt1p. The identity of 7 amino acids differs out of a total of 12 possible positions. (B) Alignment of the Loop2 region of Sec4p and Ypt1p together with other members of the Sec4p and Ypt1p groups. Amino acids considered potentially significant for Sec4p function are shaded blue and those significant for Ypt1p function are shaded red and are indicated with an asterisk; the algorithm will subsequently rank these positions according to conservation and divergence distances. (C) Qualitative overview of position ranking. The Loop2 region of Sec4p is aligned together with Sec4p-related sequences. The position of potentially significant residues is indicated with shading; of these positions, F45 and I55 (shaded blue and indicated with arrows) are strictly conserved and P47, F49, and T51 are less conserved. The algorithm will rank these residues and others over the entire Sec4p sequence according to conservation, chemical identity, and divergence from the Ypt1p sequence. (See color insert.)

4. The identification of potentially relevant residues reduces the complexity in the search for residues defining specific function. However, even for the example shown of the Loop2 region, the complexity is only reduced from 7 to 5 potentially relevant positions. The algorithm then proceeds to rank the potentially relevant residues in order of predicted importance. The complete Clustal X (1.83) alignment file includes Ypt1p and Sec4p-related sequences whose ability to functionally complement either Sec4p or Ypt1p is not known. These sequence orthologs are used to ascertain the conservation/divergence distances of the residues classified as potentially significant from the core group. This homology file is imported into the PHYLIP package in order to calculate a matrix of relatedness values between sequences using the ProtDist program (Retief, 2000). We calculate the conservation and divergence distances by sorting all of the known related proteins available in the database (identified through BLAST searches) according to their sequence homology to the first core group and by sorting a second time according to their homology to the second core group. The conservation distance is then measured for each amino acid position by identifying the first sequence in the list where that sequence and the two sequences below contain a divergent amino acid at this position (defined as the ProtDist value between this first divergent protein and any member of the first core group). Similarly, the divergence distance for each position was determined by using the sequences sorted according to their relatedness to the second core group and by finding the first sequence with a conserved amino acid (again measured as a ProtDist value). The algorithm then assigns a quantitative value to these positions by making the assumptions that a given position is likely relevant for a selective function when (1) both the conservation to the core group and divergence distance from the second core group are large, (2) when conserved residues exist at the same position in both and not only one of the core groups, and (3) if the change in amino acids between the two groups involves a change in charge or aromatic amino acid. These assumptions can be addressed by varying among input parameters assigned by the user. The ranking process is qualitatively illustrated in Fig. 3C. Residues identified as potentially significant for Sec4p function in Fig. 3B are inspected in the global alignment of Sec4p-related sequences to establish their conservation. What stands out from a perusal of the alignment is that Sec4p residues F45 and I55 are identical in all the Sec4p-related Rab protein sequences and a different amino acid is identical in every Ypt1p homolog (not shown). Even though the substitutions between Sec4p and Ypt1p in these positions would be considered conservative ones, the alignment of sequences from orthologous organisms, together with the experimental knowledge of their function, suggests that the differences are potentially significant ones.

Algorithm Implementation

Clearly manual implementation of the algorithm is possible for any particular protein pair using applications such as MatLab (Mathworks Inc.). However, a more automated implementation is desirable to reduce the tedium of individual manipulations. The algorithm has been automated with an implementation in Java (CodeWarrior, Metrowerks), the ultimate goal being to port the algorithm as an independent application on different platforms. Once coding, debugging, and testing are completed, the plan is to share the application according to established guidelines (http://www. nap.edu/books/0309088593/html/4.html), for potential use in the study of other large protein families and domains.

Algorithm Output

Figure 4 shows the algorithm output for the example of Sec4p and Ypt1p. Four out of 10 positions with the highest values are located in the Loop2, or effector region that has been previously identified to be important for Sec4p function (Brennwald and Novick, 1993; Dunn *et al.*, 1993). In addition, we did not find any residues with high specificity determining values contained within Loop7, which experimental data suggest is not important for the specific function of Sec4p (M. Nussbaum and R. Collins, unpublished data). These results indicate the predictive power of the algorithm to identify functionally important residues. However, one implication of this method is that the functionally important residues are "transplantable," namely, that these residues can be swapped between protein pairs to generate "switch-of-function" proteins, which we have not found to be the case for Sec4p and Ypt1p. The evolutionary basis of the algorithm assumes that proteins have evolved to acquire specific functionalities; however, it is also probable that some proteins have differentiated in order to lose or avoid functions; thus, transplantation to provide "switch-of-function" proteins may not work in all instances.

Comments and Cautions

When evaluating output, it is important to bear in mind that any algorithm is reliant on the quality of its assumptions, the number of parameters, and the quality of the input data. Are the Rab sequence representations in the database biased toward one of a particular experimental pair? Clearly the sequences that comprise the homology alignments of the two groups must be evenly matched. The algorithm is also extremely sensitive to the quality of the overall homology alignment. A structural

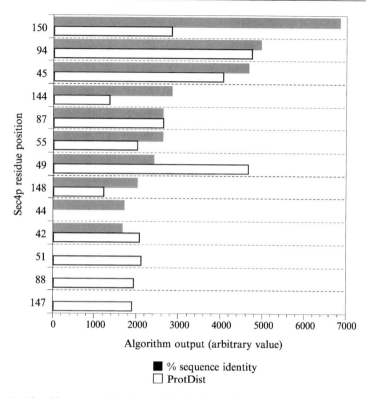

FIG. 4. Algorithm output showing predicted values of top scoring specificity determining amino acid positions generated according to sequence distances calculated either with ProtDist or % sequence identity methods.

analysis of Rab proteins reveals that conserved hydrophobic triad residues point their side chains at different angles in relation to the strands of the core (Merithew *et al.*, 2001). These angle shifts create very distinct surfaces of related GTPases. This implies that although Rabs have a related overall shape and fold, the angles of internal packing are used to create distinct recognition surfaces that may not be easily predicted through primary amino acid sequence homology-based methods. Improvements in the predictive power of methodologies such as described here will arise from incorporation of our understanding of three-dimensional structural inputs and other experimental data. Structural snapshots are also helpful to understand and provide a molecular explanation of how amino acid changes contribute to unique functionalities. Another source of variation is in the method used to generate the matrix containing values for the sequence relationships of the protein orthologs. We currently use the ProtDist program from PHYLIP to calculate these values; however, implementation of the algorithm with a different method of calculation (% identity) does yield slightly varying outputs (Fig. 4). The algorithm also ignores potential posttranslational modifications. For example, the serine residue 48 of Sec4p was not included as potentially significant even though the residue in the equivalent position for Ypt1p is aspartic acid, because other sequences with Ypt1p function also contain serine in this position. However, a hypothetical possibility is that the residue at this position is selectively phosphorylated for proteins with Ypt1p, but not Sec4p function and therefore does contribute to specific Ypt1p function. Another assumption is that conserved residues existing at the same position in both and not only one of the core groups are more significant than conserved residues within a single group. However, different, nonoverlapping regions of the protein may be dominant for different functions, making this assumption perhaps too simplistic. In summary, experimental verification of an algorithm's output is the critical (an often rate-determining) step in determining the predictive power of a particular procedure. The algorithm we describe uses a combination of homology searches and experimental inputs to provide quantitative evaluations and provides a useful starting point for experimentation.

Acknowledgments

This work is supported by grants from the U.S. National Science Foundation and U.S. National Institutes of Health. M. Nussbaum acknowledges the support of the Research Apprenticeship in the Biological Sciences program for high school students.

References

Bauer, B., Mirey, G., Vetter, I. R., Garcia-Ranea, J. A., Valencia, A., Wittinghofer, A., Camonis, J. H., and Cool, R. H. (1999). Effector recognition by the small GTP-binding proteins Ras and Ral. *J. Biol. Chem.* **274,** 17763–17770.

Brennwald, P., and Novick, P. (1993). Interactions of three domains distinguishing the Ras-related GTP-binding proteins Ypt1 and Sec4. *Nature* **362,** 560–563.

Casari, G., Sander, C., and Valencia, A. (1995). A method to predict functional residues in proteins. *Nat. Struct. Biol.* **2,** 171–178.

Clement, M., Fournier, H., de Repentigny, L., and Belhumeur, P. (1998). Isolation and characterization of the Candida albicans SEC4 gene. *Yeast* **14,** 675–680.

del Sol Mesa, A., Pazos, F., and Valencia, A. (2003). Automatic methods for predicting functionally important residues. *J. Mol. Biol.* **326,** 1289–1302.

Dietmaier, W., Fabry, S., Huber, H., and Schmitt, R. (1995). Analysis of a family of ypt genes and their products from Chlamydomonas reinhardtii. *Gene* **158,** 41–50.

Dumas, B., Borel, C., Herbert, C., Maury, J., Jacquet, C., Balsse, R., and Esquerre-Tugaye, M. T. (2001). Molecular characterization of CLPT1, a SEC4-like Rab/GTPase of the phytopathogenic fungus Colletotrichum lindemuthianum which is regulated by the carbon source. *Gene* **272,** 219–225.

Dunn, B., Stearns, T., and Botstein, D. (1993). Specificity domains distinguish the Ras-related GTPases Ypt1 and Sec4. *Nature* **362,** 563–565.

Fabry, S., Jacobsen, A., Huber, H., Palme, K., and Schmitt, R. (1993). Structure, expression, and phylogenetic relationships of a family of ypt genes encoding small G-proteins in the green alga Volvox carteri. *Curr. Genet.* **24,** 229–240.

Haubruck, H., Prange, R., Vorgias, C., and Gallwitz, D. (1989). The ras-related mouse ypt1 protein can functionally replace the YPT1 gene product in yeast. *EMBO J.* **8,** 1427–1432.

Haubruck, H., Engelke, U., Mertins, P., and Gallwitz, D. (1990). Structural and functional analysis of ypt2, an essential ras-related gene in the fission yeast Schizosaccharomyces pombe encoding a Sec4 protein homologue. *EMBO J.* **9,** 1957–1962.

Heo, W. D., and Meyer, T. (2003). Switch-of-function mutants based on morphology classification of Ras superfamily small GTPases. *Cell* **113,** 315–328.

Merithew, E., Hatherly, S., Dumas, J. J., Lawe, D. C., Heller-Harrison, R., and Lambright, D. G. (2001). Structural plasticity of an invariant hydrophobic triad in the switch regions of Rab GTPases is a determinant of effector recognition. *J. Biol. Chem.* **276,** 13982–13988.

Pertuiset, B., Beckerich, J. M., and Gaillardin, C. (1995). Molecular cloning of Rab-related genes in the yeast Yarrowia lipolytica. Analysis of RYL1, an essential gene encoding a SEC4 homologue. *Curr. Genet.* **27,** 123–130.

Retief, J. D. (2000). Phylogenetic analysis using PHYLIP. *Methods Mol. Biol.* **132,** 243–258.

Saloheimo, M., Wang, H., Valkonen, M., Vasara, T., Huuskonen, A., Riikonen, M., Pakula, T., Ward, M., and Penttila, M. (2004). Characterization of secretory genes ypt1/yptA and nsf1/nsfA from two filamentous fungi: Induction of secretory pathway genes of Trichoderma reesei under secretion stress conditions. *Appl. Environ. Microbiol.* **70,** 459–467.

Thompson, J. D., Gibson, T. J., Plewniak, F., Jeanmougin, F., and Higgins, D. G. (1997). The CLUSTAL_X windows interface: Flexible strategies for multiple sequence alignment aided by quality analysis tools. *Nucleic Acids Res.* **25,** 4876–4882.

[3] Application of Phylogenetic Algorithms to Assess Rab Functional Relationships

By Ruth N. Collins

Abstract

Researchers looking to solve biological problems have access to enormous amounts of sequence information and the desktop computational infrastructure to personally interrogate and analyze large datasets. Many powerful bioinformatics tools are available online; however, this discourages the customized analysis of data that is necessary for the experimental scientist to make maximally effective use of the information. In

METHODS IN ENZYMOLOGY, VOL. 403
0076-6879/05 $35.00
DOI: 10.1016/S0076-6879(05)03003-X

addition, a customized environment facilitates the critical evaluation of bioinformatic methods. This chapter presents a protocol developed to aid in classification of subfamilies and subclasses of a superfamily using the personal desktop computer. The visual representation of the qualitative and quantitative results of data analyses is also considered. The examples are focused on Rab GTPases but are more widely applicable to the classification of any given protein family.

Introduction

Protein sequence search algorithms are powerful tools in bioinformatics. They are used to identify functional relationships, define subgroups of a large protein families, and dissect the relationship between structure and function at the molecular level. The explosion of publically accessible databases and the revolution in desktop computing power have put the experimental biologist in the pilot seat, facilitating the collection and analysis of sequence data. For the experimental biologist who is most cognizant of the biological questions, these *in silico* methods offer valuable tools to examine hypotheses and, importantly, can also help narrow the range of experimental options. An issue commonly encountered is that to take full advantage of phylogenetic algorithms and to critically evaluate their output necessitate a certain familiarity with statistics and computational science; adequate integration of these topics remains an issue in the training of biomedical research scientists in cell and molecular biology (Bialek and Botstein, 2004). Informatic analyses of large datasets also present issues regarding the optimal way to envisage the output. For maximum clarity, a visual presentation is preferable and the example below outlines a simple method that generates a graphic representation of a large phylogenetic dataset to identify subclasses of the Rab GTPase family.

Generation of Protein Sequence Alignment

1. The first stage is to mine publically available sequence information to generate a customized database containing all the protein sequences of interest. Using a web browser, navigate to the BLAST (Altschul *et al.*, 1990) website (http://www/ncbi.nlm.nih.gov/BLAST/) and search for sequences related to the sequence of interest. Visually inspect the BLAST results and select sequences for downloading, taking care to avoid duplicates. Select sequences that are more distant to the original search sequence and use these sequences as seeds for further BLAST searching, a process termed sequence space hopping. Check both single pass methods

and pattern-based searching such as PSI-BLAST (Altschul *et al.*, 1997). The goal is to obtain as many sequences as possible in the twilight zone of 20–30% sequence similarity and avoid false positives.

2. It is not unusual for some database records to be of low accuracy and it is advisable to manually inspect each sequence and, where possible, to correlate the file information with EST databases and check the predicted splice sites of hypothetical ORFs generated from genomic sequencing.

3. Clustal X is a general purpose multiple alignment program for DNA or proteins that uses a window interface for sequence input and display (Chenna *et al.*, 2003; Thompson *et al.*, 1997). Clustal X is open source software that is available for different platforms, and a compiled version for the Mac OS X can be downloaded from http://www.embl.de/~chenna/clustal/darwin/. The quality of the alignment is critical and needs to be fine-tuned manually through the examination and modification of penalties in different regions of the alignment.

4. Once the multiple sequence alignments have been obtained, the next step is to convert the alignments into a matrix where each value is a score that represents the sequence relationship between all the sequences. As the proteins are roughly of similar overall length, one method for doing this is to create a percent identity matrix based on the alignment. MegAlign (DNAStar, Madison, WI), a commercial implementation of the Clustal W algorithm, can perform this calculation to one decimal point. Alternatively, the bioinformatics group at Cornell has created a web page where the clustal file can be uploaded to return the percent identity matrix (http://ser-loopp.tc.cornell.edu/cbsu/align_convert.htm). Generating sequence identity, however, does not take into account the chemical nature of amino acids and the detailed knowledge that has been accumulated regarding amino acid mutation frequencies. A better scoring method is to use a weighted model of amino acid replacement to create a sequence similarity table. One method for doing this is to output the Clustal alignment file in PHYLIP format, which then can be used for input to the PHYLIP ProtDist program to generate the pairwise distances between aligned sequences (http://evolution.genetics.washington.edu/phylip.html).

Principal Components Analysis of Alignment Data

5. The matrix containing the sequence relationship values is then analyzed by principal components analysis (PCA). PCA is, in general, a multivariate statistical method for finding linear combinations of variables that can be grouped together and specified by a single variable or component. Standard computer programs are available to do these types of

analyses, and the statistical toolbox from MatLab (The Mathworks, Inc.) is recommended. In theory, there are the same number of principal components as there are variables, but in practice, usually only a few of the principal components need to be identified to account for most of the data variance. The goal in this analysis is to take a large dataset with n variables and find a small number of principal components (pcp) that encompass most of the data variability.

6. Typically, the sum of the variances of the first few principal components exceeds 80% of the total variance of the original data and the pcp output can then be plotted. In the example in Fig. 1, a two-dimensional plot of the second and third principal components enables the visualization of the sequence variability among 560 unique Rab GTPase sequences, culled from a rigorous analysis of the NCBI database in early 2004. Notably, this analysis identifies 10 major groups or subclasses of Rab proteins. These subclasses contain orthologs among different species, while paralogs within a species are distributed among groups. The area of the plot where $x > 0.02$ and $y > 0.02$ shows many sequences that do not fall into clusters and is shown in greater detail in Fig. 2A. Figure 2A indicates that in the most highly studied examples, the data points in this region are derived from Rab proteins known to regulate exocytic function, as the area is bounded by subclasses of sequence groups that include Rab27, Rab3, and Rab8. This analysis might suggest that the outliers in the plot (identified in Fig. 2B) may have a common involvement in exocytic processes, whether ubiquitous, or regulated traffic with vesicles of either biosynthetic or lysosomal/endocytic origin. One implication evident from the plot is that a major evolutionary divergence in Rab sequences lies in their regulation of the exocytosis of specialized organelles. Such organelles are found in differentiated cells of higher eukaryotes (e.g., exocytosis of lung surfactant from type II alveolar cells) and in single-celled eukaryotes such as apicomplexans (e.g., rhoptries of *Toxoplasma gondii*), but not in more streamlined model eukaryotes such as *Saccharomyces cerevisiae*, whose entire complement of Rab protein sequences can be grouped within the major subclasses of Rabs identified in Fig. 1.

7. An understanding of molecular mechanism requires a detailed analysis of residues contributing to variability. Therefore, a valid question is to understand the individual residues that are responsible for the clustering of the Rab subclasses. Casari *et al.* (1995) have made use of PCA to identify such discriminant residues where the principal components are calculated for the actual amino acid identities in the homology alignment. The problem is depicted in Fig. 3, which shows an alignment of consensus sequences for the Ras, Rho, and Rab subfamilies of the Ras superfamily. The identity of conserved positions is indicated with uppercase bold

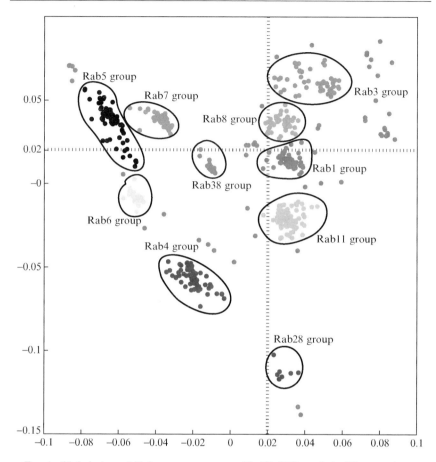

FIG. 1. Global view of Rab sequence space with 2D PCP analysis. The x and y axes represent the values of the second and third principal component, respectively. The analysis was performed on a database containing 560 individually checked and unique Rab sequences including each Rab protein identified in *S. cerevisiae*. Automatic clustering with the clusterdata function in MatLab was performed to identify groupings in the data. The 10 major groups that are color coded and named according to a representative mammalian member of the group. Dotted lines show the position of the cutoff values (0.02, −0.02) used to identify exocytic Rab sequences, shown in more detail in Fig. 2.

lettering, lowercase lettering indicates residues that vary within each subfamily, and the position of the nucleotide binding and hydrolysis motifs (universally shared among GTPases) is indicated with yellow highlighting. For reference, regions from a previous analysis predicted to define Rab family and subfamily function are indicated with red and pink highlighting,

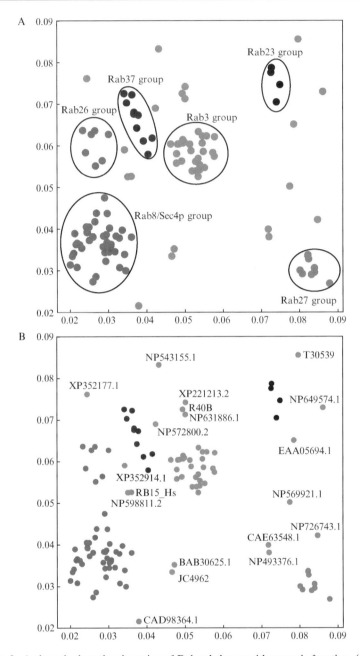

FIG. 2. A closer look at the clustering of Rab subclasses with exocytic function. (A) Rab sequences from Fig. 1 with PCP values above the cutoff of $x > 0.02$ and $y > 0.02$ have been

respectively (Pereira-Leal and Seabra, 2000). This analysis used HMM profiling and phylogenetic tree analysis to identify the major contiguous linear epitopes; each of these regions contains both residues that are conserved within the Ras and Rho families and also residues that differ among subfamilies and subclasses. From evolutionary considerations, residues that discriminate among subfamilies should be conserved within a particular subfamily, but be divergent between the subfamilies. In Fig. 3A, the residues conserved for each subfamily and shared with at least one other subfamily are highlighted with black shading. This is a reasonable assumption if the residue is conserved between all subfamilies but may be too simplistic an assumption in the light of dual specificity. Such residues would not be predicted to define an individual subfamily.

In contrast, the residues highlighted in blue shading in Fig. 3B represent conserved residues that are divergent between subfamilies and would be predicted to contain the information that defines the individual subfamilies. Those amino acids that define individual subclasses of a subfamily are expected to be divergent and not conserved in the consensus sequence. Such positions are highlighted with green shading in Fig. 3C. It is the variation of the particular residues highlighted in Fig. 3C that contributes to the subclass groupings displayed in Fig. 1. Figure 3 also shows a limitation of primary amino acid homology analysis, namely that it does not take the three-dimensional shape of the protein into account. In this representation, residues predicted to specify the function of a particular subfamily or subclass in general are distributed throughout the entire protein. However, one determinant of effector recognition for Rab proteins is the orientation of a set of three conserved residues (indicated with asterisks in Fig. 3A) that is dictated by the variation in residues at the hydrophobic core of the protein (Merithew et al., 2001). Four out of five of the core residues that participate in this variation (indicated with a § in Fig. 3C) are residues predicted by phylogenetic homology algorithms as potential subclass discriminant residues (green shaded positions in Fig. 3C). This example illustrates the predictive power of the phylogenetic homology study

---

plotted separately. These Rab sequences, where known, represent exocytic Rab function. Automatic clustering with the clusterdata function in MatLab was performed to generate groups or subclasses containing four or more members. The groups are color coded and named according to a mammalian representative of the group. Rab sequences that do not cluster into groups are collectively defined as a miscellaneous group (gray filled circles). (B) This diagram identifies the Rab proteins that are located into the space occupied by exocytic Rab proteins indicating the relative positions and accession number for each Rab protein of the miscellaneous group.

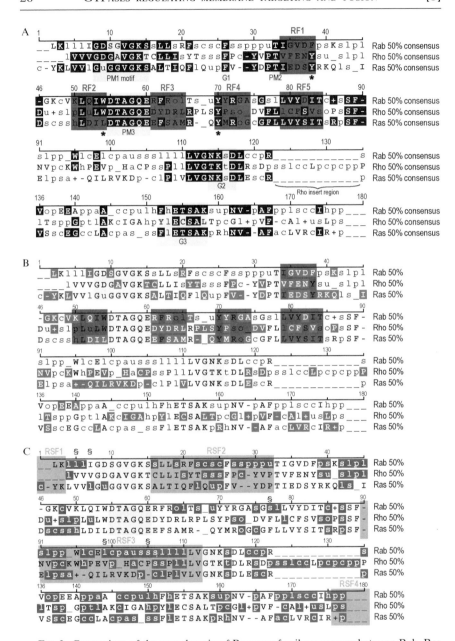

FIG. 3. Comparison of the core domain of Ras superfamily sequences between Rab, Ras, and Rho families. The core domain is aligned showing in uppercase bold letters, those residues conserved at the 50% consensus level (i.e., 50% or greater sequences) show this residue at the position indicated. Bold is also used for positions conserved for positive (+, H,

while illustrating the fact that structural information is required to interpret the molecular features identified by the homology search algorithm.

Scope and Comments

PCA is, in general, a method for finding linear combinations of variables that can be grouped together and specified by a single variable or component. One important point to note is that this method is very scalable, and many different variables can be included to make a multidimensional matrix. The variables can be theoretical such as global physicochemical parameters of the protein sequence, or raw experimental measurements such as microarray datasets (see Chapter 1). The combination of homology data with experimentally measured data is expected to create a redundancy of information that will be complementary and add reliability to the analysis. Note that the representation of the protein sequence homology data differs from the more common method of

---

K, R) or negative charge (−, D, E). In lowercase letters is shown the consensus sequence at nonconserved positions designated according to the amino acid class abbreviation; o (alcohol, S,T), l (aliphatic (I, L,V), a (aromatic, F, H,W,Y), c (charged, D,E,H,K,R), h (hydrophobic, A, C,F,G,H,I,K,L,M,R,T,V,W,Y), p (polar, C,D,E,H,K,N,Q,R,S,T), s (small, A,C,D,G,N,P,S,T, V), u (tiny, A,G,S), and t (turn-like, A,C,D,E,G,H,K,N,Q,R,S,T). The consensus sequence data were obtained from the SMART database (Schultz *et al.*, 1998) (http://smart.embl-heidelberg.de/) and are derived from 339 Ras domains, 460 Rho domains, and 1120 Rab domains. The location of the Rho insert region is marked; this insert is not contained in the Ras or Rab families. For greater clarification, the G protein-conserved sequence elements are shown highlighted in yellow. Numbering is arbitrary and intended as a descriptive guide. Highlighted in red are Rab family (RF) regions that have been proposed to uniquely distinguish the Rab subfamily of the Ras superfamily (Pereira-Leal and Seabra, 2000). Indicated below each row with an asterisk is a triad of conserved hydrophobic residues that provides structural plasticity in stabilization of the activated conformation of Rab3A and Rab5C (Merithew *et al.*, 2001). (A) Conserved positions among Rab/Rho/Ras subfamilies. In this representation, all residues that are conserved at the 50% consensus level within a subfamily and shared with at least one other subfamily member are shaded in black. An asterisk marks the positions of a triad of hydrophobic residues (position 38, 54, 70) that stabilizes the active conformation. (B) Signature motifs of Rab/Rho/Ras subfamilies. In this representation, all residues that are both conserved at the 50% consensus level within one of the subfamilies and unique within that subfamily are shaded in blue. (C) Subclass discriminant residues of Rab/Rho/Ras subfamilies. The alignment of core domains highlights nonconserved residue positions with green shading. Highlighted in pink are Rab subfamily (RSF) regions that have been proposed to identify subclasses of Rab proteins (Moore *et al.*, 1995; Pereira-Leal and Seabra, 2000). Indicated with a § above each row are the positions of amino acids that form part of the hydrophobic core between switch regions of the Rab3A and Rab5C GTPases (Merithew *et al.*, 2001) and that are predicted to be key specificity determinants as their packing in turn dictates the particular conformation of the invariant hydrophobic triad (see A). (See color insert.)

presentation in a dendrogram (Pereira-Leal and Seabra, 2001) and is complementary to such tree building methods. It should also be stressed that the results obtained with search algorithms are not static and need to be reevaluated with the ever-expanding datasets resulting from ongoing sequencing efforts.

Acknowledgments

Work in the author's laboratory is supported by the U.S. National Science Foundation and U.S. National Institutes of Health. E. Williams is thanked for helpful discussions and I. Berke for critical reading of the manuscript. This article is dedicated to P. Salvodelli.

References

Altschul, S. F., Gish, W., Miller, W., Myers, E. W., and Lipman, D. J. (1990). Basic local alignment search tool. *J. Mol. Biol.* **215,** 403–410.

Altschul, S. F., Madden, T. L., Schaffer, A. A., Zhang, J., Zhang, Z., Miller, W., and Lipman, D. J. (1997). Gapped BLAST and PSI-BLAST: A new generation of protein database search programs. *Nucleic Acids Res.* **25,** 3389–3402.

Bialek, W., and Botstein, D. (2004). Introductory science and mathematics education for 21st-century biologists. *Science* **303,** 788–790.

Casari, G., Sander, C., and Valencia, A. (1995). A method to predict functional residues in proteins. *Nat. Struct. Biol.* **2,** 171–178.

Chenna, R., Sugawara, H., Koike, T., Lopez, R., Gibson, T. J., Higgins, D. G., and Thompson, J. D. (2003). Multiple sequence alignment with the Clustal series of programs. *Nucleic Acids Res.* **31,** 3497–3500.

Merithew, E., Hatherly, S., Dumas, J. J., Lawe, D. C., Heller-Harrison, R., and Lambright, D. G. (2001). Structural plasticity of an invariant hydrophobic triad in the switch regions of Rab GTPases is a determinant of effector recognition. *J. Biol. Chem.* **276,** 13982–13988.

Moore, I., Schell, J., and Palme, K. (1995). Subclass-specific sequence motifs identified in Rab GTPases. *Trends Biochem. Sci.* **20,** 10–12.

Pereira-Leal, J. B., and Seabra, M. C. (2000). The mammalian Rab family of small GTPases: Definition of family and subfamily sequence motifs suggests a mechanism for functional specificity in the Ras superfamily. *J. Mol. Biol.* **301,** 1077–1087.

Pereira-Leal, J. B., and Seabra, M. C. (2001). Evolution of the Rab family of small GTP-binding proteins. *J. Mol. Biol.* **313,** 889–901.

Schultz, J., Milpetz, F., Bork, P., and Ponting, C. P. (1998). SMART, a simple modular architecture research tool: Identification of signaling domains. *Proc. Natl. Acad. Sci. USA* **95,** 5857–5864.

Thompson, J. D., Gibson, T. J., Plewniak, F., Jeanmougin, F., and Higgins, D. G. (1997). The CLUSTAL_X windows interface: Flexible strategies for multiple sequence alignment aided by quality analysis tools. *Nucleic Acids Res.* **25,** 4876–4882.

[4]   Application of Protein Semisynthesis for the
Construction of Functionalized Posttranslationally
Modified Rab GTPases

By ROGER S. GOODY, THOMAS DUREK, HERBERT WALDMANN,
LUCAS BRUNSVELD, and KIRILL ALEXANDROV

Abstract

Rab GTPases represent a family of key membrane traffic regulators in eukaryotic cells. To exert their function, Rab proteins must be modified with one or two geranylgeranyl moieties. This modification enables them to reversibly associate with intracellular membranes. *In vivo* the newly synthesized Rab proteins are recruited by Rab escort protein (REP) that presents them to the Rab geranylgeranyl transferase (Rab GGTase), which transfers one or two geranylgeranyl moieties to the C-terminal cysteines. Detailed understanding of the mechanism of prenylation reaction and subsequent membrane delivery of Rab proteins to the target membranes were hampered by lack of efficient technologies for the generation of preparative amounts of prenylated Rab GTPases. To circumvent this problem, we developed an approach that combines recombinant protein production, chemical synthesis of lipidated peptides with precisely designed and readily alterable structures, and a technique for peptide-to-protein ligation. Using this approach, we generated a number of semisynthetic prenylated Rab GTPases. Some of the proteins were also supplemented with fluorophores, which enabled us to develop a fluorescence-based *in vitro* prenylation assay. The approach described allows production of preparative amounts of prenylated GTPases, which was demonstrated by generation and crystallization of a monoprenylated YPT1:Rab GDI complex.

Introduction

Elucidation of protein function typically requires a combination of *in vivo* and *in vitro* approaches. In both cases only a fraction of the required information can be extracted using proteins in their native form. Commonly, modification with new functionalities is required to provide a detectable readout for the protein activity under study. Traditionally such modifications involve attachment of isotopic, fluorescent, affinity, or other functional groups to protein molecules. Although over the past century a very

METHODS IN ENZYMOLOGY, VOL. 403
0076-6879/05 $35.00
DOI: 10.1016/S0076-6879(05)03004-1

extensive toolbox of functional entities suitable for biological studies was developed, their targeted conjugation with proteins remains problematic. The problem becomes particularly acute when the desired functionalities have to be attached with a high degree of control to a protein in a site-specific manner, for instance, in the proximity of an active center or an interaction surface. These problems are related to the inapplicability of the otherwise powerful methods of organic chemistry to such large and diversely reactive molecules as proteins. The situation becomes even more dramatic when the proteins of interest undergo posttranslational modifications that are incommensurate with other, desired protein modifications or engineering since the modified proteins would not be recognizable by natural modification pathways. This situation applies to GTPases of the Rab family that function as key regulators of intracellular vesicular transport. Like most GTPases, Rab proteins are posttranslationally modified by geranylgeranyl isoprenoids covalently attached to their C-terminal cysteine(s) via a thioether linkage. This modification is essential for the ability of Rab proteins to associate with their target membranes and to exert their biological function (Pfeffer, 2001). Since geranylgeranylated GTPases are insoluble in water due to their high hydrophobicity, their production and engineering have been challenging. This chapter describes the construction of several variants of prenylated Rab GTPases either in the native form or modified with functional groups. These proteins have been used for structural studies and for the development of a novel *in vitro* prenylation assay.

Use of Expressed Protein Ligation for the Construction of Functionalized Semisynthetic GTPases

Principle

To generate preparative amounts of native or engineered monoprenylated or diprenylated Rab GTPases, we used a combination of chemical synthesis and expressed protein ligation (EPL) (Muir, 2003). Central to this method is the ability of certain protein domains (inteins) to cleave from an N-terminally fused protein of choice by combination of an N→S(O) acyl shift and a transthioesterification reaction with an added thiol reagent, thus leaving a thioester group attached to the C-terminus of the desired N-terminal protein fragment. This thioester group can then be used to couple essentially any polypeptide to the thioester-tagged protein by restoring a peptide bond in a reaction known as native chemical ligation, which is essentially a reversal of the cleavage steps (Dawson *et al.*, 1994). The only requirement for the ligation reaction is the presence of an N-terminal cysteine on the peptide or protein to be ligated.

Methods

Vector Construction, Protein Expression, Purification, and Ligation. We generated several expression vectors for C-terminal fusion of Rab7ΔC6, Rab7ΔC3, Rab3AΔC3, Rab27AΔC3, Ypt1ΔC2, Sec4ΔC2, etc. with inteins by polymerase chain reaction (PCR) amplifying the coding sequence of the GTPases with 3' oligonucleotides designed in such a way that the resulting cDNA encoded a truncated GTPase protein that could be fused to the N-terminus of the intein, which in turn was attached to a chitin-binding domain (CBD). The PCR products were subcloned into the pTYB1, pTWIN-1, or pTWIN-2 expression vectors (New England Biolabs) (Xu and Evans, 2001).

To obtain the GTPase-intein-CBD fusion protein, 1 liter of *Escherichia coli* BL21 cells transformed with the corresponding expression plasmid were grown to mid-log phase in Luria–Bertani medium and induced with 0.3 mM isopropyl-1-thio-D-galactopyranoside at 20° for 12 h. After centrifugation, cells were resuspended in 60 ml of lysis buffer (25 mM Na$_2$HPO$_4$/NaH$_2$PO$_4$, pH 7.2, 300 mM NaCl, 1 mM MgCl$_2$, 10 μM guanosine diphosphate [GDP], 1.0 mM phenylmethylsulfonyl fluoride [PMSF]) and lysed using a fluidizer (Microfluidics Corporation). After lysis, Triton X-100 was added to a final concentration of 1% (v/v). The lysate was clarified by ultracentrifugation and incubated with 9 ml of chitin beads (New England Biolabs) for 2 h at 4°. The beads were washed extensively with the lysis buffer and incubated for 14 h at room temperature with 40 ml of the cleavage buffer [25 mM Na$_2$HPO$_4$/NaH$_2$PO$_4$, pH 7.2, 300 mM NaCl, 1 mM MgCl$_2$, 10 μM GDP, and, depending on the GTPases, 50–500 mM 2-mercaptoethanesulfonic acid (MESNA)]. The supernatant containing GTPase-thioester was concentrated to a final concentration of 200 μM using Centripreps 10 (Amicon) and stored frozen at −80° until needed.

REP-1 was purified from baculovirus-infected insect cell lines as described (Alexandrov *et al.*, 1999). Yeast Rab GDI and Rab GGTase were purified as described previously (Kalinin *et al.*, 2001; Rak *et al.*, 2003).

Peptide Synthesis. Peptides C-K[5-dimethylaminonaphthalene-1-sulfonyl chloride) (Dans)]-S-C-S-C and C-K-(7-nitrobenz-2-oxa-1,3-diazol-4-yl)-C were synthesized and high-performance liquid chromatography (HPLC) purified to more than 90% purity by Thermo Hybaid (Ulm, Germany). The geranylgeranylated dipeptide H-Cys(StBu)-Cys(GerGer)-OH was synthesized using solution phase peptide chemistry (Rak *et al.*, 2003). Fmoc-Cys (S'Bu)-OH and geranylgeranylated cysteine were coupled under standard conditions employing *N*-hydroxysuccinimide and dicyclohexylcarbodiimid as coupling reagents to yield Fmoc-Cys(S'Bu)-Cys(GerGer)-OH (Brown *et al.*, 1991). Final Fmoc deprotection was performed by treatment with diethylamine/methylenechloride (1:4) to obtain the final dipeptide H_2N-Cys(S^{t}Bu)-Cys(GerGer)-OH in 70% yield. The N-terminal cysteine side chain remained protected until the ligation reaction, during which *in situ* deprotection occurred due to an excess of thiol reagent (MESNA).

For the construction of the prenylated peptides C-K(Dans)-S-C-S-C (GG), C-K(Dans)-S-C(GG)-S-C, and C-K(Dans)-S-C-(GG)-S-C(GG), we chose a block condensation strategy due to flexibility considerations (Alexandrov *et al.*, 2002; Durek *et al.*, 2004). For example, for construction of the C-K(Dans)-S-C-S-C(GG) peptide, tripeptide, Fmoc-Ser-Cys(StBu)-Ser-OAll was first deprotected at the C-terminus utilizing tetrakis(triphenylphosphine)palladium(0) with N,N'-dimethylbarbituric acid as a scavenger and coupled with the prenylated cysteine methyl ester, which was accessible via alkylation of cysteine methyl ester with geranylgeranylchloride in 2 N NH$_3$ (MeOH) (Brown *et al.*, 1991). Subsequent removal of the Fmoc-protecting group afforded tetrapeptide H-Ser-Cys(StBu)-Ser-Cys(GG)-OMe. This was then condensed with the Fmoc-Cys(StBu)-Lys (dan)-OH building block, and Fmoc removal with diethylamine finally resulted in the monogeranylgeranylated fluorescently labeled hexapeptide C-K(Dans)-S-C-S-C(GG)-OMe. A similar strategy was applied for the other peptides. Alternative approaches for the synthesis of the prenylated peptides (such as solid-phase techniques) were reported recently (Brunsveld *et al.*, 2005).

Before ligation, unprenylated peptides were dissolved to a final concentration of 50 mM in 25 mM Tris, pH 7.2, and 5% 3-[(3-cholamidopropyl)-dimethylammonio] propanesulfonate (CHAPS). The prenylated peptides were dissolved in dichloromethane/methanol (1:5) to a final concentration of ca. 50 mM.

Protein Ligation

Unprenylated Peptides. In the ligation reaction the thioester-activated Rab7 was mixed with the peptides in a buffer containing 25 mM Na$_2$HPO$_4$/NaH$_2$PO$_4$, pH 7.2, 300 mM NaCl, 500 mM MESNA, 1 mM MgCl$_2$, 5% CHAPS, and 100 μM GDP and allowed to react overnight at room temperature. The final concentrations were 200 μM and 2 mM for Rab7 and peptide, respectively. Unreacted peptide and detergent were removed by gel filtration of the reaction mixture on a Superdex-75 column (Pharmacia) equilibrated with 25 mM HEPES, pH 7.2, 40 mM NaCl, 2 mM MgCl$_2$, 10 μM GDP, and 2 mM 1,4-dithioerythritol (DTE). The extent of ligation was determined by sodium dodecyl sulfate polyacrylamide gel electrophoresis (SDS-PAGE) and mass spectrometry. For visualization of ligated fluorescent product, the reaction mixture was separated on a 15% SDS-PAGE gel and the gels were viewed in unfiltered UV light.

Prenylated Peptides. 500 μl of Rab-thioester protein (typically 20 mg/ml, ca. 500 nmol) in 25 mM Na$_2$HPO$_4$/ NaH$_2$PO$_4$, pH 7.2, 300 mM NaCl, 1 mM MgCl$_2$, and 100 μM GDP was supplemented with 50 mM cetyltrimethylammonium bromide (CTAB) and 125 mM MESNA (final concentrations). Ligation is initiated by adding 3.5–5 μmol of the respective peptide from a ca. 30 mM stock solution in dichloromethane/methanol (1:5). The reaction mixture was incubated overnight at 37° with vigorous agitation. It was then centrifuged and the supernatant removed. The pellet was washed once with 1 ml methanol, four times with 1 ml methylene chloride, four times with 1 ml methanol, and four times with 1 ml Milli-Q water at room temperature to resolubilize the contaminating peptide and unligated protein. The precipitate was dissolved in denaturation buffer (100 mM Tris-HCl, pH 8.0, 6 M guanidinium-HCl, 100 mM DTE, 1% CHAPS, 1 mM ethylenediaminetetraacetate [EDTA] to a final protein concentration of 0.5–1.0 mg/ml and incubated overnight at 4° with slight agitation. The solution was cleared by centrifugation or filtration. Protein was renatured by diluting it at least 25-fold drop wise into refolding buffer (50 mM HEPES, pH 7.5, 2.5 mM DTE, 2 mM MgCl$_2$, 10 μM GDP, 1% CHAPS, 400 mM arginine-HCl, 400 mM Trehalose, 0.5 mM PMSF, 1 mM EDTA) with gentle stirring at room temperature. The mixture was incubated for 30 min at the same temperature and was subsequently centrifuged to remove insoluble misfolded protein.

Complexation of Prenylated Rab GTPase with Rab Escort Protein-1 (REP-1) and Rab GDP Dissociation Inhibitor (Rab GDI) and Isolation of the Complexes. An equimolar amount of REP-1 or Rab GDI were added to the solution containing refolded protein, and the sample was incubated for 1 h on ice. The mixture was dialyzed over night against two 5–liter charges of dialysis buffer (25 mM HEPES, pH 7.5, 2 mM MgCl$_2$, 2 μM GDP, 2.5 mM DTE, 100 mM [NH$_4$]$_2$SO$_4$, 10% glycerol, 0.5 mM PMSF, 1 mM EDTA). The dialyzed material was concentrated to a protein concentration of 2–5 mg/ml using size-exclusion concentrators (molecular weight cutoff: 30 kDa) and loaded onto a Superdex-200 gel filtration column (Pharmacia) equilibrated with gel filtration buffer (25 mM HEPES, pH 7.5, 2 mM MgCL$_2$, 10 μM GDP, 2.5 mM DTE, 100 mM [NH$_4$]$_2$SO$_4$, 10% glycerol). Glycerol was omitted from this buffer for protein complexes intended for protein crystallization. The peak fractions containing the desired complex (judged by SDS-PAGE) were pooled, concentrated to approximately 10 mg/ml, and stored frozen at −80° in multiple aliquots. The typical recovery was 10–30% with respect to the starting Rab-thioester.

In Vitro *Prenylation of Rab7C-K(Dans)-S-C-S-C and Purification of the Rab7C-K(Dans)-S-C(GG)-S-C(GG): REP-1 Complex.* Protein complex formation and *in vitro* prenylation were performed in 1 ml of 40 mM

HEPES, pH 7.2, 150 mM NaCl, 5 mM DTE, 3 mM MgCl$_2$, and 0.3% CHAPS. A 1 ml mixture contained 50 μM REP-1, 55 μM Rab7C-K (Dans)-S-C(GG)-S-C(GG), 60 μM GST-Rab GGTase, and 500 μM geranylgeranyl pyrophosphate (GGpp). The sample was mixed, incubated at 37° for 3 min, diluted to 3 ml with the same buffer containing 5% CHAPS, and 500 μl of glutathione-Sepharose beads (Pharmacia) was added. The sample was incubated at 4° for 1 h on a rotating wheel. After the indicated period the supernatant was separated from the beads and concentrated in a Centricon 30 (Amicon) to a final volume of 300 μl. The sample was centrifuged in a bench top centrifuge for 5 min at 4° and loaded onto a 10/20 Superdex-200 gel filtration column (Pharmacia) driven by an FPLC system. The flow rate was 0.8 ml/min and fractions of 1 ml were collected and analyzed by SDS-PAGE followed by fluorescent scanning and Coomassie blue staining. Fractions containing the binary Rab7C-K(Dans)-S-C (GG)-S-C(GG):REP-1 complex were pooled, concentrated, and stored in multiple aliquots at −80°.

Fluorescence Measurements

Fluorescence measurements were performed with a Spex Fluoromax-3 spectrofluorometer (Jobin Yvon, Edison, NJ). Measurements were carried out in 1-ml quartz cuvettes (Hellman) with continuous stirring at 25°. For real-time monitoring of prenylation reactions, typically 50–100 nM of dansyl-labeled semisynthetic Rab7:REP-1 complex was mixed with an equal amount of Rab GGTase in a cuvette, containing 1 ml of buffer (50 mM HEPES, pH 7.2, 50 mM NaCl, 5 mM DTE, 2 mM MgCl$_2$, 100 μM GDP). Following a 5–min incubation at 25°, the reaction was initiated by adding GGpp to a final concentration of 10 μM. Excitation and emission monochromators were set to 280 nm and 510 nm, respectively. Data were fitted to a double exponential equation using GraFit 4.0 (Erithacus software).

Results

Choice of Expression Vector and Generation of Thioester Tagged Rab GTPases

For generation of C-terminally thioester tagged Rab GTPases, we tested pTYB-1, pTWIN-1, and pTWIN-2 vectors. In our experience, pTWIN-1 showed the best performance in terms of protein yield, protein solubility, and cleavage efficiency. In this vector a fusion protein consisting of the desired Rab GTPase and an intein-chitin binding domain assembly, fused in this order, is generated. The fusion GTPase is then isolated with chitin beads and the GTPase is cleaved off by addition of thiol reagent (MESNA in our case), which promotes the thiolysis of the polypeptide bond between the GTPase and the intein. The released protein has a thioester group at its C-terminus that can be used for the ligation reaction. Some GTPases precipitate at high concentrations of MESNA, so optimal conditions must be established experimentally for each protein.

We found that terminal Cys, Gln, and Asn residues located directly at the Rab C-terminus (former cleavage site) may lead to formation of byproducts arising from intramolecular side reactions between the thioester and the side chains of the mentioned amino acids (T. Durek, Y. Wu, and K. Alexandrov, unpublished). Therefore, mass spectrometric analysis of the protein to ascertain the presence of the intact thioester is advisable at this stage. The resulting thioester-tagged proteins can be stored for years at $-80°$.

Construction of Fluorescent Rab7CK(Dans)SC(GG)SC(GG):REP-1 Complex Using a Combination of Protein Ligation and In Vitro Prenylation

Construction of fluorescent, geranylgeranylated GTPases requires synthesis of prenylated peptides bearing the desired fluorophores and their subsequent coupling with the thioester–tagged protein. Although a very powerful approach, this has the drawback that it requires a high degree of chemical expertise and significant resources. Therefore, initially we chose a two-step procedure where first a nonprenylated fluorescent Rab GTPase was constructed using EPL and was then subjected to in vitro prenylation (Fig. 1A).

We used a fluorescently labeled peptide mimicking the truncated six amino acids of Rab7 to restore a full-length protein (Iakovenko et al., 2000). For in vitro prenylation, Rab7CK(Dans)SCSC was mixed with equimolar amounts of REP-1, GST-tagged Rab GGTase, and an approximately 10-fold molar excess of GGpp (Kalinin et al., 2001). Upon completion of the reaction, GST-tagged Rab GGTase was separated from the reaction mixture by precipitation with glutathione-Sepharose beads in the presence of 6% CHAPS. The presence of CHAPS is critical for this process, since it disrupts the interaction between Rab GGTase and the prenylated Rab7: REP-1 complex (Thoma et al., 2001a). To ensure the homogeneity of the obtained complex, as well as to separate it from the detergent, gel filtration chromatography was performed. As can be seen in Fig. 1B, Rab7CK(Dans) SC(GG)SC(GG):REP-1 eluted with a molecular mass of around 150 kDa

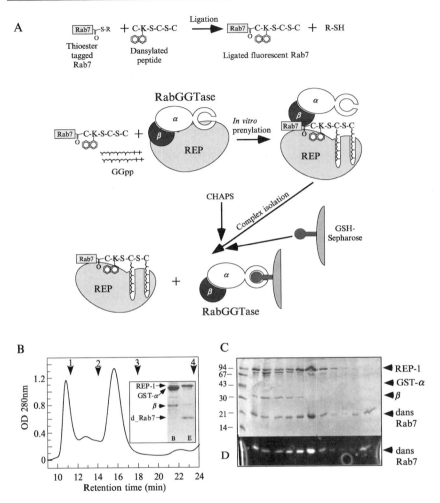

Fig. 1. (A) Schematic representation of the *in vitro* ligation and prenylation procedure for the generation of the Rab7CK(Dans)SC(GG)SC(GG):REP-1 complex. Step 1—ligation of the dansyl-containing peptide onto thioester–tagged Rab7. Step 2—*in vitro* prenylation of Rab7CK(Dans)SCSC with GST-Rab GGTase. Step 3—separation of GST-Rab GGTase using glutathione Sepharose. (B) Purification of the Rab7CK(Dans)SC(GG)SC(GG):REP-1 complex on a Superdex-200 10/20 column. Run conditions are described in Methods. Molecular weights of peaks were determined by comparison with protein standards of known molecular weight (1–670 kDa, 2–158 kDa, 3–44 kDa, 4–17 kDa) that are shown as arrowheads. The inset shows the protein distribution between the glutathione beads (B) and the eluate (E) in the loaded sample preparation. The collected fractions were subjected to SDS gel electrophoresis on 15% minigels, and proteins were visualized by fluorescent scanning (D) and Coomassie blue staining (C). Horizontal arrows denote the position of migration of REP-1, α and β subunits of Rab GGTase and Rab7 (right side), and the molecular mass markers (left side).

and was clearly separated from the minor peak of the ternary complex. The molecular weight and purity of the sample were monitored by MAL-DI-TOF mass spectroscopy. The observed molecular weight of 24191 Da closely matched the calculated mass of doubly geranylgeranylated Rab7CK(Dans)SCSC (24170 Da) (data not shown).

Construction of Monoprenylated YPT1 GTPase, Its Complexation to Yeast GDI, and Complex Crystallization

Attempts to determine the structure of the Rab:Rab GDI complex and related protein complexes have been hampered by failure to produce sufficiently large amounts of prenylated Rab GTPases, which precluded the generation of a sufficient diversity of Rab:GDI complexes necessary for successful crystallization trials. To obtain preparative amounts of preny-lated Rab GTPase, we ligated the Ypt1 protein truncated by two amino acids to a synthetic dipeptide, Cys-Cys(geranylgeranyl)-OH. This resulted in formation of native monoprenylated YPT1 GTPase. The reaction was carried out in the presence of 50 mM of the detergent CTAB, which for unknown reasons greatly facilitates the ligation of prenylated peptides (Durek *et al.*, 2004). Upon completion of the reaction, the detergent and the excess peptide have to be separated from the ligated protein. For this purpose, the reaction mixture was treated with dichloromethane, which results in complete protein precipitation and extraction of unligated pep-tide into the organic phase. The protein pellet was treated first with methanol and then with water and dissolved in 6 M guanidinium-HCl. The semisynthetic protein was refolded by diluting it in a buffer containing CHAPS, arginine-HCl, and trehalose, a procedure that is known to facili-tate the refolding process (De Bernardez *et al.*, 1999). At this stage Rab GDI was added in equimolar amounts and the complex was concentrated by ultrafiltration and further purified by gel filtration (Fig. 2). The complex between YPT1(GG):GDI eluting at the position corresponding to 50 kDa was concentrated to ca. 10 mg/ml and used for crystallization trials. Crystals were obtained at 20° using the vapor diffusion method in hanging drop setups (Rak *et al.*, 2003).

Construction of Fluorescent Monoprenylated and Diprenylated Rab7 GTPase and Development of a Fluorescence-Based Rab Prenylation Assay

Double prenylation of Rab GTPase by Rab GGTase is a multistep process that was proposed to follow a random sequential mechanism where one cysteine is somewhat preferred for the first prenylation (Shen and

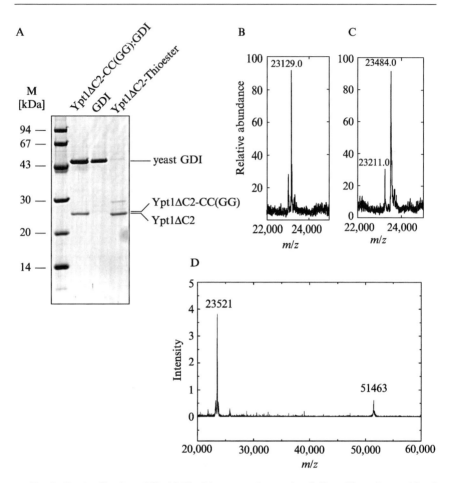

FIG. 2. *In vitro* ligation of Ypt1ΔC2 with a geranylgeranylated dipeptide and assembly of the Ypt1GG:Rab GDI complex. (A) Thioester-tagged Ypt1, yeast Rab GDI, and *in vitro* assembled Ypt1GG:Rab GDI complex resolved on an SDS-PAGE gel stained with Coomassie blue. (B) LC-ESI-MS spectrum of $^{1-204}$Ypt1Δ2-MESNA thioester ($M_{calc} = 23,131$ Da) and of (C) $^{1-204}$Ypt1Δ2-CC(GG) ($M_{calc} = 23,486$ Da). (D) MALDI-TOF-MS of the semisynthetic complex. The theoretical molecular mass of yRabGDI is 51,401 Da. The discrepancy of the determined values for the Ypt1Δ2-CC(GG) protein measured by ESI-MS (C, 23,484 Da) and MALDI-MS (D, 23,521 Da) is due to the inaccuracy of the MALDI spectrometer.

Seabra, 1996). However, in all cases monocysteine mutants of Rab proteins were used that provide only an approximation of the native situation. Moreover, the available reports dispute the exact sequence of isoprenoid addition (Shen and Seabra, 1996; Thoma *et al.*, 2001b). To clarify this point it would be necessary to generate two monoprenylated reaction intermediates and analyze the rates of their prenylation. Since *in vitro* prenylation is not a viable approach for generating these intermediates, we used EPL for generating the reaction intermediates and the double prenylated product, all carrying a dansyl label attached to the side chain of lysine 205. To this end we used a block condensation strategy for the synthesis of the three peptides: CK(Dans)SCSC(GG)-OMe, CK(Dans)SC(GG)SC-OMe, and CK(Dans)SC(GG)SC(GG)-OMe. The peptides were ligated to thioester tagged Rab7 truncated C-terminally by six amino acids (ΔC6) and the resulting protein was separated from excess peptide and detergent by extraction with organic solvents as described in the Methods section (Fig. 3A). The resulting protein was denatured in guanidinium–HCl and refolded in the presence of REP-1 protein added in stoichiometric amounts. This procedure resulted in formation of binary Rab:REP complexes that were further purified by gel filtration (Fig. 3D). MALDI-MS analysis of the resulting protein complexes revealed the expected molecular masses corresponding to monoprenylated (Fig. 3D) or diprenylated Rab7:REP-1 complexes (data not shown). We used an established *in vitro* corporation of [^{3}H]geranylgeranyl to confirm that the complexes obtained were indeed intermediates of the prenylation reaction (Seabra and James, 1998). As expected, both monoprenylated complexes could be further modified with one molecule of geranylgeranyl per molecule of Rab7 while the doubly prenylated complex showed only background incorporation of isoprenoid (Alexandrov *et al.*, 2002; Durek *et al.*, 2004).

The competence of the single prenylated semisynthetic Rab proteins to accept another isoprenoid group by Rab GGTase catalysis encouraged us to test whether the transfer of the prenyl group can be observed by fluorescence changes of the dansyl reporter group. When the single prenylated Rab7:REP-1 complexes ($\sim$75 nM) were mixed with equal amounts of transferase, formation of the ternary complex could be inferred from the strong fluorescence increase observed at 510 nm upon excitation at 280 nm (not shown). Addition of GGpp (10 μM) resulted in a time-dependent decline of the fluorescence signal. The observed reaction was essentially completed within 10 min and the time traces obtained could be fitted to a double exponential equation (Fig. 3C). Under the same conditions, the observed changes of the fluorescence amplitude of the singly prenylated Rab substrates were significantly larger than for the doubly prenylated Rab protein (Fig. 3C). The small fluorescence change observed in the latter case possibly represents low efficiency transfer of a third prenyl group onto the ligation site cysteine. The observed rate constants are in excellent agreement with values previously obtained using other assays (Thoma *et al.*, 2000).

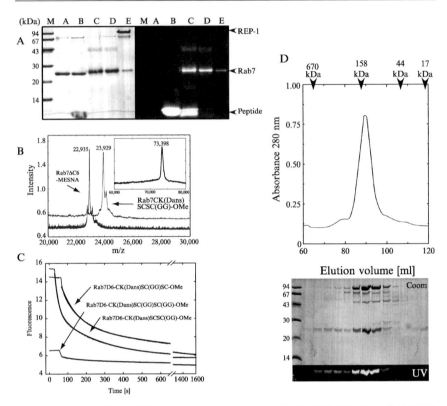

FIG. 3. Preparation of fluorescent monoprenylated Rab7. (A) Ligation of Rab7Δ6-MESNA thioester with CK(Dans)SCSC(GG)-OMe and purification of the semisynthetic Rab: REP-1 complex. SDS–PAGE gel loaded with thioester–tagged Rab7 (lane A), mixed with peptide (lane B), and following an incubation (lane C). Excess peptide was removed by washing with organic solvents (lane D). The protein was renatured, complexed to REP-1, and further purified by gel filtration (lane E). The gel was photographed while exposed to UV light (right) prior to Coomassie blue staining (left). (B) MALDI-MS analysis of $^{2-201}$Rab7Δ6-MESNA thioester (before ligation, $M_{calc} = 22,932$ Da) and the semisynthetic Rab:REP-1 protein complex [$^{2-201}$Rab7Δ6-CK(Dans)SCSC(GG)-OMe ($M_{calc} = 23,939$ Da)]. The inset shows the range from 60 to 80 kDa and the signal corresponding to REP-1 ($M_{calc} = 73,475$ Da). (C) Fluorescent *in vitro* prenylation assay: 75 nM of either Rab7Δ6-CK(Dans)SCSC (GG)-OMe, Rab7Δ6-CK(Dans)SC(GG)SC-OMe, or Rab7Δ6-CK(Dans)SC(GG)SC(66)-OMe in complex with REP-1 was incubated with 75 nM Rab GGTase. At the moment indicated by arrows 10 μM GGpp was added. (D) Preparative gel filtration of the Rab7Δ6-CK (Dans)SCSC(GG)-OMe:REP-1 complex using a Superdex-200 (26/60) column. The elution position of standard molecular weight markers is indicated by arrows. The analysis of the individual fractions by SDS-PAGE is shown below. The lower part (UV) represents the gel region corresponding to 20–30 kDa photographed in UV light prior to Coomassie blue staining (Coom).

Conclusions

Several lines of evidence suggest the physiological importance of Rab proteins in intracellular membrane transport. Nevertheless, the biochemistry of their function as well as the mechanism of their interaction with other components of the docking and fusion machinery remain largely unknown. The elucidation of Rab function requires the dissection of such interactions at the molecular level. This stresses the need for the development of sensitive biochemical assays for the study of such interactions and development of methods for production of Rab proteins in prenylated form. The semisynthesis-based methods of Rab GTPase engineering described in this chapter provide researchers with a number of tools for studying the intercommunications of Rab proteins with subunits of Rab GGTase and other molecules and provide a methodological platform for construction of fluorescent GTPases for *in vivo* experiments.

Acknowledgments

This work was supported in part by a grant of DFG number 484/5–3 to K. A. and Volkswagen Stiftung research Grant I/77 977 to R.S.G., H.W., and K.A.

References

Alexandrov, K., Heinemann, I., Durek, T., Sidorovitch, V., Goody, R. S., and Waldmann, H. (2002). Intein-mediated synthesis of geranylgeranylated Rab7 protein *in vitro*. *J. Am. Chem. Soc.* **124**(20), 5648–5649.

Alexandrov, K., Simon, I., Yurchenko, V., Iakovenko, A., Rostkova, E., Scheidig, A. J., and Goody, R. S. (1999). Characterization of the ternary complex between Rab7, REP-1 and Rab geranylgeranyl transferase. *Eur. J. Biochem.* **265**, 160–170.

Brown, M. J., Milano, P. D., Lever, D. C., Epstein, W. W., and Poulter, C. D. (1991). Prenylated proteins—a convenient synthesis of farnesyl cysteinyl thioethers. *J. Am. Chem. Soc.* **113**, 3176–3177.

Brunsveld, L., Watzke, A., Durek, T., Alexandrov, K., Goody, R. S., and Waldmann, H. (2005). Synthesis of functionalized Rab GTPases by a combination of solution- or solid-phase lipopeptide synthesis with expressed protein ligation. *Chemistry* **11**, 2756–2772.

Dawson, P. E., Muir, T. W., Clark-Lewis, I., and Kent, S. B. (1994). Synthesis of proteins by native chemical ligation. *Science* **266**, 776–779.

De Bernardez, C. E., Schwarz, E., and Rudolph, R. (1999). Inhibition of aggregation side reactions during *in vitro* protein folding. *Methods Enzymol.* **309**, 217–236.

Durek, T., Alexandrov, K., Goody, R. S., Hildebrand, A., Heinemann, I., and Waldmann, H. (2004). Synthesis of fluorescently labeled mono- and diprenylated Rab7 GTPase. *J. Am. Chem. Soc.* **126**, 16368–16378.

Iakovenko, A., Rostkova, E., Merzlyak, E., Hillebrand, A. M., Thoma, N. H., Goody, R. S., and Alexandrov, K. (2000). Semi-synthetic Rab proteins as tools for studying intermolecular interactions. *FEBS Lett.* **468**, 155–158.

Kalinin, A., Thoma, N. H., Iakovenko, A., Heinemann, I., Rostkova, E., Constantinescu, A. T., and Alexandrov, K. (2001). Expression of mammalian geranylgeranyltransferase type-II in *Escherichia coli* and its application for *in vitro* prenylation of Rab proteins. *Protein Exp. Purif.* **22,** 84–91.

Muir, T. W. (2003). Semisynthesis of proteins by expressed protein ligation. *Annu. Rev. Biochem* **72,** 249–289.

Pfeffer, S. R. (2001). Rab GTPases: Specifying and deciphering organelle identity and function. *Trends Cell Biol.* **11,** 487–491.

Rak, A., Pylypenko, O., Durek, T., Watzke, A., Kushnir, S., Brunsveld, L., Waldmann, H., Goody, R. S., and Alexandrov, K. (2003). Structure of Rab GDP-dissociation inhibitor in complex with prenylated YPT1 GTPase. *Science* **302,** 646–650.

Seabra, M. C., and James, G. L. (1998). Prenylation assays for small GTPases. *Methods Mol. Biol.* **84,** 251–260.

Shen, F., and Seabra, M. C. (1996). Mechanism of digeranylgeranylation of Rab proteins. Formation of a complex between monogeranylgeranyl-Rab and Rab escort protein. *J. Biol. Chem.* **271,** 3692–3698.

Thoma, N. H., Iakovenko, A., Kalinin, A., Waldmann, H., Goody, R. S., and Alexandrov, K. (2001a). Allosteric regulation of substrate binding and product release in geranylgeranyl-transferase type II. *Biochemistry* **40,** 268–274.

Thoma, N. H., Iakovenko, A., Owen, D., Scheidig, A. S., Waldmann, H., Goody, R. S., and Alexandrov, K. (2000). Phosphoisoprenoid binding specificity of geranylgeranyltransferase type II. *Biochemistry* **39,** 12043–12052.

Thoma, N. H., Niculae, A., Goody, R. S., and Alexandrov, K. (2001b). Double prenylation by RabGGTase can proceed without dissociation of the mono-prenylated intermediate. *J. Biol. Chem.* **276,** 48631–48636.

Xu, M. Q., and Evans, T. C., Jr. (2001). Intein-mediated ligation and cyclization of expressed proteins. *Methods* **24**(3), 257–277.

[5] Assay and Functional Properties of PrBP(PDEδ), a Prenyl-Binding Protein Interacting with Multiple Partners

By HOUBIN ZHANG, SUZANNE HOSIER, JENNIFER M. TEREW, KAI ZHANG, RICK H. COTE, and WOLFGANG BAEHR

Abstract

A 17-kDa prenyl-binding protein, PrBP(PDEδ), is highly conserved among various species from human to *Caenorhabditis elegans*. First identified as a putative regulatory δ subunit of the cyclic nucleotide phosphodiesterase (PDE6) purified from mammalian photoreceptor cells, PrBP (PDEδ) has been hypothesized to reduce activation of PDE6 by the heterotrimeric G-protein, transducin, thereby desensitizing the photoresponse. However, recent work shows that PrBP(PDEδ) interacts with

METHODS IN ENZYMOLOGY, VOL. 403
Copyright 2005, Elsevier Inc. All rights reserved.
0076-6879/05 $35.00
DOI: 10.1016/S0076-6879(05)03005-3

numerous prenylated proteins at their farnesylated or geranylgeranylated C-termini, as well as with nonprenylated proteins. These polypeptides include small GTPases such as Rab13, Ras, Rap, and Rho6, as well as components involved in phototransduction (e.g., rod and cone PDE6, rod and cone opsin kinases). Expression of PrBP(PDEδ) in tissues and organisms not expressing PDE6, the demonstration of multiple interacting partners with PrBP(PDEδ), and its low abundance in rod outer segments all argue against it being a regulatory PDE6 subunit. This raises intriguing questions as to its physiological functions. In this chapter, we review the current status of PrBP(PDEδ) and describe some of the assays used to determine these interactions in detail. In mammalian photoreceptors, the results are consistent with a role of PrBP(PDEδ) in the transport of prenylated proteins from their site of synthesis in the inner segment to the outer segment where phototransduction occurs.

Introduction

PrBP(PDEδ) Is a 17-kDa Prenyl-Binding Protein Closely Related to unc119 and RhoGDI

The crystal structure of PrBP(PDEδ) reveals an immunoglobulin-like β-fold structure in which two β-sheets form a hydrophobic pocket (Hanzal-Bayer *et al.*, 2002) into which prenyl groups can insert. The overall β-sandwich fold of PrBP(PDEδ) is similar to that of RhoGDI (Hoffman *et al.*, prenyl-binding protein involved in vesicular trafficking. Primary sequence analysis of PrBP(PDEδ) demonstrates that this protein is very highly conserved through evolution, with at least 70% sequence identity within vertebrates, and approximately 50% sequence identity when invertebrate sequences are included in the analysis. PrBP(PDEδ) is also similar to another highly conserved protein, unc119 (also referred to as RG4), whose function is uncertain (Higashide and Inana, 1999; Maduro and Pilgrim, 1995). Multiple sequence alignment of PrBP(PDEδ), unc119, and RhoGDI identifies 11 amino acid residues that are identical in all PrBP(PDEδ) and unc119 sequences. While RhoGDI sequences show very little homology with either of the other two proteins (Fig. 1), numerous residues within the geranylgeranyl-binding pocket of RhoGDI represent conservative substitutions of corresponding residues in PrBP(PDEδ). Highly conserved amino acids within each protein family and between PrBP(PDEδ) and unc119 are most frequently located within the β-sheets that comprise the hydrophobic prenyl-binding pocket. The high degree of sequence and/or structural homology of these three proteins suggests that the 17-kDa PrBP(PDEδ) is the smallest member of a superfamily of prenyl-binding proteins. While

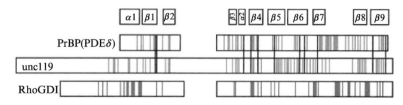

FIG. 1. Structural homology of PrBP/PDEδ, unc119, and RhoGDI. Multiple sequence alignment of PrBP(PDEδ) (8 vertebrate, 5 invertebrate sequences), unc119 (7 vertebrate, 5 invertebrate sequences), and RhoGDI (15 vertebrate, 4 invertebrate sequences) identified amino acids that are invariant within each protein family (vertical lines within boxes). Sites that are identical in all sequences of PrBP(PDEδ) and unc119 are shown as lines connecting PrBP(PDEδ) and unc119 boxes. Structural homology of the prenyl binding pocket shown at the top as α-helix and β-sheet is based on comparison of the three-dimensional structures for PrBP(PDEδ) (PDB 1KSH; (Hanzal-Bayer et al., 2002) and RhoGDI (PDB 1DOA; Hoffman et al., 2000).

PrBP(PDEδ), unc119, and RhoGDI all share the β-sandwich fold, the proteins with which each prenyl-binding protein interacts also depend on structural features distinct from the prenyl-binding pocket itself (see below). In addition to prenylated proteins, PrBP(PDEδ) can bind nonprenylated proteins through β-sheet interactions, particularly the Arf-like proteins Arl2 and Arl3. RhoGDI prefers the GDP-bound form of Rho family proteins, while PrBP(PDEδ) prefers Arl2 and Arl3 in the active form with GTP bound.

PrBP(PDEδ) Has a Wide Tissue Distribution

PrBP(PDEδ) was first characterized based on its copurification with a soluble form of rod and cone photoreceptor PDE (PDE6) from bovine retina extracts (Gillespie et al., 1989). Rod PDE6 consists of two catalytic subunits, PDEα and PDEβ (Baehr et al., 1979), and two identical inhibitory subunits PDEγ (Deterre et al., 1988; Fung et al., 1990). The catalytic α and β subunits are posttranslationally prenylated at their C-termini with farnesyl and geranylgeranyl groups (Anant et al., 1992; Qin and Baehr, 1994; Qin et al., 1992). While the bulk of rod PDE6 exists as a peripheral membrane protein, PrBP(PDEδ) can solubilize PDE6 from the rod outer segment disc membrane by direct interaction with the prenyl side chains, without affecting PDE6 catalytic activity (Florio et al., 1996; Mou et al., 1999). Based on its relative abundance in the retina (compared to other tissues) and its ability to release PDE6 from its membrane-associated state, it was hypothesized that PrBP(PDEδ) serves as a regulatory subunit of PDE6, functioning to reduce the activated lifetime of this enzyme during light adaptation of rod photoreceptors (Cook et al., 2001; Florio et al., 1996). However, unlike most proteins involved in the visual transduction pathway in rods and cones, PrBP(PDEδ) is present in nonretinal tissues that do not express photoreceptor PDE6 (Florio *et al.*, 1996; Marzesco *et al.*, 1998). PrBP(PDEδ) mRNA is found in all mammalian tissues analyzed to date, although it is present at highest levels in retina (Florio *et al.*, 1996). Furthermore, PrBP(PDEδ) orthologues are present in animals lacking a visual system, including the eyeless nematode *Caenorhabditis elegans* (Li and Baehr, 1998).

PrBP(PDEδ) Interacts with Many Proteins

Yeast two-hybrid (y2h) screening showed that PrBP(PDEδ) interacts with multiple partners. PrBP(PDEδ) was identified as a Rab13 binding partner when PrBP(PDEδ) was used as bait to screen a HeLa cDNA library (Marzesco *et al.*, 1998). The recombinant PrBP(PDEδ) has the capacity to dissociate Rab13 from membranes, most likely by interacting with the prenylated C-terminal tail of Rab13 (Marzesco *et al.*, 1998). Additional y2h screening showed that PrBP(PDEδ) was able to interact with ARF-like proteins, Arl2 and Arl3, which belong to the ARF small GTPase family (Hillig *et al.*, 2000; Linari *et al.*, 1999a). Y2h also showed that PrBP(PDEδ) can interact with the retinitis pigmentosa GTPase regulator (RPGR) (Linari *et al.*, 1999b), which is found in the connecting cilium of photoreceptors, suggesting a role in protein transport from the inner to outer segment (Zhao *et al.*, 2003). More recently, additional small GTPases of the Ras superfamily, including Ras and Rap (Nancy *et al.*, 2002), Rho6 and Rheb (Hanzal-Bayer *et al.*, 2002), and Rab8 (Norton *et al.*, 2005a), were added to the growing family of PrBP(PDEδ) interacting proteins.

PrBP(PDEδ) Interacting Proteins Identified by Yeast Two-Hybrid Screening

We prepared bovine tetina a yeast two-hybrid cDNA library in pGADT7 and screened it with a full-length bait cloned into pGBKT7. We found that a large number of colonies (>1000) were able to grow on selective medium, turning blue on selective THAL medium containing α-X-gal. We randomly picked 95 colonies, sequenced the inserts, and identified one clone encoding the C-terminal (183 amino acids) of Arl2 and three clones encoding the C-terminal of PDEα (50 amino acids). Seven clones represented partial C-terminal sequences of GRK1 (rhodopsin kinase); these seven clones corresponded to 165 (GRK1L) and 69 (GRK1S) amino acid residues at the C-terminus of GRK1, respectively. Two independent GRK1 clones encoding 165 and 69 amino acid residues of the GRK1 C-terminus, respectively, were cotransformed into yeast with bovine unc119/RG4, a retina protein that shares 23% identity with PrBP(PDEδ) in its C-terminal 153 amino acids (Higashide and Inana, 1999; Kobayashi *et al.*, 2003). We found that partial GRK1 polypeptides encoded by these two clones interacted only with PrBP(PDEδ), but not RG4/Unc119, suggesting that PrBP(PDEδ) specifically interacts with GRK1. We also cotransformed a library plasmid encoding GRK1L-SAAX (disabled CAAX box motif) into yeast with the PrBP(PDEδ) bait. The yeast cells did not grow on the selective THAL medium, suggesting that farnesylation is essential and required for the interaction between PrBP(PDEδ) and GRK1.

Methods

Construction of Bovine Retina Yeast Two-Hybrid cDNA Library. This procedure roughly follows the Two-Hybrid cDNA Library Construction kit User Manual (Clontech). In brief, fresh bovine eyes from a local vendor are dissected immediately, stored in liquid nitrogen, and bovine retina mRNAs are isolated using a FastTrack 2.0 Kit (In Vitrogen). Purified mRNA quality is estimated by 2.2 M formaldehyde/1% agarose gel electrophoresis. About 10 μg retina mRNA (1 OD$_{260nm}$ = 40 μg RNA) is used to construct the library. First-strand cDNAs are synthesized using random primers (ZAP Express cDNA Synthesis Kit, Stratagene). *Eco*RI adapters (Stratagene) are added to both ends of the double-stranded cDNAs. After the adapter ligation, the double-strand cDNAs are phosphorylated at the *Eco*RI sites (T4 polynucleotide kinase/ATP, Stratagene) and then purified using cDNA size-fractionating columns (CHROMA SPIN −400 Columns, Clontech), which remove unligated adapters, unincorporated nucleotides, and most short fragments (<300 bp). The purified, double-strand cDNAs are ligated into the *Eco*RI-digested, dephosphorylated pGADT7 vector. Transformation of the recombinant plasmids (bovine retina cDNAs in pGADT7 vector) into electrocompetent *Escherichia coli* DH5 α cells (Clontech) is performed using a Bio-Rad Gene Pulser (Bio-Rad). To determine the percentage of recombinant clones in the original library, the inserts from isolated independent clones are analyzed by *Eco*RI digestion. After amplification, the library can be stored in 25% glycerol at −80° without significant loss of titer.

Construction of PrBP(PDEδ) Bait and Library Screening. The full-length coding region of bovine PrBP(PDEδ) is amplified from the bovine retina library (above) by polymerase chain reaction (PCR) amplification. The forward primer is 5'-AAGAATTCATGTCAGCCAAGGACGAG-CG and the reverse primer is 5'-AAGGATCCTTCTCTCAAACGTA-GAAAAGCC. The PCR product is separated on 1% agarose gel. The amplified band around 450 bp is excised from the gel and purified using a gel purification kit (Qiagen). The purified DNA is digested with *Eco*RI and *Bam*HI and ligated into yeast two-hybrid bait vector pGBKT7 (Clontech) digested with *Eco*RI and *Bam*HI. The bait construct is verified by DNA sequencing. For yeast transformation, competent yeast cells are prepared using a LiAc TRAFO method (http://www.umanitoba.ca/faculties/medicine/biochem/gietz/2HS.html). An AH109 yeast colony is inoculated into 1 ml YPDA (YPD medium supplemented with 0.003% adenine hemisulfate) in a microcentrifuge tube and vortexed vigorously to disperse the cells. Then the cells are transferred to 4 ml YDPA in a test tube and grown at 30° overnight. The cell density of the overnight culture is evaluated using a hemocytometer. A proper amount of cells is transferred to 50 ml YPDA with a final cell density of 5×10^6/ml and grown at 30° until the cell density reached 2×10^7 cells/ml. The cells are harvested by centrifugation at $1500 \times g$ for 5 min at room temperature. The cells are washed once with 20 ml sterile water and resuspended in 1.0 ml 100 mM LiAC. The cells are transferred to a microcentrifuge tube and pelleted at $12,000 \times g$ for 15 s. The supernatant is discarded and the cells are resuspended in 100 mM LiAc with a final volume of 0.5 ml. Subsequently, the cells are aliquoted into 1.5-ml microcentrifuge tubes with 50 μl cells in each tube. The cells are pelleted and the LiAc was removed. A transformation mix (consisting of 240 μl of PEG 3350 is [50% w/v], 36 μl of 1.0 M LiAc, 50 μl of predenatured salmon sperm DNA [2.0 mg/ml], 5μg bovine library DNA, and 5 μg bovine PDEδ bait DNA) is added to each tube and the total volume in each tube is adjusted to 360 μl. The cells are completely dispersed by vigorously vortexing the tubes. The tubes are placed in a 30° water bath for 30 min followed by heat-shocking at 42° for 30 min. The cells are pelleted at $12,000 \times g$ for 15 s in microcentrifuge and resuspended in 300 μl in double-distilled H$_2$O and spread on synthetic dropout (SD) medium plates (150 mm) lacking tryptophan, histidine, adenine, and leucine (THAL medium). The plates are incubated at 30° for 4–5 days. The colonies growing up from the selective plates are transferred to the plates with THAL media but containing 0.2 mM α-X-gal. The colonies that turn blue are inoculated into 5 ml THAL media and grown for 5 days at 30°. The plasmid DNAs are isolated from the yeast cultures and transformed using electrocompetent *E. coli* DH5α cells. The transformants are selected by ampicillin. The plasmids isolated from the transformants are sequenced using T7 primers.

Immunoblot Detection of PrBP(PDEδ) in Retinal Homogenates and Rod Outer Segments (ROSs)

Quantitation of the PrBP(PDEδ) content of retina and ROS is determined by loading known amounts of PDE6 or rhodopsin along with purified, recombinant PrBP(PDEδ) standards on gels, and densitometrically scanning immunoreactive bands for the standards and unknowns. Using

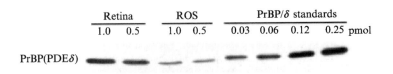

FIG. 2. Quantitative immunoblot analysis of PrBP(PDEδ) in amphibian retinal homogenates and in purified rod outer segment (ROS). Frog retinal homogenates or ROS samples containing either 1.0 or 0.5 pmol of PDE6 were compared to the indicated amounts of purified recombinant frog PrBP(PDEδ). After SDS-PAGE and transfer to PVDF, the membranes were probed with the FL antibody.

the FL antibody directed against recombinant bovine PrBP(PDEδ) protein, we observe a single 17-kDa immunoreactive band for retinal homogenates, purified ROS, and recombinant PrBP(PDEδ). The molar ratio of PrBP(PDEδ) to PDE6 in retina and in ROS can be quantified (Fig. 2). Using this approach, we find that substoichiometric levels of PrBP(PDEδ) relative to PDE6 are present in either amphibian or mammalian ROS (Norton *et al.*, 2005).

Methods

Preparation of Retinal Homogenates and Purified ROS. To prepare retinal homogenates, dark-adapted *Rana catesbeiana* retinas are isolated under infrared illumination (Cote, 2000) and homogenized directly in a homogenization buffer [50 mM Tris, pH 8.0, 150 mM NaCl, 2.0 mM dithiothreitol, 1% Triton X-100, and mammalian protease inhibitor cocktail (Sigma)] using a motor-driven pestle in a glass mortar. Following centrifugation (100,000×g, 5 min), greater than 90% of the visual pigment is recovered in the supernatant, as judged by rhodopsin difference spectroscopy (Bownds *et al.*, 1971). Retinal homogenates are added directly to Laemmli sample buffer (with DTT) for sodium dodecyl sulfate polyacrylamide gel electrophoresis (SDS–PAGE) analysis. (Freshly prepared homogenates are necessary to avoid high-molecular-weight immunoreactivity that may result from aggregation of PrBP[PDEδ] during storage.) ROS are dissociated from the retina by mechanical agitation of isolated retinas in an isotonic solution. Amphibian ROS are purified on discontinuous Percoll gradients (Cote, 2000), whereas bovine ROS are purified on discontinuous sucrose gradients (McDowell, 1993).

Immunoblot Analysis. Proteins are resolved by standard SDS–PAGE in 12% acrylamide gels and transferred to PVDF membrane (Immobilon P^{sq}, Millipore) at 60 V for 1–2 h (4°) under standard conditions (Towbin *et al.*, 1979). Membranes are blocked 1 h or overnight in 1% bovine serum albumin (BSA). (Transfer to PVDF membranes in combination with a BSA blocking buffer gives a greatly improved signal-to-noise ratio in comparison to nitrocellulose with other blocking buffers.) A rabbit poly-clonal antibody ("FL") raised to the bovine PrBP(PDEδ) protein in our laboratory (Norton *et al.*, 2005) is then incubated with the membrane for 1 h. After washing, the membrane is incubated with horseradish peroxidase-conjugated secondary antibody, washed, incubated with chemiluminescent substrate, and exposed to film. Under optimal conditions, PrBP(PDEδ) can be detected over a 100-fold range (0.01–1.00 pmol), but quantitation is limited to a ~10-fold range at a single film exposure.

Expression of GST-PrBP(PDEδ) Fusion Proteins, Purification of PrBP(PDEδ), and Binding to Immobilized PrBP(PDEδ)

Recombinant expression of PrBP(PDEδ) results in the recovery of 1–4 mg of affinity-purified protein per liter of bacterial culture. Thrombin cleavage of the GST-PrBP(PDEδ) fusion protein and removal of the GST result in a single 17-kDa band on protein gels.

Methods

Expression and Purification of GST-PrBP(PDEδ) Fusion Proteins. PrBP(PDEδ) from mouse (Zhang *et al.*, 2004), cow (Cook *et al.*, 2000), and frog (Norton *et al.*, 2005) have all been successfully subcloned into bacterial expression vectors, and the recombinant protein expressed as a GST fusion protein in *E. coli*. Expression constructs were verified by DNA sequencing prior to use.

MURINE. For expression of murine GST-PrBP(PDEδ), overnight cultures of *E. coli* ER2566 harboring the plasmid were diluted 1:100 into fresh LB and grown until the OD_{600} reached 0.6–0.8. 0.1 mM; isopropyl-β-D-thiogalacto-pyranoside (IPTG) was added, and the culture was incubated at 30° for 5 h. The cells were harvested, resuspended in phosphate-buffered saline (PBS) containing 0.5% Triton X-100, and disrupted by sonication. The lysate was cleared by centrifugation, and the GST-PrBP(PDEδ) was then purified.

BOVINE AND AMPHIBIAN. Bovine and amphibian PrBP(PDEδ) were ex-pressed as GST fusion proteins using the pGEX-KG and pGEX-2T expres-sion vectors, respectively, transformed into *E. coli* BL21(DE3) host cells. Overnight cultures were diluted with 2×TY medium and grown at 37° until the OD_{600} reached 0.6–0.8. Protein expression following addition of 1 mM IPTG proceeded for 1 h at 37°, and the cells were then pelleted, resuspended in PBS, and sonicated. The lysate was cleared by ultracentrifugation (100,000×g, 45 min).

PURIFICATION OF PrBP(PDEδ). The recombinant GST-PrBP(PDEδ) supernatant is loaded onto a glutathione-agarose column (1 ml resin/liter culture). After washing, the bound GST-PrBP(PDEδ) is eluted with 5 column volumes of 10 mM reduced glutathione. Purified GST-PrBP (PDEδ) is concentrated by ultrafiltration (Centriplus YM10), and assayed for protein content. When desired, the GST fusion partner is cleaved using a thrombin cleavage capture kit (Novagen), following the manufacturer's protocol. The cleaved PrBP(PDEδ) is separated from GST by adsorption of GST to a glutathione-agarose column, and purified PrBP(PDEδ) is recovered in the flow through. The cleaved recombinant protein is concentrated using ultrafiltration and assayed by spectrophotometry. The extinction coefficients for frog and bovine PrBP(PDEδ) are 2.1×10^4 M^{-1} cm^{-1} and 2.5×10^4 M^{-1} cm^{-1}, respectively. Recombinant PrBP (PDEδ) easily aggregates upon freezing; therefore, purified protein is routinely stored at 4°.

Immobilized PrBP(PDEδ) Pull-Down Assays for Identifying Binding Partners and Characterizing the Determinants of High-Affinity Binding to PrBP(PDEδ)

Immobilized recombinant PrBP(PDEδ) is able to bind several photo-transduction proteins from bovine retinal homogenates, including rod PDE6 catalytic subunits, rhodopsin kinase (GRK1), and the transducin β subunit (Fig. 3). In addition, Rab8, a Rab GTPase involved in rhodopsin transport at the base of the connecting cilium of rod photoreceptors (Deretic *et al.*, 1995), is also identified as a PrBP(PDEδ) binding partner. In all four instances, addition of excess soluble PrBP(PDEδ) effectively

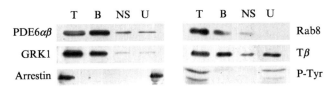

FIG. 3. PrBP(PDEδ) binds to several prenylated proteins in rod photoreceptors. Purified bovine ROS were detergent-solubilized, and incubated with PrBP(PDEδ) coupled to Sepharose beads. The beads were centrifuged and bound proteins (B) were separated from the supernatant (U). To demonstrate specificity of binding, parallel samples were incubated with an excess of PrBP(PDEδ) to control for nonspecific binding of potential binding partners with the Sepharose beads (NS). Equivalent amounts of unfractionated ROS were also examined as a measure of total immunoreactivity in the sample (T). The immunoblots were probed with antibodies specific for PDE6 (NC), GRK1 (Affinity BioReagents), visual arrestin (SCT-128), Rab8 (BD Biociences Pharmingen), transducin β subunit (Santa Cruz) and tyrosine-phosphorylated proteins (Upstate Cell Signaling Solutions).

abrogates binding, showing that these interactions are specific. PDE6 α and β subunits, GRK1, and Rab8 are all prenylated at their C-termini. Although the transducin β subunit is not prenylated, it exists in tight association with the farnesylated transducin γ subunit. In contrast, arrestin, an abundant, soluble phototransduction protein, does not bind PrBP(PDEδ). Because unc119 is structurally homologous to PrBP(PDEδ) and is a known activator of tyrosine kinases (Cen *et al.*, 2003), we also tested an antibody to tyrosine-phosphorylated proteins; in this case, we failed to detect specific interaction of tyrosine-phosphorylated proteins with PrBP(PDEδ).

We also performed pull-down assays with GST-PrBP(PDEδ) and GRK1 (expressed in Hi5 insect cells) and GRK7, a related G–protein coupled receptor kinase that was identified in several species (Chen *et al.*, 2001; Hisatomi *et al.*, 1998; Weiss *et al.*, 1998). Incubation of GST-PrBP (PDEδ) protein with the crude extract of Hi5 cells infected by GRK1– and GRK7-expressing viruses showed that both bound to GST-PrBP(PDEδ) immobilized on glutathione agarose (Fig. 4). These results demonstrate that PrBP(PDEδ) can stably bind both GRK1 (farnesylated) and GRK7 (geranylgeranylated) *in vitro*.

To ascertain the relative affinity of the prenyl moiety for PrBP(PDEδ), we performed pull-down experiments in which GST-PrBP(PDEδ) was preincubated in the presence of an excess of the synthetic isoprenoid compounds, acetyl farnesylcysteine methyl ester (AFCME) and acetyl geranylgeranylcysteine methyl ester (AGGCME) (Zhang *et al.*, 2004). Incubation with AFCME could completely abolish the binding of GST-PrBP(PDEδ) to farnesylated GRK1 and strongly compete with the binding of GST-PrBP(PDEδ) to geranylgeranylated GRK7. AGGCME also

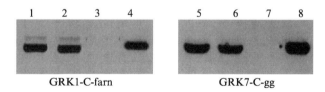

GRK1-C-farn   GRK7-C-gg

FIG. 4. GST-PrBP(PDEδ) pull–down assays of recombinant GRK1 and GRK7. Lanes 1–4, pull–down assay of GRK1. Polypeptides bound and unbound to the beads were analyzed by immunoblotting with anti-GRK1 antibody. Lane 1, supernatant following the incubation of GST with GRK1 crude extract; lane 2, supernatant following the incubation of GST-PrBP (PDEδ) with GRK1 crude extract. Note that in both lanes 1 and 2, GRK1 is in large excess. The slower moving faint band corresponds to nonprenylated GRK1. Lane 3, pellet of GST beads (no binding); lane 4, pellet of GST-PrBP(PDEδ) beads (GRK1 binds). GRK1-C-farn, farnesylated GRK1. Lanes 5–8, Pulldown assay of GRK7. Polypeptides bound and unbound to the beads were analyzed as in lanes 1–4. Lane 5, supernatant following the incubation of GST with GRK7 crude extract; lane 6, supernatant following the incubation of GST-PrBP (PDEδ) with GRK1 crude extract. Lane 7, pellet of GST beads (no binding); lane 8, pellet of GST-PrBP(PDEδ) beads. GRK7-C-gg, geranylgeranylated GRK7.

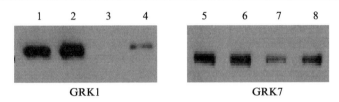

FIG. 5. Binding of GRK1 and GRK7 to GST-PrBP(PDEδ) in competition with excess synthetic prenyl compounds. GST-PrBP(PDEδ) was first incubated with one of the following: lanes 1 and 4, DMSO (solvent control); lanes 2 and 5, excess ACME (isoprenoid backbone control), lanes 3 and 7, excess AFCME, and lanes 4 and 8, excess AGGCME. Then, the GST-PrBP(PDEδ) mixture was combined with an insect cell extract expressing either GRK1 (lanes 1–4) or GRK7 (lanes 5–8). The proteins that coprecipitated with GST-PrBP(PDEδ) were analyzed by immunoblots using an anti-GRK1 antibody. Note that competition of geranylgeranylated GRK7 by AFCME is much weaker.

strongly competed with the binding of GST-PrBP(PDEδ) to GRK1 but only weakly prevented the binding of GST-PrBP(PDEδ) to GRK7 (Fig. 5). In a control experiment, ACME (the backbone of AFCME and AGGCE) did not compete (Fig. 5). These results demonstrate that the shorter farnesyl side chain (C15) has higher binding affinity for PrBP(PDEδ) than the longer geranylgeranyl chain (C20). The inability of AGGCME to chase GRK7 binding to GST-PrBP(PDEδ) suggests that additional binding sites exist on the GRK7 polypeptide that stabilize the interaction.

Methods

Identification of PrBP(PDEδ)-Interacting Proteins Using PrBP (PDEδ)-Sepharose Beads. PrBP(PDEδ) are coupled to cyanogen bromide-activated Sepharose 4B (Sigma) with a coupling efficiency of 2.5 mg of protein per 1 ml of settled beads using the manufacturer's protocol. Detergent-solubilized bovine ROS homogenates in 10 mM CHAPS, 140 mM NaCl, 0.5 mM MgCl$_2$, and 0.1 mM EDTA are incubated with PrBP (PDEδ)-Sepharose beads for 12 h at 4°. After incubation, the beads are pelleted and washed three times with TMN buffer (10 mM Tris pH 7.5, 1 mM MgCl$_2$, 300 mM NaCl, and 1 mM dithiothreitol [DTT]). The initial supernatant is collected for analysis of unbound proteins, which are precipitated with 10% trichloroacetic acid (TCA). Proteins bound to beads are eluted in SDS–PAGE sample buffer and analyzed by SDS–PAGE and immunoblot analysis. Nonspecific binding control samples are prepared by incubating the PrBP(PDEδ)-Sepharose beads with a large molar excess of GST-PrBP(PDEδ) before incubating the beads with cell extracts.

In Vitro Expression of GRK1/GRK7 for PrBP(PDEδ) Interaction Assay. GRK1 and GRK7 were each expressed in High-five (H5) insect cells (Chen *et al.*, 2001). Cultures (5 ml) are centrifuged and the cells are resuspended in 1 ml PBS containing 0.1% Triton X-100, 0.1% Tween 20, and protease inhibitor cocktail (Roche). Cell debris is pelleted by recentrifugation while the supernatant is divided between two microfuge tubes. In each tube, 20 μl of glutathione conjugated-agarose slurry and 4 μg GST protein (or an equal amount of GST-PDEδ protein) are added to supernatant. The tubes are incubated 2 h at room temperature on a rocking platform. The agarose is pelleted and washed three times with 1 ml PBS containing 0.1% Triton X-100. Bound proteins are eluted by the addition of 40 μl 2× sample buffer and boiling for 2 min. The supernatants and eluted proteins are separated by SDS–PAGE, followed by immunoblot detection with anti-GRK1 monoclonal antibody G8 (K. Palczewski, University of Washington-Seattle) or anti-GRK7 polyclonal antibody UU45APC (Chen *et al.*, 2001).

Competition Assay. For competition experiments, excess amounts of compounds in stock solution are diluted into 100 μM PBS separately and incubated with 7 μg GST-PrBP(PDEδ) overnight at room temperature. The preincubated GST-PrBP(PDEδ) is mixed with H5 cell lysate containing expressed GRK1 or GRK7 in the presence of excess AFCME or AGGCME. The mixtures are incubated for 1 h at room temperature. The pull-down assays were carried out as described above.

Conclusion: Model for PrBP(PDEδ) Involvement in Vectorial Protein Transport in Photoreceptors

Protein trafficking is particularly important for photoreceptors where high turnover of outer segments housing the phototransduction machinery requires unusually active protein transport. In approximately 100 million photoreceptors present in the human retina, an entire outer segment must be replaced once per week, an enormous task for protein transport and correct membrane targeting. As small GTPases, ARF and ARF-like proteins participate in a variety of intracellular transport processes. Arl2 and Arl3 may interact with PrBP(PDEδ) to help transport certain prenylated proteins to their destination membrane. Thus, one function of PrBP (PDEδ) may be that of a soluble transport factor interacting with a number of prenylated proteins, steered by GTP-binding proteins like Arl2/3. In Fig. 6, we picture a scenario in which PrBP(PDEδ) transport is essential for transport of rhodopsin kinase (GRK1) from the inner segment to the outer segment of rod photoreceptors. This model is consistent with preliminary experiments from a PrBP(PDEδ) knockout mouse (H. Zhang and W. Baehr, unpublished results) showing that in the absence of PrBP(PDEδ), GRK1 remains mostly localized to the inner segment where biosynthesis occurs and is not effectively transported to its outer segment destination.

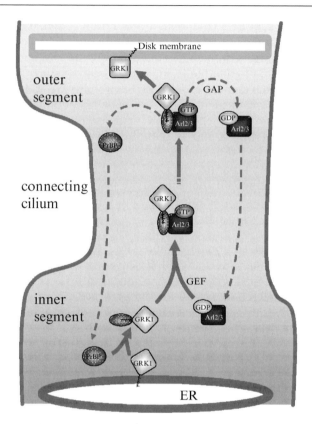

FIG. 6. Hypothetical model of PrBP(PDEδ) transport of rhodopsin kinase (GRK1) from the inner segment to the outer segment of rod photoreceptors. PrBP(PDEδ) solubilizes GRK1 from the endoplasmic reticulum (ER) membrane by binding the C-terminal farnesyl group, and the resulting soluble protein complex interacts with a trafficking protein such as Arl2/3-GDP. An unidentified guanine nucleotide exchange factor (GEF) exchanges GDP with GTP to activate Arl2/3, and the complex is then transported through the connecting cilium. When a GTPase activating protein (GAP) accelerates GTP hydrolysis on Arl2/3, the complex falls apart and GRK1 associates with the disk membrane in the outer segment. Arl2/3-GDP and PrBP(PDEδ) return to the inner segment to participate in another round of protein transport. (See color insert.)

Acknowledgments

We thank Kris Palczewski (University of Washington) for GRK antibody G8, Clay Smith (University of Florida) for the SCT-128 antibody, and Joe Beavo (University of Washington) for the bovine GST-PrBP(PDEδ) expression plasmid. This work was supported by Grants EY08123 to W.B. and EY-05798 to R.H.C., and an FFB center grant to the Department of Ophthalmology at the University of Utah.

References

Anant, J. S., Ong, O. C., Xie, H., Clarke, S., O'Brien, P. J., and Fung, B. K. K. (1992). *In vivo* differential prenylation of retinal cyclic GMP phosphodiesterase catalytic subunits. *J. Biol. Chem.* **267**, 687–690.

Baehr, W., Devlin, M. J., and Applebury, M. L. (1979). Isolation and characterization of cGMP phosphodiesterase from bovine rod outer segments. *J. Biol. Chem.* **254**, 11669–11677.

Bownds, D., Gordon-Walker, A., Gaide-Huguenin, A. C., and Robinson, W. (1971). Characterization and analysis of frog photoreceptor membranes. *J. Gen. Physiol.* **58**, 225–237.

Cen, O., Gorska, M. M., Stafford, S. J., Sur, S., and Alam, R. (2003). Identification of UNC119 as a novel activator of SRC-type tyrosine kinases. *J. Biol. Chem.* **278**, 8837–8845.

Chen, C. K., Zhang, K., Church-Kopish, J., Huang, W., Zhang, H., Chen, Y. J., Frederick, J. M., and Baehr, W. (2001). Characterization of human GRK7 as a potential cone opsin kinase. *Mol. Vis.* **7**, 305–313.

Cook, T. A., Ghomashchi, F., Gelb, M. H., Florio, S. K., and Beavo, J. A. (2000). Binding of the delta subunit to rod phosphodiesterase catalytic subunits requires methylated, prenylated C-termini of the catalytic subunits. *Biochemistry* **39**, 13516–13523.

Cook, T. A., Ghomashchi, F., Gelb, M. H., Florio, S. K., and Beavo, J. A. (2001). The delta subunit of type 6 phosphodiesterase reduces light-induced cGMP hydrolysis in rod outer segments. *J. Biol. Chem.* **276**, 5248–5255.

Cote, R. H. (2000). Kinetics and regulation of cGMP binding to noncatalytic binding sites on photoreceptor phosphodiesterase. *Methods Enzymol.* **315**, 646–672.

Deretic, D., Huber, L. A., Ransom, N., Mancini, M., Simons, K., and Papermaster, D. S. (1995). rab8 in retinal photoreceptors may participate in rhodopsin transport and in rod outer segment disk morphogenesis. *J. Cell Sci.* **108**, 215–224.

Deterre, P., Bigay, J., Forquet, F., Robert, M., and Chabre, M. (1988). cGMP phosphodiesterase of retinal rods is regulated by two inhibitory subunits. *Proc. Natl. Acad. Sci. USA* **85**, 2424–2428.

Florio, S. K., Prusti, R. K., and Beavo, J. A. (1996). Solubilization of membrane-bound rod phosphodiesterase by the rod phosphodiesterase delta subunit. *J. Biol. Chem.* **271**, 24036–24047.

Fung, B. K. K., Young, J. H., Yamane, H. K., and Griswold-Prenner, I. (1990). Subunit stoichiometry of retinal rod cGMP phosphodiesterase. *Biochemistry* **29**, 2657–2664.

Gillespie, P. G., Prusti, R. K., Apel, E. D., and Beavo, J. A. (1989). A soluble form of bovine rod photoreceptor phosphodiesterase has a novel 15-kDa subunit. *J. Biol. Chem.* **264**, 12187–12193.

Hanzal-Bayer, M., Renault, L., Roversi, P., Wittinghofer, A., and Hillig, R. C. (2002). The complex of Arl2-GTP and PDE delta: From structure to function. *EMBO J.* **21**, 2095–2106.

Higashide, T., and Inana, G. (1999). Characterization of the gene for HRG4 (UNC119), a novel photoreceptor synaptic protein homologous to unc-119. *Genomics* **57**, 446–450.

Hillig, R. C., Hanzal-Bayer, M., Linari, M., Becker, J., Wittinghofer, A., and Renault, L. (2000). Structural and biochemical properties show ARL3-GDP as a distinct GTP binding protein. *Structure. Fold. Des.* **8**, 1239–1245.

Hisatomi, O., Matsuda, S., Satoh, T., Kotaka, S., Imanishi, Y., and Tokunaga, F. (1998). A novel subtype of G-protein-coupled receptor kinase, GRK7, in teleost cone photoreceptors. *FEBS Lett.* **424**, 159–164.

Hoffman, G. R., Nassar, N., and Cerione, R. A. (2000). Structure of the Rho family GTP-binding protein Cdc42 in complex with the multifunctional regulator RhoGDI. *Cell* **100**, 345–356.

Kobayashi, A., Kubota, S., Mori, N., McLaren, M. J., and Inana, G. (2003). Photoreceptor synaptic protein HRG4 (UNC119) interacts with ARL2 via a putative conserved domain. *FEBS Lett.* **534**, 26–32.

Li, N., and Baehr, W. (1998). Expression and characterization of human PDEdelta and its Caenorhabditis elegans ortholog CEdelta. *FEBS Lett.* **440**, 454–457.

Linari, M., Hanzal-Bayer, M., and Becker, J. (1999a). The delta subunit of rod specific cyclic GMP phosphodiesterase, PDE delta, interacts with the Arf-like protein Arl3 in a GTP specific manner. *FEBS Lett.* **458**, 55–59.

Linari, M., Ueffing, M., Manson, F., Wright, A., Meitinger, T., and Becker, J. (1999b). The retinitis pigmentosa GTPase regulator, RPGR, interacts with the delta subunit of rod cyclic GMP phosphodiesterase. *Proc. Natl. Acad. Sci. USA* **96**, 1315–1320.

Maduro, M., and Pilgrim, D. (1995). Identification and cloning of unc-119, a gene expressed in the Caenorhabditis elegans nervous system. *Genetics* **141**, 977–988.

Marzesco, A. M., Galli, T., Louvard, D., and Zahraoui, A. (1998). The rod cGMP phosphodiesterase delta subunit dissociates the small GTPase Rab13 from membranes. *J. Biol. Chem.* **273**, 22340–22345.

McDowell, J. H. (1993). Preparing rod outer segment membranes, regenerating rhodopsin, and determining rhodopsin concentration. *Methods Neurosci.* **7**, 123–130.

Mou, H., Grazio, H. J., III, Cook, T. A., Beavo, J. A., and Cote, R. H. (1999). cGMP binding to noncatalytic sites on mammalian rod photoreceptor phosphodiesterase is regulated by binding of its gamma and delta subunits. *J. Biol. Chem.* **274**, 18813–18820.

Nancy, V., Callebaut, I., El Marjou, A., and de Gunzburg, J. (2002). The delta subunit of retinal rod cGMP phosphodiesterase regulates the membrane association of Ras and Rap GTPases. *J. Biol. Chem.* **277**, 15076–15084.

Norton, A. W., Hosier, S., Terew, J. M., Li, N., Dhingra, A., Vardi, N., Baehr, W., and Cote, R. H. (2005). Evaluation of the 17-kDa prenyl-binding protein as a regulatory protein for phototransduction in retinal photoreceptors. *J. Biol. Chem.* **280**, 1248–1256.

Qin, N., and Baehr, W. (1994). Expression and mutagenesis of mouse rod photoreceptor cGMP phosphodiesterase. *J. Biol. Chem.* **269**, 3265–3271.

Qin, N., Pittler, S. J., and Baehr, W. (1992). *In vitro* isoprenylation and membrane association of mouse rod photoreceptor cGMP phosphodiesterase α and β subunits expressed in bacteria. *J. Biol. Chem.* **267**, 8458–8463.

Towbin, H., Staehelin, T., and Gordon, J. (1979). Electrophoretic transfer of proteins from polyactylamide gels to nitrocellulose sheets: Procedure and some applications. *Proc. Natl. Acad. Sci. USA* **76**, 4350–4354.

Weiss, E. R., Raman, D., Shirakawa, S., Ducceschi, M. H., Bertram, P. T., Wong, F., Kraft, T. W., and Osawa, S. (1998). The cloning of GRK7, a candidate cone opsin kinase, from cone- and rod-dominant mammalian retinas. *Mol. Vis.* **4**, 27.

Zhang, H., Liu, X. H., Zhang, K., Chen, C. K., Frederick, J. M., Prestwich, G. D., and Baehr, W. (2004). Photoreceptor cGMP phosphodiesterase delta subunit (PDEdelta) functions as a prenyl-binding protein. *J. Biol. Chem.* **279**, 407–413.

Zhao, Y., Hong, D. H., Pawlyk, B., Yue, G., Adamian, M., Grynberg, M., Godzik, A., and Li, T. (2003). The retinitis pigmentosa GTPase regulator (RPGR)-interacting protein: Subserving RPGR function and participating in disk morphogenesis. *Proc. Natl. Acad. Sci. USA* **100**, 3965–3970.

[6]   Functional Assays for the Investigation of the
Role of Rab GTPase Effectors in Dense Core
Granule Release

By SÉVERINE CHEVIET, THIERRY COPPOLA, and ROMANO REGAZZI

Abstract

Rab3 and Rab27 GTPases control late events in the secretory pathway of mammalian cells including docking and fusion of secretory vesicles with the plasma membrane. The action of Rab3 and Rab27 on the exocytotic process is exerted through the activation of specific effectors. Several proteins with the capacity to interact in a GTP-dependent manner with Rab3 and Rab27 have been identified. However, for most of these potential Rab effectors a precise function in the secretory process has not yet been attributed. In this chapter we describe a series of approaches that can be applied to assess the properties of Rab3 and Rab27 effectors and their potential role in the regulation of dense core granule release.

Introduction

Rab GTPases participate in the regulation of exocytosis in neurons as well as in endocrine and exocrine glands (Seabra *et al.*, 2002). The four members of the Rab3 family (Rab3A–D) and the two members of the Rab27 family (Rab27A and Rab27B) are localized on secretory vesicles and are involved in the finals steps of the secretory process (Darchen and Goud, 2000; Fukuda, 2005; Izumi *et al.*, 2003). Despite years of intense investigation the precise role of these Rab GTPases is still not completely elucidated. Rab3 and Rab27 have been proposed to control targeting and docking of secretory vesicles to the plasma membrane (Darchen and Goud, 2000; Fukuda, 2005; Izumi *et al.*, 2003). In addition, evidence has been provided for a possible regulatory role for these Rab GTPases in vesicle fusion and in other events directly related to the exocytotic process such as cytoskeleton dynamic and endocytosis (Burns *et al.*, 1998; Kato *et al.*, 1996; Ohya *et al.*, 1998). During the past few years an increasing number of potential Rab3 and Rab27 targets have been identified (Cheviet *et al.*, 2004b; Darchen and Goud, 2000; Fukuda, 2005). To elucidate the numerous facets of Rab3 and Rab27 function a detailed characterization of these putative GTPase effectors is needed. In this chapter we describe several methodologies that allow the determination of the subcellular localization of the putative effectors, the investigation of the mechanism governing

METHODS IN ENZYMOLOGY, VOL. 403   0076-6879/05 $35.00
DOI: 10.1016/S0076-6879(05)03006-5

their interaction with Rab GTPases, and the assessment of their functional involvement in the secretory process.

Cellular Model

Since the main goal of the experiments performed in our laboratory is to elucidate the molecular mechanisms controlling exocytosis of pancreatic β-cells, the methodologies described in this chapter have been developed using the insulin-secreting cell lines INS-1E and HIT-T15. However, most of the techniques can be easily adapted to other cells performing regulated dense core granule exocytosis such as the catecholamine-secreting cell line PC12.

The INS-1E cell line was provided by Dr. C. B. Wollheim (University of Geneva). HIT-T15 cells can be obtained from the American Tissue Culture Collection (ATCC number CRL-1777). INS-1E cells are cultured in RPMI 1640 medium complemented with 5% fetal calf serum (FCS), 50 IU/ml penicillin, 50 μg/ml streptomycin, 0.1 mM sodium pyruvate, and 0.001% 2-mercaptoethanol (Asfari *et al.*, 1992). HIT-T15 cells are cultured in RPMI 1640 medium supplemented with 10% FCS, 2 mM glutamine, 32.5 μM glutathione, and 0.1 μM selenium (Regazzi *et al.*, 1990).

Determination of the Subcellular Localization of Rab Effectors

A first indication of the potential function of a protein is given by its subcellular localization. We currently use two complementary approaches to identify the compartments to which Rab effectors are associated. The first is an immunocytochemical approach coupled to the analysis of the cells by confocal microscopy. The second is a biochemical approach that involves homogenization of the cells and fractionation of the organelles on sucrose density gradients.

For immunocytochemical analysis 5×10^4 cells are seeded on glass coverslips (12 mm diameter) coated with 20 μg/ml laminin (Invitrogen, Carlsbad, CA) and 2 mg/ml poly-L-lysine (Sigma, St. Louis, MO). One or 2 days later, the coverslips are washed in phosphate-buffered saline (PBS: 138 mM NaCl, 10 mM Na$_2$HPO$_4$, 2.7 mM KCl, 1.4 mM KH$_2$PO$_4$, pH 7.4) and the cells are fixed by 20 min incubation at room temperature in 4% paraformaldehyde dissolved in PBS. After three washes in PBS, the cells are incubated for 2 h with the primary antibody diluted in PBS supplemented with 0.1% goat serum, 0.3% (v/v) Triton X-100, and 20 mg/ml bovine serum albumin (BSA). The coverslips are washed several times in PBS and incubated for 30 min with a fluorescently labeled secondary antibody (Jackson Immunoresearch Laboratories, West Grove, PA) diluted in the same buffer. After three final washes, the coverslips are mounted on glass slides with Vectashield Mounting Medium (Vector Laboratories, Burlingame, CA) and analyzed by confocal microscopy (Leica, model TCS NT, Lasertechnik, Heidelberg, Germany). To assess whether the Rab effector under investigation colocalizes with secretory organelles, labeling of the cells with a large dense core granule marker is required. Dense core granule labeling in pancreatic β-cells can be easily obtained using anti–insulin antibodies (Linco Research, St. Charles, MO). Alternatively, in β-cells, as in most cells performing dense core vesicle exocytosis, secretory granules can be labeled using antibodies directed against chromogranin A (Dako, Glostrup DK). An example of immunochemical analysis demonstrating colocalization of the Rab effector granuphilin with insulin-containing granules is shown in Fig. 1.

In some cases the protein under study is not exclusively localized on dense core granules or on the plasma membrane but displays a more complex distribution pattern. For instance, pancreatic β-cells, as well as many other secretory cells, possess small synaptic-like vesicles (Reetz et al., 1991). This and other organelles can be separated from dense core granules on sucrose density gradients (Iezzi et al., 1999; Reetz et al., 1991). For this type of analysis INS-1E cells are grown on 155-cm^2 Petri dishes. The day of the experiment about 2×10^7 cells are scraped with a rubber policeman in 500 μl of buffer A containing 5 mM HEPES (pH 7.4), 1 mM EGTA, 10 μg/ml leupeptin, 2 μg/ml aprotinin, and 0.25 M sucrose. They are then homogenized using a tight fitting glass homogenizer. This step is critical and should enable disruption of the cells without damage to cellular organelles. The best homogenization conditions should be worked out for each cell type by inspecting under the microscope lyses of the cells and preservation of intact nuclei. After homogenization the cells are centrifuged for 10 min at 3000$\times g$ to eliminate cell debris and nuclei. The supernatant is loaded onto a linear 3 ml sucrose density gradient (0.45–2.00 M). A continuous sucrose density gradient can be generated using a gradient maker (Hoefer Instruments, San Francisco, CA). Alternatively a 3 ml discontinuous sucrose density gradient (0.45–2.00 M) can be obtained by carefully layering from the bottom of the ultracentrifuge tube a series of 10 solutions (0.3 ml each) containing decreasing concentrations of sucrose. The tube is then centrifuged at 4° using a swing out rotor for 18 h at 110,000$\times g$ to separate the cellular organelles according to density. To prevent mixing of the sucrose gradient the centrifugation has to be stopped smoothly, avoiding rapid braking conditions. Fractions of 0.3 ml are collected from the top of the gradient. An aliquot of each fraction is analyzed by Western blotting using antibodies directed against markers of cellular organelles such as plasma membrane (Na$^+$/K$^+$-ATPase) and synaptic-like vesicles

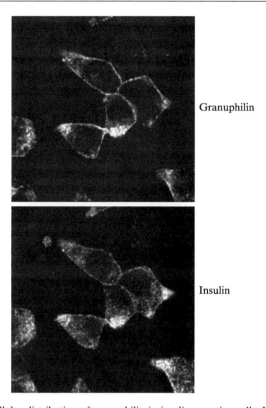

FIG. 1. Subcellular distribution of granuphilin in insulin-secreting cells. INS-1E cells were grown on glass coverslips coated with laminin and poly-L-lysine. The cells were then fixed with paraformaldehyde and incubated with a rabbit antibody against granuphilin (upper panel) and a guinea pig antibody against insulin (lower panel). The localization of granuphilin and of insulin-containing secretory granules was determined by confocal microscopy after incubation of the coverslips with an FITC-labeled anti-rabbit antibody and a Cy3-labeled anti-guinea pig antibody.

(synaptophysin) (Iezzi *et al.*, 1999). Enrichment of dense core granules is determined using an insulin radioimmunoassay. The sedimentation profile of the Rab effector determined by Western blotting is then compared to that of the markers of cellular organelles.

Measurement of the Interaction of Rab Effectors with Members of the Rab GTPase Family

Previously, each Rab GTPase was thought to possess a specific set of effectors. However, systematic screening revealed that Rab GTPases can share part of their partners with other closely related members of the Rab family (Fukuda, 2003; Kuroda *et al.*, 2002). Therefore, to elucidate the function of Rab effectors it is important to identify all their potential binding partners and to assess whether the interaction is influenced by the activation state of the GTPases. These goals can be achieved by different *in vitro* and *in vivo* methods.

An indication of the interaction between Rab GTPases and their potential effectors can be obtained by measuring the binding of bacterially produced glutathione-*S*-transferase (GST)-fusion proteins with potential partners synthetized by *in vitro* translation (Coppola *et al.*, 1999, 2001). To produce the GST-fusion proteins, the coding sequence of Rab GTPases or of the Rab effectors is cloned in frame at the 3' end of the sequence of GST into the pGEX-KG vector (Guan and Dixon, 1991). The plasmid is then introduced in the *Escherichia coli* strain BL21 (Novagen, Madison, WI). Other bacterial strains (i.e., DH5α) can also be used, but for some recombinant proteins the yield is lower than with the BL21 strain (our unpublished observations). BL21 bacteria are cultured in LB medium (GIBCO BRL, Carlsbad, CA) to a density of OD_{600} 0.8–0.9 and the expression of the fusion protein induced by addition of isopropylthio-β-galactoside (IPTG) to a final concentration of 0.1 mM. The IPTG-induced cultures are grown at 37° for 4–6 h. The cells are then harvested by centrifugation for 10 min at 5000×g. The bacterial pellet is transferred in a 1.5-ml Eppendorf tube and sonicated in an ice-cold PBS solution containing 1 mM EDTA, 0.1% 2-mercaptoethanol, and 300 μM phenylmethylsulfonyl fluoride (PMSF). Homogenization is completed by the addition of 0.5% (v/v) Triton X-100 followed by incubation of the cell suspension at 4° for 20 min. The lysate is cleared by 10 min centrifugation at 14,000 rpm in an Eppendorf microcentrifuge at 4°. The supernatant is collected, aliquoted, and stored at −80° until use. The day of the experiment, the GST–fusion proteins are immobilized by mixing 100 μl of the bacterial extract with 20 μl of glutathione-agarose beads (Sigma, St. Louis, MO). After incubation for 1 h at 4° on rotation, the beads are resuspended in buffer A (20 mM HEPES [pH 7.5], 150 mM KCl, 1 mM dithiothreitol, 5% [v/v] glycerol, 0.05% [v/v] Tween-20, and 1 mg/ml BSA). Radioactively labeled Rab GTPases or Rab effectors are freshly produced by *in vitro* translation using the TNT T7 Coupled Reticulocyte Lysate System (Promega, Madison, WI) and [^{35}S]methionine according to the manufacturer's instructions. For this purpose the cDNA of the GTPases or the effectors need to be cloned in a plasmid enabling protein expression from a T7 promoter such as pcDNA3 (Invitrogen, Carlsbad, CA). The binding reaction is performed by incubating the beads for 1 h at 4° with the ^{35}S-labeled proteins in a final volume of 30 μl. After three washes in buffer A, the beads are resuspended in 20 μl of loading buffer (250 mM Tris, pH 6.8, 46% [v/v] glycerol, 280 mM sodium dodecyl sulfate [SDS]) and the proteins remaining associated with the affinity columns are separated by

SDS–PAGE. The gel is dried and radioactively labeled proteins are visualized by autoradiography (overnight exposure of the films at $-80°$ is usually largely sufficient).

An alternative approach consists of assessing the capacity of GST-fusion proteins immobilized on agarose beads to pull down their binding partners from cell extracts. These experiments can be performed either by pulling down endogenous proteins or by using extracts of transiently transfected cells. Care should be taken if the antibody against the Rab or the effector was generated against a GST-fusion protein. In fact, analyses by Western blotting of the proteins remaining attached to the beads can be perturbed by the presence in the antiserum of antibodies directed against GST. To circumvent this problem it is possible to use extracts obtained from cells transiently expressing an epitope-tagged construct. Commercially available antibodies against c-*myc*, hemagglutinin (HA), FLAG, T7, and green fluorescent protein (GFP) do not cross-react with GST and give satisfactory results on Western blotting. Transient transfection of insulin-secreting cells can be performed either by electroporation or by lipofection. Detailed protocols have been described elsewhere (Coppola *et al.*, 1999; Waselle *et al.*, 2003). After transfection the cells (2–3×10^6 cells/point) are cultured in six multiwell plates for 2 days. They are then scraped with a rubber policeman in ice-cold buffer B (20 mM Tris–HCl, 200 mM NaCl, 10% glycerol [v/v], 1% Triton X-100 [v/v], 5 μg/ml leupeptin, and 5 μg/ml aprotinin) and lysed by sonication (3×2 s). Prior to the binding reaction, Rab GTPases can be converted to inactive or active states by exposing the homogenates for 15 min at $30°$ to either 100 μM GDP or GTPγS, respectively. The cell extracts are then incubated for 1 h at $4°$ with the GST-fusion proteins immobilized on glutathione-agarose beads. At the end of the incubation, the beads are washed in buffer B and the proteins that bind to the beads are detected by Western blotting. An illustration of the results obtained with this type of technique is shown in Fig. 2. In this case, HIT-T15 cells were transiently transfected with plasmids encoding Rab3A-GFP, Rab8-GFP, and Rab27A-GFP. After lysis, the homogenates were incubated with GST-affinity column containing the Rab-binding domain of granuphilin/Slp4. The binding of Rab3A, Rab8, and Rab27A was assessed by Western blotting with an antibody against GFP (Clontech, Palo Alto, CA).

Using an approach based on a mammalian two-hybrid system it is also possible to test the interaction of Rab GTPases with their effectors in living cells (Coppola *et al.*, 2001, 2002). For this type of analysis the cDNA of the potential Rab effector is cloned in frame with VP16 in the expression vector pACT (Promega, Madison, WI). Rab GTPases are cloned in frame with GAL4 in the expression vector pBIND (Promega, Madison, WI). This plasmid also encodes a *Renilla* luciferase under the control of a constitutive promoter that is used to normalize for the transfection efficiency. The two

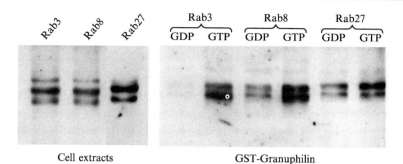

Cell extracts   GST-Granuphilin

FIG. 2. Binding of granuphilin to Rab3, Rab8, and Rab27. A GST-fusion protein containing the amino acids 1–300 of granuphilin was immobilized on glutathione-agarose beads. Lysates of HIT-T15 cells transiently transfected with GFP-tagged Rab3a, Rab8, or Rab27a were preincubated for 15 min at 37° with 100 μM GDP or GTPγS and then loaded on affinity columns containing 2 μg of GST-granuphilin 1–300. The amount of Rab3a, Rab8, and Rab27a remaining associated with the beads was determined by Western blotting using an antibody against GFP. One-tenth of the cell lysates used for the binding studies (cell extracts) was loaded on separate lanes (left panel) and served as control for the expression of the three GFP-tagged proteins.

constructs are cotransfected in insulin-secreting cells along with a third plasmid (pG5luc, Promega, Madison, WI) that encodes five GAL4 binding sites upstream of the *Firefly* luciferase reporter gene. Two days after transfection, the cells are lysed and *Firefly* and *Renilla* luciferase activities are determined using the Dual-Luciferase Reporter Assay System (Promega, Madison, WI). The interaction of Rab GTPases with their effectors causes the formation of an active VP16/GAL4 complex and leads to an increase in the synthesis of the *Firefly* luciferase. An example of positive interaction between Rab27 and granuphilin is shown in Fig. 3. This mammalian two-hybrid system presents the advantage to test protein/protein interactions within the cellular environment. However, this assay may not be appropriate to detect transient interactions occurring in restricted areas of the cells or interactions involving only a small pool of the Rab GTPase.

Functional Assessment of the Involvement in Dense Core Granule Exocytosis

To rapidly assess the functional involvement of a protein in exocytosis we take advantage of a transient transfection system originally developed by Holz and collaborators. This system is based on the observation that in cells performing regulated exocytosis ectopically expressed human growth hormone (hGH) is targeted to dense core granules and is coreleased with the endogenous secretory product (Wick *et al.*, 1993). Thus, hGH secretion can be used to selectively monitor exocytosis in transiently transfected cells

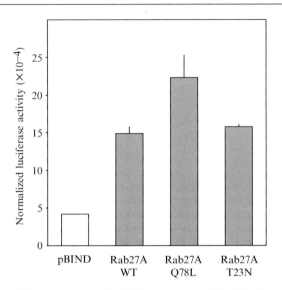

Fig. 3. Granuphilin interacts with the GTP-bound form of Rab27 in living cells. HIT-T15 cells were transiently cotransfected with a pACT plasmid directing the expression of a fusion protein of VP16 with granuphilin, a pBIND plasmid directing the expression of a fusion protein of GAL4 with the indicated mutants of Rab27, and with a plasmid containing five binding sites for GAL4 upstream of a *Firefly* luciferase reporter gene. The interaction between granuphilin and the Rab27 mutants was quantified by measuring the *Firefly* luciferase activity 2 days later. An empty pBIND vector was used as a control. Differences in transfection efficiency were corrected by normalizing the *Firefly* luciferase activity with the activity of *Renilla* luciferase expressed under the control of a constitutive promoter from the pBIND plasmid.

without the need of isolating them from untransfected cells. This assay is very sensitive and can be adapted to cell systems in which the transfection efficiency is below 10%. For this type of analysis, the cells are cotransfected with a construct leading to an alteration in the expression level of the protein under study and with pXGH5 plasmid (Nicholls, San Juan Capistrano, CA) encoding hGH. Constructs directing overexpression of the proteins investigated are generated by cloning the cDNA of the Rab effector in pcDNA3 (Invitrogen, Carlsbad, CA) or pEGFP vectors (Clontech, Palo Alto, CA). The cDNA is usually engineered to append an epitope tag (i.e., c-*myc* or GFP) either at the amino- or at the carboxy-terminus. For Rab GTPases the tags are usually integrated at the N-terminus of the protein to avoid possible perturbations in the Rab function.

Alone, overexpression experiments are often not sufficient to assess the precise role of a protein in the secretory pathway. In fact, if the quantity of the component investigated is not rate limiting, an increase in its level will not affect exocytosis. Moreover, introduction of large amounts of proteins with a positive action in the secretory process can potentially perturb the function of the secretory machinery and lead to the erroneous conclusion that the protein is an inhibitor of exocytosis. One way of solving this problem is to generate dominant negative mutants that, once introduced in the cells, selectively block the activity of the endogenous protein. However, this approach presupposes a very good knowledge of the properties of the protein under study that for most of the potential Rab effectors is not available. The discovery of the process of RNA interference (RNAi) has opened new possibilities to achieve a selective reduction in the expression level of a target gene (Hannon and Rossi, 2004; Meister and Tuschl, 2004). This powerful technique has been successfully applied to study β-cell functions (Cheviet *et al.*, 2004a; Da Silva *et al.*, 2004; Iezzi *et al.*, 2004; Waselle *et al.*, 2003). The RNAi process is elicited by the presence of short double-stranded RNA molecules called small interfering RNAs (siRNAs). siRNAs associate to an endonuclease complex (RISC) and direct a sequence-specific degradation of target mRNAs resulting in specific silencing of the selected gene (Hannon and Rossi, 2004; Meister and Tuschl, 2004). Activation of the RNAi process can be obtained by transfecting the cells with synthetic siRNAs (Da Silva *et al.*, 2004). An alternative technique consists at transfecting the cells with a plasmid directing the synthesis of a short RNA molecule that is processed by cellular enzymes to produce the siRNAs. In our laboratory we mainly use the second approach. To generate a plasmid capable of reducing the endogenous level of Rab effectors (or any other protein under investigation) we insert a 64 nucleotide sequence in the *Bg*lII and *Hin*dIII sites of the pSUPER plasmid (Oligoengine, Seattle, WA). This sequence includes 19 nucleotides derived from the target (i.e., the Rab effector) separated from its reverse 19-nucleotide complement by a short spacer (Fig. 4). Because of the presence of two complementary stretches, the RNA molecule produced from the inserted sequence folds in a hairpin structure. This secondary structure is recognized by Dicer, an endogenous enzyme that cleaves the loop and produces

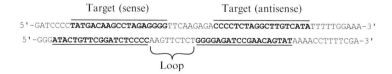

FIG. 4. Example of oligonucleotides to be inserted in the pSUPER plasmid for RNAi. A 64 nucleotide sequence including 19 nucleotides derived from the target (in this example rat NOC2) separated from its reverse 19-nucleotide complement by a short spacer (loop) is inserted in the *Bg*lII and *Hin*dIII sites of the pSUPER plasmid (Oligoengine, Seattle, WA).

the double-stranded siRNAs (Brummelkamp *et al.*, 2002). For our experiments we now routinely choose three different sequences against the target. Potential effective sequences for RNAi are identified using the siRNA Target Finder tool from Ambion (http://www.ambion.com/techlib/misc/siRNA_finder.html). We usually select sequences in the coding region, with a GC content between 42 and 47%. The presence of a stretch of four (or more) A or T will lead to premature termination of the transcript. Sequences containing these stretches should therefore be discarded. According to our experience using these simple selection rules the probability of obtaining at least one effective silencer is relatively high (about 80–90%). More sophisticated selection methods have now been developed by several companies including Dharmacon, Oligoengine, and Ambion and are available on their respective websites. To reduce the possibility of off-target effects, the specificity of each selected oligonucleotide should be verified by a search of the sequences deposited at the GenBank with the basic local alignment search tool (BLAST).

The oligonucleotides including the selected sequences are synthetized (MWG Biotech Ebersberg, Germany) and annealed in 30 mM HEPES–KOH, pH 7.4, 100 mM potassium acetate, 2 mM Mg-acetate by incubating them at 95° for 4 min and then for 10 min at 70°. The solution is successively cooled down slowly to 4°. The annealed oligonucleotides are phosphorylated by T4 PNK (Promega, Madison, WI) for 30 min at 37°. The enzyme is inactivated by 10 min incubation at 70°. The pSUPER vector is digested with *Bg*lII and *Hin*dIII and dephosphorylated with the Calf Intestinal Phosphatase (Promega, Madison, WI). Of the phosphorylated oligonucleotides 20 μg is ligated with 100 μg of linearized pSUPER. After transformation of DH5α (Invitrogen, Carlsbad, CA), the presence of the inserts is checked by screening a series of minipreps (GFX Micro Plasmid Prep kit from Amersham Biosciences, Piscataway, NJ) by *Eco*RI–*Hin*dIII digestion. In positives clones the restriction enzymes generate a fragment of about 360 pb that can be easily distinguished from the 300-pb fragment obtained from the empty vector. Screening for positive clones can also be performed by taking advantage of the fact that the *Bg*lII restriction site is lost after insertion of the oligonucleotides. Positive clones are grown in LB medium overnight, and the cDNA is prepared for transfection using the JETSTAR 2.0 Plasmid Midiprep Kit (Genomed, Löhne, Germany). The silencing activity of siRNAs is determined by transiently cotransfecting INS-1E cells with the pSUPER construct and with a vector permitting the expression of the epitope tagged target. The expression level of the target is assessed by Western blotting after 2–3 days. Under our experimental conditions the transfection efficiency is between 30 and 50%. For the most effective silencers it is possible to detect an effect on the expression level of the endogenous protein (Fig. 5A). According to our experience, at least

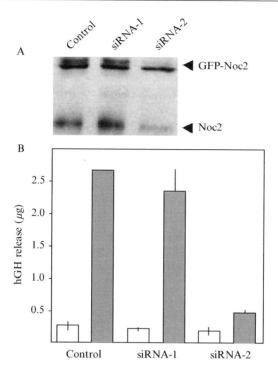

FIG. 5. Effect of Noc2 silencing on exocytosis. (A) INS-1E cells were transiently cotransfected with GFP-tagged Noc2 and with empty pSUPER vector (control) or with vectors allowing the synthesis of two different short interfering RNAs (siRNA-1 and siRNA-2) directed against rat NOC2. After 3 days the expression levels of both GFP-Noc2 and endogenous Noc2 were analyzed by Western blotting with an antibody against Noc2. (B) INS-1E cells were transiently cotransfected with a plasmid encoding hGH and with the empty pSUPER vector (control), siRNA-1 or siRNA-2. After 3 days, the cells were incubated under basal condition (open bars) or in the presence of stimulatory concentrations of glucose, K^+, forskolin, and IBMX (filled bars). After 45 min, the incubation medium was collected and the amount of hGH released was measured by ELISA.

one of the selected sequences is capable of silencing the target by more than 80%. Examples of effector silencers are given in Table I.

To assess the role of Rab effectors on insulin exocytosis 3×10^5 INS-1E cells are seeded on 24-multiwell plates and are then transiently cotransfected with the plasmid encoding hGH and with one or more silencers. Alternatively it is possible to use a plasmid encoding both the siRNA and hGH. We have engineered a plasmid (pXGH5-Sup) with these properties by introducing the sequence encoding a constitutive polymerase II promoter (metallothionine) and hGH in the pSUPER plasmid (Waselle *et al.*, 2003) (Fig. 6). This plasmid ensures 100% coexpression of hGH and

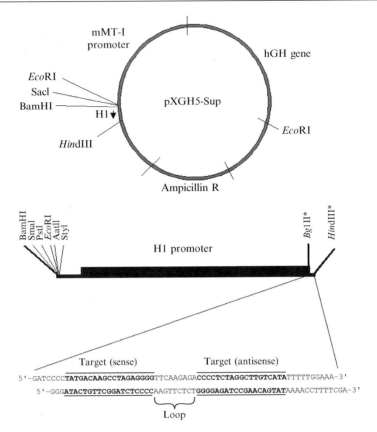

FIG. 6. Schematic representation of the plasmid permitting coexpression of siRNAs and hGH (pXGH5-Sup). A cassette containing the H1-RNA promoter and the polylinker sequence of pSUPER was amplified by PCR and subcloned within the *Xba*I and *Hin*dIII sites of pXGH5. A *Bg*lII restriction site already present in pXGH5 was mutated to allow direct insertion of the silencing sequences in the *Bg*lII and *Hin*dIII sites as in the case of the pSUPER plasmid.

siRNAs (compared to 80–90% coexpression in case of transfection of two separate plasmids). Transfection is generally performed with the Effectene reagent (Qiagen, Valencia, CA) although similar results were obtained with Lipofectamine 2000 (Invitrogen, Carlsbad, CA). Three days later, the cells are preincubated for 30 min at 37° in 20 mM HEPES (pH 7.4), 128 mM NaCl, 5 mM KCl, 1 mM MgCl$_2$, and 2.7 mM CaCl$_2$. The medium is then removed and the cells incubated for 45 min at 37° in the same buffer (basal) or in a buffer containing 20 mM HEPES (pH 7.4), 53 mM NaCl, 80 mM KCl, 1 mM MgCl$_2$, 2.7 mM CaCl$_2$, 20 mM glucose, 1 $\mu$$M$ forskolin, and 1 mM 3-isobutyl-1-methylxanthine (IBMX) (stimulated). At the end of the

TABLE I

RNAi Sequences Used in Our Laboratory to Study Rab Effector Function[a]

Target	Species	Sequence	Silencing
Rab27a	R/M	TTCATCACCACAGTGGGCA	+
MyRIP-1	R/M/H	CATCTCCACAGAAGTCCTG	+++
MyRIP-2	R/H	GAAGACGGGATCAGAAGCA	+++
NOC2-1	R/M	GTACATCTTGCCCCTGAAA	−
NOC2-2	R	TATGACAAGCCTAGAGGGG	+++
NOC2-3	R	GAACGTGATGGGCAATGGT	−

[a] The sequences shown in the table have been tested in our laboratory for their capacity to silence Rab27a and its effectors (Cheviet et al., 2004a; Waselle et al., 2003). The species in which there is a perfect match with the silencer sequence are indicated. R, rat; M, mouse; H, human. The efficacy of silencing of each sequence was evaluated by Western blotting. −, no silencing; + moderate silencing; +++, very good silencing.

incubation the supernatant is collected and analyzed by ELISA to determine the amount of hGH released in the medium (Roche Diagnostics, Alameda, CA). The effect on exocytosis caused by silencing of the Rab27 effector NOC2 is shown in Fig. 5B.

Evaluation of the role of Rab GTPases and their effectors in the secretory process is undoubtedly greatly facilitated by the opportunity to reduce the expression of most selected target genes by RNAi. Although sequence specificity of the RNAi process is clearly superior to classical antisense techniques, the possibility of generating unwanted off-target effects remains an important concern (Huppi et al., 2005). Thus, before drawing conclusions about the function of the protein under study, the effect of multiple siRNA sequences corresponding to different regions of the target mRNA is strongly suggested.

Conclusion

There is now compelling evidence for the involvement of Rab3 and Rab27 in the regulation of dense core vesicle release and for their potential role in human diseases (Kasai et al., 2005; Seabra et al., 2002; Yaekura et al., 2003). However, the mode of action of these GTPases is still not completely understood. Identification of a new group of potential partners of Rab3 and Rab27 has given a new importance to the discovery of the role of these small G proteins in the exocytotic process. Detailed characterization of the properties and functions of Rab3 and Rab27 effectors using methodologies such as the one described in this chapter will be instrumental in clarifying the mechanisms by which these GTPases modulate the release of peptide hormones, neurotransmitters, and digestive enzymes.

Acknowledgments

This work was supported by Grant 3200B0-101746 from the Swiss National Science Foundation (R.R.).

References

Asfari, M., Janjic, D., Meda, P., Li, G., Halban, P. A., and Wollheim, C. B. (1992). Establishment of 2-mercaptoethanol-dependent differentiated insulin-secreting cell lines. *Endocrinology* **130,** 167–178.

Brummelkamp, T. R., Bernards, R., and Agami, R. (2002). A system for stable expression of short interfering RNAs in mammalian cells. *Science* **296,** 550–553.

Burns, M. E., Sasaki, T., Takai, Y., and Augustine, G. J. (1998). Rabphilin-3A: A multifunctional regulator of synaptic vesicle traffic. *J. Gen. Physiol.* **111,** 243–255.

Cheviet, S., Coppola, T., Haynes, L. P., Burgoyne, R. D., and Regazzi, R. (2004a). The Rab-binding protein Noc2 is associated with insulin-containing secretory granules and is essential for pancreatic beta-cell exocytosis. *Mol. Endocrinol.* **18,** 117–126.

Cheviet, S., Waselle, L., and Regazzi, R. (2004b). Noc-king out exocrine and endocrine secretion. *Trends Cell. Biol.* **14,** 525–528.

Coppola, T., Perret-Menoud, V., Luthi, S., Farnsworth, C. C., Glomset, J. A., and Regazzi, R. (1999). Disruption of Rab3-calmodulin interaction, but not other effector interactions, prevents Rab3 inhibition of exocytosis. *EMBO J.* **18,** 5885–5891.

Coppola, T., Hirling, H., Perret-Menoud, V., Gattesco, S., Catsicas, S., Joberty, G., Macara, I. G., and Regazzi, R. (2001). Rabphilin dissociated from Rab3 promotes endocytosis through interaction with Rabaptin-5. *J. Cell Sci.* **114,** 1757–1764.

Coppola, T., Frantz, C., Perret-Menoud, V., Gattesco, S., Hirling, H., and Regazzi, R. (2002). Pancreatic beta-cell protein granuphilin binds Rab3 and Munc-18 and controls exocytosis. *Mol. Biol. Cell* **13,** 1906–1915.

Darchen, F., and Goud, B. (2000). Multiple aspects of Rab protein action in the secretory pathway: Focus on Rab3 and Rab6. *Biochimie* **82,** 375–384.

Da Silva Xavier, G., Qian, Q., Cullen, P. J., and Rutter, G. A. (2004). Distinct roles for insulin and insulin-like growth factor-1 receptors in pancreatic beta-cell glucose sensing revealed by RNA silencing. *Biochem. J.* **377,** 149–158.

Fukuda, M. (2003). Distinct Rab binding specificity of Rim1, Rim2, rabphilin, and Noc2. Identification of a critical determinant of Rab3A/Rab27A recognition by Rim2. *J. Biol. Chem.* **278,** 15373–15380.

Fukuda, M. (2005). Versatile role of Rab27 in membrane trafficking: Focus on the Rab27 effector families. *J. Biochem.* **137,** 9–16.

Guan, K. L., and Dixon, J. E. (1991). Eukaryotic proteins expressed in *Escherichia coli*: An improved thrombin cleavage and purification procedure of fusion proteins with glutathione S-transferase. *Anal. Biochem.* **192,** 262–267.

Hannon, G. J., and Rossi, J. J. (2004). Unlocking the potential of the human genome with RNA interference. *Nature* **431,** 371–378.

Huppi, K., Martin, S. E., and Caplen, N. J. (2005). Defining and assaying RNAi in mammalian cells. *Mol. Cell* **17,** 1–10.

Iezzi, M., Escher, G., Meda, P., Charollais, A., Baldini, G., Darchen, F., Wollheim, C. B., and Regazzi, R. (1999). Subcellular distribution and function of Rab3A, B, C, and D isoforms in insulin-secreting cells. *Mol. Endocrinol.* **13,** 202–212.

Iezzi, M., Kouri, G., Fukuda, M., and Wollheim, C. B. (2004). Synaptotagmin V and IX isoforms control Ca^{2+}-dependent insulin exocytosis. *J. Cell Sci.* **117,** 3119–3127.

Izumi, T., Gomi, H., Kasai, K., Mizutani, S., and Torii, S (2003). The roles of Rab27 and its effectors in the regulated secretory pathways. *Cell Struct. Funct.* **28,** 465–474.

Kasai, K., Ohara-Imaizumi, M., Takahashi, N., Mizutani, S., Zhaom, S., Kikuta, T., Kasai, H., Nagamatsu, S., Gomim, H., and Izumi, T. (2005). Rab27a mediates the tight docking of insulin granules onto the plasma membrane during glucose stimulation. *J. Clin. Invest.* **115**, 388–396.

Kato, M., Sasaki, T., Ohya, T., Nakanishi, H., Nishioka, H., Imamura, M., and Takai, Y. (1996). Physical and functional interaction of rabphilin-3A with alpha-actinin. *J. Biol. Chem.* **271**, 31775–31778.

Kuroda, T. S., Fukuda, M., Ariga, H., and Mikoshiba, K. (2002). The Slp homology domain of synaptotagmin-like proteins 1–4 and Slac2 functions as a novel Rab27A binding domain. *J. Biol. Chem.* **277**, 9212–9218.

Meister, G., and Tuschl, T. (2004). Mechanisms of gene silencing by double-stranded RNA. *Nature* **431**, 343–349.

Ohya, T., Sasaki, T., Kato, M., and Takai, Y. (1998). Involvement of Rabphilin3 in endocytosis through interaction with Rabaptin5. *J. Biol. Chem.* **273**, 613–617.

Reetz, A., Solimena, M., Matteoli, M., Folli, F., Takei, K., and De Camilli, P. (1991). GABA and pancreatic beta-cells: Colocalization of glutamic acid decarboxylase (GAD) and GABA with synaptic-like microvesicles suggests their role in GABA storage and secretion. *EMBO J.* **10**, 1275–1284.

Regazzi, R., Li, G. D., Deshusses, J., and Wollheim, C. B. (1990). Stimulus-response coupling in insulin-secreting HIT cells. Effects of secretagogues on cytosolic Ca^{2+}, diacylglycerol, and protein kinase C activity. *J. Biol. Chem.* **265**, 15003–15009.

Seabra, M. C., Mules, E. H., and Hume, A. N. (2002). Rab GTPases, intracellular traffic and disease. *Trends Mol. Med.* **8**, 23–30.

Waselle, L., Coppola, T., Fukuda, M., Iezzi, M., El-Amraoui, A., Petit, C., and Regazzi, R. (2003). Involvement of the Rab27 binding protein Slac2c/MyRIP in insulin exocytosis. *Mol. Biol. Cell* **14**, 4103–4113.

Wick, P. F., Senter, R. A., Parsels, L. A., Uhler, M. D., and Holz, R. W. (1993). Transient transfection studies of secretion in bovine chromaffin cells and PC12 cells. Generation of kainate-sensitive chromaffin cells. *J. Biol. Chem.* **268**, 10983–10989.

Yaekura, K., Julyan, R., Wicksteed, B. L., Hays, L. B., Alarcon, C., Sommers, S., Poitout, V., Baskin, D. G., Wang, Y., Philipson, L. H., and Rhodes, C. J. (2003). Insulin secretory deficiency and glucose intolerance in Rab3A null mice. *J. Biol. Chem.* **278**, 9715–9721.

[7] Analysis of Rab1 Recruitment to Vacuoles Containing *Legionella pneumophila*

By JONATHAN C. KAGAN, TAKAHIRO MURATA, and CRAIG R. ROY

Abstract

Legionella pneumophila is an intracellular bacterial pathogen that utilizes a type IV secretion system to subvert the function of the Rab1 GTPase. Bacterial proteins translocated into host cells mediate the accumulation of the Rab1 protein on vacuoles containing *Legionella pneumophila*. Assays used to investigate recruitment of Rab1 by *L. pneumophila* are described. These assays can be used to determine host and bacterial factors required for *L. pneumophila* subversion of the host secretory pathway.

METHODS IN ENZYMOLOGY, VOL. 403
Copyright 2005, Elsevier Inc. All rights reserved.
0076-6879/05 $35.00
DOI: 10.1016/S0076-6879(05)03007-7

Introduction

Intracellular pathogens have devised sophisticated strategies for subverting host cell biological processes. Many pathogens that replicate within specialized vacuoles have devised strategies for manipulation of membrane transport pathways. The bacterial pathogen *Legionella pneumophila* is able to modulate vacuole transport and fusion following uptake by phagocytic host cells (Roy and Tilney, 2002). As a result, the bacterium is able to prevent fusion of the vacuole in which it resides with lysosomes (Horwitz, 1983b; Roy *et al.*, 1998) and to subvert early secretory vesicles to create a unique compartment similar to the endoplasmic reticulum (ER) (Kagan and Roy, 2002; Kagan *et al.*, 2004). Intracellular replication of *L. pneumophila* occurs within this ER-derived organelle (Horwitz, 1983a).

To manipulate membrane transport in host cells, *L. pneumophila* employs a secretion system called the Dot/Icm transporter to translocate bacterial proteins across membranes and deliver these proteins into the cytosol of the host cell (Chen *et al.*, 2004; Luo and Isberg, 2004; Nagai *et al.*, 2002). The Dot/Icm transporter is composed of over 25 different proteins (Segal *et al.*, 1998; Vogel *et al.*, 1998). Mutations that eliminate Dot/Icm transporter function render *L. pneumophila* incapable of modulation of host membrane transport. Proteins translocated into host cells by the Dot/Icm transporter can subvert the host GTPase proteins Arf1 and Rab1. Shortly after the bacterium is internalized by a phagocytic host cell, the vacuole containing *L. pneumophila* is highly enriched for both the Arf1 and Rab1 proteins (Derre and Isberg, 2004; Kagan and Roy, 2002; Kagan *et al.*, 2004; Nagai *et al.*, 2002). Mutant bacteria defective in Dot/Icm transporter function fail to recruit Arf1 and Rab1. The RalF protein is the substrate protein translocated into host cells by the Dot/Icm transporter that mediates recruitment of Arf1 to the *L. pneumophila*-containing vacuole (Nagai *et al.*, 2002). RalF is an Arf guanine nucleotide exchange factor (GEF) that has sequence and structural similarity to host Arf GEFs containing Sec7-homology domains. The factor that recruits Rab1 to the *L. pneumophila*-containing vacuole has not been identified.

In this chapter, we describe assays that can be used to investigate Rab1 recruitment to the *L. pneumophila*-containing vacuole, and the role of Rab1 in the formation of a vacuole that supports *L. pneumophila* replication. Rab1 recruitment is used as an assay to measure *in vivo* functioning of the Dot/Icm transporter and is an indicator of subversion of early secretory vesicles to the *L. pneumophila*-containing vacuole. Dominant-interfering variants of Rab1 are used to investigate the importance of Rab1 function on recruitment and fusion of ER-derived vesicles to the *L. pneumophila*-containing vacuole.

Reagents

1. Unless otherwise stated, all reagents are obtained from Sigma-Aldrich.

2. Wild–type *L. pneumophila* strain CR39 (Lp01), the isogenic *dotA* mutant strain CR58, rabbit anti-*L. pneumophila* antibodies, pEGFP-Rab1a, pEGFP-Rab1(S25N), and pFcγRII are available from the Roy laboratory (Kagan *et al.*, 2004).

3. L929 fibroblasts (L-cells) are cultured in Dulbecco's modified Eagles medium (DMEM; Gibco) with 10% fetal bovine serum (FBS; Gemini). L-cells secrete macrophage colony-stimulating factor (MCSF). L-cell conditioned medium is prepared by growing cells in 175-cm^2 tissue culture flasks for 10 days after the cultures have reached confluency. The conditioned tissue culture medium is passed through a 0.22-μm filter and either used immediately or frozen at $-20°$ for future use.

4. Bone marrow macrophage medium consists of 50% RPMI, 20% FBS, and 30% L-cell conditioned medium.

5. Chinese hamster ovary (CHO) cells are obtained from the American Type Culture Collection. CHO cells expressing the FcγRII (CHO FcγRII cells) were constructed in the laboratory of Dr. Ira Mellman (Joiner *et al.*, 1990) and cultured in minimal Eagles media alpha (MEM; Gibco) with 10% FBS.

6. Polyclonal anti-Rab1b antibodies are obtained from Santa Cruz Biotechnology.

7. Alexa594 anti-rabbit IgG is obtained from Molecular Probes.

8. Preparation of charcoal yeast extract (CYE) agar for cultivation of *L. pneumophila* (Feeley *et al.*, 1979):

a. For 1 liter of medium add 10 g of yeast extract (Difco) and 10 g of *N*-(2-acetamido)-2-aminoethanesulfonic acid (ACES) to approximately 800 ml of distilled water and adjust to pH 6.9 using 1 *N* KOH (do not use NaOH).

b. Add 2 g of activated charcoal and 15 g of bacto-agar (Difco) to the medium and adjust the volume to 1 liter using distilled water.

c. Sterilize medium in a 2-liter flask by autoclaving.

d. Cool sterilized medium to 50° in a water bath.

e. Prepare fresh 100× filter-sterilized supplements separately. Supplements needed are L-cysteine (0.4 g/10 ml distilled water) and ferric nitrate (0.135 g [$Fe(NO_3)_3 \cdot 9H_2O$]/10 ml distilled water).

f. Add 10 ml of cysteine and 10 ml of ferric nitrate to the cooled medium from the fresh 100× stocks.

g. Dispense medium into sterile 10-cm Petri dishes and allow agar to solidify. Store plates at 4°.

9. RPMI 1640 is obtained from Gibco.
10. Fugene-6 transfection reagent is obtained from Roche.
11. Sterile phosphate-buffered saline (PBS) is obtained from Gibco.
12. A/J mice are obtained from Jackson Laboratories.

Analysis of Rab1b Recruitment to Vacuoles Containing L. pneumophila in Bone Marrow-Derived Macrophages

General Comments

This procedure is used to analyze the recruitment on endogenous Rab1 protein to vacuoles containing *L. pneumophila* in primary macrophages, which are the host cells that support bacterial growth during human infection. All procedures are performed using standard tissue culture techniques to ensure sterile conditions. Bone marrow macrophages derived from the A/J strain of mouse are used as model phagocytes for *L. pneumophila* infections. The use of macrophages prepared from other inbred mouse strains is not recommended as most inbred mouse strains restrict the intracellular growth of *L. pneumophila* (Yamamoto *et al.*, 1988), which could affect membrane transport and recruitment of host factors to the *Legionella*-containing vacuole.

Experimental Procedures

Preparation of Murine Bone Marrow-Derived Macrophages. Murine macrophages are prepared from bone marrow cells according to the procedure of Celada *et al.* (1984). After an A/J mouse is sacrificed by carbon dioxide asphyxiation, the femurs are removed and excess tissue is carefully debrided with a sterile scalpel. Using a pair of sharp sterile scissors, the two ends from each femur bone are removed. Bone marrow cells are flushed from each bone into a sterile 15-ml screw-cap tube using a 10-cm^3 syringe containing 5 ml of RPMI fitted with a 25-gauge needle. Two femurs from a single mouse should yield approximately 2×10^7 cells. The bone marrow cells are plated in 10 cm non-tissue culture-treated Petri dishes containing 20 ml of bone marrow macrophage medium. After 4 days of incubation at $37°$ in a humidified 5% CO_2 incubator, an additional 10 ml of bone marrow medium is added to each plate. Macrophages are ready to harvest 7–10 days after isolation of the bone marrow cells. Harvesting of macrophages is accomplished by vigorous washing of the adherent cells with ice-cold PBS using a 10-ml pipette, which causes the detachment of the macrophages from the dish. All washes are pooled and the cells are concentrated by centrifugation at $250 \times g$ for 10 min. Macrophages are then plated onto

12-mm sterile glass coverslips in 24-well tissue culture plates at a density of 1 × 10^5 cells/well in 0.5 ml of infection medium (85% RPMI, 10% FBS, 5% L-cell conditioned medium) and incubated for 24 h prior to use in *L. pneumophila* infections.

L. pneumophila *Infection of Bone Marrow-Derived Macrophages.* *L. pneumophila* strains CR39 and CR58 are cultured on CYE agar plates for 3–4 days at 37° in a standard incubator to obtain single colonies. A fresh single colony is then distributed evenly onto the surface a CYE agar plate and the plate is incubated at 37° for 48 h. Bacteria are transferred from the CYE agar plate to a 1.5-ml sterile Eppendorf tube containing 1 ml of water using a wooden applicator and a bacterial suspension is prepared by vortexing the tube until aggregates are no longer observed. Bacterial cells are washed by centrifugation at 12,000×*g* for 3 min at 25° and resuspended in 1 ml water. Bacterial concentrations are then determined using a spectrophotometer where an A_{600} of 1 equals approximately 10^9 bacteria ml^{-1}. Bacteria are diluted in sterile water to achieve a final solution containing 1 × 10^8 *L. pneumophila* cells ml^{-1}. *L. pneumophila* bacteria are added to the macrophages in the 24-well tissue culture plate. For most experiments, a multiplicity of infection (MOI) of 20 bacteria per macrophage is used, which is achieved by the addition of 2 μl of the final bacterial suspension to each well of macrophages. An MOI of 20 should result in roughly 10% of the macrophages being infected by a single *L. pneumophila* bacterium and very few macrophages being infected by more than 1 bacterium, which is ideal for applications requiring single cell analysis. To facilitate synchronization of infection the 24-well plates containing the infected macrophages are centrifuged at 250×*g* for 5 min at 25°. Plates are immediately transferred to a 37° water bath for 5 min to initiate bacterial internalization. Uninternalized bacteria are removed by three washes with PBS and the samples are refreshed with 0.5 ml of prewarmed infection medium and incubated for an additional 30 min at 37° in a humidified 5% CO_2 incubator.

Cell Fixation and Staining Procedures. Cells are fixed by removing the medium from each well and adding 0.5 ml of PBS containing 2% paraformaldehyde for 20 min at 25°. Fix is removed and the cells are washed twice in PBS, and then permeabilized with ice-cold methanol for 10 s. Coverslips are incubated in blocking buffer (PBS containing 2% goat serum and 50 m*M* ammonium chloride) for 30 min. Rabbit anti-Rab1b polyclonal antibody is diluted 1:100 into antibody dilution buffer (PBS containing 2% goat serum). Parafilm is stretched evenly over the back of a 24-well tissue culture plate and 40 μl of the primary antibody solution is spotted onto the parafilm. Coverslips are inverted onto a single drop of the primary antibody solution and incubated for 1 h at 25°. Coverslips are returned to the 24-well plates, washed 3× with PBS, and inverted onto a 40 μl drop of Alexa594 anti-rabbit IgG secondary antibody that has been diluted 1:250 in antibody dilution buffer. After incubation for 1 h at 25°, the coverslips are washed 3× in PBS. DNA is labeled with 4,6-diamidino-2-phenylindole (DAPI, 0.1 μg ml^{-1}) for 5 min at 25°, which will stain both the host cell nucleus and individual bacterial cells. After the coverslips are mounted on glass slides, fluorescence microscopy is used to assess Rab1 localization to vacuoles containing *L. pneumophila*.

Expected Results and Variations. Typically, 40–50% of all vacuoles containing the wild-type *L. pneumophila* strain CR39 will display Rab1 staining (Kagan *et al.*, 2004). The *dotA* mutant strain CR58 serves as a

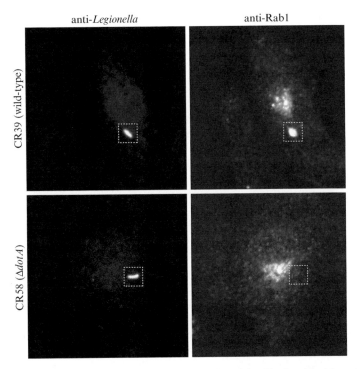

Fig. 1. Bone marrow-derived macrophages were infected for 30 min with either wild-type *L. pneumophila* (top panels) or an isogenic *eotA* mutant strain (bottom panels). The cells were double stained with a mouse anti-*L. pneumophila* antibody and a rabbit antibody specific for the Rab 1b protein, followed by fluorescein goat anti-mouse IgG and Alexa594 goat anti-rabbit IgG. Confocal microscopy was used to acquire images that show the location of *L. pneumophila* bacteria (left panels) and endogenous Rab1 localization (right panels). The boxed region in the top panels shows the location of a wild-type *L. pneumophila* bacterium that resides in a vacuole that stains positive for the Rab1 protein. The boxed region in the bottom panels shows the location of a *dotA* mutant *L. pneumophila* bacterium that resides in a vacuole that stains negative for the Rab1 protein.

negative control. Rab1 colocalization should not be observed on vacuoles containing CR58. As an alternative to using DAPI to identify bacteria, *L. pneumophila* producing the green fluorescent protein (GFP) or the red fluorescent protein (RFP) can be used to infect macrophages. Several groups have described plasmids producing GFP or RFP that can be used for localization of *L. pneumophila* by fluorescence microscopy (Chen *et al.*, 2004; Coers *et al.*, 1999). Additionally, a mouse anti-*L. pneumophila* antibody followed by a fluorescently labeled goat anti-mouse secondary can be used to identify the *L. pneumophila* cells (see Fig. 1).

Analysis of GFP-Rab1a Recruitment to Vacuoles Containing *L. pneumophila* in CHO FcγRII Cells

General Comments

Because primary macrophages are difficult to manipulate genetically and the production of these cells requires the care and use of animals, these cells are not ideal for certain applications. As an alternative, localization of GFP-Rab1a to the *Legionella*-containing vacuole in CHO FcγRII cells provides a sensitive assay that can be used to investigate the bacterial and host cell determinants important for the Rab1 recruitment process.

Transfection of CHO FcγRII Cells. The plasmid pEGFP-Rab1a produces the Rab1a protein fused to the C-terminus of GFP in the mammalian expression plasmid pEGFP-C1. This construct can be used to visualize GFP-Rab1 recruitment to the *Legionella*-containing vacuole in transfected host cells. For these studies, CHO FcγRII cells are plated onto 12-mm glass coverslips in 24-well dishes at a density of 3×10^4 cells/well in 0.5 ml of medium. After incubation for 24 h at 37° in a humidified 5% CO_2 incubator, cells are transfected with 0.5 μg of pEGFP-Rab1 using Fugene-6 according to the manufacturer's protocol. Following transfection, cells are incubated for 16 h at 37° in a humidified 5% CO_2 incubator.

L. pneumophila Infection of CHO FcγRII Cells. *L. pneumophila* strains CR39 and CR58 are grown and resuspended in sterile distilled water as described above. Prior to *L. pneumophila* infection, the culture medium for the CHO FcγRII cells is replaced with fresh medium containing a 1:1000 dilution of rabbit anti-*L. pneumophila* antisera. Because *L. pneumophila* is not able to invade nonphagocytic cells with high efficiency, opsonization with IgG permits efficient Fc receptor-mediated uptake into the CHO FcγRII cells. The route of *L. pneumophila* internalization does not affect Rab1 recruitment to the *Legionella*-containing vacuole. To initiate infection, *L. pneumophila* is added to transfected CHO FcγRII cells in 24-well plates at an MOI of 1. Samples are centrifuged for 5 min at $250 \times g$ and warmed for 5 min in a 37° water bath. Extracellular bacteria are removed by three washes with ice-cold PBS. Fresh tissue culture medium is added to each well and the plates are incubated for 30 min at 37° in a humidified 5% CO_2 incubator.

Cell Fixation and Staining Procedures. Samples are fixed in PBS containing 2% PFA for 20 min at 25°. After fixation, extracellular bacteria that have rabbit anti-*L. pneumophila* antibodies bound to their surface are labeled by inverting the coverslips onto a 40 μl drop of Alexa594-conjugated anti-rabbit IgG diluted 1:250 in antibody dilution buffer for 40 min at 25°. Coverslips are washed 3× with PBS and the bound Alexa594-conjugated antibody is immobilized by a 10 min incubation of the coverslips in PBS containing 1% PFA at 25°. Coverslips are immersed in ice-cold methanol for 10 s to permeabilize all cellular membranes. To label DNA inside of bacteria and the host cell nucleus, the cells are incubated with 0.1 μg ml^{-1} DAPI in PBS for 5 min at 25°. Coverslips are mounted onto glass slides and are visualized by fluorescence microscopy.

Expected Results and Variations. Similar to what is observed after infection of macrophages and staining for endogenous Rab1 protein, roughly 40–50% of vacuoles containing wild-type *L. pneumophila* will display GFP-Rab1 staining (Kagan *et al.*, 2004). GFP-Rab1 should not be observed on vacuoles containing *dotA* mutant *L. pneumophila*. This procedure can be adapted to analyze Rab1 recruitment to the *Legionella*-containing vacuole by time-lapse video microscopy. Cells can be infected with *L. pneumophila* producing RFP and recruitment of GFP-Rab1 can be visualized by dual color time-lapse fluorescence microscopy.

Effects of Interfering Rab1 Mutants on Replication of *L. pneumophila* in CHO FcγRII Cells

General Comments

The ability to assess *L. pneumophila* intracellular growth in cells transfected with dominant-interfering mutants of Rab1, or any other protein, is difficult because bacterial growth in untransfected cells obscures the effect the interfering protein may be having on *L. pneumophila* growth in transfected cells. Described is a relatively straightforward procedure that significantly reduces the relative frequency of *L. pneumophila* infection of untransfected cells. This approach takes advantage of the fact that *L. pneumophila* is inefficiently internalized by nonphagocytic cells. To ensure that *L. pneumophila* will be internalized preferentially by cells producing a dominant interfering Rab1 protein, nonphagocytic CHO cells are cotransfected with a plasmid encoding the FcγRII protein and a plasmid encoding the GDP-locked Rab1(S25N) protein. Because opsonized *L. pneumophila* enters FcγRII producing CHO cells with high efficiency, the majority of infected host cells will also be producing the Rab1 (S25N) protein. *L. pneumophila* colony-forming units (CFUs) are then recovered over time from the infected cells to determine the importance of Rab1 function on the intracellular growth of *L. pneumophila*.

Experimental Procedures

Cotransfection of CHO Cells. Plasmids encoding Rab1 fusion proteins with an amino-terminal GFP tag were created by ligating a cDNA fragment encoding either wild-type Rab1 or the Rab1(S25N) mutant into pEGFP-C1. CHO cells are plated onto 12-mm glass coverslips in 24-well plates at a density of 3×10^4 cells/well in 0.5 ml of medium. Triplicate samples are required for each data point. This means that six wells of transfected cells will be needed for each strain and plasmid combination to be assayed. After incubation for 24 h at 37° in a humidified 5% CO_2 incubator, cells are cotransfected with a plasmid encoding FcγRII and a plasmid encoding either GFP alone, GFP-Rab1, or GFP-Rab1(S25N) using the Fugene-6 transfection reagent. Transfected cells are incubated for 16 h at 37° in a humidified 5% CO_2 incubator. Before infection, cells are examined using an inverted fluorescence microscope, and the percent of total cells producing GFP is estimated. Transfection frequencies should be above 30% for optimal results.

L. pneumophila *Infection of Transfected CHO Cells.* L. pneumophila strains CR39 and CR58 are grown and resuspended in sterile distilled water as described above. Prior to *L. pneumophila* infection, the culture medium for the CHO FcγRII cells is replaced with fresh medium containing a 1:1000 dilution of rabbit anti-*L. pneumophila* antisera. *L. pneumophila* is then added to each well of transfected CHO cells at an MOI of 0.1. Samples are centrifuged for 5 min at $250 \times g$ and warmed for 5 min in a 37° water bath. Extracellular bacteria are removed by three washes with ice-cold PBS. Fresh medium containing the antibiotic gentamycin (20 μg ml^{-1}) is added, and the cells are incubated for 3 h at 37° in a humidified 5% CO_2 incubator. The antibiotic-containing medium is replaced with fresh tissue culture medium.

Intracellular Growth Assays. After the antibiotic-containing medium has been removed from the wells, the cells are washed 3× in PBS, and host cells in the first series of wells are lysed by the addition of 1 ml of sterile water. Vigorous pipetting of the cells in each well ensures complete lysis. The lysate is removed from each well and placed into a sterile 10-ml tube. To determine the number of *L. pneumophila* cells in each well, dilutions of the lysate are plated onto CYE agar plates and CFUs are enumerated after incubation of the plates for 3–4 days at 37°. To the remaining wells containing infected CHO cells, 1 ml of fresh medium is added, and the plates are incubated for 24 h at 37° in a humidified 5% CO_2 incubator. *L. pneumophila* is enumerated in the remaining wells by removing the tissue culture medium and placing the solution in a sterile 10-ml tube. The host cells in each well are lysed in 1 ml of sterile water and the lysate is added to the tube containing the tissue culture medium from that same well. Dilutions of the solution containing the lysate and tissue culture medium are plated onto CYE agar to determine the total number of *L. pneumophila* CFUs in each well. After *L. pneumophila* CFUs have been determined from the wells lysed immediately after antibiotic treatment and the wells lysed 24 h later, intracellular replication is calculated by determining the increase in CFUs.

Expected Results and Variations. Compared to control cells transfected with either plasmid encoding GFP or GFP-Rab1, the growth of *L. pneumophila* in cells transfected with GFP-Rab1(S25N) is typically reduced 2- or 3-fold (Kagan *et al.*, 2004). It is recommended that cells are checked to make certain that production of the GFP-Rab1(S25N) protein is having an effect on transport of secretory vesicles in transfected cells. This can be achieved using a plasmid encoding secreted alkaline phosphatase and comparing the efficiency of alkaline phosphatase secretion in the GFP-Rab1(S25N)-producing cells to cells producing GFP or GFP-Rab1. In addition to measuring *L. pneumophila* growth in the cells producing GFP-Rab1(S25N), the delivery of the v-SNARE protein Sec22b to the *L. pneumophila*-containing vacuole can be assessed using immunofluorescence microscopy. Interfering with Rab1 function delays the kinetics of Sec22b delivery to the *L. pneumophila*-containing vacuole, consistent with Rab1 being important for the transport and fusion of secretory vesicles with this organelle.

References

Celada, A., Gray, P. W., Rinderknecht, E., and Schreiber, R. D. (1984). Evidence for a gamma-interferon receptor that regulates macrophage tumoricidal activity. *J. Exp. Med.* **160,** 55–74.

Chen, J., de Felipe, K. S., Clarke, M., Lu, H., Anderson, O. R., Segal, G., and Shuman, H. A. (2004). *Legionella* effectors that promote nonlytic release from protozoa. *Science* **303,** 1358–1361.

Coers, J., Monahan, C., and Roy, C. R. (1999). Modulation of phagosome biogenesis by *Legionella pneumophila* creates an organelle permissive for intracellular growth. *Nat. Cell Biol.* **1,** 451–453.

Derre, I., and Isberg, R. R. (2004). *Legionella pneumophila* replication vacuole formation involves rapid recruitment of proteins of the early secretory system. *Infect. Immun.* **72,** 3048–3053.

Feeley, J. C., Gibson, R. J., Gorman, G. W., Langford, N. C., Rasheed, J. K., Mackel, D. C., and Blaine, W. B. (1979). Charcoal-yeast extract agar: Primary isolation medium for *Legionella pneumophila. J. Clin. Microbiol.* **10,** 437–441.

Horwitz, M. A. (1983a). Formation of a novel phagosome by the Legionnaires' disease bacterium (*Legionella pneumophila)* in human monocytes. *J. Exp. Med.* **158,** 1319–1331.

Horwitz, M. A. (1983b). The Legionnaires' disease bacterium (*Legionella pneumophila*) inhibits phagosome lysosome fusion in human monocytes. *J. Exp. Med.* **158,** 2108–2126.

Joiner, K. A., Fuhrman, S. A., Miettinen, H. M., Kasper, L. H., and Mellman, I. (1990). *Toxoplasma gondii*: Fusion competence of parasitophorous vacuoles in Fc receptor-transfected fibroblasts. *Science* **249,** 641–646.

Kagan, J. C., and Roy, C. R. (2002). *Legionella* phagosomes intercept vesicular traffic from endoplasmic reticulum exit sites. *Nat. Cell Biol.* **4,** 945–954.

Kagan, J. C., Stein, M. P., Pypaert, M., and Roy, C. R. (2004). *Legionella* subvert the functions of Rab1 and Sec22b to create a replicative organelle. *J. Exp. Med.* **199,** 1201–1211.

Luo, Z. Q., and Isberg, R. R. (2004). Multiple substrates of the *Legionella pneumophila* Dot/Icm system identified by interbacterial protein transfer. *Proc. Natl. Acad. Sci. USA* **101,** 841–846.

Nagai, H., Kagan, J. C., Zhu, X., Kahn, R. A., and Roy, C. R. (2002). A bacterial guanine nucleotide exchange factor activates ARF on *Legionella* phagosomes. *Science* **295,** 679–682.

Roy, C. R., and Tilney, L. G. (2002). The road less traveled: Transport of *Legionella* to the endoplasmic reticulum. *J. Cell Biol.* **158,** 415–419.

Roy, C. R., Berger, K., and Isberg, R. R. (1998). *Legionella pneumophila* DotA protein is required for early phagosome trafficking decisions that occur within minutes of bacterial uptake. *Mol. Microbiol.* **28,** 663–674.

Segal, G., Purcell, M., and Shuman, H. A. (1998). Host cell killing and bacterial conjugation require overlapping sets of genes within a 22-kb region of the *Legionella pneumophila* genome. *Proc. Natl. Acad. Sci. USA* **95,** 1669–1674.

Vogel, J. P., Andrews, H. L., Wong, S. K., and Isberg, R. R. (1998). Conjugative transfer by the virulence system of *Legionella pneumophila. Science* **279,** 873–876.

Yamamoto, Y., Klein, T. W., Newton, C. A., Widen, R., and Friedman, H. (1988). Growth of *Legionella pneumophila* in thioglycolate-elicited peritoneal macrophages from A/J mice. *Infect. Immun.* **56,** 370–375.

[8] Reconstitution of Rab4-Dependent Vesicle Formation *In Vitro*

By ADRIANA PAGANO and MARTIN SPIESS

Abstract

We have developed an *in vitro* assay to reconstitute the formation of endosomal recycling vesicles. To achieve specificity for endosomes as the donor organelle, cells are surface-biotinylated and allowed to endocytose for 10 min, after which the remaining surface-biotin is stripped off. The cells are then permeabilized and the cytosol washed away. Upon addition

METHODS IN ENZYMOLOGY, VOL. 403   0076-6879/05 $35.00
                          DOI: 10.1016/S0076-6879(05)03008-9

of exogenous cytosol and energy, sealed vesicles containing biotinylated recycling receptors are produced. Modification of the cytosol, for example, by immunodepletion or addition of purified proteins, allows the identification of proteins involved in vesicle formation. The results show that recycling is mediated by AP-1/clathrin-coated vesicles, requires Rab4, and is negatively regulated by rabaptin-5/rabex-5.

Introduction

Transport receptors like the transferrin receptor, the low-density lipoprotein (LDL) receptor, and the asialoglycoprotein (ASGP) receptor cycle continuously between the plasma membrane and early endosomes (Spiess, 1990; Trowbridge *et al.*, 1993). Upon internalization, endocytic vesicles fuse to sorting endosomes. The receptors exit into tubular membranes that form recycling endosomes (or endocytic recycling compartment, ERC), whereas released ligands with the main fluid volume form endosomal carrier vesicles/multivesicular bodies (ECVs/MVBs) to late endosomes (Maxfield and McGraw, 2004). There appear to be two main recycling pathways from early endosomes to the plasma membrane, directly from sorting endosomes or via recycling endosomes (Hao and Maxfield, 2000; Sheff *et al.*, 1999; van Dam *et al.*, 2002).

Rab4 and its interactor rabaptin-5 were implicated in regulating recycling to the cell surface in nonpolarized and to the apical surface in polarized cells based on overexpression and depletion studies *in vivo* (de Wit *et al.*, 2001; Deneka *et al.*, 2003; Mohrmann *et al.*, 2002; van der Sluijs *et al.*, 1992). We have reconstituted the formation of recycling vesicles *in vitro* using permeabilized cells (Pagano *et al.*, 2004). The role of candidate proteins can be tested using immunodepleted cytosol or cytosol supplemented with purified proteins or inhibitors. Since the formation of a single cohort of vesicles is analyzed, indirect effects on the organization of endosomes are less likely to affect the results than in *in vivo* studies. *In vitro* formation of endosomal vesicles containing the recycling ASGP receptor H1 was found to require Rab4 and to be inhibited by its effector rabaptin-5/rabex-5.

In Vitro Reconstitution of Endosomal Recycling Vesicles

Overview

The basic procedure of the assay is summarized schematically in Fig. 1. Since permeabilized cells, rather than purified endosomes, are used for reconstitution, the specificity for the donor organelle has to be introduced

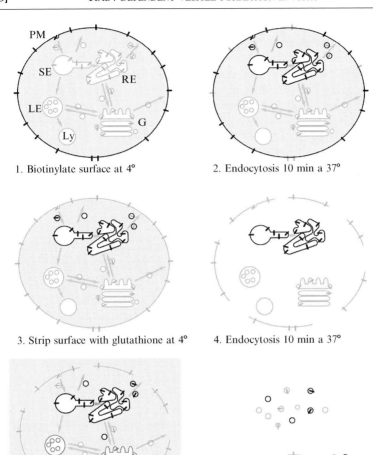

FIG. 1. Overview of the procedure to reconstitute the formation of recycling vesicles. For a description •ee the text. Black membranes contain biotinylated proteins. Recycling receptors are indicated by short lines spanning the membrane (I). G, Golgi; LE, late endosome; Ly lysosome; PM, plasma membrane; RE, recycling endosome; SE, sorting endosome. Other organelles are omitted for simplicity.

biochemically. (1) The plasma membrane of the cell is biotinylated using a membrane-impermeant reagent at 4°. (2) Labeled proteins are then allowed to internalize at 37° for 10 min, which limits access to early endosomes. (3) Back at 4°, surface biotin is released with reduced glutathione. Only endocytosed proteins in early endosomes remain labeled. (4) Cells are permeabilized by hypotonic swelling and scraping, and cytosol and small vesicles are washed away. To remove peripherally attached proteins, the membranes are washed with high-salt buffer. (5) To reconstitute vesicle formation, the permeabilized cells are incubated with exogenous cytosol and energy at 37°. (6) Newly formed vesicles are recovered in the supernatant after pelleting the broken cells and organelles. Upon membrane solubilization, biotinylated proteins, which must have originated from early endosomes, are isolated by avidin precipitation. Among this material, the protein of interest is detected by immunoblot analysis. In the case of a recycling protein like the ASGP receptor, the vesicles that contained them were destined to recycle to the plasma membrane.

Protocols

Buffer Solutions

Buffer A: 0.1 M MES/NaOH, pH 6.6, 0.5 mM MgCl$_2$, 1 mM ethyleneglycoltetraacetic acid (EGTA), 0.2 mM dithiothreitol (DTT).

PBS^{++}: phosphate-buffered saline (PBS) with 0.7 mM CaCl$_2$, 0.25 mM MgCl$_2$.

Sulfo-NHS-SS-biotin solution: 1 mg/ml sulfosuccinimidyl-2-(biotinamido)-ethyl-1,3-dithiopropionate (Pierce) in PBS^{++}, to be prepared immediately before use.

Glycine solution: 50 mM glycine in PBS^{++}.

MEM/HEPES: serum-free minimal essential medium containing 20 mM HEPES/NaOH, pH 7.2.

Glutathione solution: 50 mM reduced glutathione (Sigma), 75 mM NaCl, 75 mM NaOH, 1 mM ethylenediaminetetraacetic acid (EDTA), and 1% bovine serum albumin (Sigma). Dissolve glutathione immediately before use.

Iodoacetamide solution: 5 mg/ml iodoacetamide in PBS^{++}.

Swelling buffer: 15 mM HEPES/KOH, pH 7.2, 15 mM KCl.

Transport buffer, 10-fold concentrated stock solution: 200 mM HEPES/KOH, pH 7.2, 900 mM KOAc, 20 mM Mg(OAc)$_2$.

Stripping buffer: 20 mM HEPES/KOH, pH 7.2, 500 mM KOAc, 2 mM Mg(OAc)$_2$.

Protease inhibitor cocktail, 500-fold concentrated stock solution: 5 mg/ml benzamidine, 1 mg/ml antipain (both from Sigma), 1 mg/ml pepstatin A, 1 mg/ml leupeptin, 1 mg/ml chymostatin (all from Applichem) in 40% dimethyl sulfoxide (DMSO) and 60% ethanol.

Energy regenerating system, four separate stock solutions: 0.1 M ATP; 0.2 M GTP; 0.6 M creatine phosphate; 8 mg/ml creatine kinase (all from Roche Diagnostics).

Cytosol Preparation. Cytosol was obtained from calf brain as the high-speed supernatant after homogenization as a side product in the purification of clathrin-coated vesicles (Campbell *et al.*, 1984). The entire procedure is performed at 4°. Calf brains fresh from the slaughterhouse are cleaned of meninges using paper towels and rinsed with PBS. One volume of brain is homogenized with an equal volume of buffer A in a Waring blender (three times 8 s at medium speed) and centrifuged in a Sorvall GS3 rotor at 7000 rpm for 30 min. The supernatant is centrifuged in a Kontron TFT45.94 rotor at 40,000 rpm for 80 min. The clear supernatant is stored frozen in aliquots at −80° to be used as cytosol.

Cell Culture. Madin–Darby canine kidney (MDCK) cell line (strain II) stably expressing the ASGP receptor subunit H1 with a C-terminal myc-tag (Leitinger *et al.*, 1995) is grown in minimal essential medium supplemented with 2 mM L-glutamine, 10% fetal calf serum, 100 U/ml penicillin, 100 μg/ml streptomycin, and 0.5 mg/ml G418 sulfate (all from Life Technologies) at 37° with 7.5% CO_2. The assay was also used successfully with A431, NIH/3T3 cells, and baby hamster kidney (BHK) cells. For the experiment, 15-cm cell culture plates are coated by incubation for 30 min at 37° with 10 mg/ml poly-L-lysine (10,000–20,000; Fluka) in PBS followed by rinsing twice with PBS. Cells are seeded to be semiconfluent the next day.

Surface Biotinylation, Endocytosis, and Surface Stripping. Four 15-cm plates of semiconfluent cells are washed three times with ice-cold PBS^{++} and biotinylated at 4° for 30 min with 7 ml/plate of freshly prepared sulfo-NHS-SS-biotin solution. The reaction is quenched by rinsing the cells with cold PBS^{++}, by a 5-min incubation with glycine solution, and by another two rinses with PBS^{++}. The cells are then incubated in prewarmed MEM/HEPES for 10 min at 37° to allow endocytosis. The cells are rinsed twice with ice-cold PBS^{++}, incubated twice for 20 min with 7 ml/plate of freshly prepared glutathione solution to release surface biotin, rinsed twice with PBS^{++}, incubated for 5 min at 4° with 5 mg/ml iodoacetamide solution to quench any residual glutathione, followed by another two rinses with PBS^{++}.

Permeabilization and Cytosol Removal. The cells are permeabilized by addition of 10 ml/plate of swelling buffer at 4° for 15 min, scraped with a rubber policeman into 10 ml/plate of transport buffer, pooled, and sedimented at 800×g for 5 min. At this point essentially all cells are broken as judged by Trypan blue permeability. The pellet is resuspended twice for 10 min in 50 ml cold stripping buffer and centrifuged as before, and resuspended in transport buffer. The permeabilized cells are resuspended in 400 μl transport buffer, which should result in a protein concentration of approximately 1 μg/μl.

Vesicle Formation and Analysis. In siliconized Eppendorf tubes, 100 μl of permeabilized cells (100 μg protein) is incubated with cytosol (final protein concentration of 1.2 mg/ml) and energy-regenerating system (1.6 mM ATP, 3.2 mM GTP, 9.6 mM creatine phosphate, and 256 μg/ml creatine kinase) in a total reaction volume of 250 μl with transport buffer at 37° for 30 min. In control reactions, cytosol, energy, or both are omitted. To test the requirement for nucleotide hydrolysis, the nonhydrolyzable nucleotide analogs adenylyl imidodiphosphate (AMP-PNP; 0.16 mM) and guanylyl imidodiphosphate (GMP-PNP; 0.32 mM) are added instead of ATP and GTP. Reactions are stopped on ice and the cells sedimented at 800×g for 5 min. The supernatants are carefully aspirated and solubilized with lysis buffer (1% Triton X-100, 0.5% deoxycholate in PBS) containing protease inhibitor cocktail for 1 h at 4°. Insoluble material is removed by centrifugation in a microcentrifuge at 14,000 rpm for 10 min. Supernatants are recovered and rotated end-over-end for 1 h at 4° with 40 μl avidin-Sepharose (Pierce). The beads are washed three times with lysis buffer and boiled in SDS-gel sample buffer containing 100 mM dithiothreitol. Proteins are separated by SDS-gel electrophoresis and transferred to a polyvinylidene fluoride membrane that is then processed for immunoblot analysis using anti-myc antibody (9E10; ATCC). Antibody is detected using horseradish peroxidase-conjugated anti-mouse secondary antibody (Sigma) and the ECL kit (Amersham Biosciences).

Experimental Results

Figure 2A shows the result of a typical reconstitution experiment. No H1-containing vesicles were generated upon incubation of the permeabilized cells at 37° without additions or when incubated with nucleotides and cytosol at 4°. In the presence of ATP, GTP, and 1.2 mg/ml cytosol at 37°, typically ∼10% of biotinylated H1 in the starting material was recovered in the supernatant. With increased concentrations of up to ∼10 mg/ml cytosol, vesicle formation could be further stimulated ∼2-fold (Pagano *et al.*, 2004). Using a limiting amount of cytosol provides higher sensitivity to detect effects of depletion or addition of individual components as described below. In the absence of added cytosol, nucleotides supported a basal release of H1 into the supernatant of typically ∼20% of that in the

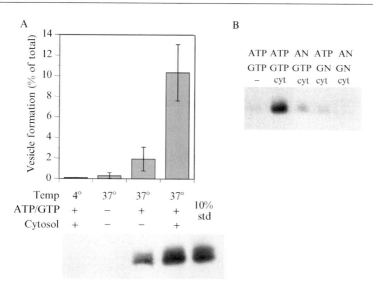

FIG. 2. Temperature-, energy-, and cytosol-dependent *in vitro* formation of endosomal vesicles containing H1. (A) Surface-biotinylated cells were incubated at 37° for 10 min to allow internalization, chilled in ice, and surface biotin was stripped with reduced glutathione. The cells were then broken and incubated for 30 min at 4° or 37°, with or without cytosol, or nucleotides and energy-regenerating system as indicated. Cells were pelleted and the supernatant was lysed and analyzed by avidin precipitation of biotinylated proteins, SDS-gel electrophoresis, and immunoblotting using anti-myc antibody. Values were normalized to the amount of total labeled H1 by analyzing 10% of a sample before centrifugation as a standard (10% std). Average and standard deviation of at least four independent determinations are shown. (B) Vesicle formation was assayed in the presence of cytosol (12.5 mg/ml) with GTP or GMP-PNP (GN), and ATP or AMP-PNP (AN), or without nucleotides. Reprinted from *Molecular Biology of the Cell* (Pagano *et al.*, 2004), with permission of The American Society for Cell Biology.

presence of cytosol (Fig. 2A). This is most likely due to residual membrane-associated proteins that had not been removed by the high-salt washes. In a time-course experiment, formation of H1-containing vesicles was already detectable within 5 min of incubation and reached a maximum after ~20 min (Pagano *et al.*, 2004). Up to 30 min, no decrease of the signal was observed, as might be expected if the vesicles would have fused efficiently with their target compartment. The integrity of the membrane vesicles containing the protein of interest can be tested by protease sensitivity, which in the case of H1 showed its cytoplasmic N-terminus to be exposed, but its transmembrane and exoplasmic domains to be protected, unless detergent was added (Pagano *et al.*, 2004).

Efficient formation of H1-containing vesicles required the presence of both ATP and GTP, suggesting the involvement of ATPase(s) and GTPase (s). The nonhydrolyzable analogues AMP-PNP and GMP-PNP did not substitute for ATP and GTP, respectively (Fig. 2B), indicating a requirement for nucleotide hydrolysis for the generation of recycling vesicles.

Vesicle Formation Using Modified Cytosol

The *in vitro* assay to reconstitute vesicle formation opens the possibility of testing the role of candidate proteins by modifying the cytosol. The most basic manipulations are the removal of a component by immunodepletion and the supplementation (or readdition) of a purified component to the cytosol. Here we present examples for these procedures.

Immunodepletion

To immunodeplete a protein from cytosol, a specific antibody may first be incubated with the cytosol before the antibody–antigen complexes are collected with protein A- or G-Sepharose, as appropriate. Alternatively, antibody may first be immobilized on protein A/G-Sepharose before incubation with cytosol. This is particularly useful when the antibody is dilute.

Protein A- and G-Sepharose (Zymed) are saturated with 5 mg/ml bovine serum albumin (fatty acid free; Sigma) overnight at 4°, washed three times, and resuspended in transport buffer as a 50% slurry. To deplete Rab4, 200 μl diluted cytosol (750 μg protein at 3.75 mg/ml) is first incubated with 15 μl of a rabbit anti-Rab4 antiserum (from Bruno Goud, Institut Curie, Paris), 35 μl transport buffer, and protease inhibitor cocktail overnight at 4°, and then mixed with 30 μl each of protein A- and G-Sepharose for 3 h at 4°. After centrifugation, the supernatant is collected and the beads are washed three times with 1 ml transport buffer and boiled with 200 μl SDS-gel loading buffer. Thirty microliters of the immunodepleted cytosol and of the initial diluted cytosol are boiled in SDS-gel loading buffer and analyzed, together with 30 μl of the material released from the beads, by SDS-gel electrophoresis and immunoblotting to determine the extent of depletion.

To deplete rabaptin-5 or Rab5, 30 μl each of protein A- and G-Sepharose are incubated overnight at 4° either with 500 μl transport buffer and 20 μl monoclonal mouse anti-rabaptin-5 antibody (Transduction Laboratories), or with 1 ml supernatant of a hybridoma-producing anti-Rab5 antibody (CL621.3; from Reinhard Jahn, Max Planck Institute, Göttingen), in the presence of protease inhibitor cocktail. The beads are then washed three times with transport buffer, mixed with 200 μl diluted cytosol (750 μg protein at 3.75 mg/ml), and incubated at 4° for 6 h to overnight. The depleted cytosol is collected and analyzed as above. Mock depletion of cytosol is performed identically with nonimmune antibody or without antibody.

To test the effect on vesicle formation, 85 μl depleted or mock-depleted cytosol is incubated with permeabilized cells (100 μg protein), energy regenerating system, and transport buffer to a total volume of 250 μl, and the released material is analyzed as described above.

Protein Purification

Rabaptin-5 is found in the cell in association with rabex-5, a Rab5 guanine nucleotide exchange factor (Horiuchi *et al.*, 1997). Purification of His$_6$-tagged rabaptin-5 in complex with rabex-5 produced in Sf9 insect cells infected with recombinant baculovirus vectors has been described in detail by Lippe *et al.* (2001). To test the effect of the purified protein, different amounts are added to standard reconstitution assays. If the buffer of the protein differs from the transport buffer, a buffer-only control needs to be included.

Experimental Results

As is shown in Fig. 3A and B, the formation of H1-containing endosomal vesicles was strongly inhibited by the removal of Rab4 from the cytosol, whereas it was not affected by depletion of more than 90% of Rab5. This confirms previous *in vivo* observations pointing to an involvement of Rab4 in receptor recycling (Deneka *et al.*, 2003; Mohrmann *et al.*, 2002; van der Sluijs *et al.*, 1992). Immunodepletion of clathrin-associated adaptor proteins similarly identified AP-1 as an essential component to generate H1-containing vesicles (Pagano *et al.*, 2004). Rabaptin-5/rabex-5 is an interesting candidate to connect Rab4 function with the AP-1 vesicle coat, because it was shown to bind to both Rab4 and the γ-ear domain of AP-1 (in addition to Rab5; de Renzis *et al.*, 2002; Deneka *et al.*, 2003; Vitale *et al.*, 1998). Upon immunodepletion of rabaptin-5/rabex-5, formation of endosomal vesicles was reproducibly stimulated (Fig. 3A and C), suggesting an inhibitory role of the complex. This was confirmed upon addition of purified, recombinantly expressed rabaptin-5/rabex-5 in amounts corresponding approximately to once or twice the amount already present in the cytosol, which clearly inhibited the production of H1-containing vesicles (Fig. 3D). The rabaptin-5/rabex-5 complex thus negatively regulates generation of recycling vesicles.

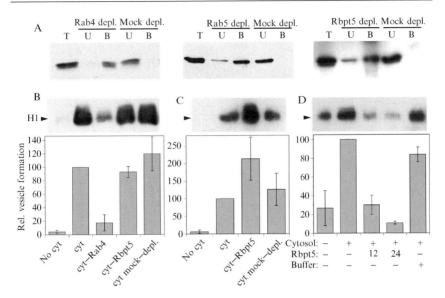

FIG. 3. Formation of recycling vesicles requires Rab4 and is inhibited by rabaptin-5/rabex-5. Bovine brain cytosol was immunodepleted of Rab4, Rab5, or rabaptin-5 (A). Total cytosol (T) and corresponding aliquots of the depleted cytosol (unbound fraction, U) and the bound material (B) were analyzed by immunoblotting with the respective antibodies. For (B) and (C), biotinylated permeabilized cells were incubated without cytosol (no cyt), with untreated cytosol (cyt), with cytosol depleted of Rab4, Rab5, or rabaptin-5 (cyt–Rab4; cyt–Rab5; cyt–Rbpt5), or with mock-depleted cytosol. For (D), the effect of supplementing 12 or 24 μg/ml purified rabaptin-5/rabex-5 or of the corresponding buffer was tested. Immunoblot analysis of biotinylated H1 in the supernatant after cell pelleting is shown for a representative experiment. Quantitation of three independent experiments (average with standard deviation) is presented below. Reprinted from *Molecular Biology of the Cell* (Pagano *et al.*, 2004), with permission of The American Society for Cell Biology.

Conclusions

Besides immunodepletion and supplementation of purified proteins, there are additional ways to manipulate the cytosol in the reconstitution reaction. Inhibitory antibodies may simply be added to the cytosol to inactivate the antigen, as was done to block clathrin function using the anti-heavy chain antibody X22 (Pagano *et al.*, 2004). It should also be possible to use cytosol extracted from cultured cells overexpressing specific wild-type or mutant proteins or depleted of a gene product by RNA interference. Since the assay focuses on a single process, the risk of indirect effects is likely to be very small. Finally, drugs may be applied to the *in vitro* reaction. For instance, we found the formation of H1-containing endosomal vesicles to be sensitive to brefeldin A (although at higher concentrations than required *in vivo* to block Arf1 nucleotide exchange factors) and insensitive to LY294002, an inhibitor of phosphatidylinositol 3-kinase involved in the fast recycling pathway from sorting endosomes (van Dam *et al.*, 2002). The results thus suggest that our procedure reconstitutes the slow recycling pathway from recycling endosomes. In contrast, in a similar *in vitro* assay, endosome-derived vesicles containing transferrin receptor and GLUT4 were generated by a brefeldin A-insensitive but neomycin-sensitive mechanism (Lim *et al.*, 2001). Since endocytic proteins had been internalized at 15°, a temperature at which transport from sorting to recycling endosomes is blocked, the starting compartment was predominantly sorting endosomes and the vesicles generated are likely to have represented the fast recycling pathway.

Acknowledgments

This work was supported by Grant 31-061579.00 from the Swiss National Science Foundation. We thank Drs. Bruno Goud, Reinhard Jahn, and Marino Zerial for reagents, and Dr. Pascal Crottet for helpful discussions.

References

Campbell, C., Squicciarini, J., Shia, M., Pilch, P. F., and Fine, R. E. (1984). Identification of a protein kinase as an intrinsic component of rat liver coated vesicles. *Biochemistry* **23**, 4420–4426.

de Renzis, S., Sönnichsen, B., and Zerial, M. (2002). Divalent Rab effectors regulate the subcompartmental organization and sorting of early endosomes. *Nat. Cell Biol.* **4**, 124–133.

de Wit, H., Lichtenstein, Y., Kelly, R. B., Geuze, H. J., Klumperman, J., and van der Sluijs, P. (2001). Rab4 regulates formation of synaptic-like microvesicles from early endosomes in PC12 cells. *Mol. Biol. Cell* **12**, 3703–3715.

Deneka, M., Neeft, M., Popa, I., van Oort, M., Sprong, H., Oorschot, V., Klumperman, J., Schu, P., and van der Sluijs, P. (2003). Rabaptin-5alpha/rabaptin-4 serves as a linker between rab4 and gamma(1)-adaptin in membrane recycling from endosomes. *EMBO J.* **22**, 2645–2657.

Hao, M., and Maxfield, F. R. (2000). Characterization of rapid membrane internalization and recycling. *J. Biol. Chem.* **275**, 15279–15286.

Horiuchi, H., Lippe, R., McBride, H. M., Rubino, M., Woodman, P., Stenmark, H., Rybin, V., Wilm, M., Ashman, K., Mann, M., and Zerial, M. (1997). A novel Rab5 GDP/GTP exchange factor complexed to Rabaptin-5 links nucleotide exchange to effector recruitment and function. *Cell* **90**, 1149–1159.

Leitinger, B., Hille-Rehfeld, A., and Spiess, M. (1995). Biosynthetic transport of the asialoglycoprotein receptor H1 to the cell surface occurs via endosomes. *Proc. Natl. Acad. Sci. USA* **92**, 10109–10113.

Lim, S. N., Bonzelius, F., Low, S. H., Wille, H., Weimbs, T., and Herman, G. A. (2001). Identification of discrete classes of endosome-derived small vesicles as a major cellular pool for recycling membrane proteins. *Mol. Biol. Cell* **12**, 981–995.

Lippe, R., Horiuchi, H., Runge, A., and Zerial, M. (2001). Expression, purification, and characterization of Rab5 effector complex, rabaptin-5/rabex-5. *Methods Enzymol.* **329,** 132–145.

Maxfield, F. R., and McGraw, T. E. (2004). Endocytic recycling. *Nat. Rev. Mol. Cell Biol.* **5,** 121–132.

Mohrmann, K., Leijendekker, R., Gerez, L., and van Der Sluijs, P. (2002). Rab4 regulates transport to the apical plasma membrane in Madin-Darby canine kidney cells. *J. Biol. Chem.* **277,** 10474–10481.

Pagano, A., Crottet, P., Prescianotto-Baschong, C., and Spiess, M. (2004). In vitro formation of recycling vesicles from endosomes requires adaptor protein-1/clathrin and is regulated by rab4 and the connector rabaptin-5. *Mol. Biol. Cell* **15,** 4990–5000.

Sheff, D. R., Daro, E. A., Hull, M., and Mellman, I. (1999). The receptor recycling pathway contains two distinct populations of early endosomes with different sorting functions. *J. Cell Biol.* **145,** 123–139.

Spiess, M. (1990). The asialoglycoprotein receptor—A model for endocytic transport receptors. *Biochemistry* **29,** 10009–10018.

Trowbridge, I. S., Collawn, J. F., and Hopkins, C. R. (1993). Signal-dependent membrane protein trafficking in the endocytic pathway. *Annu. Rev. Cell Biol.* **9,** 129–161.

van Dam, E. M., Ten Broeke, T., Jansen, K., Spijkers, P., and Stoorvogel, W. (2002). Endocytosed transferrin receptors recycle via distinct dynamin and phosphatidylinositol 3-kinase-dependent pathways. *J. Biol. Chem.* **277,** 48876–48883.

van der Sluijs, P., Hull, M., Webster, P., Male, P., Goud, B., and Mellman, I. (1992). The small GTP-binding protein rab4 controls an early sorting event on the endocytic pathway. *Cell* **70,** 729–740.

Vitale, G., Rybin, V., Christoforidis, S., Thornqvist, P., McCaffrey, M., Stenmark, H., and Zerial, M. (1998). Distinct Rab-binding domains mediate the interaction of Rabaptin-5 with GTP-bound Rab4 and Rab5. *EMBO J.* **17,** 1941–1951.

[9] Assay of Rab4-Dependent Trafficking on Microtubules

By JOHN W. MURRAY and ALLAN W. WOLKOFF

Abstract

We present an *in vitro* method to measure how Rab4 and other regulatory proteins affect microtubule-based organelle motility. The protocols utilize small-volume, disposable "microchambers" designed for epifluorescence, confocal, or other microscope platforms and into which microtubules, organelles, and primary and fluorescent secondary antibodies are added. Our work has focused on the isolation and use of endocytic vesicles from rat liver, and we present these protocols. However, the techniques can be adapted for other organelles or cell types. Multiple fluorescent probes, rapid image capture, and immunofluorescence under nonfixation conditions

METHODS IN ENZYMOLOGY, VOL. 403   0076-6879/05 $35.00
DOI: 10.1016/S0076-6879(05)03009-0

allow for measurements of the location and intensity changes of endogenous proteins upon addition of ATP or upon addition of other proteins or regulatory factors. We review measurements of microtubule-based motility as well as measurements for protein localization and protein segregation *in vitro*.

Introduction

In vitro methods to study the function of Rab proteins have the exciting potential to measure specific biochemical reactions, the concentrations of constituents, and the rates of reactions. Such measurements will lead to greater understanding of what organizational, energetic, kinetic, or solubility barriers are overcome by the presence of membrane-organizing proteins such as Rabs.

The *in vitro* assays described in this chapter employ disposable microscope chambers that allow investigation of the microtubule-based movement and protein localization of endocytic vesicles or other organelles. The assays have been developed to investigate how microtubule-based motility contributes to endosome function, which motor proteins are responsible for movement of different populations of endosomes, what proteins are associated with populations of endosomes, and how these proteins contribute to endosome motility and the related microtubule-based endosome "fission" activity. The assays are amenable to measurement of any enzyme activity that is detectable by light or fluorescence. For instance, endosome acidification has been measured by detection of the accumulation within vesicles of the fluorescent dye acridine orange (Murray *et al.*, 2002).

Isolation of Fluorescent Rat Liver Endosomes

We focus here on fluorescent endocytic vesicles prepared from rat liver (Murray *et al.*, 2000). However, we have also used Golgi and endoplasmic reticulum vesicles as well as vesicles from cultured cell lines (unpublished observations). In contrast to other systems (Blocker *et al.*, 1997; Pollock *et al.*, 1999), we have not had to add cytosol or motor proteins to the preparations to produce microtubule-based motility. The procedure takes about 7 h and the final endosomes are stored in single–use 15-μl aliquots at $-80°$.

Preparation of Fluorescent Endocytic Ligand

Endocytosis of asialoorosomucoid (ASOR) through its receptor, the asialoglycoprotein receptor (ASGPR), has been studied extensively. The receptor is abundant ($\sim 2 \times 10^5$ receptors per hepatocyte [Weigel, 1980]) and specifically expressed in hepatocytes. After endocytosis, ASOR debinds from ASGPR and is sorted to lysosomes where it is degraded, while ASGPR is recycled to the cell surface (Stockert, 1995).

Orosomucoid (α_1-acid glycoprotein, Sigma #G-9885) is desialylated to ASOR by exposure to 0.1 N H_2SO_4 for 1 h at 75° followed by neutralization and dialysis into water or appropriate buffer. The ASOR is then labeled with amine reactive fluorescent dyes (e.g., Texas Red sulfonyl chloride or Alexa Fluor 488 succinimidyl ester, Molecular Probes, Eugene, OR) according to the manufacturer's protocols to yield a molar ratio of 2:1 dye to ASOR as measured spectrophotometrically. We have measured a micromolar extinction coefficient for ASOR at JM OD_{278} of 0.0375 cm^{-1} μM^{-1} with an estimated (due to glycosylation) molecular weight of 40 kDa. Fluorescent ASOR is very stable in solution when stored at $-20°$.

Preparation of ASOR Endosomes from Rat Liver

Fifty milligrams of fluorescent ASOR is injected into the portal vein of anesthetized Sprague–Dawley rats. Use of greater than 50 mg of ASOR leads to loss of "single-wave" uptake kinetics (i.e., late endosomes will contain increased numbers of early endosomes). At specific times after injection the vena cava is cut and the liver is perfused through the portal vein with 30 ml of ice-cold phosphate-buffered saline (PBS). Subsequent procedures are performed at 4°. The liver is removed, washed in MEPS buffer plus dithiothreitol (DTT) (35 mM K_2-PIPES [note the inclusion of potassium], 5 mM ethyleneglycoltetraacetic acid [EGTA], 5 mM $MgCl_2$, 0.25 M sucrose, pH 7.1, plus 4 mM DTT), diced, and homogenized with 20 strokes in a loose Dounce homogenizer in 15 ml MEPS buffer containing 2 mM phenylmethylsulfonyl fluoride (PMSF) and 300 μl (a 1:50 dilution) of protease inhibitor cocktail (Sigma #P-8340). The homogenate is centrifuged at 3000 rpm in 15-ml conical tubes for 10 min at 4° in an Eppendorf 5810R centrifuge (radius $\sim$6 in.). The resulting postnuclear supernatant (PNS) is collected and protease inhibitors and DTT are readded at previous levels. The PNS is chromatographed on a Sephacryl S200 (Pharmacia, Uppsala, Sweden) gel filtration column equilibrated in MEPS. The opaque, yellow/white flow-through containing fluorescent endosomes is collected. Protease inhibitors and DTT are again added, and the mixture is brought to 1.4 M sucrose by addition of a 2.5 M sucrose-MEPS stock buffer. This is loaded into the bottom of a sucrose float-up step gradient consisting of 1.4, 1.2, and 0.25 M sucrose in MEPS buffer. The gradient is centrifuged at 39,000 rpm in an SW41 Ti Rotor (Beckman) for 2 h at 4°, and approximately 1.5 ml of vesicles per liver is harvested from a cloudy layer at the 1.2 M/0.25 M sucrose interface and stored at $-80°$.

Adaptation of the Isolation to Other Organelles and Cell Types

For endocytic vesicles from other cell types, the scale of the isolation should be adjusted. It is critical to have concentrated vesicles for microscopy experiments. A 10-g adult rat liver contains $\sim10^9$ cells, whereas a 100-mm Petri dish may contain 10^7 cells. For other organelles additional modifications will be required. The S200 chromatography step may be optional. We recommend against pelleting of organelles, and removal of extraneous ATPases or soluble inhibitors may be required. High concentrations of reducing agents, protease inhibitors, low temperature, and reasonably fast isolation are key to preventing motor protein inactivation. We have found that freezing vesicles as single-use aliquots yields reproducibly active plus and minus end directed vesicle motility and fission.

Preparation of Microtubule-Coated Microscopy Chambers

Disposable glass "microchambers" are constructed to view organelles and microtubules under microscopy. Similar chambers are familiar to many microscopists studying microtubule or actin-based motility (Howard and Hyman, 1993; Inoue, 1986; Pollock *et al.*, 1999; Waterman-Storer, 1998). Figure 1 shows a photograph and diagram of a chamber situated for an inverted microscope (lens is below the sample). A large coverslip (22 × 40 mm, Corning No. 290-244) forms the base of the chamber where oil and the oil immersion objective contact this coverslip. Two pieces of double-stick tape form the sides of the chamber (Scotch 3M No. 655, 0.009 cm thick), and a piece of glass slide (e.g., Fisher No. 12-550-10) forms the top and is created by scoring the slide at a width of 0.5 cm using a hand-held, wheeled glasscutter. Clear nail polish is applied to the seams to prevent the top from detaching during the washes. The internal volume is ~4 μl and buffer can be flowed in and withdrawn at either side via a Kimwipe and capillary action. Care must be taken to avoid chamber evaporation. A humidified box, excess buffer at the sides, or a small amount of glycerol to seal the chamber will prevent this.

Polymerizing Fluorescent Microtubules

Tubulin can be purified from calf brain or other sources and fluorescently labeled (Waterman-Storer, 1998). We find it convenient to purchase fluorescent tubulin. Lyophilized tubulin (e.g., #TL 238 [unlabeled], and TL 331 [rhodamine-labeled], Cytoskeleton Inc.) is brought to 10 mg/ml in cold microtubule buffer ("BRB80 buffer" [80 mM PIPES, 1 mM EGTA, 1 mM MgCl$_2$] + 3% glycerol + 1 mM GTP), clarified by centrifugation for 5 min at 14,000 rpm, and diluted to yield a molar ratio of 10:1 unlabeled to

A

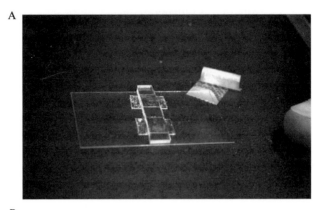

B

Assembled, DEAE-coated chamber:
Microtubules,
Block,
Suspension of organelles,
Block,
1° Antibody
Block,
2° Antibody
Anti-bleach

Coverslip

Cut glass

Tape

C

Organelles

λ 1° Antibody

✦ Fluorescent 2° Antibody

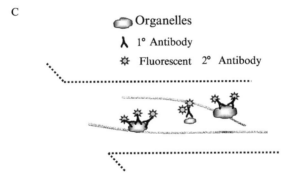

Fig. 1. Microscopy chamber and diagram. (A) Photograph shows an assembled chamber taped to a black carrying board; a fingertip indicates the scale of the image. (B) Diagram of a chamber with an overview of immunofluorescence protocol. (C) Diagram of the chamber surface with microtubules bound to glass, organelles bound to the microtubules, and primary and fluorescent secondary antibodies bound to their respective antigens.

labeled tubulin at 6 mg/ml total protein. This is brought to 37° for 10 min to polymerize the tubulin. At 10 min, a 12-fold volume of prewarmed microtubule buffer plus 20 μM Taxol is added to stabilize the microtubules. The stabilized microtubules are pelleted in a Beckman Airfuge at 15 psi for 5 min to remove unpolymerized tubulin. The pellet is resuspended in 100 μl of microtubule buffer plus 20 μM Taxol by determined up and down pipetting. The final tubulin concentration is ~1 mg/ml, good for 3 weeks in the dark, and diluted 1:70 (to 20 μg/ml) before addition to microscope chambers. The microtubules need to be long (e.g., 20 μm), free of background haze (denatured/unpolymerized tubulin), and unbundled. Haze can be removed by repelleting in the Airfuge through 30% glycerol (in microtubule buffer). "Flaky" microtubules indicate Taxol-induced polymerization. Bundled microtubules indicate that a buffer component has gone bad or is incorrect. Care should be taken to keep the unpolymerized tubulin cold, and the polymerized tubulin at 37° prior to Taxol addition.

To generate "polarity marked" microtubules (i.e., the plus end is identifiable), we grow bright fluorescent microtubules from dim, pregrown, sheared "seeds." Microtubule seeds are polymerized from clarified tubulin that has been adjusted to 80:1 unlabeled to labeled tubulin at 10 mg/ml or greater. After 5 min of polymerization the seeds are sheared by pipetting up and down 10× with the pipette tip pressed against the Eppendorf tube. Four microliters of the seeds is added to 7 μl of prewarmed microtubule buffer, and 6 μl of cold 10 mg/ml rhodamine tubulin (molar ratio of 6:1 unlabeled to labeled tubulin) is immediately added to the mixture. The microtubules are polymerized for 6 min and a 12-fold excess volume of microtubule buffer plus Taxol is added. Microtubules are centrifuged through a Beckman Airfuge for 4 min at 15 psi and resuspended in 500 μl of microtubule buffer plus Taxol. These polarity-marked microtubules are diluted 1:30 prior to addition to microscope chambers. The ratio of labeled to unlabeled tubulin should be adjusted as required for best imaging conditions. Tubulin may vary from lot to lot and polymerization conditions may need to be adjusted.

Polarity–marked microtubules should be used the same day as annealing, and breakage over time will result in loss in accuracy of the polarity mark. Typically, accuracy is 80–90%, which can be checked by microtubule gliding assays with a single species of motor protein. KRO1 (Cytoskeleton, Inc.) has worked for this purpose in our hands, though its motility is slow and light sensitive. Note that loss of accuracy will skew data toward equal plus and minus end movement. Substituting GMPCPP for GTP is reported to increase the accuracy of this procedure (Howard and Hyman, 1993). When scoring polarity, ambiguous microtubules must be eliminated from the counts, so therefore this method requires many more experiments than the standard assay to obtain significant results.

Attaching Microtubules to the Chamber

We have found that organelles, such as endosomes and nuclei, bind directly to untreated coverslips (Murray *et al.*, 2002). Microtubules generally do not bind to glass, and we therefore pretreat the coverslips with positively charged diethylaminoethyl (DEAE)-dextran (20 μg/ml) followed by extensive washing with water prior to assembling chambers. Excess DEAE-dextran will bundle microtubules. DEAE-dextran does not aggregate in solution as does polylysine but appears to be weaker at attaching the negatively charged microtubules. Consequently motor proteins can push microtubules and lead to some gliding activity. Sigma-cote can also be used to attach the microtubules (Muto *et al.*, 2005).

Microscope System

A microscope system for these experiments should have the following features: rapid switching between multiple-fluorescence channels, a full field rate of image capture of 0.5 s or less, automated X-Y-Z stage movement, low light fluorescence detection, low intensity of excitation light (to avoid photobleaching and loss of motor protein activity), deconvolution software, adaptable image analysis software, and microscope temperature control. A plexiglass enclosure for the microscope stage and hot air heater (e.g., a Sy-Airthermy, World Precision Instruments, Inc.) has given us the most stable temperatures. Thermal gradients will result in coverslip movement and loss of focus. Video capture (at 30 frames/s) is adequate for studies of spatial movement, despite poorer image quality and intensity information as compared to digital.

Laser scanning confocal microscopy is more quantitative than wide-field microscopy since out–of–focus light is eliminated. However, laser light is damaging to motor proteins. Even under reduced laser power, we have found that motility assays can be inhibited (interestingly, movement is more resistant to photodamage once it has been initiated). Deconvolution of wide-field images produces very nice looking data; however, deconvolution requires Z-series capture, is time-consuming, and can potentially introduce artifacts. Spinning disk confocal, which captures the full CCD image at once, may be the best current optical platform for these assays.

Protocol to Measure the Microtubule-Based Motility of
  Rab4-Containing Endosomes

A diagram to help visualize the procedure is provided in Fig. 1B and C.

1. Polymerize fluorescent microtubules, usable for 3 weeks.

2. Prepare DEAE-coated disposable microscope chambers for use the day of the experiment. 4–10 chambers are attached to a small, black board that can fit into an icebox.

3. Prepare assay buffer (35 mM K_2-PIPES, 5 mM $MgCl_2$, 1 mM EGTA, 0.5 mM ethylenediaminetetraacetic acid [EDTA], + 20 μM Taxol + 2 mg/ml BSA + 4 mM DTT, pH 7.4). The PIPES, $MgCl_2$, EGTA, EDTA from a 10× stock, Taxol from frozen 1 mM aliquots, the bovine serum albumin (BSA) and DTT are made fresh. Prepare MT-assay buffer (assay buffer without BSA or DTT). Prepare blocking buffer (assay buffer + 5 mg/ml casein) (the casein does not need to dissolve completely and insoluble casein can be removed by filtration after vortexing).

4. Dilute microtubules 1:70 from stock.

5. Successively add 4–5 μl of reagents followed by incubation and washing 5 × 15 μl with assay buffer or blocking buffer:

   a. Add 5 μl of microtubules, incubate 3 min at room temperature, wash with blocking buffer 3× (to reduce vesicle binding directly to glass), and wash 2× with assay buffer.

   b. Add 5 μl of Texas Red-labeled vesicles; allow binding to microtubules for 10 min at room temperature, wash with blocking buffer, and place chambers at 4° to maintain vesicle activity.

   c. Add 3 × 15 μl primary anti-Rab4 antibody (Cat. #610888 BD, Transduction Laboratories) at 5 μg/ml. Incubate 6 min and wash thoroughly in blocking buffer.

   d. Add 3 × 15 μl Cy2-anti-mouse antibody (e.g., Catalog #115-225-068 [or 115-225-166 pre-Absorbed] Jackson ImmunoResearch Laboratories, Inc.) 15 μg/ml. Incubate 6 min and wash thoroughly in assay buffer.

   e. At the microscope (heated to 37°), remove a chamber from cold and wash in assay buffer plus 2 mg/ml ascorbic acid (optional, for antibleaching).

   f. Focus on an appropriate microscope field. Avoid excess excitation light. Initiate time-series image capture of bright field, Texas Red, and Cy2 channels, 1 frame/s for 60 s and add 50 μM ATP.

Tips: Oxygen scavenging systems can be used (Kishino and Yanagida, 1988). However, we have found that catalase (or its contaminants) can bundle microtubules and inhibit motility. We have found the most consistent motility with 50 μM ATP for rat liver endosome movement. This low concentration of ATP should reduce the activation of extraneous ATPases that produce ADP and free phosphate.

Choice of Antibodies

We have assayed 50 or more primary antibodies to identify various proteins on the surface of organelles using this technique. Some antibodies do not function well for the technique, most often because of "nonspecific" staining. This means that the antibody binds all organelles brightly, binds microtubules, or binds glass itself. Antibodies can aggregate or denature in solution. Monoclonal antibodies tend to work better than polyclonal antibodies. The appearance on a Western blot often indicates how the antibody will perform. Antibodies that give multiple bands that are commonly ignored on a Western blot should be used with caution. In our experience, antibodies that work well for Western blot usually work for immunofluorescence of unfixed material. Formaldehyde or other fixatives will prevent reactivity with some antibodies.

Controls for specificity of the immunofluorescent signal include (1) the use of multiple antibodies, (2) clean reactivity by Western blot, and (3) demonstration of loss of fluorescent signal by inclusion of the antigen itself (the antigenic peptide, or bacterially expressed, purified protein) in the antibody binding step. As for Western blot or other immunofluorescence protocols, the appropriate dilution of antibody should be determined. Fluorescent signal in the absence of primary antibody is unacceptable.

Quantification of Microtubule-Based Motility Parameters

Microtubule Binding

A microtubule-coated chamber is incubated with vesicles for 10 min. After washing, multiple images are collected, and the number of vesicles per unit length of microtubules is measured.

Frequency of Movement

Microtubule-bound vesicles are selected from time-series data prior to addition of ATP and scored as moving, not moving, or "unscorable" upon addition of ATP (unscorable counts are eliminated). Motor-based movement is evident as compared to the wiggling movement seen with buffer lacking ATP. Some vesicles will release from the microtubule upon addition of ATP. We classify this as unscorable, though it can be scored as an additional parameter (e.g., nonprocessive movement). High salt will also lead to vesicle release. Excitation light, microtubule bundling, and time left in solution will reduce endosome motility. Motility remains active for 1 h or greater if vesicles are stored on ice, bound to microtubules, washed free of soluble components, and in the presence of reducing agents. Rat liver endosomes have motility frequencies of 25–40%. Vesicles from cultured cells may show lower frequency (unpublished results). Control motility assays are performed each day of the experiment to confirm the level of motility. Rat liver endosome motility, in contrast to gliding or motor-coated bead assays, arrests after 1–2 min using these techniques.

Direction of Movement with Respect to Microtubule Polarity

Frequency of movement is scored for vesicles that are bound to polarity-marked microtubules (see above). The use of purified motor proteins or motor-protein-coated beads represents an additional method to score microtubule polarity.

Run Length

The total distance that a vesicle moves before stoppage or detachment from the microtubule is scored. Kinesin-1-based movement generally has greater run length than cytoplasmic dynein-based movement (Higuchi and Endow, 2002; Wang and Sheetz, 2000).

Velocity

Organelles tend to show stop-and-start movement that is difficult to classify into absolute velocity. Gliding or bead-based assays of purified motor proteins show movement that is more continuous. For organelles, measurements of maximal velocity, pause time, and periodicity may reveal the characteristic motile state (i.e., what motors or regulatory factors are present) (Murray et al., 2000).

Fission

Early endocytic vesicles from rat liver demonstrate consistent levels (~13%) of vesicle splitting or fission, and this is significantly reduced in "late" endocytic vesicles (i.e., vesicles that have sorted from their receptor) (Bananis et al., 2004). The amount of fission is scored by observing all organelle movements over a given time period and counting those events that lead to separation into distinct fluorescence intensities. To control for the possibility that vesicles are overlapping, fission can be measured at different dilutions. An example of vesicle fission is presented in Fig. 2A. The polarity mark, which reveals the minus end of the microtubule $(-)$, is seen as a faintly staining gap in the microtubule (arrowheads). A vesicle (circle) moves toward the plus end from 5 to 20 s, at which time it undergoes fission into two vesicles (26 s), both of which move toward the plus end. This sequence demonstrates that vesicle fission can occur during plus-end-directed movement and that both daughter vesicles can contain active motors.

Figure 2B demonstrates how these parameters can be used to investigate endocytic processing. In the figure, pretreatment of early endocytic vesicles with GDP is shown to cause a substantial increase in vesicle motility as well as fission activity as compared to controls. The data suggest that the motility of endocytic vesicles, as isolated, is partially inhibited, and that GDP treatment can release this inhibition. Because of our previous results showing a high level of Rab4 staining on the endocytic vesicles (Fig. 3), as well as multiple studies indicating that Rab4 can regulate endocytic processing (McCaffrey et al., 2001; Sheff et al., 1999; van der Sluijs et al., 2001), we expressed GST-Rab4 and found that incubation of vesicles with GTPγS-Rab4 led to decreased motility, whereas incubation of vesicles with GDP-Rab4 led to a similar increase in motility as seen with GDP alone. Although GST-Rab4 purified from bacteria lacks native lipid modification, it was able to bind guanine nucleotides as well as the endocytic vesicles (Bananis et al., 2003). Additional studies with polarity-marked microtubules demonstrated that the GDP-induced increase in motility was toward the minus ends of microtubules (not shown; Bananis et al., 2003). We have also demonstrated that although dynein is present in early endocytic vesicle isolation, it is not present on the vesicles that contain ASOR, and dynein inhibitors do not affect the motility of the ASOR-containing vesicles (Bananis et al., 2000, 2003, 2004). Taken together these data lead us to the hypothesis that Rab4-GTP is inhibitory toward minus-end-directed kinesins that are present on the endocytic vesicles, and this is currently under investigation in our laboratory.

Quantification of Parameters of Protein Localization

Protein Colocalization

The colocalization of vesicle-associated fluorescent markers is assessed by counting each fluorescent vesicle in channel A and assessing for the presence of signal above background in channel B. Colocalization in the microchamber is simplified as compared to colocalization in whole cells as the signal occurs across a two-dimensional plane at discrete dots.

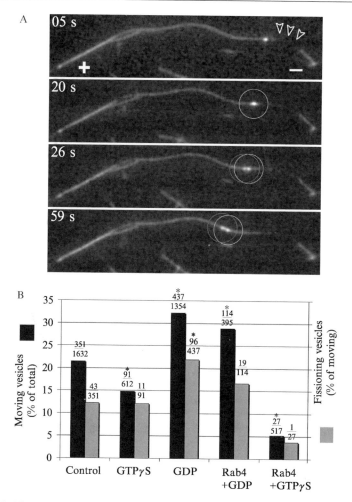

FIG. 2. Visualization and scoring of microtubule-based motility parameters. (A) A vesicle fission event observed on a polarity-marked microtubule. Fluorescent-ASOR containing mouse liver early endocytic vesicles were bound to polarity-marked microtubules in a disposable microchamber according to the given procedures. Seconds after addition of 50 μM ATP is indicated in the upper left. Arrowheads point to a dimly fluorescent microtubule "seed," which marks the minus end of the microtubule. A fluorescent vesicle (circled) moves toward the plus end and undergoes a fission event at 26 s, where upon both daughter vesicles continue to move toward the microtubule plus end. Image width, 60 μm. (B) Scoring of microtubule motility parameters. The number of moving vesicles (black bars) and vesicles undergoing fission (gray bars) were scored for a statistical sample of vesicles (vesicle number indicated above the bars, statistical difference from control indicated by the asterisk). Pretreatment of vesicles with GDP increases both motility and fission, whereas GTPγS inhibits motility and GTPγS-Rab4 further inhibits motility as well as fission. Adapted from Bananis *et al.* (2003).

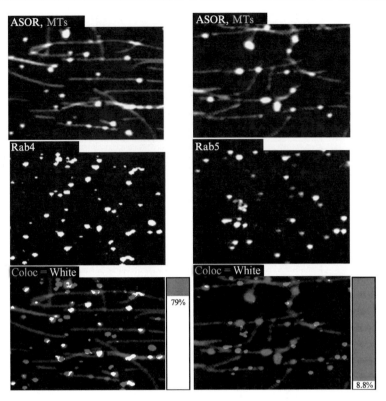

FIG. 3. Measurement of protein colocalization using the microchamber technique. Fluorescent ASOR-containing rat liver endocytic vesicles were bound to microtubules and stained for either Rab4 or Rab5 within microscope chambers. ASOR and microtubules were visualized in the rhodamine fluorescence channel and can be distinguished due to their uniform brightness (vesicles are brighter than the microtubules). Rab4 or Rab5 signal was collected in the FITC channel and merged images (bottom) indicate colocalization (white) of the Rab4 or Rab5 with ASOR. Multiple Rab4 and Rab5 experiments (>2000 vesicles) were scored and the number of ASOR vesicles containing the Rab signal is displayed in the bar alongside the merged images. Typically such images are displayed in color for clarity. Image width: 20 μm.

Figure 3 demonstrates the colocalization of Rab4 and Rab5 with early (presegregation) rat liver endosomes that were bound to microtubules. Both antibodies revealed significant immunofluorescent activity associated with vesicles within the preparation. However, only the Rab4 antibody shows significant signal overlap with the ASOR-containing endosomes (79% of all vesicles). This result suggests either that ASOR vesicles have passed through a Rab5-containing compartment after 5 min of processing in the rat liver or that they do not encounter the Rab5 compartment.

Quantitative interpretation of fluorescence intensity in terms of number of fluorescent molecules is complicated by the optical factors such as the point spread function, light scatter and/or absorption, and the size of the organelle. This can be problematic for automatic colocalization algorithms. Note that colocalization of Image A to Image B is different than colocalization of Image B to Image A.

Marker Segregation

The ability to locate proteins on organelles combined with reproducible fission activity allows for analysis of protein segregation *in vitro*. Fluorescence intensity is measured in the mother and daughter vesicles to track how markers segregate during fission. In studies of ASOR and its receptor, the asialoglycoprotein receptor, we found that a single vesicle fission event results on average of ~90% of the receptor sorting to one vesicle, whereas only 60% of the ligand sorts to this same vesicle (Bananis *et al.*, 2000). This represents a method to screen for regulatory factors that affect segregation, as putative agents can be added directly to the assay.

Binding of Exogenously Added Protein to Organelles

Fluorescent probes such as GFP or primary and fluorescent secondary antibody can be used to measure the binding of protein to organelles within the chambers. We have used this technique to estimate the equilibrium binding constant of Rab5 to endocytic vesicles (Murray *et al.*, 2002).

Conclusion

The methods described above provide a means to study endocytic motility, processing events, and protein localization in materials obtained from native tissues and cell lines without the need for transfection or overexpression.

The advantages of this technique include the following:

1. Direct access to the surface of endosomes.
2. The ability to measure before and after activities of individual endosomes.
3. The study of endogenous proteins at physiological levels of expression.
4. The ability to measure multiple probes at a time.
5. Clean optics as compared to whole cells.
6. Clear detection and quantification of protein colocalization.

The following advantages are expected for the near future:

1. Simplified approaches to estimate the number of fluorescence molecules under microscopy.
2. Large expansion in the number of optical dyes for measuring fluorescence and protein activities.
3. Increased capture rates and more deconvolution algorithms and optical modeling approaches.

The disadvantages of the technique include the following:

1. Small sample size (microscopy field) can yield high signal variation.
2. Damage induced by excitation light.
3. Potential alteration in activities induced by binding to glass.
4. Requirement for investment in microscopy systems and dedicated microscopy personnel.

As for other experimental systems, results from *in vitro* microscopy systems need to be validated. We have shown that endosomes that have taken up ASOR for 5 min contain kinesin-1 and Rab4, whereas endosomes that have taken up ASOR for 15 min are no longer associated with these proteins but instead contain dynein, kinesin-2, as well as Rab7. This has been verified by sorting endosomes through a laser-deflected flow cytometry machine, followed by Western blot analysis (Bananis *et al.*, 2004).

References

Bananis, E., Murray, J. W., Stockert, R. J., Satir, P., and Wolkoff, A. W. (2000). Microtubule and motor-dependent endocytic vesicle sorting in vitro. *J. Cell Biol.* **151,** 179–186.

Bananis, E., Murray, J. W., Stockert, R. J., Satir, P., and Wolkoff, A. W. (2003). Regulation of early endocytic vesicle motility and fission in a reconstituted system. *J. Cell Sci.* **116,** 2749–2761.

Bananis, E., Nath, S., Gordon, K., Satir, P., Stockert, R. J., Murray, J. W., and Wolkoff, A. W. (2004). Microtubule-dependent movement of late endocytic vesicles *in vitro*: Requirements for dynein and kinesin. *Mol. Biol. Cell* **15,** 3688–3697.

Blocker, A., Severin, F. F., Burkhardt, J. K., Bingham, J. B., Yu, H., Olivo, J. C., Schroer, T. A., Hyman, A. A., and Griffiths, G. (1997). Molecular requirements for bi-directional movement of phagosomes along microtubules. *J. Cell Biol.* **137,** 113–129.

Higuchi, H., and Endow, S. A. (2002). Directionality and processivity of molecular motors. *Curr. Opin. Cell. Biol.* **14,** 50–57.

Howard, J., and Hyman, A. A. (1993). Preparation of marked microtubules for the assay of the polarity of microtubule-based motors by fluorescence microscopy. *Methods Cell Biol.* **39,** 105–113.

Inoue, S. (1986). Video Microscopy. Plenum, New York.

Kishino, A., and Yanagida, T. (1988). Force measurements by micromanipulation of a single action filament by glass needles. *Nature* **334,** 74–76.

McCaffrey, M. W., Bielli, A., Cantalupo, G., Mora, S., Roberti, V., Santillo, M., Drummond, F., and Bucci, C. (2001). Rab4 affects both recycling and degradative endosomal trafficking. *FEBS Lett.* **495,** 21–30.

Murray, J. W., Bananis, E., and Wolkoff, A. W. (2000). Reconstitution of ATP-dependent movement of endocytic vesicles along microtubules *in vitro*: An oscillatory bidirectional process. *Mol. Biol. Cell* **11**, 419–433.

Murray, J. W., Bananis, E., and Wolkoff, A. W. (2002). Immunofluorescence microchamber technique for characterizing isolated organelles. *Anal. Biochem.* **305**, 55–67.

Muto, E., Sakai, H., and Kaseda, K. (2005). Long-range cooperative binding of kinesin to a microtubule in the presence of ATP. *J. Cell Biol.* **168**, 691–696.

Pollock, N., de Hostos, E. L., Turck, C. W., and Vale, R. D. (1999). Reconstitution of membrane transport powered by a novel dimeric kinesin motor of the Unc104/KIF1A family purified from Dictyostelium. *J. Cell Biol.* **147**, 493–506.

Sheff, D. R., Daro, E. A., Hull, M., and Mellman, I. (1999). The receptor recycling pathway contains two distinct populations of early endosomes with different sorting functions. *J. Cell Biol.* **145**, 123–139.

Stockert, R. J. (1995). The asialoglycoprotein receptor: Relationships between structure, function, and expression. *Physiol. Rev.* **75**, 591–609.

van der Sluijs, P., Mohrmann, K., Deneka, M., and Jongeneelen, M. (2001). Expression and properties of Rab4 and its effector rabaptin-4 in endocytic recycling. *Methods Enzymol.* **329**, 111–119.

Wang, Z., and Sheetz, M. P. (2000). The C-terminus of tubulin increases cytoplasmic dynein and kinesin processivity. *Biophys. J.* **78**, 1955–1964.

Waterman-Storer, C. M. (1998). Microtubule/organelle motility assays. *In* "Current Protocols in Cell Biology" (J. S. Bonifacino, M. Dasso, J. B. Harford, J. Lippincott-Schwartz, and K. M. Yamada, eds.), Vol. Unit 13.1. John Wiley, New York.

Weigel, P. H. (1980). Characterization of the asialoglycoprotein receptor on isolated rat hepatocytes. *J. Biol. Chem.* **255**, 6111–6120.

[10] CD2AP, Rabip4, and Rabip4': Analysis of Interaction with Rab4a and Regulation of Endosomes Morphology

By Pascale Monzo, Muriel Mari, Vincent Kaddai,
Teresa Gonzalez, Yannick Le Marchand-Brustel, and
Mireille Cormont

Abstract

In this chapter, we describe various approaches that allow us to study interactions between the small GTPase Rab4a and its two effectors, Rabip4 and CD2AP. Two complementary approaches, one using the yeast two-hybrid system and the other using a GST pull-down assay, are described. We document the studies of the localization of these proteins by cellular fractionation. Finally, we develop cellular imaging techniques to study the morphology of vesicular structures containing Rab4a. We show that the coexpression of Rab4a with its effectors affects Rab4a-containing structures, giving a clear indication of their interaction in the mammalian cellular context.

METHODS IN ENZYMOLOGY, VOL. 403   0076-6879/05 $35.00
Copyright 2005, Elsevier Inc. All rights reserved.   DOI: 10.1016/S0076-6879(05)03010-7

Introduction

Rab4a is associated with early endocytic compartments in a large number of cells (Mohrmann and van der Sluijs, 1999; van der Sluijs *et al.*, 1991). It appears to regulate sorting events from the early sorting endosomes toward numerous locations: the plasma membrane, the recycling endosomes, the late endosomes, and specialized compartments such as synaptic vesicles (Cormont *et al.*, 2003; de Wit *et al.*, 2002; McCaffrey *et al.*, 2001; Mohrmann and van der Sluijs, 1999; Zerial and McBride, 2001). The characterization of its effectors is a necessary step to understand how Rab4a controls these different pathways. Numerous effectors for Rab4a have been identified. Rabaptin-5 and Rabenosyn-5 are bivalent effectors able to interact with Rab5 and Rab4 (De Renzis *et al.*, 2002; Vitale *et al.*, 1998). These effectors are supposed to coordinate two successive traffic steps in time and space (Miaczynska and Zerial, 2002). We cloned two specific effectors of Rab4a, Rabip4 and CD2AP, using the yeast two-hybrid system (Cormont *et al.*, 2001, 2003; Mari *et al.*, 2001). A second Rabip4 isoform that is encoded by the same gene has been identified and named Rabip4' (Fouraux *et al.*, 2004; P. Monzo and M. Cormont, unpublished data). We describe some of the experiments used to characterize the interaction of Rab4a with its effectors and to study the function of Rabip4 and CD2AP.

Interaction Measurement by Yeast Two-Hybrid Experiments

Yeast Expression Vector and Yeast Strain

The yeast two-hybrid expression vectors allow for the expression of fusion proteins between the bait and the prey, fused, respectively, with the DNA-binding domain of the bacterial LexA factor and the activation domain of a compatible transcription factor as schematized in Fig. 1 (Fields and Song, 1989). The yeast strain is chosen according to the DNA-binding domain of the transcription factor used. Its genome has been modified to introduce the corresponding DNA sequence, recognized by the transcription factor, upstream of the reporter genes. It is also auxotroph for some essential amino acids, thus requiring the expression of genes present in the transformed plasmids for its growth in the absence of the corresponding amino acids. We frequently use pBTM116 (a gift from Dr. S. Fields), pACT or pGAD families vectors (BD Biosciences, Clontech), or pVP16 (a gift from Pr. S. Hollenberg) in the L40 yeast strain. pBTM116 possesses the *TRP1* gene, which allows yeast containing this plasmid to grow in minimal medium lacking tryptophan. It also contains the coding sequence for the DNA-binding domain of *Escherichia coli* LexA factor with a polylinker to insert cDNAs. The other plasmids possess the *LEU2* gene, which allows the yeasts containing this plasmid to grow in the absence of leucine. They

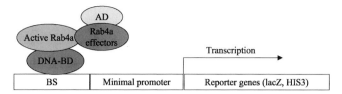

FIG. 1. Schematic representation of the yeast two-hybrid approach. Yeasts are cotransformed with plasmids allowing for the synthesis of two fusion proteins. The bait is fused with the DNA-binding domain of LexA (DNA-BD) that recognizes specific DNA-binding sequences (BS). The prey is fused to the activator domain (AD) of a transcription factor. If the prey interacts with the bait, a functional transcription factor is reconstituted and the transcription of the reporter genes cloned downstream of the sequence recognized by the DNA-binding domain of LexA is activated. In L40 yeast strain, one of the reporter genes encodes for β-galactosidase; the other allows the yeast to grow in the absence of histidine.

also contain the activation domain of the transcription factor Gal4 or VP16 followed by a polylinker to insert cDNAs. Subcloning of the cDNAs encoding for the bait, here active Rab4a (Rab4aQ67L), and the prey, in this study Rab4a effectors, is designed to preserve the reading frame between the two cDNAs. It often requires the insertion of a polymerase chain reaction (PCR) amplification of the cDNA encoding the bait and the prey to a blunted restriction site of the plasmid polylinker. To determine the interacting region on the effectors, we generated numerous constructions of truncated forms of the effectors. Stop codons are introduced at the desired position inside the coding sequence, using site-directed mutagenesis (Cormont *et al.*, 2001; Mari *et al.*, 2001). Plasmids are amplified in DH5α *Escherichia coli* strain (Invitrogen) and purified using the Qiafilter Plasmid Maxi Kit (Qiagen, 12263).

    Comments. Commercial yeast two-hybrid kits are available from companies such as BD Biosciences, Clontech (Matchmaker system), Stratagene (HybridZap systems), Invitrogen (Hybrid Hunter and ProQuest systems), and OriGene (DupLex A system). These systems generally contain the appropriate yeast strain, vectors, and controls.

Yeast Cotransformation Procedure

    The day before performing yeast cotransformation, prepare 25 ml of preculture of the adequate yeast strain (L40) in rich medium (YPD Broth, Bio101, Inc.) supplemented with 500 mg glucose. Let this culture grow overnight at 30° in a shaking incubator (150 rpm). This will allow ~15 interaction assays to be performed. Adapt the volume of preculture to your number of assays. The next day, dilute the preculture 1 to 5 in the same medium and let it grow for an additional 3 h. When the number of yeast is between 2 and 6×10^7 cells/ml, centrifuge them at $3000 \times g$ for 5 min at room temperature. The number of cells is determined using a Malassez slide. The yeast is then washed in Buffer A, resuspended in 2 ml of the same buffer, and incubated at 30° under gentle shaking for 1 h. During this incubation, prepare in a series of sterile 1.5-ml Eppendorf tubes a mix containing 1 μg of the various bait plasmids, 1 μg of the various prey plasmids, and 4 μl of fish sperm DNA (BD Biosciences, Clonetech, S0277). The concentration of the plasmid solutions should be adapted to be lower than 11 μl. Then 150 μl of yeast suspension is added in each Eppendorf tube. After a 10-min resting period at room temperature, 500 μl of Buffer C is added for a further incubation of 30 min at 30° in a shaking incubator. Cells are then transferred to 42° under continuous gentle agitation for 25 min. After centrifugation (10 s at $10,000 \times g$), cells are washed twice with 1 ml of DOB-Leu-Trp (Bio101, Inc.), resuspended in 200 μl of the same medium, and plated on selective medium made of DOBA-Leu-Trp (Bio101, Inc.). Plates are then incubated upside down in an incubator maintained at 30°, for at least 2 days.

Buffer A: LiAc 0.1 M in Tris–HCl 10 mM, pH 7.5, containing 1 mM ethylenediaminetetraacetic acid (EDTA).

Buffer B 10×: Tris–HCl 100 mM, pH 7.5, containing 10 mM EDTA.

Polyethylene glycol (PEG) solution: weight 50 g of PEG$_{3350}$ (Sigma P4338) and dissolve it in 100 ml of water.

Buffer C: mix 8 ml of PEG with 1 ml of Buffer B and 1 ml of LiAc 1 M.

Comments. The recommendation is to prepare concentrated solutions (LiAc 1 M; Tris–HCl 1 M, pH 7.5, EDTA 0.5 M), and to sterilize them, as well as the PEG solution, by filtration through 0.22-μm filters. All the solutions are kept for at least 1 month at room temperature.

Measurement of β-Galactosidase Activity

We use an "easy-to-run" 5-bromo-4-chloro-3-indolyl-β-D-galactopyranoside (X-GAL) test. Yeast colonies are patched on selective plates (DOBA-Leu-Trp). After 24 h, we perform replicates by application of a sterile Whatman 40 paper on the yeast patches. This paper is then put on a plate containing selective medium (DOBA-Leu-Trp), and yeasts grow on the paper for 1 day at 30°. For the detection of β-galactosidase activity, a Whatman 3 paper is put at the bottom of an empty plate and soaked with Buffer Z containing X-GAL (2.5 ml for a 100-mm plate). The Whatman 40 paper with the yeast patches is placed, colonies up, in liquid nitrogen for 15 s. After thawing, the Whatman 40 paper is placed on the prepared Whatman 3 paper (always colonies up). Within 30 min to 16 h, positive colonies will develop a blue coloration, indicating that the bait and the prey have interacted.

Buffer Z: Na_2HPO_4 60 mM, NaH_2PO_4 40 mM, KCl 10 mM, $MgSO_4$ 1 mM.
X-GAL stock solution: make a 2% X-GAL solution (Sigma B4252) in dimethylfluoride and store aliquots at $-20°$. For the measurement of β-galactosidase activity, prepare, for each plate, a mix with 2.5 ml of Buffer Z with 6.75 μl 2-mercaptoethanol, and 25 μl of X-GAL stock solution.

Preparation of Yeast Lysates for Analysis of the Expression of the Fusion Proteins

It is important to verify that the proteins corresponding to the bait and the prey are expressed in the yeast. For that, yeasts are grown for 2–3 days in 1.5 ml of selective medium (DOB-Leu-Trp for pBTM116 and pACT2, for example). Cells are spun down by centrifugation at 2600×g for 5 min, lysed in 1 ml 0.25 M NaOH containing 1% 2-mercaptoethanol, and incubated 10 min on ice. Proteins are then precipitated by adding 160 μl of trichloroacetic acid (50% in water) and are collected by centrifugation at 10,000×g. Pellets are washed by resuspension in 1 ml ice-cold acetone. After centrifugation at 10,000×g, the dried pellets are resuspended in 200 μl of Laemmli Buffer. Aliquots are analyzed by sodium dodecyl sulfate polyacrylamide gel electrophoresis (SDS–PAGE) and Western blots are performed using specific antibodies. Antibodies against the prey or the bait can be used when available. It is also possible to use antibodies against LexA (Santa Cruz Biotechnology, Inc.) or against the tag that is inserted in the prey (e.g., the HA tag in pACT2 vector from BD Biosciences, Clonetech).

Laemmli Buffer: Tris–HCl 70 mM, pH 4, containing 10% glycerol, 3% SDS, 0.5% bromophenol blue, and 500 mM 2-mercaptoethanol.

In Vitro Binding Assay Between Rab4a and Its Effectors (GST-Pull-Down Assay)

Expression Constructs

To investigate *in vitro* binding of Rab4a with Rabip4 and CD2AP, we generated the plasmids pGEX-2T-Rab4a, pcDNA3-myc-Rabip4, and pcDNA3-myc-CD2AP. The pGEX-2T (Amersham Biosciences) plasmid allows for the production of recombinant Rab4a in fusion with glutathione-S-transferase. The pcDNA3 plasmid (Invitrogen) possesses the T7 polymerase promoter upstream of the cDNA of interest, which allows for the production of ^{35}S-radiolabeled myc-CD2AP and myc-Rabip4 using a kit for *in vitro* transcription–translation following the instructions of the manufacturer (Promega, TNTT7 Coupled Reticulocyte Lysate System

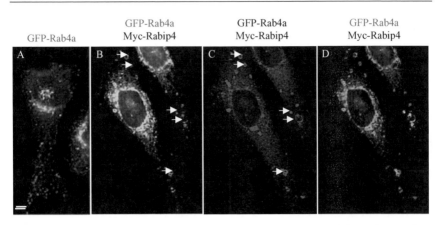

FIG. 2. Evidence for the interaction of Rab4 with Rabip4: the morphology of the Rab4a-positive endosome is changed by Rabip4 overexpression. CHO cells are transiently transfected with pEGFP-Rab4a alone (A), or together with pcDNA3-myc-Rabip4 (B–D). GFP-Rab4a labels small vesicular structures at the periphery of the cell and in a perinuclear region (A). When myc-Rabip4 is coexpressed, GFP-Rab4a is detected in enlarged vesicles (arrows) as well as in the perinuclear region (B). The same enlarged vesicles are also labeled by myc-Rabip4 (C, arrows), and the colocalization is indicated by the yellow color obtained when the green and the red images are merged (D). Cells with equal intensity in the GFP channel, i.e., that overexpressed similar amounts of Rab4a, are analyzed. Similar observations are made when myc-Rabip4' is expressed instead of myc-Rabip4. Bar = 1 μm. (See color insert.)

L4610). Use either [^{35}S]methionine or cysteine or both. We recommend using Expre^{35}S^{35}S Protein labeling mix (NEN Life Science Products). Before opening the cap of the vial, thaw the solution completely and open the cap in the hood (at hot lab) so that any vaporized materials can be trapped in the hood.

Purification of GST-Rab4a

pGEX-2T-Rab4a is transformed in *Escherichia coli* JM109 strain and purification of the protein is done by affinity chromatography (Bortoluzzi *et al.*, 1996). Transformants are grown at 37° and selected on Luria-Bertani (LB) agar plates containing 100 μg/ml ampicillin. The next day, a colony is transferred to 2 ml LB containing 100 μg/ml ampicillin and grown overnight at 37° in a shaker incubator maintained at 37°. Eight hours later the preculture is transferred into 50 ml of LB containing 100 μg/ml ampicillin and cultured overnight at 37°. The next morning, bacteria are diluted 10 X and grown until the culture reaches OD$_{600}$ ~0.6. Isopropyl

β-D-1-thiogalactopyranoside (100 μM) (Sigma, I 6758) is added for 12 h at 37°. Bacteria are harvested after centrifugation of the culture at 500×g for 20 min at 4°. The pellet is resuspended in 10 ml of ice-cold Buffer A supplemented with 20 μl of EDTA (0.5 M, pH 8) and 4 mg lysosyme (Sigma, L 7651). The bacteria homogenate is subjected to three cycles of freezing/thawing in liquid nitrogen. Then, 5 mg DNase I (Roche Diagnostics, 1284932), 100 μl MgCl$_2$ (1 M), and 150 μl sodium deoxycholate (4% in water) are added for 30 min at 4° under agitation. During this period, the homogenate, maintained at 4° on ice, is sonicated 10 X for 10 s with a 1-min interval using a Microson sonicator at an intensity of 6 (Misonix). Cell debris are pelleted by centrifugation at 4° for 30 min at 25,000×g. The supernatant is supplemented once with protease inhibitors (Complete, Roche Diagnostics) and is incubated for 1 h at 4° with 1.5 ml of glutathione–Sepharose beads (Amersham Biosciences). The mix is loaded into a plastic column with a filter at the bottom (Amersham Biosciences), washed with 20 ml of Buffer B, and then with 20 ml of Buffer C. Bound proteins are eluted with 10 ml Buffer C containing 10 mM glutathione (Sigma, G 4251). Ten fractions of 1 ml are collected. Protein concentration is determined in each fraction using the Bio-Rad protein assay (Bio-Rad Laboratories, Inc., 500-0006). Fractions containing the majority of proteins (in general fractions 2–4) are pooled and dialyzed twice against 1 liter of Buffer C. The purity of the protein estimated by SDS–PAGE analysis and staining with Coomassie Brilliant Blue (Bio-Rad Laboratories, Inc.) is normally more than 90%. The concentration of the proteins is around 1 mg/ml. Purified GST-Rab4a is snap-frozen in liquid nitrogen and stored in 200-μl aliquots at −80°.

> Buffer A: 50 mM HEPES, pH 8, 0.5 mM DTE, and protease inhibitors (Complete).
>
> Buffer B: 50 mM HEPES, pH 8, 0.5 mM DTE, and 0.6% sodium deoxycholate.
>
> Buffer C: 50 mM HEPES, pH 8, 0.5 mM DTE, 10 mM MgCl$_2$, and 100 μM GDP (Sigma G 7125).

Loading of GST-Rab4a with Guanine Nucleotides (GDP or GTPγS) and Incubation with Products of the In Vitro *Transcription–Translation Reaction*

Purified GST-Rab4a (40 μg/40 μl) is incubated with 40 μl of glutathione-Sepharose dried beads together with 50 μl of Exchange Buffer 2× and 10 μl of 10 mM GDP or 10 mM GTPγS (Sigma G 8634) for 60 min at 25°. Note that the effectors will bind only to Rab4a-GTP. Exchange buffer is aspirated (see comments) and replaced by 45 μl [^{35}S]CD2AP or [^{35}S] Rabip4 diluted in Binding Buffer. After 2 h at 4°, beads are washed twice with Washing Buffer. Then 100 μl of Elution Buffer is added for 15 min at 4° under agitation. The eluates are analyzed by SDS–PAGE. Gels are colored using 0.25% Coomassie Brilliant Blue R-250 (Bio-Rad) prepared in 50% acetic acid. The gel is destained using 7% acetic acid solution. Radioactivity emission is then enhanced using Amplify (Amersham Biosciences, NAMP100). Autoradiography is performed using Hyperfilm MP (Amersham Biosciences, RPN6K).

Comments. It is important to aspirate all the remaining buffer at each step before transition to the next step. For that, we use a 100-μl Hamilton syringe with a needle narrow enough not to aspirate the beads. Control conditions with beads incubated with glutathione-*S*-transferase purified as above from empty pGEX-2T must be performed to evaluate the binding specificity.

    Exchange Buffer 2×40 mM HEPES, pH 8,4 mM EDTA, and 2 mM DTT.

    Binding Buffer: 20 mM HEPES, pH 7.8, 100 mM NaCl, 5 mM MgCl$_2$, 1 mM DTT, and 1 mM guanine nucleotide (GDP or GTPγS). This buffer must be concentrated 2-fold and used to dilute the products of the *in vitro* translation–transduction reaction.

    Washing Buffer: 20 mM HEPES, pH 7.8, 250 mM NaCl, 1 mM DTT, and 100 μM GDP or GTPγS.

    Elution Buffer: 20 mM HEPES, pH 7.8, 1 M NaCl, 20 mM EDTA, 1 mM DTT, and 5 mM GDP.

    Stock solutions of the guanine nucleotides: Prepare 1 M solution by adding the appropriate volume of water directly in the commercial vial. Store aliquots (10–20 μl) at $-20°$ for no longer than 2 months. Dilute them in water at a concentration of 10 mM the day of the experiment.

Expression in Mammalian Cells

Expression Constructs

To analyze the cellular localization of Rab4a in terms of its effectors, plasmids are generated to encode for proteins tagged with different epitopes. We have generated fusion proteins with GFP or DsRed, and myc epitope-tagged proteins using pEGFP and pDsRed (BD Biosciences, Clontech) or pcDNA3-myc (Invitrogen). For Rab proteins, it is recommended that tag epitopes be added at the N-terminus since the C-terminus is modified by geranylgeranylation (Pereira-Leal and Seabra, 2001).

Transfection Protocol

Chinese hamster ovary (CHO) cells are grown in Ham's F12 medium (Invitrogen) with 10% fetal calf serum (Invitrogen) and penicillin-streptomycin 50 μg/ml in a humidified atmosphere (5% CO_2, 95% air) (Cormont *et al.*, 2001). Preconfluent cells are trypsinized and resuspended at a concentration of 1–2 $\times$ 10^6 cells into 400 μl of Ham's F12 medium. Cells are placed in a 0.4-cm gap cuvette (Eurogentec) along with 10–50 μg of plasmids and electroporated (260 V–1050 μF) with an Easyjet electroporator system (Equibio). Cells are resuspended in 6 ml Ham's F12 medium with 10% fetal calf serum divided into three wells of 35-mm diameter containing a glass coverslip. Two days after transfection, the cells can be used for further analysis. To obtain a 100-mm dish of transfected cells, cells from three electroporations are cultured in the same dish.

Fractionation Procedure into Cytosol and Membrane Fractions

To determine whether Rab4 effectors are cytosolic or membrane-associated proteins, it is necessary to fractionate CHO cells to separate these two fractions. Two preconfluent 100-mm dishes of CHO cells, transiently expressing myc-tagged Rab4 effectors are washed twice with ice-cold PBS. Cells are scraped in 300 μl ice-cold Homogenization Buffer and broken by 10 passages through a 22-gauge needle fixed on a 2-ml syringe. Homogenates are centrifuged at 100,000$\times g$ for 1 h at 4° (Optima X100 ultracentrifuge, Beckman). Pellets are resuspended in 300 μl of Homogenization Buffer. Samples are prepared for SDS–PAGE analysis by adding 30 μl of concentrated 10$\times$ Laemmli buffer. Equal volumes of homogenate, membrane-associated fraction, and cytosol are analyzed by SDS–PAGE and Western blotting using anti-myc antibodies (clone 9E10).

Homogenization Buffer: Tris–HCl 50 mM, pH 7.4, containing 1 mM EDTA, 250 mM sucrose, and a protease inhibitor cocktail (Complete).

Comments

To test the quality of the fractionation, the distribution of a transmembrane protein, such as the transferrin receptor, and of a cytosolic protein, such as Rho-GDI, can be controlled. For this purpose, we use commercially available specific antibodies: anti-transferrin receptor from Zymed (clone H68.4) and anti-Rho-GDI (Santa Cruz Biotechnology, Inc., N-19) (Monzo *et al.*, 2005). The pelleted fraction contains membrane-associated proteins, mitochondrial proteins, and nuclear proteins. It is possible to obtain a fraction enriched in membrane-associated proteins by centrifuging for 10 min at $500 \times g$ before the ultracentrifugation, which partly eliminates nuclei and mitochondria.

Analysis of Cellular Localization of Rab4a and Its Effectors by Fluorescence Confocal Analysis

Coverslips are fixed with 1 ml of 4% paraformaldehyde for 10 min at room temperature. Excess fixative is quenched in 1 ml of 10 mM NH$_4$Cl (10 min incubation time). After three washes with phosphate-buffered saline (PBS), coverslips are incubated in Blocking Buffer containing 0.1% Triton

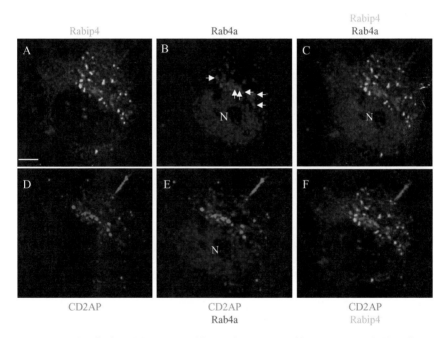

FIG. 3. Rab4a is shared between Rabip4 and CD2AP-positive structures. CHO cells are transiently cotransfected with pEGFP-Rabip4, pcDNA3-myc-Rab4a, and pDsRed-CD2AP. GFP-Rabip4 appears in green (shown in A), pDsRed-CD2AP appears in red (shown in D), and myc-Rab4a, revealed with anti-myc antibody followed by Cy5-coupled anti-mouse antibody, appears in blue (shown in B). The PhotoShop software (Adobe Systems) is used to merge the images corresponding to GFP-Rabip4 and Rab4a (C), DsRed-CD2AP and Rab4 (E), or GFP-Rabip4 and DsRed-CD2AP (F). Colocalization of Rabip4 and Rab4a in (C) results in a light blue color. Arrows in (B) point to structures containing Rab4a but not Rabip4. These structures contain in fact CD2AP as indicated by the purple color appearing in (E). CD2AP and Rabip4 do not mainly colocalize, since the merged image of CD2AP and Rabip4 (F) does not result in a yellow color. N is for nucleus with nonspecific labeling due to the Cy5-coupled anti-mouse antibodies used in this experiment. Bar = 1 μm. (See color insert.)

X-100 for 15 min before addition of the primary antibody (monoclonal antibody clone 9E10 to detect the myc-tagged proteins) diluted in the blocking buffer (Fig.2). After three washes in Blocking Buffer, coverslips are incubated with an appropriate fluorochrome-coupled anti-species antibody. When GFP and DsRed fusion proteins are analyzed together with a myc-tagged protein, we use a Cy5-coupled anti-mouse antibody from Jackson Immunoresearch Laboratories diluted to a concentration of 1 μg/ml in Blocking Buffer (see Fig. 3). The coverslips are then washed three times in Blocking Buffer, once with PBS, and finally mounted on a slide with 20 μl of Mowiol 4-88. The preparations are examined by scanning confocal fluorescence microscopy (TCS SP, Leica) with a PL APO 63 × 1.40 numeral aperture oil objective (Leica). Sequential excitations at 488, 568, and 647 nm allow us to independently detect GFP, DsRed, and Cy5 associated fluorescence, respectively. The images are then combined and merged using PhotoShop software (Adobe Systems).

Using this approach, we found that the two effectors of Rab4a, CD2AP and Rabip4 (or Rabip4'), did not mainly colocalize (Fig. 3). However, Rab4a and Rabip4 as well as CD2AP and Rab4a are colocalized, thus suggesting that Rab4a controls different trafficking pathways as a function of the effectors it binds.

> Blocking Buffer: PBS pH 7.4 containing 1% BSA (Sigma 7030) and 1% SVF.
>
> Paraformaldehyde: Paraformaldehyde (4%) is dissolved in PBS by heating at 65° for 2–3 h in a water bath. The pH is adjusted to 7.4. Boiling has to be prevented. Aliquots (2 ml) can be kept at −20° for 3 months.
>
> Mowiol 4–88: in a 50-ml plastic tube mix 6 g glycerol, 2.4 g Mowiol 4–88 (Calbiochem, 475904), 6 ml water, with 12 ml 0.2 M Tris–HCI, pH 8.5. Incubate for 6 h at 50° and vortex the mix every hour. Centrifuge 15 min at 5000×g to eliminate insoluble materials. The solution can be kept at room temperature for 1 month. For longer conservation, store aliquots at −20°. Freezing/thawing is not recommended.

Acknowledgments

This work was supported by INSERM, University of Nice/Sophia-Antipolis, the Association pour la Recherche contre le Cancer (Grant 3240), the Association Française contre la Myopathie, The Comité du Doyen Jean Lépine, and the Fondation Bettencourt-Schueller. The Fondation pour la Recherche Médicale is thanked for the support of M. Mari and V. Kaddai. P. Monzo received fellowships from the Ligue contre le Cancer, the Association pour La Recherche contre le Cancer, and the Fondation Bettencourt-Schueller.

References

Bortoluzzi, M.-N., Cormont, M., Gautier, N., Van Obberghen, E., and Le Marchand-Brustel, Y. (1996). GTPase activating protein activity for Rab4 is enriched in the plasma membrane of 3T3-L1 adipocytes. Possible involvement in the regulation of Rab4 subcellular localization. *Diabetologia* **39**, 899–906.

Cormont, M., Mari, M., Galmiche, A., Hofman, P., and Le Marchand-Brustel, Y. (2001). A FYVE-finger-containing protein, Rabip4, is a Rab4 effector involved in early endosomal traffic. *Proc. Natl. Acad. Sci. USA* **98**, 1637–1642.

Cormont, M., Meton, I., Mari, M., Monzo, P., Keslair, F., Gaskin, C., McGraw, T. E., and Le Marchand-Brustel, Y. (2003). CD2AP/CMS regulates endosome morphology and traffic to the degradative pathway through its interaction with Rab4 and c-Cbl. *Traffic* **4**, 97–112.

De Renzis, S., Sönnichsen, B., and Zerial, M. (2002). Divalent Rab effectors regulate the subcompartmental organization and sorting of early endosomes. *Nat. Cell Biol.* **4**, 124–133.

de Wit, H., Lichtenstein, Y., Kelly, R. B., Geuze, H. J., Klumperman, J., and van der Sluijs, P. (2002). Rab4 regulates formation of synaptic-like microvesicles from early endosomes in PC12 cells. *Mol. Biol. Cell* **12**, 3703–3715.

Fields, S., and Song, O.-K. (1989). A novel genetic system to detect protein-protein interactions. *Nature* **340**, 245–246.

Fouraux, M. A., Deneka, M., Ivan, V., van der Heijden, A., Raymackers, J., van Suylekom, D., van Venrooij, W. L., van des Sluijs, P., and Pruijn, G. J. (2004). Rabip4′ is an effector of rab5 and rab4 and regulates transport through early endosomes. *Mol. Biol. Cell* **15**, 611–624.

Mari, M., Macia, E., Le Marchand-Brustel, Y., and Cormont, M. (2001). Role of the FYVE-finger and the RUN domain for the subcellular localization of Rabip4. *J. Biol. Chem.* **276**, 42501–42508.

McCaffrey, M. W., Bielli, A., Cantulapo, G., Mora, S., Roberti, V., Santillo, M., Drummond, F., and Bucci, C. (2001). Rab4 affects both recycling and degradative endosomal trafficking. *FEBS Lett.* **495**, 21–30.

Miaczynska, M., and Zerial, M. (2002). Mosaic organization of the endocytic pathway. *Exp. Cell. Res.* **272**, 8–14.

Mohrmann, K., and van der Sluijs, P. (1999). Regulation of membrane transport through the endocytic pathway by rabGTPases. *Mol. Membr. Biol.* **16**, 81–87.

Monzo, P., Gauthier, N. C., Keslair, F., Loubat, A., Field, C. M., Le Marchand-Brustel, Y., and Cormont, M. (2005). Clues to CD2AP involvement in cytokinesis. *Mol. Biol. Cell*, in press.

Pereira-Leal, J. B., and Seabra, M. C. (2001). Evolution of the Rab family of small GTP-binding proteins. *J. Mol. Biol.* **313**, 889–901.

van der Sluijs, P., Hull, M., Zahraoui, A., Tavitian, A., Goud, B., and Mellman, I. (1991). The small GTP-binding protein rab4 is associated with early endosomes. *Proc. Natl. Acad. Sci. USA* **88**, 6313–6317.

Vitale, G., Rybin, V., Christoforidis, S., Thornqvist, P., McCaffrey, M., Stenmark, H., and Zerial, M. (1998). Distinct Rab-binding domains mediate the interaction of Rabaptin-5 with GTP-bound Rab4 and Rab5. *EMBO J.* **17**, 1941–1951.

Zerial, M., and McBride, H. (2001). Rab proteins as membrane organizers. *Nat. Rev. Mol. Cell. Biol.* **2**, 107–117.

[11]   Visualization of Rab5 Activity in Living Cells Using FRET Microscopy

By EMILIA GALPERIN and ALEXANDER SORKIN

Abstract

Rab5 is a member of the large family of small GTPases involved in membrane trafficking. Two genetically encoded sensors were developed to visualize Rab5 in its GTP-bound conformation in living cells. Rab5-binding fragments of Rabaptin5 or early endosomal antigen 1 (EEA.1) were fused to yellow fluorescent protein (YFP) and used in the fluorescent resonance energy transfer (FRET) assay together with Rab5-tagged cyan fluorescent protein (CFP). The presence of energy transfer between CFP-Rab5 and YFP-Rab5 binding fragments detected by sensitized FRET microscopy has validated the utility of these generated sensors to visualize the localization of GTP-bound Rab5. GTP-bound Rab5 was found in endosomes, often concentrated in distinct microdomains. Molecular architecture of the Rab5 microdomains was analyzed by three-chromophore FRET (3-FRET) microscopy, utilizing YFP, CFP, and monomeric red fluorescent proteins (mRFP.l). The results of the 3-FRET analysis suggest that GTP-bound Rab5 is capable of oligomerization and present in multiprotein complexes.

Introduction

The Rab family of small GTPases functions as compartment-specific scaffolds forming multiprotein complexes that coordinate vesicle motility, budding, and fusion. The role of Rab GTPases in membrane trafficking has been studied extensively (Szymkiewicz *et al.*, 2004; Zerial *et al.*, 2001). The activity of the Rab proteins depends of the rate of GTP hydrolysis and requires a switch between two conformations: GTP-bound or GDP-bound, respectively. The GTP-bound form is commonly considered an active conformation because in this conformation Rabs are capable of binding their interacting proteins or "effectors" (Stenmark *et al.*, 2001).

Rab5 is located on the cytoplasmic surface of early endosomes and is a key component of the protein complex responsible for homotypic fusion and cargo sorting in these organelles (Stenmark *et al.*, 1994). GTP-Loaded Rab5 interacts with several cytosolic effectors that stabilize Rab5 in its active conformation and, together with other membrane components, co-ordinates membrane docking and fusion. Q79L mutation in Rab5 results in

METHODS IN ENZYMOLOGY, VOL. 403   0076-6879/05 $35.00
DOI: 10.1016/S0076-6879(05)03011-9

significant reduction of GTP hydrolysis, as a result keeping Rab5 in a GTP-bound state. Overexpression of the Rab5(Q79L) mutant causes dramatic enlargement of early endosomes (Stenmark *et al.*, 1994). S34N mutation yields a Rab5 mutant that binds GDP with much higher affinity than GTP, thus keeping Rab5 in an inactive state (Stenmark *et al.*, 2001). The GTP-bound Rab5 interacts with several effectors, such as early endosomal autoantigen 1 (EEA.1), Rabaptin5, Rabenosin5 (Christoforidis *et al.*, 2000; Lippe *et al.*, 2001), and hVps34 (Christoforidis *et al.*, 2000). EEA.1 and Rabenosyn5 possess FYVE domains that bind to phosphatidylinositol-3-phosphate (PtdIns[3]P) (Nielsen *et al.*, 2000). Binding of the FYVE domain to PtdIns(3)P in concert with Rab5 interaction is responsible for specific targeting of these proteins to early endosomes.

Rabaptin5 is associated with the Rab5 exchange factor (Rabex5). The Rabaptin5/Rabex complex is recruited to GTP-loaded Rab5 and positively regulates Rab5 activity by slowing down GTP hydrolysis (Horiuchi *et al.*, 1997).

To visualize the active form of Rab5, and to molecularly dissect the Rab5 scaffold complex in living cells, fluorescence resonance energy transfer (FRET) microscopy has been developed. This chapter describes the design of the FRET-based sensors for Rab5 activity in living cells. It subsequently describes the 3-chromophore FRET approach to analyze multiprotein interactions within the Rab5 complex.

Description of Methods

Expression Constructs

To visualize Rab5 activity and its interactions in living cells, several fluorescently labeled fusion versions of Rab5a were prepared. Rab5 was fused to enhanced cyan (CFP) and yellow (YFP) fluorescent proteins that have been used in many FRET studies and, more recently, to monomeric red fluorescent protein (mRFP.1). To generate YFP/CFP fusion proteins, Rab5 full-length cDNAs were transferred from pcDNA-Rab5a vectors, obtained from Guthrie cDNA Resource Center (Guthrie Research Institute, Sayre, PA), using *Bam*HI and *Xho*I restriction sites, and ligated into pEYFP-C1 (Clontech) digested with *Bgl*II and *Xho*I enzymes (Fig. 1A). To generate a CFP-tagged Rab4 fusion protein, Rab4 full-length cDNAs were transferred from pcDNA-Rab4a vectors, obtained from Guthrie cDNA Resource Center (Guthrie Research Institute, Sayre, PA), using *Bam*HI and *Xho*I restriction sites, and ligated into pEYFP-C1 (Clontech) digested with *Bgl*II and *Xho*I enzymes. To generate the mRFP-Rab5 expression vector, full-length mRFP cDNA was amplified by polymerase chain

A

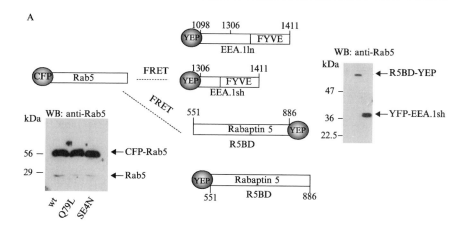

B

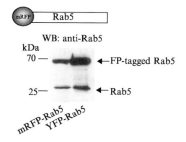

FIG. 1. Schematic representation and immunodetection of fusion proteins. (A) Depicted are Rab5, fragments of Rabaptin5, and EEA.1 proteins fused to CFP/YFP at the amino- or carboxyl-terminus. Numbers represent the amino acid residues in Rabaptin5 (R5RB) and EEA.1 (EEA.1sh) fragments according to the full-length sequences. The FYVE domain of EEA.1 is indicated. PAE cells transiently expressing CFP-Rab5 mutants ([wt], Q79L or S34N), YFP-EEA.1, or R5BD-YFP were lysed, and CFPtYFP-fusion proteins were detected by Western blotting. All fusion proteins generated in this work migrated on sodium dodecyl sulfate – polyacrylamide gel electrophoresis (SDS–PAGE) according to their predicted molecular masses. (B) PAE cells transiently expressing mRFP-Rab5 and YFP-Rab5 fusion proteins were lysed, and mRFPIYFP-fusion proteins were detected by Western blotting with anti-Rab5. All fusion proteins generated in this work migrated on SDS–PAGE according to their predicted molecular masses.

reaction (PCR) (5′GGACTTGTACAGGGCGCCGGTGGAGTGGCG3′ and 5′CCGCTAGCGGTCGCCACCATGGCCTCCTCCGAGGACG TC3′) using the pRSET-mRFP vector as a template (Campbell *et al.*, 2002) and ligated into pYFP-Rab5 (Cl) digested with *Nhe*I and *Bsr*GI enzymes, thus replacing YFP (Fig. 1B). Pfu polymerase was purchased from Stratagene (La Jolla, CA). Q79L and S34N mutants of CFP-Rab5 were also generated. Point mutations in CFP-Rab5 constructs were introduced using a QuickChange site-directed mutagenesis kit (Stratagene) (5′GGGATACAGCTGGTCTAGAACGATACCAT AG3′,5′GCTATG GTAACGTTCTAGACCAGCTGT ATCCC3′, 5′TCCGCTGTTGGC AAAAATAGCCTAGTGCTTCGTTTTG3′, and 5′CAAAACGAAG CACT AGCGT ATTTTTGCCAACAGCGGA3′).

Two fluorescent sensors of the active form of Rab5 were generated based on known interactors of Rab5, Rabaptin5, and EEAl (Fig. 1A).

Rabaptin5-Based Sensor. To generate a GTP-Rab5 sensor based on Rabaptin5, we used the previously characterized Rab5-interacting domain of Rabaptin5 (referred further as R5BD), which was mapped in *in vitro* experiments to the residues 551–862 of the carboxyl-terminus of Rabaptin5 (Vitale *et al.*, 1998). The DNA fragment corresponding to amino acid residues 551–862 of Rabaptin5 was amplified by PCR (5′GCCGCTC GAGGCCGCCATGGAAACGAGAGACCAGGTG3′ and 5′GGTACC GTCGACTGTGTCTCAGGAAGCTG3′) and ligated into pEYFP-Nl (Clontech) by using *Sal*I and *Xho*I restriction sites. YFP was also fused to the amino-terminus of R5BD (YFP-R5BD, Fig. 1A) using the pYFP-C3 vector and *Eco*RI and *Pst*I enzymes (5′GGGAATTCTATGGAAACG AGAGACCAGGTG3′ and 5′CATTGGCTGCAGTGTCTCAGGAAG CTGG3′).

EEA.1-Based Sensor. A 30-amino acid region upstream of the FYVE domain of EEA.1 was shown to be essential for Rab5 binding *in vitro*, although the functional FYVE domain is also required for efficient inter-action. Therefore, a fragment corresponding to amino acid residues 1256–1411 of EEA.1 (further referred as EEA.1sh) was amplified by PCR (5′CCCAAGCTTAAACTTACCATGCAGATTAC3′ and 5′CGG ATCCTTATCCTTGCAAGTCATTGAAAG3′) and ligated into pEYFP-C3 (Clontech) using *Hind*III and *Bam*HI restriction sites (Fig. 1A). A similar fusion protein was previously characterized and used as an early endosomal marker (Gaullier *et al.*, 1998; Lawe *et al.*, 2000). A mutat-ed version of YFP-EEA.1sh was prepared, in which histidine 1372 was mutated to tyrosine using a QuickChange site-directed mutagenesis kit (5′GT AACA GTGAGACGGCATTACTGCCGACAGTGTGG3′ and 5′CCACACTG TCGGCAGTAATGCCGTCTCACTGTTAC3′). H1372Y mutation on the PIP3 binding pocket of the FYVE domain (Kutateladze *et al.*, 2001) prevented endosomal targeting of the full-length EEA.1. Moreover, we generated a second EEA.1-based sensor (further referred to as EEA.1ln) containing a longer fragment of EEA.1. A cDNA fragment corresponding to amino acids 1098–1411 was obtained by digestion with *Eco*RI and *Bam*HI enzymes and ligated into the pYFP-C3 vector using the same restriction sites. All constructs were verified by dideoxynucleotide sequencing.

Choice of Cell Type, Transfections, and Imaging

The localization of fluorescently fused proteins was examined in several cell lines. Two cell lines, porcine aortic endothelium (PAE) and Cos-1, were chosen as preferred expression systems. These cells have minimal autofluorescence background and flattened cell shape, convenient for epifluorescent microscopy. Moreover, Cos-1 cells that express large T antigen allow pEYFP/CFP vectors, containing an SV40 origin of replication, to induce high levels of fluorescent protein expression and, therefore, better visualization of the chromophores with low brightness such as CFP. PAE cells were grown in F12 (HAM) medium (Gibco) containing 10% fetal bovine serum (HyClone) and supplemented with antibiotics and glutamine. CoS-1 cells were grown in Dulbecco's modified Eagle's medium (DMEM) (Gibco) containing 10% newborn bovine serum (HyClone), and supplemented with antibiotics and glutamine.

For fluorescent microscopy, cells were transfected using Effectene reagent (Qiagen, Hilden, Germany) in six-well plates. Cells were replated 1 day after transfection onto 25-mm untreated autoclaved glass coverslips (No. 1.5 thickness [Fisher]). Coverslips with cells were then mounted in a microscopy chamber (Molecular Probes) and imaged at room temperature in serum and phenol red-free medium.

Our fluorescence Marianas imaging workstation (Intelligent Imaging Innovation, Denver, CO) is based on an inverted Axiovert 200M Zeiss microscope equipped with a 100 plan-apo/1.4NA objective, 175W Xenon illumination source (Sutter Instruments Company, Novato, CA), Cool-SNAP HQ CCD camera (Roper Scientific, Tucson, AZ), z-step motor, independently controlled excitation and emission filter wheels (Sutter Instruments Company, Novato, CA), and a micropoint FRAP system (Photonic Instruments, Arlington Heights, IL), all controlled by SlideBook software (Intelligent Imaging Innovation, Denver, CO). The Axiovert reflector turret allows easy changing of custom reflector modules optimized for FRET experiments with different fluorochromes. Independently controlled filter wheels provide fast, automated, and efficient optical filter changing that is required in FRET experiments. A tunable dye VSL-337ND-S nitrogen ablation laser unit (Spectra-Physics, Mountain View, CA) allows FRET analysis using a method of donor fluorescence recovery after photobleaching (DFRAP) in small regions of the cell. The VSL-337ND-S laser is capable of generating 30 pulses at a repetition rate of 30 MHz and has a wavelength tunable from 337 to 590 nm using cut dyes (Photonic

Instruments). SlideBook software was used to calculate corrected FRET images and apparent FRET efficiencies. The final arrangement of images was performed using Adobe Photoshop (Adobe Systems, Mountain View, CA).

Visualization of GTP-Bound Form of Rab5 in Living Cells

The correct molecular weight of CFP/YFP and mRFP-fused Rab5 proteins was confirmed by Western blot analysis using monoclonal antibodies to Rab5 (BD Transduction Laboratories, San Diego, CA) (Fig. 1). R5BD-YFP expression was confirmed using antibodies to GFP (Zymed, San Francisco, CA) (Fig. 1A).

Fluorescently tagged proteins were expressed in Cos-1 cells, as described above. CFP-Rab5 displayed typical pattern of endosomal localization. When the R5BD-YFP fusion protein was expressed in Cos-1 cells it was found only in the cytosol rather than in endosomes (Galperin et al., 2003). The absence of R5BD-YFP binding to endogenous Rab5 is most probably related to its inability to compete with endogenous Rabaptin5 and other proteins for binding to Rab5. However, when CFP-Rab5 and R5BD-YFP were coexpressed, a significant pool of R5BD-YFP was recruited to CFP-Rab5-containing endosomes (Fig. 2A). R5BD-YFP-containing endosomes were most clearly seen in the peripheral areas of the cell, whereas large amount of cytosolic R5BD-YFP often interfered with clear visualization of endosomes in the perinuclear area (Fig. 2A). Coexpression of the CFP-Rab5(Q79L) mutant and R5BD-YFP resulted in the appearance of enlarged endosomes and massive recruitment of R5BD-YFP to Rab5-containing endosomal complexes. In contrast, coexpression of the CFP-Rab5(S34N) mutant and R5BD-YFP fusion proteins did not result in the recruitment of either R5BD-YFP or CFP-Rab5 (S34N) to endosomes. In some cells CFP-Rab5(S34N) was often located in the perinuclear area of the cell (Fig. 2A). Rabaptin5 has been reported to be a divalent Rab effector possessing distinct binding domains for two Rabs, Rab4 and Rab5 (Vitale et al., 1998). Therefore, as a control we coexpressed CFP-Rab4a and R5BD-YFP in Cos-1 cells; however, no recruitment of R5BD-YFP to Rab4 endosomes was detected (Fig. 2B). These data suggest that the R5BD protein is recruited specifically to endosomes containing overexpressed CFP-Rab5.

FRET Measurements and Calculations. To examine whether R5BD-YFP is bound to CFP-Rab5, sensitized FRET efficiencies between CFP and YFP were measured on a pixel-by-pixel basis as described below. For two-chromophore FRET (2-FRET) measurements, three images were acquired sequentially through YFP (excitation 500/20 nm, emission

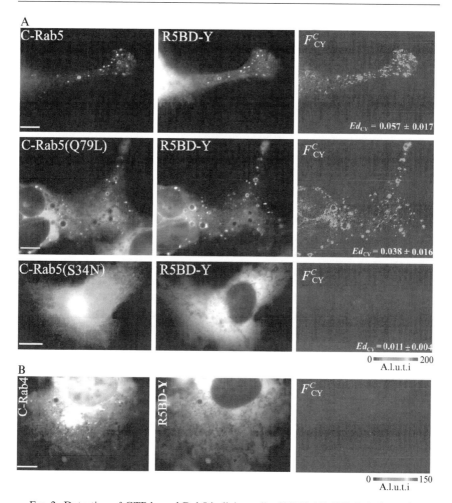

FIG. 2. Detection of GTP-bound Rab5 in living cells. CFP-Rab5, CFP-Rab5(Q79L), and CFP-Rab5(S34N) were coexpressed with R5BD-YFP (R5BD-YFP in Cos1 cells) or in Cos-1 cells (A). CFP-Rab4 was coexpressed with R5BD-YFP in Cos-1 cells (B). YFP, CFP, and FRET images were acquired from living cells at room temperature. FRET^C images were calculated as described and presented in a pseudocolor mode. Mean Ed values measured for individual endosomes of the presented cell are shown next to the corresponding image. Intensity bars are presented in arbitrary linear units of fluorescence intensity. Bar = $10\,\mu$m. A. l.u.f.i. is arbitrary linear units of fluorescence intensity. (See color insert.)

535/30 nm), CFP (excitation 436/10 nm, emission 470/30 nm), and FRET (cy) (excitation 436/10 nm, emission 535/30 nm) filter channels. An 86004BS dichroic mirror was utilized (Chroma, Inc.). Images were acquired using 2×2 binning mode and 100–250 ms integration times at room temperature. The integration time was identical for all three channels. We have noticed that at 37°, endosomes often move during image capturing through three filter channels. As a result YFP, CFP, and FRET images were often shifted. Such shifts lead to overestimation or underestimation of FRET efficiencies for cellular compartments or regions. Room temperature significantly reduced organelle movement, thus images were acquired at room temperature. For further reduction of organelle shifting, cells were treated with low concentrations of the microtubule polymerization inhibitor, nocodazole (Sigma) (20 μg/ml for 15 min at 37°). Nocodazole treatment did not alter cell shape and the morphological appearance of endosomes but dramatically reduced endosomal motility.

Prior to FRET calculations the backgrounds were subtracted from the raw images. Background values can be measured in cells that express fluorescent proteins and have been completely photobleached. Photobleaching did not affect cell autofluorescence, so images acquired after photobleaching provide correct background values. Additional images of cells that do not express fluorescent proteins have also been obtained. The average intensities of these images and images obtained after photobleaching represent the most accurate background values. However, the intensity values measured for the areas that do not contain cells were not statistically significantly different from the background values obtained using methods described above, and in most experiments with relatively bright samples, the latter method is the most practical for the analysis of multiple cells. Corrected FRET (FRET^C) was calculated for the two-chromophore FRET pairs on a pixel-by-pixel basis for the entire image, using the formula shown in Eq. (1) in the notation of Gordon *et al.* (1998):

$$FRET^C = Ff - Df(Fd/Dd) - Af(Fa/Aa) \qquad (1)$$

Df or *Af* is the fluorescence signal using the donor or acceptor filter channel, respectively, in the presence of three fluorochromes; *Ff* is the fluorescence signal through the FRET filter channel in the presence of three fluorochromes; and *Fd/Dd* and *Fa/Aa* are cross-bleed coefficients measured in cells expressing only the donor or acceptor, correspondingly.

Coefficients represent the fraction of the donor or acceptor fluorescence passing through the corresponding FRET channel. These coefficients are characteristic of the particular filter sets and do not depend on whether both YFP and CFP are present. These coefficients have been calculated as a constant proportion of the donor and acceptor bleed through the FRET filter sets. Hence, ratios *Fd/Dd* and *Fa/Aa* were determined using cells expressing either YFP-tagged or CFP-tagged proteins. Images were acquired for three channels in the presence of only one fluorochrome. These coefficients were 0.52 and 0.017 for CFP and YFP fluorescence, respectively. Bleed-through coefficients of CFP and YFP fluorescence were slightly overestimated for "safe" calculations of $FRET^C$, which could result in some underestimation of $FRET^C$ and, in several cases, negative $FRET^C$ values. Calculations were done using the "FRET" functional module of SlideBook software.

Primarily, $FRET^C$ values were calculated for the whole image and presented in pseudocolor mode. $FRET^C$ intensity is displayed stretched between the low and high renormalization values, according to a tempera-ture-based lookup table with blue (cold) indicating low values and red (hot) indicating hot values.

$FRET^C$ values were also calculated from the mean fluorescence inten-sities for each selected subregion of the image (regional analysis) contain-ing individual endosomes, ruffles, and diffuse fluorescence areas according to Eq. (1). Regional analysis allowed us to prevent overestimation and underestimation of FRET signals due to temporal shifts of organelles during image acquisition by selecting for calculations in the compartments that did not shift. "Mask" and "Image" functions of SlideBook software were utilized.

Detection of Rab5 Interaction with R5BD by FRET. As seen in Fig. 2A, $FRET^C$ images revealed energy transfer between CFP-Rab5 and R5BD-YFP, suggesting that these two proteins form a complex in early endo-somes. Positive $FRET^C$ signals were also detected in cells coexpressing CFP-Rab(Q79L) and R5BD-YFP. However, in cells expressing a GDP-bound mutant of RabS, CFP-Rab5(S34N), no $FRET^C$ signals were re-vealed (Fig. 2A). An R5BD-YFP fusion protein was not associated with Rab4-containing endosomes, and no $FRET^C$ signal was observed. These data show that R5BD-YFP may serve as a specific sensor molecule for detecting active Rab5 in living cells.

While $FRET^C$ images offer a *qualitative* indication of FRET, measure-ments of $FRET^C$ signals do not allow *quantitative* comparison of FRET efficiencies between different experimental samples. True FRET efficien-cies *(E)* can be calculated for the samples with a known stoichiometric ratio of donor and acceptor (Gordon *et al.*, 1998). In experiments where the stoichiometry of donor–acceptor interactions is unknown (Figs. 2–4), we calculated apparent FRET efficiencies *(Ed). Ed* roughly represents $FRET^C$ normalized by the donor concentration. In fact, direct comparison of different methods of calculation of FRET efficiencies and indices suggested that *Ed* is the most reliable method to calculate the apparent

FRET efficiency for samples with an unknown stoichiometry of inter-acting components (Berney *et al.*, 2003). To calculate *Ed* we used regional analysis that produces less computational noise of low-intensity pixels. The apparent FRET efficiency *Ed* for each subregion was calculated according to Eq. (2):

$$Ed = FRET^C/(Df \cdot G + FRET^C) \qquad (2)$$

where *G* is the factor relating the loss of donor signal due to FRET with the donor filter set to the increase of the acceptor emission through the FRET filter set due to FRET (Gordon *et al.*, 1998). The value of $G(cy) = 3.099$ for the CFP-YFP donor–acceptor pair was calculated, as described by Gordon *et al.* (1998). *Ed* values for the same donor–acceptor pair often vary in different experiments, which implies that *Ed* calculations can be best used in analyses of experiments with cells that express fluorescently tagged proteins at comparable levels.

Comment. Among different ways to measure FRET efficiencies, we have used a donor fluorescence recovery after photobleaching (DFRAP) approach (Wouters *et al.*, 1998). DFRAP can be easily used to calculate apparent *E* values. We have utilized this method to measure true FRET efficiencies (*E*) for the entire cell image in living cells (Galperin *et al.*, 2004). However, use of DFRAP to calculate *E* for small compartments of living cells was extremely difficult due to fast organelle movement during photo-bleaching. Therefore, we had to fix cells in order to avoid undesired endo-somal shifts. Another factor that can affect results of DFRAP analysis is partial donor photobleaching during acceptor photobleaching. Possible ex-perimental error should also be considered due to a low signal-to-noise ratio for a donor, like CFP, with low intensity brightness. The general advantages of the sensitized FRET over the DFRAB method is the substantially shorter time required for the FRET measurement and the possibility of multiple FRET measurements of the same cell using the former method.

Detection of Rab5 Interaction with EEA.1sh by FRET. To visualize GTP-bound Rab5 using an EEA.1 Rab5 binding domain, YFP-EEA.1sh protein was coexpressed with CFP-Rab5 in Cos-1 cells. Both proteins were colocalized in endosomes, and positive FRET^C signals were de-tected indicative of the interaction of Rab5 and EEA.1sh fusion proteins (Fig. 3A). On the other hand, the H1372Y mutant of YFP-EEA.1sh was not targeted to endosomes, confirming that YFP-EEA.1sh is targeted to endosomal membranes by a mechanism similar to that of full-length EEA.1 protein (data not shown). No FRET^C was observed between CFP-Rab4 and YFP-EEA.1sh proteins, although these proteins were partially colocalized in endosomes (Fig. 3A). These experiments

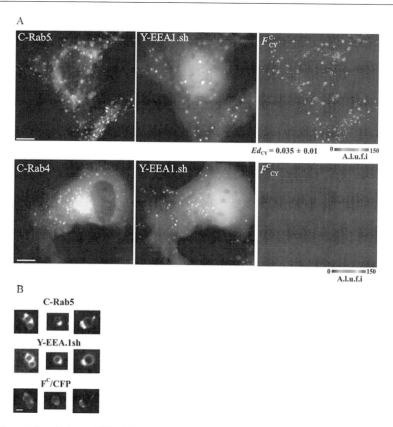

FIG. 3. Specificity of YFP-EEA.1sh sensors for Rab5. (A) CFP-Rab5 was coexpressed with YFP-EEA.1sh in Cos-1 cells. CFP-Rab4 was coexpressed with YFP-EEA.1sh in Cos-1 cells. YFP, CFP, and FRET images were acquired from living cells at room temperature. FRET^C images were calculated as described and presented in a pseudocolor mode. Mean Ed values measured for individual endosomes of the presented cell are shown next to the corresponding image. Intensity bars are presented in arbitrary linear units of fluorescence intensity. Bar $= 10$ μm. (B) Gallery of high magnification images shows individual endosomes or tethered endosomes in cells coexpressing CFP-Rab5 and YFP-EEA.1sh. FRET^C images are presented as pseudocolor intensity-modulated images (FRET^C/CFP). Bar $= 2$ μm. A.l.u.f. i. is arbitrary linear units of fluorescence intensity. (See color insert.)

validated the use of EEA.1sh as a sensor of Rab5-GTP in endosomes of living cells. However, YFP-EEA.1sh has a limited utility as a sensor of GTP-bound Rab5. The strong targeting signal of the FYVE domain directs this fragment to endosomes independently of Rab5 activity (Galperin *et al.*, 2003).

Interestingly, FRET^C signals were condensed in the clusters or "Rab5/EEA.1 microdomains" of the endosomal membrane (Fig. 3B), which is consistent with a subcompartmental organization of Rab5/EEA.1 complexes in endosomes proposed earlier (de Renzis *et al.*, 2002). To present Rab5 microdomains, we have used pseudocolor intensity modulated images (FRET^C/CFP). The CFP channel was used as a saturation channel to emphasize regions of microdomains. In these images CFP values are used as a threshold, and as a result, data higher than CFP values are displayed at full saturation, whereas data values below the low threshold are displayed with no saturation (i.e., black).

Comment. Choice of cDNA fragment used for sensor generation can greatly affect the results of FRET analysis. During this study we found that lack of FRET^C signals was not always indicative of the absence of protein–protein interaction. For instance, when YFP was fused to the amino-terminus of R5BD (YFP-R5BD, Fig. 1A), this protein was recruited to endosomes in a Rab5-dependent manner and to an extent similar to that observed for R5BD-YFP (Galperin *et al.*, 2003). However, we did not detect positive FRET^C signals despite colocalization of YFP-R5BD with either CFP-Rab5 or CFP-Rab5(Q79L) in endosomes. It is possible that the amino-terminus CFP of Rab5 is in close proximity to the carboxyl- but not the amino-terminus of Rabaptin5 within the Rab5–Rabaptin5 complex. A similar result was obtained when the YFP-EEA.1ln (Fig. 1A) construct was used as a potential sensor for Rab5 GTP binding. YFP-EEA.1ln was targeted to endosomes even more efficiently than YFP-EEA.1sh (data not shown). However, coexpression of CFP-Rab5 and YFP-EEA.1 did not result in positive FRET^C signals. In this case, the reason for the absence of FRET^C could be the long distance between YFP-CFP fluorochromes.

Detection of a Three-Protein Complex in Endosomes

The results of conventional 2-FRET analysis demonstrated the assembly of endosomal "Rab5 microdomains" in living cells. More recently we began using three-chromophore FRET (3-FRET) assay for further analysis of protein interactions participating in these domains. To this end, mRFP-Rab5, CFP-Rab5, and YFP-EEA.1sh were coexpressed in the same cells, and the 6-filter 3-FRET method was implemented to obtain three FRET^C images as described (Galperin *et al.*, 2004). As shown in Fig. 4A, all three proteins were highly colocalized in endosomes, often concentrated in confined clusters and sites of endosomal tethering.

FRET Measurements and Calculations. For 3-FRET, images were acquired sequentially through FRET(cy) (excitation 436/10 nm, emission

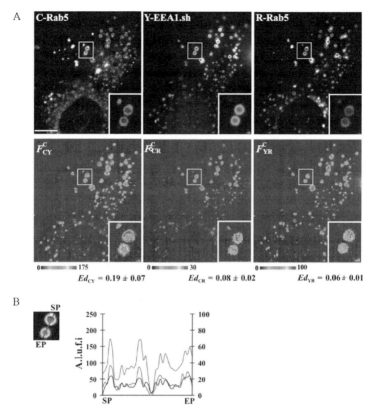

FIG. 4. 3-FRET microscopy analysis of Rab5 microdomains in single endosomes of living cells. (A) mRFP-Rab5, YFP-EEA1.sh, and CFP-Rab5 were coexpressed in Cos-1 cells, the cells were treated with nocodazole for 15 min, and six images were acquired as described in the text. FRET^C images are presented in a pseudocolor mode. Insets show an enlargement of the outlined regions of the images. Mean Ed values measured for individual endosomes of the presented cell are shown below the corresponding image inset. Bar = 10 μm. (B) Three FRET^C images in RGB color format obtained in 3-FRET experiments (FRET$^C_{CY}$ is green, FRET$^C_{YR}$ is red, and FRET$^C_{CR}$ is blue) were merged. "White" designates the overlap of red, blue, and green. The arbitrary fluorescence intensities of FRET^C signals across two endosomes were plotted. FRET$^C_{CR}$ is plotted on the right axes. SP is the starting point and EP is the end point. A.l.u.f.i. is arbitrary linear units of fluorescence intensity. (See color insert.)

535/30 nm), FRET(cr) (excitation 436/10 nm, emission 630/60 nm), FRET (yr) (excitation 492/18 nm, emission 630/60 nm), CFP (excitation 436/10 nm, emission 465/30 nm), YFP (excitation 492/18 nm, emission 535/30 nm), and mRFP (excitation 580/20 nm, emission 630/60 nm) filter channels.

A dichroic mirror #86006 (Chroma, Inc.) was used. Images were acquired under the same conditions as for the two-chromophore FRET described above. Backgrounds were subtracted from raw images prior to carrying out FRET calculations. $FRET^C$ was calculated for each of the three two-chromophore FRET pairs in the presence of three fluorescent proteins on a pixel-by-pixel basis for the entire image as described for two-chromophore FRET above, using the formula shown below as modified from Eq. (1):

$$FRET^C = Ff - Df(Fd/Dd) - Af(Fa/Aa) - Tf(Ft/Tt) \qquad (1m)$$

where Tf is a third fluorochrome and Ft/Tt is the cross-bleed coefficient for the third fluorochrome (fraction of the third fluorochrome passing through the FRET channel). We found that Ft/Tt is $<0.5\%$ for all combinations of fluorochromes and assumed these values to be zero. The cross-bleed coefficients through the FRET(yc) channel were 0.90 and 0.02 for CFP and YFP fluorescence respectively; 0.065 and 0.11 for YFP and mRFP fluorescence, respectively through the FRET(yr) channel; and 0.065 and 0.005 for CFP and mRFP fluorescence, respectively, through the FRET(cr) channel. No significant bleedthrough ($<0.5\%$) among the CFP, YFP, and mRFP filter channels was observed, so these cross-bleeds were considered effectively zero. Images were inspected for the shift of fluorescence compartments during image acquisition and discarded if such shifts occurred. The apparent FRET efficiency Ed was calculated using Eq. (2). The values of $G(cy) = 3.099$, $G(cr) = 1.290$, and $G(yr) = 0.416$ were calculated for CFP–YFP, CFP–mRFP, and YFP–mRFP donor–acceptor pairs, respectively, as described by Gordon et al. (1998). Ed was calculated using regional analysis, as was described for two-chromophore FRET analysis.

As shown in Fig. 4A, $FRET^C$ images and Ed values were indicative of the energy transfer between YFP-EEA.1sh and mRFP-Rab5, CFP-Rab5 and YFP-EEA.1sh, and CFP-Rab5 and mRFP-Rab5. Merging three $FRET^C$ images of the individual endosomes produced similar distribution of $FRET^C$ signals for three FRET pairs across the endosomes, which is consistent with the presence of three-protein complexes (Fig. 4B). Data in Fig. 4 demonstrate the detection of three-protein complexes in early endosomes of living cells. Colocalization of Rab5 and EEA.1sh proteins and maximal $FRET^C$ signals were often concentrated in microdomains within the endosomes, which likely correspond to "Rab5 microdomains." Although detection of FRET between two Rab5 proteins was somewhat surprising, this observation suggested that endosome-associated Rab5 is capable of homodimerization or oligomerization. In fact, the potential of homodimerization of Rab5 in its GTP-bound conformation has been previously demonstrated in *in vitro* experiments (Daitoku *et al.*, 2001). Hence, our data demonstrate that FRET-based sensors can be utilized for detection of the Rab5 protein complex and for analysis of the molecular architecture of this complex.

Conclusion

The method described in this chapter for visualization of Rab5 activity by fluorescent sensors using sensitized FRET microscopy is straightforward. The main advantage of genetically encoded fluorescent sensors for Rab5 activity is the detection of a GTP-bound form of Rab5 in living cells. FRET-based sensors can be useful for visualization of the cellular localization of the active Rab5 form and for elucidating the spatial–temporal relationship between Rab5 and its effectors during membrane trafficking in the living cells as well as during signaling processes. Furthermore, similar sensors can be designed to detect activity of other Rabs in living cells.

Acknowledgments

This work was supported by grants from National Cancer Institute, National Institute of Drug Abuse (A.S.), and American Cancer Society (A.S. and E.G.), and a postdoctoral fellowship from the American Heart Association (to E.G.).

References

Berney, C., and Danuser, G. (2003). FRET or no FRET: A quantitative comparison. *Biophys. J.* **84,** 3992–4010.

Campbell, R. E., Tour, O., Palmer, A. E., Steinbach, P. A., Baird, G. S., Zacharias, D. A., and Tsien, R. Y. (2002). A monomeric red fluorescent protein. *Proc. Natl. Acad. Sci. USA* **99,** 7877–7882.

Christoforidis, S., and Zerial, M. (2000). Purification and identification of novel Rab effectors using affinity chromatography. *Methods* **20,** 403–410.

Daitoku, H., Isida, J., Fujiwara, K., Nakajima, T., and Fukamizu, A. (2001). Dimerization of small GTPase Rab5. *Int. J. Mol. Med.* **8,** 397–404.

de Renzis, S., Sonnichsen, B., and Zerial, M. (2002). Divalent Rab effectors regulate the subcompartmental organization and sorting of early endosomes. *Nat. Cell. Biol.* **4,** 124–133.

Galperin, E., and Sorkin, A. (2003). Visualization of Rab5 activity in living cells by FRET microscopy and influence of plasma-membrane-targeted Rab5 on clathrin-dependent endocytosis. *J. Cell. Sci.* **116,** 4799–4810.

Galperin, E., Verkhusha, V. V., and Sorkin, A. (2004). Three-chromophore FRET microscopy to analyze multiprotein interactions in living cells. *Nat. Methods* **1,** 209–217.

Gaullier, J. M., Simonsen, A., D'Arrigo, A., Bremnes, B., Stenmark, H., and Aasland, R. (1998). FYVE fingers bind PtdIns(3)P. *Nature* **394,** 432–433.

Gordon, G. W., Berry, G., Liang, X. H., Levine, B., and Herman, B. (1998). Quantitative fluorescence resonance energy transfer measurements using fluorescence microscopy. *Biophys. J.* **74,** 2702–2713.

Horiuchi, H., Lippe, R., McBride, H. M., Rubino, M., Woodman, P., Stenmark, H., Rybin, V., Wilm, M., Ashman, K., Mann, M., and Zerial, M. (1997). A novel Rab5 GDP/GTP exchange factor complexed to Rabaptin-5 links nucleotide exchange to effector recruitment and function. *Cell* **90,** 1149–1159.

Kutateladze, T., and Overduin, M. (2001). Structural mechanism of endosome docking by the FYVE domain. *Science* **291,** 1793–1796.

Lawe, D. C., Patki, V., Heller-Harrison, R., Lambright, D., and Corvera, S. (2000). The FYVE domain of early endosome antigen 1 is required for both phosphatidylinositol 3-phosphate and Rab5 binding. Critical role of this dual interaction for endosomal localization. *J. Biol. Chem.* **275,** 3699–3705.

Lippe, R., Horiuchi, H., Runge, A., and Zerial, M. (2001). Expression, purification, and characterization of Rab5 effector complex, rabaptin-5/rabex-5. *Methods Enzymol.* **329,** 132–145.

Nielsen, E., Christoforidis, S., Uttenweiler-Joseph, S., Miaczynska, M., Dewitte, F., Wilm, M., Hoflack, B., and Zerial, M. (2000). Rabenosyn-5, a novel Rab5 effector, is complexed with hVPS45 and recruited to endosomes through a FYVE finger domain. *J. Cell. Biol.* **151,** 601–612.

Stenmark, H., and Olkkonen, V. M. (2001). The Rab GTPase family. *Genome Biol.* **2,** review S3007..

Stenmark, H., Parton, R. G., Steele-Mortimer, O., Lutcke, A., Gruenberg, J., and Zerial, M. (1994). Inhibition of rab5 GTPase activity stimulates membrane fusion in endocytosis. *EMBO J.* **13,** 1287–1296.

Szymkiewicz, I., Shupliakov, O., and Dikic, I. (2004). Cargo- and compartment-selective endocytic scaffold proteins. *Biochem. J.* **383,** 1–11.

Vitale, G., Rybin, V., Christoforidis, S., Thomqvist, P., McCaffrey, M., Stenmark, H., and Zerial, M. (1998). Distinct Rab-binding domains mediate the interaction of Rabaptin-5 with GTP-bound Rab4 and Rab5. *EMBO J.* **17,** 1941–1951.

Wouters, F. S., Bastiaens, P. I., Wirtz, K. W., and Jovin, T. M. (1998). FRET microscopy demonstrates molecular association of non-specific lipid transfer protein (nsL-TP) with fatty acid oxidation enzymes in peroxisomes. *EMBO J.* **17,** 7179–7189.

Zerial, M., and McBride, H. (2001). Rab proteins as membrane organizers. *Nat. Rev. Mol. Cell. Biol.* **2,** 107–117.

[12] Selection and Application of Recombinant Antibodies as Sensors of Rab Protein Conformation

By Clément Nizak, Sandrine Moutel,
Bruno Goud, and Franck Perez

Abstract

The existence of a conformational switch of Rabs and other small GTPases involved in intracellular transport regulation has been known for many years. This switch is superimposed on the membrane association/dissociation cycle for most of these GTPases. While these processes are key features of the dynamics of intracellular transport events, surprisingly very few previous studies have focused on the dynamics of the GDP/GTP cycle of Rab proteins in time and space. The main reason for this is the lack of tools available to dynamically probe for Rab GTPases conformation switches and membrane association/dissociation, in particular *in vivo*. We recently reported the *in vitro* selection of conformation-specific recombinant antibodies specific to the GTP-bound conformation of Rab6 proteins. These antibodies were obtained *in vitro* by phage display, a rather simple, rapid, and cheap technique. We additionally showed that these conformation-specific antibodies can be expressed in living cells to follow endogenous Rab6 in its activated conformation *in vivo*. The same strategy could be used to study other conformation switching mechanisms and, in general, to study the switching between states that antibodies can distinguish (e.g., phosphorylation, ubiquitination).

Introduction

The first step in characterizing GTPase conformation switch dynamics is to design probes that specifically detect these conformation states. One approach is to use protein domains that are naturally sensitive to the small GTPase conformation state. GDI and effectors are respectively specific for GDP or GTP conformations and subdomains could be tried and used as sensors. However, while this approach has been particularly powerful for Ras, Rho/Rac/Cdc42, or Arf proteins (Kraynov *et al.*, 2000; Schweitzer and D'Souza-Schorey, 2002), it had not yet been described for Rab proteins. An important limitation of this approach is that since the sensor is identical to natural effectors, a given GTPase cannot at the same time bind to the sensor and to its effector. This implies (1) that any GTPase engaged in a strong interaction with an effector will not be detected by the sensor and

METHODS IN ENZYMOLOGY, VOL. 403   0076-6879/05 $35.00
                   DOI: 10.1016/S0076-6879(05)03012-0

(2) that any GTPase detected by the sensor will have lost its interaction with an effector. This is particularly important when used *in vivo* since, in a way, the function of all detected GTPase molecules is necessarily altered (Kraynov *et al.*, 2000). An additional limitation of this approach is that such conformation-specific interactors are not known for many small GTPases regulating intracellular transport.

Since naturally evolved proteins seem to distinguish between different GTPases and even their conformations, we reasoned that selected antibodies may also bind specifically to a given GTPase in a particular conformational state. We have shown that it is indeed possible to isolate antibodies that specifically detect the GTP-bound conformational state of a particular Rab, Rab6 (A and A') (Nizak *et al.*, 2003b). We developed a fully *in vitro* approach based on antibody phage display (Hust and Dubel, 2004; Smothers *et al.*, 2002). A highly diverse semisynthetic library of recombinant antibodies (the Griffin.1 library, kindly provided by G. Winter, MRC, Cambridge, UK) was used. The Griffin.1 library contains 10^9 different antibodies of human origin in the form of single-chain Fv (scFv) where the V_H and the V_L of immunoglobulins are fused in a unique molecule. The library was screened for clones that specifically bind Rab6·GTP purified from bacteria and led to the identification of conformation-specific anti-Rab6 antibodies (like the scFv AA2; Fig. 1). This scFv was then successfully used to detect or purify Rab6 in its GTP-bound state by immunofluorescence or immunoprecipitation, respectively. This allowed us to show that AA2 does not bind Rab6 in competition with certain effectors. We could also confirm the membrane association of GTP-bound endogenous Rab6 proteins and follow their reaction to the overexpression of so-called dominant negative Rab6 isoforms. AA2 was then subcloned from the phage display vector into a mammalian expression vector in fusion with GFP or its spectral variants CFP and YFP. This led to the observation in real time and in living cells of the dynamics of endogenous GTP-bound Rab6 molecules and demonstrated subtle differences with the dynamics of overexpressed GFP-Rab6.

Selection of Conformation-Specific Antibodies by Phage Display

Preparation of the Target GTPase

Either a GTPγS-loaded Rab or a GTP-locked mutant can be used as antigens. For simplicity, we used the Q72L mutant of Rab6, expressed as a polyhistidine-tagged protein from a pET15b vector in BL21 *Escherichia coli* and purified the protein according to the protocol described for Rab1 and its mutants (Nuoffer *et al.*, 1995). The purity of the purified protein was assessed by sodium dodecyl sulfate, polyacrylamide gel electrophoresis (SDS–PAGE)

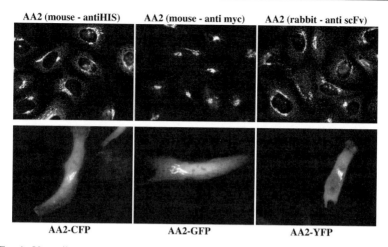

AA2 (mouse - antiHIS)   AA2 (mouse - anti myc)   AA2 (rabbit - anti scFv)

AA2-CFP   AA2-GFP   AA2-YFP

FIG. 1. Versatile usage of recombinant antibodies. In the top panel, HeLa cells were fixed in paraformaldehyde and permeabilized in saponin before being stained using the anti-Rab6 conformational scFv AA2. This illustrates that AA2 can be used as a mouse-like antibody when detected using the mouse monoclonal antibodies directed against either the HIS or the myc tag of the scFv, or as a rabbit antibody when using a rabbit anti-scFv polyclonal serum. We also successfully used goat or rabbit anti myc. This is particularly interesting when co-staining with natural antibodies, since AA2 can easily be used in conjunction with any other antibodies. In the bottom panel, HeLa cells were transfected by plasmids driving the expression of AA2 tagged with CFP, GFP or YFP and observed alive under the microscope. In this case, AA2 allows the intracellular staining and tracking of endogenous Rab6 in living cells.

examination. In addition, microinjection in HeLa cells was used to confirm that a correctly folded Rab6, in its activated conformation, was purified. Golgi enzyme redistribution to the endoplasmic reticulum was indeed observed shortly after microinjection, which is the phenotype observed upon Rab6Q72L overexpression (Echard *et al.*, 2000; Martinez *et al.*, 1997).

Comment. BL21 *E. coli* containing Rab6Q72L in a pET15b vector is inoculated into shaking cultures for 16 h at 28° without any exogenous induction: the promoter in pET15 vectors is leaky and induction with isopropylthiogalactoside (IPTG) only increases the amount of insoluble protein produced, not the production of the correctly folded soluble protein of interest (personal communication from the W. Balch laboratory concerning Sar1 and its mutants [Rowe and Balch, 1995], which we extrapolated to Rab6).

Phage Display Screen

The choice of the antibody phage display library is of course crucial. We used a semisynthetic native library from human origin: the Griffin1 library (a kind gift of Dr. G. Winter, MRC, Cambridge, UK). This library contains more than 10^9 independent clones (inserted into the pHEN2 vector), fused to myc and His$_6$ tags and displayed in fusion with the pIII protein of M13. Although often called a "phage library," the Griffin.1 is a phagemid library. This means that without helper phages, it behaves as a plasmid that confers antibiotic resistance (ampicillin) to bacteria. Only upon infection with a helper phage can phages displaying antibodies be generated. Note that the Griffin.1 library is no longer available from the MRC, but other similar libraries are freely available for academic use, in particular from the MRC.

The Griffin.1 semisynthetic naïve library is composed of scFv recombinant antibodies. We did not evaluate any Fab library. scFvs are composed of V_H and V_L chains of immunoglobulins fused together through a 15-aa-long Gly–Ser linker. Their small size (about 30 kDa compared to 150 kDa for a full IgG) in a monochain format is a huge advantage in the applications described below.

Diversity is the main criterion when choosing a library. We have exclusively used the Griffin.1 library in all our screens. With the pHEN2 vector (as well as several others), it is possible to express the scFv either as a secreted protein or fused to the M13 phage pIII surface protein for display at the phage surface. An in-frame amber stop codon inserted between the scFv sequence and the pIII gene is responsible for this: in *SupE* suppressor *E. coli* strains used for phage production (such as TG1), a fusion protein will be produced and phage particles will display scFvs at their surface. In nonsuppressor strains (like HB2151), only soluble scFvs will be produced and secreted by *E. coli* in their periplasm.

In both cases, expression can be controlled by glucose levels and IPTG. Three levels of expression are used. During bacteria amplification steps when no scFv is to be produced, 1% or 2% glucose is used to repress expression. During phage production, the ideal is to have one scFv displayed per phage particle to avoid selection by avidity but only rely on affinity. Display at the phage surface is done without induction of the promoter, only relying on its leakiness in the absence of repression. Massive production of selected scFv is done in the presence of IPTG to induce protein expression. C-terminal myc- and His$_6$-tags present just before the amber codon are used for purification and detection of the scFvs. We also generated generic polyclonal anti-scFv in rabbits and chickens.

The screens are done in suspension using biotinylated antigen and streptavidin-coated beads to recover the antigen–phage complexes. Other protocols using plastic-coated antigens exist (see Nizak *et al.*, 2003a), but this led to partial antigen denaturation. We found that scFvs directed against denatured proteins are proportionally less efficient in immunofluorescence and *in vivo* expression and this method is not adapted to the selection of conformation-sensitive scFvs.

Selection against Biotinylated Antigen

Biotin Coupling

MATERIALS

- Sulfo–NHS-biotin, or PEO–maleimide-biotin (Pierce)—stock at 10 mM
- Tween-20
- Horse radish peroxidase (HRP)-labeled streptavidin
- Anti-His$_6$ antibodies (Santa-Cruz)
- M280 streptavidin dynabeads (Dynal)
- Unlabeled streptavidin
- *Optional*: silicon-coated tubes to reduce nonspecific adsorption on tubes (necessary in the case of Rab6).

METHODS. The goal is to biotinylate and recover 100% of target antigens on streptavidin-coated magnetic beads, ideally with only one or two biotin per antigen to avoid large complex formation. While we generally use sulfo-NHS-biotin (which targets reactive amines) to modify the antigen, we failed to quantitatively modify Rab6 using this reagent. We thus used PEO-maleimide to target the carboxy-terminal Rab6 cysteines. This reagent also worked efficiently against other small GTPases. Note that it is important to use silicon-coated tubes to avoid loss of modified proteins on the tube walls. In the case of Rab6, we also used 0.5% Tween-20 to increase protein recovery. In addition, when diluted protein quantities are used, we add 0.5% of bovine serum albumin (BSA) as a stabilizer.

1. Small-scale Rab6 modification is performed in phosphate-buffered saline (PBS)-Tween-20 0.5%. A range of biotin:Rab6 protein is explored, from 1:1 to 50:1 (2 h at 4°).

2. Reaction is stopped by addition of 5× excess of free cystein.

3. Microdialysis (or microgel filtration) is then performed to remove unincorporated cysteins.

4. The percentage of modified Rab6 is calculated performing a rapid recovery test on magnetic beads (see below) using only 3 washes instead of 20. Bound and unbound fractions are analyzed by Western blotting using HRP-streptavidin (to calculate the percentage of recovered biotinylated protein, which should be close to 100%) and anti-His$_6$ antibodies (to calculate the percentage of recovered Rab6).

5. More recently, we rather use an alternative and fast way to calculate the percentage of modified protein. After biotinylation, quenching, and free biotin removal, fractions of modified proteins are mixed with a large

excess of streptavidin for 30 min at room temperature (RT). SDS-containing loading sample buffer is then added and proteins separated by SDS–polyacrylamide gel electrophoresis (PAGE) (without prior boiling), blotted and stained using either an anti-His$_6$ antibody or HRP-labeled streptavidin. Because biotin–streptavidin interaction persists in SDS, modified proteins are shifted and do not migrate at their normal size. The quantity of normally migrating proteins is proportional to the quantity of unmodified proteins.

6. In general, the quantities of modified proteins are enough to carry out the three or four cycles of selection. Otherwise, biotinylation of the antigen is repeated on a larger scale using the determined biotin:antigen ratio.

Recovery Test

MATERIALS

- M280 streptavidin Dynabeads (Dynal)
- Magnet (MPC-1 from Dynal or Magnetight separation stand from Novagen)
- PBS
- Tween-20
- Marvel fat-free milk
- Bacterial culture Falcon tubes
- *Optional*: silicon-coated tubes to reduce nonspecific adsorption on tubes (necessary in the case of Rab6).

METHODS. This step is optional. This is done to ensure that in screening condition, efficient recovery of the antigen will be achieved.

1. In an Eppendorf tube (ideally silicon-coated), add the biotinylated antigen to be tested (amount used during selection) in 1 ml PBS + 0.1% Tween-20 + 2% milk (fat-free Marvel milk. Any really low-fat milk should work). The antigen is typically diluted at 10 nM to ensure that only scFv with a good affinity will be recovered during the screen. For a 25-kDa protein, this limits the quantity needed per round of selection to 250 ng.

2. Incubate end over end for 90 min, then standing for 30 min.

3. Add 50 μl streptavidin M280 Dynabeads washed twice in PBS + T (recovered using a magnet). Incubate for another 30–60 min.

4. After isolating the beads with a magnet, remove the buffer carefully with a tip pipetting at the bottom of the tube (do not aspirate, especially not at the liquid/air interface, where beads tend to accumulate).

5. Resuspend beads gently in 1 ml PBS + T and transfer to a Falcon tube containing 7 ml PBS + T. Wash (magnet) twice more with 8 ml PBS + T and transfer to a new tube. A total of 10/20 washes (round 1 and 2/3, respectively), transferring to a new tube every three washes. Washes should be performed quickly (30–60 s total, no additional incubation time). Beads should be resuspended only when transferred to a new tube; otherwise the buffer should be added carefully so as not to disaggregate the beads, and the washing just consists in turning the tube quickly on the magnet immediately after buffer addition to allow the beads to cross the tube to the opposite side in a few seconds.

6. By SDS–PAGE/Western blotting analysis: using streptavidin–HRP and anti-His$_6$ antibodies, compare amounts of initial antigen to supernatant and washed beads. The ideal is >90% recovered, but recovery of as low as 50% is acceptable (as this was the case for Rab6).

Comment. For organelle-based screens (Nizak *et al.*, 2003a), the target antigen concentration is not controlled; it is imposed by the structure of the organelle that displays antigens in their native context.

Selection. This section describes a round of selection; the protocol is adapted from the Griffin.1 library user's manual (very similar to the Tomlinson I + J libraries protocol as well; both libraries are provided by the MRC Geneservice). We usually perform three successive rounds, and monitor the output/input phage ratio by systematic titration of phages injected in the selection tubes and phages present in the elution fraction; this yield typically increases 10- to 100-fold at the third round, indicating successful selection.

MATERIALS

- 2xYT
- TG1 *E. coli* stock (grown in nonsupplemented thiamin containing M9 minimal medium to maintain selection on the F′ episome)
- M13KO7 helper phage. Available from GE-Healthcare, but easy and cheap preparation afterward. Use 10^{12} and 10^{14}/ml stocks in PBS at $-20°$ and $-80°$
- Ampicillin, kanamycin (100 and 50 mg/ml stocks, respectively)
- 20% glucose, filter sterilized
- PEG-8000 30% + NaCl 2.5 *M*; kept at 4° to enhance phage precipitation
- M280 streptavidin Dynabeads (Dynal)
- Magnet (MPC-1 from Dynal or Magnetight separation stand from Novagen)

- PBS
- Tween-20
- Marvel fat-free milk
- Triethylamine (TEA) (Sigma)
- Tris 1 M pH 7.4
- Three 15-cm diameter and seven 10-cm diameter 2xYT + Amp + Glu plates (100 μg/ml ampicillin, 1% glucose) per round
- Bacterial culture Falcon tubes
- Optional: silicon-coated tubes (we used those for Rab6)

METHODS

1. An aliquot of the library in TG1 representing 10 times the diversity is diluted at OD = 0.05 and incubated at 37° in 2xYT + Amp + glucose 1% until midexponential growth (OD = 0.5). For the Griffin.1 library, this necessitates a 250–ml culture containing 10^{10} bacteria. *Note: We consider that 1 $OD_{600\ nm} = 8 \times 10^8$ bacteria/ml.*

2. 10^{10} bacteria (i.e., 25 ml = 1/10 of the exponential culture) are then mixed with a 20× excess of helper phage. Mixing is done inverting gently once without pipetting to prevent breaking of bacteria pili that are essential to M13 infection. Bacteria + M13 are rapidly put back at 37° and incubated standing for 30 min.

3. After centrifugation (3300×g, 20 min), infected bacteria are resuspended in 500 ml of 2xYT + Amp + Kana and incubated at 30° overnight. *Note: Kanamycin is used to select positively infected bacteria that grow more slowly than noninfected ones. Low temperature (30°) is used to help scFv folding during display.*

4. Phages are concentrated and separated from the soluble scFvs secreted by the bacteria using PEG precipitation. ScFvs are precipitated by adding one-fifth volume of PEG-8000/NaCl to the culture supernatant in an ice-cold bottle for 1 h at 4°, and centrifugation at 10,800×g for 10 min. For the first round of selection, this precipitation step is repeated after resuspension in 40 ml H_2O, addition of 8 ml PEG-8000/NaCl, and incubation for 20 min at 4°. The pellet is centrifuged once more to remove the last drops of PEG/NaCl supernatant, resuspended in 1 ml PBS, and centrifuged at 11,600×g for 10 min to remove bacterial debris. Phages remain in solution this time; the supernatant is the input phage solution.

5. Remove nonspecific binders (in particular antistreptavidin scFvs). In an Eppendorf tube (silicon-coated if necessary), add phages in 1–1.5 ml total PBS + 0.1% Tween-20 + 2% milk (PBS + T + M). We tend to use large quantities of phages (10^{14} phage particles, which is 10- to 100-fold

more than most usual phage display protocols), and this may be a critical parameter for the screening success. Add 50 μl streptavidin M280 Dynabeads washed (magnet) twice in PBS + T. Incubate end over end for 90 min, then standing for 30 min. *Optional: add "preadsorption antigen" during this step (biotinylated and tested like the "selection antigen"). In the case of Rab6, preadsorption with Rab6·GDP was not necessary.*

6. With a magnet, remove preabsorbed phages carefully with a tip at the bottom of the tube (do not aspirate, especially not at the liquid/air interface). Add preabsorbed phages to a tube containing the biotinylated antigen diluted at 10 mM (in PBS + T + M).

7. Incubate end over end for 90 min, then standing for 30 min. Add 50 μl streptavidin M280 Dynabeads (washed twice in PBS + T). Incubate for 30–60 min.

8. With a magnet, remove buffer carefully with a tip at the bottom of the tube (do not aspirate, especially not at the liquid/air interface). Resuspend beads gently in 1 ml PBS + T and transfer to a Falcon tube containing 7 ml PBS + T.

9. Wash (magnet) twice more with 8 ml PBS + T, transfer to a new tube for a total of 10 washes (first round) or 15–20 washes (subsequent rounds), transferring to a new tube every three washes. As noted before, washes should be as quick as possible. scFvs are monovalent and certain good scFvs may have high k_{off}.

10. After the last wash done in a clean tube, elute phages. We use TEA (*fresh solution* of 140 μl TEA in 10 ml H$_2$O). Then 1 ml of TEA is added and beads are incubated 10 min at RT; 500 μl of TEA is recovered (on the magnet) and neutralized in 500 μl Tris 1 M pH 7. After an additional 10 min incubation, the remaining 500 μl of TEA is recovered (magnet) and neutralized in the same Tris-containing tube. The beads are then also neutralized with 200 μl Tris 1 M pH 7 and kept separately.

11. The rescue and amplification of selected phages is achieved by infecting TG1 bacteria. Infect 10 ml TG1 (OD = 0.5) with 750 μl neutralized eluted phages. Also add 4 ml TG1 to neutralized beads, incubate 30 min at 37° standing, then pool. Save a 100 μl aliquot (to calculate the titer of eluted phage solution), spin down infected bacteria (10 min, 3300×g), and resuspend in 600 μl. Spread infected bacteria on 3 × 15-cm 2xYT Amp 1% Glu plates. It is not necessary to remove the beads before spreading bacteria onto the plates.

12. 10^{-9}, 10^{-10}, and 10^{-11} dilutions of the input phages solution are also used to infect TG1 for titration, and the infected TG1 are spread onto three separate 2xYT Amp 1% Glu plates. 10^{-1}, 10^{-2}, and 10^{-3} (10^{-2} to

10^{-4} for the third round) dilutions of a sample of the 14–ml pool of eluted phages recovered in TG1 are also spread onto three separate 2xYT Amp 1% Glu plates for output phage titration. The seventh 2xYT Amp 1% Glu plate is used for negative controls: 2xYT, input phages, TG1 *E. coli* used for infection.

13. The day after, colonies for input and output titrations are counted to determine yield. Typically we get 10^{-6} to 10^{-8} yields, with a 10- to 100-fold increase between the second and the third round. If no increase or only a poor increase is observed, we proceed to a fourth round.

14. Large plates are scraped with 3–6 ml of 2xYT + 30% glycerol to $-20°$ and the OD of 1/100 dilution is measured to determine the concentration of bacteria. The next round of selection is then started diluting bacteria to OD = 0.05 in 100 ml 2xYT Amp 1% Glu and incubated at 37° until OD = 0.5.

15. A 20-fold excess of helper phages is then added to 10 ml of this culture (80 μl of 10^{12}/ml helper phage solution) with gentle mixing and incubated standing 30 min at 37°. Bacteria are then centrifuged and resuspended in 50 ml 2xYT + Amp + Kana (**no glucose**) and incubated at 30° overnight to produce phages for another round of selection.

16. Phages are precipitated as for the first round with PEG/NaCl. The culture is first centrifuged for 10 min at $10,800 \times g$ to remove bacteria, 10 ml of ice-cold PEG/NaCl is added, and precipitation is performed for 1 h at 4°. Phages are recovered in the pellet after a 10–min $10,800 \times g$ centrifugation step that is repeated to remove the last drops of PEG/NaCl. Then the phages are resuspended in 1 ml PBS, the remaining debris are removed by spinning for 10 min at $11,600 \times g$, and the phage solution is ready for the next round of selection.

Isolation of Positive Clones

Most protocols distributed with phage display libraries recommend enzyme-linked immunosorbent assays (ELISAs) for isolation of positive clones. In an extensive assay (S. Martin and F. Perez, unpublished), we observed that ELISA cannot be used to predict whether an antibody will be usable in other methods like immunofluorescence or Western blotting. Worse, some scFvs were found negative by ELISA and positive using other methods. Thus, we systematically analyze selected scFvs using the methods we intend to use for them. In the case of Rab6, testing clones by immuno-fluorescence also gave us an immediate idea of the specificity of the antibody: perinuclear Golgi staining would indicate right away that an antibody is probably detecting Rab6 quite specifically and not all GTPases in general.

MATERIALS

- 2×YT
- Ampicillin
- 20% glucose
- 1 M IPTG
- 2–ml deep 96-well plates
- Usual immunofluorescence reagents
- scFv detection antibodies: anti-myc (9E10), anti-His$_6$ (His$_6$-1, Sigma-Aldrich)

METHODS

1. Clones are picked from the last selection round (colonies on output titration plates or obtained after spreading the last round output glycerol stock). Each colony is transferred in one well of a 2-ml deep 96-well plate with 500 μl 2×YT Amp 1% Glu, and grown overnight at 37°. Then 500 μl 2×YT 50% glycerol is added to each well to make −20° stocks.

2. Clones replicated from stock 96-well plates are grown in 2×YT Amp 1% Glu in 96-well plates (1 ml cultures in 2-ml deep wells). Cultures are started at a 1/100 dilution; 10 μl of a fresh culture is inoculated into 1 ml 2×YT Amp 0.1% Glu and grown for 2 h at 37° shaking (*note that the glucose level is reduced from 1% to 0.1% to allow for scFv expression*).

3. At this point cultures should have reached OD$_{600}$ = 0.2 roughly. Then 1 mM IPTG is added to the wells and cultures are transferred to 30° and shaken overnight (16 h).

4. After centrifugation for 5 min at 5000×g, the supernatant (culture medium containing the secreted scFvs) is transferred to a clean 96-well plate.

5. For immunofluorescence, 45 μl of the culture medium of each clone is mixed in another 96-well plate with 5 μl of a 10× solution containing 9E10 anti-myc and/or anti-His$_6$ as well as saponin when cells were fixed using paraformaldehyde (no detergent when fixed in methanol) and used as the primary antibody. Freshly fixed cells should be used (fixed cells kept overnight in PBS at 4° do not show any Rab6·GTP staining even though they still show a normal Rab6 staining with a polyclonal anti-Rab6 serum).

6. Cells are grown on coverslips, fixed, and permeabilized in saponin when needed. Coverslips are deposited onto 50 μl drops of the premixed antibodies on a parafilm and incubated in a humidified chamber for 90 min. Secondary detection is achieved with classical secondary antibodies, and coverslips are mounted using regular protocols onto slides. However, we

found that it is very important to keep washing steps very succinct for certain scFvs. Coverslips should be simply **rinsed once for 3–5 s in PBS** between primary and secondary incubations and after the secondary incubation. For instance, in the case of the GTP-restricted anti-Rab6 antibody AA2, the signal is completely abolished if coverslips are washed using the classic 3 × 5 min washes. This may be due to the high k_{off} of these monovalent scFvs.

7. Positive clones are selected and further studied after large-scale purification.

Comment. Following the classic protocol, selected clones should be transferred in the nonsuppressor strain HB2151 before characterization to increase secretion of soluble scFvs. However, we found that the soluble scFv secretion is also very efficient in TG1, the amber suppression being only partial, and, more importantly, one of our anti-Rab6·GTP scFv clones actually contains an amber codon within the scFv sequence (clone AA8), and therefore was not produced in nonsuppressor strains. The risk of missing such interesting clones convinced us to stop switching *E. coli* strains systematically after the phage display selection.

Recombinant Antibodies as Conformation Sensors *In Vitro*

Large-Scale Production and Purification of Recombinant scFvs

The miniscale production used to test clones after the screen is simply extended to larger volumes and purification is achieved according to a classical His$_6$-tag approach starting from culture medium. Because the histidine concentration in 2×YT is too high to allow for Ni-NTA-based purification, we chose to produce the scFvs in minimal medium supplemented with casamino acids. Production in 2×YT followed by ammonium sulfate precipitation or extraction from periplasm can also be used. Milligrams of scFvs are thus easily obtained overnight at a very low cost.

MATERIALS

- Supplemented M9 minimum medium
    200 ml M9 salts (5× stock)
    1.0 ml MgSO$_4$ (1 M stock)
    10 ml glycerol (20% stock)
    supplemented with + 0.1 ml vitamin B$_1$ (thiamin, 0.5% stock)
        + 5 ml casamino acids
        + 1 ml ampicillin (100 mg/ml stock)
    complete with H$_2$O to 1 liter, filter sterilize

- 2×YT
- 20% glucose
- IPTG 1 M
- PBS
- 500 mM imidazole in PBS
- Qiagen Ni-NTA or Talon resin

METHODS

1. A 500 ml 2×YT Amp 1% Glu 37° shaking culture inoculated with a positive clone at 1/100 is cultured until OD = 0.2, centrifuged for 5 min at 5000×g, and resuspended in minimal M9 medium (containing glycerol as unique carbon source, no glucose) supplemented with Amp, casamino acids, thiamin, and 1 mM IPTG. This culture is then shaken at 30° for 16 h (overnight). This induction protocol can be optimized for each clone; sometimes it is better to wait until OD = 0.6 and sometimes induction should start very early.

2. Bacteria are then centrifuged at 10,000×g for 10 min (some bacteria are lysed during this scFv secretion procedure and debris need to be removed by higher speed centrifugation). The culture medium containing secreted scFvs is then 0.2 μm filtered to remove the remaining debris, and put into a funnel adapted onto a column containing 2 ml of PBS-washed Ni-NTA or Talon resin (from then on all steps are done at 4°).

3. After two passages of the entire culture medium, the column is washed with 50 ml PBS and then eluted with 500 mM imidazole in PBS. A 250–500 μl fraction is recovered and tested for the presence of protein (almost only scFv usually) using the classic Bradford assay (Bio-Rad). After SDS–PAGE analysis, we noted that some scFvs tend to dimerize, and that these dimers become monomers in the presence of dithiothreitol (DTT) in the sample buffer.

4. The positive fractions can be pooled and either flash frozen in liquid N_2 and kept at −80° or adjusted to 50% glycerol and kept at −20°.

Comment. An additional dialysis step against PBS or another buffer prior to freezing can be performed to remove imidazole (which competes with anti-His$_6$ tag antibodies for scFv binding and thus reduces detection efficiency).

Immunofluorescence

The procedure for regular immunofluorescence is very similar to that used during the isolation of the positive clones. One important point is that washing steps should be kept as short as possible during this isolation process. Note, however, that certain scFvs can stand extended washing time.

MATERIALS

- Usual immunofluorescence materials
- scFv prep (PBS stock)
- scFv-detection antibodies: anti-myc, anti-His$_6$

METHODS

1. In the case of Rab6 staining, cells were fixed using 3% PFA for 10 min and then prepermeabilized in 0.05% saponin for 5–10 min. Again, we found that cells should be fixed just before staining for better results.
2. Coverslips are incubated for 90 min at RT with a solution of 1/100 scFv stock (1 mg/ml) mixed with 9E10 anti-myc and anti-His$_6$ antibodies in PBS + saponin.
3. Coverslips are **rinsed once for 3–5 s** in PBS and incubated with secondary antibodies in the presence of saponin. After two short rinsing steps, coverslips are mounted onto slides in mowiol.

Immunoprecipitation

The immunoprecipitation experiments carried out with anti-Rab6·GTP scFvs showed that cell lysis conditions and/or solubilization procedure are key issues. No Rab6·GTP was purified from wild-type or GFP-wtRab6 expressing HeLa cell extracts. The GTP-bound protein was efficiently purified only from GFP-Rab6Q72L expressing cells. We think either that the usual cell lysis conditions, such as the one depicted in the following protocol, do not conserve the GTP conformation state of Rabs during the experiment, or that the majority of cellular Rab6 is in a GDP conformation.

MATERIALS

- Mass culture of HeLa cells
- scFv prep (PBS stock): 50 μg per condition
- Qiagen Ni-NTA resin (Qiagen) or Talon resin (BD) or Ni-NTA magnetic beads (Qiagen)
- IP buffer: 3 mM imidazole pH 7.2, 30 mM MgCl$_2$, and protease inhibitors
- PBS

- 500 mM imidazole
- 5 M NaCl
- 2.5 M MgCl$_2$
- Triton X-100

METHODS

1. 15×10^6 HeLa cells are transfected using CaPO$_4$ with Rab6 expression plasmids (GFP-Rab6wt, GFP-Rab6Q72L, or GFP-Rab6T27N), and after 16 h, are detached from culture flasks in PBS-EDTA, centrifuged, and resuspended in 4 ml of IP buffer (from then on all steps are at 4°).

2. Cells are lysed with a ball-bearing cell cracker on ice, adjusted to 150 mM NaCl, 20 mM imidazole, and 0.5% Triton X-100, and centrifuged at $1000 \times g$ for 5 min.

3. The resulting postnuclear supernatant (PNS) is diluted with the same buffer (including NaCl, imidazole, and Triton X-100) to a final volume of 12 ml (this procedure is optimized to reduce nonspecific bead binding), and 4-ml fractions of this PNS are incubated with or without 50 μg scFv (AA2 clone) at 4° for 4 h (one fraction per condition).

4. Then 150 μl of buffer-washed Ni-NTA magnetic beads (Qiagen) is added to reactions for 1 h at 4°.

5. Beads are then washed five times in PBS, 20 mM imidazole, 30 mM MgCl$_2$, 0.5% Triton X-100 (washes are performed quickly, without resuspending the beads but just turning the tube on the magnet to allow the beads to simply cross the tube in about 1 s), and bound and unbound fractions are processed for SDS–PAGE and Western blot analysis.

Pull Down of GST-Rab6·GTP

This experiment is performed as described in Monier *et al.* (2002).
MATERIALS

- Gluthation-Sepharose beads
- GDP, GTPγS
- GST fusion of the GTPase of interest (here Rab6)
- EDTA buffer: 25 mM Tris pH 7.5, 10 mM EDTA, 5 mM MgCl$_2$
- MgCl$_2$ 2.5 M stock
- Blocking agents: BSA, casein
- Mouse PNS
- Cytosol of Rab6IP1-expressing insect cells
- scFv prep (PBS stock)

METHODS

1. Rab6-GST purified from *E. coli* is coupled to gluthatione-Sepharose beads (12.5 μg per condition), loaded with GDP or GTPγS in 10 mM EDTA buffer for 1 h at 37°, and then incubated with 0.2 μg AA2 and either 10 μg mouse PNS, 1.5 μl cytosol of Rab6IP1-expressing SF9 insect cells, 0.1% casein, 1% BSA, or no blocking agent for 90 min at RT in the presence of 10 mM MgCl$_2$.

2. Beads are washed five times and processed for SDS–PAGE. Bound and unbound fractions are analyzed by Western blotting. The AA2-Rab6·GTPγS interaction was identical in all five conditions tested.

3. βGDI (present in the mouse cytosol), Rab6IP1, and AA2 are detected with an anti-GDI antibody, an anti-Rab6IP1 antibody, and the 9E10 monoclonal anti-myc-tag antibody, respectively, followed by HRP-labeled secondary antibodies and chemiluminescence staining (Pierce).

Recombinant Antibodies as Conformation Sensors *In Vivo*

Subcloning scFvs into Mammalian Expression Vectors

We chose to introduce scFvs sequences into a widely used expression vector of the pEGFP family (Clontech/BD) but to also keep the context of scFvs in these vectors identical to that of scFvs in pHEN2. A first generation of GFP-tagging vector that we developed, and that did not keep this linker context, was inefficient for intracellular expression. The C-terminal myc and His$_6$ tags of pHEN2 were thus introduced between the scFv and the EGFP sequences (they are located between the scFv and the pIII gene in pHEN2).

We inserted a particular linker in these plasmids that allows us to subclone scFvs from pHEN2 into our modified pExFP vectors digesting the pHEN2-scFvs with *Nco*I/*Not*I and inserting into *Bbs*I/*Not*I-digested pExFP vectors; *Bbs*I digestion of these vectors creates *Nco*I cohesive extremities (the Clontech/BD pExFP vectors we use were additionally modified by removing the *Not*I site downstream of EGFP).

In Vivo *Imaging with Fluorescent scFvs*

The important issue here is the expression level of scFv–xFP fusions. The goal is to optimize the ratio of expressed scFv–xFPs to available intracellular target antigens. As with many other xFP chimeras, massively overexpressed scFv–xFP fusions tend to aggregate and accumulate near the centrosome into a so-called aggresome. This seems to occur when the scFv–xFP fusion has no binding partner in the cytoplasm. This can be due to low target antigen concentration or to the inability of the expressed scFv to bind to its target in the cytoplasm. A strong limitation of this approach is that most scFvs are expressed in the cytosol, because they are reduced in this environment and fail to fold properly. In general, we observed that about 20% of expressed scFvs can interact with their target in the cytosol.

To prevent massive overexpression, the best approach is to use "mild" transfection protocols to express scFv–xFP fusions at least at first. We have been using short–time CaPO$_4$-based transfection successfully: the expression level is very heterogeneous from cell to cell, which allows us to directly compare different expression levels in neighboring cells. Some cells will contain brightly fluorescent aggregates near the centrosome; they express scFv–xFP at too high a level compared to the availability of the target antigen. Other cells will show a weaker staining that is similar to the one observed by immunofluorescence of fixed cells stained with the same scFv.

MATERIALS

- CaCl$_2$ 2.5 M, sterilized
- HEBS, sterilized (140 mM NaCl, 1.5 mM Na$_2$HPO$_4$, 50 mM HEPES, pH 7.05)
- scFv in expression vector DNA
- Geneticin/G418 (40 mg/ml stock for HeLa cells)
- Rapid acquisition time-lapse microscopy setup (incubation chamber, etc.)

METHODS

1. Cells cultured on coverslips are transfected according to the CaPO$_4$ procedure (Jordan *et al.*, 1996) with 1 μg DNA per well (24-well plate) or 10 μg DNA (10-cm-diameter dish).

2. The medium is changed after 4 h, and cells can be observed from 12 h posttransfection on.

3. scFv–xFP expressing cells can be selected with geneticin/G418 (selection gene present in the pEGFP Clontech vectors) to obtain stable cell lines (selected at 400 μg/ml G418). Note that we have performed this successfully in the case of the anti-nmMyosinII scFv SF9 fused to YFP in HeLa cells, but failed in the case of the anti-Rab6·GTP AA2.

4. Confocal as well as conventional time-lapse microscopy setups can then be used to follow the dynamics of scFv–xFPs in live cells. Ideally, Rab6 *in vivo* dynamics should be imaged using a minimal acquisition rate of the order of one image per second, since most labeled structures move a

distance corresponding to about their size every second. Time-lapse acquisitions are usually performed for 5–10 min to obtain representative dynamics. Our experience is that confocal microscopy is better suited for these imaging experiments as less phototoxicity and photobleaching are observed, which is likely due to the short local exposure time of the scanning spot (a few microseconds) compared to 50–ms exposure times with conventional microscopy set-ups. Of course it takes at least 100–500 ms to scan a full frame with a confocal, and, therefore, the resulting image is necessarily globally distorted, but not so much locally (a given structure will be imaged correctly, but at the other end of the field the structures are not imaged simultaneously).

Comment. Photobleaching experiments such as FRAP, FLIP (etc.) can be set up to analyze the binding/unbinding cycle of GTPases from membranes, with the limitation that the fluorescence signal is linked not only to the GTPase itself but also to the probe: a cycle of scFv binding/unbinding to its target GTPase is superimposed to that of the GTPase membrane association cycle, which has to be taken into account during the analysis. We actually performed such experiments in the case of the anti-Giantin TA10-YFP intrabody (Nizak *et al.*, 2003a): Giantin is not a GTPase but a transmembrane Golgi protein, very stably anchored onto this organelle. When the cytosolic pool of TA10-YFP was bleached, we observed that the Golgi-associated pool was essentially unaffected after 15 min, which means that the intrabody remains bound to its target over time ranges that would probably be much longer than GTPase membrane association/dissociation cycles (therefore minimizing the aforementioned problem). AA2 seems to exchange more quickly, however. Note also that FRET between CFP-Rab6 and AA2-YFP may be used to quantify the interaction *in vivo*, although we have not validated this approach yet.

Acknowledgments

This work was supported by the CNRS, the Curie Institute, and the Association pour la Recherche contre le Cancer (ARC5747 and ARC5881).

References

Echard, A., Opdam, F. J., de Leeuw, H. J., Jollivet, F., Savelkoul, P., Hendriks, W., Voorberg, J., Goud, B., and Fransen, J. A. (2000). Alternative splicing of the human Rab6A gene generates two close but functionally different isoforms. *Mol. Biol. Cell* **11**, 3819–3833.

Hust, M., and Dubel, S. (2004). Mating antibody phage display with proteomics. *Trends Biotechnol.* **22**, 8–14.

Jordan, M., Schallhorn, A., and Wurm, F. M. (1996). Transfecting mammalian cells: Optimization of critical parameters affecting calcium-phosphate precipitate formation. *Nucleic Acids Res.* **24,** 596–601.

Kraynov, V. S., Chamberlain, C., Bokoch, G. M., Schwartz, M. A., Slabaugh, S., and Hahn, K. M. (2000). Localized Rac activation dynamics visualized in living cells. *Science* **290,** 333–337.

Martinez, O., Antony, C., Pehau-Arnaudet, G., Berger, E. G., Salamero, J., and Goud, B. (1997). GTP-bound forms of rab6 induce the redistribution of Golgi proteins into the endoplasmic reticulum. *Proc. Natl. Acad. Sci. USA* **94,** 1828–1833.

Monier, S., Jollivet, F., Janoueix-Lerosey, I., Johannes, L., and Goud, B. (2002). Characterization of novel Rab6-interacting proteins involved in endosome-to-TGN transport. *Traffic* **3,** 289–297.

Nizak, C., Martin-Lluesma, S. M., Moutel, S., Roux, A., Kreis, T. E., Goud, B., and Perez, F. (2003a). Recombinant antibodies selected against subcellular fractions to track endogenous protein dynamics *in vivo*. *Traffic* **7,** 739–753.

Nizak, C., Monier, S., Del Nery, E., Moutel, S., Goud, B., and Perez, F. (2003b). Recombinant antibodies to the small GTPase Rab6 as conformation sensors. *Science* **300,** 984–987.

Nuoffer, C., Peter, F., and Balch, W. E. (1995). Purification of His6-tagged Rab1 proteins using bacterial and insect cell expression systems. *Methods Enzymol.* **257,** 3–9.

Rowe, T., and Balch, W. E. (1995). Expression and purification of mammalian Sar1. *Small Gtpases Reg. Pt C* **257,** 49–53.

Schweitzer, J. K., and D'Souza-Schorey, C. (2002). Localization and activation of the ARF6 GTPase during cleavage furrow ingression and cytokinesis. *J. Biol. Chem.* **277,** 27210–27216.

Smothers, J. F., Henikoff, S., and Carter, P. (2002). Phage display: Affinity selection from biological libraries. *Science* **298,** 621–622.

[13]   Analysis of Rab Protein Function in Neurotransmitter Receptor Trafficking at Hippocampal Synapses

By NASHAAT Z. GERGES, TYLER C. BROWN, SUSANA S. CORREIA, and JOSÉ A. ESTEBAN

Abstract

Members of the Rab family of small GTPases are essential regulators of intracellular membrane sorting. Nevertheless, very little is known about the role of these proteins in the membrane trafficking processes that operate at synapses, and specifically, at postsynaptic terminals. These events include the activity-dependent exocytic and endocytic trafficking

METHODS IN ENZYMOLOGY, VOL. 403   0076-6879/05 $35.00
DOI: 10.1016/S0076-6879(05)03013-2

of AMPA-type glutamate receptors, which underlies long-lasting forms of synaptic plasticity such as long-term potentiation (LTP) and long-term depression (LTD). This chapter summarizes different experimental methods to address the role of Rab proteins in the trafficking of neurotransmitter receptors at postsynaptic terminals in the hippocampus. These techniques include immunogold electron microscopy to ultrastructurally localize endogenous Rab proteins at synapses, molecular biology methods to express recombinant Rab proteins in hippocampal slice cultures, electrophysiological techniques to evaluate the role of Rab proteins in synaptic transmission, and confocal fluorescence imaging to monitor receptor trafficking at dendrites and spines and its dependence on Rab proteins.

Introduction

The members of the Rab family of small GTPases are critical regulators of intracellular membrane trafficking and sorting in eukaryotes (Pfeffer, 2001; Zerial and McBride, 2001). This has been well established in a variety of cellular systems. However, very little is known of the functional role of Rab proteins in neurons, where polarized membrane trafficking is crucial for synaptic function and plasticity (Bredt and Nicoll, 2003; Wenthold *et al.*, 2003; Ziv and Garner, 2004). The only notable exception is Rab3, which has been shown to modulate neurotransmitter release (Geppert *et al.*, 1994, 1997; Senyshyn *et al.*, 1992) and is involved in some forms of presynaptic plasticity (Castillo *et al.*, 1997, 2002; Lonart *et al.*, 1998). This chapter focuses on experimental approaches for the study of postsynaptic functions of Rab proteins in the endocytic and exocytic trafficking of α-amino-3-hydroxy-4-isoxazolepropionic acid (AMPA)-type glutamate receptors (AMPARs) at hippocampal synapses. By mediating AMPAR synaptic trafficking, some Rab proteins, namely, Rab5, Rab8, and Rab11, have been shown to play central roles in synaptic plasticity (Brown *et al.*, 2005; Gerges *et al.*, 2004; Park *et al.*, 2004).

Ultrastructural Studies of Rab Proteins at Synaptic Terminals

Previous work has shown that multiple Rab proteins are present in axonal and dendritic regions of hippocampal neurons (de Hoop *et al.*, 1994; Fischer von Mollard *et al.*, 1990; Huber *et al.*, 1993). To determine the role of specific Rab proteins in local membrane trafficking at synapses, it is important to determine the presence and distribution of these proteins at synaptic terminals with high spatial resolution. This ultrastructural localization can be accomplished with postembedding immunogold electron microscopy (Brown *et al.*, 2005; Gerges *et al.*, 2004).

Method

Hippocampal tissue is fixed, dehydrated, and processed for osmium-free postembedding immunogold labeling, as previously described (Phend *et al.*, 1995). Thin sections are blocked with 2.5% bovine serum albumin (BSA) and 2.5% serum for 30 min at room temperature. They are then incubated with anti-Rab5 or anti-Rab8 antibodies (BD Biosciences) overnight, followed by incubation for 1 h with secondary antibodies coupled to 10-nm gold particles (Electron Microscopy Sciences). Images are acquired with a transmission electron microscope using a digital camera. Quantification of gold particles and distance measurement are performed on digital images using image analysis software.

Analysis

To interpret the synaptic distribution of endogenous Rab proteins, immunogold labeling can be binned according to its location within the synaptic terminal (Fig. 1). The following compartments are defined: presynaptic terminal (compartment "A" in Fig. 1), intracellular space underneath the postsynaptic membrane (compartment "B"), postsynaptic density (PSD) (compartment "C"), and postsynaptic plasma membrane lateral to the PSD (compartment "D"). These quantifications are limited to immunogold particles found within 600 nm from the synaptic cleft. This experimental approach has been carried out for the synaptic distribution of endogenous Rab8 and Rab5 at CA1 excitatory synapses in the

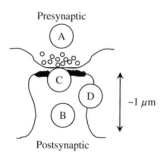

FIG. 1. Schematic representation of a synaptic terminal with the different compartments defined for quantification of immunogold labeling. (A) Presynaptic compartment. (B) Intracellular postsynaptic compartment. (C) Postsynaptic density. (D) Extrasynaptic plasma membrane lateral from the postsynaptic density.

hippocampus. The analysis indicated that Rab8 accumulates at intracellular membranes within the spine (Gerges et al., 2004), whereas Rab5 is particularly abundant at the extrasynaptic plasma membrane (Brown et al., 2005).

The distance of each gold particle to the edge of the PSD along the plasma membrane is measured for further characterization of the lateral distribution of Rab proteins at synaptic and extrasynaptic cell surfaces. These distances can be computed using image analysis software and presented as frequency histograms. This kind of analysis revealed that postsynaptic Rab5 is predominantly located outside of the PSD on lateral extrasynaptic membranes, roughly 100–300 nm away from the edge of the PSD (Brown et al., 2005).

Cloning and Expression of Recombinant Rab Proteins in Hippocampal Neurons

To perturb Rab protein function in neurons, wild-type, dominant negative, or constitutively active forms of these proteins can be overexpressed as recombinant proteins in organotypic hippocampal slice cultures. Rab protein coding sequences are cloned by standard reverse transcriptase polymerase chain reaction (RT-PCR) techniques from rat brain mRNA preparations. Rab–green fluorescent protein (GFP) fusion proteins are made by in-frame ligation of the EGFP coding sequence (Clontech) with the amino termini of the Rab protein. Single amino acid substitutions that generate dominant negative (GDP-bound) and constitutively active (GTP-bound) mutants are well established. For example, S34N and Q79L produce dominant negative and constitutively active forms of Rab5, respectively (Li and Stahl, 1993). T22N and Q67L produce equivalent phenotypes in Rab8 (Ren et al., 1996). The functionality of fluorescently tagged Rab proteins has been described in multiple previous publications (see, for instance, Sonnichsen et al., 2000). Fusion proteins of Rab5 and Rab8 with a tandem-dimer variant of the red fluorescence protein DsRed (Campbell et al., 2002) have also been generated and tested in hippocampal neurons (Brown et al., 2005; Gerges et al., 2004).

All these constructs are recloned in pSinRep5 for expression using Sindbis virus (Schlesinger and Dubensky, 1999) or in mammalian expression plasmids for biolistic delivery (Lo et al., 1994). Recombinant proteins are expressed in organotypic hippocampal slice cultures (Gahwiler et al., 1997). Briefly, hippocampal slices are prepared from young rats (postnatal days 5–7) and placed in culture on semiporous membranes. After 4–5 days in culture, the recombinant gene is delivered into the slices. For expression of single proteins, the Sindbis virus is preferable. This is a replication-deficient, low-toxicity, neurotropic virus that allows the expression of recombinant proteins exclusively in neurons upon injection of the viral solution extracellularly in the desired area of the hippocampal slice. Coexpression of several proteins can be achieved with a Sindbis virus with an intervening IRES (see, for instance, Hayashi *et al.*, 2000), or more typically, using the biolistic method with a combination of different plasmids bearing mammalian expression promoters, such as the CMV promoter. Either method leads to robust expression of the recombinant protein after a 15-h incubation (36 h expression time is typically used when expressing recombinant AMPAR subunits).

Electrophysiological Studies of Rab Protein Function

AMPA and *N*-methyl-D-aspartate (NMDA) receptors are the main ionotropic glutamate receptors at excitatory synapses in the hippocampus. Hence, the proper trafficking of these receptors is essential for synaptic function and plasticity. To study the role of Rab proteins in the targeting of AMPA and NMDA receptors into excitatory synapses, we express GFP-tagged dominant negative or constitutively active forms of these proteins in CA1 neurons from hippocampal slice cultures and monitor AMPA and NMDA receptor-mediated synaptic transmission.

Effect of Recombinant Rab Proteins on Basal Synaptic Transmission

Simultaneous double whole-cell recordings are obtained from nearby pairs of infected (expressing the recombinant protein) and uninfected (control) CA1 pyramidal neurons, under visual guidance using fluorescence and transmitted light illumination. The recording chamber is perfused with 119 mM NaCl, 2.5 mM KCl, 4 mM CaCl$_2$, 4 mM MgCl$_2$, 26 mM NaHCO$_3$, 1 mM NaH$_2$PO$_4$, 11 mM glucose, 0.1 mM picrotoxin, and 2 μM 2-chloroadenosine, at pH 7.4, gassed with 5% CO$_2$/95% O$_2$ (2-chloroadenosine is used to reduce presynaptic function and, therefore, compensate for the enhanced connectivity of the slice cultures). Patch recording pipettes (3–6 MΩ) are filled with 115 mM cesium methanesulfonate, 20 mM CsCl, 10 mM HEPES, 2.5 mM MgCl$_2$, 4 mM Na$_2$ATP, 0.4 mM Na$_3$GTP, 10 mM sodium phosphocreatine, and 0.6 mM EGTA at pH 7.25. Voltage-clamp whole-cell recordings are carried out with multiclamp 700 amplifiers (Axon Instruments, Union City, CA). Synaptic responses are evoked with bipolar electrodes using single-voltage pulses (200 μs, up to 20 V). The stimulating electrodes are placed over Schaffer collateral fibers between 300 and 500 μm from the recorded cells. Synaptic AMPA receptor-mediated responses are measured at −60 mV and NMDA receptor-mediated responses at +40 mV, at a latency when AMPA receptor responses have fully decayed (60 ms). Synaptic responses are averaged over 50–100 trials. This experimental configuration specifically addresses postsynaptic functions of Rab proteins, since the recombinant protein is always expressed in CA1 neurons and presynaptic stimulation is delivered at the Schaffer collaterals from CA3 neurons. This approach has been employed to demonstrate that Rab8, but not other exocytic Rab proteins such as Rab4 and Rab11, is required for the constitutive cycling of AMPARs at hippocampal synapses (Gerges *et al.*, 2004). Similarly, we have shown that the endocytic protein Rab5 drives the removal of AMPARs from these synapses (Brown *et al.*, 2005).

Synaptic Plasticity (LTP, LTD)

Neuronal activity continuously remodels synaptic connectivity. This process, known as synaptic plasticity, is widely thought to be the cellular correlate of learning and memory. Some of the best-studied forms of synaptic plasticity are long-term potentiation (LTP) and long-term depression (LTD) in CA1 hippocampal synapses. The involvement of specific Rab proteins in these forms of synaptic plasticity can be addressed by expressing dominant negative forms of these proteins in organotypic hippocampal slice cultures. Synaptic plasticity is then induced under whole-cell configuration on neurons expressing the recombinant protein (introduced via infection or transfection, see above) or on control neurons. LTP is induced by pairing 0 mV depolarization of the postsynaptic neuron with 3 Hz presynaptic stimulation (300 pulses). Baseline recordings before LTP induction should be limited to 2–5 min because critical factors required for LTP induction are quickly washed out in whole-cell configuration. LTD is induced by pairing 1 Hz presynaptic stimulation (500 pulses) with moderate postsynaptic depolarization (−40 mV). Using these protocols on CA1 hippocampal neurons expressing Rab8 or Rab5 dominant negative mutants, we have determined that Rab8 mediates the synaptic delivery of AMPARs during LTP (Gerges *et al.*, 2004), whereas Rab5 function is required for AMPAR internalization upon LTD induction (Brown *et al.*, 2005).

Electrophysiological Tagging

AMPARs are tetrameric molecules (Greger *et al.*, 2003; Tichelaar *et al.*, 2004) composed of different combinations of GluR1 to GluR4 subunits (Hollmann and Heinemann, 1994). In hippocampus, most of AMPA receptors are composed of GluR1/GluR2 or GluR2/GluR3 subunits (Wenthold *et al.*, 1996). These two populations reach synapses according to different pathways: GluR2/GluR3 AMPA receptors continuously cycle in and out of synapses in a manner independent of synaptic activity (Passafaro *et al.*, 2001; Shi *et al.*, 2001). In contrast, GluR1/GluR2 AMPA receptors are added into synapses in an activity-dependent manner during synaptic plasticity (Hayashi *et al.*, 2000; Passafaro *et al.*, 2001; Shi *et al.*, 2001). These two pathways have been coined as constitutive and regulated, respectively (Malinow *et al.*, 2000). On the other hand, activity-dependent removal of AMPARs during LTD seems to affect both populations of receptors (Beattie *et al.*, 2000; Brown *et al.*, 2005; Ehlers, 2000; Lee *et al.*, 2004; Lin *et al.*, 2000). Electrophysiological tagging is a powerful tool to monitor the presence of these distinct pools of receptors at the synapse, as described below.

   Most endogenous AMPA receptors in the hippocampus display a linear current–voltage relation, that is, they conduct inward currents at negative membrane potentials, and outward currents at positive ones (Fig. 2A). This is dependent on the presence of an edited GluR2 subunit (arginine 607) in the receptor (Verdoorn *et al.*, 1991). Overexpression of recombinant AMPAR subunits leads to the formation of homomeric channels that lack endogenous GluR2 subunits (Hayashi *et al.*, 2000). These channels display inward rectification, that is, they conduct inward currents at

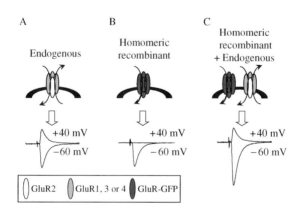

FIG. 2. Inward rectification of recombinant AMPA receptors and their use as an "electrophysiological tag." (A) Endogenous (GluR2-containing) AMPA receptors conduct inward and outward currents at negative and positive membrane potentials, respectively. (B) Recombinant AMPA receptors form homomeric channels lacking endogenous GluR2. Hence, they conduct only inward currents (inward rectification). (C) Synapses containing both endogenous and recombinant receptors display an increased ratio of inward to outward currents (rectification index) because both kinds of receptors contribute to the inward current, whereas only endogenous receptors conduct outward currents. See further explanation in the main text.

negative membrane potentials, but no outward current at positive membrane potentials (Fig. 2B). This is due to the blockade of the channel by endogenous polyamines (Bowie and Mayer, 1995; Donevan and Rogawski, 1995; Kamboj *et al.*, 1995). Thus, synaptic incorporation of homomeric recombinant receptors (either GluR1 or the unedited form of GluR2—glutamine 607) increases inward rectification of synaptic responses (Fig. 2C). This effect can be quantified as an increase in the ratio of AMPAR-mediated responses at -60 mV versus $+40$ mV (rectification index). GluR1-GFP homomeric receptors behave like GluR1/GluR2 receptors, that is, they are delivered at synapses in an activity-dependent manner. Therefore, GluR1-GFP homomers can be used as reporters for the regulated addition of AMPARs during LTP. In contrast, GluR2-GFP homomers mimic GluR2/GluR3 receptor trafficking, and, therefore, they can be used as reporters for the constitutive synaptic cycling of AMPARs (Shi *et al.*, 2001).

To test the role of Rab proteins in the synaptic delivery or removal of specific AMPA receptor populations, the Rab protein under consideration is coexpressed with either GluR1-GFP or GluR2Q607-GFP using the biolistic gene delivery method (see above). Whole-cell voltage-clamp recordings of synaptic responses are then obtained at -60 mV and $+40$ mV from transfected neurons expressing the receptor alone, or the receptor plus the Rab protein, or from control (untransfected) neurons. To isolate AMPAR-mediated responses at $+40$ mV, 0.1 mM of the NMDA receptor antagonist AP5 is added to the perfusion solution. In addition, the intracellular solution in the recording pipette is supplemented with 0.1 mM spermine, to prevent wash-out of the endogenous polyamines required for AMPAR inward rectification. The rectification index of AMPAR-mediated synaptic transmission is then calculated as the ratio between the responses at -60 mV and $+40$ mV. This method has allowed us to determine that Rab5 and Rab8 mediate, respectively, the endocytic and exocytic trafficking of both GluR1 and GluR2 populations of AMPARs at synapses (Brown *et al.*, 2005; Gerges *et al.*, 2004).

Confocal Fluorescence Imaging of Rab Protein Function in AMPA Receptor Trafficking at Dendrites and Spines

Method

To address the role of Rab proteins in the trafficking of AMPARs along dendrites and their insertion into dendritic spines, GFP-tagged AMPAR subunits (either GluR1 or GluR2) are coexpressed with Rab proteins tagged with a red fluorescence protein (RFP; see above). These proteins are cotransfected in rat hippocampal organotypic slices using the biolistic transfection system. After 1.5 days of expression, organotypic slices are processed for surface immunostaining of the GFP-tagged receptors. Briefly, slices are fixed with 4% paraformaldehyde/4% sucrose in phosphate-buffered saline (PBS) for 2 h at 4°, and then blocked in 2% serum for 1 h at room temperature. Slices are then successively incubated with anti-GFP antibody (Roche) overnight at 4°, with biotinylated anti-mouse antibody for 1 h at room temperature, and with streptavidin coupled to Cy5 (Molecular Probes) for 1 h at room temperature. All these incubations are done in the absence of detergent, and, therefore, the immunolabeling is restricted to GFP-tagged AMPARs exposed to the cell surface. Confocal fluorescence images from the GFP, RFP, and Cy5 channels are then collected using an Olympus FV500 confocal microscope with a 60× oil immersion lens. Digital images are acquired using the Fluo View software and are reconstructed and analyzed using Image J software (http://rsb.info.nih.gov/ij/). This experimental approach allows us to separately monitor the recombinant AMPAR receptor (GFP channel), its surface expression (Cy5 channel), and the coexpression of the Rab protein (RFP channel).

Analysis of Large-Scale Dendritic Trafficking

The efficiency of AMPAR transport along dendrites can be evaluated by quantifying GFP fluorescence intensity along the primary apical dendrite from neurons expressing a GFP-tagged AMPAR subunit. This is obtained by drawing a pixel-wide line along the dendrite and plotting its GFP fluorescence profile (see example in Fig. 3A). After background subtraction, the value of fluorescence intensity in each pixel is normalized to the maximum fluorescence at the soma of the neuron. This normalization accounts for variability in expression levels of the recombinant proteins. The normalized fluorescence intensity is then plotted as a function of the distance from the cell body (Fig. 3B). The effect of a recombinant Rab protein on AMPAR dendritic trafficking is then evaluated by comparing the fluorescence profile of the GFP-tagged receptor with or without coexpression of the RFP-tagged Rab protein.

Analysis of Local Dendritic Spine Trafficking

To determine AMPA receptor partition between spines and dendrites, the GFP fluorescence intensity at these two compartments is quantified using a line plot that crosses the spine head and its adjacent dendritic shaft (see Fig. 4A, left). The amount of GFP-tagged AMPAR at each compartment is then estimated from the corresponding peaks of GFP fluorescence after background subtraction (Fig. 4B, left).

A

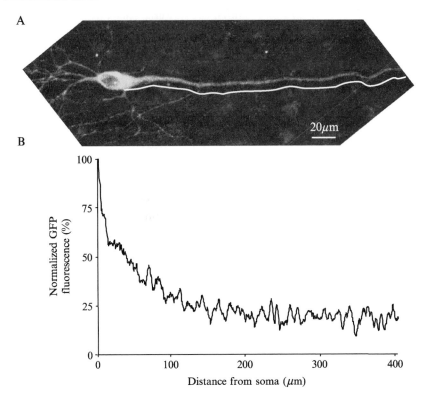

B

FIG. 3. Quantitative analysis of the large-scale distribution of GFP-tagged receptors along dendrites. (A) Representative example of a CA1 hippocampal neuron expressing GluR2-GFP. The white line represents a pixel-wide trajectory along the main apical dendrite, from which GFP fluorescence intensity is quantified. The line is shifted downward from its original position on the dendrite to facilitate visualization. (B) Quantification of GFP-fluorescence intensity along the line shown in (A). Fluorescence values are background-subtracted and normalized to the maximum at the cell soma.

To determine the surface expression of recombinant AMPA receptors, GFP (total receptor; Fig. 4A, left) and Cy5 (surface receptor; Fig. 4A, right) fluorescence intensities are quantified using line plots, as described above. Surface ratios are then calculated for spines and dendrites by dividing Cy5 and GFP fluorescence peaks after background subtraction (Fig. 4B). This method is internally normalized for immunostaining variability, since the Cy5/GFP ratios are always acquired in pairs of spine and adjacent dendrite. Additionally, spine–dendrite pairs are exclusively selected from the GFP channel, avoiding any bias with respect to their surface immunostaining.

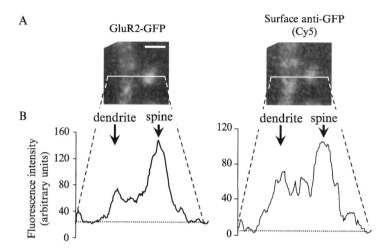

FIG. 4. Quantitative analysis of the local trafficking of GFP-tagged receptors at dendritic spines. (A) Representative confocal image of a spine and the adjacent dendritic shaft from a neuron expressing GluR2-GFP. The GFP fluorescence signal (left) represents total receptor distribution. The Cy5 signal (right) is obtained by immunostaining with anti-GFP antibodies under nonpermeabilized conditions, and, therefore, it represents the fraction of receptor exposed to the cell surface. Scale bar = 1 μm. (B) Quantification of fluorescence intensity of the GFP (left) and Cy5 (right) signals along the white lines shown in (A). The peaks of fluorescence intensity at the dendritic shaft and the spine head after background subtraction (dotted lines) are used to estimate total receptor distribution (GFP) and surface ratio (Cy5/GFP) at dendrites and spines.

This analysis is carried out from neurons expressing a GFP-tagged AMPAR subunit alone or from neurons coexpressing GFP-AMPARs together with RFP-tagged Rab proteins. To determine whether coexpression with a recombinant Rab protein alters receptor distribution between spines and dendrites, we calculate spine/dendrite ratios of either total receptor (GFP channel) or surface ratios (GFP/Cy5) for each pair of spine and dendritic shaft. For instance, in the case of the surface expression, a collection of spine/dendrite ratios would be obtained as

$$Ratio_1 = (Cy5/GFP)spine_1/(Cy5/GFP)dendrite_1$$
$$Ratio_2 = (Cy5/GFP)spine_2/(Cy5/GFP)dendrite_2$$
etc.

This collection of values can then be compared with or without coexpressed Rab protein using cumulative distributions and two-sample Kolmogorov–Smirnov tests. This experimental approach has allowed us to determine that Rab5 and Rab8 are involved, respectively, in the endocytic and exocytic trafficking of AMPARs within the spine, but not in their transport between dendritic shafts and spines (Brown *et al.*, 2005; Gerges *et al.*, 2004).

Acknowledgments

We thank Donald Backos for expert technical assistance with cloning and expression of recombinant proteins in organotypic hippocampal slice cultures. We also thank Kathryn Weiss, Dorothy Sorenson, and Chris Edwards for their help with immunogold electron microscopy. This work was supported by the National Institute of Mental Health (Grant MH070417 to J.A.E. and Grant F31-MH070205 to T.C.B.), the National Alliance for research on Schizophrenia and Depression (J.A.E.), the Alzheimer's Association (J.A.E.), and the Fundação para a Ciência e a Tecnologia (S.S.C.).

References

Beattie, E. C., Carroll, R. C., Yu, X., Morishita, W., Yasuda, H., von Zastrow, M., and Malenka, R. C. (2000). Regulation of AMPA receptor endocytosis by a signaling mechanism shared with LTD. *Nat. Neurosci.* **3**, 1291–1300.

Bowie, D., and Mayer, M. L. (1995). Inward rectification of both AMPA and kainate subtype glutamate receptors generated by polyamine-mediated ion channel block. *Neuron* **15**, 453–462.

Bredt, D. S., and Nicoll, R. A. (2003). AMPA receptor trafficking at excitatory synapses. *Neuron* **40**, 361–379.

Brown, T. C., Tran, I. C., Backos, D. S., and Esteban, J. A. (2005). NMDA receptor-dependent activation of the small GTPase Rab5 drives the removal of synaptic AMPA receptors during hippocampal LTD. *Neuron* **45**, 81–94.

Campbell, R. E., Tour, O., Palmer, A. E., Steinbach, P. A., Baird, G. S., Zacharias, D. A., and Tsien, R. Y. (2002). A monomeric red fluorescent protein. *Proc. Natl. Acad. Sci. USA* **99**, 7877–7882.

Castillo, P. E., Janz, R., Sudhof, T. C., Tzounopoulos, T., Malenka, R. C., and Nicoll, R. A. (1997). Rab3A is essential for mossy fibre long-term potentiation in the hippocampus. *Nature* **388**, 590–593.

Castillo, P. E., Schoch, S., Schmitz, F., Sudhof, T. C., and Malenka, R. C. (2002). RIM1alpha is required for presynaptic long-term potentiation. *Nature* **415**, 327–330.

de Hoop, M. J., Huber, L. A., Stenmark, H., Williamson, E., Zerial, M., Parton, R. G., and Dotti, C. G. (1994). The involvement of the small GTP-binding protein Rab5a in neuronal endocytosis. *Neuron* **13**, 11–22.

Donevan, S. D., and Rogawski, M. A. (1995). Intracellular polyamines mediate inward rectification of Ca(2+)-permeable alpha-amino-3-hydroxy-5-methyl-4-isoxazolepropionic acid receptors. *Proc. Natl. Acad. Sci. USA* **92**, 9298–9302.

Ehlers, M. D. (2000). Reinsertion or degradation of AMPA receptors determined by activity-dependent endocytic sorting. *Neuron* **28**, 511–525.

Fischer von Mollard, G., Mignery, G. A., Baumert, M., Perin, M. S., Hanson, T. J., Burger, P. M., Jahn, R., and Sudhof, T. C. (1990). rab3 is a small GTP-binding protein exclusively localized to synaptic vesicles. *Proc. Natl. Acad. Sci. USA* **87**, 1988–1992.

Gahwiler, B. H., Capogna, M., Debanne, D., McKinney, R. A., and Thompson, S. M. (1997). Organotypic slice cultures: A technique has come of age. *Trends Neurosci.* **20**, 471–477.

Geppert, M., Bolshakov, V. Y., Siegelbaum, S. A., Takei, K., De Camilli, P., Hammer, R. E., and Sudhof, T. C. (1994). The role of Rab3A in neurotransmitter release. *Nature* **369**, 493–497.

Geppert, M., Goda, Y., Stevens, C. F., and Sudhof, T. C. (1997). The small GTP-binding protein Rab3A regulates a late step in synaptic vesicle fusion. *Nature* **387**, 810–814.

Gerges, N. Z., Backos, D. S., and Esteban, J. A. (2004). Local control of AMPA receptor trafficking at the postsynaptic terminal by a small GTPase of the Rab family. *J. Biol. Chem.* **279**, 43870–43878.

Greger, I. H., Khatri, L., Kong, X., and Ziff, E. B. (2003). AMPA receptor tetramerization is mediated by Q/R editing. *Neuron* **40**, 763–774.

Hayashi, Y., Shi, S. H., Esteban, J. A., Piccini, A., Poncer, J. C., and Malinow, R. (2000). Driving AMPA receptors into synapses by LTP and CaMKII: Requirement for GluR1 and PDZ domain interaction. *Science* **287**, 2262–2267.

Hollmann, M., and Heinemann, S. (1994). Cloned glutamate receptors. *Annu. Rev. Neurosci.* **17**, 31–108.

Huber, L. A., de Hoop, M. J., Dupree, P., Zerial, M., Simons, K., and Dotti, C. (1993). Protein transport to the dendritic plasma membrane of cultured neurons is regulated by rab8p. *J. Cell Biol.* **123**, 47–55.

Kamboj, S. K., Swanson, G. T., and Cull-Candy, S. G. (1995). Intracellular spermine confers rectification on rat calcium-permeable AMPA and kainate receptors. *J. Physiol.* **486**(Pt. 2), 297–303.

Lee, S. H., Simonetta, A., and Sheng, M. (2004). Subunit rules governing the sorting of internalized AMPA receptors in hippocampal neurons. *Neuron* **43**, 221–236.

Li, G., and Stahl, P. D. (1993). Structure-function relationship of the small GTPase rab5. *J. Biol. Chem.* **268**, 24475–24480.

Lin, J. W., Ju, W., Foster, K., Lee, S. H., Ahmadian, G., Wyszynski, M., Wang, Y. T., and Sheng, M. (2000). Distinct molecular mechanisms and divergent endocytotic pathways of AMPA receptor internalization. *Nat. Neurosci.* **3**, 1282–1290.

Lo, D. C., McAllister, A. K., and Katz, L. C. (1994). Neuronal transfection in brain slices using particle-mediated gene transfer. *Neuron* **13**, 1263–1268.

Lonart, G., Janz, R., Johnson, K. M., and Sudhof, T. C. (1998). Mechanism of action of rab3A in mossy fiber LTP. *Neuron* **21**, 1141–1150.

Malinow, R., Mainen, Z. F., and Hayashi, Y. (2000). LTP mechanisms: From silence to four-lane traffic. *Curr. Opin. Neurobiol.* **10**, 352–357.

Park, M., Penick, E. C., Edwards, J. G., Kauer, J. A., and Ehlers, M. D. (2004). Recycling endosomes supply AMPA receptors for LTP. *Science* **305**, 1972–1975.

Passafaro, M., Piech, V., and Sheng, M. (2001). Subunit-specific temporal and spatial patterns of AMPA receptor exocytosis in hippocampal neurons. *Nat. Neurosci.* **4**, 917–926.

Pfeffer, S. R. (2001). Rab GTPases: Specifying and deciphering organelle identity and function. *Trends Cell Biol.* **11**, 487–491.

Phend, K. D., Rustioni, A., and Weinberg, R. J. (1995). An osmium-free method of epon embedment that preserves both ultrastructure and antigenicity for post-embedding immunocytochemistry. *J. Histochem. Cytochem.* **43**, 283–292.

Ren, M., Zeng, J., De Lemos-Chiarandini, C., Rosenfeld, M., Adesnik, M., and Sabatini, D. D. (1996). In its active form, the GTP-binding protein rab8 interacts with a stress-activated protein kinase. *Proc. Natl. Acad. Sci. USA* **93**, 5151–5155.

Schlesinger, S., and Dubensky, T. W. (1999). Alphavirus vectors for gene expression and vaccines. *Curr. Opin. Biotechnol.* **10**, 434–439.

Senyshyn, J., Balch, W. E., and Holz, R. W. (1992). Synthetic peptides of the effector-binding domain of rab enhance secretion from digitonin-permeabilized chromaffin cells. *FEBS Lett.* **309**, 41–46.

Shi, S., Hayashi, Y., Esteban, J. A., and Malinow, R. (2001). Subunit-specific rules governing AMPA receptor trafficking to synapses in hippocampal pyramidal neurons. *Cell* **105,** 331–343.

Sonnichsen, B., De Renzis, S., Nielsen, E., Rietdorf, J., and Zerial, M. (2000). Distinct membrane domains on endosomes in the recycling pathway visualized by multicolor imaging of Rab4, Rab5, and Rab11. *J. Cell Biol.* **149,** 901–914.

Tichelaar, W., Safferling, M., Keinanen, K., Stark, H., and Madden, D. R. (2004). The three-dimensional structure of an ionotropic glutamate receptor reveals a dimer-of-dimers assembly. *J. Mol. Biol.* **344,** 435–442.

Verdoorn, T. A., Burnashev, N., Monyer, H., Seeburg, P. H., and Sakmann, B. (1991). Structural determinants of ion flow through recombinant glutamate receptor channels. *Science* **252,** 1715–1718.

Wenthold, R. J., Petralia, R. S., Blahos, J., II, and Niedzielski, A. S. (1996). Evidence for multiple AMPA receptor complexes in hippocampal CA1/CA2 neurons. *J. Neurosci.* **16,** 1982–1989.

Wenthold, R. J., Prybylowski, K., Standley, S., Sans, N., and Petralia, R. S. (2003). Trafficking of NMDA receptors. *Annu. Rev. Pharmacol. Toxicol.* **43,** 335–358.

Zerial, M., and McBride, H. (2001). Rab proteins as membrane organizers. *Nat. Rev. Mol. Cell Biol.* **2,** 107–117.

Ziv, N. E., and Garner, C. C. (2004). Cellular and molecular mechanisms of presynaptic assembly. *Nat. Rev. Neurosci.* **5,** 385–399.

[14] Use of Rab GTPases to Study Lipid Trafficking in Normal and Sphingolipid Storage Disease Fibroblasts

By Amit Choudhury, David L. Marks, and Richard E. Pagano

Abstract

We describe methods for studying lipid transport in normal and sphingolipid storage disease fibroblasts. These techniques include endocytic assays with fluorescent sphingolipid analogs, expression of dominant negative (DN) Rab GTPases, and methods of manipulating cholesterol levels in intact cells and isolated cell membranes. These methods should be useful in future studies of lipid trafficking in normal and disease cell types.

Introduction

Rab GTPases play important roles in the tethering and docking of vesicles to their target membranes during intracellular vesicular transport. Rab proteins (21–25 kDa) are part of the Ras superfamily of small GTPases (Pfeffer, 2001; Stein *et al.*, 2003; Zerial and McBride, 2001).

METHODS IN ENZYMOLOGY, VOL. 403
0076-6879/05 $35.00
DOI: 10.1016/S0076-6879(05)03014-4

Human cells contain more than 60 different Rabs, each involved in specific steps in vesicle trafficking (Pfeffer, 2001; Zerial and McBride, 2001). For example, Rab5 is involved in the homotypic fusion of clathrin-derived endocytic vesicles to form early endosomes (EEs[1]), whereas Rab9 regulates the transport of membranes from late endosomes to the Golgi apparatus (Table I) (Shapiro et al., 1993; Stenmark et al., 1994).

Certain mutations in conserved guanine nucleotide-binding motifs confer Rabs with a lower affinity for GTP than for GDP (Olkkonen and Stenmark, 1997). These mutant Rabs are present mainly in a GDP-bound state and thus exert a dominant negative (DN) effect on intracellular membrane transport. DN Rabs have been useful for dissecting the intracellular vesicular transport pathways utilized by various proteins and other materials.

In this chapter, we describe methods to study the roles of Rab proteins in the intracellular sorting and transport of sphingolipids in cultured cells. We describe the use of DN Rabs in conjunction with endocytic assays using sphingolipid probes to delineate pathways utilized for the intracellular transport of sphingolipids. We also highlight the use of methods for manipulating cholesterol levels in cells and isolated membranes to investigate the regulation of intracellular transport and Rab function by cholesterol. These techniques can be readily adapted to the study of the intracellular transport of other appropriately tagged cargo molecules.

General Approach for Studying the Itinerary of Lipid Analogs or Lipid-Binding Toxins Using Mutant Rab Proteins

For studies of intracellular lipid trafficking, we have focused on the use of fluorescent sphingolipid analogs (Choudhury et al., 2002a, 2004; Martin and Pagano, 1994; Puri et al., 1999, 2001). These fluorescent lipid analogs can be integrated into cell membranes by spontaneous lipid transfer from exogenous sources (see below). Temporal changes in the intracellular distribution of the lipid analog can then be visualized in living cells by fluorescence microscopy and correlated with metabolism of the lipid, assessed using conventional lipid biochemical analyses. This approach has

---

[1] Abbreviations: BODIPY-LacCer, N-[5-(5,7-dimethyl boron dipyrromethene difluoride)-1-pentanoyl]-1′,1-lactosyl-D-erythro-sphingosine; DF-BSA, defatted bovine serum albumin; DN, dominant negative; DsRed, red fluorescent protein from Discosoma sp.; DsRed2-Nuc, DsRed protein with nuclear localization signal; EE, early endosome; EGFP, enhanced green fluorescent protein; EMEM, Eagle's minimal essential medium; FBS, fetal bovine serum; GDI, GDP dissociation inhibitor; HMEM, 10 mM HEPES-buffered minimal essential medium without indicator; HMEM + G, HMEM + glucose; HMEM − G, HMEM − glucose; HSFs, human skin fibroblasts; NP-C, Niemann–Pick type C; PM, plasma membrane; SLSD, sphingolipid storage disease; WT, wild type.

TABLE I

Rab GTPases and the Role of Dominant Negative Mutants in Endocytic Transport of Lipids

Rabs	Localization[a]	Endocytic transport of proteins	Dominant negative mutants and their effects	
			Mutation	Effect on LacCer transport[b]
Rab4a	EE, RE	EE-PM sorting or recycling	Rab4aS22N	Blocks recycling from EEs to PM
Rab5a	EE, PM, CCV	PM-EE; EE-EE fusion	Rab5aS34N; Rab5aN133I	Does not block internalization via caveolae
Rab7a	LE	EE-LE; LE-Lys	Rab7aT22N	Blocks transport to Golgi
Rab9a	LE, G	LE-G	Rab9aS21N	Blocks transport to Golgi
Rab11a	RE, G, PM	Recycling via REs and G	Rab11aS25N	Blocks recycling from REs to PM

[a] EE, early endosome; LE, late endosome; Lys, lysosome; RE, recycling endosome; PM, plasma membrane; G, Golgi apparatus.
[b] Data from Choudhury et al. (2002, 2004) and Sharma et al. (2003).

been used to study the uptake and transport of a variety of fluorescent lipids, with different phospholipid and sphingolipid analogs exhibiting different transport routes in the cell (e.g., movement along the endocytic, recycling, or secretory pathways) (Choudhury et al., 2004; Hanada and Pagano, 1995; Pagano and Sleight, 1985; Singh et al., 2003). In some cases, we have similarly utilized fluorescently tagged lipid-binding toxins (e.g., Shiga toxin [binds to globoside] and cholera toxin B subunit [binds to GM$_1$ ganglioside]) to follow the transport of endogenous sphingolipids (Choudhury et al., 2002a; Puri et al., 2001).

Using various loading and pulse/conditions, we have studied the initial endocytosis, Golgi targeting, and recycling of these lipid probes (Fig. 1). By using cells expressing a DN Rab protein that is known to block a particular step of vesicular transport, we could then determine whether that particular Rab affects the transport of a lipid probe as it moves through the cell (Choudhury et al., 2002a, 2004; Sharma et al., 2003). The DN Rab proteins that we have used are shown in Table I.

Fluorescent Sphingolipid Analogs

Mammalian cell sphingolipids consist of a fatty acid N-acylated to a long chain base and bearing a polar head group resulting in ceramide, neutral glycosphingolipids, gangliosides, or sphingomyelin. In our studies we use fluorescent sphingolipid analogs in which the natural fatty acid is replaced with a fluorescent fatty acid labeled with the NBD- or BODIPY-fluorophore (Pagano and Sleight, 1985; Pagano et al., 1991). The BODIPY-labeled analogs are particularly useful because of their higher fluorescence yield and a concentration-dependent spectral shift in their fluorescence emission spectrum (Pagano et al., 1991). This latter property allows an estimate of the molar density of the lipid analog within membranes of the living cell based on measurements of the ratio of fluorescence at red and green wavelengths, and the partitioning of sphingolipids into enriched microdomains (Chen et al., 1997; Sharma et al., 2003).

For incubation with cells, fluorescent lipid analogs are usually complexed (1:1 mol/mol) with defatted bovine serum albumin (DF-BSA; Martin and Pagano, 1994) and then diluted in a suitable culture medium (e.g., HMEM + G). Here, we highlight the use of BODIPY-lactosylceramide (BODIPY-LacCer), a glycosphingolipid analog. Using different loading and pulse/chase conditions, early endocytosis, recycling, and EE to Golgi targeting can be studied (see Fig. 1). Cells are incubated with 1–2.5 μM BODIPY-LacCer/DF-BSA for 30–60 min at 10° to allow incorporation of the lipid into the plasma membrane (PM). Cell samples are then washed and warmed to 37° for various times to allow endocytosis

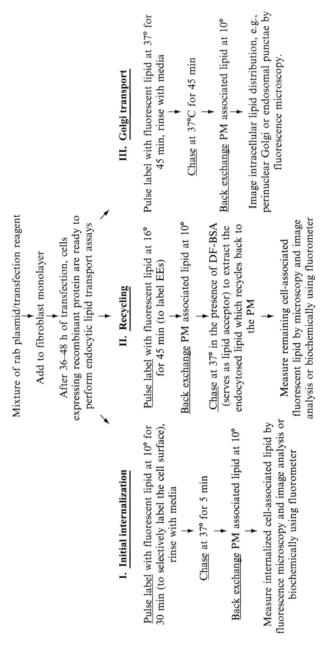

FIG. 1. Summary of assays for the study of intracellular transport of fluorescent lipid analogs. Details of the initial internalization assay were published previously (Puri *et al.*, 2001; Sharma *et al.*, 2003; Singh *et al.*, 2003). The recycling and Golgi transport assays are described in the sections on the role of Rab4 or Rab11 and role of Rab9, respectively.

and targeting to occur (see Fig. 1). Since only a small fraction of the lipid at the PM is internalized, it is usually necessary to remove any fluorescent lipid remaining at the PM prior to observation under the fluorescence microscope. This is accomplished by a procedure termed "back exchange" in which the cell samples are chilled to 4–10° and incubated multiple times with DF-BSA (Chen *et al.*, 1997; Martin and Pagano, 1994). The samples are then washed and viewed under the fluorescence microscope at green (monomer) (λ_{ex} = 450–490 nm; λ_{em} = 520–560 nm) or red (excimer) wavelengths (λ_{ex} = 450–490 nm; λ_{em} > 590 nm).

Generation of Wild-Type (WT) and DN Rab Plasmid Constructs

Full-length WT Rab cDNAs (available from UMR cDNA Resource Center, Rolla, MO) are cloned into a mammalian expression vector such as pcDNA3.1 (Invitrogen Corp., Carlsbad, CA) or fluorescent protein expression vectors (e.g., pEGFP-C or pDsRed2-C; BD Biosciences Clontech, Palo Alto, CA) (Choudhury *et al.*, 2002a; Sharma *et al.*, 2003). However, since DsRed1 and DsRed2 fusion proteins have a tendency to oligomerize, it may be advantageous to generate fusion proteins with monomeric DsRed (mRed; BD Biosciences Clontech), which does not significantly oligomerize (Campbell *et al.*, 2002). Fusion proteins were constructed with the fluorescent tag at the amino-terminus of the Rab protein and not at the carboxy-terminus to avoid interference with the Rab carboxy-terminal prenylation motif that is important for normal Rab function (Seabra *et al.*, 2002). DN mutants of Rabs are prepared by performing site-directed mutagenesis on WT Rab constructs (e.g., using the QuikChange site-directed mutagenesis kit from Stratagene [La Jolla, CA]) (Choudhury *et al.*, 2002a). Additional methods for introduction of Rabs into cells include the use of adeno (or other) viruses encoding Rab proteins (Hirosaki *et al.*, 2002) or delivery of purified recombinant Rab proteins directly into the cells by protein transduction (Narita *et al.*, 2005).

Transfection of Human Skin Fibroblasts (HSFs) with the Rab Constructs

HSFs (30–50% confluent) cultured on glass coverslips in antibiotic–free EMEM with 10% FBS are transfected with WT or DN Rab plasmid constructs using FuGene6 (Roche Applied Science, Indianapolis, IN) or other commercially available transfection reagents for 24–48 h. Before use in lipid trafficking assays, samples are assessed qualitatively for the expression of the fluorescent Rab chimeras by fluorescence microscopy using appropriate filter sets (EGFP: λ_{ex} = 489 nm; λ_{em} = 508 nm; DsRed/mRed: λ_{ex} = 558 nm; λ_{em} = 583 nm). The transfection efficiencies of HSFs are relatively low (e.g., <10% with FuGene6); however, significantly higher efficiencies can be obtained by electroporation (e.g., using the Nucleofector II device from Amaxa, Gaithersburg, MD).

When assessing the impact of a particular Rab construct on (fluorescent) lipid distribution, special precautions should be taken to minimize the effects of any spectral overlap between the fluorescent Rab chimera and the labeled lipid analog (Choudhury et al., 2002a). For BODIPY-lipids and DsRed-labeled Rab proteins, low concentrations of the BODIPY-lipid should be used to minimize excimer formation so that red (excimer) fluorescence from the lipid (and therefore spillover into the DsRed channel) is minimized. In addition, control experiments should always be carried out using "singly labeled" specimens containing either the DsRed-Rab or the BODIPY-lipid alone. By acquiring separate images of these samples at both green and red wavelengths and using identical exposure settings to those for doubly labeled specimens, the extent of this problem can be assessed. However, because of the overlap problem, only qualitative assessments of a particular Rab protein on BODIPY-lipid distribution can be made. To circumvent this problem, we use nonfluorescent chimeric Rabs expressing an HA tag at the amino-terminus (Choudhury et al., 2002a, 2004). Transfected cells can be identified by cotransfection with a reporter plasmid such as DsRed2-Nuc (BD-Clontech), which expresses DsRed2 in the nucleus of transfected cells as a result of a nuclear localization signal. Typically we use a 1:3 molar ratio of DsRed2-Nuc and the HA-Rab chimera. In control experiments we found that 90% or more of the cells exhibiting DsRed nuclear fluorescence were positive for the HA-tagged Rab protein. Using this approach, the effect of a WT or DN Rab protein on the distribution of a BODIPY-lipid can be readily assessed without any interference from a labeled Rab protein (Choudhury et al., 2004).

Use of Rab Mutants to Study Intracellular Transport of Fluorescent Sphingolipid Analogs

Roles of Rab4 and Rab11 in Membrane Recycling of BODIPY-LacCer

Rab4 regulates rapid recycling from EE, whereas Rab11 controls a slower recycling pathway from recycling endosomes to the PM (Segev, 2001; Stenmark and Olkkonen, 2001; Zerial and McBride, 2001). Using the approach described here, we showed that the dominant pathway for membrane recycling of BODIPY-LacCer was predominantly via Rab4 in normal HSFs, but in NPC1 fibroblasts, this pathway was inhibited and the Rab11-dependent pathway was dominant (see Table I, Fig. 2, and Choudhury et al. [2004]).

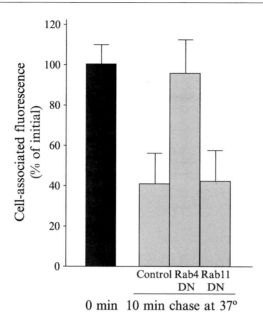

FIG. 2. Dominant negative Rab4 (Rab4DN) blocks BODIPY-LacCer recycling to the plasma membrane. Control HSFs and cells cotransfected with DsRed2-Nuc and either Rab4DN or Rab11DN were pulse labeled with LacCer at 16° to label early endosomes, followed by a chase for 10 min at 37°, and back exchange (see the section on the role of Rab4 or Rab11 and Fig. 1, Recycling). The fluorescent lipid analog remaining in control or transfected (identified by nuclear DsRed fluorescence) cells after pulse/chase was quantified by image analysis after fluorescence microscopy. Values are mean ± standard deviation from a typical experiment ($n = 30$ cells for each data point) and are expressed as a percent relative to initial LacCer fluorescence before 10 min warm up (0 min).

1. HSFs (50–60% confluent) cultured on 35-mm glass coverslips are cotransfected with an untagged Rab4a S22N mutant construct in pcDNA3.1 and pDsRed-Nuc plasmid constructs. Transfections are carried out for 48 h with 1 µg of DNA (1:3 DNA mixture of pDsRed2-Nuc plasmid and either the Rab4 or Rab11 DN construct). Coexpression of the two plasmids (Rab4 or Rab11 DN and DsRed-Nuc) is verified by immunostaining as noted above.

2. A stock solution (0.2–0.5 mM) of BODIPY-LacCer is prepared as a BSA complex (Martin and Pagano, 1994), diluted to 2.5 µM in HMEM + G.

3. Transfected cells (1–2 ml/culture dish) are incubated for 30 min at 16° with BODIPY-LacCer/BSA, washed, and further incubated in HMEM + G for an additional 15 min. Incubation at 16° allows endocytosed lipid

and proteins to accumulate in EEs as revealed by >90% colocalization with the early endosome marker, EEA1 (Choudhury *et al.*, 2004).

4. Samples are then washed in HMEM − G and back-exchanged with 5% DF-BSA in HMEM − G for 60 min at 10° to remove any fluorescent lipid that was not internalized from the PM (see above).

5. After rinsing with 5% DF-BSA/HMEM + G, samples are warmed for various times at 37°, replacing the media every 10 min with fresh, prewarmed 5% DF-BSA/HMEM + G. This incubation allows lipids to exit EEs and sort for recycling back to the PM or for transport to the Golgi apparatus. The presence of DF-BSA in the medium removes any lipid that was recycled back to the PM (Choudhury *et al.*, 2004; Sharma *et al.*, 2003).

6. Specimens are then viewed under a fluorescence microscope and digital images are acquired of transfected cells using a standard exposure time for all images. Images are acquired after various periods of chase at 37° and cell-associated BODIPY-fluorescence is then quantified by image analysis. In control untransfected cells, lipid recycling is reflected by a decrease in the cell-associated fluorescence intensity with time at 37°. Quantitation of lipid recycling is carried out by calculating the percentage of cell-associated fluorescence remaining compared to fluorescence present after 0 min of chase and expressing the values as the mean ± SD of three or more independent experiments (see Fig. 2).

Role of Rab9 in BODIPY-LacCer Transport from the PM to the Golgi Apparatus

Rab9 regulates transport from late endosomes to the Golgi apparatus (Shapiro *et al.*, 1993). Using the following approach, we demonstrated that the targeting from the PM to the Golgi complex of normal HSFs is Rab9 dependent.

1. HSFs (30–50% confluent) cultured on glass cover slips in antibiotic free EMEM with 10% FBS are transfected for 48 h with DN DsRed-Rab9 using Fugene6 (Roche) according to the manufacturer's instructions (see above).

2. A stock solution of BODIPY-LacCer (see Example 1) is diluted to 2.5 μM in EMEM with 1% FBS (EMEM/1%) and added to the cells in culture dishes (1–2 ml/dish). The cell samples are then incubated for 45 min at 37° in a CO_2 incubator.

3. The samples are then rinsed with EMEM/1% and incubated in fresh EMEM/1% for 1 h at 37° in a CO_2 incubator. Cells are then back-exchanged at low temperature to remove PM fluorescence (see Example 1 above).

4. Specimens are viewed under the fluorescence microscope. In this example, transfected cells are readily identified by the bright red *diffuse cytosolic* fluorescence of DN Rab9. BODIPY-LacCer distribution is observed in the green channel. In cells expressing empty DsRed plasmid, BODIPY-LacCer stains the Golgi apparatus, similar to the pattern seen in untransfected cells. In contrast, in cells expressing DN Rab9, BODIPY-LacCer is distributed in punctate structures, and little Golgi labeling is observed (Choudhury *et al.*, 2002a) (see Fig. 3). Control experiments can be carried out using BODIPY-ceramide, a vital stain for the Golgi apparatus (Pagano *et al.*, 1991), or immunofluorescence (e.g., using GM130), to demonstrate that DN Rab9 does not significantly alter the morphology of the Golgi apparatus (Choudhury *et al.*, 2002a).

5. The fraction of transfected cells in which Golgi targeting is disrupted is determined by visual inspection of 10–20 transfected cells in several independent experiments (see Fig. 3).

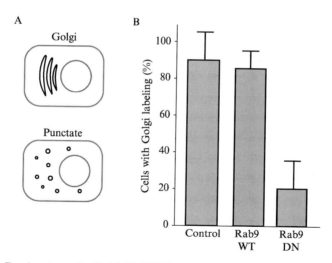

FIG. 3. Dominant negative Rab9 (Rab9DN) blocks Golgi targeting of BODIPY-LacCer. Normal HSFs were mock transfected (control) or were transfected with Rab9DN or Rab9WT. After 48 h of transfection, the cells were used for lipid transport assay (see the section on the role of Rab9 and Fig. 1, Golgi targeting). Cells were pulse-labeled with BODIPY-LacCer and chased at 37° (see text, recycling assay as described in Fig. 1). Transfected cells are identified by DsRed fluorescence and for LacCer targeting images are acquired at green wavelengths. Golgi-positive staining of LacCer (in controls) compared to punctate (Golgi-negative staining) in Rab9DN expressing cells. Cells were scored for Golgi vs. punctate appearance of fluorescent LacCer (A) and expressed as percent Golgi targeting out of total cells counted (B) Values are mean ± standard deviation from a typical experiment ($n = 30$ for each data point).

Effect of Cholesterol on Membrane Trafficking and Rab Function

Most sphingolipid storage diseases (SLSDs) are a result of a defect in lipid hydrolysis or an activator protein, resulting in an accumulation of undegraded sphingolipids (Scriver *et al.*, 2001). Cells from patients with some of these diseases also accumulate cholesterol secondary to sphingolipid storage (Puri *et al.*, 1999). Niemann–Pick type C (NP-C) disease cells also accumulate unesterified cholesterol and sphingolipids due to mutations in NPC1 or NPC2 proteins (Patterson *et al.*, 2001; Sturley *et al.*, 2004). NP-C disease appears to result from abnormal membrane trafficking, since enzyme activities are normal in these cells (Patterson *et al.*, 2001). We have shown that elevated cholesterol in SLSD fibroblasts perturbs BODIPY-LacCer Golgi targeting and recycling and more specifically affects the functions of certain Rabs (Rab4, 7, and 9) (Choudhury *et al.*, 2002a, 2004; Puri *et al.*, 1999; our unpublished observations). For these studies, a number of methods to alter free cholesterol content in both cells and isolated membranes have been useful.

Methods to Deplete Cholesterol

Cells can be cultured in media with 5% lipoprotein-deficient serum (LPDS) for 1–5 days to reduce free cholesterol. This treatment lowers cellular cholesterol by 25–50% compared to control cells grown in FBS (Martin *et al.*, 1993). *De novo* synthesis of cholesterol can be reduced by treatment of cells with inhibitors of HMG-CoA reductase (del Real *et al.*, 2004; Martin *et al.*, 1993).

Cholesterol levels can also be lowered acutely by treating live cells with methyl β-cyclodextrin (mβCD; 5–10 mM) for 30–60 min at 37°. This treatment decreases membrane cholesterol by approximately 30% of control (Troost *et al.*, 2004; our unpublished observations). Similarly, cholesterol can be extracted from isolated membranes using mβCD (Choudhury *et al.*, 2004).

Methods to Elevate Cellular Cholesterol

To chronically elevate cellular cholesterol, cells may be cultured overnight in 5% LPDS to elevate LDL receptor levels, and then incubated for 16 h with excess LDL (50–150 μg/ml) or free cholesterol (50 μg) dissolved in ethanol (Martin *et al.*, 1993; Puri *et al.*, 1999). These treatments increase cellular cholesterol to 175% compared to control cells (Martin *et al.*, 1993). In addition, several SLSD cell fibroblast types (e.g., NP-C, NP-A) are useful as natural cell models in which cholesterol is elevated relative to normal fibroblasts (Puri *et al.*, 1999). Incubation of living cells with progesterone or the hydrophobic amine U18666A has also been used to elevate intracellular cholesterol (Lebrand *et al.*, 2002).

Cholesterol can be acutely delivered to live cell membranes by incubation with a mβCD/cholesterol complex (del Real *et al.*, 2004; Sharma *et al.*, 2004). Similarly, an mβCD/cholesterol complex can be used to increase cholesterol in isolated membrane fractions up to 2- to 6-fold compared to untreated control (see the section Inhibits Rab4 Extraction by GDP Dissociation Inhibitor[G] Elevation of Cholesterol in Isolated Endosome Membranes [Choudhury *et al.*, 2004; Lebrand *et al.*, 2002]).

Methods to Measure Cellular Cholesterol

A number of methods can be used to quantify cellular cholesterol. We have used lipid extraction followed by thin-layer chromatography on silica plates in chloroform/diethyl ether/acetic acid (65:15:1) (Puri *et al.*, 2003). After drying, the plates are stained with iodine and densitometry of spots is performed (Leppimäki *et al.*, 1998). Results are expressed relative to authentic cholesterol standards included on the same chromatography plates. Cholesterol and cholesteryl esters can also be measured colorimetrically using commercially available kits (e.g., from Molecular Probes or Sigma).

An estimate of cholesterol and cholesteryl ester levels and distribution can also be obtained by staining fixed cell monolayers or tissue sections with filipin or Nile red, respectively (Choudhury *et al.*, 2002a,b; Neufeld *et al.*, 1999).

Elevation of Cholesterol in Isolated Endosome Membranes Inhibits Rab4 Extraction by GDP Dissociation Inhibitor (GDI)

GTP-bound active Rabs traffic from donor to target membrane where the GTP is then hydrolyzed to GDP. Subsequently, the GDP-bound Rab is recycled to the target membrane by the action of Rab GDI. We and others have shown that excess membrane cholesterol affects GDI-dependent cycling of selected Rab proteins by inhibiting the extraction of these Rabs from the vesicle membrane by GDI (Choudhury *et al.*, 2004; Lebrand *et al.*, 2002). The example below describes methods we used to show that *in vitro* addition of cholesterol to endosomal fractions isolated from normal fibroblasts inhibits the GDI extractability of Rab4 (Fig. 4).

Preparation of mβCD/Cholesterol Complex. This complex can be made by following the procedure described by Klein *et al.* (1995). Cholesterol (15 mg) is dissolved in 0.5 ml of a 2:1 methanol/chloroform mixture.

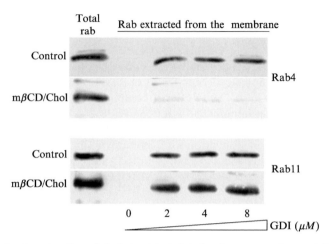

FIG. 4. *In vitro* elevation of cholesterol in isolated endosome membranes inhibits Rab4 extraction by GDI. Endosome membranes were isolated from HSFs and treated with 1 mM mβCD/cholesterol (mβCD/chol) or untreated (Control). Using normalized amounts of indicated Rabs (total Rab), the membranes were subject to GDI extraction in the presence of varying concentration of GST-GDI (see the section on Elevation of Cholesterol in Isolated Endosome Membranes). Rabs extracted from the membrane by GDI were recovered using glutathione-Sepharose 4B beads and were analyzed by SDS–PAGE followed by Western blotting using anti-Rab4 antiserum or anti-Rab11 antibody.

Dissolve 500 mg of mβCD in 10 ml of distilled water maintained at 80° while stirring continuously. Add dropwise cholesterol solution (in methanol/chloroform) to the preheated mβCD solution and continue stirring till the solution becomes clear. Snap freeze the solution in liquid nitrogen followed by freeze-drying overnight. Dissolve the powder in 0.5 ml of PBS to make the inclusion complex with 100 mM steroid concentration. In this complex the cholesterol-to-cyclodextrin molar ratio is 1:10.

Preparation of Cell Homogenate. Fibroblast monolayer cultures are scraped in homogenization buffer (HB: 38 mM potassium aspartate, 38 mM potassium glutamate, 38 mM potassium gluconate, 20 mM 3-[N-morpholino]propanesulfonic acid, pH 7.2, 5 mM sodium carbonate, and 2.5 mM magnesium sulfate) containing protease inhibitors cocktail at 4° (Lim *et al.*, 2001).

The cell suspension is homogenized by passing six times using a 1-ml syringe through a stainless steel ball–bearing homogenizer maintained at 4° (Balch and Rothman, 1985).

Preparation of Endosome-Enriched Fraction. The homogenate is clarified by centrifugation at 300g for 5 min at 4° to remove cell debris and nuclei in the pellet. The clarified supernatant is further centrifuged at 500g for 5 min and the resulting supernantant is further centrifuged for 45 min at 27,000g to generate an endosome-enriched membrane pellet.

Elevation of Membrane Cholesterol. Endosome-enriched membranes (50 μg) are suspended in 500 μl of HB with mβCD/cholesterol complex (1 mM final concentration) for 5 min at room temperature. Excess cholesterol is removed by centrifugation at 27,000g for 45 min at 4°. Elevation of cholesterol in the membrane pellet is measured by lipid extraction and TLC-based cholesterol measurement as previously described (Leppimäki *et al.*, 1998). Typically, membrane cholesterol is increased by 4- to 5-fold by this treatment (Choudhury *et al.*, 2004).

Extraction of Rab4 from Endosome-Enriched Membranes by GST-αGDI

1. Endosomal-enriched membrane pellets from untreated and mβCD/ cholesterol treated fractions are suspended in endosome buffer (30 mM HEPES, 75 mM potassium acetate, 5 mM MgCl₂ containing protease inhibitor cocktail). Protein content in various fractions is measured using Bio-Rad protein assay reagents with BSA as a standard.

2. The relative level of Rab4 in the mβCD/cholesterol-treated and -untreated endosomal fractions is measured by Western blotting.

3. GDI-mediated extraction of Rab4 from endosomes is assessed by incubating aliquots of membranes from treated and untreated samples (adjusted to equal amount of Rab4) with purified recombinant GST-GDI (0–10 μM) in extraction buffer (endosome buffer containing 100 μM ATP, 500 μM GDP) for 20 min at 30°. The sample is then centrifuged at 27,000g for 45 min at 4° to remove extracted Rab4 (which will be in the soluble supernatant fraction bound to GST-GDI) from the unextracted Rab in the membrane pellet.

4. Rab4 bound to GST-GDI is recovered on glutathione-Sepharose 4B beads (using 10 μl of a 50% slurry preequilibrated in extraction buffer) at 4°. The suspension is centrifuged for 5 min at 500g. The pellet is washed 2× with the extraction buffer (without GST-GDI).

5. Rab4 bound to GST-GDI is analyzed by SDS-gel electrophoresis (e.g., using 15% gels) followed by Western blotting using anti-Rab4 antiserum. The amount of Rab extracted from the endosome-enriched fractions containing normal vs. elevated levels of cholesterol is compared. Figure 4 shows that Rab4 extraction from membranes with elevated cholesterol was decreased 5- to 10-fold (Choudhury *et al.*, 2004) compared to normal endosomes.

Conclusion

Studies with DN Rab proteins have elucidated intracellular itineraries by which sphingolipids are sorted and transported in normal and SLSD cells. Utilization of methods to alter cholesterol in cells and isolated membranes has contributed to the conclusion that the interactions between certain Rabs and GDI are disrupted by high cholesterol levels. The techniques presented in this chapter should be useful in future investigations of membrane trafficking and Rab function.

References

Balch, W. E., and Rothman, J. E. (1985). Characterization of protein transport between successive compartments of the Golgi apparatus: Asymmetric properties of donor and acceptor activities in a cell-free system. *Arch. Biochem. Biophys.* **240,** 413–425.

Campbell, R. E., Tour, O., Palmer, A. E., Steinbach, P. A., Baird, G. S., Zacharias, D. A., and Tsien, R. Y. (2002). A monomeric red fluorescent protein. *Proc. Natl. Acad. Sci. USA* **99,** 7877–7882.

Chen, C. S., Martin, O. C., and Pagano, R. E. (1997). Changes in the spectral properties of a plasma membrane lipid analog during the first seconds of endocytosis in living cells. *Biophys. J.* **72,** 37–50.

Choudhury, A., Dominguez, M., Puri, V., Sharma, D. K., Narita, K., Wheatley, C. W., Marks, D. L., and Pagano, R. E. (2002a). Rab proteins mediate Golgi transport of caveola-internalized glycosphingolipids and correct lipid trafficking in Niemann-Pick C cells. *J. Clin. Invest.* **109,** 1541–1550.

Choudhury, A., Sharma, D. K., Wheatley, C. L., Marks, D. L., and Pagano, R. E. (2002b). Abnormal sphingolipid recycling in Niemann Pick-A (NP-A) fibroblasts. *Mol. Biol. Cell* **13S,** 93a.

Choudhury, A., Sharma, D. K., Marks, D. L., and Pagano, R. E. (2004). Elevated endosomal cholesterol levels in Niemann-Pick cells inhibit Rab4 and perturb membrane recycling. *Mol. Biol. Cell* **15,** 4500–4511.

del Real, G., Jimenez-Baranda, S., Mira, E., Lacalle, R. A., Lucas, P., Gomez-Mouton, C., Alegret, M., Pena, J. M., Rodriguez-Zapata, M., Alvarez-Mon, M., Martinez, A. C., and Manes, S. (2004). Statins inhibit HIV-1 infection by down-regulating Rho activity. *J. Exp. Med.* **200,** 541–547.

Hanada, K., and Pagano, R. E. (1995). A Chinese hamster ovary cell mutant defective in the non-endocytic uptake of fluorescent analogs of phosphatidylserine: Isolation using a cytosol acidification protocol. *J. Cell Biol.* **128,** 793–804.

Hirosaki, K., Yamashita, T., Wada, I., Jin, H. Y., and Jimbow, K. (2002). Tyrosinase and tyrosinase-related protein 1 require Rab7 for their intracellular transport. *J. Invest. Dermatol.* **119,** 475–480.

Klein, U., Gimpl, G., and Fahrenholz, F. (1995). Alteration of the myometrial plasma membrane cholesterol content with beta-cyclodextrin modulates the binding affinity of the oxytocin receptor. *Biochemistry* **34,** 13784–13793.

Lebrand, C., Corti, M., Goodson, H., Cosson, P., Cavalli, V., Mayran, N., Faure, J., and Gruenberg, J. (2002). Late endosome motility depends on lipids via the small GTPase Rab7. *EMBO J.* **21,** 1289–1300.

Leppimäki, P., Kronvist, R., and Slotte, J. P. (1998). The rate of sphingomyelin synthesis *de novo* is influenced by the level of cholesterol in cultured human skin fibroblasts. *Biochem. J.* **335,** 285–291.

Lim, S. N., Bonzelius, F., Low, S. H., Wille, H., Weimbs, T., and Herman, G. A. (2001). Identification of discrete classes of endosome-derived small vesicles as a major cellular pool for recycling membrane proteins. *Mol. Biol. Cell* **12,** 981–995.

Martin, O. C., and Pagano, R. E. (1994). Internalization and sorting of a fluorescent analog of glucosylceramide to the Golgi apparatus of human skin fibroblasts: Utilization of endocytic and nonendocytic transport mechanisms. *J. Cell Biol.* **125,** 769–781.

Martin, O. C., Comly, M. E., Blanchette-Mackie, E. J., Pentchev, P. G., and Pagano, R. E. (1993). Cholesterol deprivation affects the fluorescence properties of a ceramide analog at the Golgi apparatus of living cells. *Proc. Natl. Acad. Sci. USA* **90,** 2661–2665.

Narita, K., Choudhury, A., Dobrenis, K., Sharma, D., Holicky, E., Marks, D., Walkley, S., and Pagano, R. (2005). Protein transduction of Rab9 in Niemann-Pick cells reduces cholesterol storage. *FASEB J.* **19,** 1558–1561.

Neufeld, E. B., Wastney, M., Patel, S., Suresh, S., Cooney, A. M., Dwyer, N. K., Roff, C. F., Ohno, K., Morris, J. A., Carstea, E. D., Incardona, J. P., Strauss., III, J. F., Vanier, M. T., Patterson, M. C., Brady, R. O., Pentchev, P. G., and Blanchette-Mackie, E. J. (1999). The Niemann-Pick C1 protein resides in a vesicular compartment linked to retrograde transport of multiple lysosomal cargo. *J. Biol. Chem.* **274,** 9627–9635.

Olkkonen, V. M., and Stenmark, H. (1997). Role of Rab GTPases in membrane traffic. *Int. Rev. Cytol.* **176,** 1–85.

Pagano, R. E., and Sleight, R. G. (1985). Defining lipid transport pathway in animal cells. *Science* **229,** 1051–1057.

Pagano, R. E., Martin, O. C., Kang, H. C., and Haugland, R. P. (1991). A novel fluorescent ceramide analogue for studying membrane traffic in animal cells: Accumulation at the Golgi apparatus results in altered spectral properties of the sphingolipid precursor. *J. Cell Biol.* **113,** 1267–1279.

Patterson, M., Vanier, M., Suzuki, K., Morris, J., Carstea, E., Neufeld, E., Blanchette-Machie, E., and Pentchev, P. (2001). Niemann-Pick disease type C: A lipid trafficking disorder. *In* "The Metabolic and Molecular Bases of Inherited Disease" (C. R. Scriver, A. L. Beaudet, W. S. Sly, and D. Valle, eds.), Vol. III, pp. 3611–3634. McGraw Hill, New York.

Pfeffer, S. R. (2001). Rab GTPases: Specifying and deciphering organelle identity and function. *Trends Cell Biol.* **11,** 487–491.

Puri, V., Watanabe, R., Dominguez, M., Sun, X., Wheatley, C. L., Marks, D. L., and Pagano, R. E. (1999). Cholesterol modulates membrane traffic along the endocytic pathway in sphingolipid storage diseases. *Nature. Cell Biol.* **1,** 386–388.

Puri, V., Watanabe, R., Singh, R. D., Dominguez, M., Brown, J. C., Wheatley, C. L., Marks, D. L., and Pagano, R. E. (2001). Clathrin-dependent and -independent internalization of plasma membrane sphingolipids initiates two Golgi targeting pathways. *J. Cell Biol.* **154,** 535–547.

Puri, V., Jefferson, J. R., Singh, R. D., Wheatley, C. L., Marks, D. L., and Pagano, R. E. (2003). Sphingolipid storage induces accumulation of intracellular cholesterol by stimulating SREBP-1 cleavage. *J. Biol. Chem.* **278,** 20961–20970.

Scriver, C. R., Beaudet, A. L., Sly, W. S., and Valle, D. D. (2001). Part 16: Lysosomal enzymes. *In* "The Metabolic and Molecular Bases of Inherited Disease" (C. R. Scriver, A. L. Beaudet, W. S. Sly, and D. Valle, eds.), Vol. III, pp. 3371–3894. McGraw-Hill, New York.

Seabra, M. C., Mules, E. H., and Hume, A. N. (2002). Rab GTPases, intracellular traffic and disease. *Trends Mol. Med.* **8,** 23–30.

Segev, N. (2001). Ypt and Rab GTPases: Insight into functions through novel interactions. *Curr. Opin. Cell Biol.* **13**, 500–511.

Shapiro, A. D., Riederer, M. A., and Pfeffer, S. R. (1993). Biochemical analysis of rab9, a ras-like GTPase involved in protein transport from late endosomes to the trans Golgi network. *J. Biol. Chem.* **268**, 6925–6931.

Sharma, D. K., Choudhury, A., Singh, R. D., Wheatley, C. L., Marks, D. L., and Pagano, R. E. (2003). Glycosphingolipids internalized via caveolar-related endocytosis rapidly merge with the clathrin pathway in early endosomes and form microdomains for recycling. *J. Biol. Chem.* **278**, 7564–7572.

Sharma, D. K., Brown, J. C., Choudhury, A., Peterson, T. E., Holicky, E., Marks, D. L., Simari, R., Parton, R. G., and Pagano, R. E. (2004). Selective stimulation of caveolar endocytosis by glycosphingolipids and cholesterol. *Mol. Biol. Cell* **15**, 3114–3122.

Singh, R. D., Puri, V., Valiyaveettil, J. T., Marks, D. L., Bittman, R., and Pagano, R. E. (2003). Selective caveolin-1-dependent endocytosis of glycosphingolipids. *Mol. Biol. Cell* **14**, 3254–3265.

Stein, M.-P., Dong, J., and Wandinger-Ness, A. (2003). Rab proteins and endocytic trafficking: Potential targets for therapeutic intervention. *Adv. Drug Deliv. Rev.* **55**, 1421–1437.

Stenmark, H., and Olkkonen, V. M. (2001). The Rab GTPase family. *Genome Biol.* **2**, review S3007.

Stenmark, H., Parton, R. G., Steele-Mortimer, O., Lutcke, A., Gruenberg, J., and Zerial, M. (1994). Inhibition of rab5 GTPase activity stimulates membrane fusion in endocytosis. *EMBO J.* **13**, 1287–1296.

Sturley, S. L., Patterson, M. C., Balch, W., and Liscum, L. (2004). The pathophysiology and mechanisms of NP-C disease. *Biochim. Biophys. Acta* **1685**, 83–87.

Troost, J., Lindenmaier, H., Haefeli, W. E., and Weiss, J. (2004). Modulation of cellular cholesterol alters P-glycoprotein activity in multidrug-resistant cells. *Mol. Pharmacol.* **66**, 1332–1339.

Zerial, M., and McBride, H. (2001). Rab proteins as membrane organizers. *Nat. Rev. Mol. Cell Biol.* **2**, 107–117.

[15] Assay of Rab13 in Regulating Epithelial Tight Junction Assembly

By ANNE-MARIE MARZESCO and AHMED ZAHRAOUI

Abstract

Rab13 is recruited to tight junctions from a cytosolic pool after cell–cell contact formation. Tight junctions are intercellular junctions that separate apical from basolateral domains and are required for the establishment/maintenance of polarized transport in epithelial cells. They form selective barriers regulating the diffusion of ions and solutes between cells. They also maintain the cell surface asymmetry by forming a "fence" that prevents apical/basolateral diffusion of membrane proteins and lipids in the outer leaflet of the plasma membrane. We generate stable MDCK cell lines

METHODS IN ENZYMOLOGY, VOL. 403
0076-6879/05 $35.00
DOI: 10.1016/S0076-6879(05)03015-6

expressing inactive (T22N mutant) and constitutively active (Q67L mutant) Rab13 as GFP-Rab13 chimeras. Expression of GFP-Rab13Q67L delays the formation of electrically tight epithelial monolayers, induces the leakage of small nonionic tracers from the apical domain, and disrupts the tight junction fence diffusion barrier. It also alters the tight junction strand structure and delays the localization of the tight junction transmembrane protein, claudin1. In contrast, the inactive Rab13T22N mutant does not disrupt tight junction functions, tight junction strand architecture, or claudin1 localization. Here we describe a set of assays that allows us to investigate the role of Rab13 in modulating tight junction structure and function.

Introduction

Rab13 regulates the assembly of tight junctions in epithelial cells (Marzesco et al., 2002; Sheth et al., 2000; Zahraoui et al., 1994). Tight junctions (or zonula occludens) play a key role in the development and function of epithelia as well as endothelia. They are the most apical intercellular junctions and form a belt that completely circumvents the apex of the cells, separating apical from basolateral plasma membrane domains. Tight junctions establish a selective barrier regulating the diffusion of ions and solutes across the paracellular space (gate function). They also form a fence preventing the lateral diffusion of proteins and lipids from the apical to the lateral membrane and vice versa, thus contributing to the maintenance of epithelial cell surface asymmetry (Tsukita and Furuse, 1999; Tsukita et al., 2001). In electron micrographs of thin sections, tight junctions appear as very close contacts between the outer leaflets of the plasma membranes of neighboring cells that often appear as focal hemifusions. In freeze fracture electron microscopy replicas, tight junctions appear as a network of intramembrane strands that encircle the apex of the cells. These intramembrane strands are thought to correspond to the focal contact seen in thin sections. Several tight junction proteins are identified. Occludin and claudins, two transmembrane proteins, are thought to seal the intercellular space and to generate series of regulated channels within tight junction membranes for the passage of ions and small molecules (Tsukita and Furuse, 1999, 2000; Zahraoui et al., 2000).

Exactly how tight junctions assemble is still a matter of debate. Cytoplasmic proteins such as ZO-1 (zonula occludens 1), ZO-2, and ZO-3 link occludin and claudins to the underlying actin cytoskeleton. ZO proteins contain three PDZ (PSD95, Dlg, ZO-1) and an SH3 domain that may recruit and cluster proteins to tight junctions (Balda and Matter, 2000; Cordenonsi et al., 1999; Matter and Balda, 2003; Wittchen et al., 1999).

The expression of GFP-Rab13Q67L, but not the inactive GFP-Rab13T22N mutant, in stably transfected Madin–Darby canine kidney (MDCK) delays the formation of electrically tight epithelial monolayers as monitored by transepithelial electrical resistance (TER) and induces the leakage of small nonionic tracers from the apical domain. It also disrupts the tight junction fence diffusion barrier, induces the formation of aberrant strands along the lateral surface, and alters the distribution of the tight junction proteins, claudin1 and ZO-1. Our data reveal that Rab13 plays an important role in regulating both the structure and the function of tight junctions. We propose that Rab13, in its GTP-bound state, recruits an effector(s) that inhibits the recruitment of claudin1 to the cell surface. Conversely, Rab13T22N, which is unable to recruit the effector to the lateral membrane, may favor the assembly of tight junctions.

The methods described herein outline the different assays we used to analyze the activity of Rab13 on tight junction structure and function.

Methods

We investigate the role of Rab13 in tight junctions in the epithelial MDCK cell line grown on permeable filters. The filters are made of a porous membrane of nitrocellulose or polycarbonate, allowing these cells to grow and form a well-polarized monolayer. This enables the easy quantification of tight junction properties via the measurement of electric currents or tracer flux across monolayers. Investigation of the role of Rab13 on tight junction structure and function implies transfection of wild-type and mutant proteins. Because almost all functional tight junction assays measure properties of the entire monolayer and not of a single cell (Matter *et al.*, 2003), it is crucial that such cell populations express the transfected Rab13 proteins in a homogeneous manner. This is achieved by the selection of several clones of epithelial cells expressing Rab13 at similar levels of wild-type or mutant proteins (Fig. 1).

Cloning and Mutagenesis of Human Rab13

Rab13T22N and Q67L mutants are generated from Rab13 cDNA using a site-directed mutagenesis kit (Stratagene, La Jolla, CA). For inactive (T22N) and constitutively active (Q67L) forms of Rab13, the oligonucleotides   5'TCGGGGGTGGGCAAGAATTGTCTGATCATTCGCTT-3' and   5'-GGGACACGGCTGGCCTAGAGCGGTTCAAGACAATA-3' are used, respectively. In addition, an *Eco*RI site is introduced upstream of the ATG codon of Rab13 by polymerase chain reaction (PCR) to facilitate Rab13 cDNA cloning. *Eco*R1–*Bam*H1 inserts encoding Rab13WT,

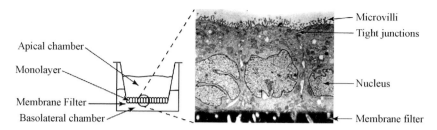

FIG. 1. Diagram of an epithelial cell monolayer grown on a permeable membrane. Electron micrograph of an MDCK cell monolayer. Arrows indicate basally located nuclei, apical microvilli, and tight junctions. The filter is seen at the base of the cells; the pores of the filter appear as white stripes.

Rab13T22N, and Rab13Q67L are fused to the C-terminus of the enhanced green fluorescent protein (GFP) and cloned into the pGFP-C3 vector (Clontech, Inc., Palo Alto, CA). All constructions are verified by sequencing.

Cell Culture and Generation of Stable MDCK Cells Expressing Wild-Type and Rab13 Mutants

MDCK strain II cells are grown in Dulbecco's modified Eagle medium (DMEM) supplemented with 10% fetal calf serum, 2 mM glutamine, 100 U/ml penicillin, and 10 mg/ml streptomycin. The cultures are incubated at 37° under a 10% CO_2 atmosphere.

Stable MDCK cell lines expressing GFP-Rab13WT, GFP-Rab13T22N, GFP-Rab13Q67L, and the mock MDCK expressing GFP are generated by transfection using the Superfect reagent kit according to the manufacturer's instructions (Qiagen GmbH, Hilden, Germany):

1. The day before transfection, seed 400,000 cells per 30-mm dish in 2 ml of DMEM, and incubate the cells at 37° under a 10% CO_2 atmosphere for 18 h.
2. Dilute 1 μg of plasmid DNA in 100 μl of DNA condensation buffer (EC buffer). Add 3.2 μl of Enhancer solution, mix by vortexing for 1 s, centrifuge a few seconds, and incubate for 10 min at room temperature.
3. Add 8 μl of Effectene reagent, mix by vortexing 10 s, centrifuge a few seconds, and incubate the mixture for 15 min at room temperature.
4. Wash the cells one time with DMEM and add 1.5 ml DMEM (with serum) to cells.

5. Add 0.5 ml of growth medium to the mixture, mix by pipeting up and down, and add the transfection reaction dropwise onto the cells in the 30-mm dishes. Swirl the dishes to distribute the transfection complex.

6. After incubation for 6 h, wash the cells two times with growth medium, and incubate the cells at 37° under a 10% CO_2 atmosphere for 18 h.

7. The cells are then trypsinized in trypsin/EDTA and plated, at a dilution of 1:10, in 10-cm dishes. Cell colonies are allowed to grow in the same medium supplemented with 1 mg/ml of G418 (Life Technologies, Inc.) for 15 days until they reach a size of 4–5 mm. Plates containing well-separated colonies are used to ensure easy isolation of individual colonies. Positive cell clones are selected using cloning rings. Colonies are grown in 48-well plates, then passed into 6-well plates, and finally cultured in 10-cm dishes. The expression of GFP, GFP-Rab13 WT, T22N, and Q67L mutants is checked by immunoblotting using a mouse monoclonal anti-GFP antibody (Roche Diagnostics GmbH, Mannheim, Germany). Stable transfected cells are maintained under selection in 500 μg/ml G418. The homogeneity of the clones expressing Rab13 wild-type T22N and Q67L mutants is checked by immunofluorescence. Two independent clones of GFP-Rab13T22N and GFP-Rab13Q67L are grown and subsequently analyzed.

Functional Analysis of Tight Junctions

In all our experiments, 600,000 cells/cm^2 are plated onto polycarbonate filters (0.4 μm pore size and 12 mm diameter; Transwell; Costar Corp., Cambridge, MA, see Fig. 1), and grown for 3–7 days. Immunoblot analysis using an anti-GFP antibody reveals that the expression of the ectopic GFP-Rab13 proteins is very weak, in particular for the GFP-Rab13T22N protein. This prompts us to treat the cells with 10 mM sodium butyrate for 15 h to stimulate the expression of the transfected cDNAs. Under these conditions, the sodium butyrate treatment does not cause changes in tight junction structure or function (Balda *et al.*, 1996; Marzesco *et al.*, 2002). Parental MDCK and MDCK expressing GFP alone are also treated with 10 mM sodium butyrate.

Localization of Tight Junction Proteins by Indirect Immunofluorescence. Analysis of tight junction assembly can easily be carried out by indirect immunofluorescence using confocal microscopy (Marzesco *et al.*, 2002). Immunofluorescence of a bona fide marker of tight junctions results in a honeycomb pattern in x–y optical sections (i.e., parallel to the monolayer),

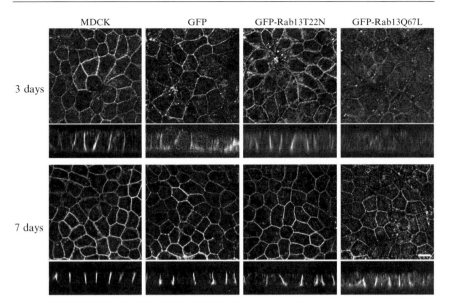

FIG. 2. Subcellular localization of claudin1 in cells expressing GFP-Rab13. Parental MDCK cells and cells stably expressing GFP, GFP-Rab13T22N, or GFP-Rab13Q67L proteins are grown on Transwell filters for 3 or 7 days. Cells are immunostained for claudin1 and analyzed by confocal laser scanning microscopy. An *x–y* optical section taken at the tight junction level and an *x–z* optical section are shown. The confocal *x–y* sections show prominent ring-like structures of claudin1. Expression of GFP-Rab13Q67L delays the recruitment of claudin1 to cell–cell junctions. Bar = 10 μm.

whereas the most apical end of the lateral membrane appears labeled in *x–z* optical sections (i.e., perpendicular to the monolayer). Actually, a large number of tight junction-associated proteins have been identified and the corresponding antibodies are available. We usually localize several transmembrane (claudins, occludin) and cytosolic (ZO-1, ZO-2) proteins to examine the assembly of tight junctions. Staining of tight junction proteins requires extraction of proteins and/or lipids, particularly for cells grown on a filter. We perform immunofluorescence staining on parental or transfected MDCK cells (Fig. 2) using the following procedure:

1. Wash cells with phosphate-buffered saline (PBS) containing 1 mM $CaCl_2$, 0.5 mM $MgCl_2$, and fix with 3% paraformaldhyde in PBS for 15 min.
2. The free aldehyde groups are quenched for 15 min with 50 mM NH_4Cl in PBS.

3. Excise the filter from the insert holder and cut a sector from the filter.
4. Permeabilize with 0.5% Triton X-100 in PBS for 15 min.
5. Block in PBS buffer containing 0.5% Triton X-100 and 0.2% bovine serum albumin (BSA). All subsequent incubations with antibodies and washes are performed with this buffer. The incubations were performed in a humid chamber, protected form light.
6. Incubate overnight at 4° with the polyclonal rabbit anti-claudin1 antibodies (Zymed Laboratories, Inc., San Francisco, CA).
7. Wash three times for 10 min each with the blocking buffer. Incubate for 45 min with affinity purified goat anti-rabbit IgG conjugated to Cy3 (Jackson ImmunoResearch Laboratories, Inc.) and diluted 1:400.
8. Wash three times with blocking buffer and three times with PBS. The filter sector is mounted in 50% glycerol-PBS, covered with a coverslip and sealed with colorless nail polish. Analyze the staining (Fig. 2) using a Leica SP2 confocal laser scanning microscope (Leica Microscopy and Systems GmbH, Heidelberg, Germany).

Assessment of Monolayer Integrity

To monitor the quality of the epithelial cell monolayer tightness, which depends upon the establishment of functional tight junctions, various approaches have been developed. One method is to visualize the integrity of the monolayers by biotinylation of the apical plasma membrane (Fig. 3).

1. Wash the intact monolayers of cells grown on a filter (six-well plate) four times with ice-cold PBS$^+$ (PBS containing 1 mM MgCl$_2$ and 0.1 mM CaCl$_2$). All of the following steps are performed on ice, in order to avoid internalization of the biotin.
2. Stock solution of EZ-Link Sulfo-NHS-LC-biotin (MW 556.58; Pierce) is prepared in dimethyl sulfoxide (DMSO) at 200 mg/ml and stored at −20°. The biotinylation solution is freshly prepared by 1:250 dilution of the stock solution in PBS$^+$ (working concentration of 0.8 mg/ml).
3. Add 250 μl of biotinylation solution to the apical chamber and 1.5 ml of PBS$^+$ to the basal chamber. Incubate on ice at 4° for 20 min. Wash the filters two times with PBS$^+$ and incubate 30 min with 0.2 M glycine.
4. Fix with 2% paraformaldhyde in PBS for 15 min. The free aldehyde groups are then quenched for 15 min with 50 mM NH$_4$Cl in PBS.
5. Excise the filter from the insert, and block in PBS buffer containing 0.5% Triton X-100 and 0.2% BSA.

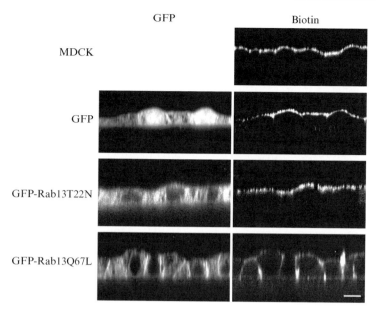

Fig. 3. Diffusion of fluorescent biotin from the apical to the lateral membrane in MDCK cells. Parental MDCK cells and cells stably expressing GFP, GFP-Rab13T22N, or GFP-Rab13Q67L proteins are grown on Transwell filters for 3 days. Biotin is loaded to the apical surface for 20 min at 0°. After washing, cells are labeled with streptavidin–Texas Red at 0°, and processed for confocal microscopy. An x–z optical section shows the distribution of fluorescent biotin. Note that biotin is restricted to the apical surface in MDCK and cells expressing GFP and GFP-Rab13T22N. However, in MDCK cells expressing GFP-Rab13Q67L, biotin is not restricted to the apical membrane but diffuses across the tight junction and stains the lateral membrane. This indicates that Rab13Q67L alters the tight junction barrier functions. Bar = 10 μm.

6. Incubate the filter for 45 min at 4° with streptavidin-Texas Red (ImmunoResearch Laboratories, Inc.) in blocking buffer.
7. After washing three times with blocking buffer and three times with PBS, the filters are mounted in 50% glycerol-PBS, covered with a coverslip, and sealed with colorless nail polish. Analyze the staining using a Leica SP2 confocal laser scanning microscope (Leica Microscopy and systems GmbH, Heidelberg, Germany).

It should be mentioned that the restriction of biotinylated proteins to the apical plasma membrane involves both the fence and gate functions of tight junctions. Therefore, to investigate the fence and the gate functions independently, other assays should be performed.

Analysis of the Junctional Fence

To investigate the tight junction fence function, we use a method that allows visualization of lipid diffusion from the apical plasma membrane to the basolateral domain. Since the diffusion fence is only efficient in the outer leaflet, a fluorescent lipid probe that does not flip-flop between the inner and outer membrane leaflet has to be used. Sphingomyelin derivatives allow efficient labeling of cells on ice, give reliable results, and are generally used. Therefore, for our studies, BODIPY R6G-sphingomyelin (550 nm fluorescent lipid) was used to label the monolayer according to the procedure described below.

Synthesis of BODIPY R6G-Sphingomyelin. BODIPY R6G-sphingomyelin is generated by coupling BODIPY R6G reactive dye (Molecular Probes, Inc., Eugene, OR) to amine-reactive sphingosylphosphorylcholine (Sigma-Aldrich Chimie GmbH, Munich, Germany) using dimethylformamide.

1. A suspension of 10.6 mg (0.0228 mmol) of sphingosylphosphorylcholine in 5 ml anhydrous dimethylformamide is stirred under inert atmosphere conditions.
2. Then 10 mg (0.0228 mmol) of BODIPY R6G (the succinimidyl ester of the 4,4-difluoro-5-phenyl-4-bora-3a,4a-diaza-*s*-indacene-3-propionic acid) is added to the mixture and incubated further for 48 h at room temperature.
3. Removal of the solvent at 50° under reduced pressure provides the wanted *N*-(4,4-difluoro-5-phenyl)-4-bora-3a,4a-diaza-*s*-indacene-3 propionyl) sphingosylphosphorylcholine as a dark powder. The purity of the fluorescent lipid is checked by thin layer chromatography.
4. BODIPY R6G-sphingomyelin/BSA complexes are obtained by adding 400 μl of BODIPY R6G-sphingomyelin stock solution (1 mM in DMSO) to 10 ml of BSA solution (0.8 mg/ml defatted BSA in 10 mM HEPES, pH 7.4, 145 mM NaCl) under vigorous vortexing.

For visualization of diffusion by confocal microscopy:

1. Wash filter-grown MDCK cells twice with cold P buffer (10 mM HEPES, pH 7.4, 145 mM NaCl, 1 mM Na-pyruvate, 10 mM glucose, 3 mM CaCl$_2$).
2. Add 250 μl of BODIPY R6G-sphingomyelin/BSA complexes to the apical chamber and incubate the cells for 10 min on ice. After washing four times with ice–cold P buffer, the cells are either left on ice for 1 h or directly mounted in P buffer.
3. Samples are prepared by cutting out the filter from the insert holder.

4. For mounting, double-sided Scotch tape is used on each side of the microscope slide to support the coverslip and avoid placing pressure on the monolayer.
5. The lateral diffusion of fluorescent lipids is analyzed by confocal microscopy within the first 10 min before internalization occurs.

Analysis of the Paracellular Gate

Tight junctions also act as a selective diffusion barrier for ions and hydrophilic nonionic molecules. The selectivity of the paracellular diffusion barrier is based on the charge and size of hydrophilic nonionic molecules and is regulated by distinct physiological and pathological stimuli. For complete characterization of the paracellular diffusion barrier, it is necessary to analyze not only the ion permeability but also the paracellular permeability to hydrophilic tracers. Measurement of the transepithelial resistance and tracer flux across the cell monolayer is used to monitor the formation of the tight junction barrier during tight junction assembly.

Measurement of Transepithelial Electrical Resistance (TER). The measurement of TER is indicative of the ion permeability of the tight junctions. Since the measurement of TER of filter-grown epithelial cells is performed in the presence of culture medium, it mostly reflects primarily Na^+ permeability. TER of filter-grown epithelial cells is originally determined in Ussing chambers and requires a complicated setup. Actually, TER measurements are performed with a voltmeter.

1. MDCK cells are plated on filters (12 mm diameter) as instant confluent monolayers and grown for 5 days.
2. The same volume of medium is added to the upper and lower chambers. The TER is temperature sensitive, so special care should be taken to avoid cooling the samples.
3. The TER value of filter–grown MDCK cells is determined by applying an AC square–wave current of 20 μA at 12.5 Hz and measuring the voltage deflection with a Ag/AgCl electrode using an Epithelial VoltOhmMeter (EVOM, World Precision Instruments).
4. The TER value is obtained by subtracting the resistance value of an empty filter with culture medium.

Paracellular Flux Assay. Paracellular permeability diffusion of hydrophilic tracers can be measured using fluorescent compounds such as dextrans or radioactively labeled compounds such as mannitol. In our laboratory, we used the following tracers: [³H]mannitol (182 Da), 4 kDa FITC-dextran, 40 kDa FITC-dextran, and 400 kDa FITC-dextran.

1. Cells are grown on filters to confluency for 3 days.
2. The stock solution of FITC-dextran (20 mg/ml) (Sigma-Aldrich Chimie GmbH, Munich, Germany) is dialyzed against P buffer (10 mM HEPES pH 7.4, 1 mM sodium pyruvate, 10 mM glucose, 3 mM CaCl$_2$, 145 mM NaCl) and diluted to 2 mg/ml in P buffer before the assay.
3. Replace the basolateral medium with 500 μl of P buffer, and the apical culture medium with either 250 μl of solution containing 2 mg/ml of 4K, 40K, or 400K FITC-dextran or 500 μl of culture medium containing 1 mM mannitol and 4 μCi/ml [^{3}H]mannitol.
4. Monolayers are incubated at 37° for either 3 h for FITC-dextran or 1 h for [^{3}H]mannitol.
5. Collect the basal chamber media. FITC-dextran is measured with a fluorometer (excitation: 392 nm; emission: 520 nm) (Perkin Elmer Applied Biosystems, Inc.). Radioactivity is counted in a liquid scintillation counter (Wallac Oy, Furky, Finland).

It should be mentioned that the TER and tracer permeability measurements are composites of the paracellular and transcellular pathways. Therefore, ultrastructural morphological studies should be undertaken to understand defects in TER and tracer diffusion.

Freeze-Fracture Electron Microscopy and Immunolabeling

This method allows visualization of tight junction strands organization and provides a measure of tight junction integrity. MDCK cells were plated in 10-cm-diameter tissue culture dishes and grown for 3 days postconfluency. The monolayers can be analyzed either by conventional freeze-fracture or by freeze-fracture immunolabeling.

Conventional Freeze-Fracture. MDCK monolayers are fixed in 2% glutaraldehyde in 0.1 M cacodylate buffer, pH 7.4, for 30 min at room temperature. Cells are scraped from the substrate with a plastic cell scraper and infiltrated with 30% glycerol for 2 h at 4°. Cell pellets are frozen by quick immersion in liquid propane (Balzers, Lichtenstein) and stored in liquid nitrogen until replicated. Freeze-fracture is performed at −130° in a Balzers freeze-fracture 301 or 400 unit (Balzers, Lichtenstein). Replicas are examined using a Philips CM12 electron microscope.

Freeze-Fracture Immunolabeling (FL). In contrast to conventional freeze-fracture, the cells are processed without fixation after a rapid wash using the cell culture medium. SDS digestion and immunolabeling of replicas with polyclonal anti-claudin1 (1:200) and polyclonal anti-occludin (1:100) antibodies followed by protein A gold (10 nm) are performed as previously described (Dunia *et al.*, 2001). Replicas are examined using a Philips CM12 electron microscope.

References

Balda, M. S., and Matter, K. (2000). Transmembrane proteins of tight junctions. *Semin. Cell Dev. Biol.* **11,** 281–289.

Balda, M. S., Whitney, J. A., Flores, C., Gonzalez, S., Cereijido, M., and Matter, K. (1996). Functional dissociation of paracellular permeability and transepithelial electrical resistance and disruption of the apical-basolateral intramembrane diffusion barrier by expression of a mutant tight junction membrane protein. *J. Cell Biol.* **134,** 1031–1049.

Cordenonsi, M., D'Atri, F., Hammar, E., Parry, D. A., Kendrick-Jones, J., Shore, D., and Citi, S. (1999). Cingulin contains globular and coiled-coil domains and interacts with ZO-1, ZO-2, ZO-3, and myosin. *J. Cell Biol.* **147,** 1569–1582.

Dunia, I., Recouvreur, M., Nicolas, P., Kumar, N. M., Bloemendal, H., and Benedetti, E. L. (2001). Sodium dodecyl sulfate-freeze-fracture immunolabeling of gap junctions. *Methods Mol. Biol.* **154,** 33–55.

Marzesco, A. M., Dunia, I., Pandjaitan, R., Recouvreur, M., Dauzonne, D., Benedetti, E. L., Louvard, D., and Zahraoui, A. (2002). The small GTPase Rab13 regulates assembly of functional tight junctions in epithelial cells. *Mol. Biol. Cell* **13,** 1819–1831.

Matter, K., and Balda, M. S. (2003). Signalling to and from tight junctions. *Nat. Rev. Mol. Cell. Biol.* **4,** 225–236.

Matter, K., Balda, M. S., Benais-Pont, G., Punn, A., Flores-Maldonado, C., Eckert, J., Raposo, G., Fleming, T. P., Cereijido, M., and Garrett, M. D. (2003). Functional analysis of tight junctions. *Methods* **30,** 228–234.

Sheth, B., Fontaine, J., Ponza, E., McCallum, A., Page, A., Citi, S., Louvard, D., Zahraoui, A., and Fleming, T. P. (2000). Differentiation of the epithelial apical junctional complex during mouse preimplantation development: A role for rab13 in the early maturation of the tight junction. *Mech. Dev.* **97,** 93–104.

Tsukita, S., and Furuse, M. (1999). Occludin and claudins in tight-junction strands: Leading or supporting players? *Trends Cell Biol.* **9,** 268–273.

Tsukita, S., and Furuse, M. (2000). Pores in the wall: Claudins constitute tight junction strands containing aqueous pores. *J. Cell Biol.* **149,** 13–16.

Tsukita, S., Furuse, M., and Itoh, M. (2001). Multifunctional strands in tight junctions. *Nat. Rev. Mol. Cell. Biol.* **2,** 285–293.

Wittchen, E. S., Haskins, J., and Stevenson, B. R. (1999). Protein interactions at the tight junction. Actin has multiple binding partners, and ZO-1 forms independent complexes with ZO-2 and ZO-3. *J. Biol. Chem.* **274,** 35179–35185.

Zahraoui, A., Joberty, G., Arpin, M., Fontaine, J. J., Hellio, R., Tavitian, A., and Louvard, D. (1994). A small rab GTPase is distributed in cytoplasmic vesicles in non polarized cells but colocalizes with the tight junction marker ZO-1 in polarized epithelial cells. *J. Cell Biol.* **124,** 101–115.

Zahraoui, A., Louvard, D., and Galli, T. (2000). Tight junction, a platform for trafficking and signaling protein complexes. *J. Cell Biol.* **151,** F31–F36.

[16] Tyrosine Phosphorylation of Rab Proteins

By JEAN H. OVERMEYER and WILLIAM A. MALTESE

Abstract

Tyrosine phosphorylation is a fundamental mechanism for regulating the functions of numerous proteins in eukaryotic cells. It has been known for some time that several members of the Rab GTPase family can undergo phosphorylation on serine or threonine residues, but the potential for tyrosine phosphorylation has been appreciated only recently, based on a single example—Rab24. Herein we describe a series of straightforward methods to facilitate an initial assessment of the potential for tyrosine phosphorylation of epitope-tagged Rab proteins transiently expressed in mammalian cells. The approach takes advantage of the availability of highly specific monoclonal antibodies against phosphotyrosine and specific chemical inhibitors for tyrosine kinases. We also describe the use of site-directed mutagenesis to identify tyrosine residues that may be targets for phosphorylation, and we discuss the possible relevance of this modification for regulating Rab function.

Introduction

Rab GTPases function as key regulators in vesicular transport pathways. Rab proteins cycle between membrane and cytoplasmic compartments, with the active, or GTP-bound version of the protein typically associating with the membrane. On the membrane, each active Rab can associate with specific proteins involved in membrane docking and fusion. Following GTP hydrolysis, the inactive, or GDP-bound Rab protein is extracted from the membrane and maintained in the cytosol through its interaction with a GDP dissociation inhibitor (GDI).

Cytosolic localization of a few members of the Rab family has been shown to be additionally regulated by phosphorylation of the Rab on Ser/ Thr residues (Fitzgerald and Reed, 1999; Gerez *et al.*, 2000). In these cases, this modification has also been demonstrated to increase the affinity of the Rab protein for GTP. For example, phosphorylation of Rab4 during mitosis increases the cytoplasmic pool of this Rab protein by 5- to 10-fold, and interestingly, these phosphorylated Rab4 molecules are in the GTP-bound form (Gerez *et al.*, 2000).

Currently, Rab24 is the only member of the Rab family that has been demonstrated to be phosphorylated on tyrosine residues (Ding *et al.*, 2003).

METHODS IN ENZYMOLOGY, VOL. 403   0076-6879/05 $35.00

The physiological function of Rab24 remains unknown. However, Olkkonen *et al.* (1993) have proposed that it plays a role in autophagy, based on its localization in the endoplasmic reticulum, *cis*-Golgi, and late endosomes. Support for this hypothesis has come from the observation that GFP-Rab24 associates with autophagosomes when cells are starved for amino acids (Munafo and Colombo, 2002). Several unusual properties have previously been reported for this GTPase (Erdman *et al.*, 2000). The majority of Rab24 is localized to the cytoplasm but is not associated with GDI and exists predominantly in the GTP-bound form. Interestingly, as reported for the Ser/Thr phosphorylation of other Rabs, the phosphorylated form of Rab24 partitions predominantly in the cytosol, leading to speculation that, in general, phosphorylation of Rab proteins may provide a mechanism for maintaining activated Rab proteins in the cytosolic compartment.

Although tyrosine phosphorylation can be detected by mass spectrometry analysis of tryptic peptides obtained from purified proteins, this approach is labor-intensive and requires specialized equipment and expertise. This chapter provides guidelines for determining if a Rab protein is phosphorylated on a tyrosine residue, using readily accessible immunochemistry reagents and techniques. We also discuss some of the potential implications of such a modification for Rab function.

Methods

Expression of the Rab Protein and Detection of Tyrosine Phosphorylation by Western Blot Analysis

To facilitate detection of the Rab protein of interest, the cDNA encoding the *rab* is inserted into a eukaryotic expression vector, such as pCMV5 (Andersson *et al.*, 1989), that is engineered to encode an in-frame, Myc epitope tag (EQKLISEEDL) at the 5′ end of the gene sequence. Previous studies (Beranger *et al.*, 1994; Brondyk *et al.*, 1993; Chen *et al.*, 1993; Overmeyer *et al.*, 1998) have shown that epitope tags do not interfere with the functional specificity, posttranslational isoprenylation, or subcellular localization of the Rab proteins. Transient overexpression of the Myc-Rab in HEK293 cells can be performed in 60-mm culture dishes using Lipofectamine-Plus reagents (Invitrogen, Carlsbad, CA) according to the manufacturer's instructions. As negative controls, Myc-tagged Rab proteins that are unmodified by tyrosine phosphorylation (Fig. 1) (Ding *et al.*, 2003) are expressed in parallel cultures. Twenty-four hours after transfection, the cells are lysed in sodium dodecyl sulfate (SDS) sample buffer (Laemmli, 1970) and aliquots from each lysate are subjected to SDS–polyacrylamide

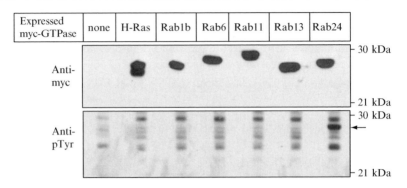

FIG. 1. Anti-phosphotyrosine (pTyr) immunoblot analysis of different Myc-tagged GTPases expressed in HEK293 cells. Aliquots of cell lysate from cultures expressing the indicated Myc-tagged proteins were subjected to SDS–PAGE and immunoblot analysis using either the 9E10 monoclonal antibody against the Myc epitope (upper panel) or the PY-Plus monoclonal antibody against pTyr (lower panel). Positions of the molecular mass standards are indicated at the right side of each panel. The arrow indicates the prominent pTyr band corresponding to Myc-Rab24. Reprinted from Ding, J., Soule, G., Overmeyer, J. H., and Maltese, W. A. (2003). Tyrosine phosphorylation of the Rab24 GTPase in cultured mammalian cells. *Biochem. Biophys. Res. Commun.* **312,** 670–675, copyright 2003 with permission from Elsevier.

gel electrophoresis (PAGE) on duplicate gels. The proteins are transferred to PVDF membranes (Millipore, Bedford, MA) in 25 mM Tris, 192 mM glycine, and 20% methanol at 100 V for 90 min at 4°. Standard immunoblotting procedures (Dugan *et al.*, 1995), using a 5.0% milk solution as a blocking agent, and the Myc monoclonal antibody (9E10, from Calbiochem, San Diego, CA) diluted 1:1000 may be used for detection of the Myc-Rab proteins. However, for the pTyr immunoblots, special care must be taken to avoid high background levels resulting from cross-reactivity with phosphorylated proteins in the blocking buffer. For this reason, milk should not be used. We have found that a commercially available blocking reagent from Zymed Laboratories (South San Francisco, CA) that contains a Tris buffer with bovine serum albumin (BSA), goat IgG, and Tween 20 works quite well to reduce the background on pTyr blots. Since no single monoclonal antibody is capable of reacting with all tyrosine-phosphorylated proteins, several different pTyr antibodies raised from different clones should be used to test the immunoreactivity of the Myc-Rab protein in question. The PY-Plus mouse anti-phosphotyrosine from Zymed is a cocktail of two different monoclonal antibodies (clones PY20 and PY-7E1) with high specificity for pTyr and no cross-reactivity with phosphoserine or phosphothreonine. Since this cocktail reacts with a broad range of pTyr proteins, it works well as an initial screen for the presence of pTyr on the Rab protein of interest (1:10,000 dilution). Other pTyr monoclonal antibodies that may be tested include 1G2, 2G8D6, and the individual PY20. All are available from Covance (Richmond, CA) as well as other suppliers. Detection of the primary antibodies is performed using horseradish peroxidase (HRP)-conjugated goat anti-mouse IgG (Pharmingen, San Diego, CA) diluted 1:3000 and ECL Western Blotting Detection Reagents (Amersham Biosciences, Piscataway, NJ).

The Myc immunoblot is used to confirm the expression levels of the epitope-tagged Rab proteins being studied. The presence of a unique pTyr signal at the same mobility on the gel as the expressed Myc-Rab protein, with no corresponding signal in the lanes containing the lysates from the cells expressing the negative control Myc-Rab proteins, suggests the possibility that the Myc-Rab protein of interest may be phosphorylated on a tyrosine residue (Fig. 1). However, due to the possibility that the pTyr signal may be originating from a distinct protein with a similar migration rate, further analysis is required to confirm this observation.

Immunoprecipitation of the Rab Protein and Detection of Tyrosine Phosphorylation by Western Blot Analysis

To verify that the pTyr signal is indeed associated with the Myc-Rab protein, the Western blot analysis described above is repeated for the Rab protein isolated by immunoprecipitation. Agarose beads conjugated to rabbit anti-Myc IgG are commercially available (Sigma, St. Louis, MO) and simplify the isolation of the epitope-tagged Rab protein. Cells grown in 100-mm dishes are transiently transfected with vector encoding the Myc-Rab protein of interest or empty vector that does not express a tagged protein. Twenty-four hours later, phosphate-buffered saline (PBS)-washed cells are homogenized in lysis buffer consisting of 10 mM HEPES, pH 7.9, 10 mM KCl, 1.5 mM MgCl$_2$, 10 mM NaF, 0.75% Nonidet P-40, and complete mini EDTA-free protease inhibitors (Roche Diagnostics, Indianapolis, IN), then centrifuged at 30,000×g for 30 min at 4° to remove insoluble material. Equal aliquots of total protein (1.5 mg) from each supernatant are mixed with 70 μl of a 50% slurry of the anti-Myc agarose beads suspended in lysis buffer. After incubation for 2 h at 4° on a rotating rack, the beads are collected by centrifugation at 2000×g for 5 min and washed three times with lysis buffer. Proteins are eluted from the beads with SDS sample buffer then separated by SDS–PAGE and analyzed by Western blotting with anti-Myc and anti-PY-Plus as described above. Due to differences in antibody affinities, we have found it useful to load proportionately more of the sample on the pTyr blot (50% of total) than on the Myc blot (10% of total). A pTyr signal associated with the immunoprecipitated Rab protein is indicative that the Rab is in fact tyrosine phosphorylated.

Mutational Analysis of the Rab Protein to Identify Specific Sites of Tyrosine Phosphorylation

Introduction of amino acid substitutions at tyrosine residues may lead to identification of specific sites that are targets for phosphorylation. To assist in identifying potential tyrosine phosphorylation sites, several software programs are available that can predict such motifs, including NetPhosK (http://www.cbs.dtu.dk/services/NetPhosK/) (Blom et al., 2004), PredPhospho (Kim et al., 2004), PHOSITE (http://www.phosite.com) (Koenig and Grabe, 2004), and GPS (Zhou et al., 2004). Identification of a unique tyrosine residue(s) that is present in the Rab protein of interest but is not found in nonphosphorylated Rabs also provides a logical target(s) for analysis.

Overlap extension polymerase chain reaction (PCR) (Higuchi, 1989) using appropriate primers and mutator oligonucleotides is one of the most convenient methods to create Rab constructs with the tyrosine residues converted to alanines. Following the transient expression of the mutated Myc-Rab proteins, immunoblot analysis of whole cell lysates or immunoprecipitated proteins with anti-Myc and anti-p-Tyr antibodies can be carried out as described above. The ECL signals are quantified on a Kodak 440 CF Image Station. A significant decrease in the pTyr/Myc ratio, when compared to that of the wild-type Myc-Rab protein, implicates the mutated tyrosine as a site for phosphorylation.

There are several potential pitfalls associated with this type of analysis. If multiple tyrosine residues on the Rab protein are phosphorylated, only a partial decrease in the pTyr/Myc ratio may be observed. Additionally, substitution of a specific Tyr may indirectly affect the phosphorylation of a distant Tyr by altering the overall protein conformation, thereby falsely implicating the substituted residue as a direct target for tyrosine kinase. In such cases, mass spectrometry or ^{32}P-phosphopeptide analyses may be required to confirm the true location of the modified residue(s).

Tyrosine Kinase Inhibitors

The use of broad-range tyrosine kinase inhibitors can confirm the presence of a phosphorylated tyrosine residue on the Rab protein. Genistein (Sigma, St. Louis, MO), when incubated with the cells for 18 h at a final concentration of 100–200 μM, is capable of inhibiting a number of receptor

TABLE I

TARGETS AND EFFECTIVE CONCENTRATIONS OF COMMONLY USED TYROSINE
KINASE INHIBITORS

Inhibitor	Final concentration in culture medium[a] (mM)	Tyrosine Kinases inhibited
Genistein[b]	0.200	EGF-R, v-Src, c-Src, v-Abl
Herbimycin A[c,d,e]	0.005	Src, Yes, Fps Rps, Abl, ErbB
PP1[f]	0.001	Src family
PP2[f]	0.010	Src family, weak inhibition of EGF-R
Tyrphostin 25[g]	0.100	EGF-R

[a] Stock solutions of all inhibitors are made in DMSO.
[b] Akiyama, T., Ishida, J., Nakagawa, S., Ogawara, H., Watanabe, S., Itoh, N., Shibuya, M., and Fukami, Y. (1987). Genistein, a specific inhibitor of tyrosine-specific protein kinases. *J. Biol. Chem.* **262,** 5592–5595.
[c] Uehara, Y., and Fukazawa, H. (1991). Use and selectivity of herbimycin A as inhibitor of protein-tyrosine kinases. *Methods Enzymol.* **201,** 370–379.
[d] Fukazawa, H., Li, P. M., Yamamoto, C., Murakami, Y., Mizuno, S., and Uehara, Y. (1991). Specific inhibition of cytoplasmic protein tyrosine kinases by herbimycin A in vitro. *Biochem. Pharmacol.* **42,** 1661–1671.
[e] Satoh, T., Uehara, Y., and Kaziro, Y. (1992). Inhibition of interleukin 3 and granulocyte-macrophage colony-stimulating factor stimulated increase of active ras.GTP by herbimycin A, a specific inhibitor of tyrosine kinases. *J. Biol. Chem.* **267,** 2537–2541.
[f] Hanke, J. H., Gardner, J. P., Dow, R. L., Changelian, P. S., Brissette, W. H., Weringer, E. J., Pollok, B. A., and Connelly, P. A. (1996). Discovery of a novel, potent, and Src family-selective tyrosine kinase inhibitor. Study of Lck- and FynT-dependent T cell activation. *J. Biol. Chem.* **271,** 695–701.
[g] Gazit, A., Yaish, P., Gilon, C., and Levitzki, A. (1989). Tyrphostins I: Synthesis and biological activity of protein tyrosine kinase inhibitors. *J. Med. Chem.* **32,** 2344–2352.

and soluble tyrosine kinases (Akiyama *et al.*, 1987). Disappearance of the anti-pTyr immunoreactive band from the Western blot analysis of the immunoprecipitated Myc-Rab after treatment with Genistein reinforces the identity of the protein modification.

On the other hand, selective inhibitors of tyrosine kinases may be employed to obtain information about the class of tyrosine kinases required for the phosphorylation. Table I lists a few of the more commonly used inhibitors, their specific targets, and the effective concentration that should be used in the medium of cultured cells.

Subcellular Distribution of the Phosphorylated Rab Protein

As Rab proteins function in various transport pathways, they successively cycle between the soluble and membrane compartments within the cell. In each of these locations, the Rab protein may interact with different subsets of effector proteins. Therefore, it may be desirable to use the immunoblot methods described above to determine if the phosphorylated Rab is localized predominantly in soluble or particulate fractions. These can be prepared by standard cell lysis and centrifugation techniques (Ding *et al.*, 2003).

Possible Effects of Tyrosine Phosphorylation on Rab Function

The three-dimensional structures of several Rab proteins have been resolved (Chattopadhyay *et al.*, 2000; Chen *et al.*, 2004; Dumas *et al.*, 1999; Esters *et al.*, 2000; Ostermeier and Brunger, 1999; Wittmann and Rudolph, 2004; Zhu *et al.*, 2003). Therefore, the knowledge of the structural and functional domains within the Rab protein, when overlaid with the location of the phosphorylated tyrosine residue(s), can be used to predict possible effects of this modification on the activity of the GTPase.

For example, Rab24 appears to be phosphorylated on at least two tyrosine residues, Y17 and Y172. The former is located in a region involved in coordination of the γ-phosphate of the bound GTP molecule. Hence, phosphorylation of this residue may contribute to the overall decreased rate of GTP hydrolysis that has been observed for Rab24. The phosphorylated tyrosine at position 172 is located within the C-terminal hypervariable domain. This region has been implicated in targeting Rab proteins to specific membrane compartments (Chavrier *et al.*, 1991), although the general applicability of this assumption has been called into question (Ali *et al.*, 2004). Additionally, the C-terminal cysteine residue(s) of Rab proteins are typically modified by geranylgeranylation, facilitating membrane association and possibly affecting guanine nucleotide binding/hydrolysis (Yang *et al.*, 1993). In light of these findings, tyrosine phosphorylation within the C-terminal domain has the potential to influence the overall function of the GTPase. In this regard, it is worth noting that Rab24 is a poor substrate for geranylgeranylation, and this deficiency is not solely related to its unique C-terminal prenylation motif (CCHH) (Erdman *et al.*, 2000). Currently it is not known whether tyrosine phosphorylation affects the ability of Rab24 to interact with Rab escort protein or geranylgeranyltransferase type II, but this clearly represents an interesting question for future study.

References

Akiyama, T., Ishida, J., Nakagawa, S., Ogawara, H., Watanabe, S., Itoh, N., Shibuya, M., and Fukami, Y. (1987). Genistein, a specific inhibitor of tyrosine-specific protein kinases. *J. Biol. Chem.* **262,** 5592–5595.

Ali, B. R., Wasmeier, C., Lamoreux, L., Strom, M., and Seabra, M. C. (2004). Multiple regions contribute to membrane targeting of Rab GTPases. *J. Cell Sci.* **117,** 6401–6412.

Andersson, S., Davis, D. L., Dahlback, H., Jornvall, H., and Russell, D. W. (1989). Cloning, structure, and expression of the mitochondrial cytochrome P-450 sterol 26-hydroxylase, a bile acid biosynthetic enzyme. *J. Biol. Chem.* **264,** 8222–8229.

Beranger, F., Cadwallader, K., Profiri, E., Powers, S., Evans, T., de Gunzberg, J., and Hancock, J. F. (1994). Determination of structural requirements for the interaction of Rab6 with RabGDI and Rab geranylgeranyltransferase. *J. Biol. Chem.* **269,** 13637–13643.

Blom, N., Sicheritz-Ponten, T., Gupta, R., Gammeltoft, S., and Brunak, S. (2004). Prediction of post-translational glycosylation and phosphorylation of proteins from the amino acid sequence. *Proteomics* **4,** 1633–1649.

Brondyk, W. H., McKiernan, C. J., Burstein, E. S., and Macara, I. G. (1993). Mutants of rab3A analogous to oncogenic ras3A analogous to oncogenic ras mutants. *J. Biol. Chem.* **268,** 9410–9415.

Chattopadhyay, D., Langsley, G., Carson, M., Recacha, R., DeLucas, L., and Smith, C. (2000). Structure of the nucleotide-binding domain of Plasmodium falciparum rab6 in the GDP-bound form. *Acta Crystallogr. D Biol. Crystallogr.* **56**(Pt. 8), 937–944.

Chavrier, P., Gorvel, J.-P., Stelzer, E., Simons, K., Gruenberg, J., and Zerial, M. (1991). Hypervariable C-terminal domain of rab proteins acts as a targeting signal. *Nature* **353,** 769–772.

Chen, L., DiGiammarino, E., Zhou, X. E., Wang, Y., Toh, D., Hodge, T. W., and Meehan, E. J. (2004). High resolution crystal structure of human Rab9 GTPase: A novel antiviral drug target. *J. Biol. Chem.* **279,** 40204–40208.

Chen, Y. T., Holcomb, C., and Moor, H. P. H. (1993). Expression and localization of two low molecular weight GTP-binding proteins, Rab8 and Rab10, by epitope tag. *Proc. Natl. Acad. Sci. USA* **90,** 6508–6512.

Ding, J., Soule, G., Overmeyer, J. H., and Maltese, W. A. (2003). Tyrosine phosphorylation of the Rab24 GTPase in cultured mammalian cells. *Biochem. Biophys. Res. Commun.* **312,** 670–675.

Dugan, J. M., deWit, C., McConlogue, L., and Maltese, W. A. (1995). The ras-related GTP binding protein, Rab1B, regulates early steps in exocytic transport and processing of β-amyloid precursor protein. *J. Biol. Chem.* **270,** 10982–10989.

Dumas, J. J., Zhu, Z., Connolly, J. L., and Lambright, D. G. (1999). Structural basis of activation and GTP hydrolysis in Rab proteins. *Structure Fold. Des.* **7,** 413–423.

Erdman, R. A., Shellenberger, K. E., Overmeyer, J. H., and Maltese, W. A. (2000). Rab24 is an atypical member of the Rab GTPase family. *J. Biol. Chem.* **275,** 3848–3856.

Esters, H., Alexandrov, K., Constantinescu, A. T., Goody, R. S., and Scheidig, A. J. (2000). High-resolution crystal structure of S. cerevisiae Ypt51(DeltaC15)-GppNHp, a small GTP-binding protein involved in regulation of endocytosis. *J. Mol. Biol.* **298,** 111–121.

Fitzgerald, M. L., and Reed, G. L. (1999). Rab6 is phosphorylated in thrombin-activated platelets by a protein kinase C-dependent mechanism: Effects on GTP/GDP binding and cellular distribution. *Biochem. J.* **342**(Pt. 2), 353–360.

Gerez, L., Mohrmann, K., van Raak, M., Jongeneelen, M., Zhou, X. Z., Lu, K. P., and van Der, S. P. (2000). Accumulation of rab4GTP in the cytoplasm and association with the peptidyl-prolyl isomerase pin1 during mitosis. *Mol. Biol. Cell* **11,** 2201–2211.

Higuchi, R. (1989). Using PCR to engineer DNA. *In* "PCR Technology" (H. A. Ehrlich, ed.), pp. 61–70. Stockton Press, New York.

Kim, J. H., Lee, J., Oh, B., Kimm, K., and Koh, I. (2004). Prediction of phosphorylation sites using SVMs. *Bioinformatics* **20,** 3179–3184.

Koenig, M., and Grabe, N. (2004). Highly specific prediction of phosphorylation sites in proteins. *Bioinformatics* **20,** 3620–3627.

Laemmli, U. K. (1970). Cleavage of structural proteins during the assembly of the head of bacteriophage T4. *Nature* **227,** 680–685.

Munafo, D. B., and Colombo, M. I. (2002). Induction of autophagy causes dramatic changes in the subcellular distribution of GFP-Rab24. *Traffic* **3,** 472–482.

Olkkonen, V. M., Dupree, P., Killisch, I., Lutcke, A., Zerial, M., and Simons, K. (1993). Molecular cloning and subcellular localization of three GTP-binding proteins of the rab subfamily. *J. Cell Sci.* **106,** 249–261.

Ostermeier, C., and Brunger, A. T. (1999). Structural basis of Rab effector specificity: Crystal structure of the small G protein Rab3A complexed with the effector domain of rabphilin-3A. *Cell* **96,** 363–374.

Overmeyer, J. H., Wilson, A. L., Erdman, R. A., and Maltese, W. A. (1998). The putative "switch 2" domain of the Ras-related GTPase, Rab1B, plays an essential role in the interaction with Rab escort protein. *Mol. Biol. Cell* **9,** 223–235.

Wittmann, J. G., and Rudolph, M. G. (2004). Crystal structure of Rab9 complexed to GDP reveals a dimer with an active conformation of switch II. *FEBS Lett.* **568,** 23–29.

Yang, C., Mollat, P., Chaffotte, A., McCaffrey, M., Cabanie, L., and Goud, B. (1993). Comparison of the biochemical properties of unprocessed and processed forms of the small GTP-binding protein, rab6p. *Eur. J. Biochem.* **217,** 1027–1037.

Zhou, F. F., Xue, Y., Chen, G. L., and Yao, X. (2004). GPS: A novel group-based phosphorylation predicting and scoring method. *Biochem. Biophys. Res. Commun.* **325,** 1443–1448.

Zhu, G., Liu, J., Terzyan, S., Zhai, P., Li, G., and Zhang, X. C. (2003). High resolution crystal structures of human Rab5a and five mutants with substitutions in the catalytically important phosphate-binding loop. *J. Biol. Chem.* **278,** 2452–2460.

[17] Assay of Rab25 Function in Ovarian and Breast Cancers

By Kwai Wa Cheng, Yiling Lu, and Gordon B. Mills

Abstract

There is a multitude of critical steps during the pathogenesis of cancer that allow cells to acquire the ability to escape from normal controls on cell growth, to avoid programmed cell death, and to become malignant. Here, we describe a molecular approach that can be broadly applied to identify drivers of genomic aberrations in cancer development. In the process, areas of genomic aberrations and genes that are dysregulated by genomic amplification are identified by array comparative genomic hybridization (CGH) and transcription profiling, respectively, with major emphasis on coordinating amplification at the CGH and RNA level and on correlation with patient's outcomes. Once candidate genes are identified, we perform

METHODS IN ENZYMOLOGY, VOL. 403   0076-6879/05 $35.00
DOI: 10.1016/S0076-6879(05)03017-X

functional genomics by manipulating levels in normal and tumor cells using RNAi or transfection, and assessing a battery of cellular functions including proliferation, anti-apoptosis, loss of contact inhibition, changes in cell signaling or transcriptional profiles, anchorage-independent growth, and *in vivo* tumor growth. We have successfully used this approach to identify the *RAB25* gene that has been implicated in the progression and aggressiveness of ovarian and breast cancers.

Introduction

Cancer is a disease of genes resulting from genetic and epigenetic changes that occur within a single cell, allowing the cell to acquire an ability to bypass the cell cycle check point, to become resistant to growth inhibitory pathways, to bypass senescence and crisis, and to become immortal (Cavanee and White, 1999; Hanahan and Weinberg, 2000). Many of the genes responsible for the development of malignancy fall into the class of oncogenes (Weinberg, 1996). The normal structure of a resident protooncogene may be converted to a dominant oncogene by mutations or chromosomal rearrangements. Occasionally, the protooncogene is not mutated but rather is expressed at higher levels, in inappropriate cells or at inappropriate times. Rab GTPases play a master role in regulating intercellular vesicle trafficking in both exocytic and endocytic pathways. (Stein *et al.*, 2003). Studies have demonstrated links between Rab GTPase dysfunction in human diseases such as Griscelli syndrome type 2 and thyroid-associated adenomas, which are caused by mutation of the *RAB27a* gene (Menasche *et al.*, 2000) and up-regulation of *RAB5a* and *RAB7* (Croizet-Berger *et al.*, 2002), respectively. In addition, dysregulation of *RAB* gene expression, in particular *RAB25*, may be a generalized component regulating the aggressiveness and potentially the outcome of human cancers, as increased *RAB25* levels have been noticed in cancers including ovarian and breast cancers (Cheng *et al.*, 2004), prostate cancer (Calvo *et al.*, 2002), transitional cell carcinoma of the bladder (Mor *et al.*, 2003), and invasive breast tumor cells (Wang *et al.*, 2002), suggesting a pathological role of Rab25 proteins in the development or progression of tumors in multiple epithelial lineages. However, the molecular mechanism by which Rab25 mediates its functions remains unknown. Here, we describe a molecular approach to examine the functional role of Rab25 in cancers.

We have instituted a generalized approach that can be broadly applied to identify drivers of genomic aberrations. In the process, areas of genomic aberrations are mapped as finely as possible using array comparative genomic hybridization (CGH) (Hodgson *et al.*, 2001; Pinkel *et al.*, 1998).

If there are many candidate regions, we select those that correlate with the patient's outcome, as these are more likely to harbor targets for therapy. To identify genes that are dysregulated by genomic amplification, we identify all of the open reading frames in the region using the University of Santa Cruz genome browser (http://genome.cse.ucsc.edu/) and utilize publicly available transcriptional profiling (Oncomine cancer microarray database) and SAGE (http://cgap.nci.nih.gov/) databases to determine levels of increases in transcripts in the tumor of interest. Any transcripts in which data are not available as well as candidates from the transcriptional profiling and SAGE analysis are assessed by quantitative polymerase chain reaction (PCR). We put a major emphasis on coordinate amplification at the CGH and RNA level and correlation with outcomes, which can provide a filter to identify genes, which are more likely to be therapeutic targets. Once candidate genes are identified, we perform functional genomics by manipulating levels in normal and tumor cells using RNAi (Milhavet *el al.*, 2003) or transfection with expression constructs, and assessing a battery of cellular functions including proliferation, anti-apoptosis, loss of contact inhibition, changes in cell signaling or transcriptional profiles, anchorage-independent growth, and *in vivo* tumor growth (Fig. 1).

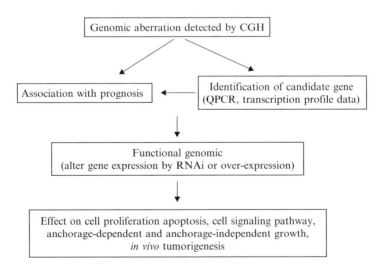

Fig. 1. Diagrammatic representation of the approach in identifying drivers of genomic aberration and its functional role. Emphasis was placed on coordinate amplification at the CGH and RNA level and correlation with outcomes, which provide a filter to identify potential therapeutic target gene(s).

Isolation and Quantitative Real-Time Polymerase Chain
Reaction (QPCR)

To determine levels for the *RAB25*, total RNA was isolated from 10
ovarian and breast cancer cell lines using Trizol reagent (Invitrogen, Carlsbad, CA) according to the manufacturer's suggested protocol. QPCR was
analyzed with an ABI PRISM 7700 Sequence Detection system from
Applied Biosystems (Foster City, CA) with a *RAB25*-specific mix of unlabeled PCR primer and Taqman MGB probe (20× concentrations) using
TaqMan One-Step RT-PCR Master Mix Reagent Kit (Applied Biosystems, Foster City, CA). The sequences of the *RAB25* forward and
reverse PCR primers are 5'-CTGAGGAGGCCCGAATGTT-3' and
5'-GGCTGAGGTCTCCAGGAAGAG-3', respectively, and the Taqman
MGB probes sequence is 5'-CGCTGAAACAATGGA-3'. The MGB
probe was labeled at the 5' end by a fluorescent FAM dye and quenched
by TAMRA dye at the 3' end. The forward and reverse primers are located
at exons 3 and 4, respectively, whereas the *RAB25* probe resides at the
boundary of exon 3 and exon 4. With this probe, one messenger RNA will
be detected. QPCR was performed by mixing 60 ng of total RNA with 25 μl
of 2× Master mix without UNG, 1.25 μl of 40× MultiScribe and RNase
Inhibitor Mix, and 2.5 μl of the 20× primer and probe to a final volume of
50 μl. The reverse transcription was carried out by incubating the mixture
at 48° for 30 min, heating at 95° for 10 min, followed by 40 PCR cycles at
95° for 15 s and 60° for 60 s. To normalize the amount of total RNA present
in each reaction, *β-actin* genes were used as an internal standard.

Preparation of *RAB25* Expression Constructs

Total RNA isolated from ovarian cancer OVCAR3 cells was used as a
template for PCR amplification to generate a 652-base pair human *RAB25*
full-length cDNA using *RAB25*-specific sense (5'-CGAAGCTTATG-
*TACCCATACGATGTTCCAGATTACGCT*GGGAATGGAACTGAG-
GAAGAT-3') and antisence (5'-GTGGATCCGAGGGGTGGACAGA-
TAAAAGAGGTATT-3') primers. *Hind*III (AAGCTT) and *Bam*HI
(GGATCC) recognition sites (underlined) were introduced to facilitate
the cloning procedure, and a hemagglutinin (HA)-tagged sequence
(*TACCCATACGATGTTCCAGATTACGCT*) was fused with the
RAB25 sequence after the ATG start codon to help in identifying the
recombinant Rab25 protein. cDNA was synthesized from total RNA using
the First Strand cDNA synthesis kit following the manufacturer's instructions (Invitrogen, Carlsbad, CA). The reaction mixture (15 μl), containing
5 μg total RNA, 5 μl bulk first-strand reaction mix, 0.2 μg oligo(dT) primer,

and 6 mM dithiothreitol (DTT), was incubated at 37° for 60 min and terminated by heating at 90° for 5 min. To amplify the *RAB25* cDNA, 5 μl reverse-transcribed cDNA was subjected to PCR amplification in a 50-μl reaction mix containing 10 mM Tris–HCl, pH 8.3, 50 mM KCl, 2.5 mM MgCl$_2$, 0.001% w/v gelatin, 10 mM each of dATP, dCTP, dTTP, and dGTP, 2.5 units of *Taq* polymerase (Invitrogen, Carlsbad, CA), and 20 pmol of each sense and antisense primer with the conditions of denaturation at 94° for 60 s, primer annealing at 55° for 65 s, and extension at 72° for 90 s. The PCR product was digested with *Hin*dIII and *Bam*HI, and ligated into pcDNA3.1 (+) vector to generate the HA-tagged Rab25 expression construct (Invitrogen, Carlsbad, CA). The insert was sequenced to confirm the presence of the correct open reading frame.

Transfection and Clonal Selection

Human epithelial ovarian carcinoma A2780, DOV13, HEY, and OCC1 cell lines were cultured in RPMI 1640 supplemented with 10% fetal bovine serum (FBS) (Atlanta Biologicals, Norcross, GA). All of the cell lines were tested to evaluate G418 (Invitrogen, Carlsbad, CA) resistance. G418 was applied in a concentration from 200 to 1000 μg/ml. The lowest concentration of G418 that killed all of the cells was applied later in the clonal selection. Stable *RAB25* expressing clones were generated by transfecting ovarian cells (1×10^5) seeded into 60-mm tissue culture plates 1 day before transfection. Two micrograms of the HA-tagged Rab25 expression construct was mixed with 6 μl of Fugene 6 reagent (Roche Molecular Biochemicals) in 100 μl of serum-free medium. The DNA mixture was incubated for 20 min at room temperature and then applied to the cells. Incubation of the cells with transfection medium was continued for approximately 24 h at 37° in 5% CO$_2$. After transfection, the cells were washed twice with culture medium and incubated with normal culture medium containing 10% FBS. *RAB25* stably–expression clones were selected for 4–6 weeks by limiting dilution in the presence of G418.

Positive clones were identified by Western blotting analysis. Cells were lysed in ice-cold lysis buffer (1% Triton X-100, 50 mM HEPES, pH 7.4, 150 mM NaCl, 1.5 mM MgCl$_2$, 1 mM EGTA, 100 mM NaF, 10 mM Na-pyrophosphate, 1 mM Na$_3$VO$_4$, 10% glycerol, 1 mM phenylmethylsulfonyl fluoride [PMSF], and 10 g/ml aprotonin) for 15 min on ice. After lysis, the samples were centrifuged at 14,000 rpm for 10 min at 4°, and the supernatants were stored at $-80°$. Proteins were diluted in sample loading buffer ($5\times$: 60 mM Tris–HCl, pH 6.8, 25% glycerol, 2% sodium dodecyl sulfate (SDS), 14.4 mM 2-mercaptoethanol, 0.1% bromophenol blue), boiled for 5 min, separated by SDS–polyacrylamide gel electrophoresis (PAGE), and transferred to

Hybond-ECL nitrocellulose membranes (Amersham Bioscience). Blots were blocked with TBST (25 mM Tris–HCl, pH 7.5, 150 mM NaCl, 0.1% Tween-20) containing 5% bovine serum albumin for at least 1 h and probed with monoclonal antibody HA.11 (1:1000 dilution; Covance, Berkeley, CA) against HA epitope overnight. The membrane was washed six times in TBST, 5 min each, and incubated with secondary antibodies coupled to horseradish peroxidase (HRP) for 1 h at room temperature. The signal was visualized by an enhanced chemiluminescence detection ECL system (Amersham Bioscience) after extensive washing in TBST.

Colony Formation Assay

For anchorage-dependent colony formation, 1×10^4 cells were transfected with either pcDNA 3.1 or *RAB25* expression vector. Forty-eight hours posttransfection, cells were trypsinized and replated in two six-wells/plate for 14 days in the presence of G418. Cells were stained with 0.1% Coomassie blue (Bio-Rad) in 30% methanol and 10% acetic acid. The number of colonies formed was counted and expressed as fold increase compared with pcDNA transfected cells.

To test the effect of *RAB25* expression on anchorage-independent colony formation, pCDNA or *RAB25* stable expressing cells were suspended at a density of 1×10^4 cells/ml in 1.5 ml of 0.3% agar (Bacto-agar; Difco, Detroit, MI) dissolved in complete medium containing 25% FBS (top layer). Cells were plated in 35-mm dishes precoated with 1.5 ml of solidified 0.6% agar base in complete medium containing 25% FBS (bottom layer). After 3 days of incubation, 200 μl fresh complete medium supplemented with 5% FBS was added to maintain humidity. Colony-forming efficiency was measured 14–18 days after plating (greater than 50 cells/colony) under a microscope and expressed as a fold increase related to control vector transfected cells.

Apoptosis Assays

Apoptosis was induced by culturing cells in 0.1% serum for 48 h, by UV radiation or by paclitaxel. The intensity of UV radiation (μJ/cm^3) and the concentrations of paclitaxel (ng/ml) required to induce apoptosis were predetermined by exposing the untransfected cells to various intensities of UV radiation (50–400 μJ/cm^3) and concentrations of paclitaxel (50–200 ng/ml). Cell were then harvested and assayed for apoptosis by flow cytometry as described below after 24 and 48 h. The lowest intensity of UV radiation and concentration of paclitaxel that induce approximately 50% of cells to undergo apoptosis were applied later to examine the effect of

RAB25 overexpression or down-regulation on cell death. For anoikis assays, cells were incubated on a rocker platform to prevent adhesion for 48 h. In each experiment, both floating and attached cells ($1–2 \times 10^6$ cells) were collected after treatment and washed twice with phosphate-buffered saline (PBS; 10 m*M* sodium phosphate, pH 7.2, 150 m*M* sodium chloride). Cells were then fixed with 1% (w/v) paraformaldehyde in PBS for 15 min on ice. Cells were collected by centrifugation at $300 \times g$ for 5 min and washed twice with PBS. After washing, cells were resuspended in 5 ml ice-cold 70% ethanol and stored in ice for at least 30 min before measuring apoptotic cells using an APO-BrdU kit (Phoenix Flow Systems, Inc., San Diego, CA) with a FACScan (Beckton Dickinson) cell sorter, air-cooled argon laser 488 nm, using CellQuest software. pcDNA transfected cells were included as control.

In Vivo Tumorigenicity Assay

To assess the impact of *RAB25* overexpression on tumorigenicity, five female BALB/c *nu/nu* mice (4 weeks old) were inoculated with either *RAB25*–overexpressing or pcDNA–transfected control cells. Animals were housed in a sterile, air-conditioned atmosphere, given food and water in standard conditions, and handled in a sterile laminar flow hood. Subcutaneous and intraperitoneal injections were performed with 5×10^6 cells in 0.2 ml PBS. Tumor development was monitored once a week. Subcutaneous tumors were measured with a digital caliper. The length (L) and width (W) of each tumor were measured to calculate tumor volume (V) as follows: $V = 2W(L/2)2$. Intraperitoneal tumor mass was measured by dissection of the tumor from the peritoneal cavity and weighing.

RNA Interference

Chemically synthesized *RAB25*–specific siRNA (5′-GGAGCUCUAU-GACCAUGCU-3′) was purchased from Xeragon, now part of Qiagen (Valencia, CA). A scramble RNAi negative control was purchased from Ambion. RNAi transfection was carried out in solution T using a Nucleofector system (Amaxa Biosystem). At the day of transfection, 1×10^6 *RAB25* stably expressed A2780 or breast MCF-7 cells were harvested and resuspended in 100 μl of solution T. Then 5 μl of *RAB25* SiRNA (20 μM) was added to the cell suspension and the mixture was transferred into an electrophoration cuvette (Amaxa Biosystem). The cuvette was inserted into a Nucleofector system and transfection was carried out using Nucleofector program R23. Immediately after transfection, 500 μl of complete medium was added to the cells and transferred into six-well plates.

Reverse-Phase Protein Lysate Array (RPPA)

RAB25-overexpressing or pcDNA-transfected control A2780 cells (5×10^5) were serum starved for 24 h before stimulation with IGF, EGF, TGF-β, and FBS for 5 min, 30 min, 2 h, and 24 h. Total cellular protein were isolated by lysis buffer as previously described. Protein concentration was determined by BCA reaction (Pierce Biotechnology, Inc., Rockford, IL) and adjusted to 1 μg/μl by diluting in 4$\times$ sodium dodecyl sulfate (SDS) buffer (250 mM Tris–HCl, pH 6.8, 35% glycerol, 8% SDS, 10% 2-mercaptoethanol) to a final 1$\times$ concentration. The samples were then boiled for 5 min and a serial 2-fold dilution was prepared using diluted lysis buffer in 4$\times$ SDS buffer. To each diluted sample, an equal amount of 80% glycerol/2$\times$ PBS solution (8 ml glycerol mixed with 2 ml of 10$\times$ PBS without Ca^{2+} and Mg^{2+}) was added. The samples were ready for printing.

Protein lysate arrays were printed on nitrocellulose-coated glass slides (FAST Slides, Schleicher & Schuell) by a GeneTAC G3 arrayer (Genomic Solution, Ann Arbor, MI) with Forty 75-μm-diameter pins arranged in a 4×10 format. Forty grids were printed at each slide with each grid contained 16 dots. Protein dots were printed in duplicate with eight concentrations (from undiluted to 1/128-fold diluted samples). Arrays were produced in batches of 10, and the occasional low-quality array (e.g., with many spot dropouts) was discarded.

Each array was incubated with a specific primary antibody and the signal was detected by using the catalyzed signal amplification (CSA) system according to the manufacturer's recommended procedure (DakoCytomation California, Inc., Carpinteria, CA). In brief, each slide was blocked with I-block (Applied Biosystems, Foster City, CA) overnight at 4°. After blocking, the slide was incubated with primary antibody and secondary antibody, diluted in DAKO antibody dilutent with background reducing compound, at room temperature for 1 h and 45 min, respectively. The slide was then incubated with streptavidin–biotin complex, biotinyl tyramide (for amplification) for 15 min, streptavidin-peroxidase for 15 min, and 3,3'-diaminobenzidine tetrahydrochloride chromogen for 5 min. Between steps, the slide was washed with TBST. The signal was scanned with an HP Scanjet 8200 scanner (Hewlett Packard, Palo Alto, CA) with a 256-shade gray scale at 600 dots per inch. Spot images were converted to raw pixel values by Microvigene version 2.0 (VigeneTech, North Billerica, MA).

Comment

In our studies, Rab25 was initially implicated in ovarian cancer by being located at a site of genomic amplification in ovarian cancer. Genomic amplification is indicative of selection for a gene or genes in the region.

Using QPCR, we measured the endogenous expression level of *RAB25* in 10 ovarian and breast cancer cell lines (Fig. 2). Normal ovarian surface epithelium was used to set the baseline for comparison. With RNA, identification of the appropriate control is a major challenge. In many cases, the cell of origin for a tumor is not known or cannot be prepared in large quantities. Further, if cancer arises in a limited population of stem cells in the precursor lineage, it may be difficult to purify sufficient cells for comparison. Identification of cancer cells with high and low endogenous *RAB25* levels provided a powerful tool in studying the functional role of *RAB25* in cancer. Altering *RAB25* levels in cell lines with *RAB25* expression constructs resulted in increasing anchorage-independent colony-forming activity (Fig. 3), cell proliferation, and cell survival under multiple stress conditions including serum starvation, anoikis (anchorage-independent stress), UV radiation and chemotherapy (paclitaxel), and *in vivo* tumor formation in murine xenografts as recently reported

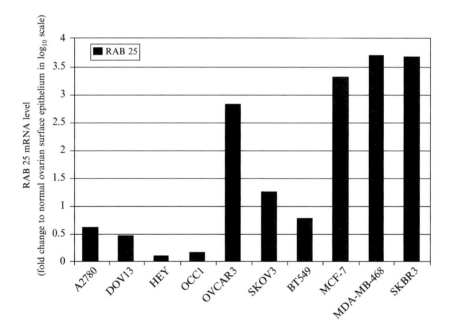

FIG. 2. Quantification of *RAB25* mRNA level in human ovarian and breast cancer cells by real-time quantitative polymerase chain reaction. Total RNA was isolated from ovarian cancer A2780, DOV13, HEY, OCC1, OVCAR3, and SKOV3, and breast cancer BT549, MCF-7, MDA-MB-468, and SKBr3 cells. The *RAB25* mRNA level in normal ovarian surface epithelium was included and set as baseline for comparison.

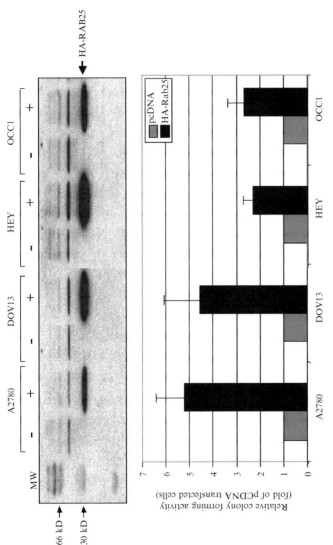

FIG. 3. *RAB25* regulates cell survival. Western blot analysis of HA-tagged RAB25 expression in ovarian cancer A2780, DOV13, HEY, and OCC1 cells (upper panel). The expression of RAB25 increases the colony-forming ability of ovarian cancer cells. Ovarian cancer cells were transfected with HA-tagged RAB25 expression construct or pcDNA control construct. The total number of colonies was counted 14 days after selection in medium containing G418, and shown as folds increase compared with pcDNA transfected cells. Results are the mean ± SD from three individual experiments.

(Cheng *et al.*, 2004). In addition, decreasing the expression of *RAB25* by RNAi transfection significantly decreased cell proliferation and markedly increased the sensitivity to apoptosis of both breast and ovarian cancer cells, confirming the role of *RAB25* in mediating cell survival (Cheng *et al.*, 2004).

A recent study has provided direct evidence that multiple signaling molecules including AKT, ERK1/2, and p38MAPK associate on endocytic vesicles and mediate their function by translocation of the vesicle to the nucleus, providing a potential mechanism by which Rab GTPases conduct cell survival signals (Delcroix *et al.*, 2003). We were able to demonstrate an increase in AKT phosphoryation in *RAB25*-overexpressed A2780 cells by Western blotting analysis (Cheng *et al.*, 2004). However, this approach is time and labor intensive. As a result, a high-throughput lysate array technology (RPPA) was introduced to dissect potential Rab25-mediated signal transduction pathways. In our hands, at least 160 samples can be examined simultaneously in one slide (Fig. 4A has 40 examples). Our data confirmed the increase in AKT phosphorylation in *RAB25*-overexpressed

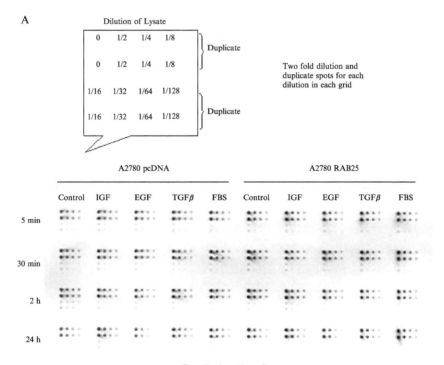

Fig. 4. *(continued)*

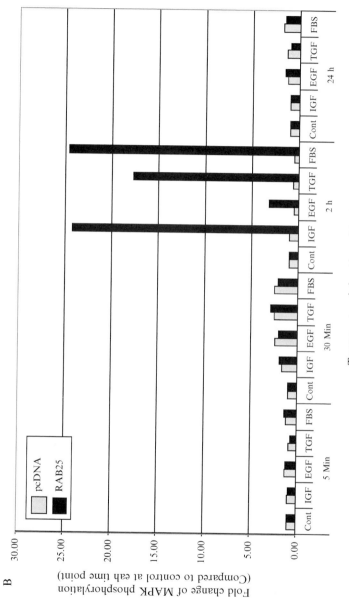

FIG. 4. Protein lysate array analysis. (A) Total cellular proteins was isolated from *RAB25* stable expressed ovarian cancer A2780 cells after treating with insulin growth factor (IGF), epithelial growth factor (EGF), transforming growth factor-β (TGFβ), and fetal bovine serum (FBS) for 5 min, 30 min, 2 h, and 24 h. Protein dots were printed in duplicate with eight concentrations (from undiluted to 1/128-fold diluted samples). A typical protein array slide was shown. (B) Effect of *RAB25* expression on MAPK phosphorylation. The change in MAPK phosphorylation was normalized with the total MAPK level and compared to the control sample at each time point.

TABLE I
ANTIBODIES USED IN PROTEIN LYSATE ARRAY STUDY

Antibody	Source	Cat. #	Dilution
Total AKT	Cell Signaling	9272	250
AKT (phospho-Ser 473)	Cell Signaling	9271	250
AKT (phospho-Thr 308)	Cell Signaling	9275	250
c-Abl	Cell Signaling	2862	500
c-Cbl	BD Transduction Lab	610441	500
EGFR	Sigma	E3138	1000
EGFR (Phospho-Tyr 992)	Cell Signaling	2234	100
EGFR (Phospho-Tyr 1068)	Cell Signaling	2235	100
JNK	Santa Cruz	SC-474	1000
JNK (Phospho-Thr 183/Tyr 185)	Cell Signaling	9251	200
Total MAPK	Cell Signaling		
MAPK (Phospho-p40/42)	Cell Signaling	9101	1000
mTOR	Cell Signaling	2972	250
mTOR (Phospho-Ser 2448)	Cell Signaling	2971	250
GAPDH	Ambion	4300	10000

cells. In addition, using robotic replication of slides and staining with multiple validated antibodies, we are able to examine the effect of Rab25 on multiple signal transduction molecules (Table I). For example, an increase in MAPK phosphorylation was observed in *RAB25*-overexpressed A2780 cells after 2 h of stimulation but not in 5 min, 30 min, and 24 h stimulation when compared to corresponding controls at each time point (Fig. 4B). These results have high reproducibility as similar signals and trends are obtained from total cellular lysate isolated from reproduced experiments. Profiling multiple signal transduction molecules by lysate array technology will certainly facilitate our understanding of the molecular mechanism by which *RAB25* regulates the aggressiveness of cancers.

Acknowledgments

K.W.C. was supported by the Odyssey Program of the Houston Endowment Scientific Achievement award from the MD Anderson Cancer Center. This work is supported by National Institutes of Health SPORE (P50-CA83639) and PPG-PO1 CA64602 to G.B.M.

References

Calvo, A., Xiao, N., Kang, J., Best, C. J., Leiva, I., Emmert-Buck, M. R., Jorcyk, C., and Green, J. E. (2002). Alterations in gene expression profiles during prostate cancer progression: Functional correlations to tumorigenicity and down-regulation of seleno-protein-P in mouse and human tumors. *Cancer Res.* **62,** 5325–5335.

Cavanee, W. K., and White, R. L. (1999). The genetic basis of cancer. *Sci. Am.* **275,** 72–79.

Cheng, K. W., Lahad, J. P., Kuo, W. L., Lapuk, A., Yamada, K., Auersperg, N., Liu, J., Smith-McCune, K., Lu, K. H., Fishman, D., Gray, J. W., and Mills, G. B. (2004). The *RAB25* small GTPase determines aggressiveness of ovarian and breast cancers. *Nat. Med.* **10**(11), 1251–1256.

Croizet-Berger, K., Daumerie, C., Couvreur, M., Courtoy, P. J., and van den Hove, M. F. (2002). The endocytic catalysts, Rab5a and Rab7, are tandem regulators of thyroid hormone production. *Proc. Natl. Acad. Sci. USA* **99**(12), 8277–8282.

Delcroix, J. D., Valletta, J. S., Wu, C., Hunt, S. J., Kowal, A. S., and Mobley, W. C. (2003). NGF signaling in sensory neurons: Evidence that early endosomes carry NGF retrograde signals. *Neuron* **39,** 69–84.

Hanahan, D., and Weinberg, R. A. (2000). The hallmark of cancer. *Cell* **100,** 57–70.

Hodgson, G., Hager, J. H., Volik, S., Hariono, S., Wernick, M., Moore, D., Nowak, N., Albertson, D. G., Pinkel, D., Collins, C., Hanahan, D., and Gray, J. W. (2001). Genome scanning with array CGH delineates regional alterations in mouse islet carcinomas. *Nat. Genet.* **29**(4), 459–464.

Menasche, G., Pastural, E., Feldmann, J., Certain, S., Ersoy, F., Dupuis, S., Wulffraat, N., Bianchi, D., Fischer, A., Le Deist, F., and de Saint Basile, G. (2000). Mutations in RAB27A cause Griscelli syndrome associated with haemophagocytic syndrome. *Nat. Genet.* **25**(2), 173–176.

Milhavet, O., Gary, D. S., and Mattson, M. P. (2003). RNA interference in biology and medicine. *Pharmacol. Rev.* **55,** 629–648.

Mor, O., Nativ, O., Stein, A., Novak, L., Lehavi, D., Shiboleth, Y., Rozen, A., Berent, E., Brodsky, L., Feinstein, E., Rahav, A., Morag, K., Rothenstein, D., Persi, N., Mor, Y., Skaliter, R., and Regev, A. (2003). Molecular analysis of transitional cell carcinoma using cDNA microarray. *Oncogene* **22,** 7702–7710.

Pinkel, D., Segraves, R., Sudar, D., Clark, S., Poole, I., Kowbel, D., Collins, C., Kuo, W. L., Chen, C., Zhai, Y., Dairkee, S. H., Ljung, B. M., Gray, J. W., and Albertson, D. G. (1998). High resolution analysis of DNA copy number variation using comparative genomic hybridization to microarrays. *Nat. Genet.* **20**(2), 207–211.

Stein, M. P., Dong, J., and Wandinger-Ness, A. (2003). Rab proteins and endocytic trafficking: Potential targets for therapeutic intervention. *Adv. Drug Deliv. Rev.* **55**(11), 1421–1437.

Wang, W., Wyckoff, J. B., Frohlich, V. C., Oleynikov, Y., Huttelmaier, S., Zavadil, J., Cermak, L., Bottinger, E. P., Singer, R. H., White, J. G., Segall, J. E., and Condeelis, J. S. (2002). Single cell behavior in metastatic primary mammary tumors correlated with gene expression patterns revealed by molecular profiling. *Cancer Res.* **62,** 6278–6288.

Weinberg, R. A. (1996). How cancer arises. *Sci. Am.* **272,** 62–70.

[18] Functional Analysis of Rab27a Effector Granuphilin in Insulin Exocytosis

By Tetsuro Izumi, Hiroshi Gomi, and Seiji Torii

Abstract

Granuphilin is specifically expressed on dense-core granules in a defined set of secretory cells such as insulin-producing pancreatic β-cells. It preferentially binds the GTP-bound form of Rab27a and regulates the exocytosis of secretory granules. Furthermore, granuphilin directly interacts with syntaxin-1a, the plasma-membrane-anchored SNARE protein, and with Munc18-1, a Sec1/Munc18 protein. We previously reported evidence that granuphilin mediates the docking of secretory granules onto the plasma membrane through these protein–protein interactions. This chapter details the methods and protocols we use to analyze the function of granuphilin with particular attention to the assays for detecting the expression, protein interactions, and effects on exocytosis of secretory granules in pancreatic β-cells and their derivative cell lines.

Introduction

Granuphilin was originally identified as one of the gene products preferentially expressed in pancreatic β-cells, as compared with α-cells, based on mRNA differential display findings (Wang et al., 1999). Although we do not know why granuphilin shows specific expression in β-cells, we have determined that it interacts with GTP-bound Rab27a located on the membrane of insulin granules (Yi et al., 2002). The domain structure of granuphilin resembles that of a Rab3a effector rabphilin-3 (Shirataki et al., 1993): it has a Rab-binding region with a zinc-finger motif at the N-terminus and C2 domains at the C-terminus (Fig. 1). Two isoforms arise from alternative splicing: granuphilin-a (76 kDa) has two C2 domains, whereas granuphilin-b (57 kDa) has only the first C2 domain of granuphilin-a with a different C-terminal sequence (Wang et al., 1999). Granuphilin also shows an affinity to GTP-bound Rab3a, as previously detected in two-hybrid assays in yeast (Yi et al., 2002) or mammalian cells (Coppola et al., 2002), in a pull-down assay using GST-fusion protein (Coppola et al., 2002), and in coprecipitation assays when overexpressed in COS-7 cells (Kuroda et al., 2002). However, there is no clear evidence for the presence of an endogenous complex between granuphilin and Rab3a in a physiological

METHODS IN ENZYMOLOGY, VOL. 403
0076-6879/05 $35.00
DOI: 10.1016/S0076-6879(05)03018-1

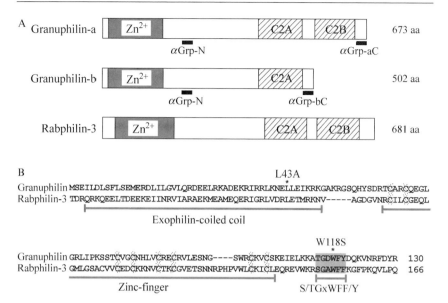

FIG. 1. Domain structures of granuphilin and rabphilin-3. (A) Rab27a effector granuphilin and Rab3a effector rabphilin-3 have a similar domain structure: a Rab-binding region containing a zinc-finger motif at the N-terminus and C2 domains that have an affinity to Ca^{2+} and/or phospholipids at the C-terminus. Two isoforms of granuphilin exist: the larger isoform, granuphilin-a, has two C2 domains, whereas the smaller isoform, granuphilin-b, contains only the first C2 domain. The amino acid numbers shown on the right are those of mouse genes. Gray box, N-terminal homologous Rab-binding region; Zn^{2+}, zinc-finger motif; dashed box, C2 domain. The black bars represent the location of peptides used to raise the anti-granuphilin antibodies. (B) The N-terminal amino acid sequence of granuphilin is aligned with that of rabphilin-3. The putative Rab-binding coiled-coil sequences (Nagashima et al., 2002) are shown. Asterisks indicate the positions of the amino acids mutated. The L43A mutant specifically loses the affinity to syntaxin-1a, whereas the W118S mutant lacks the binding activity to Rab27a (Torii et al., 2002).

state. In contrast, the endogenous complex between granuphilin and Rab27a was easily detected by coimmunoprecipitation in the extracts of a pancreatic β-cell line MIN6 (Yi et al., 2002), a corticotrope cell line AtT-20 (Zhao et al., 2002), and mouse pancreatic islets (H. Gomi, unpublished observations). Further, granuphilin and Rab27a have been shown to be codistributed in sucrose density gradient fractions of MIN6 cell extracts (Yi et al., 2002). The tissue expression pattern of granuphilin, including its absence in the brain, is highly correlated with that of Rab27a but not of Rab3a (Yi et al., 2002). Finally, the intracellular distribution of granuphilin is significantly shifted to the peripheral plasma membrane region in the pancreatic β-cells of Rab27a-mutated *ashen* mice (Kasai *et al.*, 2005). All these findings support the notion that granuphilin physiologically interacts with Rab27a, although the possibility remains that it also functions as a Rab3a effector.

This chapter describes the methods and protocols developed by our laboratory for the functional characterization of granuphilin and its interacting proteins, especially in mouse pancreatic β-cells and their derivative cell line MIN6 (Miyazaki, 1990).

Antibodies Used

Anti-granuphilin Antibodies

We produced three kinds of antibodies toward peptides synthesized from the sequence of mouse granuphilin. The peptides were coupled to *m*-malimidobenzoyl -*N*-hydroxysuccinimide ester-activated keyhole limpet hemocyanin through the cysteine sulfhydryl as described elsewhere (Izumi *et al.*, 1988). Unfortunately, the availability of these antibodies is limited because they have to be affinity purified from the sera to increase the titer and specificity. Affinity purification is performed as follows. One milligram of the peptide is crosslinked to 1 ml of Affi-Gel 10 or 15 (Bio-Rad) according to the manufacturer's instructions. Five to ten milliliters of serum is diluted 2-fold with an equal volume of phosphate-buffered saline (PBS). The diluted serum is passed two times through a column containing Affi-Gel 10 or 15. After the column is washed with buffer (50 mM HEPES [pH 7.4], 0.5 M NaCl) to remove nonspecifically bound proteins, the peptide-specific antibodies are eluted with 0.1 M glycine–HCl, pH 2.5. The fractions are collected into buffer (1 M HEPES [pH 8.0], 0.1 M NaCl) to immediately neutralize the eluate. They are then washed and concentrated three times in buffer (50 mM HEPES [pH 7.4], 0.15 M NaCl) in a Centricon 30 concentrator (Amicon) to a final concentration of 5–25 mg/ml. The relative concentrations of anti-granuphilin antibodies used for different purposes are indicated here because the absolute values vary in each batch of purified materials.

Rabbit anti-granuphilin antibody αGrp-N against a peptide corresponding to amino acids 205–220 of mouse granuphilin-a and -b, (C) SYTADSDST SRRDSLD (Yi *et al.*, 2002), recognizes both granuphilin-a and -b and is used at a 1:500~2000 dilution for immunoblotting and 1:300 for immunofluorescence microscopy. It is not suitable for immunoprecipitation.

Rabbit anti-granuphilin antibody αGrp-aC against the carboxy-terminal granuphilin-a peptide, (C)TLQLRSSMVKQKLGV (Wang *et al.*, 1999), specifically recognizes granuphilin-a and is used at a 1:2000~3500 dilution for immunoblotting and 1:100~250 for immunoprecipitation. It can be used at a 1:300 dilution for immunofluorescence analysis but is not suitable for some fixed tissues or cells.

Rabbit anti-granuphilin antibody αGrp-bC against the carboxy-terminal granuphilin-b peptide, (C)VMAKWWTGWIRLVKK (Torii *et al.*, 2002), specifically recognizes granuphilin-b and is used at a 1:1000 dilution for immunofluorescence and 1:1000 for immunoprecipitation. It is not suitable for immunoblotting.

Anti-Rab27a Antibodies

Anti-Rab27 monoclonal antibody (mouse $IgG_{2\alpha}$, 250 μg/ml) from BD Biosciences Pharmingen specifically recognizes mouse Rab27a but not Rab27b (Zhao *et al.*, 2002). It is used at a 1:1000 dilution for immunoblotting and 1:100 for immunoprecipitation. It can be used at a 1:25~300 dilution for immunofluorescence microscopy but is not suitable for some fixed tissues or cells.

Rabbit anti-Rab27a antibody against glutathione *S*-transferase (GST)-fused mouse Rab27a protein (Yi *et al.*, 2002) recognizes both mouse Rab27a and Rab27b (T. Izumi, unpublished observations). It can be used at a 1:700 dilution for immunoelectron microscopy to detect Rab27a in pancreatic β-cells (Yi *et al.*, 2002), where endogenous Rab27b is not expressed (Kasai *et al.*, 2005).

Anti-syntaxin-1a/1b (HPC-1) monoclonal antibody (mouse ascites) from Sigma is used at a 1:3000 dilution for immunoblotting and 1:500 for immunofluorescence microscopy. It is not suitable for immunoprecipitation.

Anti-Munc 18 polyclonal antibody (rabbit serum) from Synaptic Systems is used at a 1:100 dilution for immunoprecipitation. It is not suitable for immunoblotting. Anti-Munc 18 monoclonal antibody (mouse IgG1, 250 μg/ml) from BD Biosciences Pharmingen is used at a 1:300~5000 dilution for immunoblotting. It is not suitable for immunoprecipitation.

Immunohistochemistry of Granuphilin, Rab27a, and Syntaxin-1a in
    Mouse Pancreatic Islets

The adult mouse is deeply anesthetized with Nembutal (sodium pentobarbital). After washing out the blood via cardiac perfusion of physiological saline at room temperature for 2 min, the mouse is fixed with 50 ml perfusion of 4% paraformaldehyde in 0.1 *M* sodium phosphate buffer, pH 7.4, at 4° for 20 min. The pancreas is removed and postfixed with the same fixative overnight at 4° and is cut into three to four pieces, which are then equilibrated in 30% sucrose in 0.1 *M* sodium phosphate buffer, pH 7.4, overnight at 4° and frozen in Tissue-Tek O.C.T. compound (Sakura

Finetek USA) at $-80°$. Cryosections of the pancreas (20 μm thick) are prepared on a 0.1% gelatin-coated slide glass and dried in cool air overnight. On the next day, the sections are washed with PBS for 5 min and permeabilized with PBS containing 0.1% Triton X-100 for 15 min. Nonspecific reactions are blocked by an incubation with 5% normal goat serum in PBS containing 50 mM NH$_4$Cl for 30 min. Subsequently, the sections are reacted with primary antibodies in the same buffer for 1 h at room temperature and further incubated with an AlexaFluor-labeled secondary antibody (goat anti-rabbit or anti-mouse IgG, 1:1500 dilution, Molecular Probes). The immunofluorescence is viewed using a microscope equipped with an epifluorescence attachment and a charge-coupled device camera.

Immunoblot Analysis of Granuphilin, Rab27a, Syntaxin-1a, and Munc18-1 in Mouse Pancreatic Islets

To evaluate the total protein expression, the isolated pancreatic islets are directly solubilized in 1 μl per islet of 1× sodium dodecyl sulfate (SDS) sample buffer (1% sodium lauryl sulfate, 62.5 mM Tris [pH 6.8], 20% glycerol, 0.005% bromophenol blue, 5% 2-mercaptoethanol). The extracted proteins are separated by SDS–polyacrylamide gel electrophoresis (PAGE). The amount of 5 μg protein per gel lane (equivalent to 5–10 islets) should give a detectable signal in an immunoblot. The protein transferred on an Immobilon-P membrane (Millipore) is blocked with 5% dry milk dissolved in Tris-buffered saline (20 mM Tris [pH 7.4], 150 mM NaCl, 0.05% Tween 20) for 1 h with a gentle shaking. Then the membrane is reacted with primary antibodies overnight at $4°$ followed by peroxidase-labeled secondary antibodies (goat anti-rabbit or anti-mouse IgG, 1:5000 dilution, Jackson ImmunoResearch Lab) for 1 h at room temperature. Antibody detection is accomplished using enhanced chemiluminescent Western blotting detection reagents (Amersham Biosciences).

Protein Interaction Analysis

In vitro and/or *in vivo* interactions of granuphilin with Rab27a, syntaxin-la, or Munc18-1 have been previously reported (Coppola *et al.*, 2002; Torii *et al.*, 2002, 2004; Yi *et al.*, 2002) and can be investigated as described below.

Interaction of In Vitro *Translated Granuphilin with Bacterially Expressed GST-Fused Proteins*

Granuphilin cDNA is cloned in pcDNA3 plasmid (Invitrogen) with hemagglutinin (HA)-tag under the control of a T7 RNA polymerase promoter. One microgram of pcDNA3-HA-granuphilin is transcribed and translated *in vitro* (50 μl reaction volume) using the TNT Coupled Reticulocyte Lysate System (Promega). GST-fused mouse recombinant proteins (Rab27a, syntaxin-la lacking the transmembrane domain [1–264 amino acids], and Munc 18-1) are expressed in *Escherichia coli* XL-1 Blue and affinity-purified with glutathione-Sepharose 4B (Amersham Biosciences). For GST-fused Rab27a, guanine nucleotide can be exchanged on glutathione beads in buffer (50 mM Tris–HCl [pH 7.5], 150 mM NaCl, 2 mM MgCl$_2$, 5 mM EDTA, 0.5 mg of bovine serum albumin/ml) with a 1000-fold molar excess of either GTPγS or GDP at room temperature for 20 min. After the exchange reaction, MgCl$_2$ is added to a final concentration of 7 mM. Fresh sample should be used for the *in vitro* binding assay below, because mouse Rab27a is likely to be unstable with GDP (Yi *et al.*, 2002).

Purified GST-fused proteins (2 μg) immobilized on 20 μl of glutathione-Sepharose beads are incubated with various normalized amounts of *in vitro*-translated proteins ($\sim$10 μl) in 0.2 ml of binding buffer (20 mM Tris [pH 7.5], 150 mM NaCl, 2 mM MgCl$_2$, 1 mM EGTA, 0.1% Nonidet P-40, and the protease inhibitor mixture) at 4° for 1–3 h under gentle rotation. The protease inhibitor mixture described in this chapter is composed of 1 mM phenylmethylsulfonyl fluoride and 5 μg each of aprotinin, pepstatin A, and leupeptin per ml. The detergent Nonidet P-40 can be substituted with the same concentration of Triton X-100. The beads are washed four times with binding buffer and finally dissolved in SDS sample buffer. The bound proteins are analyzed by immunoblotting with rat monoclonal anti-HA (clone 3F10; Roche Diagnostics) or anti-granuphilin antibodies. In these pull-down assays, granuphilin binds to GST-Rab27a (Yi *et al.*, 2002), GST-Munc 18-1 (Coppola *et al.*, 2002; S. Torii, unpublished observations), and GST-syntaxin-1a, but not to GST-syntaxin-2 or syntaxin-3 (Torii *et al.*, 2002).

Pull-Down Assay Using GST-Fused Proteins in Cell Extracts

The interaction of GST-Rab27a with endogenous granuphilin in cell extracts is investigated as follows. MIN6 and AtT-20 cells are grown at 37° in Dulbecco's modified Eagle's medium (DMEM) supplemented with 10% fetal bovine serum, 2 mM glutamine, 100 units/ml penicillin G, and 100 units/ml streptomycin. Cells grown to confluency on 10-cm plastic dishes are lysed in 1.5 ml of lysis buffer (20 mM Tris [pH 7.5], 150 mM NaCl, 5 mM MgCl$_2$, 2 mM EGTA, 0.2% Nonidet P-40, and the protease inhibitor mixture). GST-fused Rab27a proteins (2 μg) immobilized on glutathione beads are incubated with the cell extracts (0.5 mg total proteins for MIN6 cells and 1 mg for AtT-20 cells) at 4° for 3 h with gentle rotation. The beads are washed four times with the lysis buffer. The bound proteins are analyzed by immunoblotting with anti-granuphilin antibodies. In these experiments, granuphilin specifically binds to GST-fused Rab27a and Rab27b, but not Rab3a, in the extracts of AtT-20 cells (Zhao *et al.*, 2002).

The interaction of GST-syntaxin-1a (2 μg) with endogenous granuphilin is similarly examined by incubation with MIN6 cell extracts (0.5 mg total proteins) at 4° for 12 h. In contrast to the wild type, the GST-fused L165A/E166A mutant of syntaxin-1a, which adopts a constantly open conformation *in vitro* (Dulubova *et al.*, 1999), does not bind to granuphilin (Torii *et al.*, 2002). This finding suggests that granuphilin preferentially recognizes the closed form of syntaxin-1a, although the granuphilin-binding domain of syntaxin-1a lies in the H3 region containing the SNARE motif (Torii *et al.*, 2004).

Coimmunoprecipitation in Cultured Cells

The interaction between endogenous granuphilin and Rab27a is examined as follows. MIN6 cell extracts ($\sim$2 $\times$ 10^7 cells on a 10-cm plate) are prepared in 1.5 ml of the lysis buffer described above. Approximately 20% of the cell extracts (0.5 mg protein) are incubated with anti-Rab27a monoclonal antibodies ($\sim$1 μg) at 4° for 2 h and then with 20 μl of 50% protein G-Sepharose (Amersham Biosciences) for 1 h. After the beads are washed four times with lysis buffer, immunoprecipitates are analyzed by immunoblotting using anti-granuphilin antibodies. The interaction can be easily detected because both granuphilin and Rab27a are expressed well in MIN6 and other β-cell lines (Yi *et al.*, 2002). In addition, it appears that Rab27a is stable in an active form or that the activity of guanine nucleotide exchange protein for Rab27a is high in these cells.

For the overexpression of Rab27a mutants, MIN6 cells (1–5 $\times$ 10^6) are infected with recombinant adenoviruses or transfected with expression plasmids encoding epitope (Xpress, HA, or enhanced green fluorescent protein [EGFP])-tagged Rab27a. Prior to infection or transfection, the cells are incubated for 24–48 h on 6-cm dishes. The titers of the viruses and the expression level of the proteins are examined using HEK-293 cells and MIN6 cells, respectively. Twenty microliters of virus stock diluted in 0.5 ml of medium is added to the cells at a multiplicity of infection (moi) of 5–10 plaque-forming units/cell. After incubation for 1 h with occasional agitation, 5 ml of fresh medium is added to the cells. Transfection of Rab27a plasmid is performed using LipofectAMINE 2000 reagent (Invitrogen) according to the manufacturer's instructions. Twelve to fifteen hours later, viruses or transfection reactions are removed from the cells and replaced with fresh medium. Cell extracts are prepared at 24–40 h after infection/transfection, and immunoprecipitation is performed using anti-tag antibodies (anti-Xpress mouse monoclonal: Invitrogen; anti-HA rat monoclonal: Roche Diagnostics; anti-GFP mouse monoclonal: Clontech). Granuphilin associated with each Rab27a protein is determined by immunoblotting with anti-granuphilin antibodies.

The interaction of granuphilin with endogenous syntaxin-1a (35 kDa) or Munc18-1 (68 kDa) is investigated similarly using anti-granuphilin antibody for immunoprecipitation. For detecting the interaction with Munc18-1, however, larger amounts of MIN6 cell extracts are needed (>1 mg of total proteins). In MIN6 cells overexpressing Rab27a, the interaction of granuphilin with syntaxin-1a is stimulated by the wild-type or active form (Q78L) of Rab27a, but not by the inactive form (T23N) of Rab27a (Torii *et al.*, 2002). Because anti-syntaxin-1a antibody suitable for immunoprecipitation is not available, immunoprecipitation is performed with anti-granuphilin antibody or indirectly with antibodies against epitope tags attached to Rab27a proteins. Alternatively, MIN6 cells that stably express myc-tagged syntaxin-1a are used (Torii *et al.*, 2002), in which the extracts are incubated with 3 μl of anti-c-myc mouse monoclonal antibodies (clone 9E10; Roche Diagnostics) at 4° for 12 h for immunoprecipitation.

Coimmunoprecipitation of Endogenous Proteins in Pancreatic Islets

The isolated islets are pooled and frozen at −80° until cell lysis. Protein is extracted from 400–600 islets pools in 1 μl per islet of lysis buffer containing 20 mM Tris (pH 7.5), 150 mM NaCl, 2 mM MgCl$_2$, 1 mM EGTA, 0.2~1% Triton X-100, and 1 × protease inhibitor cocktail (Complete mini; Roche Diagnostics) with vigorous pipeting for 3 min on ice. To detect the optimal interaction of granuphilin with syntaxin-1a, the lysis buffer containing 0.2% Triton X-100 should be used, although 1% Triton X-100-containing buffer can be used for the interaction with Rab27a. The protein lysate is further incubated on ice for 30 min with occasional mixing of lysate. After the insoluble materials are sedimented by centrifugation at 15,000 rpm for 30 min, the supernatant is collected and examined for the protein concentration in a small aliquot (5–10 μl) by the Coomassie brilliant blue dye method (Bio-Rad protein assay kit). Although the amount of protein depends on the islet size and the concentration of Triton X-100 in the islet lysis buffer, approximately 0.2–1.0 μg/islet protein can be recovered. Protein extracts (approximately 300–500 μg) are incubated with 30 μl of 50% slurry of protein G-Sepharose for 3–4 h at 4° under gentle rotation, and then centrifuged at 7000 rpm for 3 min. A small aliquot (20–30 μl) of the precleared lysate is taken and mixed with an equal volume of 2 × SDS sample buffer to prepare the original total protein samples. The islet extracts are incubated with affinity-purified αGrp-aC (2 μl) or

αGrp-bC (2 μl), anti-Munc18 serum (5 μl), or monoclonal anti-Rab27a antibody (5 μl) overnight at 4°. The immune complexes are captured by the addition of 20 μl of 50% protein G-Sepharose for 2–3 h at 4° under gentle rotation. The immunoprecipitates are washed four times with lysis buffer (centrifuged at 7000 rpm for 3 min). After the final wash, protein G-Sepharose and residual wash buffer adjusted to 20 μl are dissolved in 20 μl of 2 × SDS sample buffer. The immunoprecipitates and original total protein samples are analyzed by immunoblotting as previously described. The interaction between granuphilin and syntaxin-1a is specifically reduced in the islets of Rab27a-mutated *ashen* mice (Kasai *et al.*, 2005).

Secretion Assay

Recombinant adenoviruses bearing Rab27a with an Xpress-tag, granuphilin-a, granuphilin-b, and their mutant cDNAs are prepared and aliquots of viruses are stored at −80°. MIN6 cells are seeded at 5 × 10^5 cells in 12-well dishes (2.5-cm plates). Thirty to forty hours later, the cells are infected with recombinant adenoviruses (moi: 5∼10) in 150 μl of medium at 37°. After being cultured in media without viruses for an additional 24 h, the cells are incubated for 2 h in modified Krebs–Ringer buffer (120 mM NaCl, 5 mM KCl, 24 mM NaHCO$_3$, 1 mM MgCl$_2$, 2 mM CaCl$_2$, 15 mM HEPES [pH 7.4], 0.1% bovine serum albumin, 2 mM glucose). The cells are then incubated for 30 min in the same buffer or the buffer modified to include high K$^+$ (60 mM KCl–65 mM NaCl). Insulin secreted in the buffer is measured in duplicate by a radioimmunoassay (EIKEN Chemical). As shown in Fig. 2, overexpression of Rab27a significantly enhances the high K$^+$-induced insulin secretion (Yi *et al.*, 2002), whereas granuphilin overexpression profoundly inhibits high K$^+$-induced insulin secretion (Torii *et al.*, 2002). Although the precise mechanism for the opposite effects of overexpressed Rab27a and granuphilin on evoked insulin secretion remains unknown, it is likely to relate to the nature of the coupling machinery between docking and fusion in regulated exocytosis, where constitutive fusion is prohibited without an appropriate external secretagogue.

Targeting Assay

Rab27a and its effector granuphilin seem to mediate the docking of insulin granules onto the plasma membrane (Kasai *et al.*, 2005; Torii *et al.*, 2004). Overexpression of granuphilin redistributes insulin granules to the peripheral plasma membrane region in MIN6 cells (Torii *et al.*, 2004), which likely represents a physiological docking event of insulin granules

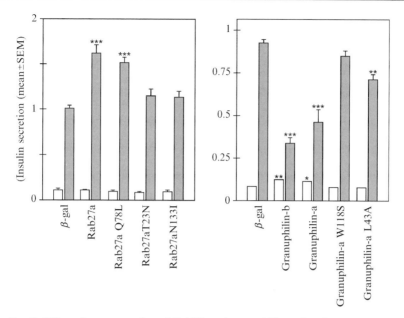

FIG. 2. Effect of overexpression of Rab27a and granuphilin on insulin secretion. MIN6 cells are infected with recombinant adenoviruses bearing β-galactosidase, Rab27a, or granuphilins cDNA. The cells are incubated for 30 min in either Krebs–Ringer buffer (open bars) or that modified to include 60 mM KCl (solid bars). Values are normalized to the release of insulin from uninfected MIN6 cells stimulated by high K$^+$. The results are given as means ± standard errors of the means. *$p < 0.05$; **$p < 0.005$; ***$p < 0.0005$, versus high K$^+$ stimulated MIN6 cells infected with the same titer of the virus bearing β-galactosidase cDNA. Overexpression of wild-type Rab27a or Rab27a Q81L enhances high K$^+$-evoked insulin secretion, whereas that of wild-type granuphilin-a or -b reduces it. Note that the inhibitory effect is either completely (W118S) or partially (L43A) lost in cells expressing mutant granuphilin. The effect of Rab27aT23N or Rab27aN133I cannot be properly assessed because these two mutants are extremely difficult to express in MIN6 cells. Figures are modified from the data of Yi *et al.* (2002) and Torii *et al.* (2002), with permission.

to the plasma membrane. This phenomenon can be observed either morphologically or biochemically.

Morphological Assay

The intracellular distribution of insulin granules is monitored in MIN6 cells stably expressing phogrin-EGFP, where phogrin-EGFP is coimmuno-localized with anti-insulin antibodies (60–90%) (Torii *et al.*, 2004). Usage of these cells has the advantage of excluding problems associated with cell permeability or immunological background. MIN6/phogrin-EGFP cells are seeded at 5×10^3 cells on eight-well Lab-Tek chamber plastic slides (Nunc). After 3 days, the cells are infected with a recombinant adenovirus bearing granuphilin cDNA ($\sim$10 moi) in 70 μl of medium. The optimum titers are preliminarily determined by immunostaining with anti-granuphilin antibodies (the expression level of exogenous granuphilin examined by immunoblotting is about 3- to 5-fold higher than those of endogenous granuphilin). The cells are cultured in DMEM without serum for 12 h and then fixed with 4% paraformaldehyde in phosphate buffer. After washing with 0.1 M sodium phosphate buffer, pH 7.4, containing 0.5 M NaCl and 0.1% Tween 20, the cells are covered with mounting medium (S3023; DAKO) and processed for microscopic analysis. Intrinsic EGFP fluorescence is observed with an epifluorescence microscope equipped with a charge-coupled device camera. As shown in Fig. 3, overexpression of granuphilin induces translocation of EGFP-labeled secretory granules toward the plasma membrane (Torii *et al.*, 2004).

A peripheral redistribution of EGFP signals is also observed in cells transfected with plasmid by conventional liposomes. MIN6/phogrin-EGFP cells on a chamber slide are transiently transfected with 0.1 μg of an expression plasmid encoding HA-granuphilin. The cells are cultured in DMEM for 24 h. They are then fixed with 4% paraformaldehyde, permeabilized with 0.05% Triton X-100 for 10 min three times at room temperature, and incubated with anti-HA rat monoclonal antibodies followed by indocarbocyanine (Cy3)-conjugated anti-rat IgG (1:500 dilution, Jackson ImmunoResearch Lab). EGFP signals reveal a linear distribution along a

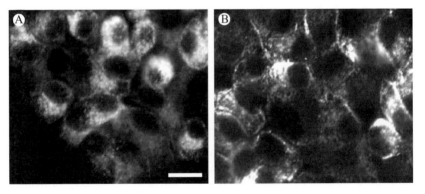

FIG. 3. Overexpression of granuphilin promotes the targeting of insulin granules to the plasma membrane. MIN6/phogrin-EGFP cells are infected with recombinant adenovirus bearing β-galactosidase (A) or granuphilin-a (B) cDNA. The cells are fixed after 12 h, and the EGFP fluorescence is observed. Note that cells infected with the adenovirus encoding granuphilin-a uniformly exhibit a prominent peripheral redistribution of EGFP-labeled granules compared with those expressing control β-galactosidase protein. Bar = 10 μm.

part of the plasma membrane and are mostly overlapped with the localization of HA-granuphilin.

The plasma membrane targeting of secretory granules promoted by wild-type granuphilin is not seen by the mutants that are defective in binding to either Rab27a (granuphilin W118S) or syntaxin-1a (granuphilin L43A) (see Fig. 1) (Torii et al., 2004). These observations suggest that interactions with both Rab27a and syntaxin-1a are required for the targeting activity of granuphilin.

Biochemical Assay

MIN6 cells (2×10^6 cells in a 6-cm dish) are infected with an adenovirus ($\sim$10 moi) bearing either β-gal or granuphilin cDNA. Twenty-four hours later, the cells are suspended in 1 ml of buffer containing 250 mM sucrose, 20 mM HEPES (pH 7.4), 2 mM MgCl$_2$, 2 mM EGTA, and the protease inhibitor mixture. The cells are then homogenized for 40 strokes by the tight-fitting Dounce homogenizer (for 5 ml) with monitoring of the cell condition under a microscope. The total homogenate (fraction T) is centrifuged at 700g for 10 min at 4° to precipitate the nuclear and intact plasma membrane fraction (P1). The resultant supernatant is then centrifuged at 12,000g for 20 min to separate the heavy organelles containing the secretory granules (P2) from the cytoplasmic materials (S). The fractionates are lysed in buffer (20 mM Tris [pH 7.5], 150 mM NaCl, 2 mM MgCl$_2$, 1 mM EDTA, 1% Triton X-100, and the protease inhibitor mixture) on ice for 30 min. Equal proportions of each lysate ($\sim$10 μg proteins in total homogenate) are subjected to SDS–PAGE and immunoblotting analysis. Monoclonal antibody against Rab27a and polyclonal antibodies toward phogrin and secretogranin III (Hosaka et al., 2004) are used to detect a marker for peripherally associated granule membrane protein, integral granule membrane protein, and granule content protein, respectively.

In control cells, the plasma membrane-associated syntaxin-1a is mainly present in the P1 fraction, whereas Rab27a and secretogranin III are distributed in the P2 fraction. Overexpression of granuphilin by the adenovirus induces significant redistributions of Rab27a, secretogranin III, and phogrin from P2 to P1, suggesting that insulin granules are moved into the heavier fraction containing the plasma membrane (Torii et al., 2004).

Comments

Through computational analyses of genome sequences, several other molecules that contain the N-terminal sequences similar to the Rab-binding region of granuphilin, called exophilins or Slp/Slac, have been identified (Izumi *et al.*, 2003; Kuroda *et al.*, 2002; Nagashima *et al.*, 2002). It is very likely that these putative Rab27a/b effectors play versatile roles in various regulated secretory pathways that are differentiated in multicellular organisms.

Acknowledgments

This work was supported by grants-in-aid for scientific research and a grant of the 21st Century Center of Excellence program from the Ministry of Education, Culture, Sports, Science, and Technology of Japan.

References

Coppola, T., Frantz, C., Perret-Menoud, V., Gattesco, S., Hirling, H., and Regazzi, R. (2002). Prancreatic β-cell protein granuphilin binds Rab3 and Munc-18 and controls exocytosis. *Mol. Biol. Cell* 13, 1906–1915.

Dulubova, I., Sugita, S., Hill, S., Hosaka, M., Fernandez, I., Südhof, T. C., and Rizo, J. (1999). A conformational switch in syntaxin during exocytosis: Role for munc18. *EMBO J.* 16, 4372–4382.

Hosaka, M., Suda, M., Sakai, Y., Izumi, T., Watanabe, T., and Takeuchi, T. (2004). Secretogranin III binds to cholesterol in the secretory granule membrane as an adapter for chromogranin A. *J. Biol. Chem.* 279, 3627–3634.

Izumi, T., Saeki, Y., Akanuma, Y., Takaku, F., and Kasuga, M. (1988). Requirement for receptor-intrinsic tyrosine kinase activities during ligand-induced membrane ruffling of KB cells: Essential sites of src-related growth factor receptor kinases. *J. Biol. Chem.* 263, 10386–10393.

Izumi, T., Gomi, H., Kasai, K., Mizutani, S., and Torii, S. (2003). Rab27 and its effectors in the regulated secretory pathways. *Cell Struct. Funct.* 28, 465–474.

Kasai, K., Ohara-Imaizumi, M., Takahashi, N., Mizutani, S., Zhao, S., Kikuta, T., Kasai, H., Nagamatsu, S., Gomi, H., and Izumi, T. (2005). Rab27a mediates the tight docking of insulin granules onto the plasma membrane during glucose stimulation. *J. Clin. Invest.* 115, 388–396.

Kuroda, T. S., Fukuda, M., Ariga, H., and Mikoshiba, K. (2002). The Slp homology domain of synaptotagmin-like proteins 1-4 and Slac2 functions as a novel Rab27a binding domain. *J. Biol. Chem.* 277, 9212–9218.

Miyazaki, J., Araki, K., Yamato, E., Ikegami, H., Asano, T., Shibasaki, Y., Oka, Y., and Yamamura, K. (1990). Establishment of a pancreatic β cell line that retains glucose-inducible insulin secretion: Special reference to expression of glucose transporter isoforms. *Endocrinology* 127, 126–132.

Nagashima, K., Torii, S., Yi, Z., Igarashi, M., Okamoto, K., Takeuchi, T., and Izumi, T. (2002). Melanophilin directly links Rab27a and myosin Va through its distinct coiled-coil regions. *FEBS Lett.* 517, 233–238.

Shirataki, H., Kaibuchi, K., Sakota, T., Kishida, S., Yamaguchi, T., Wada, K., Miyazaki, M., and Takai, Y. (1993). Rabphilin-3A, a putative target protein for *smg* p25A/*rab3A* p25 small GTP-binding protein related to synaptotagmin. *Mol. Cell. Biol.* 13, 2061–2068.

Torii, S., Zhao, S., Yi, Z., Takeuchi, T., and Izumi, T. (2002). Granuphilin modulates the exocytosis of secretory granules through interaction with syntaxin 1a. *Mol. Cell. Biol.* **22,** 5518–5526.

Torii, S., Takeuchi, T., Nagamatsu, S., and Izumi, T. (2004). Rab27 effector granuphilin promotes the plasma membrane targeting of insulin granules via interaction with syntaxin 1a. *J. Biol. Chem.* **279,** 22532–22538.

Wang, J., Takeuchi, T., Yokota, H., and Izumi, T. (1999). Novel rabphilin-3-like protein associates with insulin-containing granules in pancreatic beta cells. *J. Biol. Chem.* **274,** 28542–28548.

Yi, Z., Yokota, H., Torii, S., Aoki, T., Hosaka, M., Zhao, S., Takata, K., Takeuchi, T., and Izumi, T. (2002). The Rab27a/granuphilin complex regulates the exocytosis of insulin-containing dense-core granules. *Mol. Cell. Biol.* **22,** 1858–1867.

Zhao, S., Torii, S., Yokota-Hashimoto, H., Takeuchi, T., and Izumi, T. (2002). Involvement of Rab27b in the regulated secretion of pituitary hormones. *Endocrinology* **143,** 1817–1824.

[19] Assays for Functional Properties of Rab34 in Macropinosome Formation

By Peng Sun and Takeshi Endo

Abstract

We have shown that Rab34/Rah participates in the promotion of macropinosome formation. Here we describe procedures for the analyses of intracellular localization and some functional properties of Rab34. Rab34 lacks a consensus sequence of the fourth motif for GTP/GDP binding and GTPase activities. Indeed, GTPase assay shows that wild-type Rab34 has extremely weak GTPase activity *in vitro*. However, Rab34 exhibits appreciable GTPase activity *in vivo* probably due to the presence of specific GTPase-activating protein (GAP) activity in cells. Specific intracellular localization of Rab34 is easily detected by the expression of epitope-tagged or enhanced green fluorescent protein (EGFP)-tagged protein. It is colocalized with actin filaments to membrane ruffles and membranes of nascent macropinosomes, which are formed from the ruffles. By contrast, Rab5 is not associated with the ruffles or nascent macropinosomes but present in endosomes at later stages. The function of Rab34 in macropinosome formation is analyzed by the transfection of wild-type, constitutively active, and dominant-negative mutants of Rab34 in fibroblasts followed by treatment with platelet-derived growth factor (PDGF) or phorbol ester. These analyses indicate that Rab34 is required for efficient macropinosome formation.

METHODS IN ENZYMOLOGY, VOL. 403   0076-6879/05 $35.00
                 DOI: 10.1016/S0076-6879(05)03019-3

Introduction

Endocytosis in eukaryotic cells serves to maintain cellular and organismal homeostasis by taking up fluids and macromolecules, signaling molecules, and their receptors from the external environment (Mellman, 1996). There are at least five endocytic pathways: clathrin-dependent endocytosis mediated by clathrin-coated vesicles (100–150 nm in diameter), caveolin-dependent endocytosis mediated by caveolae (50–80 nm), clathrin/caveolin-independent endocytosis, macropinocytosis, and phagocytosis (Johannes and Lamaze, 2002; Robinson et al., 1996). Among them, macropinocytosis is carried out with relatively large macropinosomes (0.2–5 μm in diameter) formed from cell surface membrane ruffles folding back on the plasma membrane (Cardelli, 2001; Swanson and Watts, 1995). Macropinosomes are not coated with clathrin or caveolin but surrounded by actin at early stages. Macropinocytosis provides an efficient way for nonselective uptake of nutrients and solute macromolecules. It also serves for internalization of extracellular antigens by professional antigen-presenting cells such as dendritic cells. Furthermore, macropinocytosis is likely to play important roles in chemotaxis by regulating plasma membrane–actin cytoskeleton interaction and membrane trafficking. Some pathogenic bacteria, including *Salmonella typhimurium* and *Shigella flexneri*, also exploit macropinocytosis to invade the cells.

Treatment of various types of cultured cells with growth factors, cytokines, phorbol esters like phorbol 12-myristate 13-acetate (PMA), or diacylglycerol elicits a rapid and dramatic membrane ruffling and macropinocytosis (Swanson and Watts, 1995). Introduction of small GTPases, Ras or Rac1, or Tiam1, a guanine nucleotide exchange factor (GEF) for Rac1, also induces membrane ruffling and macropinocytosis in fibroblasts (Bar-Sagi and Feramisco, 1986; Michiels et al., 1995; Ridley et al., 1992). The membrane ruffling and macropinosome formation induced by growth factors, PMA, Ras, or Tiam1 are mediated by Rac1. Rac1 causes membrane ruffling by activating WAVE proteins, which activate the Arp2/3 complex involved in the formation of branched actin filament meshwork in membrane ruffles (Higgs and Pollard, 2001; Takenawa and Miki, 2001).

Rab family small GTPases play essential roles in the endocytic and exocytic processes of transport vesicle formation, motility, docking, and fusion (Rodman and Wandinger-Ness, 2000; Segev, 2001; Zerial and McBride, 2001). In mammalian cells, >60 members of the Rab family proteins have been identified. These proteins are associated with particular vesicle membrane compartments and function in specific stages of the diverse vesicle trafficking events. Among them, Rab5 is located on the membranes of clathrin-coated vesicles and early endosomes. It is involved in receptor-mediated endocytosis and fluid-phase pinocytosis. By contrast,

Rab34 is colocalized with actin to the membrane ruffles and macropinosome membrane (Sun et al., 2003). During macropinocytosis, Rab34 is associated with nascent macropinosomes and replaced by Rab5 at later stages. Overexpression of Rab34 elevates the number of macropinosomes, whereas the expression of a dominant-negative Rab34 prevents macropinosome formation induced by platelet-derived growth factor (PDGF) or PMA. On the other hand, Rab34-promoted macropinosome formation is impaired by dominant-negative mutants of Rac1 and WAVE2. Accordingly, Rab34 plays a crucial role in facilitating the formation of macropinosomes from the membrane ruffles. Here we describe the procedures for analyses of the localization and the roles of Rab34 in macropinosome biogenesis.

Expression and Purification of Recombinant Rab34

To analyze biochemical properties and cellular functions of small GTPases, constitutively active mutants, which are unable to hydrolyze bound GTP, and dominant-negative mutants, which cannot bind GTP and sequester specific GEFs in cells, are useful. These mutants are produced by introducing point mutations into cDNAs by analogy with constitutively active Ras(G12V) (Gly-12 is substituted for Val) or Ras(Q61L) and dominant-negative Ras(S17N). We can easily introduce point mutations into cDNAs in a plasmid vector by utilizing polymerase chain reaction (PCR). Wild-type (wt) or point mutated cDNAs subcloned in bacterial expression vectors are used for the preparation of recombinant proteins.

Introduction of Point Mutations into the cDNA

1. Insert the mouse Rab34 cDNA (Sun et al., 2003; DDBJ/EMBL/GenBank accession number AB082927) into the NotI–XhoI sites of pBluescript II vector (Stratagene). Introduce the constitutively active Rab34(Q111L) or the dominant-negative Rab34(T66N) mutation by PCR using high-fidelity ProofStart DNA polymerase (Qiagen). Design sense and antisense primers so that their 5' ends can attach and the PCR product will cover all the sequence of the cDNA and the plasmid (Fig. 1). Either the sense or the antisense primer contains a mutation. The primers are

Rab34(Q111L) sense primer	5'-ACGGCTGGTCTGGAAAGGTTC-3'
Rab34(Q111L) antisense primer	5'-GTCCCAAAGTTGGAGACTGAA-3'
Rab34(T66N) sense primer	5'-GTGGGGAAGAACTGTCTCATT-3'
Rab34(T66N) antisense primer	5'-AGATAGGTTCCCACAACGAT-3'

(The underlined nucleotides are substituted ones to introduce point mutations.)

2. Prepare a master mix and then add the template DNA according to the manufacturer's instruction manual of ProofStart DNA polymerase.

3. Conduct the PCR by the following step program: initial activation at 94° for 3 min; 35 cycles of three-step cycling (at 94° for 1 min, at 59° for 1 min, at 72° for 2 min); and a final extension at 72° for 10 min.

4. Because ProofStart DNA polymerase generates blunt ends, treatment of the PCR products with Klenow fragment to ensure the blunt end formation may not be required. Phosphorylate the 5′ ends of the PCR product with T4 polynucleotide kinase, and then ligate the ends of the DNA with T4 DNA ligase.

5. Transform *Escherichia coli* strain XL1-Blue with the ligated DNA and select the plasmids containing correctly mutated Rab34 cDNA.

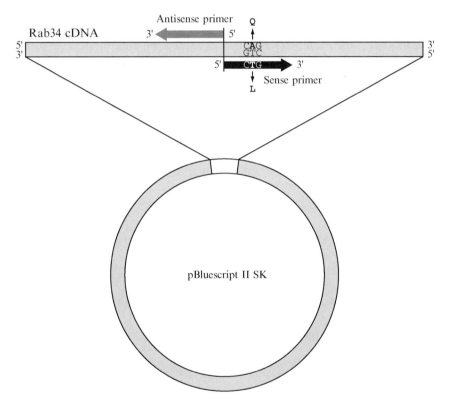

Fig. 1. A schematic representation for the introduction of a point mutation into Rab34 cDNA by PCR. The sense and the antisense primers are designed to attach at their 5′ ends. In this case, the sense primer contains a mutation. The nucleotide sequence CAG encoding Gln-111 is replaced with CTG coding for Leu in the sense primer.

Expression and Preparation of Recombinant Proteins

1. Digest Rab34 cDNAs (wt, Q111L, and T66N) in pBluescript II with *Sma*I and *Xho*I. Recover the Rab34 cDNAs by low-melting temperature agarose gel electrophoresis or other appropriate methods. Fill the protruding *Xho*I site with DNA polymerase I Klenow fragment.

2. Ligate the blunt-ended Rab34 cDNAs in the *Sma*I site of pGEX-3X (Amersham Biosciences) so that the glutathione *S*-transferase (GST) and Rab34 are in frame. Transform XL1-Blue with the pGEX/Rab34 recombinant plasmid.

3. Inoculate 500 ml of LB medium containing 50 μg/ml ampicillin in a 2-liter flask with 10 ml of overnight culture of the pGEX/Rab34-transformed *E. coli*. Culture the *E. coli* for 3 h at 37° in a bacterial shaker. To induce GST-tagged Rab34, add isopropyl β-D-1-thiogalactopyranoside (IPTG) to 10 μM (50 μl of 0.1 M stock solution), and culture for a further 12–16 h at 20–25°.

4. Collect the *E. coli* cells by centrifugation at 3000 rpm for 30 min at 4°. Resuspend the cells in wash buffer (50 mM Tris–HCl, pH 7.5, 150 mM NaCl, 5 mM MgCl$_2$), and collect the cells by centrifugation. Resuspend the collected cells in 4 ml of cold lysis buffer (50 mM Tris–HCl, pH 7.5, 150 mM NaCl, 5 mM MgCl$_2$, 1% Triton X-100, 0.5% Na-deoxycholate, 1 mM dithiothreitol [DTT], and 1 mM phenylmethylsulfonyl fluoride [PMSF]).

5. Lyse the *E. coli* cells by sonication (3 × 15 s) on ice with a Tomy Seiko UD-201 sonicator equipped with a microprobe at an output of 5. Centrifuge the lysate at 10,000 rpm for 20 min at 4°. Transfer the supernatant to a Falcon 15-ml conical tube.

6. Prewash a 150 μl bed volume of glutathione–Sepharose 4B (Amersham Biosciences) with ~10 volumes of the lysis buffer three times and keep as a 1:1 suspension. Add the suspension to the supernatant and mix the sample for 1 h at 4° with rotation. Centrifuge the sample at 500 rpm for 3 min at 4° in a swing rotor and discard the supernatant. Wash the beads with 10 ml of the lysis buffer by centrifugation three times.

7. Add 1 ml of elution buffer (50 mM Tris–HCl, pH 7.5, 150 mM NaCl, 5 mM MgCl$_2$, 1 mM DTT, 1 mM PMSF, and 10 mM glutathione), transfer the sample to a 1.5-ml microcentrifuge tube, and incubate for 10 min at 4° with rotation. Microcentrifuge the sample at 1000 rpm for 5 min at 4° in a swing rotor and collect the supernatant containing eluted GST-tagged protein. Repeat this procedure, and combine the supernatants.

8. Dialyze the supernatants against dialysis buffer (20 mM Tris–HCl, pH 7.5, 10 mM NaCl, 2 mM MgCl$_2$, and 0.1 mM DTT) for 8 h three times at 4°. Dialyzed proteins may be stored at −80° after freezing in liquid nitrogen.

GTP Binding/GTPase Assay by Thin Layer Chromatography

Because Rab34 lacks the consensus sequence of the fourth motif for GTP/GDP binding and GTPase activities (Sun *et al.*, 2003) (Fig. 2A), its GTP/GDP binding or GTPase activity is expected to be impaired. Indeed, GTPase assay *in vitro* using the bacterially expressed recombinant proteins shows that wt Rab34 binds GTP but has extremely weak GTPase activity (Fig. 2C). By contrast, wt Rab34 exhibits appreciable GTPase activity *in vivo* (Fig. 2D). This is probably due to the presence of GAP

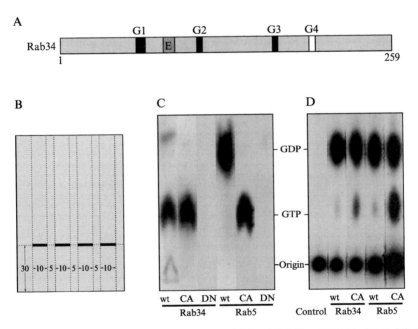

FIG. 2. Assays for GTP binding and GTPase activities of Rab34 and Rab5 by TLC. (A) A schematic drawing of Rab34 protein. G1–G4, domains for GTP/GDP binding and GTPase activities; E, core effector domain. Rab34 contains conserved G1–G3 sequences but lacks a conserved sequence for G4 (AVSA instead of the conserved sequence E*X*SA). (B) A TLC plate indicating guide vertical and horizontal lines (broken lines) and sample application lines (thick horizontal lines). Numbers indicate the length in millimeters. (C) *In vitro* GTP binding/GTPase assay. GST-tagged wt, constitutively active mutants (CA), and dominant-negative mutants (DN) of Rab34 and Rab5 are loaded with [α-^{32}P]GTP and incubated for 30 min at 37°. Bound GTP/GDP are analyzed by TLC. The positions of sample application origin, GTP, and GDP are indicated. (D) *In vivo* GTP binding/GTPase assay. Myc-tagged wt and constitutively active mutants (CA) of Rab34 and Rab5 as well as the empty vector (control) are transiently transfected to COS-1 cells. The cells are incubated with ^{32}P$_i$ for 12 h to label the proteins. Each of these proteins is immunoprecipitated with the anti-Myc mAb and the bound GTP/GDP are analyzed by TLC.

acting on Rab34 in cells. Thus, GTPase assay not only *in vitro* but also *in vivo* is required to understand correctly the regulatory mechanisms of the activity of small GTPases. Since thin-layer chromatography (TLC) can visualize the GTP/GDP binding and GTPase activities, TLC is a powerful tool for the analysis of these activities.

GTP Binding/GTPase Assay In Vitro

1. Prewash 100 μl bed volume of glutathione–Sepharose 4B with the dialysis buffer (20 mM Tris–HCl, pH 7.5, 10 mM NaCl, 2 mM MgCl$_2$, and 0.1 mM DTT) and suspend in 100 μl of the same buffer. Add ~10 μg of the affinity-purified GST-tagged protein and mix the sample for 1 h at 4° with rotation. Microcentrifuge the sample at 500 rpm for 3 min at 4° with a swing rotor and discard the supernatant. Wash the beads with 1 ml of the dialysis buffer five times by microcentrifugation.

2. Resuspend in 100 μl of incubation buffer (50 mM Tris–HCl, pH 7.5, 50 mM NaCl, 5 mM EDTA, 0.1 mM EGTA, 0.1 mM DTT, and 10 μM ATP). Add 370 kBq of [α-^{32}P]GTP (NEN, NEG-006H, or NEG-506H, 111 TBq/mmol) and incubate at 37° for 10 min with mild shaking. Wash the sample with 1 ml of ice-cold GTPase wash buffer (50 mM Tris–HCl, pH 7.5, 10 mM NaCl, 20 mM MgCl$_2$, 1 mM DTT, and 1 mg/ml bovine serum albumin) three times by microcentrifugation at 500 rpm for 3 min at 4°.

3. Add 500 μl of assay buffer (50 mM Tris–HCl, pH 7.5, 50 mM NaCl, 5 mM MgCl$_2$, 5 mM DTT, and 1 mM GTP) and allow GTPase reaction at 37° for appropriate times (e.g., 0–30 min) with mild shaking. Stop the reaction by the addition of 500 μl of the ice-cold GTPase wash buffer. Wash with 1 ml of ice-cold GTPase wash buffer three times by microcentrifugation.

4. Dissociate bound nucleotides by incubating the Sepharose beads in 20 μl of dissociation buffer (20 mM Tris–HCl, pH 7.5, 20 mM EDTA, 2% SDS, 0.5 mM GDP, and 0.5 mM GTP) at 65° for 5 min. Collect the dissociated sample by microcentrifugation at 1500 rpm for 5 min at 20°.

5. Draw vertical lines with a razor blade and a horizontal line with a soft pencil on polyethyleneimine cellulose plates (Macherey-Nagel, 20 × 20 cm) (Fig. 2B). Apply 10 μl each of the dissociated samples 10 mm wide on the horizontal line by spotting with a Pipetman P20. After applying all the samples, soak the plate in methanol and air-dry it.

6. Immerse ~1 cm of the bottom of the plate into methanol. Place the plate in a chromatography chamber containing 0.75 M KH$_2$PO$_4$-H$_3$PO$_4$ buffer (pH 3.4) at a depth of ~1 cm. Close the chamber and leave the plate until the buffer reaches the top of the plate (for 2–3 h).

7. Air-dry the plate. Wrap the plate with Saran wrap and expose a Fuji BAS imaging plate or an x-ray film to it. Visualize the radioactivity by autoradiography. Figure 2C shows GTP/GDP binding and GTPase activities of Rab34 and Rab5 *in vitro*.

GTP Binding/GTPase Assay In Vivo

1. Insert the entire coding region of the wt or mutated Rab34 cDNA into the *Bam*HI site of the pEF-BOS vector (Mizushima and Nagata, 1990) in frame with the Myc-encoding sequence (amino acid sequence, MEQKLISEEDL). Alternatively, insert the cDNA into the multiple cloning site (MCS) of the pCMV-Myc vector (BD Biosciences Clontech) in frame with the Myc-encoding sequence.

2. About 24 h before transfection, seed 5×10^5 African green monkey kidney COS-1 cells in 10 ml of Dulbecco's modified Eagle's (DME) medium containing 10% fetal bovine serum (FBS) (growth medium) on a 100-mm culture dish ($\sim$10% confluent).

3. Transfect the cells with these plasmids by using 12 μl of FuGENE 6 transfection reagent (Roche) according to the manufacturer's instruction manual.

4. Dialyze FBS against 0.15 M NaCl and 10 mM HEPES-NaOH (pH 7.9) to remove phosphate. Twelve hours after the transfection, replace the medium with 4 ml of phosphate-free DME medium (Sigma, D3656) supplemented with the dialyzed FBS at a concentration of 10%. Add 37 MBq (9.25 MBq/ml) of phosphorus-32 ($^{32}P_i$) (NEN, NEX-053H) to the medium and incubate for further 12 h.

5. Before harvesting and lysing the cells, wash a 25 μl bed volume of Protein G Sepharose 4 Fast Flow (Amersham) with 1 ml of lysis buffer (50 mM Tris–HCl, pH 7.5, 150 mM NaCl, 20 mM MgCl$_2$, 0.5% Nonidet P-40, 1 mM Na$_3$VO$_4$, and 20 μg/ml aprotinin) three times by microcentrifugation. Add 200 μl of the hybridoma culture supernatant of anti-Myc monoclonal antibody (mAb) Myc1-9E10 (American Type Culture Collection) and incubate at 4° for 60 min with mild shaking. Wash the beads with 1 ml of the lysis buffer three times by microcentrifugation.

6. Remove the COS-1 cell culture medium and wash the cells with 10 ml of ice-cold phosphate-free DME medium three times. Add 3 ml of ice-cold phosphate-free DME medium. Harvest the cells with a rubber policeman and transfer the cells to 2$\times$ microcentrifuge tubes. Collect the cells by microcentrifugation at 1500 rpm for 3 min at 4°.

7. Lyse the cells with 200 μl of the lysis buffer by pipetting with a Pipetman P200. Microcentrifuge at 3000 rpm for 3 min at 4° and collect the supernatant in another microcentrifuge tube.

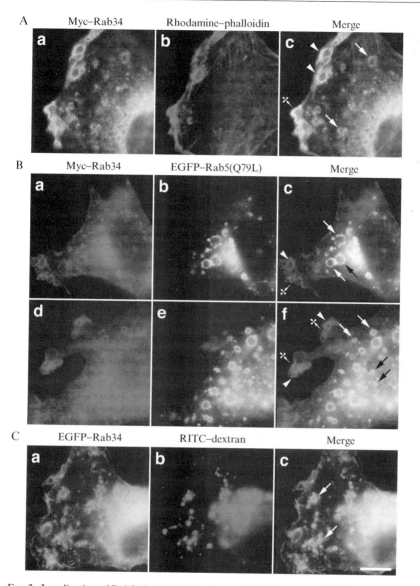

Fig. 3. Localization of Rab34 by epitope tagging and EGFP tagging. (A) Colocalization of Rab34 with actin to membrane ruffles and membranes of macropinosomes adjacent to the ruffles. (a) Myc–Rab34 detected with Alexa 488 fluorescence. (b) Actin detected by rhodamine–phalloidin staining. (c) Merged image. A dagger and arrowheads point to membrane ruffles and macropinosomes, respectively, where Rab34 and actin coexist. Rab34 remains associated with some macropinosomes (arrows) away from the ruffles, but actin is gradually dissociated from these macropinosomes. (B) Partial coexistence of Rab34 and Rab5

8. Add the anti-Myc mAb-conjugated protein G-Sepharose suspended in 200 μl of the lysis buffer to the cell lysate and incubate at 4° for 60 min with mild shaking. Wash the beads with 1 ml of the lysis buffer three times by microcentrifugation.

9. Dissociate bound nucleotides by incubating the beads in 20 μl of the dissociation buffer at 65° for 5 min. Collect the dissociated sample by microcentrifugation at 1500 rpm for 5 min at 20°.

10. Apply 10 μl each of the dissociated samples to TLC as described earlier. Visualize the radioactivity by autoradiography. Figure 2D shows GTP/GDP binding and GTPase activities of Rab34 and Rab5 *in vivo*.

Localization by Epitope Tagging and Enhanced Green Fluorescent Protein (EGFP) Tagging

Since Rab family proteins are generally involved in particular vesicle trafficking, determination of their intracellular localization is useful to infer their functions. Even if antibodies specific for particular Rab proteins are not available, epitope tagging or EGFP tagging can be used to determine their localization. In addition, application of multiple epitope tags such as Myc, hemagglutinin (HA), and FLAG or a combination of epitope tagging and EGFP tagging can determine the localization of multiple proteins in a cell. For instance, expression of Myc-tagged Rab34 and EGFP-tagged Rab5 shows that Rab34 and Rab5 coexist in a subset of macropinosome membranes but that they also locate to specific sites (Sun *et al.*, 2003) (Fig. 3A and B). Rab34 is located on the membrane ruffles or lamellipodia and nascent macropinosome membranes adjacent to the ruffles, whereas Rab5 is associated with the membranes of endosomes at later stages. These findings suggest that Rab34 and Rab5 act at early and later stages, respectively, during macropinosome dynamics and that Rab34 is replaced by Rab5 on the macropinosome membrane.

Large (several micrometers in diameter) macropinosomes are easily detected even by phase-contrast images with a 63× or a 100× objective lens. However, phase-contrast microscopy cannot correctly identify all the macropinosomes. Visualization of macropinosomes by taking advantage of

in macropinosome membrane. (a, d) Myc–Rab34 detected with Alexa 546 fluorescence. (b, e) EGFP–Rab5(Q79L). (c, f) Merged image. Daggers, arrowheads, white arrows, and black arrows point to Rab34-associated membrane ruffles, macropinosomes exclusively containing Rab34, those containing both Rab34 and Rab5, and vesicles exclusively containing Rab5, respectively. (C) Incorporation of RITC–dextran into Rab34-associated macropinosomes. (a) EGFP–Rab34. (b) Incorporated RITC–dextran. (c) Merged image. Arrows point to Rab34-associated macropinosomes incorporating RITC–dextran. Bar = 20 μm.

the incorporation of fluorescent dextran into macropinosomes is useful to detect most macropinosomes correctly in combination with the EGFP tagging (Fig. 3C).

Epitope Tagging and EGFP Tagging

1. For epitope tagging, insert the wt or mutated Rab34 or Rab5 cDNA into the *Bam*HI site of the pEF-BOS vector in frame with the Myc-encoding sequence or the 3 × HA-encoding sequence (amino acid sequence, 3 × MYPYDVPDYAGS). Alternatively, insert the cDNA into the MCS of pCMV-Myc vector or pCMV-HA vector (BD Biosciences Clontech) in frame with the Myc- or HA-encoding sequence.

2. For EGFP tagging, insert the wt or mutated Rab34 or Rab5 cDNA into the MCS of the pEGFP-C1 vector (BD Biosciences Clontech) in frame with the EGFP-encoding sequence.

3. Sterilize glass coverslips (18 × 18 mm, No. 1) in ethanol and air-dry them. Place three coverslips in a 60-mm culture dish. About 24 h before transfection, seed 2×10^5 mouse C3H/10T1/2 (10T1/2) fibroblasts in 4 ml of DME medium containing 10% FBS (growth medium) on the coverslips in the dish (~10% confluent).

4. Transfect the cells with these plasmids (e.g., cotransfection of pEF-BOS/Myc-Rab34 or pEF-BOS/HA-Rab34 together with pEGFP-C1/Rab5) by using 6 μl of FuGENE 6 transfection reagent (Roche) according to the manufacturer's instruction manual.

5. Twenty-four hours after the transfection, remove the medium and fix the cells with 4 ml of 4% paraformaldehyde in 0.1 M NaPO$_4$ buffer (pH 7.4) for 15 min at room temperature. Remove the fixative and permeabilize the cells with 4 ml of 0.2% Triton X-100 in phosphate-buffered saline (PBS) containing 0.02% NaN$_3$ (PBS/NaN$_3$) for 5 min. Remove the permeabilization solution and wash the cells by leaving for 3 min in 4 ml of PBS/NaN$_3$ four times.

6. Place the coverslip in a 35-mm dish, add 100 μl of a primary antibody (mAb Myc1-9E10 for Myc-tag staining or anti-HA polyclonal antibody [pAb] [Sigma] for HA-tag staining), and incubate for 50 min at room temperature. Remove the antibody and wash the cells by leaving for 3 min in 2 ml of PBS/NaN$_3$ four times.

7. Place the coverslip in another 35-mm dish, add 100 μl of fluorescently labeled secondary antibody (e.g., Alexa 546-conjugated goat anti-mouse IgG or anti-rabbit IgG [Molecular Probes]), and incubate for 50 min at room temperature. Remove the antibody and wash the cells as above.

7′. To see the colocalization of Rab34 and actin in the cells transfected with pEF-BOS/Myc-Rab34, add 100 μl of Alexa 488-conjugated goat anti-mouse IgG plus rhodamine (or Alexa 546)–phalloidin, and incubate for 50 min. Remove the antibody and wash the cells as described earlier.

8. Place 50 μl of mounting solution (1 mg/ml p-phenylenediamine and 20% glycerol in PBS/NaN$_3$, the pH is adjusted to 8 with 0.5 M carbonate buffer, pH 9.0) on a glass slide. (p-Phenylenediamine is an antifader. The mounting solution can be stored at $-80°$ with blocking light.) Mount the coverslip. Remove excess mounting solution by absorbing with a piece of 3MM paper. Seal the edges of the coverslip with nail polish.

9. Observe the cells with a Zeiss Axioskop microscope equipped with fluorescence optics and a 63 × or a 100 × Plan Neofluar or Plan Apochromat objective lens. Observe Alexa 488 fluorescent Myc–Rab34 with an FITC filter set and rhodamine (or Alexa 546)–phalloidin with a rhodamine filter set (Fig. 3A). Observe Alexa 546 fluorescent Myc–Rab34 with a rhodamine filter set and EGFP–Rab5 with an FITC filter set (Fig. 3B).

Detection of Macropinosomes with Fluorescent Dextran

1. Culture 10T1/2 cells on glass coverslips as described earlier. Transfect the cells with pEGFP-C1/Rab34 or pEGFP-C1/Rab5.

2. Twenty-four hours after the transfection, replace the medium with the growth medium containing 1 mg/ml rhodamine B isothiocyanate (RITC)–dextran (MW ~70,000, Sigma). If the cells are not transfected with pEGFP-C1 vector, fluorescein isothiocyanate (FITC)–dextran (MW ~70,000, Sigma) is applicable instead of RITC–dextran. Incubate for various periods of time (5–60 min).

3. Remove the medium and rinse the cells by three changes of the growth medium. Fix the cells with 4 ml of 4% paraformaldehyde in 0.1 M NaPO$_4$ buffer (pH 7.4) for 15 min. Wash the cells by three changes of 4 ml each of PBS/NaN$_3$.

4. Mount the coverslips as described earlier. Observe RITC–dextran with a rhodamine filter set and EGFP-tagged protein with an FITC filter set (Fig. 3C).

Analyses of the Role of Rab34 in Macropinosome Formation

Treatment of cultured cells with growth factors or PMA induces prominent membrane ruffling and subsequent macropinocytosis (Swanson and Watts, 1995). Introduction of Ras or Rac1 also induces membrane ruffling and macropinocytosis in fibroblasts (Bar-Sagi and Feramisco, 1986; Ridley *et al.*, 1992). Rac1 causes membrane ruffling by activating WAVE1 and

WAVE2, which induce the formation of Arp2/3 complex-mediated actin filament meshwork in membrane ruffles (Suetsugu et al., 2003; Takenawa and Miki, 2001). Because Rab34 is located to lamellipodia and macropino-some membranes, involvement of Rab34 in lamellipodia formation and macropinosome formation has been examined in correlation with the signaling by the above agents (Sun et al., 2003). Transfection of wt Rab34 or Rab34(Q111L) to 10T1/2 fibroblasts facilitates macropinosome forma-tion but does not affect lamellipodia formation. Furthermore, induction of macropinosome formation by PDGF or PMA is facilitated by Rab34 but retarded by Rab34(T66N). On the other hand, Rab34-induced macropino-some formation is markedly suppressed by the expression of dominant-negative Rac1(T17N). It is also suppressed by the expression of Ras(S17N) or WAVE2(ΔV), a dominant-negative WAVE2 mutant lacking a verprolin homology domain. Taken together, these results imply that Rab34 facil-itates the macropinosome formation triggered by Rac1–WAVE2-induced membrane ruffling and that Rac1–WAVE2-mediated actin polymerization and subsequent membrane ruffling are required for Rab34-promoted macropinosome formation.

Assay for Rab34 Function in PDGF or PMA-Treated Cells

1. Culture 10T1/2 cells on glass coverslips as described earlier. Transfect the cells with pEF-BOS/Myc-Rab34 or pEF-BOS/Myc-Rab34 (T66N).

2. Twenty-four hours after the transfection, replace the growth medium with 4 ml of DME medium containing 0.2% FBS. Culture the cells for 16 h under the serum starvation condition.

3. Add 4 μl of 3 μg/ml PDGF (Peprotech, recombinant human PDGF-AB) or 4 μl of 100 μM PMA (Sigma) to the medium (final concentration 3 ng/ml PDGF and 0.1 μM PMA). Incubate for 30 min.

4. Fix and permeabilize the cells as previously described (Epitope Tagging and EGFP Tagging [see number 5 in the section]).

5. Incubate the cells with the mAb Myc1-9E10 and wash the cells as previously described (Epitope Tagging and EGFP Tagging [see number 6 in the section]).

6. Incubate the cells with Alexa 488-conjugated goat anti-mouse IgG plus rhodamine (or Alexa 546)–phalloidin and wash the cells as previously described (Epitope Tagging and EGFP Tagging [see number 7' in the section]).

7. Mount the coverslip and observe the cells by fluorescence microscopy. Count the number of macropinosomes, whose membranes are associated with Rab34 and actin.

Assay for Rab34 Function in Correlation with Ras, Rac1, and WAVE2

1. Culture 10T1/2 cells on glass coverslips as described earlier. Cotransfect the cells with pEF-BOS/Myc-Rab34 and pEF-BOS/HA-H-Ras(S17N), pEF-BOS/HA-Rac1(T17N), or pEF-BOS/HA-WAVE2(ΔV).

2. Twenty-four hours after the transfection, fix and permeabilize the cells as previously described (Epitope Tagging and EGFP Tagging [see number 5 in the section]).

3. Incubate the cells with the mAb Myc1-9E10 plus anti-HA pAb, and wash the cells as previously described (Epitope Tagging and EGFP Tagging [see number 6 in the section]).

4. Incubate the cells with Alexa 488-conjugated goat anti-mouse IgG plus Alexa 546-conjugated goat anti-rabbit IgG, and wash the cells as previously described (Epitope Tagging and EGFP Tagging [see number 7 in the section]).

5. Mount the coverslip and observe the cells by fluorescence microscopy. Count the number of macropinosomes, whose membranes are associated with Rab34, in the cells expressing H-Ras(S17N), Rac1 (T17N), or WAVE2(ΔV).

References

Bar-Sagi, D., and Feramisco, J. R. (1986). Induction of membrane ruffling and fluid-phase pinocytosis in quiescent fibroblasts by *ras* proteins. *Science* **233**, 1061–1068.

Cardelli, J. (2001). Phagocytosis and macropinocytosis in *Dictyostelium*: Phosphoinositide-based processes, biochemically distinct. *Traffic* **2**, 311–320.

Higgs, H. N., and Pollard, T. D. (2001). Regulation of actin filament network formation through ARP2/3 complex: Activation by a diverse array of proteins. *Annu. Rev. Biochem.* **70**, 649–676.

Johannes, L., and Lamaze, C. (2002). Clathrin-dependent or not: Is it still the question? *Traffic* **3**, 443–451.

Mellman, I. (1996). Endocytosis and molecular sorting. *Annu. Rev. Cell. Dev. Biol.* **12**, 575–625.

Michiels, F., Habets, G. G. M., Stam, J. C., van der Kammen, R. A., and Collard, J. G. (1995). A role for Rac in Tiam1-induced membrane ruffling and invasion. *Nature* **375**, 338–340.

Mizushima, S., and Nagata, S. (1990). pEF-BOS, a powerful mammalian expression vector. *Nucleic Acids Res.* **18**, 5322.

Ridley, A. J., Paterson, H. F., Johnston, C. L., Diekmann, D., and Hall, A. (1992). The small GTP-binding protein rac regulates growth factor-induced membrane ruffling. *Cell* **70**, 401–410.

Robinson, M. S., Watts, C., and Zerial, M. (1996). Membrane dynamics in endocytosis. *Cell* **84**, 13–21.

Rodman, J. S., and Wandinger-Ness, A. (2000). Rab GTPases coordinate endocytosis. *J. Cell Sci.* **113**, 183–192.

Segev, N. (2001). Ypt and Rab GTPases: Insight into functions through novel interactions. *Curr. Opin. Cell Biol.* **13**, 500–511.

Suetsugu, S., Yamazaki, D., Kurisu, S., and Takenawa, T. (2003). Differential roles of WAVE1 and WAVE2 in dorsal and peripheral ruffle formation for fibroblast cell migration. *Dev. Cell* **5**, 595–609.

Sun, P., Yamamoto, H., Suetsugu, S., Miki, H., Takenawa, T., and Endo, T. (2003). Small GTPase Rah/Rab34 is associated with membrane ruffles and macropinosomes and promotes macropinosome formation. *J. Biol. Chem.* **278**, 4063–4071.

Swanson, J. A., and Watts, C. (1995). Macropinocytosis. *Trends Cell Biol.* **5**, 424–428.

Takenawa, T., and Miki, H. (2001). WASP and WAVE family proteins: Key molecules for rapid rearrangement of cortical actin filaments and cell movement. *J. Cell Sci.* **114**, 1801–1809.

Zerial, M., and McBride, H. (2001). Rab proteins as membrane organizers. *Nat. Rev. Mol. Cell Biol.* **2**, 107–117.

[20] Fluorescent Microscopy-Based Assays to Study the Role of Rab22a in Clathrin-Independent Endocytosis

By ROBERTO WEIGERT and JULIE G. DONALDSON

Abstract

The endocytic and trafficking route followed by proteins that enter cells independently of clathrin is not well understood. In HeLa cells, a distinct endocytic and recycling pathway is followed for such clathrin-independent cargo proteins, and we have characterized the role of Rab proteins in this process. Here we describe cell-based flourescence assays to examine the effects of expression and depletion of Rab22 on endosomal morphology and endocytic recycling.

Introduction

Membrane proteins, lipids, and extracellular fluids are internalized into cells from the plasma membrane (PM) via endocytosis. Endocytic pathways can be classified into two categories: clathrin dependent and clathrin independent. Clathrin-independent endocytosis is involved in the internalization of major histocompatibility complex class I protein (MHCI) and GPI-anchored proteins (Naslavsky *et al.*, 2003, 2004). These proteins are internalized in vesicles that fuse with early endosomes that contain clathrin-derived cargo proteins (e.g., transferrin, Tfn) and from there, they are either transported to the late endosome for degradation or to the endosomal recycling compartment (ERC) (Maxfield and McGraw, 2004; Naslavsky

METHODS IN ENZYMOLOGY, VOL. 403
0076-6879/05 $35.00
DOI: 10.1016/S0076-6879(05)03020-X

et al., 2003; Weigert *et al.*, 2004). From the ERC, these molecules are recycled back to the PM via tubular structures that lack clathrin-dependent recycling cargo proteins (Weigert *et al.*, 2004).

We reported that a member of the family of the Rab GTPases, namely Rab22a (Kauppi *et al.*, 2002; Mesa *et al.*, 2001; Olkkonen *et al.*, 1993), regulates the recycling of MHCI from an endosomal compartment back to the PM while recycling of cargo proteins that traffic through the clathrin-dependent pathway (i.e., Tfn) was not affected (Weigert *et al.*, 2004). Interestingly, another member of the Rab family, Rab11, regulates the recycling of MHCI, but at a different step. Specifically, activation of Rab22a is required for the formation of the recycling transport intermediates while activation of Rab11 seems to control a later step in the recycling of MHCI. Furthermore, inactivation of Rab22a was suggested to be required for the fusion of the recycling transport intermediates with the PM.

Here, we describe a series of novel approaches designed to study both qualitative and quantitative aspects of the regulation of the clathrin-independent pathway by Rab22a. These assays are based on fluorescent microscopy and rely on the use of mutants of Rab22a that are impaired in the GTP cycle and on siRNA-based depletion of Rab22a. Although our studies were extended and confirmed in various cell types, we used HeLa cells as our experimental model since the clathrin-independent and the clathrin-dependent pathways are distinct and well characterized in this cell type (Naslavsky *et al.*, 2003, 2004). Furthermore, in HeLa cells, the clathrin-independent recycling compartment is tubular, easily identifiable, and suitable for morphometric analysis. Lastly, a wide variety of markers for intracellular organelles and reagents are available and well established.

Identification of the Rab Proteins Localized on Non-Clathrin-Derived Endosomes

Due to the lack of good immunological tools to detect endogenous Rabs by immunofluorescence, our first approach was to overexpress a series of green fluorescent protein (GFP)-tagged Rabs (Rab4, Rab5, Rab7, Rab11a, Rab11b, Rab14, Rab15, Rab21, Rab22a, and Rab25) and to screen for their ability to specifically localize to the clathrin-independent endosomal system. When expressed at low levels none of the Rabs tested alters the morphology of the endosomal system. The clathrin-independent and the clathrin-dependent endosomal systems are labeled by internalizing an antibody directed against MHCI and fluorescently conjugated Tfn, respectively (see the method described later in the chapter). We find that Rab22a and Rab11a and b (hereafter referred to as Rab11 since the two isoforms behaved identically in the assays in which they were tested) are

the Rab proteins associated with the clathrin-independent pathway. Specifically, Rab22a and Rab11 are associated with vesicular endosomes located in the perinuclear area that contain both MHCI and Tfn (merge
compartment) and in tubular recycling endosomes and in vesicular endosomes clustered at the apices of the cell, beneath the PM, that contain
MHCI but are devoid of Tfn (Fig. 1A).

Using an antiserum raised against Rab22a or a Rab11a polyclonal
antibody, we find that endogenous Rab22a and Rab11a also localize to
the non-clathrin-derived endosomes (Volpicelli *et al.*, 2002; Weigert *et al.*,
2004). The localization of the endogenous Rab22a is very similar to that of
the overexpressed GFP-Rab22a with the difference that endogenous
Rab22a is only occasionally associated with the merge compartment

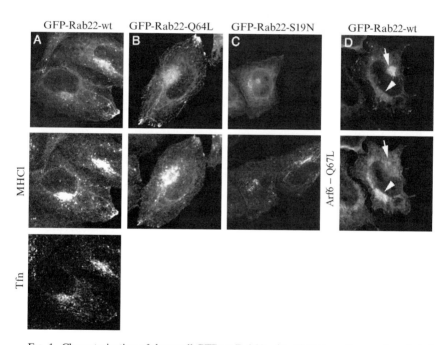

FIG. 1. Characterization of the small GTPase Rab22a. (A–C) HeLa cells were transfected
with either GFP-Rab22a-wt (A), GFP-Rab22-Q64L (B), or GFP-Rab22-S19N (C). Cells were
incubated for 30 min at 37° with an antibody against MHCI (A–C) and with 633-Tfn (A) to
allow internalization, and then fixed. To mask the MHCI antibody bound to the cell surface,
cells were probed with unconjugated goat anti-mouse antibody in the absence of saponin and
internalized MHC antibody was then visualized by probing with Alexa 594-conjugated GAM
in the presence of saponin. (D) HeLa cells were transfected with GFP-Rab22a-wt and Arf6-
Q67L, fixed, and stained for Arf6. Arrows indicate perinuclear GFP-Rab22a-labeled
endosomes and arrowheads indicate the Arf6-Q67L-induced vacuoles.

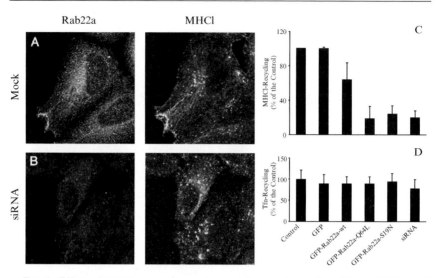

Fig. 2. Effect of Rab22a on endosomal morphology and recycling. (A and B) Loss of MHCl tubular endosomes in Rab22a-depleted cells. Mock-(A) or siRNA-treated (B) HeLa cells were allowed to internalize an antibody against MHCl for 30 min at 37°, fixed, surface MHCl antibody masked as in Fig. 1, and then stained for Rab22a and MHCl. (C, D) Recycling of MHCl and Tfn to the PM in cells transfected with Rab22a mutants or depleted of Rab22a by siRNA. HeLa cells were transfected as described in the legend to Fig. 1 or siRNA treated as described above and recycling of MHCl (C) or Tfn (D) back to the PM was measured as described in the text. As control, cells were either transfected with GFP or mock transfected (Control). SiRNA mock-treated cells are not shown since they showed behavior identical to mock-transfected cells.

(Fig. 2A). Furthermore, the MHCl-containing peripheral vesicles are more abundant and slightly larger in size (see below).

General Method to Measure Internalization of MHCI and Tfn

HeLa cells are plated on glass coverslips and used in internalization experiments 2 days later. Under these conditions the percentage of cells showing tubular recycling endosomes reaches a peak (40–60%). When cells adhere for a longer period of time or become more confluent, the percentage of cells showing tubular recycling endosomes is dramatically reduced. Cells are usually transfected in six-well dishes the day after they are plated using Fugene-6 (Roche, Indianapolis, IN). We routinely use 1 μg of each plasmid per well. After 16 h, the medium is removed, cells are washed three times with sterile phosphate-buffered saline (PBS), and then incubated with medium for an additional 4–5 h. Prior to internalization, cells are serum starved for 30 min at 37° in Dulbecco's modified Eagle's medium

(DMEM) containing 0.5% bovine serum albumin (BSA; BSA-DMEM). At the same time, 50 μl drops containing 30–50 μg/ml MHCI antibody (clone W6/32) and 5 μg/ml Alexa 633-Tfn in BSA-DMEM (Molecular Probes, Eugene, OR) are deposited on Parafilm and preequilibrated to 37° for 30 min in the cell incubator. Coverslips are placed on the drops with the cells facing the medium for 30 min at 37° to allow endocytosis of MHCI and Tfn prior to fixation with 2% formaldehyde/PBS at room temperature for 10–12 min. To mask the MHCI antibody bound to the cell surface, cells are probed with 0.1 mg/ml unconjugated goat anti-mouse antibody in the absence of saponin for 1 h, washed, and then refixed for 5 min. Internalized MHC antibody is then visualized by probing with Alexa 594-conjugated GAM (Molecular Probes, Eugene, OR) in blocking solution (PBS containing 10% fetal calf serum) plus 0.2% saponin. All images are obtained using a Zeiss 510 LSM confocal microscope (Thornwood, NJ) with a 63 × Plan Apo objective as described before (Naslavsky et al., 2003). Preparation of figures is accomplished in Adobe Photoshop 5.5.

Identification of the Endocytic Step Where Rab22a Associates with the Non-Clathrin-Derived Endosomes

We determined that Rab22a associates with the non-clathrin-derived endosomes at later stages of the endocytic process (Weigert et al., 2004). To examine when Rab22a associates with non-clathrin-internalized vesicles, we used two different approaches. First, we used the expression of the constitutively active mutant of Arf6 (Arf6-Q67L) to examine whether Rab22a associates with early endosomal structures. Expression of Arf6-Q67L promotes the fusion of the non-clathrin-derived early endosomal structures generating large vacuoles. These structures are enriched in phosphatidylinositol 4, 5-bisphosphate (PIP$_2$) and actin, and trap all the molecules trafficking through the clathrin-independent pathway (Brown et al., 2001; Naslavsky et al., 2003). Under these conditions, cargo proteins cannot reach the clathrin-derived early endosomes and do not recycle to the PM (Naslavsky et al., 2003; Weigert et al., 2004). GFP-Rab22a did not associate with the MHCI-containing vacuoles formed in cells expressing Arf6-Q67L (Fig. 1D). Notably, GFP-Rab22a is localized to perinuclear structures containing Tfn, while GFP-labeled tubules are no longer observed and the number of Rab22a-labeled peripheral vesicles is significantly decreased. Endogenous Rab22a is also not localized to the vacuoles in Arf6-Q67L-expressing cells (Weigert et al., 2004).

The second approach we used is time-lapse imaging of MHCI internalization in GFP-Rab22a-expressing cells. After 1 min of internalization, none of the MHCI-containing endosomes label with GFP-Rab22a. After

5 min, some Rab22a-labeled tubules appear to contain MHCI, as do some vesicular structures in the perinuclear area. The amount of MHCI in GFP-Rab22a tubules increases over time (Weigert *et al.*, 2004). A similar experiment can be performed internalizing 633-Tfn. Tfn-containing structures at early time points are devoid of GFP-Rab22a.

Assay to Examine the Distribution of Rab22a in Arf6-Q67L-Expressing Cells

HeLa cells are transfected as described earlier, with the exception that 0.5–0.75 μg of the plasmid encoding Arf6-Q67L is used. After 16 h the medium is exchanged with fresh media as described previously and the cells are then examined 1 h later. The amount of plasmid and the time of transfection are crucial for the visualization of the Arf6-Q67L-induced vacuoles. Prolonged times of transfection or higher amounts of plasmid induce dramatic transformations in the morphology of the cells, which tend to shrink, and furthermore, the internalization of non-clathrin-derived cargo is severely impaired (Brown *et al.*, 2001). To mask the MHCI at the cell surface, cells are incubated for 1 h at room temperature with unconjugated anti-MHCI antibody in the absence of saponin, refixed briefly, and washed 2 × 5 min with blocking solution (see previous discussion). Cells are then incubated for 1 h with an antibody against Arf6 in the presence of saponin, washed 3 × 5 min in blocking solution, and finally probed with Alexa 594-conjugated GAR and cy5-conjugated anti-MHCI to label the internal pool of MHCI directly (see below).

Preparation of Fluorescently Conjugated Anti-MHCI Antibody

Cy3-MHCI (or Cy5-MHCI) is prepared as follows: 200 μg of purified W6/32 antibody is dialyzed overnight in PBS and concentrated to 1–2 mg/ml using Microcon filters (cutoff 10 kDa, Millipore, Billerica, MA). The antibody is then conjugated with Cy3 (or Cy5) using the Cy3 (or Cy5) mAb labeling kit (Amersham, Piscataway, NJ) according to recommendations from the manufacturer. The only variation is that the ratio of dye/antibody is set to 3–4.

Time-Lapse Imaging of MHCI-Internalization in GFP-Rab22a Expressing Cells

For the time-lapse experiments, HeLa cells are plated on 2.5-cm glass coverslips and transfected as described earlier. Coverslips are mounted on a circular metal chamber, incubated with 1 ml of CO_2-independent medium

(Gibco) containing 10% fetal bovine serum (FBS) and transferred to the microscope that is prewarmed at 37° by airflow. The chamber is covered with a glass coverslip to prevent the evaporation of the medium. The entire field is scanned for transfected cells that are well spread and show clear tubular structures. Once a suitable cell is identified, Cy3-MHCI is added to the medium (20–50 μg/ml final concentration) and the recording is immediately started. Optical slices are set at 1.5 μm for each channel and the detector gain is set for the fluorescent signal to be in the linear range. Images are acquired every 6 s. Videos are generated using Metamorph (Universal Imaging, Downingtown, PA).

Effect of the Overexpression of Rab22a and Rab11 Mutants and siRNA-Mediated Depletion of Rab22a on the Morphology of the Non-Clathrin-Derived Recycling Endosomes

Overexpression of Rab22a and its mutants impaired in the GTP cycle alters the morphology of the non-clathrin-derived recycling endosomes (Weigert *et al.*, 2004). Specifically, overexpression of either GFP-Rab22a-wt or GFP-Rab22a-Q64L, a mutant defective in GTP hydrolysis, increases the number and the size of the recycling peripheral vesicles (Mesa *et al.*, 2001; Weigert *et al.*, 2004). Furthermore, expression of GFP-Rab22a-Q64L increases the percentage of cells showing MHCI-containing tubules and the number of tubules per cell (Fig. 1B). Strikingly, in cells expressing GFP-Rab22a-S19N, a mutant detective in GTP binding, tubular structures and vesicles at the cell periphery are no longer observed and MHCI and Tfn mainly localize in the perinuclear area (Mesa *et al.*, 2001; Weigert *et al.*, 2004) (Fig. 1C). The same phenotype is observed in cells depleted of Rab22a by siRNA technology (Weigert *et al.*, 2004) (Fig. 2B). On the other hand, the overexpression of GFP-Rab11-S25N, the Rab11 mutant detective in GTP binding, does not affect the percentage of cells showing MHCI-containing tubules, whereas the percentage of cells showing MHCI in peripheral vesicles is reduced (Weigert *et al.*, 2004).

Overexpression of Rab22a and Rab11 Mutants and siRNA-Mediated Depletion of Rab22a

To overexpress the various Rab mutants, cells are plated and transfected as described earlier. The amount of plasmid used to transfect the various mutants is optimized to achieve similar expression levels (1 μg for GFP-Rab22a-wt and GFP-Rab22a-Q64L and 2 μg for GFP-Rab22a-S19N and GFP-Rab11a-S25N). To knock down Rab22a levels, cells are plated on a 10-cm dish and grown in complete medium without antibiotics. After

24 h, 24 μl OligofectAmine is added to 66 μl of Opti-MEM (Invitrogen, Carlsbad, CA) and the solution is incubated for 5–10 min at room temperature (RT). This is mixed with a second solution containing 970 μl of Opti-MEM plus 300 pmol of siRNA. The mixture is incubated at RT for 15–20 min and added to the cells. After 72 h, the cells are trypsinized, diluted 1:10, and seeded in a 10-cm dish. For immunofluorescence and recycling experiments, sterile coverslips are placed into the dish. After 6 h, the second transfection is performed and cells are examined after 72 h. The target sequence for the human Rab22a (AAGGACUACGCCGACU-CUAUU) is synthesized as Option C siRNA by Dharmacon (Lafayette, CO). In HeLa cells, using this protocol, Rab22a is depleted below the detection level while other Rab proteins (namely, Rab4, Rab5, and Rab11) are not affected (Weigert et al., 2004). The siRNA-treated cells have normal morphology and can be grown for several passages in the presence of the siRNA.

Morphological Analysis of Non-Clathrin-Derived Recycling Endosomes

Internalization of MHCI is performed in either transfected or siRNA-depleted cells as described earlier. Samples are analyzed with a Zeiss Epifluorescence photomicroscope with a 63×/1.4 Plan Apo chromate objective. Transfected cells are first identified on the 488 channel and then analyzed switching to the 594 channel. Then 50–100 cells per coverslip (two coverslips per condition) are scored for the presence of MHCI-containing tubular structures and MHCI-containing peripheral vesicles defined clusters of vesicles located beneath the plasma membrane at the apices of the cell. The percentage of cells having either tubules or peripheral vesicles is calculated and expressed as percentage of the nontransfected cells.

Effect of the Overexpression of Rab22a and Rab11 Mutants and
    siRNA-Mediated Depletion of Rab22a on the Recycling to the
    PM of MHCI and Tfn

Overexpression of either GFP-Rab22a-wt, GFP-Rab22a-Q64L, or GFP-Rab22a-S19N significantly inhibits the recycling of MHCI to the PM while the recycling of Tfn is not affected (Weigert et al., 2004) (Fig. 2C and D). A similar block is observed in siRNA-depleted cells (Fig. 2C and D). Under these conditions neither the internalization of MHCI nor the fusion between MHCI-containing and clathrin-derived endosomes is affected. Notably, Tfn trafficking is not altered either (Fig. 2D). On the other hand, the overexpression of GFP-Rab11-S25N impairs both MHCI and Tfn recycling as previously reported (Powelka *et al.*, 2004; Ullrich *et al.*, 1996; Weigert *et al.*, 2004).

Assay to Measure Recycling of Cargoes from the Recycling Compartment to the PM

To quantitatively measure the effect of the expression of mutants of Rab22a and Rab11 on the recycling of cargo molecules to the PM, we decided to use a microscopy-based assay. We preferred such an approach to a more classic biochemical assay since the latter is strongly dependent on the efficiency of expression of the various constructs, which usually ranges from 10% to 50% of the cell population. The advantage of a microscopy-based assay is that such a method allows us to selectively measure recycling in transfected cells and to have an immediate correlation between the effects on the recycling and the morphology of the recycling compartment. We designed two assays based on recycling of cargo molecules accumulated in the recycling compartment: one estimating the amount of MHCI that reappears at the PM and the other estimating the disappearance of the internal pool of Tfn.

Assay to Measure Recycling of MHCI to the PM

Cells are grown on glass coverslips and incubated on ice for 30 min with 20–50 μg/ml of W6/32 antibody to allow antibody binding to MHCI at the PM. Cells are then washed with ice-cold medium to remove the unbound antibody and incubated for 30 min at 37° in the presence of 1 μM latrunculin A (LatA). We previously showed that drugs such as LatA, which disrupt the actin cytoskeleton, reversibly and selectively inhibit MHCI recycling causing MHCI to accumulate in the tubular endosomes (Radhakrishna and Donaldson, 1997). Cytochalasin D (CD) can also be used, although LatA offers the advantage of a lower toxicity. To remove the antibody that is not internalized, at the end of the internalization, cells are incubated at RT for 30–45 s with "stripping buffer" (0.5% acetic acid, 0.5 *M* NaCl, pH 3.0). Cells are rapidly rinsed twice in PBS and twice in DMEM, and then incubated at 37° for different times with complete medium to allow the recycling of MHCI. At the end of the incubations, cells are processed as follows. A set of coverslips is fixed and incubated for 1 h with 594-Alexa-conjugated antibody directed against mouse IgG (594-GAM) to determine the surface pool of MHCI. Another set of coverslips is first "stripped" for 30 s at RT and then fixed and incubated with 594-GAM in the presence of 0.2% saponin to reveal the internal pool. Each experimental condition is tested in duplicate. To determine the initial pool of internalized MHCI, a set of cells is processed immediately after the first stripping and washing steps (time 0). Cells are imaged using a Zeiss 510 LSM confocal microscope with a 40× plan Apo objective. Then 30–50 transfected cells per coverslip are randomly selected. To minimize the variability due to the differences in cell size, only cells whose area is 1000 ± 500 μm^2 (our estimated average area of HeLa cells) are included. The pinhole is completely open and all the images are taken with identical acquisition parameters, those previously optimized for the fluorescent signals to be in the dynamic range. Under these conditions, the amount of MHCI is proportional to the total fluorescence. For each channel the total fluorescence of each individual cell is measured using the Zeiss LSM image examiner (version 3.01). For each time point, the recycled MHCI is calculated as follows: first, the fluorescence measured at the surface (surface MHCI) is added to the fluorescence measured in the internal pool (internal MHCI) and termed "total MHCI"; second, the surface MHCI is expressed as a percentage of the total MHCI; finally, the recycled MHCI is calculated by subtracting the percentage of total MHCI calculated at time 0 from the percentage of total MHCI. Typically, the maximal amount of MHCI that recycles over a period of 30 min does not exceed 30–40% of the internal MHCI. Recycling can be almost completely blocked if LatA is present in the recycling medium (Weigert *et al.*, 2004).

Assay to Measure Recycling of Tfn back to the PM

To measure the recycling of Tfn, cells are serum starved for 30 min at 37° in DMEM containing 0.5% BSA, and then 5 $\mu g/ml$ of Alexa 595-Tfn (594-Tfn) is internalized for 30 min. At the end of the internalization, cells are stripped of the noninternalized Tfn as described earlier, and incubated in complete medium for different times. Cells are fixed, and the amount of Tfn inside the cell is estimated as described for MHCI and expressed as percentage of the Tfn at time 0. Recycled Tfn at a given time is calculated as the difference between the percentage of Tfn at time 0 (100%) and the percentage of internal Tfn left at a given time. Typically, the maximal amount of Tfn that recycles over a period of 30 min does not exceed 70–80% of the internal Tfn. LatA does not affect Tfn recycling (Weigert *et al.*, 2004).

Concluding Remarks

We have described a series of protocols for studying the regulation of clathrin-independent endocytosis by the small GTPase Rab22a. Most of the assays presented here rely on quantitative analysis of morphological data. These techniques are extremely powerful and can be extended to the study of other molecules and other aspects of the endocytosis like internalization or endosome fusion.

References

Brown, F. D., Rozelle, A. L., Yin, H. L., Balla, T., and Donaldson, J. G. (2001). Phosphatidylinositol 4,5-bisphosphate and Arf6-regulated membrane traffic. *J. Cell Biol.* **154,** 1007–1017.

Kauppi, M., Simonsen, A., Bremnes, B., Vieira, A., Callaghan, J., Stenmark, H., and Olkkonen, V. M. (2002). The small GTPase Rab22 interacts with EEA1 and controls endosomal membrane trafficking. *J. Cell Sci.* **115,** 899–911.

Maxfield, F. R., and McGraw, T. E. (2004). Endocytic recycling. *Nat. Rev. Mol. Cell Biol.* **5,** 121–132.

Mesa, R., Salomon, C., Roggero, M., Stahl, P. D., and Mayorga, L. S. (2001). Endocytic recycling. *J. Cell Sci.* **114,** 4041–4049.

Naslavsky, N., Weigert, R., and Donaldson, J. G. (2003). Convergence of non-clathrin- and clathrin-derived endosomes involves Arf6 inactivation and changes in phosphoinositides. *Mol. Biol. Cell* **14,** 417–431.

Naslavsky, N., Weigert, R., and Donaldson, J. G. (2004). Characterization of a nonclathrin endocytic pathway: Membrane cargo and lipid requirements. *Mol. Biol. Cell* **15,** 3542–3552.

Olkkonen, V. M., Dupree, P., Killisch, I., Lutcke, A., Zerial, M., and Simons, K. (1993). Molecular cloning and subcellular localization of three GTP-binding proteins of the rab subfamily. *J. Cell Sci.* **106**(Pt. 4), 1249–1261.

Powelka, A. M., Sun, J., Li, J., Gao, M., Shaw, L. M., Sonnenberg, A., and Hsu, V. W. (2004). Stimulation-dependent recycling of integrin b1 regulated by ARF6 and Rab11. *Traffic* **5,** 20–36.

Radhakrishna, H., and Donaldson, J. G. (1997). ARF6 regulates a novel plasma membrane recycling pathway. *J. Cell Biol.* **139,** 49–61.

Ullrich, O., Reinsch, S., Urbe, S., Zerial, M., and Parton, R. G. (1996). Rab11 regulates recycling through the pericentriolar recycling endosome. *J. Cell Biol.* **135,** 913–924.

Volpicelli, L. A., Lah, J. J., Fang, G., Goldenring, J. R., and Levey, A. I. (2002). Rab11a and myosin Vb regulate recycling of the M4 muscarinic acetylcholine receptor. *J. Neurosci.* **22,** 9776–9784.

Weigert, R., Yeung, A. C., Li, J., and Donaldson, J. G. (2004). Rab22a regulates the recycling of membrane proteins internalized independently of clathrin. *Mol. Biol. Cell* **15,** 3758–3770.

[21]   Purification and Properties of Rab3 GEP (DENN/MADD)

By TOSHIAKI SAKISAKA and YOSHIMI TAKAI

Abstract

Rab3A, a member of the Rab3 small GTP-binding protein (G protein) family, regulates Ca^{2+}-dependent exocytosis of neurotransmitter. Rab3A cycles between the GDP-bound inactive and GTP-bound active forms, and the former is converted to the latter by the action of a GDP/GTP exchange protein (GEP). We have previously purified a GEP from rat brain with lipid-modified Rab3A as a substrate. Purified Rab3 GEP is active on all the Rab3 subfamily members including Rab3A, -3B, -3C, and -3D. Purified Rab3 GEP is active on the lipid-modified form, but not on the lipid-unmodified form. Purified Rab3 GEP is inactive on Rab3A complexed with Rab GDI. The recombinant protein is prepared from the Rab3 GEP-expressed *Spodoptera frugiperda* cells (Sf9 cells). The properties of recombinant Rab3 GEP, including the requirement for lipid modifications of Rab3A, the substrate specificity, and the sensitivity to Rab GDI, are similar to those of purified Rab3 GEP. Overexpression of Rab3 GEP inhibits Ca^{2+}-dependent exocytosis from PC12 cells. On the other hand, Rab3 GEP is identical to a protein named DENN/MADD: differentially expressed in normal versus neoplastic (DENN)/mitogen-activated protein kinase-activating death domain (MADD). Here, we describe the purification method for recombinant Rab3 GEP from Sf9 cells and the functional properties of Rab3 GEP in Ca^{2+}-dependent exocytosis by use of the human growth hormone coexpression assay system of PC12 cells.

Introduction

Rab3 GDP/GTP exchange protein (GEP) is a regulator of the Rab3 family consisting of Rab3A, -3B, -3C, and -3D (Wada *et al.*, 1997). The Rab3 family members are regulated by Rab GDP dissociation inhibitor (GDI) and Rab3 GTPase-activating protein (GAP) in addition to Rab3 GEP (Fukui *et al.*, 1997; Nagano *et al.*, 1998; Sasaki *et al.*, 1990). Rab3 GEP and GAP are specific for the Rab3 family members, whereas Rab GDI is active on all the Rab family members. Of the Rab3 family members, the function and the mode of action of Rab3A have most extensively been investigated and it has been shown to be involved in Ca^{2+}-dependent

METHODS IN ENZYMOLOGY, VOL. 403
0076-6879/05 $35.00
DOI: 10.1016/S0076-6879(05)03021-1

exocytosis, particularly neurotransmitter release. The cyclical activation and inactivation of Rab3A by the action of the three regulators are essential for the action of Rab3A neurotransmitter release. A current model for the mode of action of these regulators is as follows (Takai *et al.*, 1996): (1) GDP-Rab3A forms an inactive complex with Rab GDI and stays in the cytosol of nerve terminals. (2) GDP-Rab3A released from Rab GDI is converted to GTP-Rab3A by Rab3 GEP with the help of another unidentified molecule, such as GDI displacement factor (GDF): Yip3/PRA1 for Rab5 and -9 (Pfeffer and Aivazian, 2004; Sivars *et al.*, 2003). (3) GTP-Rab3A binds effector molecules, rabphilin-3 (Shirataki *et al.*, 1993) and Rim (Wang *et al.*, 1997), which localize at synaptic vesicles and the active zone, respectively. These complexes facilitate translocation and docking of the synaptic vesicles to the active zone. (4) GTP-Rab3A is converted to GDP-Rab3A by Rab3 GAP when the vesicles fuse with the presynaptic membrane. (5) GDP-Rab3A is associated with Rab GDI and retrieved from the membrane to the cytosol with a help of Rab recycling factor (RRF): Hsp90 chaperon complex for Rab3A (An *et al.*, 2003; Sakisaka *et al.*, 2002). In Rab3A$^{-/-}$ mice, synaptic depression is increased during repetitive stimulation in the CA1 region of the hippocampus (Geppert *et al.*, 1994), and mossy fiber long-term potentiation in the CA3 region is abolished (Castillo *et al.*, 1997). In Rab3A$^{-/-}$ hippocampal neurons in culture, the quantal release per synapse is increased, whereas the size of the readily releasable pool (RRP) measured by the application of hypertonic solution is normal (Geppert *et al.*, 1997). Thus, Rab3A appears to up-regulate the steps of translocation and docking, as well as to down-regulate the step of fusion.

Rab3 GEP has originally been purified with Rab3A as a substrate from rat brain synaptic soluble fraction (Wada *et al.*, 1997). This Rab3 GEP is most active on Rab3A and -3C and partially active on Rab3B and -3D. This Rab3 GEP prefers the lipid-modified form to lipid-unmodified form. Rat Rab3 GEP consists of 1602 amino acids and shows a calculated M_r of 177,982. By Northern blot analysis, Rab3 GEP is shown to be ubiquitously expressed. The subcellular fractionation analysis in rat brain indicates that Rab3 GEP is enriched in the synaptic soluble fraction.

Studies using Rab3 GEP$^{-/-}$ mice have shown that they develop more severe phenotypes than Rab3A$^{-/-}$ and Rab GDI$^{-/-}$ mice (Ishizaki *et al.*, 2000). Rab3 GEP$^{-/-}$ mice die immediately after birth because of respiratory failure. The embryos at E18.5 show no evoked action potentials of the diaphragm and gastrocnemius muscles in response to electrical stimulation of the phrenic and sciatic nerves, respectively. The total number of the synaptic vesicles at the neuromuscular junction of Rab3 GEP$^{-/-}$

mice is remarkably reduced about 10% compared with that of wild-type (WT) mice. In the central nervous system, the release probability is markedly reduced in autapses of Rab3 GEP$^{-/-}$ hippocampal neurons in culture, whereas the size of RRP is not different between WT and Rab3 GEP$^{-/-}$ neurons, indicating that Rab3 GEP up-regulates a postdocking step of synaptic exocytosis (Yamaguchi et al., 2002). Because Rab3A reportedly down-regulates Ca^{2+}-triggered fusion of synaptic vesicles, these results provide evidence for a role of Rab3 GEP in the post-docking process distinct from Rab3A activation. Thus, Rab3 GEP is essential for neurotransmitter release in the peripheral and central nervous system and may be required for formation and trafficking of synaptic vesicles.

On the other hand, Rab3 GEP is identical to a protein named differentially expressed in normal versus neoplastic/mitogen-activated protein kinase-activating death domain (DENN/MADD) (Chow and Lee, 1996; Schievella et al., 1997). DENN/MADD is a component of a signaling protein complex that localizes to the cytosol and exerts multiple functions by using different binding partners. It is also involved in blocking the apoptosis of neuronal cells under conditions of cytotoxic stress (Del Villar and Miller, 2004). Reduced endogenous DENN/MADD expression and enhanced proapoptotic signaling have been found in the brain affected by Alzheimer's disease. Thus, Rab3 GEP has a dual role in neurotransmission and neuroprotection (Miyoshi and Takai, 2004).

We previously described the methods for purification of native Rab3 GEP from rat brain and recombinant Rab3 GEP from COS7 cells and their biochemical properties (Nakanishi and Takai, 2001). In this chapter, therefore, we describe the methods for purification of recombinant Rab3 GEP from *Spodoptera frugiperda* cells (Sf9 cells) and the functional properties of Rab3 GEP in Ca^{2+}-dependent exocytosis.

Materials

(p-Amidinophenyl)methanesulfonyl fluoride (APMSF), leupeptin, and Triton X-100 were purchased from Wako Pure Chemicals (Osaka, Japan). Bovine serum albumin (BSA) (fraction V) is from Sigma Chemical Co. (St. Louis, MO). 3-[(3-Cholamidopropyl) dimethylammonio]-1-propanesulfonic acid (CHAPS) is from Dojindo Laboratories (Kumamoto, Japan). Mono Q PC1.6/5 is from Amersham-Pharmacia Biotech. Dulbecco's modified Eagle's medium (DMEM) is from Nakalai Tesque (Kyoto, Japan). Fetal calf serum (FCS), OPTI-MEM, and LipofectAMINE reagent are from Invitrogen (Carlsbad, CA). All other chemicals are of reagent grade.

Plasmid for expression of Rab3 GEP in Sf9 cells is constructed as follows. The cDNA fragment encoding full-length rat Rab3 GEP (amino acid [aa] 1–1602) is inserted into pAcYM1 baculovirus transfer vector to construct pAcYM1-Rab3 GEP. Plasmid for expression of Rab3 GEP in PC12 cells is constructed as follows. The cDNA fragment encoding full-length rat Rab3 GEP (aa 1–1602) or C-terminal deletion mutant Rab3 GEP-ΔC (aa 1–1210) is inserted into pCMV-myc to construct pCMV-myc-Rab3 GEP or pCMV-myc-Rab3 GEP-ΔC, respectively

Methods

Purification of Recombinant Rab3 GEP from Sf9 Cells

The steps used in the purification of recombinant Rab3 GEP from Sf9 cells are as follows: (1) preparation of the crude supernatant from Sf9 cells and (2) Mono Q PC1.6/5 column chromatography.

Buffers for Purification of Recombinant Rab3 GEP from Sf9 Cells. Buffer A: 20 mM Tris–HCl at pH 7.9, 140 mM NaCl, 3 mM KCl, 1 mM CaCl$_2$, 0.5 mM MgCl$_2$, 0.9 mM Na$_2$PO$_4$, 30 μM APMSF, and 25 μM leupeptin. Buffer B: 20 mM Tris–HCl at pH 7.5, 1 mM DTT, and 0.6% CHAPS.

Preparation of the Crude Supernatant from Sf9 Cells. The following procedures are carried out at 0–4°. The Sf9 cells expressing Rab3 GEP (1 × 10^8 cells) are collected, washed three times with 30 ml of phosphate-buffered saline (PBS), and suspended with 30 ml of Buffer A. The suspension is sonicated at a setting of 60 by an ultrasonic processor (Taitec, Tokyo, Japan) on ice for 30 s four times at 1-min intervals, followed by centrifugation at 100,000×g for 1 h. The supernatant is collected and stored at −80°. The supernatant can be stored at −80° for at least 3 months.

Mono Q PC 1.6/5 Column Chromatography. The supernatant (2 ml, 4.2 mg of protein) is applied to a Mono Q PC1.6/5 column equilibrated with Buffer B containing 0.1 M NaCl. Elution is performed with a 1.5-ml linear gradient of NaCl (0.1–0.5 M) in Buffer B at a flow rate of 50 μl/min. Fractions of 50 μl each are collected. The Rab3 GEP activity appears in fractions 19–21. The active fractions (150 μl, 0.1 mg of protein) are collected and used as recombinant Rab3 GEP. This recombinant Rab3 GEP can be stored at −80° for at least 3 months.

The biochemical and physical properties of recombinant Rab3 GEP are similar to those of purified Rab3 GEP from rat brain, including the substrate specificity, the requirement of lipid modifications of Rab3A, and the ineffectiveness to Rab3A complexed with Rab GDI.

Involvement of Rab3 GEP in Ca²⁺-Dependent Exocytosis

The activity of Rab3 GEP to regulate Ca^{2+}-dependent exocytosis is assayed by measuring growth hormone (GH) release from PC12 cells cotransfected with XGH5 encoding human GH and pCMV-myc-Rab3 GEP or pCMV-myc-Rab3 GEP-ΔC. In this assay system, expressed GH is stored in dense core vesicles of PC12 cells and released in response to various agonists in an extracellular Ca^{2+}-dependent manner (Schweitzer and Kelly, 1985; Wick *et al.*, 1993).

Buffers for the GH Release Assay. Low K^+ solution: 140 mM NaCl, 4.7 mM KCl, 2.5 mM $CaCl_2$, 1.2 mM $MgSO_4$, 1.2 mM KH_2PO_4, 20 mM HEPES/NaOH at pH 7.4, and 11 mM glucose. High K^+ solution: 85 mM

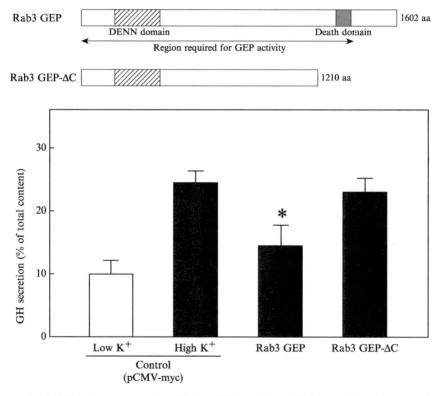

FIG. 1. Effect of overexpression of Rab3 GEP and Rab3 GEP-ΔC on release of expressed GH from PC12 cells. Data are expressed as the average percentage released of the total GH stores. The values are mean ± SE of three independent experiments. *$p < 0.01$ versus GH secretion from cells transfected with control plasmid, pCMV-myc.

NaCl, 60 mM KCl, 2.5 mM ČaCl$_2$, 1.2 mM MgSO$_4$, 1.2 mM KH$_2$PO$_4$, 20 mM HEPES/NaOH at pH 7.4, and 11 mM glucose.

Cell Culture. Stock cultures of PC12 cells are maintained at 37° in a humidified atomosphere of 10% CO$_2$ and 90% air (v/v) in DMEM containing 10% FCS, 5% horse serum (HS), penicillin (100 U/ml), and streptomycin (100 μg/ml).

Transfection. PC12 cells are plated at a density of 5×10^5 cells per 35-mm dish and incubated for 18–24 h. The cells are then cotransfected with 2 μg of pXGH and 2 μg of pCMV-myc-Rab3 GEP or pCMV-myc-Rab3 GEP-ΔC using 15 μl of LipofectAMINE reagent and 1 ml of OPTI-MEM. Six hours after the transfection, the cells were washed with DMEM and incubated in 2 ml of DMEM containing 10% FCS and 5% HS.

GH Release Experiments. GH release experiments are performed 48 h after the transfection. PC12 cells are washed with a low K$^+$ solution and incubated at 37° for 10 min with 2 ml of a high K$^+$ solution or a low K$^+$ solution. After the high K$^+$ solution or the low K$^+$ solution is removed, the cells are lysed in 2 ml of the high K$^+$ solution containing 0.5% Triton X-100 on ice. The amounts of the GH released into the high K$^+$ solution or the low K$^+$ solution and retained in the cell lysate are measured using an enzyme-linked immunosorbant (ELISA) assay according to the manufacturer's instructions (Roche).

In these experiments, the high K$^+$-induced GH release is inhibited in PC12 cells overexpressing full-length Rab3 GEP, but not Rab3 GEP-ΔC (Fig. 1) (Oishi *et al.*, 1998). As Rab3 GEP-ΔC does not show GEP activity, the GEP activity of Rab3 GEP is necessary for the inhibition of the high K$^+$-induced GH release.

Comments

Of the purification steps of recombinant Rab3 GEP, Mono Q column chromatography by use of the SMART system (Amersham Pharmacia Biotech) is the most important step to obtain a large amount of recombinant Rab3 GEP. A small total gel volume and dead volume in this system contribute to low nonspecific absorption, resulting in superior recovery. In addition, because there is less dilution in this system, it is possible to achieve sample concentration, which also increases the recovery.

References

An, Y., Shao, Y., Alory, C., Matteson, J., Sakisaka, T., Chen, W., Gibbs, R. A., Wilson, I. A., and Balch, W. E. (2003). Geranylgeranyl switching regulates GDI-Rab GTPase recycling. *Structure (Camb.)* **11,** 347–357.

Castillo, P. E., Janz, R., Sudhof, T. C., Tzounopoulos, T., Malenka, R. C., and Nicoll, R. A. (1997). Rab3A is essential for mossy fibre long-term potentiation in the hippocampus. *Nature* **388,** 590–593.

Chow, V. T., and Lee, S. S. (1996). DENN, a novel human gene differentially expressed in normal and neoplastic cells. *DNA Seq.* **6,** 263–273.

Del Villar, K., and Miller, C. A. (2004). Down-regulation of DENN/MADD, a TNF receptor binding protein, correlates with neuronal cell death in Alzheimer's disease brain and hippocampal neurons. *Proc. Natl. Acad. Sci. USA* **101,** 4210–4215.

Fukui, K., Sasaki, T., Imazumi, K., Matsuura, Y., Nakanishi, H., and Takai, Y. (1997). Isolation and characterization of a GTPase activating protein specific for the Rab3 subfamily of small G proteins. *J. Biol. Chem.* **272,** 4655–4658.

Geppert, M., Bolshakov, V. Y., Siegelbaum, S. A., Takei, K., De Camilli, P., Hammer, R. E., and Sudhof, T. C. (1994). The role of Rab3A in neurotransmitter release. *Nature* **369,** 493–497.

Geppert, M., Goda, Y., Stevens, C. F., and Sudhof, T. C. (1997). The small GTP-binding protein Rab3A regulates a late step in synaptic vesicle fusion. *Nature* **387,** 810–814.

Ishizaki, H., Miyoshi, J., Kamiya, H., Togawa, A., Tanaka, M., Sasaki, T., Endo, K., Mizoguchi, A., Ozawa, S., and Takai, Y. (2000). Role of rab GDP dissociation inhibitor alpha in regulating plasticity of hippocampal neurotransmission. *Proc. Natl. Acad. Sci. USA* **97,** 11587–11592.

Miyoshi, J., and Takai, Y. (2004). Dual role of DENN/MADD (Rab3GEP) in neurotransmission and neuroprotection. *Trends Mol. Med,* **10,** 476–480.

Nagano, F., Sasaki, T., Fukui, K., Asakura, T., Imazumi, K., and Takai, Y. (1998). Molecular cloning and characterization of the noncatalytic subunit of the Rab3 subfamily-specific GTPase-activating protein. *J. Biol. Chem.* **273,** 24781–24785.

Nakanishi, H., and Takai, Y. (2001). Purification and properties of Rab3 GDP/GTP exchange protein. *Methods Enzymol.* **329,** 59–67.

Oishi, H., Sasaki, T., Nagano, F., Ikeda, W., Ohya, T., Wada, M., Ide, N., Nakanishi, H., and Takai, Y. (1998). Localization of the Rab3 small G protein regulators in nerve terminals and their involvement in Ca2+-dependent exocytosis. *J. Biol. Chem.* **273,** 34580–34585.

Pfeffer, S., and Aivazian, D. (2004). Targeting Rab GTPases to distinct membrane compartments. *Nat. Rev. Mol. Cell Biol.* **5,** 886–896.

Sakisaka, T., Meerlo, T., Matteson, J., Plutner, H., and Balch, W. E. (2002). Rab-alphaGDI activity is regulated by a Hsp90 chaperone complex. *EMBO J.* **21,** 6125–6135.

Sasaki, T., Kikuchi, A., Araki, S., Hata, Y., Isomura, M., Kuroda, S., and Takai, Y. (1990). Purification and characterization from bovine brain cytosol of a protein that inhibits the dissociation of GDP from and the subsequent binding of GTP to smg p25A, a ras p21-like GTP-binding protein. *J. Biol. Chem.* **265,** 2333–2337.

Schievella, A. R., Chen, J. H., Graham, J. R., and Lin, L. L. (1997). MADD, a novel death domain protein that interacts with the type 1 tumor necrosis factor receptor and activates mitogen-activated protein kinase. *J. Biol. Chem.* **272,** 12069–12075.

Schweitzer, E. S., and Kelly, R. B. (1985). Selective packaging of human growth hormone into synaptic vesicles in a rat neuronal (PC12) cell line. *J. Cell Biol.* **101,** 667–676.

Shirataki, H., Kaibuchi, K., Sakoda, T., Kishida, S., Yamaguchi, T., Wada, K., Miyazaki, M., and Takai, Y. (1993). Rabphilin-3A, a putative target protein for smg p25A/rab3A p25 small GTP-binding protein related to synaptotagmin. *Mol. Cell. Biol.* **13,** 2061–2068.

Sivars, U., Aivazian, D., and Pfeffer, S. R. (2003). Yip3 catalyses the dissociation of endosomal Rab-GDI complexes. *Nature* **425,** 856–859.

Takai, Y., Sasaki, T., Shirataki, H., and Nakanishi, H. (1996). Rab3A small GTP-binding protein in Ca^{2+}-dependent exocytosis. *Genes Cells* **1,** 615–632.

Wada, M., Nakanishi, H., Satoh, A., Hirano, H., Obaishi, H., Matsuura, Y., and Takai, Y. (1997). Isolation and characterization of a GDP/GTP exchange protein specific for the Rab3 subfamily small G proteins. *J. Biol. Chem.* **272,** 3875–3878.

Wang, Y., Okamoto, M., Schmitz, F., Hofmann, K., and Sudhof, T. C. (1997). Rim is a putative Rab3 effector in regulating synaptic-vesicle fusion. *Nature* **388,** 593–598.

Wick, P. F., Senter, R. A., Parsels, L. A., Uhler, M. D., and Holz, R. W. (1993). Transient transfection studies of secretion in bovine chromaffin cells and PC12 cells. Generation of kainate-sensitive chromaffin cells. *J. Biol. Chem.* **268,** 10983–10989.

Yamaguchi, K., Tanaka, M., Mizoguchi, A., Hirata, Y., Ishizaki, H., Kaneko, K., Miyoshi, J., and Takai, Y. (2002). A GDP/GTP exchange protein for the Rab3 small G protein family up-regulates a postdocking step of synaptic exocytosis in central synapses. *Proc. Natl. Acad. Sci. USA* **99,** 14536–14541.

[22] Biochemical Characterization of Alsin, a Rab5 and Rac1 Guanine Nucleotide Exchange Factor

By Justin D. Topp, Darren S. Carney, and Bruce F. Horazdovsky

Abstract

Alsin is the gene product mutated in three juvenile-onset neurodegenerative disorders including amyotrophic lateral sclerosis 2 (ALS2). Sequence motif searches within Alsin predict the presence of Vps9, DH, and PH domains, implying that Alsin may function as a guanine nucleotide exchange factor (GEF) for Rab5 and a member of the Rho GTPase family. Procedures are presented in this chapter for the expression, purification, and biochemical characterization of the individual GEF domains of Alsin. A fractionation method is also described for the determination of Alsin's subcellular distribution. The presence of both Rac1 and Rab5 GEF activities makes Alsin a unique dual exchange factor that may couple endocytosis (via Rab5 activation) to cytoskeletal modulation (via Rac1 activation).

Introduction

Many aspects of cell growth, survival, and differentiation require the function of small monomeric GTP-binding proteins of the Ras superfamily. Small GTPases function largely as molecular switches that are regulated by their state of nucleotide binding. In the resting state, GTPases are bound to GDP. Activation via GTP exchange of GDP causes the GTPase to undergo

METHODS IN ENZYMOLOGY, VOL. 403   0076-6879/05 $35.00
                                                    DOI: 10.1016/S0076-6879(05)03022-3

a conformation change and enables it to interact specifically with a wide variety of effector proteins. The functional diversity of these effector proteins allows individual GTP-binding proteins to regulate many different cellular processes.

Activation of many GTPases is stimulated through the action of guanine nucleotide exchange factors (GEFs). GEFs act catalytically to promote the dissociation of GDP and subsequent loading of GTP on the GTPase. A multitude of proteins have been characterized that function as GEFs, most of which regulate a specific GTPase or subset of GTPases. It is becoming increasingly apparent that the activity of different exchange factors for the same GTPase can be regulated by distinct extracellular stimuli, which allows the coupling of extracellular cues to activation via specific GEFs. These aspects of specificity make activation by GEF proteins a key means of GTPase modulation. The importance of the proper activation of GTP-binding proteins is appreciated when considering the detrimental effects that mutations in exchange factors have on normal cellular function. In the case of the Rho GTPase family, mutations have been identified in Dbl and other Rho GEFs that induce transformation, presumably due to constitutive activity of the GEFs leading to Rho GTPase hyperactivation (Boettner and Van Aelst, 2002; Schmidt and Hall, 2002). In addition, mutations in Rho exchange factors that lead to neurological disorders have also been identified (Schmidt and Hall, 2002).

Over the past few years, studies have revealed that loss-of-function mutations in a gene that encodes a putative Rho and Rab guanine nucleotide exchange factor result in recessive forms of three juvenile-onset neurodegenerative diseases: amyotrophic lateral sclerosis 2 (ALS2), infantile-onset acquired hereditary spastic paraplegia (IAHSP), and primary lateral sclerosis (PLS) (Devon et al., 2003; Eymard-Pierre et al., 2002; Gros-Louis et al., 2003; Hadano et al., 2001; Yang et al., 2001). These disorders result in the progressive degeneration of motor neurons. The protein affected by these mutations, Alsin (Fig. 1), contains several intriguing domains including RCC1 (regulator of chromatin condensation-1) repeats, DH (Dbl homology) and PH (Pleckstrin homology)

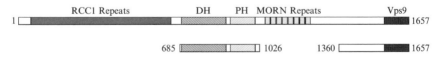

FIG. 1. Functional domains of Alsin. Shown are full-length Alsin and fragments of Alsin used for expression of DH/PH (residues 685–1026) and Vps9 (1360–1657) domains.

domains, MORN (membrane occupation and recognition nexus) repeats, and a Vps9 (vacuolar protein sorting 9) domain, implicating a role for Alsin in membrane trafficking and cytoskeletal modulation. Alsin has been shown to possess guanine nucleotide exchange activity for Rac1 and Rab5 that is mediated by the DH/PH domains and Vps9 domain, respectively (Kanekura et al., 2005; Otomo et al., 2003; Topp et al., 2004). Alsin is unique in that it is the only known protein in humans to possess both Rab5 and Rac1 GEF domains (Devon et al., 2005). All of the mutations identified are predicted to result in the premature truncation of Alsin lacking one or both of these domains, implying that the GEF activities of Alsin are crucial to motor neuron function.

This chapter presents methods used for the purification and functional analysis of the individual GEF domains of Alsin. These include demonstrations of Alsin's ability to bind specifically to and catalyze guanine nucleotide exchange on both Rac1 and Rab5. In addition, we present methods used to determine the subcellular distribution of Alsin in brain, which provide an initial framework for the purification of native Alsin protein.

Recombinant Expression and Purification of Rho and Rab GTPases, Alsin, and the GEF Domains of Alsin

Purification of Alsin and Its Individual Domains

A bacterial expression system is used to produce the Vps9 domain of Alsin (amino acids 1360–1657, Fig. 1) as a maltose-binding protein (MBP) fusion (Topp et al., 2004). The presence of the MBP tag greatly improves solubility of the Vps9 domain when compared to other tags such as His$_6$. The expression plasmid (pMBP-parallel 1) encoding the fusion is used to transform the bacterial strain HMS174 (DE3) (Novagen, Milwaukee, WI). One liter cultures of bacteria are grown at 37° to an OD$_{600}$ of 0.5–0.6, shifted to 25° for 15–30 min, and induced with 0.3 mM isopropyl-β-D-thiogalactoside (IPTG) for 10–15 h. Growth at 25° decreases overall recombinant protein production but results in an increase in soluble protein produced. After induction, cells are harvested and may be frozen at −80° for later use. A cell pellet from 1 liter of culture is resuspended in 20 ml MBP lysis buffer (20 mM Tris, pH 7.5, 200 mM NaCl, 1× protease inhibitor cocktail: N-tosyl-L-phenylalanine chloromethyl ketone, N^a-p-tosyl-L-lysine chloromethyl ketone, phenylmethylsulfonyl fluoride, leupeptin, trypsin inhibitor). All steps are carried out at 4°. Lysozyme is added to 1 mg/ml (from freshly made 10% stock in lysis buffer) and incubated for 30 min prior to probe sonication (four, 10-s pulses with a 30–45 s pause between each pulse). The homogenate is subject to centrifugation at 10,000×g for 10 min and the resulting supernatant is spun again for 1 h at 100,000×g. The supernatant from the second centrifugation is then diluted to 80 ml with MBP lysis buffer and incubated end over end for 1–2 h with 3 ml amylose resin (New England Biolabs, Beverly, MA), preequilibrated with MBP lysis buffer before adding it to the lysate. The mixture is applied to a 1-cm-diameter disposable column and the cell extract is allowed to flow through the resin by gravity. The resin is washed 10 times with 10 ml MBP lysis buffer and eluted with eight 1-ml applications of MBP elution buffer (MBP lysis buffer containing 10 mM maltose). Elution fractions are analyzed by sodium dodecyl sulfate-polyacylamide gel electrophoresis (SDS–PAGE) and Coomassie blue staining (Bio-Rad Laboratories, Hercules, CA) to check for yield and purity. Those containing the MBP-tagged Vps9 domain are pooled. Protein is then dialyzed overnight against 4 liters dialysis buffer (25 mM Tris, pH 7.5, 300 mM NaCl), concentrated, and stored at −80° for later use. Protein concentrations for all purifications described herein are determined by the Bio-Rad Bradford protein assay detection reagent (Bio-Rad Laboratories).

Full-length Alsin and the DH/PH domains of Alsin (amino acids 685–1026, Fig. 1) are prepared using a baculoviral expression system (Topp *et al.*, 2004). Recombinant baculoviruses are generated according to the manufacturer's BAC-to-BAC manual (Invitrogen, Inc., Carlsbad, CA). Briefly, bacmids encoding full-length Alsin or the DH/PH domains of Alsin as His$_6$ fusion proteins are generated and used to transfect Sf9 cells. Five days posttransfection, the cell culture medium containing the baculovirus is collected and further amplified by infecting 15 ml Sf9 cells for 5–6 days. The medium is then collected and used in all future viral infections. For protein production, Sf9 cells are infected with the appropriate baculovirus for 48–72 h and harvested by centrifugation. A time course and viral titration are recommended to determine optimal infection efficiency. In the case of Alsin, infections are performed at 1:100 for full-length and 1:500–1:1000 for individual domains, and cells are harvested after 48 h postinfection.

For the Rho GTPase binding assay described below, a cell pellet generated from a 200 ml Sf9 culture is resuspended in 10 ml His lysis buffer (50 mM NaPi, pH 8, 300 mM NaCl, 1 mM MgSO$_4$, 1% Nonidet P-40, 1× protease inhibitor cocktail) on ice. All steps to follow are carried out at 4°. Homogenates are lysed via douncing (30 strokes) and, if desired, passage through a 22-gauge needle (10 times) followed by centrifugation at 100,000×g for 1 h. The supernatant is then incubated end over end for 1–2 h with 1.5 ml preequilibrated Ni-NTA-agarose (Qiagen, Inc., Valencia, CA). The mixture is applied to a 1-cm-diameter disposable column and allowed to flow by gravity. The resin is washed five times with 10 ml His wash buffer (50 mM NaPi, pH 6, 300 mM NaCl, 1 mM MgSO$_4$, 1× protease inhibitor cocktail) and eluted in one 0.6-ml and two 1-ml fractions using His elution buffer (His wash buffer with 200 mM imidazole). For binding assays reported here, protein is used immediately following concentration, although storage at $-80°$ may be attempted for other procedures.

An enrichment in Alsin for binding assays may also be obtained by recombinant adenoviral infection in cell culture. We routinely infect SH-SY5Y cells with full-length Alsin-expressing adenovirus (see Topp *et al.* [2004] for adenovirus generation). Approximately 70–80% confluent SH-SY5Y cells (two 60-mm dishes) are infected with 10–25 plaque-forming units per cell. Cells are harvested 18–24 h later and suspended in 1 ml binding buffer (20 mM Tris, pH 7.5, 50 mM NaCl, 5% glycerol, 1 mM DTT, 10 mM EDTA, 0.1% Triton X-100, 1× protease inhibitor cocktail). Cells are lysed via 15 passages through a 25-gauge needle and centrifuged at $16,000 \times g$ for 10 min. The supernatant may be used in the binding assays below with the amount dependent on expression level; we recommend an inititial titration of supernatant.

Purification of GTPases

Plasmids (pGEX) encoding GST-Rho GTPase fusion proteins (RhoA, Cdc42, Rac1, Rac3) were kindly provided by P. Sternweis and M. Rosen (UT Southwestern Medical Center at Dallas). These vectors are transformed into HMS174 (DE3) *Escherichia coli* (Novagen). Then 500 ml cultures of bacteria are grown to an OD_{600} of 0.5–0.6, shifted to 25° for 15–30 min to equilibrate, and induced with 0.3 mM IPTG for 10–15 h. Cell pellets may be frozen at $-80°$ for later use. Cells are resuspended in 20 ml GST lysis buffer (20 mM HEPES, pH 8, 150 mM NaCl, 2 mM MgCl$_2$, 1 mM EDTA, 2 mM DTT, 1× protease inhibitor cocktail) on ice. All steps to follow are carried out at 4°. Lysozyme is added to 1 mg/ml (from freshly made 10% stock in lysis buffer) and the cell suspension is incubated for 30 min prior to sonication (as described above) and centrifugation at $10,000 \times g$ for 10 min. The supernatant is then centrifuged at $100,000 \times g$ for 1 h. The resulting supernatant is diluted 1:4 with GST lysis buffer and incubated end over end for 1–2 h with 0.5 ml preequilibrated glutathione-Sepharose (Amersham Biosciences, Piscataway, NJ). The mixture is applied to a 1-cm-diameter disposable column and allowed to flow by gravity. The resin is washed two times with 10 ml GST lysis buffer and eluted (six 0.8-ml elution steps) with GST lysis buffer containing 30 mM reduced glutathione. Yield and purity of each fraction are routinely checked by SDS–PAGE and Coomassie blue staining. Protein is then dialyzed overnight against 4 liters GTPase dialysis buffer (25 mM Tris, pH 7.5, 300 mM NaCl, 1 mM MgSO$_4$), concentrated, and stored at $-80°$ for later use.

Plasmids encoding His$_6$-Rab5 (a, b, c isoforms), His$_6$-Rab4, and His$_6$-Ypt1 (Topp *et al.*, 2004) are transformed into HMS174 (DE3) *E. coli*. Cultures (500 ml) of bacteria are grown, induced, and harvested as above. The cell pellets may be frozen at $-80°$ for later use. Cells are resuspended in 10 ml His lysis buffer (50 mM NaPi, pH 8, 300 mM NaCl, 1 mM MgSO$_4$, 1× protease inhibitor cocktail). Lysozyme is added to 1 mg/ml (from freshly made 10% stock in lysis buffer) and incubated for 30 min prior to sonication as described above. The homogenate is then pelleted at $27,000 \times g$ for 20 min and the resulting supernatant diluted 1:4 with His lysis buffer (as described above). The lysate is incubated with 2 ml preequilibrated Ni-NTA-agarose end over end for 1–2 h. The lysate/resin mixture is poured into a 1-cm-diameter disposable column and washed three times with 10 ml of His lysis buffer, three times with 10 ml His wash buffer, and eluted with His elution buffer (eight 1-ml elution steps). After an overnight dialysis against 4 liters GTPase dialysis buffer, the Rab proteins are routinely stored at $-80°$ for later use.

GTPase Binding Assays for Alsin

In Vivo *Yeast Two-Hybrid Assays*

Since other proteins that contain Vps9 domains interact with and catalyze guanine nucleotide exchange on Rab5 or the yeast homologue Vps21p (Hama *et al.*, 1999; Horiuchi *et al.*, 1997; Kajiho *et al.*, 2003; Saito *et al.*, 2002; Tall *et al.*, 2001), it was originally proposed that Alsin would also have this activity (Hadano *et al.*, 2001; Yang *et al.*, 2001). Indeed, we and others have shown that Alsin functions as a GEF for Rab5 (Otomo *et al.*, 2003; Topp *et al.*, 2004). In this and the next section, we describe experiments that may be performed to demonstrate an interaction between Alsin and Rab5. Methods such as *in vitro* binding assays or yeast two-hybrid assays may be used for these studies and both are described below.

For yeast two-hybrid assays, L40 yeast are transformed with plasmids encoding two fusion proteins: (1) the LexA DNA-binding domain fused to Rab5 ("bait") (Tall *et al.*, 2001) and (2) the Gal4 activation domain fused to the Vps9 domain of Alsin ("prey"). (Topp *et al.*, 2004). Similar experiments may be performed with full-length Alsin. Both wild-type and S34N Rab5 should be used as the S34N mutation abolishes the interaction between the γ-phosphate of GTP and Rab5 leaving the Rab in the GDP-bound or nucleotide-free state. Previous experiments have shown that this mutation in Rab5 stabilizes the interaction between Rab5 and another Vps9 domain-containing protein, Rin1 (Tall *et al.*, 2001).

Interaction between "bait" and "prey" in the L40 yeast strain drives expression of the LexA target genes, *HIS3* and *lacZ*. This interaction may be monitored by growth in the absence of histidine (L40 requires histidine to grow unless *HIS3* transcription is activated) or by scoring β-galactosidase activity in a filter assay (Vojtek *et al.*, 1993). The use of empty LexA DNA-binding domain "bait" or Gal4 activation domain "prey" plasmids provides a necessary negative control. As shown in Fig. 2, this method proves useful in demonstrating an interaction between the Vps9 domain of Alsin and Rab5 and highlights the importance of nucleotide specificity for this interaction.

In Vitro *Binding Assays*

The presence of DH and PH domains in tandem is a hallmark of proteins that possess GEF activity for GTPases of the Rho family. While the Rab5 family exchange factors can easily be predicted by computational methods, GEFs for distinct Rho family members have very similar sequences and cannot be easily differentiated. Therefore, it is necessary when studying a putative Rho GEF to first determine with which Rho GTPase the potential GEF interacts.

Binding assays are performed following the protocol described by Hart *et al.* (1994). All steps are performed at room temperature. Five micrograms of purified His_6-Rab or GST-Rho GTPases (see above) is added to 25–50 μl glutathione-Sepharose or Ni-NTA-agarose resins (preequilibrated with H_2O), brought up to 400 μl total volume with H_2O and incubated end

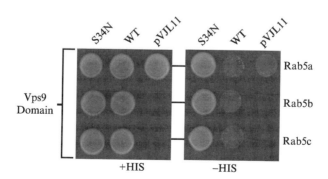

FIG. 2. Yeast two-hybrid assay with Alsin and Rab5. L40 yeast are transformed with plasmids encoding wild-type or S34N Rab5 LexA DNA-binding domain "bait" (or empty vector alone) and Alsin Vps9 domain Gal4 activation domain "prey" fusions. All strains grow on yeast supplemented with histidine but only strains in which "bait" and "prey" interact grow in the absence of histidine.

over end for 1 h. After conjugating the GTPases to the appropriate resins, the beads are pelleted and resuspended in 500 μl of binding buffer (20 mM Tris, pH 7.5, 50 mM NaCl, 5% glycerol, 1 mM DTT, 10 mM EDTA, 0.1% Triton X-100, 1× protease inhibitor cocktail). This reaction is incubated end over end for 1 h and serves to deplete the GTPase of nucleotide. This step is important as the interaction between Rho GTPases and their GEFs is thought to be enhanced when the GTPase is nucleotide free. During this incubation, Alsin protein (His$_6$-DH/PH-Alsin domains purified from Sf9 cells for Rho binding assay, or SH-SY5Y cell lysate overexpressing full-length untagged Alsin for Rab binding assay) is equilibrated in 500 μl of binding buffer. After 1 h, the GTPase/beads are pelleted, suspended in 500 μl binding buffer, and combined with the 500 μl Alsin/binding buffer solution. This mixture is incubated end over end for an additional hour. The beads are then pelleted and washed twice with 1 ml binding buffer. The protein is eluted by adding 50 μl SDS sample buffer (125 mM Tris, pH 6.8, 6% SDS, 10% α-mercaptoethanol, 20% glycerol, 0.02% bromophenol blue) and heating for 5 min at 95°. The resultant samples are analyzed by SDS-PAGE and Western blotting. With higher protein levels, the assay may be monitored by SDS-PAGE and Coomassie blue staining. Figure 3 shows an example of binding experiments in which Alsin interacts specifically with Rac1 and Rab5.

GEF Activity Assays for Alsin

Guanine nucleotide exchange factors promote the dissociation of GDP and subsequent loading of GTP on the GTPases with which they interact. Assays that measure GEF function generally monitor GDP release or GTP loading. Herein we describe *in vitro* GDP release and *in vivo* GTP loading assays to study the Rab5 and Rac1 GEF activities of Alsin, respectively.

In Vitro *Rab Nucleotide Exchange Assay*

In the *in vitro* GDP dissociation assay, Rab (100 pmol) is preincubated with 15 μM [³H]GDP for 30 min at 30° in 50 μl (total volume) Buffer A (50 mM Tris, pH 7.5, 1 mM DTT, 3 mM EDTA). After loading with [³H] GDP, an equal volume of Buffer B (50 mM Tris, pH 7.5, 1 mM DTT, 10 mM MgSO$_4$, 4 mM unlabeled GDP) containing 300 pmol of the Vps9 domain is added and the reaction returned to 30°. At 0 and 30 min, 20 μl of the sample is removed and quenched by the addition of 1 ml ice-cold Buffer C (50 mM Tris, pH 7.5, 5 mM MgSO$_4$). The sample is then applied to a nitrocellulose membrane (0.45 μM pore size) and washed twice with 5 ml ice-cold Buffer C. After drying the filter, the [³H]GDP still bound to the

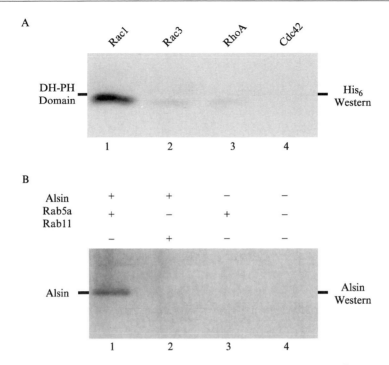

FIG. 3. Binding assays with Alsin. Purified individual domains of Alsin (Rho GTPase binding assay A) or lysates overexpressing Alsin (Rab binding assay B) are incubated with GST-Rho and His₆-Rab GTPases conjugated to the appropriate resins, and the resulting complexes are pelleted. SDS-PAGE and Western blotting reveal the specificity of Alsin's interaction with Rac1 and Rab5.

Rab is measured using a liquid scintillation counter. GEF activity stimulates release of [^{3}H]GDP from the Rab and, upon normalization to the counts at time 0, the percent Rab [^{3}H]GDP bound can be determined as shown in Fig. 4. To monitor the intrinsic rate of [^{3}H]GDP release by Rab, replace the Vps9 domain of Alsin with an equal amount of bovine serum albumin (BSA). Alternatively, a mantGDP-based fluorescence assay may be used to follow stimulated nucleotide release, as described by Davies *et al.* (this volume, Chapter 49).

In Vivo *Rac1 Nucleotide Exchange Assay*

Experiments similar to the *in vitro* exchange experiment described above for Rab5 may be performed to assay the ability of Alsin to function as a Rac1 guanine nucleotide exchange factor. However, we have been

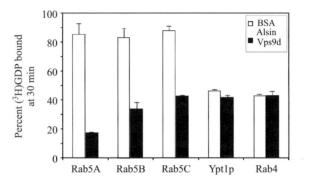

FIG. 4. *In vitro* Rab5 GDP dissociation assay. Purified Rab protein (100 pmol) is loaded with 15 μM [^{3}H]GDP, and [^{3}H]GDP release is monitored after the addition of BSA or the Vps9 domain of Alsin (300 pmol). At various 0 and 30 min time points, samples are removed, quenched, and counted by liquid scintillation counting.

unable to demonstrate Rac1 GEF activity by Alsin in these assays and have proposed that Alsin requires a posttranslational modification or a cofactor for maximal activity (Topp *et al.*, 2004). Indeed, phosphorylation or PH domain-mediated interaction with phosphoinositides has been shown previously to be necessary for the exchange activity of several Rho GEFs (discussed in Topp *et al.*, 2004).

As an alternative to the *in vitro* assays, an *in vivo* GTP loading assay has been developed to facilitate the study of Rho exchange factors. In this method, GTPase and its putative GEF are coexpressed in cells and GTP-bound protein is extracted from the cell lysate. For our studies with Alsin, we used Sf9 cells expressing Rac1 and Alsin as our model system. To isolate GTP-bound Rac1, a fusion construct consisting of the Rac1-binding domain of the effector protein p21-activated kinase (PAK) (a gift of P. Sternweis, UT Southwestern Medical Center at Dallas) fused to GST (referred to as GST-PBD) is expressed in HMS 174 (DE3) *E. coli*. Bacteria (200 ml) are grown at 37° to an OD_{600} of 0.6 and induced with 0.3 mM IPTG for 3 h at 37°. Cells are harvested and lysed in 10 ml PBD lysis buffer (20 mM HEPES, pH 7.5, 120 mM NaCl, 2 mM EDTA, 10% glycerol, 1× protease inhibitor cocktail). After sonication (twice for 15 s each), the homogenate is centrifuged for 30 min at 27,000×g. The supernatant is adjusted to 0.5% Nonidet P-40 (from a 10% stock in PBD lysis buffer) and incubated with 300 μl glutathione-Sepharose beads end over end at 4° for 1 h. The beads are then pelleted and washed five times in 10 ml PBD lysis buffer with 0.5% Nonidet P-40 and three times with 10 ml PBD lysis buffer alone. The beads are resuspended in approximately 0.5 ml PBD lysis buffer, which should result in GST-PBD concentrations of 0.5–3 mg/ml. GST-PBD can be stored at 4° for 1–2 weeks without significant functional loss.

Ten to twenty milliliter cultures of Sf9 cells are infected with Alsin (1:100; see above) and Rac1 (1:100; a gift of P. Sternweis, UT Southwestern Medical Center at Dallas) recombinant baculovirus and cells are harvested 48 h later. Cells are resuspended in 2 ml of PBD-binding buffer (50 mM Tris, pH 7.5, 500 mM NaCl, 0.5 mM MgCl$_2$, 1% Triton X-100, 0.1% SDS, 0.5% deoxycholate, 1× protease inhibitor cocktail) and lysed by douncing (30 strokes) and 10 passages through a 22-gauge needle. The lysate is centrifuged at 16,000×g for 10 min and the resulting supernatant is used in the assay. After determining cell lysate protein concentrations, 2 mg is incubated end over end with 50 μg GST-PBD at 4° for 1 h. GTP-bound Rac1 is isolated using the purified GST-PBD peptide. During this step, a portion of the cleared lysate is saved into an equal volume of SDS sample buffer and heated at 95° for 5 min, serving as the input control. After the 1 h incubation, the beads are pelleted and washed four times with 1 ml PBD wash buffer (50 mM Tris, pH 7.5, 150 mM NaCl, 0.5 mM MgCl$_2$, 1% Triton X-100, 1× protease inhibitor cocktail). The beads are then resuspended in 75 μl SDS sample buffer and boiled at 95° for 5 min. The samples are analyzed by SDS-PAGE and Western blotting with a monoclonal antibody to Rac (Upstate Cell Signaling Solutions, Waltham, MA; dilution factor 1:2000) and quantified by densitometry. It is recommended that inputs be loaded at 1:50 or 1:25 the amount loaded for the GST-PBD pulldown since typically only less than 5% of the total Rac1 is in the GTP-bound state. In the case of Alsin, negative controls for this experiment include cells expressing a form of Alsin lacking the PH/DH domains (amino acids 1–705) or no Alsin at all. An example of a typical experiment is shown in Fig. 5.

Fractionation Methods for Alsin Subcellular Distribution in Brain

Fractionation Procedure

Rab5 activation plays a key role in endosomal fusion and its impact on cell surface receptor trafficking has been well characterized. In addition, the role of Rac1 in growth factor-induced cytoskeletal modulation is well appreciated. Thus, Alsin may function to regulate endosomal or cytoskeletal dynamics through its Rab5 and Rac1 GEF domains. To begin to dissect Alsin's overall function, it is important to uncover the subcellular distribution of this protein.

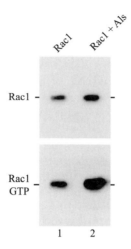

FIG. 5. *In vivo* Rac1 GTP loading assay. Sf9 cells coexpressing Alsin and Rac1 (or expressing Rac1 alone) are lysed and the GTP-bound Rac1 is isolated by incubation with the GST-PBD peptide. SDS-PAGE and Western blotting are used to visualize Rac1, and after densitometry, the GTP-bound Rac1 (pulled down with GST-PBD) is normalized to the total Rac1 in the lysate (input). Results from six experiments are averaged to provide a fold-increase with Alsin expression (3.9 ± 2.2, $p = 0.02$).

Both indirect immunofluorescence and fractionation procedures can be used to determine the compartment within the cell in which a protein resides. In some cases overexpression systems must be used to add visualization. However, overexpressed proteins may become mislocalized to other cellular locations, making interpretation difficult. Described here is a protocol for fractionating brain tissue to analyze the localization pattern of endogenous Alsin. We have used this procedure to follow Alsin localization in cerebellum, as this region of the brain has been shown to be enriched for this protein (Devon *et al.*, 2005; Hadano *et al.*, 2001; Otomo *et al.*, 2003; Yang *et al.*, 2001). Fractionations may be performed with fresh or frozen brains, as both seem to show similar Alsin distribution patterns. In addition, Alsin localization is similar when using both mouse or rat brains; a fresh rat cerebellum was the source for the data presented in Fig. 6.

One rat or three mouse cerebella are lysed in 5 ml cerebellum-lysis buffer (5 mM Tris, pH 7.5, 0.32 M sucrose, 0.5 mM CaCl$_2$, 1 mM MgCl$_2$, 1× protease inhibitor cocktail) by douncing (30 strokes) and passage through an 18-gauge needle (10–20 passages). While it is possible to use sonication to improve the lysis step, it is not recommended because this can perturb membrane structures. Centrifuge the homogenate at 500×g for 10 min,

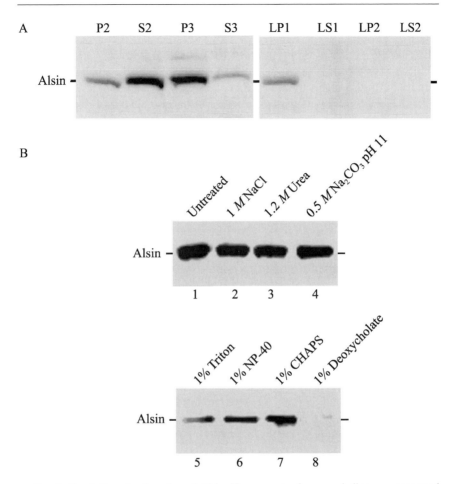

FIG. 6. Cerebellum fractionation of Alsin. Homogenates from cerebellum are generated and separated by differential centrifugation. A. Alsin is predominantly found in the P3 fraction. B. The nature of Alsin's P3 association was determined by treatment of S2 prior to the P3 spin with various reagents.

generating P1 (pellet) and S1 (supernatant) fractions. Centrifuge the resulting supernatant (S1) at 10,500×g for 15 min, generating P2 and S2 fractions. Centrifuge the resulting supernatant (S2) at 165,000×g for 2 h, generating P3 and S3 fractions. For each of these sets of samples, the pellets are suspended in SDS sample buffer, and a portion of the supernatants is added to 5 × SDS sample buffer (0.312 M Tris, pH 6.8, 10% SDS,

25% 2-mercaptoethanol, 0.05% bromophenol blue). It is recommended that the amount of material from P2 and S2 analyzed be adjusted so that equal portions of the input fraction (S1) are represented. This makes it possible to determine if protein was lost due to technical issues or protease activity at any step throughout the process.

Endosomes and endocytic vesicles, both characterized by the presence of Rab5, have been shown to separate predominantly with the P3 fraction (Devon et al., 2005; Lee et al., 2001; Topp et al., 2004). The S3 fractions contain soluble cytoplasmic proteins or proteins whose weak association with membranes is lost during the fractionation procedure. The P2 fraction contains membranous organelles such as ER, Golgi, and plasma membrane, as well as synaptosomes. Synaptosomes are synapses that have pinched off to enclose both pre- and postsynaptic membranes and their protein constituents. To further fractionate P2, resuspend it (generated from 1 ml homogenate as described above) in 50 μl cerebellum lysis buffer and hypotonically lyse by resuspending in 450 μl H$_2$O, passing through an 18- or 20-gauge needle 10 times to aid in the lysis. Centrifuge at 25,000$\times g$ for 20 min, generating LP1 (pellet) and LS1 (supernatant) fractions. Centrifuge the resulting supernatant at 165,000$\times g$ for 2 h, generating LP2 and LS2 fractions. The LS2 corresponds to soluble synaptosomal proteins while the LP1 and LP2 samples are enriched in synaptosomal membranes and synaptic vesicles, respectively. As shown in Fig. 6A, it can be easily seen that Alsin is primarily associated with the P3 cerebellar fraction.

Treatment of P3 Fraction with Agents to Determine the Nature of Alsin's Association

To determine the nature of Alsin's association with the P3 membrane fraction, incubate the fraction (either S2 or P3) with various reagents that perturb membrane association. Salts such as NaCl and urea will strip off weakly associated proteins and leave the membranes intact, while detergents will solubilize the membranes to varying degrees. For the experiments shown in Fig. 6B, 40 μl of a S2 fraction is added to 10 μl of the various 5$\times$ reagents on ice for 30 min with light vortexing several times throughout the incubation. The mixture was then spun for 2 h at 165,000$\times g$ to generate P3 and S3 fractions. The ability of a treatment to strip Alsin from the particulate fraction is shown in Fig. 6B. Only detergents were able to remove Alsin from the particulate cell fraction, indicating that at least a portion of Alsin is associated with cellular membranes. The fractionation methods described here will prove useful in developing a large-scale strategy for the purification of endogenous Alsin from brain.

References

Boettner, B., and Van Aelst, L. (2002). The role of Rho GTPases in disease development. *Gene* **286,** 155–174.

Devon, R. S., Helm, J. R., Rouleau, G. A., Leitner, Y., Lerman-Sagie, T., Lev, D., and Hayden, M. R. (2003). The first nonsense mutation in alsin results in a homogeneous phenotype of infantile-onset ascending spastic paralysis with bulbar involvement in two siblings. *Clin. Genet.* **64,** 210–215.

Devon, R. S., Schwab, C., Topp, J. D., Orban, P. C., Yang, Y. Z., Pape, T. D., Helm, J. R., Davidson, T. L., Rogers, D. A., Gros-Louis, F., Rouleau, G., Horazdovsky, B. F., Leavitt, B. R., and Hayden, M. R. (2005). Cross-species characterization of the ALS2 gene and analysis of its pattern of expression in development and adulthood. *Neurobiol. Dis.* **18,** 243–257.

Eymard-Pierre, E., Lesca, G., Dollet, S., Santorelli, F. M., di Capua, M., Bertini, E., and Boespflug-Tanguy, O. (2002). Infantile-onset ascending hereditary spastic paralysis is associated with mutations in the alsin gene. *Am. J. Hum. Genet.* **71,** 518–527.

Gros-Louis, F., Meijer, I. A., Hand, C. K., Dube, M. P., MacGregor, D. L., Seni, M. H., Devon, R. S., Hayden, M. R., Andermann, F., Andermann, E., and Rouleau, G. A. (2003). An ALS2 gene mutation causes hereditary spastic paraplegia in a Pakistani kindred. *Ann. Neurol.* **53,** 144–145.

Hadano, S., Hand, C. K., Osuga, H., Yanagisawa, Y., Otomo, A., Devon, R. S., Miyamoto, N., Showguchi-Miyata, J., Okada, Y., Singaraja, R., Figlewicz, D. A., Kwiatkowski, T., Hosler, B. A., Sagie, T., Skaug, J., Nasir, J., Brown, R. H., Jr, Scherer, S. W., Rouleau, G. A., Hayden, M. R., and Ikeda, J. E. (2001). A gene encoding a putative GTPase regulator is mutated in familial amyotrophic lateral sclerosis 2. *Nat. Genet.* **29,** 166–173.

Hama, H., Tall, G. G., and Horazdovsky, B. F. (1999). Vps9p is a guanine nucleotide exchange factor involved in vesicle-mediated vacuolar protein transport. *J. Biol. Chem.* **274,** 15284–15291.

Hart, M. J., Eva, A., Zangrilli, D., Aaronson, S. A., Evans, T., Cerione, R. A., and Zheng, Y. (1994). Cellular transformation and guanine nucleotide exchange activity are catalyzed by a common domain on the dbl oncogene product. *J. Biol. Chem.* **269,** 62–65.

Horiuchi, H., Lippe, R., McBride, H. M., Rubino, M., Woodman, P., Stenmark, H., Rybin, V., Wilm, M., Ashman, K., Mann, M., and Zerial, M. (1997). A novel Rab5 GDP/GTP exchange factor complexed to Rabaptin-5 links nucleotide exchange to effector recruitment and function. *Cell* **90,** 1149–1159.

Kajiho, H., Saito, K., Tsujita, K., Kontani, K., Araki, Y., Kurosu, H., and Katada, T. (2003). RIN3: A novel Rab5 GEF interacting with amphiphysin II involved in the early endocytic pathway. *J. Cell Sci.* **116,** 4159–4168.

Kanekura, K., Hashimoto, Y., Kita, Y., Sasabe, J., Aiso, S., Nishimoto, I., and Matsuoka, M. (2005). A Rac1/phosphatidylinositol 3-kinase/Akt3 anti-apoptotic pathway, triggered by AlsinLF, the product of the ALS2 gene, antagonizes Cu/Zn-superoxide dismutase (SOD1) mutant-induced motoneuronal cell death. *J. Biol. Chem.* **280,** 4532–4543.

Lee, S. H., Valtschanoff, J. G., Kharazia, V. N., Weinberg, R., and Sheng, M. (2001). Biochemical and morphological characterization of an intracellular membrane compartment containing AMPA receptors. *Neuropharmacology* **41,** 680–692.

Otomo, A., Hadano, S., Okada, T., Mizumura, H., Kunita, R., Nishijima, H., Showguchi-Miyata, J., Yanagisawa, Y., Kohiki, E., Suga, E., Yasuda, M., Osuga, H., Nishimoto, T., Narumiya, S., and Ikeda, J. E. (2003). ALS2, a novel guanine nucleotide exchange factor for the small GTPase Rab5, is implicated in endosomal dynamics. *Hum. Mol. Genet.* **12,** 1671–1687.

Saito, K., Murai, J., Kajiho, H., Kontani, K., Kurosu, H., and Katada, T. (2002). A novel binding protein composed of homophilic tetramer exhibits unique properties for the small GTPase Rab5. *J. Biol. Chem.* **277,** 3412–3418.

Schmidt, A., and Hall, A. (2002). Guanine nucleotide exchange factors for Rho GTPases: Turning on the switch. *Genes Dev.* **16,** 1587–1609.

Tall, G. G., Barbieri, M. A., Stahl, P. D., and Horazdovsky, B. F. (2001). Ras-activated endocytosis is mediated by the Rab5 guanine nucleotide exchange activity of RIN1. *Dev. Cell* **1,** 73–82.

Topp, J. D., Gray, N. W., Gerard, R. D., and Horazdovsky, B. F. (2004). Alsin is a Rab5 and Rac1 guanine nucleotide exchange factor. *J. Biol. Chem.* **279,** 24612–24623.

Vojtek, A. B., Hollenberg, S. M., and Cooper, J. A. (1993). Mammalian Ras interacts directly with the serine/threonine kinase Raf. *Cell* **74,** 205–214.

Yang, Y., Hentati, A., Deng, H. X., Dabbagh, O., Sasaki, T., Hirano, M., Hung, W. Y., Ouahchi, K., Yan, J., Azim, A. C., Cole, N., Gascon, G., Yagmour, A., Ben-Hamida, M., Pericak-Vance, M., Hentati, F., and Siddique, T. (2001). The gene encoding alsin, a protein with three guanine-nucleotide exchange factor domains, is mutated in a form of recessive amyotrophic lateral sclerosis. *Nat. Genet.* **29,** 160–165.

[23] Purification and Analysis of RIN Family-Novel Rab5 GEFs

By Kota Saito, Hiroaki Kajiho, Yasuhiro Araki, Hiroshi Kurosu, Kenji Kontani, Hiroshi Nishina, and Toshiaki Katada

Abstract

The small GTPase Rab5 plays important roles in membrane budding and trafficking in the early endocytic pathways, and the activation of this GTPase is mediated by several guanine nucleotide exchange factors (GEFs) at each of the transport steps. The RIN family has been identified as GEFs for Rab5 and shown to possess unique biochemical properties. The RIN family preferentially interacts with an activated form of Rab5, although it enhances guanine nucleotide exchange reaction. Moreover, biochemical analysis indicates that the RIN family functions as a tetramer. In this chapter, we describe the isolation of the recombinant RIN family via expression in *Spodoptera frugiperda* (Sf9) insect cells and in mammalian cells. In addition, functional analysis is also provided to assess the physiological properties of the RIN family.

METHODS IN ENZYMOLOGY, VOL. 403
    DOI: 10.1016/S0076-6879(05)03023-5

Introduction

The small GTPase Rab5, which cycles between GTP-bound active and GDP-bound inactive states, is involved not only in the homotypic fusion process of early endosomes but also in the budding of clathrin-coated vesicles from plasma membrane and its transport to early endosomes (Zerial and McBride, 2001). The involvement of Rab5 in the many processes indicates that the activation of Rab5 is tightly coupled with the progression of each transport step. In the inactive state, Rab5 forms a cytoplasmic complex with a regulatory protein, Rab GDP-dissociation inhibitor (RabGDI), which prevents an association with improper cellular compartments. However, Rab5 replaces GDP with GTP through its interactions with the guanine nucleotide exchange factors (GEFs) at the target membranes. Rabex-5 was first identified as a mammalian Rab5-GEF, and it contained the sequence homologous to the yeast vacuolar protein-sorting 9 (Vps9) protein (Horiuchi et al., 1997). Vps9p acts as a GEF for Vps21p, a yeast homologue of Rab5 GTPase (Hama et al., 1999). At present, several proteins containing the Vps9 domain that act as GEFs for Rab5 have been identified and considered to regulate certain transport steps in accordance with their regulation and localization (Horiuchi et al., 1997; Kajiho et al., 2003; Otomo et al., 2003; Saito et al., 2002; Tall et al., 2001; Topp et al., 2004).

We recently identified members of a novel family of Rab5-GEFs, RIN2 and RIN3, which contain an Src homology 2 (SH2) domain, proline-rich region, and Ras-association domain in addition to a Vps9 domain (Kajiho et al., 2003; Saito et al., 2002). Intriguingly, the RIN family preferentially interacts with the activated form of Rab5, although it enhances guanine nucleotide exchange reaction on Rab5. Moreover, the RIN family is composed of a homophilic tetramer and interacts with amphyphysin II, which regulates the early steps of the endocytic pathway. This chapter describes the isolation of the recombinant RIN family via expression in *Spodoptera frugiperda* (Sf9) insect cells and in mammalian cells. In addition, a functional analysis is also provided to assess the physiological properties of the RIN family.

Methods

Purification of the FLAG-RIN Family from Sf9 Cells

Construction and Selection of RIN-Containing Baculovirus. Bac-To-Bac Baculovirus Expression Systems were purchased from Invitrogen. pFastBac HTa vectors were mutagenized to FLAG-tagged vectors with polymerase chain reaction (PCR) using following primers.

Sense: 5′ AGGATGACGACGATAAGGATTACGATATCCCAA-
CGACC 3′
Antisense: 5′ TGTAATCCATGGTGGCGGTTTCGGACCGAGA-
TCCG 3′

Extended double-stranded DNAs were phosphorylated followed by ligation and subcloning. The products were verified by sequencing and renamed pFastBac FLAGa vector. Full-length cDNAs encoding human
RIN2 and RIN3 were cloned in the *Eco*RI and *Sal*I sites of the pFastBac
FLAGa vectors. These vectors were transformed into DH10Bac competent *Escherichia coli* cells so that bacmids were generated by transposition
in *E. coli* cells. The transformed cells were plated in Luria Agar plates
containing 50 μg/ml kanamycin, 7 μg/ml gentamicin, 10 μg/ml tetracycline,
100 μg/ml Bluo-gal, and 40 μg/ml isopropyl-β-D-thiogalactopyanoside
(IPTG). White colonies were picked up, and the successful transposi-
tions were verified by PCR. The bacmid DNAs were precipitated with
the standard alkali method.

Expression of the RIN Family in Sf9 Cells. Sf9 cells were grown in EX-
CELL 420 (JRH Biosciences) medium supplemented with 100 IU/ml penicillin and 100 μg/ml streptomycin at 27° in a 500-ml glass bottle with
stirring. Sf9 cells (approximately 10^6 cells) were attached in a 35-mm well
plate and transfected with bacmids using CellFECTIN reagents (Invitrogen). After 72 h, virus stocks were harvested by centrifuging at $500 \times g$ for 5
min. To amplify virus stocks, Sf9 cells attached in plates were infected with
virus at a multiplicity of infection of 0.01–0.1 for 48 h. Amplification of
virus was done twice to obtain higher-titer virus. To produce FLAG-tagged
RIN proteins, Sf9 cells were grown to a density of 5×10^6 cells/ml in a
500-ml glass bottle with stirring. The cells were infected at a multiplicity of
infection of 5–10 and incubated for 120 h. The cells were harvested and
washed twice with phosphate-buffered saline (PBS) and frozen in liquid N_2
before storage at −80°.

Purification of the RIN Family from Sf9 Cells. For purification
of FLAG-tagged RIN proteins from Sf9 cells, the cell pellet was resuspended with 25 ml of an extraction buffer consisting of 40 mM *N*-2-
hydroxyethylpiperazine-*N*′-2-ethanesulfonic acid (HEPES)-NaOH, pH 7.4,
75 mM NaCl, 15 mM NaF, 1 mM Na$_3$ VO$_4$, 10 mM Na$_4$P$_2$O$_7$, 2 mM EDTA,
1 μg/ml leupeptin, and 2 μg/ml aprotinin, and subsequently mixed with 25
ml of the extraction buffer containing 2% (w/v) Nonidet P-40 (NP-40).
The cell lysate was rotated for 15 min at 4° and then centrifuged at
$108,000 \times g$ for 30 min with an angle rotor. The supernatants were
collected and applied to a Bio-Rad open column loaded with Sepharose
CL4B (bed volume: 1 ml, Amersham Biosciences). Fractions flowed

through the column were collected and then incubated with a 500-μl bed volume of anti-FLAG M2 agarose beads (Sigma). After being rotated for 90 min at 4°, the mixture was applied to the open column. The flow-through fraction was again loaded to the column to ensure the maximum binding of RIN proteins to the column. The column was washed twice with 5 ml of a wash buffer consisting of 40 mM Tris–HCl (pH 8.0), 1 mM EDTA, 100 mM NaCl, and 0.6% (w/v) 3-[(3-cholamidopropyl) dimethylammonio]-1-propanesulfonic acid (CHAPS). The protein-bound beads were suspended in 1 ml of the wash buffer containing 100 μg/ml FLAG peptide (Sigma) and further incubated for 30 min at 4° with occasional vortexing. The eluted fraction was collected, and the beads were further eluted with 1.5 ml of the elution buffer. To remove the FLAG peptide from eluted fractions, the whole fractions were applied to the PD-10 column (Amersham Biosciences) that had been equilibrated with the wash buffer. The void-volume fractions containing RIN proteins were frozen in liquid N_2 and stored at −80°. A half-liter culture of the infected Sf9 cells yields approximately 200–300 μg of FLAG-tagged RIN proteins (Fig. 1A, top panel).

Assays for Guanine-Nucleotide Exchange and GDP-Dissociation Reactions

Materials. Prenylated Rab5b was purified from baculovirus-infected Sf9 cells according to the method described by Horiuchi *et al.* (1995, see Fig. 1A, bottom), with slight modification: GDP was added to each buffer at a final concentration of 10 μM. Purified Rab5b (180 nM) was suspended in a buffer consisting of 25 mM Tris–HCl (pH 8.0), 25 mM NaCl, 1 mM EDTA, 5 mM MgCl$_2$, 0.5 mM dithiothreitol (DTT), and 0.6% (w/v) CHAPS.

GDP-Dissociation Reaction. The dissociation of [^{3}H]GDP was assayed by measuring the decrease in the radioactivity of [^{3}H]GDP-bound Rab5b trapped on the nitrocellulose filters (ADVANTEC). [^{3}H]GDP-bound Rab5b was first made by incubating purified Rab5b (approximately 1 pmol) for 30 min at 30° in a reaction mixture (9.5 μl) consisting of 20 mM Tris–HCl (pH 8.0), 62.5 mM NaCl, 10 mM EDTA, 5 mM MgCl$_2$, 0.5 mM DTT, 0.36% (w/v) CHAPS, and 5 μM [^{3}H]GDP (10,000 cpm/pmol). After the incubation, 0.5 μl of 600 mM MgCl$_2$ was added to give a final concentration of 20 mM, and the mixture was immediately cooled on ice to prevent the dissociation of [^{3}H]GDP from Rab5b. The dissociation of [^{3}H]GDP from Rab5b was measured in a reaction mixture (65 μl) consisting of 40 mM Tris–HCl (pH 8.0), 62.5 mM NaCl, 5 mM EDTA, 15 mM MgCl$_2$, 0.5 mM DTT, 0.36% (w/v) CHAPS, 120 μM unlabeled GDP, and 40 μM GTP in the presence or absence of FLAG-RIN2 and RIN3 (approximately 20 pmol) by incubating at 30° for indicated times. The reaction was stopped

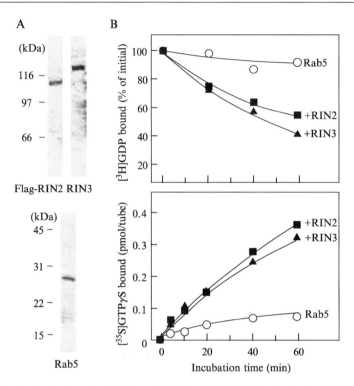

FIG. 1. Analysis of purified RINs and Rab5 by SDS-PAGE and effects of RINs on the guanine-nucleotide exchange reaction of Rab5. (A) Flag-tagged RIN2, RIN3 (top), and prenylated Rab5b (bottom) purified from baculovirus-infected Sf9 cells were separated by SDS-PAGE and stained with Coomassie brilliant blue. (B) Dissociation of [^{3}H]GDP from Rab5b (top) and [^{35}S]GTPγS binding to Rab5b (bottom) were measured in the presence or absence of RINs.

by the addition of approximately 2 ml of an ice-cold buffer consisting of 20 mM Tris–HCl (pH 8.0), 20 mM MgCl$_2$, and 100 mM NaCl, followed by rapid filtration on the nitrocellulose filters. The filters were washed five times with the same ice-cold buffer. After filtration, the radioactivity was counted. As shown in Fig. 1B (top), dissociation of [^{3}H]GDP from Rab5b was markedly stimulated in the presence of RIN2 or RIN3.

Guanine-Nucleotide Binding Reaction. The binding of [^{35}S]GTPγS was assayed by measuring the radioactivity of [^{35}S]GTPγS-bound Rab5b trapped on nitrocellulose filters. To avoid nonspecific guanine-nucleotide binding to various proteins, excess amount (50-fold to GTPγS) of ATP was added to the reaction mixture. Purified Rab5b (approximately

1 pmol) was incubated with RIN2 or RIN3 (15 pmol each) in a reaction mixture (50 μl) consisting of 40 mM Tris–HCl (pH 8.0), 62.5 mM NaCl, 5 mM EDTA, 15 mM MgCl$_2$, 0.5 mM DTT, 0.36% (w/v) CHAPS, 50 μM ATP, and 1 μM [^{35}S]GTPγS (20,000 cpm/pmol) at 30° for indicated times. The reaction was stopped by the same method as described for the GDP-dissociation reaction. As shown in Fig. 1B (bottom), [^{35}S]GTPγS binding to Rab5b was markedly stimulated in the presence of RIN2 or RIN3.

Assay for In Vitro Association between the Rab5b and RIN Family

Preparation of Nucleotide-Bound Forms of Rab5b Proteins. To prepare GTPγS- and GDP-bound forms of Rab5b, the purified protein was incubated with the nucleotides (250 μM) at 30° for 45 min in 11.2 mM Tris–HCl (pH 8.0), 50 mM HEPES-NaOH (pH 7.5), 110 mM NaCl, 0.5 mM DTT, 0.27% (w/v) CHAPS, 5 mM EDTA, and 2.2 mM MgCl$_2$. The reaction was terminated by adding MgCl$_2$ to the final concentration of 10 mM.

Expression of the FLAG-RIN Family in Mammalian Cells. COS7 or HeLa cells were maintained in Dulbecco's modified Eagle's medium (DMEM) supplemented with 10% fetal calf serum (FCS), 0.16% (w/v) NaHCO$_3$, 0.6 mg/ml L-glutamine, 100 IU/ml penicillin, and 100 μg/ml streptomycin at 37° in 95% air and 5% CO$_2$. For cell electroporation with FLAG-RIN2 or RIN3, the cells (1 × 10^7 cells) were trypsinized and washed twice with 20 ml of Opti-MEM (Invitrogen) and resuspended in 0.2 ml of Opti-MEM. The cell suspension was mixed with 10 μg of plasmids and transferred to a 0.4-cm gap cuvette (Bio-Rad) and incubated on ice for 15 min. The cells were resuspended and electroporated at 220 V, 960 μF with a Bio-Rad Gene Pulser. The cells were diluted with 20 ml of DMEM and cultured for 2 days at 37°.

Purification of the FLAG-RIN Family from COS7 Cells. The transfected COS7 cells were washed twice with PBS and solubilized with Buffer A consisting of 40 mM HEPES-NaOH (pH 7.4), 75 mM NaCl, 15 mM NaF, 1 mM Na$_3$ VO$_4$, 10 mM Na$_4$P$_2$O$_7$, 2 mM EDTA, 1 μg/ml leupeptin, 2 μg/ml aprotinin, and 1% (w/v) NP-40 by vortexing. After gently rotated for 15 min at 4°, lysates were centrifuged at 20,000×g for 15 min at 4°. Supernatants were precleared with 20 μl of anti-mouse IgG agarose beads (50% slurry, American Qualex, Inc.) and immunoprecipitated with 10 μl of the agarose beads that had been conjugated with 0.5 μg of the anti-FLAG monoclonal antibody. The beads were washed three times with 1 ml of 20 mM Tris–HCl (pH 7.5), 150 mM NaCl, and 1% (w/v) NP-40, and three more times with a buffer consisting of 50 m*M* HEPES-NaOH (pH 7.5), 100 m*M* NaCl, 7.7 m*M* MgCl$_2$, 2 m*M* EDTA, and 0.1% (w/v) NP-40. FLAG-tagged RIN was eluted from the beads with the same buffer containing 1 mg/ml of FLAG peptide.

Assay for In Vitro *Association between the Rab5 and RIN Family.* To assay the association between the Rab5 and RIN family, the GDP- or GTPγS-bound form of Rab5b was mixed with the anti-mouse IgG agarose beads (10 μl) containing FLAG-RIN2, FLAG-RIN3, or FLAG peptide alone and incubated at 25° for 60 min in 0.25 ml of 50 m*M* HEPES-NaOH (pH 7.4), 10 m*M* MgCl$_2$, 5 m*M* EDTA, 0.5 m*M* DTT, 100 m*M* NaCl, and 0.1% (w/v) NP-40. The agarose beads were washed four times with a buffer (500 μl) consisting of 50 m*M* Tris–HCl (pH 8.0), 100 m*M* NaCl, 7 m*M* MgCl$_2$, 2 m*M* EDTA, and 0.2% (w/v) NP-40. Proteins were eluted from the beads with 30 μl of the same buffer containing 1 mg/ml of FLAG peptide. After centrifugation, 24 μl of the supernatant was mixed with 8 μl of 4 × SDS sample buffer, boiled for 5 min, and followed by SDS-PAGE. Immunoblotting was performed with the anti-FLAG M2 mono-clonal antibody (Sigma) and the anti-Rab5b polyclonal antibody A-20 (Santa Cruz), respectively.

Gel Filtration Analysis of the RIN Family

Purification of the Recombinant RIN Family from HeLa cells. Trans-fected HeLa cells were washed twice with PBS and solublized with Buffer A then diluted with Buffer A not containing NP-40 to lower the concen-tration of NP-40 to 0.5% (w/v). After being gently rotated for 15 min at 4°, lysates were centrifuged at 20,000×*g* for 15 min at 4°. Supernatants were precleared with Sepharose CL4B beads (Amersham Biosciences) and im-munoprecipitated with anti-FLAG M2 agarose beads by gently rotating for 2 h at 4°. The beads, being washed three times with 0.5 ml of Tris-buffered saline/0.1% (w/v) NP-40, were further washed three more times with Buffer B consisting of 75 m*M* Tris–HCl (pH 7.5), 1 m*M* EDTA, 100 m*M* NaCl, and 0.2% (w/v) NP-40. For SDS-PAGE analysis, the beads were mixed with 24 μl of Buffer B and 12 μl of 4× sample buffer and boiled for 2 min. The samples were resolved by SDS-PAGE, and immunoblotting was performed with anti-FLAG antibody.

Gel Filtration Analysis of the RIN Family. For gel filtration analysis, the beads were washed six times with 0.5 ml of Buffer C consisting of 25 m*M* Tris–HCl (pH 7.5), 150 m*M* NaCl, and 0.2% (w/v) CHAPS, and subse-quently eluted with 0.2 ml of Buffer C containing 1 mg/ml of FLAG peptide by gently vortexing for 1 h at 4°. The eluted fraction was diluted with an equal volume of Buffer C and applied to a Superdex 200 column (HR 10/30 Amersham Biosciences) that had been equilibrated with Buffer C. Elution was carried out at room temperature at a flow rate of 0.5 ml/min with a fraction volume of 0.25 ml. The fractions were concentrated by precipitation with 10% (final) trichloroacetic acid and subjected to SDS-PAGE and silver staining. The fractions were also analyzed by SDS-PAGE and immunoblotted with anti-FLAG antibody. The elution profile of the column was calibrated with the sizing standards (Oriental Yeast Co. Ltd.) of glutamate dehydrogenase (290 kDa), lactate dehydrogenase (142 kDa), enolase (67 kDa), adenylate kinase (32 kDa), and cytochrome c (12.4 kDa).

References

Hama, H., Tall, G. G., and Horazdovsky, B. F. (1999). Vps9p is a guanine nucleotide exchange factor involved in vesicle-mediated vacuolar protein transport. *J. Biol. Chem.* **274,** 15284–15291.

Horiuchi, H., Ullrich, O., Bucci, C., and Zerial, M. (1995). Purification of posttranslationally modified and unmodified Rab5 protein expressed in Spodoptera frugiperda cells. *Methods Enzymol.* **257,** 9–15.

Horiuchi, H., Lippe, R., McBride, H. M., Rubino, M., Woodman, P., Stenmark, H., Rybin, V., Wilm, M., Ashman, K., Mann, M., and Zerial, M. (1997). A novel Rab5 GDP/GTP exchange factor complexed to Rabaptin-5 links nucleotide exchange to effector recruitment and function. *Cell* **90,** 1149–1159.

Kajiho, H., Saito, K., Tsujita, K., Kontani, K., Araki, Y., Kurosu, H., and Katada, T. (2003). RIN3: A novel Rab5 GEF interacting with amphiphysin II involved in the early endocytic pathway. *J. Cell Sci.* **116,** 4159–4168.

Otomo, A., Hadano, S., Okada, T., Mizumura, H., Kunita, R., Nishijima, H., Showguchi-Miyata, J., Yanagisawa, Y., Kohiki, E., Suga, E., Yasuda, M., Osuga, H., Nishimoto, T., Narumiya, S., and Ikeda, J. E. (2003). ALS2, a novel guanine nucleotide exchange factor for the small GTPase Rab5, is implicated in endosomal dynamics. *Hum. Mol. Genet.* **12,** 1671–1687.

Saito, K., Murai, J., Kajiho, H., Kontani, K., Kurosu, H., and Katada, T. (2002). A novel binding protein composed of homophilic tetramer exhibits unique properties for the small GTPase Rab5. *J. Biol. Chem.* **277,** 3412–3418.

Tall, G. G., Barbieri, M. A., Stahl, P. D., and Horazdovsky, B. F. (2001). Ras-activated endocytosis is mediated by the Rab5 guanine nucleotide exchange activity of RIN1. *Dev. Cell* **1,** 73–82.

Topp, J. D., Gray, N. W., Gerard, R. D., and Horazdovsky, B. F. (2004). Alsin is a Rab5 and Rac1 guanine nucleotide exchange factor. *J. Biol. Chem.* **279,** 24612–24623.

Zerial, M., and McBride, H. (2001). Rab proteins as membrane organizers. *Nat. Rev. Mol. Cell Biol.* **2,** 107–117.

[24] Purification and Functional Properties of a Rab8-Specific GEF (Rabin3) in Action Remodeling and Polarized Transport

By KATARINA HATTULA and JOHAN PERÄNEN

Abstract

Considering the large number of Rab proteins, only a few Rab-specific exchange factors have been found and characterized. Rab8 is involved in mediating polarized membrane traffic through reorganization of actin and microtubules. It is possible to use the yeast two-hybrid technique to find potential Rab activators. A human protein (Rabin8) and its rat equivalent (Rabin3) were found to bind Rab8 and function as nucleotide exchange factors for Rab8 but not for Rab3A and Rab5. Endogenous and ectopically expressed Rabin8 frequently colocalize with cortical actin. This association is increased by cytochalasin D and phorbol esters that also induced the translocation of both Rabin8 and Rab8 to lamellipodia-like structures. We also show that a GFP-fused Rabin8 behaves identically in this respect. Furthermore, coexpression of Rabin8 with the dominant negative mutant of Rab8 leads to translocation of Rabin8 onto vesicular structures enriched in cell protrusions, indicating that both Rab8 and Rabin8 are involved in mediating polarized membrane transport. This chapter presents a detailed description of the methods and protocols developed to find and characterize a Rab8-specific activator.

Introduction

Rab proteins, the largest family (> 60 members) of Ras-like small GTPases, are major regulators of membrane trafficking in eukaryotic cells (Zerial and McBride, 2001). The specific localization of each Rab protein is directly related to the transport route it regulates. The increasing amounts of effector molecules through which Rab proteins mediate membrane trafficking indicate that Rab proteins regulate not only vesicle docking and fusion but also vesicle motility (Zerial and McBride, 2001). However, much less is known about Rab-specific guanine nucleotide exchange factors (GEFs) and GTPase-activating proteins (GAPs) that regulate the activity of Rabs. There are two known Rab3-specific GEFs, GRAB and Rab3 GEP, that participate in neurotransmitter release (Luo *et al.*, 2001;

METHODS IN ENZYMOLOGY, VOL. 403
0076-6879/05 $35.00
DOI: 10.1016/S0076-6879(05)03024-7

Tanaka *et al.*, 2001; Wada *et al.*, 1997). In addition, several different GEFs (Rabex-5, Als2, RIN1, and RIN3) for Rab5 have been described that participate in endocytic membrane fusion and transport (Horiuchi *et al.*, 1997; Kajiho *et al.*, 2003; Otomo *et al.*, 2003; Tall *et al.*, 2001).

Expression of Rab8 mutants does not affect the transport kinetics of newly synthesized proteins to the plasma membrane (Ang *et al.*, 2004; Peränen *et al.*, 1996). However, Rab8 mediates polarized membrane transport through reorganization of actin and microtubules, and it has a drastic effect on cell shape (Ang *et al.*, 2003; Armstrong *et al.*, 1996; Hattula and Peränen, 2000; Peränen *et al.*, 1996). To better understand the function of Rab8 we are searching for Rab8-specific effectors and activators. To find activators we used a Rab8-GDP binding mutant (Rab8T22N) as bait in a yeast two-hybrid system (Hattula *et al.*, 2002). The most abundant interactor was the Mss4 protein that is known to bind to numerous Rab proteins along the excytosis pathway (Burton *et al.*, 1994). About 10% of the Rab8 interactors found show high identity with a rat protein called Rabin3 (Brondyk *et al.*, 1995; Hattula *et al.*, 2002). We were able to show that this protein, which we call Rabin8, promotes GDP dissociation and GTP association on Rab8, indicating that it is a true GEF. Furthermore, it was specific for Rab8, because it showed no GEF activity toward Rab3A and Rab5.

We could also demonstrate that rat Rabin3 is a Rab8-specific GEF, indicating that it is the rat equivalent of the human Rabin8 molecule. Determining the localization of Rab GEFs is of importance in understanding where Rab proteins are activated. Both endogenous Rabin8 and ectopically expressed Rabin8 were found localized to the cytoplasm and to the cortical actin. The association of Rabin8 with the cortical actin was dramatically increased by incubating cells in the presence of cytochalasin D and phorbol esters. A GFP-fused Rabin8 behaved in the same way in the presence of these reagents and may be useful when studying signals activating the Rab8 pathway (Fig. 1). Coexpression of Rabin8 with the GDP-bound Rab8 (T22N) results in translocation of Rabin8 onto membrane vesicles found in cell protrusions. The COOH-terminus of Rabin8 is essential for the association of Rabin8 with Rab8T22N-containing vesicles. This has also been shown to be the case for the Sec2p protein that is the yeast equivalent of Rabin8 (Elkind *et al.*, 2000). At present it is unclear whether Rabin8 associates with Rab8T22N-specific vesicles approaching the plasma membrane, or whether these vesicles are moving from the plasma membrane toward the cell center. Nonetheless, the close association of Rabin8 with actin elements is in accordance with the fact that activated Rab8 is involved in the reorganization of actin.

GFP-Rabin8   Phalloidin

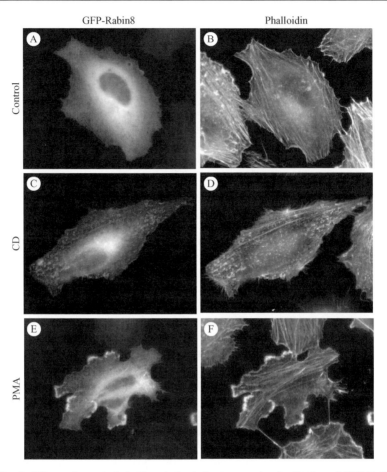

FIG. 1. Effects of cytochalasin D and phorbol esters on the distribution of GFP-Rabin8. HeLa cells were transfected overnight with a plasmid encoding GFP-Rabin8. Cells in (C) and (D) were pretreated with 0.1 μM cytochalasin D (CD) for 20 min and cells in (E) and (F) with 100 ng/ml phorbol 12-myristate 13-acetate (PMA) 30 min prior to fixation and staining with phalloidin. Untreated control cells (A and B).

This chapter describes the methods and techniques we have devised to find a Rab8-specific activator. It also shows how Rabin8 and different Rab proteins are expressed in bacteria, purified in their active form for biochemical assays and for production of antibodies. We also describe different techniques to monitor localization and behavior of endogenous and expressed Rabin8 in animal cells.

Yeast Two-Hybrid Screen

The idea behind our study was to find potential Rab8-specific activators. The yeast two-hybrid approach was chosen because it has previously been successful in isolating many Rab interacting proteins. Furthermore, the validity of the interaction can be verified by using different nucleotide-binding mutants and other members of the Ras small GTPase family.

Constructs

pAS2-Rab8bΔT22N and pAS2-Rab8bΔQ67L are constructed by inserting the *Nde*I/*Pst*I fragments of pGEM-Rab8bΔT22N and pGEM-Rab8b-ΔQ67L, respectively, into the pAS2-1 vector. The original multiple cloning sites of pGilda and pB42AD were replaced by inserting new adapters (5′-AATTCGCTAGCAGGCCTATGGCCATGGAGGCCCCGGGAGA-TC-3′, 5′-CTGAGATCTCCCGGGGCCTCCATGGCCATAGGCCTGC TAGCG-3′, 5′-AATTCATGGCCATGGAGGCCCCGGGCGGCCGCA CTAGTTCTAGAGGATCC-3′, and 5′-TCGAGGATCCTCTAGAACT AGTGCGGCCGCCCGGGGCCTCCATGGCCATG-3′) into the *Eco*-RI/*Xho* sites of pGilda and pB42AD creating plasmids pGilda-B and pB42AD-B. pGilda-Rab8Awt, pGilda-Rab8AT22N, pGilda-Rab8AQ67L, pGilda-Rab8b, pGilda-Rab8bT22N, pGilda-Rab8bQ67L, pGilda-Rab3A36N, and pGilda-Rab220N are constructed by inserting respective full-length open reading frames from pGEMmyc-Rab8A, pGEMmyc-Rab8-AT22N, pGEMmyc-Rab8Q67L, pGEM-Rab8b, pGEM-Rab8bT22N, and pGEM-Rab8bQ67L into pGilda-B (Peränen and Furuhjelm, 2000; Peränen *et al.*, 1996). pGEM-Rab3A and pGEM-Rab2 are created by amplifying the open reading frames of Rab3A and Rab2 by polymerase chain reaction (PCR) from human brain cDNA and HeLa cDNA, respectively, and cloned into pGEM-3. pGEM-Rab3AT36N and pGEM-Rab2T20N are obtained by creating indicated mutations by site-specific mutagenesis as previously described (Peränen *et al.*, 1996). pGilda-Rab3AT36N and pGilda-Rab2T20N were constructed by inserting corresponding open reading frames from pGEM-Rab3AT36N and pGEM-Rab2T20N into the *Nco*I/*Bam*HI sites of pGilda-B.

Procedure

Two different commercial yeast two-hybrid systems, the Matchmaker GAL4 Two-Hybrid System (Clontech, Palo Alto, CA) and the MATCH-MAKER LexA system (Clontech), are used to screen Rab8 interacting molecules. The initial screen is done with the Gal4-based system (Clontech, Palo Alto, CA). The pAS2-Rab8bΔT22N plasmid (a dominant negative mutant of Rab8b deleted of its lipid modification motif—CSLL) is used as a bait to screen a human brain cDNA Matchmaker GAL4 Two-Hybrid library (Clontech). Transformation is done according to the manual (Clontech). The positive clones from transformed yeast are recovered using the bacteria DH5α as a recipient. The specificity of positive clones is then tested against the original pAS2-Rab8bΔT22N, pAS2-Rab8bΔQ67L, and pAS2-laminin (Clontech). Of 10 million clones screened, 70 were true positives; 50% of the positives were Mss4, and 10% contained sequences with a high degree of homology to rat Rabin3, which we named Rabin8.

To obtain a full-length human Rabin8 open reading frame, we use a Rabin8 fragment (600 nt) obtained from the original two-hybrid clone as a probe to screen a human brain cDNA lambda Triplex human brain cDNA library (Clontech) by a PCR-based method (Israel, 1993). In the first round 64 wells with 8000 phages per well are screened. PCR on extracts from phage pools is performed and the PCR products are run on a 1% agarose gel. The agarose gel is blotted overnight to nitrocellulose (BA-85, Scleicher & Schuell, Dassel, Germany) and then probed with a Rabin8-specific oligonucleotide. One positive well was found. This is tittered and subdivided into 64 new pools, now with 150 phages per well. The screened is repeated in the same way, and one positive well was found. Now the phages from the positive well are screened by plaque lift on nitrocellulose filters according to the Triplex manual (Clontech). Positive plaque are picked and eluted overnight by incubating in lambda dilution buffer at +4°. The phage DNA is converted to plasmid form according to the Triplex manual. The inserts of two independent clones are sequenced. Full-length open reading frames were present in both clones. The open reading frame is then amplified by PCR from one of these clones and inserted into pGEM-3 to create plasmid pGEM-Rabin8. Sequence comparisons and evaluations are done with the Basic Local Alignment Search Tool (BLAST; http://www.ncbi.nlm.nih.gov). Domain and motif searches were done with the SMART tool (http://smart.embl-heidelberg.de/).

To test whether the Rabin8 retains its interacting specificity toward Rab8bΔT22N in another two-hybrid system, the Rabin8 open reading frame is cloned into pB42AD-B (Clontech; see above) of the lexA-based two-hybrid system (Clontech) and tested against different Rab GTPases: pGilda-Rab8Awt, pGilda-Rab8AT22N, pGilda-Rab8AQ67L, pGilda-Rab8b, pGilda-Rab8bT22N, pGilda-Rab8bQ67L, pGilda-Rab3A36N, and pGilda-Rab220N. To pinpoint the binding site of Rab8 on Rabin8, PCR mutagenesis is used to construct deletion of Rabin8 (1–120 aa, 1–221 aa, 1–316 aa, 101–316 aa, 222–460 aa, and 306–460 aa). These deletions are cloned into pB42AD-B, verified by sequencing, and binding toward Rab8bΔT22N is tested. The binding results showed that Rab8T22N binds to the coiled-coil region of Rabin8.

In Vitro Binding

To verify the two-hybrid interacting results between Rab8 and Rabin8, a rapid in vitro binding procedure was set up.

Constructs

The Rabin8 open reading frame is cloned into pGEX2T (Amersham Pharmacia, Piscataway, NJ) to create plasmid pGEX-Rabin8. pGEM-Rab8AQ67L and pGEM-Rab8AT22N have been previously described (Peränen et al., 1996). pGEM-Rabin8/D6 is created by amplifying an open reading frame encompassing the 1–313 aa from pGEM-Rabin8 and inserting it into pGEM-3.

Procedure

The pGEX-Rabin8 plasmid is transformed into BL21 cells, and the expression is done with 200 μM isopropylthiogalactoside (IPTG) at 15° overnight to increase the solubility of GST-Rabin8. Cells are then pelleted and resuspended in cold lysis buffer (20 mM Tris–HCl, pH 7.5, 150 mM NaCl, 1% Triton X-100, 0.4 mM phenylmethylsulfonyl fluoride [PMSF]). Cells are lysed using a French press at 1200 psi. The cell lysate is then centrifuged for 15 min at 13,000 rpm in a microcentrifuge. The supernatant fraction containing soluble GST-Rabin8 is incubated with glutathione-agarose beads (Sigma) at +4° for 1 h and then washed three times during 30 min with lysis buffer. Control beads with bound GST were done as for GST-Rabin8, but induction of GST from the pGEX vector is at 37° for 3 h instead of 15° overnight. Rab8Q67L, Rab8T22N, Rabin8 (1–316 aa), and full-length Rabin8 are translated from corresponding plasmids in vitro using a TNT Quick kit (Promega) according to the manufacturer's instructions. The in vitro translation products (two-thirds of the reaction mixture) are then incubated with GST-Rabin8 or GST-coupled glutathione agarose beads in binding buffer (50 mM Tris, pH 7.5, 150 mM NaCl, 2 mM MgCl$_2$, 1% Triton X-100, 0.4 mM PMSF) under rotation at +4° for 1 h. The beads are washed four times with binding buffer during 30 min. Bound material is eluted from the beads with 2 × Laemmli sample buffer and run on a 12% sodium dodecyl sulfate (SDS)-polyacrylamide gel. As a control, input material (one-fifth of the original reaction mix) of the in vitro translation reactions is also loaded onto the gel. The gel is dried under vacuum pressure and the protein bands are visualized by autoradiography.

Assay for GEF Activity

Contructs

Especially Rab8 is very difficult to express in a soluble form in *Escherichia coli* (Peränen *et al.*, 1996). To circumvent this we use a new expression system based on the fusion of proteins to the NusaA protein (Davis *et al.*, 1999). For expression of Rabin8 in *E. coli*, full-length Rabin8 is cloned as a *Bam*HI–*Sal*I fragment into the NusaA fusion vector pET43a (Novagen, Madison, WI). The rat Rabin3 open reading frame is amplified by PCR from PC12 cDNA, sequenced, and cloned into the *Bam*HI–*Sal*I site of pET43. Likewise, we also create pET43-Rab3A, pET43-Rab5, and pET-Rab8 by inserting the open reading frames taken from pGEM-Rab3A, pGEM-Rab5, and pGEM-Rab8 into pET43a.

Protein Expression and Purification

NusA-His-Rabin8 was expressed from the pET43-Rabin8 vector induced at 37° for 3 h with 0.5 mM IPTG. The cells pelleted from a 1-liter culture are broken by French press in lysis buffer (20 mM Tris–HCl, pH 8.5, 50 mM NaCl, PMSF), and the lysate is centrifuged at 15,000×g for 10 min. The supernatant (50 ml) is loaded onto an anion-exchange Mono Q column (Pharmacia, Biotech), and the fractions containing Rabin8 (eluted at 0.6 M NaCl) are pooled and concentrated from 20 to 2 ml. An aliquot (0.5 ml) of this concentrate is loaded onto a Superdex-75 Hiload gel exclusion chromatography column (Pharmacia Biotech) equilibrated with 20 mM Tris–HCl, pH 8.5, 100 mM NaCl, and 0.4 mM PMSF. Fractions containing Rabin8 are analyzed by SDS-PAGE and are then pooled, and concentrated in Centricon 10-kDa cutoff devices from 4 to 1 ml. This material was snap-frozen in liquid nitrogen and stored at −70°, and then used for GEF assays. The purification of NusA-His-rat Rabin3 is done exactly as described earlier for Rabin8.

Rab8-fusion proteins (GST, thioredoxin, dihydrofolate reductase [DHFR]) are mostly insoluble in all tested expression vectors and conditions, but a NusA-His-Rab8 fusion is soluble. NusA-His-Rab8 was expressed from the pET43 vector at 37° for 3 h with 0.1 mM IPTG. Cells from a 500 ml culture are pelleted and resuspended in lysis buffer (50 mM phosphate buffer, pH 7.0, 300 mM NaCl, 5 mM MgCl$_2$, 200 mM GDP, 5 mM 2-mercaptoethanol, 0.5% Triton X-100, 10% glycerol, and PMSF) and lysed by a French press (1200 psi). The lysate is run in a table centrifuge at 13,000 rpm for 10 min at 4°. The fusion protein from the supernatant fraction is bound at +4° for 30 min to Talon resin beads (Clontech). The beads are washed four times during 40 min with the same buffer not containing PMSF, and then three times in Thrombin buffer (50 mM Tris–HCl, pH 7.5, 150 mM NaCl, 5 mM MgCl$_2$, 2.5 mM CaCl$_2$). The beads with bound protein are incubated with 15 U thrombin (Sigma, St. Louis, MO) overnight at +4°. The beads and the supernatant are applied to an empty column and the eluate containing cleaved soluble wild-type Rab8 is recovered, snap-frozen in liquid nitrogen, and then stored at −70°. The yield (1 mg from a 1-liter culture) of Rab8 is considerably smaller than that expected from the total soluble fraction. We found that a large fraction of NusA-Rab8 was not bound to the Talon column, probably because the His-tag might have been masked. However, we also saw that much of the cleaved Rab8 was retained on the beads and could be eluted only by Laemmli sample buffer, indicating that this fraction of Rab8 is incorrectly folded and therefore stuck to the beads. Rab3A and Rab5 are expressed and purified in the same way as described for Rab8. It should be noted that the NusA-His-Rab3A and NusA-His-Rab5 proteins do not show the same tendency to stick to the beads as Rab8.

GDP/GTP Exchange Assays

To measure potential GDP/GTP exchange activity of Rabin8, we first followed the decrease in radioactive [³H]GDP bound to Rab3A, Rab5, and Rab8 in the presence or absence of Rabin8. The assay is initiated by pre-loading purified wild-type Rab8wt (20 pmol), Rab3A (20 pmol), and Rab5 (20 pmol) with 100 pmol [³H]GDP (10 Ci/mmol; Amersham, TRK 335) diluted in preloading buffer (20 mM HEPES, pH 7.2, 5 mM EDTA, 1 mM dithiothreitol [DTT]) for 15 min at 30°. The reaction mixtures are then transferred to ice, and MgCl$_2$ is added to a final concentration of 10 mM. The reactions are started by addition of reaction buffer (20 mM HEPES, pH 7.2, 10 mM MgCl$_2$, 1 mM GDP, 1 mM DTT) with or without purified NusA-Rabin8 (10 pmol), NusA-rat Rabin3 (10 pmol), or NusA (10 pmol) to a total reaction volume of 50 μl. The reaction mixtures are incubated at 30° for varying periods (0, 5, 10, and 30 min). Samples of 5 μl at each time point are diluted into 2 ml of ice-cold wash buffer (20 mM Tris–HCl, pH 8.0, 20 mM NaCl, 10 mM MgCl$_2$, 1 mM DTT) and directly filtered through wet) nitrocellulose filters (BA-85, Scleicher & Schuell, Dassel, Germany) that are connected to a homemade filtration apparatus. The filters were washed twice with 3 ml ice-cold wash buffer and dried before adding the scintillation fluid (Optiphase High Safe-3; Wallac, Turku, Finland).

The ability of Rabin8 to catalyze [³⁵S]GTPγS binding is assayed in the same way as the GDP release, but preloading is done with 30 pmol cold GDP nucleotide. The reaction buffer contains 1 mM ATP to prevent hydrolysis of [³⁵S]GTPγS (1117 Ci/mmol; Amersham, SJ 1320). The reaction contained 20 pmol wild-type Rab8 and 22 Ci/mmol of $[^{35}S]GTP\gamma S$. Samples are taken in the same way (0, 2, 5, 15, and 30 min) as described earlier and counted after overnight incubation in scintillation fluid.

Expression of the Rab8-Specific GEF in Cells and Tissues

Procedure

To produce specific antibodies to Rabin8, we first clone the full-length human Rabin8 from pGEM-Rabin into the pGAT-2 bacterial expression plasmid (Peränen and Furuhjelm, 2000). The His-GST-Rabin8 fusion protein is purified under denaturing conditions and prepared for immunization of rabbits as previously described for Rab8 (Peränen et al., 1996). Antibodies to Rabin8 are affinity purified by use of nitrocellulose strips containing recombinant Rabin8 (Peränen, 1992). The presence of endogenous Rabin8 in human cells is done by using commercial available cell extracts (A431, endothelial cell, HeLa, Jurkat; Trans lab) that are separated on a 12% SDS-PAGE, blotted to nitrocellulose filters, and probed by affinity purified anti-Rabin8 and monoclonal anti-Rab8 antibodies (Trans lab) as described previously (Peränen et al., 1996).

To study the expression of Rabin8 transcripts in human tissues, we use a human 12-lane Multiple Tissue Northern Blot (Clontech) that we hybridize overnight in UltraHyb solution (Ambion, Austin, TX) using standard hybridization conditions. As a probe we use a 496-bp SpeI–HindIII fragment from the full-length clone that is labeled with $[^{32}P]dCTP$ (Amersham) with a random labeling kit (Amersham). The control probe (β-actin), provided with the MTN filter, is labeled and hybridized to the stripped filter in the same way. Visualization is done either by autoradiography or by a phosphoimager.

Subcellular Localization

Constructs

Full-length Rabin8 and Rabin8D6 (1–316 aa) are provided with a myc-tag by PCR and cloned into the pEGFP-N1 vector (Clontech; J. Peränen, unpublished observations). To obtain an EGFP-fused Rabin8, the full-length Rabin8 open reading frame is taken from pGEM-Rabin8 and inserted into pEGFP-C1A (Peränen and Furuhjelm, 2000). For expression of His-Rab8T22N and His-Rab8Q67L open reading frames in mammalian cells, we cleave them out from plasmids pGEM-His-Rab8T22N and pGEM-His-Rab8Q67L and insert them into pEGFP-N1 (Hattula and Peränen, 2000).

Cell Culture and Transfections

For transfection studies we use HeLa cells that are cultured in Dulbecco's modified Eagle's medium (DMEM) supplemented with 10% fetal calf serum (FCS), 2 mM glutamine, 100 U/ml penicillin, and 100 μg/ml streptomycin. All transfection experiments are done on cells that have been grown overnight on coverslips in 35-mm plates. Cells are transiently transfected using the Fugene 6 transfection reagent according to the manufacturer (Roche, Indianapolis, IN). Alternatively, we use magnet-assisted transfection (MATra, IBA, Göttingen, Germany). Shortly, for HeLa cells grown on 35-mm plates we first add 1 μg of plasmid to 500 μl of OptiMEM (GIBCO), mix, and then 1 μl of MATra, mix, and incubate at room temperature for 20 min. We then aspirate 500 μl from the growth medium and replace it with 500 μl transfection mixture, and mix. The cell plates are then put on the magnetic plate (MATra, IBA, Göttingen, Germany) for 15 min. The cells are left in the incubator for 4 h, after which the transfection medium is replaced with fresh growth medium (DMEM, 10% FCS, 2 mM glutamine, 100 U/ml penicillin, and 100 μg/ml streptomycin). All manipulations of cells with different inhibitors are done 20 h after transfection. The inhibitors are freshly diluted in Optimem; cytochalasin D (0.1 or 1 μM) is incubated with the cells for 20 min, nocodazole (1 μg/ml) for 60 min, and phorbol 12-myristate 13-acetate (PMA; 100 ng/ml) for 30 min.

Confocal and Immunofluorescence Microscopy

Cells are prepared for immunofluorescence or confocal microscopy by fixing them with 4% paraformaldehyde, permeabilizing them with 0.1% Triton X-100, and staining them with appropriate antibodies as previously described (Peränen and Furuhjelm, 2000). Goat anti-rabbit IgG lissamine and goat anti-mouse IgG-FITC secondary antibodies are from Jackson Immunoresearch (West Grove, PA). Actin is detected in fixed cells by Alexa488-conjugated phalloidin or Texas-Red XS-conjugated phalloidin (Molecular Probes, Eugene, OR). Fluorescence of fixed cells is analyzed either with the Bio-Rad MRC-1024 confocal system (Hercules, CA) linked to a Zeiss Axiovert 135 M microscope (Thornwood, NY) or with an Olympus fluorescence microscope (Lake Success, NY).

References

Ang, A. L., Folsch, H., Koivisto, U. M., Pypaert, M., and Mellman, I. (2003). The Rab8 GTPase selectively regulates AP-1B-dependent basolateral transport in polarized Madin-Darby canine kidney cells. *J. Cell Biol.* **163,** 339–350.

Armstrong, J., Thompson, N., Squire, J. H., Smith, J., Hayes, B., and Solari, R. (1996). Identification of a novel member of the Rab8 family from the rat basophilic leukemia cell line, RBL.2H3. *J. Cell Sci.* **109**, 1265–1274.

Brondyk, W. H., McKiernan, C. J., Fortner, K. A., Stabila, P., Holz, R. W., and Macara, I. G. (1995). Interaction cloning of Rabin3, a novel protein that associates with the Ras-like GTPase Rab3A. *Mol. Cell. Biol.* **15**, 1137–1143.

Burton, J. L., Burns, M. E., Gatti, E., Augustine, G. J., and De Camilli, P. (1994). Specific interactions of Mss4 with members of the Rab GTPase subfamily. *EMBO J.* **13**, 5547–5558.

Davis, G. D., Elisee, C., Newham, D. M., and Harrison, R. G. (1999). New fusion protein systems designed to give soluble expression in *Escherichia coli*. *Biotechnol. Bioeng.* **65**, 382–388.

Elkind, N. B., Walch-Solimena, C., and Novick, P. J. (2000). The role of the COOH terminus of Sec2p in the transport of post-Golgi vesicles. *J. Cell Biol.* **149**, 95–110.

Hattula, K., and Peränen, J. (2000). FIP-2, a coiled-coil protein, links Huntingtin to Rab8 and modulates cellular morphogenesis. *Curr. Biol.* **10**, 1603–1606.

Hattula, K., Furuhjelm, J., Arffman, A., and Peranen, J. (2002). A Rab8-specific GDP/GTP exchange factor is involved in actin remodeling and polarized membrane transport. *Mol. Biol. Cell* **13**, 3268–3280.

Horiuchi, H., Lippe, R., McBride, H. M., Rubino, M., Woodman, P., Stenmark, H., Rybin, V., Wilon, M., Ashman, K., Mann, M., and Zerial, M. (1997). A novel Rab5 GDP/GTP exchange factor complexed to Rabaptin-5 links nucleotide exchange to effector recruitment and function. *Cell* **90**, 1149–1159.

Israel, D. I. (1993). A PCR-based method for high stringency screening of DNA libraries. *Nucleic Acids Res.* **21**, 2627–2631.

Kajiho, H., Saito, K., Tsujita, K., Kontani, K., Araki, Y., Kurosu, H., and Katada, T. (2003). RIN3: A novel Rab5 GEF interacting with amphiphysin II involved in the early endocytic pathway. *J. Cell Sci.* **116**, 4159–4168.

Luo, H. R., Saiardi, A., Nagata, E., Ye, K., Yu, H., Jung, T. S., Luo, X., Jain, S., Sawa, A., and Snyder, S. H. (2001). GRAB. A physiological guanine nucleotide exchange factor for Rab3A, which interacts with inositol hexakisphosphate kinase. *Neuron* **31**, 439–451.

Otomo, A., Hadano, S., Okada, T., Mizumura, H., Kunita, R., Nishijima, H., Showguchi-Miyata, J., Yanagisawa, Y., Kohiki, E., Suga, E., Yasuda, M., Osuga, H., Nishimoto, T., Narumiya, S., and Ikeda, J. E. (2003). ALS2, a novel guanine nucleotide exchange factor for the small GTPase Rab5, is implicated in endosomal dynamics. *Hum. Mol. Genet.* **12**, 1671–1687.

Peränen, J. (1992). Rapid affinity-purification and biotinylation of antibodies. *Biotechniques* **13**, 546–549.

Peränen, J., and Furuhjelm, J. (2000). Expression, purification, and properties of Rab8 function in actin cortical skeleton organization and polarized transport. *Methods Enzymol.* **329**, 188–196.

Peränen, J., Auvinen, P., Virta, H., Wepf, R., and Simons, K. (1996). Rab8 promotes polarized membrane transport through reorganization of actin and microtubules in fibroblasts. *J. Cell Biol.* **135**, 153–167.

Tall, G. G., Barbier, M. A., Stahl, P. D., and Horazdovsky, B. F. (2001). Ras-activated endocytosis is mediated by the Rab5 guanine nucleotide exchange activity of RIN1. *Dev. Cell* **1**, 73–82.

Tanaka, M., Miyoshi, J., Ishizaki, H., Togawa, A., Ohnishi, K., Endo, K., Matsubara, K., Mizoguchi, A., Nagano, T., Sato, M., Sasaki, T., and Takai, Y. (2001). Role of Rab3

GDP/GTP exchange protein in synaptic vesicle trafficking at the mouse neuromuscular junction. *Mol. Biol. Cell* **12,** 1421–1430.

Wada, M., Nakanishi, H., Satoh, A., Hirano, H., Obaishi, H., Matsuura, Y., and Takai, Y. (1997). Isolation and characterization of a GDP/GTP exchange protein specific for the Rab3 subfamily small G proteins. *J. Biol. Chem.* **272,** 3875–3878.

Zerial, M., and McBride, H. (2001). Rab proteins as membrane organizers. *Nat. Rev. Mol. Cell Biol.* **2,** 107–111.

[25] Assay and Functional Properties of SopE in the Recruitment of Rab5 on *Salmonella*-Containing Phagosomes

By Seetharaman Parashuraman and Amitabha Mukhopadhyay

Abstract

We investigated the mechanism by which *Salmonella* inhibits transport to lysosomes to survive in macrophages. We have purified the *Salmonella*-containing phagosomes from macrophages and determined the presence of different endocytic Rab proteins on the phagosomes. Our results have shown that live *Salmonella*-containing phagosomes recruit more Rab5 than dead *Salmonella*-containing phagosomes. Recruitment of Rab5 on live *Salmonella*-containing phagosomes depends on the presence of viable bacteria in the phagosomes. Subsequently, we identified an effector molecule of *Salmonella*, SopE, which specifically binds Rab5. Moreover, SopE is found to be a specific nucleotide exchange factor for Rab5 and thereby retains Rab5 in an active conformation. Activated Rab5 on *Salmonella*-containing phagosomes promotes fusion with early endosomes and thus avoids transport to the lysosomes.

Introduction

Endocytosis and phagocytosis are two primary mechanisms used by eukaryotic cells to internalize macromolecules from the extracellular milieu (Schwartz, 1990). After internalization, endocytosed materials are delivered to a common endosomal compartment where they are sorted and targeted to different intracellular destinations through a series of coordinated and specific vesicle fusion events (Balch, 1990; Gruenberg and Maxfield, 1995). Current knowledge about the intracellular transport

METHODS IN ENZYMOLOGY, VOL. 403
0076-6879/05 $35.00
DOI: 10.1016/S0076-6879(05)03025-9

of internalized ligands demonstrates that the processes of vesicle budding, docking, and fusion with acceptor compartment are regulated by the Rab family of small GTP-binding proteins (Deneka, *et al.*, 2003; Pfeffer and Aivazian, 2004; Zerial and McBride, 2001). Rab proteins execute their regulatory activity by cycling between two conformations in a nucleotide-dependent fashion (Soldati *et al.*, 1994). Membrane association and activation of Rab GTPases are regulated by Rab-GDI (Ullrich *et al.*, 1994), Rab escort protein (REP) (Alexandrov *et al.*, 1994), along with the exchange of GDP to GTP by guanine nucleotide exchange factor (GEF) (Day *et al.*, 1998). On the other hand, the switch from a GTP- to GDP-bound conformation is catalyzed through the hydrolysis of bound GTP by GTPase-activating protein (GAP) (Xiao *et al.*, 1997). Interference with these regulatory proteins by exogenous effector molecules may perturb the normal transport process and missort the cargo to a different destination.

Initially, it was thought that nascent phagosomes were destined to fuse with lysosomes and degrade their contents by the hydrolytic enzymes present in the lysosomes (Beron *et al.*, 1995). Thus, targeting the phagosomes containing intracellular pathogens to lysosomes is a major tool host cells use to protect themselves. This defense mechanism is often countered by specific pathogens through the modulation of phagosomal trafficking inside the cells (Desjardins *et al.*, 1994; Garcia-del Portillo, 1999). Given the importance of the Rab proteins along with their downstream molecules in the regulation of intracellular transport pathways, it is tempting to investigate the role of various effector molecules secreted by the intracellular pathogens that might modulate/mimic the functions of some of the transport-related regulatory molecules and thereby divert the maturation of phagosomes from the lysosomal pathway (Uchiya *et al.*, 1999). Studies with mutant organisms lacking appropriate effector molecules show impaired survival of the pathogens in the host cells (Hackstadt, 2000; VanRheenen *et al.*, 2004), suggesting that the effector molecules indeed play a major role in diverting intracellular trafficking of the pathogens. Recently, it has been shown that various intracellular pathogens like *Salmonella* and *Mycobacterium* survive in some sort of early endocytic compartment by recruiting early acting Rab5 on their phagosomes (Hashim *et al.*, 2000; Mukherjee *et al.*, 2000; Sturgill-Koszycki *et al.*, 1996; Vergne *et al.*, 2004). However, mechanisms of the recruitment of the Rab protein or their effectors on these phagosomes are not clearly established. Therefore, we focused our efforts on determining the possible role of effector molecules from *Salmonella* involved in the modulation of the trafficking related regulatory protein(s) of the host cells.

Materials

Reagents

Unless otherwise stated, all reagents were obtained from Sigma Chemical Co. (St. Louis, MO) and bacterial culture media are purchased from DIFCO. Tissue culture supplies were obtained from the Grand Island Biological Co. (Grand Island, NY). *N*-Hydroxysuccicinimidobiotin (NHS-biotin), avidin-horseradish peroxidase (Avidin-HRP), avidin, and bicinchoninic acid (BCA) reagents were purchased from Pierce Biochemicals, Rockford, IL.

Media

1. Luria broth (LB): 10 g of pancreatic digest of casein, 5 g of yeast extract, and 10 g of sodium chloride are dissolved in 1 liter of distilled water and pH is adjusted to 7.0.
2. *Salmonella* and *Shigella* agar (SS Agar): 5 g of bacto beef extract, 5 g of bacto proteose peptone, 10 g of bacto lactose, 8.5 g of bacto bile salt No. 3, 8.5 g of sodium citrate, 8.5 g of sodium thiosulfate, 10 g of ferric citrate, 13.5 g of bacto agar, 0.33 mg of brilliant green, and 0.025 g of neutral red are dissolved in 1 liter of distilled water and pH is adjusted to 7.0. The medium is boiled for 1 min and 25 ml is poured into each Petri dish (94/16 mm).
3. Internalization medium (IM): Minimum essential medium supplemented with 10 m*M* HEPES [*N*-(2-hydroxyethyl) piperazine -*N'*-(2—ethane-sulfonic acid)] and 5 m*M* glucose. pH is adjusted to 7.4 and filtered through a 0.2-μm pore filter.

Buffers

1. Phosphate-buffered saline (PBS): 10 m*M* sodium phosphate buffer, pH 7.0 containing 150 m*M* NaCl.
2. Homogenization buffer (HB): 250 m*M* sucrose, 0.5 m*M* EGTA, 20 m*M* HEPES. pH is adjusted to 7.2 with KOH.
3. Sodium dodecyl sulfate (SDS) sample buffer: 60 m*M* Tris–HCl, pH 6.8, containing 2% SDS (w/v), 10% glycerol (v/v), 3% 2-mercaptoethanol (v/v), 0.001% bromophenol blue (w/v).
4. Coating buffer: 0.1 *N* sodium carbonate buffer, pH 9.5.
5. GDP loading buffer: 20 m*M* Tris–HCl, pH 7.5, containing 5 μ*M* guanosine diphosphate, 50 m*M* NaCl, 3 m*M* MgCl$_2$, 0.1 m*M* dithiothreitol, and 0.1 m*M* EDTA.

6. Nucleotide exchange buffer: 20 mM Tris–HCl, pH 7.5, containing 5 μM [α-^{32}P]GTP (specific activity 3000 Ci/mmol), 100 mM NaCl, 10 mM MgCl$_2$, 0.5 mM dithiothreitol, and 0.5 mg/ml bovine serum albumin.

Cells

Phagocytosis is a function of specialized cells and macrophages are well characterized phagocytes in vertebrates. To understand the trafficking of *Salmonella*-containing phagosomes in macrophages, we used a J774E clone, a well-characterized mannose receptor-positive mouse macrophage cell line. Cells are grown at 37° in 5% CO$_2$ 95% air atmosphere in RPMI-1640 (Roswell Park Memorial Institute) medium supplemented with 10% (v/v) heat-inactivated fetal calf serum and gentamicin (50 μg/ml).

Salmonella

The virulent wild-type (wt) *S. typhimurium*, a clinical isolate from Lady Harding Medical College, New Delhi, India, was obtained from Dr. Vineeta Bal of the National Institute of Immunology (New Delhi, India). Bacteria are routinely grown overnight in Luria broth at 37° with constant shaking (300 rpm). The *Salmonella* are stored at $-70°$ as glycerol stock.

Recombinant Proteins

Plasmid encoding the GST-SopE$_{78-240}$ and different constructs of Rab5, namely GST-Rab5WT, GST-Rab5S34N, and GST-Rab5Q79L, along with GST-Rab7 to prepare fusion proteins, were generously provided by Prof. J. E. Galan (Yale University School of Medicine, New Haven, CT) and Prof. Philip Stahl (Washington University School of Medicine, St Louis, MO), respectively. Respective plasmids are transformed into *Escherichia coli* (DH5α) by standard technique. *E. coli* containing the respective plasmids is grown in Luria broth for 12–15 h at 37° and cells are harvested. These *E. coli*-containing respective plasmids are stored at $-70°$ as glycerol stock. For experimental purpose, cells from the glycerol stock are inoculated into LB containing 100 μg/ml of ampicillin and grown for 12–15 h at 37° with constant shaking (200 rpm). Subsequently, cells are diluted with fresh LB (1:20 v/v) and grown until the OD$_{600 \text{ nm}}$ reaches $\sim$0.6 and induced with 0.2 mM isopropyl-β-D-thiogalactopyranoside (IPTG) for 3–4 h at 37° for the expression of GST fusion proteins. Cells are harvested, lysed by sonication, and the lysate is treated with Triton X-100 (1%). Cellular debris is removed by centrifugation at $13,000 \times g$ for 10 min at $4°$. The respective fusion protein is purified from the supernatant using reduced glutathione beads by standard procedure (Smith and Corcoran, 1987).

Antibodies

Affinity-purified rabbit polyclonal anti-Rab5 and anti-Rab7 antibodies were generously provided by Dr. J. Gruenberg (EMBL, Heidelberg, Germany) and Dr. A. Wandinger-Ness (Northwestern University, Evanston, IL), respectively. Anti-*Salmonella* antibodies (anti-SopE, anti-SopB, and anti-SipC) were kindly provided by Dr. E. E. Galyov from Institute for Animal Health, Berkshire, UK. Affinity-purified rabbit polyclonal antibodies against native NSF was received as kind gift from Dr. S.W. Whiteheart (University of Kentucky, Lexington, KY). Anti-transferrin receptor antibodies were purchased from Zymed Laboratory.

Biotinylation of Rab5

Biotinylation of Rab5 is done according to procedures described earlier (Robinson and Gruenberg, 1998). Purified GST-Rab5 (800 μg) is incubated in 400 μl of 0.1 M sodium carbonate-bicarbonate buffer, pH 9.5, containing 1 mg/ml of NHS-biotin for 2 h at room temperature (RT) on an orbital shaker. Unreacted NHS-biotin is quenched by adding 40 μl of 0.2 M glycine, pH 8.0, and mixed for an additional 30 min. The reaction mixture is dialyzed against PBS containing 1 mM dithiothreitol, 5 mg/ml MgCl$_2$, and 0.1 mM phenylmethylsulfonyl fluoride with several changes for 15 h to separate the labeled protein from other reactants. Biotinylated-Rab5 is snap frozen and stored at $-80°$.

Procedure

Growing and Harvesting of Salmonella

For experimental purpose, *Salmonella* from the glycerol stock are routinely grown overnight in LB and an aliquot of cells is spread on 25 ml of SS Agar containing plate and incubated for 12 h at $37°$ to obtain *Salmonella* colonies. Subsequently, a single *Salmonella* colony is inoculated into 10 ml of LB in a 50-ml Falcon tube (30 × 115 mm) and incubated at $37°$ with constant shaking (300 rpm) to grow the culture to a cell density corresponding to an OD_{600} of about 0.9–1, which usually takes about 12 h. Finally, 0.1 ml of bacterial suspension is diluted into 10 ml of fresh LB and cells are grown under similar conditions to a cell density corresponding to an OD_{600} of about 0.5, which usually takes about 2 h.

These log phase cells are harvested by centrifugation (4000×g for 5 min) in a refrigerated Eppendorf centrifuge using an F34-6-38 rotor at 25°, washed twice with PBS, and used for further studies.

To prepare the dead *Salmonella*-containing phagosomes, 1×10^{10} *Salmonella* are resuspended in 1 ml of PBS and incubated at 65° for 45 min (Rathman *et al.*, 1996). To achieve 100% killing, cells are subsequently fixed with 1% glutaraldehyde (v/v) for 30 min at 4°. After treatment, the viability of the bacteria is checked by plating an aliquot (25 μl) of the treated cell suspension on SS agar plates. No colony appears under these conditions indicating complete loss of viability of the treated bacteria.

Preparation of Salmonella-*Containing Phagosomes*

We purified the *Salmonella*-containing phagosomes from J774E macrophages with some modification of previously described procedures of phagosome purification (Alvarez-Dominguez *et al.*, 1996). First, the J774E cells are harvested from tissue culture flasks into 10 ml of fetal calf serum (FCS) free RPMI-1640 medium by gentle scraping with a cell scraper. Cells are washed thrice with 10 ml IM by centrifugation (100×g for 6 min) in a refrigerated Eppendorf centrifuge using an F34-6-38 rotor at 4°. Subsequently, the J774E cells (5×10^8) are preincubated with 5×10^9 *Salmonella* in 700 μl of IM for 1 h at 4°. Cells are centrifuged at 1000×g for 5 min and resuspended in 1 ml of prewarmed IM to initiate phagocytosis and incubated for 5 min in a 37° water bath. Immediately, internalization of the bacteria is stopped by the addition of 10 ml of ice cold IM into the tube. Cells are washed three times with ice-cold PBS by low-speed centrifugation (100×g for 6 min) to remove uninternalized bacteria. Using this protocol, most of the free uninternalized bacteria are removed and the cell pellet predominantly contains J774E cells with internalized *Salmonella*.

To prepare the phagosomes, cells are resuspended (2×10^8 cells/ml) in HB and 1 ml of the cell suspension is homogenized in a ball-bearing homogenizer (Balch *et al.*, 1984) at 4° avoiding the formation of air bubbles. Phagosomal membranes are more prone to breakage during homogenization, so homogenization is usually carried out until 80% of the cells are broken (10–15 passages through a homogenizer), which is checked by trypan blue exclusion. Unbroken cells and nuclei are removed from the homogenates by centrifuging at 200×g for 10 min at 4°. Postnuclear supernatants (PNSs) containing phagosomes (200–300 μl aliquots) are snap frozen in liquid nitrogen and stored at −80°.

Immediately before use, an aliquot of PNS is quickly thawed and diluted with HB (1:3) and enriched phagosomal fractions are prepared by centrifugation at 12,000×g for 6 min at 4°. Subsequently, the pellet is resuspended in 100 μl of HB containing protease inhibitor cocktail (Roche Diagnostics) and loaded onto a 1 ml of 12% sucrose cushion. Purified phagosomes are recovered from the bottom of the tube (100 μl from the bottom) after centrifugation at 1700×g for 45 min at 4°. The viability of the bacteria in the phagosomes is determined by lysing an aliquot (20 μl) of respective phagosomes with 200 μl of solubilization buffer (SB; PBS containing 0.5% Triton X-100) and plating them on an SS-agar plate. Biochemical characterizations of these phagosomes by standard methods (Alvarez-Dominguez *et al.*, 1996; Bole *et al.*, 1986; Fleischer and Kervina, 1974; Ward *et al.*, 1997) show that they are free of endosome, lysosome, Golgi, and endoplasmic reticulum contamination (Hashim *et al.*, 2000).

Determination of the presence of Rab5 on Salmonella-*Containing Phagosomes*

To determine the presence of Rab5 on early live *Salmonella*-containing phagosomes, purified phagosomes are analyzed by Western blot analysis using specific antibodies. We have used dead *Salmonella*-containing phagosomes as control. Respective phagosomes (40 μg of protein estimated by bicinchoninic acid assay with bovine serum albumin as standard) are suspended in SDS sample buffer and incubated for 5 min in boiling water. Phagosomal proteins are separated by 12% SDS–polyacrylamide gel electrophoresis (SDS-PAGE) (Laemmli, 1970) (0.02 A for 70 min) and transferred onto a nitrocellulose membrane (10 V for 30 min using Bio-Rad semidry transfer apparatus). Subsequently, nitrocellulose membrane containing phagosomal proteins is probed with specific antibodies against Rab5 and Rab7 using standard protocols (Blake *et al.*, 1984) (Fig. 1). Proteins are visualized using appropriate HRP-labeled second antibody and electrochemoluminescence (ECL) (Heinicke *et al.*, 1992) (Amersham Biotech).

Recruitment of Rab5 Depends on the Presence of Viable Salmonella in the Phagosomes

To determine whether the presence of viable *Salmonella* in the phagosomes is required for the recruitment of Rab5, we purified live *Salmonella*-containing phagosomes as described in the previous section. Purified phagosomes (400 μg protein) are resuspended in 200 μl of HB and divided into two equal aliquots (100 μl). One aliquot of the phagosomes is incubated with ciprofloxacin (500 μg/ml) for 30 min at 4° to kill the resident bacteria in the phagosomes. Subsequently, treated phagosomes are washed twice with 200 μl of HB by centrifugation at 12,000×g for 6 min at 4° and finally resuspended in 100 μl of HB. The viability of the bacteria in the

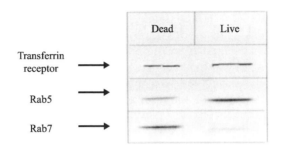

FIG. 1. Western blots showing the recruitment of more Rab5 on live *Salmonella*-containing phagosomes in comparison to dead *Salmonella*-containing phagosomes. Equivalent amount of transferrin receptor on both phagosomes indicates that both live and dead bacteria remain in the early compartment. From Mukherjee *et al.* (2000).

phagosomes is determined by lysing an aliquot (20 μl) of respective phagosomes with SB and plating them on an SS-agar plate as described previously.

Respective phagosomes (40 μg of protein) are analyzed by Western blot analysis to determine the presence of Rab5 and NSF on ciprofloxacin treated *Salmonella*-containing phagosomes using specific antibodies as described in the previous section. Under these conditions, ciprofloxacin-treated *Salmonella*-containing phagosomes do not retain Rab5 on the phagosomal membrane, suggesting that a signal from the viable bacteria is required for recruitment of Rab5 on the phagosomes (Fig. 2).

Identification of SopE, a Salmonella *Effector Protein, as a Rab5* Binding Protein

To detect the signal from live *Salmonella* required to recruit Rab5 on the phagosomes, the bacteria are grown overnight at 37° with constant shaking (300 rpm) in LB as described previously. Subsequently, 1 ml of bacterial suspension is washed twice with PBS by centrifugation (4000×g for 5 min) and resuspended in 10 ml of methionine-free RPMI-1640 medium containing 1 mCi of [^{35}S]methionine (specific activity, 1175 Ci/mmol). Cells are grown for 9 h at 37° with constant shaking (300 rpm) and subsequently washed five times with 25 ml of PBS by centrifugation (4000×g for 5 min at 4°) to remove free radioactivity. Finally, cells are resuspended in 1 ml of PBS containing protease inhibitors cocktail. Bacterial suspension is lysed by sonication on ice (10-s pulse with 10 s intervals, five times) using a sonicator (Sonics, VC 750 model) fitted with a micro tip at an amplitude of 30%. As cell lysis occurs, the suspension turns to a light

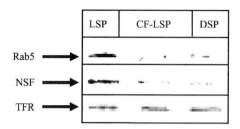

FIG. 2. Western blots showing that the recruitment of Rab5 and NSF on live *Salmonella*-containing phagosomes (LSP) depends on the presence of viable bacteria within the phagosomes. Live *Salmonella*-containing phagosomes (LSP) treated with ciprofloxacin (CF-LSP) are unable to retain Rab5 and NSF on phagosomes. The presence of an equivalent amount of transferrin receptor on different phagosomes indicates that ciprofloxacin treatment does not affect the other proteins present on phagosomes. DSP, dead *Salmonella*-containing phagosomes. From Mukherjee *et al.* (2000).

brown color and becomes more viscous. After sonication, 1% Triton X-100 (v/v) is added to the lysate and incubated in a rotator for 30 min at 4°. The lysate is centrifuged at 15,000×g for 10 min and the clear supernatant is carefully transferred into a fresh 15-ml Falcon tube.

Glutathione-agarose beads (Sigma G4510) are washed two times with several volumes of PBS containing 1% Triton X-100 (v/v). Washed gluta-thione-agarose beads (25 μl packed volume) are resuspended in 100 μl PBS containing 1% Triton X-100 (v/v) and incubated with purified recombinant GST-Rab5 (200 μg protein in 100 μl PBS containing 1% Triton X-100) for 1 h at 4° to immobilize Rab5 on the beads. Beads are washed (10,000×g for 3 min) three times to remove unbound GST-Rab5.

Finally, Rab5-immobilized glutathione-agarose beads are added to 1 ml of the metabolically labeled *Salmonella* lysate and incubated in a rotator for 1 h at 4°. Beads are washed (10,000×g for 5 min) three times to remove unbound proteins. Subsequently, the proteins are resolved by 12% SDS-PAGE. The gel is dried in a gel dryer (Hoefer slab gel dyer model GD2000) under vacuum at 80° for 1 h. Dried gel is placed in a cassette holder (8 inch × 10 inch) with an intensifying screen in the dark room. Then, an x-ray film (Amersham Hyperfilm MP) is carefully placed on the dried gel and the cassette holder is closed so that no light can penetrate. The closed cassette is kept at −70° for 12 h. After the indicated time, we have taken the film out from the cassette in the darkroom and placed it in the container containing developer solution (SoluDent developer, Solutek Corporation, Boston) with gentle agitation for about 2–5 min. Then the film is washed thoroughly under running tap water and placed in fixer solution (SoluDent fixer, Solutek Corporation, Boston) for 5–10 min. The

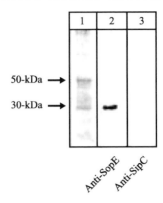

FIG. 3. Autoradiogram showing two proteins of apparent molecular weight of 50 kDa and 30 kDa from *Salmonella* lysate binds with immobilized Rab5 (Lane 1). A 30-kDa protein from *Salmonella* is identified as SopE (Lane 2) by Western blot analysis using specific antibody. From Mukherjee *et al.* (2001).

film is washed after fixing and air dried. Under these conditions, autoradiograph shows two prominent bands of about 30 and 50 kDa (Fig. 3, Lane 1). These bands do not appear when we have performed a similar experiment under identical conditions using GST immobilized beads. Similar pull out experiments are carried out with GST-Rab5 immobilized beads using unlabeled *Salmonella* lysate. GST-Rab5-bound *Salmonella* proteins are identified using antibodies against *Salmonella* secretory proteins such as SopE, SopB, and SipC by Western blot analysis as described previously. By this process, we identified the 30-kDa Rab5-binding protein from *Salmonella* as SopE (Fig. 3, Lane 2).

Determination of Binding of SopE with Rab5

To determine the ability of SopE to bind Rab5, we developed a modified enzyme-linked immunosorbent assay (ELISA) to measure the interaction between these two proteins. First, we expressed and purified both recombinant Rab5 and SopE-$_{(78-240)}$ as GST fusion proteins as described previously. To immobilize SopE onto the plastic wells of an ELISA plate, 500 ng of recombinant SopE$_{(78-240)}$ dissolved in 50 μl of coating buffer is added into the wells and incubated for 3 h at 37°. After the incubation, solution from the wells of the ELISA plates is discarded and the wells are filled with 200 μl of PBS containing 0.01% (v/v) Tween 20 (PBST) and incubated at room temperature for 5 min. These washing steps are repeated three times to remove the unbound protein. Subsequently, wells of the

ELISA plate are filled with 100 μl of blocking buffer (PBST containing 2% [w/v] of bovine serum albumin) and incubated for 1 h at 37° to block the available protein-binding sites in the well. Plates are washed three times with PBST as described previously followed by three washes with PBS. Wells containing immobilized SopE are incubated with different concentrations of biotinylated GST-Rab5WT in 50 μl of PBS for 1 h at 37°. Unbound Rab5 from the wells is removed by three washes with PBST followed by three washes with PBS.

To determine the binding of biotinylated Rab5 with SopE, wells are incubated with 50 ng of avidin-HRP dissolved in 50 μl of PBS for 30 min at 37°. Unbound avidin-HRP is removed from the wells by washing as described earlier. Subsequently, HRP activity present in each well is measured as described previously (Gruenberg *et al.*, 1989) using *o*-phenylenediamine as the chromogenic substrate to measure the binding of Rab5 with SopE. The reaction is initiated by adding 100 μl of 0.05 N sodium acetate buffer, pH 5.0, containing *o*-phenylenediamine (0.75 mg/ml) and 0.006% H_2O_2 and incubating at room temperature until a faint yellowish color develops. The reaction is terminated by adding 100 μl of 0.1 N H_2SO_4, and absorbance is measured at 490 nm in an ELISA reader to quantitate binding of Rab5 with SopE. Under these conditions, biotinylated Rab5 binds with SopE in a saturable manner (Fig. 4A). We have also measured the binding of biotinylated Rab5 with immobilized SopE in the presence of excess concentrations of unlabeled Rab5 or SopE (0–1 mg/ml)

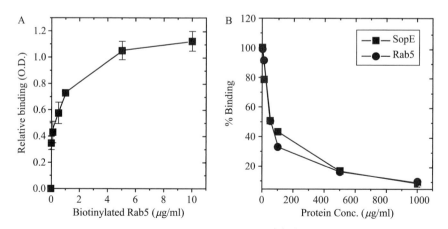

FIG. 4. *In vitro* binding of Rab5 with SopE, a *Salmonella* secretory protein (A). Unlabeled Rab5 or SopE inhibits this binding (B) indicating the specificity. From Mukherjee *et al.* (2001).

to determine the specificity. More than 90% of the binding of biotinylated Rab5 with immobilized SopE is inhibited under these conditions (Fig. 4B).

SopE Acts as a Nucleotide Exchange Factor for Rab5

Previous studies have shown that SopE acts as a guanine nucleotide exchange factor for CDC42 and Rho GTPases (Hardt *et al.*, 1998), which prompted us to investigate the possibility of SopE as a nucleotide exchange factor for Rab5. We have used different forms of Rab5 such as Rab5WT; Rab5Q79L, a GTPase-defective mutant; and Rab5S34N, a mutant locked in a GDP conformation along with Rab7 and $SopE_{(78-240)}$ as GST fusion proteins. The respective Rab protein (60 pmol) is incubated in a total volume of 20 μl of GDP loading buffer in a microfuge tube at room temperature for 30 min to load the proteins with GDP. An aliquot of 5 μl (15 pmol of Rab protein) is diluted with 200 μl of nucleotide exchange buffer with or without 15 pmol GST-$SopE_{(78-240)}$ and incubated for 30 min at room temperature to initiate the exchange reaction. To determine the amount of $[\alpha\text{-}^{32}P]GTP$ incorporated into the protein, 25 μl of the reaction mixture is blotted onto a nitrocellulose membrane and air dried. The air-dried membranes are placed in a 24-well plate and 2 ml of ice-cold wash buffer (20 mM Tris–HCl, pH 8, 100 mM NaCl, and 10 mM $MgCl_2$) is added to submerge the membrane and incubated at room temperature for 5 min on an orbital shaker. The wash buffer containing unincorporated $[\alpha\text{-}^{32}P]$ GTP is discarded and replaced with fresh ice-cold wash buffer. This washing process is repeated five times to completely remove the free radioactivity. Finally, the membranes are air dried overnight and the amount of radioactive nucleotide bound to respective Rabs is determined by scintillation counting. Under these conditions, Rab5 binds more than 2-fold $[\alpha\text{-}^{32}P]$ GTP in the presence of SopE than in the absence of SopE, indicating that SopE acts as a specific nucleotide exchange factor for Rab5 and not for Rab7 (Fig. 5).

Conclusion

We have tried to characterize the mechanism by which *Salmonella* recruit Rab5 on their phagosomes and prevent their trafficking to lysosomes in the macrophages. Our results have shown that *Salmonella*-containing early phagosomes recruit the GTP form of Rab5 from the host cells and promote fusion with the early endosomes (Mukherjee *et al.*, 2000) and thereby inhibit the transport of *Salmonella*-containing phagosomes to the lysosomes (Hashim *et al.*, 2000). Subsequently, we have shown that the bacteria regulate the recruitment of Rab5 onto the phagosomal surface

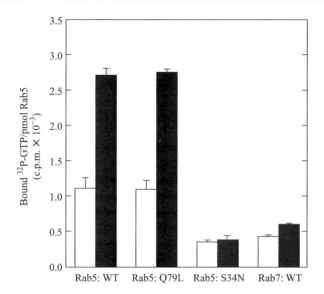

FIG. 5. Guanine nucleotide exchange assay demonstrating that SopE acts as a nucleotide exchange factor for Rab5. The presence and absence of SopE are indicated by solid and open bars, respectively. From Mukherjee *et al.* (2001).

through its type III secretory system effector, SopE. Furthermore, SopE, secreted from the *Salmonella*-containing phagosomes, acts as a nucleotide exchange factor for Rab5 converting inactive GDP-bound Rab5 to an active GTP-bound form (Mukherjee *et al.*, 2001), thereby retaining Rab5 on the phagosomal membrane in an active conformation and thus preventing its trafficking to lysosomes. In conclusion, these results suggest a plausible mechanism by which *Salmonella* survives in macrophages.

Acknowledgments

This work was supported by grants from the Department of Biotechnology, Government of India. S. Parashuraman was supported by a research fellowship from the Council of Scientific and Industrial Research, Government of India.

References

Alexandrov, K., Horiuchi, H., Steele-Mortimer, O., Seabra, M. C., and Zerial, M. (1994). Rab escort protein-1 is a multifunctional protein that accompanies newly prenylated rab proteins to their target membranes. *EMBO J.* **13,** 5262–5273.

Alvarez-Dominguez, C., Barbieri, A. M., Beron, W., Wandinger-Ness, A., and Stahl, P. D. (1996). Phagocytosed live *Listeria monocytogenes* influences rab-5 regulated *in vitro* phagosome-endosome fusion. *J. Biol. Chem.* **271,** 13834–13843.

Balch, W. E. (1990). Small GTP-binding proteins in vesicular transport. *Trends Biochem. Sci.* **15,** 473–477.

Balch, W. E., Dunphy, W. G., Braell, W. A., and Rothman, J. E. (1984). Reconstitution of the transport of protein between successive compartments of the Golgi measured by the coupled incorporation of N-acetylglucosamine. *Cell* **39,** 405–416.

Beron, W., Alvarez-Dominguez, C., Mayorga, L., and Stahl, P. D. (1995). Membrane trafficking along the phagocytic pathway. *Trends Cell Biol.* **5,** 100–104.

Blake, M. S., Johnston, K. H., Russell-Jones, G. J., and Gotschlich, E. C. (1984). A rapid, sensitive method for detection of alkaline phosphatase-conjugated anti-antibody on Western blots. *Anal Biochem.* **136,** 175–179.

Bole, D. G., Hendershot, L. M., and Kearney, J. F. (1986). Posttranslational association of immunoglobulin heavy chain binding protein with nascent heavy chains in nonsecreting and secreting hybridomas. *J. Cell Biol.* **102,** 1558–1566.

Day, G. J., Mosteller, R. D., and Broek, D. (1998). Distinct subclasses of small GTPases interact with guanine nucleotide exchange factors in a similar manner. *Mol. Cell. Biol.* **18,** 7444–7454.

Deneka, M., Neeft, M., and van der Sluijs, P. (2003). Regulation of membrane transport by Rab GTPases. *Crit. Rev. Biochem. Mol. Biol.* **38,** 121–142.

Desjardins, M., Huber, L. A., Parton, R. G., and Griffiths, G. (1994). Biogenesis of phagolysosomes proceeds through a sequential series of interactions with the endocytic apparatus. *J. Cell Biol.* **124,** 677–688.

Fleischer, S., and Kervina, M. (1974). Subcellular fractionation of rat liver. *Methods Enzymol.* **31,** 6–41.

Garcia-del Portillo, F. (1999). Pathogenic interference with host vacuolar trafficking. *Trends Microbiol.* **6,** 467–469.

Gruenberg, J., and Maxfield, F. R. (1995). Membrane transport in the endocytic pathway. *Curr. Opin. Cell Biol.* **7,** 552–563.

Gruenberg, J., Griffiths, G., and Howell, K. E. (1989). Characterization of the early endosome and putative endocytic carrier vesicle *in vivo* and with an assay of vesicle fusion *in vitro*. *J. Cell Biol.* **108,** 1301–1316.

Hackstadt, T. (2000). Redirection of host vesicle trafficking pathways by intracellular parasites. *Traffic* **1,** 93–99.

Hardt, W., Chen, L., Schuebel, K. E., Bustelo, X. R., and Galan, J. E. (1998). S. typhimurium encodes an activator of Rho GTPases that induces membrane ruffling and nuclear responses in host cells. *Cell* **93,** 815–826.

Hashim, S., Mukherjee, K., Raje, M., Basu, S. K., and Mukhopadhyay, A. (2000). Live *Salmonella* modulate expression of Rab proteins to persist in a specialized compartment and escape transport to lysosomes. *J. Biol. Chem.* **275,** 16281–16288.

Heinicke, E., Kumar, U., and Munoz, D. G. (1992). Quantitative dot-blot assay for proteins using enhanced chemiluminescence. *J. Immunol. Methods* **152,** 227–236.

Laemmli, U. K. (1970). Cleavage of structural proteins during the assembly of the head of bacteriophage T4. *Nature* **227,** 680–685.

Mukherjee, K., Siddiqi, S. A., Hashim, S., Raje, M., Basu, S. K., and Mukhopadhyay, A. (2000). Live *Salmonella* recruits N-ethylmaleimide-sensitive fusion protein on phagosomal membrane and promotes fusion with early endosome. *J. Cell Biol.* **148,** 741–753.

Mukherjee, K., Parashuraman, S., Raje, M., and Mukhopadhyay, A. (2001). SopE acts as an Rab5-specific nucleotide exchange factor and recruits non-prenylated Rab5 on

Salmonella-containing phagosomes to promote fusion with early endosomes. *J. Biol. Chem.* **276,** 23607–23615.

Pfeffer, S., and Aivazian, D. (2004). Targeting Rab GTPases to distinct membrane compartments. *Nat. Rev. Mol. Cell Biol.* **5,** 886–896.

Rathman, M., Sjaastad, M. D., and Falkow, S. (1996). Acidification of phagosomes containing *Salmonella typhimurium* in murine macrophages. *Infect. Immun.* **64,** 2765–2773.

Robinson, L. J., and Gruenberg, J. (1998). Assays measuring endocytic transport in the endocytic pathway. *In* "Cell Biology: A Laboratory Handbook" (J. E. Celis, ed.), Vol. 2, pp. 248–257. Academic Press, New York.

Schwartz, A. L. (1990). Cell biology of intracellular protein trafficking. *Annu. Rev. Immunol.* **8,** 195–229.

Smith, D. B., and Corcoran, L. M. (1987). Expression and purification of glutathione-S-transferase proteins. *In* "Current Protocols in Molecular Biology" (F. M. Ausubel *et al.,* eds.), Vol. 1, pp. 16.7.1–16.7.8. John Wiley & Sons, New York.

Soldati, T., Shapiro, A. D., Dirac Svejstrup, A. B., and Pfeffer, S. R. (1994). Membrane targeting of the small GTPase Rab9 is accompanied by nucleotide exchange. *Nature* **369,** 76–78.

Sturgill-Koszycki, S., Schaible, U. E., and Russell, D. G. (1996). Mycobacterium-containing phagosomes are accessible to early endosomes and reflect a transitional state in normal phagosome biogenesis. *EMBO J.* **15,** 6960–6968.

Uchiya, K., Barbieri, M. A., Funato, K., Shah, A. H., Stahl, P. D., and Groisman, E. A. (1999). A *Salmonella* virulence protein that inhibits cellular trafficking. *EMBO J.* **18,** 3924–3933.

Ullrich, O., Horiuchi, H., Bucci, C., and Zerial, M. (1994). Membrane association of Rab5 mediated by GDP-dissociation inhibitor and accompanied by GDP/GTP exchange. *Nature* **368,** 157–160.

VanRheenen, S. M., Dumenil, G., and Isberg, R. R. (2004). IcmF and DotU are required for optimal effector translocation and trafficking of the Legionella pneumophila vacuole. *Infect. Immun.* **72,** 5972–5982.

Vergne, I., Chua, J., Singh, S. B., and Deretic, V. (2004). Cell biology of *Mycobacterium tuberculosis* phagosome. *Annu. Rev. Cell Dev. Biol.* **20,** 367–394.

Ward, D. M., Leslie, J. D., and Kaplan, J. (1997). Homotypic lysosome fusion in macrophages: Analysis using an *in vitro* assay. *J. Cell Biol.* **139,** 665–673.

Xiao, G. H., Shoarinejad, F., Jin, F., Golemis, E. A., and Yeung, R. S. (1997). The tuberous sclerosis 2 gene product, tuberin, functions as a Rab5 GTPase activating protein (GAP) in modulating endocytosis. *J. Biol. Chem.* **272,** 6097–6100.

Zerial, M., and McBride, H. (2001). Rab proteins as membrane organizers. *Nat. Rev. Mol. Cell Biol.* **2,** 107–117.

[26] Purification and Functional Analyses of ALS2 and its Homologue

By Shinji Hadano and Joh-E Ikeda

Abstract

ALS2 is a causative gene product for a form of the familial motor neuron diseases. Computational genomic analysis identified ALS2CL, which is a novel protein highly homologous to the C-terminal region of ALS2. Both proteins contain the VPS9 domain, which is a hallmark for all known members of the guanine nucleotide exchange factors for Rab5 (Rab5GEF), and are known to act as novel factors modulating the Rab5-mediated endosome dynamics in the cells. It has also been reported that oligomerization of ALS2 is one of the fundamental features of its biochemical and physiological function involving endosome dynamics. This chapter describes methods, including purification of the recombinant ALS2 and ALS2CL, and Rab5GEF assay, which have been utilized to clarify the molecular function for ALS2 and ALS2CL.

Introduction

The small GTPases control a broad spectrum of cellular and molecular processes, such as nuclear transfer, cytoskeletal organization, various signaling cascades, and neuronal morphogenesis (Da Silva and Dotti, 2002; Etienne-Manneville and Hall, 2002; Govek *et al.*, 2005; Luo, 2000; Van Aelst and Symons, 2002). All of the small GTPases act as binary switches by cycling between an inactive (GDP-bound) and an active (GTP-bound) state, and guanine nucleotide exchange factors (GEF) are known to play a crucial role in the activation of the small GTPases by stimulating the exchange of GDP for GTP (Rossman *et al.*, 2005).

In 2001, two research groups reported that mutations in the *ALS2* gene caused a juvenile recessive form of amyotrophic lateral sclerosis (ALS), termed ALS2 (OMIM 205100), and a rare recessive form of primary lateral sclerosis (PLSJ; OMIM 606353) (Hadano *et al.*, 2001; Yang *et al.*, 2001). The *ALS2* gene encodes a novel 184-kDa protein, termed ALS2 or alsin, comprising three predicted GEF domains, i.e., regulator of chromosome condensation (RCC1) (Ohtsubo *et al.*, 1987) -like domain (RLD) (Rosa *et al.*, 1996), the Dbl homology and pleckstrin homology (DH/PH) domains (Schmidt and Hall, 2002), and a vacuolar protein sorting 9 (VPS9) domain

METHODS IN ENZYMOLOGY, VOL. 403
0076-6879/05 $35.00
DOI: 10.1016/S0076-6879(05)03026-0

(Burd *et al.*, 1996; Horiuchi *et al.*, 1997; Kajiho *et al.*, 2003; Saito *et al.*, 2002; Tall *et al.*, 2001). In addition, eight consecutive membrane occupation and recognition nexus (MORN) motifs (Takeshima *et al.*, 2000) are noted in the region between the DH/PH and VPS9 domains (Fig. 1). Recently, we identified the ALS2-associated Rab5-specific GEF activity that is mediated by the carboxy-terminal MORN/VPS9 domain of ALS2, and have also shown that ALS2 localizes preferentially onto the early endosome compartments in neuronal cells (Otomo *et al.*, 2003). Since the family of Rab GTPases has emerged as a central player for vesicle budding, motility/trafficking, and fusion (Zerial and McBride, 2001), ALS2 may be implicated in unique vesicle/membrane dynamics in neuronal cells through the activation of Rab5. Most recently, a novel *ALS2* homologous gene, termed *ALS2CL* (ALS2 C-terminal like), has been identified (Devon *et al.*, 2005; Hadano *et al.*, 2004), and its 108-kDa protein product, ALS2CL (Fig. 1), is shown to modulate the Rab5-mediated vesicle/membrane trafficking and dynamics in the cells (Hadano *et al.*, 2004).

In this chapter, we describe methods for the purification of the newly identified members of Rab5GEF family, ALS2 and ALS2CL, as well as the small GTPase Rab5, which are utilized to conduct a number of biochemical analyses including GEF activity and *in vitro* binding assays for ALS2/ALS2CL and Rab5.

Proteins Used for *In Vitro* Functional Analyses

Purification of Recombinant ALS2 and ALS2CL Proteins

Expression Constructs. The mammalian expression constructs for the N-terminally FLAG-tagged ALS2 and ALS2CL proteins are generated as follows. FLAG-linker is generated by annealing two complementary oligonucleotides (forward: 5′-CGCGAGCCACCATGGATTACAAG-GATGACGACGATAAGACGCGTGTTAACC-3′; reverse; 5′-CCGG-

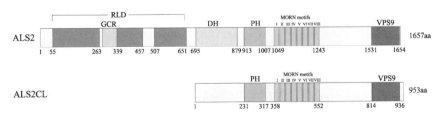

FIG. 1. Schematic representation of the domains and motifs identified in ALS2 and ALS2CL.

GGTTAACACGCGTCTTATCGTCGTCATCCTTGTAATCCATGGT-
GGCT-3′), and then inserted into the *MluI–Cfr*9I sites of pCI-neo Mammalian Expression Vector (Promega), generating the pCIneoFLAG vector. The
full-length cDNA fragments of the *ALS2/Als2* and *ALS2CL/Als2cl* genes are
amplified by reverse transcriptase polymerase chain reaction (RT-PCR) from
human or mouse brain total RNA using the following primer pairs: human
ALS2 and ALS2-L: 5′-tctcgagacgcgtATGGACT C A AAGAAGAGAA
GCTCAACAGAG-3′/ALS2-R;   5′-tatgcatcccgggtCTAGTTAAGCTTCT
CACGCTGAATCTGGTAG-3′; mouse *Als2* and mALS2-L: 5′-tctcga-
gacgcgtATGGACTCAAAGAAGAAAAGCTCAACAG-3′/ mALS2-R;
5′-tatgcatcccgggtCTAGTTAAGCTTCTCCCGCTGAATCTGGAAG-3′;
human *ALS2CL* and ALS2CL-L: 5′-atatacgcgtATGTGCAACCCTGAG-
GAGGCAGC-3′/ ALS2CL-R;   5′-atatgcggccgcCTACCAGAGCTCCC
TGGAGTGCC-3′; and mouse *Als2cl* and mALS2CL-L: 5′-atatacgcgtAT
GTCTAGCTCTGAGGAGGCAGAC-3′/ mALS2CL-R; 5′-atatgcggccgc
TCAC CAGAG-ATCTCTAGCGTCCC-3′, respectively. The resulting
ALS2/Als2 cDNAs are digested with *MluI–Cfr*9I, and inserted into the
*MluI–Cfr*9I sites of the pCIneoFLAG vector, generating pCIneoFLAG-
ALS2_L and pCIneoFLAG-mALS2_L. The *ALS2CL/Als2cl* cDNAs are
digested with *MluI–Not*I, and inserted into the *MluI–Not*I sites of the
pCIneoFLAG vector, generating pCIneoFLAG-ALS2CL and pCIneo-
FALG-mALS2CL. The DNA sequences of the insert as well as the flank-
ing regions in each clone are verified by sequencing. All other truncated
mutants for ALS2 (Otomo *et al.*, 2003) described in this chapter were
generated principally using the same method except for the primer pairs
used.

Cell Culture and Transfection. COS-7 cells are cultured in Dulbecco's
modified Eagle's medium (DMEM) supplemented with 10% heat-inactivated fetal bovine serum (Invitrogen), 100 U/ml penicillin, and 100 μg/ml
streptomycin. Cells are seeded in a T150 flask at a density of 2×10^6 cells/
flask and cultured for ∼24 h, and then transfected with 10 μg of each
plasmid construct using the Effectene Transfection Reagent (Qiagen).
After 48 h of transfection, cells are harvested by centrifugation and washed
twice with ice-cold PBS(-) by repeating the suspension and centrifugation
at +4°. The resulting cell pellets are stably stored at −80° for at least 6
months.

*Partial Purification of FLAG-Tagged ALS2/ALS2CL by Immunopre-
cipitation.* Pellets of COS-7 cells obtained from the transfection with
10 μg of pCIneoFLAG construct/T150 flask are resuspended with 1 ml of
Buffer A (50 m*M* Tris–HCl, pH 7.4, 150 m*M* NaCl, 1 m*M* EDTA, 2% [w/v]
Tween-20, 1 tablet of Complete protease inhibitor cocktail [Roche]/50 ml
of the buffer) and lysed for 3 h at 4° with a gentle rotation. Supernatants

are recovered by centrifugation at 12,000×*g* for 15 min at 4°, and are immediately subjected to immunoprecipitation using EZview Red ANTI-FLAG M2 Affinity Gel (Sigma). In brief, a 60-μl aliquot of the slurry of gel beads is washed with 2 × 1 ml of ice-cold Buffer A, mixed with an approximately 1 ml of the supernatant, and incubated for 16 h at 4° with a gentle rotation. The M2 affinity gel beads are then washed three times with ice-cold Buffer A. To estimate the approximate amount of conjugating FLAG-tagged ALS2/ALS2CL on the beads, appropriate amounts of the immunoprecipitates are analyzed by sodium dodecyl sulfate–polyacrylamide gel electrophoresis (SDS-PAGE) followed by either silver or Coomassie blue staining in combination with Western blotting analysis using an anti-FLAG antibody. The purity of the FLAG-tagged human ALS2 and its truncated mutants (silver staining) and FLAG-tagged human and mouse ALS2CL (Coomassie blue staining) prepared by this method is shown in Fig. 2. Finally, the amino-terminally FLAG-tagged ALS2/ALS2CL proteins bound to the M2 affinity gel beads are resuspended in appropriate volumes of GEF buffer (25 m*M* Tris–HCl, pH 7.4, 50 m*M*

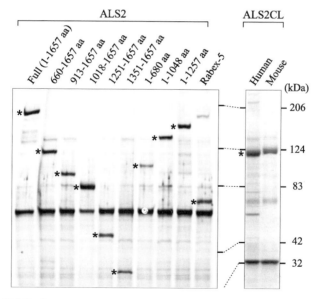

FIG. 2. SDS-PAGE analysis of the immunoprecipitated FLAG-tagged human ALS2, various truncated human ALS2 mutants, human Rabex-5 (Rab5GEF), and human and mouse ALS2CL. Asterisk indicates each immunoprecipitated protein. Left panel (ALS2, its truncated mutants, and Rabex-5), sliver staining; right panel (ALS2CL), Coomassie blue staining. Positions of size markers are shown on the right.

NaCl, 20 mM MgCl$_2$, 1 mM CHAPS) and used for GDP/GTP exchange assay *in vitro*. For longer storage, the choice of detergent in the buffer is very important. The immunoprecipitated ALS2/ALS2CL bound onto the affinity gel beads can be stored in Buffer A at 4° for at least 1 year without a significant loss of GEF activities on Rab5A. However, the buffer containing 0.1% IGEPAL CA-630 (NP-40) instead of Tween-20, for example, leads to the formation of the aggregated form of ALS2 only after several weeks, thereby decreasing their enzymatic activities.

Purification of the Recombinant Rab5A Protein

Expression Construct and Transformation. To generate the bacterial expression construct for Rab5A, the human *RAB5A* cDNA is amplified from human brain total RNA by RT-PCR, using primer pairs (Rab5A-L: 5′-atatggatccgcgaattcaATGGCTAGTCGAGGCGCAACAAGACCC-3′ and Rab5A-R: 5′-atatggatcctcgagTTAGTTACTACAACACTGATTCCTGGT TGG-3′). The amplified *RAB5A* cDNA is digested with *Bam*HI and *Xho*I and cloned into the *Bam*HI–*Xho*I sites of the pGEX-6P-2 (Amersham Pharmacia). The resulting construct, pGEX6P-Rab5A (Otomo *et al.*, 2003), is used to transform *Escherichia coli* BL21(DE3)pLysS (Novagen).

Expression and Purification of Rab5A. To express GST-fused Rab5A, 10 ml of LB medium containing 100 μg/ml of ampicillin is inoculated with a single colony of the BL21(DE3) pLysS transformant and grown to saturation overnight at 37°. Aliquots of culture (~5 ml) are then used to seed a larger culture (250 ml), and further grown at 37° until OD$_{600}$ ~1.4. The fusion protein is induced with 0.1 mM isopropyl thio-β-D-galactoside (IPTG) at 18° for 3 h. Bacteria are collected by centrifugation at 5000×g for 15 min at 4°, and frozen at −80° until use. Bacterial pellets are resuspended in 20 ml of Bacterial Lysis buffer (50 mM Tris–HCl, pH 8.0, 100 mM NaCl, 2 mM EDTA, 2 mM dithiothreitol [DTT], 10% [v/v] glycerol, 1% [w/v] CHAPS, 1 mM 4-aminophenylmethanesulfonyl fluoride [p-APMSF], 0.4 mg/ml lysozyme), and incubated for 30 min on ice, followed by the gentle sonication (1 min) on ice. The lysate is centrifuged at 10,000×g for 30 min at 4° to remove cell debris and insoluble materials, and the cleared supernatant obtained is mixed with a 1 ml slurry of the glutathione-Sepharose 4B beads (Amersham Pharmacia), which is prewashed with PBST (PBS[-], 0.5 % [v/v] Triton X-100). After mixing them for at least 3 h (or overnight) at 4°, the beads are washed four times with 10 ml of PBST and packed into a disposable column. The column is washed with 4 × 5 ml of PBST and then with 3 ml of Bacterial Lysis buffer. GST-fused Rab5A is then eluted with 9 ml of 15 mM reduced glutathione in Bacterial

Lysis buffer, snap frozen with liquid nitrogen, and stored at $-80°$ until needed.

To purify Rab5A, the GST portion of the fusion protein is removed by treatment with PreScission Protease (Amersham Pharmacia) as follows. First, to remove reduced glutathione contained in the GST-fused Rab5A sample and to replace Bacterial Lysis buffer with PSP buffer (50 mM Tris–HCl, pH 7, 100 mM NaCl, 1 mM EDTA, 1 mM DTT) appropriate for enzymatic cleavage, a 1-ml aliquot of GST-Rab5A is applied to a HiTrap Desalting column (bed volume, 5 ml, Amersham Pharmacia), which is preequilibrated with PSP buffer and eluted in 10 $\times$ 500-μl fractions of PSP buffer. Following elution, the protein concentration of each fraction is measured by a UV spectrophotometer at OD$_{280}$ and three peak fractions are pooled ($\sim$1.5 ml). Approximately 0.5 ml of PSP buffer containing 0.04% (v/v) Triton X-100 is added to this pooled fraction (final 0.01% Triton X-100 in PSP), and the resulting fraction is subjected to the digestion with PreScission Protease (20 units) overnight at $4°$. As PreScission Protease is a GST fusion protein, the cleaved GST portion of GST-Rab5A and PreScission Protease can be removed simultaneously by applying the digested samples onto a glutathione-Sepharose 4B beads column, and a follow-through fraction that contains the purified Rab5A is collected. Finally, $\sim$2 ml of the follow-through fraction is dialyzed twice using a Slide-A-Lyzer 10k Dialysis Cassette (PIERCE) against an excess volume (2 $\times$ 1 liter) of GXP loading buffer (25 mM Tris–HCl, pH 7.5, 50 mM NaCl, 10 mM EDTA, 5 mM MgCl$_2$, 1 mM DTT, 1 mM CHAPS) overnight at $4°$, and the concentration of the purified Rab5A is determined using the Bradford reaction. SDS–PAGE analysis of the expression and purification of Rab5A using the above protocol is shown in Fig. 3. Using this method, 250 ml of the starting bacterial culture is enough to obtain 1–2 mg of purified Rab5A. It is noted that although most of the other small GTPases can be purified by the same method, the yield of each small GTPase is varied.

Assays for GEF Activity

GXP Loading

The purified Rab5A (200 pmol) is loaded with [^{3}H]GDP (0.8 nmol; 370 GBq/mmol) (Amersham Pharmacia) in GXP loading buffer for 30 min at $30°$. To stabilize nucleotide binding, MgCl$_2$ is added to a final concentration of 20 mM, and the mixture is cooled to $4°$. The GDP-loaded protein sample is then applied onto the PD-10 column (Amersham Pharmacia),

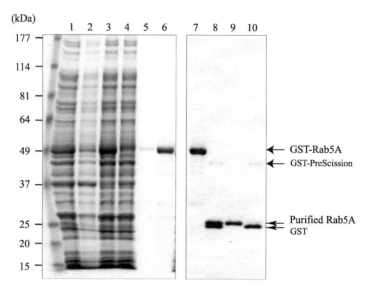

FIG. 3. Expression and purification of the recombinant Rab5A protein. Aliquots of protein sample at each purification step are analyzed by SDS-PAGE, followed by Coomassie blue staining. Lane 1, whole bacterial extract; lane 2, insoluble pellet fraction; lane 3, soluble supernatant fraction; lane 4, supernatant of soluble fraction after treatment with glutathione-Sepharose 4B beads; lane 5, proteins remaining on the beads after elution with reduced glutathione; lanes 6 and 7, eluted fraction from the beads (purified GST-Rab5A); lane 8, the digested GST-Rab5A with PreScission Protease; lane 9, the follow-through fraction (purified Rab5A); lane 10, proteins trapped onto glutathione-Sepharose 4B beads (GST and GTS-PreScission). Positions of size markers are shown on the left.

and eluted in 15 × 100 μl fractions in GEF buffer on ice. Aliquots (2.5 μl) of each fraction are counted in a liquid scintillation counter to determine the radioactivity, and six peak fractions are pooled. The loaded Rab5A is immediately used for the GDP dissociation assay, or snap frozen in liquid nitrogen and stored at −80° until needed.

GDP Dissociation Assay

Four picomoles of the [^{3}H]GDP-loaded Rab5A is preincubated for 5 min at 30°, and a nucleotide dissociation reaction was initiated by the addition of gel beads conjugating 2 pmol equivalent of the immunoprecipitated FLAG-tagged ALS2 or ALS2CL protein, and further incubated for 60 min at 30° in a final volume of 40 μl GEF buffer in the presence of 5 mM GTP. Reactions are then terminated by the addition of 2 ml of ice-cold

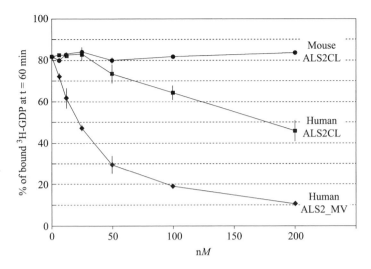

FIG. 4. The dose–response curve of the GDP dissociation activity of ALS2 and ALS2CL on Rab5A. Each value represents mean ± SD ($n = 3$) of the percentage of [^{3}H]GDP bound to Rab5A after 60 min in the presence of a given amount of either human ALS2_MV (ALS2 _1018–1657 aa) (diamonds), human ALS2CL (squares), or mouse ALS2CL (circles).

STOP buffer (25 mM Tris-HCl, pH 7.5, 100 mM NaCl, 20 mM MgCl$_2$) and filtered through BA85 nitrocellurose filters (Schleicher & Schell), followed by washing the filters with 20 ml of STOP buffer. The radioactivity trapped on the filters is counted by a scintillation counter, and the percentage of the dissociated [^{3}H]GDP is calculated. Typical results of the GDP dissociation assay for ALS2 and ALS2CL on Rab5A are shown in Fig. 4 and elsewhere (Hadano et al., 2004; Kunita et al., 2004; Otomo et al., 2003).

GTP-Binding Assay

Rab5A is preloaded with 100 mM of either GTPγS or GDP essentially in the same manner as above. Gel beads conjugating a 2 pmol equivalent of the immunoprecipitated FLAG-tagged ALS2 and 0.5 pmol of [^{35}S] GTPγS (37 TBq/mmol; Amersham Pharmacia) are preincubated for 5 min at 30°. Reactions are initiated by the addition of 4 pmol of Rab5AGDP, Rab5AGTPγS, or nucleotide-free Rab5A, and incubated for the required time, typically for 60 min, at 30° in a final volume of 40 μl GEF buffer. Reactions are then terminated and filtered, followed by scintillation counting, as indicated above (Otomo et al., 2003).

Functional Interactions of ALS2/ALS2CL

Interaction of ALS2/ALS2CL with Rab5

Purified Rab5A (4 pmol), which is preloaded with either GTPγS or GDP, or left unloaded with any nucleotides, is incubated with the FLAG-M2 affinity beads conjugating 4 pmol of the immunopurified FLAG-tagged ALS2 in 100 μl of the modified GEF buffer (25 mM Tris–HCl, pH 7.5, 100 mM NaCl, 20 mM MgCl$_2$, 1.5 mM CHAPS, 0.1% [w/v] skim milk) containing 50 μM of either GTPγS or GDP, or without nucleotides, for 2 h at 30° with a gentle shaking. After washing four times with 1 ml of the same buffer without skim milk, bound Rab5A is coeluted with the FLAG-tagged ALS2 proteins by the addition of SDS–PAGE sample buffer, and each protein is detected by Western blotting analysis using anti-Rab5 or anti-FLAG-M2 antibody. Results of the *in vitro* Rab5A binding assay are shown in Fig. 5. We have also reported the interaction between ALS2CL and Rab5A (Hadano *et al.*, 2004).

Characterization of Oliogomerized ALS2 by Gel Filtration

Recently, by immunoprecipitation and gel filtration analyses, we demonstrated that ALS2 homooligomerizes in mammalian cells (Kunita *et al.*, 2004). In this chapter, a procedure for the gel filtration of the recombinant ALS2 protein is described. In brief, FLAG-tagged ALS2

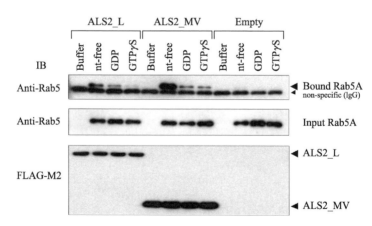

Fig. 5. *In vitro* Rab5A-binding assay. Nucleotide free (nt-free), GDP bound (GDP), or GTPγS bound (GTPγS) forms of Rab5A are incubated with the FLAG-M2 beads conjugating FLAG-tagged ALS2_L (full-length) or ALS2_1018–1657 aa (AL2_MV), or with FLAG-M2 beads alone (empty) as a control. IB, antibodies used for Western blotting analysis.

proteins on the FLAG-M2 gel beads are resuspended in Buffer B (50 mM Tris–HCl, pH 7.5, 150 mM NaCl, 0.1% IGEPAL CA-630) and subsequently eluted with Buffer B containing 500 ng/ml 3×FLAG peptide (Sigma) for 1 h at 4°. Approximately 30 pmol of FLAG-tagged ALS2 is applied to the Superdex 200 column (HR 10/30 Amersham Pharmacia), which is preequilibrated with Buffer B. Elution is carried out at 4° at a flow rate of 0.3 ml/min with a fraction volume of 0.5 ml. Fractions are subjected to Western blotting analysis with anti-ALS2 polyclonal antibody (MPF 1012–1651) (Kunita et al., 2004). The elution profile of the column is calibrated with the sizing standards (Amersham Pharmacia) of thyroglobulin (669 kDa), ferritin (440 kDa), catalase (232 kDa), aldorase (158 kDa), and ovalbumin (43 kDa). Gel filtration profiles for ALS2 have been reported by Kunita (2004).

Discussion

We have described here the procedures for the purification of ALS2, ALS2CL, and Rab5A, and the GEF activity assay as well as the in vitro binding for ALS2 and Rab5A. Using a mammalian expression system, ALS2 is highly expressed in the detergent-soluble fraction, and thus is easily enriched and purified by immunoprecipitation. Overexpressed untagged ALS2 as well as endogenous ALS2 can also be immunoprecipitated from cultured cells and/or tissue (brain) extract using the anti-ALS2 polyclonal antibody (unpublished observations). It is noteworthy that the truncated ALS2 molecules that retain Rab5GEF activity turn to a solely insoluble feature. On the other hand, ALS2CL is mostly distributed to the detergent-insoluble fraction in the cells, and thus the amount of ALS2CL in the soluble fraction is rather small. However, the expression level of the ALS2CL construct in COS-7 cells is extremely high, so that enough protein for several biochemical assays can be obtained by immunoprecipitation. Conversely, most ALS2 truncated mutants are unstable when overexpressed in the cells (Yamanaka et al., 2003) and are also insoluble in nature (unpublished observations). Therefore, to purify such proteins, a rather larger scale of culture and transfection is required when the mammalian expression system is adopted. Use of a proteasome inhibitor, such as MG132 (Calbiochem) (Yamanaka et al., 2003), is also effective in increasing the yield of proteins. Alternatively, bacterial expression (Otomo et al., 2003) and the baculovirus-Sf9 cell systems (Topp et al., 2004) can also be used for the preparation of ALS2 fragments. In fact, we have routinely used the bacterial expression system for ALS2 proteins utilizing for the antibody generation and affinity columns. However, as the homooligomerization of ALS2 is crucial for ALS2-associated Rab5GEF activity in vitro and ALS2-mediated endosome dynamics in the cells (Kunita *et al.*, 2004), a careful assessment of every experimental procedure, which may alter the properties for ALS2/ALS2CL, should be required, no matter what procedures would be used.

Acknowledgments

This work was funded by the Japan Science and Technology Agency, and in part by research grants from Research on Psychiatric and Neurological Diseases and Mental Health from the Ministry of Health, Labour and Welfare, and a Grant-in-Aide for Scientific Research from the Japan Society for the Promotion of Science.

References

Burd, C. G., Mustol, P. A., Schu, P. V., and Emr, S. D. (1996). A yeast protein related to a mammalian Ras-binding protein, Vps9p, is required for localization of vacuolar proteins. *Mol. Cell Biol.* **16**, 2369–2377.

Da Silva, J. S., and Dotti, C. G. (2002). Breaking the neuronal sphere: Regulation of the actin cytoskeleton in neuritogenesis. *Nat. Rev. Neurosci.* **3**, 694–704.

Devon, R. S., Schwab, C., Topp, J. D., Orban, P. C., Yang, Y.-Z., Pape, T. D., Helm, J. R., Davidson, T.-L., Rogers, D. A., Gros-Louis, F., Rouleau, G., Horazdovsky, B. F., Levitt, B. R., and Hayden, M. R. (2005). Cross-species characterization of the *ALS2* gene and analysis of its pattern of expression in development and adulthood. *Neuobiol. Dis.* **18**, 243–257.

Etienne-Manneville, S., and Hall, A. (2002). Rho GTPases in cell biology. *Nature* **420**, 629–635.

Govek, E.-E., Newey, S. E., and Van Aelst, L. (2005). The role of the Rho GTPases in neuronal development. *Genes Dev.* **19**, 1–49.

Hadano, S., Hand, C. K., Osuga, H., Yanagisawa, Y., Otomo, A., Devon, R. S., Miyamoto, N., Showguchi-Miyata, J., Okada, Y., Singaraja, R., Figlewicz, D. A., Kwiatkowski, T., Hosler, B. A., Sagie, T., Skaug, J., Nasir, J., Brown, R. H., Jr., Scherer, S. W., Rouleau, G. A., Hayden, M. R., and Ikeda, J.-E. (2001). A gene encoding a putative GTPase regulator is mutated in familial amyotrophic lateral sclerosis 2. *Nat. Genet.* **29**, 166–173.

Hadano, S., Otomo, A., Suzuki-Utsunomiya, K., Kunita, R., Yanagisawa, Y., Showguchi-Miyata, J., Mizumura, H., and Ikeda, J.-E. (2004). ALS2CL, the novel protein highly homologous to the carboxy-terminal half of ALS2, binds to Rab5 and modulates endosome dynamics. *FEBS Lett.* **575**, 64–70.

Horiuchi, H., Lippe, R., McBride, H. M., Rubino, M., Woodman, P., Stenmark, H., Rybin, V., Wilm, M., Ashman, K., Mann, M., and Zerial, M. (1997). A novel Rab5 GDP/GTP exchange factor complexed to Rabaptin-5 links nucleotide exchange to effector recruitment and function. *Cell* **90**, 1149–1159.

Kajiho, H., Saito, K., Tsujita, K., Kontani, K., Araki, Y., Kurosu, H., and Katada, T. (2003). RIN3: A novel Rab5 GEF interacting with amphiphysin II involved in the early endocytic pathway. *J. Cell Sci.* **116**, 4159–4168.

Kunita, R., Otomo, A., Mizumura, H., Suzuki, K., Showguchi-Miyata, J., Yanagisawa, Y., Hadano, S., and Ikeda, J.-E. (2004). Homo-oligomerization of ALS2 through its unique carboxy-terminal region is essential for the ALS2-associated Rab5 guanine nucleotide exchange activity and its regulatory function on endosome trafficking. *J. Biol. Chem.* **279,** 38626–38635.

Luo, L. (2000). Rho GTPases in neuronal morphogenesis. *Nat. Rev. Neurosci.* **1,** 173–180.

Ohtsubo, M., Kai, R., Furuno, N., Sekiguchi, T., Sekiguchi, M., Hayashida, H., Kuma, K., Miyata, T., Fukushige, S., Murotsu, T., Matsubara, K., and Nishimoto, T. (1987). Isolation and characterization of the active cDNA of the human cell cycle gene (RCC1) involved in the regulation of onset of chromosome condensation. *Genes Dev.* **1,** 585–593.

Otomo, A., Hadano, S., Okada, T., Mizumura, H., Kunita, R., Nishijima, H., Showguchi-Miyata, J., Yanagisawa, Y., Kohiki, E., Suga, E., Yasuda, M., Osuga, H., Nishimoto, T., Naurumiya, S., and Ikeda, J.-E. (2003). ALS2, a novel guanine nucleotide exchange factor for the small GTPase Rab5, is implicated in endosomal dynamics. *Hum. Mol. Genet.* **12,** 1671–1687.

Rosa, J. L., Casaroli-Marano, R. P., Buckler, A. J., Vilaró, S., and Barbacid, M. (1996). p532, a giant protein related to the chromosome condensation regulator RCC1, stimulates guanine nucleotide exchange on ARF1 and Rab proteins. *EMBO J.* **15,** 4262–4273.

Rossman, K. L., Der, C. J., and Sondek, J. (2005). GEF means go: Turning on Rho GTPases with guanine nucleotide-exchange factors. *Nat. Rev. Mol. Cell Biol.* **6,** 167–180.

Saito, K., Murai, J., Kajiho, H., Kontani, K., Kurosu, H., and Katada, T. (2002). A novel binding protein composed of homophilic tetramer exhibits unique properties for the small GTPase Rab5. *J. Biol. Chem.* **277,** 3412–3418.

Schmidt, A., and Hall, A. (2002). Guanine nucleotide exchange factors for Rho GTPases: Turning on the switch. *Genes Dev.* **16,** 1587–1609.

Takeshima, H., Komazaki, S., Nishi, M., Iino, M., and Kangawa, K. (2000). Junctophilins: A novel family of junctional membrane complex proteins. *Mol. Cell* **6,** 11–22.

Tall, G. G., Barbieri, M. A., Stahl, P. D., and Horazdovsky, B. F. (2001). Ras-activated endocytosis is mediated by the Rab5 guanine nucleotide exchange activity of RIN1. *Dev. Cell* **1,** 73–82.

Topp, J. D., Gray, N. W., Gerard, R. D., and Horazdovsky, B. F. (2004). Alsin is a Rab5 and Rac1 guanine nucleotide exchange factor. *J. Biol. Chem.* **279,** 24612–24623.

Van Aelst, L., and Symons, M. (2002). Role of Rho family GTPases in epithelial morphogenesis. *Genes Dev.* **16,** 1032–1054.

Yamanaka, K., Vande Velde, C., Eymard-Pierre, E., Bertini, E., Boespflug-Tanguy, O., and Cleveland, D. W. (2003). Unstable mutants in the peripheral endosomal membrane component ALS2 cause early-onset motor neuron disease. *Proc. Natl. Acad. Sci. USA* **100,** 16041–16046.

Yang, Y., Hentati, A., Deng, H. X., Dabbagh, O., Sasaki, T., Hirano, M., Hung, W. Y., Ouahchi, K., Yan, J., Azim, A. C., Cole, N., Gascon, G., Yagmour, A., Ben-Hamida, M., Pericak-Vance, M., Hentati, F., and Siddique, T. (2001). The gene encoding alsin, a protein with three guanine-nucleotide exchange factor domains, is mutated in a form of recessive amyotrophic lateral sclerosis. *Nat. Genet.* **29,** 160–165.

Zerial, M., and McBride, H. (2001). Rab proteins as membrane organizers. *Nat. Rev. Mol. Cell Biol.* **2,** 107–117.

[27] Polycistronic Expression and Purification of the ESCRT-II Endosomal Trafficking Complex

By Aitor Hierro, Jaewon Kim, and James H. Hurley

Abstract

Eukaryotic cells use sophisticated mechanisms to direct protein traffic between subcellular compartments. In eukaryotic cells, transmembrane proteins are delivered for degradation in the lysosome or yeast vacuole via multivesicular bodies. The sorting of proteins into lumenal vesicles within multivesicular bodies is directed by the three ESCRT protein complexes. Here we describe the expression and purification of the ESCRT-II complex using the polycistronic expression vector pST39 developed by Tan. In a modification of Tan's procedure, Pfu polymerase amplification with overlapping oligonucleotides was used to generate the translation cassettes for subcloning into pST39 expression vector in a single step. This approach reduces the number of restriction sites and subcloning steps required to express a heterooligomeric protein complex, facilitating rapid screening of multiple complexes and complex variants for crystallization or biochemical characterization.

Introduction

Eukaryotic cells are defined by internal membrane-limited organelles that delimit specific tasks. Rapid and accurate transportation and communication between these organelles are essential for cell function. Multivesicular bodies (MVBs) are intermediates in the endosomal system of eukaryotic cells that are formed by vesicle invagination from the limiting membrane of endosomes into their lumen (Gruenberg and Stenmark, 2004; Pelham, 2002; Piper and Luzio, 2001). During this vesicle-budding process transmembrane proteins and lipids are sorted into the nascent vesicles. The MVBs then fuse with the lysosome or yeast vacuole, where the vesicles and their contents are degraded by hydrolases (Futter et al., 1996). Under some circumstances, the vesicles may also be targeted for extracellular release (Blott and Griffiths, 2002), and retroviral budding from macrophages is now thought to occur by this process (Goila-Gaur et al., 2003; Martin-Serrano et al., 2003; Nguyen et al., 2003; Scarlata and Carter, 2003; Strack et al., 2003; von Schwedler et al., 2003). Proteins that are retained on the limiting membrane of the MVB are either recycled to the trans-Golgi network (TGN) or delivered to the plasma membrane or remain on the

METHODS IN ENZYMOLOGY, VOL. 403
0076-6879/05 $35.00
DOI: 10.1016/S0076-6879(05)03027-2

limiting membrane after fusion with the lysosome/vacuole. The MVB sorting pathway plays a crucial role in growth factor-receptor down-regulation (Futter *et al.*, 1996), developmental signaling (Deblandre *et al.*, 2001; Lai *et al.*, 2001; Pavlopoulos *et al.*, 2001), and regulation of the immune response (Kleijmeer *et al.*, 2001).

Cargo Recognition and Sorting

Cargo sorting into MVB is a highly regulated process and it has been shown that monoubiquitination of transmembrane proteins serves as a signal for sorting into the MVB pathway (Katzmann *et al.*, 2001; Reggiori and Pelham, 2001). The basic machinery for cargo and budding of vesicles into the lumen of MVB is a network of proteins first identified in *Saccharomyces cerevisiae* as the class E subgroup of vacuolar protein sorting (vps) mutants (Raymond *et al.*, 1992). Deletion of any of the class E vps genes in yeast results in the absence of intralumenal vesicles and the accumulation of the cargo in aberrant structures adjacent to the vacuole known as the "class E compartment" (Katzmann *et al.*, 2001; Rieder *et al.*, 1996). The yeast class E vps proteins include at least 18 members (Babst, 2005; Bowers *et al.*, 2004). These vps proteins are conserved from yeast to human, and in humans the set of class E vps homologs includes at least 26 proteins (Strack *et al.*, 2003; von Schwedler *et al.*, 2003), which comprise a network of interacting proteins. Most of the class E vps proteins are subunits of one of four large heterooligomeric complexes. These are the Vps27/Hse1 complex (Hrs/STAM in humans) and the ESCRT-I, ESCRT-II, and ESCRT-III (*E*ndosomal *S*orting *C*omplex *R*equired for *T*ranport) complexes. In working models of the ESCRT pathway, protein cargo is concentrated and moved through these complexes into the nascent vesicle of MVB in a sequential manner. First, Vps27/Hse1 is recruited to the early endosome via its FYVE domain, which binds to phosphatidylinositol 3-phosphate, and its UIM (ubiquitin interacting motif) domain, which binds to monoubiquitinated transmembrane cargo proteins in the endosomal membrane. The ESCRT-I complex (Vps23, Vps28, and Vps37) is then recruited from the cytosol by Vps27 (Katzmann *et al.*, 2003). The ESCRT-II complex (Vps22, Vps25, and Vps36) is recruited to the endosomal membrane by ESCRT-I (Babst *et al.*, 2002b; Hierro *et al.*, 2004; Teo *et al.*, 2004), and in turn is thought to recruit components of the ESCRT-III complex (Vps20, Snf7, Vps2, and Vps24) (Babst *et al.*, 2002a). The ubiquitinated cargo is sequentially transferred downstream through at least the first three of these four complexes. The last complex, ESCRT-III, is intimately associated with the invaginating vesicle. Finally, Vps4, a multimeric AAA$^+$ type ATPase (*A*TPase *a*ssociated with a variety of cellular *a*ctivities), directly binds to ESCRT-III. Following MVB vesicle formation, Vps4 catalyzes the dissociation of ESCRT protein complexes in an ATP-dependent manner, allowing further rounds of cargo sorting (Babst *et al.*, 1997, 1998). Together, the ESCRT complexes, Vps4, and related proteins form an intricate membrane-associated network.

New Approach for Expression Protein Complexes

As cargo sorting into MVB is driven by interactions between multiprotein complexes, isolation of large amounts of purified complexes for biochemical and biophysical studies is both essential for mechanistic understanding, and challenging. One method traditionally used to obtain protein complexes is to overexpress and purify each recombinant component and then reconstitute the complex *in vitro*. Some successful examples of this approach are the reconstitution of the nucleosome (Luger *et al.*, 1997), the SNARE complex (Fasshauer *et al.*, 1998), and the TFIIA/TBP/DNA complexes (Tan *et al.*, 1996). Although these examples demonstrate the viability of *in vitro* reconstitution, this method is neither infallible nor efficient. The effort of purifying each component separately and finding the correct reconstitution conditions can be tedious, and high yields of properly folded complex are not guaranteed. Another important issue for achieving higher success is the solubility of the components prior to complex reconstitution. If they are not soluble enough, they may tend to form aggregates rather than a well-folded complex, and the reconstitution approach may fail completely.

An alternative solution to *in vitro* reconstitution is what has been called *in vivo* reconstitution by coexpression of the individual components of the complex in the same cellular environment (Tan, 2001). The presence of all components simultaneously facilitates proper folding and stability. Furthermore, coexpression also offers the benefit that the complex is purified in a single preparation, minimizing the number of steps required. This not only saves effort but helps ensure the integrity of the sample by minimizing opportunities for degradation. There are a number of examples in the literature of successful coexpression of protein complexes from separate plasmids in *Escherichia coli*, for example, the myosin heavy and light chains (McNally *et al.*, 1988) or HIV-1 reverse transcriptase (Maier *et al.*, 1999), and commercial systems such as Duet (Novagen) are now available for this purpose. These systems are limited for application to complexes of many subunits. Maintaining multiple plasmids in *E. coli* requires the use of multiple antibiotics to select for the plasmids. Also, the number of copies of each plasmid may vary, giving different amounts of complex subunits after induction and generating an imbalance in expression levels (Veitia,

2003). To address this problem, Tan (2001) developed a polycistronic expression vector, pST39. In this system, each gene of interest is subcloned into a translation cassette using the vector pET3aTr, which contains the translational start signals of T7, including the translational enhancer (ε) and Shine–Dalgarno (SD) sequences. Once each component gene has been subcloned into a pET3aTr vector, additional rounds of subcloning are used to transfer these cassettes into pST39 to create a polycistronic expression plasmid. One inherent limitation is that the genes of interest, especially large genes, often contain internal restriction sites for these subcloning cassettes. Careful attention to the choice of cassettes, the order in which genes are subcloned, and sometimes the introduction of silent mutations to eliminate unwanted restriction sites is required.

One surprising result of this study is that a 1:2:1 complex of Vps22:Vps25:Vps36 was obtained (Hierro et al., 2004), despite the use of identical expression signals for all genes. The subunit stoichiometry of the complex was not known in advance. The mechanism for achieving unequal stoichiometry from equal levels of transcription and translation is unknown. It seems likely that uncomplexed Vps22 and Vps36 are destabilized and rapidly degraded. This shows that the polycistronic approach can be useful even for expression of complexes with unknown or unequal stoichiometries.

In summary, we have described a modification of Tan's approach that avoids the utilization of the pET3aTr transfer vector, together with its application to the overexpression of the heterooligomeric ESCRT-II complex. We used assembly/extension-Pfu polymerase amplification of overlapping oligonucleotides to generate in one single step the translation cassettes for subcloning into the pST39 expression vector. This method saves time and reduces the number of potential unwanted restriction sites by reducing the number of subcloning steps and the number of restriction enzymes used in the process. This system has another advantage in that it is well suited to the introduction of small affinity tags such as poly-His at the N- or C-terminus of the subunit of choice. The efficiency of the procedure makes it practical to screen several alternative placements of the affinity tag for the best results in expression and purification.

Experimental Procedures

Construction of Vps22, Vps25, and Vps36 Cassettes

The individual genes for cassette construction of Vps22, Vps25, and Vps36 were obtained by polymerase chain reaction (PCR) amplification of the DNA coding for the complex cloned into the plasmid pMB175 (Babst et al., 2002b), kindly supplied by S. D. Emr. The oligonucleotides used for

Vps25 were 5'<u>CTTTAAGAAGGAGATATACATATG</u>TCTGCATTAC-
CTCCAGTAT3'(UP) and 5'CGCGGATCCTTAAACAACCTTAATG
GCAATTACA3'(LOW) and for Vps36 were 5'<u>CTTTAAGAAGG AGA
TATACATATG</u>GAGTACTGGCATTATGTGG3'(UP) and 5'C-GCA
CGCGTTTATATATGCGAGGGCCAATATC3'(LOW). The PCR pro-
ducts were cleaned using the QIAquick PCR Purification Kit (Qiagen)
following manual instructions. It should be noted that underlined bases
of the upper primers are an extra addition to the genes for annealing with
the underlined bases of the oligonucleotide 5'CGCTCCGGAGAGCTC
TCTAGAAATAAT TTTGTTTAA<u>CTTTAAGAAGGAGATATACAT
ATG</u>3' in the second PCR round. This primer contains *Bsp*EI, *Sac*I, and
*Xba*I restriction sites in addition to the translational enhancer (ε) and SD
sequences (Figs. 1A and 2A). The lower oligonucleotides used in the
second PCR round were the same as the first round, and contain *Bam*HI
and *Mlu*I restriction sites for Vps25 and Vps36, respectively. After the
second PCR, the DNA amplifications were isolated by 1.2% agarose E-Gel
electrophoresis (Invitrogen), eluted from the gel using a Rapid Gel Extraction System (Marligen Biosciences) and digested overnight at 37° with
*Xba*I and *Bam*HI (New England) for the Vps25 cassette and *Bsp*eI and
*Mlu*I (New England) for the Vps36 cassette. After overnight digestion
DNA cassettes were cleaned using the QIAquick PCR Purification Kit
(Qiagen) following manual instructions and frozen at −20° for further
transformation into a pST39 expression vector.

For generating a Vps22 cassette, which also contains the sequence for
the 6-histidine affinity tag, spacer region, and rTEV protease cleavage site,
the approach was different. First, we constructed a dsDNA using extension-Pfu polymerase amplification of two overlapping oligonucleotides:
5'CGCTCCGGAGAGCTCTCTAGAAATAATTTTGTTTAACTTTA-

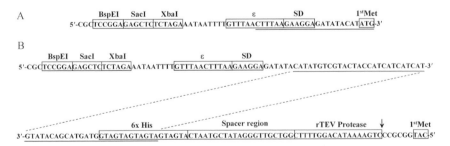

Fig. 1. Coding regions of the overlapping oligonucleotides used for (A) Vps25 and Vps36
and (B) Vps22. Restriction sites, translational enhancer (ε), Shine–Dalgarno (SD), spacer
region, and rTEV protease sequences are highlighted in boxes. Underlined nucleotides
correspond to assembly regions.

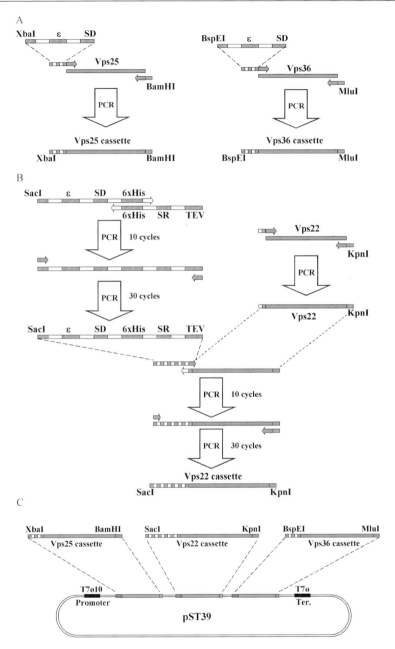

FIG. 2. Production of translation cassettes by assembly/extension-Pfu polymerase amplification of overlapping oligonucleotides. (A) Construction of cassettes for Vps25 and Vps36. (B) Construction of cassettes for Vps36. (C) Sequential insertion of Vps25, Vps22, and Vps36 into the pST39 vector.

AGAAGGAGATATA<u>CATATGTCGTACTACCATCATCATCAT</u>3′
and 5′CATGGCGCCCTGAAAATACAGGTTTTCGGTCGTTGGG-
ATATCGTAATC ATGATG<u>ATGATGATGATGGTAGTACGACA-
TATG</u>3′. The underlined nucleotides correspond to the overlap region
(Fig. 1B). After 10 extension cycles two more oligonucleotides 5′CGC-
TCCGGAGAGCTCTCTAGA3′(UP*) and 5′CATGGCGCCCTGAA-
AATACAGGTTTTCG3′(LOW) were added to the PCR reaction for
30 cycles more. The dsDNA fragment coding for the 6-histidine affinity
tag, spacer region, and rTEV protease cleavage site was isolated by
1.2% agarose E-Gel electrophoresis (Invitrogen), eluted from the gel
using a Rapid Gel Extraction System (Marligen Biosciences) and used
for another extension-Pfu polymerase amplification with the PCR product of Vps22. The oligonucleotides utilized for Vps22 amplification were
5′GAAAACCTGTATTTTCAGGGCGCC<u>ATGAAACAGTTTGGA-
CTGGCAG</u>3′(UP) and 5′CG CGGTACCTCACAGTTGCCTCG
TAAT CCAA3′(LOW*) where the underlined nucleotides correspond
to the overlapping region. After 10 extension cycles two more oligonucleotides previously used (UP* and LOW*) were added to the PCR
reaction for 30 cycles more (Fig. 2B). The dsDNA amplification was isolated
by 1.2% agarose E-Gel electrophoresis (Invitrogen), eluted from the gel
using the Rapid Gel Extraction System (Marligen Biosciences), and digested
overnight at 37° with *Sac*I and *Kpn*I restriction enzymes (New England).
After overnight digestion, the Vps22 DNA cassette containing the 6-histi-
dine affinity tag, spacer region, rTEV protease cleavage site, and Vps22 gene
was purified using the QIAquick PCR Purification Kit (Qiagen) following the
manufacturer's instructions and frozen at −20° for further transformation
into pST39 expression vector.

*Transformation of the Vps22, Vps25, and Vps36 Cassettes into pST39
  Expression Vector*

The order for transfer of the translation cassettes into pST39 was
Vps25, first, Vps22, second, and Vps36, third (Fig. 2C). This order was
chosen to avoid internal restriction sites in some cassettes. After overnight
pST39 digestion with the corresponding pair of restriction endonucleases
and alkaline phosphatase treatment, the product was purified by 1.2%
agarose E-Gel electrophoresis (Invitrogen) and eluted from the gel using
the Rapid Gel Extraction System (Marligen Biosciences). The ligation
reaction was carried out with T4 DNA ligase (Invitrogen) in a molar ratio
of 3/1 cassette/pST39 following the manufacturer's instructions. After ligation, the product was used to transform Max Efficiency DH5α competent
cells (Invitrogen). Positive colonies grown in LB/agar + 100 μg/ml ampicil-

lin were tested by PCR and sequenced to verify the presence of the cassette. The plasmid was transformed into BL21-CodonPlus-(DE3)-Ril (Stratagene) competent cells for protein complex expression.

Expression and Purification of ESCRT-II Complex

Unless otherwise stated, all chromatographic steps were performed at 4°. All buffers were repeatedly flushed with nitrogen. An aliquot of 15 ml of an overnight culture was used to inoculate 4-liter culture flasks containing 1.5 liters of terrific broth (Sigma) supplemented with 100 μg/ml ampicillin and 50 μg/ml chloramphenicol. The bacteria were grown at 37° until an OD_{600} value of 0.8 was reached and induced with 0.5 mM isopropylthiogalactoside (IPTG) at 15° for 36 h. Cells from a 9-liter culture were lysed by high-pressure homogenization in 50 mM Tris (pH 7.5), 300 mM NaCl, 5 mM 2-β mercaptoethanol (β ME), and 100 μl of protease inhibitor cocktail (Sigma) per liter of culture medium. After centrifugation the soluble fraction was adjusted to 5 mM imidazole and it was loaded into a column packed with 10 ml Co^{2+} affinity chromatography resin (Clontech). The column was washed with 2 liters 50 mM Tris (pH 7.5), 300 mM NaCl, 5 mM β ME, and 5 mM imidazole. Bound complex was eluted with the same washing buffer but with a step gradient of imidazole of 10, 20, 50, 100, and 200 mM. The fractions containing the complex were pooled and the histidine tag was removed with TEV protease, concomitant with dialysis in 50 mM Tris (pH 8.0), 150 mM NaCl, and 5 mM β ME. After overnight dialysis, the sample was loaded into column packed with 5 ml Co^{2+} affinity chromatography resin (Clontech) for removing TEV protease and the 6 His tag. The flowthrough was recovered, concentrated in a concentration cell (Amicon) to a final volume of 5 ml, and loaded into a Superdex S200 column (Amersham Biosciences) preequilibrated with 50 mM Tris (pH 7.5), 150 mM NaCl, and 5 mM β ME (Fig. 3A). The purity of the fractions analyzed by SDS-PAGE was >99% (Fig. 3B). The ESCRT-II complex was concentrated to 2 mg/ml and stored in aliquots at $-80°$ for further studies.

Concluding Remarks

Many polypeptides expressed in *E. coli* become insoluble because they might be components of obligatory complexes with predominantly hydrophobic interactions between the subunits. Soluble overexpression of these polypeptides can often be achieved if the subunits are coexpresed with their cognate partner(s). The pST39 expression system provides an effective way for constructing polycistronic expression vectors for *in vivo*

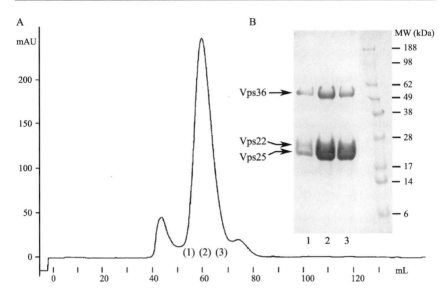

FIG. 3. Purification of the ESCRT-II complex. (A) Gel filtration (Superdex 200 16/60) profile in the last purification step showing a retention volume of 62 ml corresponding to an estimated molecular weight of 155 kDa. (B) SDS-PAGE of the collected fractions after gel filtration displaying all three components of the ESCRT-II complex.

reconstitution of heterooligomeric complexes of up to four different polypeptides. The success of this approach demands that its efficiency be optimized for the highest possible throughput. Here, we have shown how we have applied assembly/extension-Pfu polymerase amplification of overlapping oligonucleotides to generate the translation cassettes for subcloning into a pST39 expression vector in one step, with significant time savings. The utility of this system has been demonstrated with the successful cloning and reconstitution of the heterooligomeric ESCRT-II complex, and this design should also be applicable to other protein complexes.

Acknowledgments

We thank R. Grisshammer for comments on the manuscript. This work was supported by the NIDDK intramural research program and the NIH IATAP program.

References

Babst, M. (2005). A protein's final ESCRT. *Traffic* **6,** 2–9.

Babst, M., Sato, T. K., Banta, L. M., and Emr, S. D. (1997). Endosomal transport function in yeast requires a novel AAA-type ATPase, Vps4p. *EMBO J.* **16,** 1820–1831.

Babst, M., Wendland, B., Estepa, E. J., and Emr, S. D. (1998). The Vps4p AAA ATPase regulates membrane association of a Vps protein complex required for normal endosome function. *EMBO J.* **17,** 2982–2993.

Babst, M., Katzmann, D. J., Estepa-Sabal, E. J., Meerloo, T., and Emr, S. D. (2002a). Escrt-III: An endosome-associated heterooligomeric protein complex required for mvb sorting. *Dev. Cell* **3,** 271–282.

Babst, M., Katzmann, D. J., Snyder, W. B., Wendland, B., and Emr, S. D. (2002b). Endosome-associated complex, ESCRT-II, recruits transport machinery for protein sorting at the multivesicular body. *Dev. Cell* **3,** 283–289.

Blott, E. J., and Griffiths, G. M. (2002). Secretory lysosomes. *Nat. Rev. Mol. Cell Biol.* **3,** 122–131.

Bowers, K., Lottridge, J., Helliwell, S. B., Goldthwaite, L. M., Luzio, J. P., and Stevens, T. H. (2004). Protein-protein interactions of ESCRT complexes in the yeast Saccharomyces cerevisiae. *Traffic* **5,** 194–210.

Deblandre, G. A., Lai, E. C., and Kintner, C. (2001). Xenopus neuralized is a ubiquitin ligase that interacts with XDelta1 and regulates Notch signaling. *Dev. Cell* **1,** 795–806.

Fasshauer, D., Eliason, W. K., Brunger, A. T., and Jahn, R. (1998). Identification of a minimal core of the synaptic SNARE complex sufficient for reversible assembly and disassembly. *Biochemistry* **37,** 10354–10362.

Futter, C. E., Pearse, A., Hewlett, L. J., and Hopkins, C. R. (1996). Multivesicular endosomes containing internalized EGF-EGF receptor complexes mature and then fuse directly with lysosomes. *J. Cell Biol.* **132,** 1011–1023.

Goila-Gaur, R., Demirov, D. G., Orenstein, J. M., Ono, A., and Freed, E. O. (2003). Defects in human immunodeficiency virus budding and endosomal sorting induced by TSG101 overexpression. *J. Virol.* **77,** 6507–6519.

Gruenberg, J., and Stenmark, H. (2004). The biogenesis of multivesicular endosomes. *Nat. Rev. Mol. Cell Biol.* **5,** 317–323.

Hierro, A., Sun, J., Rusnak, A. S., Kim, J., Prag, G., Emr, S. D., and Hurley, J. H. (2004). Structure of the ESCRT-II endosomal trafficking complex. *Nature* **431,** 221–225.

Katzmann, D. J., Babst, M., and Emr, S. D. (2001). Ubiquitin-dependent sorting into the multivesicular body pathway requires the function of a conserved endosomal protein sorting complex, ESCRT-I. *Cell* **106,** 145–155.

Katzmann, D. J., Stefan, C. J., Babst, M., and Emr, S. D. (2003). Vps27 recruits ESCRT machinery to endosomes during MVB sorting. *J. Cell Biol.* **162,** 413–423.

Kleijmeer, M., Ramm, G., Schuurhuis, D., Griffith, J., Rescigno, M., Ricciardi-Castagnoli, P., Rudensky, A. Y., Ossendorp, F., Melief, C. J., Stoorvogel, W., and Geuze, H. J. (2001). Reorganization of multivesicular bodies regulates MHC class II antigen presentation by dendritic cells. *J. Cell Biol.* **155,** 53–63.

Lai, E. C., Deblandre, G. A., Kintner, C., and Rubin, G. M. (2001). Drosophila neuralized is a ubiquitin ligase that promotes the internalization and degradation of delta. *Dev. Cell* **1,** 783–794.

Luger, K., Rechsteiner, T. J., Flaus, A. J., Waye, M. M., and Richmond, T. J. (1997). Characterization of nucleosome core particles containing histone proteins made in bacteria. *J. Mol. Biol.* **272,** 301–311.

Maier, G., Dietrich, U., Panhans, B., Schroder, B., Rubsamen-Waigmann, H., Cellai, L., Hermann, T., and Heumann, H. (1999). Mixed reconstitution of mutated subunits of HIV-1 reverse transcriptase coexpressed in *Escherichia coli*—two tags tie it up. *Eur. J. Biochem.* **261,** 10–18.

Martin-Serrano, J., Zang, T., and Bieniasz, P. D. (2003). Role of ESCRT-I in retroviral budding. *J. Virol.* **77,** 4794–4804.

McNally, E. M., Goodwin, E. B., Spudich, J. A., and Leinwand, L. A. (1988). Coexpression and assembly of myosin heavy chain and myosin light chain in Escherichia coli. *Proc. Natl. Acad. Sci. USA* **85,** 7270–7273.

Nguyen, D. G., Booth, A., Gould, S. J., and Hildreth, J. E. (2003). Evidence that HIV budding in primary macrophages occurs through the exosome release pathway. *J. Biol. Chem.* **278,** 52347–52354.

Pavlopoulos, E., Pitsouli, C., Klueg, K. M., Muskavitch, M. A., Moschonas, N. K., and Delidakis, C. (2001). Neuralized Encodes a peripheral membrane protein involved in delta signaling and endocytosis. *Dev. Cell* **1,** 807–816.

Pelham, H. R. (2002). Insights from yeast endosomes. *Curr. Opin. Cell Biol.* **14,** 454–462.

Piper, R. C., and Luzio, J. P. (2001). Late endosomes: sorting and partitioning in multivesicular bodies. *Traffic* **2,** 612–621.

Raymond, C. K., Howald-Stevenson, I., Vater, C. A., and Stevens, T. H. (1992). Morphological classification of the yeast vacuolar protein sorting mutants: Evidence for a prevacuolar compartment in class E vps mutants. *Mol. Biol. Cell* **3,** 1389–1402.

Reggiori, F., and Pelham, H. R. (2001). Sorting of proteins into multivesicular bodies: Ubiquitin-dependent and -independent targeting. *EMBO J.* **20,** 5176–5186.

Rieder, S. E., Banta, L. M., Kohrer, K., McCaffery, J. M., and Emr, S. D. (1996). Multilamellar endosome-like compartment accumulates in the yeast vps28 vacuolar protein sorting mutant. *Mol. Biol. Cell* **7,** 985–999.

Scarlata, S., and Carter, C. (2003). Role of HIV-1 Gag domains in viral assembly. *Biochim. Biophys. Acta* **1614,** 62–72.

Strack, B., Calistri, A., Craig, S., Popova, E., and Gottlinger, H. G. (2003). AIP1/ALIX is a binding partner for HIV-1 p6 and EIAV p9 functioning in virus budding. *Cell* **114,** 689–699.

Tan, S. (2001). A modular polycistronic expression system for overexpressing protein complexes in *Escherichia coli. Protein Expr. Purif.* **21,** 224–234.

Tan, S., Hunziker, Y., Sargent, D. F., and Richmond, T. J. (1996). Crystal structure of a yeast TFIIA/TBP/DNA complex. *Nature* **381,** 127–151.

Teo, H., Perisic, O., Gonzalez, B., and Williams, R. L. (2004). ESCRT-II, an endosome-associated complex required for protein sorting: Crystal structure and interactions with ESCRT-III and membranes. *Dev. Cell* **7,** 559–569.

Veitia, R. A. (2003). Nonlinear effects in macromolecular assembly and dosage sensitivity. *J. Theor. Biol.* **220,** 19–25.

von Schwedler, U. K., Stuchell, M., Muller, B., Ward, D. M., Chung, H. Y., Morita, E., Wang, H. E., Davis, T., He, G. P., Cimbora, D. M., Scott, A., Krausslich, H. G., Kaplan, J., Morham, S. G., and Sundquist, W. I. (2003). The protein network of HIV budding. *Cell* **114,** 701–713.

[28] Analysis and Properties of the Yeast YIP1 Family of Ypt-Interacting Proteins

By Catherine Z. Chen and Ruth N. Collins

Abstract

The YIP1 family of proteins is an intriguing collection of small membrane proteins with critical roles in membrane traffic. Although their mode of action is unknown, they are receiving attention as participants in vesicle biogenesis, and as factors that may mediate the association of Rab proteins with membranes. Yeast *YIP1* is an essential gene and can be fully complemented by its human counterpart-suggesting that the essential function of Yip1p is evolutionarily conserved. This chapter presents methods for the cell biological and genetic analysis of Yip1p and other YIP1 family members in the yeast *Saccharomyces cerevisiae*.

Introduction

The Yip1p family is an evolutionarily conserved set of small integral membrane protein originally identified by two-hybrid screening with the Rab GTPases Ypt1p and Ypt31p (Yang *et al.*, 1998). YIP1 family members in the yeast *Saccharomyces cerevisiae* comprise Yip1p, Yip4p, Yip5p, and Yif1p, of which Yip1p and Yif1p provide essential cellular functions (Calero *et al.*, 2002). These proteins are capable of interaction with several partners, including other YIP1 family members, identified via both directed and global genetic/proteomic screening (Ito *et al.*, 2001; Uetz *et al.*, 2000). Evidence from other lines of investigation supports the physiological association of Yip1p with other YIP1 family members (Matern *et al.*, 2000), diprenylated Rab proteins and Yos1p, a small membrane protein of the endoplasmic reticulum (ER) (Heidtman *et al.*, 2005). The interacting capabilities of Yip1p suggest a possible role in Rab-guanine nucleotide dissociation inhibitor (GDI) displacement and Rab membrane recruitment. This hypothesis has received encouragement from the finding that human Yip3p, a protein with similar characteristics to Yip1p, has GDF activity (Sivars *et al.*, 2003), although such a role for Yip1p has yet to be demonstrated (Calero *et al.*, 2003). Yip1p is required for ER vesicle biogenesis *in vivo* and *in vitro* (Heidtman *et al.*, 2003), although the mechanism by which it participates in COPII function is not known.

METHODS IN ENZYMOLOGY, VOL. 403
Copyright 2005, Elsevier Inc. All rights reserved.
0076-6879/05 $35.00
DOI: 10.1016/S0076-6879(05)03028-4

Tagging of Yip1p

The fusion protein construct is generated by standard molecular biology techniques. An NH$_2$-terminal fusion is suggested as Yip1p retains function when NH$_2$-terminally but not COOH-terminally tagged (Fig. 1). The constructs are typically cloned under the control of the endogenous gene promoter and terminator and expressed as the sole copy of the protein in the cells. Yip4p and Yip5p constructs are transformed into their respective deletion strain. Genetic manipulations make it a straightforward procedure to test whether the tagged construct interferes with function, Yip1p and Yif1p constructs are transformed into a "tester" strain consisting of a

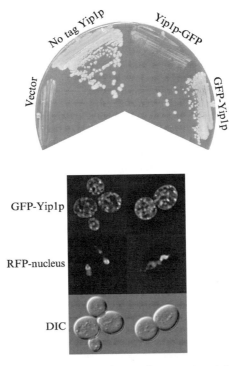

FIG. 1. Functionality and localization of green fluorescent protein (GFP)-tagged Yip1p constructs. The top panel shows GFP fused to the NH$_2$-terminus of Yip1p (GFP-Yip1p), but not the COOH-terminus (Yip1p-GFP), will complement a *yip1Δ* strain. The bottom panel shows cells expressing GFP-Yip1p as the sole source of Yip1p, analyzed by fluorescence imaging. A punctate fluorescence pattern typical of Golgi localized proteins is observed in addition to the ring-like perinuclear and subcortical structures of the ER. The inclusion of an RFP-tagged nuclear marker provides a convenient way of assessing the cell cycle stage of the imaged cell.

genomic gene deletion that is covered by an episomal URA3-containing plasmid (RCY1610, *ura3–52 leu2–3,112 YIP1ΔKAN^R* [YCP50 (pRC1245) *YIP1*] and RCY2171, *ura3Δ0 leu2Δ0 his3Δ0 YIF1ΔKAN^R* [pRS426 (pRC1037) *YIF1*]). The transformants are plated on 5-fluoroorotic acid (5-FOA) to eliminate the *URA3*-containing wild-type *YIP1* or *YIF1* plasmid, and resultant colonies now contain the new construct as the sole cellular source of Yip1p or Yif1p. The functionality assay should always be performed in comparison to yeast transformed with empty vector as a negative control and with wild-type *YIP1* as a positive control. Immunoblotting is performed with anti-tag or anti-Yip1p antibodies to determine the stability of the fusion construct.

Protocol for Immunoblotting of Yip1p and Yip1p Fusion Constructs

1. Cells in logarithmic phase growth are harvested by centrifugation, washed once, and resuspended in ice-cold TAz buffer (10 mM Tris, pH 7.5, 10 mM sodium azide) at a concentration of 10 A_{600} units/ml.

2. Of the cell suspension, 0.5 ml is transferred into a microfuge tube, the cells pelleted, and supernatant removed by aspiration.

3. The cell pellet is resuspended in 150 μl lysis buffer with protease inhibitors (4 mM phenylmethylsulfonyl fluoride [PMSF], 10 μg/ml Pepstatin A, 1 mM benzamidine, and 1 mM EDTA) added just prior to use.

4. An equal volume of glass beads is added and vortexed for 2 min in a Turbo-Beater apparatus.

5. A further 100 μl of lysis buffer is added and then 135 μl 4× sample buffer (12% w/v sodium dodecyl sulfate [SDS], 48% w/v sucrose, 150 mM Tris, pH 7.0, 6% 2-mercaptoethanol, Serva blue G).

6. The samples are mixed well and incubated at 40° for 10 min. Figure 2 shows the impact of different incubation temperatures on sample preparation.

7. The samples are microfuged at top speed for 2 min before SDS–polyacrylamide gel electrophoresis (SDS-PAGE) gel loading. Approximately 0.1 OD_{600} units are loaded for each lane.

We have found that electroblotting of the gels onto PVDF (Pall Gelman Corp., BioTrace 0.45 μm) gives good results in standard detection protocols.

Genetic Analysis of *yip1* Mutant Alleles

The mutant allele *yip1–4* is especially suited for genetic studies, as this mutant does not show revertants, is thermosensitive for growth over a wide range of temperatures, and the protein product is stable at restrictive

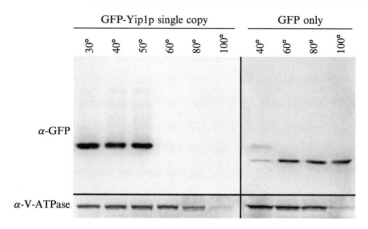

FIG. 2. Influence of sample incubation temperatures on Yip1p. Total lysates from cell expressing GFP alone or a GFP-Yip1p fusion protein were incubated in sample buffer at the temperatures indicated before SDS-PAGE and Western blotting. The V-ATPase was probed separately as an internal loading control.

temperature. For each mutant tested, a strain is created that contains a deleted genomic copy of *YIP1* together with a wild-type copy of *YIP1* on an episomal *URA3* plasmid that can be counterselected with the drug 5-FOA (Chen *et al.*, 2004). Each strain is transformed with plasmids containing either (1) *yip1*^ts allele, (2) *YIP1* wild-type plasmid, or (3) a no insert control plasmid. The resulting transformants are plated on 5-FOA-containing media to examine growth of the double mutant. The advantages of this assay are that (1) several hundred double mutant colonies can be analyzed on a single plate, rather than the few obtained with the more typical tetrad dissection analysis, (2) comparisons are made in isogenic strains where the mutant or wild-type allele is introduced on an episomal plasmid, avoiding the necessity of multiple backcrosses to compare mutations from two different strain backgrounds, (3) the assay is flexible, so the impact of several different alleles can be tested simultaneously without the need for additional crosses, and (4) the assay can potentially be made quantitative by a comparison of the growth curves and relative log phase doubling times of the single and double mutants. Figure 3A shows the synthetic interaction assay with cells grown on plates. Synthetic interactions are revealed by the presence of 5-FOA in the plate media with *vps21Δ yip1–4* double mutants as an example where no genetic interaction was observed and *ypt31Δ yip1–4* double mutants shown by comparison as an example of complete lethality between two mutants. Figure 3B demonstrates comparisons with several different *yip1* alleles simultaneously. In this example, synthetic

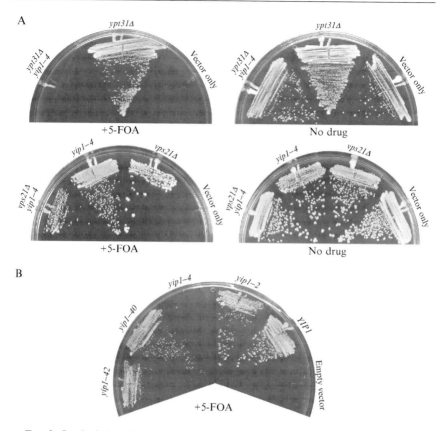

Fig. 3. Synthetic lethality assay of *yip1* alleles with Rab-encoding genes. (A) Synthetic lethality plate assay. Strains carrying thermosensitive mutations of Rab protein ORFs and various genes involved in Rab protein function and a deletion of *YIP1* together with wild-type *YIP1* on a *URA3* plasmid were transformed with vector only control (pRS315), wild-type *YIP1* control (pRC1838B), and the thermosensitive allele *yip1–4* (pRC1992). All transformants were assessed by plasmid shuffling on 5-FOA at 25°. Plates bearing *vps21Δ yip1–4* double mutants are shown as an example where no genetic interaction was observed. Similarly, *ypt31Δyip1–4* double mutants are shown as an example of complete lethality between two mutants. (B) Testing multiple alleles for synthetic lethality. Isogenic strains bearing various *yip1* conditional lethal alleles as indicated were tested for genetic interactions at 25° with the temperature-sensitive gene *ypt1*[A136D]. Wild-type *YIP1* and empty vector are included as a control for positive and negative growth, respectively.

lethality is examined between *ypt1*[A136D] and several defective *yip1* alleles; only *yip1–4* shows synthetic lethal interactions with *ypt1*[A136D]. Of the other conditional *yip1* alleles, only *yip1–42* appears to have an effect, with the double mutant appearing to be slightly sicker than the wild-type control.

The disadvantage of this type of genetic interaction assay is that the ability of cells to eliminate the URA3-containing plasmid in response to 5-FOA is variable depending on the growth rate of the strains and whether they are transferred from actively growing cultures. These factors make it imperative to include both positive and negative controls on each plate and to make comparisons only after observing the double mutant together with each of the single mutants.

Yeast Two-Hybrid Analysis of Yip1p Constructs

Yeast two-hybrid analysis was the method used in the original identification of Yip1 and Yif1p (Matern *et al.*, 2000; Yang *et al.*, 1998) and is a useful tool for identification and evaluation of YIP1 family interacting partners. The two-hybrid system results in a fusion protein with the GAL4 binding or activation domain at the NH_2-terminus of Yip1p, which does not disrupt its function (Fig. 1). We have used the yeast strain Y190 (Bai and Elledge, 1996) and the plasmids pACT2, pACTII, pAS2-CYH1, and pAS2-1 with reliable results (Chen *et al.*, 2004). One caution is that positive and negative controls must always be included with each two-hybrid experiment and each mutant allele be evaluated for its ability to transactivate in the absence of an interacting partner, because even if the wild-type protein does not transactivate, the same cannot be assumed for each mutant allele.

Acknowledgments

This work is supported by grants from the US National Science Foundation and US National Institutes of Health.

References

Bai, C., and Elledge, S. J. (1996). Gene identification using the yeast two-hybrid system. *Methods Enzymol.* **273**, 331–347.

Calero, M., Winand, N. J., and Collins, R. N. (2002). Identification of the novel proteins Yip4p and Yip5p as Rab GTPase interacting factors. *FEBS Lett.* **515**, 89–98.

Calero, M., Chen, C. Z., Zhu, W., Winand, N., Havas, K. A., Gilbert, P. M., Burd, C. G., and Collins, R. N. (2003). Dual prenylation is required for Rab protein localization and function. *Mol. Biol. Cell* **14**, 1852–1867.

Chen, C. Z., Calero, M., DeRegis, C. J., Heidtman, M., Barlowe, C., and Collins, R. N. (2004). Genetic analysis of yeast Yip1p function reveals a requirement for Golgi-localized rab proteins and rab-guanine nucleotide dissociation inhibitor. *Genetics* **168**, 1827–1841.

Heidtman, M., Chen, C. Z., Collins, R. N., and Barlowe, C. (2003). A role for Yip1p in COPII vesicle biogenesis. *J. Cell Biol.* **163**, 57–69.

Heidtman, M., Chen, C. Z., Collins, R. N., and Barlowe, C. (2005). Yos1p is a novel subunit of the Yip1p-Yif1p complex and is required for transport between the endoplasmic reticulum and the Golgi complex. *Mol. Biol. Cell* **16**, 1673–1683.

Ito, T., Chiba, T., Ozawa, R., Yoshida, M., Hattori, M., and Sakaki, Y. (2001). A comprehensive two-hybrid analysis to explore the yeast protein interactome. *Proc. Natl. Acad. Sci. USA* **98,** 4569–4574.

Matern, H., Yang, X., Andrulis, E., Sternglanz, R., Trepte, H. H., and Gallwitz, D. (2000). A novel Golgi membrane protein is part of a GTPase-binding protein complex involved in vesicle targeting. *EMBO J.* **19,** 4485–4492.

Sivars, U., Aivazian, D., and Pfeffer, S. R. (2003). Yip3 catalyses the dissociation of endosomal Rab-GDI complexes. *Nature* **425,** 856–859.

Uetz, P., Giot, L., Cagney, G., Mansfield, T. A., Judson, R. S., Knight, J. R., Lockshon, D., Narayan, V., Srinivasan, M., Pochart, P., Qureshi-Emili, A., Li, Y., Godwin, B., Conover, D., Kalbfleisch, T., Vijayadamodar, G., Yang, M., Johnston, M., Fields, S., and Rothberg, J. M. (2000). A comprehensive analysis of protein-protein interactions in Saccharomyces cerevisiae. *Nature* **403,** 623–627.

Yang, X., Matern, H. T., and Gallwitz, D. (1998). Specific binding to a novel and essential Golgi membrane protein (Yip1p) functionally links the transport GTPases Ypt1p and Ypt31p. *EMBO J.* **17,** 4954–4963.

[29] Use of Hsp90 Inhibitors to Disrupt GDI-Dependent Rab Recycling

By Christine Y. Chen, Toshiaki Sakisaka, and William E. Balch

Abstract

Guanine nucleotide dissociation inhibitor (GDI) is a central regulator of Rab GTPase family members. GDI recycles Rab proteins from the membrane and sequesters the inactive GDP-bound form of Rab in the cytosol for use in multiple rounds of transport. The balance between the membrane-bound form of Rab and the cytosolic reserve pool of the Rab–GDI complex is critical for vesicular trafficking between membrane compartments. Recycling of Rab GTPases is likely to require a membrane-bound complex of GDI, Hsp90, and Rab given that αGDI-dependent recycling of Rab3A at the synapse and neurotransmitter transmitter release is inhibited by Hsp90-specific inhibitors. Here we describe methods required for establishing the dependence of Rab recycling pathways on Hsp90 *in vitro*.

Introduction

The Rab GTPase family consists of 63 members that are distributed to different membrane compartments comprising both the exocytic and the endocytic pathway. Rab GTPases are thought to play a critical role in

METHODS IN ENZYMOLOGY, VOL. 403   0076-6879/05 $35.00
Copyright 2005, Elsevier Inc. All rights reserved.   DOI: 10.1016/S0076-6879(05)03029-6

assembling targeting complexes that promote membrane recognition and fusion (Gurkan *et al.*, 2005; Pfeffer and Aivazian, 2004). Rab GTPases, like other members of the Ras superfamily, switch between active (GTP-bound) and inactive (GDP-bound) forms. Switching from the GDP- to the GTP-bound form is regulated by guanine nucleotide exchange factors (GEFs). Activated Rab proteins recruit a variety of effectors that include tethering and fusion factors to promote interaction with target membranes (Gurkan *et al.*, 2005; Pfeffer and Aivazian, 2004). During or following membrane targeting and fusion, GTPase-activating proteins (GAPs) stimulate GTP hydrolysis that results in the dissociation of Rab from its effectors for retrieval by guanine nucleotide dissociation inhibitor (GDI). Subsequently, a new type of regulator, GDI displacement factor (GDF), is reported to play a key role in promoting the dissociation of Rab from GDI for retargeting to the membrane (Pfeffer and Aivazian, 2004; Sivars *et al.*, 2003).

Although functionally similar, GAPs and GEFs for individual Rabs share no homology at a molecular level. In contrast, GDI is highly conserved between species (Alory and Balch, 2003). Three mammalian isoforms of GDI have been identified. The α isoform is brain specific, while the β isoforms are ubiquitously expressed (Nishimura *et al.*, 1994; Shisheva *et al.*, 1994). Both α and β GDI interact with multiple Rab GTPases *in vitro* (Alory and Balch, 2003). Thus, the significance, if any, between GDI isoforms in controlling specific general Rab recycling remains unclear. Consistent with the need for only a single GDI, a GDI knockout is lethal in yeast and *Drosophila*, both of which have a single copy of GDI (Garrett *et al.*, 1994; Ricard *et al.*, 2001).

Rab GTPases are prenylated with two geranylgeranyl groups attached to cysteine residues found in their C-terminus. Prenylation is essential for membrane association and function of Rab GTPases. The structure of mammalian αGDI and yeast Gdi 1p has been solved (An *et al.*, 2003; Rak *et al.*, 2003). The primary function of GDI is to extract Rab from the membrane in a soluble form by sequestering the geranylgeranyl groups in hydrophobic pockets found in domains I and II of the protein structure (An *et al.*, 2003; Rak *et al.*, 2003). Interestingly, in the crystal structure of Rab-free GDI, the binding pocket of domain II is in a closed configuration. Following Rab binding, one of the prenyl groups is inserted into a deep hydrophobic pocket as a consequence of a major change in the conformation of domain II.

To facilitate Rab extraction, heat shock protein 90 (Hsp90) was characterized as a key recycling factor for αGDI in the synapse (Sakisaka *et al.*, 2002). Hsp90 is an abundant cytosolic chaperone that requires a variety of additional chaperones and cochaperones for steroid hormone receptor activation and the activity of >50 signaling kinase pathways (Pratt and Toft, 2003). The chaperone activity of Hsp90 is ATP dependent and sensitive to Hsp90 inhibitors including geldanamycin (GA) and radicicol. Both GA and radicicol are commonly used to analyze Hsp90 function *in vivo*.

Given the importance of Hsp90 for Rab3A recycling by GDI in the synapse and the redundancy of GDI in regulating multiple Rab GTPases, it is likely that Hsp90 plays a similar general role in recycling of other Rab GTPases (Chen *et al.*, manuscript in preparation). Below, we present two systems that apply *in vitro* protocols for analyzing the potential function of Hsp90 in the recycling of Rab GTPases through application of the Hsp90-specific inhibitors GA and radicicol.

Methods

Materials

GA and radicicol as well as other chemicals were purchased from Sigma (St. Louise, MO) unless otherwise indicated. GA was brought to a final stock concentration of 20 mM in dimethyl sulfoxide (DMSO). Radicicol was brought to a final stock concentration of 50 mM in DMSO. Stock solutions were stored at $-80°$ in small aliquots and used once after thawing. Dulbecco's minimal essential medium (DMEM) was purchased from Invitrogen/Gibco (Carlsbad, CA). Bovine growth serum was purchased from Hyclone (Logan, UT). A polyclonal antibody against Rab1 (p8868) (Plutner *et al.*, 1991) was purified by protein A chromatography. A polyclonal antibody against Rab3A was generated by immunization of rabbits with purified His$_6$-Rab3A protein. A monoclonal antibody (Cl 81.2) against αGDI used in this study was a gift from Dr. R. Jahn (The Max-Plank Institute for Biophysical Chemistry, Göttingen, Germany). Secondary antibodies conjugated with horseradish peroxidase (HRP) were purchased from Pierce (Rockford, IL). Nitrocellulose membrane for Western blotting was purchased from Schleicher and Schuell (Keene, NH). Talon resin and glutathione-Sepharose 4B resins were purchased from BD Biosciences (San Jose, CA). The Mono Q HR 5/5 column was purchased from Amersham Biosciences (Piscataway, NJ).

Recombinant Protein Purification. A construct that carried bovine αGDI in pGEX-2T vector was a gift from Y. Takai (Osaka University). pGEX-2T αGDI was transformed into *Escherichia coli* BL21 (DE3) (Invitrogen, Carlsbad) to express GST-tagged αGDI. αGDI was subcloned into a pET11d vector with 6× His linker at the N-terminal. Purification of GST-tagged αGDI or His$_6$-tagged αGDI was achieved by following the manufacturer's instructions with either glutathione-Sepharose 4B resin or Talon affinity resin. After affinity purification, recombinant proteins were further purified by a Mono Q HR 5/5 column with a linear gradient between 0 and 500 mM NaCl. Fractions containing recombinant proteins were collected and dialyzed against 25 mM HEPES-KOH (pH 7.4), 100 mM KCl, and 1 mM MgCl$_2$. Small aliquots were flash frozen in liquid nitrogen, and stored in $-80°$. GDI with either the His$_6$-tagged or the GST-tagged can be used for extracting membrane-bound Rab proteins from membranes (Cavalli et al., 2001; Peter et al., 1994; Sakisaka et al., 2002; Ullrich et al., 1993). Below we describe the use of His$_6$-tagged GDI, although similar results are observed with GST-tagged GDI.

Rab3 Extraction from Synaptic Membrane

SYNAPTOSOME PREPARATION

> Homogenizing buffer: 10 mM HEPES-KOH (pH 7.4), 1 mM EDTA, 320 mM sucrose, 250 μM dithiothreitol (DTT).
> Incubation buffer: 10 mM HEPES-KOH (pH 7.4), 5 mM NaHCO$_3$, 1.2 mM NaH$_2$PO$_4$, 5 mM KCl, 140 mM NaCl, 1 mM MgCl$_2$, 10 mM glucose.

All procedures were carried out at 0–4° unless otherwise indicated. Two rat cerebella cortex (~2.6 g) were homogenized with 20 ml of homogenize buffer with a glass homogenizer and a Teflon pestle for 10 passages. Homogenate was pelleted at 1000×g for 10 min. After centrifugation, cell debris (P1) was discarded and approximately 20 ml of supernatant (S1) was collected to layer over for a discontinuous Percoll gradient. The gradient was made with 3%, 10%, and 23% of Percoll in homogenate buffer. An equal volume of each gradient (9.6 ml of each) was overlayed with 9.6 ml of homogenate. The gradient was centrifuged at 32,500×g for 5 min. The interface between 10% and 23% Percoll was collected and diluted to 40 ml with incubation buffer followed by centrifugation at 15,000×g for 10 min. The pellet was washed with 10 ml of incubation buffer and collected by centrifugation at 15,000×g for 10 min. The pellet is now enriched with intact synaptosomes for preparing the synaptic membrane fraction.

GDI EXTRACTION ASSAY

> Reaction buffer: 50 mM HEPES-KOH (pH 7.4), 100 mM KCl, 5 mM MgCl$_2$, 10 mM EDTA, Sigma complete protease inhibitor cocktail.

The intact synaptosome pellet was resuspended with 1 ml incubation buffer and subjected to osmotic shock by diluting to 9 ml double-distilled water and stirring for 45 min at 4°. The synaptic membrane then was harvested by centrifugation at 200,000×g for 20 min and resuspended into 4 ml reaction buffer. The synaptic membrane concentration was

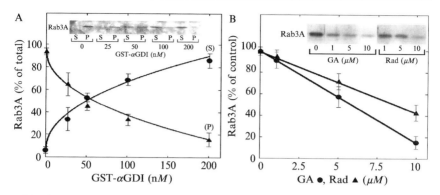

FIG. 1. Rab3A extraction is inhibited by GA and radicicol. (A) Rab3A extraction by GST-αGDI. GST-αGDI at the indicated concentrations was incubated with the synaptic membrane. Rab3A in membrane (P, closed triangles) and soluble fractions (S, closed circles) were recovered as described in the text. The recovery of Rab3A in each fraction was determined by quantitative immunoblotting. (B) Effects of GA and radicicol on Rab3A extraction by GST-αGDI. The synaptic membranes were pretreated with the indicated concentrations of drugs and subsequently incubated with 50 nM GST-αGDI. Recovery is quantitated as in (A).

determined by the BCA assay (Pierce, Rockford, IL) using a bovine serum albumin (BSA) standard. Synaptosome membrane (25 μg) and the indicated concentration of GST-αGDI were supplemented with 250 μM GDP in a 200 μl final volume of reaction buffer. After incubation for 20 min at 37°, the samples were chilled on ice and membranes pelleted by centrifugation for 20 min at 200,000×g. The membrane pellet was resuspended in a total of 80 μl sample buffer (50 mM Tris-HCl [pH 6.8], 2% SDS, 10% glycerol, 1 mM DTT). The supernatant was incubated with 50 μl of gluta-thione-Sepharose beads for 1 h at 4°. Recovered sample from beads and solublized membrane pellet were analyzed by Western blotting using anti-GDI monoclonal antibody and anti-Rab3A rabbit polyclonal antiserum (Fig. 1). To test the effect of Hsp90 inhibitors, GA, and radicicol, the synaptosome membranes were preincubated with either GA, radicicol, or DMSO for 10 min at 30° prior to homogenization and incubation with GST-αGDI.

Rab1 Extraction from Semi–intact Cells

SEMIINTACT NRK CELL PREPARATION

50/90 buffer: 50 mM HEPES-KOH (pH 7.2), 90 mM potassium acetate.

10/18 buffer: 10 mM HEPES-KOH (pH 7.2), 18 mM potassium acetate.

Normal rat kidney (NRK) cells are cultured in Dulbecco's modified Eagle's medium (DMEM) supplemented with 5% bovine growth serum, 100 units/ml penicillin, and 100 μg/ml streptomycin at 37° 5% CO_2; 5 × 10^7 cells are then seeded to a 15-cm Petri dish and incubated overnight. Cells should be at 90–100% confluence for the following step. To inactivate the abundant cytosolic Hsp90 pool with Hsp90 inhibitors, cells are washed once with 10 ml phosphate-buffered saline (PBS) followed by two sequential washes with 10 ml of serum-free DMEM. Hsp90 inhibitors or DMSO vehicle control is diluted into 5 ml of serum-free DMEM to give the final indicated concentration. Serum-free DMEM containing Hsp90 inhibitors (5 ml) is added to each dish and incubated at 37° for 1 h with occasional rocking.

To perforate pretreated cells, cells are washed three times with 10 ml of 50/90 buffer on ice and subsequently incubated with 10 ml 10/18 buffer on ice for 10 min to swell cells. Following swelling, the 10/18 buffer is replaced with 2 ml of 50/90 buffer and cells are released immediately from the dish by brusque and rapid scraping with a rubber policeman. Cells are transferred to a 15-ml Falcon centrifuge tube, and residual cells are collected by adding 2 ml of 50/90 buffer to the plate and pooled together in the same centrifuge tube. Cells are harvested by centrifugation at 1500 rpm for 5 min in a Beckman Allegra 6 R centrifuge. After centrifugation, the supernatant is discarded and the cell pellet ($\sim$80 μl/15-cm plate) is resuspended gently with 5 ml of 50/90 buffer and placed on ice for 5 min to allow release of cytosolic components. The cell suspension at this step should be examined to check the number of intact cells by light microscopy. In general, over 90% of cells should be trypan blue permeable. Cells are pelleted by centrifugation at 1500 rpm for 5 min. The pellet is resuspended to a final volume of 300 μl of 50/90 buffer per 15-cm plate of starting material. The final protein concentration is generally 2–3 mg/ml. Semiintact cells can be stored on ice for 1–2 h prior to use.

GDI EXTRACTION ASSAY The following materials and buffer are prepared prior to the assay:

   10× reaction buffer: 220 mM HEPES-KOH (pH 7.4), 200 mM Tris–HCl (pH 7.0), 1.14 M KCl, 43 mM Mg-acetate, 20 mM DTT, 1% Sigma protease inhibitor cocktail.

   Semiintact NRK cells: 2–3 mg/ml of semi–intact NRK cells in 50 mM HEPES-KOH (pH 7.4), 90 mM potassium acetate, prepared as described above.

   50 mM GDP: prepared with 20 mM HEPES-KOH (pH 7.4). Aliquots are stored at −80° and thawed on ice before use.

   Recombinant His$_6$-αGDI protein: 0.3–0.5 mg/ml (5–10 μM) in 25 mM HEPES-KOH (pH 7.4), 100 mM KCl, 1 mM MgCl$_2$. Prepared as described above. Thawed on ice before use.

Fifty microliters of semiintact cells (150–200 μg) is aliquoted to 1.5-ml Eppendorf microfuge tubes on ice. Hsp90 stock inhibitors or DMSO is diluted to 1:10 with 20 mM HEPES-KOH (pH 7.4). Then 0.5 μl of the diluted Hsp90 inhibitor and 0.5 μl of 50 mM GDP are added to each tube containing semi–intact NRK cells on ice. For multiple reactions, the above components (*semi–intact cells, GDP, and Hsp90 inhibitors*) can be pre-mixed prior to aliquoting into individual tubes. The dilution with mixing should be made immediately before use. To prepare the extraction cock-tail, 10 μl of 10× reaction buffer is mixed with double-distilled H$_2$O to achieve a final volume of 50 μl following the addition of recombinant αGDI protein (up to 1 μM). Semintact NRK cells with GDP and Hsp90 inhibitors are transferred from ice to 37° for 5 min before mixing with the extraction cocktail for an additional 20 min. After incubation, tubes are transferred to ice, and soluble and membrane associated proteins are separated by centrifugation at 16,000×g for 10 min. The supernatant (con-taining the GDI-Rab1 complex) is concentrated by the CHCl$_3$/MeOH protocol as described (Wessel and Flugge, 1984). In brief, one part of protein sample is mixed with four parts of methanol and one part of chloroform by vortexing. Three parts of sterile distilled water are added to help phase separation. After centrifuging for 1 min, protein precipitate is trapped at the interface. Liquid from aqueous layers is carefully removed and the interface is rinsed with methanol and pelleted down with centrifu-gation for 1 min. After removing the organic solvent, precipitant is dried in air and recovered into 40 μl of 50 mM Tris–HCl (pH 7.5), 1% sodium dodecyl sulfate (SDS), and 1 mM EDTA. The pellet (membrane-contain-ing fraction) is rinsed once with reaction buffer (22 mM HEPES-KOH [pH 7.4], 20 mM Tris–HCl [pH 7.0], 114 mM KCl, 4.3 mM MgOAc, 1 mM DTT, and 0.1 % Sigma protease inhibitor cocktail) and recovered by dissolving in 40 μl of 50 mM Tris–HCl (pH 7.5), 1% SDS, and 1 mM EDTA. Each sample is supplemented with 10 μl of a 5× sample loading buffer (250 mM Tris–HCl [pH 6.8], 10% SDS, 50% glycerol) and boiled for 5 min. For membrane-containing fractions, samples need to be vigorously vortexed to disrupt genomic DNA. An equal amount of recovered supernatant and pellet is analyzed by SDS-PAGE followed by quantitative Western blotting using antibody p8868 to detect Rab1.

Comments

There are several ways to prepare the perforated cells for analyzing the effect of GDI on Rab extraction *in vitro*. We demonstrated the use of a perforated membrane preparation based on osmotic shock that first ren-ders cells swollen in low salt buffer followed by shear of the cell membrane by applying a gentle physical force such as scrapping. In this case, the yield of permeable cells depends on the sensitivity of a particular cell type to osmotic shock. The second procedure that can be used utilizes digitonin. Digitonin is a mild detergent, which has been successfully used to permeabilize the plasma membrane without disturbing the intracellular membranes for the functional assays (Plutner *et al.*, 1992). Generally, most cell types can be permeabilized by incubation with 40–50 μg/ml digitonin on ice for 5–10 min. The third way to prepare perforated cells is to use the bacterial toxin streptolysin O (SLO) that forms pores in the plasma membrane by chelating cholesterol (Bhakdi *et al.*, 1985). SLO generates smaller pores (<30 nm) on the membranes than digitonin and perforation and requires a longer incubation time to equilibrate with reagents.

Several critical factors affect the successful use of GA and radicicol to measure the response of Rab recycling pathways to Hsp90. First, the final DMSO content used to deliver the drugs to cells should not exceed 0.2%, as concentrations above this value elicit nonspecific extraction. Second, Hsp90 inhibitors need to be added to cells in the presence of serum-free medium to ensure efficient uptake. Third, use of fresh membrane preparations is essential for efficient GDI extraction *in vitro*. Preincubation of cells with Hsp90 inhibitors prior to perforation significantly increases the inhibitory effect of the drugs, presumably due to inactivation of the large endogenous Hsp90 pool. Depending on the cell type and the specific Rab-dependent pathway being examined, it may be necessary to titrate the amount GDI required for extraction from control cells and to determine the concentration and time of exposure of cells to GA to inactivate the endogenous Hsp90 pool.

Acknowledgments

This work is supported by NIH grant numbers GM3301, 42336, and HL06721. TSRI manuscript number 17369-CB.

References

Alory, C., and Balch, W. E. (2003). Molecular evolution of Rab-escort-protein/guanine-nucleotide-dissociation-factor superfamily. *Mol. Biol. Cell* **14,** 3857–3867.

An, Y., Shao, Y., Alory, C., Matteson, J., Sakisaka, T., Chen, W., Gibbs, R. A., Wilson, I. A., and Balch, W. E. (2003). Geranylgeranyl switching regulates GDI-Rab GTPase recycling. *Structure (Camb.)* **11,** 347–357.

Bhakdi, S., Tranum-Jensen, J., and Sziegoleit, A. (1985). Mechanism of membrane damage by streptolysin-O. *Infect. Immun.* **47,** 52–60.

Cavalli, V., Vilbois, F., Corti, M., Marcote, M. J., Tamura, K., Karin, M., Arkinstall, S., and Gruenberg, J. (2001). The stress-induced MAP kinase p38 regulates endocytic trafficking via the GDI:Rab5 complex. *Mol. Cell* **7,** 421–432.

Garrett, M. D., Zahner, J. E., Cheney, C. M., and Novick, P. J. (1994). GDI1 encodes a GDP dissociation inhibitor that plays an essential role in the yeast secretory pathway. *EMBO J.* **13,** 1718–1728.

Gurkan, C., Lapp, H., Alory, C., Su, A. I., Hogenesch, J., and Balch, W. E. (2005). Large scale profiling of Rab GTPase trafficking neworks: The membrome. *Mol. Biol. Cell* **16,** 3847–3864.

Nishimura, N., Nakamura, H., Takai, Y., and Sano, K. (1994). Molecular cloning and characterization of two rab GDI species from rat brain: Brain-specific and ubiquitous types. *J. Biol. Chem.* **269,** 14191–14198.

Peter, F., Nuoffer, C., Pind, S. N., and Balch, W. E. (1994). Guanine nucleotide dissociation inhibitor is essential for Rab1 function in budding from the endoplasmic reticulum and transport through the Golgi stack. *J. Cell Biol.* **126,** 1393–1406.

Pfeffer, S., and Aivazian, D. (2004). Targeting Rab GTPases to distinct membrane compartments. *Nat. Rev. Mol. Cell. Biol.* **5,** 886–896.

Plutner, H., Cox, A. D., Pind, S., Khosravi-Far, R., Bourne, J. R., Schwaninger, R., Der, C. J., and Balch, W. E. (1991). Rab1b regulates vesicular transport between the endoplasmic reticulum and successive Golgi compartments. *J. Cell Biol.* **115,** 31–43.

Plutner, H., Davidson, H. W., Saraste, J., and Balch, W. E. (1992). Morphological analysis of protein transport from the ER to Golgi membranes in digitonin-permeabilized cells: Role of the P58 containing compartment. *J. Cell Biol.* **119,** 1097–1116.

Pratt, W. B., and Toft, D. O. (2003). Regulation of signaling protein function and trafficking by the hsp90/hsp70-based chaperone machinery. *Exp. Biol. Med. (Maywood)* **228,** 111–133.

Rak, A., Pylypenko, O., Durek, T., Watzke, A., Kushnir, S., Brunsveld, L., Waldmann, H., Goody, R. S., and Alexandrov, K. (2003). Structure of Rab GDP-dissociation inhibitor in complex with prenylated YPT1 GTPase. *Science* **302,** 646–650.

Ricard, C. S., Jakubowski, J. M., Verbsky, J. W., Barbieri, M. A., Lewis, W. M., Fernandez, G. E., Vogel, M., Tsou, C., Prasad, V., Stahl, P. D., Waksman, G., and Cheney, C. M. (2001). *Drosophila* rab GDI mutants disrupt development but have normal Rab membrane extraction. *Genesis* **31,** 17–29.

Sakisaka, T., Meerlo, T., Matteson, J., Plutner, H., and Balch, W. E. (2002). Rab-alphaGDI activity is regulated by a Hsp90 chaperone complex. *EMBO J.* **21,** 6125–6135.

Shisheva, A., Sudhof, T. C., and Czech, M. P. (1994). Cloning, characterization, and expression of a novel GDP dissociation inhibitor isoform from skeletal muscle. *Mol. Cell. Biol.* **14,** 3459–3468.

Sivars, U., Aivazian, D., and Pfeffer, S. R. (2003). Yip3 catalyses the dissociation of endosomal Rab-GDI complexes. *Nature* **425,** 856–859.

Ullrich, O., Stenmark, H., Alexandrov, K., Huber, L. A., Kaibuchi, K., Sasaki, T., Takai, Y., and Zerial, M. (1993). Rab GDP dissociation inhibitor as a general regulator for the membrane association of rab proteins. *J. Biol. Chem.* **268,** 18143–18150.

Wessel, D., and Flugge, U. I. (1984). A method for the quantitative recovery of protein in dilute solution in the presence of detergents and lipids. *Anal. Biochem.* **138,** 141–143.

Further Reading

Beraud-Dufour, S., and Balch, W. (2002). A journey through the exocytic pathway. *J. Cell Sci.* **115,** 1779–1780.

[30] Purification and Properties of Yip3/PRA1 as a Rab GDI Displacement Factor

By Ulf Sivars, Dikran Aivazian, and Suzanne Pfeffer

Abstract

Prenylated Rab proteins exist in the cytosol bound to guanine dissociation inhibitor (GDI). These dimeric complexes contain all of the information needed for accurate membrane delivery. We have shown that membranes contain a proteinaceous activity that is required for Rab delivery, and we named that activity GDI displacement factor (GDF). Biochemical analysis revealed that GDF activity was membrane associated and had a mass of approximately 25 kDa. We therefore used a candidate gene approach and were able to show that pure Yip3/PRA1 protein displays GDF activity. In this chapter, we review key aspects of GDF analysis: our assay and the method by which we purify Yip3/PRA1 in active form.

Discovery of GDF

About 10 years ago, we showed that purified complexes of prenylated Rab9 bound to guanine dissociation inhibitor (GDI) contain all of the information needed to target Rab9 onto late endosomes *in vitro* (Soldati *et al.*, 1994). Little delivery occurred on ice; only background membrane delivery was detected with red blood cell ghosts. A few minutes after Rab9 membrane delivery, Rab9 exchanged bound GDP for GTP. The ability of endosomes to serve as acceptors was lost upon proteinase K treatment. Importantly, the initial rate and extent of Rab recruitment were saturable. From these data, we proposed that membranes contain a proteinaceous factor that acts catalytically to displace Rab proteins from GDI and permit their association with specific membranes (Pfeffer *et al.*, 1995; Soldati *et al.*, 1994, 1995a). After many years and a great deal of effort, we succeeded in the molecular identification of the first GDI displacement factor: the integral membrane protein Yip3/PRA1 (Sivars *et al.*, 2003).

Figure 1 outlines the assay we have used to detect a membrane-associated activity that would recognize a Rab–GDI complex and catalyze the release of GDI (Dirac-Svejstrup *et al.*, 1997). We refer to this as "GDF" or GDI displacement factor activity. Rab proteins occur in the cytosol as complexes with GDI, which blocks their ability to release and exchange bound GDP. Once free from GDI, the Rab is again able to

METHODS IN ENZYMOLOGY, VOL. 403   0076-6879/05 $35.00
DOI: 10.1016/S0076-6879(05)03030-2

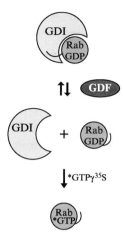

F<small>IG</small>. 1. Schematic description of the GDI displacement factor assay. See text for details.

exchange bound nucleotide at its intrinsic rate. Since GDI binds tightly to
Rab9 (Shapiro and Pfeffer, 1995), the equilibrium normally lies far in the
direction of the stable Rab9–GDI complex. However, a GDF should
facilitate the generation of free Rab9, which then exchanges bound
GDP for $[^{35}S]GTP\gamma S$. This method can detect a GDF, or alternatively,
an activity that acts on Rab–GDI complexes and stimulates that Rab's
intrinsic rate of nucleotide exchange (a GEF).

To detect GDF activity, we mix purified prenyl Rab9–GDI complexes
with protein fractions (or candidate GDF proteins) and look for an
increase in the rate of binding of $[^{35}S]GTP\gamma S$ that can occur only after
Rab dissociation from GDI. Since mammalian GDI binds only prenyl
Rabs, GDI release leads to prenyl-Rab aggregation. Therefore, assays are
carried out in the presence of detergent; a released Rab becomes
incorporated into a detergent micelle.

It was very important to establish detergent conditions that do not
destabilize prenyl Rab–GDI association (0.1% CHAPS in the *final* reaction mix). In addition, the presence of detergent is essential because GDF
activity is tightly associated with membranes and requires detergent for its
analysis. Since GDI shows a strong preference for Rabs bearing GDP,
reactions are carried out in the presence of an excess of $[^{35}S]GTP\gamma S$ to
trap the reaction product. Once released from GDI, a Rab will exchange
bound GDP for GTPγS and then be incapable of rebinding GDI.

We used this scheme to first discover a GDF activity (Dirac-Svejstrup
et al., 1997) and showed that it is not a guanine nucleotide exchange factor

because the activity did not influence the intrinsic rates of nucleotide exchange of Rabs 5, 7, or 9. Rather, GDF caused the physical release of each of these endosomal Rabs from GDI, permitting them to exchange nucleotide at their intrinsic rates. This was shown by isolating the products of a GDF reaction on a gel filtration column; at the end of the reaction, the Rab with bound GTP was well resolved from GDI (Dirac-Svejstrup *et al.*, 1997). The failure of partially purified GDF activity to act on Rab1–GDI complexes suggested that it may be specific for endosomal Rab proteins. In addition, these data suggested that there may be a distinct GDF for Rab1 complexes that function at the endoplasmic reticulum (ER) Golgi interface. We also showed that Rab1–GDI complexes can be recruited onto intact membranes under the same conditions (Dirac-Svejstrup *et al.*, 1997). We worked hard to purify this integral, endosomal membrane protein to homogeneity but were unable to identify the protein(s) involved. As described below, we had more success using a candidate gene approach that led us to Yip3 protein.

Yip3 Is a GDF for Endosomal Rabs

Given the difficulty of purifying an integral membrane protein GDF and the conservation of Rabs and GDI in yeast, we decided to take advantage of yeast genetics and look for any proteins that showed genetic interaction with yeast GDI and/or yeast Rabs. A family of proteins named Yips (Ypt-interacting proteins) and Yif (Yip1-interacting factor) emerged as strong candidates for potential GDFs. Yip1 was first discovered as an essential *Saccharomyces cerevisiae* protein that interacts with multiple Rabs in *S. cerevisiae* (Yang *et al.*, 1998). The groups of Collins and Gallwitz identified a family of Yip-related proteins in *S. cerevisiae* and human cells (Calero *et al.*, 2002; Matern *et al.*, 2000). Yip protein family members share a common topology: their amino- and carboxyl-termini face the cytosol, while the remaining bulk of their sequences are membrane embedded (Calero *et al.*, 2002). The topology of the human homologue of *S. cerevisiae* Yip3 (also known as prenylated Rab acceptor-1 [PRA1]) has been shown to include four transmembrane domains (Lin *et al.*, 2001).

In mammalian cells, YIP3/PRA1 interacts with numerous, prenylated Rabs (Bucci *et al.*, 1999; Martincic *et al.*, 1997), shows weak interaction with GDI (Hutt *et al.*, 2000), and is localized to the Golgi and endosomes (Abdul-Ghani *et al.*, 2001; Sivars *et al.*, 2003). PRA2 is ~26% identical to YIP3/PRA1 (Abdul-Ghani *et al.*, 2001). PRA2 is present in the ER, and the carboxyl-termini of the PRA proteins are responsible for their distinct localizations (Abdul-Ghani *et al.*, 2001). Because Rab9 is delivered to late endosomes, we tested whether human Yip3/PRA1 can serve as a GDF for prenyl Rab9–GDI complexes. The protein was expressed in bacteria and purified from bacterial membranes.

Yip3 Purification after Expression in Bacteria

Note: It is essential to use ultrapure CHAPS detergent and a significant excess of detergent during the purification. Otherwise, the protein is purified as an oligomer and is less active. Also, samples should not be boiled prior to sodium dodecyl sulfate–polyacrylamide gel electrophoresis (SDS-PAGE) analysis, as the hydrophobic Yip3 protein will form a large SDS-resistant aggregate. In some preparations, SDS-resistant dimers are detected (Fig. 2) and these are less active. It is important to take great care in Yip3 preparation. Also note that the protein cannot be frozen or it will lose all activity. Finally, the initial ion-exchange step is very important because it helps to remove excess lipid, detergent, and other components that interfere with the GDF assay and yield spurious results.

Materials

[^{35}S]GTPγS, 1393 Ci/mmol, 12/5 mCi/ml (New England Nuclear, Boston, MA).

Buffers and Solutions

> Lysis buffer: 25 mM HEPES, pH 7.5, 300 mM KCl, 1 mM EDTA, 1 mM dithiothreitol (DTT), 1 mM phenylmethylsulfonyl fluoride (PMSF), 1 μM pepstatin, 2.1 mM leupeptin, 0.14 TIU/ml aprotinin.
>
> Extraction buffer: 25 mM Tris–Cl, pH 8.0, 5% Triton X-100, 1 mM EDTA, 1 mM DTT, 1 mM PMSF.
>
> Q Buffer A: 25 mM Tris–Cl, pH 8.0, 10% (w/v) glycerol, 1% (w/v) Triton X-100 (150 ml needed).
>
> Q Buffer B: FFQ Buffer A + 500 mM NaCl (50 ml needed).
>
> Ni-NTA Buffer A: 25 mM HEPES/NaOH, pH 7.5, 300 mM KCl, 25 mM imidazole, and 1% CHAPS (100 ml needed).
>
> Ni-NTA Buffer B: Ni-NTA Buffer A + 400 mM imidazole (50 ml needed).
>
> Dialysis buffer: 25 mM HEPES, pH 7.5, 100 mM NaCl, 1% CHAPS (Ultrapure).
>
> GDF Buffer A (1×): 25 mM HEPES, pH 7.2, 100 mM KCl, 1.2 mM Mg(OAc)$_2$, 100 mM (NH$_4$)$_2$SO$_4$, 1 mM DTT, 1 mM PMSF, 1 μM pepstatin, 2.1 mM leupeptin, 0.14 TIU/ml aprotinin.
>
> GDF Buffer B (1×): 64 mM HEPES, pH 8.0, 2 mM EDTA, 8 mM MgCl$_2$, 100 mM NaCl.

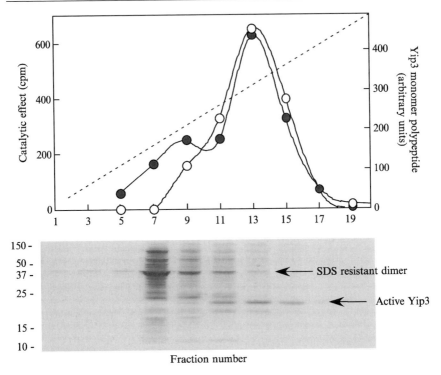

FIG. 2. Imidazole gradient elution (50–400 mM) of bacterially expressed Yip3 protein from a Nickel-NTA column. Open circles, the amount of Yip3 monomeric polypeptide detected in the fraction indicated, in arbitrary units representing quantitation of the SDS-PAGE analysis shown below. Closed circles, the amount of catalytic activity detected in each fraction. In preparations where detergent was slightly limiting, we obtained SDS-resistant, Yip3 dimers that showed significantly less activity than the monomeric form. For SDS-PAGE analysis it is important not to boil Yip3-containing samples; instead they were warmed to 37° prior to electrophoresis, to avoid generation of hydrophobic aggregates of protein.

Purification

Human Yip3 (SwissProt accession number Q9UI14) cDNA was amplified from a human testis cDNA library and cloned into pET14b (Novagen) using the *Nde*I and *Bam*HI sites. The resulting recombinant protein has an N-terminal His tag with the following sequence: *MGSSHHHHHHSSGLVPRGSH*MAAQKDQ ... LQMEPV. The protein was expressed in *Escherichia coli* BL21 (DE3) RIL (Stratagene) by induction at $OD_{600} = 0.4$ with 1 mM IPTG for 2 h at 30°. The cell pellet from a 500-ml culture was resuspended in 30 ml lysis buffer and passed through a French press at 1400 psi. The lysate was spun at 17,000 rpm in a

JA-20 rotor for 20 min at 4° (Beckman), and the supernatant removed. Membrane pellets were solubilized with Triton X-100 in 25 ml extraction buffer in a Potter-Elvehjem homogenizer (10 passes), followed by incubation at 4° on a rotator for 10 min. After a 45-min, $100,000 \times g$ spin in a cold 60Ti rotor (Beckman), the supernatant was applied onto a Q-Sepharose (Pharmacia) column (2 ml pelleted resin in a 45-ml conical tube) preequilibrated in Q Buffer A (by 2×10 ml washes of Q buffer A followed by a 2 min spin at 1000 rpm). It is important to check the conductivity of the Q buffers and the Yip3/PRA1 supernatant. Yip3 was allowed to bind the resin for 15 min with rotation in the cold. Beads are washed with 20 column volumns Q Buffer A and the wash is saved.

Yip3 was eluted using a 0–500 mM NaCl gradient (40 ml total, 1 ml fractions). Fractions were analyzed by SDS-PAGE and Western blot. Yip3-enriched fractions were pooled (9–40) and applied to an Ni-NTA agarose (Qiagen) column (1 ml) preequilibrated in Ni-NTA Buffer A (in a 45-ml conical tube with 2×10 ml washes in Ni-NTA Buffer A followed by a 2-min spin at 1000 rpm). The Yip3 was allowed to bind to the resin for 15 min in the cold with rotation. The beads were spun down and resuspended in 10 ml Ni-NTA Buffer A and poured into a column. The flow through is passed back through the column. Bound Yip3 was washed with 20 ml Ni-NTA Buffer A to remove Triton X-100, then eluted with a 50–400 mM imidazole gradient (20×1 ml) of 0–100% Ni-NTA Buffer B. Fractions were analyzed by SDS-PAGE and Western blot, and Yip3-containing fractions were pooled, in some cases, concentrated using a 10-kDa cutoff Centriprep concentrator (Amicon), and dialyzed overnight against dialysis buffer at 4°. The protein was stored on ice for up to a week.

Important cautionary note: Recombinant Yip3 loses its activity upon either freeze-thawing or heating above room temperature; its instability may reflect a cellular requirement for an additional, interaction partner. **Do not freeze**.

Assay for GDI Displacement Factor Activity

Purified nontagged prenyl Rab9–GDI complexes were used to test whether purified hYip3 could dissociate Rab–GDI complexes as described above. HYip3 stimulated the rate of nucleotide binding to Rab9 when added to prenyl Rab9–GDI complexes (Sivars *et al.*, 2003). Apparent GDF activity was proportional to the amount of Yip3 protein added and was abolished by boiling. The reaction displayed linear kinetics for 15 min at 37°. Yip3 acted catalytically because as many as $\sim$450 Rab9 molecules were released per Yip3 molecule added. After a 20-min reaction, $\sim$25% of the Rab–GDI complexes were released. Addition of more Yip3 after various times led to complete release of all Rab–GDI complexes, confirming that purified Yip3 loses activity upon incubation at 37°; nevertheless, Yip3 was only present in catalytic amounts. These data demonstrate that the GDF reaction does not reflect simple, stoichiometric binding. In control experiments, Yip3 did not enhance the intrinsic rates of nucleotide exchange of prenyl Rabs 1A, 2, 5, or 9, so it is not a GEF under these conditions (Sivars *et al.*, 2003). Gel filtration chromatography revealed that the activity observed reflects the physical displacement of Rab9 from GDI.

Yip3 was active on early endosome-associated Rab5 and late endosome-associated Rab9 and Rab7 but not on Rabs 1A and 2, Rabs of the ER and early Golgi (Sivars *et al.*, 2003). Although some weak interaction with Rab1A and Rab2 was detected that could reflect direct binding, Yip3 did not act catalytically to displace these Rabs from GDI. This is important because numerous two-hybrid screens have observed broad binding of Yip family members to a variety of Rabs (cf. Bucci *et al.*, 1999; Calero and Collins, 2002; Figueroa *et al.*, 2001). While Yip3 may bind weakly to multiple Rabs, it appears to act catalytically to displace only a subset of Rabs from GDI. A preference for endocytic Rabs was also seen previously for GDF partially purified from endosomes (Dirac-Svejstrup *et al.*, 1997).

These experiments describe the molecular identification of a protein that can act as a Rab–GDI displacement factor. We favor a model in which an ER-localized Yip protein (Yip1p or PRA-2/Yip6) drives membrane recruitment of secretory pathway Rabs, whereas the more distally localized Yip3 can recruit endosomal Rabs onto membranes (Pfeffer and Aivazian, 2004). Both of these classes of proteins may enter the secretory pathway during their biosynthesis; Yip1p and PRA-2/Yip6 may be retained in the ER while Yip3 is transported to the plasma membrane and, thereby, the endocytic pathway. Rabs could use these Yips to associate with the appropriate pathway and then become stabilized in a given compartment by interaction with specific effector proteins.

Method A: GDF Assay

The components of the GDF reaction are the Rab–GDI substrate, a test GDF sample or purified Yip3, and radiolabeled GTPγS. Rab–GDI complexes are produced using recombinant prenylated Rab proteins, purified from the membranes of expressing SF9 insect cells, and purified bovine brain GDIα. The procedure is described in detail in Soldati *et al.* (1995b). Each 100-μl reaction contains 20 pmol of Rab–GDI with as little as 1 fmol of GDF, and 2 μM (200 pmol) [^{35}S]GTPγS (0.2 μCi per reaction). A reaction mix is made. For example, for 24 reactions:

Reaction Mix:
2× GDF Buffer A, 1.25 ml
[^{35}S]GTPγS (1.0 μCi/μl), 5 μl
1 mM GTPγS, 5 μl
ddH$_2$O, 0.69 ml
Each reaction:
Reaction mix, 78.5 μl
GDF sample or buffer, in 1% CHAPS, 10 μl
0.1 mg/ml (∼200 nM) Rab–GDI stock, 11.5 μl

1. Yip3/GDF samples are diluted for analysis in a buffer containing 1% CHAPS. Rab–GDI stock is diluted in GDF Buffer B, and is added last to initiate the reactions, which are incubated at 37° for the appropriate time (usually 0–20 min).
2. Reactions are stopped by adding 500 μl of ice-cold GDF Buffer B and placing on ice.
3. The samples are applied to nitrocellulose filters (Millipore Type HA, 0.45 μm) in a vacuum manifold, where the filters have been prewet with GDF Buffer B. The samples are pulled through slowly, and the filters then washed three times with 3 ml of GDF Buffer B. After drying under a heat lamp, the filters are placed in vials containing scintillation fluid (4 ml) and counted.

It is important to determine the maximal level of Rab nucleotide exchange (and possible GDF activity) by adding 10 μl of 10% CHAPS in lieu of GDF. Rab–GDI complexes dissociate in 1% CHAPS, thus liberating the Rab to permit GTPγS binding. During the initial purification steps, it is essential to determine the background level of nucleotide binding seen without Rab–GDI complex addition because crude preparations may contain other GTP-binding constituents. In purer preparations, background is determined by inclusion of bovine serum albumin (BSA)-containing 10 μl GDF buffer containing 1% CHAPS (0.1% final) instead of GDF. The resulting signal is subtracted from the signals of samples containing GDF and 0.1% CHAPS, to obtain the amount of exchange that was due to GDF addition (catalytic effect). Finally, different BSA preparations from different suppliers yield different background signals and should be screened for this, prior to their use.

Acknowledgment

This research was funded by a grant to S.P. from the National Institutes of Health (DK37332).

References

Abdul-Ghani, M., Gougeon, P. Y., Prosser, D. C., Da-Silva, L. F., and Ngsee, J. K. (2001). PRA isoforms are targeted to distinct membrane compartments. *J. Biol. Chem.* **276,** 6225–6233.

Bucci, C., Chiariello, M., Lattero, D., Maiorano, M., and Bruni, C. B. (1999). Interaction cloning and characterization of the cDNA encoding the human prenylated Rab acceptor (PRA1). *Biochem. Biophys. Res. Commun.* **258,** 657–662.

Calero, M., and Collins, R. N. (2002). *S. cerevisiae* Pra1p/Yip3 interacts with Yip1p and Rab proteins. *Biochem. Biophys. Res. Commun.* **290,** 676–681.

Calero, M., Winand, N. J., and Collins, R. N. (2002). Identification of the novel proteins Yip4p and Yip5p as Rab GTPase interacting factors. *FEBS Lett.* **515,** 89–98.

Dirac-Svejstrup, A. B., Sumizawa, T., and Pfeffer, S. R. (1997). Identification of a GDI displacement factor that releases endosomal Rab GTPases from Rab-GDI. *EMBO J.* **16,** 465–472.

Figueroa, C., Taylor, J., and Vojtek, A. B. (2001). Prenylated Rab acceptor protein is a receptor for prenylated small GTPases. *J. Biol. Chem.* **276,** 28219–28225.

Hutt, D. M., Da-Silva, L. F., Chang, L. H., Prosser, D. C., and Ngsee, J. K. (2000). PRA1 inhibits the extraction of membrane-bound Rab GTPase by GDI1. *J. Biol. Chem.* **275,** 18511–18519.

Lin, J., Liang, Z., Zhang, Z., and Li, G. (2001). Membrane topography and topogenesis of prenylated Rab Acceptor (PRA1). *J. Biol. Chem.* **276,** 41733–41741.

Martincic, I., Peralta, M. E., and Ngsee, J. K. (1997). Isolation and characterization of a dual prenylated Rab and VAMP2 receptor. *J. Biol. Chem.* **272,** 26991–26998.

Matern, H., Yang, X., Andrulis, E., Sternglanz, R., Trepte, H. H., and Gallwitz, D. (2000). A novel Golgi membrane protein is part of a GTPase-binding protein complex involved in vesicle targeting. *EMBO J.* **19,** 4485–4492.

Pfeffer, S. R., and Aivazian, D. (2004). Targeting Rab GTPases to distinct membrane compartments. *Nat. Rev. Mol. Cell. Biol.* **5,** 1–12.

Pfeffer, S. R., Dirac-Svejstrup, A. B., and Soldati, T. (1995). Rab GDP dissociation inhibitor: Putting Rab GTPases in the right place. *J. Biol. Chem.* **270,** 17057–17059.

Shapiro, A. D., and Pfeffer, S. R. (1995). Quantitative analysis of the interactions between prenyl Rab9, GDP dissociation inhibitor-alpha, and guanine nucleotides. *J. Biol. Chem.* **270,** 11085–11090.

Sivars, U., Aivazian, D. A., and Pfeffer, S. R. (2003). Yip3 catalyses the dissociation of endosomal Rab-GDI complexes. *Nature* **425,** 856–859.

Soldati, T., Shapiro, A. D., Svejstrup, A. B., and Pfeffer, S. R. (1994). Membrane targeting of the small GTPase Rab9 is accompanied by nucleotide exchange. *Nature* **369,** 76–78.

Soldati, T., Rancano, C., Geissler, H., and Pfeffer, S. R. (1995a). Rab7 and Rab9 are recruited onto late endosomes by biochemically distinguishable processes. *J. Biol. Chem.* **270,** 25541–25548.

Soldati, T., Shapiro, A. D., and Pfeffer, S. R. (1995b). Reconstitution of the endosomal targeting of rab9 protein using purified, prenylated rab9 protein as a complex with GDI. *Methods Enzymol.* **257,** 253–259.

Yang, X., Matern, H. T., and Gallwitz, D. (1998). Specific binding to a novel and essential Golgi membrane protein (Yip1p) functionally links the transport GTPases Ypt1p and Ypt31p. *EMBO J.* **17,** 4954–4963.

[31] Purification and Analysis of TIP47 Function in Rab9-Dependent Mannose 6-Phosphate Receptor Trafficking

By ALONDRA SCHWEIZER BURGUETE, ULF SIVARS, and SUZANNE PFEFFER

Abstract

TIP47 (tail interacting protein of 47 kDa) is a cytosolic protein that is essential for the transport of mannose 6-phosphate receptors (MPRs) from endosomes to the trans-Golgi. This protein is recruited from the cytosol onto the surface of late endosomes by Rab9 GTPase, which enables TIP47 to bind to MPR cytoplasmic domains with enhanced affinity. A mutation in a deep hydrophobic cleft of TIP47 ($F^{236}C$) confers enhanced affinity binding to MPR cytoplasmic domains and stabilizes MPRs in living cells. We describe the purification of native and recombinant TIP47 proteins and assays that we use to monitor the function of this protein in MPR transport in living cells.

TIP47 in Mannose 6-Phosphate Receptor Transport

Mannose 6-phosphate receptors (MPRs) carry newly synthesized lysosomal enzymes from the Golgi to endosomes, and then return to the Golgi to pick up additional enzymes for transport (Kornfeld, 1992). Two distinct MPRs have been identified, the ~46-kDa cation-dependent (CD) MPR and the ~300-kDa cation-independent (CI) MPR. We have shown that transport of MPRs from late endosomes to the trans-Golgi requires the Rab9 GTPase (Lombardi *et al.*, 1993; Riederer *et al.*, 1994), a Rab9 effector named p40 (Diaz *et al.*, 1997), NSF, α-SNAP, and a protein named mapmodulin (Itin *et al.*, 1997, 1999).

TIP47 (tail interacting protein of 47 kDa) was identified as a protein that binds with high specificity to the cytosolic domains of the CD-MPR and CI-MPR (Diaz and Pfeffer, 1998; Krise *et al.*, 2000). As described later, TIP47 is required for the transport of these receptors from endosomes to the Golgi complex both *in vitro* and *in vivo* (Barbero *et al.*, 2002; Carroll *et al.*, 2001; Diaz and Pfeffer, 1998; Ganley *et al.*, 2004). TIP47 recognizes a Phe–Trp motif in the CD-MPR (Diaz and Pfeffer, 1998) and a proline-rich region of the CI-MPR (Orsel *et al.*, 2000). In addition, TIP47 binds directly to Rab9, which enhances TIP47's affinity for MPR cytoplasmic domains (Carroll *et al.*, 2001). In this way, Rab9 facilitates MPR cargo collection during the process of transport vesicle formation.

METHODS IN ENZYMOLOGY, VOL. 403
0076-6879/05 $35.00
DOI: 10.1016/S0076-6879(05)03031-4

We investigated the function of TIP47 by several methods. First, cellular depletion of TIP47 using antisense oligonucleotides (Diaz and Pfeffer, 1998) or siRNA (Ganley *et al.*, 2004) leads to the missorting of MPRs to lysosomes, as determined by monitoring the decrease in half-life of MPRs in metabolically labeled cells in a pulse–chase labeling experiment. Loss of TIP47 also leads to a major decrease in the half-life of Rab9 protein (Ganley *et al.*, 2004). Second, anti-TIP47 antibodies block the transport of MPRs from endosomes to the trans-Golgi in an *in vitro* system that reconstitutes this transport step, and cytosol depleted of TIP47 is inactive in this system (Diaz and Pfeffer, 1998). Third, we have generated a number of mutant TIP47 proteins and found that they inhibit the transport of MPRs from endosomes to the trans-Golgi after introduction into cultured cells. Thus, TIP47 [167]SVV-AAA was shown not to bind to the Rab9 GTPase (Hanna *et al.*, 2002) and this protein is a dominant negative inhibitor of MPR transport in living cells (Carroll *et al.*, 2001; Hanna *et al.*, 2002). Moreover, expression of this mutant protein traps MPRs in intracellular compartments and decreases the surface pool of MPRs (Barbero *et al.*, 2002). In addition, we have shown that the N-terminus of TIP47 encodes an oligomerization domain; unlike exogenously expressed, wild-type TIP47, monomeric forms of the protein (residues 152–434) are not able to stimulate the transport of MPRs from endosomes to the Golgi *in vivo* (Sincock *et al.*, 2003).

We took a chemical modification approach to map the surface of the TIP47 protein (Burguete *et al.*, 2004). We used translational misincorporation of cysteine residues to generate an ensemble of TIP47 proteins with cysteine incorporated randomly at different positions in the sequence. We analyzed mutant protein ensembles by affinity chromatography and identified mutants that either increased or decreased the affinity of TIP47 for MPR cytoplasmic domains (Burguete *et al.*, 2004). The three-dimensional structure of TIP47 residues 191–437 was reported by Hickenbottom *et al.* (2004) and is entirely consistent with our chemical modification analyses (Burguete *et al.*, 2004). The high-affinity mutant residue lies in a deep hydrophobic pocket that was proposed to represent a protein:protein interaction interface (Hickenbottom *et al.*, 2004). The low-affinity mutant lies in a connecting loop that contributes to a wall of this binding pocket (Burguete *et al.*, 2004; Hickenbottom *et al.*, 2004).

Cells expressing a TIP47 mutant protein with increased affinity for the CI-MPR cytoplasmic domain showed enhanced membrane association and stabilized CI-MPRs. TIP47 $F^{236}C$ showed a 6-fold increased affinity for the CI-MPR cytoplasmic domain ($K_d \sim 0.2\ \mu M$) with equal maximal binding to the receptor, as compared with wild-type TIP47 (Burguete *et al.*, 2004). Since MPR cytoplasmic domains are known to be important determinants

of TIP47 membrane association (Diaz and Pfeffer, 1998), a TIP47 mutant
protein with enhanced affinity for the MPR cytoplasmic domains might be
expected to show increased membrane association when expressed in living
cells. In wild-type cells, about 30% of TIP47 protein is present on membranes at steady state. A higher proportion of total TIP47 $F^{236}C$ was
present on membranes when compared with the distribution of exogenous
wild-type TIP47, expressed at comparable levels in control cells (Fig. 1A).

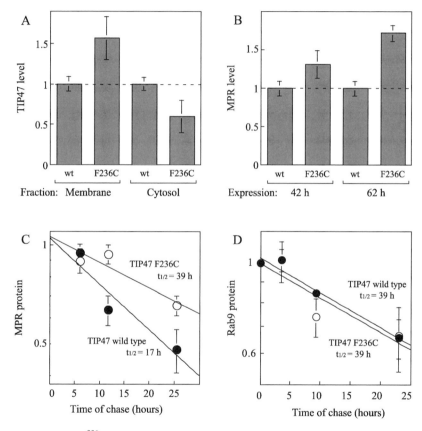

FIG. 1. TIP47 $F^{236}C$ stabilizes MPRs in living cells. (A) Membrane and cytosolic fractions
were determined by immunoblot and normalized to the amounts of EGF receptor and lamin
proteins in membrane and cytosol fractions, respectively. (B) MPRs accumulate in cells
expressing TIP47 $F^{236}C$. CI-MPR levels were determined in cells expressing wild-type and
mutant TIP47 for various times. Data were normalized to p115 protein levels. (C) CI-MPR
protein but not Rab9 GTPase (D) is stabilized in cells expressing TIP47 $F^{236}C$. Protein levels
are shown for cells expressing wild-type TIP47 (●) or TIP47 with a single amino acid
substitution at position 236 (○).

These data provide independent, *in vivo* confirmation that the mutant TIP47 F^{236}C protein binds more tightly to MPRs.

TIP47 is likely to stabilize MPRs by direct binding, because a reduction in cellular TIP47 causes a significant reduction of the CI-MPR half-life (Diaz and Pfeffer, 1998; Ganley *et al.*, 2004) and MPR mutations that block TIP47 binding also display increased MPR turnover (Schweizer *et al.*, 1997). In support of this model, MPRs accumulated in cells expressing the TIP47 F^{236}C mutant: a 30% increase in CI-MPR levels was detected 42 h after transfection and this number increased to over 50% after 62 h, as compared with cells expressing wild-type TIP47 (Fig. 1B). The accumulation observed was due to an increase in CI-MPR half-life, which more than doubled in cells expressing the TIP47 F^{236}C mutant (Fig. 1C).

Rab9 has been shown to increase the affinity with which TIP47 binds MPR cytoplasmic domains (Carroll *et al.*, 2001), likely by inducing a conformational change in TIP47. Hence, a TIP47 mutant with 6-fold increased affinity to the MPRs might be less dependent upon Rab9 for stable association with MPRs in cellular endosomes. If so, the TIP47 F^{236}C mutant might not alter the stability of the Rab9 protein. Indeed, the half-life of Rab9 was not altered by expression of the TIP47 F^{236}C mutant (Fig. 1D).

As mentioned earlier, exogenously expressed wild-type TIP47 protein has been shown to stimulate the transport of MPRs from endosomes to the Golgi complex in living cells. To determine if enhanced binding to MPR cytoplasmic domains interfered with MPR transport, we tested the consequences of mutant protein expression on this process. As shown in Fig. 2, the mutant TIP47 protein was fully capable of stimulating the transport of MPRs from endosomes to the Golgi, at levels identical to that observed for exogenously expressed wild-type TIP47 protein. Thus, TIP47 F^{236}C is functionally active both *in vitro* and in living cells, where it rescues MPRs from degradation in the lysosome and enhances their recycling to the trans-Golgi network. Taken together, all of these data support a requirement for TIP47 in the transport of MPRs from endosomes to the trans-Golgi compartment.

Purification of Native TIP47 from Bovine Kidney

Our early work revealed that TIP47 is not a monomer in cytosol; it chromatographs with an apparent mass of about 350 kDa (Diaz and Pfeffer, 1998). Recombinant, histidine-tagged TIP47 also chromatographed at that size and also at a higher molecular weight, after expression in bacteria (Diaz and Pfeffer, 1998). Subsequent cross-linking led to the conclusion that cytosolic TIP47 is a hexamer, and metabolic labeling and coimmunoprecipitation failed to identify other subunits that might contribute to its large mass (Sincock *et al.*, 2003). For this reason, we sought to purify the

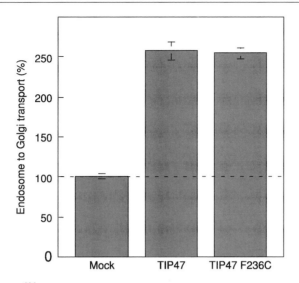

FIG. 2. TIP47 F^{236}C stimulates endosome to TGN transport of MPRs *in vivo*. Data shown represent an average of four independent experiments and are corrected for transfection efficiency (25%). Exogenously expressed TIP47 was expressed at comparable levels, within a few fold of the endogenous protein. Transport was measured as described herein.

native form of the protein to identify any potential binding partners that might be lost in an immunoprecipitate. To follow is our biochemical purification of TIP47 from bovine kidney (Sincock *et al.*, 2003).

Method

One bovine kidney (500 g), frozen by the supplier in liquid N_2 (Pel-Freez Biological, Rogers, AK), was stored at $-80°$. The tissue was incubated in liquid N_2 for 15 min and cracked into small pieces with a hammer. Chunks were homogenized in 1 liter, 10 mM HEPES/KOH, pH 7.2, 100 mM KOAc, 3 mM MgOAc, 5 mM EGTA, 1 mM phenylmethylsulfonyl fluoride, 1 mM dithiothreitol (DTT), 0.2 M sucrose, 7% (w/v) PEG 6000 (Serva), and 20 tablets of Complete EDTA free protease inhibitor (Roche, Mannheim, Germany) in a Waring blender. The homogenate was centrifuged at 10,000 rpm for 40 min in a Beckman JA10 rotor at 4°; the supernatant was filtered through two layers of cheesecloth and adjusted to pH 8.0 with Tris base. PEG 6000 was added to 18% (w/v) and stirred for 15 min before centrifugation (as above). The TIP47-containing pellet was resuspended in 1 liter 25 mM Tris–HCl, 0.2 M sucrose, pH 8.0 (resuspension buffer), and mixed with 280 ml Q Sepharose Fast-Flow (16.5 mg protein/ml resin) for 15 min on a rotating platform. The suspension was filtered through a Büchner funnel, washed with 2.8 liters of resuspension buffer, loaded into a column (53 mm), and TIP47 was step eluted in 25 mM Tris–HCl, pH 7.5, 150 mM KCl, 10% sucrose (FFQ buffer). Fractions between 140 and 700 ml were collected and mixed with 160 ml octyl-Sepharose (25 mg protein/ml resin) for 15 min on a rotating platform. The mixture was loaded onto a column (27 mm), washed with 1.6 liters FFQ buffer, and eluted with FFQ buffer containing 0.5% w/v CHAPS. Fractions eluting between 80 and 560 ml were collected and dialyzed overnight against 8 liters, 10 mM NaPO$_4$ buffer, pH 7.5, 10% glycerol, and slowly loaded onto a 30 ml hydroxyapatite Bio-Gel HT Gel (Bio-Rad, Hercules, CA, 5 mg protein/ml resin) column (27 mm). The column was washed with 300 ml 10 mM NaPO$_4$ buffer, pH 7.5, 10% glycerol, and eluted with a 600-ml gradient containing 10–100 mM NaPO$_4$ buffer, 10% glycerol. Fractions (7.5 ml) with conductivity between 0.8 and 1.6 were pooled (75 ml), dialyzed against 2 liters, 25 mM Tris–HCl, pH 8.0, 10% glycerol, and loaded to a 1 ml Mono Q column at 0.4 ml/min. The column was washed with 10 ml, 25 mM Tris–HCl, 25 mM KCl, pH 8.0, 10% glycerol, and eluted with a 20 ml, 25–200 mM KCl gradient. Fractions were analyzed for TIP47 content by immunoblot. TIP47-containing fractions were pooled and 4 ml was loaded onto a 120-ml Sephacryl S-200 HR column equilibrated in 25 mM HEPES, 150 mM KCl, pH 7.5, 10% glycerol, run at 0.4 ml/min. Fractions (1.5 ml) were analyzed by immunoblot or Coomassie blue or silver staining after concentration by speed-vac and/or TCA precipitation.

Purification of Recombinant TIP47

Full-length, recombinant TIP47 can be purified by the following procedure, but much of our work uses a monomeric form of the protein, comprised of residues 152–434 (Carroll *et al.*, 2001; Sincock *et al.*, 2003). This domain is fully capable of binding MPRs and Rab9. We have shown that residues 152–187 are required for Rab9 binding to TIP47 (Carroll *et al.*, 2001) while residues 187–434 can bind MPR cytoplasmic domain sequences (Sincock *et al.*, 2003). The Rab9 binding sequences were not present in the TIP47 fragment crystallized. TIP47 comprising amino acid residues 152–434 containing an N-terminal 6× His tag is expressed in *Escherichia coli* and purified according to the following protocol.

Materials

Resuspension Buffer: 50 mM HEPES, pH 7.5, 500 mM KCl, 25 mM imidazole, 1 mM phenylmethylsulfonyl fluoride (PMSF), 1 × complete EDTA-free Protease Inhibitor Cocktail (Roche), 10% glycerol

Elution Buffer: 50 mM HEPES, pH 7.5, 500 mM KCl, 250 mM imidazole, 10% glycerol

Dialysis Buffer: 25 mM HEPES, 150 mM KCl, 1 mM MgCl$_2$, 10% glycerol

LB plates and media

1 M isopropyl-β-D-thiogalactopyranoside (IPTG), Ni-NTA Agarose (Qiagen)

Use sterile technique throughout step 4. Keep cells on ice or in a cold room (4°) after step 5.

1. Transform BL21 DE3 *E. coli* cells with a plasmid encoding 6× His-TIP47 aa152–434.
2. Innoculate 50 ml LB media containing the appropriate selection antibiotic and grow at 37° overnight on a shaker (250 rpm).
3. Dilute the overnight culture 1:50 in 1 liter fresh LB media with the appropriate antibiotic and grow until an OD$_{600}$ of 0.3.
4. Induce with 1 mM IPTG for 3 h at 30°.
5. Pellet cells at 4000 rpm and snap freeze in liquid nitrogen. It is practical to resuspend the pellet in 20 ml LB and repellet cells in a 50-ml Falcon tube.
6. Resuspend cells in 35 ml ice-cold Resuspension Buffer.
7. French press cells at 1400 psi and transfer to a fresh tube on ice. Keep the French press cells at 4° to avoid heating the cell suspension.
8. Centrifuge the cell lysate at 17,000 rpm for 40 min at 4°.
9. Transfer the supernatant to a fresh tube and add 800 μl 50% slurry Ni-NTA agarose, prewashed in Resuspension Buffer.
10. Bind on a rotator at 4° for 40 min.
    The following purification steps are carried out at 4°.
11. Pour the slurry into a supported column and let drain by gravity without drying the bedded matrix.

Passage through the column can be accelerated by gently applying air pressure. The use of a 30-ml syringe coupled to the top of the column is useful for this purpose and allows for controlled accelerated flow in the wash steps.

12. Wash Ni-NTA beads 6× with 4 ml Resuspension Buffer without protease inhibitors.
13. Elute 3× in 250-ml Elution Buffer, incubating every elution step 3 min.
14. Dialyze in 4 liters Dialysis Buffer at 4° overnight. The buffer can be exchanged once.
15. Measure the protein concentration by standard methods, for example, a Bio-Rad protein assay, and analyze the purity of TIP47 by SDS-PAGE sodium dodecyl sulfate – polyacrylamide gel electrophoresis (SDS-PAGE), and subsequent Coomassie blue staining.

16. Snap freeze aliquots in liquid nitrogen and store at −80°. The typical yield of TIP47 protein from a 1-liter culture is ∼2 mg.

In Vivo Endosome to TGN Transport Assay

The assay makes use of the unique localization of tyrosine sulfotransferase to the trans-Golgi network and monitors the arrival of a tagged MPR from late endosomes of intact cells (Itin et al., 1997). The tag employed is comprised of a consensus sequence for tyrosine sulfation derived from cholecystokinin, a His tag for purification, and a myc tag for localization.

Chinese hamster ovary (CHO) or 293HEK cells stably expressing such a modified CD-MPR (Itin et al., 1997) are cultured in α-MEM supplemented with 7.5% fetal calf serum (FCS) and 500 μg/ml G418. Prior to assay, cells are grown in sulfate-free α-MEM supplemented with 7.5% FCS and 10 mM sodium chlorate to interfere with endogenous tyrosine sulfation. The medium is changed 24 h prior to transfection and replaced daily. If desired, cells can be transfected with plasmids encoding various forms of TIP47 (mutant or wild type) or vector alone using Fugene 6 (Roche Diagnostics). Posttransfection (43 h), 20 μg/ml cycloheximide is added for 3 h to clear any newly synthesized MPRs from the secretory pathway, prior to the addition of α-MEM containing FCS, cycloheximide, and 500 μCi sodium [^{35}S]sulfate (ICN) for 2 h at 37°. Cells are lysed in ice-cold RIPA buffer (50 mM Tris–HCl, pH 7.8, 150 mM NaCl, 1% sodium deoxycholate, 0.1% SDS, 1.5% Triton X-100, 25 mM imidazole); the lysates are then spun at 55,000×g at 4° for 15 min. Supernatants are mixed with 20 μl prewashed Ni-NTA agarose beads for 1 h at room temperature. The beads are washed 4× in RIPA; the supernatants are kept for protein assays and immunoblot analysis to confirm TIP47 expression levels. MPRs are eluted from the beads using 25 mM EDTA in RIPA and ^{35}S incorporation is quantified by scintillation counting.

Important controls include a sample that is incubated for 2 h at 4°; background incorporation can be determined by using the parental cell line that does not express tagged MPR and thus cannot incorporate sulfate into a protein collected on a nickel-NTA column. We have shown that the vast majority of the radioactivity is incorporated into the tagged MPR (Itin et al., 1997).

Cellular Depletion of TIP47 in Living Cells

HeLa and HeLaS3 cells from American Type Culture Collection are cultured at 37° and 5% CO_2 in Dulbecco's modified Eagle's media (DMEM) supplemented with 7.5% FCS, penicillin, and streptomycin. For RNA interference, HeLa and HeLaS3 cells are transfected at 50%

confluency with duplex RNA (Dharmacon Research, Inc.) using Oligofectamine (Invitrogen) according to the manufacturer and analyzed 72 h posttransfection. In some experiments we employ 293HEK cells, in which case, Lipofectamine 2000 reagent is used to improve transfection efficiency. In this case, cells are transfected at 90% confluency and analyzed after 72 h. Transfection efficiencies, determined by immunofluorescence and immunoblotting, were consistently in the range of 90%. TIP47 was targeted with the sequence AACAGAGCUACUUCGUACGUC and Rab9 with the sequence AAGUUUGAUACCCAGCUCUUC. The GL2 luciferase and lamin A/C control siRNAs were employed. Specific silencing of targeted genes was confirmed by at least three independent experiments.

Metabolic Labeling and Immunoprecipitation

Equal numbers of cells are split into 35-mm plates and grown overnight. Cells are transfected with the indicated siRNA and incubated for 24–48 h prior to labeling; MPRs have a very long half-life such that cells must be chased for an additional 24 h at minimum. For labeling, cells are washed twice with TD (25 mM Tris–HCl, pH 7.4, 5.4 mM KCl, 137 mM NaCl, 0.3 mM Na$_2$HPO$_4$) and preincubated in methionine- and cysteine-free medium containing 7.5% dialyzed FCS for 30 min. Cells were then incubated for 1 h with 100 μCi Tran[^{35}S]-label (ICN), then washed with media containing methionine and cysteine and chased for the indicated times. *Note*: When monitoring the half-life of the CI-MPR, it is important to allow the protein to leave the Golgi complex, as a significant pool of newly synthesized CI-MPR never leaves the endoplasmic reticulum and has a much shorter half-life than the fully processed form. Thus, in these experiments, the first time point is usually 3 h after the pulse, which removes confusion from the nonexported pool of protein.

Cells are then transferred to ice, washed three times with TD, and lysed in RIPA buffer containing protease inhibitors for 30 min on ice. After centrifugation for 15 min at 55,000 rpm, the supernatant is precleared with RIPA-washed protein A-agarose. Proteins to be immunoprecipitated are incubated for 2 h at room temperature with 3 μl of polyclonal antibody. Immune complexes are isolated by 30 min incubation with protein A-agarose, washed four times in RIPA, and resuspended in SDS-PAGE sample buffer. Samples are separated by SDS-PAGE and analyzed by Phosphorimager for quantification (Molecular Dynamics).

Acknowledgment

This research was supported by a grant to S.P. from the National Institutes of Health (DK37332).

References

Barbero, P., Bittova, L., and Pfeffer, S. R. (2002). Visualization of Rab9-mediated vesicle transport from endosomes to the trans Golgi in living cells. *J. Cell Biol.* **156,** 511–518.

Burguete, A. S., Harbury, P. B., and Pfeffer, S. R. (2004). *In vitro* selection and successful prediction of TIP47 protein-interaction interfaces. *Nat. Methods* **1,** 55–60.

Carroll, K. S., Hanna, J., Simon, I., Krise, J., Barbero, P., and Pfeffer, S. R. (2001). Role of the Rab9 GTPase in facilitating receptor recruitment by TIP47. *Science* **292,** 1373–1377.

Diaz, E., Schimmoller, F., and Pfeffer, S. R. (1997). A novel Rab9 effector required for endosome-to-TGN transport. *J. Cell Biol.* **138,** 283–290.

Diaz, E., and Pfeffer, S. R. (1998). TIP47: A cargo selection device for mannose 6-phosphate receptor trafficking. *Cell* **93,** 433–443.

Ganley, I., Carroll, K., Bittova, L., and Pfeffer, S. R. (2004). Rab9 regulates late endosome size and requires effector interaction for stability. *Mol. Biol. Cell* **15,** 5420–5430.

Hanna, J., Carroll, K., and Pfeffer, S. R. (2002). Identification of residues in TIP47 essential for Rab9 binding. *Proc. Natl. Acad. Sci. USA* **99,** 7450–7454.

Hickenbottom, S. J., Kimmel, A. R., Londos, C., and Hurley, J. H. (2004). Structure of a lipid droplet protein: The PAT family member TIP47. *Structure* **12,** 1199–1207.

Itin, C., Rancaño, C., Nakajima, Y., and Pfeffer, S. R. (1997). A novel assay reveals a role for alpha-SNAP in mannose 6-phosphate receptor transport from endosomes to the TGN. *J. Biol. Chem.* **272,** 27737–27744.

Itin, C., Ulitzur, N., and Pfeffer, S. R. (1999). Mapmodulin, cytoplasmic dynein and microtubules enhance the transport of mannose 6-phosphate receptors from endosomes to the trans Golgi network. *Mol. Biol. Cell* **10,** 2191–2197.

Kornfeld, S. (1992). Structure and function of the mannose 6-phosphate/insulin like growth factor II receptors. *Annu. Rev. Biochem.* **61,** 307–330.

Krise, J. P., Sincock, P. M., Orsel, J. G., and Pfeffer, S. R. (2000). Quantitative analysis of TIP47 (tail-interacting protein of 47 kD)-receptor cytoplasmic domain interactions: Implications for endosome-to-trans Golgi network trafficking. *J. Biol. Chem.* **275,** 25188–25193.

Lombardi, D., Soldati, T., Riederer, M. A., Goda, Y., Zerial, M., and Pfeffer, S. R. (1993). Rab9 functions in transport between late endosomes and the trans Golgi network *in vitro*. *EMBO J.* **12,** 677–682.

Orsel, J. G., Sincock, P. M., Krise, J. P., and Pfeffer, S. R. (2000). Recognition of the 300 K mannose 6-phosphate receptor cytoplasmic domain by TIP47. *Proc. Natl. Acad. Sci. USA* **97,** 9047–9051.

Riederer, M. A., Soldati, T., Shapiro, A. D., Lin, J., and Pfeffer, S. R. (1994). Lysosome biogenesis requires Rab9 function and receptor recycling from endosomes to the trans-Golgi network. *J. Cell Biol.* **125,** 573–582.

Schweizer, A., Kornfeld, S., and Rohrer, J. (1997). Proper sorting of the cation-dependent mannose 6-phosphate receptor in endosomes depends on a pair of aromatic amino acids in its cytoplasmic tail. *Proc. Natl. Acad. Sci. USA* **94,** 14471–14476.

Sincock, P. M., Ganley, I. G., Krise, J., Diederichs, S., Sivars, U., O'Connor, B., Ding, L., and Pfeffer, S. R. (2003). Self-assembly is important for TIP47 function in mannose 6-phosphate receptor transport. *Traffic* **4,** 18–25.

[32]   Capture of the Small GTPase Rab5 by GDI: Regulation by p38 MAP Kinase

By MICHELA FELBERBAUM-CORTI, VALERIA CAVALLI, and JEAN GRUENBERG

Abstract

The small GTPase Rab5 is one of the key regulators of early endocytic traffic and, like other GTPases, cycles between GTP- and GDP-bound states as well as between membrane and cytosol. The latter cycle is controlled by a guanine nucleotide dissociation inhibitor (GDI), which functions as a Rab vehicle in the cytosol. GDI extracts from membranes the inactive GDP-bound form of the Rab. Then, the cytosolic GDI:Rab complex is delivered to the appropriate target membrane, where the Rab protein is reloaded, presumably via a GDI displacement factor (Pfeffer and Aivazian, 2004). We previously reported that the formation of the GDI:Rab5 complex is stimulated by the mitogen-activated protein kinase p38 (Cavalli *et al.*, 2001). *Mol. Cell* **7,** 421–432.]. Selective activation of p38 MAPK increases endocytic rates *in vivo*, presumably allowing more efficient internalization of cell surface components for repair, storage, or degradation. These observations emphasize the possibility that external stimuli contribute to the regulation of membrane traffic. Here, we describe how to monitor the ability of GDI to extract Rab5 from early endosomal membranes *in vitro* and the role of p38 MAPK in this process. In addition, we detail how to investigate the possible role of p38 MAPK in the regulation of endocytosis *in vivo*.

Introduction

Internalization of cell surface receptors into animal cells occurs mainly via clathrin-dependent endocytosis, although other pathways, mediated by caveolae, rafts, or macropynosomes, have been described (Conner and Schmid, 2003). Endocytosed molecules are delivered to the early endosomes, where sorting occurs. Molecules that need to be reutilized, like receptors with housekeeping functions, are transported back to the plasma membranes. In contrast, molecules that need to be degraded, like the epidermal growth factor receptor (EGFR) and other down-regulated receptors, are sorted to late endosomes and finally delivered to lysosomes (Gruenberg, 2001).

METHODS IN ENZYMOLOGY, VOL. 403
0076-6879/05 $35.00
DOI: 10.1016/S0076-6879(05)03032-6

Endocytosis has long been considered as a mean to terminate signaling by internalizing active receptors after ligand binding. However, studies have uncovered the existence of new roles for endocytosis in the regulation of signaling events (Felberbaum-Corti and Gruenberg, 2002; Felberbaum-Corti et al., 2003; Le Roy and Wrana, 2005). In particular, targeting of ligand–receptor complexes to different endocytic compartments is proposed to directly regulate signal transduction through localized assembly of specific signaling complexes (Miaczynska et al., 2004b). The observations that after ligand addition, the majority of activated EGFRs and their downstream signaling factors such as Shc, mSos, and Grb2 are found on early endosomes led to the notion that receptor activation and thus perhaps signaling can occur intracellularly (Di Guglielmo et al., 1994). Consistently, endocytosis of activated EGFR appears necessary for full-scale mitogen-activated protein kinase (MAPK) activation (Vieira et al., 1996). Furthermore, ligand-bound EGFRs that are specifically activated in early endosomes can presumably recruit signaling molecules and elicit biological responses (Wang et al., 2002). In the same line, Teis et al. (2002) reported that the late endosomal protein p14 interacts with MP1—a scaffold protein that binds MEK1 and facilitates ERK1 activation—and that late endosomal localization of the p14/MP1–MAPK complex is required for efficient signaling in the ERK cascade after EGF stimulation. EGFR signaling might also occur from specialized endosomes primarily devoted to signaling. In fact, an endocytic structure, distinct from early endosomes and bearing Rab5 with its two effectors APPL1 and APPL2, was identified as an intermediate in signaling between the plasma membrane and the nucleus (Miaczynska et al., 2004a).

Tumor growth factor-β (TGF-β)-regulated signaling also requires targeting to early endosomes. Indeed, the interaction of TGF-β receptor with SARA, FYVE finger protein specifically recruited to early endosomes, is essential for Smad2 phosphorylation and subsequent signal propagation. A study proposes that targeting of TGF-β receptor to different endocytic pathways can influence signaling output, Clathrin-dependent internalization to early endosomes mediates signaling, while raft-dependent uptake into caveolin-positive structure leads to degradation (Di Guglielmo et al., 2003). Endosomes can also mediate transport of signaling molecules to their target sites. Particularly illustrative is the case of neurons, where signaling endosomes carrying Nerve growth factor (NGF) bound to its receptor TrkA and signaling-competent complexes undergo retrograde trafficking from the presynaptic terminals to the cell bodies that are far away located (Delcroix et al., 2003). Finally, studies in *Drosophila* indicate that morphogens can disperse through developing target tissues by trafficking through the cells rather than by free diffusion. As a result, the signaling range—that is, the distance over which the ligand can travel and signal—is controlled by adjusting the rate of recycling and degradation of the morphogens in the target cells (Gonzalez-Gaitan, 2003).

If membrane trafficking influences signal transduction, the reverse is also true. Indeed, while endocytosis has long been considered a constitutive function, evidence shows that signaling can modulate it. In particular, the small GTPase Rab5, which modulates early endocytic events, is a key regulator of this coupling. Rab5 can cycle not only between an active GTP-, and an inactive GDP-bound form, as well as between membrane and cytosol (Fig. 1). Specific factors modulate Rab5 GDP/GTP exchange (GEFs) and GTP hydrolysis (GAPs), while GDI, which is a generic Rab vehicle in the cytosol, controls the membrane–cytosol cycle. Evidence is accumulating that signaling influences each step of the Rab5 GTPase cycle, thus regulating transport.

It has long been known that binding of EGF to EGFR triggers both receptor endocytosis and fluid-phase uptake. EGF stimulation also activates Rab5, which appears essential for internalization (Barbieri et al., 2000). Activation of Rab5 is mediated, at least in part, by the ability of Ras to directly potentiate the Rab5 nucleotide exchange activity of Rin1 (Tall et al., 2001). Conversely, Rin1 probably recruits active Rab5 and Ras to the

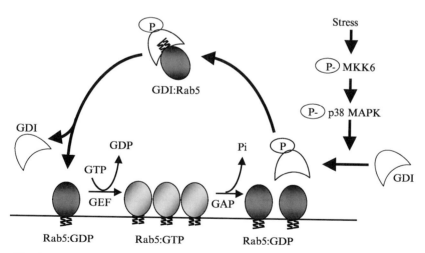

Fig. 1. Endocytosis is modulated by the p38 MAPK pathway. Activation of p38 by stress (phosphorylated p38, P-p38) leads to phosphorylation and activation of GDI, thus stimulating the formation of the cytosolic GDI:Rab5 complex. Regulation of GDI activity may provide a mechanism by which Rab5-GDP is delivered and reactivated specifically and rapidly to active regions of the membrane. The observed increase in internalization rates upon p38 MAPK activation may also reflect a direct role of Rab5 in the formation of clathrin-coated vesicles. (See color insert.)

activated receptor (Barbieri *et al.*, 2003), leading to a complex regulating both receptor signaling and trafficking. It has also been proposed that the stimulatory effect of Ras on endocytosis and Rab5 activation involves protein kinase B (PKB/Akt), a kinase involved in growth factor receptor signal transduction (Barbieri *et al.*, 1998). EGF stimulation can also negatively regulate endocytosis by modulating the GTPase activity of Rab5 through the sequential recruitment of Eps8 and RN-Tre (Lanzetti *et al.*, 2000). Eps8 is a substrate of EGFR kinase, while RN-Tre is a Rab5 GAP. Therefore, it inactivates Rab5, thus inhibiting receptor internalization and, consequently, prolonging EGFR signaling at the plasma membrane. Furthermore, RN-Tre diverts Eps8 from its Rac-activating function, resulting in attenuation of Rac signaling. Recently, the Grb2 adaptor protein was shown to mediate RN-Tre recruitment to activated receptor in an Eps8-independent manner (Martinu *et al.*, 2002). Thus, Grb2 can potentially contribute to both the activation of Rab5 (via Sos-mediated activation of Ras, leading to Rin1 recruitment) and to its inactivation (via RN-Tre).

A key component of the Rab5 cycle is Rab–GDI. GDI can be phosphorylated when complexed to Rab proteins (Steele-Mortimer *et al.*, 1993), suggesting that kinases control the Rab:GDI cycle and thus Rab activity. We found that p38 MAPK mediates GDI phosphorylation and its activation in the cytosolic cycle of Rab5, leading to enhanced formation of the GDI:Rab5 complex (Cavalli *et al.*, 2001). We also observed that p38 MAPK activation by stress stimuli modulates endocytosis, probably through a net increase in Rab5 cycling (Fig. 1). Recently, Huang *et al.* (2004) reported that p38 MAPK activation during metabotropic glutamate receptor (mGluR)-dependent long-term depression (LTD) also accelerates AMPA receptor (AMPAR) endocytosis by stimulating the formation of the GDI:Rab5 complex. Moreover, *N*-methyl-D-aspartate (NMDA) receptor-dependent LTD induction produces a rapid and transient activation of Rab5, which in turns drives the specific removal of synaptic AMPAR in a clathrin-dependent manner (Brown *et al.*, 2005). Interestingly, NMDAR opening also triggers the activation of p38 MAPK (Zhu *et al.*, 2002), which would correlate with Rab5 activation at the plasma membrane, either through regulation of the GDI:Rab5 cycle or through other unknown mechanisms.

Here, we describe the experimental system we developed to monitor *in vitro* the ability of GDI to capture Rab5 from early endosomal donor membranes and the role of p38 MAPK in this process (see Rab Capture Assay, and Figs. 2 and 3). In addition, we detail how to investigate the possible role of p38 in the *in vivo* regulation of endocytic trafficking (see Activation of p38 MAPK Modulates Endocytosis *In Vivo*, and Fig. 3).

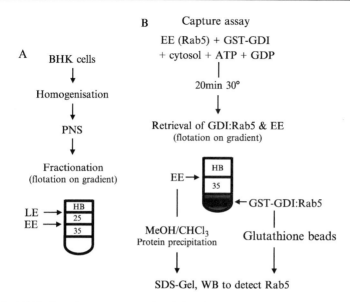

FIG. 2. (A) Early endosomes preparation. (B) Rab5 capture assay. The assay is depicted as described in (A).

Methods

Rab Capture Assay

This assay monitors the capacity of GDI to extract Rab5 from early endosomal membranes as a GDI:Rab5 complex. The formation of the complex is stimulated by a cytosolic factor that we purified and identified as p38 MAPK. Accordingly, purified recombinant forms of p38 MAPK and MKK6(E)—a constitutively active mutant of p38 MAPK upstream kinase—can substitute for the cytosol in the GDI activation process (Fig. 3A). This activation process is inhibited by a p38-specific inhibitor, SB 203580 (Fig. 3A). Our data indicate that GDI Ser-121 is necessary for p38-mediated GDI activation and thus suggests that Ser-121 is the target of p38 (Fig. 3B).

Expression and Purification of GST–GDI Protein

REAGENTS

Plasmid PGEX-2t containing the cDNA of bovine GDI (from M. Zerial, Dresden, Germany)

Escherichia coli strain BL21(DE3)plysS (Stratagene)

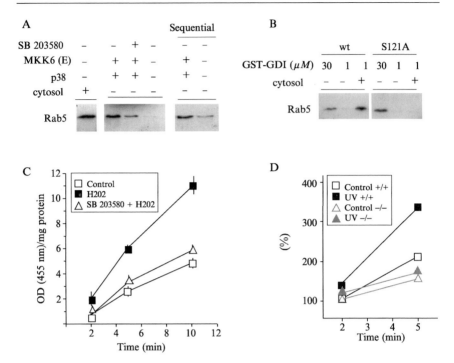

FIG. 3. (A) Rab5 capture assay with pure, recombinant proteins. The assay is supplemented with 150 nM recombinant GST-p38, 50 nM GST-MKK6(E) in the presence or absence of 10 μM SB203580. As indicated, the assay is also carried out sequentially: GST-GDI is first activated with 150 nM recombinant His-p38 and 50 nM GST-MKK6(E), and then the phosphorylated GDI is mixed with endosomes, so that it can capture membrane-associated Rab5. (B) Rab5 capture assay with GDI phosphorylation mutant. Wild-type GST-GDI or mutant are first incubated with (+) or without (−) cytosol, and then the assay is carried out sequentially. (C) Endocytosis *in vivo*. BHK cells are mock treated (open squares), treated with 50 μM H_2O_2 for 10 min (closed squares), or preincubated for 30 min with 10 μM SB203580 prior to adding 50 μM H_2O_2 for 10 min (open triangles). Then, cells are incubated for the indicated time periods with 2 mg/ml HRP as a marker of fluid-phase uptake. (D) Endocytosis in p38α ($^{-/-}$) cells *in vivo*. p38 MAPKα (+/+) (square) or (−/−) (triangle) MEFs are treated with (close symbols) or without (open symbols) UV light and then incubated with 2 mg/ml HRP. Reprinted from Cavalli *et al.* (2001). Copyright (2001), with permission from Elsevier. (See color insert.)

Luria-Bertani medium (LB, 1 liter): 10 g Bacto-tryptone, 5 g Bactoyeast extract, 5 g NaCl, pH 7.5

Ampicillin (Amp), chloramphenicol (CM), isopropyl-1-thio-D-galac-topyranoside (IPTG) Phosphate-buffered saline (PBS): 137 mM NaCl, 2.7 mM KCl, 1.5 mM KH_2PO_4, 6.5 mM Na_2HPO_4, pH 7.4

Lysis buffer: PBS, 1 mM EDTA, 1 mM dithiothreitol (DTT), protease inhibitors (10 μg/ml aprotinin, 10 μM leupeptin, 1 μM pepstatin), 1% Triton X-100
Elution buffer: 20 mM reduced L-glutathione, 120 mM NaCl, 100 mM Tris, pH 8.0 EE buffer (EEB): 75 mM KCH$_3$COO
30 mM HEPES, pH 7.4, 5 mM MgCl$_2$

PROCEDURE

1. Cultivation of *E. coli*: Bacteria are transformed with PGEX-2t-α GDI and grown overnight in LB containing 100 μg/ml Amp and 25 μg/ml CM. 400 ml LB/Amp100/CM25 is inoculated with 2 ml of the overnight culture and grown to an OD$_{600\ nm}$ of 0.2–0.3. After addition of 0.4 mM IPTG, cells are further cultured for 4 h. Cells are sedimented by centrifugation in a Sorvall GS-3 rotor for 15 min at 1500×g at 4°.

2. Cell lysis: This step and the followings are performed on ice. The bacterial pellet is resuspended in 8 ml lysis buffer. The cell suspension is sonicated 3×30-s with 30 s intervals and centrifuged at 20,000×g for 30 min in a Sorvall SS34 rotor.

3. Affinity purification: The supernatant is loaded onto 750 μl glutathione-Sepharose 4B beads (Amersham) preequilibrated in PBS and incubated on a rotating wheel for 2 h. Beads are spun down at 800×g for 2 min and washed 3× with 50 ml PBS. Beads are then packed onto a column and GST–GDI is eluted with 1.25 ml elution buffer.

4. Dialysis: The eluate is dialyzed against EEB. The purity of the preparation is analyzed by sodium dodecyl sulfate–polyacrylamide gel electrophoresis (SDS-PAGE) followed by Coomassie blue staining. Aliquots are frozen in liquid nitrogen and stored at −80°.

Generation of Recombinant GST–GDI S121A Mutant Protein. GDI S121A point mutation was carried out with the Quick-Change Site-Directed Mutagenesis kit (Stratagene). The plasmid PGEX-2t-aGDI was used as a template for the PCR reaction with 5'-GATCTACAAAGTACCA<u>GCT</u>A; CTGAGACTGAAGCC-3' sense primer and 5'-GGCTTCAGTCTCAG-TA<u>GCT</u>GGTACTTTGTAGATC-3' antisense primer. Triplets underlined in the oligonucleotide sequences encode the mutagenized alanine residue. Induction and purification of GDI mutant are as described for GST–GDI wt.

Homogenization and Preparation of a Postnuclear Supernatant (PNS). When preparing early endosomal fractions to be used in the capture assay, gentle homogenization conditions should be used to limit possible damage to endosomal elements. In addition, harsh conditions should always be avoided to limit the breakage of lysosomes and consequent proteolysis by released hydrolases.

REAGENTS

Monolayer cultures of baby hamster kidney cell line BHK-21 are grown to 80% confluence in 100-mm tissue culture dishes using G-MEM supplemented with 5% (v/v) fetal calf serum, 10% (v/v) tryptose phosphate broth, 2 mM glutamine, 100 U/ml penicillin, and 100 μg/ml streptomycin.

PBS.

Homogenization buffer (HB, 8.6% sucrose): 250 mM sucrose, 3 mM imidazole, pH 7.4. The sucrose concentration must be determined precisely with a refractometer (commercially available sucrose is hydrated).

PROCEDURE

All steps should be performed on ice, using ice-cold buffers.

1. Cell scraping: Cells are washed 2× with 10 ml PBS, and then scraped gently in 2.5 ml PBS with a rubber policeman cut at a low angle to obtain sheets of intact cells. Typically, cells from three dishes are pooled into one 15-ml tube.

2. After centrifugation for 5 min at 150×g, the cell pellet is resuspended very gently in 3 ml HB using a wide-bore pipette and centrifuged 10 min at 1200×g.

3. Homogenization: Pellets are gently resuspended in 100 μl HB/dish using a 1000-μl pipette. The homogenate is then drawn 3–10× through a 22-gauge needle attached to a 1-ml Tuberculin-type syringe and homogenization is monitored using a phase-contrast microscope. Nuclei, which appear as dark, round, or oblong structures, should remain intact and free of cellular materials.

4. The homogenate is centrifuged 10 min at 1200×g and the postnuclear supernatant (PNS) is collected.

Endosome Fractionation by Flotation in a Sucrose Gradient. Early and late endosomal fractions can easily be prepared from PNS by flotation in a step gradient of 40.6–8.6% (w/w) sucrose/H_2O (Aniento *et al.*, 1993a; Bomsel *et al.*, 1990). Early endosomes contain the transferrin receptor, Rab5, the Rab5 effector EEA1, and annexin II, while late endosomes contain Rab7, lysosomal glycoproteins (e.g., Lamp 1), and the unconventional phospholipid lysobisphosphatidic acid (LBPA).

REAGENTS

HB (8.6% sucrose)

62% sucrose solution: 2.351 M sucrose, 3 mM imidazole, pH 7.4

35% sucrose solution: 1.177 M sucrose, 3 mM imidazole, pH 7.4

25% sucrose solution: 0.806 M sucrose, 3 mM imidazole, pH 7.4

The sucrose concentration must be determined precisely with a refractometer.

PROCEDURE. All steps should be performed on ice, using ice-cold buffers.

1. Flotation gradient: The PNS is adjusted to a final sucrose concentration of 40.6% by adding ≈1.1 vol of 62% sucrose solution per volume of PNS (check with a refractometer) and transferred to a Beckmann SW60 centrifuge tube. Typically, the PNS of three dishes are loaded into a tube. The mixture is then overlaid sequentially with 1.5 ml of 35% sucrose solution, then 1 ml of 25% sucrose solution, and finally with HB to fill the tube. The gradient is centrifuged 1 h at 170,000×g.

2. Fractions collection: The interfaces should appear white. The layer of white lipids on the top of the gradient should be carefully removed. The late endosome fraction corresponds to the top interface (HB/25% interface), while the early endosome fraction (200 μl) is collected from the 25%/35% interface. The lower interface contains the plasma membrane, the endoplasmic reticulum, and the bulk of the Golgi membranes.

3. Aliquots are frozen in liquid nitrogen and stored at −80°; they should be thawed as rapidly as possible prior to use or analysis.

Cytosol Preparation

REAGENTS

HB with inhibitors cocktail protease inhibitors, 50 mM glycerol 2-phosphate—Sigma G6251, 0.1 mM sodium orthovanadate—Sigma 450243.

PROCEDURE. Cytosol from various sources can be used, but we prefer rat liver cytosol because of the high yield and protein concentration, prepared as described by Aniento *et al.* (1993b). Fresh rat liver is gently homogenized in 4 ml HB per g of liver, using a Potter homogenizer with a Teflon pestle. The homogenate is centrifuged first at 3000×g for 10 min, then at 27,000×g for 10 min in a Sorvall SS34 rotor. The supernatant is then centrifuged at 28,5000×g for 1 h in a Beckmann SW40 rotor. The supernatant from this centrifugation is the cytosol and its expected concentration is ~20 mg/ml. Aliquots should be frozen and stored in liquid nitrogen. All steps should be performed on ice, using ice-cold buffers.

Rab Capture Assay

REAGENTS

Recombinant GST–GDI, early endosomes (EE), rat liver cytosol
100 mM adenosine-5′-triphosphate (ATP, Sigma A1852), pH 7.0
100 mM guanosine-5′-diphosphate (GDP, Sigma G7127)
HB, 35% and 62% sucrose solutions
EEB with the inhibitor cocktails described earlier for cytosol preparation (EEBI)
Monoclonal antibody against Rab5 (clone 621.3 from Synaptic Systems GmbH, Göttingen, Germany; http://www.sysy.com/rab5/rab5_fs.html)

PROCEDURE A (FIG. 2B)

1. Rab5 capture by GST–GDI: In the reaction tubes placed on ice are aliquoted 15 μg GST–GDI (final 1 μM), 20 μg purified early endosomes, and EEB to 120 μl; 60 μl of EEB containing 0.3 mM ATP, 3 mM GDP, and 100 μg cytosol are added. Components are mixed briefly and incubated for 20 min at 30°, with occasional agitation.

2. Flotation gradient, EE removal: Tubes are returned to ice. The mixture is adjusted to 40.6% sucrose by adding ~1.1 vol of 62% sucrose solution (check with refractometer), transferred to a TLS55 centrifuge tube, and overlaid sequentially with 0.5 ml of 35% sucrose solution, then with HB to fill the tube. The gradient is centrifuged 1 h at 170,000×g at 4°. EE (200 μl) are collected at the HB/35% interface and total protein precipitated with MeOH/CHCl$_3$.

3. GST–GDI:Rab5 complex retrieval onto glutathione beads: The load of the gradient (~380 μl, in 40.6% sucrose) is transferred to a new tube on ice and 80 μl of a 50% glutathione-Sepharose 4B slurry preequilibrated in PBS is added. After a 30 min incubation at 4° on a rotating wheel, beads are spun down at 14,000×g for 3 min and washed 3× with 1 ml EEB. Beads are finally resuspended in sample buffer. Rab5 bound to GST–GDI and Rab5 associated to EE are analyzed by SDS-PAGE and Western blotting with anti-Rab5 antibodies.

PROCEDURE B: SEQUENTIAL

This procedure allows the separate analysis of GDI activation and Rab5 extraction.

1. GST–GDI activation: 15 μg GST–GDI, 0.5 mM ATP, 0.1 mM GDP and EEBI to 300 μl are aliquoted in the reaction tubes placed on ice. After

addition of 100 μg cytosol, components are mixed briefly and incubated for 20 min at 30°, with occasional agitation.

2. Activated GST–GDI retrieval onto glutathione beads: The tubes are returned to ice and 80 μl of a 50% glutathione-Sepharose 4B slurry preequilibrated in PBS are added. After a 30 min incubation at 4° on a rotating wheel, beads are spun down at 14,000×g for 3 min and washed 3× with 1 ml EEBI.

3. Rab5 capture by GST–GDI: Beads are finally resuspended with 20 μg purified EE, 10 μl 10 mM GDP (final 1 mM), and 10 μl EEBI (total volume ~100 μl). Components are mixed briefly and incubated for 20 min at 30° with occasional agitation.

4. Flotation gradient, EE removal: The reaction tubes are placed on ice and 80 μl of HB is added. The gradient is then prepared as in Procedure A.

5. Glutathione beads: GST–GDI:Rab5 complex analysis: The load with the beads (~380 μl) is transferred to a new tube on ice and 800 μl of EEBI is added. Beads are spun down at 14,000×g for 3 min and washed once with 1 ml EEBI. Beads and EE are analyzed as described earlier.

COMMENTS

1. *In vivo*, p38 MAPK can be activated by the upstream MAPK kinases (MAPKK) MKK6, MKK3, and MKK4. *In vitro*, the purified MKK6 S151E-T155E [MKK6(E)], a constitutive active mutant of MKK6, can phosphorylate the purified p38 MAPK. As shown in Fig. 3A, addition of both p38 and MKK6(E) recombinant proteins (from S. Arkinstall, Randolph, MA) is sufficient to activate GDI in the absence of cytosol. Activation is inhibited by SB 203580 (Calbiochem, 559389), a specific p38 inhibitor.

2. Phosphorylation of GDI, presumably on Ser-121, is required for GDI activation. Indeed, at low concentration (1 μM) the GDIS121A mutant can no longer be activated by cytosol. However, the mutant behaves like wt GDI when used at a concentration sufficiently high (30 μM) to bypass cytosol requirements, indicating that the S121A mutant is properly folded and functional (Fig. 3B).

Activation of p38 MAPK Modulates Endocytosis In Vivo

In mammalian cells the p38 MAPK cascade functions mainly in stress responses like inflammation and apoptosis. Accordingly, treatment of cul- ture cells with hydrogen peroxide (H_2O_2) or UV-C irradiation are well established ways to activate p38. To assess the physiological relevance of the p38-dependent regulation of GDI in the cytosolic cycle of Rab5, we tested whether stress modulates endocytosis. We found that activation of p38 MAPK accelerates endocytic rates *in vivo*, providing evidence that endocytosis can be regulated by the environment.

Activation of p38 MAPK by Oxidative Stress or UV Light. H_2O_2 and UV light are specific activators of the p38-dependent stress pathway and do not activate the related stress-induced kinase c-jun N-terminal kinase (JNK) when used close to physiological ranges.

REAGENTS AND PROCEDURE

1. Peroxide treatment: Cells are treated for 10 min with 50 μM H_2O_2 (Sigma, H1009). Since peroxide is rapidly metabolized by the cells, peroxide is added again after 5 min incubation.
2. UV treatment: Treatment was performed as in Tamura *et al.* (2000). Briefly, culture medium is removed and cells are irradiated for 20 s with UV-C light (254 nm, 100 J/m^2). Medium is added back to the dish and cells recover for 20 min in an incubator.
3. PNS preparation: see earlier discussion. Homogenization is performed in the presence of the inhibitor cocktail, if necessary (see Cytosol Preparation).
4. p38 MAPK activation: 100 μg PNS are analyzed by SDS-PAGE and Western blotting with anti-Phospho-p38 antibodies (Cell Signaling).

Endocytosis Measured with a Bulk Marker of the Fluid Phase. Horseradish peroxidase (HRP; Sigma P8250) is widely used as a marker for fluid-phase endocytosis, but a word of caution is needed, since solutes can be efficiently internalized by other routes in some cell types, in particular by macropinocytosis in cells with high ruffling activity. Internalized HRP is measured using *o*-dianisidine (Sigma D9154) and peroxide as substrates; the resulting brown product is quantified by measuring the absorbance at 455 nm with a spectrophotometer.

REAGENTS AND PROCEDURE. The experiment was performed as described in Cavalli *et al.* (2001).

COMMENTS

1. Internalization rates are measured over very short time periods (5 min) to limit possible indirect effects of GDI activation on other Rab proteins, hence on other transport routes.
2. As shown in Fig. 3C, peroxide causes a marked stimulation of endocytosis, which is abolished by the p38 inhibitor SB203580.
3. To further demonstrate that p38 is required for the stress-induced activation of endocytosis, the experiment is carried out in 7 p38α ($-/-$) mouse embryonic fibroblasts (MEF) (Tamura *et al.*, 2000).

UV light treatment stimulates endocytosis in control cells (+/+) but not in (−/−) cells (Fig. 3D).

4. Measuring uptake in cells overexpressing either GDI-S121A or GDI-WT allows for the monitoring of the involvement of GDI in the stress-induced activation of endocytosis.

Conclusions and Perspectives

Assays described in this chapter are useful to monitor how a stress stimulus contributes to the regulation of endocytosis. Indeed, we show that the stress-induced MAP kinase p38 regulates GDI functions in the cycle of the small GTPase Rab5 and thus modulates endocytic rates *in vivo*. Preliminary results suggest that p38 also regulates formation of a complex between GDI and Rab7, which is present on late endosomes and regulates transport to the degradative pathway, but perhaps not between GDI and all Rab proteins. Clearly, it will be important to determine to what extent this mechanism controls selective transport steps (e.g., endocytic pathway via Rab5 and Rab7) or whether its functions are more generally in membrane traffic. Conversely, other signaling pathways may regulate GDI functions. In any case, our data, together with the fact that the Rab5 cycle is regulated directly by EGF stimulation, already indicate that Rab5 may coordinate the response to various stimuli. It is attractive to speculate that Rab5 regulates both signal outcomes and endocytosis. Subsequent work is still needed to dissect the network of interactions between different signaling pathways and to characterize their precise role in the regulation of membrane traffic.

Acknowledgments

We would like to thank Marie-Hélène Beuchat and Marie-Claire Velluz for expert technical assistance. We also wish to thank Zeina Chamoun and Julien Chevallier for critical reading of the manuscript. This work was supported by the Swiss National Science Foundation and the International Human Frontier Science Program.

References

Aniento, F., Emans, N., Griffiths, G., and Gruenberg, J. (1993a). Cytoplasmic dyneindependent vesicular transport from early to late endosomes. *J. Cell Biol.* **123,** 1373–1388.

Aniento, F., Roche, E., Cuervo, A. M., and Knecht, E. (1993b). Uptake and degradation of glyceraldehyde-3-phosphate dehydrogenase by rat liver lysosomes. *J. Biol. Chem.* **268,** 10463–10470.

Barbieri, M. A., Kohn, A. D., Roth, R. A., and Stahl, P. D. (1998). Protein kinase B/akt and rab5 mediate Ras activation of endocytosis. *J. Biol. Chem.* **273,** 19367–19370.

Barbieri, M. A., Roberts, R. L., Gumusboga, A., Highfield, H., Alvarez-Dominguez, C., Wells, A., and Stahl, P. D. (2000). Epidermal growth factor and membrane trafficking. EGF receptor activation of endocytosis requires Rab5a. *J. Cell Biol.* **151,** 539–550.

Barbieri, M. A., Kong, C., Chen, P. I., Horazdovsky, B. F., and Stahl, P. D. (2003). The SRC homology 2 domain of Rin1 mediates its binding to the epidermal growth factor receptor and regulates receptor endocytosis. *J. Biol. Chem.* **278,** 32027–32036.

Bomsel, M., Parton, R., Kuznetsov, S. A., Schroer, T. A., and Gruenberg, J. (1990). Microtubule and motor dependent fusion *in vitro* between apical and basolateral endocytic vesicles from MDCK cells. *Cell* **62,** 719–731.

Brown, T. C., Tran, I. C., Backos, D. S., and Esteban, J. A. (2005). NMDA receptor-dependent activation of the small GTPase Rab5 drives the removal of synaptic AMPA receptors during hippocampal LTD. *Neuron* **45,** 81–94.

Cavalli, V., Vilbois, F., Corti, M., Marcote, M. J., Tamura, K., Karin, M., Arkinstall, S., and Gruenberg, J. (2001). The stress-induced MAP kinase p38 regulates endocytic trafficking via the GDI:Rab5 complex. *Mol. Cell* **7,** 421–432.

Conner, S. D., and Schmid, S. L. (2003). Regulated portals of entry into the cell. *Nature* **422,** 37–44.

Delcroix, J. D., Valletta, J. S., Wu, C., Hunt, S. J., Kowal, A. S., and Mobley, W. C. (2003). NGF signaling in sensory neurons: Evidence that early endosomes carry NGF retrograde signals. *Neuron* **39,** 69–84.

Di Guglielmo, G. M., Baass, P. C., Ou, W. J., Posner, B. I., and Bergeron, J. J. (1994). Compartmentalization of SHC, GRB2 and mSOS, and hyperphosphorylation of Raf-1 by EGF but not insulin in liver parenchyma. *EMBO J.* **13,** 4269–4277.

Di Guglielmo, G. M., Le Roy, C., Goodfellow, A. F., and Wrana, J. L. (2003). Distinct endocytic pathways regulate TGF-beta receptor signalling and turnover. *Nat. Cell Biol.* **5,** 410–421.

Felberbaum-Corti, M., and Gruenberg, J. (2002). Signaling from the far side. *Mol. Cell* **10,** 1259–1260.

Felberbaum-Corti, M., Van Der Goot, F. G., and Gruenberg, J. (2003). Sliding doors: Clathrin-coated pits or caveolae? *Nat. Cell Biol.* **5,** 382–384.

Gonzalez-Gaitan, M. (2003). Signal dispersal and transduction through the endocytic pathway. *Nat. Rev. Mol. Cell Biol.* **4,** 213–224.

Gruenberg, J. (2001). The endocytic pathway: A mosaic of domains. *Nat. Rev. Mol. Cell Biol.* **2,** 721–730.

Huang, C. C., You, J. L., Wu, M. Y., and Hsu, K. S. (2004). Rap1-induced p38 mitogen-activated protein kinase activation facilitates AMPA receptor trafficking via the GDI. Rab5 complex. Potential role in (S)-3,5-dihydroxyphenylglycene-induced long term depression. *J. Biol. Chem.* **279,** 12286–12292.

Lanzetti, L., Rybin, V., Malabarba, M. G., Christoforidis, S., Scita, G., Zerial, M., and Di Fiore, P. P. (2000). The Eps8 protein coordinates EGF receptor signalling through Rac and trafficking through Rab5. *Nature* **408,** 374–377.

Le Roy, C., and Wrana, J. L. (2005). Clathrin- and non-clathrin-mediated endocytic regulation of cell signalling. *Nat. Rev. Mol. Cell Biol.* **6,** 112–126.

Martinu, L., Santiago-Walker, A., Qi, H., and Chou, M. M. (2002). Endocytosis of epidermal growth factor receptor regulated by Grb2-mediated recruitment of the Rab5 GTPase-activating protein RN-tre. *J. Biol. Chem.* **277,** 50996–51002.

Miaczynska, M., Christoforidis, S., Giner, A., Shevchenko, A., Uttenweiler-Joseph, S., Habermann, B., Wilm, M., Parton, R. G., and Zerial, M. (2004a). APPL proteins link Rab5 to nuclear signal transduction via an endosomal compartment. *Cell* **116**, 445–456.

Miaczynska, M., Pelkmans, L., and Zerial, M. (2004b). Not just a sink: Endosomes in control of signal transduction. *Curr. Opin. Cell Biol.* **16**, 400–406.

Pfeffer, S., and Aivazian, D. (2004). Targeting Rab GTPases to distinct membrane compartments. *Nat. Rev. Mol. Cell Biol.* **5**, 886–896.

Steele-Mortimer, O., Gruenberg, J., and Clague, M. J. (1993). Phosphorylation of GDI and membrane cycling of rab proteins. *FEBS Lett.* **329**, 313–318.

Tall, G. G., Barbieri, M. A., Stahl, P. D., and Horazdovsky, B. F. (2001). Ras-activated endocytosis is mediated by the Rab5 guanine nucleotide exchange activity of RIN1. *Dev. Cell* **1**, 73–82.

Tamura, K., Sudo, T., Senftleben, U., Dadak, A. M., Johnson, R., and Karin, M. (2000). Requirement for p38alpha in erythropoietin expression: A role for stress kinases in erythropoiesis. *Cell* **102**, 221–231.

Teis, D., Wunderlich, W., and Huber, L. A. (2002). Localization of the MP1-MAPK scaffold complex to endosomes is mediated by p14 and required for signal transduction. *Dev. Cell* **3**, 803–814.

Vieira, A. V., Lamaze, C., and Schmid, S. L. (1996). Control of EGF receptor signaling by clathrin-mediated endocytosis. *Science* **274**, 2086–2089.

Wang, Y., Pennock, S., Chen, X., and Wang, Z. (2002). Endosomal signaling of epidermal growth factor receptor stimulates signal transduction pathways leading to cell survival. *Mol. Cell Biol.* **22**, 7279–7290.

Zhu, J. J., Qin, Y., Zhao, M., Van Aelst, L., and Malinow, R. (2002). Ras and Rap control AMPA receptor trafficking during synaptic plasticity. *Cell* **110**, 443–455.

[33]   Rab2 Purification and Interaction with Protein Kinase C ι/λ and Glyceraldehyde-3-Phosphate Dehydrogenase

By Ellen J. Tisdale

Abstract

The small GTPase Rab2 is essential for membrane trafficking in the early secretory pathway. Rab2 associates with vesicular tubular clusters (VTCs) located between the endoplasmic reticulum (ER) and the Golgi complex. VTCs function as transport intermediates and sort anterograde-directed cargo from recycling proteins. Rab2 selectively recruits atypical protein kinase C ι/λ (aPKCι/λ) and glyceraldehyde-3-phosphate (GAPDH) to VTCs where aPKCι/λ phosphorylates GAPDH. Both aPKCι/λ and GAPDH bind directly to Rab2 and this interaction ultimately results in

METHODS IN ENZYMOLOGY, VOL. 403   0076-6879/05 $35.00
DOI: 10.1016/S0076-6879(05)03033-8

COPI recruitment and the release of retrograde-directed vesicles. This chapter describes a protocol to purify recombinant Rab2 from Rab2 cDNA transformed bacteria and methods to assess recombinant Rab2 biological activity. Additionally, *in vivo* and *in vitro* assays are outlined that are employed to demonstrate Rab2 interaction with the downstream effectors aPKCι/λ and GAPDH.

Introduction

The small GTPase Rab2 is required for membrane transport in the early secretory pathway and associates with vesicular tubular clusters (VTCs) that reside between the endoplasmic reticulum (ER) and the Golgi complex (Tisdale and Balch, 1996; Tisdale *et al.*, 1992). VTCs are pleimorphic structures that serve as transport intermediates and that play an essential role in sorting anterograde-directed cargo from recycling proteins that are retrieved to the ER (Aridor *et al.*, 1995; Balch *et al.*, 1994; Horstmann *et al.*, 2002). Rab2 initiates the recruitment and directly interacts with atypical protein kinase C ι/λ (aPKCι/λ) and glyceraldehyde-3-phosphate dehydrogenase (GAPDH) on the VTC where aPKCι/λ phosphorylates GAPDH (Tisdale, 2002, 2003; Tisdale *et al.*, 2004). Moreover, GAPDH binds directly to the aPKCι/λ regulatory domain, suggesting that Rab2, aPKCι/λ, and GAPDH form a complex on the VTC (Tisdale, 2002). However, Rab2 interaction with aPKCι/λ inhibits aPKCι/λ-dependent GAPDH phosphorylation, which indicates that an unidentified downstream effector(s) recruited to the VTC modulates the inhibition imposed by Rab2 (Tisdale, 2003).

The functional relationship between Rab2 and aPKCι/λ is consistent with numerous observations indicating that intracellular membrane trafficking is regulated by phosphorylation–dephosphorylation events. In contrast, GAPDH is best known as a housekeeping enzyme that catalyzes the NAD-mediated oxidative phosphorylation of glyceraldehyde-3-phosphate to 1,3-diphosphoglycerate (Harris and Waters, 1976). Ongoing studies over the past 10 years have made it clear that GAPDH has multiple cellular activities independent of its glycolytic function (Sirover, 1999). For example, GAPDH is required for membrane transport between the ER and Golgi complex, yet elimination of GAPDH catalytic activity has no effect on this trafficking step (Tisdale *et al.*, 2004). To date, the precise function of GAPDH in the early secretory pathway is unknown. Conversely, aPKCι/λ kinase activity on the VTC is required for COPI association and subsequent formation of Rab2-mediated retrograde-directed vesicles enriched in recycling components (Tisdale, 2000).

This chapter describes a method to purify recombinant Rab2 from Rab2 cDNA-transformed bacteria and assess biochemical activity of the purified protein. In addition, a variety of protocols are outlined that can be utilized to demonstrate Rab2 interaction with the downstream effectors, aPKCι/λ and GAPDH. This "Rab2 complex" may be used as a tool in subsequent experimentation to "pull out" additional effectors essential for Rab2 function in the early secretory pathway.

Methods

Purification of Recombinant Rab2 from BL21 Transformed Bacteria

The amino-terminus of Rab2 is required for Rab2 function and the presence of epitope tags including His6, c-Myc, and hemaglutinin interferes with the inhibitory activity of Rab2 mutants when overexpressed in an *in vivo* transport assay (Tisdale and Balch, 1996). Based on this observation, a protocol was devised to generate a working stock of recombinant Rab2 for *in vitro* studies that excluded affinity chromatography. This protocol makes use of the T7 expression system described by Studier and Moffatt (1985). Human Rab2 cDNA is first subjected to polymerase chain reaction (PCR) to engineer an *Nde*I and *Bam*HI restriction site into the 5′ and 3′ ends, respectively, of the Rab2 coding sequence. The PCR product is subcloned into the *Nde*I and *Bam*HI restriction sites in pET3A (plasmid for expression of T7 RNA polymerase) and introduced into BL21 (DE3) pLysS. This bacterial strain contains the T7 RNA polymerase gene driven by the isopropyl-β-D-thiogalactopyranoside (IPTG)-inducible promoter. A 1 liter culture of the transformed bacteria is grown to 0.4–0.5 OD$_{600}$, and then induced with 0.4 mM IPTG for 3 h at 37°. The bacteria is centrifuged at 5000×g for 10 min at 4°, and the resultant bacterial pellet resuspended in 50 mM Tris (pH 7.4), 1 mM dithiothreitol (DTT), 0.1 mM phenylmethyl-sulfonyl fluoride (PMSF), 0.1 mM benzamidine, 1 mM EDTA, and 1% Triton X-100, and then homogenized by 20 passes with a Dounce tissue grinder. Lysozyme (400 μg/ml), DNase I (40 μg/ml), and 25 mM MgCl$_2$ are added and the homogenate placed on an end-over-end rocker for 30 min at 4°. The bacterial lysate is centrifuged at 22,000×g for 30 min at 4° to pellet insoluble material, and then the supernatant removed and applied to a 70 ml column containing Q Sepharose Fast Flow (Amersham Biosciences) equilibrated with Buffer A (50 mM Tris [pH 7.4], 10 mM MgCl$_2$, and 1.0 mM EDTA). The column is washed with 2 bed volumes of Buffer A and eluted with a linear NaCl gradient (0–400 mM) in Buffer A. Three-milli-liter fractions are collected and an aliquot of each fraction separated by sodium dodecyl sulfate–polyacrylamide gel electrophoresis (SDS-PAGE), transferred to nitrocellulose, and probed with a Rab2 monoclonal/polyclonal antibody (Santa Cruz) (Tisdale *et al.*, 2004). Those fractions enriched in Rab2 are pooled, concentrated by a Centriprep centrifugal filter device (Amicon), and the retentate applied to a 200 ml column containing Sephacryl S-100 (Amersham Biosciences) and eluted with Buffer A. Fractions (1.5 ml) containing Rab2 are identified by SDS-PAGE and Western blot analysis as above, pooled, dialyzed into the appropriate buffer, concentrated, and frozen ($-80°$) in aliquots.

Rab2 In Vitro *Prenylation*

The purified recombinant Rab2 (0.5 μg) is prenylated in an *in vitro* reaction (50 μl total volume) that contains 10 μg geranylgeranyl pyrophosphate (GGPP) (Sigma-Aldrich), 1 mM DTT, 25 μl rat liver cytosol (20 mg/ml), 10 mM MgCl$_2$, 1 mM ATP, 5 mM creatine phosphate, and 0.2 units of rabbit muscle creatine kinase for 1 h at 37°. The reaction is desalted through a 10-ml column of Sephadex G-25 (Amersham Biosciences) to remove incompatible reagents that may potentially inhibit subsequent activity assays, and the fraction containing prenylated Rab2 concentrated. Routinely, 40–45% of the total Rab2 is prenylated by this procedure as determined by phase separation in precycled Triton X-114 or by performing the reaction with [^{3}H]GGPP (American Radiolabeled Chemicals, Inc.) (Bordier, 1981; Tisdale and Balch, 1996).

Analysis of GDP–GTP Binding Properties of Recombinant Rab2

The ability of purified recombinant Rab2 (0.25 μg, $\sim$10 pmol) to bind GDP is determined by incubation (100 μl total volume) with 50 mM HEPES-KOH (pH 8.0), 1 mM DTT, 0.1 mg/ml bovine serum albumin (BSA), 5 mM EDTA, 4.5 or 10 mM MgCl$_2$, and 2.5 μM [^{3}H]GDP (diluted to $\sim$500 cpm/pmol with GDP) at 32° for 1 h. The reactions are then transferred to ice, diluted with ice-cold Buffer A (25 mM Tris–HCl [pH 8.0], 100 mM NaCl, 30 mM MgCl$_2$, 1 mM DTT, and 0.1 mg/ml BSA), and then immediately filtered through a prewet 0.45-μm nitrocellulose filter disc (Millipore). The filters are washed three times with Buffer A, allowed to dry, transferred to a glass vial containing scintillation fluid, and Rab2-[^{3}H]GDP quantitated by liquid scintillation counting.

The GTP-binding capacity of recombinant Rab2 can be determined by a filter binding assay (Feig *et al.*, 1986). Rab2 (10 pmol) is incubated for 60 min at 30° in 20 mM Tris–HCl (pH 7.5), 150 mM NaCl, 1 M EDTA, 2.5 mM MgCl$_2$, 1 mM DTT, 40 μg BSA, and [α-^{32}P]GTP (600 Ci/mmol) (New England Nuclear) at concentrations ranging from 1×10^{-9} to 1×10^{-4} M. To terminate binding, 2 ml of ice-cold Buffer B (25 mM Tris–HCl [pH 7.5],

20 mM MgCl$_2$, and 100 mM NaCl) is added and the reactions immediately filtered through a prewet 0.45-μm nitrocellulose filter disc (Millipore). The filters are washed three times with Buffer B, allowed to dry, transferred to a glass vial containing scintillation fluid, and Rab2-[α-^{32}P]GTP determined by liquid scintillation counting.

Determination of Recombinant Rab2-GTPase Activity

Rab2 intrinsic GTPase activity is measured by incubating recombinant Rab2 (50 nM) in 20 mM Tris–HCl (pH 7.8), 100 mM NaCl, 2.5 mM MgCl$_2$, 1 mM NaPO$_4$, 10 mM 2-mercaptoethanol, 0.1% BSA, and 1 pmol [α-^{32}P] GTP at 37° (Tisdale, 1999). Aliquots are removed at increasing times of incubation and the reaction stopped by the addition of an equal volume of 50 mM EDTA (pH 8.0), and then spotted onto polyethyleneimine-cellulose sheets (Merck). The chromatogram is developed in 0.6 M NaPO$_4$ (pH 3.5), dried, and the radioactive spots corresponding to [α-^{32}P]GDP and [α-^{32}P]GTP quantified by using a Phosphorimager (Amersham Biosciences) or excised and measured by liquid scintillation counting. GTP hydrolysis is calculated as the signal in the GDP spot relative to the total signal.

Assays to Characterize Rab2 Interaction with aPKCι/λ and GAPDH

Coimmunoprecipitation of Rab2 and Effectors

The two cell lines (HeLa and normal rat kidney cells) routinely employed in our studies contain endogenous Rab2, aPKCι/λ, and GAPDH. It is advisable to evaluate a particular cell line for aPKCι/λ expression. We have successfully used HeLa cells for Rab2 coimmunoprecipitation experiments. HeLa cells (3 × 10^6) are lysed in 50 mM Tris-buffered saline (TBS) (pH 8.0) and 1% Triton X-100 for 10 min on ice, and then the cell lysate is clarified by centrifugation at 20,000×g for 15 min at 4°. The soluble fraction is first precleared by the addition (20 μl) of settled Protein G plus/Protein A agarose (EMB Biosciences) that was washed with 50 mM TBS (pH 8.0) and 1% Triton X-100, and the cell lysate rocked end-over-end for 30 min at 4°. The agarose beads are collected by centrifugation at 4000 rpm for 5 min and the supernatant subjected to immunoprecipitation with an affinity-purified anti-Rab2 polyclonal antibody and Protein G plus/Protein A agarose. The efficiency of coprecipitation is comparable whether incubated for 4 h at room temperature or overnight at 4° (Tisdale, 2003; Tisdale *et al.*, 2004). The immune complexes are collected by centrifugation at 4000 rpm for 5 min, washed three times with 50 mM TBS (pH 8.0), 1% Triton X-100, and 100 mM NaCl, and then boiled in sample buffer. The immunoprecipitates are separated by SDS-PAGE, transferred to nitrocellulose, and then probed with reagents specific to Rab2, aPKCι/λ (BD Biosciences), and GAPDH (Chemicon).

Comment. Since there are limited reagents specific to Rab2, alternative approaches would be to (1) perform the coimmunoprecipitation with antibodies specific to either aPKCι/λ or GAPDH or (2) transfect HeLa cells with aPKCι/λ or GAPDH cDNA that contains an in-frame amino- or carboxyl-terminal epitope tag, and then perform coimmunoprecipitation experiments with antibodies made to the fusion sequences. Commercially available reagents may or may not be applicable for coimmunoprecipitation studies and should be evaluated in each system.

Mammalian Two-Hybrid Assay

The mammalian two-hybrid system is a powerful method to detect transient and weak protein–protein interaction in the cell line of choice. By employing the mammalian two-hybrid system instead of the routinely used yeast two-hybrid system, the interacting proteins are more likely to be in their native conformation and therefore posttranslationally modified, which may be required for their association.

Human Rab2 cDNA is cloned in frame to the *Eco*RI site of the GAL4 DNA-binding domain in the pM vector and aPKCι/λ cDNA or GAPDH cDNA cloned in frame to the *Eco*RI site of the activation domain in pVP16 (Clontech Laboratories). The two constructs (5 μg each) are cotransfected with the reporter vector pG5CAT (5 μg) that contains the chloramphenicol acetyltransferase (CAT) gene into HeLa cells (10^6) using a calcium phosphate transfection protocol (Chen and Okayama, 1987). We routinely obtain 40–50% transfection efficiency employing this method. The mammalian two-hybrid assay kit also includes instructions describing the appropriate control transfections to be performed in parallel. If the two fusion proteins bind *in vivo*, transcription of the CAT reporter gene is activated. The cells are collected 48–72 h posttransfection, lysed in sample buffer, and an aliquot of the cell lysate is separated by SDS-PAGE and then transferred to nitrocellulose. The blot is probed with anti-CAT polyclonal antibody (Invitrogen Life Technologies), washed, further incubated with a horseradish peroxidase (HRP)-conjugated secondary antibody, developed with enhanced chemiluminescence (ECL) (Amersham Biosciences), and then the amount of CAT protein quantitated by densitometry using the program Image Quant (Molecular Dynamics). In our studies, we have found ~15-fold increase in CAT protein expression in cells cotransfected with Rab2 and aPKCι/λ or Rab2 and GAPDH compared to nontransfected control cells, indicating that these proteins interact *in vivo* (Tisdale, 2003; Tisdale *et al.*, 2004).

Blot Overlay Assay

This simple *in vitro* assay can be used to screen for potential protein–protein interaction. For our studies, purified protein (Rab2, aPKCι/λ, or GAPDH) (5 μg) is either separated by SDS-PAGE and transferred to nitrocellulose or immobilized to nitrocellulose using a microfiltration apparatus. The recombinant Rab2 used in the assay was purified as described earlier. GAPDH purified from a variety of sources including rabbit and chicken muscle, erythrocytes, and yeast can be purchased from chemical suppliers. It is also possible to enrich for GAPDH by absorption to Blue Sepharose (Amersham Biosciences) or NAD-agarose (MP Biomedicals). To our knowledge, there is no commercially available source of aPKCι/λ and therefore the kinase is purified in the laboratory by the method outlined in the following paragraph.

After protein transfer to the membrane, the nitrocellulose is incubated in 50 mM HEPES/KOH (pH 7.2), 5 mM MgOAc, 100 mM KOAc, 10 mg/ml BSA, 0.1% Triton X-100, and 0.3% Tween 20 overnight at 4° to renature the protein. The membrane is then transferred to an overlay buffer containing 12.5 mM HEPES/KOH (pH 7.2), 1.5 mM MgOAc, 75 mM KOAc, 0.1% bovine serum albumin, 10 μM GTPγS, and 200 mM NaCl. The protein (10 μg/ml) that is being evaluated for interaction with the nitrocellulose-bound protein is added to the buffer and the blot incubated for either 4 h at room temperature or overnight at 4°. We have obtained similar results using either incubation conditions. After incubation the membrane is washed with TBS and then probed with an antibody specific to the overlay protein, washed, further incubated with an HRP-conjugated secondary reagent, and then developed with ECL.

Comment. For blot overlay assays that involve aPKCι/λ or other PKC isoforms, it is necessary to include 20 μg/ml phosphatidylserine (Avanti Polar Lipids, Inc.) in the overlay buffer. It has also been reported that protein–protein interaction can be stabilized by fixing the blot in phosphate-buffered saline (PBS)/0.5% formaldehyde for 30 min at room temperature, followed by incubation in PBS/2% glycine for 20 min at room temperature to block reactive aldehyde groups (Hyatt *et al.*, 1994). However, this fixation step may result in a high background.

Purification of His6-aPKCι/λ from Vaccinia Infected/pcDNA3-aPKCι/λ Transfected HeLa Cells. aPKCι/λ requires phosphorylation by phosphoinositide-dependent kinase 1 (PDK-1) to be active. Since PDK-1 is not expressed in bacteria, aPKCι/λ can be purified from expressing cells/tissues by standard chromatographic techniques. Alternatively, aPKCι/λ-fusion protein can be overexpressed in tissue culture cells or in baculovirus and the recombinant kinase purified by affinity chromatography.

We have generated the His6-tagged form of aPKCι/λ by PCR using a 5′ oligonucleotide primer that includes a *Bam*HI site and His6 coding sequence in tandem with the 3′ antisense oligonucleotide that encodes for an *Eco*RI site (Tisdale, 2000). The amplified product is subcloned into the *Bam*HI and *Eco*RI restriction sites in pcDNA3 (Invitrogen Life Technologies). This eukaryotic vector contains the T7 promoter and therefore can be employed in the vaccinia T7 RNA polymerase recombinant virus (vTF7-3) system to drive overexpression of the kinase (Fuerst *et al.*, 1986). HeLa cells (10^6/10-cm dish $\times$ 5) are first infected with vTF7-3 (moi of 0.1 pfu/cell) for 30 min. The cells are then transfected with 20 μg of pcDNA3-His6-aPKCλ that is premixed with 50 μl of LipofectACE (Invitrogen Life Technologies) and incubated for 4 h in 5 ml serum-free Dulbecco's modified Eagle's medium (DMEM). Five milliliters of DMEM/ 10% FBS is added and the cells returned to a 5% CO_2 incubator. The following day, cells that have rounded-up but are still attached to the tissue culture dish are dislodged by adding PBS/0.5 mM EDTA for 3 min, collected by centrifugation, and then washed once with 5 volumes of 10/18 (10 mM KOAc/18 mM HEPES, pH 7.2). The infected/transfected cells are then resuspended in 10/18 (2.5×10^6 cells/0.5 ml) containing protease inhibitors (10 μM aprotinin, 10 μM TLCK, 1 μg/ml chymotrypsin, 0.1 μM pepstatin), swollen on ice for 10 min, and homogenized by 20 passes through a 27-gauge needle after which the concentration is adjusted to 25 mM KOAc/125 mM HEPES. The nuclei are removed by centrifugation for 5 min at 500$\times g$ and the postnuclear supernatant recentrifuged at 100,000$\times g$ ($\sim$56,500 rpm) at 4° using an analytical microcentrifuge (Sorvall Discovery M120). The high-speed cytosolic fraction containing overexpressed His6-aPKCι/λ is then applied to a 2-ml column of NTA-agarose (Qiagen) equilibrated in Buffer A (10 mM HEPES [pH 7.9], 5 mM MgCl$_2$, 0.1 mM EDTA, 50 mM NaCl, and 0.8 mM imidazole). The column is washed with 10 volumes of Buffer A containing 25 mM imidazole, and His6-aPKCι/λ eluted with Buffer A supplemented with 200 mM imidazole. The purified kinase is dialyzed into 25 mM KOAc/125 mM HEPES, pH 7.2, and small aliquots snap frozen in liquid nitrogen for storage at −80°.

ELISA to Map Rab2 Interacting Domains with aPKCι/λ and GAPDH

The wells of a microtiter plate are coated with peptides synthesized to various domains in Rab2 or with purified Rab2 amino- or carboxyl-truncated recombinant proteins (1–2 μg/100 μl of 50 mM NaHCO$_3$, pH 9.6) at 4° overnight. The wells are then washed in TBS, blocked in TBS/5% FBS for 1 h at 37°, additionally washed in TBS, and then incubated in

Buffer A (50 mM Tris [pH 7.5], 5 mM MgCl$_2$, 100 mM NaCl, and 10 μM GTPγS) supplemented with 2 μg purified recombinant aPKCɩ/λ or GAPDH for 3 h at 37°. After each well is washed three times with TBS, a monoclonal antibody to aPKCɩ/λ (0.5 μg) (BD Biosciences) or a monoclonal antibody to GAPDH (0.5 μg) (Chemicon) in Buffer A is added for 2 h at 37°, washed, and further incubated with an anti-mouse alkaline phosphatase-conjugated secondary reagent (1:3000) for 1 h at 37°. The wells are again washed with TBS, developed with Sigma FAST p-Nitrophenyl phosphate (Sigma-Aldrich), and then read at 405 nm on a microplate reader.

Quantitative Membrane Binding Assay

This binding assay permits the user to quantify soluble factors recruited to membranes in response to Rab2. We have used HeLa cells and NRK cells for this assay. The cells are grown to near confluency in 10-cm tissue culture dishes, washed with ice-cold PBS by rocking the dish, and then 3 ml of 10 mM HEPES (pH 7.2) and 250 mM mannitol are added and the cells scraped off the dish with a rubber policeman. The cells are collected by centrifugation at 1000 rpm for 3 min, resuspended in 500 μl of 10 mM HEPES (pH 7.2) and 250 mM mannitol ($\sim$3 $\times$ 10^6 cells/ml), broken with 15 passes through a 27-gauge needle attached to a 3-ml syringe, and then centrifuged at 500$\times g$ for 10 min at 4°. The supernatant is removed and centrifuged at 20,000$\times g$ for 20 min at 4°. The resultant pellet containing ER, pre-Golgi, and Golgi membranes is washed with 1 M KCl in 10 mM HEPES (pH 7.2) for 10–15 min on ice to remove peripherally associated proteins, and then recentrifuged at 20,000$\times g$ for 20 min at 4°. The salt-washed membranes are resuspended (10 mg/ml total protein) in 10 mM HEPES (pH 7.2) and 250 mM mannitol. Membranes (30 μg of total protein) are added to a reaction mixture that contains 27.5 mM HEPES (pH 7.2), 2.75 mM MgOAc, 65 mM KOAc, 5 mM EGTA, 1.8 mM CaCl$_2$, 1 mM ATP, or 10 μCi [^{32}P]ATP (NEN Life Sciences), 5 mM creatine phosphate, and 0.2 units of rabbit muscle creatine kinase. Recombinant Rab2 is then added and the reaction mix incubated on ice for 10–20 min. We routinely employ a range of Rab2 concentrations (50–200 ng) to determine whether there is a dose-dependent recruitment of soluble components. Rat liver cytosol (50–75 μg total protein), which serves as the source of cytosolic factors, and 2.0 μM GTPγS are added, and the reactions shifted to 32° and incubated for 10–15 min. The binding reaction is terminated by transferring the samples to ice, and then centrifuged at 20,000$\times g$ for 15 min at 4° to obtain a pellet. The membrane pellet containing Rab2 and Rab2-recruited proteins is separated by SDS-PAGE, transferred to nitrocellulose, and then probed with reagents specific to aPKCι/λ or GAPDH. The blot is developed with ECL, and the amount of Rab2, aPKCι/λ, and GAPDH quantitated by densitometry using the program Image Quant (Molecular Dynamics).

Comments. The membranes employed in the assay can be frozen and thawed multiple times with no deleterious effect on binding activity. After identifying a downstream Rab2 effector(s), it is then possible to supplement the assay with reagents specific to those molecules (e.g., antibody or peptides) and assess recruitment of other cytosolic factors.

Acknowledgment

This work was supported by The American Heart Association-Midwest Affiliate (0030385Z) and National Institutes of Health (GMO68813).

References

Aridor, M., Bannykh, S. I., Rowe, T., and Balch, W. E. (1995). Sequential coupling between COPII and COPI vesicle coats in endoplasmic reticulum to Golgi transport. *J. Cell Biol.* **131,** 875–893.

Balch, W. E., McCaffery, J. M., Plutner, H., and Farquhar, M. G. (1994). Vesicular stomatitis virus glycoprotein is sorted and concentrated during export from the endoplasmic reticulum. *Cell* **76,** 841–852.

Bordier, C. (1981). Phase separation of integral membrane proteins in Triton X-114 solution. *J. Biol. Chem.* **256,** 1604–1607.

Chen, C., and Okayama, H. (1987). High-efficiency transformation of mammalian cells by plasmid DNA. *Mol. Cell. Biol.* **7,** 2745–2752.

Feig, L., Bin-Tao, P., Roberts, T., and Cooper, G. (1986). Isolation of ras GTP-binding mutants using an *in situ* colony-binding assay. *Proc. Natl. Acad. Sci. USA* **83,** 4607–4611.

Fuerst, T., Niles, E. G., and Studier, F. W. (1986). Eukaryotic transient-expression system based on recombinant vaccinia virus that synthesizes bacteriophage T7 RNA polymerase. *Proc. Natl. Acad. Sci. USA* **83,** 8122–8126.

Harris, J., and Waters, M. G. (1976). Glyceraldehyde-3-phosphate dehydrogenase. *In* "The Enzymes" (P. D. Boyer, ed.), Vol. XIII, pp. 1–49. Academic Press, New York.

Horstmann, H., Ng, C. P., Tang, B., and Hong, W. (2002). Ultrastructural characterization of endoplasmic reticulum–Golgi transport containers (EGTC). *J. Cell Sci.* **115,** 4263–4273.

Hyatt, S. D., Liao, L., Chapline, C., and Jaken, S. (1994). Identification and characterization of α-protein kinase C binding proteins in normal and transformed REF52 cells. *Biochemistry* **33,** 1223–1228.

Sirover, M. A. (1999). New insights into an old protein: The functional diversity of mammalian glyceraldehyde-3-phosphate dehydrogenase. *Biochim. Biophys. Acta* **1432,** 159–184.

Studier, W. F., and Moffatt, B. A. (1985). Use of bacteriophage T7 RNA polymerase to direct selective high-level expression of cloned genes. *J. Mol. Biol.* **189,** 113–130.

Tisdale, E. J. (1999). A Rab2 mutant with impaired GTPase activity stimulates vesicle formation from pre-Golgi intermediates. *Mol. Biol. Cell* **10,** 1837–1849.

Tisdale, E. J. (2000). Rab2 requires PKCι/λ to recruit β-COP for vesicle formation. *Traffic* **1,** 702–712.

Tisdale, E. J. (2002). Glyceraldehyde-3-phosphate dehydrogenase is phosphorylated by PKCι/λ and plays a role in microtubule dynamics in the early secretory pathway. *J. Biol. Chem.* **277**, 3334–3341.

Tisdale, E. J. (2003). Rab2 interacts directly with atypical protein kinase C (aPKCι/λ) ι/λ and inhibits aPKCι/λ-dependent glyceraldehyde-3-phosphate dehydrogenase phosphorylation. *J. Biol. Chem.* **278**, 52524–52530.

Tisdale, E. J., and Balch, W. E. (1996). Rab2 is essential for the maturation of pre-Golgi intermediates. *J. Biol. Chem.* **271**, 29372–29379.

Tisdale, E. J., Bourne, J. R., Khosravi-Far, R., Der, C. J., and Balch, W. E. (1992). GTP-binding mutants of Rab1 and Rab2 are potent inhibitors of vesicular transport from the endoplasmic reticulum to the Golgi complex. *J. Cell Biol.* **119**, 749–761.

Tisdale, E. J., Kelly, C., and Artalejo, C. R. (2004). Glyceraldehyde-3-phosphate dehydrogenase interacts with Rab2 and plays an essential role in endoplasmic reticulum to Golgi transport exclusive of its glycolytic activity. *J. Biol. Chem.* **279**, 54046–54052.

[34] Purification and Functional Interactions of GRASP55 with Rab2

By Francis A. Barr

Abstract

GRASP55 is a member of the GRASP family of peripheral membrane proteins thought to contribute to the organization of the Golgi apparatus. GRASPs are typically found in complex with coiled-coil protein, termed golgins, which interact with specific GTPases of the RAB family. Here I will describe the purification of native GRASP55 complexes from Golgi membranes, and the comparison with the related GRASP65 complex, standard methods for the analysis of interactions between GRASPs, golgins, and Rab GTPases using yeast two-hybrid analysis and protein biochemistry with native and recombinant proteins are also described.

Introduction

GRASPs (Golgi reassembly and stacking proteins) are a family of proteins conserved throughout evolution that were originally identified using a cell-free assay for the reassembly of the Golgi on mitotic exit as components required for the organization of Golgi cisternae into stacked structures (Barr *et al.*, 1997). They are characterized by a myristoylated membrane anchor at the amino-terminus, except for the yeast proteins, and a circularly permutated PDZ-like domain in the first 200 amino acids (Barr *et al.*, 1998). This PDZ-like domain enables GRASPs to act as adaptors for specific coiled-coil proteins of the golgin family and to link them to

METHODS IN ENZYMOLOGY, VOL. 403
0076-6879/05 $35.00
DOI: 10.1016/S0076-6879(05)03034-X

trans-membrane proteins of the Golgi apparatus such as the p24 putative cargo receptors (Barr *et al.*, 2001). GRASP65 is an adaptor for the *cis* Golgi matrix protein GM130 (Barr *et al.*, 1998), while GRASP55 is an adaptor at the medial Golgi for golgin45 (Short *et al.*, 2001). Golgin45 was identified using a combination of yeast two-hybrid screening and the biochemical purification of GRASP55 complexes from isolated Golgi membranes using an antibody affinity matrix (Short *et al.*, 2001). Like other golgins, p115 and GM130 (Allan *et al.*, 2000; Weide *et al.*, 2001), golgin45 is a binding partner for a specific Rab GTP-binding protein, Rab2 (Short *et al.*, 2001). Evidence for the importance of golgin45 for maintenance of Golgi structure and function comes from experiments in which it was depleted using RNA interference. Golgin45-depleted cells have a highly fragmented Golgi and show a strongly reduced level of protein transport (Short *et al.*, 2001). We describe methods both for the purification of GRASP–golgin complexes and for the analysis of the Rab-binding properties of golgins such as golgin45.

Purification of GRASP55 and GRASP65 Complexes from Golgi Membranes

Isolation of Rat Liver Golgi Membranes

Take two or three Sprague–Dawley rats between 150 and 200 g in weight. Kill the rats using an approved procedure and then quickly remove the livers and place them into ice-cold 0.5 M K/S (100 mM potassium phosphate, pH 6.7, 0.5 M sucrose, 5 mM MgCl$_2$). All the following steps should be carried out with ice-cold solutions in the cold room. Weigh out 18 g of liver tissue, and place into fresh 0.5 M K/S, approximately 30 ml final volume including the tissue. Cut the tissue into small pieces with a pair of sharp scissors, and then press the minced tissue through a 150-μm steel mesh sieve using the bottom of a 250-ml conical flask. Do not push too hard; gently massage the tissue through the sieve with a circular motion. Collect the homogenate and adjust to 25 ml; keep 1 ml for assay later. Prepare six gradients by placing 6.5 ml of 0.86 M K/S (100 mM potassium phosphate, pH 6.7, 0.86 M sucrose, 5 mM MgCl$_2$) in SW40 tubes (Beckman); overlay with 4 ml of homogenate in 0.5 M K/S and finally overlay with 2.5 ml of 0.25 M K/S (100 mM potassium phosphate, pH 6.7, 0.25 M sucrose, 5 mM MgCl$_2$). Balance the tubes in the rotor bucket with 0.25 M K/S, then centrifuge at 29,000 rpm for 60 min at 4°. Aspirate the lipid from the top of each tube; there should be a band of cloudy material at the 0.5 M/0.86 M interface. Collect this using a pasteur pipette; the "blood red"

0.5 M step contains cytosolic proteins. Note the volume of the collected interface and keep 100 μl for assay later. Measure the refractive index of the solution and convert this to sucrose concentration from a table of sucrose standards, then adjust to 0.25 M sucrose with 100 mM potassium phosphate, pH 6.7, and finally dilute with 0.25 M K/S to 24 ml. Place 12 ml of the diluted solution into two SW40 tubes, then slowly run 200 μl of 1.3 M K/S (100 mM potassium phosphate, pH 6.7, 1.3 M sucrose, 5 mM MgCl$_2$) down the side of the tubes so that it forms a cushion at the bottom of the tube. Balance the tubes and centrifuge at 6000 rpm for 20 min at 4°. Aspirate the straw-colored supernatant to leave the 1.3 M sucrose cushion covered by a carpet of Golgi membranes. Gently flow 5 ml of 0.25 M K/S down the wall of the tube to wash the surface of the Golgi membranes. Collect the membrane carpet using a pasteur pipette, adjust to 1.5 ml with 0.25 M K/S, and keep 50 μl for assay later. Measure the protein concentration and snap freeze 1-mg aliquots of Golgi membranes in liquid nitrogen and store at −80° until needed. Typically the concentration should be 4–5 mg/ml.

Assay the homogenate, the intermediate 0.5 M/0.86 M interface, and final Golgi fraction for protein and galactosyltransferase activity (Bretz and Staubli, 1997). Calculate the enrichment and yield of Golgi membranes over homogenate; this should be at least 100-fold.

Preparation of Affinity Columns

To prepare antibody affinity columns, large amounts of purified antibodies are required. For this reason it is advisable to use either monoclonal antibodies or an antiserum raised in a large animal such as a sheep or goat. It is also important to consider that the antibody should not be raised against a region of the protein that is suspected of forming a potential binding site for other proteins. First bind 2 mg antibody to 2 ml of either protein A or protein G Sepharose (Amersham), depending on the species and isotype of the antibody being used, for 2 h at 4°. For the sheep anti-GRASP55 antibody FBA34 (Short et al., 2001), protein G Sepharose was used. Wash the beads in a 10-ml disposable plastic column with 20 volumes of room temperature 200 mM sodium borate pH 8.0 buffer, and then transfer them to a fresh tube in 10 ml of the same buffer. Add 34 mg dimethylpimelidate (Pierce) and incubate on a tube rotator for 2 h at room temperature. Pellet the beads by centrifgation and resuspend in 200 mM ethanolamine pH 8.0 and leave overnight at 4° to quench any unreacted dimethylpimelidate. Wash the beads in a 10-ml disposable plastic column with 10 volumes of room temperature 200 mM glycine pH 2.8 to remove any noncovalently coupled antibodies, and then immediately wash with 20 volumes of phosphate-buffered saline (PBS). Resuspend the antibody beads in an equal volume of PBS, add sodium azide to 0.02% (w/v) to prevent any bacterial or fungal growth, and store at 4° until required. To control the affinity matrix, compare the starting purified antibody and the antibody bound to beads before and after cross-linking by sodium dodecyl sulfate-polyacrylamide gel electrophoresis (SDS-PAGE) on a 10% minigel. After cross-linking, only a small amount of antibody should be released from the beads by boiling in sample buffer.

Affinity Purification of Proteins from Golgi Membranes

Take a tube of Golgi membranes containing approximately 1 mg membranes from the −80° freezer, thaw rapidly at 37°, then place on ice. Add water to give 1 ml total volume, then pellet the membranes by centrifugation at $20,000 \times g$ for 10 min, 4°. This step removes the high sucrose storage buffer. Extract the membranes for 15 min on ice at a concentration of 1 mg/ml in HNT buffer (50 mM HEPES-KOH, pH 7.2, 200 mM NaCl, 0.5% [w/vl] Triton X-100) containing protease inhibitors (2 mM Pefabloc & Complete protease inhibitor cocktail [Roche Diagnostics]). Remove insoluble material by centrifugation at $20,000 \times g$ for 20 min, 4°. For affinity purification of GRASP55 complexes, add 100 μl of GRASP55 affinity matrix to 4 mg of Golgi membrane extract and incubate for 2 h at 4°. After four washes with 500 μl HNT, elute bound proteins in 100 μl of 3% (w/vl) SDS. Precipitate the eluted proteins by diluting the eluate with 900 μl water, add 7.5 μl 10% (w/vl) deoxycholate mixing well, and then add trichloroacetic acid (TCA) to a final concentration of 12% (w/vl). After 1 h on ice, recover the precipitated proteins by centrifugation at $20,000 \times g$ for 15 min, 4°. Aspirate the supernatant without disturbing the protein pellet. Add 1 ml of −20° acetone, vortex briefly, then centrifuge at $20,000 \times g$ for 5 min, 4°. The acetone washing steps help to remove the residual TCA and detergent, which otherwise cause problems for the solubilization of the precipitated protein. Remove the supernatant and repeat the acetone wash step. Add 5 μl 1 M Tris–HCl pH 8.0 followed by 20 μl of 1.5× sample buffer. Shake at room temperature until the pellets have dissolved, then heat at 99° for 3 min and analyze on 7.5% or 10% minigels.

Bound proteins can be identified following extraction from Coomassie blue-stained gel slices and digestion with sequencing-grade porcine trypsin (Promega) by peptide mass fingerprinting using a MALDI-TOF instrument (Reflex III, Bruker) and probability-based database searching (Perkins et al., 1999; Wilm et al., 1996). An example of GRASP55 and GRASP65 complexes affinity purified from rat liver Golgi membranes is shown in Fig. 1.

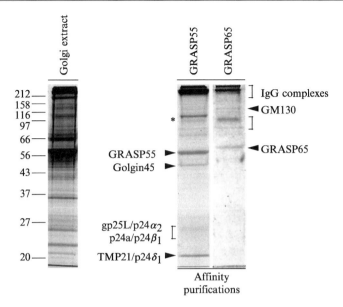

FIG. 1. Comparison of GRASP55 and GRASP65 complexes affinity purified from rat liver Golgi membranes. GRASP55 and GRASP65 complexes were affinity purified from 4 mg of Golgi membranes using the GRASP55 sheep polyclonal antibody FBA34 and the GRASP65 monoclonal antibody 7E10, respectively. The bound complexes were analyzed by SDS-PAGE on 10% minigels under nonreducing conditions.

Mapping Interaction Domains Using Directed Yeast Two-Hybrid Assays

We use a yeast two-hybrid system based on the PJ69-4A reporter yeast strain and the pGBT9 bait, and pACT2 prey vectors (James *et al.*, 1996). The advantage of this system is 2-fold. First, PJ69-4A has two independent reporters (for histidine and adenine biosynthesis) that can be used to select for interactions, as well as lacZ for blue-white colony screening. Second, the bait is expressed at a low level and can thus be used to study proteins that may be toxic if expressed at higher levels.

Yeast Growth Media

To prepare the amino acid base for synthetic complex (SC) medium, weigh out adenine 5.0 g, alanine 20 g, arginine 20 g, asparagine 20 g, aspartic acid 20 g, cysteine 20 g, glutamine 20 g, glutamic acid 20 g, glycine 20 g, inositol 20 g, isoleucine 20 g, lysine 20 g, methionine 20 g, *p*-aminobenzoic acid 2.0 g, phenylalanine 20 g, proline 20 g, serine 20 g, threonine 20 g, tyrosine 20 g, and valine 20 g. Mix thoroughly in a glass bottle on a rotor mixer overnight. Store in a cool, dark place away from moisture. To obtain SC-minus leucine and tryptophan (SC-LW), drop out supplement and weigh out 36.7 g of the amino acid base and histidine 2.0 g and uracil 2.0 g. To obtain the two-hybrid selective media (QDO quadruple drop out, lacking adenine, histidine, leucine, and tryptophan), weigh out 36.7 of the amino acid base and 2.0 g uracil. Mix thoroughly and store as above.

For 1 liter of the required SC media, weigh out 6.7 g yeast nitrogen base (without amino acids), 2.0 g of the appropriate drop-out mix, and glucose 20 g. Adjust the pH to 6.5 with NaOH, and then autoclave to sterilize. For plates, weigh 20 g of agar into a bottle, add 1 liter of the required liquid SC medium, place a stir bar in the bottom, and then sterilize by autoclaving. After autoclaving, place on a stirrer, and once cooled to approximately 55°, add 20 ml 0.2% (w/v) adenine (filter sterilized stock solution) and pour plates immediately. For two-hybrid selection plates (QDO), do not add adenine; all other media should contain adenine.

Preparation of Frozen Competent Yeast Cells

Pick several colonies from a freshly streaked plate of PJ69-4A and grow overnight in 100 ml of YPAD (autoclaved yeast extract 10 g/liter, bacto peptone 20 g/liter, sucrose 20 g/liter, adenine 100 mg/liter). Measure the OD_{600} of overnight culture and dilute to 0.15 in fresh medium. Grow at 30° to an OD_{600} of 0.5–0.6, equivalent to 1.2–1.5 $\times$ 10^7 cells. Harvest cells by centrifugation at 2000$\times g$, for 2 min at room temperature. Wash and resuspend the cells with one-half the culture volume of water and spin as before. Resuspend the pellet in one-eighth culture volume of LiSorb (filter sterilized 100 mM lithium acetate, 10 mM Tris–Cl, 1 mM EDTA, 1 M sorbitol adjusted to pH 8.0), incubate for 5 min, and spin as before. Remove supernatant and spin again to remove residual supernatant. Resuspend the cell pellet in 600 μl LiSorb per 100 ml of culture volume and add 10 μl herring sperm DNA per 100 μl yeast solution (herring sperm DNA 10 mg/ml stock [Promega] heated to 95° for 10 min before use, then kept on ice). Mix well by vortexing. Freeze cells at −80° in 100-μl aliquots; do not shock freeze, simply place a box containing cells in the freezer. Competent cells can be kept at −80° for several months.

Small-Scale Transformations for Directed Two-Hybrid Tests

Thaw cells by leaving them on the bench. Use 10 μl for each transformation; all pipetting is done on the bench at room temperature. Add 1 μl of both the bait (pGBT9) and the prey (pACT2) plasmids; we typically use 100–200 ng/μl miniprep DNA for this, and it works reproduceably with high efficiency. Add 150 μl LiPEG (filter sterilized 100 mM lithium acetate, 10 mM Tris–Cl, 1 mM EDTA, 40% [w/vl] PEG-3350 adjusted to pH 8.0) and mix well by vortexing. Incubate for 20 min at room temperature, then add 17.5 μl dimethyl sulfoxide (DMSO), and mix well by vortexing. Heat shock the cells for 15 min at 42° in a waterbath, not a heating block. Centrifuge for 2 min at 2000×g in a microcentrifuge at room temperature to pellet the cells. Carefully remove as much of the supernatant as possible. Resuspend the cell pellet in 500 μl water and plate 100 μl on appropriate selective plates, in this case synthetic complete medium lacking leucine and tryptophan (SC-Leu/Trp). Grow cells at 30° for 3 days. After this time, carefully inspect the plates. All should have evenly sized off-white/pink colonies. Large white colonies on a background of smaller pinkish colonies are almost certainly mutants. Do not pick these! Using a sterile tip or toothpick, pick up a small amount of a colony and make a small streak on SC-Leu/Trp (go back and forth three or four times), then using the same tip on QDO (synthetic complete yeast medium lacking leucine, tryptophan, adenine, and histidine). Do this for five colonies from each transformation. Grow cells at 30° for 2–3 days. After this time, carefully inspect the plates. Combinations that give rise to a two-hybrid interaction will grow on both SC-LW and QDO. Strong interactors give rise to white colonies on QDO, while weak interactors are pink due to incomplete activation of the adenine biosynthesis reporter gene. True positives are those that show bait-dependent growth and do not grow with empty bait vector control.

An example of the use of directed yeast two-hybrid assays to map the interaction sites for Rab2 and GRASP55 on golgin45 is shown in Fig. 2. This figure also illustrates the use of guanine nucleotide state-specific mutations to identify which form of the Rab GTPase interacts with the target protein. In this case, GDP-locked Rab2S20N does not interact with golgin45, while wild-type Rab2 and GTP-locked Rab2Q65L do interact, consistent with the idea that golgin45 is an effect or protein binding to the GTP form of Rab2. Deletion constructs of golgin45 map the interaction site for Rab2Q65L to amino acids 123–282 within the predicted coiled-coil region.

Rab-Effector Binding Assays

For binding assays wild-type Rabs were produced as N-terminal hexahistidine-GST-tagged fusion proteins in *Escherichia coli*, since this allows easy purification using the histidine-tag and subsequent immobilization on GST-Sepharose for binding assays. The NTA-agarose (Qiagen) used for purifying the Rabs is not suitable for binding assays, since it gives a high background binding when using mammalian cell extracts, and the presence of EDTA in a number of the binding assay buffers causes release of the Rab from the beads. The protocol described in the following subsections is modified from a previously published method (Christoforidis and Zerial, 2000) for the isolation of Rab interaction partners from Golgi membranes (Short *et al.*, 2001).

Preparation of Immobilized Rab Beads

Incubate 0.5 mg each GST-Rab protein with 50 μl packed glutathione-Sepharose (Amersham) in 1–1.5 ml total volume for 60 min at 4°. Pellet the beads by centrifugation at 2000×g for 1 min at room temperature; keep the unbound supernatant to check binding to beads. Wash the beads 3 × 500 μl with nucleotide exchange buffer (NE100: 20 mM HEPES-NaOH, pH 7.5, 100 mM NaCl, 10 mM EDTA, 0.1% [v/v] Triton X-100). Resuspend the beads in 200 μl of nucleotide loading buffer (NL100: 20 mM HEPES-NaOH, pH 7.5, 100 mM NaCl, 5 mM MgCl$_2$, 0.1% [v/v] Triton X-100) and then add 20 μl of either 10 mM GDP or GTPγS from 100 mM stocks prepared in 20 mM HEPES-NaOH, pH 7.5. Add 200 μl Golgi extract (see below) and incubate at 4° for 60 min on a tube rotator. Pellet the beads and

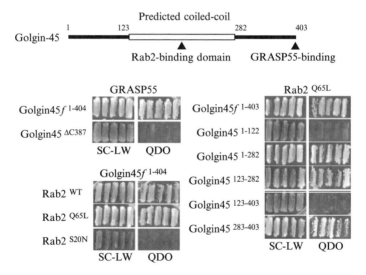

Fig. 2. Mapping the network of interactions between GRASP55, golgin45, and Rab2 using yeast two-hybrid assays. A schematic of golgin45 indicates the central coiled domain containing the binding site for the GTP form of Rab2, and the C-terminal GRASP55-binding site. Full-length GRASP55 in pFBT9 was tested against full-length golgin45 or a mutant lacking the last six amino acids (ΔC387) in pACT2. Rab2 wild type, Q65L, and S20N in pFBT9 were tested against full-length golgin45 or a series of deletion constructs in pACT2.

keep the supernatant for later analysis—20 μl is enough for one minigel. Wash the beads 3 × 500 μl with binding buffer (NB100: 20 mM HEPES-NaOH, pH 7.5, 100 mM NaCl, 5 mM MgCl$_2$, 0.1% [v/v] Triton X-100). Elute the bound effect proteins by the addition of elution buffer (NE200: 20 mM HEPES-NaOH, pH 7.5, 200 mM NaCl, 20 mM EDTA, 0.1% [v/v] Triton X-100) rotating at 4° for 5–10 min. Pellet the beads by centrifugation at 2000×g for 1 min and then transfer the supernatant to a fresh tube, avoiding the beads. An optional step at this point may sometimes be necessary to remove the Rab GST-fusion proteins from the eluate. Add 50 μl of packed glutathione Sepharose to the eluate, and then incubate for 5–10 min on ice inverting to mix. Pellet the beads by centrifugation at 2000×g for 1 min and transfer the supernatant to a fresh tube, avoiding the beads. Repeat this procedure three times.

To precipitate the eluted proteins, add 7.5 μl 10% (w/v) deoxycholate stock to the final eluate, vortex, add 300 μl 55% (w/v) TCA, vortex, and then place on ice for 40 min. Recover the precipitated proteins by centrifugation at 20,000×g for 15 min, 4°. Aspirate the supernatant without disturbing the protein pellet. Add 1 ml of −20° acetone, vortex briefly, then centrifuge at 20,000×g for 5 min, 4°. The acetone washing steps help to remove the residual TCA, which otherwise causes problems for the solubilization of the precipitated protein. Remove the supernatant and repeat the acetone wash step. Add 5 μl 1 M Tris–HCl pH 8.0 followed by 20 μl of 1.5× sample buffer. Shake at room temperature until the pellets have dissolved, then heat at 99° for 3 min and analyze on 7.5% or 10% minigels.

Preparation of Golgi Extracts for Rab-Binding Assays

Add protease inhibitors to 10 ml NL200 buffer (0.2 mM Pefabloc & Complete protease inhibitor cocktail [Roche Diagnostics]), then store on ice. Take a tube of Golgi membranes containing approximately 1 mg membranes, from the −80° freezer, thaw rapidly at 37°, then place on ice. Add water to give 1 ml total volume, then pellet the membranes by centrifugation at 20,000×g for 10 min, 4°. This step removes the high sucrose storage buffer. Remove the supernatant and add 1200 μl NL200 buffer with protease inhibitors, and then resuspend on ice with a P1000/ blue tip until all clumps are gone. Add 12 μl of 10% (v/v) Triton X-100 and pipette to mix. Leave then Golgi membranes to extract for 15 min on ice. Remove any insoluble material by centrifugation at 20,000×g for 10 min, 4°. Transfer the supernatant (this is the Golgi extract) to a fresh tube. Adjust to 1200 μl. Keep an aliquot of extract for analysis later; 20 μl is enough for a single lane of a minigel.

Analysis of Golgi Structure and Protein Transport in Golgin45-Depleted Cells

Plate HeLa cells in DME containing 10% calf serum (Invitrogen/Life Technologies) on glass coverslips at a density of 50,000 cells per well of a six-well plate and leave to attach for 24 h at 37° and 5% CO_2. RNA interference can be performed using small interfering RNA (siRNA) duplexes (Elbashir *et al.*, 2001); golgin-45 and lamin A target sequences are 5'-AATCCGAGGAGCAGGAGATGGAA-3' and 5'-AACTGGACT-TCCAGAAGAACA-3', respectively. To transfect HeLa cells, mix 3 μl preannealed siRNA duplexes (Dharmacon Research, Inc. or Qiagen) with 200 μl Optimem (Invitrogen/Life Technologies) serum-free medium, and then add 3 μl Oligifectamine (Invitrogen) and mix by pipetting. After 25 min at room temperature, add the entire mixture to one well of a six-well plate. To investigate the effects of golgin45 depletion on Golgi structure cells, fix and stain cells with antibodies to markers for different Golgi cisternae after 0, 24, 48, and 72 h of transfection (Short *et al.*, 2001).

Vesicular stomatitis virus glycoprotein (VSV-G) ts045 protein transport assays are carried out using an adaptation of a published protocol (Seemann *et al.*, 2000). After 36–48 h of siRNA, transfect the cells with a plasmid encoding green fluorescent protein (GFP)-tagged VSV-G protein. Mix 3 μl Fugene-6 (Roche Diagnostics) with 200 μl Optimem, add 1 μg plasmid DNA, and leave for 15 min at room temperature. Add this mixture to one well of a six-well plate; leave the cells for 2 h at 37° and then for 12 h at 39.5° (Toomre *et al.*, 1999). Wash the cells with ice-cold PBS and then incubate the six-well plates on ice for 1 h to promote VSV-G protein folding (Scales *et al.*, 1997). To start the transport assay, add fresh growth prewarmed medium to 31.5°. After the required chase period, fix cells with 3% (w/v) paraformaldehyde in PBS. Cell surface VSV-G is detected with a monoclonal antibody to the VSV-G lumenal domain and a donkey anti-mouse secondary coupled to CY3 (Jackson Labs), and total VSV-G by GFP fluorescence. We routinely collect images using a Zeiss Axioskop-2 with 63× Plan Apochromat oil immersion objective numerical aperture 1.4, and a 1300 × 1030 pixel cooled-CCD camera (Princeton Instruments), using Metaview software (Universal Imaging Corp.). The ratio of surface to total measured fluorescence can be used to calculate the extent of VSV-G protein transport (Seemann *et al.*, 2000).

References

Allan, B. B., Moyer, B. D., and Balch, W. E. (2000). *Science* **289,** 444–448.
Barr, F. A., Puype, M., Vandekerckhove, J., and Warren, G. (1997). *Cell* **91,** 253–262.

Barr, F. A., Nakamura, N., and Warren, G. (1998). *EMBO J.* **17,** 3258–3268.

Barr, F. A., Preisinger, C., Kopajtich, R., and Komer, R. (2001). *J. Cell Biol.* **155,** 885–891.

Bretz, R., and Staubli, W. (1977). *Eur. J. Biochem.* **77,** 191–192.

Christoforidis, S., and Zerial, M. (2000). *Methods* **20,** 403–410.

Elbashir, S. M., Harborth, J., Lendeckel, W., Yalcin, A., Weber, K., and Tuschl, T. (2001). *Nature* **411,** 494–498.

James, P., Halladay, J., and Craig, E. A. (1996). *Genetics* **144,** 1425–1436.

Perkins, D. N., Pappin, D. J., Creasy, D. M., and Cottrell, J. S. (1999). *Electrophoresis* **20,** 3551–3567.

Scales, S. J., Pepperkok, R., and Kreis, T. (1997). *Cell* **90,** 1137–1148.

Seemann, J., Jokitalo, E. J., and Warren, G. (2000). *Mol. Biol. Cell* **11,** 635–645.

Short, B., Preisinger, C., Komer, R., Kopajtich, R., Byron, O., and Barr, F. A. (2001). *J. Cell Biol.* **155,** 877–883.

Toomre, D., Keller, P., White, J., Olivo, J. C., and Simons, K. (1999). *J. Cell Sci.* **112**(Pt. 1), 21–33.

Weide, T., Bayer, M., Koster, M., Siebrasse, J. P., Peters, R., and Bamekow, A. (2001). *EMBO Rep.* **2,** 336–341.

Wilm, M., Shevchenko, A., Houthaeve, T., Breit, S., Schweigerer, L., Fotsis, T., and Mann, M. (1996). *Nature* **379,** 466–469.

[35] Purification and Properties of Rabconnectin-3

By TOSHIAKI SAKISAKA and YOSHIMI TAKAI

Abstract

Rab3A, a member of the Rab3 small GTP-binding protein (G protein) family, regulates Ca^{2+}-dependent exocytosis of neurotransmitter. The cyclical activation and inactivation of Rab3A are essential for the Rab3A action in exocytosis. GDP-Rab3A is activated to GTP-Rab3A by Rab3 GDP/GTP exchange protein (Rab3 GEP), and GTP-Rab3A is inactivated to GDP-Rab3A by Rab3 GTPase-activating protein (Rab3 GAP). We) have found a novel protein, named rabconnectin-3, that is coimmunoprecipitated with Rab3 GEP or GAP from the crude synaptic vesicle fraction of rat brain. Rabconnectin-3 constitutes a subunit structure consisting of α and β subunits and localizes at synaptic vesicles. Overexpression of the C-terminal fragment of rabconnectin-3α inhibits Ca^{2+}-dependent exocytosis from PC12 cells. We describe the purification method for native rabconnectin-3α and -3β from rat brain and the functional properties of rabconnectin-3α in Ca^{2+}-dependent exocytosis by use of human growth hormone coexpression assay system of PC12 cells.

METHODS IN ENZYMOLOGY, VOL. 403
0076-6879/05 $35.00
DOI: 10.1016/S0076-6879(05)03035-1

Introduction

Rabconnectin-3α and -3β are proteins interacting with both Rab3 GEP and GAP, which are regulators of the Rab3 family consisting of Rab3A, -3B, -3C, and -3D (Kawabe *et al.*, 2003; Nagano *et al.*, 2002). The Rab3 family members are regulated by Rab GDI in addition to Rab3 GEP and GAP (Fukui *et al.*, 1997; Nagano *et al.*, 1998; Sasaki *et al.*, 1990; Wada *et al.*, 1997). Rab3 GEP and GAP are specific for the Rab3 family members, whereas Rab GDI is active on all the Rab family members. Of the Rab3 family members, the function and the mode of action of Rab3A have been investigated most extensively and it has been shown to be involved in Ca^{2+}-dependent exocytosis, particularly neurotransmitter release. The cyclical activation and inactivation of Rab3A by the action of the three regulators are essential for the action of Rab3A neurotransmitter release. A current model for the mode of action of these regulators is as follows (Takai *et al.*, 1996): (1) GDP-Rab3A forms an inactive complex with Rab GDI and stays in the cytosol of nerve terminals. (2) GDP-Rab3A released from Rab GDI is converted to GTP-Rab3A by Rab3 GEP with the help of another unidentified molecule, such as GDI displacement factor (GDF): Yip3/PRA1 for Rab5 and -9 (Pfeffer and Aivazian, 2004; Sivars *et al.*, 2003). (3) GTP-Rab3A binds effector molecules, rabphilin-3 (Shirataki *et al.*, 1993) and Rim (Wang *et al.*, 1997), that localize at synaptic vesicles and the active zone, respectively. These complexes facilitate translocation and docking of the synaptic vesicles to the active zone. (4) GTP-Rab3A is converted to GDP-Rab3A by Rab3 GAP when the vesicles fuse with the presynaptic membrane. (5) GDP-Rab3A is associated with Rab GDI and retrieved from the membrane to the cytosol with help from Rab recycling factor (RRF): Hsp90 chaperon complex for Rab3A (An *et al.*, 2003; Sakisaka *et al.*, 2002). Thus, Rab3 GEP and GAP are presumably recruited to the vesicles when they function, but their mechanisms remain unknown.

A novel protein, which is coimmunoprecipitated with Rab3 GEP or GAP from the crude synaptic vesicle fraction of rat brain, has been isolated and named rabconnectin-3 (Kawabe *et al.*, 2003; Nagano *et al.*, 2002). Rabconnectin-3 constitutes a subunit structure consisting of α and β subunits. Human rabconnectin-3α consists of 3036 amino acids (aa) and shows a calculated M_r of 339,753. It has 12 WD domains. Human rabconnectin-3β consists of 1490 aa and shows a calculated M_r of 163,808. It has 7 WD domains. We have attempted to make recombinant proteins of full-length rabconnectin-3α in *Escherichia coli*, Sf9 cells, and COS7 cells, but we have not yet succeeded in preparing any recombinant proteins. Therefore, we describe the purification method for native rabconnectin-3α and -3β from rat brain and the functional properties of rabconnectin-3α in Ca^{2+}-dependent exocytosis.

Materials

(p-Amidinophenyl)methanesulfonyl fluoride (APMSF), leupeptin, and Triton X-100 were purchased from Wako Pure Chemicals (Osaka, Japan). Bovine serum albumin (BSA) (fraction V) is from Sigma Chemical Co. (St. Louis, MO). 3-[(3-Cholamidopropyl) dimethylammonio]-1-propane-sulfonic acid (CHAPS) is from Dojindo Laboratories (Kumamoto, Japan). Mono Q PC1.6/5 is from Amersham-Pharmacia Biotech. Dulbecco's modified Eagle's medium (DMEM) is from Nakalai Tesque (Kyoto, Japan). Fetal calf serum (FCS), OPTI-MEM, and LipofectAMINE reagent are from Invitrogen (Carlsbad, CA). All other chemicals are of reagent grade.

Plasmid for expression of various fragments of rabconnectin-3α in PC12 cells is constructed as follows. The cDNA fragment encoding rat rabconnectin-3α-fragment 1 (frg.1) (aa 1–658), -frg.2 (aa 647–1416), -frg.3 (aa 1390–2112), -frg.4 (aa 2082–2511), or -frg.5 (aa 2495–3036) is inserted into pCMV-myc to construct pCMV-myc-rabconnectin-3α-frg.1, -frg.2, -frg.3, -frg.4, or -frg.5, respectively.

Methods

Purification of Rabconnectin-3α and -3β from Rat Brain

The steps used in the purification of rabconnectin-3α and -3β from rat brain are as follows: (1) preparation of the crude synaptic vesicle fraction from rat brain; (2) sucrose density gradient ultracentrifugation; and (3) immunoprecipitation with an anti-rabconnectin-3α rabbit polyclonal antibody (pAb). All the purification procedures are carried out at 0–4°.

Buffers for Purification of Rabconnectin-3α and -3β from Rat Brain
Buffer A: 0.32 M sucrose, 1 mM NaHCO$_3$, 1 mM MgCl$_2$, 0.5 mM CaCl$_2$, and 1 μM APMSF
Buffer B: 0.32 M sucrose, 1 mM NaHCO$_3$, and 10 μg/ml leupeptin
Buffer C: 6 mM Tris–HCl, pH 8.0, and 10 μg/ml leupeptin
Buffer D: 20 mM Tris–HCl, pH 7.5, 150 mM NaCl, 1 mM EDTA, 1 mM dithiothreitol (DTT), and 1% Triton X-100

Preparation of Crude Synaptic Vesicle Fraction from Rat Brain. The crude synaptic vesicle fraction is prepared as follows (Mizoguchi *et al.*, 1989): cerebra are rapidly removed from 80 rats after decapitation. Homogenization is performed with 12 up-and-down strokes of a Potter-Elvehjem Teflon-glass homogenizer in Buffer A (cerebral tissues 10 g wet weight per 40 ml of Buffer A). The homogenates are combined, diluted to 800 ml with Buffer A, and filtrated through four layers of gauze. The homogenate is centrifuged at $1400 \times g$ for 10 min. The pellet is resuspended with three strokes of the homogenizer in 700 ml of the same buffer and centrifuged at $710 \times g$ for 10 min. The resultant pellet, designated as P1 fraction, contains nuclei and cell debris. The supernatants from the two steps of centrifugation are combined and centrifuged at $13,800 \times g$ for 10 min. The pellet is resuspended and rehomogenized in 700 ml of the same buffer and then centrifuged again at $13,800 \times g$ for 10 min. The resultant pellet, designated as P2 fraction, contains myelin, mitochondria, and synaptosomes. This pellet is resuspended with three strokes of the homogenizer in 200 ml of Buffer B. The suspension is diluted with 900 ml of Buffer C, stirred for 45 min, and centrifuged at $32,800 \times g$ for 20 min. The supernatant is further centrifuged at $78,000 \times g$ for 120 min. The final pellet is pooled as the crude synaptic vesicle fraction. This crude synaptic vesicle fraction can be stored at $-80°$ for at least 3 months.

Sucrose Density Gradient Ultracentrifugation. The proteins extracted from the crude synaptic vesicle fraction (500 μl, 1.25 mg of protein) with Buffer D are layered onto a 10.5 ml linear 10–50% sucrose density gradient in Buffer D and subjected to centrifugation at $207,000 \times g$ for 18 h. Fractions of 500 μl each are collected. An aliquot of each fraction is subjected to sodium dodecyl sulfate–polyacrylamide gel electrophoresis (SDS-PAGE), followed by immunoblotting with the anti-rabconnectin-3α and -3β pAbs. Thyroglobulin (19S), catalase (11.3S), γ-globulin (7.4 S), and bovine serum albumen (BSA) (4.6S) are used as *S*-value markers on a parallel gradient. Rabconnectin-3α and -3β are mostly recovered in fractions 7–10, which are numbered from the bottom (around 19S) (Kawabe *et al.*, 2003).

Immunoprecipitation. The fractions, which are reacted with the anti-rabconnectin-3α pAb, are collected and diluted with three times volume of Buffer D to dilute sucrose concentration. The collected fraction is incubated at 4° overnight with the anti-rabconnectin-3α pAb immobilized on protein A Sepharose beads (20 μl of wet volume). After the beads are extensively washed with Buffer D, the bound proteins are eluted by boiling the beads in an SDS sample buffer (60 mM Tris–HCl, pH 6.7, 3% SDS, 2% [v/v] 2-mercaptoethanol, and 5% glycerol). The sample is subjected to SDS-PAGE, followed by protein staining with silver or by immunoblotting with the anti-rabconnectin-3α and -3β pAbs (Kawabe *et al.*, 2003).

The sum of calculated molecular masses of rabconnectin-3α and -3β is about 480 kDa. Rabconnectin-3α and -3β are mostly recovered in the fractions whose molecular mass is about 570 kDa estimated by sucrose density gradient ultracentrifugation. The molecular mass of the peak is larger than the sum of their molecular masses. One possible explanation is that their protein conformation may change when they form a complex.

Another explanation is that the complex may contain other unidentified molecules. When the fractions are immunoprecipitated with rabconnectin-3α pAb, rabconnectin-3α and -3β are coimmunoprecipitated at a molar ratio of about 1:1. Taken together, it is likely that rabconnectin-3 constitutes a subunit structure consisting of the α and β subunits.

Involvement of Rabconnectin-3α in Ca^{2+}-Dependent Exocytosis

The activity of rabconnectin-3α to regulate Ca^{2+}-dependent exocytosis is assayed by measuring growth hormone (GH) release from PC12 cells cotransfected with XGH5 encoding human GH and pCMV-myc-rabconnectin-3α-frg.1, -frg.2, -frg.3, -frg.4, or -frg.5. In this assay system, expressed GH is stored in dense core vesicles of PC12 cells and released in response to various agonists in an extracellular Ca^{2+}-dependent manner (Schweitzer and Kelly, 1985; Wick *et al.*, 1993).

Buffers for the GH Release Assay

Low K^+ solution: 140 mM NaCl, 4.7 mM KCl, 2.5 mM CaCl$_2$, 1.2 mM MgSO$_4$, 1.2 mM KH$_2$PO$_4$, 20 mM HEPES/NaOH, pH 7.4, and 11 mM glucose

High K^+ solution: 85 mM NaCl, 60 mM KCl, 2.5 mM CaCl$_2$, 1.2 mM MgSO$_4$, 1.2 mM KH$_2$PO$_4$, 20 mM HEPES/NaOH, pH 7.4, and 11 mM glucose

Cell Culture. Stock cultures of PC12 cells are maintained at 37° in a humidified atmosphere of 10% CO$_2$ and 90% air (v/v) in DMEM containing 10% FCS, 5% horse serum (HS), penicillin (100 U/ml), and streptomycin (100 μg/ml).

Transfection. PC12 cells are plated at a density of 5×10^5 cells per 35-mm dish and incubated for 18–24 h. The cells are then cotransfected with 2 μg of pXGH and 2 μg of pCMV-myc-rabconnectin-3α-frg.1, -frg.2, -frg.3, -frg.4, or -frg.5 using 15 μl of LipofectAMINE reagent and 1 ml of OPTI-MEM. Six hours after the transfection, the cells are washed with DMEM and incubated in 2 ml of DMEM containing 10% FCS and 5% HS.

GH Release Experiments. GH release experiments are performed 48 h after the transfection. PC12 cells are washed with a low K^+ solution and incubated at 37° for 10 min with 2 ml of a high K^+ solution or a low K^+ solution. After the high K^+ solution or the low K^+ solution is removed, the cells are lysed in 2 ml of the high K^+ solution containing 0.5% Triton X-100 on ice. The amounts of the GH released into the high K^+ solution or the low K^+ solution and retained in the cell lysate are measured using an enzyme–linked immunosorbent assay (ELISA) assay according to the manufacturer's instructions (Roche).

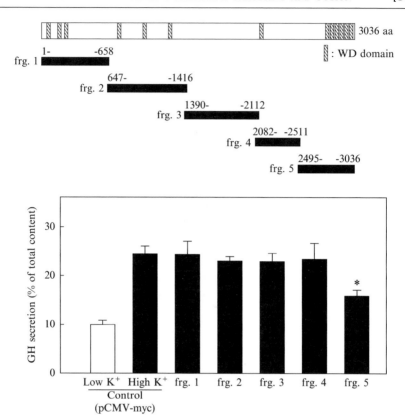

FIG. 1. Effect of overexpression of various fragments of rabconnectin-3α on release of expressed GH from PC12 cells. Data are expressed as the average percentage released of the total GH stores. The values are mean ±ISE of three independent experiments. $*p < 0.01$ versus GH secretion from cells transfected with control plasmid, pCMV-myc.

In these experiments, the high K^+-induced GH release is inhibited in PC12 cells overexpressing rabconnectin-3α-frg.5, not but other fragments (Fig. 1).

Comments

Of the purification steps of rabconnectin-3α and -3β, sucrose density gradient ultracentrifugation is the most important step to obtain a large amount of rabconnectin-3α and -3β. There is less dilution in this system than in gel filtration. In addition, rabconnectin-3α and -3β are very sticky in the absence of detergent. Therefore, rabconnectin-3α and -3β should be stored at $-80°$ with detergent (1% Triton X-100).

References

An, Y., Shao, Y., Alory, C., Matteson, J., Sakisaka, T., Chen, W., Gibbs, R. A., Wilson, I. A., and Balch, W. E. (2003). Geranylgeranyl switching regulates GDI-Rab GTPase recycling. *Structure (Camb.)* **11,** 347–357.

Fukui, K., Sasaki, T., Imazumi, K., Matsuura, Y., Nakanishi, H., and Takai, Y. (1997). Isolation and characterization of a GTPase activating protein specific for the Rab3 subfamily of small G proteins. *J. Biol. Chem.* **272,** 4655–4658.

Kawabe, H., Sakisaka, T., Yasumi, M., Shingai, T., Izumi, G., Nagano, F., Deguchi-Tawarada, M., Takeuchi, M., Nakanishi, H., and Takai, Y. (2003). A novel rabconnectin-3-binding protein that directly binds a GDP/GTP exchange protein for Rab3A small G protein implicated in Ca^{2+}-dependent exocytosis of neurotransmitter. *Genes Cells* **8,** 537–546.

Mizoguchi, A., Ueda, T., Ikeda, K., Shiku, H., Mizoguti, H., and Takai, Y. (1989). Localization and subcellular distribution of cellular ras gene products in rat brain. *Mol. Brain Res.* **5,** 31–44.

Nagano, F., Sasaki, T., Fukui, K., Asakura, T., Imazumi, K., and Takai, Y. (1998). Molecular cloning and characterization of the noncatalytic subunit of the Rab3 subfamily-specific GTPase-activating protein. *J. Biol. Chem.* **273,** 24781–24785.

Nagano, F., Kawabe, H., Nakanishi, H., Shinohara, M., Deguchi-Tawarada, M., Takeuchi, M., Sasaki, T., and Takai, Y. (2002). Rabconnectin-3, a novel protein that binds both GDP/GTP exchange protein and GTPase-activating protein for Rab3 small G protein family. *J. Biol. Chem.* **277,** 9629–9632.

Pfeffer, S., and Aivazian, D. (2004). Targeting Rab GTPases to distinct membrane compartments. *Nat. Rev. Mol. Cell. Biol.* **5,** 886–896.

Sakisaka, T., Meerlo, T., Matteson, J., Plutner, H., and Balch, W. E. (2002). Rab-alphaGDI activity is regulated by a Hsp90 chaperone complex. *EMBO J.* **21,** 6125–6135.

Sasaki, T., Kikuchi, A., Araki, S., Hata, Y., Isomura, M., Kuroda, S., and Takai, Y. (1990). Purification and characterization from bovine brain cytosol of a protein that inhibits the dissociation of GDP from and the subsequent binding of GTP to smg p25A, a ras p21-like GTP-binding protein. *J. Biol. Chem.* **265,** 2333–2337.

Schweitzer, E. S., and Kelly, R. B. (1985). Selective packaging of human growth hormone into synaptic vesicles in a rat neuronal (PC12) cell line. *J. Cell Biol.* **101,** 667–676.

Shirataki, H., Kaibuchi, K., Sakoda, T., Kishida, S., Yamaguchi, T., Wada, K., Miyazaki, M., and Takai, Y. (1993). Rabphilin-3A, a putative target protein for smg p25A/rab3A p25 small GTP-binding protein related to synaptotagmin. *Mol. Cell. Biol.* **13,** 2061–2068.

Sivars, U., Aivazian, D., and Pfeffer, S. R. (2003). Yip3 catalyses the dissociation of endosomal Rab-GDI complexes. *Nature* **425,** 856–859.

Takai, Y., Sasaki, T., Shirataki, H., and Nakanishi, H. (1996). Rab3A small GTP-binding protein in Ca^{2+}-dependent exocytosis. *Genes Cells* **1,** 615–632.

Wada, M., Nakanishi, H., Satoh, A., Hirano, H., Obaishi, H., Matsuura, Y., and Takai, Y. (1997). Isolation and characterization of a GDP/GTP exchange protein specific for the Rab3 subfamily small G proteins. *J. Biol. Chem.* **272,** 3875–3878.

Wang, Y., Okamoto, M., Schmitz, F., Hofmann, K., and Sudhof, T. C. (1997). Rim is a putative Rab3 effector in regulating synaptic-vesicle fusion. *Nature* **388,** 593–598.

Wick, P. F., Senter, R. A., Parsels, L. A., Uhler, M. D., and Holz, R. W. (1993). Transient transfection studies of secretion in bovine chromaffin cells and PC12 cells. Generation of kainate-sensitive chromaffin cells. *J. Biol. Chem.* **268,** 10983–10989.

[36]   Physical and Functional Interaction of Noc2/Rab3 in Exocytosis

By TADAO SHIBASAKI and SUSUMU SEINO

Abstract

Rab, monomeric small Ras-like GTPase, regulates intracellular membrane trafficking in eukaryotic cells. Rab3 is involved in the exocytotic process in a variety of secretory cells including neuronal, neuroendocrine, endocrine, and exocrine cells. Noc2, originally identified as a molecule homologous to Rabphilin-3, is a putative effector of Rab3. Noc2 interacts with the active (GTP-bound) form of Rab3 and regulates hormone secretion in neuroendocrine and endocrine cells and enzyme release in exocrine cells. This chapter describes two kinds of interaction assay by which the association of Noc2 with Rab3 is analyzed: a yeast two-hybrid assay to detect the interaction of Noc2 with the active form of Rab3 in intact cells and a pull-down assay using GST-fused Noc2 protein to ascertain the physical interaction of Noc2 and Rab3 *in vitro*. Thus, the Noc2 knockout mouse is a useful model for studying the functional consequences of disruption of the interaction.

Introduction

Rab belongs to the Ras superfamily of small GTP-binding protein and regulates different steps of membrane traffic within the cell (Pfeffer, 2001; Segev, 2001; Zerial and McBride, 2001). Rab3 is found on secretory granules in a variety of cell types and is involved in regulated exocytosis (Darchen and Goud, 2000; Sudhof, 2004). In mammals, the Rab3 subfamily includes four structurally related isoforms (Rab3A, B, C, and D). The GTP/GDP exchange cycle of Rab3 is required in regulated exocytosis (Takai *et al.*, 2001). The interaction of the GTP-bound form of Rab3 with its effector proteins plays an important role in the exocytotic process (Darchen and Goud, 2000; Sudhof, 2004; Takai *et al.*, 2001). Rabphilin-3 (Shirataki *et al.*, 1993), Noc2 (Kotake *et al.*, 1997), Rims (Rim1 and Rim2) (Ozaki *et al.*, 2000; Wang *et al.*, 1997, 2000), and granuphilin (Wang *et al.*, 1999) have been identified as putative Rab3 effectors. Noc2, Rim2, and granuphilin, all of which are expressed in neuroendocrine and endocrine cells, are involved in hormone secretion (Kotake *et al.*, 1997; Ozaki *et al.*, 2000; Wang *et al.*, 1999).

METHODS IN ENZYMOLOGY, VOL. 403
0076-6879/05 $35.00
DOI: 10.1016/S0076-6879(05)03036-3

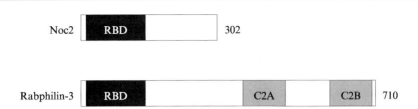

FIG. 1. Comparison of functional domains of Noc2 and Rabphilin-3. Both Noc2 and Rabphilin-3 share a conserved Rab-binding domain (RBD) in the amino-terminal region. Rabphilin-3 also has two C2-domains (C2A and C2B) in the carboxyl-terminal region.

Noc2 (no C2-domain) is a protein of 302 amino acids that is expressed in neuroendocrine and endocrine tissue and their derived cell lines such as pituitary and pancreatic β cells (Kotake *et al.*, 1997). Noc2 has 77.9% similarity to the amino-terminal Rab-binding domain of Rabphilin-3 but lacks the two C2 domains that are present in the carboxyl-terminal half of Rabphilin-3 (Fig. 1) (Kotake *et al.*, 1997). We found in Noc2 knockout (Noc2$^{-/-}$) mice that insulin secretion is markedly impaired under stressful conditions (Matsumoto *et al.*, 2004) but is restored by inhibition of pertussis toxin (PTX)-sensitive trimeric GTP-binding protein $G_{i/o}$, which blocks insulin secretion in pancreatic β cells. Noc2$^{-/-}$ mice also exhibit marked accumulation of secretory granules and dysfunction of enzyme release in pancreatic exocrine cells (Matsumoto *et al.*, 2004). In epithelial cells, Noc2 is proposed to control cell surface transport of membrane protein vesicular stomatitis virus glycoprotein (VSV-G) via its interaction with the active form of Rab3B (Manabe, 2004). In addition to Rab3, Noc2 interacts with Rab27 and Rab8 (Coppola *et al.*, 2002; Fukuda, 2003), both of which are involved in exocytosis. Thus, formation of the Noc2/Rab3 complex is required in normal regulated exocytosis in both endocrine and exocrine cells.

Materials

Phenylmethylsulfonyl fluoride (PMSF), glutathione-agarose beads, reduced glutathione, anti-FLAG M2 antibody, GDPβS, GTPγS, Dulbecco's modified Eagle's medium (DMEM), dithiothreitol (DTT), 3-[(3-cholamidopropyl) dimethylammonio]propanesulfonic acid (CHAPS), polyethylene glycol (PEG), 2-mercaptoethanol, and pFLAG-CMV2 were purchased from Sigma (Madison, WI). The ECL detection kit and pGEX4T are from Amersham Biosciences (Uppsala, Sweden). Agar, lithium acetate, *o*-nitrophenyl-β-D-galactopyranoside (ONPG), dimethyl sulfoxide (DMSO), ethylenediamine-*N,N, N',N'*-tetraacetic acid (EDTA),

2-[4-(2-hydroxyethyl)-1-piperazinyl]ethanesulfonic acid (HEPES), and *N*-tris(hydroxymethyl) methyl-2-aminoethanesulfonic acid (Tris) are from Nacalai Tesque (Kyoto, Japan). Isopropylthiogalactoside (IPTG) is from Takara Bio Inc. (Otsu, Japan). Sodium pyruvate, Opti-MEM, Lipofect-AMINE, PLUS reagent, and RPMI medium are from Invitrogen Corp. (Carlsbad, CA). Polyvinylidene difluoride (PVDF) is from Millipore Corp. (Bedford, MA). Yeast extract, Bacto Peptone, and yeast nitrogen base without amino acid are from Difco (Detroit, MI). The adenoviral expression system is from Stratagene (La Jolla, CA). The insulin radioimmunoassay kit is from Eiken Chemical (Tokyo, Japan).

Assay for Binding of Noc2 to Rab3A, B, C, and D Using Yeast Two-Hybrid System

Construct

Wild-type Noc2 cDNA (Kotake, 1997) is subcloned into expression vector pVP16. Noc2 is expressed as a VP16 LexA-activating domain-fused protein in yeast. pVP16 carries a gene involved in leucine synthesis. Wild-type Rab3A, Rab3A(Q81L), wild-type Rab3B, Rab3B(Q81L), wild-type Rab3C, Rab3C(Q81L), wild-type Rab3D, and Rab3D(Q81L) cDNAs are subcloned into expression plasmid pBTM116. Rab3 is expressed as a LexA DNA-binding domain-fused protein in yeast. pBTM116 carries a gene involved in tryptophan synthesis. In Rab3A(Q81L), Rab3B(Q81L), Rab3C(Q81L), and Rab3D(Q81L) mutants, glutamine (amino acid residue 81) is replaced with leucine. These mutants are dominant active Rab3 isoforms, which have low GTPase activity and should stabilize the GTP-bound form (Brondyk *et al.*, 1993).

Media

YPAD medium contains 10 g of yeast extract, 20 g of Bacto-Peptone, 2% (w/v) glucose, and 20 g of agar for plates, per 1 liter. Yc-Leu-Trp media contains 1.2 g of Bacto-yeast nitrogen base without amino acids, $(NH_4)_2SO_4$, 10 g of succinic acid, 6 g of NaOH, 100 ml of 10 $\times$ dropout mix, and 20 g of agar for plates, per 1 liter; 10 $\times$ dropout mix contains 1 g of adenine, 1 g of arginine, 1 g of cysteine, 1 g of leucine, 1 g of lysine, 1 g of threonine, 1 g of tryptophan, 1 g of uracil, 0.5 g of asparatic acid, 0.5 g of histidine, 0.5 g of isoleucine, 0.5 g of methionine, 0.5 g of phenylalanine, 0.5 g of serine, 0.5 g of tyrosine, and 0.5 g of valine without the appropriate supplement, per 1 liter.

Solution

> TE: 10 mM Tris–HCl (pH 7.5), 1 mM EDTA
> TE/lithium acetate: 10 mM Tris–HCl (pH 7.5), 1 mM EDTA, and 100 mM lithium acetate
> PEG/TE/lithium acetate: 40% PEG (avg. 3350), 10 mM Tris–HCl (pH 7.5), 1 mM EDTA, and 100 mM lithium acetate
> Z-buffer: 62 mM Na_2HPO_4, 42.8 mM NaH_2PO_4, 10.1 mM KCl, and 1.0 mM $MgSO_4$ (adjusted to pH 7.0)
> Z-buffer/2-mercaptoethanol solution: 1 ml of Z-buffer and 2.7 μl of 2-mercaptoethanol (prepare daily)
> 10 mg/ml sheared and denatured salmon sperm DNA
> ONPG solution: 4 mg of ONPG in 1 ml of Z-buffer (prepare daily)
> Stop solution: 1 M Na_2CO_3

Yeast Transformation

Yeast strain L40 (*Mata, his3D200, trp1-901, Leu2–3, 112, ade2*) stored at $-80°$ is inoculated on a YPAD plate using an inoculating needle, and cultured for a few days at 30°. A single colony is inoculated into 5 ml of YPAD and grown overnight at 30° with shaking at 230 rpm. The overnight preculture is transferred into 50 ml of YPAD and grown further at 30° with shaking at 230 rpm. After the culture reached OD_{600} of 0.5~0.6, the cells are harvested by centrifugation at $1500 \times g$ for 5 min at room temperature. The pellet is washed with 40 ml of TE. The washed pellet is suspended with 2 ml of TE/lithium acetate and incubated for 10 min at room temperature. The cells obtained are competent for transformation. One hundred microliters of the cells is mixed with 0.5~2.0 μg of plasmid DNA (pBTM116 and pVP16) and 10 μl of 10 mg/ml salmon sperm DNA. The mixture is suspended with 700 μl of PEG/TE/lithium acetate and incubated for 30 min at 30° with shaking at 230 rpm. The cells are suspended with 88 μl of DMSO and subjected to heat shock treatment for 7 min at 42°. They are collected by centrifugation for 30 s. After removal of the supernatant, the cells are suspended with 100 μl of TE, plated on Yc-Leu-Trp medium, and cultured at 30° until colonies grow.

Measurement of β-Galactosidase Activity by Liquid Culture

A single colony is inoculated into 4 ml of Yc-Leu-Trp and grown overnight. Three milliliters of the overnight culture is transferred into a 50-ml tube containing 7 ml of YPAD medium and cultured until OD_{600} of 0.6~0.9 is reached. An aliquot (1.5 ml) of the culture is transferred into a

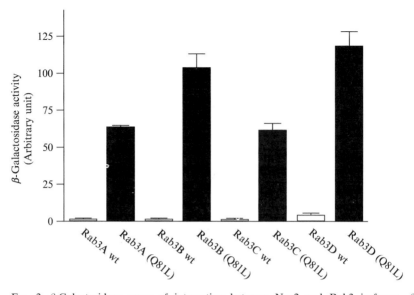

FIG. 2. β-Galactosidase assay of interaction between Noc2 and Rab3 isoforms. β-Galactosidase activity was measured with ONPG as described in the text. Noc2 bound to all dominant active mutants of Rab3 isoforms but not to wild type. Open columns, wild-type Rab3s (Rab3 wt); solid columns, Rab3 (Q81L) mutants. The values are mean ± SEM of two independent experiments ($n = 6$).

1.5-ml tube and centrifuged. The pellet is washed with 1.5 ml of Z-buffer and suspended with 600 μl of Z-buffer. Four hundred microliters of the suspension is used for calculation of β-galactosidase activity to adjust the cell density (OD_{600}) for each cell. One hundred microliters of the suspension is also frozen in liquid nitrogen for 10 s and quickly thawed at 30°. The lysate is mixed with 700 μl of Z-buffer/2-mercaptoethanol solution and 160 μl of ONPG solution (Fig. 2). The mixture is incubated for a period from 5 min to 24 h at 30° until a yellow color develops. The reaction is stopped by addition of 400 μl of 1 M Na$_2$CO$_3$. Cell debris is discarded by centrifugation. The supernatant is subjected to measurement of OD_{420}. β-Galactosidase is calculated using the following formula:

$$\beta\text{-galactosidase activity} = 1000 \times (OD_{420} \text{ of cultured cells} - OD420 \text{ of blank})/(t \cdot v \cdot OD_{600})$$

where t = time (minutes or hour), v = the volume (ml) of cultured cells, and OD_{600} is the density of cultured cells.

The reaction time of β-galactosidase depends generally on the intensity of the protein–protein interaction. In the cells expressing Noc2 and dominant active mutants of Rab3 isoforms, yellow color develops within 30 min. After addition of stop solution, the intensity of the color does not change.

Assay for *In Vitro* Binding of Noc2 and Rab3 Isoforms

Constructs

Both wild-type Noc2 and Noc2AAA mutant cDNAs are subcloned into expression vector pGEX4T. In the Noc2AAA mutant, tryptophan, phenylalanine, and tyrosine (amino acid residues 154, 155, and 156, respectively) are replaced with alanines, and the mutant does not bind to Rab3A *in vitro* (Haynes *et al.*, 2001).

Buffers

Buffer A: 50 mM Tris–HCl (pH 7.4), 50 mM NaCl, 1 mM EDTA, 2 mg/ml lysozyme, 1% Triton X-100, and 1 mM PMSF

Buffer B: 50 mM Tris–HCl (pH 7.4), 50 mM NaCl, 1 mM EDTA, and 1% Triton X-100

Buffer C: phosphate-buffered saline (PBS) and 1 mM EDTA

Buffer D: 100 mM Tris–HCl (pH 7.4), 120 mM NaCl, and 20 mM reduced glutathione

Buffer E: 20 mM HEPES-NaOH (pH 7.4), 200 mM NaCl, 1 mM DTT, 5 mM MgCl$_2$, 1 mM ATP, and 0.26% CHAPS

Purification

Both the pGEX4T wild-type Noc2 and Noc2AAA mutants are transformed in *Escherichia coli* strain DH5α. The transformants are cultured overnight at 37° in 50 ml of LB medium containing 100 μg/ml of ampicillin. The culture is diluted with 500 ml of LB/ampicillin medium. After the culture reaches OD$_{600}$ 0.5~0.9, IPTG is added to the final concentration of 200 μM and cultured for 3 h at 37°. The cells are collected by centrifugation at 6000×g for 15 min. The pellet is suspended with 20 ml of ice-cold Buffer A and incubated on ice for 15 min. The suspension is sonicated twice for 1 min on ice at a 30-s interval. The homogenate is centrifuged at 20,000×g for 30 min at 4°. The supernatant is transferred into a 1.5-ml tube containing 0.5 ml of glutathione-agarose beads equilibrated with Buffer B. After agitation of the mixture for 2 h at 4°, the complex of glutathione S-transferase (GST)-fused Noc2 with glutathione-agarose beads is washed three times with Buffer C. The GST-Noc2 is eluted with 0.5 ml of Buffer D. The elute is dialyzed against Buffer E. Purified GST-Noc2 is stored at $-80°$.

Expression of Rab3 Isoforms and Rab5 Protein

Rab3A, B, C, D, and Rab5 are subcloned into expression vector pFLAG-CMV2. COS-1 cells are maintained at $37°$ in a humidified atmosphere of 5% CO_2 and 95% air (v/v) in growth medium (DMEM containing 10% fatal bovine serum, 1 mM sodium pyruvate). COS-1 cells to be transfected with pFLAG-CMV2-Rab3A, B, C, D, and Rab5 are plated at a density of 50–80% confluence per 60-cm dish the day before transfection. Three micrograms of expression plasmid is diluted with 250 μl of Opti-MEM containing 8 μl of PLUS reagent, followed by incubation for 15 min at room temperature. After incubation, the mixture is combined with 250 μl of Opti-MEM containing 12 μl of LipofectAMINE reagent, allowed to be incubated for 15 min at room temperature, and diluted with 2 ml of Opti-MEM. COS-1 cells are washed with Opti-MEM and cultured in the solution containing plasmid DNA, PLUS reagent, and LipofectAMINE reagent at $37°$. After 3 h, the solution is replaced with 5 ml of growth medium. Forty-eight hours after transfection, the cells are subjected to *in vitro* binding assay.

In Vitro *Binding Assay*

One microgram of GST-wild type Noc2 or Noc2AAA mutant is immobilized onto 20 μl of glutathione-agarose beads by incubation in Buffer E supplemented with 50 mM GDPβS or GTPγS for 1 h at $4°$ (Fig. 3). The cells expressing FLAG-tagged Rab3A, B, C, D, and Rab5 are washed three times with ice-cold PBS, collected using a cell scraper, and transferred into a centrifuge tube and centrifuged at $500 \times g$ for 3 min. The cell pellet is solubilized with 0.5 ml of Buffer E supplemented with 50 μM GDPβS or GTPγS. The lysate is sonicated three times for 5 s and centrifuged at 15,000 $\times g$ for 30 min on ice. The supernatant is agitated together with GST-wild type Noc2 or Noc2AAA mutant immobilized on glutathione-agarose beads for 2 h $4°$. The beads are washed five times with 500 μl of GDPβS or GTPγS-containing Buffer E. Bound protein is eluted by 20 μl of Laemmli sample buffer. Each elute is subjected to sodium dodecyl sulfate, polyacylamide gel electrophoresis (SDS-PAGE) (15%) and electrophoretically transferred to PVDF membrane for immunoblotting. The membrane is processed using an ECL detection kit to detect FLAG-tagged Rab3 isoforms and Rab5 with anti-FLAG M2 antibody (1:5000).

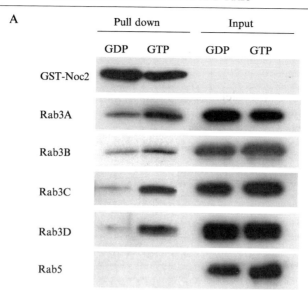

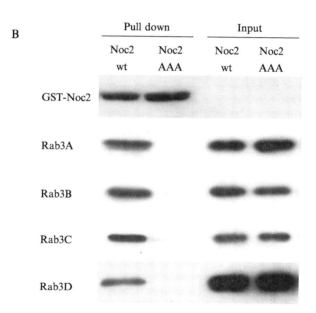

FIG. 3. *In vitro* binding of Noc2 to Rab3 isoforms. (A) Lysate from COS-1 expressing Rab proteins was incubated with GST-Noc2 in the presence of GDPβS or GTPγS. GST-wild-type Noc2 bound preferably to the GTPγS-bound form of Rab3 isoforms. GST-wild-type Noc2 did not bind to Rab5. (B) Lysate incubated with GST-Noc2 or GST-Noc2AAA. Bound Rab proteins were detected by immunoblotting with anti-FLAG M2 antibody. The Noc2AAA mutant did not bind to any isoform of Rab3. Reprinted from Matsumoto *et al.* (2004). *Proc. Natl. Acad. Sci. USA* **101,** 8313–8318. Copyright (2004), National Academy of Sciences, U.S.A.

Insulin Secretion in Pancreatic Islets Isolated from Noc2$^{-/-}$ Mice

Buffer

HEPES-Krebs buffer/BSA: 10 mM HEPES, 118.4 mM NaCl, 4.7 mM KCl, 1.2 mM KH$_2$PO$_4$, 2.4 mM CaCl$_2$, 1.2 mM MgSO$_4$, 20 mM NaHCO$_3$, 2.8 mM glucose, supplemented with 0.2% BSA

Batch Incubation of Isolated Pancreatic Islets

Pancreatic islets are isolated by the collagenase digestion method as described (Wollheim *et al.*, 1990). Batch incubation is performed by the method described by Okamoto *et al.* (1992) with slight modification. Isolated pancreatic islets are cultured overnight at 37° in RPMI 1640 medium containing 10% fetal bovine serum. The cultured pancreatic islets (10 in each tube) are washed three times with HEPES-Krebs buffer/BSA and preincubated for 30 min at 37° in HEPES-Krebs buffer/BSA. The incubated pancreatic islets are washed once with HEPES-Krebs buffer/BSA and incubated in HEPES-Krebs buffer/BSA containing various stimuli of insulin secretion. Insulin released into the medium for 30-min incubation is measured using a radioimmunoassay kit.

Infection of Pancreatic Islets Isolated from Noc2$^{-/-}$ Mice with Wild-Type Noc2 or Noc2AAA cDNA

Recombinant adenovirus carrying LacZ (Ad-LacZ), wild-type Noc2 (Ad-wild type Noc2), or Noc2AAA (Ad-Noc2AAA) (Fig. 4) is generated according to the manufacturer's instructions. Isolated pancreatic islets are transferred into a well of a 48-well tissue culture plate (30 islets/200 μl of infection medium/well) and infected with recombinant adenovirus (multiplicity of infection, 100). The infection medium contains RPMI 1640 medium, 10% fetal bovine serum, and recombinant adenovirus. The pancreatic islets are incubated for 1 h at 37° in a humidified atmosphere of 5% CO$_2$ and 95% air (v/v) and agitated every 20 min during incubation. Three hundred microliters of RPMI 1640 medium containing 10% fetal bovine serum is added to each well and cultured for 48 h at 37°. The pancreatic islets infected with Ad-LacZ, Ad-wild type Noc2, or Ad-Noc2AAA are washed three times with HEPES-Krebs buffer/BSA in a well of a six-well tissue culture plate and incubated in 2 ml of the same buffer at 37°. After 30 min, five size-matched islets are collected, washed once with HEPES-Krebs buffer/BSA, and incubated in 500 μl of HEPES-Krebs buffer/BSA containing 60 mM KCl. Insulin released into the medium for 30-min incubation is measured using a radioimmunoassay kit.

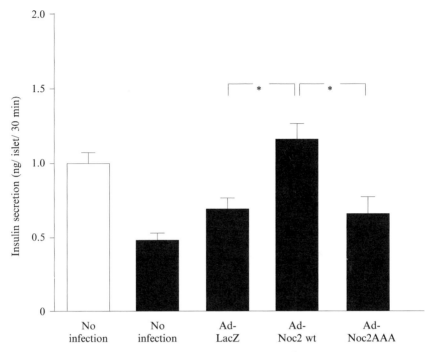

Fig. 4. Rescue of impaired insulin secretion in Noc2$^{-/-}$ mice by Noc2 transgene. Pancreatic islets isolated from Noc2$^{-/-}$ mice were infected with recombinant adenoviruses carrying LacZ, wild-type Noc2, or Noc2AAA cDNA. Ad-LacZ, Ad-Noc2 wt, and Ad-Noc2AAA indicate adenovirus carrying LacZ, wild-type Noc2, and Noc2AAA, respectively. Open column, wild-type mice; solid columns, Noc2$^{-/-}$ mice. Ca^{2+}-triggered insulin secretion in pancreatic islets isolated from Noc2$^{-/-}$ mice was completely restored by wild-type Noc2 gene transfer, whereas the Noc2AAA mutant, which did not bind to any Rab3 isoform, had no effect. Ad-LacZ, 0.69 ± 0.07 ng/islet/30 min; Ad-Noc2 wt, 1.20 ± 0.11 ng/islet/30 min; and Ad-Noc2AAA, 0.65 ± 0.11 ng/islet/30 min. Values are mean ± SEM ($n = 9$, *$p < 0.01$). Reprinted from Matsumoto *et al.* (2004). *Proc. Natl. Acad. Sci. USA* **101,** 8313–8318. Copyright (2004), National Academy of Sciences, U.S.A.

Concluding Remarks

From the study of Noc2$^{-/-}$ mice (Matsumoto, 2004), it is clear that Noc2 participates in regulated exocytosis in both endocrine and exocrine cells. Although the effects of Noc2 on exocytosis require interaction with Rab3, other molecules such as Rab27 and Rab8, both of which interact with Noc2, may also be involved in the exocytotic process. Thus, biochemical and functional assays of Noc2/Rab interactions should be helpful in clarifying the mechanism of the intracellular signal that couples Noc2 to Rab-mediated exocytosis.

Acknowledgment

This work was supported by Grant-in-Aid from the Ministry of Education, Science, Sports, Culture, and Technology.

References

Brondyk, W. H., McKiernan, C. J., Burstein, E. S., and Macara, I. G. (1993). Mutants of Rab3A analogous to oncogenic Ras mutants. Sensitivity to Rab3A-GTPase activating protein and Rab3A-guanine nucleotide releasing factor. *J. Biol. Chem.* **268**, 9410–9415.

Coppola, T., Frantz, C., Perret-Menoud, V., Gattesco, S., Hirling, H., and Regazzi, R. (2002). Pancreatic β-cell protein granuphilin binds Rab3 and Munc-18 and controls exocytosis. *Mol. Biol. Cell* **13**, 1906–1915.

Darchen, F., and Goud, B. (2000). Multiple aspects of Rab protein action in the secretory pathway: Focus on Rab3 and Rab6. *Biochimie* **82**, 375–384.

Fukuda, M. (2003). Distinct Rab binding specificity of Rim1, Rim2, Rabphilin-3, and Noc2. Identification of a critical determinant of Rab3A/Rab27A recognition by Rim2. *J. Biol. Chem.* **278**, 15373–15380.

Haynes, L. P., Evans, G. J., Morgan, A., and Burgoyne, R. D. (2001). A direct inhibitory role for the Rab3-specific effector, Noc2, in Ca^{2+}-regulated exocytosis in neuroendocrine cells. *J. Biol. Chem.* **276**, 9726–9732.

Kotake, K., Ozaki, N., Mizuta, M., Sekiya, S., Inagaki, N., and Seino, S. (1997). Noc2, a putative zinc finger protein involved in exocytosis in endocrine cells. *J. Biol. Chem.* **272**, 29407–29410.

Manabe, S., Nishimura, N., Yamamoto, Y., Kitamura, H., Morimoto, S., Imai, M., Nagahiro, S., Seino, S., and Sasaki, T. (2004). Identification and characterization of Noc2 as a potential Rab3B effector protein in epithelial cells. *Biochem. Biophys. Res. Commun.* **316**, 218–225.

Matsumoto, M., Miki, T., Shibasaki, T., Kawaguchi, M., Shinozaki, H., Nio, J., Saraya, A., Koseki, H., Miyazaki, M., Iwanaga, T., and Seino, S. (2004). Noc2 is essential in normal regulation of exocytosis in endocrine and exocrine cells. *Proc. Natl. Acad. Sci. USA* **101**, 8313–8318.

Okamoto, Y., Ishida, H., Taminato, T., Tsuji, K., Kurose, T., Tsuura, Y., Kato, S., Imura, H., and Seino, Y. (1992). Role of cytosolic Ca^{2+} in impaired sensitivity to glucose of rat pancreatic islets exposed to high glucose *in vitro*. *Diabetes* **41**, 1555–1561.

Ozaki, N., Shibasaki, T., Kashima, Y., Miki, T., Takahashi, K., Ueno, H., Sunaga, Y., Yano, H., Matsuura, Y., Iwanaga, T., Takai, Y., and Seino, S. (2000). cAMP-GEFII is a direct target of cAMP in regulated exocytosis. *Nat. Cell Biol.* **2**, 805–811.

Pfeffer, S. R. (2001). Rab GTPases: Specifying and deciphering organelle identity and function. *Trends Cell Biol.* **11**, 487–914.

Segev, N. (2001). Ypt and Rab GTPases: Insight into functions through novel interactions. *Curr. Opin. Cell Biol.* **13**, 500–511.

Shirataki, H., Kaibuchi, K., Sakoda, T., Kishida, S., Yamaguchi, T., Wada, K., Miyazaki, M., and Takai, Y. (1993). Rabphilin-3A, a putative target protein for smg p25A/rab3A p25 small GTP-binding protein related to synaptotagmin. *Mol. Cell. Biol.* **13**, 2061–2068.

Sudhof, T. C. (2004). The synaptic vesicle cycle. *Annu. Rev. Neurosci.* **27**, 509–547.

Takai, Y., Sasaki, T., and Matozaki, T. (2001). Small GTP-binding proteins. *Physiol. Rev.* **81**, 153–208.

Wang, J., Takeuchi, T., Yokota, H., and Izumi, T. (1999). Novel rabphilin-3-like protein associates with insulin-containing granules in pancreatic β cells. *J. Biol. Chem.* **274**, 28542–28548.

Wang, Y., Okamoto, M., Schmitz, F., Hofmann, K., and Sudhof, T. C. (1997). Rim is a putative Rab3 effector in regulating synaptic-vesicle fusion. *Nature* **388,** 593–859.
Wang, Y., Sugita, S., and Sudhof, T. C. (2000). The RIM/NIM family of neuronal C2 domain proteins. Interactions with Rab3 and a new class of Src homology 3 domain proteins. *J. Biol. Chem.* **275,** 20033–20044.
Wollheim, C. B., Meda, P., and Halban, P. A. (1990). Isolation of pancreatic islets and primary culture of the intact microorgans or of dispersed islet cells. *Methods Enzymol.* **192,** 188–223.
Zerial, M., and McBride, H. (2001). Rab proteins as membrane organizers. *Nat. Rev. Mol. Cell. Biol.* **2,** 107–117.

[37] Functional Analysis of Slac2-a/Melanophilin as a Linker Protein between Rab27A and Myosin Va in Melanosome Transport

By TARUHO S. KURODA, TAKASHI ITOH, and MITSUNORI FUKUDA

Abstract

Slac2-a/melanophilin regulates melanosome transport in mammalian skin melanocytes by linking melanosome-bound Rab27A and an actin-based motor protein, myosin Va. Slac2-a consists of an N-terminal Slp homology domain (SHD), which has been identified as a specific GTP-Rab27-binding domain, a myosin Va-binding domain (MBD) in the middle region, and an actin-binding domain (ABD) at the C-terminus. Mutations in the *slac2-a/mlph* gene cause the abnormal pigmentation (i.e., perinuclear melanosome aggregation in melanocytes) in human Griscelli syndrome type III and in *leaden* mice because of the inability to form the tripartite protein complex consisting of Rab27A, Slac2-a, and myosin Va. In this chapter we describe the methods, including *in vivo* melanosome distribution assay combined with dominant-negative approaches and RNA interference technology, that have been used to analyze the function of Slac2-a in melanosome transport in melanocytes.

Introduction

Slac2-a (synaptotagmin-like protein [Slp] lacking C2 domains-a) was originally identified as a protein that contains an N-terminal Slp homology domain (SHD), the same as Slp family members (Slp1/JFC1, Slp2-a, Slp3-a, Slp4/granuphilin, and Slp5), but lacks C2 domains (Fukuda, 2002a, 2005a,b; Fukuda *et al.*, 2001; Izumi *et al.*, 2003). Since the SHD consists of two

METHODS IN ENZYMOLOGY, VOL. 403
0076-6879/05 $35.00
DOI: 10.1016/S0076-6879(05)03037-5

conserved potential α-helical regions (named SHD1 and SHD2) that exhibit weak homology with a previously characterized Rab3A-binding domain of rabphilin, Rim, and Noc2, the SHD was expected to act as the binding domain of certain Rab protein(s) other than Rab3A. Biochemical screening for specific interactions between the SHD of Slp/Slac2 and Rab proteins has revealed that the entire SHD binds directly to the GTP-bound active form of Rab27A/B (Kuroda *et al.*, 2002a,b). In addition, the SHD1 of Slac2-a is responsible for the interaction with the switch II region of the GTP-bound form of Rab27A (Fukuda, 2002b).

Slac2-a has been independently identified as melanophilin, a product of the *leaden* gene, whose mutations in mice cause lighter coat color (Matesic *et al.*, 2001). Genetic analyses of coat-color mutant mice have indicated that the *leaden* gene product functions in melanosome transport in collaboration with the products of the *ashen* and *dilute* genes (Moore *et al.*, 1988). *Ashen* encodes Rab27A (Wilson *et al.*, 2000), a small Ras-like GTPase that is a member of the Rab family of membrane transport proteins (Zerial and McBride, 2001), and *dilute* encodes myosin Va (Mercer *et al.*, 1991), an unconventional actin-based motor that moves toward the plus end of actin filaments and functions in intracellular organelle transport. Myosin Va is recruited to Rab27A on mature melanosomes in melanocytes, but the two proteins do not directly interact with each other (Bahadoran *et al.*, 2001; Hume *et al.*, 2001; Wu *et al.*, 2001), suggesting the existence of a linker protein between Rab27A and myosin Va in melanocytes. Biochemical and cell biological analyses have revealed that Slac2-a functions as the "missing link" between Rab27A and myosin Va: the middle region of Slac2-a directly binds both the exon F and the adjacent globular tail of myosin Va and the SHD of Slac2-a binds GTP-Rab27A on melanosomes (Fukuda and Kuroda, 2004; Fukuda *et al.*, 2002; Hume *et al.*, 2002; Nagashima *et al.*, 2002; Provance *et al.*, 2002; Strom *et al.*, 2002; Wu *et al.*, 2002). The tripartite protein complex of Rab27A, Slac2-a, and myosin Va maintains normal melanosome distribution at the cell periphery and in the dendrites of melanocytes by capturing melanosomes on the peripheral actin filaments (Kuroda *et al.*, 2003; Wu *et al.*, 1998, 2002). Defects in the formation of the Rab27A-Slac2-a-myosin Va complex result in perinuclear melanosome aggregation because of failure of melanosome transfer from the microtubules to actin filaments, as seen in the melanocytes of coat-color mutant mice (*ashen, leaden,* and *dilute*) and the three types of human Griscelli syndrome patients (Ménasché *et al.*, 2000, 2003; Pastural *et al.*, 1997). Slac2-a also contains an actin-binding domain (ABD) at the C-terminus that is required for the efficient transfer of melanosomes from microtubules to actin filaments (Fukuda and Kuroda, 2002; Kuroda *et al.*, 2003), and the association between Slac2-a and actin is stimulated by cyclic AMP

(Passeron *et al.*, 2004). This chapter describes methods we have used to investigate the functions of Slac2-a in melanosome transport in melan-a mouse melanocytes.

Assay for Melanosome Transport in Melan-a Cells: Dominant Negative Approaches

Melanosomes are specialized organelles that produce and store melanin pigments, and they are excellent experimental material for the study of organelle transport because they can be seen through a light microscope without staining (Marks and Seabra, 2001; Nascimento *et al.*, 2003). Under normal conditions melanosomes are mainly distributed at the cell periphery and in the dendrites of melanocytes (Fig. 1C). Dysfunction of the Rab27A-Slac2-a-myosin Va complex caused by mutations in either the *rab27a, slac2-a,* or *myosin Va* gene causes perinuclear melanosome aggregation in the melanocytes of human Griscelli syndrome patients and in coat-color-mutant mice (Fig. 1F). The same perinuclear aggregation phenotype can be produced by overexpression of a recombinant protein(s) in melanocytes that disrupts the formation of the Rab27A-Slac2-a-myosin Va complex (Fukuda and Kuroda, 2004; Hume *et al.*, 2001; Kuroda and Fukuda, 2004; Kuroda *et al.*, 2003; Strom *et al.*, 2002; Wu *et al.*, 2002). In this section we provide detailed descriptions of the protocols for the "melanosome distribution assay" that has been used to determine the precise functions of Slac2-a in melanosome transport in melan-a cells (Kuroda *et al.*, 2003).

Culture of Melan-a Mouse Melanocytes and Transfection

Melan-a cells are an immortal line of pigmented mouse melanocytes that was originally derived from normal embryonic skin of inbred C57BL mice, and were generously donated for use in our studies by Bennett *et al.* (1987). Melan-a cells are cultured according to the methods of Bennett and colleagues with slight modifications. Melan-a cells are cultured in RPMI-1640 medium (catalogue number R8758; Sigma Chemical Co., St. Louis, MO) supplemented with 10% fetal bovine serum, 100 U/ml penicillin G, 100 μg/ml streptomycin, 7.5 μg/ml phenol red, and 0.1 mM 2-mercaptoethanol. It should be noted that pigmentation of melan-a cells requires a lower pH ($\sim$pH 6.9) that can be achieved by growing them under 10% CO_2 at 37° and by adding 2.7 mM HCl to the medium. Immediately prior to use, 200 nM phorbol 12-myristate 13-acetate (PMA, also known as 12-*O*-tetradecanoyl phorbol 13-acetate [TPA]; Sigma Chemical Co.) is added to the medium to maintain melan-a cell proliferation.

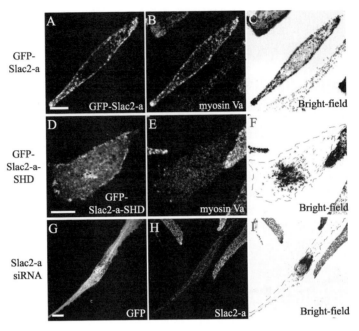

FIG. 1. Normal peripheral distribution and abnormal perinuclear aggregation of melano-somes in melan-a mouse melanocytes. Melan-a cells were transfected with pEGFP-C1-Slac2-a (A–C), pEGFP-C1-Slac2-a-SHD (D–F), and pSilencer-Slac2-a plus pEGFP-C1 (G–I). Cells were fixed, permeabilized, and immunostained with anti-myosin Va rabbit IgG (B and E) and anti-Slac2-a rabbit IgG (H). Fluorescence and bright-field images were acquired with a confocal laser-scanning microscope. The bright-field images show the melanosome distribution in the cells (C, F, and I). Bars in A, D, and G represent 10 μm. The wild-type GFP-Slac2-a (A) expressed in melan-a cells supported normal distribution of melanosomes at the cell periphery (C) and recruited myosin Va to the melanosomes (B), whereas expression of GFP-Slac2-a-SHD (D) induced perinuclear aggregation of melanosomes (F) and attenuation of myosin Va immunoreactivity (E). Expression of the Slac2-a siRNA, together with GFP (G), induced the depletion of Slac2-a protein (H) and perinuclear melanosome aggregation (I), the same as in the melanocytes of Griscelli syndrome patients and *leaden* mice.

Melan-a cells are subcultured by removing them with 0.025% trypsin and 0.1 mM EDTA·4Na in phosphate-buffered saline (PBS) (lower trypsin concentration than used for subculture of most cell types). Melan-a cells should be suspended gently during subculture; vigorous suspension should be avoided, because it often reduces the attachment of melan-a cells to the dish, especially when melan-a cells are used for immunofluorescence analysis. Culture medium supplemented with 5% dimethyl sulfoxide is used for frozen storage of melan-a cells. Melan-a cells between the 25th passage (p25) and p30 should be used for melanosome distribution assays, because melan-a cells younger than p25 often exhibit low transfection efficiency, and those older than p30 tend to be less pigmented.

Melanosome Distribution Assay

Melan-a cells (7.5×10^4 cells, the day before transfection) seeded on 3.5-cm glass-bottom dishes (MatTek Corp., Ashland, MA) are transfected with $2 \mu g$ of pEGFP-C1 vector (BD Clontech, Palo Alto, CA) or its derivatives by using 3 μl of FuGENE6 reagent (Roche, Basel, Switzerland) according to the manufacturer's notes. At 48–72 h after transfection, the cells are washed twice with PBS for 5 min, fixed with 4% paraformaldehyde (PFA) (catalogue number 168–20955; Wako Pure Chemicals, Osaka, Japan) in 0.1 M phosphate buffer, pH 7.4, for 20 min with gentle agitation every 5 min, and then washed with 1 ml of PBS for 5 min three times. Fluorescence and bright-field images of the transfected cells are acquired with a confocal laser-scanning microscope (Fluoview FV500; Olympus, Tokyo, Japan) equipped with a $60\times$ oil immersion objective lens (NA = 1.40; Olympus) and three lasers of different wavelengths (488, 543, and 633 nm; Melles Griot, Carlsbad, CA). At least 150 cells (more than 50 cells/dish, three independent dishes for each plasmid) are acquired at random, and the images are processed with Adobe Photoshop software (version 7.0). Cells in which more than half of the melanosomes are located around the nucleus are classified as "perinuclear aggregation" cells. The results are expressed as the percentage of cells bearing green fluorescence protein (GFP) fluorescence that show perinuclear melanosome aggregation. Under these experimental conditions, expression of GFP alone and GFP-tagged wild-type Slac2-a in melan-a cells induces a basal level of perinuclear melanosome aggregation of approximately 5% and 20%, respectively (Kuroda *et al.*, 2003), and most of the cells expressing GFP-Slac2-a show normal melanosome distribution at the cell periphery, with GFP-Slac2-a being present on melanosomes (Fig. 1A–C). Note that the level of expression of endogenous and exogenous proteins in melan-a cells should be compared by immunoblot analysis, because melanosome distribution is highly sensitive to the amount of protein expressed: the dominant negative effect of expressed proteins is most prominent when their expression level in melan-a cells is very high.

Immunofluorescence Analysis of Endogenous Rab27A and Myosin Va in Melan-a Cells

Melan-a cells seeded on glass-bottom dishes are fixed with 4% PFA for 20 min and washed with PBS as described earlier. Longer fixation time or higher concentrations of PFA should be avoided, because such intense fixation conditions often reduce the immunoreactivity of proteins, especially of Slac2-a and Slp2-a, in melan-a cells. The cells are permeabilized with 1 ml of 0.3% Triton X-100 in PBS for 2 min, washed with 1 ml of blocking buffer (1% bovine serum albumin, fraction V [catalogue number A-7906; Sigma Chemical Co.] and 0.1% Triton X-100 in PBS) for 5 min three times, blocked with 1 ml of the blocking buffer for 1 h, and immunostained for 1 h with the first antibody in 100 μl of the blocking buffer at the following concentrations: anti-Rab27A mouse IgG 5.0 μg/ml (BD Transduction Laboratories, Lexington, KY) and anti-myosin Va rabbit IgG 6.9 μg/ml (Fukuda et al., 2002). After washing with 1 ml of the blocking buffer for 5 min three times, cells are reacted with secondary IgG conjugated with Alexa-Fluor 568 or 633 (1/5000 dilution in 100 μl of the blocking buffer; Molecular Probes, Eugene, OR) followed by washing with 1 ml of the blocking buffer for 5 min three times. Fluorescence images of GFP, Rab27A, and myosin Va are sequentially acquired with a confocal laser-scanning microscope using three wavelength lasers as described earlier. Under normal conditions, immunostaining signals of endogenous Rab27A and myosin Va are detected on melanosomes in melan-a cells (Fig. 1B), but the endogenous myosin Va signal is often reduced and/or segregated from the signal of endogenous Rab27A when perinuclear melanosome aggregation is induced by a dominant negative effect of recombinant proteins expressed in melan-a cells (Fig. 1E).

Assay for Interaction between Slac2-a and Rab27A, Myosin Va, and Actin In Vivo

Slac2-a is composed of three functional domains: an N-terminal SHD (amino acids 1–153), a myosin Va exon-F-binding domain (MBD) in the middle of the molecule (amino acids 241–405), and a C-terminal ABD (amino acids 401–590) (Fig. 2). Since a single domain of Slac2-a, SHD, MBD, and ABD alone is necessary and sufficient for the interaction with Rab27A, myosin Va, and actin, respectively, these domains are suitable for developing tools to assess the interaction between Slac2-a and Rab27A, myosin Va, and actin in vivo. To do so we constructed pEGFP-C1 vectors encoding Slac2-a-SHD, -MBD, and -ABD as a trapper of Rab27A, myosin Va, and actin, respectively, and prepared their binding-defective mutants (SHD[E14A], MBD[EA], and ABD[KA]) by Ala-based site-directed mutagenesis as negative controls (Kuroda et al., 2003). The SHD(E14A) contains a Glu-to-Ala substitution at amino acid position 14 and completely lacks Rab27A-binding activity. The loss of Rab27A-binding activity by SHD(E14A) is demonstrated by cotransfection and coimmunoprecipitation assays (Kuroda et al., 2003; see also Chapter 39, this volume).

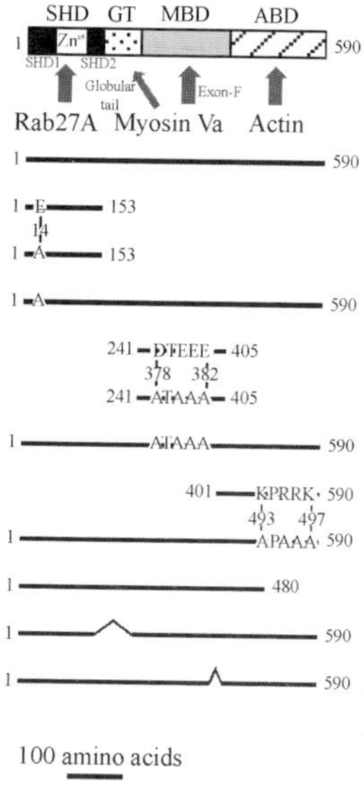

	Rab27A binding	Myosin Va binding	Actin binding	Perinuclear melanosome aggregation
Slac2-a				
Wild type	+	+	+	<20%
SHD	+	−	−	~90%
SHD(E14A)	−	−	−	<20%
Slac2-a(E14A)	−	+	+	~90%
MBD	−	+	−	~90%
MBD(EA)	−	−	−	<20%
Slac2-a(EA)	+	−	+	~90%
ABD	−	−	+	~20%
Slac2-a(KA)	+	+	±	~90%
ΔABD	+	+	−	~90%
ΔGT	+	+*	+	~90%
ΔPEST	+	+	+	~40%
GFP only	−	−	−	~5%

FIG. 2. Summary of melanosome distribution assays. A schematic representation of the domain structure of mouse Slac2-a is shown at the upper left: closed boxes, the SHD1 and SHD2; Zn^{2+}, two zinc finger motifs; dotted box, globular tail-binding site; shaded box, the MBD; hatched box, the ABD. Constructs of Slac2-a mutants are depicted. As an example, SHD(E14A) encodes amino acids 1–153 of Slac2-a and carries a substitution of alanine for glutamic acid at amino acid position 14. Amino acid numbers are stated on both sides. The Rab27A-, myosin Va-, and actin-binding activity of each construct (−, ± or +) and the results of the melanosome distribution assay (approximate percentage of cells exhibiting perinuclear melanosome aggregation) are summarized on the right (Fukuda and Itoh, 2004; Fukuda and Kuroda, 2004; Kuroda et al., 2003). Slac2-a-ΔGT (a mutant lacking the myosin Va-globular tail-binding region) strongly induced perinuclear aggregation in melan-a cells but retained myosin Va-binding activity through the high-affinity myosin-Va binding site (asterisk).

Similarly, the MBD(EA) containing D378A, E380A, E381A, and E382A substitutions and the ABD(KA) containing K493A, R495A, R496A, and K497A substitutions lack myosin Va-binding activity and actin-binding activity, respectively. When GFP-tagged SHD or MBD is expressed in melan-a cells as described above, a high percentage of cells (more than 85%) exhibit perinuclear melanosome aggregation due to the occupation of endogenous Rab27A and myosin Va, respectively (Fig. 1D–F), whereas expression of the GFP-SHD(E14A) or -MBD(EA) mutant in melan-a cells has no inhibitory effect on melanosome transport. Although expression of GFP-ABD in melan-a cells does not inhibit distribution of melanosomes at the cell periphery, it does induce excess formation of actin bundles and the exclusion of melanosomes from peripheral actin bundles (Kuroda et al., 2003).

We have also prepared full-length GFP-Slac2-a carrying E14A, EA, and KA mutations (i.e., Slac2-a[E14A], Slac2-a[EA], or Slac2-a[KA]) to analyze the function of Slac2-a in melanosome transport *in vivo*. Expression of these full-length mutants in melan-a cells also causes approximately 90% of the transfected cells to exhibit perinuclear aggregation of melanosomes, indicating that all three ligand-binding abilities of Slac2-a are indispensable for melanosome transport in melanocytes (Kuroda *et al.*, 2003). The effects of expression of Slac2-a point and deletion mutants on melanosome transport and their Rab27A-, myosin Va-, and actin-binding activity are summarized in Fig. 2.

Assay for Weak Interaction between Slac2-a and the Globular Tail of Myosin Va In Vivo

The association between Slac2-a and myosin Va is mediated by two modes of domain–domain interactions: the strong interaction between Slac2-a-MBD and myosin Va-exon-F (i.e., high-affinity site) and the weak interaction between amino acids 147–200 of Slac2-a (between the SHD and MBD) and the myosin Va-globular tail (i.e., low-affinity site) (Fukuda and Kuroda, 2004; Kuroda *et al.*, 2003; Wu *et al.*, 2002). A mutant Slac2-a-ΔGT lacking amino acids 147–200 is prepared to investigate the significance of the weak interaction of Slac2-a with the globular tail of myosin Va. Although the binding affinity between Slac2-a-ΔGT and the myosin Va tail is slightly lower than between wild-type Slac2-a and the myosin Va tail, and Slac2-a-ΔGT can form a complex with Rab27A and the myosin Va tail, expression of GFP-Slac2-a-ΔGT in melan-a cells strongly inhibits melanosome transport in melan-a cells (Fig. 2), indicating that the low-affinity myosin Va-binding site in Slac2-a is also required for melanosome transport (Fukuda and Kuroda, 2004). Consistent with this, missense mutations in the globular tail of myosin Va that lose low-affinity Slac2-a-binding activity cause abnormal pigmentation in *dilute* mice (Huang *et al.*, 1998).

Assay for the Function of PEST-Like Sequences in the C-terminus of Slac2-a

Slac2-a contains multiple PEST-like sequences, putative signals for rapid protein degradation, in the MBD and ABD at the C-terminus, and they are highly sensitive to various proteases *in vitro* and endogenous Ca^{2+}-dependent calpains in melan-a cells (Fukuda and Itoh, 2004). Because of the PEST-like sequences, the Slac2-a molecule is readily degraded during the process of immunoprecipitation in the absence of protease inhibitors. Slac2-a-ΔPEST lacks one of the PEST-like sequences (STSSEDET; amino acids 399–405) located between the MBD and the ABD and is more resistant to proteases than the wild-type Slac2-a. Although Slac2-a-ΔPEST normally interacts with Rab27A, myosin Va, and actin, the same as the wild-type Slac2-a, expression of GFP-Slac2-a-ΔPEST in melan-a cells induces perinuclear melanosome aggregation more frequently (more than 40%) than wild-type GFP-Slac2-a expression (Fig. 2), suggesting that Slac2-a degradation is involved in proper melanosome distribution in melanocytes (Fukuda and Itoh, 2004).

Assay for Melanosome Transport by RNA Interference (RNAi) Technology

In the previous section we described methods of analyzing the function of Slac2-a in melanosome transport by overexpression of point or deletion mutants of Slac2-a. However, since it is also important to analyze the function of endogenous Slac2-a molecules in melan-a cells, we developed small interfering RNA (siRNA) targeted against Slac2-a and siRNA-resistant Slac2-a (Kuroda and Fukuda, 2004). Knockdown of endogenous Slac2-a by the specific siRNA induces perinuclear melanosome aggregation in melan-a cells, and expression of siRNA-resistant Slac2-a can rescue this phenotype. This section describes experimental protocols for studies on the function of Slac2-a in melanosome transport by RNAi technology.

Assay for Depletion of Endogenous Slac2-a in Melan-a Cells by Specific siRNA

The following nucleotides with a 19-base target site (in bold) and a 9-base loop (underlined) are used to obtain pSilencer-Slac2-a: Slac2-a siRNA(+) primer (5'-**GAAGGAAATGGAGACAGTG**TTCAAGAGA-CACTGTCTCCATTTCCTTCTTTTTT-3') and Slac2-a siRNA(-) primer (5'-AATTAAAAAA**GAAGGAAATGGAGACAGTG**TCTCTTGAAC ACTGTCTCCATTTCCTTCGGCC-3'). These nucleotides are mixed, denatured at 94° for 2 min, annealed at 72° for 1 min, gradually cooled to 4° for 2 h, and then subcloned into the *Apa*I and *Eco*RI sites of the pSilencer 1.0-U6 vector (Ambion, Austin, TX), which expresses short hairpin RNA under the control of the mouse U6 promoter. Knockdown efficiency is evaluated by cotransfection of pEF-T7-Slac2-a (or pEGFP-C1-Slac2-a) and pSilencer-Slac2-a into NIH/3T3 cells (or COS-7 cells) (Kuroda and Fukuda, 2004).

Melan-a cells seeded on 3.5-cm glass-bottom dishes are cotransfected with 0.75 μg of pSilencer-Slac2-a and 0.25 μg of pEGFP-C1vector, as a marker of transfected cells, with 1 μl of FuGENE6 reagent. At 48–72 h after transfection, cells are subjected to immunofluorescence analysis using anti-Slac2-a rabbit IgG (Imai *et al.*, 2004). Under the above experimental conditions, more than 90% of cells expressing GFP and the Slac2-a siRNA exhibited perinuclear melanosome aggregation because of down-regulation of endogenous Slac2-a protein (Fig. 1G–I) (Kuroda and Fukuda, 2004). Expression of siRNA directed against Rab27A (target site; 5′-AA-GAGAGTGGTGTACAGAG-3′) also frequently induces perinuclear melanosome aggregation, whereas expression of siRNA directed against Slp2-a (target site; 5′-CACCTTTGTCTCATTATCA-3′), another Rab27A-binding protein in melanocytes, induces loss of melanosomes at the cell periphery, rather than inducing perinuclear melanosome aggregation (Kuroda and Fukuda, 2004).

Protocols for the Rescue Experiments by Expression of siRNA-Resistant Slac2-a

cDNA of s̲iRNA-r̲esistant Slac2-a (referred to as Slac2-a^{SR}) is prepared by the conventional polymerase chain reaction (PCR) using SR primer (5′-A̲A̲G̲C̲T̲T̲GGA**GG**AGGGTAACGGTGATA**GAGA**GCAGACTGAT-GA-3′, sense) with a restriction enzyme site (underlined) and substituted nucleotides (in bold). The resulting cDNA fragment contains seven nucleotide substitutions in the target site of Slac2-a siRNA without amino acid substitution and is subcloned into the pEGFP-C1 vector (named pEGFP-C1-Slac2-a^{SR}). The siRNA resistance of Slac2-a^{SR} cDNA is confirmed by a cotransfection assay using COS-7 cells. COS-7 cells are transfected with pEGFP-C1-Slac2-a^{SR} (or pEGFP-C1-Slac2-a) and pSilencer-Slac2-a (or a control vector). The cell lysate is prepared with 1% SDS and analyzed by 7.5% SDS-polyacrylamide gel electrophoresis followed by immunoblotting with anti-GFP rabbit IgG (1/1000 dilution; MBL Co., Ltd., Nagoya, Japan).

Melan-a cells seeded on 3.5-cm glass-bottom dishes are cotransfected with 1 μg of Slac2-a siRNA expression vector and 1 μg of pEGFP-C1-Slac2-a^{SR} (or pEGFP-C1 alone) by using 2 μl of FuGENE6 reagent. At 48–72 h after transfection, cells are fixed and analyzed with a confocal laser-scanning microscope as described previously. Under these experimental conditions, more than 80% of the cells expressing the Slac2-a siRNA and GFP exhibit perinuclear aggregation of melanosomes, and simultaneous expression of GFP-Slac2-a^{SR} with the Slac2-a siRNA completely rescues the perinuclear melanosome aggregation phenotype to the basal level (< 20%) (Kuroda and Fukuda, 2005).

Concluding Remarks

We have described methods of analyzing the function of Slac2-a in melanosome transport by dominant negative approaches and RNAi technology. By using these methods, we reported finding that Slp2-a, a previously uncharacterized Rab27A-binding protein in melanocytes, controls melanosome distribution in the cell periphery and regulates the morphology of melanocytes through the phospholipid-binding activity of its C2 domains (Kuroda and Fukuda, 2004). That study shed light on the differential and sequential roles of Rab27A-binding proteins in melanosome transport in melanocytes: first, Slac2-a mediates the transfer of melanosomes from microtubules to peripheral actin filaments; and second, Slp2-a functions in the attachment of melanosomes to the plasma membrane.

Acknowledgments

This work was supported in part by grants from the Ministry of Education, Culture, Sports, and Technology of Japan.

References

Bahadoran, P., Aberdam, E., Mantoux, F., Buscà, R., Bille, K., Yalman, N., de Saint-Basile, G., Casaroli-Marano, R., Ortonne, J. P., and Ballotti, R. (2001). Rab27a: A key to melanosome transport in human melanocytes. *J. Cell Biol.* **152**, 843–850.

Bennett, D. C., Cooper, P. J., and Hart, I. R. (1987). A line of non-tumorigenic mouse melanocytes, syngeneic with the B16 melanoma and requiring a tumour promoter for growth. *Int. J. Cancer* **39**, 414–418.

Fukuda, M. (2002a). Slp and Slac2, novel families of Rab27 effectors that control Rab27-dependent membrane traffic. *Recent Res. Dev. Neurochem.* **5**, 297–309.

Fukuda, M. (2002b). Synaptotagmin-like protein (Slp) homology domain 1 of Slac2-a/melanophilin is a critical determinant of GTP-dependent specific binding to Rab27A. *J. Biol. Chem.* **277**, 40118–40124.

Fukuda, M. (2005a). Versatile role of Rab27 in membrane trafficking: Focus on the Rab27 effector families. *J. Biochem.* **137**, 9–16.

Fukuda, M. (2005b). Slp homology domain: A novel protein motif that specifically binds small GTPase Rab27. *Recent Res. Dev. Biochem.* **6**, 13–29.

Fukuda, M., and Itoh, T. (2004). Slac2-a/melanophilin contains multiple PEST-like sequences that are highly sensitive to proteolysis. *J. Biol. Chem.* **279**, 22314–22321.

Fukuda, M., and Kuroda, T. S. (2002). Slac2-c (synaptotagmin-like protein homologue lacking C2 domains-c), a novel linker protein that interacts with Rab27, myosin Va/VIIa, and actin. *J. Biol. Chem.* **277**, 43096–43103.

Fukuda, M., and Kuroda, T. S. (2004). Missense mutations in the globular tail of myosin-Va in dilute mice partially impair binding of Slac2-a/melanophilin. *J. Cell Sci.* **117**, 583–591.

Fukuda, M., Saegusa, C., and Mikoshiba, K. (2001). Novel splicing isoforms of synaptotagmin-like proteins 2 and 3: Identification of the Slp homology domain. *Biochem. Biophys. Res. Commun.* **283**, 513–519.

Fukuda, M., Kuroda, T. S., and Mikoshiba, K. (2002). Slac2-a/melanophilin, the missing link between Rab27 and myosin Va: Implications of a tripartite protein complex for melanosome transport. *J. Biol. Chem.* **277**, 12432–12436.

Huang, J. D., Mermall, V., Strobel, M. C., Russell, L. B., Mooseker, M. S., Copeland, N. G., and Jenkins, N. A. (1998). Molecular genetic dissection of mouse unconventional myosin-VA: Tail region mutations. *Genetics* **148**, 1963–1972.

Hume, A. N., Collinson, L. M., Rapak, A., Gomes, A. Q., Hopkins, C. R., and Seabra, M. C. (2001). Rab27a regulates the peripheral distribution of melanosomes in melanocytes. *J. Cell Biol.* **152,** 795–808.

Hume, A. N., Collinson, L. M., Hopkins, C. R., Strom, M., Barral, D. C., Bossi, G., Griffiths, G. M., and Seabra, M. C. (2002). The *leaden* gene product is required with Rab27a to recruit myosin Va to melanosomes in melanocytes. *Traffic* **3,** 193–202.

Imai, A., Yoshie, S., Nashida, T., Shimomura, H., and Fukuda, M. (2004). The small GTPase Rab27B regulates amylase release from rat parotid acinar cells. *J. Cell Sci.* **117,** 1945–1953.

Izumi, T., Gomi, H., Kasai, K., Mizutani, S., and Torii, S. (2003). The roles of Rab27 and its effectors in the regulated secretory pathways. *Cell Struct. Funct.* **28,** 465–474.

Kuroda, T. S., and Fukuda, M. (2004). Rab27A-binding protein Slp2-a is required for peripheral melanosome distribution and elongated cell shape in melanocytes. *Nat. Cell Biol.* **6,** 1195–1203.

Kuroda, T. S., and Fukuda, M. (2005). Functional analysis of Slac2-c/MyRIP as a linker protein between melalosomes and myosin Vlla. *J. Biol. Chem.* **280,** 28015–28022.

Kuroda, T. S., Fukuda, M., Ariga, H., and Mikoshiba, K. (2002a). The Slp homology domain of synaptotagmin-like proteins 1–4 and Slac2 functions as a novel Rab27A binding domain. *J. Biol. Chem.* **277,** 9212–9218.

Kuroda, T. S., Fukuda, M., Ariga, H., and Mikoshiba, K. (2002b). Synaptotagmin-like protein 5: A novel Rab27A effector with C-terminal tandem C2 domains. *Biochem. Biophys. Res. Commun.* **293,** 899–906.

Kuroda, T. S., Ariga, H., and Fukuda, M. (2003). The actin-binding domain of Slac2-a/melanophilin is required for melanosome distribution in melanocytes. *Mol. Cell. Biol.* **23,** 5245–5255.

Marks, M. S., and Seabra, M. C. (2001). The melanosome: Membrane dynamics in black and white. *Nat. Rev. Mol. Cell Biol.* **2,** 738–748.

Matesic, L. E., Yip, R., Reuss, A. E., Swing, D. A., O'Sullivan, T. N., Fletcher, C. F., Copeland, N. G., and Jenkins, N. A. (2001). Mutations in *Mlph*, encoding a member of the Rab effector family, cause the melanosome transport defects observed in *leaden* mice. *Proc. Natl. Acad. Sci. USA* **98,** 10238–10243.

Ménasché, G., Pastural, E., Feldmann, J., Certain, S., Ersoy, F., Dupuis, S., Wulffraat, N., Bianchi, D., Fischer, A., Le Deist, F., and de Saint Basile, G. (2000). Mutations in *RAB27A* cause Griscelli syndrome associated with haemophagocytic syndrome. *Nat. Genet.* **25,** 173–176.

Ménasché, G., Ho, C. H., Sanal, O., Feldmann, J., Tezcan, I., Ersoy, F., Houdusse, A., Fischer, A., and de Saint Basile, G. (2003). Griscelli syndrome restricted to hypopigmentation results from a melanophilin defect (GS3) or a *MYO5A* F-exon deletion (GS1). *J. Clin. Invest.* **112,** 450–456.

Mercer, J. A., Seperack, P. K., Strobel, M. C., Copeland, N. G., and Jenkins, N. A. (1991). Novel myosin heavy chain encoded by murine *dilute* coat colour locus. *Nature* **349,** 709–713.

Moore, K. J., Swing, D. A., Rinchik, E. M., Mucenski, M. L., Buchberg, A. M., Copeland, N. G., and Jenkins, N. A. (1988). The murine dilute suppressor gene *dsu* suppresses the coat-color phenotype of three pigment mutations that alter melanocyte morphology, *d, ash* and *ln. Genetics* **119,** 933–941.

Nagashima, K., Torii, S., Yi, Z., Igarashi, M., Okamoto, K., Takeuchi, T., and Izumi, T. (2002). Melanophilin directly links Rab27a and myosin Va through its distinct coiled-coil regions. *FEBS Lett.* **517,** 233–238.

Nascimento, A. A., Roland, J. T., and Gelfand, V. I. (2003). Pigment cells: A model for the study of organelle transport. *Annu. Rev. Cell Dev. Biol.* **19,** 469–491.

Passeron, T., Bahadoran, P., Bertolotto, C., Chiaverini, C., Buscà, R., Valony, G., Bille, K., Ortonne, J. P., and Ballotti, R. (2004). Cyclic AMP promotes a peripheral distribution of melanosomes and stimulates melanophilin/Slac2-a and actin association. *FASEB J.* **18**, 989–991.

Pastural, E., Barrat, F. J., Dufourcq-Lagelouse, R., Certain, S., Sanal, O., Jabado, N., Seger, R., Griscelli, C., Fischer, A., and de Saint Basile, G. (1997). Griscelli disease maps to chromosome 15q21 and is associated with mutations in the myosin-Va gene. *Nat. Genet.* **16**, 289–292.

Provance, D. W., Jr., James, T. L., and Mercer, J. A. (2002). Melanophilin, the product of the *leaden* locus, is required for targeting of myosin-Va to melanosomes. *Traffic* **3**, 124–132.

Strom, M., Hume, A. N., Tarafder, A. K., Barkagianni, E., and Seabra, M. C. (2002). A family of Rab27-binding proteins: Melanophilin links Rab27a and myosin Va function in melanosome transport. *J. Biol. Chem.* **277**, 25423–25430.

Wilson, S. M., Yip, R., Swing, D. A., O'Sullivan, T. N., Zhang, Y., Novak, E. K., Swank, R. T., Russell, L. B., Copeland, N. G., and Jenkins, N. A. (2000). A mutation in *Rab27a* causes the vesicle transport defects observed in *ashen* mice. *Proc. Natl. Acad. Sci. USA* **97**, 7933–7938.

Wu, X., Bowers, B., Rao, K., Wei, Q., and Hammer, J. A., III (1998). Visualization of melanosome dynamics within wild-type and dilute melanocytes suggests a paradigm for myosin V function in vivo. *J. Cell Biol.* **143**, 1899–1918.

Wu, X., Rao, K., Bowers, M. B., Copeland, N. G., Jenkins, N. A., and Hammer, J. A., III (2001). Rab27a enables myosin Va-dependent melanosome capture by recruiting the myosin to the organelle. *J. Cell Sci.* **114**, 1091–1100.

Wu, X. S., Rao, K., Zhang, H., Wang, F., Sellers, J. R., Matesic, L. E., Copeland, N. G., Jenkins, N. A., and Hammer, J. A., III (2002). Identification of an organelle receptor for myosin-Va. *Nat. Cell Biol.* **4**, 271–278.

Zerial, M., and McBride, H. (2001). Rab proteins as membrane organizers. *Nat. Rev. Mol. Cell Biol.* **2**, 107–117.

[38] Identification and Biochemical Analysis of Slac2-c/MyRIP as a Rab27A-, Myosin Va/VIIa-, and Actin-Binding Protein

By TARUHO S. KURODA and MITSUNORI FUKUDA

Abstract

Slac2-c/MyRIP is a specific Rab27A-binding protein that contains an N-terminal synaptotagmin-like protein (Slp) homology domain (SHD, a newly identified GTP-Rab27A-binding motif), but in contrast to the Slp family proteins, it lacks C-terminal tandem C2 domains. *In vitro* Slac2-c simultaneously directly interacts with both Rab27A and an actin-based motor protein, myosin Va, via its N-terminal SHD and middle region, respectively, consistent with the fact that the overall structure of Slac2-c is similar to that of Slac2-a/melanophilin, a linker protein between Rab27A and myosin Va in the melanosome transport in melanocytes. Unlike

METHODS IN ENZYMOLOGY, VOL. 403
Copyright 2005, Elsevier Inc. All rights reserved.

0076-6879/05 $35.00
DOI: 10.1016/S0076-6879(05)03038-7

Slac2-a, however, the middle region of Slac2-c interacts with two types of myosins, myosin Va and myosin VIIa. In addition, the most C-terminal part of both Slac2-a and Slac2-c functions as an actin-binding domain: it directly interacts with globular and fibrous actin *in vitro*, and the actin-binding domain of Slac2-a and Slac2-c colocalizes with actin filaments when it is expressed in living cells (i.e., PC12 cells and mouse melanocytes). In this chapter we describe the methods that have been used to analyze the protein–protein interactions of Slac2-c, specifically with Rab27A, myosin Va/VIIa, and actin.

Introduction

The Slac2 (synaptotagmin-like protein [Slp] homologue lacking C2 domains) family is a group of proteins that contains an N-terminal Slp homology domain (SHD, known as the specific Rab27A-binding motif) lacking C-terminal tandem C2 domains, and three members have been reported in mammals to date: Slac2-a/melanophilin, Slac2-b, and Slac2-c/MyRIP (El-Amraoui *et al.*, 2002; Fukuda, 2002, 2005; Fukuda and Kuroda, 2002; Fukuda *et al.*, 2001; Kuroda *et al.*, 2002a; Matesic *et al.*, 2001). Slac2-c was originally identified by EST database searches as a homologue of Slac2-a, a linker protein between Rab27A and myosin Va (Fukuda, *et al.*, 2002; Kuroda *et al.*, 2003; Strom *et al.*, 2002; Wu *et al.*, 2002). The N-terminal SHD is highly conserved between Slac2-a and Slac2-c at the amino acid level (> 40% identity), but their C-terminal regions containing a putative myosin Va-binding site are less well conserved (about 20% identity) (Fukuda and Kuroda, 2002). Slac2-c has also been independently identified as MyRIP (myosin VIIa- and Rab-interacting protein) by yeast two-hybrid screening of a human retina cDNA library with the tail domain of myosin VIIa as the bait (El-Amraoui *et al.*, 2002).

Sequence comparison between Slac2-a and Slac2-c and systematic deletion analyses have revealed that both Slac2-a and Slac2-c contain three functional domains (Fukuda and Kuroda, 2002; Kuroda *et al.*, 2003). (1) The N-terminal SHD specifically binds the GTP-bound form of Rab27A/B, the same as the SHD of Slp proteins (Kuroda *et al.*, 2002a,b); (2) the middle domain of Slac2-a and Slac2-c interacts with the tail domain of myosin Va, but only Slac2-c is capable of interacting with the tail domain of myosin VIIa (El-Amraoui *et al.*, 2002; Fukuda and Kuroda, 2002); and (3) the most C-terminal part (approximately 200 amino acids) of Slac2-c and Slac2-a functions as a novel actin-binding motif (Desnos *et al.*, 2003; Fukuda and Kuroda, 2002; Kuroda *et al.*, 2003; Passeron *et al.*, 2004; Waselle *et al.*, 2003). Based on these observations, together with the fact that Slac2-c is present on the melanosomes of mouse retinal pigment epithelium (RPE) cells (El-Amraoui *et al.*, 2002), it has been suggested that Slac2-c controls retinal melanosome transport in RPE cells in a manner analogous to the function of Slac2-a (or the Rab27A-Slac2-a-myosin Va complex) in melanosome transport in skin melanocytes. It has also been demonstrated that Slac2-c regulates secretory granule exocytosis in some secretory cells through interaction with actin filaments independently of the function of myosin Va/VIIa (Desnos *et al.*, 2003; Imai *et al.*, 2004; Waselle *et al.*, 2003).

This chapter describes experimental methods that we have used to characterize the three functional domains of Slac2-c: the N-terminal SHD, the myosin-binding domain (MBD), and the C-terminal actin-binding domain (ABD).

Specific Interaction between the SHD of Slac2-c and Rab27A/B

The SHDs of Slp and Slac2 proteins are composed of two highly conserved potential α-helical regions, referred to as SHD1 ($\sim$45 amino acids) and SHD2 ($\sim$35 amino acids) (Fig. 1A). The SHD of Slp3-a, Slp4/granuphilin, Slp5, Slac2-a, and Slac2-c contains two zinc finger motifs between two α-helices ($\sim$80 amino acids), whereas the SHD of Slp1, Slp2-a, and Slac2-b lacks such zinc finger motifs, and their SHD1 and SHD2 are linked together (Fukuda, 2002, 2005; Fukuda *et al.*, 2001; Kuroda *et al.*, 2002a). To characterize SHD as a specific Rab27-binding domain, we have established panels of Rab proteins in COS-7 cells (Fukuda, 2003; Kuroda *et al.*, 2002a) (see also Chapter 39, this volume) and tested recombinant Slac2-c for interaction with a series of Rab proteins. The results clearly have indicated that SHD functions as a novel Rab27-binding motif. In this section we describe several protocols for evaluating the specific interaction between Slac2-c and GTP-Rab27A *in vitro*.

Expression and Purification of GST-Slac2-c-SHD

pGEX-4T-3-Slac2-c-SHD vector (Amersham Biosciences, Buckinghamshire, UK), which encodes the SHD of mouse Slac2-c (amino acids 1–145; Fig. 1A), is constructed by standard molecular cloning techniques (Kuroda *et al.*, 2002a), and the plasmid is transformed into *Escherichia coli* JM109. A single bacterial colony is transferred into 20 ml of LB medium containing 100 μg/ml of ampicillin (LA) in a 50-ml tube, and precultured by incubation at 37° overnight (with constant shaking at 200 rpm). The bacterial culture is then diluted into 400 ml of LA in a 1-liter flask and incubated for 1 h at 30°. Isopropylthlogalactoside (IPTG; final concentration 1 m*M*) is then added to the culture to induce expression of GST-Slac2-c-SHD

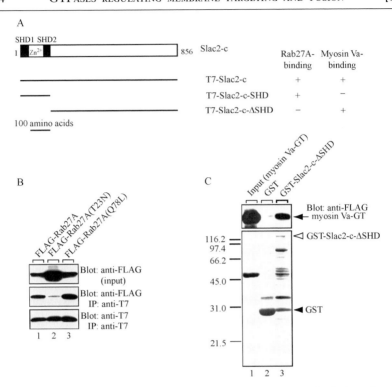

FIG. 1. Slac2-c interacts with Rab27A and myosin Va through distinct domains. (A) Schematic representation of the mouse T7-Slac2-c deletion mutants in (B) and (C) below. Slac2-c contains an N-terminal SHD composed of two potential α-helical regions (SHD1 and SHD2; black boxes) separated by two zinc finger motifs (indicated by Zn^{2+}). The Rab27A- and myosin Va-binding activity of each Slac2-c deletion mutant is indicated on the right. (B) Specific interaction between Slac2-c and the GTP-bound form of Rab27A. T7-Slac2-c-SHD was coexpressed in COS-7 cells together with FLAG-Rab27A, the T23N mutant (constitutive inactive form), and the Q78L mutant (constitutive active form). The binding experiment was performed as described in the text. Proteins bound to the T7-Slac2-c-SHD beads were analyzed by 12.5% SDS-PAGE and visualized with HRP-conjugated anti-FLAG tag antibody (middle panel) and HRP-conjugated anti-T7 tag antibody (bottom panel). The top panel indicates the total expressed FLAG-Rab27A proteins used for immunoprecipitation (input). (C) Direct interaction between Slac2-c-ΔSHD and the globular tail of myosin Va. A binding experiment using purified components of GST-Slac2-c-ΔSHD (or GST alone) and FLAG-myosin Va-GT was performed as described in the text. The myosin Va-GT protein trapped by the beads was detected with HRP-conjugated anti-FLAG tag antibody (arrow in top panel), and GST-Slac2-c-ΔSHD (open arrowhead) and GST alone (closed arrowhead) coupled with the beads were confirmed by Amido Black staining (bottom panel). Molecular mass markers ($\times 10^{-3}$) are shown on the left. Reproduced with permission from Fukuda and Kuroda (2002).

protein. After incubation at 30° for 3.5 h, the bacterial pellets are resuspended in 30 ml of a resuspension buffer (20 mM phosphate buffer, pH 7.0, 150 mM NaCl) containing 0.1 mM phenylmethylsulfonylfluoride (PMSF). The bacterial cells are then disrupted by sonication for 1 min on ice five times at 5-min intervals, and GST-Slac2-c-SHD protein is solubilized by gentle agitation with 1% Triton X-100 for 15 min at 4°. After removing insoluble material by centrifuging at $11,050 \times g$ for 15 min at 4° and filtering through a 0.45-μm disk filter (Millipore Corp, Bedford, MA), the supernatant is applied to a glutathione-Sepharose 4B column (packed volume 1 ml; Amersham Biosciences) preequilibrated with phosphate-buffered saline (PBS) containing 1% Triton X-100 in an Evergreen column (Los Angeles, CA). After washing the column with 10 ml of PBS, GST-Slac2-c-SHD is eluted from the beads with a buffer containing 5 mM glutathione and 50 mM Tris–HCl, pH 8.0 (1 ml × 7). The eluate is dialyzed overnight against 500 ml of PBS (changed three times) and concentrated to less than 500 μl with Centriprep10 (Millipore Corp.). The concentration of the GST-Slac2-c-SHD protein obtained is determined with the Bio-Rad (Hercules, CA) protein assay kit using bovine serum albumin (BSA) as a reference. The entire procedure is carried out at low temperature (4°) to the extent possible.

Assay for Interaction between GST-Slac2-c-SHD and FLAG-Rab Proteins

Panels of FLAG-Rab proteins (i.e., diluted COS-7 cell lysates containing equivalent amounts of each FLAG-Rab protein) are prepared as described in Chapter 39 (this volume) (Fukuda, 2003; Kuroda *et al.*, 2002a). Purified GST-Slac2-c-SHD (1 μg) is coupled with glutathione-Sepharose 4B beads (wet volume 20 μl) by gentle agitation for 30 min at 4° in a washing buffer (50 mM HEPES-KOH, pH 7.2, 150 mM NaCl, 0.2% Triton X-100, 0.1 mM PMSF, 10 μM leupeptin, and 10 μM pepstatin A; "protease inhibitors" refers to these three inhibitors throughout the text). After washing the beads three times with the washing buffer, the GST-Slac2-c-SHD beads are incubated for 2 h at 4° with 1 ml of the diluted cell lysates containing almost equal amounts of each FLAG-Rab protein. Note that addition of 0.5 mM GTPγS [guanosine 5'-O-(3-thiotriphosphate)] and 1 mM MgCl$_2$ to the reaction mixtures is recommended to stabilize the interaction between the Slac2-SHD and Rab. After washing the beads five times with 1 ml of the washing buffer, proteins trapped by the beads are analyzed by 12.5% sodium dodecyl sulfate–polyacrylamide gel electrophoresis (SDS-PAGE) followed by immunoblotting with horseradish peroxidase (HRP)-conjugated anti-FLAG tag (M2) antibody (1/10,000 dilution; Sigma Chemical Co., St. Louis, MO) and enhanced chemiluminescence

(ECL) detection (Amersham Biosciences) as described previously (Fukuda *et al.*, 1999; Kuroda *et al.*, 2002a). The same blots are stripped and then reprobed with HRP-conjugated anti-GST antibody (1/10,000 dilution; Santa Cruz Biotechnology, Inc., Santa Cruz, CA) to ensure that the same amounts of GST-Slac2-c-SHD are loaded (see also Chapter 40, this volume). Since GST-SHD proteins prepared from bacteria are often contaminated by degradation products, GST-SHD occasionally nonspecifically binds Rabs other than Rab27A/B under the above experimental conditions. Cotransfection assay in COS-7 cells (T7-Slac2/Slp and FLAG-Rab) has also been developed to avoid such nonspecific binding (Fukuda and Kuroda, 2002; Fukuda *et al.*, 2002; Kuroda *et al.*, 2002a) and protocols are described in detail in Chapter 39 (this volume).

Assay for Specific Interaction between the Slac2-c SHD and the GTP-Bound Form of Rab27A

Constitutive active (Q78L, a Glu-to-Leu substitution at amino acid position 78) and constitutive inactive (T23N) mutants of Rab27A are constructed by site-directed mutagenesis, and mutant cDNA fragments are subcloned into the eukaryotic expression vector pEF-FLAG (modified from pEF-BOS) by standard molecular cloning techniques to obtain pEF-FLAG-Rab27A(Q78L) and (T23N) (Fukuda *et al.*, 1999; Kuroda *et al.*, 2002a; Mizushima and Nagata, 1990). Cotransfection of pEF-T7-Slac2-c-SHD and pEF-FLAG-Rab27A(T23N) or -Rab27A(Q78L) in COS-7 cells and coimmunoprecipitation assay are performed as described in Chapter 39 (this volume). Under these experimental conditions, T7-Slac2-c specifically interacts with Rab27A(Q78L), which mimics the GTP-bound form, not with Rab27A(T23N), which mimics the GDP-bound form (Fig. 1B) (Fukuda and Kuroda, 2002).

Guanine-nucleotide-selective interaction of Rab27A with Slp/Slac2 is also tested by incubation of T7-tagged Slp/Slac2 with wild-type FLAG-Rab27A in the presence of 0.5 mM GTPγS (fixed in an active form) or 1 mM GDP (fixed in an inactive form) (Kuroda *et al.*, 2002a). However, GTP-dependent Rab27A binding to Slp/Slac2 is demonstrated more clearly when Rab27A(Q78L) and Rab27A(T23N) mutants are used.

Slac2-c Directly Interacts with Myosin Va and Myosin VIIa

Since Slac2-a has been shown to function as a myosin Va receptor via the middle region of Slac2-a (i.e., MBD) (Fukuda, *et al.*, 2002; Kuroda *et al.*, 2003; Nagashima *et al.*, 2002; Strom *et al.*, 2002; Wu *et al.*, 2002) and Slac2-c contains a putative MBD in the middle part of the molecule,

Slac2-c and myosins have been tested for interaction *in vitro*. The results have shown that Slac2-c interacts with the tail domain of both myosin Va and myosin VIIa, whereas Slac2-a interacts with the tail domain of myosin Va alone, and not with the tail domain of myosin VIIa (Desnos *et al.*, 2003; Fukuda and Kuroda, 2002). This section describes several experimental protocols that are used to evaluate the interaction between Slac2-c and myosin Va/VIIa *in vitro* and in cultured melanocytes.

Assay for Direct Interaction between GST-Slac2-c-ΔSHD and the Globular Tail of Myosin Va

Construction of pEF-T7-GST-Slac2-c-ΔSHD and pEF-T7-GST-FLAG-myosin Va-GT (encoding the globular tail of mouse myosin Va) is performed as described previously (Fukuda and Kuroda, 2002; Fukuda *et al.*, 2002). COS-7 cells (7.5×10^5 cells/10-cm dish, the day before transfection) are transfected with 4 μg of pEF-T7-GST-Slac2-c-ΔSHD or pEF-T7-GST-FLAG-myosin Va-GT by using LipofectAmine Plus reagent (Invitrogen, Carlsbad, CA) according to the manufacturer's notes. Three days after transfection, cells are harvested and homogenized in 1 ml of a buffer containing 50 mM HEPES-KOH, pH 7.2, 250 mM NaCl, and protease inhibitors in a glass-Teflon Potter homogenizer with 10 strokes at 900–1000 rpm. After solubilization with 1% Triton X-100, insoluble material is removed by centrifugation at $17,360 \times g$ for 10 min at 4°. The GST fusion proteins expressed are affinity purified on glutathione-Sepharose beads (wet volume 20 μl) according to the manufacturer's notes. After extensively washing the T7-GST-FLAG-myosin Va-GT-coupled beads with 10 mM HEPES-KOH, pH 7.2, 100 mM NaCl, 0.2% Triton X-100, and protease inhibitors, they are digested with thrombin (1 unit; Sigma Chemical Co.) on the same column at 25° for 1 h. The eluate containing FLAG-myosin Va-GT protein is then incubated with benzamidine-Sepharose 6B (wet volume 20 μl; Amersham Biosciences) to remove thrombin. Protein concentrations are estimated by 10% SDS-PAGE or determined with a Bio-Rad protein assay kit using BSA as a reference.

The purified FLAG-myosin Va-GT protein is incubated with glutathione-Sepharose beads (wet volume 20 μl) either coupled with T7-GST-Slac2-c-ΔSHD or T7-GST alone in 50 mM HEPES-KOH, pH 7.2, 100 mM NaCl, 1 mM MgCl$_2$, and 0.2% Triton X-100 for 1 h at 4°. After washing three times with 1 ml of the binding buffer without recombinant proteins, proteins trapped by the beads are analyzed by 10% SDS-PAGE followed by immunoblotting with HRP-conjugated anti-FLAG tag antibody (1/10,000 dilution) and Amido Black staining. Under the above experimental conditions, myosin Va-GT directly interacts with Slac2-c-Δ

SHD (Fig. 1C) and Slac2-a-ΔSHD (Fukuda *et al.*, 2002). Note that the interaction between myosin Va and Slac2-c (or Slac2-a) is stronger when the whole tail domain containing the exon F of myosin Va is used for the binding assays (Fukuda and Kuroda, 2002, 2004) (see also below).

Cotransfection Assay for Interaction between Slac2-c and the Tail Domain of Myosin Va and Myosin VIIa

The tail domain of mouse myosin Va has three alternative splicing regions, namely, exon B, exon D, and exon F. Brain-type (BR) myosin Va contains the exon B but lacks the exon D and exon, F, whereas melanocyte-type (MC) myosin Va contains the exon D and exon F instead of the exon B (Huang *et al.*, 1998; Wu *et al.*, 2002). The mouse BR and MC myosin Va tail cDNAs and the mouse myosin VIIA cDNA are amplified by polymerase chain reactions and subcloned into the eukaryotic expression vector pEF-FLAG as described previously (Fukuda and Kuroda, 2002). pEF-T7-Slac2-c and pEF-FLAG-MC-myosin Va tail (-FLAG-BR-myosin Va tail, or -FLAG-myosin VIIa tail) are cotransfected into COS-7 cells as described earlier. The cells are homogenized in 1 ml of a buffer containing 50 mM HEPES-KOH, pH 7.2, 250 mM NaCl, and protease inhibitors, and solubilized with 1% Triton X-100 at 4° for 1 h. The proteins expressed are immunoprecipitated at 4° for 1 h with anti-T7 tag antibody-conjugated agarose (wet volume 10 μl; Novagen, Madison, WI). The beads are then washed three times with 1 ml of 50 mM HEPES-KOH, pH 7.2, 150 mM NaCl, 0.2% Triton X-100, and protease inhibitors, and resuspended in SDS sample buffer. Coimmunoprecipitated FLAG-myosins are detected first with HRP-conjugated anti-FLAG tag antibody (1/10,000 dilution). The same blots are then stripped and reprobed with HRP-conjugated anti-T7 tag antibody (1/10,000 dilution) to ensure that the same amounts of T7-Slac2-c protein are loaded. Under these experimental conditions, Slac2-c interacts with MC-myosin Va in preference to BR-myosin Va, the same as Slac2-a (Fukuda and Kuroda, 2002; 2004), and the affinity of Slac2-c for myosin VIIa is much greater than its affinity for myosin Va (Kuroda and Fukuda, 2005).

Assay for In Vitro Formation of a Tripartite Protein Complex of Rab27A, Slac2-a/c, and Myosin Va from Purified Components

pEF-GST-FLAG-myosin Va-GT (4 μg of plasmid) is transfected into COS-7 cells as described earlier. After purification of the expressed GST-fusion protein from the COS-7 cell lysate with glutathione-Sepharose beads, recombinant FLAG-myosin Va-GT protein is prepared by cleavage of GST by means of thrombin digestion as described earlier.

The mouse Rab27A cDNA is subcloned into the eukaryotic expression vector pEF-HA (named pEF-HA-Rab27A) as described previously (Fukuda and Kuroda, 2004). pEF-T7-Slac2-c and pEF-HA-Rab27A (2 μg of each plasmid) are cotransfected into COS-7 cells, and the T7-Slac2-c-HA-Rab27A complex is affinity purified with anti-T7 tag antibody-conjugated agarose beads as described earlier. The beads are then incubated for 1 h at 4° with purified FLAG-myosin Va-GT proteins (see previous discussion) in a buffer containing 50 mM HEPES-KOH, pH 7.2, 100 mM NaCl, 1 mM MgCl$_2$ 0.5 mM GTPγS, and 0.2% Triton X-100. After washing three times with 1 ml of the binding buffer without recombinant proteins, proteins trapped by the beads are analyzed by 10% SDS-PAGE followed by immunoblotting with HRP-conjugated anti-FLAG tag antibody, anti-HA tag (YPYDVPDYA) antibody (1/10,000 dilution; Sigma Chemical Co.), and anti-T7 tag antibody as described earlier. *In vivo* formation of the tripartite protein complex of Rab27A-Slac2-a-myosin Va in mouse melanocytes is also demonstrated by coimmunoprecipitation assay with anti-Slac2-a specific antibody (Fukuda *et al.*, 2002). The protocols for the coimmunoprecipitation assay are described in detail in Chapter 39 (this volume).

Interaction between the C-Terminus of Slac2-a/c and Actin

Although the minimal myosin Va-binding domain is located in the middle of the Slac2-a molecule (241–400 amino acids) and the most C-terminal part (approximately 200 amino acids) of Slac2-a is not essential for myosin Va-binding activity, we noted relatively high similarity (21.5% identity) between the most C-terminal part of Slac2-a and Slac2-c. To characterize the C-terminal part of Slac2-c biochemically, we searched for the C-terminal region-binding protein and succeeded in isolating a single protein with an approximate molecular mass of 45 kDa that was subsequently identified as actin with a specific anti-actin antibody. In this section we describe the experimental methods used to characterize the C-terminus of Slac2-c (or Slac2-a) as a novel actin-binding motif, both *in vitro* and in intact cells.

Assay for Determination of the Actin-Binding Domain of Slac2-c

The actin-binding domain of Slac2-a/c is mapped by performing a binding experiment between ectopically expressed T7-Slac2-a/c deletion mutants and endogenous actin in COS-7 cells. Four deletion mutants of mouse Slac2-c are constructed and subcloned into the pEF-T7 vector as described previously (Fukuda and Kuroda, 2002): Slac2-c-SHD (amino acids 1–145), Slac2-c-ΔSHD (amino acids 146–856), Slac2-c-Δ146/494 (amino acids

146–494), and Slac2-c-495/856 (amino acids 495–856; ABD) (Fig. 2A). These pEF-T7-Slac2-c deletion mutants are expressed in COS-7 cells, and the T7-tagged proteins are affinity purified with anti-T7 tag antibody-conjugated agarose beads as described previously. Copurified proteins are

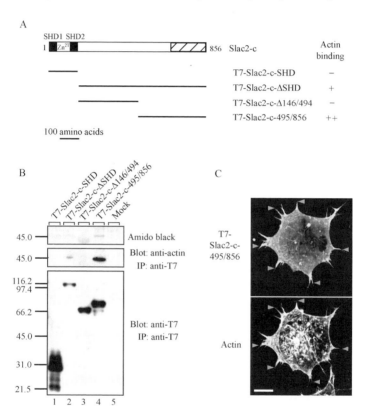

FIG. 2. Characterization of the actin-binding domain of Slac2-c. (A) Schematic representation of the mouse T7-Slac2-c deletion mutants used in (B) and (C) below. The putative ABD of Slac2-c is indicated by the hatched box. The actin-binding activity of each mutant is shown on the right. (B) Mapping of the actin-binding domain of Slac2-c. T7-Slac2-c deletion mutants were expressed in COS-7 cells and affinity purified with anti-T7 tag antibody-conjugated agarose beads as described in the text. Coimmunoprecipitated actin is detected by Amido Black staining (top panel) and immunoblotting with anti-actin antibody (middle panel). The bottom panel shows immunoprecipitated T7-Slac2-c mutants visualized with HRP-conjugated anti-T7 tag antibody. Molecular mass markers ($\times 10^{-3}$) are shown on the left. (C) PC12 cells expressing T7-Slac2-c-495/856 were fixed, permeabilized, and immunostained with anti-T7 tag mouse monoclonal antibody (upper panel). Actin filaments were visualized with Texas Red-conjugated phalloidin (lower panel). Arrowheads indicate colocalization of Slac2-c-495/856 with actin filaments and the inhibited neurite outgrowth. The bar in the lower panel indicates 10 μm. Reproduced with permission from Fukuda and Kuroda (2002).

visualized by Amido Black staining or with anti-actin goat antibody (1/200 dilution; Santa Cruz Biotechnology, Inc.). The immunoprecipitated T7-Slac2-c mutants are visualized with HRP-conjugated anti-T7 tag antibody. Under the above experimental conditions, the interaction between full-length Slac2-a/c and actin is much weaker than the interaction between the ABD of Slac2-a/c and actin (Fig. 2B) (Fukuda and Kuroda, 2002).

Assay for Direct Interaction between Slac2-a and Globular and Fibrous Actin

The cDNA encoding the ABD of mouse Slac2-a (amino acids 401–590) is subcloned into the pGEX-4T-3 vector by standard molecular cloning techniques, and GST-Slac2-a-ABD (or GST alone) is expressed and purified as described earlier. Globular actin (G-actin) or fibrous actin (F-actin) (catalogue number A-3653; Sigma Chemical Co.) from bovine muscle is prepared as described previously (Perelroizen et al., 1996). A 100 μg quantity of actin is mixed with 1 μg of GST-Slac2-a-ABD (or GST alone as a control) in 400 μl of G-buffer (5 mM Tris–HCl, pH 7.5, 1 mM dithiothreitol, 0.2 mM ATP, and 0.1 mM CaCl$_2$) or F-buffer (5 mM Tris–HCl, pH 7.5, 1 mM dithiothreitol, 0.2 mM ATP, 0.1 mM CaCl$_2$, 2 mM MgCl$_2$, and 0.1 M KCl). After incubation for 1 h at 4° with gentle agitation, the actin–ABD complex in the reaction mixtures is immunoprecipitated for 1 h at 4° with anti-actin mouse monoclonal antibody (C-2)-conjugated agarose (wet volume 10 μl; Santa Cruz Biotechnology, Inc.). After washing with 1 ml of the G-buffer or F-buffer five times, proteins bound to the beads are analyzed by 10% SDS-PAGE followed by immunoblotting with HRP-conjugated anti-GST antibody (1/2000 dilution; Santa Cruz Biotechnology, Inc.) and ECL detection as described earlier. The blot is then reprobed with anti-actin goat polyclonal antibody (I-19; 1/200 dilution; Santa Cruz Biotechnology, Inc.) to ensure that the same amount of actin has been loaded. Under the above experimental conditions, the ABD of Slac2-a directly interacts with both G-actin and F-actin (Fukuda and Kuroda, 2002).

Localization of the Slac2-c ABD on Actin Filaments in PC12 Cells

PC12 cells are cultured at 37° in Dulbecco's modified Eagle's medium supplemented with 10% fetal bovine serum, 10% horse serum, 100 U/ml penicillin G, and 100 μg/ml streptomycin, under 5% CO$_2$. Transfection of pEF-T7-Slac2-c-495/856 (4 μg of plasmid) into PC12 cells (0.5~1.0 × 10^5 cells, the day before transfection/3.5-cm dish [MatTek Corp., Ashland, MA] coated with collagen type IV) is performed by using Lipofect-AMINE 2000 reagent (8 μl; Invitrogen, Carlsbad, CA) according to the manufacturer's notes. One day after transfection, cells are stimulated to extend neurites with β-nerve growth factor (NGF) (100 ng/ml; Merck, Darmstadt, Germany). Three days after transfection, cells are fixed with 4% paraformaldehyde for 20 min, permeabilized with 0.3% Triton X-100 for 2 min, and blocked with blocking buffer (1% BSA and 0.1% Triton X-100 in PBS) for 1 h. Next, the cells are immunostained with anti-T7 tag mouse monoclonal antibody (1/4000 dilution; Novagen) for 1 h and then incubated with Alexa-Fluor 488-labeled anti-mouse secondary antibody (1/5000 dilution; Molecular Probes, Inc., Eugene, OR) and Texas Red-conjugated phalloidin (1/5000 dilution; Molecular Probes, Inc.) for 1 h to visualize T7-Slac2-c-495/856 protein and F-actin, respectively. The transfected cells are examined for fluorescence with a confocal laser-scanning microscope (Fluoview FV500; Olympus, Tokyo, Japan), and the images are processed with Adobe Photoshop software (version 7.0) (Fig. 2C). The method of immunofluorescence analysis is described in detail in Chapter 40 (this volume).

Under these experimental conditions, the ABD of both Slac2-a or Slac2-c expressed in PC12 cells colocalizes with actin filaments in the cell periphery and inhibits NGF-induced neurite outgrowth of PC12 cells, probably due to the inhibition of actin filaments remodeling during neurite outgrowth (Fukuda and Kuroda, 2002) (Fig. 2C). In addition, mouse melanocyte melan-a cells expressing the Slac2-a ABD exhibit excess formation of actin bundles in the cell periphery and altered distribution of melanosomes (Kuroda et al., 2003). It is noteworthy that low expression levels of green fluorescence protein (GFP)-tagged Slac2-a are usually present on the melanosomes in melan-a cells, whereas high expression levels of GFP-Slac2-a are mainly present on actin filaments, rather than on melanosomes, and often cause perinuclear aggregation of melanosomes (Fukuda and Itoh, 2004).

Concluding Remarks

We have described several methods of analyzing the three-domain structure of Slac2-c in in vitro and in cultured cells. Consistent with the fact that the overall structure of Slac2-c is similar to that of Slac2-a (Fukuda and Kuroda, 2002), Slac2-c simultaneously interacts with Rab27A via the N-terminal SHD, myosin Va via the middle region, and actin via the ABD, the same as Slac2-a (Kuroda et al., 2003). The most striking difference between Slac2-a and Slac2-c is that only Slac2-c is capable of interacting with myosin VIIa (El-Amraoui et al., 2002; Fukuda and Kuroda, 2002). Several physiological functions of Slac2-c have been proposed based on these biochemical properties of Slac2-c. The most likely function of Slac2-c is as a linker protein between retinal melanosome-bound Rab27A and myosin VIIa in RPE cells in retinal melanosome transport (El-Amraoui *et al.*, 2002), the same as the function of Slac2-a in skin melanosome transport, although no precise functional analysis of Slac2-c in retinal melanosome transport has yet been performed. Slac2-c has been shown to be involved in the control of secretion/motion of secretory granules in pancreatic β-cell lines (Waselle *et al.*, 2003), PC12 cells (Desnos *et al.*, 2003), and parotid acinar cells (Imai *et al.*, 2004). It should be noted, however, that the control of secretory granule exocytosis by Slac2-c depends on its actin- and Rab27A-binding ability but is independent of interaction with myosin Va/VIIa in these secretory cells.

Acknowledgments

This work was supported in part by grants from the Ministry of Education, Culture, Sports, and Technology of Japan.

References

Desnos, C., Schonn, J. S., Huet, S., Tran, V. S., El-Amraoui, A., Raposo, G., Fanget, I., Chapuis, C., Ménasché, G., de Saint Basile, G., Petit, C., Cribier, S., Henry, J. P., and Darchen, F. (2003). Rab27A and its effector MyRIP link secretory granules to F-actin and control their motion towards release sites. *J. Cell Biol.* **163**, 559–570.

El-Amraoui, A., Schonn, J. S., Kussel-Andermann, P., Blanchard, S., Desnos, C., Henry, J. P., Wolfrum, U., Darchen, F., and Petit, C. (2002). MyRIP, a novel Rab effector, enables myosin VIIa recruitment to retinal melanosomes. *EMBO Rep.* **3**, 463–470.

Fukuda, M. (2002). Slp and Slac2, novel families of Rab27 effectors that control Rab27-dependent membrane traffic. *Recent Res. Dev. Neurochem.* **5**, 297–309.

Fukuda, M. (2003). Distinct Rab binding specificity of Rim1, Rim2, rabphilin, and Noc2: Identification of a critical determinant of Rab3A/Rab27A recognition by Rim2. *J. Biol. Chem.* **278**, 15373–15380.

Fukuda, M. (2005). Versatile role of Rab27 in membrane trafficking: Focus on the Rab27 effector families. *J. Biochem.* **137**, 9–16.

Fukuda, M., and Itoh, T. (2004). Slac2-a/melanophilin contains multiple PEST-like sequences that are highly sensitive to proteolysis. *J. Biol. Chem.* **279**, 22314–22321.

Fukuda, M., and Kuroda, T. S. (2002). Slac2-c (synaptotagmin-like protein homologue lacking C2 domains-c), a novel linker protein that interacts with Rab27, myosin Va/VIIa, and actin. *J. Biol. Chem.* **277**, 43096–43103.

Fukuda, M., and Kuroda, T. S. (2004). Missense mutations in the globular tail of myosin-Va in *dilute* mice partially impair binding of Slac2-a/melanophilin. *J. Cell Sci.* **117**, 583–591.

Fukuda, M., Kanno, E., and Mikoshiba, K. (1999). Conserved N-terminal cysteine motif is essential for homo- and heterodimer formation of synaptotagmins III, V, VI, and X. *J. Biol. Chem.* **274**, 31421–31427.

Fukuda, M., Saegusa, C., and Mikoshiba, K. (2001). Novel splicing isoforms of synaptotagmin-like proteins 2 and 3: Identification of the Slp homology domain. *Biochem. Biophys. Res. Commun.* **283**, 513–519.

Fukuda, M., Kuroda, T. S., and Mikoshiba, K. (2002). Slac2-a/melanophilin, the missing link between Rab27 and myosin Va: Implications of a tripartite protein complex for melanosome transport. *J. Biol. Chem.* **277,** 12432–12436.

Huang, J. D., Mermall, V., Strobel, M. C., Russell, L. B., Mooseker, M. S., Copeland, N. G., and Jenkins, N. A. (1998). Molecular genetic dissection of mouse unconventional myosin-VA: Tail region mutations. *Genetics* **148,** 1963–1972.

Imai, A., Yoshie, S., Nashida, T., Shimomura, H., and Fukuda, M. (2004). The small GTPase Rab27B regulates amylase release from rat parotid acinar cells. *J. Cell Sci.* **117,** 1945–1953.

Kuroda, T. S., and Fukuda, M. (2005). Functional analysis of Slac2-c/MyRIP as a linker protein between melanosomes and myosin VIIa. *J. Biol. Chem.* **280,** 28015–28022.

Kuroda, T. S., Fukuda, M., Ariga, H., and Mikoshiba, K. (2002a). The Slp homology domain of synaptotagmin-like proteins 1–4 and Slac2 functions as a novel Rab27A binding domain. *J. Biol. Chem.* **277,** 9212–9218.

Kuroda, T. S., Fukuda, M., Ariga, H., and Mikoshiba, K. (2002b). Synaptotagmin-like protein 5: A novel Rab27A effector with C-terminal tandem C2 domains. *Biochem. Biophys. Res. Commun.* **293,** 899–906.

Kuroda, T. S., Ariga, H., and Fukuda, M. (2003). The actin-binding domain of Slac2-a/melanophilin is required for melanosome distribution in melanocytes. *Mol. Cell. Biol.* **23,** 5245–5255.

Matesic, L. E., Yip, R., Reuss, A. E., Swing, D. A., O'Sullivan, T. N., Fletcher, C. F., Copeland, N. G., and Jenkins, N. A. (2001). Mutations in *Mlph*, encoding a member of the Rab effector family, cause the melanosome transport defects observed in *leaden* mice. *Proc. Natl. Acad. Sci. USA* **98,** 10238–10243.

Mizushima, S., and Nagata, S. (1990). pEF-BOS, a powerful mammalian expression vector. *Nucleic Acids Res.* **18,** 5322.

Nagashima, K., Torii, S., Yi, Z., Igarashi, M., Okamoto, K., Takeuchi, T., and Izumi, T. (2002). Melanophilin directly links Rab27a and myosin Va through its distinct coiled-coil regions. *FEBS Lett.* **517,** 233–238.

Passeron, T., Bahadoran, P., Bertolotto, C., Chiaverini, C., Buscà, R., Valony, G., Bille, K., Ortonne, J. P., and Ballotti, R. (2004). Cyclic AMP promotes a peripheral distribution of melanosomes and stimulates melanophilin/Slac2-a and actin association. *FASEB J.* **18,** 989–991.

Perelroizen, I., Didry, D., Christensen, H., Chua, N.-H., and Carlier, M.-F. (1996). Role of nucleotide exchange and hydrolysis in the function of profilin in action assembly. *J. Biol. Chem.* **271,** 12302–12309.

Strom, M., Hume, A. N., Tarafder, A. K., Barkagianni, E., and Seabra, M. C. (2002). A family of Rab27-binding proteins: Melanophilin links Rab27a and myosin Va function in melanosome transport. *J. Biol. Chem.* **277,** 25423–25430.

Waselle, L., Coppola, T., Fukuda, M., Iezzi, M., El-Amraoui, A., Petit, C., and Regazzi, R. (2003). Involvement of the Rab27 binding protein Slac2c/MyRIP in insulin exocytosis. *Mol. Biol. Cell* **14,** 4103–4113.

Wu, X. S., Rao, K., Zhang, H., Wang, F., Sellers, J. R., Matesic, L. E., Copeland, N. G., Jenkins, N. A., and Hammer, J. A., III (2002). Identification of an organelle receptor for myosin-Va. *Nat. Cell Biol.* **4,** 271–278.

[39] Analysis of the Role of Rab27 Effector Slp4-a/ Granuphilin-a in Dense-Core Vesicle Exocytosis

By MITSUNORI FUKUDA and EIKO KANNO

Abstract

Slp4-a/granuphilin-a is a member of the synaptotagmin-like protein (Slp) family and consists of an N-terminal Slp homology domain (SHD) and C-terminal tandem C2 domains. Slp4-a is specifically localized on secretory granules in some endocrine and exocrine cells through its SHD, and it attenuates Ca^{2+}-dependent dense-core vesicle (DCV) exocytosis when transiently expressed in endocrine cells. Although the SHD of Slp4-a interacts with three distinct Rab species (Rab3A, Rab8A, and Rab27A) *in vitro*, in contrast to other Slp members, which only recognize Rab27 isoforms, Slp4-a functions as a Rab27A effector during DCV exocytosis under physiological conditions. This chapter describes various approaches that have been used to characterize the function of Slp4-a as a Rab27A effector, rather than a Rab3A or Rab8A effector, both in *in vitro* and in neuroendocrine PC12 cells. Specifically, the methods that have been used to analyze (1) the physical interaction between Slp4-a and Rab27A, including pull-down assay and cotransfection assay in COS-7 cells; (2) the localization of Slp4-a-Rab27A complex on DCVs in PC12 cells; and (3) the involvement of Slp4-a and Rab27A in DCV exocytosis by neuropeptide Y (NPY) cotransfection assay combined with site-directed mutagenesis are described.

Introduction

Granuphilin-a was originally described as a rabphilin-like protein that is abundantly expressed on insulin-containing granules in pancreatic β-cells (Wang *et al.*, 1999). It was subsequently reported to be the fourth member of the synaptotagmin-like protein (Slp) family (i.e., Slp4-a), which is distinct from the rabphilin/Doc2 family (Fukuda, 2003a; Fukuda and Mikoshiba, 2001), and Slp4-a has also been found on dense-core vesicles (DCVs) in other endocrine cells (e.g., chromaffin cells, PC12 cells, and AtT-20 cells) and amylase-containing granules in parotid acinar cells (Imai *et al.*, 2004). The Slp family consists of five members (Slp1/JFC1, Slp2-a, Slp3-a, Slp4-a/ granuphilin-a, and Slp5) in mammals and one member (dm-Slp/Btsz) in fruit flies, and an Slp protein is defined as a protein having a putative

METHODS IN ENZYMOLOGY, VOL. 403   0076-6879/05 $35.00
          DOI: 10.1016/S0076-6879(05)03039-9

Rab-binding domain (called Slp homology domain; SHD) at the N-terminus and two putative Ca^{2+}- and/or phospholipid-binding motifs at the C-terminus (called the C2A domain and the C2B domain) (Fukuda, 2002a, 2005; Fukuda *et al.*, 2001; Izumi *et al.*, 2003; Kuroda *et al.*, 2002b). The SHD of Slp family members have been shown to function as an effector domain for specific Rab, a small GTP-binding protein believed to be essential for membrane trafficking in eukaryotic cells (Zerial and McBride, 2001). All SHDs directly interact with the GTP-bound form of Rab27A and Rab27B *in vitro* (Fukuda, 2002b, 2003b; Kuroda *et al.*, 2002a,b; Strom *et al.*, 2002), but the Slp4-a SHD is exceptional because it is also capable of interacting with Rab8 isoforms (Rab8A/B) and Rab3 isoforms (Rab3A// B/C/D) (Coppola *et al.*, 2002; Fukuda, 2003c; Fukuda *et al.*, 2002a; Yi *et al.*, 2002), whereas the others recognize only Rab27 isoforms (El-Amraoui *et al.*, 2002; Fukuda and Kuroda, 2002; Fukuda *et al.*, 2002d; Kuroda and Fukuda, 2004). Thus, functional assays are essential to determine which of the Rab-Slp4-a complexes is involved in the control of regulated granule exocytosis *in vivo*.

We have developed several functional assays to identify Slp4-a-binding proteins and shown that Slp4-a preferentially interacts with Rab27A on DCVs in neuroendocrine PC12 cells and regulates their exocytosis through specific interaction with Rab27A. In this chapter we describe the methods that we have used in the study of Slp4-a as a Rab27A effector in PC12 cells.

Biochemical Characterization of the Rab-Binding Partners of Slp4-a *In Vitro*

Since the SHD of Slp shows little similarity to the previously described Rab3A-binding domain of rabphilin, Rim, and Noc2, it has been proposed that it functions as a binding domain of some Rab other than Rab3 (Fukuda *et al.*, 2001; Kuroda *et al.*, 2002a). Although several commonly used methods (e.g., yeast two-hybrid screening and pull-down assay) are available to identify the Rab that binds the SHD, it is of paramount importance to determine whether the interaction between SHD and the Rab occurs in a Rab species (or subfamily)-specific manner, because more than 60 Rabs (approximately 40 Rab subfamilies) are present in mammals (Pereira-Leal and Seabra, 2001). To do so, we cloned almost all of the mouse Rab clones that belong to distinct Rab subfamilies (i.e., Rab1–40) and established panels of FLAG-tagged Rab proteins in COS-7 cells. The Rab panels have subsequently proved to be powerful and useful tools for determining the Rab-binding specificity of SHD *in vitro* (Fukuda, 2003b; Fukuda *et al.*, 2002d; Kuroda *et al.*, 2002a,b), and SHD is now generally

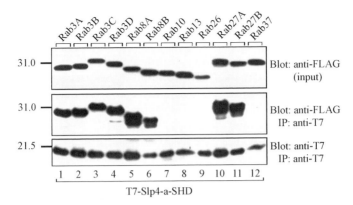

FIG. 1. Interaction between Slp4-a-SHD and three distinct Rab species (Rab3, Rab8, and Rab27). The T7-Slp4-a beads were incubated with the COS-7 cell lysates containing the FLAG-Rab proteins indicated (top panel; 1/80 volume of the reaction mixture used for immunoprecipitation). Proteins trapped by the beads were analyzed by 10% SDS-PAGE followed by immunoblotting with HRP-conjugated anti-FLAG tag antibody (middle panel; Blot: anti-FLAG; IP: anti-T7). The same blots were then stripped and reprobed with HRP-conjugated anti-T7 tag antibody to ensure that similar amounts of T7-tagged proteins had been loaded (bottom panel; Blot: anti-T7; IP: anti-T7). Note that the Slp4-a SHD interacted with Rab3A/B/C/D, Rab8A/B, and Rab27A/B. The positions of the molecular mass markers ($\times 10^{-3}$) are shown on the left.

thought to be a specific Rab27-binding motif. The only exception is the Slp4-a SHD, because it is also capable of interacting with FLAG-tagged Rab3 isoforms and Rab8 isoforms *in vitro* (Fig. 1), the same as rabphilin and Noc2 (Fukuda, 2003c; Fukuda *et al.*, 2002a, 2004). In this section we describe several experimental protocols for evaluating the interaction between Slp4-a and Rabs *in vitro*.

Expression of FLAG-Rab1–40 in COS-7 Cells and Generation of FLAG-Rab Panels

The mouse Rab1–40 cDNAs are amplified from Marathon-Ready adult mouse brain or heart cDNA (BD Clontech, Palo Alto, CA) by polymerase chain reaction (PCR) and subcloned into the eukaryotic expression vector pEF-FLAG (modified from pEF-BOS) by standard molecular cloning techniques to obtain pEF-FLAG-Rab1–40 (Fukuda *et al.*, 1999; Kuroda *et al.*, 2002a; Mizushima and Nagata, 1990). Plasmid DNA for transfection into mammalian cell cultures is prepared by using Qiagen (Chatsworth, CA) maxiprep kits according to the manufacturer's notes. COS-7 cells are cultured in Dulbecco's modified Eagle's medium (DMEM) supplemented with 10% fetal bovine serum, 100 U/ml penicillin G, and 100

μg/ml streptomycin, at 37° under 5% CO_2. A 4 μg amount of each pEF-FLAG-Rab plasmid is transfected into COS-7 cells (7.5 × 10^5 cells, the day before transfection/10-cm dish) by using Lipofect Amine Plus reagent (Invitrogen, Carlsbad, CA) according to the manufacturer's notes. Three days after transfection, cells (from one to several 10-cm dishes) are harvested in 1.5-ml microtubes and cell pellets are stored at −80° until used (for at least 6 months). Before use, cells are thawed on ice and homogenized in a buffer containing 500 μl of 50 mM HEPES-KOH, pH 7.2, 150 mM NaCl, 1 mM MgCl$_2$, 0.5 mM GTPγS (guanosine 5′- O-[3-thiotriphosphate), and protease inhibitors (0.1 mM phenylmethylsulfonyl fluoride, 10 μM leupeptin, and 10 μM pepstatin A; "protease inhibitors" refers to these three inhibitors throughout the text) in a glass-Teflon Potter homogenizer with 10 strokes at 900–1000 rpm. Proteins are solubilized with 1% Triton X-100 with gentle agitation at 4° for 1 h, and insoluble material is removed by centrifugation at 15,000 rpm (TOMY MX-100 high-speed refrigerated microcentrifuge; Tokyo, Japan) for 10 min. Since the levels of expression of recombinant FLAG-tagged Rab proteins in total cell lysates vary among the members of the Rab family, it is necessary to dilute cell lysates containing high levels of Rab proteins with the above homogenization buffer containing 1% Triton X-100 so that the amounts of FLAG-Rab proteins in the diluted lysates of COS-7 cells are similar. In brief, 10 μl of total cell lysates is analyzed by 10% sodium dodecyl sulfate–polyacrylamide gel electrophoresis (SDS–PAGE) followed by immunoblotting with horseradish peroxidase (HRP)-conjugated anti-FLAG tag (M2) mouse monoclonal antibody (1/10,000 dilution) (Sigma Chemical Co., St. Louis, MO) and enhanced chemiluminescence (ECL) detection (Amersham Biosciences, Buckinghamshire, UK) as described previously (Fukuda et al., 1999). The original cell lysates are appropriately diluted based on the intensity of the immunoreactive bands on x-ray film quantified by Lane Analyzer (version 3.0) (ATTO Corp., Tokyo, Japan), so that the amounts of FLAG-Rab proteins in the diluted samples are similar, which is subsequently confirmed by immunoblotting with HRP-conjugated anti-FLAG tag antibody (Fukuda, 2003b; Fukuda et al., 2002d; Kuroda et al., 2002a). Diluted samples can be stored at −80° for at least several months, but repeated freezing and thawing should be avoided (no more than three times).

Assay for Interaction between T7-Slp4-a and FLAG-Rabs In Vitro

The mouse Slp4-a cDNA is amplified from the mouse brain cDNA by PCR and subcloned into the eukayotic expression vector pEF-T7 (named pEF-T7-Slp4-a) as described previously (Fukuda et al., 1999; Kuroda et al., 2002a). Plasmid DNA is prepared with Qiagen maxiprep kits. Transfection of pEF-T7-Slp4-a or pEF-T7-Slp4-a-SHD (4 μg of plasmid) into COS-7 cells and preparation of total cell lysates are performed as described previously, except that the composition of the homogenization buffer is 50 mM HEPES-KOH, pH 7.2, 250 mM NaCl, 1 mM MgCl$_2$, and protease inhibitors. The total cell lysates (400 μl) are incubated with anti-T7 tag antibody-conjugated agarose beads (wet volume 20 μl; Novagen, Madison, WI) with gentle agitation at 4° for 1 h, and the T7-Slp4-a-bound beads are then washed once with 1 ml of 50 mM HEPES-KOH, pH 7.2, 150 mM NaCl, 1 mM MgCl$_2$, and protease inhibitors. Note that T7-tagged proteins are more efficiently immunoprecipitated with anti-T7 tag antibody-conjugated agarose in the presence of 250 mM NaCl than in the presence of 150 mM NaCl. The T7-Slp4-a beads are then incubated with 400 μl of the above diluted FLAG-Rab-containing cell lysates with gentle agitation at 4° for 1 h. The beads are washed three times with 1 ml of 50 mM HEPES-KOH, pH 7.2, 150 mM NaCl, 1 mM MgCl$_2$, 0.2% Triton X-100, and protease inhibitors. After removing the washing buffer, SDS sample buffer is added to the beads, and SDS samples are boiled for 3 min. Coimmunoprecipitated FLAG-Rab proteins and immunoprecipitated T7-Slp4-a are analyzed by 10% SDS-PAGE followed by immunoblotting with HRP-conjugated anti-FLAG tag antibody and anti-T7 tag antibody (1/10,000 dilution; Novagen), respectively (Fukuda *et al.*, 1999; Kuroda *et al.*, 2002a). T7-Slp4-a interacts with three distinct Rab isoforms, Rab3, Rab8, and Rab27, and not with any other Rab isoforms (middle panel of Fig. 1).

GST (Glutathione S-Transferase)-Fusion Protein Pull-Down Assay: Assay for Interaction between GST-Slp4-a-SHD and FLAG-Rabs

GST-Slp4-a-SHD (amino acid residues 1–150) is expressed in *Escherichia coli* JM109 and prepared as described in Chapter 38 (this volume). Note that GST-Slp4-a-SHD contains a large amount of degradation products even after affinity purification (Kuroda *et al.*, 2002a). Glutathione-Sepharose beads (wet volume 20 μl; Amersham Biosciences) coupled with 1 μg of the purified GST-Slp4-a-SHD are incubated with 400 μl of the above diluted FLAG-Rab-containing cell lysates with gentle agitation at 4° for 1 h. Beads are washed five times with 1 ml of 50 mM HEPES-KOH, pH 7.2, 150 mM NaCl, 1 mM MgCl$_2$, 0.2% Triton X-100, and protease inhibitors. After removing the washing buffer, SDS sample buffer is added to the beads. The FLAG-Rab proteins bound to the beads and GST-Slp4-a-SHD are analyzed by 10% SDS-PAGE followed by immunoblotting with HRP-conjugated anti-FLAG tag antibody and anti-GST antibody (1/10,000 dilution; Santa Cruz Biotechnology, Inc., Santa Cruz, CA), respectively.

Cotransfection Assay in COS-7 Cells

pEF-T7-Slp4-a and pEF-FLAG-Rab (2 μg of each plasmid) are cotransfected into COS-7 cells (7.5×10^5 cells, the day before transfection/ 10-cm dish) by using LipofectAmine Plus reagent. Three days after transfection, total cell lysates in 50 mM HEPES-KOH, pH 7.2, 250 mM NaCl, 1 mM MgCl$_2$, 0.5 mM GTPγS, 1% Triton X-100, and protease inhibitors are obtained as described earlier. Coimmunoprecipitation of T7-Slp4-a-FLAG-Rab complex with anti-T7 tag antibody-conjugated agarose beads and immunoblotting with HRP-conjugated anti-FLAG tag and anti-T7 tag antibodies are performed as described previously. Under the above experimental conditions, T7-Slp4-a specifically interacts with three Rab isoforms: Rab3, Rab8, and Rab27 (Fukuda, 2003b; Fukuda *et al.*, 2002a; Kuroda *et al.*, 2002a).

Both Slp4-a and Rab27A are Present on Dense-Core Vesicles in PC12 Cells

Since endocrine PC12 cells endogenously express all four proteins (Rab3A, Rab8, Rab27A, and Slp4-a) and are often utilized for studies on DCV exocytosis, PC12 cells are an appropriate choice of cells to determine which Rabs are *in vivo* binding partners of Slp4-a. Although the level of expression of Rab27A in PC12 cells is more than four times less than that of Rab3A and more than 10 times less than that of Rab8, the majority of Slp4-a proteins form a complex with Rab27A on DCVs (Fukuda *et al.*, 2002a). This section describes several protocols, including coimmunoprecipitation, GST pull-down, subcellular fractionation, and immunofluorescence, that are used to evaluate the presence of the Rab27A-Slp4-a complex on DCVs in PC12 cells

Antibody Production

pGEX-Slp4-a-C2B vector (Amersham Biosciences), which encodes the C2B domain of mouse Slp4-a (amino acids 489–673), is prepared by standard molecular cloning techniques (Fukuda *et al.*, 2002a). GST-Slp4-a-C2B is expressed and purified as described in Chapter 38 (this volume). New Zealand white rabbits are subcutaneously immunized with purified GST-Slp4-a-C2B (or GST-Slp4-a-SHD) using a RIBI adjuvant system (Hamilton, MT), and anti-Slp4-a-C2B (or anti-Slp4-a-SHD) antibody is affinity purified by exposure to antigen-bound Affi-Gel 10 beads (Bio-Rad, Hercules, CA) as described previously (Fukuda and Mikoshiba, 1999). The specificity of the antibody is checked by immunoblotting with recombinant T7-tagged Slp1, Slp2-a, Slp3-a, Slp4-a, Slp5, rabphilin, and Syt I expressed in COS-7 cells (Fukuda *et al.*, 2002a,c; Imai *et al.*, 2004). Due to the high sequence similarity between Slp4-a and Slp5, anti-Slp4-a C2B antibody weakly recognizes Slp5, but cross-reactive components can be easily removed by passage through glutathione-Sepharose beads coupled with >1 mg of GST-Slp5-C2B. The purified anti-Slp4-a antibody recognizes endogenous Slp4-a in PC12 cells (Fukuda *et al.*, 2002a).

Coimmunoprecipitation of Endogenous Slp4-a Rab27A Complex from PC12 Cell Lysates by Anti-Slp4-a Antibody

PC12 cells are cultured in DMEM supplemented with 10% fetal bovine serum, 10% horse serum, 100 U/ml penicillin G, and 100 μg/ml streptomycin, at 37° under 5% CO_2. PC12 cells (one confluent 10-cm dish) are homogenized in a buffer containing 500 μl of 50 mM HEPES-KOH, pH 7.2, 150 mM NaCl, 1 mM MgCl$_2$, 0.5 mM GTPγS, and protease inhibitors in a glass-Teflon Potter homogenizer with 10 strokes at 900–1000 rpm, and the proteins are solubilized with 1% Triton X-100 at 4° for 1 h. After removing of insoluble material by centrifugation at 15,000 rpm for 10 min, the supernatant is incubated at 4° for 1 h with 30 μl (wet volume) of protein A-Sepharose beads (Amersham Biosciences) for preabsorption. The resultant supernatant is incubated with either anti-Slp4-a-C2B IgG or control rabbit IgG (10 μg/ml) at 4° for 1 h, and then with protein A-Sepharose beads (wet volume 15 μl) at 4° for 1 h. After washing the beads five times with 10 mM HEPES-KOH, pH 7.2, 150 mM NaCl, 1 mM MgCl$_2$, 0.2% Triton X-100, and protease inhibitors, the immunoprecipitates are analyzed by 10% SDS-PAGE followed by immunoblotting with anti-Rab3A (1/500 dilution; BD Transduction Laboratories, Lexington, KY), anti-Rab8 (1/1000 dilution; BD Transduction Laboratories), anti-Rab27A (1/250 dilution; BD Transduction Laboratories), or anti-Slp4-a antibody (1 μg/ ml dilution) as described previously (Fukuda, 2003c; Fukuda *et al.*, 1999, 2002a). The anti-Slp4-a antibody efficiently coimmunoprecipitates endogenous Rab27A rather than Rab8, and hardly any Rab3A is detected in the immunoprecipitates (Fukuda *et al.*, 2002a). Under the same experimental conditions, the anti-Slp4-a antibody also coimmunoprecipitates endogenous syntaxin 1A (t-SNARE) and Munc18-1, but not SNAP-25 or VAMP-2 (Coppola *et al.*, 2002; Fukuda, 2003c; Torii *et al.*, 2002). Immunoreactive bands are visualized with ECL.

Assay for Interaction between GST-Slp4-a and Endogenous Rab27A in PC12 Cells

The mouse Slp4-a cDNA is subcloned into the eukaryotic expression vector pEF-T7-GST (named pEF-T7-GST-Slp4-a) as described previously (Fukuda *et al.*, 2002a,d). Plasmid DNA is prepared with Qiagen maxiprep kits. Transfection of pEF-T7-GST-Slp4-a or pEF-T7-GST (8 μg of plasmid) into PC12 cells (3 $\times$ 10^6 cells, the day before transfection/6-cm dish coated with collagen type I) is performed by using LipofectAMINE 2000 reagent (Invitrogen) according to the manufacturer's notes. Three days after transfection, cells are harvested and homogenized in 1 ml of the homogenization buffer (50 mM HEPES-KOH, pH 7.2, 150 mM NaCl, 1 mM MgCl$_2$, 0.5 mM GTPγS, and protease inhibitors). After solubilization with 1% Triton X-100, insoluble material is removed by centrifugation at 15,000 rpm, and the expressed GST fusion proteins are affinity purified on glutathione-Sepharose beads (wet volume 20 μl). After washing the beads five times with 10 mM HEPES-KOH, pH 7.2, 150 mM NaCl, 1 mM MgCl$_2$, 0.2% Triton X-100, and protease inhibitors, proteins bound to the beads are analyzed by 10% SDS-PAGE, followed by immunoblotting with anti-Rab3A, anti-Rab8, and anti-Rab27A antibodies as described previously. T7-GST fusion proteins are detected with HRP-conjugated anti-T7 tag antibody (1/10,000 dilution). T7-GST-Slp4-a, but not T7-GST alone, preferentially interacts with endogenous Rab27A (Fukuda *et al.*, 2002a).

Subcellular Fractionation and Immunofluorescence Analyses

Subcellular fractionation of PC12 cells with a linear sucrose gradient (0.6–1.8 M) is performed as described previously (Saegusa *et al.*, 2002). The distribution of DCVs and synaptic-like microvesicles is determined by using anti-synaptotagmin IX antibody (1 μg/ml) (Fukuda *et al.*, 2002c) and anti-synaptophysin antibody (1/4000 dilution; Sigma Chemical Co.), respectively. Immunofluorescence analysis of PC12 cells is also performed as described previously (Fukuda and Mikoshiba, 1999).

Inhibition of Dense-Core Vesicle Exocytosis by Slp4-a through Specific Interaction with Rab27A in PC12 Cells

We and others have previously shown that expression of Rab27A and Slp4-a in endocrine cells has opposite effects on DCV exocytosis (Coppola *et al.*, 2002; Fukuda *et al.*, 2002a; Yi *et al.*, 2002). Expression of Rab27A, but not Rab8A, promotes DCV exocytosis in PC12 cells, whereas expression of Slp4-a diminishes DCV exocytosis in PC12 cells (Fukuda *et al.*, 2002a). Since the inhibitory effect of Slp4-a on DCV exocytosis depends on the N-terminal SHD (i.e., deletion of the SHD reverses the inhibitory effect), it is important to determine which Rab·Slp4-a complexes are involved in the inhibition of DCV exocytosis *in vivo*. To do so, we developed a neuropeptide Y (NPY) cotransfection assay in PC12 cells that makes it possible to evaluate the effect of various Slp4-a mutants on DCV exocytosis (Fukuda,

2003c; Fukuda *et al.*, 2002a). This section describes the experimental protocols used to compare the contribution of three Rab-Slp4-a complexes to controlling DCV exocytosis.

Site-Directed Mutagenesis and Construction of Slp4-a Mutant Plasmids

Mutant Slp4-a plasmids carrying an Ile-to-Ala substitution at amino acid position 18 (named Slp4-a[I18A]), a V21A substitution, or a TGDWFY-to-AGAAAY substitution at amino acid positions 115–120 (named Slp4-a[A4]) are obtained by conventional PCR techniques as described previously (Fukuda, 2003c; Fukuda *et al.*, 1995, 2002a). pEF-T7-Slp4-a-ΔSHD and pEF-T7-Slp4-a(I18A), (V21A), or (A4) are constructed as described previously (Fukuda, 2003c; Fukuda *et al.*, 1999, 2002a). We have shown by cotransfection assay in COS-7 cells that deletion of the SHD completely abrogates the Rab-binding activity of Slp4-a, whereas the I18A mutation and V21A mutation selectively abrogate Rab3A- and Rab8A-binding activity and Rab3A-binding activity, respectively (Fukuda *et al.*, 2002a). The Slp4-a(A4) mutant specifically interacts with Rab27A(Q78L) (mimics the GTP-bound form of Rab27A), not with Rab27A(T23N) (mimics the GDP-bound form of Rab27A), whereas the wild-type Slp4-a is capable of interacting with both Rab27A(Q78L) and (T23N), unlike other Slp members (Fukuda, 2003c).

NPY Release Assay

The NPY cDNA we used was a kind gift of Dr. Wolfhard Almers (Vollum Institute, Portland, OR). The addition of the C-terminal T7-GST tag (GGSGGTGG**MARMTGGQQMGR** + GST; Gly linker, underlined; T7 tag, in bold) to NPY is essentially performed by PCR, and the resulting NPY-T7-GST fragments are subcloned into the *Not*I site of the modified pShooter vector (Invitrogen) (named pShooter-NPY-T7-GST) as described previously (Fukuda *et al.*, 2002b,e). PC12 cells (1.5 × 10^6 cells, the day before transfection/6-cm dish) are cotransfected with 4 μg each of pShooter-NPY-T7-GST and pEF-FLAG-Rab (or pEF-T7-Slp4-a mutants) by using LipofectAMINE 2000 reagent according to the manufacturer's notes. Three days after transfection, cells are washed with prewarmed low-KCl buffer (5.6 mM KCl, 145 mM NaCl, 2.2 mM CaCl$_2$, 0.5 mM MgCl$_2$, 5.6 mM glucose, and 15 mM HEPES-KOH, pH 7.4) and then stimulated with either low-KCl buffer or high-KCl buffer (56 mM KCl, 95 mM NaCl, 2.2 mM CaCl$_2$, 0.5 mM MgCl$_2$, 5.6 mM glucose, and 15 mM HEPES-KOH, pH 7.4) for 10 min at 37°. Released NPY-T7-GST is recovered by incubation with glutathione-Sepharose beads and analyzed by immunoblotting with

HRP-conjugated anti-T7 tag antibody. The intensity of the immunoreactive bands on x-ray film is quantified by Lane Analyzer (version 3.0) and normalized by total expressed NPY-T7-GST. Total cell lysates are obtained by incubation with a lysis buffer (10 mM Tris–HCl, pH 8.0, 150 mM NaCl, 1 mM EDTA, and 1% Nonidet P-40, and protease inhibitors). Expressed FLAG-Rab and T7-Slp4-a in the PC12 cell lyastes can be detected with HRP-conjugated anti-FLAG tag antibody and anti-T7 tag antibody, respectively (inset in Fig. 2). Under these experimental conditions, NPY-T7-GST is targeted to DCVs, and approximately 5–10% of total NPY-T7-GST is released in a high-KCl-dependent manner. Note that expression of the Rab3A/8A-binding-defective Slp4-a mutant (I18A/V21A) still attenuates NPY secretion, the same as the wild-type protein, whereas expression of the Rab27A-binding-defective mutant (ΔSHD) or of the GDP-Rab27A-binding-defective mutant (A4) had no effect on NPY secretion (Fig. 2).

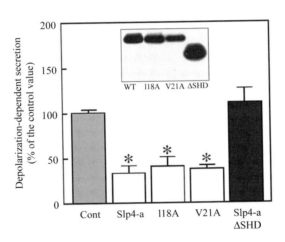

FIG. 2. Effect of expression of Slp4-a mutants on DCV exocytosis in PC12 cells. The NPY-T7-GST secretion assay was performed as described in the text. The results are expressed as percentages of NPY-T7-GST secretion in control samples (cont; shaded bar) without expression of recombinant proteins. Bars indicate the means ± SE of three determinations. The results shown are representative of three independent experiments. The inset shows expressed recombinant Slp4-a proteins visualized with anti-T7 tag antibody. Note that expression of Slp4-a-ΔSHD had no effect on high-KCl-dependent NPY secretion (closed bar), whereas expression of Slp4-a(I18A) (Rab27A specific) and Slp4-a(V21A) (Rab8A and Rab27A specific) inhibited NPY secretion, the same as the wild-type protein (open bars). *$p < 0.01$, Student's t-test. Reproduced with permission from Fukuda et al. (2002a).

Summary of Results

We have described several methods of analyzing Slp4-a interaction with specific Rab, including *in vitro* binding assays and functional studies in PC12 cells. *In vitro* binding assays and coimmunoprecipitation assays have shown that Slp4-a preferentially interacts with Rab27A, rather than Rab3A or Rab8A, and NPY cotransfection assays combined with site-directed mutagenesis confirmed that Slp4-a functions as a Rab27A effector during DCV exocytosis in PC12 cells. Expression of Slp4-a was found to inhibit DCV exocytosis through specific interaction with the GDP-bound form of Rab27A in PC12 cells, and Slp4-a has been shown to interact with syntaxin 1A (Torii *et al.*, 2004) as well as Munc18-1 (Coppola *et al.*, 2002; Fukuda, 2003c), which may be involved in the docking of DCVs with the plasma membrane in endocrine cells.

Acknowledgments

This work was supported in part by grants from the Ministry of Education, Culture, Sports, and Technology of Japan.

References

Coppola, T., Frantz, C., Perret-Menoud, V., Gattesco, S., Hirling, H., and Regazzi, R. (2002). Pancreatic β-cell protein granuphilin binds Rab3 and Munc-18 and controls exocytosis. *Mol. Biol. Cell* **13,** 1906–1915.

El-Amraoui, A., Schonn, J. S., Kussel-Andermann, P., Blanchard, S., Desnos, C., Henry, J. P., Wolfrum, U., Darchen, F., and Petit, C. (2002). MyRIP, a novel Rab effector, enables myosin VIIa recruitment to retinal melanosomes. *EMBO Rep.* **3,** 463–470.

Fukuda, M. (2002a). Slp and Slac2, novel families of Rab27 effectors that control Rab27-dependent membrane traffic. *Recent Res. Dev. Neurochem.* **5,** 297–309.

Fukuda, M. (2002b). Synaptotagmin-like protein (Slp) homology domain 1 of Slac2-a/melanophilin is a critical determinant of GTP-dependent specific binding to Rab27A. *J. Biol. Chem.* **277,** 40118–40124.

Fukuda, M. (2003a). Molecular cloning, expression, and characterization of a novel class of synaptotagmin (Syt XIV) conserved from *Drosophila* to humans. *J. Biochem.* **133,** 641–649.

Fukuda, M. (2003b). Distinct Rab binding specificity of Rim1, Rim2, rabphilin, and Noc2: Identification of a critical determinant of Rab3A/Rab27A recognition by Rim2. *J. Biol. Chem.* **278,** 15373–15380.

Fukuda, M. (2003c). Slp4-a/granuphilin-a inhibits dense-core vesicle exocytosis through interaction with the GDP-bound form of Rab27A in PC12 cells. *J. Biol. Chem.* **278,** 15390–15396.

Fukuda, M. (2005). Versatile role of Rab27 in membrane trafficking: Focus on the Rab27 effector families. *J. Biochem.* **137,** 9–16.

Fukuda, M., and Kuroda, T. S. (2002). Slac2-c (synaptotagmin-like protein homologue lacking C2 domains-c), a novel linker protein that interacts with Rab27, myosin Va/VIIa, and actin. *J. Biol. Chem.* **277,** 43096–43103.

Fukuda, M., and Mikoshiba, K. (1999). A novel alternatively spliced variant of synaptotagmin VI lacking a transmembrane domain: Implications for distinct functions of the two isoforms. *J. Biol. Chem.* **274**, 31428–31434.

Fukuda, M., and Mikoshiba, K. (2001). Synaptotagmin-like protein 1-3: A novel family of C-terminal-type tandem C2 proteins. *Biochem. Biophys. Res. Commun.* **281**, 1226–1233.

Fukuda, M., Kojima, T., Aruga, J., Niinobe, M., and Mikoshiba, K. (1995). Functional diversity of C2 domains of synaptotagmin family: Mutational analysis of inositol high polyphosphate binding domain. *J. Biol. Chem.* **270**, 26523–26527.

Fukuda, M., Kanno, E., and Mikoshiba, K. (1999). Conserved N-terminal cysteine motif is essential for homo- and heterodimer formation of synaptotagmins III, V, VI, and X. *J. Biol. Chem.* **274**, 31421–31427.

Fukuda, M., Saegusa, C., and Mikoshiba, K. (2001). Novel splicing isoforms of synaptotagmin-like proteins 2 and 3: Identification of the Slp homology domain. *Biochem. Biophys. Res. Commun.* **283**, 513–519.

Fukuda, M., Kanno, E., Saegusa, C., Ogata, Y., and Kuroda, T. S. (2002a). Slp4-a/granuphilin-a regulates dense-core vesicle exocytosis in PC12 cells. *J. Biol. Chem.* **277**, 39673–39678.

Fukuda, M., Katayama, E., and Mikoshiba, K. (2002b). The calcium-binding loops of the tandem C2 domains of synaptotagmin VII cooperatively mediate calcium-dependent oligomerization. *J. Biol. Chem.* **277**, 29315–29320.

Fukuda, M., Kowalchyk, J. A., Zhang, X., Martin, T. F. J., and Mikoshiba, K. (2002c). Synaptotagmin IX regulates Ca^{2+}-dependent secretion in PC12 cells. *J. Biol. Chem.* **277**, 4601–4604.

Fukuda, M., Kuroda, T. S., and Mikoshiba, K. (2002d). Slac2-a/melanophilin, the missing link between Rab27 and myosin Va: Implications of a tripartite protein complex for melanosome transport. *J. Biol. Chem.* **277**, 12432–12436.

Fukuda, M., Ogata, Y., Saegusa, C., Kanno, E., and Mikoshiba, K. (2002e). Alternative splicing isoforms of synaptotagmin VII in the mouse, rat and human. *Biochem. J.* **365**, 173–180.

Fukuda, M., Kanno, E., and Yamamoto, A. (2004). Rabphilin and Noc2 are recruited to dense-core vesicles through specific interaction with Rab27A in PC12 cells. *J. Biol. Chem.* **279**, 13065–13075.

Imai, A., Yoshie, S., Nashida, T., Shimomura, H., and Fukuda, M. (2004). The small GTPase Rab27B regulates amylase release from rat parotid acinar cells. *J. Cell Sci.* **117**, 1945–1953.

Izumi, T., Gomi, H., Kasai, K., Mizutani, S., and Torii, S. (2003). The roles of Rab27 and its effectors in the regulated secretory pathways. *Cell Struct. Funct.* **28**, 465–474.

Kuroda, T. S., and Fukuda, M. (2004). Rab27A-binding protein Slp2-a is required for peripheral melanosome distribution and elongated cell shape in melanocytes. *Nat. Cell Biol.* **6**, 1195–1203.

Kuroda, T. S., Fukuda, M., Ariga, H., and Mikoshiba, K. (2002a). The Slp homology domain of synaptotagmin-like proteins 1-4 and Slac2 functions as a novel Rab27A binding domain. *J. Biol. Chem.* **277**, 9212–9218.

Kuroda, T. S., Fukuda, M., Ariga, H., and Mikoshiba, K. (2002b). Synaptotagmin-like protein 5: A novel Rab27A effector with C-terminal tandem C2 domains. *Biochem. Biophys. Res. Commun.* **293**, 899–906.

Mizushima, S., and Nagata, S. (1990). pEF-BOS, a powerful mammalian expression vector. *Nucleic Acids Res.* **18**, 5322.

Pereira-Leal, J. B., and Seabra, M. C. (2001). Evolution of the Rab family of small GTP-binding proteins. *J. Mol. Biol.* **313**, 889–901.

Saegusa, C., Fukuda, M., and Mikoshiba, K. (2002). Synaptotagmin V is targeted to dense-core vesicles that undergo calcium-dependent exocytosis in PC12 cells. *J. Biol. Chem.* **277**, 24499–24505.

Strom, M., Hume, A. N., Tarafder, A. K., Barkagianni, E., and Seabra, M. C. (2002). A family of Rab27-binding proteins: Melanophilin links Rab27a and myosin Va function in melanosome transport. *J. Biol. Chem.* **277,** 25423–25430.

Torii, S., Zhao, S., Yi, Z., Takeuchi, T., and Izumi, T. (2002). Granuphilin modulates the exocytosis of secretory granules through interaction with syntaxin 1a. *Mol. Cell. Biol.* **22,** 5518–5526.

Torii, S., Takeuchi, T., Nagamatsu, S., and Izumi, T. (2004). Rab27 effector granuphilin promotes the plasma membrane targeting of insulin granules via interaction with syntaxin 1a. *J. Biol. Chem.* **279,** 22532–22538.

Wang, J., Takeuchi, T., Yokota, H., and Izumi, T. (1999). Novel rabphilin-3-like protein associates with insulin-containing granules in pancreatic beta cells. *J. Biol. Chem.* **274,** 28542–28548.

Yi, Z., Yokota, H., Torii, S., Aoki, T., Hosaka, M., Zhao, S., Takata, K., Takeuchi, T., and Izumi, T. (2002). The Rab27a/granuphilin complex regulates the exocytosis of insulin-containing dense-core granules. *Mol. Cell. Biol.* **22,** 1858–1867.

Zerial, M., and McBride, H. (2001). Rab proteins as membrane organizers. *Nat. Rev. Mol. Cell. Biol.* **2,** 107–117.

[40] Assay and Functional Interactions of Rim2 with Rab3

By MITSUNORI FUKUDA

Abstract

Rim was originally identified as a protein that contains a putative Rab3A-effector domain at the N-terminus, the same as rabphilin, and two forms of Rim, Rim1 and Rim2, have been reported in mammals. The putative Rab3A-binding domain (RBD) of Rim consists of two α-helical regions (named RBD1 and RBD2) separated by two zinc finger motifs, and several alternative splicing events occur in the RBD1 of both Rims that result in the production of long forms and short forms of RBD. The short forms of Rim2 RBD are capable of interacting with Rab3A with high affinity *in vitro*, and it is recruited to dense-core vesicles (DCVs) in neuroendocrine PC12 cells through interaction with endogenous Rab3A, whereas the long forms of Rim2 RBD show dramatically reduced Rab3A-binding activity *in vitro* (more than a 50-fold decrease in affinity compared with the short forms of Rim2 RBD), and it is mainly present in the cytoplasm and nucleus. Expression of the shortest form of Rim2 RBD, but not its Rab3A binding-defective mutant (E36A/R37S), promotes high-KCl-dependent neuropeptide Y secretion from PC12 cells, suggesting that the Rim2 containing the short forms of RBD functions as a Rab3A effector during DCV

METHODS IN ENZYMOLOGY, VOL. 403
0076-6879/05 $35.00
DOI: 10.1016/S0076-6879(05)03040-5

exocytosis. In this Chapter, I describe several assay methods that have been used to determine the physiological significance of the alternative splicing event in the RBD1 of Rim2, including assays for the *in vitro* interaction between Rim2 RBD and Rab3A and for the localization of Rim2-RBD on DCVs in PC12 cells.

Introduction

Rim1 was originally identified as a protein that contains a putative Rab3A-effector domain at the N-terminus (Wang *et al.*, 1997), the same as rabphilin (Shirataki *et al.*, 1993), and it has been shown to be involved in the regulation of secretory vesicle exocytosis and synaptic plasticity (reviewed in Südhof, 2004). Rim2 has been reported as a second isoform of Rim that shares the same domain structures as Rim1: a putative Rab3A-binding domain (RBD) that consists of two α-helical regions (named RBD1 and RBD2) separated by zinc finger motifs at the N-terminus, a PDZ domain in the middle region, and two C2 domains at the C-terminus (Ozaki *et al.*, 2000; Wang *et al.*, 2000). Although both Rim isoforms are coexpressed in brain, the RBDs of Rim1 and Rim2 are structurally different because of alternative splicing (Fig. 1) (Fukuda, 2003a, 2004a). The most abundant form of Rim1 in brain contains a 50-amino acid insertion in the RBD1 (i.e., the longest form; simply designated Rim1 RBD below), and the Rim1Δ83–105 form (i.e., long form; a 23-amino acid insertion) and the Rim1Δ56–105 form (i.e., short form; no insertion) are less abundant. By contrast, the most abundant form of Rim2 in brain lacks such an insertion (short form; simply designated Rim2 RBD below), and hardly any of the long form of Rim2 RBD (named Rim2 RBD[+40A]) is detected. It is of great interest that the intracellular localization of Rim1 with the long RBD and Rim2 with the short RBD is completely different in pancreatic β-cells, where Rim1 is associated with the plasma membrane, possibly through a PDZ domain or C2 domains, and is not present on secretory granules (Iezzi *et al.*, 2000), where Rim2 is associated with insulin-containing granules through the N-terminal RBD (Shibasaki *et al.*, 2003), despite the fact that the RBD of Rim1 and Rim2 is capable of interacting with Rab3A *in vitro* (Fukuda 2003a, 2004a; Iezzi *et al.*, 2000; Ozaki *et al.*, 2000; Sun *et al.*, 2001; Wang *et al.*, 1997, 2000, 2001). The reason for this distinct subcellular localization of Rim1 RBD and Rim2 RBD has now been explained by the discovery that insertion of Rim1 and Rim2 into the RBD1 dramatically reduces Rab3A-binding activity, and it has been proposed that the Rab3A-effector function of mammalian Rim is regulated by alternative splicing in the RBD1 of Rims (Fukuda, 2004a). This chapter describes the methods used to investigate

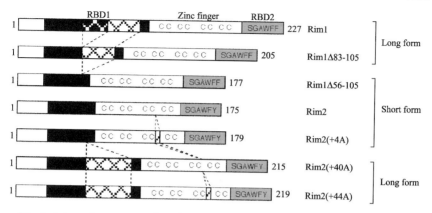

FIG. 1. Schematic representation of the alternative splicing isoforms of mouse Rim1 and Rim2. The RBD of Rim1 and Rim2 consists of two α-helical regions (RBD1 and RBD2; black boxes and shaded boxes, respectively) that are separated by two zinc finger motifs (indicated by "Cs"), the same as the RBD of rabphilin (Ostermeier and Brunger, 1999). The conserved SGAWF(Y/F) motif in RBD2 is essential for Rab3A binding by rabphilin (Ostermeier and Brunger, 1999). Amino acid insertions (or deletions) by alternative splicing are found in the C-terminus of the RBD1 and the second zinc finger motif (indicated by cross-hatched and hatched boxes, respectively). At least three forms of the Rim1 RBD and four forms of the Rim2 RBD can be produced by alternative splicing, but the longest form of Rim1 and the shortest form of Rim2 are predominant in mouse brain (Fukuda, 2003a). The amino acid positions are indicated on both sides. Reproduced with permission from Fukuda (2004a).

the functional relationship between alternative splicing in the RBD1 of Rim2 and Rab3A-binding activity both in *in vitro* and in intact PC12 cells.

Characterization of the Rab3A-Binding Activity of the Long Forms and Short Forms of Rim2 RBD *In Vitro*

Although rabphilin, Rim, and Noc2 were previously believed to be specific Rab3A effectors, a systematic analysis of their Rab-binding specificity by using panels of FLAG-Rab1-40 in COS-7 cells has indicated that rabphilin and Noc2 are different from Rim in terms of Rab27A-binding activity (Cheviet *et al.*, 2004; Fukuda, 2003a, 2004a,b; Fukuda *et al.*, 2004; Kuroda *et al.*, 2002) (see Chapter 41, this volume). Both rabphilin and Noc2 interact with Rab27A in preference to Rab3A, the same as the synaptotagmin-like protein homology domain (SHD) of Slp4-a (Fukuda, 2003b, 2005; Fukuda *et al.*, 2002), whereas Rim does not interact with Rab27A at all because of the presence of an acidic cluster (e.g., Glu-50, Glu-51, and Glu-52 of Rim2) in the middle of RBD1 (Fukuda, 2003a). Systematic deletion analysis has further indicated that the RBD1 of Rim2 is a central Rab3A-binding domain, although the RBD1 alone is insufficient for high-affinity Rab3A binding (Fukuda, 2004a; Wang et al., 2001). Thus, the alternative splicing events that occur in the RBD1 of Rim1 and Rim2 are of great interest with regard to the regulation of Rab3A (Fig. 1). I developed in vitro binding assays to compare the relative Rab3A-binding affinity of Rim RBDs and found that the short forms of Rim RBD (Rim1Δ56–105, Rim2, Rim2[+4A]) bind Rab3A with higher affinity than the long forms of Rim RBD do (Rim1, Rim1Δ83–105, Rim2[+40A], Rim2[+44A]) (Fukuda, 2004a). In this section I describe several experimental protocols for evaluating differences between the Rab3A binding affinity of the short forms and long forms of Rim2 RBD.

Methods of Identification of Alternative Splicing Isoforms of Rim1 and Rim2

cDNA encoding an RBD of mouse Rim1 and Rim2 is amplified from brain cDNA and testis cDNA (mouse MTC panel I; BD Clontech, Palo Alto, CA), respectively, by polymerase chain reaction (PCR) using the following pairs of oligonucleotides with restriction enzyme sites (under-lined) or stop codons (boldface) as described previously (Fukuda et al., 1999): 5′-C<u>GGATCC</u>ATGTCCTCGGCCGTGGGGCC-3′ (Rim1-Met primer; sense) and 5′-**TCA**CACCTCAGATCCAGCACCTG-3′ (Rim1-RBD-3′ primer; antisense); 5′-C<u>GGATCC</u>ATGTCGGCTCCGCTCG GGCC-3′ (Rim2-Met primer; sense) and 5′-**TCA**GGCTTCCTCATTTC-GAAGCC-3′ (Rim2-RBD-3′ primer; antisense). The purified PCR products are directly inserted into the pGEM-T Easy vector (Promega, Madison, WI) and verified by DNA sequencing as described previously (Fukuda et al., 1999). Three different Rim1 RBDs (no deletion type [simply referred to as Rim1 RBD below], Δ83–105 [deletion of amino acid residues 83–105 in Rim1] and Δ56–105 [deletion of amino acid residues 56–105 in Rim1]) are produced by alternatively splicing at the RBD1 (GenBank accession numbers AB162895–AB162897, respectively), and the longest form of Rim1 RBD is predominant in mouse brain (Fukuda, 2003a). By contrast, two alternative splicing sites are present in the Rim2 RBD (see Fig. 1): a 40-amino acid insertion in the Rim2 RBD1, the same as in Rim1, and a 4-amino acid insertion (i.e., EDKV) in the second zinc finger motif. As a result of these alternative splicing events, four different types of Rim2 RBD are produced: Rim2, Rim2(+4A), Rim2(+40A), and Rim2(+44A) (GenBank accession numbers AB162898–AB162901, respectively), and the shortest form of Rim2 is dominant in all mouse tissues (Fukuda, 2004a).

Cotransfection Assay in COS-7 Cells

The cDNA of Rim2-RBD, Rim2(+4A)-RBD, Rim2(+40A)-RBD, and Rim2(+44A)-RBD is subcloned into the eukaryotic expression vector pEF-T7 (modified from pEF-BOS) by standard molecular cloning techniques to obtain pEF-T7-Rim2-RBDs (Fukuda, 2004a; Fukuda *et al.*, 1994, 1999; Mizushima and Nagata, 1990). pEF-FLAG-Rab3A is prepared as described previously (Kuroda *et al.*, 2002). Plasmid DNA for transfection into mammalian cell cultures is prepared by using Qiagen (Chatsworth, CA) maxiprep kits according to the manufacturer's notes. COS-7 cells are cultured in Dulbecco's modified Eagle's medium (DMEM) supplemented with 10% fetal bovine serum, 100 U/ml penicillin G, and 100 μg/ml streptomycin, at 37° under 5% CO_2. Cotransfection of pEF-T7-Rim2-RBDs (1 μg of plasmid) and pEF-FLAG-Rab3A (0.5 μg of plasmid) into COS-7 cells (7.5 × 10^5 cells, the day before transfection/10-cm dish) is carried out by using LipofectAmine Plus reagent (Invitrogen, Carlsbad, CA) according to the manufacturer's notes. Because of the extremely high level of expression of FLAG-Rab3A in COS-7 cells, a smaller amount of plasmids should be used for cotransfection assay than usual (2 μg of each plasmid) to highlight the difference in the Rab3A-binding capacity of Rim2-RBDs (Fukuda, 2004a).

Three days after transfection, cells are homogenized in a buffer containing 500 μl of 50 mM HEPES-KOH, pH 7.2, 250 mM NaCl, 1 mM $MgCl_2$, 0.5 mM GTPγS (guanosine 5'-O- [3-thiotriphosphate]), and protease inhibitors (0.1 mM phenylmethylsulfonyl fluoride, 10 μM leupeptin, and 10 μM pepstatin A; "protease inhibitors" refers to these three inhibitors throughout the text) in a glass-Teflon Potter homogenizer with 10 strokes at 900–1000 rpm. Note that the addition of the nonhydrolyzable GTP analog GTPγS to the reaction mixtures is an essential step, because the intrinsic GTPase activity of Rab3A is higher than that of other Rabs (e.g., Rab27A) (Larijani *et al.*, 2003). Proteins are solubilized by gentle agitation with 1% Triton X-100 at 4° for 1 h, and insoluble material is removed by centrifugation at 15,000 rpm (TOMY MX-100 high speed refrigerated micro centrifuge; Tokyo, Japan) for 10 min. T7-Rim2-RBDs are immunoprecipitated from total cell lysates (400 μl) by gentle agitation with anti-T7 tag antibody-conjugated agarose beads (wet volume 20 μl; Novagen, Madison, WI) at 4° for 1 h. After washing the beads three times with 1 ml of a buffer consisting of 50 mM HEPES-KOH, pH 7.2, 150 mM NaCl, 1 mM $MgCl_2$, 0.2% Triton X-100, and protease inhibitors ("washing buffer" refers to this buffer throughout the text), sodium dodecyl sulfate (SDS) sample buffer is added to the beads, and SDS samples are boiled for 3 min.

The proteins bound to the beads are analyzed by 10% SDS–polyacryl-amide gel electrophoresis (PAGE) and transferred to a polyvinylidene difluoride (PVDF) membrane (Millipore Corp., Bedford, MA) by electro-blotting. Blots are blocked with 1% skim milk and 0.1% Tween-20 in phosphate-buffered saline (PBS) for 1 h at room temperature and then incubated with horseradish peroxidase (HRP)-conjugated anti-FLAG tag (DYKDDDDK) mouse monoclonal antibody (M2) (1/10,000 dilution with 1% skim milk in PBS containing 0.1% Tween-20; Sigma Chemical Co., St. Louis, MO) for 1 h at room temperature. After washing the membrane with PBS containing 0.1% Tween-20 (PBS-T) for 5 min. three times, immunoreactive bands are detected by enhanced chemilumines-cence (Amersham Biosciences, Buckinghamshire, UK). The same blots are incubated with HRP-conjugated anti-T7 tag (MASMTGGQQMG) antibody (1/10,000 dilution with 1% skim milk/PBS-T; Novagen) for an additional 1 h at room temperature, and then subjected to enhanced chemiluminescence (ECL) detection (Fukuda et al., 1999). Since the apparent molecular masses of T7-Rim2-RBD(+40A) and FLAG-Rab3A are close to each other, it is necessary to remove the anti-T7 tag antibody from the membrane by incubation with a stripping buffer (100 mM 2-mercaptoethanol, 2% SDS, 62.5 mM Tris–HCl, pH 6.8) for 30 min at 50°. The same blots are blocked again with 1% skim milk in PBS-T for 1 h at room temperature, and then incubated with HRP-conjugated anti-FALG tag antibody and subjected to ECL detection. Interaction between the short forms of Rim2-RBD and Rab3A is readily observed under these experimental conditions, whereas hardly any interaction between the long forms of Rim2-RBD and Rab3A is detected (Fukuda, 2004a). Similar results are obtained when the full-length Rim2(+4A) and Rim2(+44A) are used for cotransfection assay (Fukuda, 2004a).

Assay for Competition Experiments

pEF-FLAG-Rab3A and pEF-T7-Rim2-RBDs (4 μg of each plasmid) are separately transfected into COS-7 cells as described previously. Three days after transfection, cells (one 10-cm dish) are homogenized in a buffer containing 2 ml of 50 mM HEPES-KOH, pH 7.2, 150 mM NaCl, 1 mM MgCl$_2$, 0.5 mM GTPγS, and protease inhibitors in a glass-Teflon Potter homogenizer with 10 strokes at 900–1000 rpm. Proteins are solubilized by gentle agitation with 1% Triton X-100 at 4° for 1 h, and insoluble material is removed by centrifugation at 15,000 rpm. The supernatant containing FLAG-Rab3A is divided into four portions (400 μl each in 1.5-ml micro-tubes), and the FLAG-Rab3A proteins in each supernatant are affinity-purified with anti-FLAG tag (M2) antibody-conjugated agarose beads (wet volume 70 μl; Sigma Chemical Co.). The agarose beads coupled with FLAG-Rab3A are then divided into seven equal portions (10 μl each in 1.5-ml microtubes; a total of 4 × 7 samples).

Total cell lysates (10 μl) containing T7-tagged Rim2-RBD, Rim2 (+4A)-RBD, Rim2(+40A)-RBD, or Rim2(+44A)-RBD proteins are analyzed by 10% SDS-PAGE followed by immunoblotting with HRP-conjugated anti-T7 tag antibody and ECL detection. The intensity of the immunoreactive bands on x-ray film is quantified with Lane Analyzer (version 3.0) (ATTO Corp. Tokyo, Japan), and original cell lysates are diluted with sufficient homogenization buffer containing 1% Triton X-100 to obtain diluted samples containing the same amounts of T7-RBDs.

The FLAG-Rab3A beads (wet volume 10 μl) are incubated with 400 μl of a solution containing a long form of Rim2 RBD (e.g., Rim2 RBD[+40A] or Rim2 RBDp[+44A]) and a short form of Rim2 RBD (e.g., Rim2 RBD or Rim2 RBD[+4A]) in the following proportions (long/short: 400 μl/0 μl, 360 μl/40 μl, 300 μl/100 μl, 200 μl/200 μl, 100 μl/300 μl, 40 μl/360 μl, 0 μl/400 μl) for 1 h at 4° in 50 mM HEPES–KOH, pH 7.2, 150 mM NaCl, 1 mM MgCl$_2$, 1% Triton X-100, and protease inhibitors. Note that the amount of T7-tagged proteins in the solution should be much greater than the amount of the FLAG-Rab3A protein on the beads, so that competition between the long and short form of Rim2 RBD occurs efficiently. After washing the beads three times with 1 ml of the washing buffer, the proteins bound to the beads are analyzed by 12.5% SDS-PAGE followed by immunoblotting with HRP-conjugated anti-T7 tag antibody and anti-FLAG tag antibody as described earlier (Fukuda et al., 1999). Under these experimental conditions, Rab3A preferentially interacts with the Rim2 RBD, even when a 10 times greater amount of the Rim2 RBD(+40A) than the Rim2 RBD is present in the reaction mixtures (Fukuda, 2004a).

Assay for Comparison of Rab3A-Binding Affinity by Long Forms and Short Forms of Rim2 RBD In Vitro

FLAG-Rab3A beads (wet volume 10 μl) in 1.5-ml microtubes (×7 samples) are prepared as described previously. Equal amounts of T7-Rim2-RBD, T7-Rim2(+4A)-RBD, T7-Rim2(+40A)-RBD, and T7-Rim2(+44A)-RBD proteins in 50 mM HEPES-KOH, pH 7.2, 150 mM NaCl, 1 mM MgCl$_2$, 0.5 mM GTPγS, 1% Triton X-100, and protease inhibitors are also prepared as described earlier, and they are systematically diluted thus: 1, 1/4, 1/10, 1/40, 1/100, 1/400, and 1/1000. The diluted samples are incubated with beads coupled with FLAG-Rab3A protein. After washing the beads three times with 1 ml of the washing buffer, proteins bound to the beads are analyzed by 12.5% SDS-PAGE and visualized with HRP-conjugated anti-T7 tag antibody and anti-FLAG tag antibody (Fig. 2A). Immunoreactive bands of T7-Rim2 RBDs on x-ray film are captured and quantified with Lane Analyzer (Fig. 2B). Experiments are repeated at least three times, and EC_{50} values for the RBD-Rab3A interaction are calculated with GraphPad PRISM software (version 4.0) (Fukuda, 2004a). Under these experimental conditions, amino acid insertion into the RBD1 of Rim2 dramatically reduces Rab3A binding

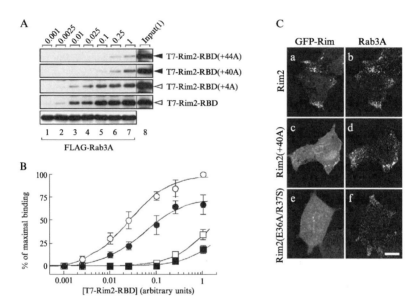

FIG. 2. Distinct Rab3A-binding activity of the long forms and short forms of Rim2 RBD in *in vitro* and in PC12 cells. (A) T7-Rim2-RBDs were systematically diluted as indicated, and diluted samples were incubated with beads coupled with FLAG-Rab3A protein. Proteins bound to the beads were analyzed by 12.5% SDS-PAGE and visualized with HRP-conjugated anti-T7 tag antibody (top four panels). The bottom panel shows the FLAG-Rab3A protein visualized with HRP-conjugated anti-FLAG tag antibody. The results shown are representative of three independent experiments. (B) The immunoreactive bands of Rim2 RBDs on x-ray film in (A) were captured and quantified as described previously (Fukuda, 2004a). The EC_{50} values calculated with GraphPad PRISM software (version 4.0) were 0.025 for the Rim2·Rab3A interaction (open circles) and 0.057 for the Rim2(+4A)–Rab3A interaction (solid circles). Bars indicate the means ± SE of three independent experiments. (C) Rab3A-dependent recruitment of the short form of the Rim2 RBD to DCVs in PC12 cells. PC12 cells expressing GFP-Rim2-RBD (a and b), Rim2-RBD(+40A) (c and d), and Rim2-RBD(E36A/R37S) (e and f) were fixed, permeabilized, and stained with anti-Rab3A mouse monoclonal antibody (b, d, and f). Note that the short form of the Rim2 RBD overlapped well with Rab3A, whereas the long form of the Rim2 RBD was mostly found in the cytoplasm and in the nucleus (compare a and b, or c and d). Note that the Rab3A-binding-defective mutant of Rim2(E36A/R37S) was also localized in the cytoplasm. Scale bar in f indicates 10 μm. Reproduced with permission from Fukuda (2004a).

affinity (50-fold decrease in affinity; compare open circles and open squares in Fig. 2B), and insertion of four amino acids into the second zinc finger motif of Rim2 also reduces Rab3A-binding affinity (2-fold decrease in affinity; compare open and solid circles in Fig. 2B).

High-Affinity Rab3A-Binding Activity Is Required for Dense-Core Vesicle Localization of Rim2 in PC12 Cells

Although the long forms of Rim2 RBD show relatively weak Rab3A-binding capacity *in vitro* compared with the short forms of Rim2 RBD, the Rab3A-binding activity of the long forms of Rim2 RBD may still be sufficient to recognize the endogenous expression level of Rab3A in living cells. To investigate this possibility, I expressed green fluorescence protein (GFP)-tagged Rim2 RBDs in PC12 cells, where Rab3A is abundantly expressed, and compared the localization of GFP-Rim2-RBDs and endogenous Rab3A by immunofluorescence analysis (Fukuda, 2004a). This section describes the experimental protocols used to determine (1) the DCV localization of Rim2 RBDs in PC12 cells and (2) the involvement of the short forms of Rim2 RBD in DCV exocytosis.

Site-Directed Mutagenesis and Construction of GFP-Rim2-RBD Plasmids

Mutant Rim2 plasmid carrying Glu-to-Ala and Arg-to-Ser substitutions at amino acid positions 36 and 37 (E36A/R37S) is obtained by two-step PCR techniques as described previously (Fukuda, 2004a; Fukuda *et al.*, 1995). The mutant cDNA is subcloned into pEF-T7 expression vector as described previously (Fukuda, 2004a; Fukuda *et al.*, 1999). Plasmid DNA is prepared using Qiagen maxiprep kits. Lack of Rab3A-binding activity of the Rim2 RBD(E36A/R37S) mutant is confirmed by *in vitro* binding assays as described earlier.

Immunofluorescence Analysis of GFP-Tagged Rim2 RBDs

The Rim-RBD fragments are subcloned into the *Bgl*II/*Sal*I site of pEGFP-Cl plasmids (BD Clontech). PC12 cells are cultured in DMEM supplemented with 10% fetal bovine serum, 10% horse serum, 100 U/ml penicillin G, and 100 μg/ml streptomycin, at 37° under 5% CO_2. Transfection of pEGFP-C1-Rim2-RBDs (4 μg of plasmid) into PC12 cells (0.5~1.0 $\times$ 10^5 cells, the day before transfection/3.5-cm dish [MatTek Corp., Ashland, MA] coated with collagen type IV) is performed by using LipofectAMINE 2000 reagent (Invitrogen) according to the manufacturer's notes. Three days after transfection, cells are fixed with 4% paraformaldehyde in 0.1 *M* sodium phosphate buffer for 20 min at room temperature and then washed with 0.1 M glycine. The fixed cells are permeabilized with 0.3% Triton X-100 in PBS for 2 min and immediately washed with the blocking solution (1% bovine serum albumin fraction V [Sigma Chemical Co.] and 0.1% Triton X-100 in PBS) three times for 5 min each (Fukuda and Mikoshiba, 1999). The cells are incubated in blocking solution for 1 h at room temperature and then incubated with anti-Rab27A mouse monoclonal antibody (1/100 dilution) for 1 h at room temperature. After washing out the primary antibody with blocking solution three times for 5 min each, the cells are incubated with Alexa Fluor 568-labeled anti-rabbit IgG (1/5000 dilution; Molecular Probes, Inc., Eugene, OR) for 1 h at room temperature. After washing out the secondary antibody with blocking solution five times for 5 min each, immunoreactivity is analyzed with a confocal fluorescence microscope (Fluoview; Olympus, Tokyo, Japan), and the images are processed with Adobe Photoshop software (version 7.0). Under these experimental conditions, GFP-tagged short forms of Rim2 RBD (Rim2-RBD and Rim2-RBD[+4A]) colocalized well with endogenous Rab3A (Fig. 2C[a, b]), whereas the GFP-tagged long forms of Rim2 RBD (Rim2-RBD[+40A] and Rim2-RBD[+44A]) are mostly localized in the cytoplasm, in the nucleus, or near the plasma membrane (Fig. 2C[c, d]). GFP-Rim2-RBD(E36A/R37S), which completely lacks Rab3A-binding activity, is distributed throughout the cytoplasm, and no colocalization of the Rim2 mutant with Rab3A is detected (Fig. 2C[e, f]).

NPY Release Assay

PC12 cells are cotransfected with 4 μg each of pShooter-NPY-T7-GST and pEF-T7-Rim2-RBD (or pEF-T7-Rim2-RBD[E36A/R37S] mutant) by using LipofectAMINE 2000 reagent, and NPY release assay is performed as described in Chapter 39 (this volume). Under these experimental conditions, expression of T7-Rim2-RBD attenuates high-KCl-dependent NPY secretion from PC12 cells, whereas expression of the Rab3A binding-defective mutant T7-Rim2-RBD(E36A/R37S) has no effect on secretion (Fukuda, 2004a).

Summary of Results

We have described several methods of analyzing the Rab3A-binding properties of the long and short forms of Rim2 RBD in *in vitro* and in intact PC12 cells. Rim2 containing the 40-amino acid insertion in the RBD1 (i.e., Rim2-RBD[+40A]) exhibited very low Rab3A-binding capacity *in vitro*, in contrast to the Rim2 containing RBD1 without any insertion. Functional studies in PC12 cells have demonstrated that the Rab3A-binding activity of

Rim2-RBD(+40A) is insufficient to assess the endogenous expression level of Rab3A in PC12 cells, and thus only the shortest form of the Rim2 RBD, which is the predominant isoform in all mouse tissues tested by RT-PCR analysis, functions as a Rab3A effector *in vivo*. I have also tested the interaction between invertebrate (*Caenorhabditis elegans* and *Drosophila*) Rim RBD and Rab3 by *in vitro* binding assays as described previously. Unexpectedly, however, the Rim RBD of these animals was unable to interact with invertebrate Rab3 (Fukuda, 2004a), indicating that the Rab3-effector function of Rim during secretory vesicle exocytosis has not been retained during evolution.

Acknowledgments

This work was supported in part by grants from the Ministry of Education, Culture, Sports, and Technology of Japan.

References

Cheviet, S., Coppola, T., Haynes, L. P., Burgoyne, R. D., and Regazzi, R. (2004). The Rab-binding protein Noc2 is associated with insulin-containing secretory granules and is essential for pancreatic β-cell exocytosis. *Mol. Endocrinol.* **18,** 117–126.

Fukuda, M. (2003a). Distinct Rab binding specificity of Rim1, Rim2, rabphilin, and Noc2: Identification of a critical determinant of Rab3A/Rab27A recognition by Rim2. *J. Biol. Chem.* **278,** 15373–15380.

Fukuda, M. (2003b). Slp4-a/granuphilin-a inhibits dense-core vesicle exocytosis through interaction with the GDP-bound form of Rab27A in PC12 cells. *J. Biol. Chem.* **278,** 15390–15396.

Fukuda, M. (2004a). Alternative splicing in the first α-helical region of the Rab-binding domain of Rim regulates Rab3A binding activity: Is Rim a Rab3 effector protein during evolution? *Genes Cells* **9,** 831–842.

Fukuda, M. (2004b). Rabphilin and Noc2 function as Rab27 effectors that control Ca^{2+}-regulated exocytosis. *Recent Res. Dev. Neurochem.* **7,** 57–69.

Fukuda, M. (2005). Versatile role of Rab27 in membrane trafficking: Focus on the Rab27 effector families. *J. Biochem.* **137,** 9–16.

Fukuda, M., and Mikoshiba, K. (1999). A novel alternatively spliced variant of synaptotagmin VI lacking a transmembrane domain: Implications for distinct functions of the two isoforms. *J. Biol. Chem.* **274,** 31428–31434.

Fukuda, M., Aruga, J., Niinobe, M., Aimoto, S., and Mikoshiba, K. (1994). Inositol-1,3,4,5-tetrakisphosphate binding to C2B domain of IP4BP/synaptotagmin II. *J. Biol. Chem.* **269,** 29206–29211.

Fukuda, M., Kojima, T., Aruga, J., Niinobe, M., and Mikoshiba, K. (1995). Functional diversity of C2 domains of synaptotagmin family: Mutational analysis of inositol high polyphosphate binding domain. *J. Biol. Chem.* **270,** 26523–26527.

Fukuda, M., Kanno, E., and Mikoshiba, K. (1999). Conserved N-terminal cysteine motif is essential for homo- and heterodimer formation of synaptotagmins III, V, VI, and X. *J. Biol. Chem.* **274,** 31421–31427.

Fukuda, M., Kanno, E., Saegusa, C., Ogata, Y., and Kuroda, T. S. (2002). Slp4-a/granuphilin-a regulates dense-core vesicle exocytosis in PC12 cells. *J. Biol. Chem.* **277,** 39673–39678.

Fukuda, M., Kanno, E., and Yamamoto, A. (2004). Rabphilin and Noc2 are recruited to dense-core vesicles through specific interaction with Rab27A in PC12 cells. *J. Biol. Chem.* **279,** 13065–13075.

Iezzi, M., Regazzi, R., and Wollheim, C. B. (2000). The Rab3-interacting molecule RIM is expressed in pancreatic β-cells and is implicated in insulin exocytosis. *FEBS Lett.* **474,** 66–70.

Kuroda, T. S., Fukuda, M., Ariga, H., and Mikoshiba, K. (2002). The Slp homology domain of synaptotagmin-like proteins 1–4 and Slac2 functions as a novel Rab27A binding domain. *J. Biol. Chem.* **277,** 9212–9218.

Larijani, B., Hume, A. N., Tarafder, A. K., and Seabra, M. C. (2003). Multiple factors contribute to inefficient prenylation of Rab27a in Rab prenylation diseases. *J. Biol. Chem.* **278,** 46798–46804.

Mizushima, S., and Nagata, S. (1990). pEF-BOS, a powerful mammalian expression vector. *Nucleic Acids Res.* **18,** 5322.

Ostermeier, C., and Brunger, A. T. (1999). Structural basis of Rab effector specificity: Crystal structure of the small G protein Rab3A complexed with the effector domain of rabphilin-3A. *Cell* **96,** 363–374.

Ozaki, N., Shibasaki, T., Kashima, Y., Miki, T., Takahashi, K., Ueno, H., Sunaga, Y., Yano, H., Matsuura, Y., Iwanaga, T., Takai, Y., and Seino, S. (2000). cAMP-GEFII is a direct target of cAMP in regulated exocytosis. *Nat. Cell Biol.* **2,** 805–811.

Shibasaki, T., Sunaga, Y., Fujimoto, K., Kashima, Y., and Seino, S. (2003). Interaction of ATP sensor, cAMP sensor, Ca^{2+} sensor, and voltage-dependent Ca^{2+} channel in insulin granule exocytosis. *J. Biol. Chem.* **279,** 7956–7961.

Shirataki, H., Kaibuchi, K., Sakoda, T., Kishida, S., Yamaguchi, T., Wada, K., Miyazaki, M., and Takai, Y. (1993). Rabphilin-3A, a putative target protein for smg p25A/rab3A p25 small GTP-binding protein related to synaptotagmin. *Mol. Cell. Biol.* **13,** 2061–2068.

Südhof, T. C. (2004). The synaptic vesicle cycle. *Annu. Rev. Neurosci.* **27,** 509–547.

Sun, L., Bittner, M. A., and Holz, R. W. (2001). Rab3a binding and secretion-enhancing domains in Rim1 are separate and unique: Studies in adrenal chromaffin cells. *J. Biol. Chem.* **276,** 12911–12917.

Wang, X., Hu, B., Zimmermann, B., and Kilimann, M. W. (2001). Rim1 and rabphilin-3 bind Rab3-GTP by composite determinants partially related through N-terminal α-helix motifs. *J. Biol. Chem.* **276,** 32480–32488.

Wang, Y., Okamoto, M., Schmitz, F., Hofmann, K., and Südhof, T. C. (1997). Rim is a putative Rab3 effector in regulating synaptic-vesicle fusion. *Nature* **388,** 593–598.

Wang, Y., Sugita, S., and Südhof, T. C. (2000). The RIM/NIM family of neuronal C2 domain proteins: Interactions with Rab3 and a new class of Src homology 3 domain proteins. *J. Biol. Chem.* **275,** 20033–20044.

[41] Assay of the Rab-Binding Specificity of Rabphilin and Noc2: Target Molecules for Rab27

By MITSUNORI FUKUDA and AKITSUGU YAMAMOTO

Abstract

Rabphilin and Noc2 were originally described as Rab3A effector proteins involved in the regulation of secretory vesicle exocytosis in neurons and certain endocrine cells. Both proteins share the conserved N-terminal Rab-binding domain (RBD) that consists of two α-helical regions separated by two zinc finger motifs. However, the RBD of rabphilin and Noc2 has been shown to bind Rab27A (the closest homologue of Rab3 isoforms) in preference to Rab3A, both *in vitro* and *in vivo*. Rabphilin and Noc2 are recruited to dense-core vesicles (DCVs) in neuroendocrine PC12 cells and regulate their exocytosis through interaction with Rab27A rather than with Rab3A. Rab3A-binding-defective mutants of rabphilin(E50A) and Noc2 (E51A) retain the ability to target DCVs in PC12 cells, the same as the wild-type proteins, whereas Rab27A-binding-defective mutants of rabphilin(E50A/I54A) and Noc2(E51A/I55A) do not (i.e., they are present throughout the cytoplasm). Expression of the wild-type or the E50A mutant of rabphilin-RBD, but not the E50A/I54A mutant of rabphilin-RBD, in PC12 cells significantly attenuated DCV exocytosis monitored by high-KCl-stimulated neuropeptide Y secretion. In this chapter we describe various assay methods that have been used to characterize the RBD of rabphilin and Noc2 as "RBD27 (Rab-binding domain for Rab27)."

Introduction

The Rab3 subfamily is evolutionarily conserved from nematodes to humans, and Rab3 subfamily proteins are thought to control Ca^{2+}-regulated exocytosis of secretory granules through specific interaction with effector molecule(s) (Geppert and Südhof, 1998; Takai *et al.*, 1996). The first such effector molecule identified was rabphilin, which specifically binds the GTP-bound activated form of Rab3A on synaptic vesicles (Li *et al.*, 1994; Shirataki *et al.*, 1993). Rabphilin belongs to the C-terminal type tandem C2 proteins (Fukuda and Mikoshiba, 2001) and consists of an N-terminal Rab-binding domain (RBD) (i.e., two α-helical regions [named RBD1 and RBD2] and zinc finger motifs) (McKiernan *et al.*, 1996; Ostermeier and Brunger, 1999; Yamaguchi *et al.*, 1993) and C-terminal tandem C2 domains that bind Ca^{2+} and phospholipids (Chung

METHODS IN ENZYMOLOGY, VOL. 403
0076-6879/05 $35.00
DOI: 10.1016/S0076-6879(05)03041-7

et al., 1998; Fukuda *et al.*, 1994; Yamaguchi *et al.*, 1993). Although it has been proposed that the Rab3A-rabphilin system is important for the control of Ca^{2+}-regulated exocytosis based on the results of overexpression and peptide-fragment-injection experiments (Burns *et al.*, 1998; Komuro *et al.*, 1996), evidence has indicted that rabphilin functions independently of Rab3A during Ca^{2+}-regulated exocytosis (Chung *et al.*, 1999; Joberty *et al.*, 1999; Schluter *et al.*, 1999, 2004; Staunton *et al.*, 2001). More recently, the RBD of rabphilin and Noc2, another putative Rab3A effector (Haynes *et al.*, 2001; Kotake *et al.*, 1997), has been found to have sequence similarity with the specific Rab27A/B-binding domain (called [the Slp homology domain] [SHD]) of the Slp (synaptotagmin-like protein) family and Slac2 (Slp homologue lacking C2 domains) family, and to interact with Rab27 isoforms in preference to Rab3A *in vitro* (Fukuda, 2002a, 2003a, 2005; Fukuda and Kuroda, 2002; Fukuda *et al.*, 2001, 2002b; Kuroda *et al.*, 2002a,b; Strom *et al.*, 2002). Consistent with these findings, rabphilin and Noc2 are recruited to dense-core vesicles (DCVs) through interaction with Rab27A and regulate their Ca^{2+}-dependent exocytosis in endocrine cells (Cheviet *et al.*, 2004; Fukuda *et al.*, 2004). In addition, we have shown that only rabphilin-Rab27 interaction, not rapphilin-Rab3 interaction, is phylogenetically retained in *Caenorhabditis elegans* and *Drosophila*, indicating that rabphilin functions as a Rab27 effector across phylogeny (Fukuda, 2004a; Fukuda *et al.*, 2004).

In this chapter we describe several experimental protocols that have been used to compare the Rab3A- and Rab27A-binding properties of RBD of rabphilin and Noc2 *in vitro* and in intact PC12 cells.

Rab-Binding Domain of Rabphilin and Noc2 Interacts with Three Distinct Rab Species *In Vitro* and in PC12 Cells

Although rabphilin was originally reported as a specific Rab3A-binding protein on synaptic vesicles (Li *et al.*, 1994; Shirataki *et al.*, 1993), the Rab-binding specificity of rabphilin was determined by using only a small number of Rabs (e.g., Rab11) or other small GTPases (e.g., Ki-Ras and Rho) (Shirataki *et al.*, 1993). Thus, it is possible that rabphilin (or another putative Rab3A effector Noc2) interacts with other Rabs in preference to Rab3 isoforms (the same possibility is true of other Rab effectors identified thus far, except the Rab27A effectors, Slp and Slac2) (Fukuda, 2003a). However, because of the large number of Rab proteins (>60 Rabs) in mammals (Pereira-Leal and Seabra, 2001), this possibility had not been systematically investigated until we established panels of FLAG-tagged Rab proteins (i.e., Rab1–40) in COS-7 cells (Fukuda, 2003a; Kuroda *et al.*, 2002a). Use of the Rab panels revealed that rabphilin and Noc2 interact with three distinct Rab isoforms (Rab3A/B/C/D, Rab8A, and Rab27A/B) *in vitro*, the same as the SHD of Slp4-a/granuphilin-a (Fukuda, 2003b; Fukuda *et al.*, 2002a; Kuroda *et al.*, 2002a). We have further shown that rabphilin and Noc2 bind Rab27A with more than 3 and with more than 10 times higher affinity, respectively, than they bind Rab3A (Fukuda, 2004a). In this section we describe several experimental protocols that are used to analyze the Rab-binding specificity of the RBD of rabphilin and Noc2 with special attention to their relative affinity for Rab27A instead of Rab3A *in vitro* and in intact PC12 cells.

Assay for Interaction between T7-rabphilin or T7-Noc2 and FLAG-Rabs In Vitro

cDNA encoding the full open reading frame of mouse rabphilin is obtained by screening of a mouse cerebellum oligo(dT)-primed cDNA library constructed in λgt11 (Fukuda *et al.*, 1994), and cDNA encoding the full open reading frame of mouse Noc2 is amplified by polymerase chain reaction (PCR) from Marathon-Ready adult mouse brain cDNA (BD Clontech, Palo Alto, CA) as described previously (Fukuda *et al.*, 1994, 2004). The mouse rabphilin or Noc2 cDNA is subcloned into the eukaryotic expression vector pEF-T7 (modified from pEF-BOS) by standard molecular cloning techniques to obtain pEF-T7-rabphilin and pEF-T7-Noc2 (Fukuda *et al.*, 1999, 2004; Kuroda *et al.*, 2002a; Mizushima and Nagata, 1990). Plasmid DNA for transfection into mammalian cell cultures is prepared by using Qiagen (Chatsworth, CA) maxiprep kits according to the manufacturer's notes. COS-7 cells are cultured in Dulbecco's modified Eagle's medium (DMEM) supplemented with 10% fetal bovine serum, 100 U/ml penicillin G, and 100 μg/ml streptomycin, at 37° under 5% CO_2. Transfection of pEF-T7-rabphilin or pEF-T7-Noc2 (4 μg of plasmid) into COS-7 cells (7.5 × 10^5 cells, the day before transfection/10-cm dish) is carried out by using LipofectAmine Plus reagent (Invitrogen, Carlsbad, CA) according to the manufacturer's notes. Three days after transfection, cells are homogenized in a buffer containing 500 μl of 50 mM HEPES-KOH, pH 7.2, 250 mM NaCl, 1 mM $MgCl_2$, 0.5 mM GTPγS (guanosine 5'-O-[3-thiotriphosphate]), and protease inhibitors (0.1 mM phenylmethylsulfonyl fluoride, 10 μM leupeptin, and 10 μM pepstatin A; "protease inhibitors" refers to these three inhibitors throughout the text) in a glass-Teflon Potter homogenizer with 10 strokes at 900–1000 rpm. Proteins are solubilized by gentle agitation with 1% Triton X-100 at 4° for 1 h, and insoluble material is removed by centrifugation at 15,000 rpm (TOMY MX-100 high-speed refrigerated microcentrifuge; Tokyo, Japan) for 10 min. T7-rabphilin (or T7-Noc2) protein is immunoprecipitated from the total cell lysates (400 μl) by gentle agitation with anti-T7 tag antibody-conjugated agarose beads (wet volume 20 μl; Novagen, Madison, WI) at 4° for 1 h, and the T7-rabphilin (or T7-Noc2)-bound beads are washed once with 1 ml of 50 mM HEPES-KOH, pH 7.2, 150 mM NaCl, 1 mM MgCl$_2$, 0.2% Triton X-100, and protease inhibitors.

The T7-rabphilin (or T7-Noc2) beads are probed with FLAG-Rab panels prepared as described in Chapter 39 (this volume). In brief, the beads are incubated by gentle agitation with 400 μl of the FLAG-Rab-containing cell lysates at 4° for 1 h, and the beads are then washed three times with 1 ml of 50 mM HEPES-KOH, pH 7.2, 150 mM NaCl, 1 mM MgCl$_2$, 0.2% Triton X-100, and protease inhibitors. The proteins bound to the beads are analyzed by 10% sodium dodecyl sulfate–polyacrylamide gel electrophoresis (SDS–PAGE) followed by immunoblotting with horseradish peroxidase (HRP)-conjugated anti-FLAG tag mouse monoclonal antibody (M2) (1/10,000 dilution; Sigma Chemical Co., St. Louis, MO) and HRP-conjugated anti-T7 tag (1/10,000 dilution) (see Chapter 40 [this volume] for details). Immunoreactive bands are detected by enhanced chemiluminescence (ECL) (Amersham Biosciences, Buckinghamshire, UK). Under these experimental conditions, both T7-rabphilin and T7-Noc2 (or their RBD alone) interact with three distinct Rab isoforms: Rab3, Rab8, and Rab27 (Fukuda, 2003b; Kuroda *et al.*, 2002a).

Similar results are obtained by cotransfection assay in COS-7 cells (i.e., 2 μg each of pEF-T7-rabphilin and pEF-FLAG-Rab is cotransfected into COS-7 cells) (see Chapter 39 [this volume] for details) or by pull-down assay using glutathione S-transferase (GST)-rabphilin-RBD (see Chapter 37 [this volume] for details).

Assay for Competition Experiments

cDNA of the mouse rabphilin-RBD (amino acid residues 1–186) or Noc2-RBD (amino acid residues 1–180) is subcloned into the eukaryotic expression vector pEF-T7 by conventional PCR techniques (named pEF-T7-rabphilin-RBD or pEF-T7-Noc2-RBD, respectively) (Fukuda, 2003a). T7-rabphilin-RBD (or Noc2-RBD) and FLAG-Rab27A (or Rab3A) proteins are separately expressed in COS-7 cells (one 10-cm dish) as described previously. Preparation of total cell lysates (400 μl) containing T7-rabphilin-RBD (or Noc2-RBD) proteins is also performed as described earlier. T7-tagged proteins are affinity-purified by anti-T7 tag antibody-conjugated agarose (wet volume 140 μl), and the agarose beads are divided into seven portions (20 μl each in 1.5-ml microtubes).

COS-7 cells expressing FLAG-Rab27A (or FLAG-Rab3A) are homogenized in a buffer containing 1.5 ml of 50 mM HEPES-KOH, pH 7.2,

150 mM NaCl, 1 mM MgCl$_2$, 0.5 mM GTPγS, and protease inhibitors in a glass-Teflon Potter homogenizer with 10 strokes at 900–1000 rpm. Proteins are solubilized by gentle agitation with 1% Triton X-100 at 4° for 1 h, and insoluble material is removed by centrifugation at 15,000 rpm. Total cell lysates (10 μl) containing FLAG-Rab proteins are analyzed by 10% SDS–PAGE followed by immunoblotting with HRP-conjugated anti-FLAG tag antibody and ECL detection. The intensity of the immunoreactive bands on x-ray film is quantified by Lane Analyzer (version 3.0) (ATTO Corp., Tokyo, Japan), and, if necessary, original cell lysates are appropriately diluted with the homogenization buffer containing 1% Triton X-100 so that the amounts of FLAG-Rab proteins in the diluted samples are similar. Under these experimental conditions, however, the expression levels of FLAG-Rab3A and FLAG-Rab27A are almost the same.

The T7-rabphilin-RBD (or T7-Noc2-RBD) beads (wet volume 20 μl) are incubated with 400 μl of a solution containing FLAG-Rab3A and FLAG-Rab27A in various proportions (Rab3A/Rab27A: 400 μl/0 μl, 360 μl/40 μl, 300 μl/100 μl, 200 μl/200 μl, 100 μl/300 μl, 40 μl/360 μl, 0 μl/ 400 μl; see Fig. 1) for 1 h at 4° in 50 mM HEPES-KOH, pH 7.2, 150 mM NaCl, 1 mM MgCl$_2$, 1% Triton X-100, and protease inhibitors. Note that the

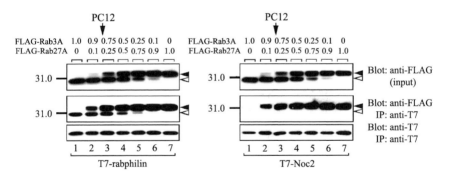

Fig. 1. Rabphilin and Noc2 interact with Rab27A in preference to Rab3A *in vitro*. Competition experiments revealed that rabphilin and Noc2 preferentially interact with Rab27A rather than Rab3A. T7-rabphilin-RBD beads (or T7-Noc2-RBD beads) were incubated with solutions containing Rab3A and Rab27A in the proportions indicated. After washing the beads, proteins bound to the beads were analyzed by 12.5% SDS–PAGE followed by immunoblotting with HRP-conjugated anti-FLAG tag antibody (Blot, anti-FLAG; IP, anti-T7; middle panels) and anti-T7 tag antibody (Blot, anti-T7; IP, anti-T7; bottom panels). Input means 1/80 volume of the reaction mixture used for immunoprecipitation (top panels). The open and closed arrowheads indicate the position of FLAG-Rab3A and FLAG-Rab27A, respectively. The arrows indicate the proportion of Rab3A (0.8) to Rab27A (0.2) in PC12 cells as reported previously (Fukuda *et al.*, 2002a). The positions of the molecular mass markers ($\times$ 10^{-3}) are shown on the left. Reproduced with permission from Fukuda *et al.* (2004).

amount of FLAG-Rab proteins in the solution should be much greater than the amount of T7-tagged proteins on the beads, so that competition between Rab3A and Rab27A occurs efficiently. After washing the beads three times with 1 ml of 50 mM HEPES-KOH, pH 7.2, 150 mM NaCl, 1 mM MgCl$_2$, and 0.2% Triton X-100, the proteins bound to the beads are analyzed by 12.5% SDS–PAGE followed by immunoblotting with HRP-conjugated anti-T7 tag antibody and anti-FLAG tag antibody as described previously (Fukuda *et al.*, 1999). Under these experimental conditions, Noc2 interacts with Rab27A alone, even when a 10 times greater amount of Rab3A than Rab27A is present in the reaction mixture (right middle panel of Fig. 1), whereas rabphilin interacts with both Rabs, but preferentially interacts with Rab27A rather than Rab3A (left middle panel of Fig. 1). Since the ratio between expression level of Rab3A and Rab27A in PC12 cells is about 0.8:0.2 (= 4:1) (Rab3A/Rab27A) (Fukuda *et al.*, 2002a), Noc2 should function as a specific Rab27A effector in intact PC12 cells, and rabphilin should interact with both Rab3A and Rab27A (arrows in Fig. 1).

Assay for Determining Relative Rab-Binding Affinity of Rabphilin and Noc2 In Vitro

T7-rabphilin-RBD (or T7-Noc2-RBD) beads (wet volume 20 μl) in 1.5-ml microtubes ($\times$ 7 samples) are prepared as described earlier. Equal amounts of FLAG-Rab27A and FLAG-Rab3A in 50 mM HEPES-KOH, pH 7.2, 150 mM NaCl, 1 mM MgCl$_2$, 0.5 mM GTPγS, 1% Triton X-100, and protease inhibitors are also prepared as described previously, and they are systematically diluted thus: 1, 1/4, 1/10, 1/40, 1/100, 1/400, and 1/1000. The diluted samples are incubated with beads coupled with T7-rabphilin-RBD (or T7-Noc2-RBD) protein. After washing the beads three times with 1 ml of 20 mM HEPES-KOH, pH 7.2, 150 mM NaCl, 1 mM MgCl$_2$, and 0.2% Triton X-100, proteins bound to the beads are analyzed by 12.5% SDS–PAGE and visualized with HRP-conjugated anti-FLAG tag antibody and anti-T7 tag antibody. Immunoreactive bands of FLAG-Rabs on x-ray film are captured and quantified by Lane Analyzer. Experiments are repeated at least three times, and EC$_{50}$ values for the RBD-Rab27A interaction or RBD-Rab3A interaction are calculated with GraphPad PRISM software (version 4.0) (Fukuda, 2004a,b).

Assay for Interaction between GST-Rabphilin-RBD (or GST-Noc2-RBD) and Endogenous Rab27A in PC12 Cells

The mouse rabphilin-RBD cDNA (or Noc2-RBD) is subcloned into the eukaryotic expression vector pEF-T7-GST (named pEF-T7-GST-rabphilin-RBD or -Noc2-RBD) as described previously (Fukuda *et al.*, 2002b,

2004). All procedures, including preparation of plasmids, DNA transfection into PC12 cells, GST pull-down, and immunoblotting, are described elsewhere in detail (see Chapter 39, this volume). Consistent with the *in vitro* binding assays in COS-7 cells, T7-GST-rabphilin interacts with both endogenous Rab27A and Rab3A in PC12 cells, whereas T7-GST-Noc2-RBD specifically interacts with endogenous Rab27A alone (Fukuda *et al.*, 2004).

Assay for Coimmunoprecipitation of Endogenous Rabphilin-Rab27A or Noc2-Rab27A Complex from PC12 Cell Lysates with Specific Antibodies

Antibody against the C-terminus of mouse Noc2 (amino acid residues 181–302) (i.e., GST-Noc2-ΔRBD) is prepared as described previously (Fukuda and Mikoshiba, 1999; Fukuda *et al.*, 2004). Anti-rabphilin mouse monoclonal antibody is obtained from BD Transduction Laboratories (Lexington, KY). All other procedures, including preparation of PC12 cell lysates, immunoprecipitation with specific IgGs (10 μg/ml), and immunoblotting, are described elsewhere in detail (see Chapter 39, this volume). Both the anti-rabphilin antibody and the anti-Noc2 antibody efficiently coimmunoprecipitate endogenous Rab27A from the PC12 cell lysates, whereas only anti-rabphilin antibody coimmunoprecipitates endogenous Rab3A (Fukuda *et al.*, 2004).

Rabphilin and Noc2 Are Recruited to Dense-Core Vesicles and Regulate Their Exocytosis through Interaction with Rab27A in PC12 Cells

The RBD of rabphilin and Noc2 consists of two α-helical regions (RBD1 and RBD2) separated by two zinc finger motifs, and systematic deletion analysis has shown that the RBD1 of rabphilin and Noc2 is a central Rab27A-binding domain, the same as the SHD1 of Slac2-a and Slp4-a (Fukuda, 2002b, 2003b; Fukuda *et al.*, 2004). Although the RBD1 of rabphilin and Noc2 is also necessary for interaction with Rab3A (Wang *et al.*, 2001), another α-helical region RBD2 is required for high-affinity Rab3A recognition, indicating different mechanisms of Rab27A/Rab3A recognition by rabphilin and Noc2 (Fukuda *et al.*, 2004). Consistent with this, we have succeeded in creating a loss-of-function-type rabphilin and Noc2 (i.e., Rab3A-binding-defective mutant or Rab27A-binding-defective mutant) by Ala-based site-directed mutagenesis focused on the amino acids conserved among rabphilin, Noc2, Slp4-a, and Slac2-a (Fukuda *et al.*, 2004). For instance, a single mutation (a Glu-to-Ala substitution at amino acid position 51 [E51A] of Noc2 or E50A of rabphilin) results in loss of Rab3A-binding activity without affecting Rab27A-binding activity at all

(lanes 2 and 5 in the middle panel of Fig. 2A). By contrast, double mutations (E51A/I55A of Noc2 and E50A/I54A of rabphilin) dramatically
reduced Rab27A-binding activity (lane 6 in the middle panel of Fig. 2A).
These loss-of-function-type mutants are powerful tools for investigating the
Rab27A effector function of rabphilin and Noc2 *in vivo*. This section
describes the experimental protocols that are used to analyze the function
and localization of loss-of-function-type rabphilin and Noc2 mutants in
PC12 cells.

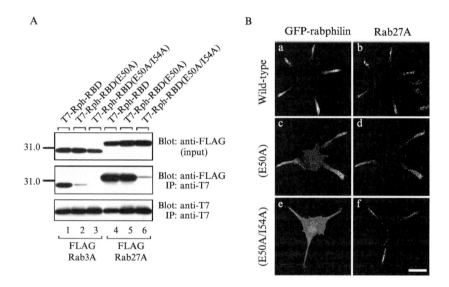

FIG. 2. Characterization of Rab3A–binding-defective (E50A) and Rab27A–binding-
defective (E50A/I54A) mutants of rabphilin *in vitro* and in PC12 cells. (A) pEF-T7-
rabphilin-RBD point mutants and pEF-FLAG-Rab were cotransfected into COS-7 cells, and
associations between the T7-rabphilin-RBD mutant and FLAG-Rab were evaluated by
coimmunoprecipitation assay as described previously (Fukuda *et al.*, 1999, 2004). Coimmunoprecipitated FLAG-Rab and the immunoprecipitated (IP) T7-rabphilin mutant were
visualized with HRP-conjugated anti-FLAG tag antibody (Blot, anti-FLAG; IP, anti-T7;
middle panel) and anti-T7 tag antibody (Blot, anti-T7; IP, anti-T7; bottom panel),
respectively. Input means 1/80 volume of the reaction mixture used for immunoprecipitation
(top panel). The positions of the molecular mass markers ($\times 10^{-3}$) are shown on the left. (B)
PC12 cells expressing GFP-tagged rabphilin-RBD mutants were fixed, permeabilized, and
stained with anti-Rab27A mouse monoclonal antibody (b, d, and f). Note that the wild-type
(a) and the E50A mutant (c) that lacks Rab3A-binding activity are colocalized with Rab27A
in the distal portion of the neurites, where DCVs are enriched, whereas the E50A/I54A
mutant, which lacks Rab27A-binding activity, is distributed throughout the cytoplasm. Scale
bar in f indicates 50 μm. Reproduced with permission from Fukuda *et al.* (2004).

Site-Directed Mutagenesis and Construction of Rabphilin and Noc2 Mutant Plasmids

Mutant rabphilin plasmids carrying an E50A or E50A/I54A substitution and mutant Noc2 plasmids carrying an E51A or E51A/I55A substitution are obtained by two-step PCR techniques (Fukuda *et al.*, 1995, 2004). The mutant cDNAs are subcloned into a pEF-T7 or pEGFP-C1 (BD Clontech) expression vector as described previously (Fukuda *et al.*, 1999, 2004). Plasmid DNA is prepared using Qiagen maxiprep kits. The Rab–binding specificity of each mutant is confirmed by *in vitro* binding assays for interaction between T7-RBD and FLAG-Rab as described previously (Fig. 1A).

Immunofluorescence Analysis of Green Fluorescent Protein-Tagged Rabphilin and Noc2

pEGFP-C1-rabphilin-RBD or -Noc2-RBD (4 μg of plasmid) is transfected into PC12 cells (0.5~1.0 × 10^5 cells, the day before transfection/3.5-cm dish coated with collagen type IV) by using LipofectAMINE 2000 reagent (Invitrogen) according to the manufacturer's notes. Three days after transfection, cells are fixed, permeabilized with 0.3% Triton X-100, and blocked with the blocking solution (1% bovine serum albumin [BSA] and 0.1% Triton X-100 in phosphate-buffered saline [PBS]) as described in Chapter 40 (this volume). The cells are incubated with anti-Rab27A mouse monoclonal antibody (1/100 dilution) for 1 h at room temperature and then with Alexa Fluor 568-labeled anti-rabbit IgG (1/5000 dilution; Molecular Probes, Inc., Eugene, OR) for 1 h at room temperature. Immunoreactivity is analyzed with a confocal fluorescence microscope (Fluoview; Olympus, Tokyo, Japan), and the images are processed with Adobe Photoshop software (version 7.0). Under these experimental conditions, Rab3A-binding-defective mutants of GFP-rabphilin-RBD and GFP-Noc2-RBD are still colocalized with endogenous Rab27A on DCVs in PC12 cells, the same as the wild-type protein, whereas Rab27A-binding-defective mutants are present throughout the cytoplasm (Fig. 2B).

Immunoelectron Microscopic Analysis

Localization of rabphilin and Noc2 is analyzed immunoelectron microscopically using the preembedding colloidal gold-silver enhancement method. GFP-rabphilin-RBD or GFP-Noc2-RBD is expressed in PC12 cells (cultured on collagen type IV-coated plastic coverslips; Sumitomo Bakelite, Tokyo, Japan) as described earlier. Three days after transfection, cells are fixed in 4% paraformaldehyde (catalogue number 261–26; Nakalai Tesque, Kyoto, Japan) in 0.1 *M* sodium phosphate buffer (PB), pH 7.4, for

2 h. After washing in the same buffer three times for 5 min, the fixed cells are incubated for 15 s in PB containing 14% glycerol and 35% sucrose, and permeabilized by freezing and thawing in liquid nitrogen. In some cases cells are permeabilized by incubating for 30 min in PB containing 0.25% saponin and 5% BSA. When cells are frozen and thawed in liquid nitrogen, preservation of the ultrastructure is better in the former than in the latter, while saponin treatment enhances permeability of antibody into the inside of vesicular structure in the cytoplasm. The cells are washed in PB once for 3 min and then incubated for 30 min for blocking in PB containing 0.005% saponin, 10% BSA, 10% normal goat serum, and 0.1% cold water fish skin gelatin. The cells are then exposed to anti-GFP rabbit polyclonal antibody (1/500 dilution; BD Clontech) in the blocking solution overnight. After washing in PB containing 0.005% saponin six times for 10 min, the cells are incubated for 2 h with the Fab' fragment of goat anti-rabbit IgG that has been conjugated to colloidal gold (1.4-nm diameter; 1/500 dilution; Nanoprobes, Inc., Stony Brook, NY) in the blocking solution. The cells are then washed with PB six times for 10 min and fixed with 1% glutaraldehyde in PB for 10 min. After washing in 50 mM HEPES buffer, pH 5.8, three times for 3 min, and once in Milli-Q water for 1 min, the gold labeling is intensified with a silver enhancement kit (HQ silver; Nanoprobes, Inc.) for 6 min at 20° in the dark according to the manufacturer's notes. After washing in distilled water three times for 1 min, the cells are postfixed in 0.5% OsO_4 in PB for 90 min at 4° and washed in PB three times for 3 min and once in distilled water for 3 min. The cells are dehydrated in a series of graded ethanol solutions once with 30, 50, 70, and 90% ethanol for 10 min each, twice in 100% ethanol, and embedded in epoxy resin. In some cases cells are stained en bloc with 2% uranyl acetate in 70% ethanol for 1 h between 70 and 90% ethanol to improve the ultrastructure of the cells. After hardening of the resin, the plastic coverslip is removed from the epoxy resin. Ultrathin sections are cut horizontally to the cell layer, doubly stained with uranyl acetate and lead citrate, and examined with an H7600 electron microscope (Hitachi, Tokyo, Japan).

NPY Release Assay

PC12 cells are cotransfected with 4 μg each of pShooter-NPY-T7-GST and pEF-T7-rabphilin-RBD mutants by using LipofectAMINE 2000 reagent, and an NPY release assay is performed as described in Chapter 39 (this volume). Under these experimental conditions, expression of T7-rabphilin-RBD(E50A) attenuates high-KCl-dependent NPY secretion from PC12 cells, the same as the wild-type protein, whereas expression of T7-rabphilin-RBD(E50A/I54A) has no effect on secretion (Fukuda *et al.*, 2004).

Summary of Results

We have described several methods of comparing the Rab27A- and Rab3A-binding activities of rabphilin and Noc2, including *in vitro* binding assays and functional studies in PC12 cells combined with site-directed mutagenesis. Both rabphilin and Noc2 were found to be present on the DCVs in PC12 cells and to regulate their exocytosis through interaction with Rab27A, rather than with Rab3A.

Acknowledgments

This work was supported in part by grants from the Ministry of Education, Culture, Sports, and Technology of Japan.

References

Burns, M. E., Sasaki, T., Takai, Y., and Augustine, G. J. (1998). Rabphilin-3A: A multifunctional regulator of synaptic vesicle traffic. *J. Gen. Physiol.* **111,** 243–255.

Cheviet, S., Coppola, T., Haynes, L. P., Burgoyne, R. D., and Regazzi, R. (2004). The Rab-binding protein Noc2 is associated with insulin-containing secretory granules and is essential for pancreatic β-cell exocytosis. *Mol. Endocrinol.* **18,** 117–126.

Chung, S. H., Song, W. J., Kim, K., Bednarski, J. J., Chen, J., Prestwich, G. D., and Holz, R. W. (1998). The C2 domains of Rabphilin3A specifically bind phosphatidylinositol 4,5-bisphosphate containing vesicles in a Ca^{2+}-dependent manner: *In Vitro* characteristics and possible significance. *J. Biol. Chem.* **273,** 10240–10248.

Chung, S. H., Joberty, G., Gelino, E. A., Macara, I. G., and Holz, R. W. (1999). Comparison of the effects on secretion in chromaffin and PC12 cells of Rab3 family members and mutants: Evidence that inhibitory effects are independent of direct interaction with Rabphilin3. *J. Biol. Chem.* **274,** 18113–18120.

Fukuda, M. (2002a). Slp and Slac2, novel families of Rab27 effectors that control Rab27-dependent membrane traffic. *Recent Res. Dev. Neurochem.* **5,** 297–309.

Fukuda, M. (2002b). Synaptotagmin-like protein (Slp) homology domain 1 of Slac2-a/melanophilin is a critical determinant of GTP-dependent specific binding to Rab27A. *J. Biol. Chem.* **277,** 40118–40124.

Fukuda, M. (2003a). Distinct Rab binding specificity of Rim1, Rim2, rabphilin, and Noc2: Identification of a critical determinant of Rab3A/Rab27A recognition by Rim2. *J. Biol. Chem.* **278,** 15373–15380.

Fukuda, M. (2003b). Slp4-a/granuphilin-a inhibits dense-core vesicle exocytosis through interaction with the GDP-bound form of Rab27A in PC12 cells. *J. Biol. Chem.* **278,** 15390–15396.

Fukuda, M. (2004a). Rabphilin and Noc2 function as Rab27 effectors that control Ca^{2+}-regulated exocytosis. *Recent Res. Dev. Neurochem.* **7,** 57–69.

Fukuda, M. (2004b). Alternative splicing in the first α-helical region of the Rab-binding domain of Rim regulates Rab3A binding activity: Is Rim a Rab3 effector protein during evolution? *Genes Cells* **9,** 831–842.

Fukuda, M. (2005). Versatile role of Rab27 in membrane trafficking: Focus on the Rab27 effector families. *J. Biochem.* **137,** 9–16.

Fukuda, M., and Kuroda, T. S. (2002). Slac2-c (synaptotagmin-like protein homologue lacking C2 domains-c), a novel linker protein that interacts with Rab27, myosin Va/VIIa, and actin. *J. Biol. Chem.* **277,** 43096–43103.

Fukuda, M., and Mikoshiba, K. (1999). A novel alternatively spliced variant of synaptotagmin VI lacking a transmembrane domain: Implications for distinct functions of the two isoforms. *J. Biol. Chem.* **274,** 31428–31434.

Fukuda, M., and Mikoshiba, K. (2001). Synaptotagmin-like protein 1–3: A novel family of C-terminal-type tandem C2 proteins. *Biochem. Biophys. Res. Commun.* **281,** 1226–1233.

Fukuda, M., Aruga, J., Niinobe, M., Aimoto, S., and Mikoshiba, K. (1994). Inositol-1,3,4,5-tetrakisphosphate binding to C2B domain of IP4BP/synaptotagmin II. *J. Biol. Chem.* **269,** 29206–29211.

Fukuda, M., Kojima, T., Aruga, J., Niinobe, M., and Mikoshiba, K. (1995). Functional diversity of C2 domains of synaptotagmin family: Mutational analysis of inositol high polyphosphate binding domain. *J. Biol. Chem.* **270,** 26523–26527.

Fukuda, M., Kanno, E., and Mikoshiba, K. (1999). Conserved N-terminal cysteine motif is essential for homo- and heterodimer formation of synaptotagmins III, V, VI, and X. *J. Biol. Chem.* **274,** 31421–31427.

Fukuda, M., Saegusa, C., and Mikoshiba, K. (2001). Novel splicing isoforms of synaptotagmin-like proteins 2 and 3: Identification of the Slp homology domain. *Biochem. Biophys. Res. Commun.* **283,** 513–519.

Fukuda, M., Kanno, E., Saegusa, C., Ogata, Y., and Kuroda, T. S. (2002a). Slp4-a/granuphilin-a regulates dense-core vesicle exocytosis in PC12 cells. *J. Biol. Chem.* **277,** 39673–39678.

Fukuda, M., Kuroda, T. S., and Mikoshiba, K. (2002b). Slac2-a/melanophilin, the missing link between Rab27 and myosin Va: Implications of a tripartite protein complex for melanosome transport. *J. Biol. Chem.* **277,** 12432–12436.

Fukuda, M., Kanno, E., and Yamamoto, A. (2004). Rabphilin and Noc2 are recruited to dense-core vesicles through specific interaction with Rab27A in PC12 cells. *J. Biol. Chem.* **279,** 13065–13075.

Geppert, M., and Südhof, T. C. (1998). RAB3 and synaptotagmin: The yin and yang of synaptic membrane fusion. *Annu. Rev. Neurosci.* **21,** 75–95.

Haynes, L. P., Evans, G. J., Morgan, A., and Burgoyne, R. D. (2001). A direct inhibitory role for the Rab3-specific effector, Noc2, in Ca^{2+}-regulated exocytosis in neuroendocrine cells. *J. Biol. Chem.* **276,** 9726–9732.

Joberty, G., Stabila, P. F., Coppola, T., Macara, I. G., and Regazzi, R. (1999). High affinity Rab3 binding is dispensable for Rabphilin-dependent potentiation of stimulated secretion. *J. Cell Sci.* **112,** 3579–3587.

Komuro, R., Sasaki, T., Orita, S., Maeda, M., and Takai, Y. (1996). Involvement of rabphilin-3A in Ca^{2+}-dependent exocytosis from PC12 cells. *Biochem. Biophys. Res. Commun.* **219,** 435–440.

Kotake, K., Ozaki, N., Mizuta, M., Sekiya, S., Inagaki, N., and Seino, S. (1997). Noc2, a putative zinc finger protein involved in exocytosis in endocrine cells. *J. Biol. Chem.* **272,** 29407–29410.

Kuroda, T. S., Fukuda, M., Ariga, H., and Mikoshiba, K. (2002a). The Slp homology domain of synaptotagmin-like proteins 1–4 and Slac2 functions as a novel Rab27A binding domain. *J. Biol. Chem.* **277,** 9212–9218.

Kuroda, T. S., Fukuda, M., Ariga, H., and Mikoshiba, K. (2002b). Synaptotagmin-like protein 5: A novel Rab27A effector with C-terminal tandem C2 domains. *Biochem. Biophys. Res. Commun.* **293,** 899–906.

Li, C., Takei, K., Geppert, M., Daniell, L., Stenius, K., Chapman, E. R., Jahn, R., De Camilli, P., and Südhof, T. C. (1994). Synaptic targeting of rabphilin-3A, a synaptic vesicle $Ca^{2+}/$phospholipid-binding protein, depends on rab3A/3C. *Neuron* **13,** 885–898.

McKiernan, C. J., Stabila, P. F., and Macara, I. G. (1996). Role of the Rab3A-binding domain in targeting of rabphilin-3A to vesicle membranes of PC12 cells. *Mol. Cell. Biol.* **16,** 4985–4995.

Mizushima, S., and Nagata, S. (1990). pEF-BOS, a powerful mammalian expression vector. *Nucleic Acids Res.* **18,** 5322.

Ostermeier, C., and Brunger, A. T. (1999). Structural basis of Rab effector specificity: Crystal structure of the small G protein Rab3A complexed with the effector domain of rabphilin-3A. *Cell* **96,** 363–374.

Pereira-Leal, J. B., and Seabra, M. C. (2001). Evolution of the Rab family of small GTP-binding proteins. *J. Mol. Biol.* **313,** 889–901.

Schluter, O. M., Schnell, E., Verhage, M., Tzonopoulos, T., Nicoll, R. A., Janz, R., Malenka, R. C., Geppert, M., and Südhof, T. C. (1999). Rabphilin knock-out mice reveal that rabphilin is not required for rab3 function in regulating neurotransmitter release. *J. Neurosci.* **19,** 5834–5846.

Schluter, O. M., Schmitz, F., Jahn, R., Rosenmund, C., and Südhof, T. C. (2004). A complete genetic analysis of neuronal Rab3 function. *J. Neurosci.* **24,** 6629–6637.

Shirataki, H., Kaibuchi, K., Sakoda, T., Kishida, S., Yamaguchi, T., Wada, K., Miyazaki, M., and Takai, Y. (1993). Rabphilin-3A, a putative target protein for smg p25A/rab3A p25 small GTP-binding protein related to synaptotagmin. *Mol. Cell. Biol.* **13,** 2061–2068.

Staunton, J., Ganetzky, B., and Nonet, M. L. (2001). Rabphilin potentiates soluble N-ethylmaleimide sensitive factor attachment protein receptor function independently of rab3. *J. Neurosci.* **21,** 9255–9264.

Strom, M., Hume, A. N., Tarafder, A. K., Barkagianni, E., and Seabra, M. C. (2002). A family of Rab27-binding proteins: Melanophilin links Rab27a and myosin Va function in melanosome transport. *J. Biol. Chem.* **277,** 25423–25430.

Takai, Y., Sasaki, T., Shirataki, H., and Nakanishi, H. (1996). Rab3A small GTP-binding protein in Ca^{2+}-dependent exocytosis. *Genes Cells* **1,** 615–632.

Wang, X., Hu, B., Zimmermann, B., and Kilimann, M. W. (2001). Rim1 and rabphilin-3 bind Rab3-GTP by composite determinants partially related through N-terminal α-helix motifs. *J. Biol. Chem.* **276,** 32480–32488.

Yamaguchi, T., Shirataki, H., Kishida, S., Miyazaki, M., Nishikawa, J., Wada, K., Numata, S., Kaibuchi, K., and Takai, Y. (1993). Two functionally different domains of rabphilin-3A, Rab3A p25/smg p25A-binding and phospholipid- and Ca^{2+}-binding domains. *J. Biol. Chem.* **268,** 27164–27170.

[42] Functional Properties of the Rab-Binding Domain of Rab Coupling Protein

By ANDREW J. LINDSAY, NICOLAS MARIE, and MARY W. MCCAFFREY

Abstract

Rab-Coupling Protein (RCP) is an approximately 80-kDa, hydrophilic protein that belongs to a recently identified family of proteins that is characterized by its ability to interact with Rab11 via a highly homologous Rab-binding domain positioned at its carboxy-termini. Five members of

METHODS IN ENZYMOLOGY, VOL. 403
0076-6879/05 $35.00
DOI: 10.1016/S0076-6879(05)03042-9

this family have been identified; however, a number of these Rab11-FIPs, including RCP and Rip11, have several splice isoforms. RCP is involved in regulating transport of membrane-bound vesicles from the endocytic recycling compartment to the plasma membrane, and it can be found at both locations within the cell.

Introduction

The Rab GTPases comprise the largest family of the Ras superfamily of small GTPases. Approximately 60 Rab proteins identified in humans (Bock *et al.*, 2001) are involved in regulating specific membrane transport pathways within the cell. Rab4 and Rab11 are involved in regulating the transport of vesicles from the sorting endosome (SE) and the endocytic recycling compartment (ERC), respectively (Maxfield and McGraw, 2004). To help elucidate the functions of Rab4, we used the yeast two-hybrid technique to identify Rab4 effector proteins. One of the clones "fished" out of this screen encoded a protein that we called Rab Coupling Protein (RCP). Surprisingly, we found that in addition to interacting with Rab4, RCP displayed a stronger interaction with Rab11 (Lindsay *et al.*, 2002). We suggested that RCP serves as a "molecular link" between Rab4-regulated and Rab11-regulated membrane transport.

Database searches revealed that RCP belongs to a family of proteins characterized by the presence of a highly homologous, approximately 20-amino acid domain at its carboxy-termini. We and others demonstrated that this domain mediates binding to Rab11 (Hales *et al.*, 2001; Lindsay and McCaffrey, 2002, 2004b; Prekeris *et al.*, 2001; Wallace *et al.*, 2002). This domain was termed the Rab11-binding domain (RBD) and proteins that possess it belong to the Rab11 *family* of *interacting proteins* (Rab11-FIPs). The Rab11-FIPs can be subdivided into distinct classes—the class I Rab11-FIPs, which include RCP (Lindsay *et al.*, 2002), Rip11 (Prekeris *et al.*, 2000), and Rab11-FIP2 (Hales *et al.*, 2001; Lindsay and McCaffrey, 2002), all possess a homologous C2 domain near their amino-termini. The class II Rab11-FIPs, which include Rab11-FIP3/arfophilin-1 and Rab11-FIP4/arfophilin-2, lack a C2 domain but possess EF-hand and ERM motifs (Wallace *et al.*, 2002).

RCP is localized predominantly to the ERC and plays a role in regulating transport from this compartment to the plasma membrane. The C2 domains of the class I Rab11-FIPs mediate their targeting to the plasma membrane by binding to PtdIns(3,4,5)P_3 and phosphatidic acid (Lindsay and McCaffrey, 2004a). We believe that certain signaling pathways trigger the release of Rab11 FIP-regulated transport vesicles from the ERC, followed by their subsequent delivery to regions of the plasma membrane that are enriched in PtdIns(3,4,5)P_3 and phosphatidic acid. A key event in the regulation of Rab11-FIP function is the recruitment by Rab11 from the cytosol to the ERC.

This chapter describes the generation, by site-directed mutagenesis, of a series of RCP constructs that have key residues in the RBD mutated. The chapter then describes the methods we use to determine the effect of these mutations on the interaction of RCP with Rab4 and Rab11 and on the intracellular localization of RCP.

Materials

The template used for the site-directed mutagenesis of RCP has been described (Lindsay and McCaffrey, 2004b). Essentially, the approximately 2.5-kb fragment containing the RCP open reading frame, minus the first methionine, and possessing *Bam*HI sites at the 5' and 3' ends was amplified by polymerase chain reaction (PCR). The product was digested with *Bam*HI and inserted into *Bam*HI-cut pEGFP-C3 (Clontech). This construct allows for the expression of RCP with green fluorescent protein (GFP) fused to its amino-terminus, under the control of the cytomegalovirus (CMV) promoter.

Methods

Site-Directed Mutagenesis

The Rab11-binding domain of RCP is an approximately 20-amino acid motif located near its carboxy-terminus. It is highly homologous to the same region of the other members of the Rab11-FIP family (Lindsay and McCaffrey, 2004b). We had previously narrowed the Rab4- and Rab11-binding regions of RCP to their final 65 amino acids (Lindsay *et al.*, 2002), and Prekeris *et al.* (2001) determined that the minimal region of Rip11 required for Rab11 binding was this homologous 20-amino acid domain. Analysis of the primary sequence of RCP using an algorithm that predicts the likelihood of it forming coiled-coils (Lupas *et al.*, 1991) revealed that there is a putative coiled-coil domain near the carboxy-terminus. Coiled-coil domains are bundles of α-helices that wind into a superhelix and are important in mediating oligomerization and protein–protein interactions (Burkhard *et al.*, 2001). This putative coiled-coil domain encompasses the RBD. To determine the effect that single amino acid changes have on the potential for coiled-coil formation, individual residues were changed and the sequence run through the Lupas algorithm. Using this method it was determined that changing the aliphatic isoleucine at position 621 to

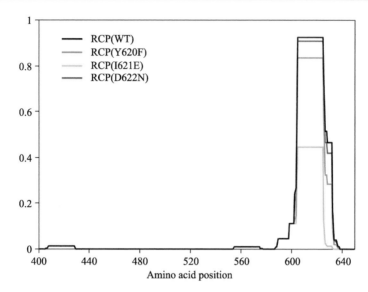

FIG. 1. Probability of RCP and mutants forming an α-helical coiled coil. Amino acids 400–649 of RCP, with or without the indicated amino acid changes, were run through Lupas' algorithm. Changing the isoleucine at position 621 to glutamic acid dramatically reduces the probability of the carboxy-terminus of RCP forming a coiled-coil.

glutamic acid would dramatically reduce the potential for coiled-coil formation (to below threshold) (Fig. 1). In contrast, changing the amino acids on either side of the isoleucine had less effect on the potential for coiled-coil formation. To determine experimentally whether these residues are important for mediating the Rab4 and Rab11 interactions, site-directed mutagenesis was used to change the candidate amino acids. The effect of these changes was determined using yeast two-hybrid, biochemical, and cell biology techniques.

The site-directed mutagenesis of RCP was performed using two primers, both containing the desired mutation, with each primer binding to the same sequence on complementary strands of the template. Upon binding to the template the primers are extended, without primer displacement, using a high-fidelity DNA polymerase. This generates a mutated plasmid containing staggered nicks. After a number of rounds of cycling the parental template plasmid is digested with the restriction enzyme *Dpn*I. *Dpn*I digests methylated and hemimethylated DNA; therefore, it will selectively digest the bacterially purified template. The nicked DNA containing the required mutation is subsequently transformed into competent *Escherichia coli* cells. This QuikChange technique (Stratagene) is highly efficient and due to the small amount of template and low number of cycles required reduces the chances of undesirable mutations.

The RCP mutagenic primers were all approximately 40 bases in length with the mutation centrally located (Table I). All primers were purified by polyacrylamide gel electrophoresis (Proligo, France). Each mutagenesis reaction contains 10 ng of pEGFP-C3 RCP template, 125 ng each of the sense and antisense primers, 10 mM dNTPs, and 1 μl (2.5 U/μl) PfuTurbo (Stratagene). The reaction volume was made up to 50 μl with water. The thermal cycling was performed on a PTC-200 Thermal Cycler (MJ Research) using the parameters indicated in Table II. Once the thermal cycling is finished 1 μl DpnI (10 U/μl) is added to the reaction, and incubated at 37° for a further 1 h. When digestion of the methylated template is complete, 2 μl of the reaction is used to transform $E. coli$ XL-1 cells. Transformants are "spread-plated" onto Luria broth (LB) agar containing

TABLE I
RCP SITE-DIRECTED MUTAGENESIS PRIMERS[a]

RCP(Y620F) sense	5'-GTCCGCGAGCTGGAAGAC**TTC**ATTGACAACCTGCTTGTC-3'
RCP(Y620F) antisense	5'- GACAAGCAGGTTGTCAAT**GAA**GTCTTCAGCTCGCGGAC-3'
RCP(I621E) sense	5'-CGCGAGCTGGAAGACTAC**GAG**GACAACCTGCTTGTCAGG-3'
RCP(I621E) antisense	5'-CCTGACAAGCAGGTTGTC**CTC**GTAGTCTTCCAGCTCGCG-3'
RCP(D622N) sense	5'-GCTGGAAGACTACATT**AAC**AACCTGCTTGTCAGGG-3'
RCP(D622N) antisense	5'-CCCTGACAAGCAGGTT**GTT**AATGTAGTCTTCCAGC-3'

[a] Changes are indicated in bold.

TABLE II
CYCLING PARAMETERS FOR THE SITE-DIRECTED MUTAGENESIS OF RCP

Cycles	Temperature	Time
1	95°	1 min
16	95°	50 s
	60°	50 s
	68°	15 min
1	68°	7 min

the appropriate selective antibiotic and incubated overnight at 37°. Individual colonies are then picked and cultured overnight in LB. Purified plasmid DNA is digested and the open reading frame is subjected to DNA sequencing analysis to ensure that the desired mutation has been incorporated and that no random mutations have been introduced. Each of the RBD mutants was subsequently subcloned from the pEGFP-C3 vector into pTrcHisB (Invitrogen) for prokaryotic expression and pGADGH for yeast two-hybrid analysis.

Note: We find that this method of site-directed mutagenesis is rapid and highly efficient.

Protein–Protein Interactions: Yeast Two-Hybrid Technique

Plasmids, Strains. Plasmids used were RCP, RCP(Y620F), RCP (I621E), and RCP(D622N) subcloned into the *Bam*HI–*Eco*RI sites of pGADGH (see previous discussion). The proteins are expressed in yeast with an amino-terminal GAL4 activation domain (AD) fusion. The pLex-Rab11Q70L and pVJL10-Rab4Q67L plasmids express constitutively active forms of both Rab proteins, fused at their amino-termini to the DNA-binding domain (BD) of the GAL4 transcription factor (Lindsay *et al.*, 2002). The yeast reporter strain was *Saccharomyces cerevisiae* L40 (MATa *trp1 leu2 his3::lexA-His3 URA3::lexA-lacZ*).

Transformation. L40 yeast from a fresh overnight culture is inoculated into sterile YPD (1% yeast extract, 2% Bacto-peptone, and 2% glucose) and incubated at 30° until the OD_{600} is approximately 0.4. The yeast cells are centrifuged, washed with 0.1 *M* lithium acetate/TE, and subsequently incubated in 0.1 *M* lithium acetate/TE for a further 1 h with shaking at 30°. Meanwhile, approximately 0.4 μg each of the AD and BD plasmids are mixed together with 40 μg of salmon sperm DNA in a 1.5-ml tube. Of the yeast suspension 150 μl is then added to each transformation reaction and incubated for 10 min at 30°. Addition of 500 μl of 0.1 *M* lithium acetate/TE/ 50% polyethylene glycol and further incubation at 30° for 1 h are followed by a heat-shock step at 42° for 25 min. The tubes should be inverted regularly during these steps, as the yeast tends to settle at the bottom of the tube. After the heat shock the transformations are centrifuged at $1200 \times g$ for 5 min and the PEG removed. The cells are washed twice with 1 ml of sterile dH_2O and finally resuspended in 100 μl of dH_2O. Each transformation is then plated onto drop-out medium lacking tryptophan (W$^-$) and leucine (L$^-$), and placed in a 30° incubator for 3 days.

Note: It is important to wash the cells well after the heat shock, as the presence of polyethylene glycol can severely reduce the transformation efficiency.

Interaction Assay. To assay for an interaction between the AD- and BD-fusion proteins, we use of the HIS3 reporter gene. This is assayed by growth of yeast double-transformants on media lacking tryptophan (W⁻), leucine (L⁻), and histidine (H⁻). Single well-spaced, medium-sized, colonies are picked from the transformation plate and diluted in 500 μl of sterile *d*H₂O and vortexed thoroughly. Of each dilution, 5 μl is "spotted" in parallel onto W⁻L⁻ and W⁻L⁻H⁻ media. The plates are then incubated at 30° for 2–3 days. All transformants should grow on the W⁻L⁻ media, but only the transformants that have their GAL4 AD and BD brought together in close proximity, due to the interaction of the two fusion proteins, will activate transcription of the HIS3 gene (necessary for histidine biosynthesis) and therefore can grow on media lacking histidine. RCP(I621E) fails to interact with both Rab4 and Rab11, RCP(D622N) interacts with Rab11 but not Rab4, whereas wild-type RCP and RCP(Y620F) interact with both Rab GTPases (Fig. 2).

Note: This HIS3 reporter assay is straightforward and highly reproducible. It can also be adapted to give quantitative results by measuring the rate of growth of transformants in liquid W⁻L⁻H⁻ media (Rous *et al.*, 2002).

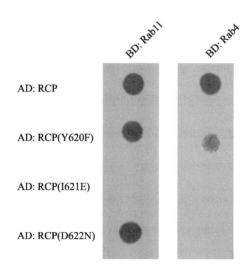

FIG. 2. Protein–protein interactions: Yeast two-hybrid technique. *S. cerevisiae* L40 cotransformants expressing the indicated activation domain (AD) and DNA-binding domain (BD) fusion proteins spotted on to media lacking histidine. Growth on this selective media indicates an interaction.

Protein–Protein Interactions: Far-Western Technique

The Far-Western technique is a simple and rapid method for determining whether two recombinant proteins interact directly. It does not require much optimization of buffer conditions and is highly reproducible. It is also ideal for simultaneously testing for ability of a large number of proteins to interact with a specific "probe." We have used this technique to test, *in vitro*, the ability of RCP, and the RBD mutants of RCP, to interact with GST-fused Rab4 and Rab11 "probes."

Protocol. A fresh overnight culture of *E. coli* containing pTrcHisB RCP or mutants is used to inoculate LB. The bacteria are grown at 37° to an $OD_{600} = 0.5$ and expression of His-tagged fusion protein is induced with 0.3 mM isopropylthiogalactose (IPTG). After a further 4 h at 37° the cells are spun-down and resuspended in $1\times$ sample buffer (25% glycerol, 6.6% sodium dodecyl sulfate [SDS], 160 mM Tris, pH 6.8, 0.025% bromophenol blue). Equal amounts of overexpressed His-tagged RCP, and RCP mutants, are separated on a 10% SDS–PAGE gel and transferred to nitrocellulose. The nitrocellulose is then "blocked" overnight at 4° with Basic buffer (BB: 20 mM HEPES, pH 7.5, 50 mM KCl, 10 mM MgCl$_2$, 1 mM dithiothreitol [DTT], and 0.1% NP-40) plus 5% nonfat dried milk. Prior to overlay the GST-tagged Rab4 or Rab11 probes are loaded with the nonhydrolyzable GTP analogue, GTPγS. This is achieved by incubating 10 μg of the GST-fusion protein with 100 μM GTPγS in Loading buffer (20 mM HEPES, pH 7.5, 4 mM EDTA) for 30 min at 37°. MgCl$_2$ is then added to a final concentration of 10 mM. The probes are subsequently diluted in Interaction buffer (BB plus 1% nonfat dried milk), overlayed onto the nitrocellulose, and rocked at 4° for 4 h. The nitrocellulose is washed with TBS-T (10 mM Tris, pH 7.5, 150 mM NaCl, 0.1% Tween-20). Bound probe is detected with an anti-GST antibody (Sigma) followed by incubation with a horseradish peroxidase (HRP)-conjugated anti-rabbit IgG (Sigma) secondary antibody. This biochemical assay confirms the results from the yeast two-hybrid assay, in that RCP(I621E) does not interact with Rab4 or Rab11, RCP(D622N) does not interact with Rab4, and RCP(WT) and RCP(Y620F) interact with both Rab GTPases (Fig. 3).

Intracellular Localization of Mutants

Expression of GFP-tagged versions of RCP mutants and the analysis of their localization by fluorescence microscopy revealed that the RCP (Y620F) and RCP(D622N) mutants displayed a vesicular pattern very similar to that of wild-type RCP. In contrast, RCP(I621E) appeared to be completely cytosolic (Lindsay and McCaffrey, 2004b). To confirm this

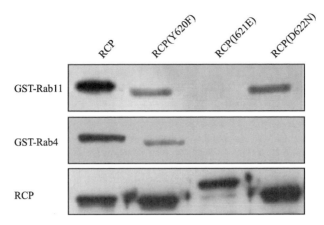

FIG. 3. Protein–protein interactions: Far-Western technique. *E. coli* lysates expressing the indicated His-tagged fusion proteins were separated by SDS–PAGE, transferred to nitrocellulose, and overlayed with either GTPγS-loaded GST-Rab4 or GST-Rab11. After several washes bound Rab fusion protein was detected with an anti-GST antibody. Levels of protein expression were determined by probing with an anti-RCP antibody (bottom panel). Note that the I621E mutation causes a shift in the mobility of RCP.

biochemically, cytosol and membrane fractions were prepared from cells expressing each of the mutants.

Subcellular Fractionation. HeLa cells seeded in 100-mm dishes to a density such that they will be 60–70% confluent the next day are transfected with the GFP-fusion constructs using Effectene transfection reagent (Qiagen) according to the manufacturer's instructions. Twenty-four hours posttransfection the cells are resuspended in Fractionation buffer (FB: 50 mM Tris, pH 7.4, 250 mM sucrose, 1 mM EDTA) plus protease inhibitors (Roche). Lysis of the cells is achieved by three freeze–thaw cycles in liquid nitrogen and passage through a 26-gauge needle. A postnuclear supernatant (PNS) is generated by centrifugation at 5000×g for 5 min at 4°. The PNS is subsequently separated by centrifugation at 200,000×g for 30 min at 4° in an MLA-130 rotor (Optima Ultracentrifuge, Beckman). The cytosol (high-speed supernatant) was removed and the membrane fraction (high-speed pellet) is resuspended in an equal volume of FB plus protease inhibitors. The same volumes of cytosol and membrane fractions are separated on a 10% SDS–PAGE gel and transferred to nitrocellulose. An anti-GFP antibody (Abcam) is used to detect the fusion proteins. RCP(I621E) is present exclusively in the cytosolic fraction, whereas the other mutants and wild-type RCP are found in both the membrane and the cytosol (Fig. 4). This suggests that RCP must be able to bind Rab11 for it to localize to the correct intracellular compartment.

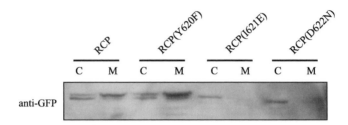

FIG. 4. Subcellular fractionation of RCP mutants. HeLa cells expressing the indicated GFP-tagged RCP fusion protein were separated into cytosol (C) and membrane (M) fractions. Equal volumes of each fraction were separated by SDS–PAGE and transferred to nitrocellulose. GFP-RCP and its mutants were detected with an anti-GFP antibody.

Note: We routinely obtain 40–50% transfection efficiency of the GFP-RCP plasmid constructs in HeLa cells using Effectene. It is important to omit any detergent from the fractionation buffer, as this will solubilize the membrane compartments. To monitor the fractionation procedure we routinely use TfnR and β-tubulin as markers for the membrane and cytosol fractions, respectively.

Acknowledgments

We are grateful to C. Horgan, A. Zhdanov, T. Zurawski, and S. Hanscom for useful discussions and assistance during this work. The work was supported by Science Foundation Ireland (SFI) (Investigator Grant 02/IN.1/B070) and the European Union (Marie Curie Development Host fellowship, Grant HPMD-CT-2000-00024).

References

Bock, J. B., Matern, H. T., Peden, A. A., and Scheller, R. H. (2001). A genomic perspective on membrane compartment organization. *Nature* **409,** 839–841.

Burkhard, P., Stetefeld, J., and Strelkov, S. V. (2001). Coiled coils: A highly versatile protein folding motif. *Trends Cell Biol.* **11,** 82–88.

Hales, C. M., Griner, R., Hobdy-Henderson, K. C., Dorn, M. C., Hardy, D., Kumar, R., Navarre, J., Chan, E. K., Lapierre, L. A., and Goldenring, J. R. (2001). Identification and characterization of a family of Rab11-interacting proteins. *J. Biol. Chem.* **276,** 39067–39075.

Lindsay, A. J., and McCaffrey, M. W. (2002). Rab11-FIP2 functions in transferrin recycling and associates with endosomal membranes via its COOH-terminal domain. *J. Biol. Chem.* **277,** 27193–27199.

Lindsay, A. J., and McCaffrey, M. W. (2004a). The C2 domains of the class I Rab11 family of interacting proteins target recycling vesicles to the plasma membrane. *J. Cell. Sci.* **117,** 4365–4375.

Lindsay, A. J., and McCaffrey, M. W. (2004b). Characterisation of the Rab binding properties of Rab coupling protein (RCP) by site-directed mutagenesis. *FEBS Lett.* **571,** 86–92.

Lindsay, A. J., Hendrick, A. G., Cantalupo, G., Senic-Matuglia, F., Goud, B., Bucci, C., and McCaffrey, M. W. (2002). Rab coupling protein (RCP), a novel Rab4 and Rab11 effector protein. *J. Biol. Chem.* **277**, 12190–12199.

Lupas, A., Van Dyke, M., and Stock, J. (1991). Predicting coiled coils from protein sequences. *Science* **252**, 1162–1164.

Maxfield, F. R., and McGraw, T. E. (2004). Endocytic recycling. *Nat. Rev. Mol. Cell Biol.* **5**, 121–132.

Prekeris, R., Klumperman, J., and Scheller, R. H. (2000). A Rab11/Rip11 protein complex regulates apical membrane trafficking via recycling endosomes. *Mol. Cell* **6**, 1437–1448.

Prekeris, R., Davies, J. M., and Scheller, R. H. (2001). Identification of a novel rab11/25 binding domain present in eferin and rip proteins. *J. Biol. Chem.* **276**, 38966–38970.

Rous, B. A., Reaves, B. J., Ihrke, G., Briggs, J. A., Gray, S. R., Stephens, D. J., Banting, G., and Luzio, J. P. (2002). Role of adaptor complex AP-3 in targeting wild-type and mutated CD63 to lysosomes. *Mol. Biol. Cell* **13**, 1071–1082.

Wallace, D. M., Lindsay, A. J., Hendrick, A. G., and McCaffrey, M. W. (2002). The novel Rab11-FIP/Rip/RCP family of proteins displays extensive homo- and hetero-interacting abilities. *Biochem. Biophys. Res. Commun.* **292**, 909–915.

[43] Purification and Functional Properties of Rab11-FIP2

By Andrew J. Lindsay and Mary W. McCaffrey

Abstract

Rab11-FIP2 is a 512-amino acid protein that was first identified in a screen for Rab11 interacting proteins. Database analysis revealed that it belongs to a family of proteins characterized by the presence of a highly homologous domain located at its carboxy-termini. This family was termed the Rab11 family of interacting proteins (Rab11-FIPs), as all members have been demonstrated to interact with Rab11. The Rab11-FIPs can be further subdivided into two classes. Rab11-FIP2 belongs to the class I Rab11-FIPs due to the presence of a C2 domain near its amino-terminus. RCP and Rip11 are the other class I family members. A number of proteins that interact directly with Rab11-FIP2, in addition to Rab11, have been identified. These include the EH domain-containing protein Reps1, the AP-2 subunit α-adaptin, the actin-based motor protein myosin Vb, and the chemokine receptors CXCR2 and CXCR4. It is hypothesized that Rab11-FIP2 functions to transport cargo, such as chemokine receptors and the EGF receptor, from the endocytic recycling compartment (ERC) to the plasma membrane.

METHODS IN ENZYMOLOGY, VOL. 403
0076-6879/05 $35.00
DOI: 10.1016/S0076-6879(05)03043-0

Introduction

Rab11 belongs to the Rab family of small GTPases, of which there are approximately 60 members in the human genome (Bock et al., 2001). Rab11 is found primarily at the endocytic recycling compartment (ERC) but has also been localized to the trans-Golgi network (TGN) (Urbe et al., 1993). It functions to regulate the transport of cargo from the ERC to the plasma membrane and between the ERC and the TGN (Ullrich et al., 1996). The Rab11 subfamily is composed of three members—Rab11a, Rab11b, and Rab25. Rab11a and Rab11b seem to be ubiquitously expressed, whereas Rab25 expression is restricted to epithelial tissue (Goldenring et al., 1993). The primary sequence of Rab25 is the most divergent of the Rab11 subfamily (Fig. 1A). One of the methods researchers have used to further understand the function of Rab11 and to dissect the specific roles of each isoform is to identify effector proteins through which it mediates its function. To this end, a number of laboratories identified Rab11-FIP2 by virtue of its homology to other members of the Rab11-FIP family (Hales et al., 2001; Lindsay and McCaffrey, 2002; Prekeris et al., 2001). Rab11-FIP2 was initially isolated as part of a project to clone large cDNA open reading frames and was designated KIAA0941 (Nagase et al., 1999). Primary sequence analysis revealed the presence of an approximately 20-amino acid Rab-binding domain (RBD) at its carboxy-terminus that mediates the interaction with Rab11. Rab11-FIP2 has a C2 domain near its amino-terminus that is highly homologous to the C2 domains of RCP and Rip11 and binds the phospholipids, phosphatidic acid, and PtdIns$(3,4,5)P_3$ (Lindsay and McCaffrey, 2004).

The central region of the class I Rab11-FIPs that links the C2 domain and the RBD is highly divergent and therefore likely to mediate the specific functions of these related proteins. This could either be by targeting the protein for posttranslational modification, or via interaction with other protein(s). For example, Rab Coupling Protein (RCP) possesses three PEST domains that target it for calpain-specific proteolysis (Marie et al., 2005). Neither Rip11 nor Rab11-FIP2 contains PEST motifs. In contrast, Rab11-FIP2 has three asparagine, proline, phenylalanine (NPF) repeats in the "linker" region (Fig. 1B). NPF repeats are binding motifs for EH domain-containing proteins. Indeed, the EH domain protein Reps1 was found to be a binding partner of Rab11-FIP2 (Cullis et al., 2002).

Rab11-FIP2 has also been demonstrated to interact with the actin-based motor protein myosin Vb and is likely to be involved in regulating the movement of transport vesicles along actin filaments (Hales et al., 2002). One set of cargo molecules that is recycled in a Rab11-FIP2-dependent manner is the chemokine receptors CXCR2 and CXCR4. In fact,

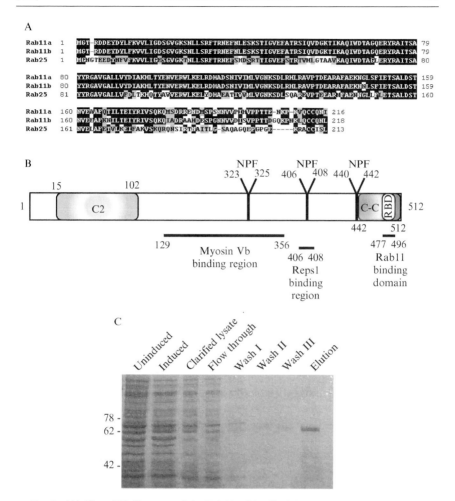

FIG. 1. (A) ClustalW alignment of the Rab11 subfamily. Identities are highlighted in black and similarities are highlighted in gray. (B) Schematic of Rab11-FIP2. Numbers correspond to amino acid positions. Regions of Rab11-FIP2 that mediate the interactions with its binding partners are indicated. RBD, Rab11-binding domain; C-C, coiled-coil domain; C2, phospholipid-binding C2 domain; NPF, asparagine–proline–phenylalanine repeat. (C) Purification of His-Rab11-FIP2. Coomassie blue-stained SDS–PAGE gel loaded with samples taken from each stage of the purification procedure. A polypeptide of approximately 64 kDa corresponding to His-tagged Rab11-FIP2 is eluted at the final step.

Rab11-FIP2 forms a protein complex with myosin Vb and CXCR2 (Fan *et al.*, 2004).

In this chapter the methodology we have used to study Rab11-FIP2 function is described. This includes the purification of recombinant protein and the development of a fluorescent-based trafficking assay to study the effects of overexpression of Rab11-FIP2 truncation mutants on transferrin recycling.

Materials

Reagents for bacterial growth were purchased from Difco. Isopropyl-β-D-thiogalactopyranoside (IPTG) and dithiothreitol (DTT) are from Melford. Alexa594-transferrin is from Molecular Probes. Iron-saturated holotransferrin and desferrioxamine are from Sigma. Effectene transfection reagent and Ni-NTA Agarose are from Qiagen. Dulbecco's modified Eagle's medium (DMEM), L-glutamine, penicillin/streptomycin, trypsin-EDTA, and fetal bovine serum were purchased from Cambrex. The $6\times$ His-tagged fusion vector pTrcHisC was purchased from Invitrogen and pEGFP-C1 was from Clontech.

A plasmid containing the complete open reading frame of Rab11-FIP2 (KIAA0941) was kindly supplied by the Kazusa Human cDNA Project (Kazusa, Japan) (Nagase *et al.*, 1999). This was used as the template in a polymerase chain reaction (PCR) that amplified the open reading frame, minus the first methionine, and attached *Eco*RI sites at the 5' and 3' ends. The resulting product was digested with *Eco*RI and inserted into the *Eco*RI sites of pTrcHisC and pEGFP-C1. The PCR products were sequenced to ensure that no random mutations had been incorporated. The dominant-negative Rab11-FIP2 carboxy-terminal green fluorescent protein (GFP) fusion construct was generated by inserting the approximately 200-bp *Pst*I fragment encoding amino acids 446–512 into the *Pst*I site of pEGFP-C1.

Methods

Batch Purification of 6×His-Rab11 FIP2

Rab11-FIP2 is a 512-amino acid protein with a predicted molecular weight of 58 kDa. The $6\times$ histidine tag was placed at the amino-terminus to avoid any interference with the RBD. BL21 (DE3) cells transformed with pTrcHisC Rab11-FIP2 are grown at 37° in 1 liter of Luria broth (LB) containing 50 μg/ml ampicillin to an OD_{600} of approximately 0.4. Expression of the recombinant protein is induced with 0.1 mM IPTG and the culture is incubated for a further 4 h, this time at 30° to reduce the potential of high-level expression resulting in the formation of inclusion bodies. The cells are harvested by centrifugation and resuspended in 20 ml of Buffer A (50 mM sodium phosphate, pH 8.0, 300 mM NaCl, and 10 mM imidazole), snap-frozen, and stored at $-80°$. To prepare a crude lysate, the suspension is rapidly thawed at $37°$ and sonicated at a medium setting for three 30–s intervals (MSE Soniprep 150). All further steps should be performed at $4°$. Triton X-100 is added to a final concentration of 0.1% and the lysate is rotated at $4°$ for 30 min. The lysate is clarified by centrifugation at $14,000 \times g$ for 20 min. Meanwhile, 1 ml of 50% Ni-NTA agarose is preequilibrated in Buffer A plus 0.1% Triton X-100 for 30 min on ice. The clarified lysate is added to the agarose and the mixture is rotated for 2 h. The agarose is subsequently pelleted by spinning at $1000 \times g$ and washed three times with 5 ml of Buffer B (50 mM sodium phosphate, pH 8.0, 300 mM NaCl, and 40 mM imidazole). The His-Rab11-FIP2 recombinant protein is eluted by incubating the agarose beads in 500 μl of Buffer C (50 mM sodium phosphate, pH 8.0, 300 mM NaCl, and 250 mM imidazole) for 5 min at room temperature. This elution step is repeated twice more and the samples are pooled and dialyzed overnight against phosphate-buffered saline (PBS). Purified His-Rab11-FIP2 is aliquoted and stored at $-80°$.

Note: Samples should be taken at each step of the purification and checked by sodium dodecyl sulfate–polyacrylamide gel electrophoresis (SDS–PAGE) (Fig. 1C). Prepared under the conditions described above Rab11-FIP2 is largely soluble and the majority of the recombinant protein will partition in the clarified lysate.

Fluorescent-Based Recycling Assay

To determine if Rab11-FIP2 functions to regulate the movement of cargo along the endocytic-recycling pathway, a transport assay was developed. The effect of the overexpression of wild-type Rab11-FIP2, or a truncation mutant containing the RBD, on the trafficking of transferrin was assessed. The constitutive recycling pathway of transferrin is well defined and has thus proved to be a useful tool in characterizing proteins involved in endocytosis.

To determine the effect of overexpression of wild-type and truncation mutants of Rab11-FIP2 on the trafficking of transferrin, HeLa or A431 cells are seeded onto 10-mm coverslips in 24-well plates, at a density such that they will be 60–70% confluent the next day. The cells are cultured in DMEM plus 10% fetal bovine serum, 2 mM glutamine, and penicillin/streptomycin. The plasmids pEGFP-C1 Rab11-FIP2 and pEGFP-C1 Rab11-FIP2$_{(446-512)}$ are purified from *Escherichia coli* XL-1 bacteria using the Wizard PureFection Plasmid DNA Purification System from Promega. The day after seeding 0.4 μg of each plasmid is transfected into separate wells using the Effectene transfection reagent, according to the manufacturer's instructions. Approximately 8 h posttransfection the medium containing the liposome:DNA complexes is removed, cells are washed once with PBS, and fresh medium is added. Expression of the GFP-fusion proteins is allowed to proceed a further 16 h, for a total of 24 h. The cells are then washed with PBS and depletion media (DMEM, plus 2 mM glutamine, penicillin/streptomycin, minus serum) is added for 1 h. Once the "starvation" period is complete, uptake medium (depletion medium supplemented with 25 μg/ml Alexa594-transferrin) is added and the cells are allowed to internalize the transferrin for 45 min at 37°.

To assess the effect that the expression of the exogenous proteins has on the internalization of transferrin, the cells are washed with ice-cold PBS and fixed immediately with 3% paraformaldehyde. Alternatively, to determine whether the overexpression of these constructs has an effect on the recycling of transferrin back out of the cell, the cells are washed to remove excess transferrin and chase medium (depletion medium, 250 μg/ml unlabeled iron-saturated holotransferrin, and 50 μM desferrioxamine) is added and the cells are incubated for a further 60 min at 37°. Each well is then washed with ice-cold PBS and fixed immediately.

In control cells fixed directly after the 45 min uptake, either nontransfected cells or cells expressing GFP on its own, the Alexa594-transferrin will label the entire recycling pathway including the sorting endosomes, ERC, and intermediate transport vesicles. In cells fixed after the chase, the majority of the transferrin will have recycled out, yielding a dramatically reduced fluorescence signal. If expression of the GFP-fusion protein has an effect on internalization of cargo, the pattern of Alexa594-Tfn in the uptake samples should be different from nontransfected cells in the same field and from the pattern of cells expressing GFP alone. Similarly, if the recombinant protein functions along the recycling step, expression of truncation mutants should alter the transferrin distribution in the chase sample. Full-length Rab11-FIP2 or the carboxy-terminal truncation mutant has no effect on transferrin internalization; however, the overexpression of GFP-Rab11-FIP2$_{(446-512)}$ inhibits the rate of recycling of transferrin. In cells transfected with this mutant, the Alexa594-Tfn is trapped in an intricate tubulovesicular structure after the chase, whereas nontransfected cells in the same field have recycled the fluorescent transferrin back into the medium (Fig. 2). Expression of full-length Rab11-FIP2 does not inhibit the rate of transferrin recycling.

Note: The desferrioxamine is added to the chase medium to chelate free iron in the medium, thus preventing the recycled transferrin from binding iron, which could then be reinternalized. This assay can be adapted for use with any construct suspected of involvement in the receptor recycling pathway.

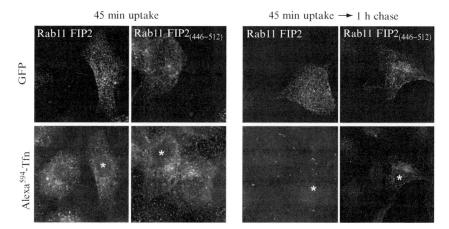

Fig. 2. Effect of Rab11-FIP2 expression on transferrin trafficking. HeLa cells were transfected with pEGFP-Rab11-FIP2 or pEGFP-Rab11-FIP2$_{(446-512)}$. Twenty-four hours posttransfection the cells were serum starved and then allowed to internalize 25 μg/ml Alexa594-Tfn for 45 min at 37°. After the internalization step, the cells were either fixed immediately or the fluorescent transferrin was chased with excess unlabeled transferrin for 60 min and then fixed. Expression of Rab11-FIP2 or its truncation mutant had no effect on the internalization of transferrin; however, Rab11-FIP2$_{(446-512)}$ inhibited the rate of transferrin recycling. Transfected cells are indicated by an asterisk.

Quantification of Transferrin Internalization and Recycling

The effect of the expression of GFP-Rab11-FIP2, and mutants, on the internalization and recycling of transferrin can be quantified by measuring the intensity of Alexa594-Tfn fluorescence in cells fixed after internalization of the ligand or fixed after the chase. Measuring the fluorescence intensity in a large number of cells from multiple experiments can yield statistically significant results.

We acquired images of multiple fields, under each condition, from several experiments, using a Nikon Eclipse E600 fluorescence microscope fitted with a Hamamatsu Chilled CCD (C5985) camera. Two images from each field were captured, one using a green filter (Nikon BA515–555) to identify transfected cells and one image captured with a red filter (Nikon BA600–660) to visualize the transferrin within the cell. All fields were acquired under the same magnification, exposure, and gain settings. Care was taken to ensure that the fluorescence signal was not saturated. The recorded conditions included wild-type and truncated Rab11-FIP2 mutant expressing cells fixed immediately after the uptake, or fixed after the chase. To measure the intensity of the Alexa-Tfn in these transfected cells, we used the NIH Image Version 1.63 software run on an Apple Mac Power PC. The images are imported into the software and a uniform box size is overlaid onto each transfected cell. The average fluorescence intensity of the Alexa-Tfn within the box is measured. A threshold background fluorescence is calculated from cells that have not been allowed to internalize transferrin. This threshold value accounts for the intrinsic autofluorescence of the cell and is subtracted from each experimental value. In this way, a large number of cells can be efficiently measured. We observed that Rab11-FIP2 or Rab11-FIP2$_{(446-512)}$ had no effect on the internalization of transferrin; however, there was over a 2-fold increase in the amount of Alexa-Tfn retained in cells expressing Rab11-FIP2$_{(446-512)}$ after the chase (Lindsay and McCaffrey, 2002). This indicates that Rab11-FIP2 plays a role in the recycling of transferrin back to the plasma membrane.

Note: This fluorescence-based transport assay can be modified to measure an effect on the time course of ligand internalization or recycling. For the former, Alexa-Tfn is incubated with the cells at 4°. This allows the transferrin to bind to its receptors on the cell surface but prevents the endocytosis of the receptor–ligand complex. The excess ligand is then washed off and the medium is warmed to 37° to initiate internalization. The cells can then be fixed at various times after warming. The majority of transferrin will enter the sorting endosomes after 5 min and have reached the ERC by 20 min. To analyze the kinetics of recycling, the cells are allowed to internalize Alexa-Tfn at 16°. At this temperature the ligand can enter the sorting endosome, but onward transport from this compartment is blocked. To allow recycling to initiate, after washing away excess ligand, the medium is warmed to 37°, and the cells can be fixed at various times after.

Acknowledgments

We are grateful to N. Marie, C. Horgan, A. Zhdanov, and S. Hanscom for useful discussions and assistance during this work. The work was supported by Science Foundation Ireland (SFI) (Investigator Grant 02/IN.1/B070).

References

Bock, J. B., Matern, H. T., Peden, A. A., and Scheller, R. H. (2001). A genomic perspective on membrane compartment organization. *Nature* **409,** 839–841.

Cullis, D. N., Philip, B., Baleja, J. D., and Feig, L. A. (2002). Rab11-FIP2, an adaptor protein connecting cellular components involved in internalization and recycling of epidermal growth factor receptors. *J. Biol. Chem.* **277,** 49158–49166.

Fan, G. H., Lapierre, L. A., Goldenring, J. R., Sai, J., and Richmond, A. (2004). Rab11-family interacting protein 2 and myosin Vb are required for CXCR2 recycling and receptor-mediated chemotaxis. *Mol. Biol. Cell* **15,** 2456–2469.

Goldenring, J., Shen, K., Vaughan, H., and Modlin, I. (1993). Identification of a small GTP-binding protein, Rab25, expressed in the gastrointestinal mucosa, kidney, and lung. *J. Biol. Chem.* **268,** 18419–18422.

Hales, C. M., Griner, R., Hobdy-Henderson, K. C., Dorn, M. C., Hardy, D., Kumar, R., Navarre, J., Chan, E. K., Lapierre, L. A., and Goldenring, J. R. (2001). Identification and characterization of a family of Rab11-interacting proteins. *J. Biol. Chem.* **276,** 39067–39075.

Hales, C. M., Vaerman, J. P., and Goldenring, J. R. (2002). Rab11 family interacting protein 2 associates with myosin Vb and regulates plasma membrane recycling. *J. Biol. Chem.* **277,** 50415–50421.

Lindsay, A. J., and McCaffrey, M. W. (2002). Rab11-FIP2 functions in transferrin recycling and associates with endosomal membranes via its COOH-terminal domain. *J. Biol. Chem.* **277,** 27193–27199.

Lindsay, A. J., and McCaffrey, M. W. (2004). The C2 domains of the class I Rab11 family of interacting proteins target recycling vesicles to the plasma membrane. *J. Cell Sci.* **117,** 4365–4375.

Marie, N., Lindsay, A. J., and McCaffrey, M. W. (2005). Rab coupling protein is selectively degraded by calpain in Ca 2+-dependent manner. *Biochem. J.* **389,** 223–231.

Nagase, T., Ishikawa, K., Suyama, M., Kikuno, R., Hirosawa, M., Miyajima, N., Tanaka, A., Kotani, H., Nomura, N., and Ohara, O. (1999). Prediction of the coding sequences of unidentified human genes. XIII. The complete sequences of 100 new cDNA clones from brain which code for large proteins *in vitro*. *DNA Res.* **6,** 63–70.

Prekeris, R., Davies, J. M., and Scheller, R. H. (2001). Identification of a novel rab11/25 binding domain present in eferin and rip proteins. *J. Biol. Chem.* **276,** 38966–38970.

Ullrich, O., Reinsch, S., Urbe, S., Zerial, M., and Parton, R. G. (1996). Rab11 regulates recycling through the pericentriolar recycling endosome. *J. Cell Biol.* **135,** 913–924.

Urbe, S., Huber, L. A., Zerial, M., Tooze, S. A., and Parton, R. G. (1993). Rab11, a small GTPase associated with both constitutive and regulated secretory pathways in PC12 cells. *FEBS Lett.* **334,** 175–182.

[44] Purification and Functional Properties of Rab11-FIP3

By Conor P. Horgan, Tomas H. Zurawski, and Mary W. McCaffrey

Abstract

The Rab family of small GTPases are key regulators of membrane trafficking in eukaryotic cells. Rab11, one member of this family, plays a role in regulating various cellular functions such as plasma membrane recycling, phagocytosis, and cytokinesis. A family of Rab11-binding proteins has been identified and termed the Rab11 family interacting proteins or Rab11-FIPs. Rab11-FIP3, a member of this Rab11-binding protein family, in addition to interacting with Rab11, is also capable of interaction with members of the ADP-Ribosylation Factor (ARF) GTPase family. Here we describe

METHODS IN ENZYMOLOGY, VOL. 403
0076-6879/05 $35.00
DOI: 10.1016/S0076-6879(05)03044-2

the purification of Rab11-FIP3 and report its biological properties in eukaryotic cells as visualized by immunofluorescence microscopy.

Introduction

Proteins belonging to the Rab and ARF GTPase families have long been known to regulate various membrane trafficking events. Two such proteins, namely Rab11 and ARF6, have been specifically implicated in various cellular processes. This multiplicity of function suggests that a variety of effector proteins may exist and are involved in controlling these processes. Rab11-FIP3 is one such candidate effector protein.

Rab11-FIP3 (also known as Arfophilin and Eferin) was originally reported as a cDNA clone (KIAA0665) isolated from human brain (Nagase et al., 1999). Subsequent to this, it was reported as an ARF-binding protein and named Arfophilin (Shin et al., 1999, 2001). As a protein capable of interaction with Rab11, Rab11-FIP3 was then reported by two independent groups, in one case named Eferin (Hales et al., 2001; Prekeris et al., 2001). It has been demonstrated that Rab11-FIP3 binds all the Rab11 subfamily members (Rab11a, Rab11b, and Rab25) as well as class II and class III ARFs in a GTP-dependent manner (Hales et al., 2001; Shin et al., 1999, 2001). The function of Rab11-FIP3 is still not fully understood, but it is believed to be implicated in the delivery, targeting, and/or fusion of recycling endosomes with cleavage furrow/midbody during cell division (Horgan et al., 2004; Wilson et al., 2005).

To elucidate the function of Rab11-FIP3, the need for recombinant protein arose. This valuable reagent can be utilized in several important assays such as antibody production and characterization, identification of interacting partners using pull-down techniques, and investigation of *in vitro* protein–protein interactions utilizing a Far-Western overlay technique.

Methods

Expression and Purification of Recombinant Rab11-FIP3

Expression of hexahistidine-fused recombinant full-length protein in a prokaryotic cell system resulted in production of the protein with very low solubility, and purification by conventional methods was unsatisfactory. Attempts to decrease the insolubility problems through optimization of protein expression conditions (altering the bacterial strain, lowering the bacterial incubation temperature, varying the concentration of isopropyl-β-D-thiogalactopyranoside [IPTG], inducing at higher cell density for a shorter period of time, increasing aeration) proved ineffective.

Following purification under either native or denaturing conditions, this system did not appear to be very effective in terms of yield of the recombinant protein. With a view to obtaining better protein yields, a decision to express and purify Rab11-FIP3 as a glutathione *S*-transferase (GST) fusion was made. GST-fused proteins were expressed and purified more successfully, but the yield of full-length protein remained somewhat unsatisfactory.

Due to the poor yield obtained for the full-length protein and in addition to the requirement of a specific antigen for antibody production, we decided to purify a carboxy-terminally truncated form of the protein. Rab11-FIPs display minimal amino acid homology at their amino-termini, and so this region, if used as an immunogen, would be more likely to yield an antibody that would specifically detect Rab11-FIP3.

Plasmid Construction. For purification of hexahistidine-fused recombinant Rab11-FIP3, a 2.4-kb *Eco*RI fragment encoding full-length Rab11-FIP3$_{(2–756)}$ was subcloned from the previously described pGADGH/Rab11-FIP3$_{(2–756)}$ construct (Wallace *et al.*, 2002) into pTrcHisC plasmid (Invitrogen). To express and purify the amino-terminal region of Rab11-FIP3 as a hexahistidine fusion, a 0.8-kb *Bam*HI fragment corresponding to amino acids 2–246 was subcloned into a pTrcHis2C plasmid (Invitrogen) generating a pTrcHis2C/Rab11-FIP3$_{(2–246)}$ construct. To purify GST-fused full-length recombinant Rab11-FIP3, a 2.4-kb *Eco*RI fragment from the pTrcHisC/Rab11-FIP3$_{(2–756)}$ construct was subcloned into a pGEX-3X plasmid (Amersham Biosciences). pGEX-3X/Rab11-FIP3$_{(2–246)}$ has been previously described (Horgan *et al.*, 2004).

Hexahistidine-Fused Recombinant Protein Purification. Escherichia coli (XL-1) cells are freshly transformed with pTrcHis2C/Rab11-FIP3$_{(2–246)}$ or pTrcHisC/Rab11-FIP3$_{(2–756)}$. A single colony transformant is picked and cultured overnight in a temperature-controlled orbital shaking incubator at 37° in 5 ml of sterile Luria–Bertani (LB) broth (10 g/liter of Bacto-tryptone, 5 g/liter of Bacto-yeast extract, and 10 g/liter NaCl) including 0.1 mg/ml ampicillin (Sigma). On the following morning, these starter cultures are diluted 1/100 into 500 ml of sterile LB broth containing ampicillin at 0.1 mg/ml and cultured with agitation at 37° until the OD$_{600}$ reaches 0.6–1.0. Protein expression is then induced with 0.1 m*M* IPTG (Qiagen) for 1 h at 30°. Bacterial cells are pelleted by centrifugation of the culture (7000×*g* for 15 min at 4°). The pellets are washed with ice-cold phosphate-buffered saline (PBS; 137 m*M* NaCl, 2.7 m*M* KCl, 1.8 m*M* KH$_2$PO$_4$, 8.1 m*M* Na$_2$HPO$_4$, pH 7.4) and resuspended in 5 ml of ice-cold Lysis buffer (50 m*M* NaH$_2$PO$_4$, 300 m*M* NaCl, 10 m*M* imidazole, 1 m*M* dithiothreitol [DTT], pH 7.9). Resuspended pellets are snap frozen in liquid nitrogen and stored at −80° overnight. Pellets are thawed on ice and Triton X-100

(Sigma) is added to a final concentration of 0.5% together with 1 mg/ml lysozyme (Sigma) and protease inhibitor cocktail (Sigma). Cells are then lysed for 30 min under rotation (12 rpm) at 4°. While cells are being lysed, 50% (v/v) Ni^{2+}-NTA agarose slurry (Qiagen) is equilibrated on ice in Lysis buffer including identical supplements as the sample. The cell lysate is clarified by centrifugation ($15,000 \times g$ for 15 min at 4°). The supernatant protein suspension is collected and incubated under rotation at 4° for 1 h with preequilibrated Ni^{2+}-NTA agarose slurry (1 ml/100 ml of lysate). The resulting nickel beads with bound hexahistidine-recombinant protein are centrifuged at low speed ($500 \times g$ for 5 min at 4°) and the supernatant containing unbound proteins is removed. Pelleted beads are washed twice by repeated low-speed centrifugation ($500 \times g$ for 5 min at 4°) followed by gentle resuspension in 10 times the volume of the pellet of ice-cold Washing buffer 1 and 2 (Lysis buffer containing an increasing concentration of imidazole [WB1 = 20 mM, WB2 = 40 mM imidazole]). Bound proteins are eluted from the beads using Elution buffer (Lysis buffer + 300 mM imidazole) by repeated low-speed centrifugation and gentle resuspension in Elution buffer. Fractions from each step of purification are collected and analyzed by sodium dodecyl sulfate–polyacrylamide gel electrophoresis (SDS-PAGE) and Western blotting utilizing anti-Xpress antibody (Invitrogen) and rabbit-polyclonal anti-Rab11-FIP3 antibody (Horgan et al., 2004). Eluted peak fractions are pooled and dialyzed against PBS before snap freezing in liquid nitrogen and storage at −80°. The protein concentration is quantified by standard Bradford assay. This protocol results in the purification of <0.1 mg of the full-length protein and of ~1.0 mg of His_{6X}-Rab11-FIP3$_{(2–246)}$ from 500 ml of culture (purity >80%). Due to these very poor yields, we proceeded to purify the protein as a GST fusion.

GST Recombinant Protein Purification. E. coli (XL-1) cells are freshly transformed with pGEX-3X/Rab11-FIP3$_{(2–246)}$ or pGEX-3X/Rab11-FIP3$_{(2–756)}$. A single colony transformant is then cultured overnight. These overnight starter cultures are then diluted, cultured, induced, and pelleted as described earlier for hexahistidine-fused protein purification. The pellets are washed with ice-cold PBS, snap frozen in liquid nitrogen, and stored at −80° overnight. Frozen pellets are thawed on ice and resuspended in 10 ml of ice-cold PBS containing 1 mg/ml lysozyme and protease inhibitor cocktail. Cells are then lysed for 30 min under rotation (12 rpm) at 4°. Cell lysate is clarified by centrifugation and the supernatant-containing protein suspension is incubated under rotation at 4° for 1 h with preequilibrated glutathione-Sepharose beads 50% (v/v) agarose slurry (Sigma), as described for purification of hexahistidine-fused protein. The beads with bound GST-recombinant protein are centrifuged ($500 \times g$ for 5 min at 4°) and supernatant containing unbound proteins is removed. Pelleted beads are washed three times by repeated low-speed centrifugation and gentle resuspension in ice-cold PBS (10 volumes of pellet). Bound proteins are eluted from the beads using Elution buffer (20 mM reduced glutathione in 50 mM Tris–HCl, pH 8.0) by repeated low-speed centrifugation and gentle resuspension in Elution buffer. Fractions from each step of purification are collected and analyzed by SDS-PAGE and Western blotting utilizing anti-GST antibody (Sigma) and our rabbit polyclonal anti-Rab11-FIP3 antibody. Eluted peak fractions are pooled and dialyzed against PBS before snap freezing in liquid nitrogen and storage at −80°. The protein concentration is quantified by standard Bradford assay. This protocol leads to the purification of a somewhat greater but still insufficient amount of full-length protein (<0.2 mg from 500 ml of culture; purity 85%) (Fig. 1A). On the other hand, expression and purification of the truncated form GST-Rab11-FIP3$_{(2–246)}$ resulted in a yield of >2.5 mg from 500 ml of culture (purity >90%) (Fig. 1B). Empty pGEX-3X plasmid was expressed and purified as a positive control using the same purification conditions. The yield was >5 mg (purity >90%).

Antibody Generation and Characterization

Epitope Prediction. Studies on Rab11-FIP3 and its biological significance have relied heavily on the overexpression of the protein, which alters the morphology of the endosomal–recycling compartment (ERC) as previously demonstrated (Horgan *et al.*, 2004). Consequently, we considered it imperative to examine Rab11-FIP3 at the endogenous level. To obtain an antibody specific to Rab11-FIP3, we pursued an anti-peptide approach. A number of parameters were taken into consideration in identifying the peptide to be used for the immunization. Initially, we identified the regions of the Rab11-FIP3 amino acid sequence that are unique to this protein. This was achieved by utilizing BioEdit Sequence Alignment Editor software (Hall, 1999) to align the amino acid sequences of all the currently known Rab11-FIP family members by ClustalW (Thompson *et al.*, 1994), followed by identification of the regions of Rab11-FIP3 that are not conserved in the other family members. Subsequent to this, the resultant polypeptide sequences were analyzed by BLAST analysis/search (http://www.ncbi.nlm.nih.gov/BLAST/) and those found to be unique to Rab11-FIP3 were further analyzed as follows.

Peptide amino acid sequences identified as unique to Rab11-FIP3 and in the size range of 10–20 amino acids were examined to determine whether they would successfully elicit an immune response. These predictions were based on a combination of three distinct parameters: antigenicity, accessibility, and hydrophilicity. Predictions were made using a

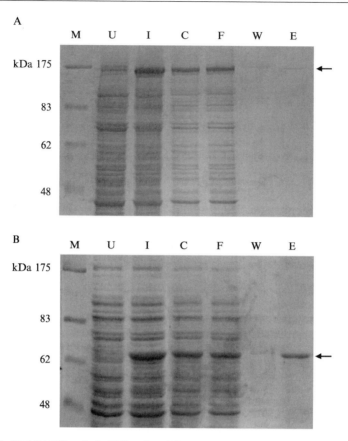

FIG. 1. SDS-PAGE analysis (10% gel) and Coomassie blue staining of protein purification fractions from *E. coli* (Xl-1) cells containing recombinant plasmid expressing GST-fusion protein. (A) Purification of GST-Rab11-FIP3$_{(2-756)}$. (B) Purification of GST-Rab11-FIP3$_{(2-246)}$. Lanes U and I contain equivalent amounts of total cellular protein from uninduced and induced samples, respectively. Volumes of clarified lysate (C) and flow through (F) were normalized. Lane (W) contains washes and lane (E) contains eluted protein. Arrows point to purified protein.

combination of sequence analysis tools, which included Peptide Companion, MacVector 7.2, and ExPASy Proteomics tools available online at http://us.expasy.org/tools/.

Based on predictions outlined previously, we identified the polypeptide sequence corresponding to Rab11-FIP3 amino acid residues 80–97 (GGPRD PGPSAPPPRSGPR) as the polypeptide most likely to successfully yield antibodies specific to this protein.

Immunization. A synthetic peptide was conjugated to keyhole limpet hemocyanin (KLH) via its amino-terminus. This conjugate was used to immunize New Zealand white rabbits for antibody generation. The resulting immune serum was affinity purified against 3 mg of the epoxy-immobilized synthetic peptide and yielded approximately 1.5 mg of affinity-purified IgG as quantified by standard Bradford assay. The peptide synthesis and conjugation, in addition to the immunization and affinity purification, were performed by Davids Biotechnologie (http://www.dabio.de/).

Characterization of the Rabbit Polyclonal Anti-Rab11-FIP3 Affinity-Purified Antibody. We confirmed the reactivity and specificity of our Rab11-FIP3 antibody using a combination of immunofluorescence and immunoblotting assays, which we described in considerable detail in Horgan *et al.* (2004). Briefly these included determination of the ability of the antibody to detect exogenously expressed Rab11-FIP3, but not other Rab11-FIPs *in vivo* by immunofluorescence; the ability of the antibody to detect purified recombinant Rab11-FIP3, but not other Rab11-FIPs by immunoblotting; and a preadsorption assay, which involves incubation of the antibody with either purified recombinant protein or the immunizing peptide, and resulting in complete abolition of antibody immunoreactivity. We determined the optimal working concentrations for the affinity-purified antibody to detect endogenous Rab11-FIP3 to be 0.86 μg/ml (working dilution 1/150) for immunofluorescence and 0.65 μg/ml (working dilution 1/200) for immunoblotting.

Analysis of Subcellular Distribution of Endogenous Rab11-FIP3

Cell Culture and Transfection. Human cervical epithelial carcinoma cells (HeLa cell line) are cultured in Dulbecco's modified Eagle's medium supplemented with 10% (v/v) fetal bovine serum (FBS), 2 mM L-glutamine, and 100 U/ml penicillin-streptomycin (all from BioWhittaker) and grown in 5% CO_2 at 37°. For overexpression studies, cells are transfected with the indicated plasmid constructs using Effectene (Qiagen) as a transfection reagent according to the manufacturer's instructions. Sixteen to eighteen hours posttransfection the cells are processed for immunofluorescence microscopy as described in the following paragraph.

Immunofluorescence Microscopy. HeLa cells are grown on 10-mm round glass coverslips in 24-well plates and allowed to reach 70–80% confluency. Samples are then washed twice with 1 ml of PBS for 5 min *in situ* before fixation with 3% paraformaldehyde for 15 min at room temperature (RT). Samples are then washed twice with PBS, following which free aldehyde groups are quenched with 50 mM NH$_4$Cl for 15 min at RT and washed twice with PBS. Cells are permeabilized with

Permeabilization buffer (0.05% [w/v] saponin, 0.2% [w/v] bovine serum albumin [BSA] in PBS) for 5 min at RT. Samples are then incubated with primary antibodies diluted in 5% (v/v) FBS/PBS containing 0.05% (w/v) saponin for 45 min at RT and then washed twice with 0.5 ml of PBS before incubation with appropriate secondary antibodies for 45 min at RT. Cover-slips are mounted in MOWIOL (CalBiochem), which contains 100 mg/ml DABCO (1,4-diazabicyclo[2,2,2]octane, Sigma) and images are recorded using a Zeiss LSM 510 META confocal microscope fitted with a 63×/1.4 plan apochromat lens. Samples labeled with blue fluorophores are excited at 405 nm and emissions detected with a 420- to 480-nm bandpass filter. Green fluorophores are excited at 488-nm and emissions detected with a 505- to 550-nm bandpass filter. Red fluorophores are excited at 543 nm, and emissions detected with a 561- to 753-nm long-pass filter. Images are processed using Zeiss LSM Image Browser, version 3,1,0,99 (CarlZeiss), and Adobe Illustrator, version 10 (Adobe), software.

Mitotic Synchronization and Rab11-FIP3 Subcellular Localization. To examine the localization of endogenous Rab11-FIP3 in interphase and mitotic cells we performed mitotic synchronization experiments. Mitotic synchronization is achieved as follows: HeLa cells at 80–90% confluency in a standard 75-cm^2 tissue culture flask are incubated in media containing 1 μM nocodazole (Sigma) for 15 h at standard conditions (5% CO_2, 37°). Following this incubation, mitotic cells (~90% of the population) display a rounded morphology and detach easily into suspension with gentle tapping of the flask, thus facilitating the harvesting of the population of the cells blocked at the G_2/M stage of the cell cycle. The resulting media (mitotic cell suspension) is then collected, centrifuged (200×g for 5 min), washed with PBS at 4°, resuspended in nocodazole-free media, and seeded onto poly-L-lysine-(Sigma) coated coverslips. The synchronized population of cells is allowed to progress into mitosis until it is fixed at various time points following release from the mitotic block. Within approximately 120 min following release from nocodazole arrest, most cells have completed cytokinesis, as evidenced by the central spindle remnant present between dividing cells. Samples are then processed for immunofluorescence microscopy and mitotic progression is assessed by monitoring α-tubulin (Sigma) and Hoechst 33258 (Molecular Probes) staining by (immuno)fluorescence techniques.

Utilizing our affinity-purified antibody (Horgan et al., 2004) we examined by immunofluorescence microscopy the localization of Rab11-FIP3 in HeLa cells that had been either left untreated or fixed in various stages of mitosis. We found that during interphase Rab11-FIP3 localizes to vesicular structures that are dispersed throughout the cell (Fig. 2). Interestingly, we found that in mitotic cells, or, more specifically, in cells undergoing cytokinesis, a substantial proportion of the Rab11-FIP3 protein localized to cleavage furrow/midbody region (Fig. 2, arrow).

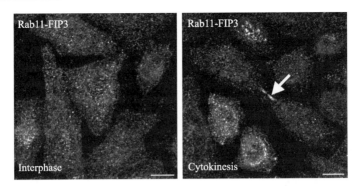

FIG. 2. Subcellular localization of endogenous Rab11-FIP3 in interphase and mitotic HeLa cells as visualized by immunofluorescence microscopy. The arrow is pointing to the cleavage furrow/midbody region. Bar = 10 μm.

Drug Treatments. To further elucidate the subcellular distribution of Rab11-FIP3, we examined its localization in cells that have been treated with cytoskeletal depolymerizing agents and brefeldin A (BFA), as the dynamics of endocytic proteins in the presence of these agents can reveal features of their native localization and trafficking patterns. Drug treatment experiments are performed as follows. Nocodazole (microtubule-depolymerizing agent) and cytochalasin D (actin filament-depolymerizing agent; Sigma) are solubilized in dimethyl sulfoxide (DMSO, Sigma) and BFA (Sigma) is solubilized in methanol. HeLa cells growing on 10-mm glass coverslips are incubated with either 10 μM nocodazole or 10 μM cytochalasin D for 30 min at 37° or with 20 μg/ml BFA for 15 min at 37° prior to fixation and immunostaining with anti-Rab11-FIP3 antibody. As a control, we treat cells with the same concentration of drug solvents (0.2% [v/v] DMSO or 1% [v/v] methanol) alone to monitor solvent effects.

Utilizing our polyclonal anti-Rab11-FIP3 antibody, we previously reported that Rab11-FIP3 subcellular localization is dependent on both microtubule and actin cytoskeletal integrity (Horgan *et al.*, 2004). Additionally, we found that while the localization of other Rab11-FIP members is affected by BFA treatment, BFA does not perturb the localization of Rab11-FIP3 (Horgan *et al.*, 2004).

Analysis of Exogenously Expressed Wild-Type and Mutant Rab11-FIP3

Plasmid Construction. pEGFP-C1/Rab11-FIP3$_{(2–756)}$, pARF6/GFP^2-N3, and pEGFP-C3/Rab11-FIP3$_{(244–756)}$ constructs have been previously described (Horgan *et al.*, 2004). pcDNA3.1HisB/Rab11-FIP3$_{(2–756)}$ was constructed by subcloning the 2.4-kb *Eco*RI fragment from the previously

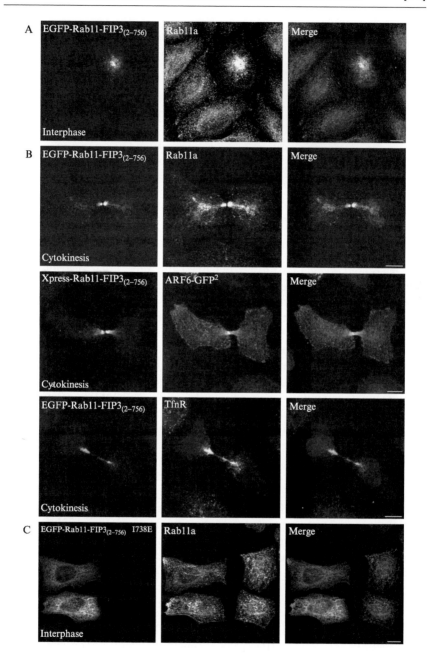

described pGADGH/Rab11-FIP3$_{(2-756)}$ construct (Wallace *et al.*, 2002) into pcDNA3.1HisB (Invitrogen). pEGFP-C1/Rab11-FIP3$_{(2-756)}$ I738E was generated by site-directed mutagenesis using the QuikChange Site-Directed Mutagenesis Kit (Stratagene) utilizing a sense primer *FWD I738E* (5′-CGCCTGCAGGACTACGAAGACAGGATCATCGTGGCC -3′) and an antisense primer *REV I738E* (5′-GGCCACGATGATCCTG TCTTCGTAGTCCTGCAGGCG-3′), with pEGFP-C3/Rab11-FIP3$_{(2-756)}$ as template. Constructs generated by PCR were confirmed to be correct by double-strand sequencing.

Analysis of Exogenous Expression of Rab11-FIP3. To investigate Rab11-FIP3 function, we transiently transfected HeLa cells with a construct encoding EGFP-tagged Rab11-FIP3$_{(2-756)}$ and examined its localization by immunofluorescence microscopy. In interphase HeLa cells, we found that the exogenously expressed protein concentrates near the nucleus (Fig. 3A). Previously, we have shown that this concentration of protein surrounds the centrosome as evidenced by γ-tubulin staining (Horgan *et al.*, 2004). Additionally, in some cells, an elaborate tubular or tubulovesicular staining pattern emanating from the centrosome is observed (data not shown). Notably, the phenotype that we observe for the exogenously expressed protein differs significantly from what we observe for the endogenous protein. A comparison of Fig. 2 and Fig. 3A demonstrates this. In Fig. 2 endogenous Rab11-FIP3 localizes to vesicular structures that are dispersed throughout the cell, while the overexpressed Rab11-FIP3 protein predominantly concentrates in a pericentrosomal location as illustrated in Fig. 3A. Immunostaining HeLa cells expressing the EGFP-tagged Rab11-FIP3$_{(2-756)}$ with an anti-Rab11a antibody (Zymed) revealed extensive colocalization between the exogenously expressed protein and Rab11a (Fig. 3A). Interestingly, when Rab11-FIP3 is overexpressed, it alters the Rab11a staining pattern as it condenses the Rab11a staining pattern into the Rab11-FIP3-positive compartment near the

---

FIG. 3. Localization of exogenously expressed EGFP-fused Rab11-FIP3 and Rab11-FIP3 mutants. (A and B, upper and lower panels) HeLa cells were transfected with pEGFP-C1/ Rab11-FIP3$_{(2-756)}$ (green in merge). Sixteen to eighteen hours posttransfection the cells were processed for immunofluorescence as described in Methods and immunostained with antibodies to Rab11a or TfnR (red in merge). (B, middle panel) HeLa cells were double transfected with pcDNA3.1HisB/Rab11-FIP3$_{(2-756)}$ (red in merge) and pARF6/GFP^2-N3 (green in merge). Sixteen hours posttransfection the cells were immunostained with anti-Xpress antibody (red in merge). (C) HeLa cells were transfected with pEGFP-C1/Rab11-FIP3$_{(2-756)}$ I738E (green in merge). Sixteen hours posttransfection the cells were immunostained with an antibody to Rab11a (red in merge). Hoechst dye (blue) was used to visualize the nuclei. Bar = 10 μm. (See color insert.)

centrosome (compare transfected versus non-transfected cells in Fig. 3A). Furthermore, as reported in Horgan *et al.* (2004), we observed a similar albeit less dramatic perturbation on the localization of two further endo-somal-recycling markers, the transferrin receptor (TfnR) and Rab coupling protein (RCP).

Taken together, these data demonstrate that exogenously expressed Rab11-FIP3 localizes predominantly to a pericentrosomal Rab11a-positive ERC during interphase, and that its overexpression alters the ERC morphology.

We observed, on examination of HeLa cells exogenously expressing Rab11-FIP3 and undergoing mitosis, that the Rab11-FIP3 protein was localized to the cleavage furrow/midbody region (Fig. 3B). This correlates with what we observed for the endogenous protein (Fig. 2, arrow). Since Rab11-FIP3 interacts with Rab11 and ARF6 (Hales *et al.*, 2001; Shin *et al.*, 2001), two small GTPases that have been implicated in cell division (Cheng *et al.*, 2002; Finger and White, 2002; Horgan *et al.*, 2004; Riggs *et al.*, 2003; Schweitzer and D'Souza-Schorey, 2002; Skop *et al.*, 2001), we were inter-ested in examining the localization of these two proteins in mitotic cells expressing Rab11-FIP3. We found that within the cleavage furrow of mitotic cells exogenously expressing Rab11-FIP3, a high level of colocali-zation is observed between Rab11-FIP3 and Rab11a (Fig. 3B, upper pan-el). In the case of ARF6, a protein known to localize to the cleavage furrow and midbody during cell division (Schweitzer and D'Souza-Schorey, 2002), we found in cells exogenously expressing both Rab11-FIP3 (detected with anti-Xpress antibody) and ARF6, that these two proteins localize in close proximity of each other with the ARF6 protein concentrated at the mid-body and Rab11-FIP3 concentrated within the intracellular bridge between the dividing cells (Fig. 3B, middle panel). Since both Rab11 and ARF6 are involved in endosomal recycling, we were interested in investigating if the TfnR, the classic marker of endosomal recycling, was associated with a Rab11-FIP3-positive structure localized to the cleavage furrow in mitotic cells. Indeed, we observed a high degree of colocalization between the TfnR (Zymed) and Rab11-FIP3 in mitotic cells (Fig. 3B, lower panel). These data suggest that Rab11-FIP3 functions in the delivery of recycling endosomes to the cleavage furrow/midbody during cell division.

Analysis of the Functional Consequences of Rab11-FIP3 Mutant Expression. We performed site-directed mutagenesis to mutate a hydro-phobic isoleucine residue contained within the Rab-binding domain (RBD) of Rab11-FIP3 to hydrophilic glutamic acid to generate a mutant of Rab11-FIP3 that has been reported to be deficient in Rab11 binding (Wilson *et al.*, 2005). This mutant of Rab11-FIP3 has been previously shown to localize to the midbody independent of Rab11 (Wilson *et al.*,

2005). Furthermore, it has been demonstrated that its expression increases the number of binucleate cells, suggesting that Rab11/Rab11-FIP3 complex formation is required for completion of cytokinesis (Wilson *et al.*, 2005). We examined the localization of this mutant in HeLa cells by fluorescence microscopy. We found that in cells expressing this construct (pEGFP-C1/Rab11-FIP3$_{(2-756)}$ I738E), the protein predominantly localized to the cytosol, with a minor proportion of the protein found at the cell periphery and forming a reticular type of structure around the nucleus (Fig. 3C). Moreover, in contrast to cells overexpressing wild-type Rab11-FIP3, in cells transfected with this Rab11-binding deficient mutant and counterstained for Rab11a, we observed no condensation of the Rab11a immunoreactivity into the centrosomal region of the cell (compare Fig. 3A and C).

To examine the functional consequences of expression of an amino-terminally truncated mutant of Rab11-FIP3 (Rab11-FIP3$_{(244-756)}$) in transferrin (Tfn) trafficking we performed a Tfn uptake and chase assay, described in detail in Chapter 43 (this volume). As reported in Horgan *et al.* (2004), we observed no discernible effect on either Tfn uptake or recycling in HeLa cells expressing this truncated mutant of Rab11-FIP3. This indicates that unlike other Rab11-FIP members, Rab11-FIP3 is not involved in Tfn trafficking.

Acknowledgments

The authors are grateful to N. Marie and S. Hanscom for their invaluable assistance during the execution of this work. This work was supported by Science Foundation Ireland (SFI) (Investigator Grant 02/1N.1/B070) and the European Union (Marie Curie Development Host Fellowship, Grant HPMD-CT-2000-00024) to M.M.C., and an Irish Research Council for Science, Engineering and Technology (IRCSET) (Grant RS/2002/202-6) grant to C.H.

References

Cheng, H., Sugiura, R., Wu, W., Fujita, M., Lu, Y., Sio, S. O., Kawai, R., Takegawa, K., Shuntoh, H., and Kuno, T. (2002). Role of the Rab GTP-binding protein Ypt3 in the fission yeast exocytic pathway and its connection to calcineurin function. *Mol. Biol. Cell* **13**, 2963–2976.

Finger, F. P., and White, J. G. (2002). Fusion and fission: Membrane trafficking in animal cytokinesis. *Cell* **108**, 727–730.

Hales, C. M., Griner, R., Hobdy-Henderson, K. C., Dorn, M. C., Hardy, D., Kumar, R., Navarre, J., Chan, E. K., Lapierre, L. A., and Goldenring, J. R. (2001). Identification and characterization of a family of Rab11-interacting proteins. *J. Biol. Chem.* **276**, 39067–39075.

Hall, T. A. (1999). BioEdit: A user-friendly biological sequence alignment editor and analysis program for Windows 95/98/NT. *Nucl. Acids Symp. Ser.* **41**, 95–98.

Horgan, C. P., Walsh, M., Zurawski, T. H., and McCaffrey, M. W. (2004). Rab11FIP3 localises to a Rab11-positive pericentrosomal compartment during interphase and to the cleavage furrow during cytokinesis. *Biochem. Biophys. Res. Commun.* **319**, 83–94.

Nagase, T., Ishikawa, K., Kikuno, R., Hirosawa, M., Nomura, N., and Ohara, O. (1999). Prediction of the coding sequences of unidentified human genes. XV. The complete sequences of 100 new cDNA clones from brain which code for large proteins *in vitro*. *DNA Res.* **6,** 337–345.

Prekeris, R., Davies, J. M., and Scheller, R. H. (2001). Identification of a novel Rab11/25 binding domain present in Eferin and Rip proteins. *J. Biol. Chem.* **276,** 38966–38970.

Riggs, B., Rothwell, W., Mische, S., Hickson, G. R., Matheson, J., Hays, T. S., Gould, G. W., and Sullivan, W. (2003). Actin cytoskeleton remodeling during early Drosophila furrow formation requires recycling endosomal components nuclear-fallout and Rab11. *J. Cell Biol.* **163,** 143–154.

Schweitzer, J. K., and D'Souza-Schorey, C. (2002). Localization and activation of the ARF6 GTPase during cleavage furrow ingression and cytokinesis. *J. Biol. Chem.* **277,** 27210–27216.

Shin, O. H., Ross, A. H., Mihai, I., and Exton, J. H. (1999). Identification of arfophilin, a target protein for GTP-bound class II ADP-ribosylation factors. *J. Biol. Chem.* **274,** 36609–36615.

Shin, O. H., Couvillon, A. D., and Exton, J. H. (2001). Arfophilin is a common target of both class II and class III ADP-ribosylation factors. *Biochemistry* **40,** 10846–10852.

Skop, A. R., Bergmann, D., Mohler, W. A., and White, J. G. (2001). Completion of cytokinesis in *C. elegans* requires a brefeldin A-sensitive membrane accumulation at the cleavage furrow apex. *Curr. Biol.* **11,** 735–746.

Thompson, J. D., Higgins, D. G., and Gibson, T. J. (1994). CLUSTAL W: Improving the sensitivity of progressive multiple sequence alignment through sequence weighting, position-specific gap penalties and weight matrix choice. *Nucleic Acids Res.* **22,** 4673–4680.

Wallace, D. M., Lindsay, A. J., Hendrick, A. G., and McCaffrey, M. W. (2002). The novel Rab11-FIP/Rip/RCP family of proteins displays extensive homo- and hetero-interacting abilities. *Biochem. Biophys. Res. Commun.* **292,** 909–915.

Wilson, G. M., Fielding, A. B., Simon, G. C., Yu, X., Andrews, P. D., Hames, R. S., Frey, A. M., Peden, A. A., Gould, G. W., and Prekeris, R. (2005). The FIP3-Rab11 Protein Complex Regulates Recycling Endosome Targeting to the Cleavage Furrow during Late Cytokinesis. *Mol. Biol. Cell* **16,** 849–860.

[45]   Class I FIPs, Rab11-Binding Proteins That Regulate Endocytic Sorting and Recycling

By Elizabeth Tarbutton, Andrew A. Peden, Jagath R. Junutula, and Rytis Prekeris

Abstract

Rab11 GTPase is an important regulator of endocytic membrane traffic. In the GTP-bound form Rab GTPases interact with effector proteins and each Rab–effector complex is proposed to regulate a unique trafficking step/event such as vesicle docking, budding, transport, or fusion. At least six Rab11 effectors (family of Rab11 interacting proteins, FIPs) have been

METHODS IN ENZYMOLOGY, VOL. 403   0076-6879/05 $35.00
              DOI: 10.1016/S0076-6879(05)03045-4

identified and shown to interact with Rab11. Based on the sequence homology FIPs are divided in class I and class II subfamilies. Class I FIPs have been hypothesized to regulate the recycling of plasma membrane receptors. In contrast, class II FIPs have been implicated in regulating membrane traffic during more specialized cellular functions, such as cytokinesis. This chapter reviews the background and methodology required for characterizing interactions between FIPs and Rab11, as well as understanding their role in regulating endocytic membrane traffic.

Introduction

Cell surface proteins perform a variety of functions and their levels at the plasma membrane must be tightly regulated. Such regulation in part is achieved by endocytosis, where molecules are internalized, sorted, and either returned to the cell surface or degraded. Sorting and recycling of endocytosed proteins are required for proper cellular function, yet the mechanisms involved in regulating these processes are still not fully understood. At least some of the sorting occurs at the level of early endosomes (EE) via generation of EE tubular extensions that give rise to recycling endosomes (RE) and ensure the delivery of proteins back to the plasma membrane (Mellman, 1996; Robinson *et al.*, 1996). In addition to the RE-dependent recycling, proteins can also be transported directly from EE to plasma membrane, via the "fast" recycling pathway (Mellman, 1996; Robinson *et al.*, 1996). These functionally distinct endocytic compartments are often distinguished by the presence of different small monomeric GTPases, known as Rabs. At least seven Rabs (4, 5, 7, 9, 11, 14, and 15) serve to mediate vesicular traffic in the endocytic recycling pathway and may regulate distinct, although probably overlapping, transport steps (Chavrier and Goud, 1999; Ullrich *et al.*, 1996; van der Sluijs *et al.*, 1992).

In the past few years Rab11 GTPases have emerged as key regulators of endocytic recycling (Prekeris, 2003; Ullrich *et al.*, 1996). The Rab11 family is composed of three closely related proteins: Rab11a, Rab11b, and Rab25. Rab11a and Rab11b are ubiquitously expressed, while Rab25 is present only in epithelial cells and is thought to regulate apical protein targeting (Goldenring *et al.*, 1993, 2001). Rab11a was originally cloned as a GTPase that regulates constitutive protein recycling; however, data suggest that Rab11 family members are involved in a variety of different membrane trafficking pathways such as phagocytosis (Cox *et al.*, 2000), apical targeting in epithelial cells (Wang *et al.*, 2000), insulin-dependent glucose transporter 4 (GLUT4) transport to the plasma membrane (Kessler *et al.*, 2000), and protein transport to and from endosomes to the *trans*-Golgi network (Wilcke *et al.*, 2000).

Identification of the Rab11 Family Interacting Proteins (FIPs)

Cycling between GTP- and GDP-bound forms of Rab GTPases regulates the recruitment of various effector proteins to cellular membranes, thereby affecting the targeting and fusion of transport vesicles (Gonzalez and Scheller, 1999). Because the ability of Rabs to interact with several different effector molecules could be the basis for the specific function of Rabs, much effort has been invested in trying to identify effector proteins for Rab11. As a result, several proteins have been shown to interact with the GTP form of Rab11, including myosin V and the exocyst complex (Lapierre *et al.*, 2001; Zhang *et al.*, 2004). In addition, we identified Rip11/Rab11-FIP5, a novel Rab11-interacting protein (Prekeris *et al.*, 2000), and subsequently five other related proteins have been found (Hales *et al.*, 2001; Hickson *et al.*, 2003; Lindsay *et al.*, 2002; Prekeris *et al.*, 2001). All together they form the Rab11 family of interacting proteins (FIPs) (Fig. 1) (Hales *et al.*, 2001; Prekeris, 2003; Prekeris *et al.*, 2001). Based on sequence homology FIPs can be divided into two main classes (Fig. 1). Class I FIPs (Rip11, Rab-coupling protein [RCP], and FIP2) contain a C2 domain at the N-terminus of the protein (Prekeris, 2003). Class II FIPs (FIP3 and FIP4) contain two EF hands (Prekeris, 2003). In addition, class II FIPs have been shown to interact with Arf GTPases (Hickson *et al.*, 2003; Shin *et al.*, 1999). Arf6 and Rab11 GTPases bind to distinct domains in FIP3, thus allowing FIP3 to play a role as an Arf6 and Rab11 cross-linking protein (Fielding *et al.*, 2005).

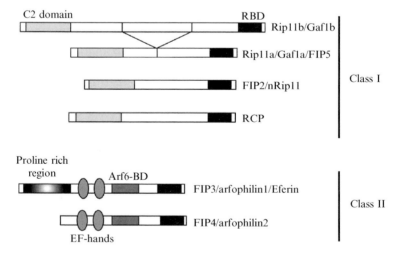

FIG. 1. Schematic representation of the FIP protein family. RBD, Rab11-binding domain; Arf6-BD, Arf6-binding domain.

Despite the identification of FIPs as Rab11-binding proteins, we are only beginning to understand the properties of Rab11-FIP interactions. Interestingly, some work suggests that FIP2 and RCP may bind to Rab11 in a GTP-independent manner (Lindsay and McCaffrey, 2002; Lindsay et al., 2002). In addition, little information is available on the affinities of Rab11 interactions with the different FIP family members. Thus, we have analyzed the properties of FIP binding to Rab11 using isothermal titration calorimetry (ITC) (Junutula et al., 2004). The ITC analysis demonstrated that FIPs bind to Rab11a and Rab11b with a similar affinity (Table I). Indeed, it has been previously reported that siRNA-based down-regulation of Rab11a or Rab11b alone had no effect on the subcellular distribution of FIP proteins, indicating that Rab11b can compensate for Rab11a and vice versa. In contrast, down-regulation of both Rab11 isoforms resulted in redistribution of FIPs to the cytosol (Junutula et al., 2004). Thus, while it has been reported that Rab11a and Rab11b can have distinct, although overlapping, cellular functions, they do not display any differences in the ability to bind various FIPs (Junutula et al., 2004).

All FIP isoforms interact with Rab11 in a GTP-dependent manner in vitro (Table I). This is not surprising, as the Rab11-binding domain in all FIP isoforms is highly conserved (Meyers and Prekeris, 2002) (Fig. 2). Interestingly, while FIPs interact strongly with GTP-bound Rab11 ($\sim$50–100 nM), weaker ($\sim$1000–1500 nM) GDP-dependent interactions are also observed (Table I) (Junutula et al., 2004). This raises the possibility that FIPs might be able to interact with Rab11-GDP in vivo. The cellular

TABLE I
THE AFFINITIES OF INTERACTIONS BETWEEN Rab11 AND FIPs[a]

Rab11-FIP complex	K_d (nM)
Rip11 with Rab11a-GDP	928
Rip11 with Rab11a-Gpp(NH)p	54.4
Rip11 with Rab11b-GDP	797
Rip11 with Rab11b-Gpp(NH)p	45.9
FIP2 with Rab11a-GDP	1317
FIP2 with Rab11a-Gpp(NH)p	40
FIP2 with Rab11b-GDP	1106
FIP2 with Rab11b-Gpp(NH)p	44
RCP with Rab11a-GDP	1178
RCP with Rab11a-Gpp(NH)p	107
RCP with Rab11b-GDP	1287
RCP with Rab11b-Gpp(NH)p	173

[a] All of the ITC experiments were carried out as described in the text. Each value is an average of three independent experiments.

	625			628	629	630				633					638						
Rip11	L	E	S	Y	I	D	R	L	L	V	R	I	M	E	T	S	P	T	L	L
FIP2	L	E	D	Y	I	D	N	L	L	V	R	V	M	E	E	T	P	S	I	L
RCP	L	E	D	Y	I	D	N	L	L	V	R	V	M	E	E	T	P	N	I	L

FIG. 2. The alignment of RBD domains from Rip11, FIP2, and RCP proteins. Boxed regions mark highly conserved RBD regions. Asterisks mark the amino acid residues that form a "hydrophobic patch."

concentrations of FIPs range between 200 and 500 nM (Junutula *et al.*, 2004). Therefore, in the cell, GDP-dependent interactions will not be favored due to their much weaker affinity, although it is possible that relatively low affinity interactions between Rab11-GDP may be compensated by high concentrations of these proteins on the endocytic membranes.

Isothermal Titration Calorimetry (ITC)

ITC measures the heat generated or absorbed during molecular interaction and has been routinely used to study antigen–antibody, protein–ligand, and protein–protein interactions (Ladbury and Chowdhry, 1996; Leavitt and Freire, 2001). To determine the affinities of Rab11 binding to different FIPs, we performed ITC analysis. ITC offers several advantages over glutathione bead pull-down or surface plasmon resonance assays. First, the measurements are performed in solution, thus eliminating artifacts caused by the attachment of proteins to surfaces (bead or "chip"). Second, it does not require any modifications of the proteins such as addition of fluorophores or tags, thus allowing the study of native protein interactions. Third, ITC measurements provide a complete thermodynamic profile, including binding affinity (K_a), binding stoichiometry (n), enthalpy (ΔH), and entropy (ΔS). The following section provides a protocol that can be used to characterize Rab11 binding to FIP proteins by ITC analysis. All experiments were done using a VP-ITC calorimeter (Microcal LLC).

Expression and Purification of the Recombinant Rab11 and FIPs

1. RCP, FIP2, Rip11, Rab11a, and Rab11b are cloned into pGEX-KG (Pharmacia) and transformed in BL21-Codon Plus (DE3)-RIL *Escherichia coli* cells.

2. *E. coli* cells are inoculated into 100 ml of Terrific Broth media and grown overnight at 37°.

3. The next morning cells are diluted to 0.1 OD and grown at 37° until they reach 0.6 OD. The GST-fusion protein expression is then induced

with 1 mM isopropyl-β-D-thiogalactopyranoside (IPTG) and the cells are grown for a further 2.5 h at 37°.

4. The cells are sedimented and resuspended in lysis buffer (100 mM phosphate buffer, pH 7.4, with 0.1% Tween, 1 mM dithiothreitol [DTT], 1 μM phenylmethylsulfonyl fluoride [PMSF]).

5. Cells are lysed via French pressing and centrifuged at 30,000×g for 10 min to sediment cell debris and collect lysate.

6. The resulting lysate is incubated with glutathione beads for 1 h at 4°. Beads are washed three times with 50 ml of lysis buffer, followed by two washes with ITC titration buffer (phosphate-buffered saline [PBS], pH 7.4, with 5 mM MgCl$_2$).

7. To obtain homogeneous preparation of either GppNHp-bound Rab11 or GDP-bound Rab11, glutathione beads are washed with nucleotide exchange buffer (PBS buffer, pH 7.4, 5 mM MgCl$_2$, 10 mM EDTA with 0.5 mM GppNHp or GDP) followed by a wash with nucleotide stabilization buffer (PBS buffer, pH 7.4, 5 mM MgCl$_2$, with 0.5 mM GppNHp or GDP).

8. Protein is eluted using thrombin (1 h at room temperature or overnight at 4°). The purity of the protein is analyzed by sodium dodecyl sulfate–polyacrylamide gel electrophoresis (SDS-PAGE). If needed, the protein is further purified by gel-filtration chromatography. The resulting protein was quantified using Bradford protein assay and used for ITC analysis.

ITC Analysis

1. Recombinant Rab11a or Rab11b protein (8 μM) in PBS containing 5 mM MgCl$_2$ and 0.5 mM GppNHp or GDP is loaded in the sample cell.

2. Recombinant Rip11, RCP, or FIP2 (200 μl in the same buffer) is titrated into the sample cell by 5 μl injections, up to a total of 30–40 injections. The titration is performed while samples are stirred at 300 rpm at 25°. An interval of 4 min between each injection is allowed for the baseline to stabilize.

3. To estimate a background, a blank titration is performed by injecting the corresponding FIP into buffer alone. The data from the blank run are subtracted from the corresponding Rab11-FIP titration.

4. The data are fitted via the one-set-of-sites model to calculate the binding constant using origin software (Micocal LLC).

5. The ITC experiments for Rab-FIP interaction are repeated at various temperatures to determine the change in heat capacity (ΔC_p).

Additional Comments. It is important to make sure that the buffer solutions used to reconstitute the proteins are identical or large variations in baseline will be observed. The molar concentration of protein/ligand in the ITC cell can be 3–1000 times greater than the dissociation constant of interaction. Care should be taken that the titration concentration reaches at least a 3-fold molar excess over the protein/ligand present in the ITC cell at the end of the experiment. If possible, avoid using unstable reducing agents such as DTT and 2-mercaptoethanol, as they will rapidly oxidize and cause large fluctuations in the baseline. These reducing agents can be replaced with more stable ones, such as TCEP. There are several disadvantages with ITC analysis. First, it cannot measure subnanomolar affinities accurately. Second, it requires milligram quantities of protein. Third, for precise measurement, each ITC injection should have an average of at least 3–5 μcal of heat absorbed or evolved into a 1.3-ml cell; any signal less than this cannot be studied (VP-ITC MicroCalorimeter User's Manual).

Identification and Characterization of Rab11-Binding Domain (RBD)

The common feature of all FIPs is the presence of a highly conserved motif at the C-terminus of the protein that is necessary and sufficient for the binding of Rab11 (RBD) (Figs. 1 and 2) (Prekeris et al., 2001). The RBD is part of the larger C-terminal α-helical domain (Meyers and Prekeris, 2002). When the C-terminus of FIPs is plotted in the α-helical conformation, all hydrophobic residues of the RBD form a hydrophobic patch (Meyers and Prekeris, 2002). Substitution of the central isoleucine with glutamic acid (I629E) completely abolishes Rab11 binding, while the conservative substitution (I629V) has little effect on Rip11 and Rab11 interaction (Fig. 2 and Table II) (Junutula et al., 2004). Mutations of other amino acids (L625A and L633A) within the hydrophobic patch also inhibit Rab11 binding (Fig. 2 and unpublished observations). Consistent with the

TABLE II
MUTATIONAL ANALYSIS OF RBD DOMAIN[a]

Rab11-GTP and Rip11 complex	K_d (nM)
Rip11 wild type	54.4
Rip11-I629V	160
Rip11-I629E	30,000
Rip11-D630A	38
Rip11-E638A	50
Rip11-Y628A	1900
Rip11-Y628F	530

[a] All of the ITC experiments were carried out as described in the text. Each value is an average of three independent experiments.

role of the hydrophobic patch in Rab11 binding, ITC analysis shows negative ΔC_p values for both GDP-(-0.84 kcal/mol/K) and GTP-dependent (-0.97 kcal/mol/K) Rab11–Rip11 interactions (Junutula *et al.*, 2004).

In addition to the conserved hydrophobic patch, the RBD domain also contains a conserved tyrosine residue (Y628) as well as several conserved negatively charged residues (E626, D630, and E638) (Fig. 2). Mutation of the tyrosine (Y628A) results in the inhibition of Rip11 and Rab11 binding (Table II), suggesting that Y628 may also participate in Rab11 binding (Junutula *et al.*, 2004). Surprisingly, mutations of the conserved negatively charged residues (D630A and E638A) have no effect on Rip11 interactions with Rab11 (Table II) (Junutula *et al.*, 2004). However, even if these residues do not directly participate in Rab11 binding, they could still be important in interactions between FIPs and other proteins, such as myosin Vb.

Role of RCP and Rip11 in TfR Recycling and Degradation

Data characterizing Rab11 and FIP interactions demonstrate that all FIPs bind to Rab11 GTPases with similar affinity. Given that most mammalian cells express several FIPs, each Rab11-FIP complex might be involved in distinct endocytic transport events. Consistent with this hypothesis, it has been shown that Rab11 forms mutually exclusive complexes with each of the different FIPs (Meyers and Prekeris, 2002). Furthermore, FIPs also display distinct, though partially overlapping, localization patterns within the cell (Meyers and Prekeris, 2002). To gain insight into the function of these various FIPs, their respective RBD domains have been overexpressed to produce dominant-negative effects that are assumed to reflect FIP isoform-specific functions. These studies have implicated FIP2 and RCP in regulating epidermal growth factor receptor (EGFR) transport and TfR recycling, respectively (Lindsay and McCaffrey, 2002; Lindsay *et al.*, 2002). Interestingly, the dominant-negative effects on endocytic transport are similar for all the FIP RBDs (Peden *et al.*, 2004). For instance, overexpression of RCP-RBD, FIP2-RBD, and Rip11-RBD results in the tubulation of recycling endosomes where many endocytic markers accumulate (EGFR, TfR, chemokine receptors, e-cadherin, and LDLR) (Cullis *et al.*, 2002; Lindsay and McCaffrey, 2002; Meyers and Prekeris, 2002; Peden *et al.*, 2004 and unpublished observations). While it is possible that all FIPs regulate the same membrane trafficking pathway, it is more likely that isoform-specific function is mediated by domains outside of the RBD and that overexpression of any of the RBDs results in general inhibition of Rab11/FIP function.

To gain an insight into the function of the various FIPs, we used RNA-interference (RNAi) to disrupt their expression. All the members of the FIP family can be easily down-regulated using RNAi (Peden *et al.*, 2004;

Wilson *et al.*, 2004). The use of fluorescence-activated cell sorting (FACS) assays, in combination with RNAi, provides a powerful tool for the indentification FIP protein function (Peden *et al.*, 2004). Using RNAi studies, it has been shown that class I FIPs appear to be involved in regulating the recycling of various plasma membrane receptors and the class II FIPs were shown to regulate more specialized cell processes, such as cytokinesis and cell motility (Wilson *et al.*, 2004 and unpublished observations).

Consistent with the involvement of RCP in regulation of transferrin receptor (TfR) recycling, knock-down of RCP inhibits Tf uptake in HeLa cells (Fig. 3C) (Peden *et al.*, 2004). Surprisingly, knock-down of RCP decreased the cell surface (Fig. 3B) and total levels of RCP, while having little effect on TfR recycling to the plasma membrane (Fig. 3D) (Peden *et al.*, 2004). Furthermore, the decrease in cellular levels of TfR can be blocked by lysosomal inhibitors. In contrast to RCP, knock-down of Rip11 stimulated Tf uptake (Fig. 3C) (Peden *et al.*, 2004) and resulted in the slight increase in cellular TfR levels (unpublished observations).

In summary, class I FIPs appear to regulate distinct Rab11-dependent steps of endocytic sorting and traffic. For instance, RCP appears to regulate transport of TfR to the recycling pathway, while Rip11 may be required for apical protein targeting (Prekeris *et al.*, 2000). The function of the FIP2, the third member of class I FIPs, remains to be analyzed by RNAi and FACS. However, data suggest that FIP2 may regulate yet another endocytic pathway, such as traffic of EGFR and/or chemokine receptors (Cullis *et al.*, 2002; Fan *et al.*, 2003).

Knock-Down of FIPs by RNA-Interference (RNAi)

To determine the role of the FIPs in endocytic membrane traffic, we used RNAi to down-regulate the expression of individual FIPs. The following section provides a protocol for the effective knock-down of Rip11 and RCP levels in HeLa cells. The siRNA to either RCP or Rip11 was obtained from Qiagen based on human RCP and Rip11 sequences (see Peden *et al.*, 2004, for sequences).

1. siRNA is reconstituted in 1 ml of 25 mM phosphate buffer, pH 7.5. The siRNA is heated at 90° for 1 min. It is then incubated in a 37° water bath for 1 h. This procedure disrupts higher aggregates that can form during the siRNA lyophilization process. The reconstituted siRNA is then stored at −20°.

2. One day before transfection, 7.5×10^4 HeLa cells are seeded onto a 6-cm plate in normal growth media and incubated at 37° overnight.

3. For each condition, 40 μl of Plus reagent (Invitrogen) is added to 250 μl of Opti-Mem (Invitrogen). To this solution, 8 μl of siRNA is added. We have found that using multiple siRNAs enhances the knock-down of the FIPs. This mixture is incubated at room temperature for 15 min. We

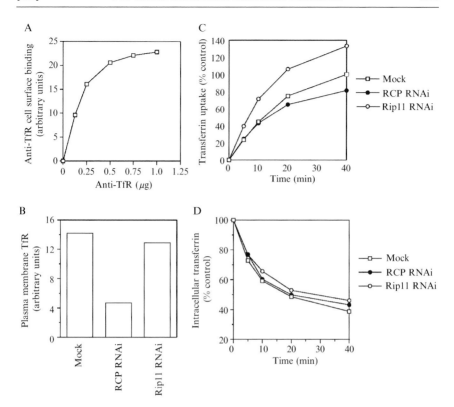

FIG. 3. Effect of RCP and Rip11 knock-down on endocytic TfR transport. (A) Increasing amounts of anti-TfR-Alexa647 antibody were incubated with HeLa cells for 30 min at 4°. Cells are then washed and the amount of bound antibody analyzed by flow cytometry. (B) HeLa cells were incubated with anti-TfR-PE antibody for 30 min at 4°. The amount of plasma membrane bound antibody was analyzed by flow cytometry. (C) HeLa cells were incubated for varying amounts of time at 37° in the presence of Tf-Alexa488. Cells were then washed and the amount of internalized Tf-Alexa488 was analyzed by flow cytometry. (D) HeLa cells were incubated for 30 min at 37° in the presence of Tf-Alexa488. Cells were then washed and incubated at 37° for varying amounts of time in the presence of unlabeled Tf. The level of cell-associated Tf-Alexa488 was determined by flow cytometry.

have found that use of the Plus reagent enhances cell survival after RNAi and reduces the amount of Lipofectamine 2000 needed.

4. While the siRNA mixture is incubating, 10 μl of Lipofectamine 2000 (Invitrogen) is added to 250 μl of Opti-Mem (for each condition). It is important to mix the Lipofectamine 2000 well to ensure even distribution of liposomes between samples.

5. After the 15-min incubation period, the Lipofectamine mixture is added to the siRNA mixture, mixed gently, and incubated at room temperature for 15 min to allow proper formation of liposomes.

6. During the 15-min incubation period, the normal growth medium is aspirated off of the cells in a sterile cell culture hood. The medium is replaced with 2 ml of prewarmed Opti-Mem (37°).

7. The siRNA-containing liposome mixture is added to the plate of cells and allowed to incubate at 37° for 3 h.

8. After 3 h, the liposome mixture is aspirated off of the cells and replaced with 5 ml of prewarmed normal growth media. Cells are allowed to recover for 48–72 h after the RNAi before use in FACS assays.

9. To quantify the extent of knock-down, cells are harvested and washed in 1× PBS. Cells are resuspended in 500 μl of PBS with 1 mM protease inhibitors (we use PMSF), 5 mM EDTA, and 1% Triton X-100. Cells are incubated on ice for 30 min. To sediment insoluble material, cells are centrifuged at 5000×g for 5 min at 4°. The supernatant is moved to a clean tube and protein content is assayed by Bradford protein analysis. To quantify the level of down-regulation, samples are separated on SDS–PAGE gels and probed with either anti-Rip11 or anti-RCP antibodies.

Flow Cytometry-Based Transport Assays

We and others have previously shown that disruption of Rab11 function leads to defects in the trafficking of the TfR (Peden *et al.*, 2004). To determine the role of Rab11 effectors on this pathway, we established flow cytometry-based assays for measuring the trafficking of the TfR. We prefer flow cytometry-based assays over more conventional ^{125}I-labeled Tf assays for several reasons. First, flow cytometry allows one to analyze the transport of several molecules simultaneously (e.g., TF-Alexa488 and anti-EGF-R-Alexa647 can be measured on the FL1 and FL4 channels of a 2 laser BD FACSCalibur). Gating can be used to specifically analyze the transfected cells (e.g., if CFP, YFP, and GFP-tagged constructs are used) (Junutula *et al.*, 2004; Peden *et al.*, 2004). Second, flow cytometry-based assays are far less prone to errors caused by differences in cell number, transfection efficiency, and carryover from unbound label in washing steps. The reduced sensitivity of FACS-based assays relative to ^{125}I-based assays is usually not an issue unless studying receptors with very low levels.

Before flow cytometry assays can be used on adherent cell lines, it is prudent to perform several control experiments. The first control that should be carried out is to establish the most appropriate method for detaching the cells from the tissue culture dish. This can be determined by comparing the FACS signal obtained from cells detached using trypsin versus cells detached using cell dissociation buffer (Sigma). We have found that most markers appear to be resistant to trypsinization (TfR, EGFR, CD63, and Lamp1), and that trypsin-treated cells have greater viability and are less likely to form clumps during incubations (cell viability can be determined by PI exclusion).

Titration of antibody or ligand also needs to be performed to determine the optimal concentration of reagent required to provide a satisfactory signal. For many antibodies 1 μg/million cells is a good starting point. In some cases the signal may be weak and it may be necessary to switch the fluorophore (e.g., anti-LDLR antibodies give almost no signal when conjugated to Alexa488 but when changed to Alexa647 provide acceptable signal). In most FACS machines the FL4 channel (Cy5, APC, and Alexa647) gives approximately 10-fold more signal than the FL1 channel (FITC, Cy3, and Alexa488) due to suboptimal excitation and emission filters used for the FL1 channel.

Finally, if a bivalent antibody is being used to follow the trafficking of the receptor, it is important to try and establish that the antibody does not alter the trafficking of the molecule being studied. This can be determined by pretreating the cells with antibody and determining whether there has been a change in steady distribution of the marker (alteration in cell surface levels or localization within the cell, detected using a second noncompetitive antibody). In some cases this issue can be avoided by using high-affinity FABs or by using the antibody to measure steady-state levels of the receptor of interest (cell surface or total levels).

The following is a protocol designed to measure cell surface levels of TfR and the kinetics of receptor uptake and recycling.

1. Mock or siRNA-treated HeLa cells are harvested via trypsin digestion. Cells are counted on a hemacytometer, and 500,000 cells are placed in a tube for each condition.

2. Cells are pelleted by spinning at 1000 rpm for 5 min and resuspended in 500 μl of ice-cold serum-free media (DMEM) that is supplemented with 20 mM HEPES, pH 7.4.

3. Two microliters of 2.5 mg/ml Tf-Alexa488 or 2 μl of 1 mg/ml anti-TfR-PE is added to each tube of the cells. The control cells do not receive transferrin or anti-TfR antibody. Cells are then incubated on ice for 30 min (for cell surface measurements, the cells are washed and then analyzed as below).

4. For the uptake assays, cells are moved to a 37° water bath and incubated for various time points. For recycling assays, all cells were incubated at 37° for 30 min. Unbound Tf-Alexa488 is then washed and cells incubated with excess unlabeled Tf (50 μg/ml) for varying amounts of time at 37°.

5. After incubation, cells are washed two times with ice-cold media and placed on ice until used for FACS analysis (the cells can be resuspended in 2–4% paraformaldehyde for later analysis).

6. Cells are analyzed using flow cytometry. We use an unlabeled control and single labeled controls to optimize the parameters of the assay. We follow 5000 events to obtain our data. Data are normalized to our control and plotted on a line graph.

Due to the limitations in sensitivity, the uptake relies on the continuous presence of Tf to obtain a significant signal. Thus, this assay measures the multiple cycles of uptake and recycling of TfR (the first 5 min of the assay should reflect the internalization kinetics more accurately). If more accurate internalization rates are required, an anti-TfRAlexa488 antibody quenching system should be used or acid stripping experiments should be performed (Austin *et al.*, 2004).

References

Austin, C. D., De Maziere, A. M., Pisacane, P. I., van Dijk, S. M., Eigenbrot, C., Sliwkowski, M. X., Klumperman, J., and Scheller, R. H. (2004). Endocytosis and sorting of Erb2 and the site of action of cancer therapeutics trastuzumab and geldanamycin. *Mol. Biol. Cell* **15,** 5268–5282.

Chavrier, P., and Goud, B. (1999). The role of ARF and Rab GTPases in membrane transport. *Curr. Opin. Cell Biol.* **11,** 466–475.

Cox, D., Lee, D. J., Dale, B. M., Calafat, J., and Greenberg, S. (2000). A Rab11-containing rapidly recycling compartment in macrophages that promotes phagocytosis. *Proc. Natl. Acad. Sci. USA* **97,** 680–685.

Cullis, D. N., Philip, B., Baleja, J. D., and Feig, L. A. (2002). Rab11-FIP2, an adaptor protein connecting cellular components involved in internalization and recycling of epidermal growth factor receptors. *J. Biol. Chem.* **277,** 49158–49166.

Fan, G. H., Lapierre, L. A., Goldenring, J. R., and Richmond, A. (2003). Differential regulation of CXCR2 trafficking by Rab GTPases. *Blood* **101,** 2115–2124.

Fielding, A. B., Schonteich, E., Yu, X., Matheson, J., Wilson, G., Xinzi, Y., Hickson, G. R. X., Srivastava, S., Baldwin, S. A., Prekeris, R., and Gould, G. W. (2005). Rab11-FIP3 and Rab11-FIP4 interact with ArF6 and Exocyst to control membrane traffic during cytokinesis. *EMBO J.* In press.

Goldenring, J. R., Shen, K. R., Vaughan, H. D., and Modlin, I. M. (1993). Identification of a small GTP-binding protein, Rab25, expressed in the gastrointestinal mucosa, kidney, and lung. *J. Biol. Chem.* **268,** 18419–18422.

Goldenring, J. R., Aron, L. M., Lapierre, L. A., Navarre, J., and Casanova, J. E. (2001). Expression and properties of Rab25 in polarized Madin-Darby canine kidney cells. *Methods Enzymol.* **329,** 225–234.

Gonzalez, L., Jr., and Scheller, R. H. (1999). Regulation of membrane trafficking: Structural insights from a Rab/effector complex. *Cell* **96,** 755–758.

Hales, C. M., Griner, R., Hobdy-Henderson, K. C., Dorn, M. C., Hardy, D., Kumar, R., Navarre, J., Chan, E. K., Lapierre, L. A., and Goldenring, J. R. (2001). Identification and characterization of a family of Rab11-interacting proteins. *J. Biol. Chem.* **276,** 39067–39075.

Hickson, G. R. X., Matheson, J., Riggs, B., Maier, V. H., Fielding, A. B., Prekeris, R., Sullivan, W., Barr, F. A., and Gould, G. W. (2003). Arfophilins are dual Arf/Rab11 binding proteins that regulate recycling endosome distribution and are related to Drosophila nuclear fallout. *Mol. Biol. Cell* **14,** 2908–2920.

Junutula, J. R., Schonteich, E., Wilson, G. M., Peden, A. A., Scheller, R. H., and Prekeris, R. (2004). Molecular characterization of Rab11 interactions with members of the family of Rab11-interacting proteins. *J. Biol. Chem.* **279,** 33430–33437.

Kessler, A., Tomas, E., Immler, D., Meyer, H. E., Zorzano, A., and Eckel, J. (2000). Rab11 is associated with GLUT4-containing vesicles and redistributes in response to insulin. *Diabetologia* **43,** 1518–1527.

Ladbury, J. E., and Chowdhry, B. Z. (1996). Sensing the heat: The application of isothermal titration calorimetry to thermodynamic studies of biomolecular interactions. *Chem. Biol.* **3,** 791–801.

Lapierre, L. A., Kumar, R., Hales, C. M., Navarre, J., Bhartur, S. G., Burnette, J. O., Provance, D. W., Jr., Mercer, J. A., Bahler, M., and Goldenring, J. R. (2001). Myosin vb is associated with plasma membrane recycling systems. *Mol. Biol. Cell* **12,** 1843–1857.

Leavitt, S., and Freire, E. (2001). Direct measurement of protein binding energetics by isothermal titration calorimetry. *Curr. Opin. Struct. Biol.* **11,** 560–566.

Lindsay, A. J., and McCaffrey, M. W. (2002). Rab11-FIP2 functions in transferrin recycling and associates with endosomal membranes via its COOH-terminal domain. *J. Biol. Chem.* **277,** 27193–27199.

Lindsay, A. J., Hendrick, A. G., Cantalupo, G., Senic-Matuglia, F., Goud, B., Bucci, C., and McCaffrey, M. W. (2002). Rab coupling protein (RCP), a novel Rab4 and Rab11 effector protein. *J. Biol. Chem.* **277,** 12190–12199.

Mellman, I. (1996). Endocytosis and molecular sorting. *Annu. Rev. Cell Dev. Biol.* **12,** 575–625.

Meyers, J. M., and Prekeris, R. (2002). Formation of mutually exclusive Rab11 complexes with members of the family of Rab11-interacting proteins regulates Rab11 endocytic targeting and function. *J. Biol. Chem.* **277,** 49003–49010.

Peden, A. A., Schonteich, E., Chun, J., Junutula, J. R., Scheller, R. H., and Prekeris, R. (2004). The RCP-Rab11 complex regulates endocytic protein sorting. *Mol. Biol. Cell* **15,** 3530–3541.

Prekeris, R. (2003). Rabs, Rips, FIPs, and endocytic membrane traffic. *Scientific World Journal* **3,** 870–880.

Prekeris, R., Klumperman, J., and Scheller, R. H. (2000). A Rab11/Rip11 protein complex regulates apical membrane trafficking via recycling endosomes. *Mol. Cell* **6,** 1437–1448.

Prekeris, R., Davies, J. M., and Scheller, R. H. (2001). Identification of a novel Rab11/25 binding domain present in Eferin and Rip proteins. *J. Biol. Chem.* **276,** 38966–38970.

Robinson, M. S., Watts, C., and Zerial, M. (1996). Membrane dynamics in endocytosis. *Cell* **84,** 13–21.

Shin, O. H., Ross, A. H., Mihai, I., and Exton, J. H. (1999). Identification of arfophilin, a target protein for GTP-bound class II ADP-ribosylation factors. *J. Biol. Chem.* **274,** 36609–36615.

Ullrich, O., Reinsch, S., Urbe, S., Zerial, M., and Parton, R. G. (1996). Rab11 regulates recycling through the pericentriolar recycling endosome. *J. Cell Biol.* **135,** 913–924.

van der Sluijs, P., Hull, M., Webster, P., Male, P., Goud, B., and Mellman, I. (1992). The small GTP-binding protein rab4 controls an early sorting event on the endocytic pathway. *Cell* **70,** 729–740.

Wang, X., Kumar, R., Navarre, J., Casanova, J. E., and Goldenring, J. R. (2000). Regulation of vesicle trafficking in Madin-Darby canine kidney cells by Rab11a and Rab25. *J. Biol. Chem.* **275,** 29138–29146.

Wilcke, M., Johannes, L., Galli, T., Mayau, V., Goud, B., and Salamero, J. (2000). Rab11 regulates the compartmentalization of early endosomes required for efficient transport from early endosomes to the trans-Golgi network. *J. Cell Biol.* **151,** 1207–1220.

Wilson, G. M., Fielding, A. B., Simon, G. C., Yu, X., Andrews, P. D., Peden, A. A., Gould, G. W., and Prekeris, R. (2004). The FIP3-Rab11 protein complex regulates recycling endosome targeting to the cleavage furrow during late cytokinesis. *Mol. Biol. Cell* **16,** 849–860.

Zhang, X. M., Ellis, S., Sriratana, A., Mitchell, C. A., and Rowe, T. (2004). Sec15 is an effector for the Rab11 GTPase in mammalian cells. *J. Biol. Chem.* **279,** 43027–43034.

[46]   Expression and Properties of the Rab4, Rabaptin-5α, AP-1 Complex in Endosomal Recycling

By Ioana Popa, Magda Deneka, and Peter van der Sluijs

Abstract

We previously showed that the small GTPase Rab4 regulates formation of recycling vesicles from early endosomes. To understand how Rab4 accomplishes this task, we started to identify the Rab4 effector protein network. In this chapter, we describe experiments leading to the characterization of a complex consisting of Rab4GTP, its effector Rabaptin-5α, and the adaptor protein complex AP-1, which regulates recycling from endosomes.

Introduction

Early endosomes are an important sorting station where proteins and lipids that are destined for late endocytic compartments are sorted away from cargo molecules that are recycled back to the plasma membrane. Early endosomes receive material from the plasma membrane by endocytosis as well as from the *trans*-Golgi network (TGN) via exocytic routes. Exit routes from early endosomes lead to the cell surface and intracellular compartments such as late endosomes and the TGN. The massive flow of membrane through early endosomes needs to be tightly controlled to maintain the structural and architectural integrity of this compartment (see Gruenberg, 2001). Indeed the small GTPases Rab5 and Rab4 have been shown to be important regulators that act sequentially in controlling entry into and recycling from early endosomes, respectively (see Deneka *et al.*, 2003b). They accomplish this feat through the spatial and temporal recruitment of a structurally diverse group of cytoplasmic effector proteins that bind to their active GTP forms. Effector proteins generally have been found to cooperate with their Rab protein partners at multiple stages in a transport pathway that include vesicle formation, transport along the cytoskeleton, and tethering and docking on a target membrane (see Zerial and McBride, 2001). Interestingly, a subgroup of the Rab5 effectors, including rabaptin-5, rabaptin-5α, rabaptin-5β as well as some of the FYVE domain proteins such as rabenosyn-5 and rabip4' also interact with Rab4 in a separate binding site (de Renzis *et al.*, 2002; Fouraux *et al.*, 2003; Nagelkerken *et al.*, 2000; Vitale *et al.*, 1998). The bivalent nature of these effectors is thought to coordinate the activities of Rab5 and Rab4 in distinct but overlapping

METHODS IN ENZYMOLOGY, VOL. 403   0076-6879/05 $35.00
DOI: 10.1016/S0076-6879(05)03046-6

microdomains on early endosomes (de Renzis *et al.*, 2002; Sönnichsen *et al.*, 2000). We found before that Rab4 controls recycling vesicle formation from endosomes and that Rab4 and brefeldin A act in the same recycling pathway (de Wit *et al.*, 2001; Mohrmann *et al.*, 2002). Because the hetero-tetrameric adaptor complex AP-1 localizes to early endosomes (Stoorvogel *et al.*, 1996) and is involved in transport vesicle formation (Odorizzi and Trowbridge, 1997), it is important to investigate the functional relationships of the Rab4, rabaptin-5α, and AP-1 adaptor in endosomal recycling.

Methods

Yeast Two-Hybrid Interaction Assay of Rabaptin-5 Constructs and AP Complex Subunits

To assess interactions between rabaptins and AP-1 subunits we used initially Gal4-based two-hybrid assays since this format is well suited for rapid testing of direct interactions (Fig. 1A and B).

Expression Constructs. pGBT9-rabaptin-5α, pGBT9-rabaptin-5, pGBT9-rabaptin-5α(1–390), pGBT9-rabaptin-5α(509–830), pGBT9-rabaptin-5α(509–863), pGBT9-rabaptin-5α(1–592), pGBT9-rabaptin-5α(301–830), pGBT9-rabaptin-5α(301–863), pGBT9-rabaptin-5α(301–592), pGBT9-rabaptin-5α(301–449), +pGBT9-rabaptin-5α(449–592), pACT2-α-adaptin, pACT2-β2-adaptin, pACT2-μ1-adaptin, pACT2-μ2-adaptin, pGADH-γ1-adaptin, pACT2-δ-adaptin, and pACT2-σ1-adaptin have been described (Deneka *et al.*, 2003). The two-hybrid plasmids pGBT9, pACT2, and pGADGH were obtained from Clontech.

Reagent. SC medium: 6.7 g yeast nitrogen base lacking amino acids (Difco), 0.75 g Complete Supplement Medium without leucine, tryptophan, and histidine (Bio 101), 20 g glucose (Sigma), 0.1 g adenine hemisulfate (Sigma), brought to 1 liter with ddH$_2$O. The pH of the medium is 5.8.

Protocol. YGH1 (*MATa trp1 leu2 his3 LYS2::GAL1-HIS3 URA3:: GAL1-LacZ*) cells (Spaargaren and Bischoff, 1994) were transformed with 3–6 μg DNA and grown at 28° on SC medium lacking tryptophan and leucine. Colonies were patched on SC plates lacking leucine, tryptophan, and histidine (growth selection) and SC plates without leucine and trypto-phan for β-galactosidase assay. For the latter, plates were overlaid with Hydrobond-N nylon membrane (Amersham Pharmacia Biotech) under light pressure. Filters are quickly lifted and immersed for 20 s in liquid nitrogen. After a brief warming, filters are overlaid (cells on top) with Whatman 3 paper soaked in 2 ml 60 mM NaH$_2$PO$_4$, pH 7, 10 mM KCl, 1 mM MgSO$_4$, containing 50 mM α-mercaptoethanol and 0.02% X-gal (both freshly added).

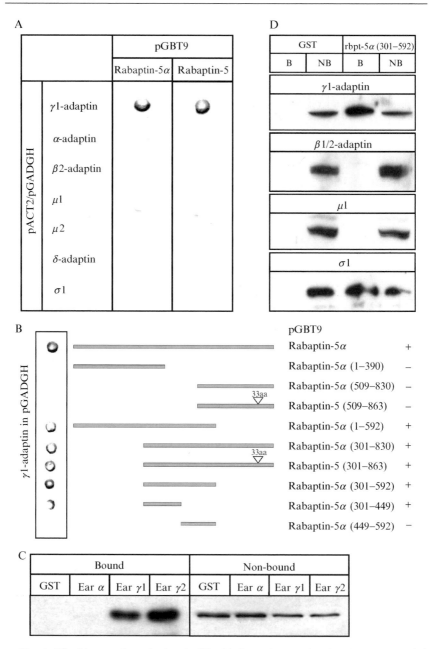

FIG. 1. The hinge region of rabaptin-5/5α binds to the ear domain of γ1-adaptin (A) β-galactosidase reporter activation of YGH1 cells cotransformed with pGBT9-rabaptin-5α or pGBT9-rabaptin-5 and pGADH-γ1-adaptin. None of the other tested subunits of AP-1

Comments. Over the past several years a number of rabaptin-5 variants have been identified that with the exception of rabaptin-5β are thought to arise by alternative splicing (Gournier *et al.*, 1998; Korobko *et al.*, 2002; Nagelkerken *et al.*, 2000). Rabaptin-5α contains a C-terminal 33-amino acid internal deletion compared to rabaptin-5, immediately adjacent to the Rab5-binding site. So far we have not found functional differences between rabaptin-5 and rabaptin-5α, and indeed binding of γ1-adaptin to rabaptin-5 and rabaptin-5α is identical and occurs in a completely conserved region. Korobko *et al.* (2005), however, observed that rabaptin-5δ, which contains an in-frame deletion in the Rab4-binding site, lost the ability to interact with Rab4 while Rab5 binding is not affected.

Binding Assay of Rabaptin-5α to Glutathione S-transferase (GST) Fusions of Adaptor Ears

An independent biochemical method was developed to confirm the results of the yeast two-hybrid interaction trap. The method is based on the immobilization of recombinant adaptin ear domains as GST fusions on GSH beads, followed by the retrieval of rabaptin-5α from detergent lysates and detection by Western blot (Fig. 1C).

Reagents

Buffer A: 25 mM Tris–HCl, pH 7.5, 0.5 mM ethylenediaminetetrascetic acid (EDTA), 10% sucrose, 1 mM dithiothreitol (DTT)

Buffer B: 25 mM Tris–HCl, pH 7.5, 0.5 mM EDTA, 1 mM DTT

Buffer C: 25 mM Tris–HCl, pH 7.5, 0.5 mM EDTA, 1% Triton X-100, 1 mM DTT

Buffer D: 50 mM Tris–HCl, pH 8.0, 25 mM GSH

Protease inhibitors stocks: 5 mg/ml leupeptin, 10 mg/ml aprotinin, 1 mg/ml pepstatin, 100 mM phenylmethylsulfonyl fluoride (PMSF) (all stored at $-20°$)

1 M DTT stock (stored at $-20°$)

250 mM reduced glutathione (GSH, stored at $-80°$)

(μ1- and σ1-adaptin), AP-2 (α-, β2-, and μ2-adaptin), or AP-3 (δ-adaptin) bound to rabaptins. (B) Binding of rabaptin-5α from a brain extract to GST fusions containing the ear domain of γ1-adaptin and γ2-adaptin. (C) Summary of the two-hybrid assays with rabaptin-5α truncations showing that the domain required for γ1-adaptin binding is between amino acids 301 and 449; binding is indicated by +. (D) Pull-down assay showing binding of AP-1 from a HeLa cell extract to GST-rabaptin-5α. The Western blot of bound (B) and nonbound (NB) fractions was probed with antibodies against γ1-, β1,2-, σ1-, and μ1-adaptin. Since rabaptin-5 and rabaptin-5α behave identically in the assays and have complete sequence identity in the hinge region, we performed the other experiments with rabaptin-5α (Adapted with permission from Deneka *et al.*, 2003a).

Expression and Isolation of GST Adaptin Ear Fusion Proteins.
pGEX4T3, pGEX1λT ear α-adaptin, pGEX4T3 ear γ1-adaptin, and
pGEX4T3 ear γ2-adaptin are transformed in *Escherichia coli* BL21(DE3)
and bacteria are grown at 37° on Luria–Bertani (LB) agar plates containing
50 μg/ml ampicillin. The next day a colony is transferred to 2.5 ml LB
containing 50 μg/ml ampicillin (medium) and grown for 5 h. The cultures
are subsequently diluted to 125 ml and grown overnight at 30°. The next
morning suspension cultures are diluted four times with medium and
maintained at 30° until OD_{600} reached 0.6. Freshly dissolved isopropylthiogalactoside (IPTG) is then added to 1 mM (final concentration) and the
cells are grown for 3 h at 30°. Bacteria are harvested by centrifugation
for 10 min at 5000 rpm in a Sorvall SLA-3000 rotor. Pellets are washed
in cold phosphate-buffered saline (PBS), combined, and recentrifuged for
10 min at 4500 rpm in a Beckman table centrifuge. Bacterial pellets are
snap-frozen in liquid nitrogen and stored at −80°. All steps of this protocol
are performed at 4°. Glutathione (GSH) Sepharose beads are washed twice
with Buffer B and twice with Buffer C before use in protein isolation.
Bacterial pellets corresponding to 2 liter culture volume are resuspended in
20 ml Buffer A containing 2 mg/ml lysozyme and kept for 15 min on ice.
The suspension is then diluted with 80 ml Buffer A containing protease
inhibitors (5 μg/ml leupeptin, 10 μg/ml aprotinin, 1 μg/ml pepstatin, 1 mM
PMSF), sonicated two times 1 min on ice, and centrifuged for 15 min at
15,000 rpm in a Sorvall SS-34 rotor. Fresh protease inhibitors and 1%
Triton X-100 (final concentration) are added to the supernatant, which is
subsequently incubated with 1.5 ml washed GSH beads by end-over-end
rotation for 2 h. Beads are washed with Buffer C and four times with Buffer
B in batch mode. Fusion protein is then eluted by end-over-end rotation
with 1.5 ml Buffer D. After 30 min, beads are centrifuged and supernatant
containing fusion protein is saved. The last incubation is repeated
once with the same volume of Buffer D, and the eluates are combined,
aliquotted, snap-frozen in liquid nitrogen, and stored at −80°.

*Preparation of Pig Brain Extract and Binding Assay with GST-Ear
Adaptins.* All steps of this protocol are performed at 4°. Fresh pig brain is
obtained from a local slaughterhouse and 100 g of brain is homogenized with
a Waring blender at maximal speed in two volumes phosphate-buffered
saline (PBS) containing 5 μg/ml leupeptin, 10 μg/ml aprotinin, 1 μg/ml pepstatin, and 1 mM PMSF (protease inhibitors). NP-40 is added to 0.1% (final
concentration) and the mixture is stirred slowly on ice. After 30 min, the
detergent extract is centrifuged at 10,000 rpm in a Sorvall SLA1500 rotor
for 40 min. The supernatant is subsequently retrieved and recentrifuged
for 60 min at 33,000 rpm in a Beckman Ti45 rotor. The pellet is discarded
and the supernatant is snap-frozen in liquid nitrogen and stored at −80°.

For binding experiments, brain extract is thawed on ice and centrifuged for 15 min at 50,000 rpm in a Beckman TLA100.2 rotor. Aliquots (7.5 mg protein) of extract are precleared by end-over-end incubation with 100 μg GST and 200 μl GSH beads. In the meantime GST adaptin ear fusion proteins are defrosted and centrifuged as described earlier. In a typical binding assay 200 μl precleared brain extract (1.5 mg protein) is incubated with 20 μg clarified fusion protein and 60 μl GSH beads by end-over-end rotation. After 2 h, the beads are washed five times with PBS containing 0.1% NP-40. Bound material is boiled off the beads in 20 μl reducing Laemmli buffer. The proteins are resolved on a 10% sodium dodecyl sulfate–polyacrylic acid (SDS-PAA) gel and transferred to a PVDF membrane for 1.5 h at 225 mA. Bound rabaptin-5α is then assayed by Western blot using a rabbit antibody as previously described (van der Sluijs *et al.*, 2000).

Comment. We found that rabaptin-5α binds equally well to the ear domains of both γ1-adaptin and γ2-adaptin. The γ2-adaptin variant, however, appears to have a more distinct function than γ1-adaptin or is expressed in different tissues because a γ1-adaptin knockout cannot be compensated for by γ2-adaptin (Meyer *et al.*, 2000).

Binding Assay of AP-1 to Recombinant Rabaptin-5α Hinge Fragment

To investigate whether other subunits of AP-1 bind to rabaptin-5α, we slightly modified the biochemical binding assay. This alternative binding assay capitalizes on the observation made in the two-hybrid assay that the hinge region (aa 301–592) of rabaptin-5α contains the information required to interact with γ1-adaptin. In contrast to full-length rabaptin-5α, which expresses poorly in *E. coli*, the hinge fragment can be obtained readily as a GST fusion protein (Fig. 1D).

Preparation of HeLa Cell Lysate and Binding Assay. pRP269 and pRP269-rabaptin-5α(301–592) are transformed in *E. coli* BL21(DE3) and proteins are isolated and stored as described previously for the adaptin ear fusion proteins. Five confluent 10-cm dishes of HeLa cells are washed twice with 10 ml ice-cold PBS and lysed in 1 ml PBS containing 0.5% NP-40 for 30 min on a rocking platform. Dishes are scraped and lysates are shaken for 20 min in a cold room and centrifuged for 15 min at 14,000 rpm in a microcentrifuge. The HeLa cell lysate (5 ml) is precleared for 2 h by incubation with 200 μl GSH beads and 100 μg GST. Precleared lysate is divided in two equal aliquots and incubated with 20 μg fusion protein and 60 μl GSH beads for 2 h under rotation. The beads are washed three times with 1 ml PBS containing 0.1% NP-40. Bound proteins are eluted in 70 μl reducing Laemmli buffer, resolved on 7.5% SDS-PAA gels, and transferred to a PVDF membrane for 50 min at 400 mA. Bound AP-1 subunits are analyzed by Western blot using antibodies against γ1-adaptin (Sigma) and σ1-adaptin (generously provided by Linton Traub). Dilutions of antibodies are made in PBS containing 5% skimmed milk and 0.2% Tween-20.

Comments. To detect the small σ1-subunit and medium μ1-subunit, we use 12% (instead of 7.5%) SDS-PAA gels. Initially we could detect only γ1-adaptin and the σ1-subunit in the bound protein fraction (Deneka *et al.*, 2003a). More recently we found the entire set of AP-1 subunits in the eluted protein pool using a freshly prepared extract. Possibly this reflects differences between freshly prepared and stored detergent extracts.

Analysis of the Rabaptin-5α/AP-1 Complex by Fluorescence Microscopy

Transfection and Immunolabeling Protocol. The pcDNA3-rabaptin-5α, pcDNA3.1His-rabaptin-5α, pEYFP-Rab4, and pEGFP-Rab5 constructs used in the following experiments are generated as described before (Deneka *et al.*, 2003a). None of the available Rab4 and rabaptin-5α antibodies allows for the detection of the endogenous protein by immunofluorescence microscopy, thus requiring ectopic expression for this type of analysis. We generally employ HeLa cells as recipient since they can be easily transfected and contain a convenient number of transferrin (Tf) receptors that serve as markers for early endosomes. Typically, HeLa cells are grown to 40% confluency on 10-mm #1 coverslips in 6-cm dishes. Six microliters Fugene 6 (Roche) is mixed with 94 μl serum-free minimal essential medium (SFMEM). After 5 min at room temperature, 3 μg expression construct is added, and after 15 min the mixture is added to the cells. After 24 h transfected cells are processed for further analysis. Cells are washed once with PBS, fixed with 3% paraformaldehyde for 30 min, washed once with PBS, and incubated for 5 min with PBS containing 50 mM NH$_4$Cl. Fixed cells are washed once with PBS and incubated for 1 h in PBS containing 0.5% bovine serum albumin (BSA) and 0.1% saponin (block buffer). Incubation with primary antibodies is done for 1 h and is followed by three 5-min washes with block buffer. Staining with appropriately labeled secondary antibodies is done for 1 h and is followed by three 5-min washes with block buffer and two washes with PBS. Coverslips are mounted on objective glasses with 3 μl Moviol (Hoechst).

Assay for Rabaptin-5α and γ1-Adaptin Colocalization on Rab4-Containing Endosomes

HeLa cells transfected with pcDNA3.1His-rabaptin-5α are grown for 24 h on coverslips and incubated for 30 min in SFMEM, washed once with uptake medium (SFMEM, 20 mM HEPES, pH 7.4, 0.1% BSA), and incubated with 15 μg/ml human Alexa594-Tf (Molecular Probes) for 60 min at 37° in uptake medium. Cells are fixed for 15 min with 2.5% PFA in PBS at 37°, followed by a fixation for 30 min at room temperature. Cells are subsequently labeled with a rabbit antibody against rabaptin-5α followed by Alexa488-conjugated IgG. So far the antibodies against rabaptin-5α and γ1-adaptin do not allow for a triple label experiment. We therefore perform double label immunofluorescence in a parallel dish with cells that are not exposed to Alexa594-Tf, with a rabbit antibody against rabaptin-5α and a monoclonal antibody against γ1-adaptin. Staining is done with Alexa488 goat anti-rabbit IgG and Cy3 goat anti-mouse IgG (Fig. 2A and B). γ1-Adaptin localizes to both the *trans*-Golgi network (TGN) and endosomes, although we find γ1-adaptin and rabaptin-5α complex only on endosomes. This suggests that an endosome-specific factor is required for the unique localization of the complex, which could be either Rab5 or Rab4, of which rabaptin-5α is an effector. This can be tested in coexpression studies of either pEYFP-Rab4 or pEGFP-Rab5 in combination with pcDNA3-rabaptin-5α (Fig. 2C). Cells are then labeled separately with a monoclonal antibody against γ1-adaptin or a rabbit antibody against rabaptin-5α followed by secondary Cy3 goat anti-rabbit IgG or Cy3 goat anti-mouse IgG.

Comment. The enlarged endosomes that are seen by overexpression of rabaptin-5α look similar to those observed after overexpression of a GTP-hydrolysis-deficient Rab5 mutant (Stenmark *et al.*, 1995). Experiments with rabaptin-5α truncation mutants lacking the Rab5 binding site rule out an involvement of Rab5 (Deneka *et al.*, 2003a).

Assay for the Interaction of Rab4 with Endogenous Rabaptin-5α–AP-1 Complex

To investigate whether Rab4 can interact with a rabaptin-5α–AP-1 complex we use pulldown assays with recombinant GST-Rab4 and brain cytosol (Fig. 2D).

Buffers

> Nucleotide exchange buffer (NE buffer): 20 mM HEPES, pH 7.5, 100 mM NaCl, 10 mM EDTA, 5 mM MgCl$_2$, 1 mM DTT
> Nucleotide stabilization buffer (NS buffer): 20 mM HEPES, pH 7.5, 100 mM NaCl, 5 mM MgCl$_2$, 1 mM DTT
> Wash buffer: 20 mM HEPES, pH 7.5, 250 mM NaCl, 1 mM DTT
> Elution buffer: 20 mM HEPES, pH 7.5, 1.5 M NaCl, 20 mM EDTA, 1 mM DTT, 5 mM GDP

Isolation of GST-Rab Fusion Protein and Guanine Nucleotide Loading. The expression constructs Rab4-pGEX2T, Rab5-pGEX1λT, and

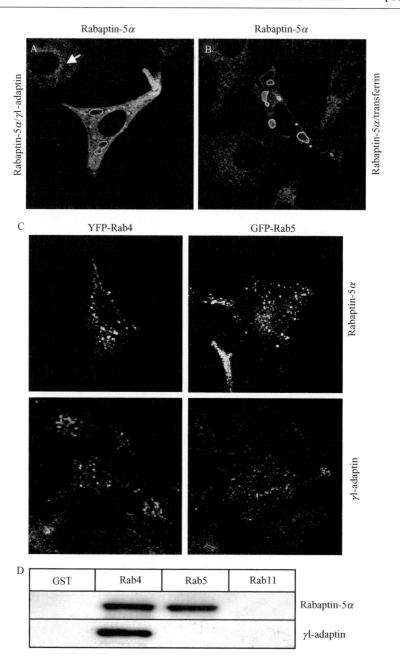

FIG. 2. Rab4-dependent localization of the rabaptin-5α/AP-1 complex to endosomes. (A) HeLa cells were transfected with pcDNA3.1His-rabaptin-5α. The cells are labeled for rabaptin-5α (green) and endogenous γ1-adaptin (red). Note the colocalization of rabaptin-5α

Rab11-pGEX5X3 have been described (Deneka *et al.*, 2003a), and expression of GST, GST-Rab4, GST-Rab5, and GST-Rab11 in *E. coli* is done as for the GST ear adaptin constructs. Bacterial pellets of 1-liter cultures are resuspended in 20 ml PBS containing 5 m*M* MgCl$_2$, 5 m*M* α-mercaptoethanol, 2 mg/ml lysozyme, 10 μg/ml RNase, 10 μg/ml DNase, protease inhibitors (cf. above), and 0.2 m*M* GDP. Lysates are sonicated two times 1 min on ice and centrifuged for 1 h at 33,000 rpm in a Beckman Ti45 rotor. The supernatants are incubated with 0.6 ml GSH beads (washed three times with PBS, 5 m*M* MgCl$_2$, and 5 m*M* α-mercaptoethanol prior to use) for 2 h by end-over-end rotation. Beads are washed four times with PBS, 5 m*M* MgCl$_2$, 5 m*M* α-mercaptoethanol, and 0.1 m*M* GDP, and stored overnight at 4° in PBS, 5 m*M* MgCl$_2$, 5 m*M* α-mercaptoethanol, 0.1 m*M* GDP, and 0.02% sodium azide. To establish the fusion protein concentration on the beads, 10-μl beads are boiled in 50 μl reducing Laemmli buffer and analyzed on a 12.5% SDS-PAA gel that is also loaded with a concentration range of BSA. This also allows for normalization of the amount of beads to be used in binding experiments. Empty GSH beads are added to have the same amount of beads in the assay. Beads are washed with NE buffer containing 10 μ*M* GTPγS, and incubated for 30 min under rotation with NE buffer containing 1 m*M* GTPγS at room temperature. The wash and incubation steps are repeated twice to ensure optimal nucleotide loading. Beads were washed once with NS buffer containing 10 μ*M* GTPγS and further incubated with NS buffer containing 1 m*M* GTPγS for 20 min at room temperature and immediately used in binding assays.

Pig Brain Cytosol Preparation and Binding Assay

Pig brain (200 g) is homogenized on ice (as described earlier for the preparation of brain extract) in 350 ml NS buffer containing 1 m*M* DTT and protease inhibitors. The homogenate is centrifuged for 40 min at

---

and γ1-adaptin in transfected cells and the distinct localization of γ1-adaptin in nontransfected cells (arrow). (B) HeLa cells were incubated with Alexa594-Tf for 60 min at 37° and subsequently labeled with anti-rabaptin-5α antibody and Alexa488-IgG, showing that the rabaptin-5α/AP-1 complex localizes to early endosomes. (C) Localization of γ1-adaptin to endosomes depends on Rab4. HeLa cells expressing His$_6$-rabaptin-5α in combination with YFP-Rab4 or GFP-Rab5 and labeled with anti-γ1-adaptin antibody and anti-rabaptin-5α antibody, detected with Cy3-labeled secondary antibodies. Note that GFP-Rab5 colocalizes with rabaptin-5α, but not with γ1-adaptin. (D) Rab4 recruits rabaptin-5α/AP-1 complex. GST-Rab4, GST-Rab5, and GST-Rab11 were isolated on GSH beads, loaded with GTPγS and incubated with pig brain cytosol. Bound fractions were immunoblotted with antibodies against rabaptin-5α and γ1-adaptin. Note that GST-Rab4 and GST-Rab5 both retrieve rabaptin-5α, but that only GST-Rab4 bound material includes γ1-adaptin (Adapted with permission from Denaka *et al.*, 2003d). (See color insert.)

10,000 rpm in a Sorvall SLA1500 rotor, and the supernatant is recentrifuged for 1 h at 33,000 rpm in a Beckman Ti45 rotor. The cytosol is dialyzed overnight against NS buffer to remove free nucleotides, and used immediately in binding experiments. Beads (250 μl) (corresponding to 2 mg protein) are rotated with 8 ml cytosol for 2.5 h at 4° in the presence of 100 μM GTPγS. Beads are next washed twice with 2.5 ml NS buffer in the presence of 10 μM GTPγS, twice with 2.5 ml NS containing 250 mM NaCl, 10 μM GTPγS, and finally washed once with 1.5 ml wash buffer. Bound proteins are eluted with 150 μl elution buffer for 30 min at room temperature. Eluates are either resolved immediately on 10% and 12% SDS-PAA gels and analyzed by Western blot with antibodies against rabaptin-5α and γ1-adaptin or snap-frozen and stored at −80° until further analysis.

Comments. Instead of the nonhydrolyzable GTP analog GTPγS, we now routinely use guanosine-5'-(β,γ-imido)triphosphate (GMP-PNP) and obtain the same results with the considerably cheaper alternative. The beads with bound GST-Rab proteins can be stored several days at 4°. In our hands, best results are obtained when storage time is kept to a minimum.

Analysis of Rabaptin-5α Function in Recycling from Endosomes

To determine the role of rabaptin-5α, we analyze transferrin recycling using a pulse–chase protocol in cells transfected with rabaptin-5α. In principle it is also possible to perform functional experiments in HeLa cells in which rabaptin-5α is knocked down with RNAi. Rab5-mediated endocytosis is, however, critically dependent on rabaptin-5α, and knock down of the effector precludes the analysis of recycling since insufficient amounts of transferrin are internalized (Deneka *et al.*, 2003a) (Fig. 3).

Pulse–Chase Protocol for Fluorescently Labeled Transferrin. HeLa cells on coverslips are transfected with pcDNA3.1His-rabaptin-5α alone or in combination with pcDNA3-γ1-adaptin (1.5 μg DNA of each/6-cm dish). After 24 h, cells are depleted of endogenous transferrin during a 30-min incubation in bicarbonate, free SFMEM, 20 mM HEPES, pH 7.4, 0.1% BSA (uptake medium). The cells are then incubated with uptake medium containing 15 μg/ml Alexa488-transferrin (single transfectants) or Alexa594-transferrin (double transfectants) at 16°, which causes accumulation of the endocytic tracer in early endosomes (de Wit *et al.*, 2001). Medium is aspirated after 30 min, and cells are briefly washed with 16° uptake medium. Recycling is initiated by adding 37° uptake medium containing 100 μM desferal. The iron chelator serves to inhibit reinternalization of recycled transferrin. After different periods of time, the cells are fixed and processed for double label fluorescence microscopy with anti-Xpress (against His$_6$ rabaptin) followed by Cy3-labeled anti-mouse IgG (single transfectants) or

Alexa488-labeled anti-mouse IgG (double transfectants). A semiquantitative estimate of the effect of rabaptin-5 on Tf kinetics can be made by selecting five rabaptin-5α-positive endosomes (per cell and per chase time) and count how many of those contain Tf. This analysis is performed on 10 cells per time point and in duplicate transfections.

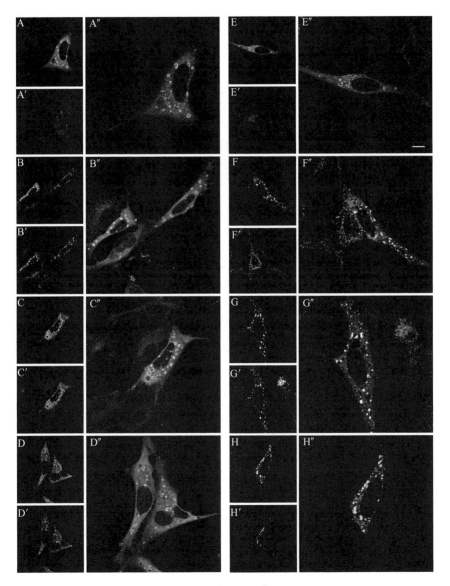

FIG. 3. (*continued*)

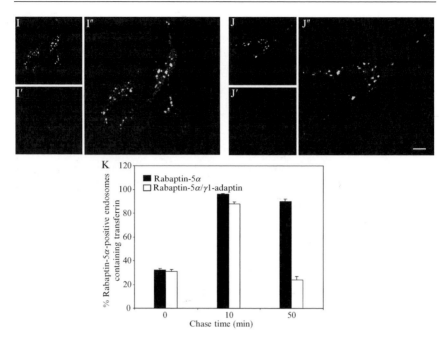

FIG. 3. Transfected rabaptin-5α delays transferrin recycling. HeLa cells were transfected with pcDNA3.1 His-rabaptin-5α (A–E) alone or in combination with pcDNA3-γ1-adaptin (F–J). The cells were incubated with 15 μg/ml Alexa488-Tf (A–E) or Alexa594-Tf (F–J) at 16° for 30 min, subsequently chased at 37° for 0 min (A′, F′), 10 min (B′,G′), 20 min (C′, H′), 40 min (D′, I′), and 50 min (E′, G′) and labeled with anti-Xpress followed by Cy3-labeled anti-mouse IgG (A–E) or Alexa488-labeled anti-mouse IgG (F–J). Merged images are shown in A″–J″. Bar = 10 μm. (K) Quantitation of the fraction of rabaptin-5α-positive endosomes containing Alexa-Tf at 0, 10, and 50 min of chase. Error bars denote the standard deviation ($n = 10$). Note that entry of Alexa-Tf into early endosomes is the same in single rabaptin-5α transfectants as in the double transfectants, showing that Rab5-dependent internalization is not affected. Recycling of Alexa488-Tf in the pcDNA3.1 His-rabaptin-5α transfectants, however, is strongly diminished compared to pcDNA3.1 His-rabaptin-5α/pcDNA3-γ1-adaptin transfected (or control) cells. Others reported a similar negative regulatory function of rabaptin-5 in an *in vitro* endosome recycling assay (Pagano *et al.*, 2004) (Adapted with permission from Denaka *et al.*, 2003a). (See color insert.)

Comments. We perform the pulse on a 16° waterbath and the chase on a 37° waterbath. This allows for better temperature control and allows precise chase times to be obtained, which is especially important for short periods of chase.

Concluding Remarks

Rabaptin-5 and rabaptin-5α belong to a rapidly expanding group of accessory proteins that bind with low affinity to the heterotetrameric AP-1 and monomeric GGA adaptors (Hirst *et al.*, 2001; Mattera *et al.*, 2003; Mills *et al.*, 2003; Neubrand *et al.*, 2005; Ritter *et al.*, 2004; Shiba *et al.*, 2002). It is reasonable to assume that they assist in AP-1 and GGA-dependent formation of clathrin-coated vesicles from intracellular membranes. To understand, the precise mechanism of their activity, however, needs further investigations, especially since some of these acessory proteins also bind to the plasma membrane adaptor AP-2. The indirect link between the active form of the small GTPase Rab4 and AP-1 suggests an extra layer of control of AP-1 function and testifies to the complexity of the mechanisms controlling membrane dynamics between endosomes, the plasma membrane, and the TGN. The assays described in this chapter provide tools that might contribute to our understanding of regulatory factors in transport pathways from endosomes.

Acknowledgments

    This research was supported by grants from NWO/MW and NWO-ALW (to PvdS). We thank Marc van Peski for help with preparation of the figures.

References

de Renzis, S., Sönnichsen, B., and Zerial, M. (2002). Divalent rab effectors regulate the sub-compartmental organization and sorting function of early endosomes. *Nature Cell Biol* **4**, 124–133.

de Wit, H., Lichtenstein, Y., Kelly, R. B., Geuze, H. J., Klumperman, J., and van der Sluijs, P. (2001). Rab4 regulates formation of synaptic-like microvesicles from early endosomes in PC12 cells. *Mol Biol Cell* **12**, 3703–3715.

Deneka, M., Neeft, M., Popa, I., van Oort, M., Sprong, H., Oorschot, V., Klumperman, J., Schu, P., and van der Sluijs, P. (2003a). rabaptin-5α/rabaptin-4 serves as a linker between rab4 and γ1-adaptin in membrane recycling from endosomes. *EMBO J* **22**, 2645–2657.

Deneka, M., Neeft, M., and van der Sluijs, P. (2003b). Rab GTPase switch in membrane dynamics. *CRC Rev Biochem Mol Biol* **38**, 121–142.

Fouraux, M., Deneka, M., Ivan, V., van der Heijden, A., Raymackers, J., van Suylekom, D., van Venrooij, W. J., van der Sluijs, P., and Pruijn, G. J. M. (2003). rabip4′ is an effector of rab5 and rab4 and regulates transport through early endosomes. *Mol Biol Cell* **15**, 611–624.

Gournier, H., Stenmark, H., Rybin, V., Lippe, R., and Zerial, M. (1998). Two distinct effectors of the small GTPase rab5 cooperate in endocytic membrane fusion. *EMBO J* **17**, 1930–1940.

Gruenberg, J. (2001). The endocytic pathway: a mosaic of domains. *Nature Revs Mol Cell Biol* **2**, 721–730.

Hirst, J., Lindsay, M. R., and Robinson, M. S. (2001). Golgi-localized, γ-ear-containing, ADP-ribosylation factor-binding proteins: roles of the different domains and comparison with AP-1 and clathrin. *Mol Biol Cell* **12**, 3573–3588.

Korobko, E., Kiselev, S., Olsnes, S., Stenmark, H., and Korobko, I. (2005). The rab5 effector rabaptin-5 and its isoform rabaptin-5δ differ in their ability to interact with the small GTPase rab4. *FEBS J* **272**, 37–46.

Korobko, E. V., Kiselev, S. L., and Korobko, I. V. (2002). Multiple rabaptin-5-like transcripts. *Gene* **292**, 191–197.

Mattera, R., Arighi, C. N., Lodge, R., Zerial, M., and Bonifacino, J. S. (2003). Divalent interaction of the GGAs with the rabaptin-5-rabex-5 complex. *EMBO J* **22**, 78–88.

Meyer, C., Zizioli, D., Lausmann, S., Eskelinen, E. L., Hamann, J., Saftig, P., von Figura, K., and Schu, P. (2000). μ1A-adaptin-deficient mice: lethality, loss of AP-1 binding and rerouting of mannose 6-phosphate receptors. *EMBO J* **19**, 2193–2203.

Mills, I. G., Praefcke, G. J. K., Vallis, Y., Peter, B. J., Olesen, L. E., Gallop, J. L., Butler, P. J. G., Evans, P. R., and McMahon, H. T. (2003). Epsin R: an AP1/clathrin interacting protein involved in vesicle trafficking. *J Cell Biol* **160**, 213–222.

Mohrmann, K., Leijendekker, R., Gerez, L., and van der Sluijs, P. (2002). Rab4 regulates transport to the apical plasma membrane in madin-darby canine kidney cells. *J Biol Chem* **277**, 10474–10481.

Nagelkerken, B., van Anken, E., van Raak, M., Gerez, L., Mohrmann, K., van Uden, N., Holthuizen, J., Pelkmans, L., and van der Sluijs, P. (2000). Rabaptin4, a novel effector of rab4a, is recruited to perinuclear recycling vesicles. *Biochem J* **346**, 593–601.

Neubrand, V. E., Will, R. D., Mobius, W., Poustka, A., Wiemann, S., Schu, P., Dotti, C. G., Pepperkok, R., and Simpson, J. C. (2005). gamma-BAR, a novel AP-1 interacting protein involved in post-Golgi trafficking. *EMBO J* **24**, 1122–1133.

Odorizzi, G., and Trowbridge, I. S. (1997). Structural requirements for basolateral sorting of the human transferrin receptor in the biosynthetic and endocyte compartments of MDCK cells. *J Cell Biol* **137**, 1255–1267.

Pagano, A., Crottet, P., Prescianotto-Baschong, C., and Spiess, M. (2004). In vitro formation of recycling vesicles from endosomes requires AP-1/clathrin and is regulated by rab4 and the connector rabaptin-5. *Mol Biol Cell* **15**, 4990–5000.

Ritter, B., Denisov, A. Y., Philie, J., Deprez, C., Tung, E. C., Gehring, K., and McPherson, P. S. (2004). Two WXXF based motifs in NECAPs define the specificity of accessory protein binding to AP-1 and AP-2. *EMBO J* **23**, 3701–3710.

Shiba, Y., Takatsu, H., Shin, H. W., and Nakayama, K. (2002). γ-adaptin interacts directly with rabaptin-5 through its ear domain. *J Biochem* **131**, 327–336.

Spaargaren, M., and Bischoff, J. R. (1994). Identification of the guanine nucleotide dissociation stimulator for ral as a putative effector molecule for r-ras, H-ras, K-ras, Rap. *Proc Natl Acad Sci USA* **91**, 12609–12613.

Stenmark, H., Vitale, G., Ullrich, O., and Zerial, M. (1995). Rabaptin-5 is a direct effector of the small GTPase rab5 in endocytic membrane fusion. *Cell* **83**, 423–432.

Stoorvogel, W., Oorschot, V., and Geuze, H. J. (1996). A novel class of clathrin coated vesicles budding from endosomes. *J Cell Biol* **132**, 21–34.

Sönnichsen, B., De Renzis, S., Nielsen, E., Rietdorf, J., and Zerial, M. (2000). Distinct membrane domains on endosomes in the recycling pathway visualized by multicolor imaging of rab4, rab5, and rab11. *J Cell Biol* **149**, 901–913.

van der Sluijs, P., Mohrmann, K., Deneka, M., and Jongeneelen, M. (2000). Expression and properties of rab4 and its effector rabaptin-4 in endocytic recycling. *Meth Enzymol* **329**, 111–119.

Vitale, G., Rybin, V., Christoforidis, S., Thornqvist, P. O., McCaffrey, M., Stenmark, H., and Zerial, M. (1998). Distinct rab-binding domains mediate the interaction of rabaptin-5 with GTP-bound rab4 and rab5. *EMBO J* **17**, 1941–1951.

Zerial, M., and McBride, H. (2001). Rab proteins as membrane organizers. *Nature Rev Mol Cell Biol* **2**, 107–117.

[47]   Measurement of the Interaction of the p85α Subunit of Phosphatidylinositol 3-Kinase with Rab5

By M. DEAN CHAMBERLAIN and DEBORAH H. ANDERSON

Abstract

During endocytosis of the activated platelet-derived growth factor (PDGF) receptor, phosphatidylinositol 3-kinase (PI3K) remains associated with the receptor. We found that the p85α subunit of PI3 kinase binds directly to Rab5 and possesses GTPase activating protein (GAP) activity toward Rab5. Rab5 is a small monomeric GTPase involved in regulating vesicle fusion events during receptor-mediated endocytosis. We used two methods to characterize the direct binding between Rab5 in various nucleotide-bound states and the p85 protein. In the first assay, the ability of p85 to bind to Rab5 is measured using an enzyme-linked immunosorbent assay (ELISA). The second assay is a glutathione *S*-transferase (GST) pull-down approach in which GST-Rab5 proteins in various nucleotide-bound states are allowed to bind p85. In both instances, bound p85 is detected using anti-p85 antibodies.

Introduction

Down-regulation of signal transduction pathways activated by receptor tyrosine kinases, such as the platelet-derived growth factor (PDGF) receptor, involves endocytosis of the activated receptor complex and receptor degradation (Kapeller *et al.*, 1993; Sorkin and Waters, 1993). Receptor-mediated endocytosis involves multiple vesicle fusion events that effectively deliver the receptor-signaling complex to the early endosome. This complex is then disassembled and the receptor is either recycled back to the plasma membrane or sorted to the late endosome and lysosome for degradation (Nilsson *et al.*, 1983; Rosenfeld *et al.*, 1984).

Rab5 is a small monomeric GTPase involved in early endosomal fusion events such as the fusion of early endocytic vesicles (containing activated receptors undergoing endocytosis) with the early/sorting endosomes (Armstrong, 2000; Stenmark and Olkkonen, 2001). During endocytosis of the activated PDGF receptor, a class-I phosphatidylinositol 3-kinase (PI3K) consisting of an 85-kDa regulatory subunit (p85α) and a 110-kDa catalytic subunit (p110α or β) remains associated with the receptor (Kapeller *et al.*, 1993). PI3K activity is important for activating downstream signaling pathways such as the PDK1/Akt cell survival pathway (Cantley, 2002). Results

METHODS IN ENZYMOLOGY, VOL. 403   0076-6879/05 $35.00
                   DOI: 10.1016/S0076-6879(05)03047-8

from our laboratory now show that the p85 protein can also interact directly with Rab5 and stimulate Rab5 GTPase activity (Chamberlain *et al.*, 2004). This chapter describes the methods used to characterize the direct binding between Rab5 and the p85 protein.

Expression and Purification of Rab5, Glutathione S-Transferase (GST), and p85 Proteins

GST-Rab5 Wild Type and Mutants GST-Rab5Q79L and GST-Rab5S34N

Plasmids encoding human Rab5 (wild-type and mutants Q79L and S34N) were generously provided by Dr. G. Li (University of Oklahoma [Liu and Li, 1998]). The Q79L mutation inactivates the Rab5 GTPase activity such that this mutant is locked in a GTP-bound active conformation. The S34N mutation causes Rab5 to preferentially bind GDP such that this mutant is in a GDP-bound inactive conformation. Each of these Rab5 inserts was provided in a pGEX3X plasmid (Amersham Biosciences), and therefore, induction of protein expression resulted in the production of GST-Rab5 fusion proteins. A detailed protocol for the induction of the GST-Rab5 proteins, isolation on glutathione–Sepharose beads, and factor Xa cleavage from GST appears in Anderson and Chamberlain, Chapter 48 (this volume).

For the enzyme-linked immunosorbent assay (ELISA)-based binding assay, the Rab5 protein is further purified using gel filtration chromatography on a Sephacryl HR-200 column in phosphate buffer (50 mM NaPO$_4$, pH 7.0, 150 mM NaCl). Aliquots of the resulting column fractions are resolved by sodium dodecyl sulfate–polyacrylamide gel electrophoresis (SDS-PAGE) (15%) and stained with Coomassie blue. Fractions containing pure Rab5 are pooled, concentrated, and buffer exchanged into Rab5 storage buffer (50 mM HEPES, pH 7.5, 50 mM NaCl, 1 mM EDTA) using an Amicon ultracentrifugal filter (molecular weight cutoff of 5 kDa). The concentration of the pure Rab5 protein is determined using a Bradford assay, with a bovine serum albumin (BSA) standard curve. Glycerol is added to 20% and Rab5 samples are flash frozen in aliquots and stored at −80° (Fig. 1, lane 2).

GST Protein

GST protein (BL21/pGEX2T; Amersham Biosciences) is induced (100 ml) and purified on glutathione-Sepharose beads (100 μl of 50% slurry), as described for GST-Rab5 (Anderson and Chamberlain, Chapter 48,

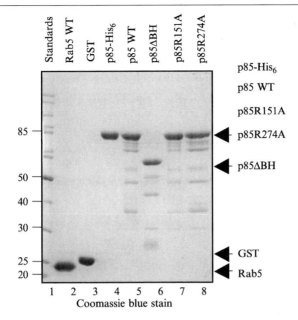

FIG. 1. SDS-PAGE (12%) and Coomassie blue staining of purified proteins (1 μg each) used for binding experiments. GST-Rab5 wild=type (WT) was isolated on glutathione Sepharose beads, cleaved from GST with factor Xa. The resulting Rab5 protein was further purified by gel filtration chromatography. GST protein was purified using glutathione Sepharose beads and eluted off the beads. GST-p85-His$_6$ was bound to glutathione-Sepharose beads, cleaved from GST with PreScission protease, and further purified by binding and elution from Talon beads. GST-p85 wild type (WT) and mutants GST-p85ΔBH, GST-p85R151A, and GST-p85R274A were isolated on glutathione-Sepharose beads and cleaved from GST with thrombin. The sizes of the molecular weight markers (Standards; Fermentas) are 200, 150, 120, 100, 85, 70, 60, 50, 40, 30, 25, and 20 kDa.

this volume). The GST protein is eluted (twice) from the beads by incubating with 500 μl reduced glutathione (15 mM) in 10 mM Tris–HCl, pH 8.0, for 10 min at room temperature. The beads are centrifuged at 2000×g and the supernatant containing the GST protein is concentrated using an Amicon ultracentrifugal filter (molecular weight cutoff of 5 kDa). Protein concentration is determined using a Bradford assay. Samples are aliquoted, flash frozen, and stored at −80° (Fig. 1, lane 3).

p85 Wild Type, and Mutants p85ΔBH, p85R151A, and p85R274A

The cDNA encoding full-length wild-type bovine p85α was subcloned into pGEX2T (Amersham Biosciences) (King *et al.*, 2000) and is expressed in BL21 cells, as described for pGEX3X-Rab5 (Anderson and Chamberlain,

Chapter 48, this volume). The p85ΔBH mutant was generated by piecing together the p85 sequences encoding the p85 SH3 domain (amino acids 1–83) and the p85 (N+C)SH2 (amino acids 314–724) as described (Chamberlain *et al.*, 2004). The point mutations (R151A and R274A) were generated using the Quikchange mutagenesis method (Stratagene). DNA sequencing to ensure that the inserts did not contain additional mutations verified all p85 sequences. Both the p85ΔBH and the p85R274A mutant proteins possess little or no Rab5 GAP activity, whereas the p85R151A protein has 65% of the Rab5 GAP activity of the wild-type p85 protein (Chamberlain *et al.*, 2004).

Induction (1 liter) and isolation of GST-p85 proteins on glutathione-Sepharose beads are carried out as described for GST-Rab5 proteins (Anderson and Chamberlain, Chapter 48, this volume). The p85 proteins are cleaved from GST using 100 NIH units of thrombin (Sigma T6884) in a total volume of 1 ml phosphate buffered saline (PBS) (137 mM NaCl, 2.7 mM KCl, 4.3 mM sodium phosphate, 1.4 mM potassium phosphate, pH 7.3) overnight at room temperature. The supernatant and three washes are pooled and buffer exchanged into protein storage buffer (20 mM Tris–HCl, pH 8, 100 mM NaCl, 1 mM EDTA, 1 mM dithiothreitol [DTT]) using an Amicon ultracentrifugal filter (molecular weight cut-off of 10 kDa). The concentrations of the pure p85 proteins are determined using a Bradford assay. Glycerol is added to 10% and p85 samples are flash frozen in aliquots and stored at −80° (Fig. 1, lanes 5–8).

For the ELISA-based binding assay, wild-type p85 is expressed from a modified pGEX6P1 plasmid (Amersham Biosciences) such that it contains both an N-terminal GST-tag and a C-terminal His$_6$-tag. This dual-tagged expression system allows for the purification of p85-His$_6$ protein containing no residual GST proteins. This is accomplished by isolating the GST-p85-His$_6$ on glutathione-Sepharose beads, cleaving the p85-His$_6$ protein from the GST sequences using PreScission (Amersham Biosciences), and further purifying the p85-His$_6$ protein on Talon beads (Clontech).

The altered pGEX6P1 plasmid, pGEX6P-His, was generated by inserting a pair of oligonucleotides encoding six histidine residues and a stop codon, between the *Eco*RI and *Sal*I sites such that the sequence of this region is now (5′-GAA TTC CAT CAT CAC CAT CAC CAT TGA GTC GAC-3′). The p85 encoding insert (not including the stop codon) was amplified by polymerase chain reaction and subcloned into the *Bam*HI and *Eco*RI sites of pGEX6P-His.

The expression of GST-p85-His$_6$ is induced (1 liter) and bound to glutathione-Sepharose beads (Anderson and Chamberlain, Chapter 48, this volume). The beads are washed three times in 20 ml PreScission buffer (50 mM Tris–HCl, pH 7.0, 150 mM NaCl, 1 mM EDTA, 1 mM DTT). The p85-His$_6$ protein is cleaved from GST in a total volume of 1 ml PreScission buffer with 35 units of PreScission protease (Amersham Biosciences) for 16 h at +4°. The supernatant from the cleavage reaction and four washes (1 ml each in PreScission buffer) are pooled. The EDTA and DTT present in the PreScission buffer are not compatible with the His$_6$-tag-binding Talon beads (Clontech) used next. Therefore, the p85-His$_6$ protein samples are buffer exchanged into Talon extraction/wash buffer (50 mM NaPO$_4$, 300 mM NaCl) using an Amicon ultracentrifugal filter (molecular weight cut-off of 10 kDa). The p85-His$_6$ protein is bound to 1 ml of a 50% slurry of Talon beads that have been previously washed three times in Talon extraction/wash buffer. Samples are mixed end over end for 20 min at room temperature. Beads are recovered by centrifugation at $800 \times g$ for 5min and washed in 20 ml Talon extraction/wash buffer three times, recovering the washed beads by centrifugation as before. The Talon beads are transferred into a small column using additional Talon extraction/wash buffer. The p85-His$_6$ protein is eluted from the beads by the addition of 1 ml of Talon elution buffer (50 mM NaPO$_4$, 300 mM NaCl, 150 mM imidazole) three times. The p85-His$_6$ protein is diluted in Talon extraction/wash buffer and concentrated using a new Amicon ultracentrifugal filter (molecular weight cutoff of 10 kDa). The protein concentration is determined using a Bradford assay. Glycerol is added to 20% and p85-His$_6$ samples are flash frozen in aliquots and stored at $-80°$ (Fig. 1, lane 4).

ELISA-Based Binding Assay

A simple and sensitive method to assay the direct binding between purified Rab5 and p85 was developed using an ELISA-based procedure. The Rab5 protein is preloaded with different nucleotides and bound to an ELISA plate. Increasing concentrations of p85 protein are added and bound p85 protein is quantified using an anti-p85 antibody.

Nucleotide Loading of the Rab5 Protein

Rab5 is diluted to 40 μg/ml in PBS + 2 mM EDTA for 20 min at room temperature to remove bound nucleotide. To prepare Rab5 with no bound nucleotide, an equal volume of 20 mM MgCl$_2$ in PBS is added. To prepare Rab5 with bound GDP, an equal volume of 20 mM MgCl$_2$ + 200 μM GDP (Sigma) in PBS is added. To prepare Rab5 with bound GTPγS (a non-hydrolyzable form of GTP), an equal volume of 20 mM MgCl$_2$ + 200 μM GTPγS (Sigma) in PBS is added. Each is incubated for a further 20 min at room temperature to load each of the nucleotides onto the Rab5 protein.

General. Immulon 4 flat-bottomed 96-well ELISA plates (VWR) are used. All incubations are carried out at room temperature, unless otherwise indicated, in a humidified plastic container (i.e., damp paper towels in the bottom of a plastic container with a lid). Determinations are carried out at least in duplicate. The wells around the perimeter of the plate are avoided since they show larger well-to-well differences in absorbance (at 450 nm) even when empty (0.036–0.042), as compared to the interior wells (0.036–0.039).

1. Bind Rab5 protein that has been preloaded with nucleotide to 12 wells for each test protein (20 μg/ml in PBS, 50 μl per well; 1 μg total) overnight is best, but 2 h may suffice. For controls, prepare four wells with just PBS.

2. Empty out wells into sink and bang out excess liquid.

3. *Blocking*: Fill each well with blocking buffer (5% Carnation skim milk, 0.2% Tween-20 in PBS) and incubate for 2 h.

4. Empty blocking as in step 2 and wash four times with distilled water. To *"wash"* fill a plastic basin one-half to three-quarters full of distilled water, and with the ELISA plate at a 45° angle, submerge the plate to fill the wells with water. Repeat, tipping a different edge of the plate down each time until all four edges have been done. Be sure to bang out the excess water onto a paper towel.

5. *Add p85-His$_6$ test protein-containing solution*: Make duplicate wells (50 μl per well) and use 2-fold serial dilutions of protein in blocking (e.g., 250, 125, 62.5, 31.25, 15.6 nM), as well as duplicate 0 nM wells. A set using GST at the same concentrations is also made (to control for background nonspecific binding). As an additional control, the highest concentration of p85-His$_6$ or GST test protein (e.g., 250 nM) is added to two wells each that received PBS instead of Rab5 protein in step 1. Incubate for 2 h.

6. Wash 16 times in distilled water: four times for one leading edge, then change the water and repeat for each of the four edges of the plate. Bang out the excess water as before.

7. *Add primary antibody—anti-p85* (affinity purified rabbit antibody directed against p85 amino acids 314–724; King *et al.* [2000]) for wells where p85-His$_6$ has been added (including 0 nM wells that did not receive any test protein), 0.25 μg/ml in blocking buffer; 100 μl per well. *Or add anti-GST* (Santa Cruz Biotechnology sc-138 mouse mAb) for wells where GST has been added (including 0 nM wells), 0.0625 μg/ml in blocking buffer; 100 μl per well. Incubate for 1 h.

8. *Washes* as in step 6.

9. *Add secondary antibody—for anti-p85 wells*, anti-rabbit horseradish peroxidase (Santa Cruz Biotechnology sc-2004), 0.125 μg/ml in blocking buffer; 100 μl per well. *Or for anti-GST wells*, add anti-mouse horseradish peroxidase (Santa Cruz Biotechnology sc-2005), 0.25 μg/ml in blocking buffer; 100 μl per well. Incubate for 1 h.

10. *Washes* as in step 6.

11. *TMB reaction*: Add 100 μl per well of 3,3′5,5′-tetramethylbenzidine (TMB) (Kirkegaard & Perry Labs; mix equal volumes of two solutions immediately prior to use: TMB Peroxidase substrate & Peroxidase solution B). Incubate for 30 min.

12. Add 100 μl 1M H$_3$PO$_4$ per well. Read the absorbance at 450 nm.

Comments. Prior to carrying out this ELISA-based assay, several different anti-p85 antibodies were tested. Three commercial anti-p85 antibodies (Transduction Laboratories P13020, tested at 1 μg/ml; Upstate Biotechnology #05 212, tested at 1 μg/ml; Upstate Biotechnology #05 217, tested at 1:1000 dilution) also showed a good signal (i.e., ability to detect p85-His$_6$ protein bound to wells) and low background (did not nonspecifically bind to wells containing no p85-His$_6$ protein). For each of these mouse antibodies, anti-mouse horseradish peroxidase was used at 0.25 μg/ml. Of these three commercial anti-p85 antibodies, the Upstate Biotechnology #05 212 was the most sensitive in detecting p85-His$_6$ associated with Rab5 bound in the well.

After these initial tests for the suitability of the particular anti-p85 antibody, titrations of the primary (anti-p85, generated in our laboratory; 1, 0.5, 0.25, 0.125, 0.0625 μg/ml) and secondary (anti-rabbit horseradish peroxidase; 0.25, 0.125, 0.0625 μg/ml) antibody were carried out using p85-His$_6$ protein bound to the wells. The lowest concentrations of antibody that still gave maximal sensitivity were selected and these are the ones used in the above method. Similar antibody titrations are recommended if the mouse anti-p85 antibody from Upstate Biotechnology #05 212 and the mouse horseradish peroxidase-conjugated secondary antibody are to be used instead. Similar antibody titrations have been carried out for the anti-GST antibody (Fang *et al.*, 2002).

In the absence of Rab5 bound to the plate, the wells with added 250 nM p85-His$_6$ typically have an absorbance of 0.06–0.09, while those with added 250 nM GST show an absorbance of 0.05. When Rab5 proteins are bound to the plate, the "background absorbance" in the absence of p85-His$_6$ protein (i.e., 0 nM test protein in step 5) is relatively small (for Rab5 it is 0.06–0.07; for Rab5-GDP it is 0.09–0.11; for Rab5-GTPγS it is 0.11). The corresponding background absorbance in the absence of GST protein is 0.06–0.08. Each experiment is carried out at least twice, with duplicate determinations. After subtracting the "background absorbance" from each experimental value, these four (or more) results are averaged and plotted using Prism software (GraphPad Software, Inc., San Diego, CA) $\pm$ standard error from the mean (Fig. 2).

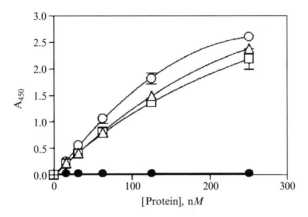

FIG. 2. ELISA-based binding assay showing that the p85α protein binds directly to Rab5, Rab5-GDP, and Rab5-GTPγS. Wells containing Rab5 in the absence of nucleotide (○, ●), Rab5-GDP (△), or Rab5-GTPγS (□) were blocked and incubated with increasing concentrations of purified p85 protein (○, △, □) or control GST protein (●). Bound p85 (or GST) protein was detected using anti-p85 (or anti-GST) antibodies, followed by a secondary antibody conjugated to horseradish peroxidase and quantified by measuring the absorbance at 450 nm of the acidified product. Reproduced with permission from Chamberlain *et al.* (2004).

Pull-Down Assay for GST-Rab5 and p85

A convenient method to detect the association between two proteins is described. This method is frequently used to map binding domains capable of mediating protein–protein interactions. The Rab5 protein is expressed in bacteria as a GST-Rab5 fusion protein and is immobilized on glutathione-Sepharose beads. The purified p85 protein is added to the immobilized GST-Rab5 and bound p85 protein is detected using an immunoblot analysis with anti-p85 antibodies.

Preparation of Immobilized GST Fusion Proteins

GST, GST-Rab5WT, GST-Rab5S34N, and GST-Rab5Q79L are each induced (50 ml culture each) and bound to glutathione-Sepharose beads (100 μl of a 50% slurry) as described (Anderson and Chamberlain, Chapter 48, this volume). The GST-fusion proteins are left immobilized on the beads (50 μl packed volume) in a total volume of 1 ml PBS. An aliquot (2 μl) of each suspension, together with known amounts of bovine serum albumin standards, is resolved by SDS-PAGE (12%) and stained with Coomassie blue, to assess the relative amounts of each GST-fusion protein (Fig. 3). If the relative amounts of the fusion proteins vary too much, the suspension volumes can be adjusted accordingly and a second gel can be

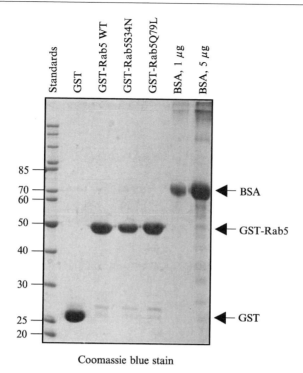

Coomassie blue stain

FIG. 3. SDS-PAGE (12%) and Coomassie blue staining of GST and GST-Rab5 fusion proteins that have been immobilized on glutathione-Sepharose beads in preparation for GST pull-down assays. Aliquots (2 μl) of the indicated immobilized GST, and GST-Rab5 fusion proteins before normalization. The sizes of the molecular weight markers (Standards; Fermentas) are 200, 150, 120, 100, 85, 70, 60, 50, 40, 30, 25, and 20 kDa.

used to normalize the amounts more accurately. Five micrograms of each GST-fusion protein is used in subsequent pull-down experiments.

GST Pull-Down Assay

1. An aliquot of suspension for each different immobilized GST-fusion protein (GST, GST-Rab5) corresponding to 5 μg is transferred to a 1.5-ml microcentrifuge tube. An additional 20 μl of 50% glutathione-Sepharose beads is added to ensure that there are sufficient beads to see easily during the wash steps. Beads are centrifuged at 2000×g for 1 min and all liquid is removed.

2. The GST-Rab5 proteins are loaded with nucleotide:

  a. To each sample, 500 μl Buffer A (20mM Tris–HCl, pH 7.5, 50 mM NaCl, 1 mM DTT, 5% glycerol, 0.1% Triton X-100, 10 μg/ml aprotinin, 10 μg/ml leupeptin) + 10 mM EDTA is added to first remove nucleotide for 20 min at room temperature. Beads are centrifuged at 2000×g for 1 min and all liquid is removed.

b. One of the following is then added:

i. For no nucleotide, 400 μl Buffer A + 10 mM EDTA.

ii. For Mg/GTPγS, 400 μl Buffer A + 10 mM MgCl$_2$ + 200 μM GTPγS.

iii. For Mg/GDP, 400 μl Buffer A + 10 mM MgCl$_2$ + 200 μM GDP.

iv. For Mg/GDP-AlF$_4^-$ (the transition state analogue), 400 μl Buffer A + 10 mM MgCl$_2$ + 1 mM AlCl$_3$ + 10 mM NaF + 1 mM GDP.

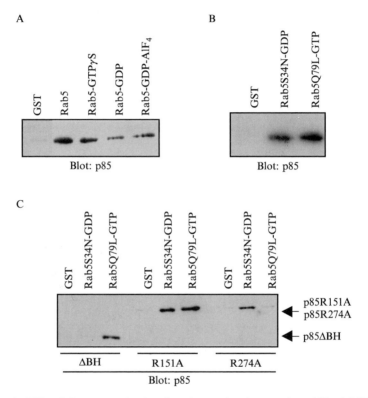

Fig. 4. GST pull-down assay showing direct interactions between immobilized GST-Rab5 proteins in various nucleotide bound states and purified p85. (A) GST and GST-Rab5 were immobilized on glutathione-Sepharose beads and loaded with the indicated nucleotide. Purified thrombin-cleaved wild-type p85 was added and bound p85 was detected after washing, using an immunoblot analysis with anti-p85 antibodies. (B) GST and GST-Rab5 mutants known to selectively bind GDP (S34N) or GTP (Q79L) were used in a pull-down assay with wild-type p85 protein. (C) The indicated p85 mutants (ΔBH, R151A, R274A) were tested for their abilities to bind to GST (control), GST-Rab5S34N-GDP, and GST-Rab5Q79L-GTP. Reproduced with permission from Chamberlain *et al.* (2004).

Samples are allowed to bind nucleotide for 30 min at room temperature. Beads are centrifuged at 2000×g for 1 min and all liquid is removed.

3. The p85 test protein (10 μg; Fig. 1, lanes 5–8) is added in a volume of 50 μl of the corresponding buffer used in step 2b + 1% Carnation skim milk powder. Samples are incubated for 1–2 h at +4°.

4. Beads are centrifuged at 2000×g for 1 min and washed three times with 500 μl wash buffer (50 mM Tris–HCl, pH 7.5, 150 mM NaCl, 1% Nonidet P-40).

5. Proteins associated with the washed beads are resolved by SDS-PAGE (10%), and the gel is soaked in transfer buffer (48 mM Tris, 39 mM glycine, 0.0375% SDS, 20% methanol) for 10 min. The proteins are transferred to a nitrocellulose membrane (VWR; Catalog #CA27376-991) using a semidry blotting apparatus at 400 mA constant current for 15 min.

6. The nitrocellulose is blocked in blotto (5% Carnation skim milk powder in TBST [100 mM Tris–HCl, pH 8, 150 mM NaCl, 0.05% Tween-20]) for 1 h at room temperature or overnight at +4°.

7. The blot is probed with anti-p85 antibody (1:200 in blotto; Upstate Biotechnology #05 217) for 1 h at room temperature and washed three times in TBST for 5 min each.

8. The membrane is next probed with anti-mouse horseradish peroxidase-conjugated secondary antibody (1:2000 in blotto; Santa Cruz Biotechnology sc-2005) for 1 h at room temperature and washed as before.

9. The nitrocellulose is incubated for 1 min in Western Lightning reagent (equal volumes of the two chemiluminescence reagents in PerkinElmer kit NEL105), wrapped in Saran Wrap, and exposed to X-Omat Blue XB-1 film (Kodak) for up to 5 min in a darkroom (Fig. 4).

Acknowledgments

We are grateful to G. Li and M. Waterfield for providing plasmids encoding the Rab5 and p85 proteins, respectively. We thank A. Hawrysh and K. James for expert technical assistance. M.D.C. is supported by a doctoral scholarship from the Canadian Institutes of Health Research Regional Partnership Program. This work was supported by grants from the Natural Sciences and Engineering Research Council and the Saskatchewan Cancer Agency Research Fund.

References

Armstrong, J. (2000). How do Rab proteins function in membrane traffic? *Int. J. Biochem. Cell Biol.* **32**, 303–307.

Cantley, L. C. (2002). The phosphoinositide 3-kinase pathway. *Science* **296**, 1655–1657.

Chamberlain, M. D., Berry, T. R., Pastor, M. C., and Anderson, D. H. (2004). The p85{α} subunit of phosphatidylinositol 3'-kinase binds to and stimulates the GTPase activity of Rab proteins. *J. Biol. Chem.* **279,** 48607–48614.

Fang, Y., Johnson, L. M., Mahon, E. S., and Anderson, D. H. (2002). Two phosphorylation-independent sites on the p85 SH2 domains bind A-Raf kinase. *Biochem. Biophys. Res. Commun.* **290,** 1267–1274.

Kapeller, R., Chakrabarti, R., Cantley, L., Fay, F., and Corvera, S. (1993). Internalization of activated platelet-derived growth factor receptor–phosphatidylinositol-3' kinase complexes: Potential interactions with the microtubule cytoskeleton. *Mol. Cell. Biol.* **13,** 6052–6063.

King, T. R., Fang, Y., Mahon, E. S., and Anderson, D. H. (2000). Using a phage display library to identify basic residues in A-Raf required to mediate binding to the Src homology 2 domains of the p85 subunit of phosphatidylinositol 3'-kinase. *J. Biol. Chem.* **275,** 36450–36456.

Liu, K., and Li, G. (1998). Catalytic domain of the p120 Ras GAP binds to Rab5 and stimulates its GTPase activity. *J. Biol. Chem.* **273,** 10087–10090.

Nilsson, J., Thyberg, J., Heldin, C. H., Westermark, B., and Wasteson, A. (1983). Surface binding and internalization of platelet-derived growth factor in human fibroblasts. *Proc. Natl. Acad. Sci. USA* **80,** 5592–5596.

Rosenfeld, M. E., Bowen-Pope, D. F., and Ross, R. (1984). Platelet-derived growth factor: Morphologic and biochemical studies of binding, internalization, and degradation. *J. Cell. Physiol.* **121,** 263–274.

Sorkin, A., and Waters, C. M. (1993). Endocytosis of growth factor receptors. *Bioessays* **15,** 375–382.

Stenmark, H., and Olkkonen, V. M. (2001). The Rab GTPase family. *Genome Biol.* **2,** Review S3007.

[48] Assay and Stimulation of the Rab5 GTPase by the p85α Subunit of Phosphatidylinositol 3-Kinase

By Deborah H. Anderson and M. Dean Chamberlain

Abstract

Rab5 is a small monomeric GTPase involved in regulating vesicle fusion events during receptor-mediated endocytosis. During endocytosis of the activated platelet-derived growth factor receptor, phosphatidylinositol 3-kinase (PI3K) remains associated with the receptor. We have found that the p85α subunit of PI3K binds directly to Rab5 and possesses GTPase-activating protein (GAP) activity toward Rab5. We describe two methods used to characterize the GAP activity of p85 toward the Rab5 protein. The first method is a steady-state GAP assay, used to show that the p85α protein has GAP activity toward Rab5. The second method is a single turnover GAP assay and measures changes in the catalytic rate of Rab5 GTP hydrolysis with or without the p85α protein.

METHODS IN ENZYMOLOGY, VOL. 403   0076-6879/05 $35.00
Copyright 2005, Elsevier Inc. All rights reserved.   DOI: 10.1016/S0076-6879(05)03048-X

Introduction

The Rab family of small monomeric G proteins contains more than 60 known family members and is involved in intracellular vesicle trafficking, specifically exocytosis and endocytosis (Deneka *et al.*, 2003; Maxfield and McGraw, 2004; Seabra and Wasmeier, 2004). These Rab proteins have two states: an active GTP-bound form and an inactive GDP-bound form. The best-studied example of Rab protein function is that of Rab5. In the inactive state, Rab5 is located in the cytosol and is bound to a guanine dissociation inhibitor (GDI) (Pfeffer and Aivazian, 2004). In the active state, Rab5 is associated with early endosomal membranes and recruits the early endosomal autoantigen 1 (EEA1) to the vesicle surface (Christoforidis *et al.*, 1999; Simonsen *et al.*, 1998). EEA1 tethers the membranes of two vesicles together by binding to Rab5-GTP on one vesicle and phosphatidylinositol 3-phosphate (PI3P) on the other vesicle (Patki *et al.*, 1997; Simonsen *et al.*, 1998). This allows the SNARE proteins on the two vesicles to interact, which facilitates the fusion of the two vesicles (Dietrich *et al.*, 2003; Fasshauer, 2003).

The cycle between the inactive and active states of a Rab is critical for its proper function. Two classes of proteins, the guanine nucleotide exchange factors (GEFs) and the GTPase-activating proteins (GAPs), govern the change between the two states. GEFs exchange the GDP with GTP to activate the Rab protein, whereas GAPs increase the intrinsic GTPase activity of the Rab, stimulating the hydrolysis of the bound GTP to GDP. We have recently shown that the p85α subunit of PI3K can bind directly to Rab5 and that p85 also has GAP activity toward Rab5 (Chamberlain *et al.*, 2004).

PI3K is a lipid kinase that is involved in cellular signaling downstream from several receptor tyrosine kinases (Cantley, 2002). PI3K generates the phospholipid second messengers, phosphatidylinositol 3,4-bisphosphate and phosphatidylinositol 3,4,5-trisphosphate. These phospholipids recruit and activate several downstream signaling proteins to the plasma membrane, including PDK1 and Akt, that promote cell proliferation and anti-apoptotic signals. We now find that in addition to its role in cell signaling, the p85 subunit of PI3K also regulates vesicle fusion events during receptor endocytosis through its Rab GAP activity (Chamberlain *et al.*, 2004). This chapter describes the methods used to assay the stimulation of Rab5 GTPase activity by the p85 protein.

Expression and Purification of Proteins

Rab5 Wild Type

The plasmid encoding human Rab5 was generously provided by G. Li (University of Oklahoma). The plasmid provided was in the pGEX3X vector (Amersham Biosciences), which produces a glutathione S-transfer-ase (GST)-Rab5 fusion protein. The GST tag is cleavable with factor Xa (Amersham Biosciences) to produce Rab5. For purification purposes the pGEX3X-Rab5 plasmid is expressed in *Escherichia coli* BL21 cells [F$^-$ *ompT hsdS$_B$* (r$_B^-$ m$_B^-$) *gal dcm*; Novagen]. These cells are protease deficient to minimize protein degradation during purification.

Luria-Bertani (LB; 100 ml) containing 100 μg/ml ampicillin is inocu-lated with *E. coli* BL21/pGEX3X-Rab5 cells and grown overnight at 37°. The next day 1 liter of LB containing 100 μg/ml ampicillin is inoculated with the entire 100 ml overnight culture. The culture is grown at 37° until an OD$_{600}$ of 0.5–0.7 is reached ($\sim$ 1 h). Isopropylthiogalactoside (IPTG; Sigma) is added to a concentration of 0.1 m*M* to induce protein expression. The culture is grown at 37° for 4 h and the cells are harvested by centrifu-gation at 4200×*g* for 15 min at 4°. The cell pellet can be frozen at −20° or lysed immediately.

For lysis, the cell pellet is resuspended in 10 ml phosphate-buffered saline (PBS; 137 m*M* NaCl, 2.7 m*M* KCl, 4.3 m*M* sodium phosphate, 1.4 m*M* potassium phosphate, pH 7.3) containing freshly added aprotinin (10 μg/ml; Sigma), leupeptin (10 μg/ml; Sigma), and phenylmethylsulfonyl fluoride (PMSF; 1 m*M*; Sigma). The cells are lysed by three bursts of sonification for a duration of 30 s at a setting of 2.5 using a Model 250/450 Sonifier (Branson Ultrasonics) with 2 min of chilling on ice between bursts. Triton X-100 (Sigma) is added to a concentration of 1% (to minimize protein–protein interactions and reduce contamination of the sample) and the sample is centrifuged at 12,000×*g* for 30 min at 4° to remove cell debris. The supernatant is filtered through a 0.45-μm cellulose acetate membrane (Nalgene) and mixed with 1 ml of a 50% slurry of glutathione-Sepharose beads (Amersham Biosciences). The lysate and beads are in-cubated together for 1 h at room temperature with rocking. The beads are recovered by centrifugation at 800×*g* for 5 min and the beads are washed three times with 50 ml of ice-cold PBS.

For cleavage of the GST tag, the beads are resuspended to a 50% slurry (total volume of 1 ml) with the addition of 50 units of factor Xa (Amersham Biosciences). The cleavage reaction is carried out for 4 h at room temperature. Longer cleavage times can result in proteolysis of the Rab5 protein. The beads are centrifuged at 2000×*g* and the supernatant containing the purified Rab5 is collected. The beads are washed four times with 1 ml of PBS to recover any residual Rab5 trapped between the beads. The samples are pooled and concentrated using an ultracentrifugal filter (Amicon) with a molecular weight cutoff of 5000 Da. As the Rab5 is concentrated, it is also buffer exchanged into protein storage buffer (50 m*M* HEPES, pH 7.5, 50 m*M* NaCl, 1 m*M* EDTA). Glycerol is added

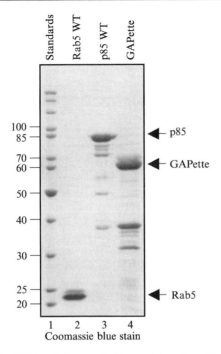

FIG. 1. SDS-PAGE (12%) and Coomassie blue staining of purified proteins (1 μg each) used for GAP assays. GST-Rab5–wild type (WT) was isolated on glutathione–Sepharose beads, cleaved from GST with Factor Xa. The wild-type p85 protein (p85 WT) and GAPette proteins were induced as GST-fusion proteins, isolated on glutathione-Sepharose beads, and cleaved from GST with thrombin. The sizes of the molecular weight markers (Standards; Fermentas) are 200, 150, 120, 100, 85, 70, 60, 50, 40, 30, 25, and 20 kDa.

to 40% and the concentration of the purified Rab5 is determined using a Bradford assay. A sample is resolved by sodium dodecyl sulfate-polyacrylamide gel electrophoresis (SDS-PAGE; 12%) and stained with Coomassie blue to determine purity (Fig. 1, lane 2). The purified Rab5 is flash-frozen in liquid nitrogen and stored at −80°.

p85 Wild-Type and the p85R274A Mutant

The cDNA encoding full-length wild-type bovine p85 was subcloned into pGEX2T (Amersham Biosciences) and is expressed in *E. coli* BL21 cells, as described previously for pGEX3X-Rab5. The GST-p85 is bound to glutathione-Sepharose beads using the same method as described earlier for GST-Rab5. For cleavage of the GST tag from the p85 protein, 100 NIH units of thrombin (Sigma, T6884) is used in a total volume of 1 ml PBS. The cleavage reaction is performed overnight at room temperature. The purified p85 protein is harvested the same as described earlier and buffer exchanged into protein storage buffer using an ultracentrifugal filter (Amicon) with a molecular weight cutoff of 10,000 Da. Glycerol is added to a concentration of 10%. The protein concentration is determined by Bradford assay and the purity of the p85 protein is assessed by SDS-PAGE (12%) with Coomassie blue staining (Fig. 1, lane 3). The p85 samples are flash-frozen and stored at $-80°$.

A mutant p85 protein containing a single point mutation within the domain encoding the GAP activity (p85R274A) was prepared similarly.

GAPette

GAPette is the catalytic domain of the p120 Ras GAP that has been shown to bind to and stimulate Rab5 (Liu and Li, 1998). It can be used as a positive control in the GAP assays. We used a clone of human p120 RasGAP that we had previously obtained from F. McCormick (University of California, San Francisco) to generate the GAPette clone. The pUC101a-p120 RasGAP plasmid was digested with ScaI and EcoRI and gel purified on a 0.8% low melt agarose gel to obtain the sequences encoding the catalytic domain (GAPette). The insert was ligated into pGEX2T digested with SmaI and EcoRI. This clone encoded a GST-GAPette protein containing amino acids 450–1047 of human p120 RasGAP, and after thrombin cleavage, the GST portion is ~60 kDa in size (Fig. 1, lane 4). The protocol for producing purified GAPette is the same as that for the p85 protein.

GAP Assay

This assay measures the hydrolysis of GTP by the Rab5 protein by determining the amount of GDP produced. Other GAP assays measure the amount of phosphate released (Self and Hall, 1995). The assay has three steps. The first is the loading of the Rab5 protein with $[\alpha\text{-}^{32}P]$GTP. The second step is the hydrolysis reaction converting the bound $[\alpha\text{-}^{32}P]$GTP to $[\alpha\text{-}^{32}P]$GDP. The last step is the separation of the $[\alpha\text{-}^{32}P]$GTP and $[\alpha\text{-}^{32}P]$GDP by thin-layer chromatography (TLC). There are two different variations of the GAP assay described in this chapter that were adapted from the method used by Liu and Li (1998). The first is the Steady–State GAP Assay. The second is the Single Turnover GAP Assay.

Steady-State GAP Assay

This assay measures the amount of GDP generated for a fixed time but with increasing concentrations of the GAP protein, p85. This assay has the possibility of new $[\alpha\text{-}^{32}P]$GTP loading onto the Rab5 protein and therefore cannot be used to determine the catalytic rate. It is used to determine the appropriate concentration range to use the p85 protein.

Preparation for the GAP Assay

The TLC chamber should be equilibrated with the developing solvent (0.75 M KH$_2$PO$_4$) for 1–24 h in advance. The PEI Cellulose F plates (VWR Canlab, Catalog #CAM05725-01) are labeled using a pencil. A line is drawn 2 cm from the bottom of the plate, and across this line, the lanes are marked with a small "x" 1.5 cm apart and 2 cm from the outside edge of the plate. Labeled TLC plates can be dried overnight at ~75° or for at least 1 h.

Nucleotide Loading of the Rab Protein. The Rab5 protein is diluted in loading buffer (20 mM Tris–HCl, pH 8.0, 2 mM EDTA, 1 mM dithiothreitol [DTT]) to a concentration of 200 nM. (α-^{32}P)GTP (Perkin Elmer, Catalog #BLU506H; 3000 Ci/mmol; 0.51 μl) is added to the Rab5 protein to a concentration of 85 nM, with the total reaction volume of the Rab5 protein plus the [α-^{32}P]GTP of 20 μl per sample. The Rab5 and nucleotide mix can, however, be made in a master mix for the number of samples plus one. The reaction mix is incubated at room temperature for 30 min, which allows the formation of Rab5[α-^{32}P]GTP. A negative control of nucleotide ([α-^{32}P]GTP) without any Rab5 protein should be prepared as well.

Hydrolysis Reaction. As the [α-^{32}P]GTP/Rab5 mix incubates, tubes of GAP protein (p85 or GAPette) are prepared. The hydrolysis reaction is carried out in a total volume of 30 μl: 9 μl p85 protein (diluted in loading buffer; sufficiently concentrated to give a final concentration of 0–35 μM in the 30 μl assay volume), 1 μl MgCl$_2$ (300 mM made fresh; final concentration of 10 mM), and 20 μl of the [α-^{32}P]GTP/Rab5 (prepared as described previously). For the positive control of GAPette, a final concentration of 0.8 μM in the 30 μl assay volume is used, and for the negative control, loading buffer alone is used. Therefore, a set of tubes containing 10 μl of the p85 and MgCl$_2$ (controls: GAPette/MgCl$_2$; buffer/MgCl$_2$) is prepared that will give the appropriate final concentrations of p85 and MgCl$_2$ after the 20 μl of [α-^{32}P]GTP/Rab5 is added.

After the Rab5 protein and the nucleotide mixture are finished incubating, 20 μl of this mix is added to the p85 protein plus MgCl$_2$, as well as to each of the control tubes. This is quickly mixed and incubated for 10 min at room temperature. The MgCl$_2$ locks the [α-^{32}P]GTP into the [α-^{32}P]GTP-Rab5 complex. After 10 min the reaction is stopped by the addition of 6 μl of elution buffer (1% SDS, 25 mM EDTA, 25 mM GDP, 25 mM GTP) and heated to 65° for 2 min.

Separation of [α-^{32}P]GTP/ [α-^{32}P]GDP by TLC. The samples are spotted onto the TLC plates 5 μl at a time until the entire sample is applied, allowing the spots to dry between additions. After the entire sample has

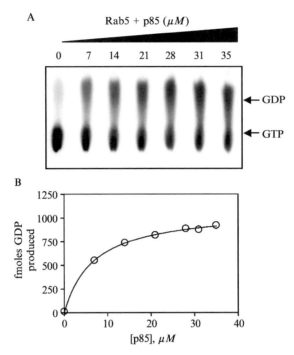

FIG. 2. Steady-state GAP assay for p85 and the Rab5 GTPase. (A) Rab5 was loaded with $[\alpha\text{-}^{32}P]GTP$, and hydrolysis to $[\alpha\text{-}^{32}P]GDP$ was assayed for 10 min in the absence and presence of increasing concentrations of p85 as indicated. Nucleotides were resolved by thin-layer chromatography and visualized using a Phosphorimager. (B) The results in (A) were quantified (see text) and the average of at least three independent determinations is shown. Reproduced with permission from Chamberlain *et al.* (2004).

been added, the plate is allowed to dry. When inserted into the TLC chamber, the developing solvent should be several millimeters below the line marking the lanes so that the developing solvent does not smear the spots. The samples are allowed to migrate in the TLC chamber until the solvent front is ~1 cm from the top of the plate, which takes ~2 h. After the plate is removed and dried, the separated $[\alpha\text{-}^{32}P]GTP$ and $[\alpha\text{-}^{32}P]$ GDP are visualized using a phosphorimager (Molecular Imager FX Pro Plus; Bio-Rad, Catalog #170–7850) and quantified using Quantity One software (Bio-Rad) (Fig. 2A). The amount of GTP hydrolyzed is calculated as femtomoles of $[\alpha\text{-}^{32}P]GDP$ produced or as a relative GAP activity using the ratio of $[\alpha\text{-}^{32}P]GDP$ to $[\alpha\text{-}^{32}P]GTP$ on the TLC plate and the starting amount of $[\alpha\text{-}^{32}P]GTP$ used (Fig. 2B). The value from the nucleotide alone sample is subtracted from the experimental values since it is the background of $[\alpha\text{-}^{32}P]GDP$ present in the absence of Rab5. The relative GAP activity is determined by normalizing the data to the amount of $[\alpha\text{-}^{32}P]$ GDP produced by the Rab5 protein in the absence of a GAP protein. The results are graphed and statistical analysis is performed using Prism software (GraphPad Software, Inc., San Diego, CA).

Single Turnover GAP Assay

This assay measures the rate of Rab5 GTP hydrolysis over time with a fixed concentration of p85 protein. There is no possibility of $[\alpha\text{-}^{32}P]GTP$ reloading due to the presence of a large excess of unlabeled GTP. Using this assay, it is possible to measure the catalytic rate of GTP hydrolysis by the Rab5 protein and how that rate is changed by the addition of the p85 protein.

The single turnover GAP assay is performed the same way as the steady-state GAP assay with the following changes. There is 1.7 mM unlabeled GTP added to the tube containing the p85 protein and the MgCl$_2$. Also the concentration of the p85 protein is held constant and the time of incubation is varied from 0 to 25 min at 5-min intervals (Fig. 3). When choosing a concentration to use for the p85 protein, it is ideal to choose a concentration that is submaximal in the steady-state GAP assay. This allows for both increases or decreases in rates to be observed easily. For the proposes of our experiments, we used 10 μM p85.

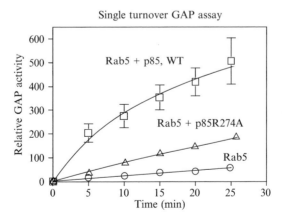

FIG. 3. Single turnover GAP assay showing that the p85 protein stimulates the catalytic rate of Rab5 GTP hydrolysis. Rab5 was loaded with $[\alpha\text{-}^{32}P]GTP$ and hydrolysis to $[\alpha\text{-}^{32}P]GDP$ was assayed for different times in the presence of a large excess of unlabeled GTP (1.7 mM), either alone ($\bigcirc$), or in the presence of wild-type p85 (10 μM, $\square$), or the mutant p85R274A (10 μM; $\triangle$). Reproduced with permission from Chamberlain et al. (2004).

A p85 mutant (R274A) with a single point mutation within the domain encoding GAP activity is also assayed for GAP activity. This mutation severely decreases the Rab5 GAP activity of the p85 protein (Fig. 3).

Comments

It should be noted that high concentrations of glycerol adversely affect the GAP assay. Thus, if the concentrations of the p85 or Rab5 proteins are low, it may be necessary to add less glycerol to the proteins prior to freezing. We found that with the 40% glycerol in the Rab5 protein, there was no experimental effect from the glycerol if the Rab5 was diluted 1:100 with loading buffer before using it for nucleotide loading. For p85, we found that 10% glycerol could be added without a significant effect on the GAP assay even though the p85 was not always diluted before addition into the assay reaction mixture. We also noted that different preparations of the proteins have slightly different specific activities, suggesting that it would be best to use the same batch of Rab5 and p85 proteins for a particular set of experiments.

Acknowledgments

We are grateful to G. Li, M. Waterfield, and F. McCormick for providing plasmids encoding the Rab5, p85, and p120 RasGAP proteins, respectively. We thank A. Hawrysh and K. James for expert technical assistance. M.D.C. is supported by a doctoral scholarship from the Canadian Institutes of Health Research Regional Partnership Program. This work was supported by grants from the National Sciences and Engineering Research Council and the Saskatchewan Cancer Agency Research Fund.

References

Cantley, L. C. (2002). The phosphoinositide 3-kinase pathway. *Science* **296,** 1655–1657.

Chamberlain, M. D., Berry, T. R., Pastor, M. C., and Anderson, D. H. (2004). The p85{α} subunit of phosphatidylinositol 3′-kinase binds to and stimulates the GTPase activity of Rab proteins. *J. Biol. Chem.* **279,** 48607–48614.

Christoforidis, S., Miaczynska, M., Ashman, K., Wilm, M., Zhao, L., Yip, S. C., Waterfield, M. D., Backer, J. M., and Zerial, M. (1999). Phosphatidylinositol-3-OH kinases are Rab5 effectors. *Nat. Cell. Biol.* **1,** 249–252.

Deneka, M., Neeft, M., and van der Sluijs, P. (2003). Regulation of membrane transport by rab GTPases. *Crit. Rev. Biochem. Mol. Biol.* **38,** 121–142.

Dietrich, L. E., Boeddinghaus, C., LaGrassa, T. J., and Ungermann, C. (2003). Control of eukaryotic membrane fusion by N-terminal domains of SNARE proteins. *Biochim. Biophys. Acta.* **1641,** 111–119.

Fasshauer, D. (2003). Structural insights into the SNARE mechanism. *Biochim. Biophys. Acta.* **1641,** 87–97.

Liu, K., and Li, G. (1998). Catalytic domain of the p120 Ras GAP binds to Rab5 and stimulates its GTPase activity. *J. Biol. Chem.* **273**, 10087–10090.

Maxfield, F. R., and McGraw, T. E. (2004). Endocytic recycling. *Nat. Rev. Mol. Cell. Biol.* **5**, 121–132.

Patki, V., Virbasius, J., Lane, W. S., Toh, B. H., Shpetner, H. S., and Corvera, S. (1997). Identification of an early endosomal protein regulated by phosphatidylinositol 3-kinase. *Proc. Natl. Acad. Sci. USA* **94**, 7326–7330.

Pfeffer, S., and Aivazian, D. (2004). Targeting Rab GTPases to distinct membrane compartments. *Nat. Rev. Mol. Cell. Biol.* **5**, 886–896.

Seabra, M. C., and Wasmeier, C. (2004). Controlling the location and activation of Rab GTPases. *Curr. Opin. Cell. Biol.* **16**, 451–457.

Self, A. J., and Hall, A. (1995). Measurement of intrinsic nucleotide exchange and GTP hydrolysis rates. *Methods Enzymol.* **256**, 67–76.

Simonsen, A., Lippe, R., Christoforidis, S., Gaullier, J. M., Brech, A., Callaghan, J., Toh, B. H., Murphy, C., Zerial, M., and Stenmark, H. (1998). EEA1 links PI(3)K function to Rab5 regulation of endosome fusion [see comments]. *Nature* **394**, 494–498.

[49] Ubiquitin Regulation of the Rab5 Family GEF Vps9p

By Brian A. Davies, Darren S. Carney, and Bruce F. Horazdovsky

Abstract

To maintain cellular homeostasis, the levels of transmembrane receptors found on the plasma membrane must be tightly regulated. Endocytosis of activated receptors and the eventual degradation of these transmembrane proteins in the lysosome serve a vital role in maintaining the plasma membrane receptor levels as well as attenuating the downstream signaling pathways. Two processes that regulate this receptor trafficking are the covalent modification of the receptor with ubiquitin (ubiquitylation) and the activation of the Rab5 family of small GTPases. Activation of Rab5 family proteins has been shown to be critical for early steps of the endocytic pathway including delivery of activated receptors to the early endosome, while ubiquitylation of activated receptors has been shown to be involved in receptor internalization, delivery to the endosome, and sorting into the multivesiclar body. In yeast, the guanine nucleotide exchange factor Vps9p serves to integrate the activation of a Rab5 protein (Vps21p) via the Vps9 domain with ubiquitin binding via the CUE domain to facilitate the delivery of ubiquitylated receptors to the endosome. Here we provide detailed protocols for the study of Vps9p *in vivo* and *in vitro* with regard to Vps21p activation, ubiquitin binding, and Vps9p ubiquitylation.

METHODS IN ENZYMOLOGY, VOL. 403   0076-6879/05 $35.00
Copyright 2005, Elsevier Inc. All rights reserved.   DOI: 10.1016/S0076-6879(05)03049-1

Introduction

The trafficking of material through the endocytic system is a complex and tightly regulated process (Aguilar and Wendland, 2005). In addition to internalization, this pathway is responsible for the appropriate and timely delivery of activated cell surface receptors through a series of compartments whereon compartment-specific signal transduction events may occur (early endosome; endocytosis-dependent signaling), wherein the receptor is sequestered away from the cytoplasm (late endosome; multivesicular body [MVB] sorting), or within which the activated receptor is finally degraded (lysosome). Disregulation of this trafficking system results in aberrant signal transduction activity, leading to either uncontrolled growth and cancer as seen with mutation of the protooncogene c-Cbl or motor neuron degeneration as seen with mutations in Alsin (Carney et al., 2005; Shtiegman and Yarden, 2003). Two critical means of regulating traffic within the endocytic system are the covalent modification of proteins with ubiquitin (ubiquitylation) and the activation of the Rab5 family of GTPases.

The covalent modification of proteins with ubiquitin has traditionally been associated with polyubiquitylation and proteosomal degradation. More recently, however, monoubiquitylation has been demonstrated to serve as a sorting determinant in receptor trafficking as well as a modulator of protein function. Several proteins of the endocytic pathway have been shown to bind ubiquitin via a series of conserved domains, including the ubiquitin interaction motif (UIM), ubiquitin conjugating enzyme E2 variant (UEV), ubiquitin associated (UBA), Npl4 zinc finger (NZF), and CUE domains (sequence conserved with Cue1p) (Hicke and Dunn, 2003). Receptor ubiquitylation drives recognition by the internalization machinery via interaction with Ede1p (UBA) and the epsins (UIM). Subsequently, interaction with Vps9p (CUE) facilitates delivery of the ubiquitylated receptor to the endosome (Carney et al., 2005). And finally, interactions with Vps27p (UIM), Hse1p (UIM), Vps23p (UEV), and Vps36p (NZF) facilitate sequestration of the ubiquitylated receptor away from the cytosol through sorting into the MVB (Babst, 2005). Together, these various interactions serve to deliver the activated receptor to the lumen of the yeast vacuole/lysosome for degradation.

Activation of the Rab5 family of GTPases is also a critical step in the delivery of receptors to the lysosome/yeast vacuole (Carney et al., 2005). Rab proteins serve as molecular switches through binding GDP in the inactive state or GTP in the active state. The conversion from the active to inactive states occurs through the intrinsic weak GTP hydrolysis activity of the Rab, which can be stimulated by the GTPase-activating proteins

(GAPs). The conversion of the Rab to the active GTP-bound state occurs through promoting GDP release to permit GTP binding; this activation is accomplished by guanine nucleotide exchange factors (GEFs). Rab5 activation is promoted by the Vps9 domain, the structure of which has been determined by Delprato et al. (2004) (Fig. 1A). In mammalian systems, 10 proteins have been identified that contain this Vps9 domain, a subset of which is depicted in Fig. 1B. In addition to the Vps9 domain, these GEFs contain a variety of signaling domains, including Src homology 2 (SH2) and Ras association (RA) domains (Rin proteins), ankyrin repeats (Ankyrin Repeat Domain 27), Ras GAP (hRME-6), and Rac1 GEF domains (Alsin) (Carney et al., 2005). This observation suggests that these Vps9 domain-containing proteins serve to integrate signal transduction pathways with endocytic flux via Rab5 activation.

In yeast, activation of the Rab5 ortholog Vps21p is mediated by Vps9p (Hama et al., 1999). Both Vps21p and Vps9p function in endosomal delivery of cargo derived from the trans-Golgi network (biosynthetic pathway) as well as from the plasma membrane (endocytic pathway) (Burd et al., 1996; Gerrard et al., 2000; Horazdovsky et al., 1994; Singer-Kruger et al., 1994, 1995). The catalytic domain responsible for Vps21p activation comprises the middle portion of Vps9p. In addition, Vps9p contains a ubiquitin-binding domain (CUE) in its carboxy-terminus (Davies et al., 2003; Donaldson et al., 2003; Ponting, 2000; Shih et al., 2003). Structural determination of the Vps9p CUE domain in complex with ubiquitin revealed that Vps9p forms a dimer to bind ubiquitin with high affinity (Fig. 2) (Prag et al., 2003). This high-affinity binding is required for efficient delivery of ubiquitylated receptors to the vacuole as well as for ubiquitylation of Vps9p itself (Davies et al., 2003; Prag et al., 2003; Shih et al., 2003). These observations indicate that ubiquitin represents an important regulator of Vps9p in the endocytic pathway and that the CUE domain integrates ubiquitin binding with Rab activation to regulate flux through the endocytic pathway. We hope that through understanding this regulation of Vps9p by ubiquitin we will gain insight into the roles mammalian Vps9 domain proteins play integrating signal transduction pathways with endocytic flux. Some of the techniques we employ are described below.

Analysis of VPS9 Function in the Biosynthetic Pathway

vps9 was originally uncovered in genetic selections for defects in the delivery of soluble proteases to the vacuole (Bankaitis et al., 1986; Rothman and Stevens, 1986). Cloning of VPS9 by complementation, subsequent deletion of the candidate gene, and characterization of the Δvps9

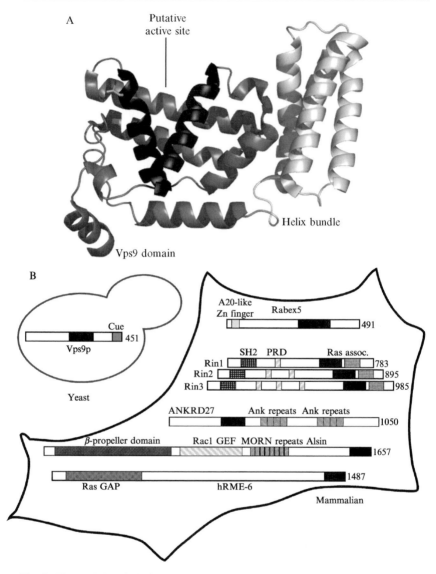

FIG. 1. The Vps9 domain activates Rab5 family proteins. (A) Structure of the Vps9 domain as determined by Delprato *et al.* (2004). The domain (gray) and the preceding helix bundle (light gray) are predominantly α-helical. The helices form a scaffold on which a hydrophobic grove (black) is positioned. This grove is the putative active site for Rab5 activation. (B) The Vps9 domain (black) is found in conjunction with a number of other signaling domains in mammalian systems, including SH2, proline rich, and Ras association domains in the Rin proteins, ankyrin repeats in ANKRD27, MORN repeats and a Rac1 GEF domain in Alsin, and a Ras GAP domain in hRME-6. In yeast, Vps9p contains a CUE domain that binds ubiquitin.

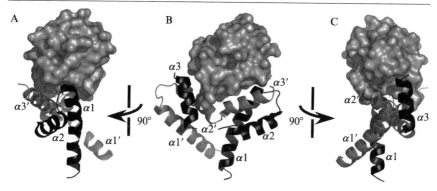

FIG. 2. The CUE domain forms an asymmetric dimer to bind ubiquitin with high affinity. One interaction surface, shown in (C) and the right portion of (B), is formed by residues from the first helix of one subunit (α1) and the third helix of the second subunit (α3′); this interaction surface is similar to the interaction observed between ubiquitin and the monomeric CUE domains of Cue2p or with the UBA domain. The second interaction surface, shown in (A) and the left portion of (B), is formed by residues from the second helix of the second subunit (α2′).

strain conclusively demonstrated that Vps9p is required for vacuolar delivery of the soluble proteases carboxypeptidase Y (CPY) and proteinase A (PrA) as well as carboxypeptidase S (CPS) and alkaline phosphatase (ALP), two enzymes that traffic to the vacuole as transmembrane zymogens (Burd *et al.*, 1996). In addition, $\Delta vps9$ cells exhibit defects in endocytosis and the enlarged vacuole morphology characteristic of the Class D *vps* mutants (Banta *et al.*, 1988; Raymond *et al.*, 1992). While quantitative assessment of the endocytic defects in *vps9* hypomorphs is possible and will be discussed later in this chapter, the most convenient quantitative assay for *VPS9* function is the analysis of CPY maturation. Briefly, following pulse–chase labeling with [^{35}S]methionine and -cysteine, yeast lysates are generated under denaturing conditions, and CPY is immunoprecipitated from the cleared lysate and resolved by sodium dodecyl sulfate–polyacrylamide gel electrophoresis (SDS-PAGE). The delivery of CPY to the vacuole results in activation via cleavage of the sorting signal (Bryant and Stevens, 1998); thus, vacuolar sorting can be measured by following conversion of the Golgi-modified CPY precursor (p2CPY) to the mature form (mCPY). The transport of CPY to the vacuole is dependent on the Rab5 family protein Vps21p as well as its GEF Vps9p. The *in vivo* activation of Vps21p by Vps9p can therefore be assessed by quantitating CPY sorting using the procedure described below.

Buffer/Materials

> Minimal media: 0.67% Difco yeast nitrogen base without amino acids, 2% glucose, 25mM potassium phosphate, pH 5.4[1], supplemented with appropriate amino acids for the particular strain under examination[2]
>
> 25× chase: 5% yeast extract, 125 mM methionine, 25 mM cysteine[3]
>
> Boiling buffer: 50 mM Tris, pH 7.5, 1 mM EDTA, 1% SDS
>
> Tween-20 IP buffer: 50 mM Tris, pH 7.5, 150 mM NaCl, 0.5% Tween-20, 0.1 mM EDTA
>
> Protein A Sepharose slurry: 0.4 g protein A Sepharose 4B CL (Amersham Biosciences) resuspended in 10 mM Tris, pH 7.5, 1 mg/ml bovine serum albumin (BSA) Fraction V, 1 mM NaN_3[4]
>
> Tween-20 urea buffer: 100 mM Tris, pH 7.5, 200 mM NaCl, 2 M urea, 0.5% Tween-20 Tris-buffered saline (TBS): 50 mM Tris, pH 7.5, 150 mM NaCl
>
> 5× Laemmli sample buffer: 0.312 M Tris, pH 6.8, 10% SDS, 25% 2-mercaptoethanol, 0.05% bromophenol blue.

Procedure

1. Grow cultures overnight in minimal media at 30°. In the morning, dilute to 0.2 OD_{600}/ml and continue incubations at 30° until OD_{600} of 0.6–0.8 is achieved.

2. Harvest 5 OD_{600} and pellet in a Falcon 2059 tube. Resuspend in 1 ml minimal media and incubate in a 30° agitating waterbath for 5–10 min.

3. Initiate metabolic labeling by the addition of 50–100 μCi ^{35}S-Promix (Amersham Biosciences) and return to the agitating waterbath for 10 min.

4. Add 40 μl 25× chase to prevent further labeling and continue incubation in an agitating waterbath for 30 min.

5. Transfer cultures to 1.5-ml Eppendorf tubes and add 100 μl trichloroacetic acid (TCA) to precipitate the samples. Incubate on ice for 10 min.

6. Spin 1–2 min in a microfuge at 13,000×g. Aspirate the supernatant and resuspend the pellets in 1 ml cold acetone. Use of a sonicating waterbath, such as the Branson Ultrasonic Cleaner 2510, seems to be the most

---

[1] Buffering the media may help eliminate a spurious processing of secreted p2CPY.

[2] We utilize 50× amino acid stocks containing 0.2% histidine, 0.3% leucine, 0.3% lysine, 0.2% tryptophan, 0.2% adenine, and 0.2% uracil as the complete mixture for addition to minimal media, or leave out particular amino acids to select for plasmid retention.

[3] Aliquots are stored in −20° and thawed for use.

[4] Following the initial resuspension and overnight swelling, replace the buffer. Store at 4°.

effective method for resuspension. Then pellet and repeat the acetone wash procedure. Finally, pellet the samples and aspirate the acetone. Dry the pellets in an evacuated centrifuge (speedvac) for 5–60 min without heating.

7. Resuspend the dried TCA pellets in 100 μl Boiling buffer by sonicating. Then add 150 μl 0.5-mm glass beads (BioSpec Products, Inc.) using a 200-μl Eppendorf tube as a scoop and a minifunnel generated by trimming a P1000 pipette tip[5]. Vortex the samples thoroughly—1–2 min if individually agitated or 10 min if using a multitube vortexer.

8. Heat the samples at 95° for 4 min. Then add 1 ml Tween-20 IP buffer, 10 μl 100 mg/ml BSA, and vortex briefly.

9. Centrifuge the samples at 13,000×g for 10 min at room temperature and transfer 1 ml of the samples to a second tube. Repeat the spin on this second tube and transfer 950 μl to a third tube to generate the cleared lysate[6]. Use a small portion of the samples (2–5 μl) to determine labeling efficiency; 2×10^6 to 2×10^7 cpm/OD$_{600}$ are reasonable findings.

10. Add 2 μl of anti-CPY antiserum (U1345, generated by P. Marshall), and incubate the samples for at least 3 h, but preferably overnight, at 4° with gentle rocking. Add 100 μl of protein A Sepharose slurry and continue incubations for 90 min at 4°.

11. Spin the samples for 5 s at 13,000×g to pellet the protein A beads. Aspirate the supernatant and add 1ml Tween-20 IP buffer. Invert to mix, pellet, and resuspend in 1ml Tween-20 Urea buffer. Pellet again and wash with 1 ml Tris-buffered saline (TBS). Pellet a fourth time and aspirate all but 40 μl of the supernatant.

12. Add 10 μl 5× Laemmli sample buffer and heat the samples at 95° for 4 min. After spinning the samples at 13,000×g for 1 min, resolve 10 μl of each sample by SDS-PAGE on a 10% acrylamide gel. Use a prestained ladder to monitor and facilitate the resolution of proteins in the 50–70 kDa range.

13. Fix the gel in 10% acetic acid, 10% methanol for 10 min with gentle agitation. Then perform two 10-min washes with water prior to incubating the gel in 1 M sodium salicylate, 10% glycerol for 10 min. Dry the gel on filter paper using a gel dryer (1 h at 80°). Then load in a cassette with X-ray film, and incubate at −80° for optimal signal detection. Exposure times (12 h to 2 weeks) will vary based on labeling efficiency.

---

[5] Avoid spilling glass beads around the collar of the Eppendorf tube as these beads may compromise tube closure and lead to contamination of the work area.

[6] It is important to avoid transferring any precipitated material or the glass beads, as they will obscure the relevant CPY immunoprecipitation signal.

14. Develop the film and quantitate the relative amounts of mature CPY (69 kDa) and Golgi-modified CPY (61 kDa) to determine the percent CPY sorting.

From these experiments, the ability of Vps9p to activate Vps21p *in vivo* for biosynthetic pathway vacuolar protein sorting can be assessed. However, defects in Vps9p ubiquitin binding do not perturb this biosynthetic pathway as measured by this CPY sorting assay. The CPY sorting assay does serve as a positive control for general function of Vps9p alleles defective for ubiquitin binding.

Analysis of *Vps9* Function in the Endocytic Pathway

In contrast to the biosynthetic pathway phenotypes, Vps9p mutants defective for ubiquitin binding exhibit deficits in endocytic pathway transport to the vacuole (Davies *et al.*, 2003; Donaldson *et al.*, 2003). Both the Rab Vps21p and the GEF Vps9p are required for fusion of vesicles with the endosome, a step subsequent to receptor internalization yet prior to sorting into the MVB, and defects in Vps9p ubiquitin binding affect the efficiency of this transport process. To study transport through the endocytic pathway, a variety of approaches have been utilized, including analysis of radiolabeled α-factor uptake and degradation, cyclohexamide-chase experiments to determine turnover rates for α-factor or α-factor receptors (Ste2p and Ste3p, respectively; refer to Katzmann and Wendland, 2005), or microscopic analysis of GFP-tagged Ste2p or Ste3p. We have utilized both cyclohexamide-chase experiments with Western blotting for endogenous or HA-tagged Ste3p turnover (Davies *et al.*, 2003) as well as microscopic analysis of Ste3p-GFP (Prag *et al.*, 2003). We find the microscopic analysis to be most useful, as it distinguishes between defects in internalization and subsequent trafficking events, and thus the procedure is described as follows.

Buffer/Materials

pRS304 Ste3GFP: The promoter and coding region of *STE3* were cloned in-frame to a carboxy-terminal GFP fusion coding sequence. The *STE3::GFP* was cloned into the *Xho*I and *Sac*I sites of pRS304 (Sikorski and Hieter, 1989) to generate pRS304 Ste3GFP. This vector contains a single *Eco*RV recognition site, present within the *TRP2* portion of the plasmid
Minimal media: as previously described
YPD: 1% yeast extract, 2% peptone, 2% glucose

FM4-64 media: 16 μM FM4-64 (T3166, Molecular Probes) in YPD, generated by diluting the 1.6 mM FM4-64 stock solution (in dimethyl sulfoxide [DMSO]) 1:100 in YPD

Cyclohexamide solution: 1 mg/ml cyclohexamide (C7698, Sigma) in 95% ethanol, generated from a 10 mg/ml stock (in ethanol, stored at $-20°$).

Procedure

1. The $\Delta vps9$ strain (CBY1; Burd *et al.*, 1996) was transformed with the *Eco*RV digested pRS304 Ste3GFP plasmid (to drive integration at *TRP2*), and yeast harboring the integrated reporter were selected for growth on synthetic media lacking tryptophan. The resultant BHY93 strain (Prag *et al.*, 2003) was then transformed with yeast replicating shuttle vectors (2 μ or CEN) containing *vps9* alleles. Analysis of the ensuing strains was performed as follows.

2. Grow cultures at $30°$ overnight in minimal media to select for plasmid retention. In the morning, dilute to 0.2 OD$_{600}$/ml in minimal media and continue incubations at $30°$ until OD$_{600}$ of approximately 0.5 is achieved.

3. Harvest 1 OD$_{600}$ and resuspend in 5 ml YPD. Culture at $30°$ for 1 h, and then pellet 1 OD$_{600}$ in a Falcon 2059 tube.

4. Resuspend the samples in 50 μl FM4-64[7] media; this should be done for all samples at the same time. Incubate 30 min in a $30°$ agitating waterbath to label with FM4-64, then add 1 ml YPD, and chase the FM4-64 to the vacuole with an incubation greater than 1 h.

5. Fifteen minutes after the addition of the YPD chase, add 2 μl cyclohexamide solution to the first sample; stagger subsequent cyclohexamide additions to other samples by 15 min.

6. Incubate the samples for 45 min at $30°$ following cyclohexamide addition to eliminate the pool of Ste3pGFP reporter trafficking through the secretory pathway to the plasma membrane and to allow specific examination of the endocytic pathway trafficking.

7. Staggering the samples by 15 min, harvest 500 μl of sample in a 1.5-ml Eppendorf tube via a $13,000 \times g$ spin for 1 min. Resuspend the pellet in 50 μl minimal media lacking tryptophan with 2 μg/ml cyclohexamide. Spot 2 μl on a slide and apply a coverslip.

8. Using an Olympus IX70 inverted microscope with fluorescein isothiocyanate (FITC) and rhodamine filter sets and a Photometrix digital camera, record the fluorescence of the Ste3pGFP reporter (FITC, 2 s exposure) and the FM4-64-labeled vacuoles (rhodamine, 0.2 s exposure)

---

[7] FM4-64 is a lipophilic fluorescent dye that is transported to, and thus assists in visualization of, the vacuolar limiting membrane.

throughout the yeast cell using 0.2-μm vertical sections (12–16 sections per field of yeast). Also record the visible field for each layer.[8]

9. At the end of the experiment, deconvolve the image stacks using the DeltaVision package (Applied Precision), and count the number of perivacuolar puncta per cell.

From this procedure, we were able to uncover a deficit in the efficiency of Ste3p trafficking to the vacuole in yeast harboring Vps9p alleles defective for ubiquitin binding.

Vps9p Expression and Purification

To directly examine the ability of Vps9p to activate Vps21p, we utilize a GDP release assay with purified proteins. Expression of both proteins is performed in bacteria, and purification is accomplished via two to four steps. While Vps21p expression and purification have always been straightforward, Vps9p purification has been significantly optimized over the years. Initial expression and purification were performed using a hexahistidine-tagged Vps9p expression vector (pQE31-Vps9p, Hama *et al.*, 1999). Examination of protein solubility indicated that only 10–20% of the expressed protein was soluble; however, Ni^{2+}-affinity purification of the soluble material yielded sufficient protein for *in vitro* analyses (Davies *et al.*, 2003; Hama *et al.*, 1999). At the suggestion of Steve Sprang (UTSW), we generated a maltose-binding protein-Vps9p fusion (MBP-Vps9p) and found that this protein was much more soluble (>90%), suggesting that the MBP epitope facilitates Vps9p folding. We subsequently generated a hexahistidine-MBP fusion vector (pET28MBP, Fig. 3) and used this vector for Vps9p expression (pET28MBP Vps9). The His_6MBP-Vps9p exhibits good solubility when expressed at 25° and offers a simple purification strategy for the isolation of high amounts of pure protein. Briefly, the His_6MBP-Vps9p fusion is purified by Ni^{2+} affinity chromatography, and then the His_6MBP tag is removed by digestion with a hexahistidine-tagged tobacco etch virus protease (His_6TEV; Invitrogen). Ni^{2+} affinity chromatography is then used to remove both the His_6MBP tag and His_6TEV from Vps9p. Vps9p is then further purified by ion-exchange chromatography (Bioscale Q2 or HiTrap Q FF 5 ml) and/or gel filtration chromatography (Superdex-75 or -200) to yield a sample of greater than 95% purity. Typical yields for Vps9p from a 1-liter culture are 20–30 mg of protein. We have also used this His_6MBP tag and purification procedure with great success for isolation of human Rabex5 as well as the Vps9 domain of Rabex5. The detailed

---

[8] Since you have only 15 min to record images for each sample, you will need to work quickly. Alternatively, you can increase the stagger time.

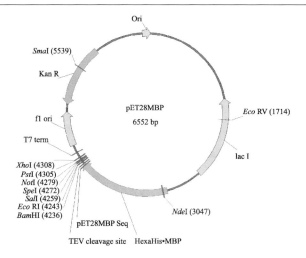

FIG. 3. His₆MBP expression vector. The maltose-binding protein (MBP) coding sequence was inserted into the pET28b bacterial expression vector to generate pET28MBP. To sequence inserts, we use the pET28MBP Seq oligo (5'ACAACAACCTCGGGATCG) and T7 Terminator oligo (5'ATGCTAGTTATTGCTCAGCGGTGG). pET28MBP is used for Vps9p expression and purification. The MBP epitope facilitates protein folding of Vps9p in bacteria, and the His₆-tag permits Ni²⁺ affinity purification. The His₆MBP tag is then cleaved with His₆TEV treatment, and both the epitope and protease are removed by Ni²⁺ affinity chromatography to yield purified Vps9p.

procedure for Vps9p purification using the pET28MBP-Vps9 vector is described below.

Buffer/Materials

pET28MBP: The maltose-binding protein and carboxy-terminal TEV protease cleavage site coding sequences were amplified from pMBP parallel 1 (Sheffield *et al.*, 1999) using the oligos MBP-3 (5'AGGACCATAGATTATGAAAATCGAAGAAGGTAAAC) and MBP-4 (5'TACCGCATGCCTCGAGACTGCAGGCT). The polymerase chain reaction (PCR) product was digested with *Nde*I and *Xho*I and cloned into the *Nde*I, *Xho*I sites of pET28b (Novagen) to yield the pET28MBP vector. The vector map and multicloning site are shown in Fig. 3.

pET28MBP-Vps9: *VPS9* had previously been cloned into the *Bam*HI, *Sal*I sites of pQE31 (Qiagen) (Hama *et al.*, 1999). This *Bam*HI, *Sal*I fragment was subcloned into the *Bam*HI, *Sal*I sites of pET28MBP.

HMS174 DE3: bacterial expression strain obtained from Novagen

LB: 1% tryptone, 0.5% yeast extract, 1% NaCl

Ni^{2+} lysis buffer: 50 mM sodium phosphate, pH 7.5, 300 mM NaCl, 0.1 mM AEBSF, protease inhibitors[9]

Lysozyme solution: 10 mM Tris, pH 8.0, 10 mg/ml lysozyme

Ni^{2+} Buffer A: 50 mM sodium phosphate, pH 7.5, 300 mM NaCl

Ni^{2+} Buffer B: 50 mM sodium phosphate, pH 7.5, 300 mM NaCl, 500 mM imidazole

Q Buffer A: 25 mM Tris, pH 7.5

Q Buffer B: 25 mM Tris, pH 7.5, 1 M NaCl

Superdex buffer: 25 mM Tris, pH 7.5, 150 mM NaCl.

Procedure

1. Transform the HMS174 DE3 strain with the pET28MBP Vps9 plasmid, and grow colonies overnight at 37° on LB Kan plates (50 μg/ml kanamyacin).

2. Start four 5 ml LB Kan (50 μg/ml) cultures in the morning from fresh colonies (pick multiple colonies per tube). Incubate at 37° until an OD_{600} of approximately 0.5 is achieved (usually 2–3 h). Dilute the overday cultures into 1 liter LB Kan media in a 2.8-liter Fernbach flask and continue incubation at 37° until an OD_{600} of 0.5–0.8 is achieved. Shift the culture to a 25° incubator for 30 min before adding 500 μl of 1 M isopropyl-β-D-thiogalactopyransodie (IPTG). Continue inductions overnight for a total of 14–24 h.

3. Harvest the culture and resuspend in 10 ml Ni^{2+} lysis buffer with protease inhibitors freshly added. Add 1 ml lysozyme solution and incubate 30 min at 4° with gentle rocking. Flash freeze in liquid nitrogen; alternatively, the sample can be stored at −80° until proceeding with the purification.

4. Thaw the sample and pass three times through a French pressure cell to complete lysis and shear the DNA[10]. Clear the lysate by centrifugation at 40,000 rpm in a Beckmann 70Ti rotor at 4°. Filter the lysate through a 0.45-μm syringe filter and load into a 50-ml Superloop.

---

[9] Protease inhibitors: *N*-tosyl-L-phenlalanine-chloromethyl ketone (TPCK), Nα-p-tosyl-L-lysine-chloromethyl ketone (TLCK), leupeptin, trypsin inhibitor.

[10] We have also successfully used simple DNase I treatments or ultrasonication with a probe sonicator, but the French press is our preferred method.

5. Perform Ni^{2+} affinity purification with an AKTA FPLC machine using a 5-ml HiTrap Chelating HP column (Amersham-Pharmacia) charged with $NiSO_4$ per the manufacturer's instructions and the Ni^{2+} Buffer A and Buffer B with a flow rate of 2.5 ml/min collecting 5-ml fractions. Following application of the sample, the column is washed with 2 column volumes Buffer A and then 5 column volumes 4% Buffer B (20 mM imidazole). Protein is then eluted via a 4–50% Buffer B gradient (20–250 mM imidazole) over 20 column volumes. The column is then washed with 5 column volumes Buffer B and reequilibrated in 5 columns Buffer A. Use the UV trace to identify candidate fractions and analyze by SDS-PAGE and Coomassie staining. His_6MBP-Vps9 elutes at approximately 50–125 mM imidazole.

6. Pool fractions containing His_6MBP-Vps9 and concentrate to less than 5 ml using a Vivaspin 20 ml 30,000 MWCO concentrator. Transfer the sample to a Falcon 2059 tube and add dithiothreitol (DTT) to 1 mM and EDTA to 0.5 mM. Add approximately 250 U His_6TEV protease (Invitrogen) and incubate the sample overnight at 16°.

7. Replace the sample buffer with Ni^{2+} Buffer A using a HiPrep 26/10 Desalting column (Amersham Pharmacia) at a flow rate of 2 ml/min collecting 1-ml fractions. Pool protein-containing fractions, as determined by UV trace, and load into a 50-ml Superloop.

8. Remove His_6TEV and His_6MBP by Ni^{2+} affinity purification with the AKTA FPLC using the 5-ml HiTrap Chelating HP column (charged with $NiSO_4$) and the Ni^{2+} Buffer A and Buffer B with a flow rate of 1 ml/min collecting 1-ml fractions. Following application of the sample (Vps9p should be in the flowthrough), the column is washed with 10 column volumes Buffer A, and then a 5 column volume 0–10% Buffer B gradient (0–50 mM imidazole) is initiated. The column is then washed with 5 column volumes Buffer B and reequilibrated in 5 column volumes Buffer A. Use the UV trace to identify candidate fractions of the flowthrough and early wash and analyze by SDS-PAGE and Coomassie staining.

9. Pool Vps9p fractions and concentrate in a Vivaspin 30,000 MWCO concentrator to less than 1 ml. Dilute sample 10-fold in Q Buffer A and load into a 10-ml Superloop.

10. Perform ion-exchange purification with the AKTA FPLC using the HiTrap Q FF 5-ml column and the Q Buffer A and Buffer B with a flow rate of 2.5 ml/min collecting 2.5-ml fractions. Following application of the sample, the column is washed with 5 column volumes Buffer A, and then a 20 column volume 0–50% Buffer B gradient (0–500 mM NaCl) is initiated. The column is then washed with 5 column volumes Buffer B and reequilibrated in 5 columns Buffer A. Use the UV trace to identify

candidate Vps9p fractions and analyze by SDS-PAGE and Coomassie staining. Vps9p elutes in the 200–300 mM NaCl range.

11. Pool Vps9p fractions and concentrate in a Vivaspin 20 ml 30,000 MWCO concentrator to less than 1 ml. Load the sample into a 1-ml loop.

12. Perform gel filtration purification with the AKTA FPLC using the HiLoad 16/60 Superdex 75 column and the Superdex buffer with a flow rate of 1 ml/min collecting 1-ml fractions. Use the UV trace to identify candidate Vps9p fractions and analyze by SDS-PAGE and Coomassie staining. Vps9p fractionates as a monomer by this procedure.

13. Pool Vps9p fractions and concentrate in a Vivaspin 2 ml 10,000 MWCO concentrator to less than 1 ml. The sample is then divided into 100-μl aliquots, flash frozen in liquid nitrogen, and stored in the $-80°$ freezer until needed.

While this protocol includes both the ion-exchange and gel filtration chromatography procedures, use of either separation technique in conjunction with the two rounds of Ni^{2+} affinity purification is likely sufficient to produce recombinant Vps9p suitable for most *in vitro* analyses. The assays we have utilized include glutathione *S*-transferase (GST) pulldown experiments with carboxy-terminally tagged ubiquitin or Vps21p, gel filtration analyses of Vps9p•Ub and Vps9p•Vps21p complexes, and the GDP release assay with Vps21p, described below.

Analysis of Vps9p-Stimulated Vps21p Nucleotide Release

Vps21p, the yeast Rab5 ortholog, is preloaded with labeled GDP and then incubated with excess GDP or GTP in the presence of exchange factor to facilitate release of the labeled nucleotide. Initial experiments by Hama *et al.* (1999) utilized [^{3}H]GDP-loaded Vps21p and filter binding experiments, but more recently we have employed mantGDP-loaded Vps21p in fluorescence-based assays. These fluorescence assays can either be configured to take advantage of the increased mantGDP fluorescence when bound to Vps21p using excitation and emission wavelengths of 340 and 440 nm, respectively, or as a fluorescence resonance energy transfer (FRET) between Vps21p and mantGDP when the nucleotide is bound (ex, 290 nm; em, 440 nm). For Vps9p with Vps21p, we find that the FRET experiment gives less background fluorescence upon nucleotide release, so we utilize 290 nm as the excitation wavelength. However, some of the other Vps9 domain-containing proteins are more amenable to direct analysis of mantGDP fluorescence changes; therefore, we would recommend checking both excitation conditions for a particular GEF/Rab nucleotide release assay.

Vps21p is expressed in bacteria as a hexahistidine-fusion protein using the pET28b Vps21 vector. His$_6$-Vps21p is purified by Ni^{2+} affinity chromatography using the initial steps described for His$_6$MBP-Vps9p purification; the hexahistidine tag is then removed by thrombin cleavage[11], and the protein is fractionated by ion-exchange chromatography similar to the procedure described for Vps9p[12]. The purified protein is then stored at −80°. Yields of Vps21p tend to be fairly robust with approximately 75–100 mg from 1-liter inductions. Vps21p is loaded with mantGDP by the following procedure.

Buffer/Materials

Vps21p: expressed in bacteria, purified by Ni^{2+} affinity and ion-exchange chromatography, and stored at −80° in 5-mg aliquots at 37.5 mg/ml (133 μl)
mantGDP: 5 mM N-methylanthraniloyl-GDP (M-12414, Molecular Probes)
mantGDP release assay buffer: 20 mM HEPES, pH 7.5, 5 mM MgSO$_4$
PD10 desalting column (17-0851-01, Amersham Biosciences)
Bio-Rad Protein Assay (500-0006, Bio-Rad).

Procedure

1. Thaw 5 mg Vps21p aliquot on ice.
2. Combine the 133 μl 1.6 mM Vps21p with 316 μl Tris, pH 8.0, 1 μl 1 M DTT, 4 μl 500 mM EDTA, and 46 μl 5 mM mantGDP (approximately equimolar with Vps21p). Incubate 30 min at 37°; during incubation, equilibrate PD10 desalting column with 25 ml mantGDP release assay buffer.
3. Add 4 μl of 1 M MgSO$_4$ to terminate the loading reaction, and apply loading reaction and 2 ml mantGDP release assay buffer to the PD10 desalting column (discard the eluate).
4. Apply 3.5 ml mantGDP release assay buffer to the column and collect the eluate in seven 500-μl fractions. Use the Bio-Rad Protein Assay to identify Vps21p-containing fractions. Combine Vps21p fractions, determine the protein concentration, and divide into 100-μl aliquots for storage at −80°.

---

[11] His$_6$-Vps21p can be utilized in the nucleotide release assays; however, better activity is observed using Vps21p.
[12] Vps21p elutes between 200 and 300 mM NaCl using 25 mM Tris, pH 7.5, buffers and the Bioscale Q2 or HiTrap Q FF columns.

Although the labeled Vps21p generated via this procedure represents, at best, half of the Vps21p pool, this concentration of mantGDP•Vps21p is sufficient for strong fluorescence using the FRET assay outlined below (Fig. 4).

Buffer/Materials

    mantGDP release assay buffer as previously described
    mantVps21: mantGDP•Vps21p stored at −80° in mantGDP release
        buffer; 100-μl aliquots at 137 μM
    Vps9p: stored in −80° in TBS; 125-μl aliquots at 460 μM
    1 mM GTP: 100 mM GTP diluted 1:100 in mantGDP release assay
        buffer
    Photon Technology International Quanta Master 2001 fluorometer
    1 ml Quartz fluorometer cuvette.

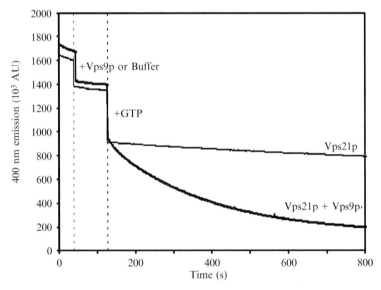

Fig. 4. FRET-based guanine nucleotide exchange activity assay. FRET between mantGDP and Vps21p is measured with the PTI Quanta Master 2001 fluorometer (em, 290 nm; ex, 440 nm). Vps21p loaded with mantGDP is added to the fluorometer cuvette in 800 μl total. Vps9p or buffer (100 μl) is then added to the sample (first dashed line). Following a reequilibration period, excess unlabeled GTP is added (100 μl 1 mM GTP) for exchange onto Vps21p (second dashed line). Data points are collected every second, and the rate of mantGDP release is determined.

Procedure

1. To cuvette, add 782 μl mantGDP release assay buffer and 18.25 μl mantVps21[13]. Place the cuvette with the microstir bar in the fluorometer and stir at 700 rpm.

2. Using the PTI Felix32 Software Package, initiate a 30-min kinetic experiment[14] with excitation at 290 nm and emission at 440 nm, taking timepoints every second (slits set at 4). Initially record for 30–60 s (same for all samples) to ensure consistent mantVps21 addition.

3. Pause the experiment and quickly add 100 μl of mantGDP release assay buffer or Vps9p diluted in mantGDP release assay buffer.[15] Continue recording the fluorescence. The dilution effect should cause an approximately 10% drop in fluorescence detected. Allow the signal to stabilize for 1–2 min (same for all samples).

4. Pause the experiment, quickly add 100 μl of 1 mM GTP, resume data collection as fast as possible, and continue for the remainder of the experiment. The addition of GTP will substantially diminish the fluorescence distinct from dilution effects; this effect is consistent throughout the buffer alone or Vps9p-containing samples, indicating that this decrease is also independent of mantGDP release from Vps21p. We therefore take the fluorescence immediately following GTP addition as the value for normalization between runs.

5. Data are transferred to Microsoft Excel, normalized to initial fluorescence after GTP addition, and the rate of nucleotide release is determined.

From these experiments, the relative activity of various Vps9p mutants can be quantitatively compared to the wild-type protein.

Analysis of Vps9p Ubiquitylation *In Vivo*

In addition to binding ubiquitin via the CUE domain, Vps9p is subject to covalent modification with ubiquitin. Vps9p can be isolated from yeast via affinity purification or immunoprecipitation, and the presence of the ubiquityl moiety is confirmed by Western blotting for endogenous ubiquitin conjugation or blotting for the HA epitope when using yeast expressing HA-tagged ubiquitin (using pGPD416 HA-Ub plasmid). Interestingly, the ubiquityl group on Vps9p appears somewhat stable as the inclusion of 10 mM N-ethylmaleimide (NEM) is not required to permit isolation of

---

[13] Yields 2.5 μM Vps21p in 1 ml reaction volume.
[14] Figure 4 presents a shortened experiment of 800 s.
[15] Final Vps9p concentrations of 100 nM to 40 μM have been examined.

Ub-Vps9p from lysates; this observation contrasts with the requirement for NEM in isolation of ubiquitylated CPS (D. Katzmann, personal communication). Thus, confirmation of Vps9p ubiquitylation is fairly straightforward through the isolation and Western blotting techniques. However, a more simple approach is to examine Vps9p in crude yeast lysates by Western blotting with the Vps9p antiserum (Burd et al., 1996). Through this procedure, Vps9p and Ub-Vps9p can be discerned as bands migrating around 55 and 63 kDa, respectively. This whole cell Western procedure is described below.

Buffer/Materials

YPD or minimal media as previously described
$5\times$ Laemmli sample buffer as previously described
Glass beads: 0.5-mm glass beads (BioSpec Products, Inc.)
TBS: 50 mM Tris, pH 7.5, 150 mM NaCl
TBST: TBS with 0.05% Tween-20
TBSTM: TBST with 5% dry milk
Vps9 antiserum: raised in rabbits against a TrpE-Vps9 fusion protein containing the carboxy-terminal 138 residues, including the CUE domain (Burd et al., 1996).

Procedure

1. Culture strains overnight at 30° in 5ml YPD or appropriate minimal media for plasmid selection, and dilute back in the morning to 0.2 OD_{600}/ml. Continue incubations at 30° until the OD_{600} is 0.5–0.8.
2. Harvest 1 OD_{600} and pellet in a 1.5-ml Eppendorf tube.
3. Resuspend the sample in 100 μl $5\times$ Laemmli sample buffer and add 150 μl glass beads.
4. Vortex for 10 min in a multitube shaker on the highest setting.
5. Heat for 4 min at 95° and spin for 1 min at 13,000 rpm.
6. Resolve 10 μl by SDS-PAGE using 10% acrylamide gels. We find use of the 25 μl Hamilton syringe for gel loading helps prevent accidental transfer of glass beads. Gels are run long enough to permit good separation in the 40–80 kDa range.
7. Transfer protein to nitrocellulose using the Bio-Rad mini transblot cell; 70 V for 2 h is sufficient.
8. Block in TBSTM for 30 min at room temperature.
9. Incubate with Vps9 antiserum (1:2000 in TBSTM) either at 4° overnight or at room temperature for 2–3 h.
10. Wash three times with TBST.

11. Incubate with horseradish peroxidase-conjugated anti-rabbit anti-serum (1:3000 in TBSTM; Amersham Biosciences) for 60–90 min at room temperature.
12. Wash three times with TBST and two times with TBS.
13. Develop the fluorescence signal using the SuperSignal West Femto maximum sensitivity substrate diluted 1:1 in TBS and the UVP gel documentation system.

Using this straightforward assay, the percentage of Vps9p present in the ubiquitylated form can be determined. This percent Ub-Vps9p correlates with the ability of Vps9p to bind ubiquitin as demonstrated by analysis of Vps9p CUE domain mutants (Prag *et al.*, 2003). In addition, this ubiquity-lation is dependent *in vivo* on *RSP5* (Davies *et al.*, 2003; Shih *et al.*, 2003).

Analysis of Vps9p Ubiquitylation *In Vitro*

In vitro ubiquitylation of Vps9p has been reconstituted and is also dependent on Rsp5p, a HECT domain E3 ubiquitin ligase. Initial ubiquitin modification requires three enzymes: the ubiquitin-activating enzyme (E1), the ubiquitin-conjugating enzyme (E2), and the ubiquitin-ligating enzyme (E3). It is possible to reconstitute Vps9p ubiquitylation by incubating purified Vps9p with ATP, ubiquitin (His$_6$HA-Ub), purified human E1 (His$_6$hE1), a yeast E2 (MBP-Ubc5p), and extracts generated from wild-type yeast or extracts generated from *rsp5^{ts}* yeast but supplemented with purified Rsp5p (MBP-Rsp5p), using the procedure below (Fig. 5).

Buffer/Materials

YPD as previously described
Glass beads as previously described
TBS as previously described
His$_6$HAUb: epitope-tagged ubiquitin expressed in bacteria from the pET28-HA-Ub plasmid and purified by Ni^{2+} affinity purification; stored in $-80°$ at 190 μM
His$_6$hE1: ubiquitin-activating enzyme (E1) expressed in SF9 cells and purified by Li Deng and Zhijian Chen (UTSW); stored in $-80°$ at 0.65 mg/ml
MBP-Ubc5p: ubiquitin-conjugating enzyme (E2) expressed in bacteria and purified by amylose affinity purification; stored in $-80°$ at 0.52 mg/ml
MBP-Rsp5p: ubiquitin ligase (E3) expressed in bacteria and purified by amylose affinity purification; stored in $-80°$ at 0.54 mg/ml

MYY290 Extract: +   −   +   −   +   −   +   −
MYY808 Extract: −   +   −   +   −   +   −   +
MBP-Rsp5p: −   −   +   +   −   −   +   +

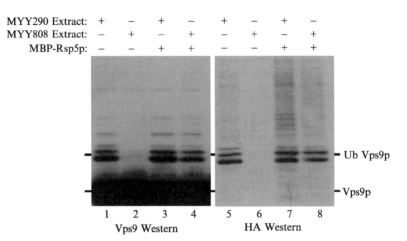

1   2   3   4   5   6   7   8
Vps9 Western   HA Western

Ub Vps9p

Vps9p

Fig. 5. *In vitro* reconstitution of Vps9p ubiquitylation. Extracts from wild-type (MYY290) or *rsp5^{ts}* (MYY808) strains, generated by a glass bead lysis method, are combined with Vps9p (substrate), HA-ubiquitin, purified E1 and E2, and ATP (lanes 1, 2, 5, and 6). Recombinant Rsp5p is added back to the reactions in lanes 3, 4, 7, and 8. Western blotting was performed with Vps9 antiserum (lanes 1–4) or the HA.11 monoclonal antibody (lanes 5–8) to detect ubiquitylated Vps9p.

Vps9p: expressed in bacteria, purified as described; stored in −80° at 1.37 mg/ml (25.3 μM)

10× Ubiquitylation assay buffer: 250 mM Tris, pH 8.0, 1.25 M NaCl, 20 mM MgCl$_2$, with protease inhibitors previously described

50 mM ATP

5× Laemmli sample buffer as previously described

Vps9 antiserum as previously described

HA.11 monoclonal antibody (MMS-101R, Covance Research Products).

Procedure

1. Grow 10 ml overnight cultures of wild-type (MYY290) and *rsp5^{ts}* (MYY808) yeast strains at 25°. Harvest approximately 50 OD$_{600}$ and wash with water. Resuspend in 100 μl cold TBS with protease inhibitors and combine with 150-μl glass beads in a 1.5-ml Eppendorf tube. Vortex samples 6 × 30 s to lyse with incubations on ice between. Add 400 μl TBS with protease inhibitors and vortex briefly. Spin samples 10 min at 13,000 × g at 4°, and divide into ten 50 μl aliquots (approximately 8 mg/ml) for storage at −80°.

2. Thaw lysates and purified proteins in ice.

3. Set up 20 μl ubiquitylation reactions (final concentration in parentheses) with 2 μl 10× Ubiquitylation assay buffer, 1 μl 50 mM

ATP (2.5 mM), 0.5 μl Vps9p (0.34 mg/ml), 0.4 μl His₆HAUb, 0.1 μl His₆hE1 (3.25 μg/ml), 0.6 μl MBP-Ubc5p (0.8 mg/ml), yeast MYY290 or MYY808 lysates (0.8 mg/ml) with or without 1 μl MBP-Rsp5p (27 μg/ml), and water to make up the remaining volume.

4. Incubate 1 h at 25°.

5. Add 10 μl 5× Laemmli sample buffer and heat at 95° for 4 min.

6. Spin down samples and resolve 10 μl by SDS–PAGE on two 10% acrylamide gels. Run the gels such that the free ubiquitin is run off the gel. Transfer to nitrocellulose and perform Western blotting with Vps9p antiserum (1:2000 in TBSTM, with anti-rabbit:HRP secondary at 1:3000 in TBSTM) or HA.11 monoclonal antibody (1:2000 in TBSTM, with anti-mouse:HRP secondary at 1:3000 in TBSTM).

7. Develop fluorescence signal using the SuperSignal West Femto maximum sensitivity substrate diluted 1:4 in TBS.

Using this protocol, *in vitro* reconstitution of Vps9p can be done in a manner dependent on the presence of Rsp5p, either supplied in a crude yeast lysate or as recombinant protein; however, further work is needed to optimize and clarify this crude *in vitro* assay system.

Concluding Remarks

Vps9p can both bind ubiquitin and be ubiquitylated; however, the precise means by which these two events regulate Vps9p is not yet known. Further use of the assays described above with Vps9p mutants unable to bind ubiquitin, unable to undergo ubiquitylation, or unable to bind and activate Vps21p should shed light on the mechanism by which the Vps9 domain and the CUE domain coordinate to facilitate appropriate Vps21p activation.

Acknowledgments

We would like to acknowledge a number of individuals who assisted in the development or optimization of these protocols, including David Katzmann, Beverly Wendland, Marina Ramirez-Alverado, Stephen Sprang, and James Chen, as well as members of the Horazdovsky Laboratory past and present.

References

Aguilar, R. C., and Wendland, B. (2005). Endocytosis of membrane receptors: Two pathways are better than one. *Proc. Natl. Acad. Sci. USA* **102**, 2679–2680.

Babst, M. (2005). A protein's final ESCRT. *Traffic* **6**, 2–9.

Bankaitis, V. A., Johnson, L. M., and Emr, S. D. (1986). Isolation of yeast mutants defective in protein targeting to the vacuole. *Proc. Natl. Acad. Sci. USA* **83**, 9075–9079.

Banta, L. M., Robinson, J. S., Klionsky, D. J., and Emr, S. D. (1988). Organelle assembly in yeast: Characterization of yeast mutants defective in vacuolar biogenesis and protein sorting. *J. Cell. Biol.* **107**, 1369–1383.

Bryant, N. J., and Stevens, T. H. (1998). Vacuole biogenesis in *Saccharomyces cerevisiae*: Protein transport pathways to the yeast vacuole. *Microbiol. Mol. Biol. Rev.* **62**, 230–247.

Burd, C. G., Mustol, P. A., Schu, P. V., and Emr, S. D. (1996). A yeast protein related to a mammalian Ras-binding protein, Vps9p, is required for localization of vacuolar proteins. *Mol. Cell. Biol.* **16**, 2369–2377.

Carney, D. S., Davies, B. A., and Horazdovsky, B. F. (2005). Rab5 guanine nucleotide exchange factors: Regulators of receptor trafficking from yeast to neurons. *Trends in Cell Biol.* in press.

Davies, B. A., Topp, J. D., Sfeir, A. J., Katzmann, D. J., Carney, D. S., Tall, G. G., Friedberg, A. S., Deng, L., Chen, Z., and Horazdovsky, B. F. (2003). Vps9p CUE domain ubiquitin binding is required for efficient endocytic protein traffic. *J. Biol. Chem.* **278**, 19826–19833.

Delprato, A., Merithew, E., and Lambright, D. G. (2004). Structure, exchange determinants, and family-wide rab specificity of the tandem helical bundle and Vps9 domains of Rabex-5. *Cell* **118**, 607–617.

Donaldson, K. M., Yin, H., Gekakis, N., Supek, F., and Joazeiro, C. A. (2003). Ubiquitin signals protein trafficking via interaction with a novel ubiquitin binding domain in the membrane fusion regulator, Vps9p. *Curr. Biol.* **13**, 258–262.

Gerrard, S. R., Bryant, N. J., and Stevens, T. H. (2000). VPS21 controls entry of endocytosed and biosynthetic proteins into the yeast prevacuolar compartment. *Mol. Biol. Cell* **11**, 613–626.

Hama, H., Tall, G. G., and Horazdovsky, B. F. (1999). Vps9p is a guanine nucleotide exchange factor involved in vesicle-mediated vacuolar protein transport. *J. Biol. Chem.* **274**, 15284–15291.

Hicke, L., and Dunn, R. (2003). Regulation of membrane protein transport by ubiquitin and ubiquitin-binding proteins. *Annu. Rev. Cell Dev. Biol.* **19**, 141–172.

Horazdovsky, B. F., Busch, G. R., and Emr, S. D. (1994). VPS21 encodes a rab5-like GTP binding protein that is required for the sorting of yeast vacuolar proteins. *EMBO J.* **13**, 1297–1309.

Katzmann, D. J., and Wendland, B. (2005). Analysis of ubiquitin-dependent protein sorting within the endocytic pathway in *Saccharomyces cerevisiae*. *Methods Enzymol.* in press.

Ponting, C. P. (2000). Proteins of the endoplasmic-reticulum-associated degradation pathway: Domain detection and function prediction. *Biochem. J.* **351**(Pt. 2), 527–535.

Prag, G., Misra, S., Jones, E. A., Ghirlando, R., Davies, B. A., Horazdovsky, B. F., and Hurley, J. H. (2003). Mechanism of ubiquitin recognition by the CUE domain of Vps9p. *Cell* **113**, 609–620.

Raymond, C. K., Howald-Stevenson, I., Vater, C. A., and Stevens, T. H. (1992). Morphological classification of the yeast vacuolar protein sorting mutants: Evidence for a prevacuolar compartment in class E vps mutants. *Mol. Biol. Cell* **3**, 1389–1402.

Rothman, J. H., and Stevens, T. H. (1986). Protein sorting in yeast: Mutants defective in vacuole biogenesis mislocalize vacuolar proteins into the late secretory pathway. *Cell* **47**, 1041–1051.

Sheffield, P., Garrard, S., and Derewenda, Z. (1999). Overcoming expression and purification problems of RhoGDI using a family of "parallel" expression vectors. *Protein Expr. Purif.* **15**, 34–39.

Shih, S. C., Prag, G., Francis, S. A., Sutanto, M. A., Hurley, J. H., and Hicke, L. (2003). A ubiquitin-binding motif required for intramolecular monoubiquitylation, the CUE domain. *EMBO J.* **22,** 1273–1281.

Shtiegman, K., and Yarden, Y. (2003). The role of ubiquitylation in signaling by growth factors: Implications to cancer. *Semin. Cancer Biol.* **13,** 29–40.

Sikorski, R. S., and Hieter, P. (1989). A system of shuttle vectors and yeast host strains designed for efficient manipulation of DNA in Saccharomyces cerevisiae. *Genetics* **122,** 19–27.

Singer-Kruger, B., Stenmark, H., Dusterhoft, A., Philippsen, P., Yoo, J. S., Gallwitz, D., and Zerial, M. (1994). Role of three rab5-like GTPases, Ypt51p, Ypt52p, and Ypt53p, in the endocytic and vacuolar protein sorting pathways of yeast. *J. Cell Biol.* **125,** 283–298.

Singer-Kruger, B., Stenmark, H., and Zerial, M. (1995). Yeast Ypt51p and mammalian Rab5: Counterparts with similar function in the early endocytic pathway. *J. Cell Sci.* **108**(Pt. 11), 3509–3521.

[50] Analysis of the Interaction between GGA1 GAT Domain and Rabaptin-5

By Guangyu Zhu, Peng Zhai, Nancy Wakeham, Xiangyuan He, and Xuejun C. Zhang

Abstract

GGAs are a family of adaptor proteins involved in vesicular transport. As an effector of the small GTPase Arf, GGA interacts using its GAT domain with the GTP-bound form of Arf. The GAT domain is also found to interact with ubiquitin and rabaptin-5. Rabaptin-5 is, in turn, an effector of another small GTPase, Rab5, which regulates early endosome fusion. The interaction between GGAs and rabaptin-5 is likely to take place in a pathway between the *trans*-Golgi network and early endosomes. This chapter describes *in vitro* biochemical characterization of the interaction between the GGA1 GAT domain and rabaptin-5. Combining with the complex crystal structure, we reveal that the binding mode is helix bundle-to-helix bundle in nature.

Introduction

Golgi-associated, γ-adaptin-related, Arf-binding proteins (GGAs) are a family of cytosolic proteins that mediate clathrin-coated vesicle assembly at the *trans*-Golgi network (TGN) membrane in the presence of small GTPase ADP ribosylation factor (Arf) (Bonifacino, 2004; Nakayama and Wakatsuki, 2003). There exist three isoforms of GGA, termed GGA1–3, in humans. All of them consist of multiple domains including the N-terminal

METHODS IN ENZYMOLOGY, VOL. 403
0076-6879/05 $35.00
DOI: 10.1016/S0076-6879(05)03050-8

VHS (VPS27, Hrs, and STAM homology domain, residues 1–150 in GGA1), GAT (GGA and TOM1 homology domain, residues 170–302), hinge region, and the C-terminal GAE (γ-adaptin ear homolog domain, residues 494–639) (Bonifacino, 2004; Nakayama and Wakatsuki, 2003). Important functions have been assigned to each of these domains. In particular, the GAT domain has been identified as the structural entity that interacts with the membrane-associated GTP-bound form of Arf. The GAT domain has also been shown to interact with rabaptin-5 (Mattera *et al.*, 2003) and ubiquitin (Bilodeau *et al.*, 2004; Puertollano and Bonifacino, 2004; Shiba *et al.*, 2004). Previous crystallography studies illustrate that the GGA1 GAT domain contains two distinct regions: an N-terminal helix–loop–helix motif (residues 170–209) and a C-terminal three-helix bundle motif (residues 210–302) (Collins *et al.*, 2003; Shiba *et al.*, 2003; Suer *et al.*, 2003; Zhu *et al.*, 2003). The helix–loop–helix motif is responsible for Arf binding.

Rabaptin-5, consisting of 862 amino acid residues, is an effector of Rab5, which functions as a regulator in homotypic early endosome fusion and heterotypic fusion between early endosomes and clathrin-coated vesicles (Zerial and McBride, 2001). Rabaptin-5 is present mainly in the cytosol, while a small pool is also found at the cell membrane (Stenmark *et al.*, 1995). Based on sequence analysis, rabaptin-5 is predicted to contain multiple coiled-coil regions. It directly binds with many partners including small GTPases Rab5 and Rab4 (Stenmark *et al.*, 1995; Vitale *et al.*, 1998), rabex-5 (Horiuchi *et al.*, 1997), rabphilin (Ohya *et al.*, 1998), γ-adaptin of the AP-1 adaptor complex (Shiba *et al.*, 2002), and GGAs. These bindings have been postulated to connect different steps and pathways of membrane transport. For example, the binding between rabaptin-5 and GGAs is involved in a pathway between TGN and the early endosomes. Interestingly, the interaction between rabaptin-5 and GGAs is divalent: GGA GAE domains recognize an FGPLV sequence (residues 439–443) in a predicted random coil region of rabaptin-5, while GGA GAT domains bind to a C-terminal coiled-coil region of rabaptin-5 (i.e., residues 551–661) (Mattera *et al.*, 2003). The first binding site has been illustrated structurally (Miller *et al.*, 2003). Here we describe the experiments to characterize the second interaction.

Methods

Materials

Unless otherwise stated, all reagents used were of the highest quality and obtained from Fisher Scientific (Pittsburgh, PA) and Sigma (St. Louis, MO); restriction enzymes were from New England Biolabs (Beverly, MA); LB culture media (Catalog #1.10285.5007) was from EMD (Gibbstown, NJ) supplemented with 100 $\mu g/mL$ ampicillin; lysates were supplemented with Complete protease inhibitor tablet without ethylenediaminetetraacetic acid (EDTA, Catalog #1–873–580) and DNase I (Catalog #1–873–580) from Roche (Indianapolis, IN); chromatography and sodium dodecyl sulfate–polyacrylamide gel electrophoresis (SDS-PAGE) equipment and supplies were from Amersham Biosciences (Piscataway, NJ); samples were conditioned and/or concentrated as appropriate using Spectra/Por dialysis (MWCO: 6000–8000) membrane (Catalog #132650) from Spectrum (Rancho Dominquez, CA) and Amicon Ultra-4 (MWCO: 5000) centrifugal filter devices (Catalog #UFC800596) from Millipore (Bedford, MA), respectively.

Purification of Recombinant GGA GAT Fragments

Construction of GGA GAT Domain Fragments. The cDNAs encoding GGA1 regions of residues 170–302 (i.e., the GAT domain; GenBank accession number AF233521), 148–308 (i.e., the GAT domain with an extra N-terminal peptide), 141–326 (i.e., the GAT domain with extra peptides on both ends), 142–210 (i.e., the helix–loop–helix motif with an extra N-terminal peptide), 210–302, (i.e., the helix–loop–helix motif), and 210–326 (i.e., the three-helix bundle motif with an extra C-terminal peptide), GGA2 region of residues 157–342 (equivalent to GGA1 141–326; accession number AF233522), and GGA3 region of residues 147–313 (equivalent to GGA1 141–326; accession number AF219138) were inserted into Amersham Biosciences pGEX6P1 vector (Catalog #27-4597-01) between *Bam*HI and *Eco*RI sites for GGA1 or GGA2, and at the *Eco*RI site for GGA3 in constructing glutathione *S*-transferase (GST)–GAT fusion proteins (Table I). Point mutations were introduced into the GST–GGA1 141–326 parental construct in the pGEX6P1 plasmid using synthesized oligo primers and polymerase chain reaction (PCR)-based mutagenesis. All constructs were verified by DNA sequencing.

Expression of GGA GAT Domain Fragments. Each GGA GAT fragment or variant was transformed into BL21 Star (DE3) *Escherichia coli* cells (Catalog #C6010-03) from Invitrogen (Carlsbad, CA) and expressed as a soluble recombinant protein using standard procedures. Bacterial cells grown at $37°$ as $6\times$ 1-L cultures were induced at 0.6 OD_{600} with 0.1 mM isopropyl-β-D-thiogalactoside (IPTG) for 3 h after which they were harvested, resuspended in 200 mL of $1\times$ phosphate-buffered saline (PBS, pH 7.4), lysozyme (2.5 mg/L of cell culture), Complete protease inhibitor tablet, and frozen at $-20°$ overnight. Upon thawing, DNase I (1200 units) was added, and the lysate stirred for 1 h at $4°$. It was then clarified by centrifugation at $65,000\times g$ for 30 min at $4°$, and the supernatant containing the recombinant protein was further purified.

TABLE I

Schemata for Analyzing Interaction between GAT Domain and Rabaptin-5

Constructs	aa sequence	Expression source	Sample preparation	GST-mediated pull-down[a]	Competition, BIACORE and crystallization[a]
GGA GAT fragments					
GST-GGA1	141–326	pGEX6P1/BL21	Soluble/purified	a, b, c	e, f, g
GST-GGA2	157–342	pGEX6P1/BL21	Soluble/purified	a	
GST-GGA3	147–313	pGEX6P1/BL21	Soluble/purified	a	
GST-GGA1	170–302	pGEX6P1/BL21	Soluble/purified	b	
GST-GGA1	148–308	pGEX6P1/BL21	Soluble/purified	b	
GST-GGA1	142–210	pGEX6P1/BL21	Soluble/purified	b	
GST-GGA1	210–302	pGEX6P1/BL21	Soluble/purified	b	g
GST-GGA1	210–326	pGEX6P1/BL21	Soluble/purified	b	
GST-GGA1 141–326	Variants[b]	pGEX6P1/BL21	Soluble/purified	d	
GST-GGA3 147–313	Variants[b]	pGEX6P1/BL21	Soluble/purified	d	
Rabaptin-5 and fragments					
Rabaptin-5	1–862	HEK293	Lysate	a, b, d	e
His6–rabaptin-5	551–862	pET15b/BL21	Soluble/purified	c	f
Rabaptin-5	551–661	pET11a/BL21	Refolded/purified	c	e, f, g
Rabaptin-5 551–661	Variants[b]	pET11a/BL21	Soluble/lysate	c	

[a] Components used within a particular assay are represented by letters: "a–d," GST-mediated pull-down, where "a" was to analyze rabaptin-5 specificity to different GGA GAT domains (Zhai et al., 2003), "b" to identify the motif in GAT that binds with rabaptin-5 (Zhai et al., 2003; Zhu et al., 2003), and "c" and "d" to define key residues in the interface on the rabaptin-5 and GAT sides, respectively (Zhai et al., 2003; Zhu et al., 2004); "e," GST-mediated competition pull-down (Zhu et al., 2004); "f," BIACORE (Zhai et al., 2003); and "g," crystallization trial (Zhu et al., 2003, 2004).

[b] Point mutations in GGA1 141–326 include K198E, K213E, R214E, K226E, K250E, R258E, P261R, P261E, F264R, R265E, D274R, L277E, E279R, N284S, D285R, Q289R, N292R, and Q296T; point mutations in GGA3 147–313 include S283N and S283D; and point mutations in rabaptin-5 551–661 include Q561A, M563R, Q566A, Q566R, N568A, and N568R.

Purification of GGA GAT Domain Fragments. The protein was purified by gravity affinity chromatography using glutathione Sepharose 4 Fast Flow (10 mL) beads (GSH Seph 4 FF, Catalog #17-5132-01) and eluting with 50 mM Tris–HCl (pH 8.0) containing 10 mM reduced glutathione (GSH). The eluted fraction was dialyzed into 10 mM Tris–HCl (pH 8.5) and run over a Resource Q ion-exchange (6 mL) column (Catalog 17-1179-01) connected to an ÄKTA FPLC system and eluted with a 0–1 M NaCl gradient. Fractions containing the protein were pooled and dialyzed against 10 mM Tris–HCl (pH 8.5) and 0.1% (v/v) 2-mercaptoethanol (2ME), and stored on ice. Typically, 10–30 mg of recombinant GST-fusion proteins can be purified per liter of cell culture. To obtain an untagged protein, the GAT domain was separated from gel-immobilized GST following cleavage with PreScission Protease (Catalog #27-0843-01) from Amersham Biosciences following instructions from the manufacturer. The sample was further purified by Resource Q chromatography with the protein eluting between 150 and 250 mM NaCl. Protein concentration was based on the $OD_{280\ nm}$ and extinction coefficient calculated from the amino acid sequence entered into VectorNTI computer software (InforMax, Inc., Frederick, MA). The correct secondary structures of selected GAT point mutation variants were verified by circular dichroism analysis. GAT variants and wild-type samples (0.5 mg/ml) dialyzed into ultrapure water were subjected to a 190–300 nm wavelength scan using a peltier-type temperature controlled J-715 spectropolarimeter (Jasco, Inc., Tokyo, Japan). The secondary structure compositions were determined using CDPro software (Sreerama and Woody, 2000).

Purification of Recombinant Rabaptin-5 Fragments and Variants

Construction, Expression, and Purification of Rabaptin-5 551–862 Fragment. To express the recombinant human rabaptin-5 551–862 fragment (accession number X91141), rabaptin-5 cDNA corresponding to this region flanked by *Nde*I and *Xho*I sites was amplified and cloned into Novagen vector pET15b (Catalog #69661-3) and expressed in BL21 cells as described before. Ni-IDA immobilized metal affinity chromatography (IMAC) was used to purify the recombinant His-tag fusion protein. A 5-ml Hi-trap Chelating HP column (Catalog #17-0409-01) was charged with 0.1 mM NiCl$_2$ and 0.02% NaN$_3$, and equilibrated; the protein was then bound in *binding buffer* (50 mM Tris–HCl [pH 8.5], 200 mM NaCl, and 5 mM imidazole) and eluted with *binding buffer* containing 0.2 M imidazole. The collected protein sample was further purified by Resource Q chromatography as described before. The protein eluted at ~250 mM NaCl was identified by SDS-PAGE stained with Coomassie Brilliant blue

(Catalog #17-0518-01) using a PhastSystem (Amersham Biosciences), pooled, dialyzed against 10 mM Tris–HCl (pH 8.0) and 0.1% (v/v) 2ME, and stored on ice. Because of the lack of tryptophan residues in this fragment, the protein concentration was determined using the Bio-Rad Bradford assay (Catalog #500-0006), which yielded ~2.5 mg/L of cell culture.

Construction, Expression, and Purification of Rabaptin-5 551–661 Fragment. A shorter construct encoding rabaptin-5 residues 551–661 was subcloned into a pET11a vector (Novagen, Catalog #69436-3) between *Nde*I and *Bam*HI restriction sites, and overexpressed as inclusion bodies (IBs) in BL21 cells at 37° with 1 mM IPTG induction for 3 h. Protein released from solubilized IBs was refolded using the rapid dilution method (Lin *et al.*, 1994). Cells from a 6-liter culture were harvested and lysed as described previously. The IB pellet was separated from supernatant by centrifugation at 18,000×g for 30 min, then resuspended and washed three times with 1× PBS containing 0.1% Triton X-100, and dissolved in 60 mL *solubilization buffer* (8 M urea, 2 mM EDTA, 1 mM glycine, 100 mM Tris [pH 10.5], and 5 mM 2ME). Any residual insoluble material was removed by centrifugation at 35,000×g for 1 h, and the cleared supernatant was diluted to 200 mL ~5.0 OD$_{280}$ with *solubilization buffer* containing 300 mg dithiothreitol (DTT), 60 mg reduced GSH, and 12 mg oxidized GSH. The sample was rapidly diluted dropwise (2 ml/min using a peristaltic pump) into 4 liters of stirring room temperature *refolding buffer* (20 mM Tris [pH 10.5], 1% glycerol, and 2 mM EDTA). The pH is adjusted to 9.0 and it is stored at 4°. Over the course of 2 days the pH was progressively adjusted to pH 8.0. The protein solution was concentrated to ~100 mL in a Millipore Pellicon 2 tangential flow ultrafiltration system fitted with two ultrafiltration filters (MWCO: 10,000) and further concentrated to ca. 20 ml in a nitrogen-pressurized Amicon stirred cell. The refolded protein was loaded directly onto a Sephacryl S-200 HR column (3.2 × 100 cm, Catalog #17-0584-01) controlled by a GradiFrac system, first equilibrated with 20 mM Tris (pH 8.0) and 0.4 M urea. Finally, fractions identified by nonreducing SDS-PAGE as the correctly refolded monomeric protein form were combined and further purified by Resource Q chromatography with the protein eluting between ~100 and 200 mM NaCl. The yield of this short rabaptin-5 construct was ~3 mg/L of cell culture.

Construction and Expression of Rabaptin-5 551–661 Variants. Point mutations were introduced into the construct of rabaptin-5 551–661 using PCR-based site-directed mutagenesis, expressed in BL21 cells at 30° with 0.1 mM IPTG induction for 5 h; the cells were then harvested and lysed as described previously. Each lysate containing a soluble rabaptin-5 551–661 variant was used directly in subsequent pull-down assays.

Expression of Full-Length Rabaptin-5 in HEK293 Cells. A pCDNA3.1
expression vector containing cDNA of full-length human rabaptin-5 was
transfected into HEK293 cells using the Lipofectamin2000 transfection
reagent (Invitrogen, Catalog #R795–07 and 11668–027, respectively). The
transfected cells were grown in a T-25 culture flask for 36 h, resuspended in
buffer A ($1\times$ PBS, 1 mM EDTA, 1 mM MgCl$_2$, 1 mM DTT, 1% [v/v] NP40
detergent, supplemented with Calbiochem [San Diego, CA] Protease Inhibitor Cocktail Set I [Catalog #539131]), then broken by repeated ejection
of the lysate through a 21-gauge needle. Supernatant from the homogenate,
first collected from centrifugation at 10,000$\times g$ for 10 min and then at
120,000$\times g$ for 1 h at 4°, was used for the binding assay.

Biochemical Characterization of the Interaction between GGA GAT Domain and Rabaptin-5

Pull-Down Assay for GGA–Rabaptin-5 Interaction. An aliquot of cell
lysate containing $\sim$30 μg of a fusion protein of GST–GAT variant was
incubated for 30 min at 22° with 20 μL of 50% slurry of GSH Seph 4 FF
beads in a final volume of 500 μl of *binding buffer* ($1\times$ PBS, 0.2% [v/v]
Triton X-100, 0.2% [v/v] 2ME, and Complete protease inhibitor tablet).
Beads containing immobilized GST fusion protein were washed three
times by alternating centrifugation (3 min at 2000$\times g$) and resuspension in
binding buffer. Washed beads were incubated for 2 h at 4° with 30 μg of
purified rabaptin-5 551–862 in 1 ml *binding buffer* and washed again three
times with 1 ml *binding buffer*. After the last centrifugation, 30 μl of
2$\times$ SDS sample buffer was added, and the sample was heated for 5 min
at 90°, followed by separating the supernatant from the beads for SDS–
PAGE analysis. The binding affinities between rabaptin-5 551–661 variants
in crude cell lysate and purified GST–GGA1 141–326 were analyzed in a
similar manner.

*Competition Analysis Emphasizing the Interaction between GGA1
GAT Domain and Full-Length Rabaptin-5.* An aliquot of full-length ra-
baptin-5 lysate (200 μl containing $\sim$100 μg of total protein) was incubated
in *buffer A* containing 20 μg of purified GST–GGA1 141–326 fusion
protein (prebound to 40 μL of 50% slurry of GSH Seph 4 FF beads) in
the presence of varied concentrations (0–20 μM) of rabaptin-5 551–661 in a
400 μl final volume for 2 h at 22° with gentle rotation. The beads were
washed three times with 500 μl of *buffer A*, 25 μl reducing 2$\times$ SDS sample
buffer was added, and bound proteins were applied to a Novex (San Diego,
CA) 10% Tris–tricine SDS-PAGE and blotted onto Immobilon-P^{SQ}
PVDF transfer membrane (Millipore, Catalog #ISEQ 07850). The blot
incubated sequentially with mouse monoclonal anti-rabaptin-5 antibody

(1:500 dilution, Catalog #610676) supplied by BD Biosciences (San Jose, CA), which shows no cross-reactivity to the rabaptin-5 551–661 fragment, and Sigma peroxidase-conjugated anti-mouse IgG (1:5000 dilution, Catalog #A9044) was detected by film using ECL Plus chemiluminescence reagent (Amersham Biosciences, Catalog #RPN2132). The presence of rabaptin-5 551–661 in the complex was confirmed by SDS-PAGE.

Biomolecular Interaction Analysis between GGA1 GAT and Rabaptin-5 Fragments Using Surface Plasmon Resonance. Binding affinity between GGA1 GAT and rabaptin-5 fragments was quantitatively determined by surface plasmon resonance (SPR) using a BIACORE 3000 biosensor (BIACORE, Inc., Piscataway, NJ). Recombinant GGA1 141–326 with the N-terminal GST tag removed, as previously described, was prepared at a concentration of 5 μg/ml in 10 mM sodium acetate (pH 5.0) and immobilized in random orientation onto the surface of a BIACORE carboxymethylated dextran matrix (CM5, Catalog #BR-1000–12) sensor chip. Sequential injections were applied to achieve a target average surface density of 500 RU (response unit). Analyses were performed in *running buffer* (10 mM HEPES [pH 7.2], 150 mM NaCl, 0.1 mM DTT, and 0.005% [v/v] P-20 surfactant) at 25°. Purified recombinant proteins of rabaptin-5 551–661 and 551–862 fragments in varied concentrations (75–800 nM) were injected individually over the sensor surface at 0.5 μL/s for a 180 s association phase, followed by a 180 s *running buffer* only dissociation period. Interactions were recorded in the form of sensorgrams with nonspecific refraction subtracted automatically and concurrently. After each cycle, the sensor surface was regenerated by injecting 5 μl of 2 M MgCl$_2$ and 1 mM DTT followed by a 120 s equilibration in the *running buffer*. First the dissociation rate constant (k_{off}) and then the association rate constant (k_{on}) were calculated using homogeneous kinetics for dissociation and association type-1 models through BIAevaluation software (version 3.1). Apparent equilibrium dissociation constants were calculated using the equation $K_d = k_{off}/k_{on}$, based on a model assuming an average of one GAT binding site per rabaptin-5 monomer.

Cocrystallization of the GGA1–Rabaptin-5 Complex. A crystal form of GGA1 GAT and rabaptin-5 fragment complex was obtained by cross-seeding the crystallization drop with a crystal of GGA1 210–302 alone. This GGA1 210–302 crystal seed was obtained in 1.0 M sodium citrate and 0.1 M HEPES (pH 7.5). It belongs to a tetragonal space group with cell parameters $a = b = 115$ Å and $c = 61$ Å, and diffracts to about 3.8 Å resolution. The complex crystallized in a completely different crystal form of the $P3_221$ space group with cell parameters of $a = b = 155$ Å and $c = 53$ Å. Crystals of this form were obtained at 20° using the hanging drop method under the following condition: protein samples of GGA1 210–302

(20 mg/mL) and rabaptin-5 551–661 (8 mg/mL) were first mixed 1:2 (v/v), then this sample was mixed 2:1 (v/v) with the *reservoir solution* (1.1 M [NH$_4$]$_2$SO$_4$, 0.1 M Tris [pH 9.0], and 0.1% [v/v] 2ME). The hanging drop was equilibrated with *reservoir solution* for 48 h, and then seeded with either the GGA1 210–302 crystal or the complex one. A few complex crystals grew to a size of ∼0.8 × 0.4 × 0.4 mm from a 4.5-μl drop in ∼10 days.

Concluding Remarks

GGA proteins function predominantly in the vesicle trafficking pathways that connect to Golgi networks, while rabaptin-5 is recognized as an essential regulator of Rab5 functioning in endocytosis. The experiments described here contribute to our understanding of involvement of these elements and their connection to other steps in regulating vesicle trafficking between TGN and endosomes. In particular, the interaction mode between GGA GAT and rabaptin-5 is found to be helix bundle-to-helix bundle in nature, and it involves the C-terminal three-helix bundle motif of the GAT domain and a homodimer region within the rabaptin-5 551–661 fragment (Zhai *et al.*, 2003; Zhu *et al.*, 2004).

Acknowledgments

This work was supported by a grant from the Oklahoma Center for the Advancement of Science and Technology to X.C.Z.

References

Bilodeau, P. S., Winistorfer, S. C., Allaman, M. M., Surendhran, K., Kearney, W. R., Robertson, A. D., and Piper, R. C. (2004). The GAT domains of clathrin-associated GGA proteins have two ubiquitin binding motifs. *J. Biol. Chem.* **279**, 54808–54816.

Bonifacino, J. S. (2004). The GGA proteins: Adaptors on the move. *Nat. Rev. Mol. Cell Biol.* **5**, 23–32.

Collins, B. M., Watson, P. J., and Owen, D. J. (2003). The structure of the GGA1-GAT domain reveals the molecular basis for ARF binding and membrane association of GGAs. *Dev. Cell* **4**, 321–332.

Horiuchi, H., Lippe, R., McBride, H. M., Rubino, M., Woodman, P., Stenmark, H., Rybin, V., Wilm, M., Ashman, K., Mann, M., and Zerial, M. (1997). A novel Rab5 GDP/GTP exchange factor complexed to Rabaptin-5 links nucleotide exchange to effector recruitment and function. *Cell* **90**, 1149–1159.

Lin, X. L., Lin, Y. Z., and Tang, J. (1994). Relationships of human immunodeficiency virus protease with eukaryotic aspartic proteases. *Methods Enzymol.* **241**, 195–224.

Mattera, R., Arighi, C. N., Lodge, R., Zerial, M., and Bonifacino, J. S. (2003). Divalent interaction of the GGAs with the Rabaptin-5-Rabex-5 complex. *EMBO J.* **22**, 78–88.

Miller, G. J., Mattera, R., Bonifacino, J. S., and Hurley, J. H. (2003). Recognition of accessory protein motifs by the gamma-adaptin ear domain of GGA3. *Nat. Struct. Biol.* **10,** 599–606.

Nakayama, K., and Wakatsuki, S. (2003). The structure and function of GGAs, the traffic controllers at the TGN sorting crossroads. *Cell Struct. Funct.* **28,** 431–442.

Ohya, T., Sasaki, T., Kato, M., and Takai, Y. (1998). Involvement of Rabphilin3 in endocytosis through interaction with Rabaptin5. *J. Biol. Chem.* **273,** 613–617.

Puertollano, R., and Bonifacino, J. S. (2004). Interactions of GGA3 with the ubiquitin sorting machinery. *Nat. Cell Biol.* **6,** 244–251.

Shiba, Y., Takatsu, H., Shin, H. W., and Nakayama, K. (2002). Gamma-adaptin interacts directly with Rabaptin-5 through its ear domain. *J. Biochem. (Tokyo)* **131,** 327–336.

Shiba, T., Kawasaki, M., Takatsu, H., Nogi, T., Matsugaki, N., Igarashi, N., Suzuki, M., Kato, R., Nakayama, K., and Wakatsuki, S. (2003). Molecular mechanism of membrane recruitment of GGA by ARF in lysosomal protein transport. *Nat. Struct. Biol.* **10,** 386–393.

Shiba, Y., Katoh, Y., Shiba, T., Yoshino, K., Takatsu, H., Kobayashi, H., Shin, H. W., Wakatsuki, S., and Nakayama, K. (2004). GAT (GGA and Tom1) domain responsible for ubiquitin binding and ubiquitination. *J. Biol. Chem.* **279,** 7105–7111.

Sreerama, N., and Woody, R. W. (2000). Estimation of protein secondary structure from circular dichroism spectra: Comparison of CONTIN, SELCON, and CDSSTR methods with an expanded reference set. *Anal. Biochem.* **287,** 252–260.

Stenmark, H., Vitale, G., Ullrich, O., and Zerial, M. (1995). Rabaptin-5 is a direct effector of the small GTPase Rab5 in endocytic membrane fusion. *Cell* **83,** 423–432.

Suer, S., Misra, S., Saidi, L. F., and Hurley, J. H. (2003). Structure of the GAT domain of human GGA1: A syntaxin amino-terminal domain fold in an endosomal trafficking adaptor. *Proc. Natl. Acad. Sci. USA* **100,** 4451–4456.

Vitale, G., Rybin, V., Christoforidis, S., Thornqvist, P., McCaffrey, M., Stenmark, H., and Zerial, M. (1998). Distinct Rab-binding domains mediate the interaction of Rabaptin-5 with GTP-bound Rab4 and Rab5. *EMBO J.* **17,** 1941–1951.

Zerial, M., and McBride, H. (2001). Rab proteins as membrane organizers. *Nat. Rev. Mol. Cell Biol.* **2,** 107–117.

Zhai, P., He, X., Liu, J., Wakeham, N., Zhu, G., Li, G., Tang, J., and Zhang, X. C. (2003). The interaction of the human GGA1 GAT domain with rabaptin-5 is mediated by residues on its three-helix bundle. *Biochemistry* **42,** 13901–13908.

Zhu, G., Zhai, P., He, X., Terzyan, S., Zhang, R., Joachimiak, A., Tang, J., and Zhang, X. C. (2003). Crystal structure of the human GGA1 GAT domain. *Biochemistry* **42,** 6392–6399.

Zhu, G., Zhai, P., He, X., Wakeham, N., Rodgers, K., Li, G., Tang, J., and Zhang, X. C. (2004). Crystal structure of human GGA1 GAT domain complexed with the GAT-binding domain of Rabaptin5. *EMBO J.* **23,** 3909–3917.

[51]   Purification and Properties of Rab6
Interacting Proteins

By SOLANGE MONIER and BRUNO GOUD

Abstract

A crucial step in the characterization of novel partners of Rab proteins is the confirmation that they indeed interact together by techniques other than the yeast two-hybrid assay used to discover them. Some methods and clues that would help to discriminate between putative interactors are summarized. Pull-down, coimmunoprecipitation, and gel filtration experiments are described as ways of checking protein–protein interaction *in vitro* and *in vivo*.

Introduction

This chapter describes a general approach used for the study of Rab interacting proteins. After the initial identification of putative Rab interactors by a yeast two-hybrid screen of a library, the major issue that remains to be clarified is the specificity and the validity of the interaction. Both proteins with completely unknown functions and proteins sharing identified homology domains, must after their discovery, be confirmed as genuine Rab interactors before their involvement in a Rab-regulated pathway can be further investigated. We describe here a few clues to proceed and hopefully to define *bona fide* interactors of our protein of interest.

The Golgi-associated Rab6 protein is involved in the retrograde transport from the endosomal recycling compartment to the Golgi and from there to the endoplasmic reticulum (ER) (Girod *et al.*, 1999; White *et al.*, 1999). This was assessed by using Rab6 mutant overexpression: overexpression of the GTPase-deficient mutant of Rab6, Rab6Q72L, results in a massive redistribution of Golgi resident proteins to the ER. This process is microtubule dependent (Young *et al.*, 2005). Whether and how Rab6 effectors such as rabkinesin 6 (Echard *et al.*, 1998) or the dynein/dynactin motor complex (BicD2, p150^{glued}, p50) (Matanis *et al.*, 2002; Short *et al.*, 2002) regulate this pathway is still under debate. Identification of more Rab6 partners may help in understanding how complex pathways that make intracellular traffic are modulated.

Rab proteins cycle between an inactive GDP-bound cytosolic form and the active membrane-bound GTP-associated form. These two conformations confer to Rab proteins the ability to bind specific sets of proteins. The

METHODS IN ENZYMOLOGY, VOL. 403
0076-6879/05 $35.00
DOI: 10.1016/S0076-6879(05)03051-X

best characterized proteins that specifically bind Rabs in their GDP form are REP, GDI, and PRA1 (Martincic *et al.*, 1997; Seabra and Wasmeier, 2004; Zhang, 2003), which are involved in the movement of Rab toward or off the membranes; in addition, guanine nucleotide exchange factors (GEFs) are essential partners that activate Rab by mediating exchange of GDP by GTP. Many proteins have been characterized, which specifically interact with the active GTP-bound membrane-associated form of Rab, and are supposed to be involved in the various steps of vesicular transport. Among them, GAPs promote Rab inactivation and allow a new cycle. Both Rab configurations can be obtained *in vitro* and thereby specific inter-actors of both forms can be tested by the pull-down experiment technique. Next, the possible interaction of Rab partners *in vivo* can be tested by a coimmunoprecipitation experiment. Further characterization is provided by gel filtration. Here we illustrate the approach of the characterization of Rab6 partners with R6IP2, a protein involved in endosome-to-Golgi transport (Monier *et al.*, 2002).

Materials

Guanosine 5′-(3-*O*-thio)triphosphate (GTPγS) is obtained from Roche, GDP from Sigma, and glutathione Sepharose (GSH) 4B beads, Superdex200 PC3.2/30 column from Pharmacia.

GST-Rabs are produced in bacteria, in their nonprenylated form.

Bovine brain extract and postnuclear supernatants (PNSs) from tissue and cells are prepared as previously described (Monier *et al.*, 2002).

Buffers and Solutions

GDP is solubilized to 0.1 M stock solution in 0.1 M Tris–HCl, pH 7.5

Loading buffer: 25 mM Tris–HCl, pH 7.5, 10 mM EDTA, 0.05% CHAPS, 5 mM MgCl$_2$

Interacting buffer: 25 mM Tris–HCl, pH 7.5, 10 mM MgCl$_2$, 50 mM NaCl, 0.1% Triton X-100

Immunoprecipitation buffer: 20 mM Tris–HCl, pH 7.5, 150 mM NaCl, 10 mM MgCl$_2$, 0.5% Triton X-100

Gel filtration buffer: 50 mM Tris–HCl, pH 8, 0.1 M NaCl

Pull-Down Experiment

GST-tagged Rab protein is immobilized on GSH-coupled Sepharose 4B beads, loaded with GDP or with GTPγS in the presence of low Mg^{2+} concentration, then incubated in the presence of interactor, either provided as a protein purified from bacteria (small protein, < 50 kDa) or from Sf9/ baculovirus system (for larger proteins or for processed proteins), or synthesized *in vitro* using an SP6/T7 transcription/translation kit, or as a protein present in a tissue or cell extract, either endogenously or overexpressed using appropriate vector.

Of GST-Rab ten μg is incubated in a volume higher than 500 μl on a rotating wheel at 4° for 90 min with 30 μl GSH-Sepharose 4B previously washed in loading buffer and resuspended as indicated by the manufacturer. Beads are centrifuged at 700×g for 5 min and resuspended in the same buffer containing 200 μM GDP or GTPγS, incubated at 37° for 1 h. At the end of incubation, a final concentration of 20 mM MgCl$_2$ is added, before samples are centrifuged at 4° and incubated in the presence of interacting buffer containing the interactor of interest. Incubation is performed at room temperature for 90 min under rotation. Samples are then centrifuged and washed four times with the same buffer. The bead-associated proteins are solubilized and submitted to SDS-PAGE. Proteins associated with Rab:GDP or Rab:GTPγS are detected by Western blot or autoradiography.

Figure 1 shows the interaction *in vitro* of Rab6 with R6IP2, a Rab6 effector involved in ERC to Golgi transport (Monier *et al.*, 2002), and with AA2, a recombinant antibody recognizing Rab6:GTP (Nizak *et al.*, 2003). GST-Rab6A (lanes 1 and 2) and GST alone (lane 3) were incubated in the presence of either mouse brain PNS (upper and middle panel) or recombinant antibody AA2 (lower panel). Rab6 in its GTP conformation specifically binds R6IP2 and AA2 (lane 2). In constrast, only Rab6:GDP interacts with GDIα and β (lane 1).

GDI binding proves to be an essential and useful control of Rab loading; indeed, GDI is abundant in cell or tissue extract, and binds with very high specificity to Rab:GDP (even to unprenylated Rab), in particular to Rab6, Rab4, and Rab33 (when the anti-GDI antibody must be incubated in the presence of 0.04% Triton X-100 instead of 0.1% Tween-20 for immunodetection). However, Rab11, Rab24, Rab3, and Rab5 do not bind GDI in the same conditions.

The robustness of interaction can be tested by increasing the stringency of the interacting buffer: up to 500 mM NaCl and 0.5% Triton X-100 can indicate the interaction strength, or decrease the background.

1. GST protein is the usual recommended control for a pull-down assay. However, the use of other GST-Rab fusion proteins constitutes the best alternative since its structure is very similar to the Rab of interest.

2. A major issue is the source of Rab interactor used. Indeed, although other techniques such as the yeast two-hybrid assay strongly suggest direct

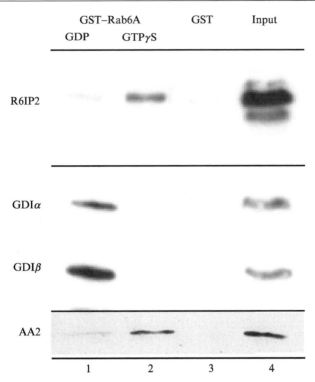

FIG. 1. Specific interactors of Rab6 in its GDP and GTP conformation. GST-Rab6A loaded with GDP (lane 1) or GTPγS (lane 2) was incubated with mouse brain PNS (upper and middle panel) or AA2 (lower panel). A fifth of the amount used in the interaction step was loaded as "input" (lane 4). On the left are indicated the protein identified using a specific antibody (anti-R6IP2 or anti-GDIβ), or using antibodies against the tagged protein (anti-His + anti-Myc for AA2).

interaction between one Rab and its interactor of interest, it often appears difficult to strictly confirm it using a pull-down experiment, especially when extracts are used rather than a purified compound. A recommendation is to use various sources, such an *in vitro* synthesized protein together with extract.

3. Depending on the source of interactor used and on the affinity of interactors for Rab proteins, washes may become very critical steps. Indeed when using the recombinant antibody AA2 as shown in Fig. 1, interaction is readily lost after the third wash. Conversely, background can be significantly reduced by additional wash when using PNS instead of cytosolic cell or tissue fractions.

Coimmunoprecipitation Experiment

To demonstrate that our proteins of interest are also able to interact in living cells is the next step to prove that the interaction observed in a pull-down experiment is relevant.

Endogenous Rab:GTP protein has proven to be fairly impossible to coimmunoprecipate with any of its partners. The reason is unknown and is possibly linked to the Rab:GTP instability or its low abundance in a cell lysate.

One way to overcome this situation is to overexpress the Rab:GTP mutant as well as its putative partner and to perform immunoprecipation by using tagged proteins.

Gel Filtration

Separation of cytosolic proteins according to their size indicates whether they are present as a monomer or in a complex.

Eighty microliters of brain extracts containing about 1 mg of protein is ultracentrifuged and the supernatant loaded on an analytical Superdex-200 column. Fractions of 50 μl eluted from the column are analyzed for their protein content by SDS-PAGE and Western blot. Figure 2 shows the separation of cytosolic proteins present in bovine brain extract. R6IP2 is found at its expected molecular mass size ($\sim$110 kDa), in fractions also

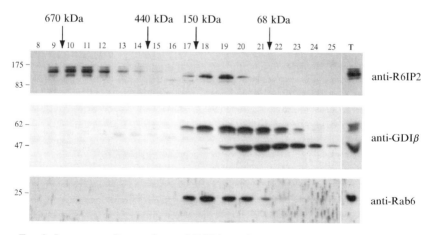

FIG. 2. Large cytosolic complexes of R6IP2. Bovine brain cytosol was submitted to gel filtration. On the top of the figure are indicated the size markers used to calibrate the column, on the left markers for the gel; the antibodies used to reveal the protein are indicated on the right; unfractionated brain cytosol was loaded in lane T.

containing Rab6 proteins and GDIα and β. It is unlikely that R6IP2 is associated with any other protein given its molecular weight and its retention volume. However, as already known, Rab6 and GDI are likely to be found in the same complex (Rab6 elutes around 150 kDa, while its molecular weight is about 20 kDa). In addition, R6IP2 is also found in fractions eluted at very high molecular mass (∼ 670 kDa, Fig. 2). It is important to note that this technique allows detection of protein pools that are less represented than the major forms of the protein. Indeed, it appears that the R6IP2 isoform found in large complexes is slightly different from the form retained with the 150-kDa marker.

We also found that GAPcenA (Cuif *et al.*, 1999) and R6IP1 (unpublished results), two other Rab6 partners mostly cytosolic, are present in the form of very high molecular weight complexes (data not shown), which only partly overlap, indicating that they likely represent independent complexes.

Whether these proteins can form complexes can be further studied by coimmunoprecipitation experiments.

Protein complexes are, however, thought to essentially associate to membranes, especially as far as vesicular traffic is concerned. The challenge for such studies is to find the appropriate detergent that preserves protein interaction while solubilizing the membrane lipids.

References

Cuif, M. H., Possmayer, F., Zander, H., Bordes, N., Jollivet, F., Couedel-Courteille, A., Janoueix-Lerosey, I., Langsley, G., Bornens, M., and Goud, B. (1999). Characterization of GAPCenA, a GTPase activating protein for Rab6, part of which associates with the centrosome. *EMBO J.* **18,** 1772–1782.

Echard, A., Jollivet, F., Martinez, O., Lacapere, J. J., Rousselet, A., Janoueix-Lerosey, I., and Goud, B. (1998). Interaction of a Golgi-associated kinesin-like protein with Rab6. *Science* **279,** 580–585.

Girod, A., Storrie, B., Simpson, J. C., Johannes, L., Goud, B., Roberts, L. M., Lord, J. M., Nilsson, T., and Pepperkok, R. (1999). Evidence for a COP-I-independent transport route from the Golgi complex to the endoplasmic reticulum. *Nat. Cell Biol.* **1,** 423–430.

Martincic, I., Peralta, M. E., and Ngsee, J. K. (1997). Isolation and characterization of a dual prenylated Rab and VAMP2 receptor. *J. Biol. Chem.* **272,** 26991–26998.

Matanis, T., Akhmanova, A., Wulf, P., Del Nery, E., Weide, T., Stepanova, T., Galjart, N., Grosveld, F., Goud, B., De Zeeuw, C. I., Barnekow, A., and Hoogenraad, C. C. (2002). Bicaudal-D regulates COPI-independent Golgi-ER transport by recruiting the dynein-dynactin motor complex. *Nat. Cell Biol.* **4,** 986–992.

Monier, S., Jollivet, F., Janoueix-Lerosey, I., Johannes, L., and Goud, B. (2002). Characterization of novel Rab6-interacting proteins involved in endosome-to-TGN transport. *Traffic* **3,** 289–297.

Nizak, C., Monier, S., del Nery, E., Moutel, S., Goud, B., and Perez, F. (2003). Recombinant antibodies to the small GTPase Rab6 as conformation sensors. *Science* **300,** 984–987.

Seabra, M. C., and Wasmeier, C. (2004). Controlling the location and activation of Rab GTPases. *Curr. Opin. Cell Biol.* **16,** 451–457.

Short, B., Preisinger, C., Schaletzky, J., Kopajtich, R., and Barr, F. A. (2002). The Rab6 GTPase regulates recruitment of the dynactin complex to Golgi membranes. *Curr. Biol.* **12,** 1792–1795.

White, J., Johannes, L., Mallard, F., Girod, A., Grill, S., Reinsch, S., Keller, P., Tzschaschel, B., Echard, A., Goud, B., and Stelzer, E. H. (1999). Rab6 coordinates a novel Golgi to ER retrograde transport pathway in live cells. *J. Cell Biol.* **147,** 743–760.

Young, J., Stauber, T., del Nery, E., Vernos, I., Pepperkok, R., and Nilsson, T. (2005). Regulation of microtubule-dependent recycling at the trans-Golgi network by Rab6A and Rab6A'. *Mol. Biol. Cell* **16,** 162–177.

Zhang, H. (2003). Binding platforms for Rab prenylation and recycling: Rab escort protein, RabGGT, and RabGDI. *Structure (Camb.)* **11,** 237–239.

[52] Affinity Purification of Ypt6 Effectors and Identification of TMF/ARA160 as a Rab6 Interactor

By Symeon Siniossoglou

Abstract

Rab/Ypt GTPases are key regulators of intracellular traffic in eukaryotic cells. One important function of Rab/Ypts is the nucleotide-dependent recruitment of downstream effector molecules onto the membrane of organelles. In budding yeast Ypt6 is required for recycling of membrane proteins from endosomes back to the Golgi. A biochemical approach based on the affinity purification of Ypt6:GTP-interacting proteins from yeast cytosol led to the identification of two conserved Ypt6 effectors, the tetrameric VFT complex and Sgm1. The mammalian homolog of Sgm1, TMF/ARA160, contains a short conserved coiled-coil motif that is sufficient for the binding to the three mammalian orthologs of Ypt6, Rab6A, Rab6A', and Rab6B.

Introduction

Rab/Ypt GTPases are members of the Ras superfamily of monomeric GTP-binding proteins that function in vesicle formation, motility, tethering, and fusion with target membranes (Pfeffer and Aivazian, 2004; Zerial and McBride, 2001). Rab/Ypts (11 members in budding yeast and more than 60 in humans) localize to different intracellular compartments and regulate membrane transport by cycling between a GDP- and a GTP-bound state. A basic property of activated, GTP-loaded Rab/Ypts is the

METHODS IN ENZYMOLOGY, VOL. 403   0076-6879/05 $35.00
DOI: 10.1016/S0076-6879(05)03052-1

recruitment of effectors, soluble proteins or multisubunit complexes that mediate docking of donor and acceptor membranes. Therefore, selective recruitment of effectors onto the different intracellular organelles by Rab/Ypt GTPases is a key event that contributes to the specificity of membrane transport. Moreover, a number of effectors bind components of the core factors mediating the lipid bilayer fusion, SNAREs, suggesting that they could also play a more direct role in membrane fusion (Segev, 2001).

Ypt6, the yeast homolog of mammalian Rab6, is associated with the Golgi complex and is implicated in the recycling of proteins from endosomes back to the late Golgi (or *trans*-Golgi network in mammalian cells). In *ypt6* mutants the sorting receptor Vps10p is mislocalized to the vacuole and its cargo, the vacuolar protease carboxypeptidase Y, is secreted as a result (Bensen *et al.*, 2001; Tsukada *et al.*, 1999). Moreover, removal of Ypt6 blocks recycling of the late Golgi SNAREs Tlg1p and Tlg2p and the exocytic SNARE Snc1p from endosomes back to the Golgi (Siniossoglou and Pelham, 2001).

Fusion of endosome-derived vesicles with the late Golgi depends on a cascade of protein–protein interactions, initiated by the activation of Ypt6 to its GTP-bound form catalyzed by the heterodimeric exchange factor Ric1–Rgp1 (Siniossoglou *et al.*, 2000), a peripheral membrane protein complex that localizes to the Golgi. Ypt6:GTP then recruits Sgm1, a conserved coiled-coil protein, and VFT (also called GARP by Conibear *et al.*, 2003), a tetrameric complex that binds independently the SNARE Tlg1 (Siniossoglou and Pelham, 2001, 2002). This could provide a mechanism by which the asymmetric localization of the exchange factor results in the docking of Tlg1p-containing vesicles onto Golgi membranes.

Mapping of the Ypt6:GTP-binding domain of Sgm1 led to the identification of a short 100-residue coiled-coil motif found in a wide range of eukaryotes. The mammalian homolog containing this motif is TMF/ARA160, a protein previously identified in various screens as a putative transcription or chromatin remodeling factor (Garcia *et al.*, 1992; Hsiao and Chang, 1999; Mori and Kato, 2002). TMF/ARA160 localizes to the Golgi complex, binds the three known Rab6 isoforms (Rab6A, Rab6A', and Rab6B), and is required for proper Golgi morphology (Fridmann-Sirkis *et al.*, 2004). These data, together with the fact that mammalian cells contain putative homologs of Ric1 and VFT, suggest that the Ypt6 Golgi targeting machinery is evolutionarily conserved.

This chapter describes the procedures to (1) identify Ypt6 effectors from yeast cytosol, based on the method used by Christoforidis and Zerial (2000) to purify Rab5 effectors from brain extracts, and (2) assay the binding of TMF/ARA160 to Rab6. The ease of genetic manipulation in yeast allows the rapid preparation of cytosol(s) containing any tagged wild-type or mutant protein(s) one would like to assay for binding to a given Ypt GTPase. The results can then be validated by *in vivo* assays such as localization of effector-GFP fusions in *ypt* mutants.

Identification of Ypt6 Effectors from Yeast Cytosol

Preparation of GST-Ypt6 Beads and Loading with GTPγS

   Solutions

      Bacterial lysis buffer: phosphate-buffered saline (PBS) consisting of 137 mM NaCl, 2.7 mM KCl, 10 mM Na$_2$HPO$_4$, 2 mM KH$_2$PO$_4$ pH 7.4, 1% Triton X-100
      Binding buffer: 20 mM Tris–HCl, pH 8.0, 110 mM KCl, 5 mM MgCl$_2$, 1 mM dithiothreitol (DTT)
      NE buffer: binding buffer containing 10 mM EDTA, 10 μM GTPγS
      NS buffer: binding buffer containing 10 μM GTPγS

Procedure. To construct a GST–Ypt6 fusion, the open reading frame of *YPT6* is cloned into the expression vector pGEX6P2 (Amersham Biosciences) so that its amino-terminal end is fused in frame to the GST tag sequence in the vector. The plasmid containing the fusion gene is then transformed into the *Escherichia coli* strain *MC1061* (available from ATCC, http:/www.atcc.org) and one transformant expressing high levels of GST–Ypt6 is inoculated in 10 ml 2 × TY medium containing 100 μg/ml ampicillin and grown overnight at 37°. Next morning, dilute the cells into 800 ml of medium and grow to OD$_{600}$ 0.6. Add isopropyl-β-D-thiogalacto-pyranoside (IPTG) to 0.2 mM and grow the cells at 37° for 3 h. Collect the cells by centrifuging at 6000×g for 10 min. Wash the cells once with water, freeze in liquid nitrogen, and store the pellet at −80°. To purify the fusion protein, thaw the cells in ice and resuspend them in 24 ml bacterial lysis buffer containing 1 tablet of EDTA-free protease inhibitors (Roche) and 50 μg/ml phenylmethylsulfonyl fluoride (PMSF, Sigma). Dounce the cell suspension until it becomes homogeneous and then sonicate on ice for 5 × 30 s with 2 min of cooling in between. Centrifuge the extract at 30,000×g for 20 min. Collect the supernatant and add 0.6 ml glutathione Sepharose beads (Amersham Biosciences) preequilibrated with bacterial lysis buffer. Incubate at 4° for 30 min under rotation. Spin down the beads at 4000×g for 3 min and wash them once with 30 ml bacterial lysis buffer and once with 10 ml binding buffer. Recombinant GST fusions are loaded with GTPγS as described in Christoforidis and Zerial (2000) with minor modifications: 0.3-ml beads are transferred into a 1-ml Mobicol column

(Mobitec) containing a 90-μm pore size filter and washed with 1 ml NE buffer. Columns are spun at 500×g for 1 min to remove the buffer and then beads are incubated on a turning wheel with 1 ml NE buffer containing 1 mM GTPγS for 30 min at room temperature. This wash/incubation step is repeated twice. The beads are then washed once with 1 ml NS buffer and incubated with 1 ml NS buffer containing 1 mM GTPγS for 30 min at room temperature. The excess of Mg^{2+} in the NS buffer stabilizes the GTPase in the GTP-bound form. Control beads loaded with GDP are prepared as described earlier, with the exception that the NE and NS buffers contain GDP instead of GTPγS.

Preparation of yeast cytosol

Solutions

Alkaline DTT buffer: 100 mM Tris–HCl, pH 9.4, 10 mM DTT
Spheroplasting buffer: 1.2 M Sorbitol, 20 mM K$_2$HPO$_4$/KH$_2$PO$_4$, pH 7.4
Binding buffer: see prior description

Procedure. Grow an overnight culture of wild-type yeast (strain SEY6210, *Mat*α *ura3 his3 leu2 trp1 suc2 lys2*) (Robinson *et al.*, 1988) in 100 ml YPD medium (1% yeast extract, 2% peptone, 2% D-glucose) at 30°. Next morning, dilute the preculture into 2 liters of fresh medium to OD$_{600}$ 0.1 and let it grow at 30° to OD$_{600}$ 1.0. Harvest the cells by centrifuging at 4000×g for 5 min and wash them once with water. Resuspend the cells in 100 ml alkaline DTT buffer and incubate for 10 min at room temperature under moderate shaking. Centrifuge as described earlier, weigh the cell pellet, and resuspend it in 100 ml of spheroplasting buffer. Add 5 mg of Zymolyase 20T per gram of cell pellet, mix gently, and incubate under gentle shaking at room temperature for 15 min. Monitor spheroplasting by diluting 50 μl of the yeast suspension into 1 ml water. Loss of turbidity indicates cell lysis. Most wild-type yeast strains will need 15–20 min for an efficient spheroplasting. Centrifuge at 3000×g, and without resuspending the pellet, rinse it once with spheroplasting buffer. Spin again, pour off the supernatant, snap-freeze the spheroplasts in liquid nitrogen, and store them at −80°. To lyse spheroplasts, the frozen cell pellet is thawed in ice and resuspended in 3.5 ml of binding buffer per gram of cells, containing EDTA-free protease inhibitors (Roche). All the following steps take place at 4°. Dounce with 10 strokes using a Dounce homogenizer. Centrifuge the lysate at 100,000×g for 30 min. Collect the supernatant and dialyze it against 3 liters of lysis buffer to remove any endogenous nucleotide that could interfere with the GST-Ypt/Rab beads. Dialyzed cytosol is then precleared by incubation for 1 h with 300 μl of glutathione-Sepharose beads preequilibrated in lysis buffer. This step helps reduce background due to the unspecific binding of cytosolic proteins to the Sepharose beads. The beads are removed by centrifugation at $4000 \times g$ for 3 min and the cytosol is stored on ice.

For the preparation of cytosol containing protein A-tagged versions of the VFT complex (Vps52–PtA, Vps53–PtA, Vps54–PtA), Sgm1 (Sgm1–PtA), and the Sgm1 Ypt6-binding domain (Sgm1[597–707]–PtA), grow the corresponding strains (*vps52Δ* + YCplac111–*VPS52*–PtA, *vps53Δ* + YCplac111–*VPS53*–PtA, *vps54Δ* + YCplac111–*VPS54*–PtA, *sgm1Δ* + YCplac111–*SGM1*–PtA, *sgm1Δ* + YEplac181–*SGM1[597–707]*–PtA) in selective medium (0.67% Yeast Nitrogen Base with ammonium sulfate, 2% D-glucose supplemented with amino acid drop out lacking leucine) and process them as described above.

Identification of Effector Proteins Binding to GST-Ypt6:GTPγS

Solutions

Binding buffer: see previous description
Wash buffer I: 20 mM Tris–HCl, pH 8.0, 220 mM KCl, 5 mM MgCl$_2$, 1 mM DTT
Wash buffer II: 20 mM Tris–HCl, pH 8.0, 220 mM KCl, 5 mM MgCl$_2$, 1 mM DTT
Elution buffer: 20 mM Tris–HCl, pH 8.0, 1.5 M KCl, 20 mM EDTA, 1 mM DTT

Procedure. The beads containing GST–Ypt6:GTPγS are transferred into a Poly-prep chromatography column (Bio-Rad). Yeast cytosol (10 ml) prepared as described earlier and containing 100 μM GTPγS is added into the column and the slurry is incubated for 2 h at 4° under rotation. The columns are then opened and the cytosol is removed by gravity flow. Beads are washed sequentially with 3 ml of binding buffer containing 10 μM GTPγS, 3 ml of wash buffer I containing 10 μM GTPγS, and 1 ml of wash buffer II, and then transferred into a Mobicol column. The buffer from the last wash is drained out by centrifugation at $1000 \times g$ for 10 s and GST-Ypt6:GTPγS-interacting proteins are eluted by incubating the beads with 0.3 ml of elution buffer containing 5 mM GDP for 15 min at room temperature under rotation. The GST-Ypt6:GDP eluate was obtained as for the GST-Ypt6:GTPγS with the difference that all buffers contained GDP instead of GTPγS and the elution buffer contained –1 mM GTPγS instead of GDP. The eluates are concentrated by ultrafiltration (Nanosep, Pall-Gellman) to 50 μl. Half of each eluate is loaded on a 10% SDS-PAGE and stained using Coomassie BB.

Comments

SPECIFICITY. In the experiment shown in Fig. 1, specificity is demonstrated by (1) the preferential binding of the effectors to the GTPγS-bound form of Ypt6 (although we found that within the same experiment, Sgm1 binds reproducibly more efficiently to Ypt6:GTPγS than the VFT complex, probably reflecting the different properties of these effectors), (2) the lack

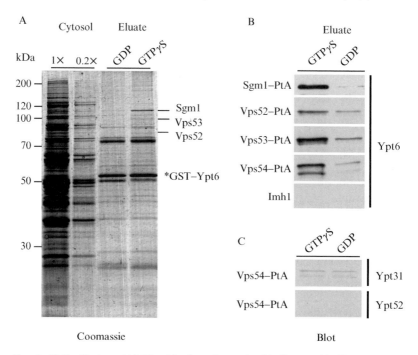

FIG. 1. Ypt6 effectors. (A) Identification of proteins binding specifically to GST–Ypt6:GTPγS. Cytosol from spheroplasts of a wild-type strain was prepared as described in the text ("1×," 2 μl or "0.2×," 0.4 μl) and incubated with GST–Ypt6:GTPγS or GST–Ypt6:GDP containing beads. Bound proteins were eluted and analyzed by SDS-PAGE followed by Coomassie staining as described in the text. Molecular mass markers (in kDa) are shown on the left. Proteins bound preferentially to GST–Ypt6:GTPγS are indicated on the right. The identity of each band was determined by mass spectrometry and verified by Western blot analysis (see B). The asterisk indicates GST–Ypt6 that leaks from the glutathione beads during the elution. (B) Western blot analysis of fractions eluted from GST–Ypt6:GTPγS or GST–Ypt6:GDP beads following incubation with cytosol of yeast strains expressing Sgm1–PtA, Vps52–PtA, Vps53–PtA, or Vps54–PtA fusion proteins, using anti-protein A antibodies (DAKO). Bound fractions from cytosol containing Vps52–PtA were also probed with an antibody recognizing Imh1p (Tsukada et al., 1999). (C) Western blot analysis of eluates obtained from GST–Ypt31 and GST–Ypt52 beads, preloaded with GTPγS or GDP as described above for Ypt6 and incubated with cytosol from a strain expressing Vps54–PtA. Blots were incubated with anti-protein A antibodies.

of binding of another Golgi-associated coiled coil protein, Imh1, to Ypt6: GTPγS, and (3) the lack of binding of Vps54–PtA to two other GTPases, the Golgi-associated Ypt31 and the endosomal Ypt52, both of which were purified and loaded with nucleotides as Ypt6.

CYTOSOL PREPARATION. By obtaining cytosol from various yeast mutants, the described procedure can be used to map GTPase-interaction domains of effector molecules. For example, by preparing cytosol from *vps52/53/54* deletion mutants that impair the assembly of the complex, we assessed the binding properties of the single proteins and showed that Vps52 is the subunit that mediates binding to Ypt6 (Siniossoglou and Pelham, 2002). Similarly, binding of cytosols derived from *sgm1Δ* cells expressing various fragments of Sgm1 with Ypt6 led to the identification of a short coiled-coil domain (corresponding to residues 597–707) that is sufficient and necessary for binding to Ypt6-GTPγS (Fridmann-Sirkis *et al.*, 2004, see also Fig. 2A). Finally the presence of the TEV-protease cleavage site-protein A tag at the C-terminus of the VFT subunits can allow the

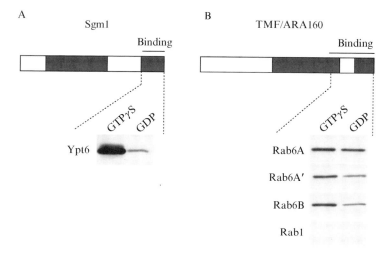

FIG. 2. Binding of Sgm1 and TMF/ARA160 to Ypt6 and Rab6 through a conserved coiled-coil motif. Schematic representation of the primary structure of yeast Sgm1 (A) and mouse TMF/ARA160 (B). Gray boxes indicate the coiled-coil domains. (A) Cytosol from a *sgm1Δ* yeast strain expressing a carboxy-terminal fragment of Sgm1 (amino acid residues 597–707) tagged with protein A was incubated with GST-Ypt6:GTPγS or GST-Ypt6:GDP as in Fig. 1. Eluted fractions were loaded onto SDS-PAGE, transferred to nitrocellulose, and probed with anti-protein A antibodies. (B) A lysate from bacteria expressing the 310 carboxy-terminal residues of mouse TMF/ARA160 as a His$_6$-tagged fusion was incubated with the indicated recombinant Rab6 and Rab1 as in (A). Eluted fractions were loaded onto SDS-PAGE, transferred to nitrocellulose, and probed with an anti-penta-His antibody.

rapid purification and elution of the native complex from yeast cytosol to analyze its direct binding on Ypt6.

Binding of the Conserved Ypt6-GTP Interacting Domain of TMF/ARA160 to Rab6

Procedure

The amino-terminal ends of the full-length cDNAs of human Rab6A, Rab6A′, Rab6B, and Rab1 are fused in frame to GST into pGEX6P2 and the resulting fusion proteins are expressed in MC1061, purified, and loaded with GTPγS or GDP as described for Ypt6. All recombinant Rabs are expressed equally well in bacteria at 37° with the exception of GST-Rab6A′, whose yield is reduced by 50%. The 310 C-terminal residues of mouse TMF/ARA160 are fused in frame to the His$_6$ tag sequence in the expression vector pET30a (Novagen). The plasmid is then transformed into the *Escherichia coli* strain BL21(DE3). One transformant expressing high levels of His$_6$-TMF-C is inoculated in 10 ml 2 × TY medium containing 40 μg/ml kanamycin and grown overnight at 30°. The next morning, dilute the cells into 100 ml of medium and grow the cells at 30° to OD$_{600}$ 0.6. Add IPTG to 0.5 mM and grow the cells at 30° for 2 h. Lyse the cells in 10 ml binding buffer, centrifuge the extract, and dialyze the supernatant against the binding buffer as described previously. Prepare eight 1-ml Mobicol columns, each containing 400 μl binding buffer, 10 μg BSA, and 50 μl glutathione-Sepharose beads, each with bound GST-Rab6A, GST-Rab6A′, GST-Rab6B, and GST-Rab1, preloaded with GTPγS or GDP. Transfer 10 μl of the supernatant containing His$_6$-TMF-C into each column. Binding reactions, washes, and elution were performed as described previously. One-fifth of each eluate was loaded on a 10% SDS-PAGE, transferred onto nitrocellulose membranes, and probed with the penta-His antibody (Qiagen).

Acknowledgments

I am grateful to Hugh Pelham for advise and stimulating discussions during the course of this work in his laboratory. I thank Sean Munro for the Rab6A and Rab1 cDNAs and helpful comments on the manuscript and Uri Nir for TMF plasmids. This work was supported by a long-term postdoctoral fellowship from the Human Frontiers Science Program to SS.

References

Bensen, E. S., Yeung, B. G., and Payne, G. S. (2001). Ric1p and the Ypt6p GTPase function in a common pathway required for localization of *trans*-Golgi network membrane proteins. *Mol. Biol. Cell* **12**, 13–26.

Christoforidis, S., and Zerial, M. (2000). Purification and identification of novel Rab effectors using affinity chromatography. *Methods* **20,** 403–410.

Conibear, E., Cleck, J. N., and Stevens, T. H. (2003). Vps51p mediates the association of the GARP (Vps52/53/54) complex with the late Golgi t-SNARE Tlg1p. *Mol. Biol. Cell* **14,** 1610–1623.

Fridmann-Sirkis, Y., Siniossoglou, S., and Pelham, H. R. (2004). TMF is a golgin that binds Rab6 and influences Golgi morphology. *BMC Cell Biol.* **5,** 18.

Garcia, J. A., Ou, S.-H. I., Wu, F., Lusis, A. J., Sparkes, R. S., and Gaynor, R. B. (1992). Cloning and chromosomal mapping of a human immunodeficiency virus 1 "TATA" element modulatory factor. *Proc. Natl. Acad. Sci. USA* **89,** 9372–9376.

Hsiao, P.-W., and Chang, C. (1999). Isolation and characterization of ARA160 as the first androgen receptor N-terminal-associated coactivator in human prostate cells. *J. Biol. Chem.* **274,** 22373–22379.

Mori, K., and Kato, H. (2002). A putative nuclear receptor coactivator (TMF/ARA160) associates with hbrm/hSNF2 alpha and BRG-1/hSNF2 beta and localizes in the Golgi apparatus. *FEBS Lett.* **520,** 127–132.

Pfeffer, S., and Aivazian, D. (2004). Targeting Rab GTPases to distinct membrane compartments. *Nat. Rev. Mol. Cell Biol.* **5,** 886–896.

Robinson, J. S., Klionsky, D. J., Banta, L. M., and Emr, S. D. (1988). Protein sorting in *Saccharomyces cerevisiae* — isolation of mutants defective in the delivery and processing of multiple vacuolar hydrolases. *Mol. Cell. Biol.* **8,** 4936–4948.

Segev, N. (2001). Ypt and Rab GTPases: Insight into functions through novel interactions. *Curr. Opin. Cell Biol.* **13,** 500–511.

Siniossoglou, S., and Pelham, H. R. (2001). An effector of Ypt6p binds the SNARE Tlg1p and mediates selective fusion of vesicles with late Golgi membranes. *EMBO J.* **20,** 5991–5998.

Siniossoglou, S., and Pelham, H. R. (2002). Vps51p links the VFT complex to the SNARE Tlg1p. *J. Biol. Chem.* **277,** 48318–48324.

Siniossoglou, S., Peak-Chew, S. Y., and Pelham, H. R. (2000). Ric1p and Rgp1p form a complex that catalyses nucleotide exchange on Ypt6p. *EMBO J.* **19,** 4885–4894.

Tsukada, M., Will, E., and Gallwitz, D. (1999). Structural and functional analysis of a novel coiled-coil protein involved in Ypt6 GTPase-regulated protein transport in yeast. *Mol. Biol. Cell* **10,** 63–75.

Zerial, M., and McBride, H. (2001). Rab proteins as membrane organizers. *Nat. Rev. Mol. Cell Biol.* **2,** 107–117.

[53] Assay and Properties of Rab6 Interaction with Dynein–Dynactin Complexes

By Evelyn Fuchs, Benjamin Short, and Francis A. Barr

Abstract

RAB GTPases help to maintain the fidelity of membrane trafficking events by recruiting cytosolic tethering and motility factors to vesicle and organelle membranes. In the case of Rab6, it recruits the dynein-dynactin

METHODS IN ENZYMOLOGY, VOL. 403
Copyright 2005, Elsevier Inc. All rights reserved.

0076-6879/05 $35.00
DOI: 10.1016/S0076-6879(05)03053-3

complex to Golgi-associated vesicles via an adaptor protein of the Bicaudal-D family. Here we describe methods for the identification of Rab6-binding partners in cell extracts. We then focus on the biochemical analysis of interactions with the dynein-dynactin complex and the adaptor proteins Bicaudal-D1 and -D2. Standard protocols for yeast two-hybrid analysis, and biochemical assays for the analysis of the interactions between Rab6, Bicaudal-D, and the subunits of the dynein-dynactin complex are outlined.

Introduction

Rab6 has been implicated in microtubule-mediated membrane traffic events between endosomes and the *trans*-Golgi network and in retrograde trafficking within the Golgi and from the Golgi back to the endoplasmic reticulum (Echard et al., 1998; Mallard et al., 2002; Young et al., 2005). A number of effector proteins for mammalian Rab6 have been described that may explain how Rab6 function relates to microtubules. First, a kinesin-9 family motor protein Rabkinesin-6/Rab6-KIFL was proposed to be a Rab6 effector based on a yeast two-hybrid screen (Echard et al., 1998). However, recent findings suggest that Rabkinesin-6/Rab6-KIFL is a mitosis-specific motor protein required for the transport of Aurora and Polo family protein kinases during cytokinesis (Fontijn et al., 2001; Gruneberg et al., 2004; Hill et al., 2000; Neef et al., 2003). Based on these findings and close sequence relationship between Rabkinesin-6/Rab6-KIFL and the mitotic kinesins of the MKlp1/kinesin-9 family, we have proposed that it be reclassified as a mitotic kinesin with the name MKlp2 (mitotic kinesin-like protein 2) (Neef et al., 2003). A second link between Rab6 and microtubules comes from the findings that Rab6 can interact with a coiled-coil protein of the *trans*-Golgi Bicaudal-D (BicD) (Hoogenraad et al., 2001; Matanis et al., 2002; Short et al., 2002). Human cells have two BicD proteins, BicD1 and Bicd2, which can interact with both the dynein–dynactin motor complex (Hoogenraad et al., 2001) and Rab6 (Matanis et al., 2002; Short et al., 2002). Additionally, p150[glued], a core component of the dynactin complex, can also interact with Rab6, suggesting that Rab6 directly interacts with dynactin through p150[glued] and indirectly via BicD proteins (Short et al., 2002). Rab6 together with BicD proteins may therefore be involved in the dynein–dynactin-dependent capture and movement of Golgi-associated vesicles along microtubules (Vaughan et al., 2002). In this chapter, we describe methods for the analysis of interactions between Rab6, p150[glued], and BicD proteins and for the isolation of Rab6-binding partners from cell extracts.

Directed Yeast Two-Hybrid Analysis of Rab6–Effector Interactions

We use a yeast two-hybrid system based on the PJ69-4A reporter yeast strain and the pGBT9 bait, and pACT2 prey vectors (James *et al.*, 1996). The advantage of this system is 2-fold. First, PJ69-4A has two independent reporters (for histidine and adenine biosynthesis) that can be used to select for interactions, as well as lacZ for blue-white colony screening. Second, the bait is expressed at a low level and can thus be used to study proteins that may be toxic if expressed at higher levels. We have observed that a number of small coiled-coil fragments display yeast two-hybrid interactions with Rab GTP-binding proteins, while the corresponding full-length proteins do not. This highlights the importance of verifying interactions using other methods such as binding assays with recombinant proteins, or cell extracts. Another consideration for Rab-binding partners is that many have extensive coiled-coil regions; coiled-coil proteins are often self-activating when placed in the two-hybrid bait vector and for this reason it is better to perform screens or tests with the Rab in the bait vector and the coiled-coil protein in the prey vector.

In the examples shown in Fig. 1, we tested the interaction of the wild-type, GDP- and GTP-locked forms of Rab6 with BicD1 and BicD2. This analysis shows that GTP-locked Rab6^{Q72L} exhibits a clear interaction with both BicD proteins (Fig. 1A). An equivalent series of Rab1 constructs shows no interaction with BicD proteins, but GTP-locked Rab1Q70L does interact with its effector p115 as expected (Fig. 1A). Deletion constructs can be used to map the interaction sites, in this case the binding sites for BicD1 and BicD2 are mapped to a carboxy-terminal region conserved in both proteins (Fig. 1A). A similar approach used for p150^{glued} shows that it specifically interacts with the GTP form of Rab6 and identifies a region containing a number of coiled-coil segments between amino acids 540 and 1278 as the Rab6-binding site.

Yeast Growth Media

To prepare the amino acid base for synthetic complex medium (SC), weigh out adenine 5.0 g, alanine 20 g, arginine 20 g, asparagine 20 g, aspartic acid 20 g, cysteine 20 g, glutamine 20 g, glutamic acid 20 g, glycine 20 g, inositol 20 g, isoleucine 20 g, lysine 20 g, methionine 20 g, *p*-amino-benzoic acid 2.0 g, phenylalanine 20 g, proline 20 g, serine 20 g, threonine 20 g, tyrosine 20 g, and valine 20 g. Mix this thoroughly in a glass bottle on a rotor mixer overnight and store in a cool, dark place away from moisture. To obtain SC-minus leucine and tryptophan (SC-LW) drop out the supplement and weigh out 36.7 g of the amino acid base and histidine 2.0 g and uracil 2.0 g. To obtain the two-hybrid selective media (QDO quadruple

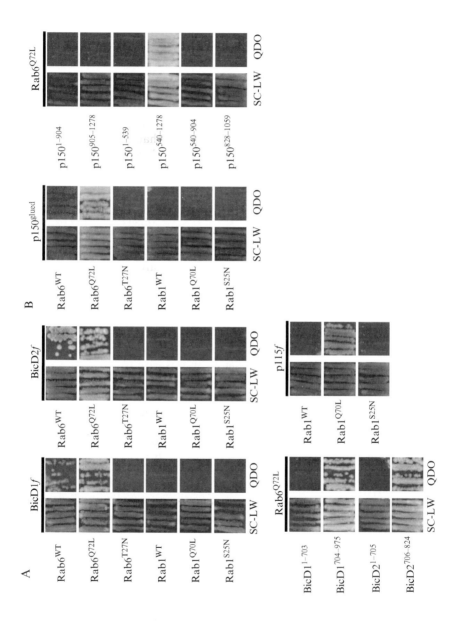

drop out, lacking adenine, histidine, leucine, and tryptophan) weigh out 36.7 of the amino acid base and 2.0 g of uracil. Mix this thoroughly and store as stated previously.

For 1 liter of the required SC media weigh out 6.7 g yeast nitrogen base (without amino acids), 2.0 g of the appropriate drop out mix, and glucose 20 g. Adjust the pH to 6.5 with NaOH, and then autoclave to sterilize. For plates, weigh 20 g of agar into a bottle, add 1 liter of the required liquid SC medium, place a stir bar in the bottom, and then sterilize by autoclaving. After autoclaving place it on a stirrer, and once cooled to approximately 55° add 20 ml 0.2% (w/v) adenine (filter sterilized stock solution) and pour the plates immediately. For two-hybrid selection plates (QDO) do not add adenine; all other media should contain adenine.

Preparation of Frozen Competent Yeast Cells

Select several colonies from a freshly streaked plate of PJ69-4A and grow them overnight in 100 ml of YPAD (autoclaved yeast extract 10 g/liter, bacto peptone 20 g/liter, sucrose 20 g/liter, adenine 100 mg/liter). Measure the OD_{600} of the overnight culture and dilute to 0.15 in fresh medium. Grow at 30° to an OD_{600} of 0.5–0.6, equivalent to 1.2–1.5×10^7 cells. Harvest the cells by centrifugation at $2000 \times g$, for 2 min, at room temperature. Wash and resuspend the cells with one-half the culture volume of water and spin as before. Resuspend the pellet in 1/8th culture volume of LiSorb (filter sterilized 100 mM lithium acetate, 10 mM Tris–Cl, 1 mM ethylenediaminetetraacidic acid [EDTA], 1 M sorbitol adjusted to pH 8.0), incubate for 5 min, and spin as before. Remove the supernatant and spin again to remove residual supernatant. Resuspend the cell pellet in 600 μl LiSorb per 100 ml of culture volume, and add 10 μl herring sperm DNA per 100 μl yeast solution (herring sperm DNA 10 mg/ml stock [Promega]

FIG. 1. BicD1, BicD2, and p150 interact with Rab6 in Y2H. (A) Full-length BicD1 and BicD2 show interaction with Rab6 but not with Rab1. The interaction with wild-type Rab6 ($Rab6^{WT}$) is weaker than with the GTP-locked Rab6 mutant ($Rab6^{Q72L}$); note that Rab6Q72L colonies are less pink on QDO. The inactive GDP-locked Rab6 ($Rab6^{T27N}$) does not bind any of the effector proteins. As a positive control for Rab1, the interaction with the cis-Golgi tethering protein p115 and known Rab1 effector is shown. Note that the GTP-locked Rab1^{Q72L} but not the wild-type Rab1 shows an interaction in the yeast two-hybrid assay. To map the binding region for Rab6 various constructs of BicD1 and BicD2 were tested against Rab6^{Q72L}. BicD1 and BicD2 bind Rab6 via the C-terminal quarter of the protein (aa 704–975 in BicD1, aa 706–824 in BicD2). (B) Full-length p150^{glued} shows a strong interaction with activated Rab6^{Q72L} and a weaker interaction with wild type Rab6, but no interaction with Rab1. The minimal binding region in p150 covers the C-terminal half of the protein (aa 540–1278). (See color insert.)

heated to 95° for 10 min before use, then kept on ice). Mix well by vortexing. Freeze the cells at −80° in 100-μl aliquots. Do not shock freeze; simply place a box containing the cells in the freezer. Competent cells can be kept at −80° for several months.

Small-Scale Transformations for Directed Two-Hybrid Tests

Thaw the cells by leaving them on the bench; use 10 μl for each transformation. All pipetting is done on the bench at room temperature. Add 1 μl of both bait (pGBT9) and prey (pACT2) plasmids; we typically use 100–200 ng/μl miniprep DNA for this, and it works reproducibly with high efficiency. Add 150 μl LiPEG (filter sterilized 100 mM lithium acetate, 10 mM Tris–Cl, 1 mM EDTA, 40% [w/v] PEG-3350 adjusted to pH 8.0) and mix well by vortexing. Incubate for 20 min at room temperature, then add 17.5 μl dimethyl sulfoxide (DMSO), and mix well by vortexing. Heat shock the cells for 15 min at 42° in a waterbath, not a heating block. Centrifuge for 2 min at 2000×g in a microcentrifuge at room temperature to pellet the cells. Carefully remove as much of the supernatant as possible. Resuspend the cell pellet in 500 μl water and plate 100 μl on appropriate selective plates, in this case synthetic complete medium lacking leucine and tryptophan (SC-Leu/Trp). Grow cells at 30° for 3 days. After this time, carefully inspect the plates. All should have evenly sized off-white/pink colonies. Large white colonies on a background of smaller pinkish colonies are almost certainly mutants. Do not select these! Using a sterile tip or toothpick, pick up a small amount of a colony and make a small streak on SC-Leu/Trp (go back and forth three or four times), then use the same tip on QDO (synthetic complete yeast medium lacking leucine, tryptophan, adenine, and histidine). Do this for five colonies from each transformation. Grow the cells at 30° for 2–3 days. After this time, carefully inspect the plates. Combinations that give rise to a two-hybrid interaction will grow on both SC-LW and QDO. Strong interactors give rise to white colonies on QDO, whereas weak interactors are pink because of incomplete activation of the adenine biosynthesis reporter gene. True positives show bait-dependent growth and do not grow with empty bait vector control.

Biochemical Analysis of Rab6-Binding Partners

Binding assays using recombinant proteins are a simple way to analyze Rab6-binding partners *in vitro*. Briefly, recombinant Rab6 expressed in and purified from bacteria can be immobilized on a solid support and loaded with the appropriate nucleotide. An additional level of specificity comes from the use of nucleotide-state-specific mutants of the Rab protein—in the case of Rab6 the T27N GDP-locked and Q72L GTP-locked mutants, respectively. The potential binding partners to be tested are then added to the immobilized Rab proteins and eluted by the addition of EDTA at the end of the binding period. Chelation of magnesium strips the bound nucleotide from the Rab and hence the release of the bound nucleotide and any nucleotide-state-specific binding partners. Other proteins nonspecifically adhering to the beads are unlikely to be eluted by this treatment, in contrast to elution with salts or chaotropic agents. An example of such a binding assay is illustrated by the binding of recombinant BicD2 to the GTP but not the GDP form of Rab6 (Fig. 2A). Note that carboxy-terminally truncated BicD2 (asterisk) fails to bind to Rab6, consistent with the yeast two-hybrid mapping data (Fig. 1A). In contrast to previous reports, we have been unable to find any interaction between Rabkinesin-6/Rab6-KIFL/Mklp2 and either of the two splice variants of Rab6a (Fig. 2B) under the same conditions used for the binding of BicD proteins to Rab6.

A similar approach can be used to study the interaction of Rab6 with protein complexes in cell extracts, to confirm two-hybrid screening data or recombinant protein-binding assays. In the example shown, the dynactin complex subunit p150^{glued} and BicD1 and BicD2 could be isolated from HeLa cell extracts on GTP-loaded Rab6^{WT} or Rab6^{Q72L} beads but not GDP-loaded Rab6^{T27N} beads (Fig. 3). Scaling up such binding assays can generate sufficient amounts of material to identify novel Rab-binding proteins (Short *et al.*, 2002).

Expression and Purification of Recombinant Rab6

Hexahistidine-GST-tagged Rab6 is expressed in *Escherichia coli* BL21 (pRIL) using the bacterial expression vector pGAT2 after induction from an OD$_{600}$ of 0.5 with 0.5 mM isopropyl-β-D-thiogalactoside (IPTG) overnight at 18°. Hexahistidine-tagged BicD1 and BicD2 in pQE32 (Qiagen) can be expressed by induction from an OD$_{600}$ of 0.5 at 37° for 3 h in *E. coli* JM109. Cell pellets can be frozen at $-20°$ until required or processed immediately. Thaw the cell pellets at room temperature by the addition of 10 ml lysis buffer per liter of original culture (IMAC5: 20 mM Tris–HCl, pH 8.0, 300 mM NaCl, 5 mM imidazole, 0.2% [w/v] Triton X-100, plus 0.5 mg/ml lysozyme and Complete protease inhibitor cocktail [Roche Diagnostics]). Resuspend using a 10-ml pipette and pool the lysate in one tube; at this point it will be extremely viscous. Sonicate on ice four times for 30 with a 30-s rest period to shear the DNA; the lysate will become much less viscous. Keep a 20-μl aliquot of the lysate. Centrifuge at 28,000 rpm in an SW28 rotor for 30 min at 4° to pellet the cell debris. Decant the

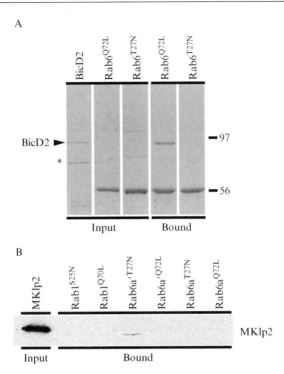

Fig. 2. Rab6-binding assays with recombinant BicD2 and MKlp2. (A) Recombinant 6×His-tagged BicD2 and GST-tagged Rab6 were used in a direct binding assay using glutathione-Sepharose beads. The eluates of the binding reaction (bound) were analyzed on a CBB gel and compared to the recombinant proteins alone. A degradation product in the BicD2 protein preparation is visible in the load (asterisk), but only full-length BicD2 (arrowhead, 90 kDa) bound to the active form of Rab6 (Rab6^{Q72L}). As a control the GDP-locked Rab6 (Rab6^{T27N}) was used. (B) Recombinant His-tagged MKlp2 and GST-tagged Rab6a, Rab6a', and Rab1 were used in a direct binding assay using glutathione-Sepharose beads. The eluates of the binding reaction (bound) were analyzed on a Western blot using antibodies to MKlp2 to detect the bound protein. MKlp2 was detected by the antibody in the load (20% of MKlp2 input), but could not be detected in the bound fraction with either isoform of Rab6a or Rab1 under these conditions.

supernatant into a clean tube and store on ice; this is the cleared bacterial lysate; keep a 20-μl aliquot. Take 0.5 ml of NTA-agarose per liter of original culture, wash in IMAC5, and then add it to the cleared bacterial lysate. The amount of resin can be decreased if the protein is expressed at a low level; 1 ml of resin binds approximately 10 mg of protein. Rotate at 4° for 60–120 min to allow the protein to bind to the resin. Wash the resin four times for 5 min at room temperature in a batch with 10 ml IMAC20AT

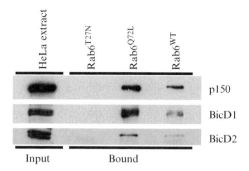

FIG. 3. Rab6 pulls down endogenous p150, BicD1, and BicD2 from HeLa extract. GST-tagged Rab6 was bound to glutathione-Sepharose beads, loaded with GTP (Rab6^{WT} and Rab6^{Q72L}) or GDP (Rab6^{T27N}) and incubated with HeLa extract made in RIPA buffer. Eluates from the reaction were analyzed on Western blot and probed with mouse antibodies to p150^{glued}, and rabbit antibodies to BicD1 or BicD2. All three Rab6 effector proteins could be detected in the pull-down with Rab6^{WT} and the GTP-locked form of Rab6 (Rab6^{Q72L}) but not with the inactive Rab6 mutant (Rab6^{T27N}). Consistent with theory the binding is more efficient using the active mutant Rab6^{Q72L} than with Rab6^{WT}.

(20 mM Tris–HCl, pH 8.0, 300 mM NaCl, 20 mM imidazole, 1 mM ATP, 1 mM MgCl$_2$, 0.2% [w/v] Triton X-100). Pellet the resin by centrifugation at 1000×g for 5 min, 4°. Keep a 20-μl aliquot of the first supernatant/ unbound fraction. Transfer the resin to a clean column, then wash with 20 resin volumes of IMAC20 (no ATP or Triton X-100). Elute the column with 20 ml IMAC200 (20 mM Tris–HCl, pH 8.0, 300 mM NaCl, 200 mM imidazole) and collect 0.5–1.0 ml fractions depending on the bed volume of the column. Analyze 2-μl aliquots of the lysate, flow-through, and wash, and 5–10 μl of the fractions on minigels; a single band should be visible in the IMAC200 elution corresponding to the recombinant Rab protein. Dialyze or desalt the protein into the appropriate buffer immediately; we typically use either phosphate-buffered saline (PBS) or 20 mM Tris–HCl, pH 8.0, 150 mM NaCl containing 1 mM dithiothreitol (DTT), then aliquot and store at −80°. Do not freeze proteins in the presence of imidazole–containing buffers, since they often precipitate on freezing!

While Rab proteins express well with yields up to 10 mg/liter culture and are soluble at concentrations up to 10 mg/ml, coiled-coil proteins such as BicD1 and BicD2 are expressed at lower levels and typically give lower yields of 1 mg/liter culture. For this reason it is often better to express smaller domains corresponding to the Rab-binding sites identified by yeast two-hybrid analysis. Rabkinesin-6/Rab6-KIFL/MKlp2 cannot be expressed as a full-length soluble protein in bacteria but can be obtained using the baculovirus Sf9 insect cell expression system.

Rab6-Binding Assays with Cell Extracts

HeLa cells grown in 3×15-cm-diameter dishes until confluence were washed three times with PBS, then pipetted or scraped from the dish in PBS and the cells pelleted by centrifugation at $600 \times g$ for 2 min. The cell pellets were extracted for 15 min on ice with RIPA buffer (10 mM Tris–HCl, pH 7.5, 150 mM NaCl, 1% [w/v] Triton X-100, 0.1% [w/v] Na-SDS, 1% [w/v] sodium deoxycholate containing 2 mM Pefabloc and Complete protease inhibitor cocktail [Roche Diagnostics]). Cell debris was removed by centrifugation at $20,000 \times g$ for 15 min at $4°$, and the cell lysate stored on ice until required. While binding assays can be done with frozen extract, it is better to use fresh extract to ensure that protein complexes are not disrupted by the freeze-thawing associated with storage.

For effector binding assays from cell extracts 500 μg GST-Rab6 was bound to 50 μl glutathione Sepharose 4B (Amersham) in a total volume of 1 ml in buffer NE100 (20 mM HEPES-NaOH, pH 7.5, 100 mM NaCl, 10 mM EDTA, 0.1% [v/v] Triton X-100) on a roller for 60 min at $4°$. The beads were pelleted by centrifugation at $600 \times g$ for 1 min at $4°$ and washed three times with 1 ml buffer NE100 to remove bound nucleotide from the Rab. For the incubation with extract the loaded beads were resuspended in buffer NL100 (20 mM HEPES-NaOH, pH 7.5, 100 mM NaCl, 5 mM MgCl$_2$, 0.1% [v/v] Triton X-100) and transferred to a 15-ml tube. The beads were incubated in a total volume of 6 ml NL100 for 60 min on a roller at $4°$ with 40 mg protein from a RIPA extract of HeLa cells (14 mg/ml protein). To load the Rab with nucleotide GTP (for Rab6^{WT} and Rab6^{Q72L}) or GDP (for Rab6^{T27N}) was added to a final concentration of 50 μM from 100 mM stocks in NL100. The beads were pelleted by centrifugation and washed three times with 1 ml NL100. Bound proteins were eluted in 500 μl buffer NE200 (20 mM HEPES-NaOH, pH 7.5, 200 mM NaCl, 20 mM EDTA, 0.1% [v/v] Triton X-100), rotating at $4°$ for 5 min. For trichloroacetic acid (TCA) precipitation the eluate was diluted with water to 1 ml and sodium deoxycholate was added to a final concentration of 0.02% (w/v). Proteins were precipitated with a final TCA concentration of 12% (w/v) for 1 h on ice, and pelleted by centrifugation at $20,000 \times g$ for 15 min at $4°$. Pellets were washed two times with cold acetone and dissolved in $1.5 \times$ SDS sample buffer.

Rab6 Effector-Binding Assays with Purified Proteins

For effector binding assays 4 μg GST-Rab6 was immobilized on 15 μl glutathione Sepharose 4B (Amersham) in a total volume of 400 μl in buffer NE100 (20 mM HEPES-NaOH, pH 7.5, 100 mM NaCl, 10 mM EDTA, 0.1% [v/v] Triton X-100) on a roller for 60 min at $4°$. The beads were pelleted by centrifugation at $600 \times g$ for 1 min at 4 ° and washed three times with 1 ml buffer NE100 to remove bound nucleotide from the Rab. For the binding reaction the loaded beads were resuspended in buffer NL100 (20 mM HEPES-NaOH, pH 7.5, 100 mM NaCl, 5 mM MgCl$_2$, 0.1% [v/v] Triton X-100). The beads were incubated in a total volume of 400 μl NL100 for 60 min on a roller at 4° with 10 μg His-BicD2. To load the Rab with nucleotide either GTP (for Rab6^{WT} and Rab6^{Q72L}) or GDP (for Rab6^{T27N}) was added to a final concentration of 50 μM from 100 mM frozen stocks in NL100. The beads were pelleted by centrifugation and washed three times with 1 ml NL100. Bound protein was eluted directly in 1.5$\times$ SDS sample buffer and analyzed on a 7.5% minigel. Because of the limited amount of insect cell purified His-tagged Mklp2, only 1 μg recombinant protein was used for the binding reaction, followed by an analysis of the eluates on Western blot.

References

Echard, A., Jollivet, F., Martinez, O., Lacapere, J. J., Rousselet, A., Janoueix-Lerosey, I., and Goud, B. (1998). Interaction of a Golgi-associated kinesin-like protein with Rab6. *Science* **279,** 580–585.

Fontijn, R. D., Goud, B., Echard, A., Jollivet, F., van Marle, J., Pannekoek, H., and Horrevoets, A. J. (2001). The human kinesin-like protein RB6K is under tight cell cycle control and is essential for cytokinesis. *Mol. Cell. Biol.* **21,** 2944–2955.

Gruneberg, U., Neef, R., Honda, R., Nigg, E. A., and Barr, F. A. (2004). Relocation of Aurora B from centromeres to the central spindle at the metaphase to anaphase transition requires MKlp2. *J. Cell. Biol.* **166,** 167–172.

Hill, E., Clarke, M., and Barr, F. A. (2000). The Rab6-binding kinesin, Rab6-KIFL, is required for cytokinesis. *EMBO J.* **19,** 5711–5719.

Hoogenraad, C. C., Akhmanova, A., Howell, S. A., Dortland, B. R., De Zeeuw, C. I., Willemsen, R., Visser, P., Grosveld, F., and Galjart, N. (2001). Mammalian Golgi-associated Bicaudal-D2 functions in the dynein-dynactin pathway by interacting with these complexes. *EMBO J.* **20,** 4041–4054.

James, P., Halladay, J., and Craig, E. A. (1996). Genomic libraries and a host strain designed for highly efficient two-hybrid selection in yeast. *Genetics* **144,** 1425–1436.

Mallard, F., Tang, B. L., Galli, T., Tenza, D., Saint-Pol, A., Yue, X., Antony, C., Hong, W., Goud, B., and Johannes, L. (2002). Early/recycling endosomes-to-TGN transport involves two SNARE complexes and a Rab6 isoform. *J. Cell. Biol.* **156,** 653–664.

Matanis, T., Akhmanova, A., Wulf, P., Del Nery, E., Weide, T., Stepanova, T., Galjart, N., Grosveld, F., Goud, B., De Zeeuw, C. I., Barnekow, A., and Hoogenraad, C. C. (2002). Bicaudal-D regulates COPI-independent Golgi - ER transport by recruiting the dynein - dynaction motor complex. *Nat. Cell. Biol.* **4,** 986–992.

Neef, R., Preisinger, C., Sutcliffe, J., Kopajtich, R., Nigg, E. A., Mayer, T. U., and Barr, F. A. (2003). Phosphorylation of mitotic kinesin-like protein 2 by polo-like kinase 1 is required for cytokinesis. *J. Cell. Biol.* **162,** 863–875.

Short, B., Preisinger, C., Schaletzky, J., Kopajtich, R., and Barr, F. A. (2002). The Rab6 GTPase regulates recruitment of the dynactin complex to Golgi membranes. *Curr. Biol.* **12,** 1792–1795.

Vaughan, P. S., Miura, P., Henderson, M., Byrne, B., and Vaughan, K. T. (2002). A role for regulated binding of p150(Glued) to microtubule plus ends in organelle transport. *J. Cell. Biol.* **158,** 305–319.

Young, J., Stauber, T., Del Nery, E., Vernos, I., Pepperkok, R., and Nilsson, T. (2005). Regulation of microtubule-dependent recycling at the trans-Golgi network by Rab6A and Rab6A'. *Mol. Biol. Cell* **16,** 162–177.

[54] Assay and Functional Properties of Rabkinesin-6/Rab6-KIFL/MKlp2 in Cytokinesis

By Rüdiger Neef, Ulrike Grüneberg, and Francis A. Barr

Abstract

Here we describe methods for the characterization of the kinesin-9 family motor protein Rabkinesin-6/Rab6-KIFL/MKlp2 in cytokinesis. Here we outline biochemical assays for studying the interaction of Rabkinesin-6/Rab6-KIFL/MKlp2 with microtubules, and its regulation and interaction with the mitotic polo-like kinase 1 using recombinant proteins expressed in and purified from insect cells. Protocols for the *in vivo* functional analysis of Rabkinesin-6/Rab6-KIFL/MKlp2 using depletion with small interfering RNA duplexes are described.

Introduction

Rabkinesin-6/Rab6-KIFL is a vertebrate-specific kinesin-9 family motor protein originally identified in a yeast two-hybrid screen as an interaction partner of the Golgi-localized GTP-binding protein Rab6 (Echard *et al.*, 1998). However, like other mitotic kinesin-like proteins of the kinesin-9 family such as MKlp1, Rabkinesin-6/Rab6-KIFL is under tight cell cycle regulation at both the level of transcription and protein stability (Fontijn *et al.*, 2001; Hill *et al.*, 2000). It is thus essentially absent from interphase cells and accumulates in G_2 cells prior to entry into mitosis (Hill *et al.*, 2000), and for this reason we now refer to Rabkinesin-6/Rab6-KIFL as MKlp2 (mitotic kinesin-like protein 2) (Neef *et al.*, 2003). In G_2 cells MKlp2 is sequestered in the nucleus, then released on nuclear envelope breakdown, and remains cytoplasmic until the end of metaphase. At the metaphase-to-anaphase transition, MKlp2 is recruited to the central spindle, an anti-parallel array of microtubules that forms between the segregating chromosomes, and becomes restricted to a narrow band in the center of this structure during anaphase and telophase, where it remains until cytokinesis

METHODS IN ENZYMOLOGY, VOL. 403
0076-6879/05 $35.00
DOI: 10.1016/S0076-6879(05)03054-5

occurs. Interference with the function of MKlp2 by microinjection of inhibitory antibodies or depletion using RNA interference results in the failure of cytokinesis (Fontijn *et al.*, 2001; Hill *et al.*, 2000; Neef *et al.*, 2003). Consistent with its localization and involvement in cytokinesis, MKlp2 is required for central spindle localization of two mitotic kinases of the Polo and Aurora families also involved in cytokinesis (Gruneberg *et al.*, 2004; Neef *et al.*, 2003). MKlp2 is a substrate and binding partner for the human Polo-like kinase 1 (Plk1). This interaction is mediated by the carboxy-terminal Polo box domain of Plk1 that forms a phosphopeptide-binding domain capable of binding to phosphorylated MKlp2 (Elia *et al.*, 2003a,b; Neef *et al.*, 2003). Thus, Plk1 is restricted to the central spindle in anaphase cells by the action of MKlp2. Aurora B is a component of the passenger protein complex that is required for the bipolar attachment to, and correct segregation of chromosomes by, the mitotic spindle. In metaphase cells, it is present on kinetochores and then relocates to the central spindle in anaphase cells by a microtubule-dependent transport process. In cells depleted of MKlp2, Aurora B fails to relocate to the central spindle and remains at kinetochores, implying that MKlp2 is the motor protein responsible for mediating this transport event (Gruneberg *et al.*, 2004).

This chapter describes the purification of recombinant MKlp2 for use in microtubule-binding studies, and as a substrate and binding partner for Plk1. We also describe the use of RNA interference to study the function on MKlp2 in cytokinesis in cells.

Purification of Recombinant Proteins

It has not proven possible to obtain soluble active full-length MKlp2 or Plk1 from bacterial expression systems, and we therefore produce these proteins in Sf9 insect cells using the baculovirus expression vector system (Becton Dickinson). Recombinant viruses are produced according to the manufacturer's protocol by cotransfecting the desired construct together with the viral helper DNA (Baculogold, Pharmingen/Becton Dickinson). The third amplification, or P3, is used for infection of insect cells for protein expression.

Purification of MKlp2 and Plk1 from Insect Cells

Sf9 insect cells are grown in suspension at 27° in spinner flasks stirring at 50 rpm (Cellspin, Integra Biosciences) containing 500 ml TC100 medium plus 10% fetal calf serum (FCS) (Invitrogen/Life Technologies). Once the culture has reached a density of 1.0×10^6 cells/ml harvest the cells by centrifugation for 8 min at $400 \times g$ at room temperature using sterile 200-ml centrifuge tubes. Resuspend the Sf9 cells in 30 ml of medium and infect with 10 ml supernatant of the P3-amplified virus. After 1 h of incubation on a roller at room temperature, transfer the infected cells back to the original spinner flask and add growth medium to obtain a cell density of 1.0×10^6 cells/ml. Grow the cells for 48–72 h at 27° with 50 rpm rotation, and monitor the status of the infection under the light microscope.

Harvest the infected cells by centrifugation for 8 min at $400 \times g$ and room temperature in 200-ml centrifuge tubes. Wash the cell pellets once in 200 ml phosphate-buffered saline (PBS), and pellet as before. Combine and resuspend the cell pellets in 20 ml cold lysis buffer (20 mM Tris–HCl, pH 8.0, 300 mM NaCl, 5 mM imidazole, 0.1% [w/v] Triton X-100, 0.2 mM phenylmethylsulfonyl fluoride [PMSF], Complete protease inhibitor cocktail [Roche Diagnostics]) and incubate for 30 min on ice (keep a 10-μl aliquot on ice). Centrifuge the lysate for 30 min with 27,000 rpm at 4° in a Beckmann SW40 rotor; keep a 10-μl aliquot of the supernatant on ice. During this time, equilibrate 500 μl of the Nickel-NTA beads (Qiagen) by washing 3× with 1 ml of lysis buffer; spin down for 2 min at $3000 \times g$ in a microfuge. Incubate the insect cell extract with the Ni-NTA beads for 2 h on a roller at 4° to allow binding of His-tagged proteins. Recover the beads by centrifugation for 5 min at $1000 \times g$ and 4°, washed once with 20 ml lysis buffer. Keep 10-μl aliquots of the unbound fraction and the first wash on ice. Transfer the beads to a column and wash with 20 ml of IMAC20T (20 mM Tris–HCl pH 8.0, 300 mM NaCl, 20 mM imidazole, 0.1% [w/v] Triton X-100), and four times with 20 ml IMAC20 (IMAC20T without Triton X-100). Elute 10 times with 500 μl IMAC 200 and collect the fractions. Take an aliquot of 10 μl of every fraction and analyze for expression together with the aliquots of the purification on an appropriate Coomassie gel. Dialyze or desalt the protein into the appropriate buffer immediately; we typically use either PBS or 20 mM Tris–HCl, pH 8.0, 150 mM NaCl containing 1 mM dithiothreitol (DTT), then aliquot and store at −80°.

Purification of Proteins from Bacteria

The hexahistidine-GST-tagged Polo box domain of Plk1, amino acids 303–603, can be expressed in *Escherichia coli* BL21(pRIL) using the T7 polymerase bacterial expression vector pGAT2 after induction from an OD_{600} of 0.5 at 37° for 3 h in the *E. coli* strain BL21(DE3). Cell pellets can be frozen at −20° until required or processed immediately. They should be thawed at room temperature by the addition of 10 ml lysis buffer per liter of original culture (IMAC5: 20 mM Tris–HCl, pH 8.0, 300 mM NaCl, 5 mM imidazole, 0.2% [w/v] Triton X-100, plus 0.5 mg/ml lysozyme and

Complete protease inhibitor cocktail [Roche Diagnostics]). Resuspend them using a 10-ml pipette and pool the lysate in one tube; at this point it will be extremely viscous. Sonicate the lysate on ice, four times 30 s with a 30-s rest period to shear the DNA; the lysate will become much less viscous. Keep a 20-μl aliquot of the lysate. Centrifuge it at 28,000 rpm in an SW28 rotor for 30 min at 4° to pellet the cell debris. Decant the supernatant into a clean tube and store on ice; this is the cleared bacterial lysate; keep a 20-μl aliquot. Take 0.5 ml of NTA-agarose per liter of original culture, wash in IMAC5, and then add it to the cleared bacterial lysate. The amount of resin can be decreased if the protein is expressed at a low level; 1 ml of resin binds approximately 10 mg protein. Rotate the mixture at 4° for 60–120 min to allow the protein to bind to the resin. Wash the resin four times for 5 min at room temperature in batch with 10 ml IMAC20AT (20 mM Tris–HCl, pH 8.0, 300 mM NaCl, 20 mM imidazole, 1 mM ATP, 1 mM MgCl$_2$, 0.2% [w/v] Triton X-100). Pellet the resin by centrifugation at 1000×g for 5 min, 4°. Keep a 20-μl aliquot of the first supernatant/ unbound fraction. Transfer the resin to a clean column, then wash with 20 resin volumes of IMAC20 (no ATP or Triton X-100). Elute the column with 20 ml IMAC200 (20 mM Tris–HCl, pH 8.0, 300 mM NaCl, 200 mM imidazole,); collect 0.5–1.0 ml fractions depending on the bed volume of the column. Analyze 2-μl aliquots of the lysate, flow-through, and wash, and 5–10 μl of the fractions on minigels; a single band should be visible in the IMAC200 elution corresponding to the Polo box domain. Dialyze or desalt the protein into the appropriate buffer immediately; we typically use either PBS or 20 mM Tris–HCl, pH 8.0, 150 mM NaCl containing 1 mM DTT; then aliquot and store at −80°.

Interaction of MKlp2 with Microtubules

MKlp2 is a kinesin family motor protein and therefore is expected to interact with microtubules. To study this interaction, recombinant MKlp2 purified from insect cells is incubated with microtubules made from purified tubulin. A simple pelleting assay can be used to determine the extent of microtubule binding (Fig. 1). With this assay it is important to perform a binding assay in the absence of microtubules to control for precipitation of the protein and to control for the recovery of microtubules on a Coomassie brilliant blue-stained gel. MKlp2 has potent microtubule-bundling activity, and if fluorescently labeled tubulin is used then this can be visualized under the microscope (Fig. 2). The effect of MKlp2 phosphorylation by mitotic kinases such as Plk1 and Cdk-cyclin B on microtubule binding can also be investigated with this protocol (Neef *et al.*, 2003).

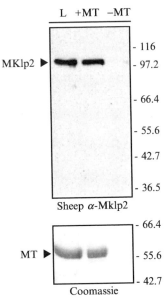

FIG. 1. Binding of MKlp2 to microtubules. MKlp2 was incubated in the presence (+MT) or absence of microtubules (−MT) for 20 min at room temperature. The microtubules and bound proteins were pelleted by centrifugation, then the washed pellets and a sample equivalent to the input material (L) were analyzed by SDS-PAGE and Coomassie brilliant blue staining to check the recovery of microtubules and Western blotting with antibodies to detect MKlp2. (See color insert.)

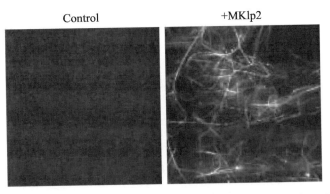

FIG. 2. MKlp2 is a microtubule-bundling protein. Rhodamine-labeled microtubules incubated with either buffer (control) or in the presence of MKlp2 for 20 min at room temperature were dropped on glass slides and covered with a glass coverslip. Microtubules and microtubule bundles were visualized by rhodamine fluorescence using a 63× oil immersion objective.

Microtubule Polymerization

Tubulin can either be purified from pig or bovine brain and modified with fluorescent dyes such as rhodamine using published methods (Hyman *et al.*, 1991) or be purchased from a commercial supplier (Cytoskeleton, Inc.). To polymerize tubulin into microtubules, incubate 25 μl pig or bovine brain tubulin (10 mg/ml) for 30 min at 37° in a total of 70 μl BRB80 (80 mM PIPES pH 6.8, 1 mM MgCl$_2$, 1 mM EGTA) containing 4 mM MgCl$_2$, 1 mM GTP, and 4.8% (v/v) dimethyl sulfoxide (DMSO). For microtubule bundling assays, use tubulin conjugated to a fluorescent dye such as rhodamine. To recover the polymerized tubulin, centrifuge the reaction for 15 min at 35° through a 75-μl cushion of 40% (v/v) glycerol in BRB80 at 55,000 rpm in a TLA100 rotor (Beckmann). Wash the microtubule pellet once with 50 μl BRB80 containing 50 μM taxol, then centrifuge again at 55,000 rpm in a TLA100 rotor for 5 min at 37°. Resuspend the microtubule pellet in 50 μl BRB80 containing 50 μM taxol. The final concentration of tubulin incorporated into microtubules is 50 μM (5 mg/ml).

Microtubule-Binding and -Bundling Assays

For microtubule-binding assays, incubate the recombinant protein, typically 10–250 ng, with 2-μl microtubules for 20 min at room temperature in a total volume of 20 μl BRB80 containing 10 μM taxol. To recover the microtubules and bound proteins, centrifuge the reaction for 20 min at 25° through a 75-μl cushion of 40% (v/v) glycerol in BRB80 at 55,000 rpm in a TLA100 rotor (Beckmann). Wash the microtubule pellet once with 50 μl BRB80 containing 50 μM taxol, then centrifuge again at 55,000 rpm in a TLA100 rotor for 5 min at 37°. Resuspend the pellet of microtubules and bound proteins in 30 μl of sample buffer; load 80% on 10% minigel for Coomassie blue staining and 20% on a 10% minigel for Western blotting.

For microtubule-bundling assays, incubate the recombinant protein, typically 10–250 ng, with 2-μl microtubules prepared from rhodamine-labeled tubulin for 20 min at room temperature in a total volume of 20 μl BRB80 containing 10 μM taxol. Directly after incubation the bundling reaction should be spotted onto a glass slide, covered by dropping a 15-mm-diameter glass coverslip onto the drop of liquid, and then analyzed under a microscope; for this purpose use a Zeiss Axioskop-2 with a 63× oil immersion lens of numerical aperture 1.4, and a 1300 by 1030 pixel cooled-CCD camera (Princeton Instruments), using Metaview software (Universal Imaging Corp.). Under these conditions individual microtubules form regular fine filaments 20–30 μm in length, while microtubule bundles are longer and thicker structures greater than 100 μm long.

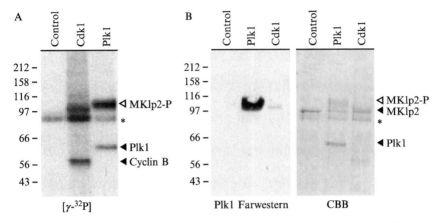

FIG. 3. Interaction of MKlp2 with Plk1. (A) MKlp2 was incubated with buffer (Control), Plk1, or Cdk1-cyclin B (Cdk1) for 30 min at 30° in the presence of [γ-^{32}P]ATP, then analyzed by SDS-PAGE on 7.5% minigels. The dried gel was exposed to film. Arrowheads indicate phosphorylated proteins, and the asterisk marks a 90-kDa phosphorylated protein contaminating the MKlp2 preparation. (B) MKlp2 was incubated with buffer (Control), Plk1, or Cdk1-cyclin B (Cdk1) for 30 min at 30°, then analyzed by far Western blotting with the Polo-box domain of human Plk1. A Plk1 farwestern blot and a corresponding Coomassie brilliant blue (CBB)-stained gel are shown. Arrowheads indicate MKlp2 and Plk1, and the asterisk marks a 90-kDa phosphorylated protein contaminating the MKlp2 preparation.

Regulation of MKlp2 by Plk1

MKlp2 is a substrate for two mitotic kinases, Plk1 and Cdk1-cyclin B (Neef *et al.*, 2003). This can be reconstituted *in vitro* using recombinant kinases and MKlp2 purified from the insect cell expression system (Fig. 3A). MKlp2 phosphorylated in this way can be used for a number of different assays to test its microtubule-binding and -bundling properties, or interaction with the phospho-binding Polo-box domain of Plk1. Using a far Western blotting approach, it is possible to demonstrate the interaction of Plk1 with MKlp2 phosphorylated by Plk1 but not other mitotic kinases such as Cdk1-cyclin B (Fig. 3B). This approach can also be used to study the interaction of other mitotic phosphoproteins with Plk1 (Preisinger *et al.*, 2005).

Phosphorylation of MKlp2 by Plk1

To phosphorylate MKlp2 with Plk1, take 100 ng of recombinant MKlp2, 10–50 ng recombinant Plk1, mM ATP, 0.1 μl [γ-^{32}P] 0.5 mM ATP (3000 Ci/mmol and 10 mCi/ml, Amersham) in a total volume of 20 μl BRB80 for 30 min at 30°. For assays with microtubules, add 2 μl of polymerized tubulin prepared as described previously to the reaction. If phosphorylated MKlp2 is used for microtubule-binding or -bundling assays, use 500 ng of MKlp2 and omit the radioactive ATP. Analyze the entire mixture by sodium dodecyl sulfate polyacrylamide gel electrophoresis (SDS–PAGE) on 7.5% minigels, and then dry and expose the gel to film or a phospho-imaging system.

Farwestern Ligand Blots

For ligand blots, incubate 500 ng recombinant MKlp2 with or without 50 ng of recombinant Plk1 for 45 min at 37° in BRB80 containing 500 μM ATP and 1 mM DTT. After incubation denature the samples by boiling in SDS sample buffer. Load 10% on 10% SDS-PAGE minigel for Western blotting, and the remaining 90% on a separate 10% minigel for Coomassie blue staining. Any standard Western blotting system can be used for the transfer; we use a semidry transfer system (Bio-Rad Laboratories). Incubate the blotted membrane overnight at 4° in TBSTM (50 mM Tris–HCl, pH 7.4, 137 mM NaCl, 0.1% [v/v] Tween-20, 4% [w/v] skim milk powder). After this renaturation step, probe the membrane with 1.5 μg/ml GST-Polo box in TBSTM for 6 h at 4°. Wash the membrane four times for 5 min with TBSTM at room temperature. To detect the bound GST-Polo box, incubate the membrane with primary antibodies to GST, and the appropriate horseradish peroxidase (HRP)-coupled secondary antibodies. Reveal the signal using a chemiluminescence detection system (ECL, Amersham).

Studying MKlp2 Function in Cytokinesis

MKlp2 is highly suited to study by RNA interference since it has to be newly synthesized prior to each cell division (Neef *et al.*, 2003). HeLa cells transfected with short interfering RNA (siRNA) specific for MKlp2 are depleted of the protein within 24–30 h. From 24 h onward, an accumulation of binucleated cells is observed, and by 48 h when all cells have gone through at least one cell cycle, greater than 90% of cells on a coverslip are binucleated (Fig. 4). Lamin A is a good control since this protein is not essential for growth in cell culture, and cells depleted of lamin A show less than 2% binucleation 48 h after transfection with siRNA. MKlp2 is required to localize the mitotic kinases Aurora B and Plk1 to the central spindle and staining for these proteins can be used as a read out of the MKlp2 depletion phenotype (Gruneberg *et al.*, 2004; Neef *et al.*, 2003). In Fig. 5 the staining of Aurora B in lamin A and MKlp2-depleted anaphase cells is shown. The failure of Aurora B to relocate from the kinetochores to the central spindle can be seen by the two bars of Aurora B lying over the chromosomes in MKlp2-depleted cells compared to the

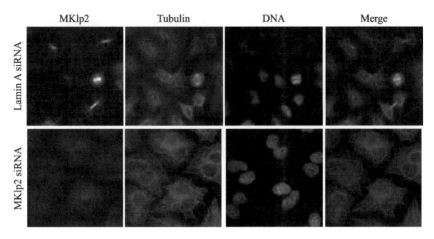

FIG. 4. Depletion of MKlp2 causes a failure of cytokinesis and the accumulation of binucleated cells. HeLa cells transfected with siRNA duplexes for Lamin A and MKlp2 were fixed after 48 h with 3% (w/v) paraformaldehyde and stained with a sheep antibody to MKlp2, mouse monoclonal antibodies to α-tubulin (DM1a, Sigma-Aldrich), and DAPI for DNA. In the merged image MKlp2 is red, α-tubulin is green, and DNA is blue.

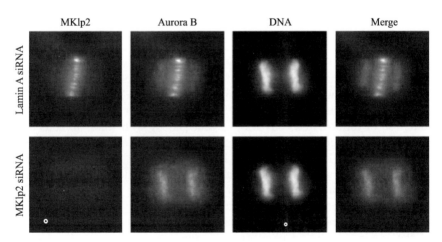

FIG. 5. Aurora B is unable to relocate from kinetochores to the central spindle in MKlp2-depleted cells. HeLa cells were transfected with siRNA duplexes for Lamin A or MKlp2 for 24 h, fixed with 3% (w/v) paraformaldehyde, and stained with a sheep antibody to MKlp2 (Hill et al., 2000), mouse monoclonal antibodies to Aurora B (AIM-1, Becton Dickinson), and DAPI for DNA. In the merged image MKlp2 is green, Aurora B is red, and DNA is blue. (See color insert.)

single bar falling between the segregating chromosomes in the lamin A control cells (Fig. 5).

Depletion of MKlp2 by RNA Interference

Plate HeLa cells in DME containing 10% calf serum (Invitrogen/Life Technologies) on glass coverslips at a density of 50,000 cells per well of a six-well plate and leave them to attach for 24 h at 37° and 5% CO_2. RNA interference can be performed using siRNA duplexes (Elbashir *et al.*, 2001); MKlp2 and lamin A were targeted with siRNA duplexes designed from the sequences 5′-aagatcagggttgtgtccgtatt-3′and 5′-aaggcttctaggcgtgag-gagtt-3′, respectively. To transfect HeLa cells, mix 3 μl preannealed siRNA duplexes (Dharmacon Research Inc. or Qiagen) with 200 μl Optimem (Invitrogen/Life Technologies) serum-free medium, and then add 3 μl Oligofectamine (Invitrogen) and mix by pipetting. After 25 min at room temperature, add the entire mixture to one well of a six-well plate. To investigate the effects of MKlp2 depletion, fix and stain cells with antibodies to proteins such as Plk1, Aurora B, α-tubulin, and other proteins involved in cytokinesis after 0, 24, 48, and 72 h of transfection (Gruneberg *et al.*, 2004; Neef *et al.*, 2003). We routinely collect images using a Zeiss Axioskop-2 with 63× Plan Apochromat oil immersion objective numerical aperture 1.4, and a 1300 by 1030 pixel cooled-CCD camera (Princeton Instruments), using Metaview software (Universal Imaging Corp.).

References

Echard, A., Jollivet, F., Martinez, O., Lacapere, J. J., Rousselet, A., Janoueix-Lerosey, I., and Goud, B. (1998). Interaction of a Golgi-associated kinesin-like protein with Rab6. *Science* **279,** 580–585.

Elbashir, S. M., Harborth, J., Lendeckel, W., Yalcin, A., Weber, K., and Tuschl, T. (2001). Duplexes of 21-nucleotide RNAs mediate RNA interference in cultured mammalian cells. *Nature* **411,** 494–498.

Elia, A. E., Cantley, L. C., and Yaffe, M. B. (2003). Proteomic screen finds pSer/pThr-binding domain localizing Plk1 to mitotic substrates. *Science* **299,** 1228–1231.

Elia, A. E., Rellos, P., Haire, L. F., Chao, J. W., Ivins, F. J., Hoepker, K., Mohammad, D., Cantley, L. C., Smerdon, S. J., and Yaffe, M. B. (2003). The molecular basis for phosphodependent substrate targeting and regulation of Plks by the Polo-box domain. *Cell* **115,** 83–95.

Fontijn, R. D., Goud, B., Echard, A., Jollivet, F., van Marle, J., Pannekoek, H., and Horrevoets, A. J. (2001). The human kinesin-like protein RB6K is under tight cell cycle control and is essential for cytokinesis. *Mol. Cell. Biol.* **21,** 2944–2955.

Gruneberg, U., Neef, R., Honda, R., Nigg, E. A., and Barr, F. A. (2004). Relocation of Aurora B from centromeres to the central spindle at the metaphase to anaphase transition requires MKlp2. *J. Cell. Biol.* **166,** 167–172.

Hill, E., Clarke, M., and Barr, F. A. (2000). The Rab6-binding kinesin, Rab6-KIFL, is required for cytokinesis. *EMBO J.* **19,** 5711–5719.

Hyman, A., Drechsel, D., Kellogg, D., Salser, S., Sawin, K., Steffen, P., Wordeman, L., and Mitchison, T. (1991). Preparation of modified tubulins. *Methods Enzymol.* **196,** 478–485.

Neef, R., Preisinger, C., Sutcliffe, J., Kopajtich, R., Nigg, E. A., Mayer, T. U., and Barr, F. A. (2003). Phosphorylation of mitotic kinesin-like protein 2 by polo-like kinase 1 is required for cytokinesis. *J. Cell. Biol.* **162,** 863–875.

Preisinger, C., Korner, R., Wind, M., Lehmann, W. D., Kopajtich, R., and Barr, F. A. (2005). Plk1 docking to GRASP65 phosphorylated by Cdk1 suggests a mechanism for Golgi checkpoint signalling. *EMBO J.* **24,** 753–765.

[55] Interaction and Functional Analyses of Human VPS34/p150 Phosphatidylinositol 3-Kinase Complex with Rab7

By Mary-Pat Stein, Canhong Cao, Mathewos Tessema, Yan Feng, Elsa Romero, Angela Welford, and Angela Wandinger-Ness

Abstract

The Rab7 GTPase is a key regulator of late endocytic membrane transport and autophagy. Rab7 exerts temporal and spatial control over late endocytic membrane transport through interactions with various effector proteins. Among Rab7 effectors, the hVPS34/p150 phosphatidylinositol (PtdIns) 3-kinase complex serves to regulate late endosomal phosphatidylinositol signaling that is important for protein sorting and intraluminal vesicle sequestration. In this chapter, reagents and methods for the characterization of the interactions and regulation of the Rab7/hVPS34/p150 complex are described. Using these methods we demonstrate the requirement for activated Rab7 in the regulation of hVPS34/p150 PtdIns 3-kinase activity on late endosomes *in vivo*.

Introduction

Rab7 is a small GTPase with critical functions in late endocytic membrane transport, protein sorting, and autophagy (Cantalupo *et al.*, 2001; Dong *et al.*, 2004; Feng *et al.*, 1995, 2001; Gutierrez *et al.*, 2004; Jager *et al.*, 2004; Press *et al.*, 1998; Stein *et al.*, 2003b; Vitelli *et al.*, 1995, 1997). Rab7 belongs to a large family of GTPases that serves to provide the observed specificity of membrane trafficking between compartments (Somsel Rodman and Wandinger-

0076-6879/05 $35.00
DOI: 10.1016/S0076-6879(05)03055-7

Ness, 2000; Stein *et al.*, 2003b). In their function as "master" regulators, Rab GTPases do not function independently, but rather appear to orchestrate the temporal and spatial regulation of events needed for vesicular transport by interfacing with a variety of interacting partners.

A number of Rab7 effectors have been identified. These include modulators of endosome acidification (rabring7) and transport on the cytoskeleton (RILP) (Cantalupo *et al.*, 2001; Feng *et al.*, 2001; Jordens *et al.*, 2001; 2003; Mizuno *et al.*, 2003). Our group has concentrated on two further Rab7 effectors. The first is XAPC7, an α proteasome subunit discussed in Chapter 56 (Mukherjee *et al.*, this volume) (Dong *et al.*, 2004). The second is the hVPS34/p150 phosphatidylinositol (PtdIns) 3-kinase complex and is the focus of this chapter (Feng *et al.*, 2001; Stein *et al.*, 2003b). Rab7 functions in the control of localized phosphatidylinositol 3-phosphate [PtdIns(3)P] biosynthesis by regulating the late endosomal targeting and activity of hVPS34/p150 PtdIns 3-kinase activity. Localized PtdIns(3)P synthesis in turn serves to recruit proteins with PtdIns(3)P recognition motifs, which include another lipid kinase, PIKfyve, implicated in intraluminal vesicle formation (Ikonomov *et al.*, 2001, 2002, 2003; Sbrissa *et al.*, 1999; Shisheva, 2001), as well as myotubularin lipid phosphatases to control endosomal levels of PtdIns(3)P (Blondeau *et al.*, 2000; M.P. Stein, unpublished observations). The hVPS34/p150 kinase complex is also present on early endosomes where it is regulated by Rab5 (Christoforidis *et al.*, 1999b; Murray *et al.*, 2002). It therefore remains of interest to establish how the activity of the hVPS34/p150 kinase complex is coordinated between two Rab GTPases and to understand more precisely how the levels of PtdIns(3)P, PtdIns(3,5)P$_2$, and PtdIns(5)P are spatially and temporally controlled through the interactions of lipid kinases and phosphatases (Fig. 1).

Here we describe the reagents along with the morphological and biochemical assays we have developed to monitor the functions of Rab7 and the hVPS34/p150 phosphatidylinositol 3-kinase complex. These assays have been used to demonstrate that active Rab7 and hVPS34/p150 cooperatively promote transport from early to late endosomes (Stein *et al.*, 2003b). Furthermore, our data suggest that active Rab7 is required to activate the hVPS34/p150 complex, but the complex subsequently dissociates to permit downstream downregulation by the myotubularin phosphatidylinositol 3-phosphatases. Thus, the methodologies provide the framework for dissecting the interactions between Rab7 and the hVPS34/p150 kinase complex and monitoring the regulation of PtdIns(3)P biosynthesis that is important for late endocytic membrane transport.

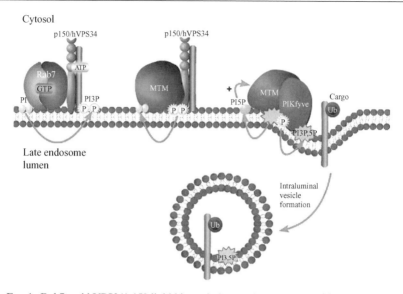

FIG. 1. Rab7 and hVPS34/p150 lipid kinase in late endosomal PtdIns(3)P regulation. The figure summarizes our working model for PtdIns(3)P regulation on late endosomes. GTP-bound Rab7 is responsible for the activation of the hVPS34/p150 lipid kinase and leads to the formation of PtdIns(3)P microdomains. Rab7 is known to bind to the WD40 domain of p150. These domains in turn recruit proteins with PtdIns(3)P binding domains such as further lipid kinases (PIKfyve) that participate in luminal vesicle formation and lipid phosphatases such as the myotubularins (MTM1 and MTMR2). The action of the myotubularins regulates late endosomal PtdIns(3)P and PtdIns(3,5)P_2 levels produced by the lipid kinases. The product of PtdIns(3,5)P_2 degradation, PtdIns(5)P, is an allosteric activator of the myotubularins (Schaletzky *et al.*, 2003).

Methods

General Reagents. All reagents unless otherwise specified were from Sigma-Aldrich (http://www.sigmaaldrich.com/Local/SA_Splash.html).

Construction of Wild-Type and Mutant hVPS34 Expression Vectors and PtdIns 3-Kinase Activity Assay

Constructs. The full-length cDNA of human VPS34 (hVPS34) (GenBank accession number AA081286) was obtained from Genome Systems (St. Louis, MO) and subcloned into the *Eco*RI/*Xho*I sites of pcDNA3 (Invitrogen Corporation, Carlsbad, CA). Kinase inactive hVPS34 (hVPS34kd) mutants were prepared by site-directed mutagenesis of critical residues in the ATP binding pocket. Three different mutants were constructed. K636R or R744P mutants in the kinase domain were prepared by polymerase chain reaction (PCR)-mediated site-directed mutagenesis. A double mutant in the kinase domain of hVPS34 D743A/N748I was constructed as described for the rat protein (Row *et al.*, 2001) using the QuikChange site-directed mutagenesis system according to the manufacturer's instruction (Stratagene, La Jolla, CA) and the forward primer: GGAGTTGGAGC-CAGGCACCTGGATATCCTTGTGCTA (modified bases underlined). All mutants were confirmed by DNA sequencing and screened for *in vitro* kinase activity (see below and Row *et al.* [2001]). The double mutant was used as the kinase dead variant in our published work (Stein *et al.*, 2003b). N-terminal (P5 produced by *KpnI, PstI* digest) and C-terminal (P3 produced by *XhoI, PstI* digest) fragments of hVPS34 were subcloned into a customized 6 × His-tagged pRSET (Invitrogen Corporation) prokaryotic expression vector with an inverted multiple cloning site (pTESR). Recombinant hVPS34 protein expressed in *Escherichia coli* BL21 was purified and used for raising antibodies to hVPS34 as detailed in the section Generation of Specific Antibody Reagents later in the chapter. Wild-type hVPS34 and the K636R or R744P mutants were also successfully cloned into baculovirus expression vectors pBlueBac (Invitrogen Corporation) as *Bam*H1/*Eco*RI (blunt end) and *Xho*1 fragments. pBlueBac hVPS34 was expressed in *Spodoptera frugiperda* (Sf9) cells infected at 5 or 10 moi after 48–90 h of infection (Fig. 2). For further information regarding baculovirus-mediated expression of rab GTPases, see Christoforidis *et al.* (2001).

Canine Rab7 was subcloned into pGEX 5X-2 (Amersham Biosciences, Piscataway, NJ) and used to purify recombinant GST-Rab7 for the *in vitro*

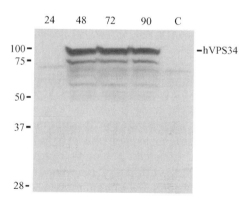

Fig. 2. Recombinant baculovirus to express hVSP34 in Sf9 insect cells. Sf9 cells were infected at 5 moi of recombinant baculovirus encoding hVPS34. Lysates were prepared after 24–90 h of infection, resolved by SDS-PAGE, and immunoblotted for hVPS34 using our rabbit polyclonal diluted 1:10,000. Lysates from uninfected Sf9 cells served as a negative control (C).

interaction studies detailed in the section on *in vitro* protein interaction Assays later in this chapter. The GTP-binding activity of wild-type and all mutant forms of Rab7 was confirmed by GTP-overlay blotting as previously described (Feng *et al.*, 1995) or by solution binding as described in the section on *in vitro* protein interaction assays.

Constructs encoding the full-length human V5-tagged p150 adaptor protein, as well as various deletion constructs, were as previously described by Murray *et al.* (2002).

Together, the described plasmid expression vectors may be used for numerous *in vivo* and *in vitro* binding assays to study the interactions between Rab7 and the lipid kinase complex, some of which are described in the following sections. All recombinant DNA molecules were prepared in accordance with NIH guidelines and our research is judged exempt from all USDA/CDC select agents defined by 42CFR part 73, 9 CFR part 121, and 7 CFR part 331 under the auspices of the UNM Biosafety Office.

PtdIns 3-Kinase Activity Assay. Kinase activities of the wild-type and mutant hVPS34 proteins were scored using an *in vitro* PtdIns 3-kinase assay (Panaretou *et al.*, 1997; Whitman *et al.*, 1985). For this purpose the hVPS34 proteins were overexpressed in BHK21 cells (as detailed in the section Cell Line, Protein Expression, and Morphological Analysis later in this chapter) and cell lysates were prepared in lysis buffer (137 mM NaCl, 20 mM Tris, pH 8.0, 5 mM MgCl$_2$, 1 mM CaCl$_2$, 10% glycerol, 1% NP40). Lysates were clarified by centrifugation at 15,000 rpm in a microfuge for 15 min. The hVPS34 proteins were immunoprecipitated with our specific antibody (detailed below) and the immunoprecipitates were washed twice with lysis buffer, once with 500 mM LiCl, 100 mM Tris, pH 7.4, and three times with 10 mM Tris, pH 7.4, 100 mM NaCl, 1 mM EDTA. Immunoprecipitates were mixed with an equal volume of 2-fold concentrated reaction mix (40 mM Tris, pH 7.6, 200 mM NaCl, 10 mM MgCl$_2$, 0.4 mg/ml phosphatidylinositol, 0.8 mCi/ml [γ-^{32}P]ATP, 100 μM ATP) and incubated on ice for 10 min. MnCl$_2$ may be used in lieu of MgCl$_2$ and is reported to enhance the specificity of the assay for hVPS34 activity (Panaretou *et al.*, 1997). The reactions were then incubated at room temperature for 30 min; 100 μl of 1 M HCl was added to stop the reaction. Phospholipids were extracted into the organic phase by adding 200 μl of 1:1 CH$_3$OH:CHCl$_3$. The extract was washed once with 80 μl HCl and the phosphatidylinositol products and free [γ-^{32}P]ATP in the organic phase were resolved by thin-layer chromatography using Whatman Silica60 plates (LK6F, Whatman International Ltd., UK) and 9:7:2 CHCl$_3$:CH$_3$OH:4 M NH$_4$OH as the solvent. The radioactive PtdIns(3)P was detected by autoradiography or

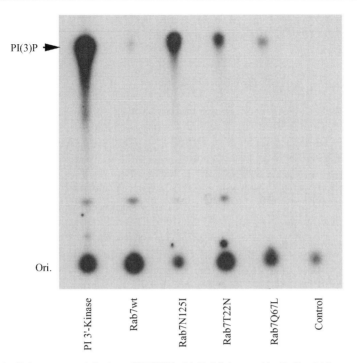

PI(3)P →

Ori.

PI 3'-Kinase | Rab7wt | Rab7N125I | Rab7T22N | Rab7Q67L | Control

FIG. 3. Coimmunoprecipitation of hVPS34 with Rab7 detected by PtdIns 3-kinase activity. Mock transfected BHK21 cells and BHK21 cells infected/transfected and overexpressing hVPS34 alone or with Rab7 wild-type (wt), dominant-negative Rab7N125I or Rab7T22N, or dominant-active Rab7Q67L. Lysates were immunoprecipitated with anti-hVPS34 (PtdIns 3-kinase and control samples) or with anti-Rab7 (all Rab7 samples) antibodies. The PtdIns 3-kinase activity associated with each immunoprecipitate was measured as described in the text, and products were resolved by thin-layer chromatography and visualized by autoradiography. Based on immunoblot analysis the amount of Rab7 protein precipitated was similar in all cases, but the amounts of hVPS34 protein varied (Stein *et al.*, 2003b). Therefore, the amount of PtdIns(3)P activity closely reflects the amount of coprecipitated kinase.

by phosphoimage analysis and positively identified by comparison with PtdIns(3)P produced by purified hVPS34.

Figure 3 illustrates the utility of the assay for monitoring hVPS34 activity associated with Rab7 by coimmunoprecipitation. A composite analysis of coprecipitated kinase activity, IP/Westerns, and coprecipitation of metabolically labeled proteins suggests that active Rab7 binds and activates hVPS34 but readily dissociates. The nucleotide-free Rab7N125I protein on the other hand binds very tightly to hVPS34/p150 and may represent a transition state that binds the kinase complex very tightly, impeding PtdIns(3)P signaling and accounting for the severe inhibition of late endocytic transport imposed by this mutant (Feng *et al.*, 2001; Stein *et al.*, 2003b).

Generation of Specific Antibody Reagents

Chicken Anti-Rab7 Antibody. We have found antibodies raised in chickens to be ideal for morphological studies. A 50:50 mix of two synthetic peptides (KQETEVELYNEFPEPIKLDKNDRAKTSAESCSC and KQ-ETEVELYNEFPEPIK) corresponding to the hypervariable C-terminus of Rab7 was conjugated to keyhole limpet hemocyanin (KLH) (Calbiochem/EMD Biosciences, San Diego, CA) using recrystallized benzoquinone (stored in the dark). KLH (10 mg/ml) was dissolved in water (total volume 2 ml) and adjusted to 150 mM NaCl and 100 mM sodium phosphate, pH 7.4. Fifteen milligrams benzoquinone was dissolved in 0.6 ml ethanol and added dropwise while vortexing to the KLH solution. The mixture was incubated in the dark at room temperature for 15 min. The activated KLH was purified on a 1.2 × 12 cm Sephadex G-50 column equilibrated with phosphate-buffered saline (PBS). Two peaks were observed during chromatography. The first peak was pinkish and contained 95% of the activated KLH protein, while the second ochre brown peak contained the benzoquinone and side reaction products. Four milligrams of the activated KLH (1.5 ml) was immediately mixed with 2.4 mg of the Rab7 peptides (dissolved in 0.24 ml PBS) and 200 μl sodium carbonate buffer pH 9.0 was added and the mixture was incubated overnight in the dark at room temperature. The KLH–peptide conjugate can be stored frozen and was used to immunize two hens (100–125 μg peptide/injection) and isolate the IgY from the eggs (performed by Aves Labs, Inc., Tigard, OR). The resulting immune IgY (300 mg/40 ml) was of extremely high titer based on immunofluorescence assays (see the section on morphological assay for VSVG protein endocytosis later in this chapter) and was useful for detecting both endogenous (1:200) (Fig. 4) and overexpressed (1:800) (Stein *et al.*, 2003b) Rab7 protein. The chicken antibody is not suited for immunoblotting or immunoprecipitation. For biochemical experiments we rely on several rabbit polyclonal antibodies raised against the C-terminus of Rab7 (Press *et al.*, 1998).

Rabbit Anti-hVPS34 Antibody. Two rabbit polyclonal antibodies were raised against the recombinant His-tagged, N- (P5) and C-terminal (P3) domains of hVSP34 purified from *E. coli*. Briefly, plasmids encoding the His-tagged domains of hVPS34 were used to transform *E. coli* BL21. Individual colonies from fresh transformants were used to inoculate 500 ml T-broth cultures containing ampicillin and cultured overnight at 37°. Cells were extracted with 1% Triton X-100, and subsequently the cell

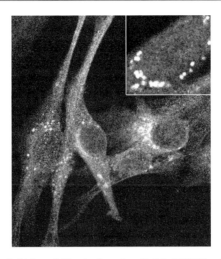

FIG. 4. Specificity of chicken IgY raised against Rab7. BHK21 cells were processed for immunofluorescence staining as described in the text. Endogenous Rab7 was visualized with our chicken anti-Rab7 (1:200) and an FITC-conjugated anti-chicken secondary antibody. Samples were imaged on a Zeiss LSM 510. Inset shows a 2× zoom to illustrate the large, coarse granular Rab7 staining pattern.

pellets containing the insoluble protein were treated with 10 ml 6 M guanidine-HCl, 0.1 M sodium phosphate, 10 mM Tris, pH 8.0, for 1 h at room temperature with stirring. The extract was clarified by centrifugation for 15 min at 15,000 rpm in a Sorvall SS34 rotor. The supernatants were incubated with 4 ml of a 50% slurry Ni-NTA resin (QiaExpress, Qiagen, Valencia, CA). The resin was sequentially washed: once with 6 M guanidine HCl, 0.1 M sodium phosphate, 10 mM Tris, pH 8.0; twice with 8 M urea, 0.1 M sodium phosphate, 10 mM Tris, pH 8.0; and twice with 8 M urea, 0.1 M sodium phosphate, 10 mM Tris, pH 6.3. Subsequently, the proteins were eluted by incubating the resin in 8 M urea, 250 mM imidazole overnight. Purity was confirmed by sodium dodecyl sulfate–polyacrylamide gel electrophoresis (SDS-PAGE) and yield quantified (1 mg P3; 5 mg P5). Proteins were dialyzed against three changes of 20 mM HEPES, pH 7.2, 100 mM NaCl, 2 mM MgCl$_2$, 5 mM 2-mercaptoethanol, 0.1% Triton X-100, concentrated to 1 mg/ml protein using a Centricon spin concentrator (Millipore Corporation, Billerica, CA) and aliquots quick frozen and stored at −80° were used for antibody production. P3 and P5 fractions were used to inject one rabbit each (performed by HRP, Inc., Denver, CO). The antiserum (NW259) raised against the P3 C-terminal domain had a higher titer than the antiserum (NW260) raised against the P5 N-terminal

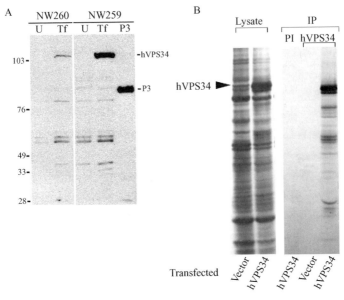

FIG. 5. Specificity of rabbit polyclonal antisera raised against hVPS34. (A) Immunoblot. His-tagged N-terminal (P5) and C-terminal (P3) fragments of hVPS34 were expressed and purified from *E. coli* and used to immunize rabbits as detailed in the text. Lysates (1/10 volume from a 3.5-cm dish) from untransfected (U) BHK21 or from cells infected/transfected to overexpress hVPS34 (Tf) were resolved by SDS-PAGE and immunoblotted with NW260 (raised against P5) or with NW259 (raised against P3) using a 1:5000 dilution of the primary antiserum. Approximately 80 ng of purified P3 was used as a positive control. Migration of standards is as shown. (B) Immunoprecipitation. BHK21 cells grown in 3.5-cm dishes were mock treated or infected/transfected with pcDNA3 hVPS34 radiolabeled with 100 μCi/ml Trans^{35}S-Label for 2 h and lysed in 1 ml RIPA buffer as detailed in the text. Of the supernatant 300 μl was incubated with 10 μl preimmune (PI) or immune NW259 (anti-hVPS34) at 4° with rotation and immunoprecipitates were collected using protein A-Sepharose. Samples were resolved on 10% polyacrylamide gels. Migration of hVPS34 is indicated by an arrow.

domain (Fig. 5A) and was used without purification for immunoblotting (1:5000), immunoprecipitation (10 μl antiserum/300 μl metabolically labeled lysate in RIPA buffer; Fig. 5B), and immunofluorescence staining (1:250; see Fig. 6). The antibody was not reactive against endogenous BHK21 hVPS34 by immunoblotting (Fig. 5A), but detected endogenous hVPS34 in human cells (HEK293A, HepG2, HeLa) and in bovine brain cytosol (1:5000, not shown).

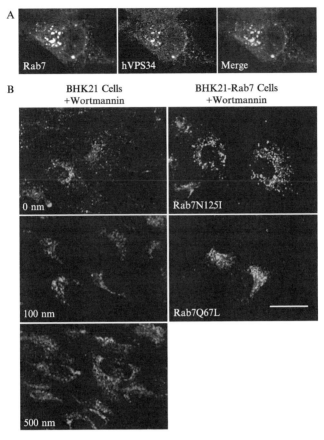

FIG. 6. Colocalization of hVPS34 and Rab7 by immunofluorescence microscopy. (A) BHK21 cells were infected/transfected as detailed in the text to overexpress hVPS34 and Rab7. Samples were processed for immunofluorescence staining. Overexpressed Rab7 was visualized with our chicken anti-Rab7 (1:800) and overexpressed hVPS34 was detected with our rabbit anti-hVPS34 NW259 (1:250). FITC-conjugated donkey anti-chicken and Texas Red-conjugated donkey anti-rabbit were used as secondary detection antibodies. Samples were imaged on a Zeiss LSM510. Merge shows the overlap between the two channels. (B) Parental BHK21 cells were left untreated or treated with 100 or 500 nm wortmannin for 30 min. In comparison, stable cell lines expressing dominant-negative (Rab7N125I) or dominant-active (Rab7Q67L) Rab7 were treated with 100 nM wortmannin for 30 min. Cells were fixed and stained with rabbit pAb against mannose 6-phosphate receptor (red) (kind gift of A. R. Robbins, NIDDK, Bethesda, MD) and mouse mAb 4A1 against lgp120 (green) (kind gift of J. Gruenberg, University of Geneva, Switzerland) and imaged on a Zeiss LSM410. Note the increased dilation of endosomes in wortmannin-treated cells expressing the dominant-negative Rab7N125I protein (comparable to control cells treated with a 5-fold higher concentration of wortmannin) versus the normal appearance of endosomes (comparable to untreated control cells) in cells expressing the dominant-positive Rab7Q67L protein. (See color insert.)

Cell Line, Protein Expression, and Morphological Analyses

For all analyses, baby hamster kidney (BHK21) cells obtained from American Type Culture Collection (Rockville, MD) were used due to the ease of transfection and their well-defined endocytic transport pathways. For overexpression of recombinant proteins, BHK21 cells were infected with recombinant vaccinia virus (vTF7.3) encoding T7 RNA polymerase (Fuerst *et al.*, 1986) and then transfected with Lipofectamine according to the manufacturer's instructions (Invitrogen Corporation) for 5–6 h with plasmids containing Rab7 or hVPS34 cDNA under the control of the bacteriophage T7 promoter, as previously described (Feng *et al.*, 1995). Hydroxyurea was added to a final concentration of 10 m*M* during the transfection period to block viral replication and minimize cytopathic effects. **Note**: In cotransfection experiments the ratios of individual plasmids must be carefully titrated to achieve optimal coexpression of all proteins, while not exceeding the maximal optimal DNA:lipofectamine ratio; e.g., the optimal ratio for Rab7/hVPS34 cotransfection experiments was 1.2 μg pGEM Rab7:0.2 μg pcDNA3 hVPS34. Despite the presence of a CMV promoter, pcDNA3 hVPS34 could not be used for expression by transfection alone. The safe use of vaccinia virus requires primary and booster vaccination against smallpox.

Immunofluorescence and Immunoelectron Microscopy Colocalization Studies to Document hVPS34/p150 PtdIns 3-Kinase Function on the Late Endocytic Pathway

Immunofluorescence. For immunofluorescence colocalization studies, BHK21 cells were seeded in 35-mm dishes containing 20-mm square glass coverslips. Cells were infected/transfected for 5–6 h (as described in the section Cell Line, Protein Expression, and Morphological Analysis earlier in this chapter). Prior to fixation, cells were prepermeabilized with 0.05% (w/v) saponin in 80 m*M* PIPES buffer, pH 6.8, and subsequently fixed in 3% PFA prepared in PBS. (*Note*: saponin permeabilization is useful for extracting cytosolic contents and allows for improved staining of membrane components. Saponin concentration may need to be increased up to 0.5% or extraction time increased depending on cell type used and cell confluence.) Cells were washed with PBS/0.05% saponin, quenched with NH$_4$Cl, and incubated with blocking solution (0.2% fishskin gelatin or 2% normal goat serum and 2% normal calf serum containing 0.05% saponin). Samples were incubated with primary antibodies diluted in blocking solution for 30–90 min and washed three times with PBS/0.05% saponin. Samples were incubated with secondary antibodies diluted in blocking solution for 30 min, washed three times with PBS/0.05% saponin, and mounted in Mowiol mounting medium (6 g glycerol, 2.4 g Mowiol 4–88 [Calbiochem/EMD Biosciences] added while vortexing, add 6 ml distilled water and leave at room temperature for 2 h, add 12 ml 0.2 M Tris, pH 8.5, and incubate at 53° overnight until Mowiol is dissolved, clarify by centrifugation at 4000 rpm for 20 min, and store as 1-ml aliquots at −20°). Care was taken that all secondary antibodies were raised in donkey (Jackson ImmunoResearch, West Grove, PA) to minimize secondary antibody cross-reactivities. Figure 6A illustrates the colocalization of Rab7 and hVPS34. Similarly, Rab7 could be shown to colocalize with p150 (Stein *et al.*, 2003). Limited colocalization was observed between overexpressed Rab5 and hVPS34 except when Rab5Q79L was expressed. Treatment of stable cell lines expressing wild-type or mutant forms or Rab7 (Press *et al.*, 1998) with the PtdIns 3-kinase inhibitor wortmannin first revealed a synergistic functional relationship between Rab7 and the PtdIns 3-kinase that was later confirmed by the biochemical experiments detailed in the following sections (Fig. 6B). Since hVPS34/p150 is a downstream effector of both Rab5 and Rab7 GTPases, it is of interest to determine how the kinase complex is coordinately regulated by these two GTPases, e.g., using quantitative flow cytometry bead-based GST-binding assays described in the section on *in vitro* interaction assays later in this chapter.

Immunoelectron Microscopy. In preparation for immunoelectron microscopy, BHK21 cells were grown on 10-cm dishes and proteins were overexpressed by infection/transfection. BHK21 cells overexpressing Rab7 or hVPS34 were released from the culture dishes with 20 mM EDTA in PBS (5 ml), gently pelleted, resuspended, and fixed initially for 15 min in 8% paraformaldehyde/0.01% glutaraldehyde in 0.1 M cacodylate buffer, pH 7.4. Cells were pelleted by centrifugation at 1200 rpm. Cells were then further fixed in 4% paraformaldehyde/0.01% glutaraldehyde in 0.1 M cacodylate buffer, pH 7.4, for an additional 45 min. Fixatives and cacodylate buffer were from EMD Chemicals, Inc. (formerly EM Science, Gibbstown, NJ). The pellet was rinsed with 0.1 M cacodylate buffer three times then allowed to sit in cryoprotectant (20% [w/v] polyvinylpyrrolidone, 1.86 M sucrose in PBS) 24–48 h before freezing in liquid nitrogen. Seventy-nanometer-thin frozen sections were produced on a Reichert Ultracut S microtome equipped with a Leica EM FCS for cryosectioning, collected with a 1:1 2.3 M sucrose to 2% (w/v) methylcellulose drop, and placed onto Formvar-coated gold clad grids. The grids were floated on 1% normal donkey serum (Jackson Immunoresearch Laboratories, Inc., West Grove, PA) in PBS until immunolabeled.

Immunolabeling was carried out using 1% normal donkey serum in PBS as the diluent for the primary antibodies against hVPS34 or p150

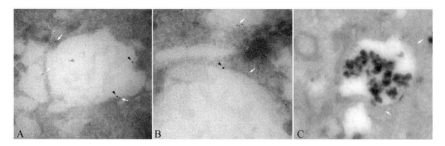

FIG. 7. Colocalization of hVPS34, p150, and Rab7 by immunoelectron microscopy. BHK21 cells overexpressing dominant-active Rab7Q67L and V5-tagged p150 or V5-tagged hVPS34 were processed for immunoelectron microscopy as described in the text. (A, B) Rab7 was detected with polyclonal anti-Rab7 antibody raised in rabbit and 6 nm gold-conjugated donkey anti-rabbit antibody (white arrows). V5-tagged p150 was detected with anti-V5 mAb and 12 nm gold-conjugated donkey anti-mouse antibody (black arrowheads). (C) V5-tagged hVPS34 was detected with anti-V5 mAb and 12 nm gold-conjugated donkey anti-mouse antibody. Images were taken on a Hitachi 7500 at 80 kV and at magnifications of (A) ×70,000, (B) ×120,000, and (C) ×50,000. Panel A adapted from Stein *et al.*, 2003b.

V5-tagged (mAb directed against V5 epitope, Santa Cruz Biotechnologies, Santa Cruz, CA) and Rab7 (rabbit anti-Rab7). Secondary antibodies conjugated to gold (12 nm donkey anti-mouse and 6 nm donkey anti-rabbit) were obtained from commercial sources (Jackson Immunoresearch Laboratories, Inc.). Labeled sections were fixed in 2% glutaraldehyde in PBS, counterstained with freshly made uranyl acetate (0.1 g/2.5 ml distilled water), and then embedded in polyvinyl alcohol. Sections were examined and images collected on a Hitachi 7500 transmission electron microscope at 80 kV equipped with a 1-megapixel digital camera. Figure 7 shows the colocalization of p150 and Rab7 on late endosomes and hVPS34 on multivesicular structures.

Morphological Assay for VSVG Protein Endocytosis to Monitor Late Endocytic Membrane Transport and Identify the Functions of hVPS34 and Rab7

Background. Vesicular stomatitis virus (VSV)G protein has been widely used for tracing both exocytic and endocytic membrane transport (Fuller *et al.*, 1984; Gruenberg *et al.*, 1989). Our laboratory has established both morphological and biochemical assays for tracing VSVG protein transport along the endocytic pathway (Feng *et al.*, 1995, 2001; Stein *et al.*, 2003b). As an endocytic tracer, VSVG protein has the benefit that it is synchronously and preferentially transported to late endosomes following appropriate manipulations and it undergoes defined cleavages that can distinguish transit from early to late endosomes and from late endosomes to lysosomes (Feng *et al.*, 2001; Stein *et al.*, 2003b).

Morphological Evaluation of Late Endocytic Transport. BHK21 were infected with vaccinia virus and transfected with plasmids encoding VSVG protein (pAR-G) (Whitt *et al.*, 1989), Rab7, and hVPS34 wild-type or hVPS34kd (kinase dead). After 6 h of transfection, cells were washed with ice-cold Glasgow's modified Eagle's medium (GMEM) containing low bicarbonate (0.35 g/liter), 60 mg/liter KH_2PO_4, and 10 mM HEPES, pH 7.2, to provide appropriate buffering under atmospheric conditions (GMEM buffer). The cells were incubated for 30 min on ice to prevent internalization with excess mAb 8G5F11 or I1 (culture supernatant diluted 1:30 in GMEM buffer) directed against the ectoplasmic domain of VSVG protein (Lefrancois and Lyles, 1982).

To allow synchronous internalization to early endosomes, cells were incubated at 15° for 45 min in a heating, circulating waterbath placed in the cold room and set to 15°. Cells were washed two times with GMEM and transferred to warm GMEM and incubated at 37° for 0–60 min. Transport to late endosomes is observed within 30 min, is complete by 60 min, and is confirmed by colocalization with the late endosomal marker mannose 6-phosphate receptor (Stein *et al.*, 2003b). Using this assay we showed that both Rab7 dominant-negative mutant proteins and hVSP34kd inhibit late endosomal delivery of VSVG and that the two proteins act synergistically to regulate transport (Stein *et al.*, 2003b).

Coimmunoprecipitation Experiments to Document Protein–Protein Interactions

We have found coimmunoprecipitation assays to be invaluable for characterizing *in vivo* protein–protein interactions. For this purpose, BHK21 cells were infected and cotransfected with plasmids encoding Rab7, V5-tagged p150, and hVPS34 as described in the section on Cell Line, Protein Expression, and Morphological Analyses earlier in this chapter. Approximately 5 h posttransfection, cells were starved in Dulbecco's modified Eagle's medium (DMEM) lacking cysteine and methionine for 30 min. After starvation, cells were incubated in DMEM lacking cysteine and methionine supplemented with 100 mCi/ml Trans[^{35}S]-Label (MP [formerly ICN] Biomedicals, Irvine, CA) for 45 min. Cells were washed once with PBS and to enhance detection of membrane-associated complexes, cells were prepermeabilized in 0.05% saponin in PIPES buffer (80 mM PIPES, pH 6.8, 5 mM EGTA, 1 mM $MgCl_2$ in PBS) for 5 min, then washed once again with PBS. Samples were either stored at −80° until further processing or lysed immediately in RIPA buffer (50 mM Tris, pH 7.4, 150 mM NaCl,

1% [v/v] Nonidet P-40, 0.5% [w/v] deoxycholate, and 0.1% [w/v] SDS) containing chymostatin, leupeptin, aprotinin, pepstatin, and 4-(2-aminoethyl)benzenesulfonyl fluoride (AEBSF) on ice for 20 min. Samples were collected and clarified by centrifugation for 15 min at 14,000 rpm in an Eppendorf microfuge at 4°. Supernatants were recovered and approximately 1/10 volume was removed for SDS-PAGE analysis of total labeled protein. Antiserum against Rab7, V5, or hVPS34 was added and samples were incubated at 4° rotating for 1 h. For monoclonal antibodies, a rabbit anti-mouse IgG linker antibody (Jackson Immunoresearch Laboratories, Inc.) was added for 30 min prior to addition of protein A–Sepharose. Samples were incubated an additional 30 min while rotating at 4°, washed three times with RIPA buffer, and once in RIPA buffer without detergents prior to solubilization in Laemmli sample buffer (Laemmli, 1970).

Immunoprecipitated proteins were resolved by SDS-PAGE on 4–16.5% gradient gels and quantified using a Molecular Dynamics Storm Phosphoimager and Molecular Dynamics ImageQuant software (Sunnyvale, CA). Values were quantified as ratios of coimmunoprecipitated hVPS34 or V5-p150 to Rab7 and an average ratio from three to six independent experiments was calculated along with the standard error of the mean using GraphPad Prism software (San Diego, CA). Based on such coimmunoprecipitation studies, we established that wild-type Rab7 is associated with the hVPS34/p150 PtdIns 3-kinase complex *in vivo* (Stein *et al.*, 2003b). The Rab5 GTPase could also be coprecipitated with the hVPS34/p150 PtdIns 3-kinase complex, but Rab9, another late endosomal GTPase, was not found associated with the PtdIns 3-kinase complex (not shown).

In Vitro *Protein Interaction Assays and Mapping of Protein–Protein Interaction Sites with GST-Protein–Protein Binding Assays*

Background. In vitro protein–protein interaction studies are useful for identifying binding affinities, direct interactions, and specific interaction domains. Our laboratory has used glutathione *S*-transferase (GST)-immobilized proteins to detect interactions with *in vitro* translated proteins as detailed below. Purified unlabeled proteins or crude cell lysates may also be used depending on the application of interest (Christoforidis and Zerial, 2001; Christoforidis *et al.*, 1999a).

GST-Rab7 Fusion Protein Expression and Purification. Recombinant Rab7 was expressed as an N-terminal GST-fusion protein in *E. coli* BL21 (for vector description, see the section Construction of Wild-Type and Mutant hVPS34 Expression Vectors). Four-hundred milliliters of transformed bacterial culture was grown to $A_{600} = 0.5$ in Luria-Bertani (LB) medium containing 100 μg/ml ampicillin in a 2-liter flask. Cells were chilled on ice for 20 min. Protein expression was then induced overnight at room temperature by the addition of isopropyl-β-thiogalactopyranoside (IPTG) to a final concentration of 50 μM. (Taylor *et al.*, 1997). Room temperature induction is particularly important for larger proteins that are insoluble in cultures induced at 30°. Bacteria were harvested by centrifugation at 4000 rpm for 10 min at 4°, suspended in 10 ml of GST lysis buffer (25 mM Tris, pH 7.4, 2 mM EDTA, 137 mM NaCl, 2.6 mM KCl, 1 mM dithiothreitol (DTT), 1 mM phenylmethylsulfonyl fluoride, 10 μg/ml each chymostatin, leupeptin, antipain, and pepstatin), quick frozen, and stored at $-80°$ in 1-ml aliquots. Aliquots were thawed just before use and lysed by sonication (Sonics & Materials, Inc., Danbury, CT) five times, 13 s (50% duty cycle, micro-tip, and output control 5) and on ice between cycles. Triton X-100 was added to a final concentration of 1% (v/v). The lysate was incubated at 4° rotating for 30 min, pelleted by centrifugation at 14,000 rpm for 10 min. The GST-fusion protein was affinity purified using glutathione-Sepharose 4B (Amersham Biosciences). A 50% glutathione-Sepharose slurry (40 μl) was incubated with lysate (400 μl) for 1 h at 4°. Glutathione-Sepharose was washed three times with GST Lysis Buffer and the GST fusion protein was eluted by incubation with 60 μl of 10 mM glutathione elution buffer (10 mM reduced glutathione in 50 mM Tris, pH 8.0) at 4° for 60 min. Protein concentration was measured using Bio-Rad DC Protein Assay (Bio-Rad Laboratories, Inc., Hercules, CA) according to manufacturer's instruction. The estimated yield from a 400-ml culture is $\sim$1–2 mg pure GST-Rab7.

Monitoring GTP-Binding Activity of GST-Rab7. GST-Rab7 was expressed and purified from *E. coli* BL21 cells as described earlier. To test the nucleotide-binding activity of the purified proteins prior to being used for protein–protein interaction studies, the GST-Rab7 protein (5 μg) was bound to 100 μl of a 50% slurry of glutathione-Sepharose 4B (Amersham Biosciences) in PBS. The GST-Rab7 loaded beads were washed with cold binding buffer (50 mM Tris, pH 7.5, 5 mM MgCl$_2$, 10 mM EDTA, 1 mM DTT, 1 mg/ml bovine serum albumin [BSA]), collected by centrifugation for 5 min at 500g and resuspended in 50 μl binding buffer containing 25 μCi $[\gamma$-^{32}P]GTP (800 Ci/mmol, MP Biomedicals) on ice. Negative control samples also contained 1 mM GTP. All samples were incubated at 30° with gentle agitation for 30 min. Beads were collected by centrifugation and washed three times with wash buffer (50 mM Tris, pH 7.5, 20 mM MgCl$_2$, 1 mg/ml BSA). Bound radioactivity was measured by scintillation counting (Fig. 8). Radioactive GTP binding and comparison of protein binding to

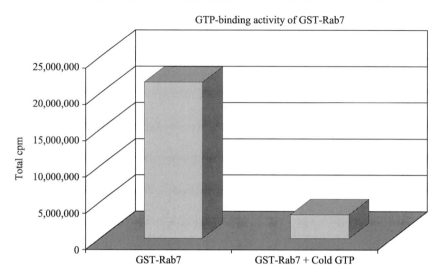

FIG. 8. *In vitro* GTP-binding activity of GST-Rab7. Purified GST-Rab7 was bound to glutathione-Sepharose 4B and GTP-binding activity assessed using [γ-^{32}P]GTP as detailed in the text. Plotted are the counts per minute [γ-^{32}P]GTP bound to immobilized GST-Rab7 relative to a control sample containing excess unlabeled GTP in the reaction mix.

GDP versus GTPγS-loaded GST-Rab7 is important to ensure the specificity of Rab GTPase/effector interactions (Christoforidis *et al.*, 1999a).

GST-Rab7/p150-Binding Assay. Full-length (p150-WT) deletion mutants or isolated p150 domains (Murray *et al.*, 2002) were synthesized *in vitro* by TNT Quick Coupled Transcription/Translation Systems (Promega, Madison, WI). Briefly, the reaction components (40 μl TNT Quick Master Mix, 2 μl of 10 mCi/ml Trans[^{35}S]-Label [MP [ICN] Biomedicals), 1 μg p150 plasmid DNA template, nuclease-free water to 50 μl final volume] were combined and incubated at 30° for 90 min. *In vitro* translated p150 (10 μl) was added to equimolar GST-Rab7 or GST immobilized on glutathione Sepharose beads and incubated at 4° for 2 h. Beads were washed three times with GST lysis buffer and boiled in SDS–PAGE sample buffer. Proteins were resolved by SDS-PAGE and quantified on a Molecular Dynamics Phosphoimager. Figure 9 illustrates the utility of the assay for assessing the interaction between Rab7 and p150. Using p150 deletion mutants the WD40 domain of p150 was found to be responsible for Rab7 binding (data not shown), analogous to what is reported for Rab5 (Murray *et al.*, 2002). No direct interaction between Rab7 and hVPS34 was observed (data not shown).

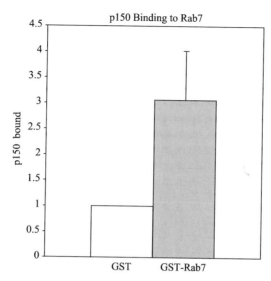

FIG. 9. Rab7 and p150 interaction detected by *in vitro* binding assay. GST or GST-Rab7 was immobilized on glutathione-Sepharose 4B and *in vitro* translated full-length p150 was allowed to bind as described in the text. Graph shows the binding of p150 to GST-Rab7 relative to GST control and data are representative of four independent experiments.

Flow Cytometry-Based Glutathione Bead Assay. The major drawbacks of most *in vitro* binding assays (including GST-protein–protein binding assays) are (1) the large amount of purified proteins required to perform the assays and (2) the lack of quantitative and order of binding information. To address these problems, we are developing a GSH-bead (glutathione-labeled dextran microspheres)-based flow cytometry assay that can generate reliable quantitative data using small amounts of protein. Such sensitive bead-based flow cytometry assays have already been successfully applied to studies of other multipartite complexes including the assembly of heterotrimeric G-proteins, G-coupled receptors, and ligands (Buranda *et al.*, 2003; Simons *et al.*, 2003, 2004). The assay has utility for drug discovery and/or measuring kinetics of association and complex stability. In preliminary experiments with GST-tagged GFP, we could accurately quantify the concentration of GST-GFP/bead and precisely determine the binding and dissociation constants (K_d) using as little as 1–10 nM of the purified GST-tagged protein. We anticipate using the system to immobilize GST-Rab7 and/or GST-p150 for purposes of obtaining accurate binding and dissociation constants, as well as the order of assembly between Rab7 and its interacting partners as a function of nucleotide bound status or mutation.

Monitoring PtdIns(3)P Formation In Vivo with $2\times$-FYVE^{Hrs}-myc

Background. PtdIns(3)P is specifically recognized by proteins containing FYVE domains. The crystal structure of the FYVE domain reveals a binding pocket that closely interfaces with the PtdIns(3)P head group. Stenmark and colleagues pioneered the use of a synthetic peptide with two sequential FYVE domains derived from hepatocyte growth factor-regulated tyrosine kinase substrate (Hrs) fused to the myc epitope tag to specifically detect intracellular PtdIns(3)P pools (Gillooly *et al.*, 2000).

PtdIns(3)P Formation In Vivo. BHK21 cells were infected and cotransfected with plasmids expressing wild-type or mutant Rab7 and the $2\times$-FYVE^{Hrs}-myc construct (Gillooly *et al.*, 2000) to visualize intracellular PtdIns(3)P. Cells were washed with PBS, prepermeabilized with 0.05% saponin in PIPES buffer, washed again with PBS, and fixed in 3% PFA. Rab7 was visualized using our chicken anti-Rab7 antibody while anti-myc tag mAb 9E10 (Santa Cruz Biotechnology, Inc., Santa Cruz, CA) was used to monitor $2\times$-FYVE^{Hrs}-myc binding to PtdIns(3)P *in vivo*.

The amount of $2\times$-FYVE^{Hrs}-myc binding was quantified in a minimum of 30 cells in two or three different experiments using the Zeiss LSM software (Carl Zeiss, Inc., Thornwood, NY). Average values of total $2\times$-FYVE^{Hrs}-myc staining per cell were determined and values were expressed as a percent of the total $2\times$-FYVE^{Hrs}-myc binding in cells expressing wild-type Rab7. Decreased $2\times$-FYVE^{Hrs}-myc binding in cells expressing mutant versus wild-type Rab7 demonstrated that the activity of Rab7 plays an important role in the formation of PtdIns(3)P *in vivo* (Stein *et al.*, 2003b). The requirement for active Rab7 in PtdIns(3)P formation on late endosomes was further evidenced by the fact that wild-type and active forms of Rab7 colocalized with $2\times$-FYVE^{Hrs}-myc, in stark contrast to the lack of $2\times$-FYVE^{Hrs}-myc staining on endosomes bearing inactive Rab7.

Conclusions

PtdIns(3)P signaling on endosomes provides temporal and spatial control over membrane fusion and protein sorting via intraluminal vesicle formation. The formation of PtdIns(3)P on late endosomes is coordinated by the Rab7 GTPase and the hVPS34/p150 PtdIns 3-kinase complex. The interaction of Rab7 with the PtdIns 3-kinase is mediated via binding of the WD40 domain on p150. Intraluminal vesicle formation depends on the recruitment of a second lipid kinase, PIKfyve, to the locally produced PtdIns(3)P, while down-regulation of late endosomal PtdIns(3)P signaling is regulated by the myotubularins whose membrane recruitment is also dictated by FYVE domain–PtdIns(3)P interactions. A combination of genetic, morphological, and biochemical approaches has led to a basic understanding of the role of PtdIns(3)P signaling in endocytic membrane transport. In this chapter we have detailed assays useful for dissecting these processes in mammalian cells. Future challenges remain in defining how the PtdIns3-kinase complex is coordinately regulated on early and late endosomes through the actions of Rab5 and Rab7 and identifying the mechanisms governing the temporal assembly and disassembly of the regulatory molecules involved in PtdIns(3)P signaling and its down-regulation.

Acknowledgments

A.W.N. gratefully acknowledges support from NSF MCBQ446179. M.P.S. was supported by a University of New Mexico CRTC fellowship while at the University of New Mexico and is currently supported by NRSA 1F32AI063911. We are grateful to Melanie Lenhart for expert technical assistance and Ms. Valorie Bivins for outstanding administrative support. We thank Dr. Jonathan Backer (Albert Einstein College of Medicine, New York, NY) for the p150 constructs, Dr. Harald Stenmark (Norwegian Radium Hospital, Oslo, Norway) for the 2x-FYVE^{Hrs}-myc construct, and Dr. Jianbo Dong (University of Texas Southwestern, Dallas, TX) for numerous helpful scientific discussions. Confocal images were acquired using the University of New Mexico CRTC Fluorescence Microscopy Facility (http://kugrserver.health.unm.edu:16080/microscopy/).

References

Blondeau, F., Laporte, J., Bodin, S., Superti-Furga, G., Payrastre, B., and Mandel, J. L. (2000). Myotubularin, a phosphatase deficient in myotubular myopathy, acts on phosphatidylinositol 3-kinase and phosphatidylinositol 3-phosphate pathway. *Hum. Mol. Genet.* **9,** 2223–2229.

Buranda, T., Lopez, G. P., Simons, P., Pastuszyn, A., and Sklar, L. (2003). Detection of epitope-tagged proteins in flow cytometry: Fluorescence resonance energy transfer-based assays on beads with femtomole resolution. *Anal. Biochem.* **298,** 151–162.

Cantalupo, G., Alifano, P., Roberti, V., Bruni, C. B., and Bucci, C. (2001). Rab-interacting lysosomal protein (RILP): The Rab7 effector required for transport to lysosomes. *EMBO J.* **20,** 683–693.

Christoforidis, S., and Zerial, M. (2001). Purification of EEA1 from bovine brain cytosol using Rab5 affinity chromatography and activity assays. *Methods Enzymol.* **329,** 120–132.

Christoforidis, S., McBride, H. M., Burgoyne, R. D., and Zerial, M. (1999a). The Rab5 effector EEA1 is a core component of endosome docking. *Nature* **397,** 621–625.

Christoforidis, S., Miaczynska, M., Ashman, K., Wilm, M., Zhao, L., Yip, S. C., Waterfield, M. D., Backer, J. M., and Zerial, M. (1999b). Phosphatidylinositol-3-OH kinases are Rab5 effectors. *Nat. Cell Biol.* **1,** 249–252.

Dong, J., Chen, W., Welford, A., and Wandinger-Ness, A. (2004). The proteasome alpha subunit XAPC7 interacts specifically with Rab7 and late endosomes. *J. Biol. Chem.* **279,** 21334–21342.

Feng, Y., Press, B., and Wandinger-Ness, A. (1995). Rab7: An important regulator of late endocytic membrane traffic. *J. Cell Biol.* **131,** 1435–1452.

Feng, Y., Press, B., Chen, W., Zimmerman, J., and Wandinger-Ness, A. (2001). Expression and properties of Rab7 in endosome function. *Methods Enzymol.* **329,** 175–187.

Fuerst, T. R., Niles, E. G., Studier, F. W., and Moss, B. (1986). Eukaryotic transient-expression system based on recombinant vaccinia virus that synthesizes bacteriophage T7 RNA polymerase. *Proc. Natl. Acad. Sci. USA* **83,** 8122–8126.

Fuller, S., von Bonsdorff, C. H., and Simons, K. (1984). Vesicular stomatitis virus infects and matures only through the basolateral surface of the polarized epithelial cell line, MDCK. *Cell* **38,** 65–77.

Gillooly, D. J., Morrow, I. C., Lindsay, M., Gould, R., Bryant, N. J., Gaullier, J. M., Parton, R. G., and Stenmark, H. (2000). Localization of phosphatidylinositol 3-phosphate in yeast and mammalian cells. *EMBO J.* **19,** 4577–4588.

Gruenberg, J., Griffiths, G., and Howell, K. E. (1989). Characterization of the early endosome and putative endocytic carrier vesicles *in vivo* and with an assay of vesicle fusion *in vitro*. *J. Cell Biol.* **108,** 1301–1316.

Gutierrez, M. G., Master, S. S., Singh, S. B., Taylor, G. A., Colombo, M. I., and Deretic, V. (2004). Autophagy is a defense mechanism inhibiting Bcg and mycobacterium tuberculosis survival in infected macrophages. *Cell* **119,** 753–766.

Ikonomov, O. C., Sbrissa, D., and Shisheva, A. (2001). Mammalian cell morphology and endocytic membrane homeostasis require enzymatically active phosphoinositide 5-kinase PIKfyve. *J. Biol. Chem.* **276,** 26141–26147.

Ikonomov, O. C., Sbrissa, D., Mlak, K., Kanzaki, M., Pessin, J., and Shisheva, A. (2002). Functional dissection of lipid and protein kinase signals of PIKfyve reveals the role of PtdIns 3,5-P2 production for endomembrane integrity. *J. Biol. Chem.* **277,** 9206–9211.

Ikonomov, O. C., Sbrissa, D., Foti, M., Carpentier, J. L., and Shisheva, A. (2003). PIKfyve controls fluid phase endocytosis but not recycling/degradation of endocytosed receptors or sorting of procathepsin D by regulating multivesicular body morphogenesis. *Mol. Biol. Cell* **14,** 4581–4591.

Jager, S., Bucci, C., Tanida, I., Ueno, T., Kominami, E., Saftig, P., and Eskelinen, E. L. (2004). Role for Rab7 in maturation of late autophagic vacuoles. *J. Cell Sci.* **117,** 4837–4848.

Jordens, I., Fernandez-Borja, M., Marsman, M., Dusseljee, S., Janssen, L., Calafat, J., Janssen, H., Wubbolts, R., and Neefjes, J. (2001). The Rab7 effector protein RILP controls lysosomal transport by inducing the recruitment of dynein-dynactin motors. *Curr. Biol.* **11,** 1680–1685.

Jordens, I., Marsman, M., Kuijl, C., Janssen, L., and Neefjes, J. (2003). Il-24 the small GTPase Rab7 and its effector protein RILP regulate lysosomal transport. *Pigment Cell Res.* **16,** 583.

Laemmli, U. K. (1970). Cleavage of structural proteins during the assembly of the head of bacteriophage T4. *Nature* **227,** 680–685.

Lefrancois, L., and Lyles, D. S. (1982). The interaction of antibody with the major surface glycoprotein of vesicular stomatitis virus. Ii. Monoclonal antibodies of nonneutralizing and cross-reactive epitopes of Indiana and New Jersey serotypes. *Virology* **121,** 168–174.

Mizuno, K., Kitamura, A., and Sasaki, T. (2003). Rabring7, a novel Rab7 target protein with a ring finger motif. *Mol. Biol. Cell* **14,** 3741–3752.

Murray, J. T., Panaretou, C., Stenmark, H., Miaczynska, M., and Backer, J. M. (2002). Role of Rab5 in the recruitment of hVps34/p150 to the early endosome. *Traffic* **3,** 416–427.

Panaretou, C., Domin, J., Cockcroft, S., and Waterfield, M. D. (1997). Characterization of p150, an adaptor protein for the human phosphatidylinositol (PtdIns) 3-kinase. Substrate presentation by phosphatidylinositol transfer protein to the p150/PtdIns 3-kinase complex. *J. Biol. Chem.* **272,** 2477–2485.

Press, B., Feng, Y., Hoflack, B., and Wandinger-Ness, A. (1998). Mutant Rab7 causes the accumulation of cathepsin D and cation-independent mannose 6-phosphate receptor in an early endocytic compartment. *J. Cell Biol.* **140,** 1075–1089.

Row, P. E., Reaves, B. J., Domin, J., Luzio, J. P., and Davidson, H. W. (2001). Overexpression of a rat kinase-deficient phosphoinositide 3-kinase, Vps34p, inhibits cathepsin D maturation. *Biochem. J.* **353,** 655–661.

Sbrissa, D., Ikonomov, O. C., and Shisheva, A. (1999). PIKfyve, a mammalian ortholog of yeast Fab1p lipid kinase, synthesizes 5-phosphoinositides. Effect of insulin. *J. Biol. Chem.* **274,** 21589–21597.

Schaletzky, J., Dove, S. K., Short, B., Lorenzo, O., Clague, M. J., and Barr, F. A. (2003). Phosphatidylinositol-5-phosphate activation and conserved substrate specificity of the myotubularin phosphatidylinositol 3-phosphatases. *Curr. Biol.* **13,** 504–509.

Shisheva, A. (2001). PIKfyve: The road to PtdIns 5-P and PtdIns 3,5-P(2). *Cell Biol. Int.* **25,** 1201–1206.

Simons, P. C., Shi, M., Foutz, T., Cimino, D. F., Lewis, J., Buranda, T., Lim, W. K., Neubig, R. R., McIntire, W. E., Garrison, J., Prossnitz, E., and Sklar, L. A. (2003). Ligand-receptor-G-protein molecular assemblies on beads for mechanistic studies and screening by flow cytometry. *Mol. Pharmacol.* **64,** 1227–1238.

Simons, P. C., Biggs, S. M., Waller, A., Foutz, T., Cimino, D. F., Guo, Q., Neubig, R. R., Tang, W. J., Prossnitz, E. R., and Sklar, L. A. (2004). Real-time analysis of ternary complex on particles: Direct evidence for partial agonism at the agonist-receptor-G protein complex assembly step of signal transduction. *J. Biol. Chem.* **279,** 13514–13521.

Somsel Rodman, J., and Wandinger-Ness, A. (2000). Rab GTPases coordinate endocytosis. *J. Cell Sci.* **113**(Pt 2), 183–192.

Stein, M.-P., Dong, J., and Wandinger-Ness, A. (2003a). Rab proteins and endocytic trafficking: Potential targets for therapeutic intervention. *Adv. Drug Deliv. Rev.* **55,** 1421–1437.

Stein, M.-P., Feng, Y., Cooper, K. L., Welford, A. M., and Wandinger-Ness, A. (2003b). Human Vps34 and p150 are Rab7 interacting partners. *Traffic* **4,** 754–771.

Taylor, G. S., Liu, Y., Baskerville, C., and Charbonneau, H. (1997). The activity of Cdc14p, an oligomeric dual specificity protein phosphatase from Saccharomyces cerevisiae, is required for cell cycle progression. *J. Biol. Chem.* **272,** 24054–24063.

Vitelli, R., Chiariello, M., Bruni, C. B., and Bucci, C. (1995). Cloning and expression analysis of the murine Rab7 cDNA. *Biochim. Biophys. Acta* **1264,** 268–270.

Vitelli, R., Santillo, M., Lattero, D., Chiariello, M., Bifulco, M., Bruni, C. B., and Bucci, C. (1997). Role of the small GTPase Rab7 in the late endocytic pathway. *J. Biol. Chem.* **272,** 4391–4397.

Whitman, M., Kaplan, D. R., Schaffhausen, B., Cantley, L., and Roberts, T. M. (1985). Association of phosphatidylinositol kinase activity with polyoma middle-T competent for transformation. *Nature* **315,** 239–242.

Whitt, M. A., Chong, L., and Rose, J. K. (1989). Glycoprotein cytoplasmic domain sequences required for rescue of a vesicular stomatitis virus glycoprotein mutant. *J. Virol.* **63,** 3569–3578.

[56]   Functional Analyses and Interaction of the XAPC7 Proteasome Subunit with Rab7

By Sanchita Mukherjee, Jianbo Dong, Carrie Heincelman, Melanie Lenhart, Angela Welford, and Angela Wandinger-Ness

Abstract

Proteasomes have long been known to mediate the degradation of polyubiquitinated proteins in the cytoplasm and the nucleus. Additionally, proteasomes have been identified as participating in cellular degradative pathways involving the endomembrane system. In conjunction with the endoplasmic reticulum, proteasomes serve as a quality control mechanism for disposing of malfolded newly synthesized proteins, while on the endocytic pathway they serve to facilitate the degradation of key signaling and nutrient receptors as well as the destruction of phagocytosed pathogens. Our laboratory has identified a direct interaction between the late endocytic Rab7 GTPase and the α-proteasome subunit, XAPC7, thus providing the first molecular link between the endocytic trafficking and cytosolic degradative machineries. In this chapter reagents and methods for studying the regulation and interactions between XAPC7, the 20S proteasome, and Rab7 are described.

Introduction

The Rab7 GTPase coordinates late endocytic membrane transport and protein degradation in conjunction with the α-proteasome subunit XAPC7 (Dong et al., 2004; Feng et al., 1995). Our data provided the first molecular link between the cytosolic and lysosomal degradative pathways that had previously only been suggested by proteasome inhibitor studies (van Kerkhof et al., 2001). The involvement of the proteasome at the late endosome is intriguing because of the importance of ubiquitin signals in cargo sorting by the phosphatidylinositol 3-phosphate (PtdIns[3]P) binding protein Hrs and associated ESCRT complexes (Clague and Urbe, 2003; Conibear, 2002). Together, the proteasome, a lipid kinase, hVPS34/p150, and lipid phosphatases, myotubularins, are hypothesized to regulate late endocytic phosphatidylinositol lipid dynamics and control the sorting of ubiquitinated membrane receptors in late endosomes (Fig. 1). This postulate is being tested using the reagents and methods described in this chapter and

METHODS IN ENZYMOLOGY, VOL. 403
0076-6879/05 $35.00
DOI: 10.1016/S0076-6879(05)03056-9

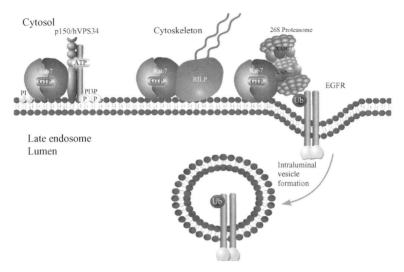

FIG. 1. Rab7 and associated factors regulate late endocytic events. The schematic depicts known Rab7 interacting proteins and their potential functions in endocytosis. Rab7 binds to the hVPS34/p150 phosphatidylinositol 3-phosphate [PtdIns(3)P] kinase via the p150 adaptor protein and serves to regulate PtdIns(3)P synthesis on late endosomes in conjunction with the myotubularin lipid phosphatase (Stein *et al.*, 2003). Rab7 has also been shown to interact with Rab7 interacting lysosome protein (RILP), which controls late endosome motility on microtubules (Cantalupo *et al.*, 2001; Jordens *et al.*, 2001) and Rab7-interacting RING finger protein (Rabring7), which controls acidification (Mizuno *et al.*, 2003). In addition, our laboratory has identified a specific interaction with the 26S proteasome mediated through the interaction with the α subunit XAPC7 (Dong *et al.*, 2004). Proteasome activity is known to be important for growth factor receptor degradation, though further specific mechanistic details are unknown and are currently under investigation.

an accompanying chapter (Chapter 55) in this volume on the hVPS34/p150 PtdIns 3-kinase complex.

Here we describe morphological and biochemical assays we have developed to monitor the activity and assembly of XAPC7 with Rab7 to regulate late endocytic membrane transport. The assays have been used to demonstrate that normal proteasome activity is crucial for the transport of cargo such as the epidermal growth factor receptor (EGFR) from early to late endosomes, as well as its subsequent degradation (Dong *et al.*, 2004; S. Mukherjee, unpublished observations). Furthermore, our data suggest that XFP-tagged variants of XAPC7 are useful probes for studying the temporal membrane recruitment of proteasomes. This chapter describes methodologies useful for dissecting the interactions between Rab7, XAPC7, and the 20S proteasome and establishing their functions in late endocytic membrane transport of key growth factor receptors.

Methods

Construction of Wild-Type and XFP-Tagged XAPC7 Vectors

To construct expression vectors encoding the full-length XAPC7 clone under the control of the T7 promoter, we obtained human expressed sequence tag (EST) clone (IMAGE:84457, Stratagene, La Jolla, CA) and confirmed that it encoded the full-length cDNA of proteasome α-subunit XAPC7 by nucleotide sequencing (Dong et al., 2004). We amplified the cDNA using polymerase chain reaction (PCR) amplification and then subcloned it into pGEM3 by restriction digest with EcoR1 and XhoI. An epitope-tagged variant of XAPC7 was generated in pcDNA3 by incorporating the TDIEMNRLG sequence derived from VSVG protein in-frame onto the N-terminus of XAPC7 using a PCR-based strategy. Green fluorescent protein (GFP)-XAPC7 chimeric protein under the control of a human cytomegalovirus (CMV) promoter was prepared by subcloning the cDNA fragments from the corresponding pGEM-XAPC7 constructs 3′ in frame into the pEGFP-C2 vector (Clontech, Palo Alto, CA). XAPC7-DsRed chimeric protein under the control of a CMV promoter was generated by PCR amplification of XAPC7 from pGEM-XAPC7 and subcloned 5′ in Frame into DsRed-pcDNA3 by restriction digest with BamH1 and XhoI. The cDNA for monomeric DsRed was kindly provided by Dr. Roger Tsien (University of California, San Diego, CA).

Mammalian expression vectors encoding canine Rab7 (wild-type and various mutant forms) under the control of the bacteriophage T7 promoter are as previously described (Feng et al., 1995). A Rab7/Rab11 chimera was constructed by replacing the 33 C-terminal residues of Rab7 with the 43 C-terminal residues of Rab11 using a PCR-based strategy to yield the Rab7-(1–174)Rab11-(160–202) chimera (Dong et al., 2004). The chimera was subcloned into GFP-C1 (BD Biosciences, San Jose, CA).

All recombinant DNA molecules were prepared in accordance with National Institutes of Health (NIH) guidelines and our research is judged exempt from all USDA/CDC select agents defined by 42CFR part 73, 9 CFR part 121, and 7 CFR part 331 under the auspices of the University of New Mexico Biosafety Office.

Generation of Specific Antibody Reagents

Rabbit Anti-XAPC7 Antibody. A rabbit polyclonal antiserum reactive against XAPC7, an α-type subunit of the 26S proteasome, was raised using clone C7 (108 C-terminal amino acids of canine XAPC7 isolated in our two-hybrid screen) expressed in pGEX 5X-2 (Amersham Biosciences,

Piscataway, NJ) and purified from *Escherichia coli* BL21 on glutathione agarose as the antigen. More details about our procedures for expressing and purifying GST-fusion proteins is given in Chapter 55 in this volume on Rab7 and hVPS34/p150. Antibody production in rabbits was contracted to HRP, Inc. (Denver, CO).

The polyclonal XAPC7 antibody specifically recognized the XAPC7 α-subunit (tested against hamster, canine, and human XAPC7), but not other α-subunits in immunoprecipitation (60 μl/200 μl lysate from a 3.5-cm dish), immunoblotting (1:1000), and immunofluorescence (1:400) assays.

Immunofluorescence and Immunoelectron Microscopy Colocalization Studies to Document XAPC7 Function on the Late Endocytic Pathway

Background. Proteasome α-subunit XAPC7 was found localized on late endosomal membranes. The membrane localization of XAPC7 was determined through fractionation and colocalization studies (Dong *et al.*, 2004). BHK21 cells on cotransfection with XAPC7 and Rab7 revealed their colocalization on late endosomes. In contrast, no colocalization was observed with Rab5 on early endosomes or wild-type Rab11 on recycling endosomes.

Immunofluorescence. To document the late endosomal recruitment of XAPC7 by Rab7, a chimeric Rab protein construct was used to determine whether binding to Rab7 was principally responsible for the specific late endosomal membrane localization of XAPC7. Membrane localization of Rab proteins is in part determined by their hypervariable C termini and may in some cases be documented by swapping C-terminal sequences (Ali *et al.*, 2004; Stenmark *et al.*, 1994). A Rab7-(1–174)Rab11-(160–202) chimera was constructed where the C-terminal hypervariable domain of Rab7-(175–207) was replaced with the counterpart of Rab11-(160–202) as detailed in the section Construction of Wild-Type and XFP-Tagged XAPC7 Vectors on the previous page. BHK21 cells were grown and processed for immunofluorescence staining. Cells were prepermeabilized with 0.05% (v/v) saponin in PIPES buffer (80 mM PIPES-KOH, pH 6.8, 5 mM EGTA, 1 mM MgCl$_2$) for 2–3 min, to obtain optimal visualization of membrane structures and minimize cytosolic protein staining, and fixed with 3% (w/v) paraformaldehyde in phosphate-buffered saline (PBS). All incubations with primary and detecting antibodies were conducted at room temperature for 30 min with antibodies diluted in PBS/0.05% saponin. The coverslips were washed four times with PBS/saponin and mounted on glass slides in Mowiol 4–88 (see Chapter 55, this volume, for details). Cells were viewed on a Zeiss LSM510 confocal microscope equipped with a 63× oil immersion lens. BHK21 cells cotransfected with this chimera and vesicular stomatitis virus G protein (VSVG)-tagged XAPC7 caused both proteins to change their distribution from disperse Rab7-positive vesicles to recycling endosomes (Dong *et al.*, 2004). This result confirmed that XAPC7 interacts with Rab7 (residues 1–174) independent of the extreme carboxy-terminus domain and XAPC7 is recruited to late endosomal membranes through an interaction with Rab7.

Immunoelectron Microscopy. The presence of Rab7 and XAPC7 together on the membranes of multivesicular late endosomes was established by immunoelectron microscopy. For immunoelectron microscopy, cells in suspension were lightly fixed, pelleted, then fixed in 4% paraformaldehyde and 0.01% glutaraldehyde in 0.1 M sodium cacodylate buffer, pH 7.4, for 1 h. Fixatives and cacodylate buffer were from EMD Chemicals, Inc. (formerly EM Science, Gibbstown, NJ). Immunolabeling was carried out using 1% normal donkey serum/PBS as the diluent for the primary antibodies against XAPC7 (rabbit) and Rab7 (chicken), followed by gold-conjugated secondary antibodies (6 nm donkey anti-rabbit and 12 nm donkey anti-chicken from Jackson Immunoresearch Laboratories, Inc., West Grove, PA). Labeled sections were fixed in 2% glutaraldehyde in PBS, contrast stained with freshly made uranyl acetate, then embedded in polyvinyl alcohol. For additional details regarding general electron-immunomicroscopy procedures, see our chapter on hVPS34/p150 and rab7.

Assays for Epidermal Growth Factor Receptor Endocytosis to Monitor Late Endocytic Membrane Transport and Identify the Functions of XAPC7 and Rab7

Background. Rab7 is a key regulator of endocytosis from early to late endosomes and proteasome inhibitors are known to interfere with EGFR sorting on the endocytic pathway (Longva *et al.*, 2002). Given the colocalization of XAPC7 with wild-type Rab7 on late endosomes, it was of interest to determine whether the two proteins function on the same pathway. In our published studies, we describe a morphological assay for tracing the endocytic transport of VSVG protein, as well as of EGFR (Dong *et al.*, 2004). Based on these morphological assays, the overexpression of XAPC7 was shown to interfere with normal delivery to late endosomes. To confirm the morphological data, we also established a biochemical assay to monitor XAPC7 function as detailed in this section. The inhibitory effect of XAPC7 expression suggests endosome-associated proteasome may have both negative and positive roles in the transport process. For example, excess proteasomes may inactivate Rab7 by impairing its interaction with other effectors because XAPC7 is present in two of the heptameric rings constituting the 20S core of the proteasome (Kopp *et al.*, 1997) (Fig. 1). Our current work is directed toward identifying more precisely the function of the proteasome on endosomal membranes through some of the assays indicated in the following subsection.

EGFR Degradation Assay. Epidermal growth factor (EGF) binds to and activates the EGFR by dimerization to initiate multiple signal transduction events that trigger cell proliferation. Simultaneously, activation of EGFR also accelerates its endocytosis through clathrin-coated pits, resulting in lysosomal targeting and down-regulation of EGFR protein levels to attenuate receptor signaling. We have chosen A431 cells for measuring EGFR degradation, as this cell line expresses nearly 1–2×10^6 EGF receptors at the cell surface, allowing for ease of analysis (Gamou *et al.*, 1984; Gullick *et al.*, 1984). Upon stimulation of A431 cells with EGF, there is a dramatic increase in EGFR tyrosine phosphorylation and the receptors are synchronously endocytosed.

In preparation for the assay, A431 cells were cultured in Dulbecco's modified Eagle's medium (DMEM) with 10% fetal bovine serum, antibiotics, and glutamine for 24 h and serum starved for 3 h to maximize cell surface EGFR expression. Cells were washed and incubated with EGF (50–100 ng/ml) and cycloheximide (CHX) (25 μg /ml) to inhibit new protein synthesis for 0–4 h at 37°. Controls were not stimulated with EGF. Cells were lysed with sodium dodecyl sulfate (SDS) sample buffer (2% SDS, 50 mM Tris, pH 6.8, with HCl, 2.5 mM EDTA, 7% [w/v] glycerol, 0.01% bromophenol blue, and 5% [v/v] 2-mercaptoethanol). Lysates were heated to 95° for 10 min and samples were resolved by sodium dodecyl sulfate polyacrylamide gel electrophoresis (SDS-PAGE) on 8% gels and immunoblotted for EGFR using mAb against epidermal growth factor receptor, clone F4 (Sigma-Aldrich). Under normal culture conditions EGFR has a half-life of 10 h, but in the presence of EGF this value is reduced to about 1 h (Fig. 2). EGF stimulation resulted in a visible degradation of EGFR as

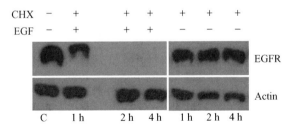

FIG. 2. Degradation of EGFR upon incubation with EGF. A431 cells were incubated with cycloheximide (CHX, 25 μg/ml) and ligand (EGF, 100 ng/ml) at 37° for the indicated time periods. The cells were lysed and subjected to SDS-PAGE. EGFR and actin were detected by immunoblotting as detailed in the text. C, control cells not incubated with ligand or CHX.

early as 1 h. Further incubation with EGF for 2–4 h resulted in complete degradation of EGFR, whereas cells without EGF ligand stimulation showed no obvious receptor turnover at the comparable time point. Actin was used as a control for equal protein loading and was detected with mouse monoclonal antibody (mAb) clone C4 (ICN Biomedicals, Inc., Costa Mesa, CA).

Using this assay, we found that normal proteasome activity is crucial for the degradation of cargo such as EGFR. When A431 cells were transfected with plasmid XAPC7-DsRed using LipofectAMINE-2000 reagent to overexpress XAPC7, EGFR degradation was inhibited even after extended EGF stimulation (data not shown). Parallel morphological studies using A431 cells overexpressing XAPC7-DsRed also showed visible accumulation of EGFR primarily at the plasma membrane in unstimulated cells and exclusively in endosomes on ligand stimulation (data not shown). Morphological assays to monitor the fate of EGFR were conducted as described for the biochemical experiments except that cells were fixed and processed for immunofluorescence staining as detailed in the section Immunofluorescence and Immunoelectron Microscopy. EGFR was stained with a mouse mAb F4 (Sigma-Aldrich) and XAPC7 was stained with our rabbit polyclonal antibody.

Coimmunoprecipitation Experiments to Document Protein–Protein Interactions

Metabolically Labeled Lysates. The interaction between Rab7 and XAPC7 observed by a yeast two-hybrid assay was further confirmed by coimmunoprecipitation experiments (Dong et al., 2004). BHK21 cells cultured in 3.5-cm dishes were infected with a recombinant vaccinia virus encoding T7 RNA polymerase (Fuerst et al., 1986), and then transfected (using LipofectAMINE [Invitrogen Corporation, Carlsbad, CA] according to the manufacturer's instruction) with plasmids encoding full-length human XAPC7 in combination with wild-type and mutant forms of Rab7 or wild-type Rab5. (**Note:** The safe use of vaccinia virus requires primary and booster vaccination against smallpox.) After 5 h, the cells were starved 30 min for cysteine and methionine and metabolically labeled for 30–90 min with Trans[^{35}S]-Label (MP [formerly ICN] Biomedicals, Irvine, CA). Cell lysates were prepared in RIPA buffer (1% [v/v] Nonidet P-40, 0.5% [w/v] deoxycholate, 0.1% [w/v] SDS, 50 mM Tris, pH 7.4, and 150 mM NaCl) containing protease inhibitor cocktail (1 mM phenylmethylsulfonyl fluoride, 1 mM benzamidine,1 μg/ml chymostatin, leupeptin, antipain, and pepstatin). Lysates were precleared by two treatments with 80 μl Pansorbin (EMD Biosciences/Calbiochem, San Diego, CA) for 20 min each at 4° and clarified by centrifugation in a microfuge at 15,000 rpm for 15 min. Cleared lysates were immunoprecipitated with an antibody directed against Rab7 (2 μl/ml) or XAPC7 (4 μl/ml) at 4° for 1 h. Pansorbin (80 μl) was used to absorb the immune complexes and the immunoprecipitated samples were resolved by SDS-PAGE on 12.5% gels and quantified as described (Dong *et al.*, 2004). Endogenous and overexpressed XAPC7 was specifically and proportionately coimmunoprecipitated with all forms of Rab7 tested. Quantification of these data shows that there is a much stronger interaction between XAPC7 and all forms of Rab7 than with Rab5 (Dong *et al.*, 2004).

Immunoprecipitation and Western Blot. BHK21 cells were grown on 100-mm dishes and infected/transfected to overexpress XAPC7 and Rab7, as described earlier, and lysates prepared in 150 μl RIPA buffer. Samples were kept on ice for 15 min and then diluted to 1 ml with IP incubation buffer (0.5% [v/v] Triton X-100, 15 mM Tris, pH 8.0, 150 mM NaCl, 4 mM EDTA, containing protease inhibitor cocktail [1 mM phenylmethylsulfonyl fluoride, 1 mM benzamidine,1 μg/ml chymostatin, leupeptin, antipain, and pepstatin]). Lysates were clarified by centrifugation at 15,000 rpm in a microcentrifuge tube for 15 min at 4°. Samples were immunoprecipitated using our polyclonal rabbit antibodies directed against Rab7 (Press *et al.*, 1998) or XAPC7 (Dong *et al.*, 2004) at 2–4 μl/ml and incubated for 1 h on a rotator at 4°. Protein A-Sepharose (Sigma-Aldrich) was added and incubation was for another 1 h at 4° with rotation. Protein A-Sepharose beads were washed two times with IP wash buffer (50 mM Tris, pH 8.0, 4 mM EDTA, 150 mM NaCl, 0.1% [v/v] Triton X-100, 0.1% [w/v] SDS) and two times with the same buffer without Triton X-100. Protein A-Sepharose beads were boiled in SDS sample buffer and resolved by SDS-PAGE and transferred to nitrocellulose Hybond-C (Sigma-Aldrich). Immunoblot analysis was performed with a monoclonal antibody for XAPC7 (Affiniti Research Products Ltd., Maunhead, UK) at a 1:1000 dilution and our rabbit polyclonal antibody was directed against Rab7 (NW112, 1:500).

Monitoring Proteasome Assembly and Activity

Background. Proteasome α-subunit XAPC7 colocalizes with Rab7 on endosomal membranes and a significant fraction of the total endogenous or overexpressed XAPC7 was not limited to the cytoplasm but detected in the membrane pool (Dong *et al.*, 2004). Proteasome α-subunit XAPC7 is assembled into proteasomes (Tipler *et al.*, 1997). Thus, the binding of XAPC7 to Rab7 may provide for temporally and spatially regulated recruitment of proteasomes to late endosomes. Here we describe assays for monitoring proteasome assembly on endosomal membranes using two

XFP-tagged variants of XAPC7 and a combination of immunofluorescence and immunoprecipitation assays, along with a novel phagocytosis recruitment assay.

Characterization of XFP-XAPC7 Variants by Immunofluorescence Localization. Two fluorescent protein variants of XAPC7 were constructed as detailed in the section on Construction of Wild-Type and XFP-Tagged XAPC7 Vectors. A Ds-Red variant of XAPC7 (XAPC7-DsRed) tagged at the C-terminus leaves the critical N-terminus free for proteasome assembly (Zwickl *et al.*, 1994), while a GFP-XAPC7 variant is N-terminally tagged. Both XFP-XAPC7 chimeras were found to colocalize with Rab7 analogous to the untagged protein (Fig. 3A and data not shown).

Proteasome Assembly Assay. Two independent measures for purifying intact proteasome and monitoring the integration of the XFP-XAPC7 chimeras into the 20S proteasomes were used. The first entailed a coimmunoprecipitation strategy and was based on cofractionation.

To purify 20S proteasome by immunoprecipitation, BHK21 cells were grown on two 100-mm dishes and were cotransfected with GFP-Rab7 and XAPC7-DsRed or GFP-XAPC7, respectively. Cells were lysed by suspending the pellet in three pellet volumes of homogenization buffer (20 mM HEPES, pH 7.4, 2 mM CaCl$_2$) and passage through a 22-gauge syringe 10–12 times. Complete lysis was monitored by microscopy and considered complete when nuclei were intact and ∼80% cells appeared disrupted. Cell debris and nuclei were pelleted by two successive low-speed centrifugations at 1000 rpm in an Eppendorf microcentrifuge at 4° for 5 min each. Mitochondria were sedimented by centrifugation at 1950×g for 20 min at 4°. The resulting supernatant was overlaid on a 200 μl 10% (w/v) sucrose cushion and subjected to centrifugation in a Beckman Optima TL ultracentrifuge using a swinging bucket TLA100.2 rotor at 100,000 rpm for 1 h at 4°. The supernatant was reserved as the cytosolic fraction and the sucrose cushion was carefully removed. The resulting crude membrane pellet fraction was resuspended in 150 μl in RIPA buffer and incubated for 1 h with a rabbit polyclonal antibody reactive against 20S proteasome from diverse species (PW8155 Affiniti Research Products Ltd.). We also found a mouse mAb (PW8100) directed against α6 (HC2) to be effective for immunoprecipitation of 20S proteasome from HeLa cells. The immune complexes were collected using Protein A-Sepharose, as detailed in the section on Coimmunoprecipitation Experiments, resuspended in SDS-PAGE sample buffer, heated for 10 min, and resolved by SDS-PAGE on 12% gels. A mouse mAb directed against XAPC7 (Affiniti Research Products Ltd.) was used to identify the presence of both XFP-XAPC7 chimeras along with endogenous XAPC7 in the immunopreciptated fractions (Fig. 3B). Immunoblotting for the β3 subunit HC10 served as a

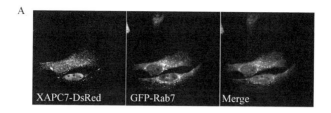

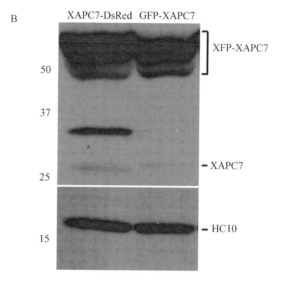

FIG. 3. Proteasome incorporation of XFP-variants of XAPC7. (A) Human cervical carcinoma (HeLa) cells were cotransfected with plasmids encoding XAPC7-DsRed and GFP-Rab7, treated with 0.05% saponin, fixed with 3% paraformaldehyde in PBS as detailed in the text, and imaged using a Zeiss LSM510 fitted with a 63× objective. (B) BHK21 cells were transfected with plasmids encoding GFP-XAPC7 or XAPC7-DsRed and cells were grown in complete media for 18 h. 20S proteasome preparations were isolated by immunoprecipitation with a rabbit polyclonal antibody directed against the 20S complex, resolved by SDS-PAGE on 12% gels, and immunoblotted with a mouse monoclonal antibody against XAPC7 (upper panel) and a mouse monoclonal antibody against the β3(HC10) subunit of the 20S proteasome (lower panel). The bands corresponding to the chimeric and endogenous XAPC7 are indicated. An additional band is observed in the XAPC7-DsRed-transfected sample that is thought to be a degradation fragment. Prestained kaleidoscope molecular weight standards were from Bio-Rad (Hercules, CA) and migration is as indicated. (See color insert.)

positive control for the recovery of 20S protesome in the initial immuno-precipitate (Fig. 3B lower panel).

To purify 20S proteasome by sedimentation, BHK21 cells were separately transfected with plasmids encoding XFP-XAPC7 chimeras and 24 h posttransfection cells were washed once with PBS, were scraped with a rubber policeman into 0.75 ml PBS containing 1 mM dithiothreitol (DTT) and 1 mM adenosine triphosphate (ATP), and were pelleted by centrifugation at 800×g for 5 min. The inclusion of ATP stabilizes the intact proteasome. The supernatant was discarded and the pellet was resuspended in three pellet volumes of Buffer A (20 mM Tris adjusted with HCl to pH 7.5, 5 mM MgCl$_2$, 1 mM DTT, 1 mM ATP, and 250 mM sucrose) (Hu *et al.*, 1999). The suspension was passed through 22-gauge syringe 10–12 times on ice. The lysate was subjected to centrifugation at 5000 rpm for 10 min and the supernatant was removed to a fresh tube and subjected to centrifugation at 15,000 rpm all at 4° in an Eppendorf microcentrifuge. The supernatant was collected and subjected to centrifugation at 100,000×g in a Beckman Optima TL ultracentrifuge using a swinging bucket TLA100.2 rotor for 1 h at 4°. The supernatant following ultracentrifugation was collected and layered on top of a 10% (w/v) sucrose cushion (in Buffer A) and then subjected to ultracentrifugation at 200,000×g in a TLA100.2 rotor for 5 h at 4°. The final pellet enriched in proteasomes was resuspended in Buffer A containing 10% glycerol. The resuspended fraction was treated with SDS sample buffer, resolved by SDS-PAGE on 12% gels, and transferred to nitrocellulose. Western blotting revealed the presence of XFP-XAPC7 along with the other proteasome subunits, providing independent confirmation for chimeric XAPC7 incorporation into 20S proteasome (data not shown).

XAPC7 Function Measured by Phagocytosis. Proteasomes are important for late endosomal/phagosomal protein processing events (Lee *et al.*, 2005; Vieira *et al.*, 2002). To study the temporal parameters of proteasome recruitment to membranes, we coexpressed GFP-Rab7 and XAPC7- Ds-Red in macrophages and monitored whether the two proteins are coordinately recruited to latex bead phagosomes via live cell imaging of maturing phagosomes. For this purpose Raw264.7 mouse macrophages (American Type Culture Collection, Rockville, MD) were plated on 25-mm-diameter coverslips in six-well tissue culture dishes and cultured overnight in DMEM with 4 mM L-glutamine adjusted to contain 1.5 g/liter sodium bicarbonate and 4.5 g/liter glucose, and 10% fetal bovine serum. Subconfluent cells were transfected with plasmids encoding XAPC7-DsRed and GFP-Rab7 by nucleofection using a Cell Line Nucleofector kit R solution with the T-20 program and following the manufacturer's protocol (Amaxa Biosystems, Inc., Gaithersburg, MD). Both the chimeric proteins GFP-Rab7 and XAPC7-DsRed were coexpressed and colocalized on endosomes in RAW 264.7 cells (not shown). Phagocytosis recruitment assays were conducted 18 h posttransfection.

Phagocytosis of latex beads was performed as described (Chua and Deretic, 2004). Briefly, streptavidin-conjugated polystyrene beads (Sigma-Aldrich) were washed with PBS and then opsonized with 100% fetal bovine serum while incubating at 37° for 15 min. Cells were washed two times in cold Hank's buffered saline solution (HBSS) or DMEM without phenol red and left in 1 ml of the same solution on ice. The bead suspension of 25 μl (1 μl /0.2 ml) was added to the coverslips in DMEM or HBSS and the tissue culture plates were subjected to centrifugation in a Sorvall RT6000D centrifuge (GMI, Inc., MN) at 800 rpm for 5 min at 4°. To initiate phagocytosis, coverslips were transferred immediately to a perfusion chamber (Harvard Apparatus) and incubated in prewarmed phenol-free DMEM at a constant temperature of 37°. Phagocytosis was monitored by live cell microscopy on a Nikon TE2000 equipped with a Bio-Rad Radiance 2100 confocal scanhead. Transfected cells chosen for observation were selected based on the presence of moving (Brownian motion) particles to be phagocytosed by adjacent macrophages. The latex beads were identified using the DIC optics, while fluorescent protein chimeras were

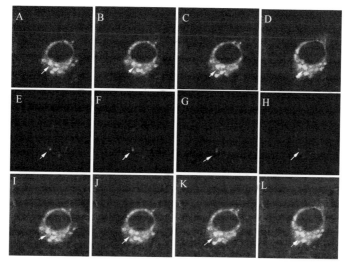

Fig. 4. XAPC7 is involved in phagocytosis. Raw 264.7 mouse macrophages were cotransfected with expression plasmids encoding GFP-Rab7 and XAPC7-DsRed. Live cells were imaged at 30-s intervals for 10 min with a digital camera. In the sequential frames shown in (A)–(D) Rab7 (green) and (E)–(H) XAPC7 (red) are seen to incorporate into latex bead phagosomes (arrows). The time frame for incorporation is 7–10 min. In (I)–(L) a merged image shows the DIC image with the fluorescence images for both proteins superimposed. (See color insert.)

imaged using argon and green HeNe 1 lasers and images were captured by confocal scanning selected areas at 30-s intervals over a time period of 10 min. As evidenced by the live cell imaging results Ds-Red XAPC7 and GFP-Rab7 were coordinately recruited to latex bead phagosomes within 7–10 min (Fig. 4). The timeframe for Rab7 recruitment is as reported by others (Vieira *et al.*, 2002). The need for proteasome activity and ubiquitinylated membrane proteins in phagosome maturation was demonstrated using proteasome inhibitors (Lee *et al.*, 2005). Thus, the convergence of Rab7-positive late endosomes with phagosomes may provide a source of the requisite factors for phagosome maturation including proteasomes.

Conclusions

Down-regulation of growth factors is critical for normal cell growth control and may be dysregulated during malignant transformation due to altered sorting and processing on the endocytic pathway. The Rab7 GTPase and its interacting partner XAPC7, an α-subunit of the 26S proteasome, are shown to be important regulators of EGFR degradation through the morphological and biochemical analyses detailed in this chapter. In future studies it will be of interest to establish how disease-causing mutations in the Rab7 GTPase impact the interactions with XAPC7 as well as other known partners and assess the effects on cell growth control.

Acknowledgments

A.W.N. gratefully acknowledges support from NSF MCB0446179. C.H. is an undergraduate at Washington University, St. Louis, and was supported by two NSF REU supplements (2004, 2005). We thank Dr. Roger Tsien for the monomeric DsRed construct. We are grateful to Elsa Romero for expert technical assistance and Ms. Valorie Bivins for outstanding administrative support. Confocal images were acquired using the University of New Mexico CRTC Fluorescence Microscopy Facility (http://kugrserver.health.unm.edu:16080/microscopy/).

References

Ali, B. R., Wasmeier, C., Lamoreux, L., Strom, M., and Seabra, M. C. (2004). Multiple regions contribute to membrane targeting of rab GTPases. *J. Cell Sci.* **117**, 6401–6412.

Cantalupo, G., Alifano, P., Roberti, V., Bruni, C. B., and Bucci, C. (2001). Rab-interacting lysosomal protein (RILP): The rab7 effector required for transport to lysosomes. *EMBO J.* **20**, 683–693.

Chua, J., and Deretic, V. (2004). Mycobacterium tuberculosis reprograms waves of phosphatidylinositol 3-phosphate on phagosomal organelles. *J. Biol. Chem.* **279**, 36982–36992.

Clague, M. J., and Urbe, S. (2003). Hrs function: Viruses provide the clue. *Trends Cell Biol.* **13**, 603–606.

Conibear, E. (2002). An ESCRT into the endosome. *Mol. Cell* **10**, 215–216.

Dong, J., Chen, W., Welford, A., and Wandinger-Ness, A. (2004). The proteasome alpha subunit XAPC7 interacts specifically with Rab7 and late endosomes. *J. Biol. Chem.* **279**, 21334–21342.

Feng, Y., Press, B., and Wandinger-Ness, A. (1995). Rab 7: An important regulator of late endocytic membrane traffic. *J. Cell Biol.* **131**, 1435–1452.

Fuerst, T. R., Niles, E. G., Studier, F. W., and Moss, B. (1986). Eukaryotic transient-expression system based on recombinant vaccinia virus that synthesizes bacteriophage T7 RNA polymerase. *Proc. Natl. Acad. Sci. USA* **83**, 8122–8126.

Gamou, S., Kim, Y. S., and Shimizu, N. (1984). Different responses to EGF in two human carcinoma cell lines, A431 and UCVA-1, possessing high numbers of EGF receptors. *Mol. Cell. Endocrinol.* **37**, 205–213.

Gullick, W. J., Downward, D. J., Marsden, J. J., and Waterfield, M. D. (1984). A radioimmunoassay for human epidermal growth factor receptor. *Anal. Biochem.* **141**, 253–261.

Hu, Z., Zhang, Z., Doo, E., Coux, O., Goldberg, A. L., and Liang, T. J. (1999). Hepatitis B virus X protein is both a substrate and a potential inhibitor of the proteasome complex. *J. Virol.* **73**, 7231–7240.

Jordens, I., Fernandez-Borja, M., Marsman, M., Dusseljee, S., Janssen, L., Calafat, J., Janssen, H., Wubbolts, R., and Neefjes, J. (2001). The rab7 effector protein RILP controls lysosomal transport by inducing the recruitment of dynein-dynactin motors. *Curr. Biol.* **11**, 1680–1685.

Kopp, F., Hendil, K. B., Dahlmann, B., Kristensen, P., Sobek, A., and Uerkvitz, W. (1997). Subunit arrangement in the human 20S proteasome. *Proc. Natl. Acad. Sci. USA* **94**, 2939–2944.

Lee, W. L., Kim, M. K., Schreiber, A. D., and Grinstein, S. (2005). Role of ubiquitin and proteasomes in phagosome maturation. *Mol. Biol. Cell* **16**, 2077–2090.

Longva, K. E., Blystad, F. D., Stang, E., Larsen, A. M., Johannessen, L. E., and Madshus, I. H. (2002). Ubiquitination and proteasomal activity is required for transport of the EGF receptor to inner membranes of multivesicular bodies. *J. Cell Biol.* **156**, 843–854.

Mizuno, K., Kitamura, A., and Sasaki, T. (2003). Rabring7, a novel Rab7 target protein with a ring finger motif. *Mol. Biol. Cell* **14**, 3741–3752.

Press, B., Feng, Y., Hoflack, B., and Wandinger-Ness, A. (1998). Mutant Rab7 causes the accumulation of cathepsin D and cation-independent mannose 6-phosphate receptor in an early endocytic compartment. *J. Cell Biol.* **140**, 1075–1089.

Stein, M. P., Feng, Y., Cooper, K. L., Welford, A. M., and Wandinger-Ness, A. (2003). Human vps34 and p150 are Rab7 interacting partners. *Traffic* **4**, 754–771.

Stenmark, H., Valencia, A., Martinez, O., Ullrich, O., Goud, B., and Zerial, M. (1994). Distinct structural elements of Rab5 define its functional specificity. *EMBO J.* **13**, 575–583.

Tipler, C. P., Hutchon, S. P., Hendil, K., Tanaka, K., Fishel, S., and Mayer, R. J. (1997). Purification and characterization of 26S proteasomes from human and mouse spermatozoa. *Mol. Hum. Reprod.* **3**, 1053–1060.

van Kerkhof, P., Alves dos Santos, C. M., Sachse, M., Klumperman, J., Bu, G., and Strous, G. J. (2001). Proteasome inhibitors block a late step in lysosomal transport of selected membrane but not soluble proteins. *Mol. Biol. Cell* **12**, 2556–2566.

Vieira, O. V., Botelho, R. J., and Grinstein, S. (2002). Phagosome maturation: Aging gracefully. *Biochem. J.* **366**, 689–704.

Zwickl, P., Kleinz, J., and Baumeister, W. (1994). Critical elements in proteasome assembly. *Nat. Struct. Biol.* **1**, 765–770.

[57]   Expression, Assay, and Functional Properties of RILP

By Anna Maria Rosaria Colucci, Maria Rita Spinosa, and Cecilia Bucci

Abstract

Rab proteins are master regulators of vesicular membrane traffic of endocytic and exocytic pathways. They basically serve to recruit proteins and lipids required for vesicle formation, docking, and fusion. Each Rab protein is able to recruit one or more effectors, and, through the action of effectors, it drives its specific downstream functions. The Rab interacting lysosomal protein (RILP) is a common effector of Rab7 and Rab34, two Rab proteins implicated in the biogenesis of lysosomes. RILP is recruited onto late endosomal/lysosomal membranes by Rab7-GTP where it induces the recruitment of the dynein–dynactin motor complexes. Therefore, through the timed and selective dynein motor recruitment onto late endosomes and lysosomes, Rab7 and RILP control transport to endocytic degradative compartments. A similar role for Rab7 and RILP has been demonstrated also for phagosomes. Indeed, RILP recruits dynein–dynactin motors on Rab7-GTP-positive phagosomes and the recruitment not only displaces phagosomes centripetally, but also promotes the extension of phagosomal tubules toward late endocytic compartments. RILP is therefore a key protein for the biogenesis of lysosomes and phagolysosomes. This chapter describes how to express wild-type or mutated RILP in mammalian cells, and how to test the effects caused by RILP dysfunction. In particular, we report assays to monitor the interaction between RILP and Rab7, morphology and distribution of endosomes, and to measure degradation of endocytic markers.

Introduction

Rab proteins, the largest family of monomeric GTPases, regulate vesicular transport. A single activated Rab protein can selectively bind to several effector proteins, and, through a series of sequential interactions, it is responsible for different aspects of a single transport step (Hammer and Wu, 2002; Pfeffer and Aivazian, 2004; Seabra and Wasmeier, 2004). Indeed, Rab proteins are now considered fundamental for spatial and temporal coordination of vesicular transport. Several Rab effector proteins have been identified and characterized. Analysis of these effectors revealed that most of them are unrelated proteins with diverse functions. This

METHODS IN ENZYMOLOGY, VOL. 403   0076-6879/05 $35.00

striking finding supports the idea that Rab GTPases play a crucial role in determining the specificity of each vesicular transport step.

The Rab interacting lysosomal protein (RILP) has been first identified as a Rab7 effector protein (Bucci *et al.*, 2001; Cantalupo *et al.*, 2001). The Rab7 GTPase controls late endocytic traffic and it is a key protein for biogenesis of lysosomes and phagosomes (Bucci *et al.*, 2000; Harrison *et al.*, 2003; Press *et al.*, 1998; Vieira *et al.*, 2003; Vitelli *et al.*, 1997). Moreover, Rab7 is important for maturation of late autophagic vacuoles (Gutierrez *et al.*, 2004; Jager *et al.*, 2004). RILP is able to interact also with the small GTPase Rab34, a Rab protein associated primarily with the Golgi apparatus (Wang and Hong, 2002). RILP interacts preferentially with the GTP-bound form of Rab7 and Rab34, and it is important in transport to and maintenance of degradative endocytic compartments. Indeed, the active form of Rab7 (GTP bound) recruits RILP onto Rab7-positive compartments, and RILP then induces recruitment of dynein–dynactin motor complexes to Rab7-containing organelles (Cantalupo *et al.*, 2001; Harrison *et al.*, 2003; Jordens *et al.*, 2001). Recruitment of motors causes the movement of Rab7-containing vesicles/organelles toward the microtubule-organizing center clearing these structures from the cell periphery. Expression of an RILP deletion mutant, lacking 216 aa at the N-terminus, containing only the Rab7 but not the dynein–dynactin binding domain, causes inhibition of protein degradation, inhibition of transport to lysosomes, dispersion of late endosomes/lysosomes at the cell periphery, and inhibition of phagolysosome maturation. In addition, Rab7 and/or RILP are involved in the maturation of pathogen-containing vacuoles (Beron *et al.*, 2002; Clemens *et al.*, 2000; Deretic *et al.*, 1997; Guignot *et al.*, 2004; Harrison *et al.*, 2004; Marsman *et al.*, 2004; Roy *et al.*, 1998; Rzomp *et al.*, 2003).

Recently, mutations in the Rab7 gene have been associated with hereditary sensory and motor neuropathies (Houlden *et al.*, 2004; Verhoeven *et al.*, 2003). However, how Rab7 mutations act in the development of peripheral neuropathies is still unknown. One of the mutations leads to the virtual absence of expression of RILP, implying that the pathogenetic mechanism for this neuropathy may be through a pathway involving RILP (Houlden *et al.*, 2004).

Expression in Mammalian Cells

Construction of Recombinant Expression Vectors

Epitope tagging is a general method for tracking recombinant protein. We have added to the N-terminus of RILP two different tags: HA and GFP. The HA tag is derived from an epitope of the influenza hemagglutinin protein, which has been extensively used in expression vectors since it can be efficiently recognized by commercially available monoclonal and polyclonal antibodies. The green fluorescent protein (GFP) is a 27-kDa protein that generates a green fluorescence visible in both living and fixed samples. In the past years GFP has become a valuable tool for studying complex biological processes since it can be fused to the N- or C-terminus of other proteins, generally without altering their localization or function (Gerdes and Kaether, 1996).

To obtain HA-RILP, we amplified cDNA coding for RILP by polymerase chain reaction (PCR) using the upstream primer RILPf 5'-AAG-GAATTCATGGAGCCCAGGAGG-3' and the downstream primer RILPr 5'-AGAGAATTCTCAGGCCTCTGGG-3'. Because RILP has a high GC content, we used 10% dimethyl sulfoxide (DMSO) in the PCR reactions to facilitate DNA denaturation. The *Eco*RI-restricted amplified fragment was then inserted into the mammalian expression pCEFL-HA vector, a modified pCDNAIII expression vector containing the elongation factor 1α promoter driving the expression of the in-frame N-terminal HA tag of nine amino acids (Marinissen *et al.*, 1999). To obtain HA-RILPC33, cDNA coding for the C-terminal 185 aa of RILP was amplified by PCR with the following oligonucleotides: RILPC33f 5'-ATGAATT-CACCGCCGGCGCAGGCG-3' and RILPC33r 5'-ATAAGAATGCGG CCGCTCAGGCCTCTGGGG-3'. The fragment was restricted with *Eco*RI *Not*I and then inserted into pCEFL-HA.

To obtain GFP-RILP, we cloned a *Sal*I–*Bam*HI (partial digestion) fragment into *Sal*I–*Bgl*II sites of the pEGFPC1 vector (Clontech). For GFP tagging of RILPC33, an *Eco*RI–*Sal*I fragment was cloned into the pEGFP-C1 vector.

Rab7 and RILP constructs in pGEM under the control of the T7 promoter have been described (Cantalupo *et al.*, 2001; Vitelli *et al.*, 1997).

All recombinant expression plasmids were purified using Qiagen plasmid kits and verified by double-stranded sequencing. Expression of these constructs revealed that epitope tags at the N-terminus of RILP or of RILPC33 did not alter subcellular localization or functional properties.

Cell Culture and Transient Transfection

HeLa cells (ATCC, CCL-2) were grown routinely at $37°$ in 5% CO_2 in Dulbecco's modified Eagle's medium supplemented with heat-inactivated 10% fetal bovine serum (FBS), 50 U/ml penicillin, and 50 μg/ml streptomycin.

For transient transfection, cells were grown on 11-mm round glass coverslips to 80% confluence and then transfected with the appropriate plasmids, using either DOSPER or Fugene (Roche) strictly following the manufacturer's instructions. Expression was allowed for 24–36 h; subsequently cells were processed for immunofluorescence or biochemical assays. Under these conditions, 30–50% of the cells were detected as transfected.

Alternatively, we expressed RILP using the T7 RNA polymerase recombinant vaccinia virus as described (Bucci *et al.*, 1992; Fuerst *et al.*, 1986; Stenmark *et al.*, 1995). HeLa cells were washed once with serum-free medium and then infected for 30 min with the vT7 vaccinia virus in serum-free medium containing 20 m*M* HEPES, pH 7.2, and 20 μg/ml soybean trypsin inhibitor. The virus was then aspirated and the cells were transfected as described earlier in the presence of 10 m*M* hydroxyurea to avoid new viral production. Using this protocol, the cDNA to be expressed has to be cloned downstream of a T7 promoter, and cells can be processed for immunofluorescence or biochemical assays after 4–7 h. This system allows a high level of expression over a short period of time. Under these conditions 50–60% of cells were transfected.

Assays

Molecular Interaction of RILP

Several *in vivo* and *in vitro* systems are available to test RILP interactions. RILP has been first isolated as a 33-kDa truncated form lacking its N-terminal half (RILP-C33) by using the yeast two-hybrid system in an attempt to search for novel Rab7 interacting proteins. The interaction then has to be confirmed in HeLa cells using affinity chromatography and protein overlay (Cantalupo *et al.*, 2001). The yeast system remains the method of choice to search for and to test RILP interactions. This system, combined with the use of altered Rab7 proteins generated by *in vitro* mutagenesis, has been fundamental to elucidate functional properties of RILP (Cantalupo *et al.*, 2001). In particular, Rab7I41M and Rab7Q67L are GTPase-defective mutant proteins that behave as Rab7-activating mutants, whereas Rab7T22N and Rab7N125I behave as Rab7-dominant negative mutants.

To detect interaction by the two-hybrid system, the *Saccharomyces cerevisiae* AH109 strain was transformed with the appropriate constructs; transformants were plated on synthetic medium lacking leucine and tryptophan. After 2 days of growth at 30°, colonies were selected and assayed for growth at 30° on synthetic medium lacking histidine, leucine, adenine, and tryptophan and containing 5 m*M* 3-amino-1,2,4-triazole (a competitive inhibitor of the His3 protein). Alternatively, fresh colonies were transferred onto nitrocellulose filters and assayed for β-galactosidase activity, using 5-bromo-4-chloro-3-indolyl β-D-galactopyranoside (X-Gal) as a substrate. Nitrocellulose filters were dipped in liquid nitrogen for 10 s to permeabilize yeast cells, allowed to thaw at room temperature, and then colonies side up, were placed onto filter paper soaked with Z buffer (60 mM Na_2HPO_4, 40 mM NaH_2PO_4, 10 mM KCl, 1 mM $MgSO_4$), to which 2.7 μl/ml of 2-mercaptoethanol and 0.4 mg/ml of X-Gal had been added. Usually the blue color developed within 1 h at 30°. To have quantitative data, colonies were grown in synthetic liquid medium lacking leucine and tryptophan to exponential phase ($OD_{600} = 0.5$–0.8). The exact OD_{600} when the cells were harvested was recorded. Cells were concentrated five times by centrifugation and 200 μl of concentrated yeast cells suspension was added to 800 μl of Z buffer with 2-mercaptoethanol. Then 50 μl of chloroform and 50 μl of sodium dodecyl sulfate (SDS) 0.1% were added, vortexed, and 160 ml of fresh o-nitrophenyl-β-D-galactoside (ONPG) solution (4 mg/ml in Z buffer) was added. After mixing, samples were incubated at 30° until color development; 400 μl of 1 M Na_2CO_3 was added to terminate the reaction and the elapsed time was recorded in minutes (t). After centrifugation (10 min at 12,000 rpm) to pellet debris, the OD_{420} of the upper phase was determined. β-Galactosidase activity was calculated as $1000 \times OD_{420}/(OD_{600} \times V \times t)$ where V is the volume of yeast cells suspension used in the reaction multiplied by the concentration factor (Bartel *et al.*, 1993; Guarente, 1983).

In the two-hybrid system RILP-C33 interacts with Rab7 wt, Rab7I41M, Rab7Q67L, and Rab7N125I, but only weakly (~40-fold less than to Rab7 wt) with Rab7T22N (Table I). The interaction is weaker (3- to 5-fold less)

TABLE I
INTERACTION BETWEEN RAB7 WT OR RAB7 MUTANTS AND RILP-C33

Construct	β-Galactosidase (relative units)[a]
Rab7 wt	247 ± 38
Rab7T22N	7.1 ± 0.5
Rab7I41M	288 ± 32
Rab7Q67L	339 ± 47
Rab7N125I	320 ± 21
Rab7ΔC	42 ± 3
Rab7Q67LΔC	54 ± 5

[a]β-Galactosidase activities of RILP-C33/Rab7 double transformants were measured. Data are expressed as arbitrary relative units and represent mean values $\pm$ SEM of six independent transformants.

if the last three amino acids are removed from Rab7 wt or Rab7Q67L proteins, suggesting a role for C-terminal prenylation in the interaction. The strong interaction with the Rab7N125I (a dominant-negative mutant in functional assays) can be explained considering that this mutant weakly binds both nucleotides (GTP and GDP) and therefore is considered to have a high turnover. Similar results can be obtained using full-length RILP (data not shown).

To prove direct interaction between RILP and its Rab interactors, a protein overlay assay can be performed (Cantalupo *et al.*, 2001). The method requires purified proteins that may be obtained as GST fusions from *Escherichia coli* expression systems. Briefly, 5 μg of the protein of interest is separated by sodium dodecyl sulfate-polyacrylamide gel electrophoresis (SDS-PAGE) and transferred to nitrocellulose. The filter is then incubated with gentle agitation in 50 mM HEPES, pH 7.2, 5 mM Mg $(CH_3COO)_2$, 100 mM CH_3COOK, 3 mM dithiothreitol (DTT), 10 mg/ml bovine serum albumin (BSA), 0.1% Triton X-100, and 0.3% Tween. The purified Rab protein was, in the meantime, loaded with radiolabeled GTP at a working concentration of 0.1 μM in 20 mM HEPES, pH 7, 4 mM EDTA for 1 h at 37°. Then 10 mM $MgCl_2$ was added to stabilize the Rab–nucleotide complex. The protein was then added in 10 ml of binding buffer (12.5 mM HEPES, pH 7.2, 1.5 mM Mg[CH_3COO]$_2$, 75 mM CH_3COOK, 1 mM DTT, 2 mg/ml BSA, 0.05% Triton X-100) and applied to the renatured blot for 2 h at room temperature. The filter was then washed three times for 10 min with 20 mM Tris-Cl, pH 7.4, 100 mM NaCl, 20 mM $MgCl_2$, 0.005% Triton X-100, and then exposed to X-ray film.

Confocal Fluorescence Microscopy

After transfection, cells were washed once with phosphate-buffered saline (PBS) and were fixed with 3% paraformaldehyde for 15 min at room temperature. Free aldehyde groups were quenched with 50 mM NH_4Cl in PBS for 10 min. Cells were then permeabilized with 0.5% saponin for 10 min. Antibodies were diluted in 5% horse serum, 0.1% saponin in PBS. Coverslips were washed twice in PBS containing 0.1% saponin, incubated for 20 min with primary antibodies, washed twice with 0.1% saponin in PBS, and incubated for 20 min with secondary antibodies. Coverslips were then washed twice in 0.1% saponin in PBS, once in PBS, and dipped very briefly in water. Coverslips were then mounted onto slides with Mowiol (Sigma). Anti-lysosomal-associated membrane protein 1 (anti-LAMP-1) monoclonal antibodies (H4A3) were from the Developmental Studies Hybridoma Bank (Iowa City, IA) while monoclonal anti-hemagglutinin (anti-HA) antibody was purchased from Sigma. The secondary antibodies used were donkey anti-mouse conjugated to Texas Red. All antibodies were used at a 1:500 dilution.

When cytosolic signal may interfere with detection of the membrane-bound fraction of the protein to be visualized, it is better to permeabilize the cells before fixation. This is especially true when strong expression systems are used causing accumulation of the expressed protein in the cytosol. Indeed this is the case of Rab proteins and RILP, which cycle from cytosol to membranes. Permeabilization was accomplished with 0.1% saponin (Sigma) in 80 mM PIPES, pH 6.8, 1 mM MgCl$_2$, and 5 mM EGTA for 5 min at room temperature prior to paraformaldehyde fixation. This procedure washed out the majority of cytosolic proteins. To label acidic compartments we used LysoTracker Red DND-99 from Molecular Probes. Cells were exposed to LysoTracker Red (50 nM) in Dulbecco's modified Eagle's medium (DMEM) for 2 h prior to fixation. Cells were analyzed by confocal microscopy using an LSM 510 laser scanning confocal microscope (Zeiss) with a 100× oil immersion objective. All recorded images were processed using Adobe Photoshop software.

Estimation of [^{125}I]LDL Degradation

Lipoprotein-deficient serum (LDS) was prepared from serum of normolipidemic subjects by ultracentrifugation at a density of 1.210 g/ml. LDS was then extensively dialyzed for 24 h at 4° against 0.15 M NaCl, 0.24 mM disodium EDTA, pH 7.4. Cells were incubated for 24 h in medium complemented with human LDS before transfection. After transfection with the appropriate plasmid, 20 μg/ml of ^{125}I low-density lipoprotein (LDL) (Amersham Biosciences) was added to the medium. Cells were allowed to internalize and process [^{125}I]LDL for 5 h. Subsequently the amount of [^{125}I]LDL surface bound, internalized, and degraded was estimated as described (Brown and Goldstein, 1975; Vitelli et al., 1997). Briefly, to determine the amount of [^{125}I]LDL degraded, the medium was collected and treated with trichloroacetic acid. Then samples were extracted with chloroform and hydrogen peroxide to remove free iodine, and an aliquot of the aqueous phase was counted with a γ-counter. This acid-soluble material is mainly [^{125}I]iodotyrosine, the product of LDL degradation. To determine [^{125}I]LDL binding and internalization, after the 5 h incubation with [^{125}I]LDL, cells were washed six times with 0.1% BSA in PBS, once with PBS, and then treated with 3 mg/ml Pronase in serum-free medium containing 10 mM HEPES, pH 7.3, for 1 h at 0 °. In these conditions cells were detaching from the wells and were recovered by centrifugation at 3000 rpm for 3 min. Radioactivity present in pellet and supernatant was detected with a γ-counter. The radioactivity present in the Pronase-treated supernatant represented the [^{125}I]LDL surface bound while the cell-associated radioactivity represented the [^{125}I]LDL internalized.

Functional Properties

RILP Promotes Centripetal Migration of Rab7-Positive Organelles

RILP is a ubiquitously expressed cytosolic protein that is recruited on late endosomal, lysosomal, and phagosomal membranes by Rab7-GTP (Cantalupo *et al.*, 2001; Harrison *et al.*, 2003). Indeed, the portion of RILP on membranes colocalizes with late endosomal/lysosomal markers as Lamp1, Lamp2, and CathD but not with early endosomal, ER, or Golgi markers (data not shown). We overexpressed RILP and its deletion mutant RILP-C33 in HeLa cells and examined the effects caused on endocytic acidic compartments by confocal immunofluorescence microscopy. Indeed, these compartments, labeled by LysoTracker, were profoundly affected (Fig. 1).

Overexpression of RILP caused a high degree of clustering in the peri-nuclear region (Fig. 1A–C) while overexpression of RILP-C33 caused dispersion of these compartments (Fig. 1D–F). These effects were much

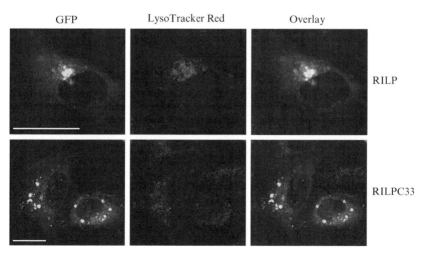

FIG. 1. Confocal fluorescence analysis of HeLa cells expressing GFP-RILP or GFP-RILPC33. Cells were transfected with the appropriate constructs as indicated. Two hours before fixation cells were incubated with LysoTracker Red. Bars = 10 μm. (See color insert.)

TABLE II
Distance of Lamp1-Labeled Structures from γ-Tubulin Immuno-
stained MTOC in Control or HA-Tagged Rab7Q67L, Rab7T22N,
RILP, or RILPC33 HeLa Expressing Cells

Transfected construct	Distance[a]
Rab7Q67L	2.9 ± 0.4
Rab7T22N	8.1 ± 0.5
RILP	1.3 ± 0.3
RILP-C33	7.3 ± 0.6
Control	4.1 ± 0.5

[a] Values are expressed in micrometers and are means of three independent experiments performed in duplicate.

stronger if RILP and RILP-C33 were expressed using the vaccinia T7 expression system, probably due to the high expression level accomplished in a short time (Cantalupo et al., 2001). The RILP-promoted perinuclear clustering of the late endosomal/lysosomal compartment was reversed by nocodazole, indicating that the process was microtubule dependent (Cantalupo et al., 2001).

In Table II the mean distance of Lamp1-labeled structures from γ-tubulin immunostained microtubule organizing center (MTOC) in control or HA-tagged Rab7Q67L, Rab7T22N, RILP, or RILPC33 expressing cells is reported. The results indicate that interaction between RILP and Rab7 promotes the centripetal migration of late endosomal/lysosomal organelles.

Similar effects were observed for Latex beads-containing phagosomes in macrophage cell lines where RILP promotes extension of phagosomal tubules toward endocytic compartments (Harrison et al., 2003).

RILP Is Required for Transport to Lysosomes

The analysis of kinetics of uptake and degradation of endocytic markers represents a useful tool to measure the efficiency of the endocytic/lysosomal route. No effects on kinetics of internalization of various endocytic markers were observed when RILP or RILP-C33 was overexpressed in HeLa cells, indicating that RILP and Rab7 as well are not involved in the early steps of endocytosis. In contrast, overexpression of RILP-C33, but not of RILP, caused dramatic inhibition of endocytic marker degradation, comparable with that produced by overexpression of the dominant-negative mutant Rab7T22N (Cantalupo et al., 2001; Vitelli et al., 1997). Moreover, overexpression of RILP, but not RILP-C33, was able to reverse/

prevent the inhibitory effect of Rab7T22N. In contrast, overexpression of the Rab7-activating mutant Rab7Q67L, locked in the GTP-bound form, was not able to reverse/prevent the inhibitory effect of RILP-C33, indicating that Rab7 and RILP work sequentially in the regulation of the late endocytic pathway.

Conclusions

RILP is a key protein for the biogenesis of lysosomes and phagolysosomes. Indeed, it is responsible for transport to these compartments and movement of these organelles toward the MTOC through the recruitment of dynein–dynactin motor complexes. However, no data on direct interaction of RILP with these motors are yet available. Therefore, further work will be necessary to address in more detail the molecular mechanism of action of this protein and how its function is regulated.

References

Bartel, P., Chien, C., Sternglanz, R., and Fields, S. (1993). Elimination of false positives that arise in using the two-hybrid system. *Biotechniques* **14,** 920–924.

Beron, W., Gutierrez, M., Rabinovitch, M., and Colombo, M. (2002). Coxiella burnetii localizes in a Rab7-labeled compartment with autophagic characteristics. *Infect. Immun.* **70,** 5816–5821.

Brown, M., and Goldstein, J. (1975). Regulation of the activity of the low density lipoprotein receptor in human fibroblasts. *Cell* **6,** 307–316.

Bucci, C., Parton, R., Mather, I., Stunnenberg, H., Simons, K., Hoflack, B., and Zerial, M. (1992). The small GTPase rab5 functions as a regulatory factor in the early endocytic pathway. *Cell* **70,** 715–728.

Bucci, C., Thomsen, P., Nicoziani, P., McCarthy, J., and van Deurs, B. (2000). Rab7: A key to lysosome biogenesis. *Mol. Biol. Cell* **11,** 467–480.

Bucci, C., De Gregorio, L., and Bruni, C. B. (2001). Expression analysis and chromosomal assignment of PRA1 and RILP genes. *Biochem. Biophys. Res. Commun.* **286,** 815–819.

Cantalupo, G., Alifano, P., Roberti, V., Bruni, C., and Bucci, C. (2001). Rab-interacting lysosomal protein (RILP): The Rab7 effector required for transport to lysosomes. *EMBO J.* **20,** 683–693.

Clemens, D., Lee, B., and Horwitz, M. (2000). Mycobacterium tuberculosis and Legionella pneumophila phagosomes exhibit arrested maturation despite acquisition of Rab7. *Infect. Immun.* **68,** 5154–5166.

Deretic, V., Via, L., Fratti, R., and Deretic, D. (1997). Mycobacterial phagosome maturation, rab proteins, and intracellular trafficking. *Electrophoresis* **18,** 2542–2547.

Fuerst, T., Niles, E., Studier, F., and Moss, B. (1986). Eukaryotic transient-expression system based on recombinant vaccinia virus that synthesizes bacteriophage T7 RNA polymerase. *Proc. Natl. Acad. Sci. USA* **83,** 8122–8126.

Gerdes, H., and Kaether, C. (1996). Green fluorescent protein: Applications in cell biology. *FEBS Lett.* **389,** 44–47.

Guarente, L. (1983). Yeast promoters and LacZ fusions designed to study expression of cloned genes in yeast. *Methods Enzymol.* **101,** 181–189.

Guignot, J., Caron, E., Beuzon, C., Bucci, C., Kagan, J., Roy, C., and Holden, D. (2004). Microtubule motors control membrane dynamics of Salmonella-containing vacuoles. *J. Cell Sci.* **117,** 1033–1045.

Gutierrez, M., Munafo, D., Beron, W., and Colombo, M. (2004). Rab7 is required for the normal progression of the autophagic pathway in mammalian cells. *J. Cell Sci.* **117,** 2687–2697.

Hammer, J. A., 3rd., and Wu, X. S. (2002). Rabs grab motors: Defining the connections between Rab GTPases and motor proteins. *Curr. Opin. Cell Biol.* **14,** 69–75.

Harrison, R., Bucci, C., Vieira, O., Schroer, T., and Grinstein, S. (2003). Phagosomes fuse with late endosomes and/or lysosomes by extension of membrane protrusions along microtubules: Role of Rab7 and RILP. *Mol. Cell. Biol.* **23,** 6494–6506.

Harrison, R., Brumell, J., Khandani, A., Bucci, C., Scott, C., Jiang, X., Finlay, B., and Grinstein, S. (2004). Salmonella impairs RILP recruitment to Rab7 during maturation of invasion vacuoles. *Mol. Biol. Cell* **15,** 3146–3154.

Houlden, H., King, R., Muddle, J., Warner, T., Reilly, M., Orrell, R., and Ginsberg, L. (2004). A novel RAB7 mutation associated with ulcero-mutilating neuropathy. *Ann. Neurol.* **56,** 586–590.

Jager, S., Bucci, C., Tanida, I., Ueno, T., Kominami, E., Saftig, P., and Eskelinen, E. (2004). Role for Rab7 in maturation of late autophagic vacuoles. *J. Cell Sci.* **117,** 4837–4848.

Jordens, I., Fernandez-Borja, M., Marsman, M., Dusseljee, S., Janssen, L., Calafat, J., Janssen, H., Wubbolts, R., and Neefjes, J. (2001). The Rab7 effector protein RILP controls lysosomal transport by inducing the recruitment of dynein-dynactin motors. *Curr. Biol.* **11,** 1680–1685.

Marinissen, M., Chiariello, M., Pallante, M., and Gutkind, J. (1999). A network of mitogen-activated protein kinases links G protein-coupled receptors to the c-jun promoter: A role for c-Jun NH2-terminal kinase, p38s, and extracellular signal-regulated kinase 5. *Mol. Cell. Biol.* **19,** 4289–4301.

Marsman, M., Jordens, I., Kuijl, C., Janssen, L., and Neefjes, J. (2004). Dynein-mediated vesicle transport controls intracellular Salmonella replication. *Mol. Biol. Cell* **15,** 2954–2964.

Pfeffer, S., and Aivazian, D. (2004). Targeting Rab GTPases to distict membrane compartments. *Nat. Rev. Mol. Cell Biol.* **5,** 886–896.

Press, B., Feng, Y., Hoflack, B., and Wandinger-Ness, A. (1998). Mutant Rab7 causes the accumulation of cathepsin D and cation-independent mannose 6-phosphate receptor in an early endocytic compartment. *J. Cell Biol.* **140,** 1075–1089.

Roy, C., Berger, K., and Isberg, R. (1998). Legionella pneumophila DotA protein is required for early phagosome trafficking decisions that occur within minutes of bacterial uptake. *Mol. Microbiol.* **28,** 663–674.

Rzomp, K., Scholtes, L., Briggs, B., Whittaker, G., and Scidmore, M. (2003). Rab GTPases are recruited to chlamydial inclusions in both a species-dependent and species-independent manner. *Infect. Immun.* **71,** 5855–5870.

Seabra, M., and Wasmeier, C. (2004). Controlling the location and activation of Rab GTPases. *Curr. Opin. Cell. Biol.* **16,** 451–457.

Stenmark, H., Bucci, C., and Zerial, M. (1995). Expression of Rab GTPases using recombinant vaccinia viruses. *Methods Enzymol.* **257,** 155–164.

Verhoeven, K., De Jonghe, P., Coen, K., Verpoorten, N., Auer-Grumbach, M., Kwon, J., FitzPatrick, D., Schmedding, E., De Vriendt, E., Jacobs, A., Van Gerwen, V., Wagner, K., Hartung, H., and Timmerman, V. (2003). Mutations in the small GTPase late endosomal protein RAB7 cause Charcot-Marie-Tooth type 2B neuropathy. *Am. J. Hum. Genet.* **72,** 722–727.

Vieira, O., Bucci, C., Harrison, R., Trimble, W., Lanzetti, L., Gruenberg, J., Schreiber, A., Stahl, P., and Grinstein, S. (2003). Modulation of Rab5 and Rab7 recruitment to phagosomes by phosphatidylinositol 3-kinase. *Mol. Cell. Biol.* **23,** 2501–2514.

Vitelli, R., Santillo, M., Lattero, D., Chiariello, M., Bifulco, M., Bruni, C., and Bucci, C. (1997). Role of the small GTPase Rab7 in the late endocytic pathway. *J. Biol. Chem.* **272,** 4391–4397.

Wang, T., and Hong, W. (2002). Interorganellar regulation of lysosome positioning by the Golgi apparatus through Rab34 interaction with Rab-interacting lysosomal protein. *Mol. Biol. Cell* **13,** 4317–4332.

[58] Assay and Functional Properties of Rab34 Interaction with RILP in Lysosome Morphogenesis

By Tuanlao Wang *and* Wanjin Hong

Abstract

We have recently characterized Rab34 as a new member of the Rab GTPase family based on its ability to regulate lysosomal morphology. Rabbit polyclonal antibody raised against recombinant Rab34 reveals that Rab34 is a 29-kDa protein present both in the cytosol and in the Golgi apparatus. A GTP overlay assay shows that a wild-type and GTP-restricted mutant form of recombinant Rab34 bind GTP *in vitro*. Yeast two-hybrid interaction screens identify Rab7-interacting lysosomal protein (RILP) as a partner of Rab34. Both GST pull-down experiments and direct binding assays *in vitro* demonstrate that RILP interacts selectively with the wild-type and GTP-restricted but not GDP-restricted form of Rab34. A key residue (K82) of Rab34 is necessary for interaction with RILP. Expression of EGFP-tagged Rab34 wild-type or GTP-restricted forms in mammalian cells results in redistribution of clustered lysosomes to the peri-Golgi region and this property depends on K82, suggesting that Rab34 regulates lysosome distribution via interaction with RILP. These results suggest that RILP is a common effector shared by Rab7 and Rab34. We describe the methods used in our study.

Introduction

Rab small GTPases participate in vesicular trafficking and regulate vesicle budding, movement, docking, and fusion processes via interacting with myriad effectors (Chavrier and Goud, 1999; Olkkonen and Stenmark, 1997;

METHODS IN ENZYMOLOGY, VOL. 403
0076-6879/05 $35.00
DOI: 10.1016/S0076-6879(05)03058-2

Pfeffer, 2001; Segev, 2001; Somsel and Wandinger-Ness, 2000; Zerial and McBride, 2001). Around 60 members of the Rab GTPase family are identified in mammalian cells (Wennerberg *et al.*, 2005). The partial cDNA of Rab34 was originally identified as Rah (ras-related homolog) (Morimoto *et al.*, 1991), and we have shown that Rab34 is a new Rab protein with an N-terminal extension larger than that of most of the other Rab proteins (Wang and Hong, 2002). Rab34 is present in the Golgi apparatus and cytosol and it interacts with Rab7-interacting lysosomal protein (RILP) to regulate the morphology and spatial distribution of lysosomes (Wang and Hong, 2002). RILP, originally identified as a Rab7 effector (Cantalupo *et al.*, 2001), is involved in the regulation of lysosome morphogenesis through cooperation with the dynein–dynactin complex (Jordens *et al.*, 2001). We recently identified two proteins related to RILP and defined a unique 62-residue region in RILP that is necessary for regulation of lysosomal morphology via interaction with Rab7 and Rab34 (Wang *et al.*, 2004). We describe some assays used to investigate the functional properties of Rab34 and its interaction with RILP in regulating lysosome morphogenesis.

Expression Constructs and Mutagenesis

Our characterization of the full-length sequence of Rah (Morimoto *et al.*, 1991) suggests that it is a new member of the Rab family. It was therefore renamed Rab34. Since the biological function of Rab34 is unknown, we first explored the strategy of overexpressing Rab34 in transiently transfected cells. EGFP-Rab34 is a chimeric protein with enhanced green fluorescent protein (EGFP) fused to the N-terminus of Rab34 and it is constructed as follows. The coding region of mouse Rab34 is retrieved by polymerase chain reaction (PCR) with primer 1 (5-AAT TCT CGA GTG AAC ATT CTG GCG CCC GTG CGG AGG-3) and primer 2 (5-GC GGA TCC TCA GGG ACA ACA TGT GGC CTT CTT-3) from a mouse EST clone (GenBank accession number AW 742422, clone ID 2780131). The PCR products are digested with restriction enzymes *Xho*I and *Bam*HI and resolved by agarose gel electrophoresis. The desired DNA fragment is purified and ligated into the *Xho*I/*Bam*HI sites of the pEGFP-C1 vector (BD Clontech).

Generation of different mutant forms is a conventional strategy for functional analysis of Rab proteins. Some conserved domains/motifs are critical for nucleotide binding or hydrolysis of Rab proteins. For example, substitution of T/S with N in the GX4GK(T/S) motif results in restriction of Rab preferentially in the GDP-bound inactive form, while substitution of Q with L in the WDTAGQE motif will significantly decrease GTPase activity and result in restriction of Rab in the GTP-bound active state

(Olkkonen and Stenmark, 1997). A Rab34T66N mutant is generated that serves as a dominant negative form, while the Rab34Q111L mutant acts as a dominant active form. PCR-based point mutagenesis is carried out as below (use Rab34T66N as an example): primer 1 and primer 3 (5-GAA CCT ATT AAT GAG ACA <u>ATT</u> CTT CCC CAC AGA TAG GTC-3) are used to generate PCR fragment 1, whereas fragment 2 is generated by PCR using primer 4 (GAC CTA TCT GTG GGG AAG <u>AAT</u> TGT CTC ATT AAT AGG TTC-3) and primer 2. Gel-purified fragment 1 and fragment 2 are annealed and used as templates for another PCR using primer 1 and primer 2 to generate full-length DNA containing the point mutation. After digestion with *Xho*I and *Bam*HI, the PRC fragments are resolved by agarose gel. The desired fragment is purified and subcloned into the *Xho*I/*Bam*HI sites of the pEGFP-C1 vector. Rab34K82Q, Rab34Q111L, and Rab34F132Y mutants, as well as the Golgi targeting chimera Rab34Q111L-GS15, Rab34K82Q-GS15, and Rab34Q111L-GRIP, are generated using this approach and subcloned into a pEGFP-C1 vector so that all mutants are expressed as a fusion protein with N-terminal EGFP. All constructs are confirmed by DNA sequencing.

Generation and Purification of Rabbit Polyclonal Antibody against Rab34

Rab34 has a unique N-terminal extension as compared to other Rab proteins. This provides the ability to generate Rab34-specific antibodies. The coding region for the N-terminal residues 1–58 is amplified by PCR using primer 5 (5-GG GGA TCC ATG AAC ATT CTG GCG CCC GTG -3) and primer 6 (5-GAGA CTC GAG CAC AAC GAT GAC CTT GGA TAT-3). After digestion with *Bam*HI and *Xho*I, the PCR fragments are resolved by agarose gel. The desired fragment is purified and subcloned into *Bam*HI/*Xho*I sites of the pGEX-KG vector. The recombinant plasmid, pGEX-KG/Rab34(1–58), will express Rab34(1–58) fused to the C-terminus of glutathione *S*-transferase (GST). GST-Rab34(1–58) is expressed by transforming the plasmid into *Escherichia coli* DH5α cells. A bacterial colony with isopropyl β-D-thiogalactopyranoside (IPTG)-inducible expression of GST-Rab34(1–58) is expanded and the cells are grown to OD_{600} of 0.6–0.8 at 37° in LB containing 100 µg/ml ampicillin. IPTG is added to 0.1 mM (final concentration) and the culture is incubated at room temperature overnight. Bacterial cells are harvested by centrifugation for 10 min at 5000 rpm in a Sorvall GS3 rotor. The cell pellet is resuspended in *lysis buffer* (phosphate-buffered saline [PBS] containing 20 mM Tris–HCl, pH 8.0, 0.1% Triton-100, 1 mM dithiothreitol [DTT], 1 mM $MgCl_2$, 0.1 mM phenylmethylsulfonyl fluoride [PMSF], 0.1 mg/ml lysozyme) on ice for 30 min.

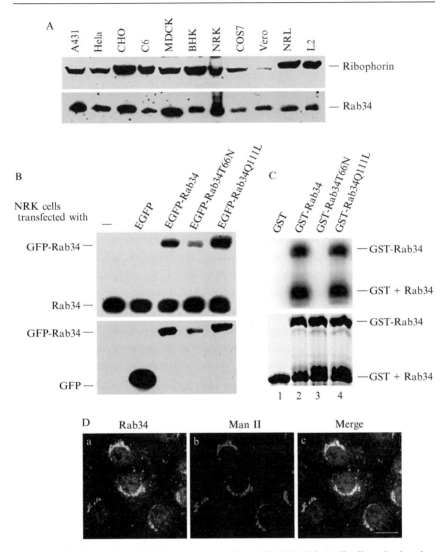

FIG. 1. Biochemical and cellular characterizations of Rab34. (A) Antibodies raised against GST-Rab34(1–58) recognize a 29-kDa protein in different cells. About 30 μg of total cell lysate derived from the indicated cell lines is resolved by SDS-PAGE and transferred to a filter. The filter is probed with anti-Rab34 antibodies or antibodies against ribophorin (kindly provided by D. I. Meyer, University of California, Los Angeles). (B) Characterization of EGFP-Rab34 fusion proteins. Control NRK cells and cells transfected with constructs for expressing EGFP, EGFP-Rab34, GDP-restricted EGFP-Rab34T66N, and GTP-restricted EGFP-Rab34Q111L as indicated are analyzed by Western blot using anti-Rab34 antibodies (upper panel) or anti-EGFP antibodies (lower panel). In addition to endogenous 29-kDa Rab34 polypeptide, anti-Rab34 also detects the fusion proteins of about 59 kDa.

The cells are sonicated and then incubated on ice for another 30 min. The cell lysate is clarified of cell debris by centrifugation for 30 min at 14,000 rpm in a Sorvall SS34 rotor. The cell lysate is incubated with glutathione-Sepharose 4B beads (Amersham) for 1 h at 4° and the sample is loaded onto a column followed by washing three times with cold *GST buffer* (PBS containg 20 mM Tris–HCl, pH 8.0, 1 mM MgCl$_2$). GST-Rab34(1–58) is eluted using GST buffer containing 3 mg/ml of reduced glutathione. The purified GST-Rab34(1–58) is dialyzed in PBS at 4° overnight and concentrated to 1 μg/μl using Centricon (Millipore).

As described previously (Wong *et al.*, 1999), 500 μg antigen, GST-Rab34(1–58), is mixed with an equal volume of Freund's adjuvant (Life Technologies, Inc.) and injected into female New Zealand white rabbits at 2-week intervals. After the third injection, antiserum is collected 10 days after each boost injection. For affinity purification, serum is diluted twice with PBS and incubated with cyanogen bromide-activated Sepharose 4B beads (Amersham) coupled with the GST-Rab34(1–58) according to the manufacturer's protocol. Specific antibodies are eluted from the beads with ImmunoPure IgG elution buffer (Pierce) and neutralized with 1 M Tris–HCl, pH 9 (300 μl for 10 ml elution buffer). The antibody is then dialyzed in PBS, pH 7.3. The antibody specifically detects Rab34 in different cell lysates as a 29-kDa band, or EGFP-Rab34 in transfected cell lysate as a 59-kDa band by Western blot analysis (Fig. 1A and B). This antibody is used to reveal that Rab34 is primarily associated with the Golgi apparatus and cytosol in NRK cells (Fig. 1D).

GTP Overlay Assay

GST-Rab34wt, -Rab34T66N, and -Rab34Q111L are generated in a manner similar to that described earlier for GST-Rab34(1–58). A GTP overlay assay is performed as described previously (Bucci *et al.*, 1992) with minor modifications. The fusion proteins are resolved by SDS-PAGE. The resulting gel is soaked in 50 mM Tris–HCl, pH 7.5, with 20% glycerol

---

(C) Wild-type and Rab34Q111L but not Rab34T66N bind GTP as revealed by a GTP overlay assay. The indicated proteins resolved by SDS–PAGE are either stained with Coomassie blue (lower panel) or transferred to a filter and then incubated with [^{32}P]GTP (upper panel). In addition to the GST-Rab34, autocleavage of fusion proteins into GST and Rab34 is observed. (D) Rab34 is present in the Golgi apparatus and cytosol. NRK cells are fixed, permeabilized, and incubated with anti-Rab34 antibodies (a) and monoclonal antibody against Golgi mannosidase II (Man II) (b). The merged image is also shown (c). As shown, Rab34 colocalizes with the Golgi apparatus marked by Man II. Bar = 10 μm. This figure is adopted from Wang and Hong (2002) with copyright permission of *Mol. Biol. Cell* of the American Society of Cell Biology. (See color insert.)

for 2 h at room temperature and the proteins are transferred onto the nitrocellulose filter using buffer containing 10 mM NaHCO$_3$ and 3 mM Na$_2$CO$_3$, pH 9.8. The filter is rinsed twice for 10 min in PBS containing 10 mM MgCl$_2$, 2 mM DTT, and 0.3% Tween-20, pH 7.3, and then incubated with 0.1 mCi/ml of [α-^{32}P]GTP for 2 h. After extensive washing, the filter is analyzed by PhosphorImager (Fig. 1C).

Yeast Two-Hybrid Interaction Assay

The yeast two-hybrid system is a genetic tool used to study protein–protein interaction *in vivo*. When a bait protein fused to a Gal4 DNA-binding domain interacts with a prey protein fused to a Gal4 DNA activation domain, a reporter gene (such as the β-galactosidase gene) will be transcriptionally activated and provides a simple genetic assay. It is often more sensitive than commonly used *in vitro* assays such as coimmunoprecipitation (Fields and Sternglanz, 1994; Miller and Stagljar, 2004). We have used the yeast two-hybrid system (MATCHMAKER Two-Hybrid System 3 kits, BD Clontech) to identify Rab34 partners. We used the active form of Rab34Q111L as the bait by subcloning the coding region into *Nde*I/*Bam*HI sites of pGBKT7 in the same reading frame as the Gal4 DNA binding domain (Gal-BD). The construct is transformed into yeast strain AH109 cells. The transformants are selected on SD/-Trp agar plates, and the expression of the Gal-BD-Rab34Q111L is confirmed by Western blot of the cells' lysate. For screening, AH109 cells expressing Gal-BD-Rab34Q111L are mated with a pool (about 5.5 × 10^7 independent colonies) of Y187 yeast cells pretransformed with a human kidney cDNA library fused to the Gal4 activation domain (AD) in the pACT2 vector (BD Clontech) in 50 ml of 2 × YPD medium with gentle swirling (30–50 rpm). After 20 h of mating, the mating culture is checked for the presence of zygotes. The mating is continued for another 4 h until the mating efficiency is >5%. The mating culture is spun down at 1000 rpm for 10 min in a 50-ml tube and washed once with sterilized water. Cells are resuspended in 10 ml 0.5 × YPD medium, and plated on SD/-Trp/-Leu/-His/-Ade (QDO) agar plates to select for interacting clones. The Gal4-AD-cDNA in the pACT2 vector is extracted from positive yeast clones and recovered by transformation into *Escherichia coli* DH5α cells. The identity of interacting protein is revealed by sequencing the Gal4-AD-cDNA recovered from the bacteria cells.

Seven positive clones are identified in the screen at the highest stringency and three of them contain partial cDNA encoding RILP (103–401 aa). To confirm the interaction, the recovered DNA is retransformed into Y187 to perform a yeast mating assay with AH109 cells expressing various bait proteins, and the diploids are selected on QDO plates. As shown in Fig. 2A, RILP(103–401 aa) interacts with Rab34 and Rab34Q111L but not Rab34T66N. Furthermore, mutation of Lys82 (K82) to Q in the switch I region abolishes the interaction.

GST Pull-Down Experiment and Immunoblotting Analysis

To further confirm the interaction between Rab34 and RILP, a GST pull-down assay is used in our experiments. The coding region for residues 103–394 of RILP is amplified by PCR with primers 7 (5-AAGA ATT CGC GCG GGG CCA CAG GAG GAG CGC -3) and 8 (5-AAG GGC GGC CGC CCC CAG ACA AAG GTG TTC GTG GAG -3) and then cloned into *EcoRI/NotI* sites of the pGEX-4T-1 vector (Amersham). As described earlier, the GST fusion protein is expressed in *E. coli* DH5α cells. The purified GST-RILP(103–394) is coupled to the GST-Sepharose 4B beads for an *in vitro* binding assay. For pull-down experiments, HeLa cells are transfected to express EGFP-Rab34, Rab34T66N, Rab34K82Q, Rab34Q111L, and Rab34F132Y, respectively. Cells are lysed in *lysis buffer* (20 mM HEPES, pH 7.4, 100 mM NaCl, 5 mM MgCl$_2$, 1% Triton-100, 0.1 M PMSF, and EDTA-free proteinase inhibitor cocktail [Roche]) for 1 h at 4°. The lysates are spun down using a TLA-100 rotor at 55,000 rpm for 30 min. The supernatants are incubated with 25 μg of GST-RILP(103–394) beads in the presence of 100 μM GTPγS (or GDP for EGFP-Rab34T66N). After overnight incubation, the beads are washed sequentially with *Buffer 1* (20 mM HEPES, pH 7.4, 100 mM NaCl, and 5 mM MgCl$_2$, 1% Triton-100), *Buffer 2* (20 mM HEPES, pH 7.4, 100 mM NaCl, and 5 mM MgCl$_2$, 0.5% Triton-100), and *Buffer 3* (20 mM HEPES, pH 7.4, 100 mM NaCl, and 5 mM MgCl$_2$); proteins retained on the beads are then processed for SDS-PAGE and Western blot.

For Western blot, the samples are resolved by 10% SDS-PAGE gel and the proteins are transferred to a Hybon-C extra nitrocellulose filter. The filter is blocked with 5% skim milk in PBS overnight at 4°, and then incubated with monoclonal antibody against EGFP (B.D. Clontech) for 1 h at room temperature. The filter is washed with PBS and incubated with horseradish peroxidase-conjugated goat anti-mouse IgG (Pierce) for 30 min at room temperature. The blots are detected using a chemiluminescence detection kit (Pierce). As shown in Fig. 2B, EGFP-Rab34, EGFP-Rab39Q111L, and EGFP-Rab34F132Y but not others are efficiently retained (about 5–10% of total input) by immobilized GST-RILP(103–394) but not GST. Less than 1% of GDP-restricted EGFP-Rab34T66N is retained by the GST-RILP(103–394) beads, which is similar to the results from yeast two-hybrid assays. Since EGFP-Rab34K82Q is not retained by GST-RILP(103–394), it confirms the

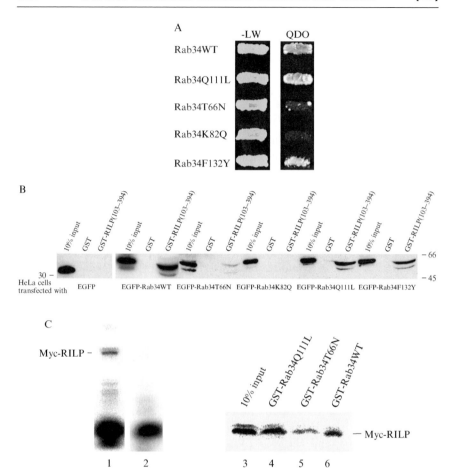

FIG. 2. Rab34 interacts with RILP. (A) Yeast two-hybrid assay showing interaction of Rab34 with RILP. AH109 yeast cells transformed with pGBKT7-based constructs expressing Rab34WT, Rab34Q111L, Rab34T66N, Rab34K82Q, or Rab34F132Y are each mated with Y187 yeast cells expressing RILP(103–401) in a pACT2 vector. The diploids are grown on either SD/-Leu/-Trp (-LW) or SD/-Leu/-Trp/-His/-Ade (QDO) plates. (B) GST pull-down assay showing K82-dependent interaction of RILP with a wild-type and GTP-restricted form of Rab34. HeLa cells are transfected to express EGFP, EGFP-Rab34T66N, EGFP-Rab34K82Q, EGFP-Rab34Q111L, or EGFP-Rab34F132Y. The resulting cell lysates are incubated with immobilized GST-RILP(103–394) or GST in the presence of 100 μM GTPγS (or GDP for Rab34T66N). Proteins retained on the beads are resolved by SDS-PAGE (along with 10% of starting materials) and then processed for Western blot using a monoclonal antibody against EGFP. (C) Rab34 binds directly to RILP. ^{35}S-Met-labeled Myc-RILP is generated by *in vitro* translation (lane 1, 10 μl of *in vitro* translation reaction). After immunoprecipitation with Myc antibodies, labeled RILP is efficiently depleted (lane 2, 10 μl of *in vitro* translation reaction after immunoprecipitation). The immunopurified RILP is

conclusion derived from the yeast two-hybrid assay that K82 is necessary for Rab34 to interact with RILP.

In Vitro Direct Binding of ^{35}S-Labeled RILP with GST-Rab34

To demonstrate a direct interaction between RILP and Rab34, the full-length RILP cDNA is cloned into *Eco*RI/*Not*I sites of the pDMYCneo vector (Xu *et al.*, 2001). ^{35}S-Met-labeled Myc-RILP is generated by TNT quick coupled transcription/translation systems according to the manufacturer's protocol (Promega). The reaction contains 80 μl TNT Quick Master Mix, 2 μl [^{35}S]methionine (2000 Ci/mmol), 1 μg pDMYCneo-RILP plasmid, and 17 μl nuclease-free water and is incubated at 30° for 90 min. ^{35}S-Met-labeled Myc-RILP is affinity purified by immunoprecipitation from the reaction with polyclonal anti-myc tag antibody (Upstate) precoupled to protein A-Sepharose CL-4B beads (Amersham). After washing several times with *binding buffer* (20 mM HEPES, pH 7.4, 100 mM NaCl, 5 mM MgCl$_2$, and EDTA-free proteinase inhibitor cocktail), Myc-RILP is eluted twice with ImmunoPure IgG elution buffer (50 μl each) (Pierce). The eluate is neutralized as described earlier and diluted to 500 μl using binding buffer, which is then incubated with immobilized GST-Rab34 preloaded with GTPγS or GDP. After overnight incubation, the beads are washed sequentially with binding buffer containing 1%, 0.5%, and 0.1% Triton-100. The samples are then resolved by 10% SDS-PAGE gel. The gel is dried and the amounts of ^{35}S-Met-labeled Myc-RILP retained are revealed by PhosphorImager. As shown in Fig. 2C, RILP alone is efficiently bound to GTP-restricted GST-Rab34Q111L (lane 4) and GST-Rab34WT (lane 6); a smaller amount of RILP is retained by immobilized GST-Rab34T66N (lane 5). These results suggest that Rab34 can directly interact with RILP.

Rab34 Regulates Lysosome Positioning by Interacting with RILP

To study the function of Rab34, constructs for expressing various EGFP-tagged forms of Rab34 are transfected transiently in NRK or HeLa cells. Cells are grown on coverslips in Dulbecco's modified Eagle's medium (DMEM) supplemented with 10% fetal bovine serum (Gibco) in a 5% CO$_2$ incubator at 37° and transfected with the constructs using

---

incubated with immobilized GST-Rab34Q111L (lane 4), GST-Rab34T66N (lane 5), or GST-Rab34wt (lane 6). The amounts of ^{35}S-Met labeled Myc-RILP retained by these beads, together with 10% input (lane 3), are resolved by SDS-PAGE and revealed by PhosphorImager. This figure is adopted from Wang and Hong (2002) with copyright permission of *Mol. Biol. Cell* of the American Society of Cell Biology.

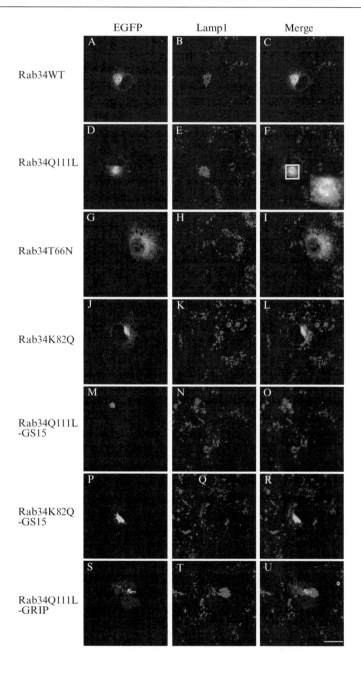

LipofectAMINE according to the manufacturer's instructions (Invitrogen). Cells are processed for indirect immunofluorescence microscopy analysis as described previously (Lu *et al.*, 2001). Cells (control or transfected) grown on coverslips are washed twice with PBSCM (PBS containing 1 mM CaCl$_2$ and 1 mM MgCl$_2$) and fixed for 15–30 min with 3% paraformaldehyde in PBSCM at 4°. After sequential washing with PBSCM containing 50 mM NH$_4$Cl and then PBSCM, cells are permeabilized with 0.1% saponin (Sigma) in PBSCM for 15 min at room temperature. The permeabilized cells are incubated with primary antibodies in fluorescence dilution buffer (FDB) (PBSCM containing 5% normal goat serum, 5% fetal bovine serum, and 2% bovine serum albumin) for 1 h at room temperature. After extensive washing, cells are incubated with Texas Red-conjugated secondary antibodies (Jackson ImmunoResearch) in FDB for 1 h at room temperature. Cells are washed and mounted with Vectashield (Vector Laboratories). Confocal microscopy was performed using a Zeiss Axioplan II microscope equipped with a Bio-Rad MRC-1024 confocal scanning laser. The EGFP signal is revealed directly.

As shown in Fig. 3, expression of EGFP-Rab34 (Fig. 3A–C) and EGFP-Rab34Q111L (Fig. 3D–F) dramatically alters the distribution of lysosomes marked by Lamp1 in NRK cells by clustering lysosomes to the peri-Golgi region. EGFP-Rab34T66N does not significantly affect the distribution of lysosomes (Fig. 3G–I). More than 90% of cells expressing EGFP-Rab34 or EGFP-Rab34Q111L have altered distribution of lysosomes, while lysosomes in 90% of EGFP-Rab34T66N-expressing cells are not altered. Similar results are obtained in HeLa cells (Wang and Hong, 2002). Since RILP is an interacting protein of Rab34 and has previously been implicated in lysosomal distribution, one possibility is that Rab34 alters lysosomal distribution via its direct interaction with RILP. Consistent with this, expression of EGFP-Rab34K82Q fails to redistribute lysosomes to the peri-Golgi region (Fig. 3J–L), suggesting that interaction of Rab34 with RILP is essential for its effect on lysosomal distribution.

Fig. 3. Rab34 regulates lysosomal distribution. NRK cells are transfected to express EGFP-Rab34wt (A–C), EGFP-Rab34Q111L (D–F), EGFP-Rab34T66N (G–I), EGFP-Rab34K82Q (J–L), EGFP-Rab34Q111L-GS15 (M–O), EGFP-Rab34K82Q-GS15 (P–R), or EGFP-Rab34Q111L-GRIP (S–U). Transfected cells are fixed, permeabilized, and then labeled with monoclonal antibody against lysosomal marker Lamp1. As shown, expression of a wild-type GTP-restricted but not GDP-restricted form of Rab34 specifically shifts lysosomes from the periphery to the peri-Golgi region. This effect is dependent on K82 and can occur when Rab34 is tethered to the Golgi apparatus by fusion to GS15 or to the *trans*-Golgi network by fusion with the GRIP domain of Golgin-97. Bar = 10 μm. This figure is adopted from Wang and Hong (2002) with copyright permission of *Mol. Biol. Cell* of the American Society of Cell Biology. (See color insert.)

As Rab34 is distributed both in the cytosol and the Golgi apparatus, it is important to examine whether Golgi-localized Rab34 can execute an effect on lysosomal distribution. Three mutants of Rab34 are created: EGFP-Rab34Q111L-GS15 and EGFP-Rab34K82Q-GS15 are created by replacing the C-terminal prenylation signal (CCP motif) of Rab34Q111L and EGFP-Rab34K82Q, respectively, with the coding region of GS15, a SNARE of the Golgi apparatus (Xu *et al.*, 1997), while EGFP-Rab34Q111L-GRIP is constructed by replacing the C-terminal CCP motif of Rab34Q111L by the GRIP domain (residue 670–748) of Golgin-97, which is sufficient for targeting to the *trans*-Golgi network (Kjer-Nielsen *et al.*, 1999; Munro and Nichols, 1999). Rab34Q111L-GS15 is still able to redistribute the lysosomes to the peri-Golgi region (Fig. 3M–O). This effect is dependent on Rab34's interaction with RILP as Golgi-anchored EGFP-Rab34K82Q-GS15 does not affect lysosomal positioning (Fig. 3P–R). Rab34Q111L-GRIP also has potent effects on lysosomal distribution (Fig. 3S–U). These results suggest that Golgi-localized Rab34 regulates lysosomal distribution via interaction with RILP.

References

Bucci, C., Parton, R. G., Mather, I. H., Stunnenberg, H., Simons, K., Hoflack, B., and Zerial, M. (1992). The small GTPase Rab5 functions as a regulatory factor in the early endocytic pathway. *Cell* **70**, 715–728.

Cantalupo, G., Alifano, P., Roberti, V., Bruni, C. B., and Bucci, C. (2001). Rab-interacting lysosomal protein (RILP): The Rab7 effector required for transport to lysosomes. *EMBO J.* **20**, 683–693.

Chavrier, P., and Goud, B. (1999). The role of ARF and Rab GTPases in membrane transport. *Curr. Opin. Cell Biol.* **11**, 466–475.

Fields, S., and Sternglanz, R. (1994). The two-hybrid system: An assay for protein-protein interactions. *Trends Genet.* **10**, 286–292.

Jordens, I., Fernandez-Borja, M., Marsman, M., Dusseljee, S., Janssen, L., Calafat, J., Janssen, H., Wubbolts, R., and Neefjes, J. (2001). The Rab7 effector protein RILP controls lysosomal transport by inducing the recruitment of dynein-dynactin motors. *Curr. Biol.* **11**, 1680–1685.

Kjer-Nielsen, L., Teasdale, R. D., van Vliet, C., and Gleeson, P. A. (1999). A novel Golgi-localisation domain shared by a class of coiled-coil peripheral membrane proteins. *Curr. Biol.* **9**, 385–388.

Lu, L., Horstmann, H., Ng, C., and Hong, W. (2001). Regulation of Golgi structure and function by ARF-like protein 1 (Arl1). *J. Cell Sci.* **114**, 4543–4555.

Miller, J., and Stagljar, I. (2004). Using the yeast two-hybrid system to identify interacting proteins. *Methods Mol. Biol.* **261**, 247–262.

Morimoto, B. H., Chuang, C. C., and Koshland, D. E., Jr. (1991). Molecular cloning of a member of a new class of low-molecular-weight GTP-binding proteins. *Genes Dev.* **5**, 2386–2391.

Munro, S., and Nichols, B. J. (1999). The GRIP domain—a novel Golgi-targeting domain found in several coiled-coil proteins. *Curr. Biol.* **9**, 377–380.

Olkkonen, V. M., and Stenmark, H. (1997). Role of Rab GTPases in membrane traffic. *Int. Rev. Cytol.* **176,** 1–85.

Pfeffer, S. R. (2001). Rab GTPases: Specifying and deciphering organelle identity and function. *Trends Cell Biol.* **11,** 487–491.

Segev, N. (2001). Ypt and Rab GTPases: Insight into functions through novel interactions. *Curr. Opin. Cell Biol.* **13,** 500–511.

Wang, T., and Hong, W. (2002). Interorganellar regulation of lysosome positioning by the Golgi apparatus through Rab34 interaction with Rab-interacting lysosomal protein. *Mol. Biol. Cell* **13,** 4317–4332.

Wang, T., Wong, K. K., and Hong, W. (2004). A unique region of RILP distinguishes it from its related proteins in its regulation of lysosomal morphology and interaction with Rab7 and Rab34. *Mol. Biol. Cell* **15,** 815–826.

Wennerberg, K., Rossman, K. L., and Der, C. J. (2005). The Ras superfamily at a glance. *J. Cell. Sci.* **118,** 843–846.

Wong, S. H., Xu, Y., Zhang, T., Griffiths, G., Lowe, S. L., Subramaniam, V. N., Seow, K. T., and Hong, W. (1999). GS32, a novel Golgi SNARE of 32 kDa, interacts preferentially with syntaxin 6. *Mol. Biol. Cell* **10,** 119–134.

Xu, Y., Wong, S. H., Zhang, T., Subramaniam, V. N., and Hong, W. (1997). GS15, a 15kilodalton Golgi soluble N-ethylmaleimide-sensitive factor attachment protein receptor (SNARE) homologous to rbet1. *J. Biol. Chem.* **272,** 20162–20166.

Xu, Y., Hortsman, H., Seet, L. F., Wong, S. H., and Hong, W. (2001). SNX3 regulates endosomal function via its PX domain-mediated interaction with PtdIns(3)P. *Nat. Cell Biol.* **3,** 658–666.

Zerial, M., and McBride, H. (2001). Rab proteins as membrane organizers. *Nat. Rev. Mol. Cell Biol.* **2,** 107–117.

Further Reading

Somsel Rodman, J., and Wandinger-Ness, A. (2000). Rab GTPases coordinate endocytosis. *J. Cell Sci.* **113,** 183–192.

[59] Rabring7: A Target Protein for Rab7 Small G Protein

By KOUICHI MIZUNO, AYUKO SAKANE, and TAKUYA SASAKI

Abstract

Rab7, a member of the Rab family of small G proteins, has been shown to regulate late endocytic traffic and lysosome biogenesis, but the exact roles and the mode of actions of Rab7 are still undetermined. Accumulating evidence suggests that each Rab protein has multiple target proteins and works together with them to coordinate the individual step of vesicle traffic. Rabring7 (*Rab7*-interacting RING finger protein) is a Rab7 target protein that has been isolated using a CytoTrap system. This protein shows

METHODS IN ENZYMOLOGY, VOL. 403
0076-6879/05 $35.00
DOI: 10.1016/S0076-6879(05)03059-4

no homology with RILP, which has been reported as another Rab7 target protein. Rabring7 is recruited efficiently to late endosome/lysosome by the GTP-bound form of Rab7. Exogenous expression of Rabring7 not only affects epidermal growth factor degradation but also induces the perinuclear aggregation of lysosomes and the increased acidity in the lysosomes. This chapter describes the procedures for the isolation of Rabring7 with a CytoTrap system, the analysis of the Rab7–Rabring7 interactions, and the properties of Rabring7.

Introduction

The Rab family of small G proteins appears to be a key regulator of intracellular vesicle traffic, including exocytosis and endocytosis (Pfeffer, 2001; Segev, 2001; Takai et al., 2001; Zerial and McBride, 2001). There are more than 60 members in mammalian cells and at least 13 Rab proteins have been identified in the endocytic pathway (Stein et al., 2003). Some of them involved in the early steps of endocytosis have been studied extensively and the molecular mechanisms of these steps are well understood. Rab5 regulates the first step of internalization and subsequent fusion of vesicles with early endosomes. Rab5 is also involved in the homotypic fusion of early endosomes. To date, Rabaptin-5, EEA1, p110β, hVPS34/ p150, Rabenosyn-5, and APPL1/2 have been identified as Rab5 target proteins, which specifically interact with the GTP-bound active form of Rab5 (Callaghan et al., 1999; Christoforidis et al., 1999a,b; Miaczynska et al., 2004; Murray et al., 2002; Stenmark et al., 1995). Analysis of these target proteins suggests that Rab5 can regulate several steps through these multiple target proteins and that these target proteins act cooperatively to coordinate the various functions of Rab5 in early endocytic traffic.

In contrast, less well-characterized Rab proteins are involved in late endocytic traffic. Rab7 localizes to the late endocytic structures, late endosome/lysosome. Previous data obtained with a dominant-negative mutant and a constitutive active mutant of Rab7 suggest that Rab7 regulates vesicle traffic from early to late endosomes and from late endosome to lysosome (Feng et al., 1995; Meresse et al., 1995; Press et al., 1998; Vitelli et al., 1997). Moreover, Rab7 has also been implicated in the lysosome biogenesis, by regulation of the heterotypic fusion between late endosomes and lysosomes and the homotypic fusion of lysosomes (Bucci et al., 2000). Thus, Rab7 is involved in several steps of late endocytic traffic, but the exact roles and the mode of actions of Rab7 are still unknown.

A Rab7 target protein, Rab-interacting lysosomal protein (RILP), has been identified and characterized (Bucci et al., 2001; Cantalupo et al., 2001; Jordens et al., 2001). RILP recruits a minus-end directed dynein–dynactin motor to the late endosome/lysosome and consequently to stimulate the transport of these compartments toward the perinuclear region along microtubules (Jordens *et al.*, 2001). We have identified another Rab7 target protein, named Rabring7, which contains an H2 type RING finger motif at the C-terminus, using a CytoTrap system (Mizuno *et al.*, 2003). There is no homology between RILP and Rabring7, but both target proteins have similar properties: both are recruited to late endosome/lysosome by the GTP-bound form of Rab7, exogenous expression of these proteins causes perinuclear aggregation of late endosomes/lysosomes, and they affect epidermal growth factor (EGF) degradation. The unique feature of Rabring7 is that exogenous expression of Rabring7 shows increased acidity in the lysosome, whereas RILP does not seem to affect it. It has been reported that a dominant-negative mutant of Rab7, Rab7T22N, decreases lysosomal acidity (Bucci *et al.*, 2000). Based on these observations, Rabring7 might play a specific role in the Rab7-mediated lysosomal acidification in addition to the common function with RILP.

This chapter first describes the procedures for the isolation of Rabring7 with the CytoTrap system and the analysis of Rab7–Rabring7 interactions. It then describes the properties of Rabring7.

Materials

Anti-GFP and anti-6 × His monoclonal antibodies are purchased from Clontech (Palo Alto, CA). An anti-Xpress monoclonal antibody is from Invitrogen (Carlsbad, CA). The anti-HA monoclonal antibody comes from Roche (Basel, Switzerland) and anti-Sos monoclonal antibody from Transduction Laboratories (Lexington, KY). Horseradish peroxidase-linked anti-mouse IgG antibody is from Bio-Rad (Hercules, CA). The enhanced chemiluminescence (ECL) is from Amersham Bioscience (Piscataway, NJ). ^{125}I-labeled human EGF (>27.8 TBq/mmol, 3.7 MBq/ml) is from Amersham Bioscience (Piscataway, NJ). Reduced glutathione and isopropyl-β-D-thiogalactopyranoside (IPTG) are from Wako Pure Chemicals (Osaka, Japan). GST-expression vector, pGEX-5X, glutathione-Sepharose 4B, and rProtein A-Sepharose Fast Flow are from Amersham Bioscience (Piscataway, NJ). FuGENE 6 is from Roche (Basel, Switzerland). Dulbecco's modified Eagle's medium (DMEM) and fetal bovine serum (FBS) are from Sigma-Aldrich (St. Louis, MO) and Invitrogen (Carlsbad, CA), respectively. Alexa488- or TexasRed-conjugated secondary antibody, LysoTracker Red DND-99, and TexasRed-transferrin are from Molecular Probe (Eugene, OR). All other chemicals are of reagent grade.

Plasmids for expression of *murine* Rab proteins (Rab4, Rab7, and Rab8) are constructed as follows. The cDNA encoding full open reading frames of the *murine* Rab proteins (Rab4, Rab7, and Rab8) is amplified from a *murine* immature B cell line (WEHI231) cDNA library by polymerase chain reaction (PCR). The PCR products are purified from agarose gel and directly inserted into the pGEM-T Easy vector (Promega, Madison, WI). A site-directed mutagenesis is carried out by a PCR-based method. A deletion mutant is also generated by a PCR-based method. The pSos/Rab7 plasmids are made by subcloning the Rab7wt, Q67L, T22N, N125I, wtΔC, Q67LΔC, T22NΔC, and N125IΔC cDNA into pSos (Stratagene, La Jolla, CA) from the pGEM/Rab7 plasmids, respectively. The pEGFP/Rab7 plasmids are made by subcloning the Rab7wt, Q67L, T22N, and N125I cDNA into pEGFP-C3 (Clontech, Palo Alto, CA) from the pSos/Rab7 plasmids, respectively.

Plasmids for expression of full-length Rabring7 and its truncated mutants are constructed as follows. The Rabring7 cDNA is amplified from the WEHI231 cDNA library by PCR. The truncated mutants, Rabring7N (amino acids 1–145) and Rabring7C (amino acids 146–305), are generated by PCR. The PCR products are subcloned into pcDNA4/HisMaxC (Invitrogen, Carlsbad, CA) to obtain the pcDNA/Rabring7 plasmids. The Rabring7, Rabring7N, and Rabring7C are also inserted into pCIneoHA (Promega, Madison, WI) and pGEX-5X (Amersham Bioscience, Piscataway, NJ) to produce the pCIneoHA/Rabring7 and pGEX/Ranring7 plasmids, respectively.

Methods

Isolation of Rabring7 with a CytoTrap System

Strain and Media. The yeast strain cdc25Hα used in a CytoTrap system has the following genotype: *MATα ura3, lys2, leu2, trp1, his200, ade101, cdc25-2, GAL+* (Aronheim *et al.*, 2001). A point mutation of the *cdc25* gene in cdc25Hα prevents host growth at a restrictive temperature (37°), but not at a permissive temperature (24°), because this mutation leads to the inactivation of the yeast Ras GEP, cdc25p, at 37°. The strain is grown in YPAD medium or Burkholder's minimum media (BMM) with appropriate supplements at 24°. For selection of transformants, glucose base BMM lacking leucine and uracil (BMM/glucose −leu, ura) is used. For induction of protein expression from pMyr (Stratagene, La Jolla, CA), galactose base BMM lacking leucine and uracil (BMM/galactose −leu, ura) is used.

CytoTrap Screening. A human Ras GEP, Sos-tagged GTPase-deficient mutant of Rab7, Rab7Q67L, is used as bait. The final three amino acids of Rab7 are deleted in this mutant to prevent the C-terminal posttranslational prenylation, because such a modification causes a high background in this system. The yeast strain cdc25Hα is transformed with pSos/Rab7Q67LΔC by the standard lithium acetate method (Gietz *et al.*, 1992). The yeast containing pSos/Rab7Q67LΔC is then transformed with a WEHI231 cDNA library that is constructed by ligation of the oligo(dT)-primed cDNA fragments derived from WEHI231 cells to pMyr. The transformants are grown on a BMM/galactose −leu, ura plate at 24° for 3 days and then grown at 37°. The colonies grown at 37° are selected and tested for growth on BMM/glucose −leu, ura and BMM/galactose −leu, ura plates at 37°. The prey is expressed as a fusion protein with a myristylation sequence that anchors the fusion protein onto the plasma membrane. If the bait protein physically interacts with the prey protein, Sos is recruited to the membrane, thereby activating the yeast RAS-signaling pathway and allowing cdc25Hα to grow at 37°. The library plasmids are isolated from the clones that exhibit galactose-dependent growth at 37° and are retransformed into the cdc25Hα cells either with pSos/Rab7Q67LΔC or with pSos/collagenase IV as a negative control. Only plasmids that suppress the cdc25Hα phenotype in the presence of pSos/Rab7Q67LΔC are sequenced. Using this system, Rabring7 is isolated as one of the clones that specifically interact with Rab7Q67LΔC. This clone also interacts with the wild-type Rab7 lacking the C-terminal three amino acids, Rab7wtΔC, but not with two mutants lacking the C-terminal three amino acids, the dominant-negative mutant locked in the GDP-bound form, Rab7T22NΔC, or the nucleotide-empty mutant, Rab7N125IΔC. These results suggest that Rabring7 is a Rab7 target protein.

Assay for the Rab7–Rabring7 Interactions

To confirm the interaction between Rab7 and Rabring7 observed in the CytoTrap system, a GST pull-down assay and the coimmunoprecipitation assay are performed.

GST Pull-Down Assay

Purification of GST-Rabring7. GST and GST-tagged full-length and truncated mutants of Rabring7 are expressed in the *BLR* bacteria strain after being induced by IPTG at a final concentration of 0.1 m*M* and are then purified on a glutathione Sepharose 4B column.

Culture. COS1 cells are maintained at 37° in a humidified atmosphere of 5% CO_2 and 95% air (v/v) in DMEM containing 10% FBS, 100 U/ml penicillin, and 100 μg/ml streptomycin.

EXPRESSION OF EGFP-TAGGED RAB PROTEINS IN COS1 CELLS. Rab proteins are expressed as N-terminal EGFP-tagged proteins (EGFP-Rabs) by transfection of pEGFP/Rabs in COS1 cells. The COS1 cells are plated at a density of 1×10^6 cells/10-cm dish and are incubated at 37° for 24 h. Five micrograms of pEGFP/Rabs and 15 μl of FuGENE 6 are mixed with 485 μl of DMEM. The mixed solution is then allowed to be incubated at room temperature for 15 min and is added to each dish. The cells are then incubated at 37° for 48 h.

GST PULL-DOWN ASSAY. Fourty-eight hours after transfection, the COS1 cells expressing EGFP-Rabs are scraped from the dishes in phosphate-buffered saline (PBS) and washed with the same buffer twice. The cells are lysed in 1 ml of a lysis buffer (20 mM HEPES [pH 7.5], 100 mM NaCl, 0.1% NP-40, 5 mM EDTA, 10 mM MgCl$_2$ and 1 mM dithiothreitol [DTT]) for 10 min at 4°. The cell lysates are centrifuged at 16,100$\times g$ for 10 min at 4° and the supernatants are used as the cell extracts. The cell extracts are incubated with GST or GST-tagged proteins immobilized on glutathione-Sepharose 4B beads. The beads are washed extensively with the same buffer three times. Comparable amounts of the samples are subjected to sodium dodecyl sulfate–polyacrylamide gel electrophoresis (SDS–PAGE) and transferred to a polyvinylidene difluoride (PVDF) membrane. After blocking, the membrane is treated with the anti-GFP antibody and immunoreactive proteins are detected according to the ECL protocol using the horseradish peroxidase-linked secondary antibody. In these experiments, GST-Rabring7 interacts with EGFP-Rab7wt, but not with EGFP-Rab4wt. Very weak interaction is observed with EGFP-Rab8wt. GST alone does not interact with any EGFP-Rabs. GST-Rabring7 also interacts with EGFP-Rab7Q67L, although there is little interaction of GST-Rabring7 with the two mutants, EGFP-Rab7T22N and EGFP-Rab7-N125I. Moreover, GST-Rabring7N interacts with EGFP-Rab7Q67L as well as GST-Rabring7, but GST-Rabring7C does not.

Coimmunoprecipitation Assay

CELL CULTURE AND TRANSFECTION. COS1 cells are cultured and plated as described previously. After 24 h, 2.5 μg of pEGFP/Rabs, 2.5 μg of pCIneoHA/Rabring7, and 15 μl of FuGENE 6 are mixed with 485 μl of DMEM. The mixed solution is then allowed to be incubated at room temperature for 15 min and is added to the dish. The cells are then incubated at 37° for 48 h.

IMMUNOPRECIPITATION. Immunoprecipitation is performed at 48 h after the transfection. The cell extracts are prepared as described earlier and incubated with the anti-HA antibody or the control mouse IgG for 2 h at 4°, followed by treatment with rProtein A–Sepharose FF beads. The beads are washed extensively with a lysis buffer three times. Comparable amounts of the samples are subjected to SDS-PAGE and are transferred to a PVDF membrane. After blocking, the membrane is treated with the anti-HA antibody and the immunoreactive proteins are detected according to the ECL protocol using the horseradish peroxidase-linked secondary antibody. After stripping, the PVDF membranes are reprobed by the anti-GFP antibody and the immunoreactive proteins are detected as described earlier.

In these experiments, EGFP-tagged Rab7Q67L and Rab7wt are coimmunoprecipitated with HA-Rabring7, although EGFP-tagged Rab7T22N and Rab7N125I are not, in the cells coexpressing the respective proteins. This result indicates that Rabring7 interacts with the GTP-bound form of Rab7 in COS1 cells. The Rab specificity is also observed in the coimmunoprecipitation assays. These indicate that Rabring7 binds the GTP-bound form of Rab7 more preferentially than the GDP-bound form, which supports the idea that Rabring7 is a Rab7 target protein.

Properties of Rabring7

Assay for Recruitment of Rabring7 to Late Endosome/Lysosome by the GTP-bound Form of Rab7. BHK cells are maintained at 37° in a humidified atmosphere of 5% CO_2 and 95% air (v/v) in DMEM containing 10% FBS, 100 U/ml penicillin, and 100 μg/ml streptomycin. The cells cultured on cover glasses are transfected with the various plasmids using FuGENE 6 as described previously. After 48 h, the cells are fixed with 3.7% formaldehyde in PBS for 15 min at room temperature. The cells are washed three times with PBS and are treated with 0.1% Triton X-100 in PBS for 5 min at room temperature. After blocking, the cells are incubated with the first antibodies for 1 h at room temperature and are washed three times with PBS, followed by incubation with the Alexa488- or TexasRed-conjugated secondary antibody for 30 min at room temperature. The cells are washed three times with PBS and are mounted on slide glasses. Fluorescence is visualized through a Bio-Rad Radiance 2000 confocal laser scanning microscope. In these experiments, when BHK cells are transiently transfected with the N-terminal 6 × His- and Xpress-tagged full-length Rabring7 (His-Rabring7) expression plasmid, His-Rabring7 is diffusely observed in the cytosol. EGFP-Rab7wt and Rab7Q67L are mostly concentrated in the perinuclear region corresponding to late endosome/lysosome, whereas EGFP-Rab7T22N and Rab7N125I are distributed throughout the cytoplasm. Coexpression of EGFP-Rab7wt or Rab7Q67L with His-Rabring7 causes changes in the distribution of His-Rabring7, and His-Rabring7 is accumulated in the perinuclear region where Rab7wt or Rab7Q67L and Rabring7 are colocalized. EGFP-Rab7T22N and Rab7N125I do not affect the subcellular distribution of Rabring7.

Assay for the Effect of Rabring7 on Subcellular Distribution and Acidity of Lysosome. BHK cells are cultured on cover glasses and are transfected as described earlier. After 48 h, the cells are incubated with 100 nM LysoTracker Red DND-99 for 30 min at 37° for labeling of lysosome. The cells are then fixed and immunofluorescence microscopy is performed as described previously. In these experiments, LysoTracker-accumulating vesicles are dispersed throughout the cytoplasm of the untransfected cells, whereas they are aggregated in the perinuclear region of the Rabring7- and Rabring7N-expressing cells. In addition, these vesicles show enhanced acidity, as shown by the very strong fluorescent signal relative to the untransfected cells. These effects are not observed in the Rabring7C-expressing cells. In the transferrin uptake assay, the transfected cells are incubated with 50 μg/ml TexasRed-transferrin for 30 min at 37°. Rabring7 does not change the distribution of internalized transferrin in BHK cells.

EGF Degradation Assay. The EGF degradation assay is performed as previously described (Cantalupo *et al.*, 2001). HeLa cells cultured in six-well plates are transfected with various expression plasmids as described earlier. After 24 h, the cells are incubated overnight with serum-starved medium. They are incubated further with ^{125}I-labeled EGF (37 kBq/ml) on ice for 1 h. Unbound ligand is washed off and the cells are allowed to internalize and to degrade the EGF at 37°. After 2 h, the extracellular media are collected and treated with trichloroacetic acid (TCA). The extracellular TCA-soluble ^{125}I-labeled EGF is quantitated as a degraded EGF fraction by gamma counter. In these experiments, Rab7N125I inhibits EGF degradation, as previously reported (Bucci *et al.*, 2000; Feng *et al.*, 1995). Overexpression of Rabring7 significantly inhibits EGF degradation, although it does not affect the uptake or recycling of EGF.

Comments

The target proteins for small G proteins have been isolated by several biochemical and molecular biological methods: affinity column chromatography, blot overlay, and yeast two-hybrid system. Among these methods, the yeast two-hybrid system is most popular, because this method is easy and reproducible. However, if the target proteins belong to transcriptional activators or repressors, proteins that require posttranslational modification, and proteins with large molecular weights, it is almost impossible to isolate them by this method, because this exploits the modular nature of eukaryotic transcription factors. In such cases, the CytoTrap two-hybrid system should be used instead of the conventional two-hybrid system. The CytoTrap system is based upon generating fusion proteins whose interaction in the yeast cytoplasm activates the RAS-signaling pathway, resulting in cell growth. These properties of the CytoTrap system enable us to detect protein–protein interactions that cannot be assayed by the conventional two-hybrid system. As to Rabring7, the RING finger motif can give a false positive result in the conventional two-hybrid system. In addition to Rabring7, we have succeeded in isolating Munc13–4 as a Rab27 target protein using the CytoTrap system (Goishi *et al.*, 2004).

References

Aronheim, A. (2001). Ras signaling pathway for analysis of protein-protein interactions. *Methods Enzymol.* **332**, 260–270.

Bucci, C., Thomsen, P., Nicoziani, P., McCarthy, J., and van Deurs, B. (2000). Rab7: A key to lysosome biogenesis. *Mol. Biol. Cell* **11**, 467–480.

Bucci, C., De Gregorio, L., and Bruni, C. B. (2001). Expression analysis and chromosomal assignment of PRA1 and RILP genes. *Biochem. Biophys. Res. Commun.* **286**, 815–819.

Callaghan, J., Nixon, S., Bucci, C., Toh, B. H., and Stenmark, H. (1999). Direct interaction of EEA1 with Rab5b. *Eur. J. Biochem.* **265**, 361–366.

Cantalupo, G., Alifano, P., Roberti, V., Bruni, C. B., and Bucci, C. (2001). Rab-interacting lysosomal protein (RILP): The Rab7 effector required for transport to lysosomes. *EMBO J.* **20**, 683–693.

Christoforidis, S., McBride, H. M., Burgoyne, R. D., and Zerial, M. (1999a). The Rab5 effector EEA1 is a core component of endosome docking. *Nature* **397**, 621–625.

Christoforidis, S., Miaczynska, M., Ashman, K., Wilm, M., Zhao, L., Yip, S. S., Waterfield, M. D., Backer, J. M., and Zerial, M. (1999b). Phosphatidylinositol-3-OH kinases are Rab5 effectors. *Nat. Cell Biol.* **1**, 249–252.

Feng, Y., Press, B., and Wandinger-Ness, A. (1995). Rab7: An important regulator of late endocytic membrane traffic. *J. Cell Biol.* **131**, 1435–1452.

Gietz, D., St. Jean, A., Woods, R. A., and Schiestl, R. H. (1992). Improved method for high efficiency transformation of intact yeast cells. *Nucleic Acids Res.* **20**, 1425.

Goishi, K., Mizuno, K., Nakanishi, H., and Sasaki, T. (2004). Involvement of Rab27 in antigen-induced histamine release from rat basophilic leukemia 2H3 cells. *Biochem. Biophys. Res. Commun.* **324**, 294–301.

Jordens, I., Fernandez-Borja, M., Marsman, M., Dusseljee, S., Janssen, L., Calafat, J., Janssen, H., Wubbolts, R., and Neefjes, J. (2001). The Rab7 effector protein RILP controls lysosomal transport by inducing the recruitment of dynein-dynactin motors. *Curr. Biol.* **11**, 1680–1685.

Meresse, S., Gorvel, J. P., and Chavrier, P. (1995). The rab7 GTPase resides on a vesicular compartment connected to lysosomes. *J. Cell Sci.* **108**, 3349–3358.

Miaczynska, M., Christofordis, A., Giner, A., Shevchenko, A., Uttenweiler-Joseph, S., Habermann, B., Wilm, M., Parton, B. G., and Zerial, M. (2004). APPL proteins link Rab5 to nuclear signal transduction via endosomal compartment. *Cell* **116**, 445–456.

Mizuno, K., Kitamura, A., and Sasaki, T. (2003). Rabring7, a novel Rab7 target protein with a RING finger motif. *Mol. Biol. Cell* **14**, 3741–3752.

Murray, J. T., Panaretou, C., Stenmark, H., Miaczynska, M., and Baker, J. M. (2002). Role of Rab5 in the recruitment of Vps34/p150 to the early endosome. *Traffic* **3**, 2219–2228.

Pfeffer, S. R. (2001). Rab GTPases: Specifying and deciphering organelle identity and function. *Trends Cell Biol.* **11**, 487–491.

Press, B., Feng, Y., Hoflack, B., and Wandinger-Ness, A. (1998). Mutant Rab7 causes the accumulation of cathepsin D and cation-independent mannose 6-phosphate receptor in an early endocytic compartment. *J. Cell Biol.* **140,** 1075–1089.

Segev, N. (2001). Ypt and Rab GTPases: Insight into functions through novel interactions. *Curr. Opin. Cell Biol.* **13,** 500–511.

Stein, M.-P., Dong, J., and Wandinger-Ness, A. (2003). Rab proteins and endocytic trafficking: Potential targets for therapeutic intervention. *Adv. Drug Deliv. Rev.* **55,** 1421–1437.

Stenmark, H., Vitale, G., Ullrich, O., and Zerial, M. (1995). Rabaptin-5 is a direct effector of the small GTPase Rab5 in endocytic membrane fusion. *Cell* **83,** 423–432.

Takai, Y., Sasaki, T., and Matozaki, M. (2001). Small GTP-binding proteins. *Physiol. Rev.* **81,** 153–208.

Vitelli, R., Santillo, M., Lattero, D., Chiariello, M., Bifulco, M., Bruni, C. B., and Bucci, C. (1997). Role of the small GTPase Rab7 in the late endocytic pathway. *J. Biol. Chem.* **272,** 4391–4397.

Zerial, M., and McBride, H. (2001). Rab proteins as membrane organizers. *Nat. Rev. Mol. Cell Biol.* **2,** 107–117.

[60] Analysis of Potential Binding of the Recombinant Rab9 Effector p40 to Phosphoinositide-Enriched Synthetic Liposomes

By Diego Sbrissa, Ognian C. Ikonomov, and Assia Shisheva

Abstract

The transport factor p40 is thought to assist Rab9 in mediating late-endosome-to-TGN vesicular transport. p40 was recently identified as an associated protein of the PtdIns 5-P/PtdIns 3,5-P2-producing kinase PIK-Fyve. Moreover, p40 recovery in membrane fractions appeared totally dependent on the presence of an intact enzymatic activity of PIKFyve, implying a mechanism of p40 membrane association dependent on membrane PtdIns 5-P and/or PtdIns 3,5-P2. Here we have evaluated plausible interaction of recombinant p40 with the PIKFyve products PtdIns 5-P and PtdIns 3,5-P2 by a synthetic-liposome binding assay.

Introduction

In the biosynthetic pathway, newly synthesized lysosomal hydrolases are sorted away from secreted proteins through binding to one of the two mannose 6-phosphate receptors (MPR) that recognize the posttranslationally acquired mannose 6-phosphate marker (Ghosh *et al.*, 2003; Gruenberg, 2001; Luzio *et al.*, 2003). Sequestered into clathrin-coated vesicles at the *trans*-Golgi network (TGN), the enzymes arrive in late endocytic/prelysosomal compartments, where they dissociate from MPR and are ultimately delivered to lysosomes. To maintain the forward flow of the enzymes to

METHODS IN ENZYMOLOGY, VOL. 403
0076-6879/05 $35.00
DOI: 10.1016/S0076-6879(05)03060-0

lysosomes, MPR must be efficiently recycled back to the TGN for reuse. The small GTPase Rab9, shown to localize primarily on late endosomes, is suggested to be important in this transport step (Pfeffer, 2001, 2003). Video microscopy in live cells demonstrating the fusion of Rab9-positive transport vesicles with TGN, along with other lines of functional data, is consistent with this hypothesis (Barbero *et al.*, 2002; Pfeffer, 2001, 2003). In addition to Rab9 GTPase, the late-endosome-to-TGN transport requires the protein TIP47 that binds to the cytoplasmic domain of MPRs and likely serves as a coat of these transport/carrier vesicles (Diaz and Pfeffer, 1998; Pfeffer, 2001, 2003). In addition, a protein transport factor p40 has been found to directly interact and synergize with Rab9 in stimulating the *in vitro* late endosome-to-TGN transport of MPR (Diaz *et al.*, 1997). The p40 sequence is composed almost entirely of six internally repeated motifs of ~50 residues in length, known as kelch repeats, predicted to form a four-stranded β-sheet corresponding to a single blade of a β-propeller (Adams *et al.*, 2000).

We have identified that p40 interacts with the conserved chaperonin domain of PIKfyve in yeast two-hybrid screens and GST pull-down assays (Ikonomov *et al.*, 2003a). PIKfyve is a large enzyme with dual specificity to phosphorylate, both phosphatidylinositol (PtdIns) or PtdIns 3-P on position D-5, and proteins (Ikonomov *et al.*, 2001, 2003b; Sbrissa *et al.*, 2000, 2002b; Shisheva, 2001). Of these activities, the cellular role of PIKfyve-catalyzed PtdIns 3,5-P2 synthesis in the context of the late endocytic pathway has been best characterized. Thus, expression of PIKfyve point mutants selectively deficient in PtdIns 3,5-P2-producing activity was found to elicit enormous dilation of endosomes and formation of gross cytoplasmic vacuoles (Ikonomov *et al.*, 2002). This severe phenotype could be corrected by high levels of PIKfyve^{WT} or microinjected PtdIns 3,5-P2 but not by PtdIns 3-P or PtdIns 5-P, implying a selective role for PtdIns 3,5-P2 in endosome trafficking. Intriguingly, biochemical fractionation demonstrated that p40 membrane association requires the enzymatic activity of PIKfyve, since the high-speed membrane fraction, derived from a cell line stably expressing the kinase-deficient dominant-negative PIKfyve^{K1831E} mutant, was profoundly depleted in immunoreactive p40 that was recovered in the soluble pool (Fig. 1; Ikonomov *et al.*, 2003a). This observation led us to suggest that p40, bound to PIKfyve through the enzyme's chaperonin-like domain, is recruited to membranes where a subsequent interaction with PtdIns 3,5-P2 produced by PIKfyve stabilizes the p40 membrane association and its Rab9-dependent function in late endosome-to-TGN membrane retrieval. Perturbation in this step due to p40 membrane release should result in late-endosome membrane enlargement that could explain, at least in part, the observed enormous dilation of endosome structures on expression of enzymatically inactive PIKfyve mutants (Ikonomov *et al.*, 2001, 2002).

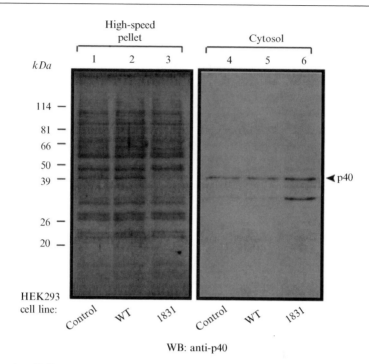

FIG. 1. p40 dissociation from membranes upon expression of dominant-negative kinase-deficient PIKfyve[K1831E]. HEK293 stable cell lines induced to express PIKfyve[WT] or the dominant-negative mutant PIKfyve[K1831E] were homogenized and centrifuged to obtain the high-speed pellet at 200,000×g and soluble fraction (cytosol) as detailed in Ikonomov et al. (2003). Aliquots were processed by SDS-PAGE and immunoblotting with anti-p40 antibodies. Note the p40 ablation from the pellet fraction and an increase in the soluble pool (the arrow).

The hypothesis that p40 could interact with membrane PtdIns 3,5-P2 has two additional lines of support. First, the p40 protein sequence displays two clusters of basic residues positioned prior to the first kelch repeat ([12]KPRP[15] and [46]KRGK[49]), consistent with a plausible interaction with acidic phospholipids (Ikonomov et al., 2003a). Second, findings identify a family of conserved proteins in both yeast and mammals, harboring WD40 repeats, that displays a PtdIns 3,5-P2-binding specificity (Dove et al., 2004; Jeffries et al., 2004). An interaction with PtdIns 5-P and PtdIns 3-P has also been demonstrated for the mammalian WD40 repeat protein WIPI49 (Jeffries et al., 2004). Intriguingly, although unrelated in primary structure to kelch repeat proteins, WD40 repeat proteins are also predicted to fold to form β-propellers (Gettemans et al., 2003). Here we describe a liposome-based assay to determine possible interactions of recombinant p40 with PtdIns 3,5-P2, PtdIns 5-P, and PtdIns 3-P.

General Considerations

Currently, common methods for studying protein–lipid interaction primarily include surface plasmon resonance (SPR) (James *et al.*, 1996), protein overlay on phosphoinositides (PIs) immobilized on nitrocellulose membrane filters (Dowler *et al.*, 2002), and centrifugation-based liposome binding assays (Klarlund and Czech, 2001; Schiavo *et al.*, 1996). The method described here for monitoring p40 Rab9 effector protein binding to liposomes has been previously used for determination of FYVE-domain peptide-specific binding to PtdIns 3-P (Sbrissa *et al.*, 2002a). It is a centrifugal assay, in which synthetic liposomes incorporating radiolabeled [^{3}H]phosphatidylcholine with and without various PIs are tested for their ability to bind to glutathione *S*-transferase (GST)-p40 fusion protein immobilized on glutathione (GSH)-agarose beads. The interaction allows a direct quantitation by measuring the radioactivity of radiolabeled liposomes bound to the bead-immobilized protein. In addition to radiolabeled lipid, a scintillation counter, and a bath sonicator, the assay makes use of equipment and supplies readily available to most investigators, is relatively rapid and economical, and is versatile (e.g., use of membrane filtration instead of centrifugation, or fusion proteins immobilized via different tag/bead chemistries is possible) and therefore can likely be successfully adapted to screen other proteins for lipid interactions.

Experimental Procedures

Expression and Purification of Recombinant GST-p40 Fusion Protein

Construction of pGEX5X-1-p40 cDNA. Preparation of GST-p40 fusion is conducted as previously reported (Ikonomov *et al.*, 2003a). Briefly, full-length human p40 obtained as pQE31-p40 cDNA (a kind gift by Dr. Susan Pfeffer, Department of Biochemistry, Stanford University, Stanford, CA) is fused in-frame to the GST C-terminus by first amplifying the coding sequence of p40 by PCR using primers with engineered *Eco*RI or *Xho*I restriction sites: sense 5′-CGGC<u>GAATTC</u>ATGAAGCAAC-3′ and antisense 5′-CAGC<u>CTCGAG</u>TTAGTCCACTAC-3′; restriction sites are underlined. The PCR product is digested with *Hind*III, and the resulting two fragments of 180 bp (nt 150–329 of p40 sequence GI:5032014) and 968 bp (nt 329–1297) are subsequently digested with *Eco*RI and *Xho*I, respectively, and ligated into the *Eco*RI/*Xho*I digest of pGEX-5X-1 (Amersham Biosciences Corp., Piscataway, NJ; www.amershambiosciences.com). The expected organization of the construct is confirmed by restriction endonuclease digestion and sequencing.

Growth of Bacterial Cultures. GST fusion proteins are produced in the XA90 *Escherichia coli* strain transformed with the above construct or empty pGEX5X vector, for GST alone, which is used as a negative control in the liposome binding assay. The confirmed positive clones encoding GST-p40 are grown overnight in 50 ml of Luria-Bertani (LB) medium (Sambrook *et al.*, 1989) plus 50 µg/ml ampicillin at 37°. The next day, the culture is diluted into 500 ml of fresh LB plus ampicillin for 1.5 h of growth before being induced with 0.1 mM isopropyl-1-thio-β-D-galactopyranoside for an additional 6 h. Cells are harvested by centrifugation at 5000×g for 20 min and resuspended in 10 ml of phosphate–buffered saline (PBS) containing 1 mM phenylmethylsulfonyl fluoride (PMSF) by repeatedly pipeting up and down with a 10-ml pipette. Resuspended cells are then frozen in 0.5-ml aliquots in 1.5-ml microfuge tubes at −80°. An aliquot of the cells is processed for lysis as outlined in the following step and subjected to sodium dodecyl sulfate-polyacrylamide gel electrophoresis (SDS-PAGE) analysis for estimating the concentration and quality of the purified proteins (see below). Frozen aliquots are similarly processed for lysis in small amounts as needed on the day of the liposome assay, by small-scale affinity purification of GST-fusion protein on GSH-agarose beads as outlined in the following subsection.

Affinity Purification of GST-p40-Fusion Protein on GSH-Agarose. All procedures for preparing bacterial lysates are conducted on ice (4°). Aliquots (as many as needed for 0.1 nmol fusion protein per liposome-binding assay sample) of frozen cells prepared as directed above are lysed by repeated freezing and thawing with 1 mg/ml lysozyme in PBS plus 1 mM PMSF and subsequent incubation for 30 min on ice. Cell lysates are treated with DNase (0.1 mg/ml) in the presence of 10 mM MgCl$_2$ for 10 min or until no longer viscous, and cleared at 10,000×g for 15 min at 4°. Cleared lysates are incubated for 1 h at 4° with 10-µl packed GSH-agarose beads (Sigma Chemical Co., St. Louis, MO; www.sigma-aldrich.com) per aliquot. The beads are washed several times with PBS plus 1 mM PMSF and subsequently washed two more times with liposome-binding assay buffer (20 mM HEPES, pH 7.5, 100 mM KCl, 0.2 mM dithiothreitol [DTT]) and used directly for the binding assay with liposomes (see below).

Quality and Quantity Check for GST-Fusion Protein Expression. Prior to the liposome-binding assay, an aliquot of the protein affinity purified on GSH-agarose beads as outlined in the preceding step is assessed for purity and yield of the desired GST fusion product. For this purpose, an aliquot of the beads is boiled with ∼ 30–35 µl of 1 × Laemmli sample buffer (Laemmli, 1970) for 3 min. The concentration and quality of the purified proteins bound to the beads are then determined electrophoretically by reducing discontinuous SDS-PAGE (Laemmli, 1970) on a 10% (w/v) acrylamide/0.8% (w/v) bis-acrylamide minigel along with 10 µl of Benchmark

Pre-Stained Protein Ladder molecular weight markers (Invitrogen Corporation, Carlsbad, CA; www.invitrogen.com) and several lanes run with 1, 5, and 10 μg of bovine serum albumin (BSA) standard (Pierce Biotechnology, Inc., Rockford, IL; www.piercenet.com) in 1 × SB. The gel is stained in 0.02% (w/v) Coomassie Brilliant Blue R (Sigma Chemical Co.) in 20% methanol–7% acetic acid (v/v), and the quantity of the Coomassie Blue-stained protein bands is estimated by comparing their intensities to those of the BSA standards. The GST-p40 migrates as an ~70-kDa band.

Preparation of Phosphoinositide-Enriched Liposomes

Preparation and Storage of Phospholipid Stocks. Stock solutions containing 1 mM of the following lipids (Avanti Polar Lipids, Inc., Alabaster, AL; www.avantilipids.com) are prepared in chloroform, or chloroform with a few drops of methanol for those not dissolving in chloroform alone, sealed under N_2 in glass tubes with Teflon-lined screw caps, and stored at −20°: brain phosphatidylcholine (PtdCho), phosphatidylserine (PtdSer), phosphatidylethanolamine (PtdEth), and phosphatidylinositol (PtdIns). Stock solutions of 0.5 mM dipalmitoyl PIs (Echelon Biosciences, Inc., Salt Lake City, UT; www.echelon-inc.com) are prepared in 9:9:2 (v/v) chloroform–methanol–water, requiring some heating and sonication for complete dissolution, and are aliquoted (20–100 μl) and stored in small glass vials under N_2 at −80°. Tritiated PtdCho ([choline-*methyl*-^{3}H]-L-α-dipalmitoyl PtdCho, 76 μCi/nmol) is obtained as a 1:1 (v/v) toluene–ethanol solution from PerkinElmer Life Sciences, Inc. (Boston, MA; www.perkinelmer.com/lifesciences) and stored at −20°. The reader may find some useful technical information for lipid handling and liposome preparation in the "Practical Approach Series" (New, 1990) and in the product catalog from Avanti ("Avanti Polar Lipids, Inc., Products Catalog Edition VI," pp. 159–172. Avanti Polar Lipids, Inc., Alabaster, AL).

Preparing and Drying the Lipid Mixture. To prepare a suspension of 660 μl (enough for three duplicate liposome samples or six in total) of liposomes incorporating 0/2 mol% of PI, first a ^{3}H-labeled lipid mixture is made, containing the following mol% of lipids: 63/65% PtdCho, 20% PtdSer, 15% PtdEth, and 0/2% PtdIns. This is done by mixing 94 μl of 1 mM PtdCho stock (94 nmol), 60 μl of 0.5 mM PtdSer stock (30 nmol), 23 μl of 1 mM PtdEth stock (23 nmol), and 0.55 μl of [^{3}H]PtdCho stock (0.55 μCi) together in a glass tube and diluting the mixture to a convenient volume with 1:1 (v/v) chloroform–methanol or benzene–ethanol. This ^{3}H-labeled lipid mixture (150 nmol of total lipid) is then divided into several equal aliquots in microfuge tubes, depending on the number of different PIs to be studied; for example, if liposomes with two PIs are needed then the mixture is divided into three equal parts (50 nmol of total lipid in each) and 1 μl of 1 mM PtdCho stock is added to one aliquot to obtain PI-free liposomes, or 2 μl of 0.5 mM dipalmitoyl PI stock, to the other aliquots for the other two 2%PI-enriched liposomes. In this way, all of these different lipid mixtures should have a more uniform amount of both radiolabel and total lipid. These ^{3}H-labeled lipid mixtures are then dried down under blowing N$_2$ with a Model 10 N-EVAP nitrogen evaporator (Organomation Associates, Inc., Berlin, MA; www.organomation.com) at a bath temperature of 35–40° for 15 min. The residue is redissolved in 200–300 μl of ethanol mixed with a few drops of benzene and evaporated again to remove residual solvent and traces of water. After a dry film is obtained, drying is continued with N$_2$ for a further 15 min to ensure complete dryness of the lipid film.

Hydration and Dispersion of Lipid and Final Liposome Purification. Next, each aliquot of dried ^{3}H-labeled lipid mixture (50 nmol of total lipid) is rehydrated in 220 μl of binding assay buffer (20 mM HEPES, pH 7.5, 100 mM KCl, 0.2 mM DTT) at 35–40° for 0.5 h and agitated vigorously on a vortex mixer occasionally to disperse lipid and promote formation of multi-lamellar vesicles (MLVs), giving a final total lipid concentration of ~230 μM and 0.83 μCi ^{3}H/ml after hydration. The suspension of MLVs is then subjected to ultrasonic disintegration at 35–40° for 5 min in a bath sonicator (Laboratory Supplies Company, Inc., Hicksville, NY) to reduce them to more homogeneous, small unilamellar vesicles, followed by purification by centrifugation at 15,000×g for 10 min to remove undispersed lipid aggregates, large MLVs, and foreign particles. The clear supernatant containing the ^{3}H-labeled liposomes is then removed and transferred to a clean tube, care being taken not to disturb the lower 10- to 20-μl bottom layer to be discarded. A small aliquot (5–10 μl) of the liposome suspension is saved for scintillation counting on a Model 4430 Packard TRI–CARB liquid scintillation counter (Packard Instrument Co., Inc., Meriden, CT; www.packardinst.com) to determine whether the various liposomes are uniformly labeled and allow for small corrections to normalize binding data from the different liposome preparations.

Liposome-Binding Assay

All procedures are conducted at room temperature (20°). Duplicate samples of freshly prepared GST-p40 fusion protein (0.1 nmol) bound to 10-μl packed GSH-agarose beads, along with other proteins at equal amounts that will serve as a negative (GST) or a positive control (GST-FYVE finger domain peptide derived from PIKfyve; Sbrissa *et al.*, 2002a), are washed twice with binding assay buffer (see previous section) and incubated with 90 μl of the cleared ^{3}H-labeled liposome suspension (prepared as described previously) for exactly 0.5 h on an end-over-end mixer or at low speed on an Eppendorf 5432 vortex mixer (Eppendorf, Hamburg, Germany; www.eppendorf.com). The binding is terminated by dilution with 1 ml of assay buffer containing 0.01% (v/v) Nonidet P-40 and beads are collected at $3000 \times g$ (6500 rpm) for 30 s in an IEC Micromax benchtop centrifuge (International Equipment Company, Needham Hts., MA; www.thermo.com). Supernatants are removed and discarded, whereas the beads are quickly washed three more times the same way. The bound liposomes remaining on the washed beads are solubilized with 320 μl of 10% (w/v) SDS for 5 min on a vortexer. The samples are centrifuged at $12,000 \times g$ for 30 s, and 300 μl of the supernatant is counted by liquid scintillation for 5 min along with 300 μl of 10% (w/v) SDS for counting background. Inclusion of GST by itself is useful as a negative control as is another irrelevant/non-lipid-binding GST-fusion protein for a nonspecific binding control (ideally a mutant version of the protein that does not bind lipids would be used as a negative control). One or more fusion peptides that bind positively to one of the tested PIs (Fig. 2) is also recommendable. After

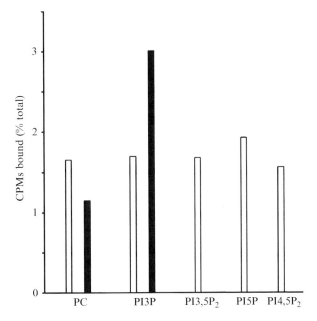

FIG. 2. Binding of various phosphoinositides to GST-fusion proteins of p40 and PIKfyve FYVE-finger domain as determined by the liposome-binding assay described herein. Note that whereas GST-p40 (open bars) does not bind specifically to any of the tested liposomes, binding of PtdIns 3-P-enriched liposomes to the GST-PIKfyve FYVE-finger domain (solid bars) was demonstrated and served as a positive control.

subtracting the background sample from the raw counts, the percentage of total lipid bound is calculated and plotted both for the PI-free liposome control and the PI-enriched liposome sample for the protein of interest. The same is done for the negative and positive GST-based control proteins.

Applications

A representative p40 binding assay is shown in Fig. 2. Unexpectedly, we could not demonstrate binding by GST-p40 to the PtdIns 3,5-P2 and/or PtdIns 5-P that is predicted here. Binding to other PI was also not documented. By contrast, the positive control used in this set of experiments— the GST-fusion of the PIKfyve N-terminus containing the FYVE finger domain, predicted to interact with PtdIns 3-P (Sbrissa *et al.*, 2002a)—binds nicely to PtdIns 3-P-enriched liposomes. It should be concluded that although p40 has a predicted structure of a β-propeller, shown to bind PtdIns 3,5-P2, PtdIns 5-P, and PtdIns 3-P (Adams *et al.*, 2000; Diaz *et al.*, 1997; Dove *et al.*, 2004; Jeffries *et al.*, 2004), GST-p40 is not specifically interacting with these lipids in a liposome-binding assay.

Acknowledgments

This work was supported by National Institutes of Health Grant DK58058 and an American Diabetes Association Research grant (to A.S).

References

Adams, J., Kelso, R., and Cooley, L. (2000). The kelch repeat superfamily of proteins: Propellers of cell function. *Trends Cell Biol.* **10**, 17–24.
"Avanti Polar Lipids, Inc., Products Catalog Edition VI" from Avanti Polar Lipids, Inc., Alabaster, AL 35007, USA, pp. 159–172.
Barbero, P., Bittova, L., and Pfeffer, S. R. (2002). Visualization of Rab9-mediated vesicle transport from endosomes to the trans-Golgi in living cells. *J. Cell Biol.* **156**, 511–518.
Diaz, E., and Pfeffer, S. R. (1998). TIP47: A cargo selection device for mannose 6-phosphate receptor trafficking. *Cell* **93**, 433–443.
Diaz, E., Schimmöller, F., and Pfeffer, S. R. (1997). A novel Rab9 effector required for endosome-to-TGN transport. *J. Cell Biol.* **138**, 283–290.
Dove, S. K., Piper, R. C., McEwen, R. K., Yu, J. W., King, M. C., Hughes, D. C., Thuring, J., Holmes, A. B., Cooke, F. T., Michell, R. H., Parker, P. J., and Lemmon, M. A. (2004). SVP 1p defines a family of phosphatidy limositol 3,5-bis phosphate effectors. *EMBO J.* **23**, 1922–1933.
Dowler, S., Kular, G., and Alessi, D. R. (2002). Protein lipid overlay assay. *Sci. STKE* **129**, PL6.
Gettemans, J., Meerschaert, K., Vanderkerckhove, J., and De Corte, V. (2003). A kelch beta propeller featuring as a G beta structural mimic: Reinventing the wheel? *Sci. STKE* **191**, PE27.
Ghosh, P., Dahms, D. M., and Kornfeld, S. (2003). Mannose 6-phosphate receptors: New twists in the tale. *Mol. Cell. Biol.* **4**, 202–212.
Gruenberg, J. (2001). The endocytic pathway: A mosaic of domains. *Rev. Mol. Cell. Biol.* **2**, 721–730.

Ikonomov, O. C., Sbrissa, D., and Shisheva, A. (2001). Mammalian cell morphology and endocytic membrane homeostasis require enzymatically active phosphoinositide 5-kinase PIKfyve. *J. Biol. Chem.* **276,** 26141–26147.

Ikonomov, O. C., Sbrissa, D., Mlak, K., Kanzaki, M., Pessin, J., and Shisheva, A. (2002). Functional dissection of lipid and protein kinase signals of PIKfyve reveals the role of PtdIns 3,5-P2 production for endomembrane integrity. *J. Biol. Chem.* **277,** 9206–9211.

Ikonomov, O. C., Sbrissa, D., Mlak, K., Deeb, R., Fligger, J., Soans, A., Finley, R. L., Jr., and Shisheva, A. (2003a). Active PIKfyve associates with and promotes the membrane attachment of the late endosome-to-trans-Golgi network transport factor Rab9 effector P40. *J. Biol. Chem.* **278,** 50863–50870.

Ikonomov, O. C., Sbrissa, D., Foti, M., Carpentier, J.-L., and Shisheva, A. (2003b). PIKfyve controls fluid phase endocytosis but not recycling/degradation of endocytosed receptors or sorting of procathepsin D by regulating multivesicualr body morphogenesis. *Mol. Biol. Cell* **14,** 4581–4591.

James, S. R., Downes, C. P., Gigg, R., Grove, S. J. A., Holmes, A. B., and Alessi, D. R. (1996). Specific binding of the Akt-1 protein kinase to phosphatidylinositol 3,4,5-trisphosphate without subsequent activation. *Biochem. J.* **315,** 700–713.

Jeffries, T. R., Dove, S. K., Michell, R. H., and Parker, P. J. (2004). PtdIns-specific MPR pathway association of a novel WD40 repeat protein, WIPI49. *Mol. Biol. Cell* **15,** 2652–2663.

Klarlund, J. K., and Czech, M. P. (2001). Isolation and properties of GRP1, an ADP-ribosylation factor (ARF)-guanine nucleotide exchange protein regulated by phosphatidylinositol 3,4,5-trisphosphate. *Methods Enzymol.* **329,** 279–289.

Laemmli, U. K. (1970). Cleavage of structural proteins during the assembly of the head of bacteriophage T4. *Nature* **227,** 680–685.

Luzio, J. P., Poupon, V., Lindsay, M. R., Mullock, B. M., Piper, R. C., and Pryor, P. R. (2003). Membrane dynamics and the biogenesis of lysosomes. *Mol. Membr. Biol.* **20,** 141–154.

New, R. R. C. (ed.) (1990). "Liposomes: A Practical Approach." Oxford University Press, New York.

Pfeffer, S. R. (2001). Rab GTPases: Specifying and deciphering organelle identity and function. *Trends Cell Biol.* **11,** 487–491.

Pfeffer, S. (2003). Membrane domains in the secretory and endocytic pathways. *Cell* **112,** 507–517.

Sambrook, J., Fritsch, E. F., and Maniatis, T. (1989). *In* "Molecular Cloning: A Laboratory Manual," 2nd Ed., pp. A.1 and A.4. Cold Spring Harbor Laboratory Press, Cold Spring Harbor, NY.

Sbrissa, D., Ikonomov, O. C., and Shisheva, A. (2000). PIKfyve lipid kinase is a protein kinase: Downregulation of 5′-phosphoinositide product formation by autophosphorylation. *Biochemistry* **39,** 15980–15989.

Sbrissa, D., Ikonomov, O. C., and Shisheva, A. (2002a). Phosphatidylinositol 3-phosphate-interacting domains in PIKfyve. Binding specificity and role in PIKfyve. Endomembrane localization. *J. Biol. Chem.* **277,** 6073–6079.

Sbrissa, D., Ikonomov, O. C., Deeb, R., and Shisheva, A. (2002b). Phosphatidylinositol 5-phosphate biosynthesis is linked to PIKfyve and is involved in osmotic response pathway in mammalian cells. *J. Biol. Chem.* **277,** 47276–47284.

Schiavo, G., Gu, Q.-M., Prestwich, G. D., Söllner, T. H., and Rothman, J. E. (1996). Calcium-dependent switching of the specificity of phosphoinositide binding to synaptotagmin. *Proc. Natl. Acad. Sci. USA* **93,** 13327–13332.

Shisheva, A. (2001). PIKfyve: The road to PtdIns 5-P and PtdIns 3,5-P(2). *Cell Biol. Int.* **25,** 1201–1206.

[61]   Assessment of Rab11-FIP2 Interacting Proteins *In Vitro*

By NICOLE A. DUCHARME, MIN JIN, LYNNE A. LAPIERRE, and JAMES R. GOLDENRING

Abstract

Members of the Rab family of small GTPases are involved in multiple trafficking events in both endocytotic and biosynthetic pathways. To understand more fully the regulation of these events, a concerted effort is underway to ascertain the binding partners and regulators of Rabs. Here, we describe methods to assess binding of Rab11a with Rab11-FIP2 and other Rab11-FIPs utilizing a modified far-Western approach. We then broaden this application to assess binding of Rab11-FIP2 with myosin Vb and homodimerization of Rab11-FIP2.

Introduction

Members of the Rab family of small GTPases play diverse roles in membrane trafficking events. Rab11a, a member of the Rab11 subfamily, participates in the regulation of recycling endosomal trafficking. Plasma membrane recycling is critical in maintaining proper membrane protein expression in response to stimuli for such diverse events as nutrient internalization and the recycling of ion channels and receptors. Rab11a is involved in trafficking such diverse cargoes as GLUT4 (Kessler *et al.*, 2000; Zeigerer *et al.*, 2002), respiratory syncytial virus (Brock *et al.*, 2003), the chemokine receptor CXCR2 (Fan *et al.*, 2003), polymeric IgA receptor (Casanova *et al.*, 1999), and transferrin receptor (Green *et al.*, 1997; Ullrich *et al.*, 1996), suggesting it as a critical component of the intracellular membrane recycling system.

The function of Rab proteins is determined by their interaction with specific upstream and downstream regulators. In 1999, two laboratories independently isolated a Rab11a interacting protein, designated Rabphilin-11 or Rab11 binding protein (Rab11BP) (Mammoto *et al.*, 1999; Zeng *et al.*, 1999). Rab11BP preferentially binds to the GTP bound form of Rab11 and colocalizes at the perinuclear region and along microtubules. Because parietal cells contain an amplified apical recycling system, we utilized a parietal cell yeast two-hybrid library to screen for Rab11a interacting

METHODS IN ENZYMOLOGY, VOL. 403
Copyright 2005, Elsevier Inc. All rights reserved.

0076-6879/05 $35.00
DOI: 10.1016/S0076-6879(05)03061-2

proteins. This screen yielded both myosin Vb and a family of interacting proteins (Hales *et al.*, 2001; Lapierre *et al.*, 2001).

The Rab11a family of interacting proteins (FIPs) was discovered in the past 5 years: Rab11-FIP1, Rab11-FIP2, Rab11-FIP3 (Hales *et al.*, 2001), Rab11-FIP4 (Wallace *et al.*, 2002b), pp75/Rip11 (Prekeris *et al.*, 2000), and Rab-coupling protein (RCP) (Lindsay *et al.*, 2002). They each interact with the carboxyl-terminus of Rab11 through predicted coiled-coil regions in what is termed the Rab-binding domain (RBD) (Hales *et al.*, 2001). One hypothesis suggests that FIPs compete for binding to Rab11 and through this competition recruit Rab11 to various intracellular locations (Meyers and Prekeris, 2002). For example, Rab11-FIP2 localizes to recycling endosomes linking myosin Vb with Rab11-positive vesicles (Hales *et al.*, 2002) while Rab11-FIP4 localizes to early endosomes (Wallace *et al.*, 2002b). Most members of this family contain other binding motifs either with EF domains or with C2 domains (Rizo and Sudhof, 1998). In addition, two Rab11-FIPs contain ERM (ezrin, radixin, and moesin) domains that are thought to bind to the actin cytoskeleton and plasma membrane (Tsukita and Yonemura, 1999).

Rab11-FIP2 interacts with both GDP- and GTP-bound Rab11 as well as myosin Vb (Lapierre *et al.*, 2001). The McCaffrey laboratory reported that Rab11-FIP2 (FIP2) homodimerizes in parallel and anti-parallel conformations (Wallace *et al.*, 2002a). FIP2 suppresses epidermal growth factor receptor (EGFR) uptake and binds α-adaptin, which associates with clarthrin-coated pits (Cullis *et al.*, 2002), implying a critical role in early endocytosis. Myosin Vb and Rab11 are involved in the trafficking of the M4 muscarinic acetylcholine receptor (Volpicelli *et al.*, 2002), which suggests a potential role for FIP2 in this interaction. Given these data, it is possible that FIP2 serves as the linker and regulator of Rab11/myosin Vb interaction.

The molecular motor myosin Vb is a critical regulator of trafficking through plasma membrane recycling systems (Lapierre *et al.*, 2001). We have demonstrated previously that Rab11-FIP2 binds to both myosin Vb and Rab11a (Lapierre *et al.*, 2001), suggesting a role for a ternary complex of Rab11-FIP2, myosin Vb, and Rab11a in regulating recycling. The tail region of myosin Vb contains differentiable Rab11-FIP2 (Hales *et al.*, 2002) and Rab11a binding domains (Lapierre *et al.*, 2001; J.R. Goldenring, Hobdy-Henderson, and Roland, unpublished observations). Expression of the tail region lacking a motor domain prominently inhibits the recycling of transferrin receptor in nonpolarized cells (Lapierre *et al.*, 2001) and apical recycling and transcytosis of pIgAR in polarized Madin–Darby canine kidney (MDCK) cells. Expression of myosin Vb tail can also block ligand-induced recycling of M_4-muscarinic acetylcholine receptor (Volpicelli *et al.*, 2002). Furthermore, we have found that ligand-induced CXCR2 internalization and recycling are dependent on both Rab11-FIP2 and myosin Vb interaction and function (Fan *et al.*, 2004). CXC-L1 binding induced the association of myosin Vb and Rab11-FIP2 with CXCR2 in an immunoprecipitable complex. Thus, the regulation of the components of the Rab11a/myosin Vb/Rab11-FIP2 complex may serve as a critical point for modulation of recycling traffic and may provide insight into the poorly understood regulation of myosin-Rab trafficking.

More recently, we have begun to focus on the ability of Rab11-FIP2 to bind with Rab11 recycling components utilizing a modified far-Western approach. This approach allows us to ascertain requirements for biochemical protein–protein interactions, which are often difficult to study using conventional methodologies. For example, antibodies against Rab11a do not immunoprecipitate Rab11a interacting proteins. The results garnered from the far-Western approach can then be confirmed *in situ* using model systems.

Methods

General Methods

Antibodies. Rab11 monoclonal antibody 8D3 was raised against a recombinant human Rab11a sequence. The antibody recognizes Rab11a, Rab11b, and Rab25 but not other Rab proteins. Rabbit anti-Rab11a (VU57) antibodies were developed against the amino-terminus of human Rab11a and were specific for Rab11a versus Rab11b and Rab25 (L.A. Lapierre, in preparation). Chicken anti-myosin Vb was a gift from Dr. Richard Cheney (University of North Carolina). S-protein horseradish peroxidase (HRP) conjugate was from Novagen. Peroxidase-conjugated secondary antibodies were from Jackson Immuno Research Laboratories (West Grove, PA).

Recombinant Protein Production. For recombinant protein production constructs in pET-30a vectors were retransformed into BL21(DE3)pLysS bacteria. Bacteria were grown to log phase and then induced with iso-propyl-β-D-thiogalactopyranoside (IPTG) (400 ng/ml) for 3 h at $37°$. To harvest protein, bacteria were pelleted at $2000 \times g$ and then resuspended in lysis buffer (50 mM sodium phosphate buffer, pH 8.0, 300 mM NaCl with protease inhibitors [Protein Buffer], and 10 mM imidazole). Protein was harvested according to the manufacturer's protocol (Novagen). Briefly, the bacteria were then sonicated four times for 20 s at maximum potency on ice. The lysate was extracted with 0.1% Triton X-100 for 5 min on ice. The extracted lysate was cleared by centrifugation at $15,000 \times g$, and the resulting supernatant was incubated with nickel-affinity resin at $4°$ (His-Bind, Novagen). The beads and protein were washed in Protein Buffer with 20 mM imidazole. The bound protein was eluted overnight at $4°$ with Elution Buffer (Protein Buffer with 250 mM imidazole).

GST-tagged proteins were grown in a similar manner. The pelleted bacteria were resuspended in phosphate-buffered saline (PBS) with protease inhibitors. Following extraction and lysis, the cleared supernatant was incubated with glutatione-Sepharose 4B beads (Amersham Biosciences) at $4°$. The beads with bound protein were washed with PBS plus protease inhibitors three times. The bound protein was eluted with 10 mM glutathione in 50 mM Tris, pH 8.0.

Analysis of Rab11a Binding

Traditionally, Rab11 interactions have been studied utilizing γ-^{35}S-GTP Rab11 overlays (Hales *et al.*, 2001). This methodology requires the use of radioactive materials, which can often be a limiting factor in many laboratories. We have developed an approach using nonradioactive GTPγS and antibodies against Rab11a.

1. One microgram of His-tagged or 20 μg of total cell lysate protein was resolved on 10% sodium dodecyl sulfate-polyacrylamide gel electrophoresis (SDS-PAGE) gels. Proteins were transferred to nitrocellulose for 2 h at 250 mA or overnight at 100 mA with SDS in transfer buffer.
2. The blots were blocked in 5% nonfat dry milk in TBS-Tween (0.05%) for a minimum of 30 min.
3. For generalized binding: 10 μg of Rab11a protein was loaded with 10 μM GTPγS for 1 h at $30°$ in the presence of 0.5 mM MgCl$_2$ and protease inhibitors.
4. The reaction was stopped with the addition of 20 mM MgCl$_2$ (final concentration).
5. For overlay of Rab11-FIP2, unloaded recombinant Rab11a was incubated with monoclonal anti-Rab11 8D3 supernatant (1:1) for 1 h at room temperature in 2.5% nonfat dry milk in TBS-Tween-20. For overlay of RCP, the GTPγS preloaded Rab11a was incubated with rabbit anti-Rab11a (1:1000) for 1 h at room temperature in 2.5% nonfat dry milk in TBS-Tween-20. In both cases, the mixtures were added to the blots and incubated overnight at $4°$.
6. Blots were washed three times with TBS-Tween (0.05%).

7. Blots were incubated with the appropriate secondary antibody for 1 h at room temperature.
8. Blots were washed three times with TBS-Tween (0.05%) and visualized by ECL.

We have utilized this methodology to analyze the interaction of Rab11a with Rab11-FIP proteins. Figure 1 demonstrates that Rab11a interacts with recombinant Rab11-FIP2. Rab11a did not interact with a truncated mutant Rab11-FIP2 lacking the Rab11 binding domain (Fig. 1). In a lysate of MDCK cells, Rab11a overlay detected a major 55-kDa protein, compatible with endogenous Rab11-FIP2. The mixture also recognizes endogenous Rab11 in the cell lysate as expected. In overlays of Rab11-FIP2, we did not include GTP because Rab11 binds to Rab11-FIP2 in either its GTP-bound or its

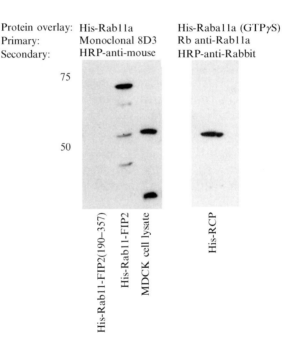

Fig. 1. Rab11a interacts with known interacting proteins *in vitro*. Recombinant proteins and MDCK cell lysates were resolved by SDS-PAGE and transferred to nitrocellulose. Left: Recombinant His-Rab11a was incubated with monoclonal antibody 8D3 and then overlayed onto the blot overnight. The binding of Rab11a was assessed with an anti-mouse IgG secondary antibody. The recombinant protein missing the Rab11 binding domain is not recognized. Full-length recombinant Rab11-FIP2 is bound by Rab11a. Rab11a also binds to a Rab11-FIP2 found in cell lysate. Right: Recombinant Rab11a was preloaded with GTPγS, incubated with rabbit anti-Rab11a, and overlayed onto a blot with recombinant RCP. The binding of Rab11a was assessed with an anti-rabbit IgG secondary antibody.

GTP-bound state. Thus, to assess whether this is a generalizable methodology, we overlayed GTPγS loaded Rab11a onto blots with recombinant RCP, another member of the Rab11-FIPs. As shown in Fig. 1, Rab11a binding is also seen for RCP. These results indicate that the far-Western approach is a feasible, reliable method to use when assessing binding partners of Rab11a.

Analysis of Mysoin Vb Tail Binding with Rab11-FIP2

We have previously demonstrated by yeast two-hybrid analysis that Rab11-FIP2 interacts with the tail region of myosin Vb lacking the motor domain. We have now used overlays and a far-Western approach to assess binding *in vitro* between myosin Vb tail and Rab11-FIP2.

1. One microgram of His-tagged proteins was resolved on 10% SDS-PAGE gels.
2. Proteins were transferred to nitrocellulose for 2 h at 250 mA or overnight at 100 mA with SDS in transfer buffer.
3. The blots were blocked in 5% nonfat dry milk in TBS-Tween (0.05%) for a minimum of 30 min.
4. GST-myosin Vb tail (5 μg/ml) was incubated with a chicken anti-myosin Vb antibody (1:500) for 1 h at room temperature in 2.5% nonfat dry milk in TBS-Tween (0.05%).
5. This mixture was added to the blot and incubated overnight at 4°.
6. Blots were washed three times with TBS-Tween (0.05%).
7. Blots were incubated with HRP-conjugated anti-chicken antibody for 1 h at room temperature.
8. Blots were washed three times with TBS-Tween (0.05%) and visualized by ECL.

Figure 2A demonstrates that the myosin Vb tail interacted with Rab11-FIP2 in far-Western blot overlays. The multiple bands observed reflect the rapid breakdown of recombinant Rab11-FIP2 following isolation. Association of myosin Vb tail with Rab11 FIP2 breakdown fragments supports its interaction with discrete regions of Rab11-FIP2.

Analysis of Rab11-FIP2 Binding to Rab11-FIP2

Rab11-FIP2 homodimerizes in parallel and antiparallel conformations (Wallace *et al.*, 2002a). We therefore sought to evaluate Rab11-FIP2 dimerization using blot overlays.

1. GST-tagged proteins (1 μg) were resolved on 10% SDS-PAGE gels.
2. Proteins were transferred for 2 h at 250 mA or overnight at 100 mA with SDS in transfer buffer.

A   B

Protein overlay: GST-myosin Vb tail   His-Rab11-FIP2

Primary: Chicken anti-myosin Vb   S-tag HRP

Secondary: HRP anti-chicken

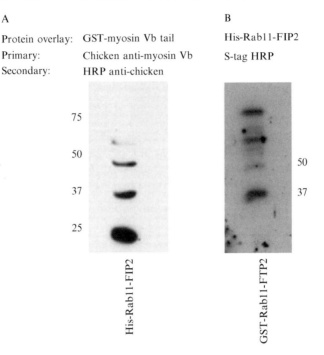

FIG. 2. Rab11-FIP2 interacts with myosin Vb and self-dimerizes *in vitro*. Recombinant Rab11-FIP2 was resolved by SDS-PAGE and transferred to nitrocellulose. (A) GST-myosin Vb tail was incubated with a chicken anti-myosin Vb antibody and then incubated with the blot overnight. The interaction was assessed by anti-chicken secondary antibodies. (B) His-Rab11-FIP2 was incubated with the blot overnight. The dimerization was assessed by the recognition of the S-tag on the His-tagged protein.

3. The blots were blocked in 5% nonfat dry milk in TBS-Tween (0.05%) for a minimum of 30 min.

4. His-Rab11-FIP2 (50 μg) was diluted in 2.5% nonfat dry milk in TBS-Tween (0.05%).

5. This mixture was added to the blot and incubated overnight at 4°.

6. Blots were washed three times with TBS-Tween (0.05%).

7. Blots were incubated with S-Tag HRP (1:5000) for 1 h at room temperature.

8. Blots were washed three times with TBS-Tween (0.05%) and visualized by ECL.

Figure 2B demonstrates that the Rab11-FIP2 homodimerizes in blot overlays. As noted earlier, Rab11-FIP2 dimerized with multiple breakdown products present in the recombinant GST-Rab11-FIP2 preparation.

Comments

An understanding of the interactions of proteins associated with recycling system function is critical for an appreciation of the mechanisms that regulate aspects of vesicle trafficking. Present studies support a role for the assembly of multiprotein complexes coordinated around Rab11a. The present methodologies allow for the dissection of protein interactions using protein overlays and far-Western methodologies without employing radioactivity. These techniques for far-Western methodologies using bacterial expression protein rely on the ability of protein association domains to fold into their active conformation following resolution on SDS-PAGE gels and transfer to nitrocellulose membranes. In the case of the Rab11a binding motif in Rab11-FIP2, the amphipathic α-helix responsible for this interaction should assemble with a large high free energy, thus accounting for the ability to perform successful protein overlays. Delineation of similar interactions between Rab11a and myosin Vb has not been successful, indicating that the binding motifs on myosin Vb may be more complex. Similarly, the ability of Rab11-FIP2 to dimerize and interact with myosin Vb tail in protein overlays also implies a high free energy for reconstitution of binding motifs. Although the binding site for dimerization appears to require both the C2 domain and the carboxyl-terminal region of the protein, the precise structural requirements for dimerization or heterooligomerization with other Rab11-FIP proteins remains unclear.

Acknowledgments

These investigations were supported by Grants AR49311 and DK48370 from the National Institutes of Health.

References

Brock, S. C., Goldenring, J. R., and Crowe, J. E., Jr. (2003). Apical recycling systems regulate directional budding of respiratory syncytial virus from polarized epithelial cells. *Proc. Natl. Acad. Sci. USA* **100**, 15143–15148.

Casanova, J. E., Wang, X., Kumar, R., Bhartur, S. G., Navarre, J., Woodrum, J. E., Altschuler, Y., Ray, G. S., and Goldenring, J. R. (1999). Association of Rab25 and Rab11a with the apical recycling system of polarized Madin-Darby canine kidney cells. *Mol. Biol. Cell* **10,** 47–61.

Cullis, D. N., Philip, B., Baleja, J. D., and Feig, L. A. (2002). Rab11-FIP2, an adaptor protein connecting cellular components involved in internalization and recycling of epidermal growth factor receptors. *J. Biol. Chem.* **277,** 49158–49166.

Fan, G. H., Lapierre, L. A., Goldenring, J. R., and Richmond, A. (2003). Differential regulation of CXCR2 trafficking by Rab GTPases. *Blood* **101,** 2115–2124.

Fan, G.-H., Lapierre, L. A., Goldenring, J. R., Sai, J., and Richmond, A. (2004). Rab11-family interacting protein 2 and myosin Vb are required for CXCR2 recycling and receptor-mediated chemotaxis. *Mol. Biol. Cell.* **15,** 2456–2469.

Green, E. G., Ramm, E., Riley, N. M., Spiro, D. J., Goldenring, J. R., and Wessling-Resnick, M. (1997). Rab11 is associated with transferrin-containing recycling compartments in K562 cells. *Biochem. Biophys. Res. Commun.* **239,** 612–616.

Hales, C. M., Griner, R., Hobdy-Henderson, K. C., Dorn, M. C., Hardy, D., Kumar, R., Navarre, J., Chan, E. K., Lapierre, L. A., and Goldenring, J. R. (2001). Identification and characterization of a family of Rab11-interacting proteins. *J. Biol. Chem.* **276,** 39067–39075.

Hales, C. M., Vaerman, J. P., and Goldenring, J. R. (2002). Rab11 family interacting protein 2 associates with myosin Vb and regulates plasma membrane recycling. *J. Biol. Chem.* **277,** 50415–50421.

Kessler, A., Tomas, E., Immler, D., Meyer, H. E., Zorzano, A., and Eckel, J. (2000). Rab11 is associated with GLUT4-containing vesicles and redistributes in response to insulin. *Diabetologia* **43,** 1518–1527.

Lapierre, L. A., Kumar, R., Hales, C. M., Navarre, J., Bhartur, S. G., Burnette, J. O., Provance, D. W., Jr., Mercer, J. A., Bahler, M., and Goldenring, J. R. (2001). Myosin vb is associated with plasma membrane recycling systems. *Mol. Biol. Cell* **12,** 1843–1857.

Lindsay, A. J., Hendrick, A. G., Cantalupo, G., Senic-Matuglia, F., Goud, B., Bucci, C., and McCaffrey, M. W. (2002). Rab coupling protein (RCP), a novel Rab4 and Rab11 effector protein. *J. Biol. Chem.* **277,** 12190–12199.

Mammoto, A., Ohtsuka, T., Hotta, I., Sasaki, T., and Takai, Y. (1999). Rab11BP/rabphilin-11, a downstream target of rab11 small G protein implicated in vesicle recycling. *J. Biol. Chem.* **274,** 25517–25524.

Meyers, J. M., and Prekeris, R. (2002). Formation of mutually exclusive Rab11 complexes with members of the family of Rab11-interacting proteins regulates Rab11 endocytic targeting and function. *J. Biol. Chem.* **277,** 49003–49010.

Prekeris, R., Klumperman, J., and Scheller, R. H. (2000). A Rab11/Rip11 protein complex regulates apical membrane trafficking via recycling endosomes. *Mol. Cell* **6,** 1437–1448.

Rizo, J., and Sudhof, T. C. (1998). C2-domains, structure and function of a universal Ca2+-binding domain. *J. Biol. Chem.* **273,** 15879–15882.

Tsukita, S., and Yonemura, S. (1999). Cortical actin organization: Lessons from ERM (Ezrin/Radixin/Moesin) proteins. *J. Biol. Chem.* **274,** 34507–34510.

Ullrich, O., Reinsch, S., Urbe, S., Zerial, M., and Parton, R. G. (1996). Rab11 regulates recycling through the pericentriolar recycling endosome. *J. Cell Biol.* **135,** 913–924.

Volpicelli, L. A., Lah, J. J., Fang, G., Goldenring, J. R., and Levey, A. I. (2002). Rab11a and myosin Vb regulate recycling of the M4 muscarinic acetylcholine receptor. *J. Neurosci.* **22,** 9776–9784.

Wallace, D. M., Lindsay, A. J., Hendrick, A. G., and McCaffrey, M. W. (2002a). The novel Rab11-FIP/Rip/RCP family of proteins displays extensive homo- and hetero-interacting abilities. *Biochem. Biophys. Res. Commun.* **292,** 909–915.

Wallace, D. M., Lindsay, A. J., Hendrick, A. G., and McCaffrey, M. W. (2002b). Rab11-FIP4 interacts with Rab11 in a GTP-dependent manner and its overexpression condenses the Rab11 positive compartment in HeLa cells. *Biochem. Biophys. Res. Commun.* **299,** 770–779.

Zeigerer, A., Lampson, M. A., Karylowski, O., Sabatini, D. D., Adesnik, M., Ren, M., and McGraw, T. E. (2002). GLUT4 retention in adipocytes requires two intracellular insulin-regulated transport steps. *Mol. Biol. Cell* **13,** 2421–2435.

Zeng, J., Ren, M., Gravotta, D., De Lemos-Chiarandini, C., Lui, M., Erdjument-Bromage, H., Tempst, P., Xu, G., Shen, T. H., Morimoto, T., Adesnik, M., and Sabatini, D. D. (1999). Identification of a putative effector protein for rab11 that participates in transferrin recycling. *Proc. Natl. Acad. Sci. USA* **96,** 2840–2845.

[62] Interactions of Myosin Vb with Rab11 Family Members and Cargoes Traversing the Plasma Membrane Recycling System

By Lynne A. Lapierre and James R. Goldenring

Abstract

Myosin Vb interacts with Rab11 family members and is the major motor protein identified in association with plasma membrane recycling systems in nonpolarized and polarized cells. Using yeast two-hybrid binary screens, we demonstrated that dominant active Rab25S21V fails to interact with myosin Vb, but does interact with Rab11-FIP2. Transfection of DsRed2-myosin Vb tail in MDCK cell lines stably transfected with wild-type or dominant active forms of Rab11a or Rab25 demonstrated that the distribution of Rab25S21V is only partially altered by expression of the myosin Vb tail. Finally, we demonstrate EGF-dependent sequestration of internalized EGF receptor by EGFP-myosin Vb in A431 cells. Expression of myosin Vb tail represents a powerful provocative test for the trafficking of cargoes through Rab11-containing plasma membrane recycling systems.

Introduction

Members of the family of myosin V proteins are plus end actin motors that move vesicles toward the cell periphery. There are three members of this family, myosin Va, Vb, and Vc. Myosin Va interacts with Rab27a and is involved with regulation of melanosome and secretory lysosome trafficking (Wu *et al.*, 2001). Myosin Vc also interacts with trafficking vesicles, but

METHODS IN ENZYMOLOGY, VOL. 403
0076-6879/05 $35.00
DOI: 10.1016/S0076-6879(05)03062-4

there identity is unclear (Rodrigues and Cheney, 2001). Myosin Vb interacts with Rab11a and is a critical regulator of vesicle trafficking through the plasma membrane recycling system in both nonpolarized and polarized cells (Lapierre et al., 2001). The interaction of Rab11a and myosin Vb was originally identified through the screening of a rabbit gastric parietal cell yeast two-hybrid library with a bait for dominant active Rab11aS20V (Lapierre et al., 2001). Subsequently, transfection of the motor-deficient tail region of myosin Vb has served as a powerful dominant negative inhibitor of trafficking through plasma membrane recycling systems (Fan et al., 2004; Volpicelli et al., 2002). In this chapter we discuss a combination of in situ association assays using both yeast two-hybrid assays and cultured cell transfection methods to analyze putative interactions of myosin Vb with Rab11 family members as well as to assess the trafficking of a specific cargo, the epidermal growth factor (EGF) receptor, through Rab11a-containing plasma membrane systems.

Use of Yeast Two-Hybrid Assays to Identify Interacting Proteins for the Myosin Vb Tail

Yeast two-hybrid methods have served as critical discovery tools to elucidate the interactions of myosin Vb with Rab11a as well as with Rab11-FIP2 (Hales et al., 2002; Lapierre et al., 2001). In addition to discovery of interactions, binary screening of yeast two-hybrid interactions between defined bait and target constructs allows simple evaluation of protein–protein interactions in a system where expressed proteins achieve more reliable folding within yeast cells. We have utilized binary yeast two-hybrid assays to evaluate the interaction of myosin Vb with wild-type and mutant forms of Rab11 family members.

Constructs

The rabbit myosin Vb tail construct was isolated from a rabbit gastric parietal cell cDNA yeast two-hybrid library constructed in the pADGAL (Stratagene) by its interaction with a Rab11a dominant active mutant (Rab11aS20V) (Lapierre et al., 2001). This truncation consists of the last 588 amino acids of myosin Vb and encodes the globular tail and the coiled-coil domain but does not contain the motor domain. Wild-type Rab11a, Rab11b, and Rab25 were amplified with Pfu polymerase using sense and antisense primers containing restriction sites (EcoRI and SalI for Rab11a and Rab25 and SalI for Rab11b) and the amplified products were ligated into pBDGAL(CaM) (Stratagene). The Rab11aS20V, Rab11aS25N, Rab11aN124I, Rab11bS20V, Rab11bS25N, Rab11bN124I, Rab25S21V, and RabT26N mutations were constructed by single base-pair replacements in the pBDGAL(CaM) wild-type constructs using a two complementary oligonucleotide method with Pfu polymerase (Stratagene). The Rab11-FIP constructs in pBDGAL(Cam) were as previously described (Hales *et al.*, 2001).

Yeast Two-Hybrid Binary Screening

Yeast two-hybrid binary screens were performed as follows. The yeast strain Y190 was grown overnight in YPD medium (1% yeast extract, 2% peptone, 2% dextrose). The next day the yeast were harvested by centrifugation at $1000 \times g$ for 5 min, washed once with sterile water, then resuspended in LA/TE solution (100 mM lithium acetate, 10 mM Tris–HCl, 1 mM EDTA, pH 7.5) at a concentration of 20 μl of LA/TE for every 2 ml of yeast culture and incubated for 5 min at room temperature (RT). The reaction mixtures were set up in microcentrifuge tubes consisting of 200 ng each of both the pBDGAL(CaM) and the pADGAL construct plasmids, 20 μg of heat-denatured salmon sperm DNA, 120 μl of LA/TE/PEG solution (LA/TE solution with 40% polyethylene glycol [PEG] 4000), and 20 μl of the yeast/LA/TE solution. All solutions were made fresh the day of use. The yeast/DNA mixture was mixed well by inversion and placed horizontally at 30° for 30 min with shaking. The yeast were heat shocked at 42° for 15 min and pelleted at $1000 \times g$ for 30 s. The yeast were resuspended in 100 μl of sterile water and plated onto tryptophan- and leucine-deficient medium (yeast nitrogen base supplemented with all amino acids except for tryptophan and leucine). The plates were incubated for 3 days at 30°.

β-Galactosidase Assay

Colonies were lifted onto 75-mm filter paper discs, frozen twice in liquid nitrogen allowing them to completely thaw in between. A second piece of filter paper was placed into a 100-mm Petri dish and presoaked with 1.5 ml of fresh X-Gal solution 1.67 ml of X-Gal stock (50 mM 5-bromo-4-chloro-3-indolyl β-D-galactoside) diluted in 100 ml of Z buffer (60 mM Na$_2$HPO$_4$, 40 mM NaH$_2$PO$_4$, 10 mM KCl, 1 mM MgSO$_4$). The yeast lifted filter paper was laid onto the presoaked paper colony side up and incubated at RT; a blue color was observed to develop from 1 h to overnight. Positive interactions were indicated by the development of blue color and classified as +++ for 1–2 h, ++ for 3–4 h, + for 5–18 h, and the results were negative if there was no blue after 18 h.

Table I shows interactions between the myosin Vb tail and members of the Rab11 family of small GTPases. Myosin Vb tail interacts strongly (indicated by +++) with the wild-type of Rab11a, Rab11b, and Rab25, as well as the GTPase-deficient mutation (DA; dominant active) of Rab11a and Rab11b (S20V mutants); interestingly this mutation in Rab25 (S21V) did not interact with the myosin Vb tail. Two classes of dominant-negative forms of the Rabs were used: GDP-bound (Rab11aS25N, Rab11bS25N, and Rab25T26N) and the nucleotide-free (Rab11aN124I and Rab11b N124I) mutants. While both classes of DN mutants of Rab11a and the nucleotide-free mutant of Rab11b (Rab11bN124I) interacted to some degree with myosin Vb tail, neither the Rab11bS25N nor the Rab25T26N did.

A group of related proteins has been shown to interact with the Rab11 family, the Rab11 family of interacting proteins (Rab11-FIPs), pp75/RIP11, and rab coupling protein (RCP). Myosin Vb tail interacted only with Rab11-FIP2 of the Rab11-FIP family and did not interact with pp75/RIP11 (Hales *et al.*, 2002). We have looked at Rab11-FIP2's ability to interact with the Rab11 family of small GTPases and their mutants (Table I). All interacted strongly with Rab11-FIP2 except for two of the GDP-bound mutants, Rab11aS25N, which had a medium interaction (++), and Rab25T26N, which had a very weak interaction (+).

Assay of Myosin Vb Tail Interactions with Rab11 Family Members in Transfected Madin–Darby Canine Kidney (MDCK) Cells

In some cases, interactions observed in yeast two-hybrid assays are not confirmed *in situ*. The results in yeast two-hybrid studies indicated that myosin Vb did not interact with the constitutively GTP-bound Rab25S21V. Thus, to provide a second assay of myosin Vb interactions with Rab11

TABLE I

Yeast Two-Hybrid Interaction of Myosin Vb Tail with Rab11a, Rab11b, Rab25, and Their Mutants

	Myosin Vb tail[a]	Rab11-FIP2[a]
Rab11a	+++	+++
Rab11aS20V	+++	+++
Rab11aS25N	++	++
Rab11aN124I	+++	+++
Rab11b	+++	+++
Rab11bS20V	+++	+++
Rab11bS25N	−	+++
Rab11bN124I	+++	+++
Rab25	+++	+++
Rab25S21V	−	+++
Rab25T26N	−	+

[a] The time of blue development is as indicated +++, 1–2 h; ++, 3–4 h; +, 5–18 h; and −, no blue after 18 h.

family members, we assessed the interaction of the myosin Vb tail with the wild-type and DA mutants of Rab11 family members in MDCK cell lines stably expressing the wild-type and DA mutants of Rab11a and Rab25. We transiently transfected the four cell lines with a DsRed2 chimera of myosins Vb tail. Rab11b-expressing lines were not used because EGFP-Rab11b artifactually displaces endogenously expressed Rab11a from recycling system vesicles compartment (Lapierre *et al.*, 2003).

EGFP-Rab11a, EGFP-Rab11aS20V, and EGFP-Rab25 all localized to a perinuclear region of the cell and all colocalized with the transfected DsRed2-myosin Vb tail (Fig. 1). Indeed, the colocalized EGFP-Rab

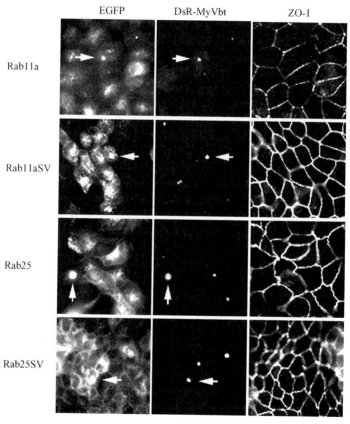

Fig. 1. Overexpression of the myosin Vb tail causes redistribution of EGFP-Rab-11a, EGFP-Rab11aS20V, and EGFP-Rab25, but only partial redistribution of EGFP-Rab25S21V. Stable MDCK cell lines of the EGFP-tagged Rabs were transiently transfected with DsRed2-myosin Vb tail (DsR-MyVbT). Fixed cells were stained with antibodies against ZO-1. The arrows indicate the positions of dual transfected cells.

chimeras in the DsRed2-myosin Vb tail-transfected cells were much brighter than in the neighboring non-dual-transfected cells. EGFP-Rab25S21V localized to vesicles in the perijunctional region of the cell. In cells cotransfected with the DsRed2-myosin Vb tail, a large proportion of the EGFP-Rab25S21V remained out at the periphery. Some EGFP-Rab25S21V colocalized with the DsRed2-myosin Vb tail in the perinuclear region. However, in the case of the other Rab constructs, DsRed2-myosin Vb tail appeared to sequester the vast majority of EGFP-Rab-containing vesicles into the central tubular nidus. These results provide *in situ* confirmation of the yeast two-hybrid studies, suggesting an altered interaction of myosin Vb tail with the Rab25S21V.

Assessment of EGF Receptor Trafficking Through the Rab11a and Myosin Vb-Containing Recycling System

The myosin Vb tail provides a powerful dominant-negative influence on trafficking through the Rab11a-containing plasma membrane recycling system. Thus, recycling cargoes can enter the tubulovesicular recycling system, but they cannot exit. Rab11a appears to identify a slow recycling pathway through which cargoes traverse back to the plasma membrane over a 60–90 min period. Other recycling pathways likely exist including those for rapid recycling of internalized cargoes. These other pathways are not regulated by Rab11a and myosin Vb, so expression of myosin Vb tail is a powerful provocative test for the assessment of the pathway responsible for the recycling of particular cargoes. We have used transfection of fluorescent chimeras of myosin Vb to determine that a number of recycling cargoes including transferrin receptors in nonpolarized cells, polymeric IgA receptors (Lapierre *et al.*, 2001), M4-muscarinic receptors (Volpicelli *et al.*, 2002), and CXCR-2 (Fan *et al.*, 2004) all traffic through the plasma membrane recycling system regulated by Rab11a and myosin Vb. A number of investigations have suggested that EGF receptor recycles to the plasma membrane following ligand-dependent internalization (for review see Harris *et al.* [2003]). To assess the pathway for recycling we evaluated the effects of EGFP-myosin Vb tail expression on EGF receptor recycling in A431 cells.

EGFP-Rab Constructs and Cell Lines

The wild-type and mutant Rab11a and Rab25 inserts were removed from the pBDGAL(CaM) vector using *Eco*RI and *Sal*I and cloned into pEGFP-C2 (Clontech). MDCK cells were transfected in suspension with Effectine (Qiagen) following the manufacturer's protocol for a 10-cm dish. The cells were allowed to recover and the next day the cells were resuspended, serially diluted, and replated in 15-cm dishes supplanted with 0.5 mg/ml of G418. The cells were allowed to grow for 5–8 days until colonies were observed. Individual EGFP-expressing colonies were identified under fluorescence microscopy, harvested, and transferred to 24-well plates for expansion. Three independent cell lines for each Rab were characterized. Cell lines were grown in 10% fetal bovine serum (FBS), Dulbecco's modified Eagle's medium (DMEM) supplemented with G418 (0.25 mg/ml).

Myosin Vb Tail Constructs and Transfections

Myosin Vb tail was amplified with Pfu polymerase using sense and antisense primers containing (*Bam*HI and *Sal*I) restriction sites and the amplified products were ligated into pDsRed2-C1 (Clontech). The pDsRed2-myosin Vb tail vectors were transiently transfected into the GFP-Rab cell lines in suspension using Effectine, following the manufacturer's protocol for 12-well plates. The transfections were plated onto 12-mm, 0.4-mm pore clear Transwells (Costar), refed the next day, and allowed to grow and polarize for 4 days.

Myosin Vb tail was amplified with Pfu polymerase using sense and antisense primers containing (*Bam*HI and *Sal*I) restriction sites and the amplified products were ligated into pEGFP-C1 (Clontech) (Lapierre *et al.*, 2001). A431 cells were seeded onto 25-mm coverslips and grown in 10% FBS DMEM to 80% confluence. They were then transiently transfected with Effectine using the manufacturer's protocol for a six-well plate; after 5 h, the medium was removed and replaced with 10% FBS DMEM; after a further 19 h (total of 24 h after transfection), EGF (stock 2.5 μg/ml; Peprotech, Rocky Hill, NJ) was added to the wells at a final concentration of 10 ng/ml. Cells were washed and fixed at either 0 or 60 min after EGF addition.

Immunofluorescence

Cells were washed in phosphate-buffered saline (PBS), fixed in 4% paraformaldehyde-PBS for 20 min at RT, then blocked and extracted by incubating in 10% normal donkey serum, 0.3% Triton X-100 PBS for 30 min at RT. Cells were stained with Rat anti-ZO-1 (1:200, Chemicon) to visualize tight junctions (Rab11 cell lines) or rabbit anti-EGF receptor (1:100; Upstate) and mouse anti-Rab11a (purified IgG 1:25 [Goldenring *et al.*, 1996]) (A431 cells), followed by species-specific secondary antibodies, Cy5-labeled donkey anti-rat or Cy5-labeled donkey anti-mouse and Cy3-labeled donkey anti-rabbit (all 1:200 Jackson Immunoresearch) and

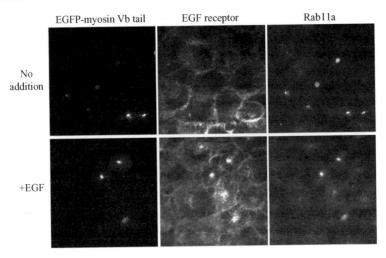

FIG. 2. Overexpression of the myosin Vb tail sequesters recycling EGF receptor. A413 cells were transiently transfected with EGFP-myosin Vb tail, then treated with 10 ng/ml of EGF for 0 (no addition) or 60 min. The cells were fixed and costained for the EGF receptor and Rab11a. EGF treatment caused internalization and sequestration of EGF receptor with EGFP-myosin Vb tail and Rab11a.

mounted in Prolong anti-fade (Molecular Probes). All cells were visualized with a Zeiss Axiophot epifluorescence microscope.

In serum-starved A431 cells, the endogenous Rab11a colocalized with the transfected myosin Vb tail, while the EGF receptor was present on the plasma membrane (Fig. 2). Incubation with EGF caused internalization of the EGF receptor, and after 60-min the internalized EGF receptor colocalized with EGFP-myosin Vb tail and Rab11a in a compacted perinuclear tubular compartment. These results indicate that in A431 cells the EGF receptor traffics through a pathway that is controlled by Rab11a and uses the myosin Vb motor similar to the CXCR2 receptor (Fan *et al.*, 2003, 2004).

Acknowledgments

These investigations were supported by Grants AR49311 and DK48370 from the National Institutes of Health.

References

Fan, G.-H., Lapierre, L. A., Goldenring, J. R., and Richmond, A. (2003). Differential regulation of CXCR2 trafficking by Rab GTPases. *Blood* **101**, 2115–2124.

Fan, G.-H., Lapierre, L. A., Goldenring, J. R., Sai, J., and Richmond, A. (2004). Rab11-family interacting protein 2 and myosin Vb are required for CXCR2 recycling and receptor-mediated chemotaxis. *Mol. Biol. Cell* **15,** 2456–2469.

Goldenring, J. R., Smith, J., Vaughan, H. D., Cameron, P., Hawkins, W., and Navarre, J. (1996). Rab11 is an apically located small GTP-binding protein in epithelial tissues. *Am. J. Physiol.* **270**(33), G515–G525.

Hales, C. M., Griner, R., Dorn, M. C., Hardy, D., Kumar, R., Navarre, J., Chan, E. K. C., Lapierre, L. A., and Goldenring, J. R. (2001). Identification and characterization of a family of Rab11 interacting proteins. *J. Biol. Chem.* **276,** 39067–39075.

Hales, C. M., Vaerman, J.-P., and Goldenring, J. R. (2002). Rab11 family interacting protein 2 associates with myosin Vb and regulates plasma membrane recycling. *J. Biol. Chem.* **277,** 50415–50421.

Harris, R. C., Chung, E., and Coffey, R. J. (2003). EGF receptor ligands. *Exp. Cell Res.* **284,** 2–13.

Lapierre, L. A., Kumar, R., Hales, C. M., Navarre, J., Bhartur, S. G., Burnette, J. O., Provance, J. D. W., Mercer, J. A., Bahler, M., and Goldenring, J. R. (2001). Myosin Vb is associated with and regulates plasma membrane recycling systems. *Mol. Biol. Cell* **12,** 1843–1857.

Lapierre, L. A., Dorn, M. C., Zimmerman, C. F., Navarre, J., Burnette, J. O., and Goldenring, J. R. (2003). Rab11b resides in a vesicular compartment distinct from Rab11a in parietal cells and other epithelial cells. *Exp. Cell Res.* **290,** 322–331.

Rodrigues, O. C., and Cheney, R. E. (2001). Human myosin-Vc is a novel class V myosin expressed in epithelial cells. *J. Cell Sci.* **115,** 991–1004.

Volpicelli, L. A., Lah, J. J., Fang, G., Goldenring, J. R., and Levey, A. I. (2002). Rab11a and myosin Vb regulate recycling of the M4 muscarinic acetylcholine receptor. *J. Neurosci.* **22,** 9776–9784.

Wu, X., Rao, K., Bowers, M. B., Copeland, N. G., Jenkins, N. A., and Hammer, J. A. I. (2001). Rab27a enables myosin Va-dependent melanosome capture by recruiting the myosin to the organelles. *J. Cell Sci.* **114,** 1091–1100.

[63] Properties of Rab13 Interaction with Protein Kinase A

By AHMED ZAHRAOUI

Abstract

The small GTPase Rab13 localizes to tight junctions in epithelial cells and regulates the recruitment of claudin1 and ZO-1, two proteins required for the assembly of functional tight junctions. Rab13 directly binds to the α-catalytic subunit of protein kinase A (PKA α cat) and reversibly inhibits PKA-dependent phosphorylation of vasodilator-stimulated phosphoprotein (VASP), a key actin cytoskeletal remodeling protein. The inhibition of VASP phosphorylation abolishes the targeting of VASP to cell–cell junctions, which in turn leads to a delay in the recruitment of claudin1

METHODS IN ENZYMOLOGY, VOL. 403   0076-6879/05 $35.00

and ZO1 into tight junctions. Consequently, tight junctions formed in epithelial cells expressing the GTP-bound Rab13 are structurally disorganized and functionally leaky for small molecules (A. M. Marzesco *et al.* [2002]. *Mol. Biol. Cell* **13**, 1819–1831; K. Kohler *et al.* [2004]. *J. Cell Biol.* **165**, 175–180). Our data provide the first direct link between activation of small GTPases and the recruitment of cytoskeletal modulators into tight junctions. Here, we describe different procedures we used to demonstrate that Rab13 interacts with PKA and reversibly controls phosphorylation and recruitment of VASP.

Introduction

Small GTPase Rab proteins act in conjuction with a variety of protein effectors to regulate different steps of exocytic and endocytic pathways. They regulate specific membrane trafficking events including vesicle formation, motility via kinesins or myosins, tethering, and fusion (Pfeffer, 2001; Zerial and McBride, 2001). The small GTPase Rab13 is closely related to mammalian Rab8 and Rab10 and to the yeast Sec4 involved in polarized secretion during the budding process (Guo, 1999; Zahraoui *et al.*, 1994). Rab13 is recruited from a cytoplasmic pool to tight junctions at an early stage during tight junction assembly (Sheth *et al.*, 2000). Tight junction proteins are involved in cell–cell adhesion and cell polarity, proliferation, and differentiation (Matter and Balda, 2003; Zahraoui *et al.*, 2000). The tight junction transmembrane proteins, claudins, are required for the establishment of tight junction diffusion barriers (Tsukita and Furuse, 2000). Scaffolding proteins such as ZO-1 (zonula occludens 1), a PDZ (*PSD95, D*lg. ZO-1)-containing protein, connect transmembrane proteins to the underlying actin cytoskeleton and recruit cytosolic proteins, such as kinases, and transcription factors (Balda *et al.*, 2003; Itoh *et al.*, 1999; Wittchen *et al.*, 1999). Vasodilator-stimulated phosphoprotein (VASP), a protein required for actin polymerzation, is also necessary for epithelial tight junction assembly (Lawrence *et al.*, 2002). VASP is a substrate for both protein kinase A (PKA) and PKG, and phosphorylation of VASP abolishes its interaction with actin and inhibits actin polymerization (Harbeck *et al.*, 2000). However, the regulatory pathways that link actin rearrangement to tight junctions are still unclear.

Expression of the active (Rab13Q67L) mutant of Rab13 in epithelial Madin–Darby canine kidney (MDCK) cells delays the recruitment of claudin1 and ZO-1, two proteins required for the assembly of tight junctions. Interestingly, expression of Rab13Q67L does not impair the recruitment of E-cadherin (an adherens junction protein) to the lateral membrane. In contrast, the inactive Rab13T22N mutant does not alter tight junction assembly. Our results suggest that Rab13 plays an important role in coordinating the recruitment of claudin1/ZO-1 to specific domains on the lateral membrane (Marzesco *et al.*, 2002). Strikingly, expression of the GTP-bound Rab13 inhibits PKA-dependent phosphorylation and tight junction recruitment of VASP, an actin remodeling protein. Rab13GTP directly binds to PKA and inhibits its activity. Activation of PKA abolishes the inhibitory effect of Rab13 on the recruitment of VASP, claudin1, and ZO-1 to cell–cell junctions. Rab13 is therefore the first GTPase that controls PKA activity required for tight junction assembly (Kohler *et al.*, 2004). We present here different techniques used to show that Rab13 interacts with the PKA α-catalytic subunit and inhibits its activity both *in vitro* and *in vivo*.

Methods

Anti-VASP monoclonal antibody is from BD Transduction Laboratories (San Diego, CA) and anti-PKA α-cat antibodies from Santa Cruz Biotechnology (Santa Cruz, CA). Secondary antibodies coupled to Cy3 are from Jackson Immunoresearch Laboratories (West Grove, PA). Purified PKA α-catalytic subunit and the PKA peptide inhibitor, PKI, are purchased from Sigma.

Cell Culture

MDCK cells (clone II) are cultured in Dulbecco's modified Eagle's medium (DMEM) supplemented with 10% fetal calf serum, 2 mM glutamine, 100 U/ml penicillin, and 10 mg/ml streptomycin. The cells are incubated at 37° under 10% CO_2 atmosphere. Isolation of MDCK cells stably expressing GFP, GFP-Rab13T22N, or GFP-Rab13Q67L is described in Chapter 62 (this volume). For the calcium switch experiments, cells are cultured in Eagle's minimum essential medium (EMEM, from Bio-Whittaker, Belgium).

Calcium Switch Experiments

Epithelial cells plated at high density establish tight junctions in 12–15 h through a process initiated by the recruitment to the lateral membrane of E-cadherin, a calcium-binding transmembrane protein. In the absence of calcium, epithelial cells make no cell–cell junctions. Subsequent restoration of a physiological level of calcium results in the synchronous *de novo* assembly of adherens and tight junctions with faster kinetics. This assay, called a calcium switch protocol, can be used as a cellular model system to study cell–cell junction assembly. To examine the role of Rab13 on the recruitment of tight junction proteins during the assembly of cell–cell junctions, we perform the calcium switch experiments as described in the following:

1. Plate MDCK cells at 500,000 cells/cm^2 and incubate them in EMEM (a medium without calcium) for 15 h. Under these conditions, cells attach to the substrate but do not form cell–cell junctions.
2. Cells are then rinsed with normal DMEM, and incubated in DMEM at 37° under a 10% CO_2 atmosphere for 0, 4, or 6 h.
3. Cells are then washed in PBS containing 0.5 mM $MgCl_2$ and 1 mM $CaCl_2$ and analyzed by immunofluorescence or immunoblotting (Figs. 1 and 2).

Immunofluorescence Microscopy

Immunofluorescence is performed as previously described (Marzesco *et al.*, 2002). Cells are fixed with 3% paraformaldehyde for 15 min. They are permeabilized with 0.5% Triton X-100 in phosphate-buffered saline (PBS) for 15 min and then blocked in PBS buffer containing 0.5% Triton

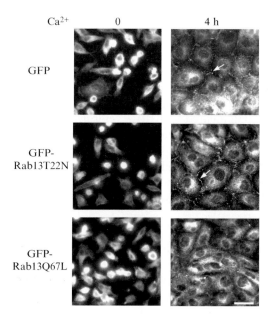

Fig. 1. The recruitment of VASP at 0 and 4 h after induction of cell–cell contacts. In cells expressing GFP, GFP-Rab13T22N, and GFP-Rab13Q67L mutants, the absence of Ca^{2+} (time 0 h) results in the absence of cell–cell junctions and the distribution of VASP in the cytoplasm. Within 4 h after the addition of Ca^{2+}, VASP is detected at cell–cell junctions (arrows) in MDCK cells expressing GFP and GFP-Rab13T22N, but not in those expressing GFP-Rab13Q67L mutant. This indicated that Rab13Q67L inhibits the recruitment of VASP to sites of cell–cell contacts during junction assembly.

Ca²⁺   0   6 h

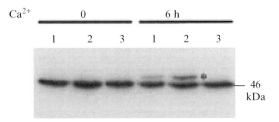

FIG. 2. PKA-dependent VASP phosphorylation induces a conformational change in VASP and a shift in SDS-PAGE mobility from 46 to 50 kDa. MDCK cells expressing (1) GFP, (2) GFP-Rab13T22N, or (3) GFP-Rab13Q67L are subjected to calcium switch experiments. Cell lysates are prepared at 0 and 6 h after addition of calcium in the medium and analyzed by immunoblot using the monoclonal anti-VASP antibody recognizing both the phosphorylated (50 kDa) and the nonphosphorylated (46 kDa) form of VASP. In the absence of calcium, only the nonphosphorylated VASP is detected, indicating that cell–cell junction formation is required for VASP phosphorylation. After addition of calcium, both nonphosphorylated and phosphorylated VASP (indicated by an asterisk) are detected in cells expressing GFP and GFP-Rab13T22N. In contrast, phosphorylated VASP is absent in cells expressing GFP-Rab13Q67L, indicating that GTP-bound Rab13 inhibits phosphorylation of VASP.

X-100 and 0.2% bovine serum albumin (BSA). All subsequent incubations with antibodies and washes are performed with this buffer. Cells are incubated 1 h at 4° with the anti-VASP antibodies, rinsed three times for 10 min with the blocking buffer, and then incubated with affinity purified secondary antibodies raised in goat and conjugated to Cy3. After 45 min cells are washed three times with PBS buffer containing 0.5% Triton X-100 and 0.2% BSA and three times with PBS. Cells are analyzed with a fluorescence microscope (Zeiss, Germany) and further processed with Adobe Photoshop Software (Adobe Systems, Mountain View, CA).

Immunoblotting

A total of 500,000 cells expressing GFP, GFP-Rab13T22N, or GFP-Rab13T22N are grown on 3.5-cm-diameter culture plates for 16 h, washed three times with ice-cold PBS, and extracted in 0.5% Triton, 10 mM Tris–HCl, pH 7.6, 120 mM NaCl, 25 mM KCl, 1.8 mM CaCl$_2$, 1 mM sodium vanadate, 50 mM NaF, and a mixture of protease inhibitors (Sigma) on a rocker platform for 30 min at 4°. Solubilized material is recovered by pelleting at 18,000×g for 15 min at 4°. Supernatants are collected and protein concentrations determined using the Bio-Rad protein assay kit (Bio-Rad Laboratories, Hercules, CA). Proteins are separated by sodium dodecyl sulfate polyacrylamide gel electrophoresis (SDS-PAGE) and transferred electrophoretically to immobilon filters. The filters are probed with polyclonal anti-VASP antibodies and revealed by enhanced chemiluminescent (ECL) substrate detection for horseradish peroxidase (HRP) (Pierce, Rockford, IL).

Incubation with Kinase Inhibitors or Activators

Three protein kinases, PKA, PKC, and PKG, can phosphorylate VASP. To determine which kinase is implicated, we use specific kinase inhibitors or activators and examine their effect on the Rab13-dependent phosphorylation of VASP. For this purpose, 500,000 cells are plated onto 3.5-cm-diameter culture plates overnight and treated with cell-permeable inhibitors/activators of protein kinases at 37° as follows: MDCK cells or cells expressing GFP and GFP-Rab13T22N are incubated 1 h with 30 μM PKA inhibitor H-89, 30 min with 10 μM PKC inhibitor, Rö-32–0432, and with 5 μM PKG inhibitor, KT5823 (Calbiochem, USA). MDCK cells expressing GFP-Rab13Q67L are incubated 30 min with 100 μM cAMP or cGMP (activator of PKG), or 20 min with 10 μM forskolin (Sigma); cAMP and forskolin activate PKA. Cells are washed three times in PBS and analyzed either by immunofluorescence or by immunoblotting procedures. In the calcium switch experiments, forskolin and H89 are added to cell culture at time 0 after addition of Ca^{2+}.

The treatment of cells expressing GDP- or GTP- bound Rab13 with inhibitors/activators also reveals that Rab13 reversibly controls VASP phosphorylation *in vivo*.

Immunoprecipitation of VASP

Epithelial MDCK cells expressing Rab13Q67L are grown for 3 days on 10-cm dishes and washed with PBS–1 mM Ca^{2+}, 0.5 mM MgCl$_2$. Cells are extracted in IP buffer (50 mM NaCl, 25 mM Tris, pH 8, 1 mM EDTA, 0.25% Triton, 1 mM sodium vanadate, 50 mM NaF, and protease inhibitors). After centrifugation at 18,000×g for 15 min at 4°, supernatants are recovered and incubated with 5 μg of anti-VASP antibody overnight at 4°. Protein G agarose beads are added for 2 h at 4° and the beads washed three times with IP buffer. Equal amounts of proteins are separated by SDS-PAGE and transferred electrophoretically to nitrocellulose filters. Filters are probed with anti-VASP antibodies before ECL detection according to the manufacturer's protocols (Pierce, Rockford, IL). Immunoprecipitated VASP is used in the kinase assay (see later discussion).

Expression of Rab13 in Escherichia coli and Purification

Two oligonucleotides, 5'-GAATTCCATGGCCAAAGCCTACGAC-3' and 5'-GTCGACTCAGCCCAGGGAGCACTT6-3', are used to create by polymerase chain reaction an *Eco*RI site upstream of the ATG initiator codon of Rab13 coding sequence and an *Sal*I downstream Rab13 stop codon, respectively. The *Eco*RI–*Sal*I fragment containing the complete Rab13 coding sequence is cloned into a pGEX-4T-3 expression vector. This vector carries the sequence encoding glutathione *S*-transferase (GST) adjacent to the cloning site and allows the expression of a GST-Rab13 fusion protein. It also contains a thrombin site that allows cleavage and elution of Rab13 protein. In addition, the procedure of purification of the GST fusion protein is rapid and leads to a highly pure protein (>95%).

GST-Rab13 fusion protein is expressed in *E. coli* and purified according to the manufacturer's protocol (Amersham Pharmacia, Uppsala, Sweden).

1. Inoculate a colony of recombinant pGEX-4T plasmid into 50 ml LB/ampicillin medium and grow overnight at 37° in a shaking incubator.
2. Dilute at 1:50 into 2 liters fresh LB/ampicillin medium and grow for 3 h until an OD of 0.6. Add 0.5 mM of isopropyl-β-D-thiogalactopyranoside (IPTG) to the culture and incubate for 1.5 h.
3. Collect bacteria by centrifugation for 10 min at 5000×g. Resuspend the pellet in 20 ml PBS on ice. Lyse cells by sonication, add 1% Triton X-100 to the lysate, and incubate on a rocker platform for 15 min. Centrifuge 10 min at 10,000×g.
4. Add the supernatant to 1 ml of 50% slurry glutathione beads that have been previously equilibrated in PBS. Rotate gently for 30 min. After washing three times with 100 ml cold PBS, the purity of the fusion protein retained on the beads can be analyzed by SDS-PAGE. Store the protein beads at 4° until utilization (for 1 mo).

GST Pull-Down Assay

Epithelial MDCK cells are grown on 10-cm dishes for 3 days, washed in PBS, and extracted in 10 mM Tris–HCl, pH 7.6, 120 mM NaCl, 25 mM KCl, 1.8 mM CaCl, 1 mM sodium vanadate, 50 mM NaF, 1% NP40, and a mixture of protease inhibitors (Sigma) on a rocker platform for 30 min at 4°. Solubilized material is recovered by pelleting at 18,000×g for 15 min at 4°. Supernatants are collected and protein concentration determined using the Bio-Rad protein assay kit (Bio-Rad Laboratories, Hercules, CA). Purified GST-Rab13 bound to glutathione beads are loaded with 1 mM GDP or GTPγS (a poorly hydrolyzable GTP analogue) for 90 min at room temperature in incubation buffer (100 mM NaCl, 20 mM Tris, 10 mM EDTA, 5 mM MgCl$_2$, and 1 mM dithiothreitol [DTT], pH 7.6). After washing with the same buffer, beads are incubated overnight with 10 mg of MDCK cell extracts at 4°. Rab13 beads are extensively washed and the retained proteins are subjected to SDS–PAGE and immunoblotted with antibodies against the α catalytic subunit of PKA. As a negative control, we use the GST-Rab6 beads (gift from B. Goud, Institut Curie). To prove the direct interaction of Rab13 (Fig. 3) with PKA, 100 μg of purified PKA catalytic subunit from bovine heart (Sigma) is incubated with a 10-fold molar excess of purified GST, GST-Rab13-GDP, or GST-Rab13-GTPγS proteins in the incubation buffer overnight at 4°. GST-Rab6 is used as negative control. After washing, the binding of PKA is determined by SDS-PAGE and Western blotting using polyclonal anti-PKA α-catalytic subunit. As a negative control, the filter is probed with anti-PKCζ antibodies, a protein kinase associated with tight junctions.

In Vitro *Kinase Assay*

cAMP-dependant protein kinase (PKA) is ubiquitous serine/threonine kinase. In the absence of cAMP, PKA is an inactive tetramer composed of two regulatory subunits and two catalytic subunits. Binding of cAMP to regulatory subunits induces dissociation of regulatory subunits from catalytic subunits and concomitant activation of catalytic subunits. Free catalytic subunits of PKA can phosphorylate a variety of cellular target proteins (Francis and Corbin, 1994). Different isoforms of regulatory and catalytic subunits have been identified. The recombinant α-catalytic subunit of 41 kDa corresponds to the predominant α-isoform, has a broad tissue distribution, and can be utilized for *in vitro* kinase assays.

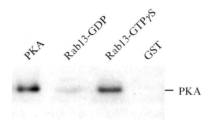

Fig. 3. Rab13 directly interacts with PKA. The purified α-catalytic subunit of PKA is incubated with GST or GST-Rab13 fusion protein preloaded with GDP or GTPγS. Proteins retained on GST beads are analyzed by SDS-PAGE and immunoblotted with antibodies against the α-catalytic subunit of PKA.

In Vitro *Phosphorylation of VASP Protein by the PKA* α *Catalytic Subunit*

Given that Rab13Q67L inhibits PKA-dependant VASP phosphorylation *in vivo* (Fig. 1), we immunoprecipitate nonphosphorylated VASP from epithelial MDCK cells expressing Rab13Q67L and use it as a physiologically relevant substrate for PKA activity. The PKA α-catalytic subunit phosphorylates VASP on a preferential serine 157 residue. This phosphorylation induces a conformational change in VASP and a shift in SDS-PAGE mobility from 46 to 50 kDa. This VASP property is used to monitor VASP phosphorylation in our assay.

Ten micrograms of purified GST or GST-Rab13 fusion protein loaded with GTPγS and immunoprecipitated VASP is resuspended in 50 μl of PKA kinase buffer (100 mM NaCl, 20 mM Tris, pH 7.5, 10 mM MgCl₂, 1 mM DTT, 1 mM ATP). The reaction is started by addition of 10 units of purified PKA α-catalytic subunit and incubated for 30 min at 30°. A PKA assay is performed in the presence of 1 μg of PKI, a specific peptide inhibitor of PKA (Sigma). The reaction is terminated by boiling in SDS-Laemmli sample buffer, separated on SDS-PAGE, and VASP phosphorylation determined by Western blot using the anti-VASP antibodies (Kohler *et al.*, 2004).

References

Balda, M. S., Garrett, M. D., and Matter, K. (2003). The ZO-1-associated Y-box factor ZONAB regulates epithelial cell proliferation and cell density. *J. Cell Biol.* **160,** 423–432.

Francis, S.H, and Corbin, J. D. (1994). Structure and function of cyclic nucleotide-dependent protein kinases. *Annu. Rev. Physiol.* **56,** 237–272.

Guo, W., Roth, D., Walch-Solimena, C., and Novick, P. (1999). The exocyst is an effector for Sec4p, targeting secretory vesicles to sites of exocytosis. *EMBO J.* **18,** 1071–1080.

Harbeck, B., Huttelmaier, S., Schluter, K., Jockusch, B. M., and Illenberger, S. (2000). Phosphorylation of the vasodilator-stimulated phosphoprotein regulates its interaction with actin. *J. Biol. Chem.* **275,** 30817–30825.

Itoh, M., Morita, K., and Tsukita, S. (1999). Characterization of ZO-2 as a MAGUK family member associated with tight as well as adherens junctions with a binding affinity to occludin and alpha catenin. *J. Biol. Chem.* **274,** 5981–5986.

Kohler, K., Louvard, D., and Zahraoui, A. (2004). *Rab13* regulates PKA signaling during tight junction assembly. *J. Cell Biol.* **165,** 175–180.

Lawrence, D. W., Comerford, K. M., and Colgan, S. P. (2002). Role of VASP in reestablishment of epithelial tight junction assembly after Ca2+ switch. *Am. J. Physiol. Cell Physiol.* **282,** C1235–C1245.

Marzesco, A. M., Dunia, I., Pandjaitan, R., Recouvreur, M., Dauzonne, D., Benedetti, E. L., Louvard, D., Galli, T., and Zahraoui, A. (2002). The small GTPase Rab13 regulates assembly of functional tight junctions in epithelial cells. *Mol. Biol. Cell* **13,** 1819–1831.

Matter, K., and Balda, M. S. (2003). Signalling to and from tight junctions. *Nat. Rev. Mol. Cell. Biol.* **4,** 225–236.

Pfeffer, S. R. (2001). Rab GTPases: Specifying and deciphering organelle identity and function. *Trends Cell Biol.* **11,** 487–491.

Sheth, B., Fontaine, J., Ponza, E., McCallum, A., Page, A., Citi, S., Louvard, D., Zahraoui, A., and Fleming, T. P. (2000). Differentiation of the epithelial apical junctional complex during mouse preimplantation development: A role for rab13 in the early maturation of the tight junction. *Mech. Dev.* **97,** 93–104.

Tsukita, S., and Furuse, M. (2000). Pores in the wall: Claudins constitute tight junction strands containing aqueous pores. *J. Cell Biol.* **149,** 13–16.

Wittchen, E. S., Haskins, J., and Stevenson, B. R. (1999). Protein interactions at the tight junction. Actin has multiple binding partners, and ZO-1 forms independent complexes with ZO-2 and ZO-3. *J. Biol. Chem.* **274,** 35179–35185.

Zahraoui, A., Joberty, G., Arpin, M., Fontaine, J. J., Hellio, R., Tavitian, A., and Louvard, D. (1994). A small rab GTPase is distributed in cytoplasmic vesicles in non polarized cells but colocalizes with the tight junction marker ZO-1 in polarized epithelial cells. *J. Cell Biol.* **124,** 101–115.

Zahraoui, A., Louvard, D., and Galli, T. (2000). Tight junction, a platform for trafficking and signaling protein complexes. *J. Cell Biol.* **151,** F31–F36.

Zerial, M., and McBride, H. (2001). Rab proteins as membrane organizers. *Nat. Rev. Mol. Cell Biol.* **2,** 107–117.

[64] Functional Properties of Rab15 Effector Protein in Endocytic Recycling

By LISA A. ELFERINK and DAVID J. STRICK

Abstract

Receptor recycling has emerged as an important regulatory mechanism for cell surface composition, pathogen invasion, and for control over the intensity and duration of receptor signaling in multiple cell types. In the case of the transferrin receptor, receptor recycling is an important step for facilitating iron uptake into the cell, by regulating the availability of the receptor at the cell surface. Following internalization into clathrin-coated pits, the transferrin receptor first enters peripheral sorting endosomes. Here, internalized transferrin receptor is either sorted for recycling back to the cell surface directly, or targeted to a slower route of recycling through a perinuclear population of endosomes termed the endocytic recycling compartment. This chapter describes methodologies to examine the fast and slow modes of transferrin receptor recycling, with a particular emphasis on the function of the novel protein Rab15 effector protein.

METHODS IN ENZYMOLOGY, VOL. 403
0076-6879/05 $35.00
DOI: 10.1016/S0076-6879(05)03064-8

Introduction

Early endosomes can be divided into two mechanistically distinct classes—the Rab4-positive sorting endosome (SE) and the Rab11-positive endocytic recycling compartment (ERC) (Deneka et al., 2003; Maxfield and McGraw, 2004). Following internalization into SEs, the transferrin receptor (TfR) and its ligand transferrin (Tfn) have been shown to recycle using two distinct pathways. One route involves rapid receptor recycling to the cell surface, directly from the SEs (Bottger et al., 1996; de Renzis et al., 2002; van der Sluijs et al., 1992). The second is a slower pathway, involving receptor trafficking through the ERC (Ren et al., 1998; Sheff et al., 1999; Ullrich et al., 1996). Rab15 differs from Rab 4 and 11 in that it distributes between the SEs and the ERC when expressed in a variety of cells (Zuk and Elferink, 1999, 2000), suggesting that Rab15 and/or its interacting partners may link these distinct recycling compartments. Consistent with this idea, overexpression of constitutively inactive Rab15 mutants differentially regulated transport through SEs and the ERC (Zuk and Elferink, 2000). Recycling of the TfR was promoted from both SEs and the ERC in cells expressing GDP-bound Rab15T22N. Conversely, overexpression of the guanine nucleotide–free mutant Rab15N121I increased the indirect or slower mode of TfR recycling from the ERC, without affecting rapid recycling from SEs, suggesting that unique Rab15-effector interactions could discriminate between these transport steps.

To understand how Rab15 differentially regulates receptor trafficking through SEs and the ERC, we searched for Rab15 binding partners. Using a yeast two-hybrid approach, we identified the novel protein Rab15 effector protein (REP15) as a binding partner for Rab15 (Strick and Elferink, 2005). REP15 is compartment specific, colocalizing with Rab15 and Rab11 on the ERC, but not with Rab15 or Rab4 on the SEs. Consistent with its localization to the ERC, REP15 overexpression and short interfering RNA (siRNA)-mediated knockdown blocked TfR recycling from this compartment. Tfn uptake into SEs, rapid receptor recycling to the plasma membrane, and receptor delivery to the ERC were unaltered by REP15 overexpression. We describe here some of the recycling assays used to distinguish the function of REP15 in the fast and slow routes of TfR recycling.

Cell Surface Biotinylation Assays to Measure TfR Recycling

To study REP15 function in TfR recycling from the ERC, we modified an established cell surface biotinylation scheme (Schmidt et al., 1997; Tanowitz and von Zastrow, 2003) for measuring receptor endocytosis.

This assay offers several advantages over traditional methods using radiolabeled, fluorescent, or enzyme-linked agonists. First, direct examination of receptor trafficking using covalently bound biotin is independent of ligand binding, eliminating concerns that the ligand could dissociate from the receptor while in transit through endosomal compartments. Second, ligand labeling may reduce its affinity for the receptor, possibly through steric hindrance. Third, the availability of some ligands may be limiting, or unknown as in the case of orphan receptors. Finally, cell surface biotinylation makes it possible to follow the trafficking of several receptors simultaneously.

Cell Surface Biotinylation Assay for Measuring ERC Recycling

An outline of the assay for measuring TfR recycling from the ERC is shown in Fig. 1A. Cells are surface biotinylated at 4° with sulfo-NHS-SS-Biotin, which is cleaved by washing the cells with the impermeant reducing agent 2-mercaptoethanesulfonic acid sodium salt (MesNa). Following biotinylation, the cells were incubated for 10 min at 37° to allow biotinylated proteins to endocytose and traffic to the SEs, followed by a primary MesNa wash to remove biotin from noninternalized and recycled TfR. The cells are then subjected to successive incubations at 37° to promote TfR trafficking, followed by ice-cold MesNa washes to specifically load the ERC (and not the SEs) with biotinylated TfR. One plate is maintained at 4° to measure the amount of total biotinylated receptor loaded into the ERC. The remaining plates are chased for 10, 20, or 30 min at 37° in media lacking Tfn, to promote receptor recycling from the ERC to the cell surface. All cells are subjected to a final MesNa wash, and biotinylated proteins remaining in the ERC are isolated from cell lysates using streptavidin agarose. Internalized TfR was identified by Western analysis, and the relative amount of recycled TfR was measured as the percentage of the ERC-associated TfR.

Buffers and Reagents

MesNa buffer: 50 mM 2-mercaptoethanesulfonic acid sodium salt (MesNa, Catalog No. M1511, Sigma Aldrich, St. Louis, MO), 150 mM NaCl, 1 mM EDTA, 0.2% bovine serum albumin (BSA), 20 mM Tris, pH 8.6

PBS^{2+}: 138 mM NaCl, 2.7 mM KCl, 8.2 mM Na$_2$HPO$_4$, 1.5 mM KH$_2$PO$_4$, 1 mM CaCl$_2$, 1 mM MgCl$_2$

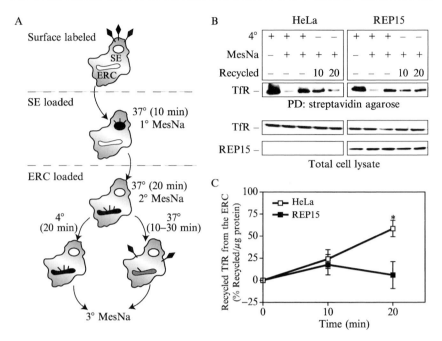

FIG. 1. REP15 overexpression decreases TfR recycling from the ERC. (A) Schematic diagram of the cell surface biotinylation assay for TfR recycling. (B) HeLa cells transiently expressing REP15 or mock-transfected cells (HeLa) were surface biotinylated and then incubated at $4°$ (+) to retain TfR on the cell surface or at $37°$ (−) to promote TfR internalization and recycling as described in the text. Noninternalized TfR was stripped of biotin (+) with MesNa washes. Control plates were washed in washing buffer lacking MesNa (−) to measure total biotinylated TfR. Biotinylated proteins were isolated by streptavidin pull downs (PD) and analyzed by Western analysis as indicated. Representative blots from three independent experiments are shown. (C) Values represent the mean ± SE of recycled TfR from three independent assays and are expressed as a percentage of the total (ERC loaded) TfR (ANOVA, $*p < 0.01$).

Wash buffer: 150 mM NaCl, 1 mM EDTA, 0.2% BSA, 20 mM Tris, pH 8.6

Lysis buffer: 150 mM NaCl, 10 mM HEPES, pH 7.4, 1 mM EDTA, 0.5% NP-40, and 0.5% Triton X-100, 2 μg/ml pepstatin, 2 μg/ml aprotinin, 2 μg/ml leupeptin, 1 mM phenylmethylsulfonyl fluoride (PMSF)

Sulfo NHS-SS-Biotin (Catalog No. 21331, Pierce, Rockford, IL)

Streptavidin agarose (Catalog No. S1638, Sigma Aldrich, St. Louis, MO)

Cell Culture Conditions

The success of this assay is cell type dependent, since it uses multiple washes that could affect cell adherence to the tissue culture substrate. Additional care is therefore required for weakly adhering cells that could detach during the washes. We routinely use HeLa cells that are cultured at 37° in 5% CO_2 in Dulbecco's modified Eagle's medium (DMEM) containing penicillin streptomycin and 10% cosmic calf serum (Hyclone, Logan, UT). For cell surface biotinylation studies, HeLa cells were grown on 100-mm dishes to 70–80% confluency. Greater cell density could adversely affect the efficiency of the biotinylation step and/or subsequent MesNa washes. Cells were transiently transfected as required using Lipofect-AMINE (Invitrogen Life Technologies, Carlsbad, CA). Prior to biotinylation, the transfected cells were routinely incubated in serum-free medium for an empirically determined amount of time, to deplete the endogenous ligand. This is particularly important for signaling receptors and for studies using labeled ligands.

Assay Protocol

Semiconfluent plates of HeLa cells were washed once in PBS^{2+} at 37° for 10 min, followed by two ice-cold washes in PBS^{2+}. Cells were surface biotinylated at 4° using 0.250 mg/ml of Sulfo NHS-SS-Biotin in PBS^{2+} with gentle rocking for 30 min. It is critical that the monolayer of cells is completely covered by the biotinylation medium; we typically use 40 μl/cm^2 of growth area in a tissue culture dish. The reaction was quenched by washing the cells three times for 5 min each with PBS^{2+} containing 50 mM glycine, followed by another three washes 5 min each in PBS^{2+}. To initiate receptor internalization into and recycling from the peripheral SEs, the cells were then incubated in prewarmed DMEM containing 2 mg/ml BSA at 37° for 10 min. Receptor internalization and recycling were halted by placing the cells on ice, followed by three washes in ice–cold wash buffer for 10 min at 4° with gentle rocking. Biotin at the cell surface was stripped by a 1-h wash at 4°, using freshly prepared MesNa buffer. After three ice–cold rinses with PBS^{2+}, the cells were chased a second time at 37° for 20 min in prewarmed DMEM containing 2 mg/ml BSA, to load biotinylated TfR into the ERC and complete receptor recycling from the SEs. The cells were washed in MesNa as described earlier, to remove residual cell surface biotin. After three ice–cold rinses with PBS^{2+}, one plate of cells was incubated at 4° to measure total biotinylated TfR loaded into the ERC (time = 0 min). The other plates of cells were chased at 37° in DMEM containing 2 mg/ml BSA for increasing time (10–20 min) to promote recycling from the ERC. The cells were subjected to a final

MesNa wash to strip the residual cell surface biotin from receptors that recycled from the ERC to the cell surface. The cells were washed three times with ice–cold PBS^{2+} to remove the residual MesNa, and then lysed in 500 μl of lysis buffer and stored at $-80°$. Typically, 250 μg of protein in a cell lysate at a concentration of 1 μg/μl is optimal for pull downs using streptavidin agarose.

For pull downs, the thawed lysates were cleared by centrifugation at $14,000\times g$ for 10 min. A 50% slurry of streptavidin agarose was prewashed three times with ice–cold lysis buffer, and 30 μl of the washed slurry incubated with 250 μg of protein overnight at $4°$, with gentle rocking end over end. A key step is to ensure that equal amounts of streptavidin agarose are dispensed. Differences in bead volume could result in differences in the efficiency of the pull downs, and compromise the reproducibility of the results. After incubation with the lysates, the beads were pelleted by centrifugation at $1000\times g$ for 5 min and washed three times in lysis buffer, followed by two washes in 50 mM Tris–Cl, pH 7.5. The lysates should be retained and examined by immunoprecipitation and Western analysis for biotinylaed TfR, to ensure the efficiency of the pull downs. The beads were resuspended in 30 μl of SDS sample buffer, boiled for 5 min at $100°$, and 20 μl of the sample was analyzed by Western analysis. The resulting images were generated using a Fluorochem 8200 AlphaImager and data were quantified using the AlphaEase v.3.1.2 spot densitometry function (Alpha Innotech Corp. San Diego, CA). To avoid saturation of the Western blot images, multiple exposures within the linear range are required.

Under these conditions, increased chase times correspond to increased recycling of the TfR from the ERC in mock-transfected HeLa cells (Fig. 1B and C). Conversely, transient overexpression of REP15 resulted in a 50% decrease in the level of TfR recycled from the ERC, relative to mock-transfected cells. Western analysis of total protein in the cell lysates confirmed that the differences in surface levels of TfR were not a consequence of discernible differences in the expression level of the receptor, consistent with a role for REP15 in regulating the endocytic recycling of the TfR from the ERC.

Additional Comments

It is necessary to empirically optimize the biotinylation reaction and to monitor the efficiency and sensitivity of the cells to the MesNa washes. We recommend starting with wash buffers containing 10–25 mM MesNa, for three 45-min washes at $4°$. Higher MesNa concentrations can be used (e.g., 50 mM), if the incubation time is decreased accordingly (e.g., twice for 60 min each). In our hands, optimal MesNa washes require use of freshly prepared buffer. Including an avidin quench using PBS^{2+} containing 0.2% BSA, 50 μg/ml avidin after the final MesNa wash and prior to cell lysis could prove useful in decreasing background surface biotin levels. However, care must be taken to remove excess avidin. One observed disadvantage of this assay is that it is not readily amenable for studying the kinetics of rapid receptor recycling from the SEs. These types of assays are often performed using a combination of low temperature and 37° steps, which would alter the kinetics of membrane trafficking. Finally, the high amount of protein required for cell surface biotinylation studies renders these assays expensive when using siRNA technologies. Furthermore, this assay is not readily adaptable to high–throughput screening, such as assaying for recycling defects.

ELISA for Distinguishing TfR Recycling from the SEs and the ERC

As an alternative to cell surface biotinylations, we describe a simple and rapid procedure to measure both the fast and the slow modes of Tfn recycling. The assay uses an enzyme-linked immunosorbant assay (ELISA) format to detect recycled biotinylated-transferrin (B-Tfn) captured on antibody-coated plates. This assay offers several advantages over the standard cell surface biotinylation approach described earlier. First, technical challenges make it difficult to accurately measure rapid TfR recycling from the SEs using cell surface biotinylation assays. Second, using ligand-specific antibodies, the ELISA can be adapted to measure the recycling of multiple receptor–ligand complexes. Third, it has the potential to be used for high-throughput screens. Fourth, this assay is conducive to studies using siRNAs to knockdown protein function, since it requires relatively small samples of material.

Buffers and Reagents

> Phosphate-buffered saline (PBS) pH 4.2: 138 mM NaCl, 2.7 mM KCl, 8.2 mM Na_2HPO_4, 1.5 mM KH_2PO_4, 25 mM glacial acetic acid, pH 4.2
> PBS–BSA: 138 mM NaCl, 2.7 mM KCl, 8.2 mM Na_2HPO_4, 1.5 mM KH_2PO_4, 0.2% BSA
> Wash buffer: 50 mM Tris, pH 8.0, 0.14 M NaCl, and 0.05% Tween-20
> Blocking buffer: 50 mM Tris, pH 8.0, 0.14 M NaCl, 1% BSA, pH 8.0
> Maxisorp ELISA plates: (Catalog No. 437111, Nalgene-Nunc, Rochester, NY)
> Human Transferrin ELISA Quantitation Kit: E80-128, Bethyl Laboratories, Montgomery, TX

Streptavidin coupled horseradish peroxidase (HRP), 0.5 mg/ml:
Catalog No. 016-030-084, Jackson Immuno Research Laboratories,
West Grove, PA

Goat–anti-mouse coupled HRP: Catalog No. 115-035-003, Jackson
Immuno Research Laboratories, West Grove, PA

Quanta Blu fluorogenic substrate: Catalog No. 15169, Pierce,
Rockford, IL

Transferrin (Tfn): Catalog No. T4132, Sigma, St. Louis, MO

Biotinylated-Tfn: Catalog No. T23363, Molecular Probes, Eugene, OR

Assay Protocol

Triplicate sets of HeLa cells were routinely cultured in 24-well dishes to
70–80% confluency and transiently transfected as required. The cells were
washed three times in PBS–BSA, and then incubated for 2 h at 37° in
serum-free DMEM to deplete endogenous Tfn in the culture media. Endocytosis at reduced temperatures (16–20°) has been shown to selectively
load the SEs with Tfn, by preventing the exit of the TfR from this compartment (Dunn *et al.*, 1980; Ren *et al.*, 1998; Schmidt *et al.*, 1997). The
temperature-induced block in TfR trafficking is reversible, since subsequent chases at 37° result in direct receptor recycling as well as receptor
transport to the ERC. Thus, to maximize the loading of B-Tfn into the
SEs, the cells were incubated at 16° for 1 h in DMEM containing B-Tfn.
Cells were washed three times, alternatively in PBS, pH 4.2, and PBS–BSA
at 4° (acid wash), to remove surface bound B-Tfn. The cells were then
incubated at 37° for 20 min in DMEM containing 2 mg/ml BSA, 500 μg/ml
unlabeled Tfn to promote trafficking to the ERC and recycling from the
SE. The cells were placed on ice to halt trafficking, and B-Tfn that recycled
from the SEs was removed using alternate ice-cold PBS, pH 4.2, and
PBS–BSA washes as described above. One set of cells was maintained
at 4° as a background control for recycling. The remaining sets of cells
were chased at 37° for 1–20 min in 300 μl of DMEM containing 2 mg/ml
BSA, 500 μg/ml Tfn to promote receptor recycling from the ERC. The cells
were shifted to 4° and the chase medium was removed to measure recycled
levels of B-Tfn. The cells were extensively washed a third time with
alternate ice-cold PBS, pH 4.2, and PBS–BSA washes, and then lysed in
200 μl of lysis buffer overnight at 4° for protein determination.

B-Tfn recycled into the chase media was measured using a human Tfn
ELISA. In this assay, a Nalgene-Nunc Maxisorp ELISA plate was coated
with an anti-Tfn antibody diluted 1:100 in 0.05 M sodium carbonate
buffer, pH 9.6, for 1 h at room temperature. One set of wells was coated
with goat–anti-mouse IgG coupled to HRP, diluted 1:10,000 in 0.05 M

sodium carbonate buffer, pH 9.6, as a control for consistency between plate assays. The plate was washed three times with wash buffer, and the wells blocked by incubation with 200 μl of blocking buffer for 1 h at room temperature. Excess blocking buffer was removed by aspiration, washed three times in wash buffer, and each Tfn antibody-coated well incubated with 300 μl of the chase medium for 1 h at room temperature. The wells were washed six times in wash buffer, then incubated with 100 μl of streptavidin-conjugated HRP (Av-HRP) diluted 1:1000 in blocking buffer for 1 h at room temperature. Excess Av-HRP was removed using six washes of wash buffer, and HRP activity detected using QuantaBlu fluorogenic substrate (Pierce, Rockford, IL) according to the manufacturer's instructions. HRP activities were measured at the excitation wavelength of 320 nm and the emission wavelength of 420 nm using a Molecular Devices Gemini Fluorescent Plate Reader. It is critical to measure ligand recycling with respect to the relative amount of starting material (i.e., microgram of protein), especially when the experimental treatments affect cell growth and/or viability. Thus, protein concentrations of the initial lysates were determined using a BCA kit (Pierce) and the level of HRP activity normalized to the protein concentration of the cell lysates. In cells overexpressing REP15, recycling of B-Tfn from the ERC was decreased $\sim$3-fold relative to mock-transfected HeLa cells (Fig. 2), consistent with the surface biotinylation assays (see Fig. 1C).

We adapted the protocol to measure the direct mode of TfR recycling from SEs (Fig. 3A). In this assay, the cells were loaded with B-Tfn at 16°, and noninternalized B-Tfn was removed using alternate ice-cold PBS, pH 4.2, and PBS–BSA washes as outlined earlier. The cells were then shifted to 37° for increasing periods of time (5–10 min) to induce receptor recycling from the SEs. The appearance of B-Tfn in the resulting chase media was measured using the Tfn ELISA as described previously. Since comparable amounts of B-Tfn were recycled from the SEs in both mock-transfected and cells transiently expressing REP15 (Fig. 3B), the data confirm that REP15 functions to regulate receptor recycling from the ERC and not SEs.

Additional Comments

Many factors could affect the efficiency of the Tfn-ELISA, including the level of TfR expression in a cell type, variations in cell density, and the sensitivity of the cells to the acid washes. Hence, assay reliability depends on standardizing these variables between experiments. This assay can be readily modified to measure the recycling of other ligand–receptor complexes. However, researchers should consider variables that are specific and important to a given ligand–receptor interaction. For example, this

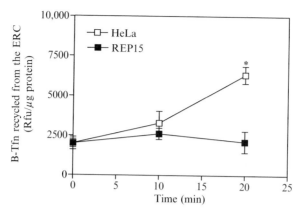

FIG. 2. A B-Tfn ELISA to measure TfR recycling from the ERC. The relative amount of B-Tfn recycled from the ERC in HeLa cells overexpressing REP15 and mock-transfected cells (HeLa) was quantified using a B-Tfn ELISA assay. All values are the mean of triplicate values and are expressed as relative fluorescent units (Rfu) per μg protein. Statistical differences between experimental and control sets were observed (ANOVA, $^*p < 0.01$).

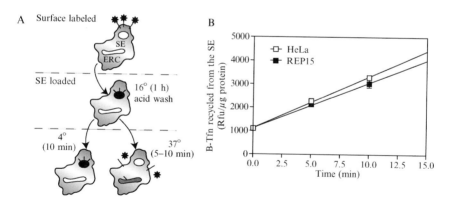

FIG. 3. REP15 does not regulate rapid TfR recycling from the SEs. (A) Schematic of the TfR recycling assay from the ERC. (B) HeLa cells overexpressing REP15 (REP15) and mock-transfected cells (Con) were loaded with B-Tfn (*) at 16° for 1 h, acid washed with PBS pH 4.2, and chased for 5 and 10 min at 37° in media containing unlabeled Tfn. The amount of B-Tfn recycled into the media from SEs was measured using a B-Tfn ELISA. Values represent the mean ± SE of three independent assays and are expressed as relative fluorescent units (Rfu) per μg protein. No significant differences (one-way ANOVA and a Newman–Keuls posttest) were observed between experimental and control conditions.

assay is dependent on the close association of the ligand with its receptor during endocytic trafficking through SEs and the ERC. Similarly, the assay is dependent on the availability of a suitable ligand antibody.

Concluding Remarks

It remains a major challenge to identify new compartment-specific components that govern distinct recycling pathways from SEs and the ERC. The basic protocols presented here possess a flexibility that allows the assays to be adapted to future applications, examining the compartment-specific function of REP15 in receptor recycling, as well as endocytic trafficking in general.

Acknowledgments

This work was supported in part by the National Science Foundation Grant IBN-0343739 to L.A.E. We thank Patricia Gazzoli for assistance in the preparation of the manuscript and Ning Li for helpful comments.

References

Bottger, G., Nagelkerken, B., and van der Sluijs, P. (1996). Rab4 and Rab7 define distinct nonoverlapping endosomal compartments. *J. Biol. Chem.* **271**, 29191–29197.

Deneka, M., Neeft, M., and van der Sluijs, P. (2003). Regulation of membrane transport by rab GTPases. *Crit. Rev. Biochem. Mol. Biol.* **38**, 121–142.

de Renzis, S., Sonnichsen, B., and Zerial, M. (2002). Divalent Rab effectors regulate the sub-compartmental organization and sorting of early endosomes. *Nat. Cell Biol.* **4**, 124–133.

Dunn, W. A., Hubbard, A. L., and Aronson, N. N., Jr. (1980). Low temperature selectively inhibits fusion between pinocytic vesicles and lysosomes during heterophagy of 125I-asialofetuin by the perfused rat liver. *J. Biol. Chem.* **255**, 5971–5978.

Maxfield, F. R., and McGraw, T. E. (2004). Endocytic recycling. *Nat. Rev. Mol. Cell Biol.* **5**, 121–132.

Ren, M., Xu, G., Zeng, J., De Lemos-Chiarandini, C., Adesnik, M., and Sabatini, D. D. (1998). Hydrolysis of GTP on rab11 is required for the direct delivery of transferrin from the pericentriolar recycling compartment to the cell surface but not from sorting endosomes. *Proc. Natl. Acad. Sci. USA* **95**, 6187–6192.

Schmidt, A., Hannah, M. J., and Huttner, W. B. (1997). Synaptic-like microvesicles of neuroendocrine cells originate from a novel compartment that is continuous with the plasma membrane and devoid of transferrin receptor. *J. Cell Biol.* **137**, 445–458.

Sheff, D. R., Daro, E. A., Hull, M., and Mellman, I. (1999). The receptor recycling pathway contains two distinct populations of early endosomes with different sorting functions. *J. Cell Biol.* **145**, 123–139.

Strick, D. J., and Elferink, L. A. (2005). Rab15 effector protein: A novel protein for receptor recycling from the endocytic recycling compartment. Submitted.

Tanowitz, M., and von Zastrow, M. (2003). A novel endocytic recycling signal that distinguishes the membrane trafficking of naturally occurring opioid receptors. *J. Biol. Chem.* **278,** 45978–45986.

Ullrich, O., Reinsch, S., Urbe, S., Zerial, M., and Parton, R. G. (1996). Rab11 regulates recycling through the pericentriolar recycling endosome. *J. Cell Biol.* **135,** 913–924.

van der Sluijs, P., Hull, M., Webster, P., Male, P., Goud, B., and Mellman, I. (1992). The small GTP-binding protein rab4 controls an early sorting event on the endocytic pathway. *Cell* **70,** 729–740.

Zuk, P. A., and Elferink, L. A. (1999). Rab15 mediates an early endocytic event in Chinese hamster ovary cells. *J. Biol. Chem.* **274,** 22303–22312.

Zuk, P. A., and Elferink, L. A. (2000). Rab15 differentially regulates early endocytic trafficking. *J. Biol. Chem.* **275,** 26754–26764.

[65] Assays for Interaction between Rab7 and Oxysterol Binding Protein Related Protein 1L (ORP1L)

By MARIE JOHANSSON and VESA M. OLKKONEN

Abstract

ORP1L belongs to the recently described family of human oxysterol binding protein homologues. We have previously shown that ORP1L localizes to late endosomes. In this chapter we describe methods that have been used to investigate the functional link of ORP1L with the protein machinery regulating late endosomal membrane trafficking. Coimmunoprecipitation, COS cell two-hybrid, and pull-down assays were applied to demonstrate a physical interaction between ORP1L and the late endosomal small GTPase Rab7. With these methods we were able to map the Rab7-binding determinant of ORP1L to the amino-terminal ankyrin repeat region (aa 1–237) and show that the interaction is preferentially with the GTP-bound form of Rab7. Furthermore, we describe approaches based on transient transfection and confocal immunofluorescence microscopy, which were employed to study the effect of this amino-terminal ORP1L fragment on late endosome morphology. The ankyrin repeat fragment induces juxtanuclear clustering of late endosomes, dependent on an intact microtubule network. When it is coexpressed with the dominant inhibitory Rab7 mutant T22N, the clustering is inhibited, suggesting that the effect involves interaction of the fragment with active Rab7.

METHODS IN ENZYMOLOGY, VOL. 403   0076-6879/05 $35.00
Copyright 2005, Elsevier Inc. All rights reserved.   DOI: 10.1016/S0076-6879(05)03065-X

Introduction

Oxysterol binding protein (OSBP) is a cytoplasmic protein with affinity for a number of oxysterols, oxygenated derivatives of cholesterol (Dawson *et al.*, 1989; Taylor *et al.*, 1984). Overexpression of OSBP in Chinese hamster ovary (CHO) cells causes alterations in cholesterol and sphingomyelin metabolism (Lagace *et al.*, 1997, 1999), and the protein has been suggested to function in lipid trafficking from the endoplasmic reticulum (ER) to the Golgi apparatus (Wyles *et al.*, 2002). Furthermore, OSBP was identified as a regulator of the extracellular signal-regulated kinase (ERK) signaling pathways (Wang *et al.*, 2005). Families of proteins homologous to OSBP have been discovered in eukaryotic organisms from yeasts to humans (reviewed by Lehto and Olkkonen, 2003). These proteins, denoted OSBP-related proteins (ORP), have been implicated in diverse aspects of cell regulation, including sterol metabolism, vesicle transport, and cell signaling. However, their function has remained largely enigmatic. The ORPs are cytosolic but associate peripherally with distinct subcellular membrane compartments. The human ORP family consists of 12 members, which can be divided into six subfamilies (Jaworski *et al.*, 2001; Lehto *et al.*, 2001). ORP1 is present in human tissues as two major variants, ORP1L and ORP1S. The former carries, in addition to the C-terminal OSBP-related ligand-binding domain, a pleckstrin homology domain and an N-terminal extension with three ankyrin repeats (Fig. 1). We have shown that ORP1L localizes to late endosomes (LEs) (Johansson *et al.*, 2003).

The small Rab GTPases are central regulators of intracellular membrane trafficking. They are found at specific locations on organelles that constitute the biosynthetic and endocytic pathways of eukaryotic cells, and function in transport vesicle formation, vesicle movement along cytoskeletal tracks, as well as in the tethering/docking of vesicles at the target membrane (reviewed by Pfeffer, 2001; Zerial and McBride, 2001). Rab7 is present on LEs (Chavrier *et al.*, 1990; Meresse *et al.*, 1995) and controls membrane trafficking from early endosomes to late LEs (Feng *et al.*, 1995;

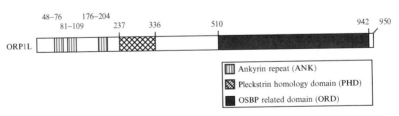

Fig. 1. Schematic presentation of ORP1L structure. The numbers indicate amino acid positions.

Press *et al.*, 1998). Rab7 also regulates the microtubule-dependent motility of late endocytic compartments (Cantalupo *et al.*, 2001; Harrison *et al.*, 2003; Jordens *et al.*, 2001; Lebrand *et al.*, 2002) and has been connected to a spectrum of LE functions (Edinger *et al.*, 2003; Jager *et al.*, 2004; Marsman *et al.*, 2004). We observed that ORP1L interacts physically with Rab7, which creates a novel connection between the OSBP-related proteins and the Rab-based machineries that control membrane trafficking (Johansson *et al.*, 2005). In this chapter we describe approaches used to demonstrate the interaction between ORP1L and Rab7.

Colocalization of ORP1L with Rab7 Assessed by Confocal Immunofluorescence Microscopy

The intracellular localization of ORP1L was studied by transiently transfecting CHO-K1 cells with cDNA encoding the full-length protein (aa 1–950). The expressed ORP1L protein was shown to have a perinuclear organellar staining pattern (Johansson *et al.*, 2003). In cells with high ORP1L expression levels, these organelles appeared clustered and enlarged. The organelles were identified as LEs by the extensive colocalization of the EGFP–ORP1L fusion protein and the endogenous late endosomal marker Rab7 (Fig. 2).

The full-length ORP1L cDNA was generated by reverse transcribing 1 μg of total RNA from human fetal brain (Stratagene, La Jolla, CA) using the Superscript II enzyme (Invitrogen, Carlsbad, CA) according to the

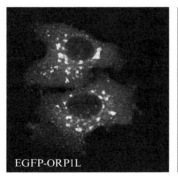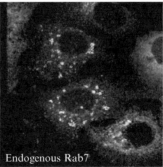

FIG. 2. Colocalization of ORP1L with Rab7 assessed by confocal immunofluorescence microscopy (see the section on colocalization of ORP1L with Rab7). CHO-K1 cells were transiently transfected with an EGFP-ORP1L expression plasmid, fixed, and stained with anti-Rab7 antibodies. The bound antibodies were visualized with an anti-chicken Alexa Fluor secondary antibody conjugate. Images were obtained with a Leica TCS SP1 laser scanning confocal microscope.

manufacturer's instructions. Overlapping ORP1L 5'-end and 3'-end cDNA fragments were generated by polymerase chain reaction (PCR) carried out with 2.5 U of Pfu Turbo DNA-polymerase (Stratagene) in a reaction volume of 20 μl containing 2 μl of template cDNA, 10× cloned Pfu DNA polymerase reaction buffer, 200 μM dNTPs, 10 pmol of each primer, and 10% dimethyl sulfoxide. The thermal cycling protocol applied for the reactions consisted of denaturation at 94° for 5 min followed by 35 cycles of denaturation at 94° for 30 s, annealing at 60° for 30 s, extension at 72° for 3 min, and a final extension at 72° for 10 min. The N-terminal fragment was subsequently ligated, using a unique XbaI site in the overlap region, into the C-terminal construct, generating the full-length ORP1L cDNA (Johansson et al., 2003). Oligonucleotide primers for generation of the cDNA constructs are listed in Table I.

For the purpose of intracellular localization studies, the full-length insert was subcloned into the BamHI site of the mammalian expression vector pEGFP-C1 (BD Biosciences Clontech, Palo Alto, CA). Plasmid DNA used for subcloning and transfection was purified using the Qiagen Maxi kit (Valencia, CA).

CHO-K1 cells (ATCC CCL-61) were seeded onto coverslips 1 day prior to transfection and grown to 70% confluency in 500 μl of Iscove's modified Dulbecco's medium (IMDM; Sigma-Aldrich, St. Louis, MO), 10% fetal bovine serum (FBS), 100 U/ml penicillin, and 100 μg/ml streptomycin. Two microliters of LipofectAMINE 2000 (Invitrogen) was added to 48 μl serum-free medium, let stand for 5 min at room temperature, and combined with 1.0 μg plasmid DNA diluted in 50 μl serum-free medium. The transfection complex was allowed to form for 25 min, after which the mix (100 μl) was added dropwise onto the cells in complete medium. Cells were incubated for 24 h at 37° in the presence of the transfection complexes.

After transfection, cells were washed twice with ice-cold phosphate-buffered saline (PBS) and fixed with 4% paraformaldehyde, 250 mM HEPES, pH 7.4, for 30 min. Free aldehyde groups were quenched by incubation with 50 mM NH$_4$Cl for 10 min. Cells were permeabilized for 25 min with 0.05% Triton X-100 in PBS. Nonspecific binding of antibodies was blocked with 10% FBS/PBS for 20 min at 37°. Chicken anti-Rab7 antibody (Dong et al., 2004; Stein et al., 2003), a kind gift from Dr. Angela Wandinger-Ness (Department of Pathology, University of New Mexico Health Sciences Center, Albuquerque, NM), was diluted (1:50) in 5% FBS/PBS, and cells were incubated for 30 min at 37°. After washing 3 × 10 min with PBS, Alexa Fluor secondary antibody conjugate (Molecular Probes, Eugene, OR) was added (1:150 in 5% FBS/PBS) and incubated for 30 min at 37°. Cells were washed 3 × 10 min with PBS and mounted in Mowiol (Calbiochem, San Diego, CA) containing

TABLE I
OLIGONUCLEOTIDE PRIMER SEQUENCES FOR GENERATION OF cDNA CONSTRUCTS

cDNA construct	Forward primer (5' → 3')	Reverse primer (5' → 3')
5' end of ORP1L	ACTCGGATCCATG-AACACAGAAGCG-GAGCAACA	TCACGAATTCGGCCAGCG-TCTCCAGTGCTTC
3' end of ORP1L	CCAGCATCCTTAGC-GAGGACGGA	ACTCGGATCCGTAGATTA-GCCAAACACCCTGAC
ANK	ACTCGGATCCTTATG-AACACAGAAGCGGA-GCAACA	CTCGGATCCTATCGTTTCAA-TGCTTTGTAGATGAC
ORD + PHD	TCACGGATCCAAACCTC-TTGACCTTGCCCAG	ACTCGGATCCTACAGCTGG-TCCTGGGAACAGTA
wtRab7 (XhoI)	AGACTCCTCGAGATGA-CCTCTAGGAAGAAA-GTGTTG	AGACTCCTCGAGTCAGCAA-CTGCAGCTTTCCGCT
wtRab7 (HindIII)	AGACTAAGCTTAATGA-CCTCTAGGAAGAAAGT	AACATCAAGCTTTCAGCAA-CTGCAGCTTCCG
Rab7T22N	CTGGAGTTGGTAAGAACT-CACTCATGAACCAG	CTGGTTCATGAGTGAGTTC-TTACCAACTCCAG
Rab7Q67L	ACACAGCAGGCCTGGAA-CGGTTCCAG	CTGGAACCGTTCCAGGCC-TGCTGTGT

50 mg/ml of the antifading agent 1,4-diazocyclo[2,2,2]octane (Sigma-Aldrich). The specimens were analyzed with a TCS SP1 laser scanning confocal microscope (Leica, Wetzlar, Germany). A representative image obtained with this method is shown in Fig. 2.

Using the same techniques, we also detected colocalization of transfected EGFP-ORP1L with the LE markers Rab7, Rab9, and Lamp-1 in HeLa cells. Furthermore, the endogenous HeLa cell ORP1L was detected on LEs with this methodology using an affinity-purified antibody prepared against a fusion protein carrying aa 428–553 of the protein (M. Johansson et al., 2005). For double immunostaining, the primary antibodies were added simultaneously on the coverslips. Similarly, the Alexa Fluor-conjugated secondary antibodies were added together on the specimens.

Approaches to Studying the Physical Interaction of ORP1L and Rab7

The changes in late endosomal morphology caused by overexpression of ORP1L suggested an interference with the machinery involved in late endosomal tethering, docking, or fusion. This prompted us to investigate a possible physical interaction between Rab7, a protein controlling late endosomal dynamics, and ORP1L. For this purpose, we carried out coimmunoprecipitation experiments with the overexpressed and the endogenous cellular proteins. The preferential binding of ORP1L to either the GDP- or GTP-bound Rab7 was studied with a two-hybrid assay based on a luciferase reporter system. Finally, the interacting region of ORP1L was mapped by a pull-down approach, in which *in vitro* translated ^{35}S-labeled fragments of ORP1L were incubated with immobilized glutathione-*S*-transferase (GST)-Rab7 or plain GST.

Immunoprecipitation

Rab7 and ORP1L cDNA inserts were subcloned into mammalian expression vectors pcDNA4HisMaxC (Invitrogen) and pcDNA3.1(−) (Invitrogen), respectively. The pcDNA4HisMaxC vector encodes an N-terminal Xpress epitope, for which a monoclonal antibody is commercially available (Invitrogen). The full-length ORP1L cDNA insert was directly transferred into the *Bam*HI site of pcDNA3.1(−), while the Rab7 insert was amplified using an EGFP-Rab7 construct kindly provided by Dr. Angela Wandinger-Ness (Department of Pathology, University of New Mexico Health Sciences Center, Albuquerque, NM) as template, and cloned into the *Xho*I site of pcDNA4HisMaxC. Oligonucleotide primers for the generation of the cDNA construct are shown in Table I. Amplification by polymerase chain reaction (PCR) was carried out as described in the previous section, except that no dimethyl sulfoxide was added to the reaction, the annealing temperature in the thermal cycling protocol was 55°, and 125 pg of plasmid template was used per reaction.

HeLa cells were seeded onto 6-cm dishes and grown to 90% confluency in 2 ml of Dulbecco's modified Eagle's medium (DMEM; Sigma-Aldrich), 10% FBS, 20 mM HEPES, pH 7.4, 100 U/ml penicillin, and 100 μg/ml streptomycin 1 day prior to transfection. Fifteen microliters of Lipofect-AMINE 2000 (Invitrogen) was added to 85 μl serum-free medium, let stand for 5 min, and combined with 2.5 μg plasmid DNA encoding untagged ORP1L and 2.5 μg plasmid DNA encoding Xpress epitope-tagged Rab7, diluted in 100 μl serum-free medium. The transfection complex was allowed to form for 25 min, after which the mix (200 μl) was added dropwise to the cells in complete medium. Cells were transfected for 24 h at 37°.

The cells were washed twice with ice-cold PBS, and 500 μl of lysis buffer (10 mM HEPES, pH 7.6, 150 mM NaCl, 0.5 mM MgCl$_2$, 10% glycerol, 0.5% Triton X-100) with complete EDTA-free protease inhibitor cocktail (Roche Diagnostics, Indianapolis, IN) was added per dish. Cells were scraped and kept on ice for 15 min, centrifuged for 15 min at 16,000×g in a microcentrifuge at +4°, and the supernatant was collected and kept on ice. Protein G-Sepharose 4 fast flow (Amersham Biosciences AB, Uppsala Sweden) (60 μl of 50% slurry) was washed 4 times with 400 μl of lysis buffer. The Sepharose was centrifuged at 400×g for 1 min between the washes. The supernatant was preadsorbed with the protein G-Sepharose matrix for 30 min on a roller at +4°. The beads were centrifuged as before, and the supernatant was incubated overnight at +4° with Xpress or control antibodies. The lysate–antibody mixture was added to fresh protein G-Sepharose and incubated for 4 h on a roller at +4°. The matrix was washed four times with lysis buffer and boiled for 5 min in 30 μl of sodium dodecyl sulfate-polyacrylamide gel electrophoresis (SDS-PAGE) loading buffer.

Proteins were resolved on 12.5% SDS-polyacrylamide gels and transferred to Hybond-C extra nitrocellulose (Amersham Pharmacia Biotech, Piscataway, NJ). Nonspecific binding of antibodies was blocked with, and all antibody incubations were carried out in 5% fat-free powdered milk in 10 mM Tris–HCl, pH 7.4, 150 mM NaCl, 0.05% Tween-20. The ORP1L protein was detected with polyclonal rabbit anti-ORP1L affinity-purified antibody (1:500) against aa 428–950 (Johansson et $al.$, 2003) while Rab7 was detected with monoclonal anti-Xpress antibody (1:5000). The bound primary antibodies were visualized with horseradish peroxidase-conjugated goat-anti-rabbit or goat-anti-mouse IgG (Bio-Rad Laboratories, Hercules, Ca) and the enhanced chemiluminescence system ECL (Amersham Pharmacia Biotech). Specific coimmunoprecipitation of ORP1L with Xpress-tagged Rab7 is illustrated in Fig. 3.

The immunoprecipitation of endogenous proteins was accomplished by applying the same protocol with minor modifications and using a polyclonal Rab7 antibody (Santa Cruz Biotechnology, Santa Cruz, CA). The lysis buffer composition was 20 mM HEPES, pH 7.6, 150 mM NaCl, 2.0 mM MgCl$_2$, 10% glycerol, 0.5% Triton X-100, 1 mM dithiothreitol (DTT), with complete EDTA-free protease inhibitor cocktail added prior to use. After preadsorption with protein G Sepharose, samples were incubated with the respective antibodies for 2 h at +4°, after which the lysate–antibody mixture was added to fresh protein G Sepharose beads and incubated overnight at +4° on a roller. The endogenous ORP1L protein was detected by Western blotting with affinity-purified rabbit anti-ORP1L (1:500) against aa 428–553, while Rab7 was detected with rabbit anti-Rab7 (1:1000) (Santa Cruz Biotechnology).

Two-Hybrid Luciferase Assay

The two-hybrid luciferase assay used is based on measuring the activity of firefly luciferase, an enzyme that catalyzes the reaction where the chemical energy of luciferin oxidation is converted through an electron

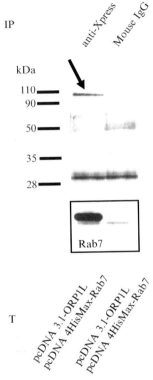

FIG. 3. Coimmunoprecipitation of ORP1L with Rab7 (see the section on immunoprecipitation). ORP1L and Xpress epitope-tagged Rab7 constructs were transfected into HeLa cells (T indicates the transfected constructs), and immunoprecipitation of the expressed proteins was carried out using Xpress-antibody or irrelevant mouse IgG (IP). The precipitates were Western blotted with anti-ORP1L. The arrow in the left lane indicates the ORP1L band. The inset shows the corresponding lanes blotted with Xpress antibody, detecting epitope-tagged Rab7.

transition into oxyluciferin and emitted light. The luciferase activity can be used as an indirect measure of its expression. In this assay, the expression of the luciferase gene in vector pG5luc-1 (Promega, Madison, WI) is under the control of the transcription factor Gal4 regulatory element. The Gal4 DNA binding domain is encoded in vector pM (BD Biosciences Clontech). In the two-hybrid assay, transcription of the luciferase gene requires an interaction of the Gal4 DNA binding domain with a transcriptional activation domain derived from herpes simplex virus protein 16, encoded in vector pVP16 (BD Biosciences Clontech). The constructs of interest are cloned into pM or pVP16 and are expressed as fusion proteins with the

respective domains encoded by the vectors. When the two proteins interact, the hybrid binds the *GAL4* DNA element in the pG5luc-1 reporter
plasmid and the luciferase gene is transcribed. The luciferase activity
is therefore dependent on the formation of a hybrid complex. The principle
of the system is presented in Fields and Song (1989) and Sadowski
et al. (1988).

For the assay ORP1L cDNA was cloned into the *Bam*HI site of a vector
derived from pM, pM2 (Dr. Ivan Sadowski, Department of Biochemistry
and Molecular Biology, Faculty of Medicine, Vancouver, B.C. Canada),
which has a reading frame compatible with cloning into the *Bam*HI site.
Three different Rab7 inserts were cloned into the *Hin*dIII site of vector
pVP16. The inserts encoded the wild-type protein, a dominant inhibitory
mutant, and a constitutively active mutant. The dominant inhibitory mutant carries a mutation of threonine (T) at amino acid position 22 to
asparagine (N). This mutation exchanges a side chain hydroxyl group for
an amide group and leads to a protein that is preferentially GDP bound. In
the constitutively active mutant, glutamine (Q) at position 67 is mutated to
leucine (L), which results in a GTPase-deficient, constitutively GTP-bound
Rab7. The mutants were generated using the Quikchange site-directed
mutagenesis kit from Stratagene. The kit provides all the necessary reagents except the oligonucleotide primers and the cDNA template. Oligonucleotide primers for generation of the Rab7 mutant constructs are listed
in Table I. Primers were designed according to the manufacturer's guidelines and reaction conditions were as instructed in the kit manual. Wild-
type Rab7 cDNA was used as template for PCR with mutagenic forward
and reverse primers. After the PCR step, amplification products were
digested with *Dpn*I, which digests the methylated template DNA, leaving
the mutated nonmethylated products intact, which can then be transformed
into competent *Escherichia coli.* A pSV-β-galactosidase vector from Promega was used to correct the luciferase activity data for the transfection
frequency.

COS-1 cells were seeded on 12-well plates the day prior to transfection
and grown to 90% confluency in 1 ml of Eagle's medium (DMEM; Sigma-
Aldrich), 10% FBS, 20 mM HEPES, pH 7.4, 100 U/ml penicillin, and 100 μg/
ml streptomycin. In this assay we used the FuGENE 6 transfection reagent
(Roche Diagnostics). We have noted that this reagent results in a higher
transfection frequency in COS-1 cells than LF 2000. Transfection mixtures
were prepared consisting of 400 ng of wt or mutant pVP16-Rab7, 400 ng of
pM2-ORP1L, 200 ng of pG5luc, and 100 ng of pSV-β-galactosidase. Two
control mixtures were also prepared, one consisting of empty pVP16 vector,
pM2-ORP1L, pG5luc, and pSV-β-galactosidase, and the other consisting
of empty pVP16 vector, empty pM2, pG5luc, and pSV-β-galactosidase.

Mixtures were prepared in triplicate. 2 μl of FuGENE 6 were diluted in 48 μl of serum-free medium and let stand for 5 min. The plasmid DNA was added directly to the mix, and the complex was allowed to form for an additional 20 min. The transfection mixture was added dropwise onto the cells in complete medium. Cells were transfected for 24 h at 37°.

Cells were washed twice with ice-cold PBS and 100 μl of 1× reporter lysis buffer (Promega) was added to each well. The cells were transferred to −20° and allowed to freeze. They were then thawed at room temperature and collected by scraping with a rubber policeman on ice. The lysates were centrifuged for 3 min at 16,000×g and the supernatants transferred to new tubes. Samples (20 μl) were pipetted onto a white microtiter plate (Thermo Electron Corporation, Vantaa, Finland). Luciferase assay substrate (100 μl) (Promega) was mixed with the samples and luminescence was measured instantly. Because the luminescence fades extremely fast, we used a multichannel pipet for addition of substrate and rapid mixing. No more than 12 samples were measured at a time. The luminescence was measured using a Victor 1420 Multilabel Counter (Wallac, Turku, Finland).

To correct the data for differences in transfection frequency between the samples, the β-galactosidase activity in the lysates was measured. Sample lysates (10 μl) were pipetted in duplicate onto a F96 cert maxisorp immuno-plate (Nunc A/S, Roskilde, Denmark). Water (10 μl) was used as a 0-sample for measuring background. The reaction mixture (65 μl) consisting of 1 mM MgCl$_2$, 45 mM α-mercaptoethanol, 1 mg/ml *ortho*-nitrophenyl-β-D-galactopyranoside (ONPG; Sigma-Aldrich), and 0.1 M Na-phosphate, pH 7.0, was mixed with the samples. The hydrolysis of ONPG produces the yellow compound *ortho*-nitrophenol. Samples were incubated at +37° for 15–30 min until the yellow color developed, and the reaction was stopped by adding 125 μl of 1 M Na$_2$CO$_3$. The absorbance was measured at 420 nm. Figure 4 shows the results from a single representative two-hybrid experiment.

Pull-Down Assay

Our pull-down assay is based on interaction of *in vitro*-translated ^{35}S-labeled protein fragments with GST-Rab7 or plain GST (negative control) immobilized on glutathione-Sepharose. The full-length ORP1L/pcDNA3.1 plasmid was used as template for the amplification of truncated constructs. ORP1L fragments were cloned into vector pcDNA4 HisMaxC (Invitrogen), which contains the promoter for bacteriophage T7 RNA

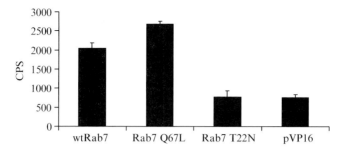

FIG. 4. A two-hybrid luciferase assay for ORP1L interaction with Rab7 carried out in COS-1 cells (see the section on Two-Hybrid Luciferase Assay). ORP1L expression plasmid was transfected into cells together with a plasmid encoding wtRab7, the GTPase-deficient mutant Rab7 Q67L, the dominant inhibitory mutant Rab7 T22N, or empty vector pVP16 as a negative control. The catalytic activity of the luciferase enzyme, detected as production of light (CPS), is a measure of the transcriptional activity of the reporter plasmid, which in turn is dependent on formation of the two-hybrid complex. The diagram shows the results from a single representative experiment.

polymerase. Wild-type Rab7 cDNA was cloned into the *Xho*I site of vector pGEX4T1 (Amersham Pharmacia Biotech) for production of the GST fusion protein. Oligonucleotide primers for the generation of the cDNA constructs are listed in Table I.

Sulfur-35-labeled full-length ORP1L and ORP1L protein fragments were synthesized using the TnT T7-coupled reticulocyte lysate system (Promega). This kit combines transcription and translation in a single tube reaction. Reaction mixes (25 μl) were prepared consisting of 6.5 μl water treated with 0.1% diethylpyrocarbonate (DEPC), 1.0 μl TnT reaction buffer, 12.5 μl rabbit reticulocyte lysate, 0.5 μl amino acid mixture minus methionine or cysteine (1 mM), 1.0 μl Redivue Promix ^{35}S cell labeling mix (14.3 mCi/ml, >1000 Ci/mmol) (Amersham), 2.5 μl plasmid DNA (200ng/ μl), 0.5 μl RNasin RNase inhibitor (Promega), and 0.5 μl T7 polymerase. The rabbit reticulocyte lysate, the TnT T7 polymerase, the reaction buffer, and the amino acid mixtures are provided in the kit. Reaction mixtures were incubated for 1.5 h at 30° and placed on ice. A small aliquot (2 μl) of each reaction was transferred to another tube, boiled with 10 μl of SDS sample buffer, and stored at −20°.

Sixty microliters of glutathione-Sepharose 4B slurry (Amersham Biosciences AB) was washed four times with coupling buffer (PBS, 2 mM MgCl$_2$, 1 mM DTT). The beads were centrifuged at 400×g for 1 min between each wash. Purified GST-Rab7 (150 μg) or plain GST (150 μg) was added to the washed beads and incubated on a roller overnight at $+4°$. The beads with bound proteins were washed once with coupling buffer and once with loading buffer (20 mM HEPES, 100 mM KAc, 0.5 mM MgCl$_2$, 1 mM DTT, 2 mM EDTA, pH 7.2) and incubated with 10 μM guanosine-5'-O-(3-thiotriphosphate) (GTPγS; Sigma-Aldrich), a nonhydrolyzable GTP analogue, in loading buffer (300 μl) for 1 h at $+37°$. The MgCl$_2$ concentration was adjusted to 10 mM to stabilize the GTPγS–Rab7 complex, and *in vitro* translated ORP1L fragments (11.5 μl) and albumin (10 mg/ml) were added. The beads were kept on a roller for 1 h at room temperature and washed three times with wash buffer (20 mM HEPES, 100 mM KAc, 5 mM MgCl$_2$, 1 mM DTT, pH 7.2). The coupled proteins were released from the beads with 30 μl of 15 mM glutathione in wash buffer for 20 min at room temperature. The supernatant was recovered and boiled with 30 μl SDS-PAGE loading buffer. Half of the sample was resolved in a 12.5% SDS-polyacrylamide gel. The gel was fixed with 10% acetate, 25% methanol, for 30 min, and then incubated with Amplify (Amersham Biosciences, Buckinghamshire, England) for 30 min, dried, and exposed on Kodak X-ray film at $-70°$. The 2-μl sample taken directly after the *in vitro* translation reaction was exposed for 2 days, while the eluted samples were exposed for 1 week. Figure 5 shows specific pull-down

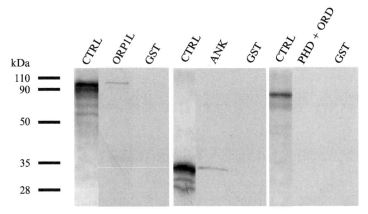

Fig. 5. Pulldown assay for interaction of ORP1L fragments with GST-Rab7 (see the section on Pulldown Assay). Full-length ORP1L or truncated fragments (ANK, aa 1–237; PHD + ORD, aa 211–950) were ^{35}S labeled during *in vitro* translation. The labeled proteins were incubated with GST-Rab7 or plain GST (negative control) immobilized on glutathione-Sepharose. Full-length ORP1L and the N-terminal ankyrin repeat region containing fragment were specifically pulled down by GST-Rab7. The CTRL lanes show the products of each *in vitro* translation reaction (2 μl loaded directly onto the gel) (Adapted with permission from Johansson *et al.*, 2005).

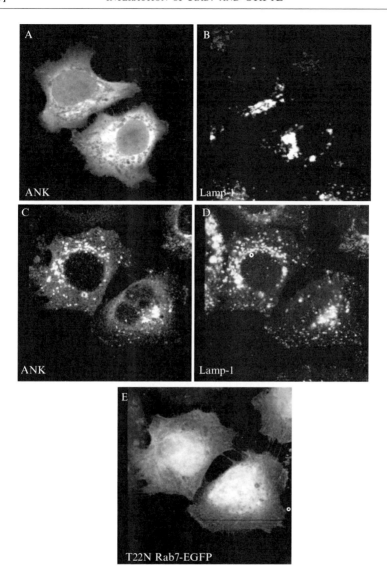

FIG. 6. Immunofluorescence microscopy of HeLa cells overexpressing the ankyrin repeat region of ORP1L. The ankyrin repeat region causes clustering of late endocytic compartments (A and B). The ANK fragment (aa 1–237) was overexpressed and cells were double immunostained with antibodies against ANK and endogenous Lamp-1 (the method is described in the section on colocalization of ORP1L with Rab7). The clusters are partially dissolved when the dominant inhibitory Rab7 T22N is coexpressed with the ANK fragment (C–E).

of the full-length ORP1L and the N-terminal ankyrin repeat-containing (ANK) fragment (aa 1–237) by GST-Rab7. When these N-terminal residues of ORP1L were absent, the result was negative. The assay demonstrates that the N-terminal region of aa 1–237 contains the determinant for interaction with Rab7.

The Ankyrin Repeat Region of ORP1L Affects the Microtubule-Dependent Localization of Late Endosomes

The results from the pull-down assay (see the section on pull-down assay) showed that the N-terminal ankyrin repeat region (aa 1–237) of ORP1L binds Rab7. When this fragment is overexpressed in HeLa cells, a marked change in the morphology of Lamp-1-positive compartments is evident. The compartments appear as large clusters in the perinuclear region (Fig. 6). Treatment of the cells with the microtubule depolymerizing agent nocodazole (10 μM, for 30 min at $+37°$) showed that this clustering is dependent on a functional microtubule network. The nocodazole treatment caused dispersal of the Lamp-1-positive clusters toward the cell periphery, partially dissolving them (not shown).

ORP1L preferentially binds active Rab7, as evidenced by the results from the two-hybrid assay (see the section Two-Hybrid Luciferase Assay). Overexpression of wild-type Rab7 or the GTPase-deficient Q67L mutant induces formation of enlarged late endosomes in the perinuclear region. In contrast, when the dominant inhibitory mutant Rab7 T22N is overexpressed late endosomes are scattered in the cell periphery (Bucci et al., 2000). Rab7 has been shown to play a role in the microtubule-dependent mobility of late endosomes (Cantalupo et al., 2001; Harrison et al., 2003; Jordens et al., 2001; Lebrand et al., 2002). When the ANK fragment of ORP1L was expressed together with the Rab7 T22N mutant, the late endosomal clusters were dispersed as in the nocodazole-treated cells, suggesting that the ANK-induced late endosomal clustering involves interaction with active Rab7 (Fig. 6).

Acknowledgments

We are grateful to Seija Puomilahti and Pirjo Ranta for skillful technical assistance. M.J. is a member of the Helsinki Graduate School of Biotechnology and Molecular Biology. Dr. Angela Wandinger-Ness is thanked for kindly providing reagents. This study was supported by the Academy of Finland (Grants 54301 and 206298 to V.M.O.), the Finnish Foundation for Cardiovascular Research (V.M.O.), and the Sigrid Juselius Foundation (V.M.O.).

References

Bucci, C., Thomsen, P., Nicoziani, P., McCarthy, J., and van Deurs, B. (2000). Rab7: A key to lysosome biogenesis. *Mol. Biol. Cell* **11**, 467–480.

Cantalupo, G., Alifano, P., Roberti, V., Bruni, C. B., and Bucci, C. (2001). Rab-interacting lysosomal protein (RILP): The Rab7 effector required for transport to lysosomes. *EMBO J.* **20**, 683–693.

Chavrier, P., Parton, R. G., Hauri, H. P., Simons, K., and Zerial, M. (1990). Localization of low molecular weight GTP binding proteins to exocytic and endocytic compartments. *Cell* **62**, 317–329.

Dawson, P. A., Van der Westhuyzen, D. R., Goldstein, J. L., and Brown, M. S. (1989). Purification of oxysterol binding protein from hamster liver cytosol. *J. Biol. Chem.* **264**, 9046–9052.

Dong, J., Chen, W., Welford, A., and Wandinger-Ness, A. (2004). The proteasome alpha-subunit XAPC7 interacts specifically with Rab7 and late endosomes. *J. Biol. Chem.* **279**, 21334–21342.

Edinger, A. L., Cinalli, R. M., and Thompson, C. B. (2003). Rab7 prevents growth factor-independent survival by inhibiting cell-autonomous nutrient transporter expression. *Dev. Cell* **5**, 571–582.

Feng, Y., Press, B., and Wandinger-Ness, A. (1995). Rab 7: An important regulator of late endocytic membrane traffic. *J. Cell Biol.* **131**, 1435–1452.

Fields, S., and Song, O. (1989). A novel genetic system to detect protein-protein interactions. *Nature* **340**, 245–246.

Harrison, R. E., Bucci, C., Vieira, O. V., Schroer, T. A., and Grinstein, S. (2003). Phagosomes fuse with late endosomes and/or lysosomes by extension of membrane protrusions along microtubules: Role of Rab7 and RILP. *Mol. Cell. Biol.* **23**, 6494–6506.

Jager, S., Bucci, C., Tanida, I., Ueno, T., Kominami, E., Saftig, P., and Eskelinen, E. L. (2004). Role for Rab7 in maturation of late autophagic vacuoles. *J. Cell Sci.* **117**, 4837–4848.

Jaworski, C. J., Moreira, E., Li, A., Lee, R., and Rodriguez, I. R. (2001). A family of 12 human genes containing oxysterol-binding domains. *Genomics* **78**, 185–196.

Johansson, M., Bocher, V., Lehto, M., Chinetti, G., Kuismanen, E., Ehnholm, C., Staels, B., and Olkkonen, V. M. (2003). The two variants of oxysterol binding protein-related protein-1 display different tissue expression patterns, have different intracellular localization, and are functionally distinct. *Mol. Biol. Cell* **14**, 903–915.

Johansson, M., Lehto, M., Tanhuanpää, K., Cover, T. L., and Olkkonen, V. M.. (2005). The oxysterol binding protein homologue ORP1L interacts with Rab7 and alters functional properties of late endocytic compartments. *Mol. Cell. Biol.* In press.

Jordens, I., Fernandez-Borja, M., Marsman, M., Dusseljee, S., Janssen, L., Calafat, J., Janssen, H., Wubbolts, R., and Neefjes, J. (2001). The Rab7 effector protein RILP controls lysosomal transport by inducing the recruitment of dynein-dynactin motors. *Curr. Biol.* **11**, 1680–1685.

Lagace, T. A., Byers, D. M., Cook, H. W., and Ridgway, N. D. (1997). Altered regulation of cholesterol and cholesteryl ester synthesis in Chinese-hamster ovary cells overexpressing the oxysterol-binding protein is dependent on the pleckstrin homology domain. *Biochem. J.* **326**, 205–213.

Lagace, T. A., Byers, D. M., Cook, H. W., and Ridgway, N. D. (1999). Chinese hamster ovary cells overexpressing the oxysterol binding protein (OSBP) display enhanced synthesis of sphingomyelin in response to 25-hydroxycholesterol. *J. Lipid Res.* **40**, 109–116.

Lebrand, C., Corti, M., Goodson, H., Cosson, P., Cavalli, V., Mayran, N., Faure, J., and Gruenberg, J. (2002). Late endosome motility depends on lipids via the small GTPase Rab7. *EMBO J.* **21,** 1289–1300.

Lehto, M., and Olkkonen, V. M. (2003). The OSBP-related proteins: A novel protein family involved in vesicle transport, cellular lipid metabolism, and cell signalling. *Biochim. Biophys. Acta* **1631,** 1–11.

Lehto, M., Laitinen, S., Chinetti, G., Johansson, M., Ehnholm, C., Staels, B., Ikonen, E., and Olkkonen, V. M. (2001). The OSBP-related protein family in humans. *J. Lipid Res.* **42,** 1203–1213.

Marsman, M., Jordens, I., Kuijl, C., Janssen, L., and Neefjes, J. (2004). Dynein-mediated vesicle transport controls intracellular Salmonella replication. *Mol. Biol. Cell* **15,** 2954–2964.

Meresse, S., Gorvel, J. P., and Chavrier, P. (1995). The rab7 GTPase resides on a vesicular compartment connected to lysosomes. *J. Cell Sci.* **108,** 3349–3358.

Pfeffer, S. R. (2001). Rab GTPases: Specifying and deciphering organelle identity and function. *Trends Cell Biol.* **11,** 487–491.

Press, B., Feng, Y., Hoflack, B., and Wandinger-Ness, A. (1998). Mutant Rab7 causes the accumulation of cathepsin D and cation-independent mannose 6-phosphate receptor in an early endocytic compartment. *J. Cell Biol.* **140,** 1075–1089.

Sadowski, I., Ma, J., Triezenberg, S., and Ptashne, M. (1988). GAL4-VP16 is an unusually potent transcriptional activator. *Nature* **335,** 563–564.

Stein, M. P., Feng, Y., Cooper, K. L., Welford, A. M., and Wandinger-Ness, A. (2003). Human VPS34 and p150 are Rab7 interacting partners. *Traffic* **4,** 754–771.

Taylor, F. R., Saucier, S. E., Shown, E. P., Parish, E. J., and Kandutsch, A. A. (1984). Correlation between oxysterol binding to a cytosolic binding protein and potency in the repression of hydroxymethylglutaryl coenzyme A reductase. *J. Biol. Chem.* **259,** 12382–12387.

Wang, P. Y., Weng, J., and Anderson, R. G. (2005). OSBP is a cholesterol-regulated scaffolding protein in control of ERK 1/2 activation. *Science* **307,** 1472–1476.

Wyles, J. P., McMaster, C. R., and Ridgway, N. D. (2002). Vesicle-associated membrane protein-associated protein-A (VAP-A) interacts with the oxysterol-binding protein to modify export from the endoplasmic reticulum. *J. Biol. Chem.* **277,** 29908–29918.

Zerial, M., and McBride, H. (2001). Rab proteins as membrane organizers. *Nat. Rev. Mol. Cell Biol.* **2,** 107–117.

[66]   Characterization of Rab23, a Negative Regulator of Sonic Hedgehog Signaling

By TIMOTHY M. EVANS, FIONA SIMPSON,
ROBERT G. PARTON, and CAROL WICKING

Abstract

The hedgehog signaling pathway is indispensable in embryogenesis, being responsible for the development of a wide array of vertebrate organs. Given its importance in embryogenesis, the precise regulation of hedgehog signaling is crucial. Aberrant activation of this pathway in postnatal life has been associated with a number of tumor types, reinforcing the role of developmental signaling pathways in tumorigenesis. The small GTPase Rab23 acts as a negative regulator of the hedgehog signaling pathway, most notably in the vertebrate neural system. By analogy with studies of other Rab proteins, analysis of the localization of wild-type and constitutively active and inactive forms of Rab23 provides the potential to shed light on the role of Rab23 at the cellular level. We previously produced expression constructs encoding these proteins for analysis in mammalian cell cultures at both the light and the electron microscopy level. This revealed that both wild-type and active Rab23 localizes to the plasma membrane and to endocytic vesicles (T. M. Evans *et al.* [2003] *Traffic* **4,** 869–884). We describe the methods used to design and make the Rab23 expression constructs, and to assess their localization relative to key hedgehog pathways and endocytic markers in both transiently and stably transfected cell cultures.

Introduction

The hedgehog signaling pathway is pivotal in embryogenesis, being implicated in the development of a wide array of vertebrate organs (Ingham and McMahon, 2001). Aberrant activation of this pathway in postnatal life has also been linked with a number of tumor types, reinforcing the key role of developmental signaling pathways in tumorigenesis (reviewed in Beachy *et al.*, 2004; Wicking and McGlinn, 2001). In the prevailing model for hedgehog signaling, reception of hedgehog ligand at the plasma membrane serves to alleviate static repression of the pathway, ultimately leading to a change in transcriptional output of target genes. In the absence of hedgehog ligand, the activity of smoothened, a

METHODS IN ENZYMOLOGY, VOL. 403   0076-6879/05 $35.00
DOI: 10.1016/S0076-6879(05)03066-1

7-transmembrane molecule, is inhibited by the 12-transmembrane hedge-hog receptor, patched. Binding of hedgehog to patched relieves this inhibition on smoothened, thus facilitating the transmission of the signal to the nucleus. While initial details of this pathway emerged largely from genetic–based studies in *Drosophila*, the past decade has seen an increase in our understanding of the details regulating the reception and transduction of the vertebrate hedgehog signal at the cellular level. An emerging theme from these studies has been the importance of the role that cholesterol levels and intracellular trafficking events play in the precise regulation and coordinated movement of key pathway members within the cell. An important milestone was the discovery that the small GTPase Rab23 acts as a negative regulator of the hedgehog signaling pathway, most notably in the vertebrate neural system (Eggenschwiler *et al.*, 2001).

The initial connection between hedgehog signaling and Rab23 came from analysis of the phenotype of two independent alleles of the mouse mutant *open-brain (opb)*, one that arose spontaneously and one as a result of ethyl-nitrosylurea (ENU) mutagenesis. Phenotypically the *opb* embryos resemble a partial loss of function of the Sonic hedgehog receptor Patched, particularly evident in the spinal cord where dorsal cell types fail to develop (Eggenschwiler and Anderson, 2000). This is in contrast to *Sonic hedgehog (Shh)* null embryos that lack ventral cell types in the spinal cord (Chiang *et al.*, 1996). These data, combined with genetic studies in *Shh/opb* double mutant mice, led to the conclusion that the product of the *opb* gene acts distal to Shh as a negative regulator of hedgehog signaling (Eggenschwiler *et al.*, 2001). Positional cloning revealed that the *opb* locus encodes Rab23, a relatively uncharacterized member of the Rab GTPase family of vesicle transport proteins (Eggenschwiler *et al.*, 2001).

Rab proteins localize to specific membrane compartments and coordinate stages of membrane transport, such as vesicle formation, vesicle and organelle motility, and tethering of vesicles to their target compartment (reviewed in Zerial and McBride, 2001). Rab proteins cycle between a cytosolic inactive GDP-bound and a membrane-associated active GTP-bound form, a process tightly regulated by a number of accessory proteins (reviewed in Alory and Balch, 2001; Olkkonen and Stenmark, 1997). A full appreciation of the role of Rab23 in embryogenesis will be enhanced by an understanding of its function at the cellular level. By analogy with studies of other Rab proteins, we made expression constructs of wild-type, constitutively active and inactive forms of Rab23. Analysis of localization of these constructs in mammalian cell cultures at both the light and the electron microscopy level revealed that both wild-type and active Rab23 localizes to the plasma membrane and to endocytic vesicles (Evans *et al.*, 2003). In this chapter we describe the methods used to design and make the Rab23 constructs, and to assess their localization relative to key hedgehog pathway and endocytic markers in both transiently and stably transfected cell cultures.

Detection of *Rab23* Expression by RT-PCR

Primer Design

The detection of *Rab23* transcripts by reverse transcriptase polymerase chain reaction (RT-PCR) in cell lines derived from different species is reliant on thoughtful primer design. The cell lines analyzed for *Rab23* expression were C3H/10T1/2 from mouse, A431 and HeLa from human, and BHK-21 from hamster (Fig. 1). To assist in primer design, sequence comparisons were performed using the Clustal W algorithm (Thompson *et al.*, 1994; Figs. 2 and 3). Primers were designed to regions unique to Rab23 sequences (encoding amino acids 1–9 and 178–205; Fig. 2), but having complete identity between human and mouse DNA sequences (Fig. 3). The hamster *Rab23* sequence is not represented in GenBank and was therefore not included in the alignment. Nonetheless, on the basis of mice and hamsters sharing the same subgroup of the rodent evolutionary tree, it was rationalized that the hamster *Rab23* sequence would be closely related to that of the mouse. The primers designed meeting these criteria are detailed in Table I.

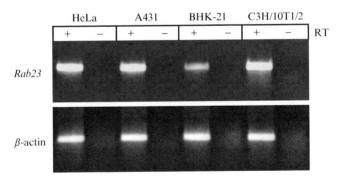

FIG. 1. RT-PCR for *Rab23* transcripts. *Rab23* transcripts were detected in the human cervical adenocarcinoma cell line HeLa, human epidermoid carcinoma cell line A431, the Syrian golden hamster kidney fibroblast cell line BHK-21, and the sonic hedgehog responsive mouse mesodermal cell line C3H/10T1/2. To detect *Rab23*, PCR was performed for 35 cycles with an annealing temperature of 55°. For β-actin, PCR was performed for 30 cycles with an annealing temperature of 60°. No bands were amplified from control reactions where reverse transcriptase (RT) was excluded (−). Reproduced with permission from Evans *et al.* (2003).

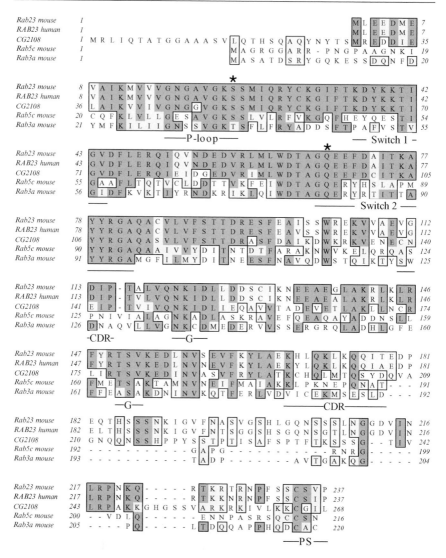

FIG. 2. Alignment of Rab23 proteins. Mouse and human Rab23 proteins, the *Drosophila* Rab23 orthologue CG2108, mouse Rab5c, and mouse Rab3a proteins were aligned using the Clustal W algorithm (Thompson *et al.*, 1994). The GenBank accession numbers for each protein are Rab23 mouse, NP_033025; Rab23 human, NP_899050; CG2108, NP_649574; Rab5c mouse, NP_077776; Rab3a mouse, NP_033027. Functional domains as defined in other Rab family proteins are indicated below the sequence. P-loop, phosphate-binding loop; Switch 1 and Switch 2, conformational switch regions; CDR, Rab effector-binding regions; G, guanine nucleotide binding residues; PS, prenylation site. Amino acids mutated by site-directed mutagenesis for analysis in this study were S23N and Q68L, both indicated by an asterisk above the amino acid sequence.

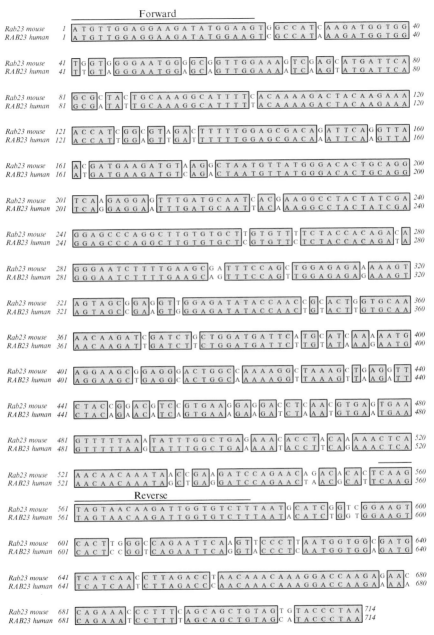

FIG. 3. Alignment of mouse and human Rab23 coding sequences. Mouse *Rab23* and human *RAB23* coding sequences were aligned using the Clustal W algorithm (Thompson *et al.*, 1994) to assist in the design of RT-PCR primers. Primers designed were to regions unique to Rab23 but having total identity between human and mouse DNA sequences as indicated.

TABLE I

Oligonucleotide Primers Used for RT-PCR of Rab23 Transcripts, and the
Cloning and Site-Directed Mutagenesis of Rab23 Coding Sequence in
Mammalian Expression Vectors[a]

Primers	
RT-PCR	
Forward	5′-GAGAAGCTTCGATGTTGGAGGAAGATATGGAAG-3′
Reverse	5′-GGTCTAGATTTAGGGTACACTACAGCTGCTG-3′
Cloning	
Forward	5′-GAG<u>AAGCTT</u>CGATGTTGGAGGAAGATATGGAAG-3′
Reverse	3′-GG<u>TCTAGA</u>TTTAGGGTACACTACAGCTGCTG-3′
Site-directed mutagenesis	
S23N	
Forward	5′-GGGCGGTTGGAAAG**AAC**AGCATGATTCAG- CGCTACTG-3′
Q68L	
Reverse	3′-CAGTAGCGCTGAATCATGCT**GTT**CTTTCCAACCGCCCC-3′
Forward	5′-GTTATGGGACACTGCAGGTC**T**AGAGGAGTTTGATG-3′
Reverse	5′-CATCAAACTCCTCT**A**GACCTGCAGTGTCCCATAAC-3′

[a] Underlined sequences represent restriction sites incorporated into the primers. The bases changed in the mutagenesis reaction are indicated in bold.

Additionally, designing RT-PCR primers to target transcript regions spanning intron/exon boundaries can improve specificity of primers for transcript sequences thus decreasing the chance of amplification from contaminating genomic DNA. Comparison of human *RAB23* (ENSEMBL gene ID: ENSG00000112210) and mouse *Rab23* (ENSEMBL gene ID: ENSMUSG00000004768) genomic sequence and structure revealed that target regions for each RT-PCR primer designed are present on separate exons.

Reagents

1–3 μg total RNA from cell lines

$10 \times$ *Taq* polymerase PCR buffer (100 mM Tris–HCl, pH 8.4, 500 mM KCl)

90 OD U/ml pd(N)$_6$

25 mM dNTPs

Moloney murine leukemia virus (MMLV) reverse transcriptase

25 mM MgCl$_2$

Taq polymerase

7.5 mM MgCl$_2$

RT-PCR primers as above

Diethylpyrocarbonate (DEPC)-treated Milli Q water

Protocol

1. RT reactions should be prepared to contain the following in a 30 μl final volume:
   - 2 μg of total RNA
   - 4.5 μl of 10 $\times$ *Taq* polymerase PCR buffer
   - 3 mM MgCl$_2$
   - 1 μl of 25 mM dNTPs
   - 3 U/ml pd(N)$_6$
   - 100 U of MMLV reverse transcriptase
2. Incubate reactions at room temperature for 10 min followed by 42° for 1 h and 95° for 5 min to stop the reaction.
3. PCR reactions should be prepared to contain the following in a 20 μl final volume:
   - 6 μl of RT reaction above
   - 1.1 μl of 10 $\times$ *Taq* polymerase PCR buffer
   - 1.6 μl of 7.5 mM MgCl$_2$
   - 2.5 ng/μl PCR primers
   - 1 U *Taq* DNA polymerase
4. Overlay the reaction mixture with mineral oil and cycle in a thermal cycler as follows: 94° for 2 min, followed by 35 cycles of 94° for 30 s, 60° for 45 s, 72° for 30 s.

Cloning of Rab23 for Transient Expression

The strategy to study the localization of Rab23 transiently expressed as green fluorescent protein (GFP) fusion proteins has been used to analyze numerous other Rab proteins (reviewed in Olkkonen and Stenmark, 1997). Based on published open reading frame predictions (Olkkonen *et al.*, 1994), the mouse *Rab23* coding sequence was amplified by PCR amplification from a *Rab23* cDNA clone using the primers detailed in Table I. The resulting fragment was cloned into pEGFP-C1 expression construct to allow expression of GFP-tagged Rab23 protein (pEGFP-Rab23, Fig. 4A). The Rab23 coding sequence was also cloned into a mammalian expression vector 3′ to a hemagglutinin (HA) epitope tag (pcHA-Rab23, Fig. 4A). For both constructs transient expression of protein of the predicted size was confirmed by standard Western blotting techniques (Sambrook *et al.*, 1989; Fig. 4B).

To facilitate the analysis of Rab23 cellular distribution, site-directed mutagenesis was performed on the wild-type Rab23-GFP at conserved residues found in the GTP/GDP binding motifs of all Ras superfamily GTPases (Feig, 1999; Olkkonen and Stenmark, 1997). Mutations in the GDP/GTP binding regions in p21-ras were shown to influence guanine

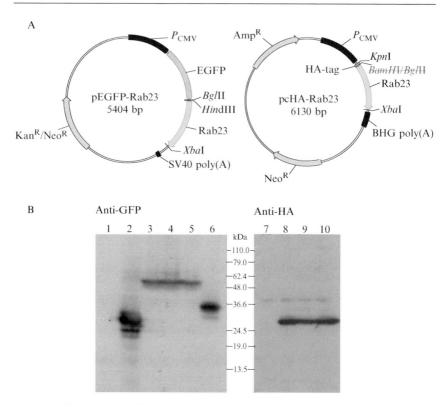

FIG. 4. Cloning of Rab23 for mammalian expression. (A) Schematic representation of Rab23 expression constructs, pEGFP-Rab23 and pcHA-Rab23, as used in this study. Kan^R, kanamycin resistance; Neo^R, neomycin resistance; Amp^R, ampicillin resistance; P_{CMV}, cytomegalovirus promoter; EGFP, enhanced green fluorescent protein; SV40 poly(A), simian virus 40 polyadenylation signal; BHG poly(A), bovine growth hormone polyadenylation signal. (B) Protein expression from Rab23 constructs analyzed by Western blotting according to standard techniques (Sambrook *et al.*, 1989). BHK-21 cells were transfected with the following: 1 and 7, untransfected control; 2, pEGFP-C1; 3, wild-type pEGFP-Rab23; 4, pEGFP-Rab23S23N; 5, pEGFP-Rab23Q68L; 6, pEGFP-Rab23R80X; 8, wild-type pcHA-Rab23; 9, pcHA-Rab23S23N; 10, pcHA-Rab23Q68L. Lanes 1–6 were probed with rabbit anti-GFP antibody and lanes 7–10 were probed rabbit anti-HA antibody followed by anti-rabbit HRP antibody. The sizes of detected bands are as predicted: GFP, ∼28 kDa; Rab23-GFP fusion, ∼54 kDa; Rab23R80X-GFP, ∼36 kDa; Rab23-HA fusion proteins, ∼27 kDa. (See color insert.)

nucleotide binding and GTP hydrolysis, and these were subsequently applied to the study of Rab proteins, initially to Rab5 (Li and Stahl, 1993; Stenmark *et al.*, 1994). The alignment of Rab23, Rab5, and Rab3 illustrates the conservation of certain amino acids critical to Rab function (Fig. 2). It was predicted that the mutation S23N would confer a loss of function resulting from a decreased affinity for GTP, and the mutation Q68L would confer a gain of function resulting from a loss of GTPase activity. The amino acid changes S23N and Q68L were introduced into the Rab23 sequence of the plasmid pEGFP-Rab23 using site-directed mutagenesis as described below. All constructs generated were sequenced for incorporation of the desired mutation. The subcellular distribution of wild-type active and inactive forms of Rab23 protein was analyzed.

Site-Directed Mutagenesis

Primer Design. A primer pair for each site-directed mutagenesis should be designed according to the following parameters. The mutation to be made should be in the middle of both primers with approximately 10–15 bases of correct sequence on both sides and should be between 25 and 45 bases in length. The melting temperature (T_m) should be greater than or equal to 78° as calculated by the following formula:

$$T_m = 81.5 + 0.41(\%GC) - 675/N - \%\text{mismatch}$$

where N is the primer length in bases and where the values for %GC and %mismatch are whole numbers.

Primers should have one or more C or G bases at their 3′ ends. Primers designed according to these parameters are frequently the reverse complement of each other. Primers purified by high-performance liquid chromatography (HPLC) or polyacrylamide gel electrophoresis (PAGE) are preferred, although desalted primers perform adequately. Primers successfully used for site-directed mutagenesis of Rab23 coding sequence are shown in Table I.

Reagents

10 × *Pfu* polymerase buffer (20 mM Tris–HCl, pH 8.8, 10 mM [NH$_4$]$_2$SO$_4$, 10 mM KCl, 0.1% Triton X-100, 0.1 mg/ml bovine serum albumin [BSA], 2 mM MgSO$_4$)
2 mM dNTPs
Pfu polymerase
Primers designed as above to incorporate desired mutation
*Dpn*I restriction enzyme
Competent *Escherichia coli* cells such as DH5α or XL1 Blue

Protocol

1. Reactions should be prepared to contain the following in a 50 μl final volume:
   • 1 × *Pfu* polymerase reaction buffer

- 5–50 ng of plasmid DNA template
- 125 ng each of primer #1 and primer #2
- 0.2 mM dNTP
- 2.5 U of *Pfu* polymerase

2. Overlay the reaction mixture with mineral oil and cycle in a thermal cycler as follows: 95° for 30 s, followed by 12–16 cycles of 95° for 30 s, 55° for 1 min, 68° for ≥2 min/kb of plasmid length.

3. (Optional) Run 10 μl of reaction on a 1% (w/v) agarose gel. Successful reactions usually have a visible band at a size corresponding to the linearized parent plasmid.

4. To digest parent (nonmutated) plasmid, add 1 μl of *Dpn*I restriction enzyme and incubate the reaction at 37° for 1 h.

5. Transform 1 μl of the reaction into a strain of competent *E. coli* cells suitable for the propagation and maintenance of plasmid DNA such as DH5α or XL1 Blue. Competent cells having transformation efficiency of greater than 1×10^7 cfu/μg DNA are preferred.

6. Propagate the resulting colonies to recover the mutated plasmid and sequence the insert to ensure integrity.

Endotoxin Removal from Plasmid Maxi-prep DNA

Lipopolysaccharide (LPS) or endotoxin is a major component of Gram-negative bacteria cell walls and is a common contaminant in plasmid DNA preparations. Removal of LPS from plasmid DNA preparations used for transfection of mammalian cells gives increased transfection rates and decreased apparent toxicity upon transfection (Cotten *et al.*, 1994). Methods were adapted from those used to remove LPS from protein samples (Aida and Pabst, 1990) and are subsequent to purification of cleared lysates by column purification as in Genomed (Lohne, Germany), Jetstar maxi, and Qiagen (Valencia, CA) plasmid maxi kits.

Reagents

TE buffer (pH 8.0)
3 M sodium acetate solution (pH 5.2)
Triton X-114
70% and 100% (v/v) ethanol solutions

Protocol

1. Plasmid DNA in 500 μl of TE buffer (pH 8.0) should be adjusted to 0.3 M sodium acetate (pH 5.2).

2. Add Triton X-114 to a final concentration of 1%. Vortex the samples and incubate on ice for 5 min with intermittent mixing to ensure a homogeneous solution.
3. Heat chilled samples to 37° for 5 min followed by centrifugation in a benchtop microfuge for 1 min at maximum speed at room temperature to allow two phases to form.
4. Remove the upper aqueous phase (approximately 450 μl) immediately so as not to disrupt the detergent phase in the form of an oily droplet at the bottom of the tube, as this phase contains the LPS.
5. Add 900 μl of 100% ethanol to the collected aqueous phase to precipitate the plasmid DNA. The DNA is then pelleted by centrifugation.
6. Wash the pelleted DNA with 70% ethanol and then resuspend in TE buffer (pH 8.0). To determine the concentration of plasmid DNA, dilute the samples 1 in 300 in water and analyze spectrophotometrically as per standard techniques (Sambrook *et al.*, 1989). Plasmid DNA samples should be stored at a concentration of 1 μg/μl at $-20°$ until use.

Transfection of Mammalian Cells

Reagents

Cell culture medium: Dulbecco's modified Eagle's medium (DMEM) containing 5% (v/v) serum, with glutamine but without penicillin/ streptomycin
Rab23 in mammalian expression construct purified as described earlier
LipofectAMINE 2000 (Invitrogen Corp., Carlsbad, CA)
Opti-MEM (Invitrogen)

Protocol

1. On the day prior to transfection, seed the cells to be transfected from a flask of subconfluent cells to achieve 30% confluency on the day of transfection. Cells should be plated in cell culture medium without penicillin/streptomycin. On the day of transfection the media should be replaced, again omitting penicillin/streptomycin.
2. Dilute the DNA to be transfected into the suggested volume of Opti-MEM (Table II) without serum in a microfuge tube or 15-ml Falcon tube. It is important to note that DNA purity is an essential factor in achieving the best transfection efficiency with the least cell

TABLE II
Suggested DNA, LF2000, and Opti-MEM Mixes for Transfection

Plate size	Plate surface area (approx.) (cm²)	Plating volume (ml)	DNA per well (μg)	Opti-MEM volume for DNA dilution (μl)	Volume of LF2000 reagent (μl)	Opti-MEM for LF2000 dilution (μl)
24-well	2.0	0.5	0.7	50	2	50
6-well/35 mm	9.6	1.5	2	150	6	150
60 mm	21.1	3	4	300	12	300
100 mm	58.1	9	12	900	36	900

death. LPS removal using Triton X-114 is suggested and is described earlier in this chapter.

3. Dilute the LipofectAMINE 2000 (LF2000) into the suggested volume of Opti-MEM (Table II) without serum in a 1.5-ml microfuge tube or 15-ml Falcon tube and incubate at room temperature for 5 min (30 min maximum). Decreased transfection efficiencies are observed with longer incubations. If DMEM is used in place of Opti-MEM then this incubation should be no longer than 5 min.

4. Combine the diluted DNA with the diluted LF2000. Mix gently by pipetting and incubate for 20 min at room temperature.

5. Add the DNA-LF2000 complexes to the media on the cells and mix briefly with gentle rocking. This mixing is crucial to minimize localized toxicity of high LF2000 concentration.

6. Change the media on the transfected cells 3–4 h after transfection for fresh cell culture medium without penicillin/streptomycin.

7. Incubate the cells at 37° for 16–24 h prior to analysis.

Preparation of Stable Cell Lines

Reagents

Rab23 in mammalian expression construct with a neomycin resistance cassette

Selective medium: cell culture medium (including serum) containing an apropriate concentration of the selective drug geneticin or G418 (Invitrogen)

Cell culture medium including serum

Phosphate-buffered saline (PBS)

Cloning rings: prepared by cutting the upper 6 mm of blue (1 ml) pipette tips and sterilized by autoclaving

Grease, such as O-ring grease (Catalog No. 6101351: Techne, Inc., Burlington NJ), autoclaved in a glass petri dish

Trypsin-EDTA solution (Invitrogen)

Protocol

1. To determine the appropriate concentration of selective agent, incubate the cells to be transfected (at approximately 30% confluency) with increasing concentrations of selective agent. For G418, a concentration range 100–800 μg/ml in cell culture medium is appropriate. Choose the lowest concentration that efficiently kills all cells after 7 days as the working concentration for the selection.

2. Seed the cells to be transfected from a flask of subconfluent cells to achieve 80% confluency on the day of transfection.

3. Transfect the cells with Rab23 mammalian expression construct as above using LF2000.

4. After transfection, incubate the cells for approximately 16–20 h.

5. Trypsinize the transfected cells and prepare 1:10, 1:100 dilutions of cells in 10 ml of selective medium, in triplicate. Transfer these into 10-cm tissue culture dishes containing selective medium.

6. Incubate the cells for 10–14 days at 37° until colonies of approximately 5 mm in diameter appear. The media should be changed every 3 days or whenever dead cells appear.

7. Select the dilution of cells with an appropriate number of isolated colonies on each plate (approximately 10–15). Wash once with PBS.

8. Dip the cloning rings in sterile grease and place over the colonies. Add 100 μl of trypsin-EDTA into each cloning ring. An alternative to using trypsin-EDTA is to apply 100 μl of PBS into each cloning ring and pipette to dislodge the cells.

9. When the cells have detached, transfer the cells into 1 ml of selective medium in a 24-well plate.

10. When the cells are confluent, expand them into 6-well plates and further as required in the presence of the selective drug. Cells can be assayed for stable protein expression by immunofluorescence microscopy or Western blotting.

Transferrin Internalization

Cells to be used for transferrin internalization experiments should express the transferrin receptor. If the cells to be used do not endogenously express the transferrin receptor then human transferrin receptor expressed from a mammalian expression vector can be transiently transfected into the required cell line. It should also be noted that ferritinized transferrin is endocytosed more efficiently.

Reagents

PBS
Biotin- or fluorescein isothiocyanate (FITC)-conjugated human transferrin (Molecular Probes, Eugene, OR)
Hanks' buffered saline solution (Invitrogen)
7.5% (w/v) NaHCO$_3$ solution
1 M HEPES (pH 7.4)
DMEM supplemented with 5% (v/v) serum heated to 37°
1 M acetic acid stock solution
4% (w/v) paraformaldehyde in PBS

Protocol

1. Transiently transfect BHK-21 cells as described earlier, with an expression construct containing the human transferrin receptor the day prior to the experiment.
2. Wash the BHK-21 cells expressing the human transferrin receptor thoroughly with PBS.
3. Place the cells on ice and add 25 μg/ml human biotin-or FITC-conjugated human transferrin in Hanks' buffered saline solution supplemented with 0.75 g/liter NaHCO$_3$ and 10 mM HEPES (pH 7.4). Incubate on ice for 30 min.
4. Wash the cells thoroughly with PBS. Internalization of bound transferrin is initiated by the incubation of the cells at 37° in DMEM supplemented with 5% serum. Cells are then incubated for specific time points over a 60 min time course. Cells should be collected at 2, 5, 15, 30, and 60 min.
5. Prior to fixation remove the surface-bound transferrin by washing the coverslips twice in cold PBS, once in cold PBS containing 25 mM acetic acid, then twice in cold PBS.
6. Fix the coverslips in 4% paraformaldehyde prior to immunofluorescence.
7. Immunofluorescence methods are described below. FITC-conjugated transferrin can be viewed directly while biotin-conjugated transferrin is probed with Alexa- or FITC-conjugated streptavidin for imaging.

Immunofluorescence and Colocalization

To analyze colocalization of Rab23 or disruption of molecular trafficking pathways of Patched (PTCH) or Smoothened (Smo-HA) in cells co-transfected with Rab23 wild-type and mutant constructs, standard immunofluorescence analysis was used (Fig. 5).

Reagents

PBS
4% (w/v) paraformaldehyde in PBS
0.1% (v/v) Triton X-100 in PBS
2% (w/v) bovine serum albumin in PBS (blocking solution)
Primary antibodies such as chicken anti-PTCH antibody (Evans *et al.*, 2003) and rabbit anti-HA antibody (Cell Signaling Technology, Beverly, MA)
Secondary antibodies such as goat–anti-chicken Alexa fluor 594 and goat–anti-rabbit Alexa fluor (Molecular Probes)
Mowiol mounting solution: prepared by mixing 6 g glycerol, 2.4 g Mowiol 4–88 (Calbiochem, San Diego, CA), 6 ml dH$_2$O, and 12 ml 0.2 M Tris–HCl, pH 8.5, incubating overnight at 50° or until soluble; the solution is clarified by centrifugation at 4000–5000 rpm, sterile filtered, and stored at −20° until use

Protocol

1. Wash the cells cultured on glass coverslips in a 24-well plate that has been transfected as described earlier three times in PBS. Add 4% paraformaldehyde in PBS and fix the cells in this solution for at least 30 min at room temperature. Paraformaldehyde should be used in a fume cupboard and recommended safety precautions followed.

2. After fixation wash the cells twice in PBS and add PBS containing 0.1% Triton X-100. Incubate the cells at room temperature for 10–20 min. This treatment permeabilizes the cells.

3. Wash the cells twice in PBS. Cover the slips with blocking solution for 15 min. During this time stock solutions of primary antibodies to be used for immunofluorescence labeling should be centrifuged at 15,000 rpm at 4° for 15 min to pellet any precipitates.

4. Dilute the primary antibody to be used in blocking solution. To reduce the amount of antibody used, 25–30 μl drops of primary antibody in blocking solution can be placed on a sheet of Parafilm on the bench. The coverslips are then placed on the drops. Care should be taken to ensure

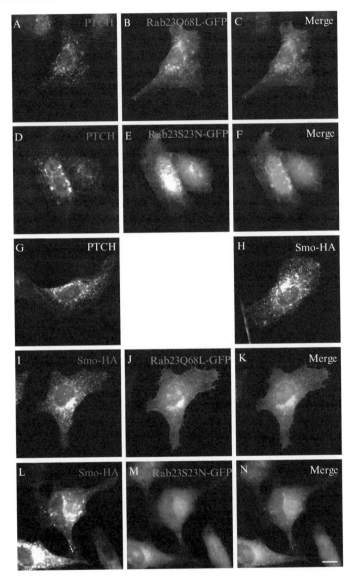

FIG. 5. Patched or smoothened distribution is unaltered in the presence of mutant Rab23. Human patched, PTCH, was transfected with Rab23Q68L-GFP (A–C), Rab23S23N-GFP (E–F), or alone (G). Rat smoothened, Smo-HA, was transfected with Rab23Q68L-GFP (I–K), Rab23S23N-GFP (L–N), or alone (H). Cells were labeled for PTCH with chicken anti-PTCH followed by anti-chicken Alexa fluor 594 (A, D, G, C red, F red) antibodies or labeled for Smo-HA with rabbit anti-HA followed by anti-rabbit Alexa fluor 594 antibodies (H, I, L, K red, N red). Rab23Q68L-GFP fluorescence (B, C green, J, K green) and Rab23S23N-GFP

that the side of the coverslip that has the cells on it is placed onto the liquid drop. Cells on coverslips should not be left dry at any point to prevent salt precipitation on the slips. Incubate the coverslips with the primary antibody for 1 h at room temperature.

5. Wash the coverslips at least three times in blocking solution over a 30-min time period. During this time the secondary antibody should be centrifuged to remove precipitates as described for the primary antibody in step 3.

6. Dilute the secondary antibody to the required concentration in blocking solution. Incubate the coverslips on drops of secondary antibody diluent as described in step 4. Incubate for 1 h at room temperature.

7. Wash the coverslips with at least three changes of blocking solution over a 30-min period, followed by two washes with PBS. (Optional: The coverslips can be immersed in dH$_2$O very rapidly and instantly blotted briefly by setting the edge of the coverslip against a piece of Whatman 3 M paper. Care should be taken not to allow drying of the slip and not to leave the slip in water for any time. This step can be used to reduce salt precipitation formed from PBS.)

8. Place a drop of Mowiol mounting solution on a clean slide. Mount the immunofluorescence coverslips by placing them cell side down on the Mowiol mounting solution. Ensure the coverslips are not moved after mounting and allow the Mowiol to harden prior to imaging.

Electron Microscopy

Cells transfected with Rab23-GFP were fixed with 8% (w/v) paraformaldehyde in PBS and then processed for frozen sectioning according to published methods (Liou *et al.*, 1996). Thawed sections were single-labeled with polyclonal rabbit anti-GFP followed by protein A-gold (10 nm). Plasma membrane sheets were prepared from Rab23-GFP transfected cells as previously described (Parton and Hancock, 2001) and labeled as described earlier. Here we describe the immunolabeling procedure.

Immunolabeling

Protocol. Antibody incubations and washing steps can be performed on a clean surface of the Parafilm held in place by the surface tension of a drop of water.

(E, F green, M, N green) fluorescence is shown. The distribution of PTCH or Smo-HA remains unaltered in the presence of these Rab23 mutants compared with cells transfected with PTCH or Smo-HA alone. Bar = 10 μm. Reproduced with permission from Evans *et al.* (2003). (See color insert.)

1. Prepare methyl cellulose (Sigma Chemical Co., St. Louis, MO) several days prior to the experiment. Make up a 2% (w/v) solution of methyl cellulose by adding the powder to cold distilled water. Methyl cellulose is slow to dissolve and is best dissolved in cold water at 4° for 2–3 days. When fully dissolved centrifuge at high speed (60,000 rpm in a Beckman 60 Ti or 70 Ti rotor) for 60 min at 4°. The solubilized methyl cellulose should be removed without disturbing the pellet and then stored at 4°.

2. Nonspecific binding is reduced by incubating grids on a blocking solution of 0.2% fish skin gelatin/0.1% BSA in PBS containing 50 mM glycine (to quench free aldehyde groups) for 10 min at room temperature.

3. The antibody, diluted in blocking solution, should be centrifuged for 5 min at 13,000 rpm in a bench-top centrifuge to remove any aggregates formed during storage.

4. Place 5 μl drops of the antibody on the Parafilm surface and float the grids, with sections down, on the grids for 30 min at room temperature.

5. Wash the grids in PBS five times for 5 min each. Remove the grids from the antibody with forceps and place onto a large drop of PBS. Transfer the grids from drop to drop with forceps or a large wire loop. Take care not to dry the surface of the grid.

6. Dilute the protein A-gold (University of Utrech, The Netherlands) in blocking solution (do not centrifuge prior to use).

7. Place the grids section down on 5 μl drops of protein A-gold for 30 min.

8. Repeat the wash step as for step 4 but washing six times for 5 min each.

9. Wash the grids by floating them on large drops of water. This step is very **important** as it removes all phosphate molecules prior to incubation with uranyl acetate. Phosphates present in the sections will precipitate the uranyl salts.

10. Contrasting and drying. The general procedure is as follows: mix the methyl cellulose with a 3% (w/v) aqueous uranyl acetate solution to give a final concentration of 0.3% (w/v) uranyl acetate (9 parts methyl cellulose to 1 part uranyl acetate).

11. Put drops of this solution onto a clean surface (Parafilm on a plastic platform) on ice and float the grids section down on the drops for 10 min.

12. Remove each grid individually from the methyl cellulose-uranyl acetate solution.

13. Remove excess liquid from the loop using filter paper.

References

Aida, Y., and Pabst, M. J. (1990). Removal of endotoxin from protein solutions by phase separation using Triton X-114. *J. Immunol. Methods* **132,** 191–195.

Alory, C., and Balch, W. E. (2001). Organization of the Rab-GDI/CHM superfamily: The functional basis for choroideremia disease. *Traffic* **2,** 532–543.

Beachy, P. A., Karhadkar, S. S., and Berman, D. M. (2004). Tissue repair and stem cell renewal in carcinogenesis. *Nature* **432**, 324–331.

Chiang, C., Litingtung, Y., Lee, E., Young, K. E., Corden, J. L., Westphal, H., and Beachy, P. A. (1996). Cyclopia and defective axial patterning in mice lacking Sonic hedgehog gene function. *Nature* **383**, 407–413.

Cotten, M., Baker, A., Saltik, M., Wagner, E., and Buschle, M. (1994). Lipopolysaccharide is a frequent contaminant of plasmid DNA preparations and can be toxic to primary human cells in the presence of adenovirus. *Gene Ther.* **1**, 239–246.

Eggenschwiler, J. T., and Anderson, K. V. (2000). Dorsal and lateral fates in the mouse neural tube require the cell-autonomous activity of the open brain gene. *Dev. Biol.* **227**, 648–660.

Eggenschwiler, J. T., Espinoza, E., and Anderson, K. V. (2001). Rab23 is an essential negative regulator of the mouse Sonic hedgehog signalling pathway. *Nature* **412**, 194–198.

Evans, T. M., Ferguson, C., Wainwright, B. J., Parton, R. G., and Wicking, C. (2003). Rab23, a negative regulator of hedgehog signaling, localizes to the plasma membrane and the endocytic pathway. *Traffic* **4**, 869–884.

Feig, L. A. (1999). Tools of the trade: Use of dominant-inhibitory mutants of Ras-family GTPases. *Nat. Cell Biol.* **1**, E25–E27.

Ingham, P. W., and McMahon, A. P. (2001). Hedgehog signaling in animal development: Paradigms and principles. *Genes Dev.* **15**, 3059–3087.

Li, G., and Stahl, P. D. (1993). Structure-function relationship of the small GTPase rab5. *J. Biol. Chem.* **268**, 24475–24480.

Liou, W., Geuze, H. J., and Slot, J. W. (1996). Improving structural integrity of cryosections for immunogold labeling. *Histochem. Cell Biol.* **106**, 41–58.

Olkkonen, V. M., and Stenmark, H. (1997). Role of Rab GTPases in membrane traffic. *Int. Rev. Cytol.* **176**, 1–85.

Olkkonen, V. M., Peterson, J. R., Dupree, P., Lutcke, A., Zerial, M., and Simons, K. (1994). Isolation of a mouse cDNA encoding Rab23, a small novel GTPase expressed predominantly in the brain. *Gene* **138**, 207–211.

Parton, R. G., and Hancock, J. F. (2001). Caveolin and Ras function. *Methods Enzymol.* **333**, 172–183.

Sambrook, J., Fritsch, E. F., and Maniatis, T. (1989). "Molecular Cloning a Laboratory Manual." Cold Spring Harbour Laboratory, Cold Spring Harbour, New York.

Stenmark, H., Parton, R. G., Steele-Mortimer, O., Lutcke, A., Gruenberg, J., and Zerial, M. (1994). Inhibition of rab5 GTPase activity stimulates membrane fusion in endocytosis. *EMBO J.* **13**, 1287–1296.

Thompson, J. D., Higgins, D. G., and Gibson, T. J. (1994). CLUSTAL W: Improving the sensitivity of progressive multiple sequence alignment through sequence weighting, position-specific gap penalties and weight matrix choice. *Nucleic Acids Res.* **22**, 4673–4680.

Wicking, C., and McGlinn, E. (2001). The role of hedgehog signalling in tumorigenesis. *Cancer Lett.* **173**, 1–7.

Zerial, M., and McBride, H. (2001). Rab proteins as membrane organizers. *Nat. Rev. Mol. Cell Biol.* **2**, 107–117.

[67] Purification and Functional Analysis of a Rab27 Effector Munc13-4 Using a Semiintact Platelet Dense-Granule Secretion Assay

By RYUTARO SHIRAKAWA, TOMOHITO HIGASHI, HIROKAZU KONDO, AKIRA YOSHIOKA, TORU KITA, and HISANORI HORIUCHI

Abstract

We have demonstrated that small GTPase Rab27 regulates dense-granule secretion in platelets. Using Rab27a affinity chromatography, we purified Munc13-4 as a novel Rab27a interacting protein from platelet cytosol. This chapter describes the purification of Munc13-4 and an *in vitro* assay system analyzing the mechanism of dense-granule secretion in platelets. The activity of Munc13-4 is tested in this assay.

Introduction

Rab27a was originally identified as a ras-related gene in a cDNA library of megakaryocytes (Nagata *et al.*, 1990). Subsequent discoveries of Rab27a as the responsible gene for Griscelli syndrome (Menasche *et al.*, 2000), an immunodefective disease with hypopigmentation, and its mouse model *ashen* (Wilson *et al.*, 2000) revealed the essential roles of Rab27a in cytotoxic T lymphocytes (CTLs) and melanocytes. In CTLs and melanocytes, Rab27a regulates lytic granule secretion (Stinchcombe *et al.*, 2001) and peripheral distribution of pigment granules (Hume *et al.*, 2001), respectively.

To date, several Rab27 effector molecules, including Slp1~5 and Slac2-a~c, have been documented (Kuroda *et al.*, 2002a,b). All these effectors interact with the GTP-bound form of Rab27 through their N-terminal conserved region, termed the Slp homology domain (SHD) (Kuroda *et al.*, 2002a,b). Evidence suggests that Rab27 interacts with different effectors in different types of cells to regulate distinct steps of membrane traffic. In melanocytes, for example, Rab27a interacts with its effector Slac2-a/melanophilin to regulate melanosome trafficking to the cell periphery, and another effector Slp2 for the tethering of melanosomes to the plasma membrane (Kuroda and Fukuda, 2004).

Rab27a and its close relative Rab27b are abundant Rab proteins in platelets, and they are shown to be associated with dense granules (Barral *et al.*, 2002; Shirakawa *et al.*, 2004). Since thrombogenic mediators, such as

METHODS IN ENZYMOLOGY, VOL. 403   0076-6879/05 $35.00

ADP and serotonin, are contained in these granules, the release of dense granules from activated platelets contributes to hemostasis and thrombosis. In a recent study, we have shown that Rab27 regulates Ca^{2+}-induced dense-granule secretion in permeabilized platelets (Shirakawa *et al.*, 2004). Furthermore, we performed Rab27a affinity chromatography and successfully identified Munc13-4 as a novel GTP-bound Rab27a interacting protein in platelet cytosol (Shirakawa *et al.*, 2004). In contrast to Slps and Slac2s, Munc13-4 lacks the conserved Rab27-binding domain (SHD), indicating that Munc13-4 is a distinct type of Rab27 effector. Munc13-4 is a nonneuronal homologue of Munc13-1, a presynaptic protein essential for neurotransmitter release. This structural similarity implies a critical role of Munc13-4 in platelet granule secretion.

Recently, it has been reported that mutations in Munc13-4 cause familial hemophagocytic lymphohistiocytosis type 3 (FHL3), an immunodefective disease with impaired lymphocyte cytotoxicity (Feldmann *et al.*, 2003; Ishii *et al.*, 2005). In CTLs from FHL3 patients, lytic granule secretion is impaired at the prefusion step as observed in Rab27a-defective *ashen* mice (Feldmann *et al.*, 2003). Taken together with our finding that Munc13-4 is a Rab27-binding protein (Shirakawa *et al.*, 2004), this study indicates that Munc13-4 could act as a Rab27a effector in the secretion in not only platelets but also CTLs.

In this chapter, we describe the purification of Munc13-4 from platelet cytosol using Rab27a affinity chromatography. We also describe an *in vitro* secretion assay we developed to analyze the molecular mechanism of dense-granule secretion. The activity of Munc13-4 was evaluated in this assay using recombinant Munc13-4.

Purification of Munc13-4 from Platelet Cytosol Using Rab27a Affinity Chromatography

To identify Rab27 effector proteins in platelets, we performed Rab27a affinity chromatography using platelet cytosol as the source for interacting proteins. This experiment includes three principal steps: (1) preparation of platelet cytosol, (2) purification of glutathione *S*-transferase (GST)-tagged Rab27a, and (3) purification of interacting proteins from platelet cytosol using Rab27a affinity chromatography.

Solution

Buffer A: 50 mM HEPES/KOH, pH 7.4, 78 mM KCl, 4 mM MgCl$_2$, 2 mM EGTA, 0.2 mM CaCl$_2$, 1 mM dithiothreitol (DTT)

Preparation of Platelet Cytosol

Human platelets outdated for transfusion are provided by the Kyoto Red Cross Blood Center. All steps of the preparation are performed at 4°. Platelet pellet (~4 ml prepared from approximately 4 liters of blood) is resuspended in 40 ml Buffer A containing protease inhibitor cocktail (Sigma) and sonicated five times each for 10 s with a Branson probe sonicator at an output level 5 under cooling. The cell lysate is centrifuged for 60 min at 50,000 rpm in a Beckman 60 Ti rotor. The supernatant is dialyzed three times against 1 liter Buffer A each for 2 h. Then the sample was centrifuged for 60 min at 50,000 rpm to remove potential aggregates. The supernatant is used as platelet cytosol. This procedure typically yields 40 ml of cytosol at a protein concentration of 5 mg/ml determined with the method of Bradford (Bio-Rad) using bovine serum albumin (BSA) as a standard. The cytosol is used for affinity purification immediately after the preparation since platelet cytosol is prone to aggregate. Part of the cytosol is stored at −80° and used for the reconstitution of the secretion in the functional assay described later in this chapter.

Preparation of GST-Rab27a

Rab27a cDNA is subcloned into the *Bam*HI site of pGEX-2T (Amersham Biosciences) to produce GST-tagged Rab27a. *Escherichia coli* strain BL21 (DE3) cells transformed with the plasmid are precultured in 100 ml Luria-Bertani (LB) medium containing ampicillin at 50 μg/ml (LB/ampicillin) and grown overnight. The next day, the confluent culture is diluted 10 times with LB/ampicillin (1 liter total) and grown to an OD_{600} of 0.6 at 37° with vigorous shaking. Protein expression is induced with 0.3 mM isopropylthiogalactopyranoside (IPTG). After 3 h induction, cells are harvested by centrifugation at 5000 rpm for 10 min in a Beckman J2–21 rotor. The bacterial pellet is washed once with cold phosphate-buffered saline and stored at −80°.

Frozen cells are thawed, resuspended in 40 ml Buffer A containing protease inhibitor cocktail, and disrupted by sonication five times each for 30 s. The cell lysate is centrifuged at 50,000 rpm for 1 h at 4° in a Beckman 60 Ti rotor. The supernatant is collected and incubated with 0.5 ml glutathione-Sepharose beads (Amersham Biosciences) in a 50-ml conical tube for 1 h at 4° on a rotator. The beads are then collected by centrifugation and washed with 40 ml Buffer A four times in the same tube. The beads are transferred into a 10-ml Poly-Prep chromatography column (Bio-Rad), washed with 10 ml Buffer A, and bound GST-Rab27a is eluted with 2.5 ml Buffer A containing 10 mM glutathione. The eluate is dialyzed against 1 liter Buffer A, followed by centrifugation at 100,000 rpm for 10 min in a Beckman 100.2 rotor. The supernatant is recovered, aliquoted, and stored at −80° until use. The purity of GST-Rab27a is examined by sodium dodecyl sulfate-polyacrylamide gel electrophoresis (SDS-PAGE) followed by Coomassie blue staining, and the concentration determined with the method of Bradford using bovine serum albumin (BSA) as a standard. This procedure typically yields 5 mg GST-Rab27a at a purity >90%.

Affinity Chromatography

To prepare GTP-bound active and GDP-bound inactive Rab27a-coated beads, 50 μg of GST-Rab27a is bound to 50 μl glutathione-Sepharose beads in each 1.5-ml tube. GST-Rab27a-coated beads are then resuspended in 500 μl Buffer A containing 10 mM EDTA and 1 mM GTPγS, a slowly hydrolyzable analogue of GTP, or 1 mM GDP and incubated for 90 min at 25° under gentle rotation. Chelating Mg^{2+} with EDTA serves to accelerate the rate of nucleotide exchange. The nucleotide exchange reaction is stopped by the addition of MgCl$_2$ at a final concentration of 14 mM. Beads are further incubated for 20 min to stabilize the nucleotide state of Rab27a.

These GST-Rab27a:GTPγS- and GST-Rab27a:GDP-coated beads are washed once with ice-cold Buffer A, mixed with 10 ml platelet cytosol prepared as described earlier in 15-ml conical tubes, and incubated at 4° under gentle rotation. As a negative control, GST-coated beads are also prepared and treated identically. After 1 h incubation, the beads are collected by centrifugation at 740×g for 2 min and transferred to 1.5-ml tubes. The beads are quickly washed with 1 ml Buffer A five times and bound proteins are eluted by boiling in 100 μl SDS sample buffer. Samples (20 μl each) are analyzed on a 4–20% gradient SDS-polyacrylamide gel (Daiichi Chemical, Japan) with Coomassie blue staining. As shown in Fig. 1A, a protein band migrating at 120-kDa is specifically detected in the GST-Rab27a:GTPγS lane. Mass spectrometry analysis of the band reveals that the 120-kDa protein is Munc13-4. The identity of Munc13-4 is confirmed by Western blotting analysis using anti-Munc13-4 antibody (Fig. 1B).

Expression and Purification of Recombinant Munc13-4

For characterization of the role and significance of Munc13-4 in platelet granule secretion, recombinant Munc13-4 protein was first produced. Since Munc13-4 is a large protein (120-kDa) and hardly purified as a soluble protein in bacterial cells, we used a baculovirus expression system. Using recombinant Munc13-4, we confirmed the direct interaction with Rab27:

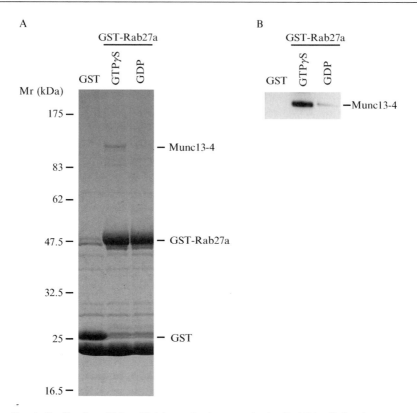

Fig. 1. Purification of Munc13-4 from platelet cytosol using Rab27a affinity chromatography. (A) GST-Rab27a was expressed in bacteria, bound to glutathione-Sepharose beads, and the corresponding nucleotide forms were prepared. After incubation with platelet cytosol, proteins bound to beads coated with either form of Rab27a and GST were eluted. The eluates (each 20 μl) were separated on a 4–20% gradient SDS-polyacrylamide gel and the gel was stained with Coomassie blue. Mass spectrometry analysis revealed that the 120-kDa protein is Munc13-4. (B) Western blotting analysis of the eluate (2 μl) obtained from the Rab27a affinity chromatography using anti-Munc13-4 antibody.

GTP and examined possible involvement of Munc13-4 in the regulation of dense-granule secretion in platelets (Shirakawa *et al.*, 2004). The method of recombinant Munc13-4 production is described in the following subsection.

Solution

Buffer B: 50 mM HEPES/KOH, pH 7.4, 150 mM KCl, 4 mM MgCl$_2$, 1 mM DTT

Expression of Recombinant Munc13-4 in Insect Cells

Hexahistidine (His_6)-tagged Munc13-4 is expressed in insect cells using the Bac-to-Bac baculovirus expression system (Invitrogen) according to the manufacturer's instructions. Briefly, Munc13-4 cDNA is subcloned into a pDEST10 donor vector to produce recombinant viruses encoding His_6-Munc13-4. Next, DH10Bac competent cells are transformed with the plasmid and plated on a selection plate containing 50 $\mu g/ml$ kanamycin, 7 $\mu g/ml$ gentamicin, 10 $\mu g/ml$ tetracycline, 100 $\mu g/ml$ X-Gal, and 50 $\mu g/ml$ IPTG. After a 48-h incubation, a white colony harboring the recombinant bacmid is inoculated into a 2-ml culture, and the recombinant viral genome is extracted.

For virus propagation and recombinant protein expression, *Spodoptera frugiperda* (Sf9) insect cells are used. Sf9 cells are cultured in Grace's insect medium (Gibco) supplemented with 5% fetal calf serum and penicillin/streptomycin. Sf9 cells seeded on a 10-cm^2 dish are transfected with the recombinant bacmid by liposome-mediated transfection using CellFEC-TIN (Invitrogen). After a 72-h incubation, medium containing recombinant viruses (P1) is collected, and Sf9 cells in a 75-cm^2 flask are infected with the P1 virus stock to amplify the virus titer. After 2 days, the medium is collected (P2), and virus amplification is repeated to obtain high-titer virus stocks. For recombinant protein expression, Sf9 cells in twenty 175-cm^2 flasks are prepared and infected with the high-titer virus stock (usually 1 ml of P4 per flask). After a 3-day incubation, cells are harvested, washed, snap-frozen in liquid nitrogen, and stored at $-80°$. The titer of the virus is examined by Munc13-4 protein expression. The high-titer virus stocks can be stored at $4°$ with light-shielding.

Purification of His_6-Munc13-4

Frozen cells ($\sim$2 ml) are thawed on ice and resuspended in 20 ml Buffer B containing protease inhibitor cocktail. The cells are disrupted by sonication and centrifuged for 1 h at 50,000 rpm in a Beckman 60 Ti rotor. Approximately 70% of His_6-Munc13-4 is recovered in the supernatant. The supernatant is collected and added with imidazole at a final concentration of 20 mM. The addition of a low concentration of imidazole prevents nonspecific binding of proteins to Ni-NTA agarose beads (Qiagen). The supernatant is incubated in batch with 0.5 ml Ni-NTA agarose beads for 1 h on a rotator. The beads are collected by centrifugation, washed three times with 40 ml Buffer B containing 20 mM imidazole, and loaded into a 10-ml chromatography column. After washing the beads with 10 ml of Buffer B, bound His_6-Munc13-4 is eluted with 2 ml Buffer B containing

250 mM imidazole. Following dialysis three times against 1 liter Buffer A each for 2 h, the eluted sample is centrifuged at 100,000 rpm for 10 min at 4° in a Beckman 100.2 rotor. The supernatant is collected, snap-frozen, and stored at −80° in aliquots. This procedure typically yields 0.8 mg of His$_6$-Munc13-4.

Functional Analysis of Munc13-4 Using Permeabilized Platelets

The major obstacle to biochemical studies of regulated exocytosis is inaccessibility to the secretory apparatus in intact cells. Several permeabilization techniques, including mechanical disruption, electropermeabilization, and use of detergents, have been employed to access the intracellular environment of secretory cells. We have used Streptolysin-O (SLO) (Palmer *et al.*, 1998), a bacterial pore-forming toxin, to establish a semiintact assay system analyzing dense-granule secretion in platelets (Shirakawa *et al.*, 2000).

This assay essentially involves four steps (Fig. 2): (1) isolation of platelets and labeling of dense granules with [^{3}H]serotonin, (2) permeabilization of the labeled platelets with SLO, (3) addition of platelet cytosol and ATP, and (4) stimulation with Ca^{2+} and counting of released [^{3}H]serotonin. It is well known that agonists induce platelet granule secretion by increasing the intracellular calcium ion concentration. Because [Ca^{2+}] inside the platelets would be the same as that outside following permeabilization, we used calcium chloride as a stimulus. Therefore, we analyzed the secretion mechanism triggered by increased [Ca^{2+}]. In this assay, secretion appears to be physiological with a time course and Ca^{2+} sensitivity similar to those in

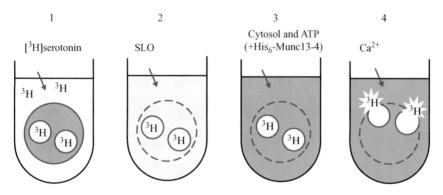

Fig. 2. Schematic representation of semiintact dense-granule secretion assay. (1) Isolate platelets from blood and label dense granules with [^{3}H]serotonin. (2) Permeabilize labeled platelets with SLO. (3) Add platelet cytosol and ATP. (4) Add Ca^{2+} and count released [^{3}H] serotonin.

intact platelets. The activity of Munc13-4 is tested in this assay system using the recombinant protein purified as described earlier.

Solutions

ACD buffer: 41.6 mM citric acid, 75 mM sodium citrate, 136 mM dextrose

Isotonic citrate (IC) buffer: 50 mM sodium citrate, 100 mM NaCl, 138 mM dextrose, pH 6.2

Stimulation buffer: 50 mM HEPES/KOH, pH 7.4, 78 mM KCl, 4 mM MgCl$_2$, 2 mM EGTA, 18 mM CaCl$_2$, 1 mM DTT

Stop buffer: 50 mM HEPES/KOH, pH 7.4, 78 mM KCl, 4 mM MgCl$_2$, 9 mM EGTA, 0.2 mM CaCl$_2$, 1 mM DTT

Isolation of Platelets

All the isolation steps are carried out at room temperature. Blood from a healthy donor is collected by venipuncture into one-sixth volume of ACD buffer. Citrate-anticoagulated blood is centrifuged at 200×g for 15 min to prepare platelet-rich plasma. Then, platelet-rich plasma is centrifuged at 1160×g for 10 min to sediment the platelets. The supernatant is discarded and the platelet pellet is gently resuspended in 10 ml IC Buffer containing 20 nM prostaglandin E$_1$ (PGE$_1$) (Sigma). Treatment with PGE$_1$ prevents activation of platelets through the preparation. After a 10-min incubation, platelets are centrifuged at 1160×g for 10 min and resuspended in 1 ml IC buffer.

Labeling of Dense Granules with [^{3}H]serotonin

To label dense granules with [^{3}H]serotonin, the platelet suspension (1 ml) is incubated with 5 μCi of [^{3}H]serotonin (10–20 Ci/mmol, Amersham Biosciences) at 30° for 5 min. Since platelets actively incorporate extracellular serotonin into dense granules through serotonin transporters that reside in the plasma membrane and dense granule membrane, this short incubation time enables efficient labeling of the organelle. Labeled platelets are centrifuged at 600×g for 5 min and resuspended in 1 ml ice-cold Buffer A containing 4 mg/ml BSA (Buffer A/BSA) and 0.6 μg/ml SLO obtained from Dr. S. Bhakdi (Mainz University, Germany) (Palmer *et al.*, 1998). The mixture is incubated for 10 min at 4°. At this temperature, SLO binds to the plasma membrane as monomers but does not oligomerize to form pores. Then, platelets are centrifuged at 600×g for 5 min to remove unbound SLO and resuspended in 1 ml Buffer A/BSA. Next, SLO-bound platelets are warmed to 30° and incubated for 5 min. This step allows SLO to oligomerize into ring-shaped structures surrounding pores of ∼35 nm diameter (Palmer *et al.*, 1998). This diameter permits free access of macromolecules. Permeabilized platelets are transferred onto ice and further incubated for 15 min to release cytosolic proteins and small molecules, such as ATP. This step results in more than 90% leakage of cytosol as determined by the release of lactate dehydrogenase, a cytosolic protein marker. Following incubation, platelets are centrifuged at $600 \times g$ for 5 min to remove the leaked cytosol and resuspended again in Buffer A/BSA. At this stage, platelets are rendered unresponsive to Ca^{2+} stimulation due to the leakage of ATP and cytosolic factors (Yoshioka *et al.*, 2001b). In spite of the high degree of permeability achieved, the intracellular ultrastructure is well preserved (Yoshioka *et al.*, 2001a).

Serotonin Secretion Assay

Standard assay is performed as described later in the chapter. Permeabilized platelets (approximately 10^8 platelets, 50,000 cpm/assay) in Buffer A/BSA (calculated free $[Ca^{2+}]$ at 20 nM [Fabiato and Fabiato, 1979]) are

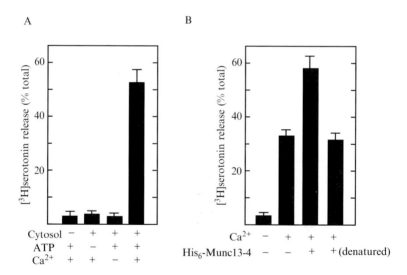

Fig. 3. Characterization of dense-granule secretion assay and functional analysis of Munc13-4 in the assay. (A) Permeabilized platelets are incubated with or without ATP and cytosol at 1.5 mg protein/ml, and stimulated with 20 μM free $[Ca^{2+}]$ or 20 nM $[Ca^{2+}]$ as indicated on the figure. After 1 min incubation, released $[^3H]$serotonin is counted. (B) Permeabilized platelets are incubated with ATP and cytosol (0.8 mg protein/ml) in the absence or presence of His_6-Munc13-4 (0.5 μM) and heat-denatured His_6-Munc13-4 (0.5 μM) as indicated on the figure. Released $[^3H]$serotonin after 1 min stimulation with 20 μM $[Ca^{2+}]$ is measured. The extent of secretion is expressed as percentage of total. The data shown are expressed as means $\pm$ SE ($n = 3$).

mixed with platelet cytosol at 1.5 mg protein/ml, ATP regeneration system (8 mM creatine phosphate, 50 μg/ml creatine phosphokinase, and 1 mM ATP), and other regents, and incubated on ice for 15 min in a total volume of 90 μl. Then, the reaction mixture is incubated at 30° for 5 min, and finally stimulated with Ca^{2+} by addition of 10 μl Stimulation Buffer, which gives 20 μM free Ca^{2+} (Fabiato and Fabiato, 1979). After 1 min stimulation, 400 μl of ice-cold Stop Buffer is added to quench the reaction. Platelets are removed by centrifugation at 2000×g for 5 min and the released [^{3}H] serotonin is recovered in the supernatant. The radioactivity of [^{3}H]seroto-nin in 100 μl of the supernatant is counted in 3 ml Clear-sol II (Nacalai Tesque, Japan) using a liquid scintillation counter (Beckman). In this system, dense-granule secretion depends on exogenously added cytosol and ATP (Fig. 3A). The maximum extent of released serotonin under optimal conditions (with platelet cytosol at 1.5 mg protein/ml) represents 50–60% of the total intracellular serotonin pool, whereas the extent of the control (without Ca^{2+} stimulation) is ~5% (Fig. 3A).

The activity of Munc13-4 is tested in this assay system. As shown in Fig. 3B, addition of recombinant His$_6$-Munc13-4 in the reaction mixture significantly enhances (Shirakawa *et al.*, 2004) [^{3}H]serotonin secretion, whereas heat-denatured His$_6$-Munc13-4 does not.

Acknowledgments

We thank Dr. Y. Nozawa for Rab27a cDNA, the Kyoto Red Cross Blood Center for providing outdated platelets, and T. Matsubara for excellent technical assistance. R.S. is a recipient of Fellowship of Japan Society for the Promotion of Science. This work was supported by the Japan Ministry of Education, Culture, Sports, Science, and Technology Research Grants 16–1244 (to R.S.), 15590740 (to H. H.), and 16209031 (to T. K.), by Health and Labour Sciences Research Grant H14-Chouju-012 from the Ministry of Health Labour and Welfare (to T. K. and H. H.), and in part by grants from the Takeda Science Foundation (to H.H).

References

Barral, D. C., Ramalho, J. S., Anders, R., Hume, A. N., Knapton, H. J., Tolmachova, T., Collinson, L. M., Goulding, D., Authi, K. S., and Seabra, M. C. (2002). Functional redundancy of Rab27 proteins and the pathogenesis of Griscelli syndrome. *J. Clin. Invest.* **110,** 247–257.

Fabiato, A., and Fabiato, F. (1979). Calculator programs for computing the composition of the solutions containing multiple metals and ligands used for experiments in skinned muscle cells. *J. Physiol. (Paris)* **75,** 463–505.

Feldmann, J., Callebaut, I., Raposo, G., Certain, S., Bacq, D., Dumont, C., Lambert, N., Ouachee-Chardin, M., Chedeville, G., Tamary, H., Minard-Colin, V., Vilmer, E., Blanche, S., Le Deist, F., Fischer, A., and de Saint Basile, G. (2003). Munc13-4 is essential for cytolytic granules fusion and is mutated in a form of familial hemophagocytic lymphohistiocytosis (FHL3). *Cell* **115**, 461–473.

Hume, A. N., Collinson, L. M., Rapak, A., Gomes, A. Q., Hopkins, C. R., and Seabra, M. C. (2001). Rab27a regulates the peripheral distribution of melanosomes in melanocytes. *J. Cell Biol.* **152**, 795–808.

Ishii, E., Ueda, I., Shirakawa, R., Yamamoto, K., Horiuchi, H., Ohga, S., Furuno, K., Morimoto, A., Imayoshi, M., Ogata, Y., Zaitzu, M., Sako, M., Koike, K., Sakata, A., Takada, H., Hara, T., Imashuku, S., Sasazuki, T., and Yasukawa, M. (2005). Genetic subtypes of familial hemophagocytic lymphohistiocytosis: Correlations with clinical features and cytotoxic T lymphocyte/natural killer cell functions. *Blood* **105**, 3442–3448.

Kuroda, T. S., and Fukuda, M. (2004). Rab27A-binding protein Slp2-a is required for peripheral melanosome distribution and elongated cell shape in melanocytes. *Nat. Cell Biol.* **6**, 1195–1203.

Kuroda, T. S., Fukuda, M., Ariga, H., and Mikoshiba, K. (2002a). The Slp homology domain of synaptotagmin-like proteins 1–4 and Slac2 functions as a novel Rab27A binding domain. *J. Biol. Chem.* **277**, 9212–9218.

Kuroda, T. S., Fukuda, M., Ariga, H., and Mikoshiba, K. (2002b). Synaptotagmin-like protein 5: A novel Rab27A effector with C-terminal tandem C2 domains. *Biochem. Biophys. Res. Commun.* **293**, 899–906.

Menasche, G., Pastural, E., Feldmann, J., Certain, S., Ersoy, F., Dupuis, S., Wulffraat, N., Bianchi, D., Fischer, A., Le Deist, F., and de Saint Basile, G. (2000). Mutations in RAB27A cause Griscelli syndrome associated with haemophagocytic syndrome. *Nat. Genet.* **25**, 173–176.

Nagata, K., Satoh, T., Itoh, H., Kozasa, T., Okano, Y., Doi, T., Kaziro, Y., and Nozawa, Y. (1990). The ram: A novel low molecular weight GTP-binding protein cDNA from a rat megakaryocyte library. *FEBS Lett.* **275**, 29–32.

Palmer, M., Harris, R., Freytag, C., Kehoe, M., Tranum-Jensen, J., and Bhakdi, S. (1998). Assembly mechanism of the oligomeric streptolysin O pore: The early membrane lesion is lined by a free edge of the lipid membrane and is extended gradually during oligomerization. *EMBO J.* **17**, 1598–1605.

Shirakawa, R., Yoshioka, A., Horiuchi, H., Nishioka, H., Tabuchi, A., and Kita, T. (2000). Small GTPase rab4 regulates Ca2+-induced alpha-granule secretion in platelets. *J. Biol. Chem.* **275**, 33844–33849.

Shirakawa, R., Higashi, T., Tabuchi, A., Yoshioka, A., Nishioka, H., Fukuda, M., Kita, T., and Horiuchi, H. (2004). Munc 13-4 is a GTP-Rab27-binding protein regulating dense core granule secretion in platelets. *J. Biol. Chem.* **279**, 10730–10737.

Stinchcombe, J. C., Barral, D. C., Mules, E. H., Booth, S., Hume, A. N., Machesky, L. M., Seabra, M. C., and Griffiths, G. M. (2001). Rab27a is required for regulated secretion in cytotoxic T lymphocytes. *J. Cell Biol.* **152**, 825–834.

Wilson, S. M., Yip, R., Swing, D. A., O'Sullivan, T. N., Zhang, Y., Novak, E. K., Swank, R. T., Russell, L. B., Copeland, N. G., and Jenkins, N. A. (2000). A mutation in Rab27a causes the vesicle transport defects observed in ashen mice. *Proc. Natl. Acad. Sci. USA* **97**, 7933–7938.

Yoshioka, A., Horiuchi, H., Shirakawa, R., Nishioka, H., Tabuchi, A., Higashi, T., Yamamoto, A., and Kita, T. (2001a). Molecular dissection of alpha- and dense-core granule secretion of platelets. *Ann. NY Acad. Sci.* **947**, 403–406.

Yoshioka, A., Shirakawa, R., Nishioka, H., Tabuchi, A., Higashi, T., Ozaki, H., Yamamoto, A., Kita, T., and Horiuchi, H. (2001b). Identification of protein kinase Calpha as an essential, but not sufficient, cytosolic factor for Ca2+-induced alpha- and dense-core granule secretion in platelets. *J. Biol. Chem.* **276**, 39379–39385.

[68] Analysis of hVps34/hVps15 Interactions with Rab5 *In Vivo* and *In Vitro*

By James T. Murray and Jonathan M. Backer

Abstract

The hVps34 phosphatidylinositol (PI) 3-kinase plays an important role in the regulation of vesicular trafficking in the endosomal system. hVps34 associates with a myristylated protein kinase, hVps15. The two proteins are targeting to early endosomal membranes by interactions between hVps15 and activated (GTP-bound) Rab5. This leads to the production of the hVps34 product, PI(3)P, in the endosomal membrane, and subsequent recruitment of FYVE and PX domain-containing effector proteins. This chapter describes the analysis of hVps34/hVps15 interactions with Rab5 in tissue culture cells and *in vitro*.

Introduction

Mammalian cells express multiple phosphatidylinositol (PI) 3-kinases that are regulated by different mechanisms and produce different 3-phosphoinositide products (Fruman *et al.*, 1998; Vanhaesebroeck *et al.*, 2001). Class I PI 3-kinases (p85/p110 and PI3Kγ) produce primarily phosphatidylinositol-3,4,5-trisphosphate, (PI[3,4,5]P$_3$) in intact cells (Auger *et al.*, 1989). Class II PI 3-kinases produce phosphatidylinositol-3,4-bisphosphate (PI[3,4]P$_2$) and phosphatidylinositol 3-phosphate (PI[3]P) *in vitro*, but do not produce PI(3,4,5)P$_3$ (Vanhaesebroeck *et al.*, 2001). Finally, Class III PI 3-kinases, the yeast Vps34 and its human homolog, hVps34, produce only PI(3)P (Herman *et al.*, 1992; Volinia *et al.*, 1995).

Vps34 was first described as a gene essential for the sorting of carboxypeptidase Y (CPY) to the yeast vacuole (Herman and Emr, 1990). The discovery that Vps34 was a PI 3-kinase (Schu *et al.*, 1993) led to a number of studies demonstrating a role for PI 3-kinases, and hVps34 in particular, in mammalian endocytic trafficking (reviewed in Backer, 2000). hVps34 has been implicated in endosomal sorting (Siddhanta *et al.*, 1998), early endosomal fusion and motility (Christoforidis *et al.*, 1999b; Nielsen *et al.*, 1999), multivesicular body formation (Futter *et al.*, 2001), polarized sorting in hepatocytes (Tuma *et al.*, 2001), phagosome maturation (Fratti *et al.*, 2001; Vieira *et al.*, 2001), and mast cell degranulation (Windmiller and Backer, 2003). A key finding in elucidating the mechanism of hVps34

METHODS IN ENZYMOLOGY, VOL. 403
Copyright 2005, Elsevier Inc. All rights reserved.

0076-6879/05 $35.00
DOI: 10.1016/S0076-6879(05)03068-5

signaling in the endocytic system was the identification of EEA1 as a regulator of endosomal fusion (reviewed in Corvera *et al.*, 1999; Wurmser *et al.*, 1999). EEA1 contains an N-terminal zinc finger domain, the FYVE domain, that binds specifically to the product of hVps34, PI(3)P (Corvera *et al.*, 1999; Wurmser *et al.*, 1999). FYVE domains are found in a number of proteins that regulate endosomal sorting. A second PI(3)P-binding domain, the Phox homology or PX domain, has also been described (Wishart *et al.*, 2001).

The recruitment of PX and FYVE domain-containing proteins to endosomal compartments requires the production of PI(3)P at appropriate sites. Thus, the targeting of hVps34 is an important aspect of its signaling specificity. hVps34 is recruited to sites of early endosome fusion by the small GTPase, Rab5 (Christoforidis *et al.*, 1999b; Murray *et al.*, 2002). Recruitment of hVps34 by Rab5 is indirect and requires the adaptor protein hVps15, which binds to and regulates the lipid kinase activity of hVps34 (Murray *et al.*, 2002; Panaretou *et al.*, 1997). Analysis of domain deletion mutants of hVps15 has revealed that the C-terminal WD repeats are responsible for binding to active Rab5:GTP (Murray *et al.*, 2002). Following the recruitment of hVps34/hVps15 complexes to sites of early endosome docking, hVps34 catalyzes the production of PI(3)P, which in conjunction with active Rab5 recruits proteins that regulate endosome docking and fusion (Christoforidis *et al.*, 1999a). In this chapter we describe the methods used to analyze Rab5-mediated targeting of hVps15/hVps34 *in vivo* and Rab5 binding hVps15 *in vitro*.

Analysis of Rab5-Mediated Targeting of hVps15 and hVps34 by Immunofluorescence

The effect of wild-type and mutant Rab5 on the targeting of hVps15 and hVps34 to intracellular membranes can be analyzed by indirect immunofluorescence. Unfortunately, antibodies suitable for the detection of endogenous hVps34 and hVps15 are not commercially available. Thus, these studies will be described using epitope-tagged hVps34 and hVps15. The detection of myc-tagged hVps34 in cells expressing WT or Q79L Rab5 is shown in Fig. 1.

Buffers

Blocking buffer: 3% immunoglobulin (IgG)-free bovine serum albumin (BSA) (Jackson ImmunoResearch, West Grove, PA) in phosphate-buffered saline (PBS).

α-Rab5   α-hVps34

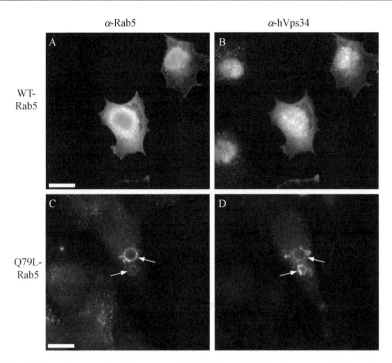

FIG. 1. Recruitment of epitope-tagged hVps34 to Rab5 endosomes. HeLa cells were microinjected with cDNAs encoding myc-tagged hVps34 with either wild-type (A and B) or Q79L-Rab5 (C and D). Cells were incubated for 24 h, fixed, and stained with antibodies against Rab5 (A and C) or hVps34 (B and D). Arrows show examples of colocalization. Reprinted with permission from Murray *et al.* (2002). Role of Rab5 in the recruitment of hVps34/p150 to the early endosome. *Traffic* **3**, 416–427, Blackwell Publishing, Oxford, UK.

Method

Transient Transfection of HeLa Cells Seeded on Glass Coverslips. HeLa cells are seeded onto 13-mm round coverslips (No. 0, VWR, West Chester, PA) in six-well tissue culture dishes and cultured to 50–60% confluence. The cells are transfected with cDNA encoding epitope-tagged hVps34 or hVps15 plus wild-type or Q79L-Rab5 (provided by Dr. Marino Zerial, Max Planck Institute, Dresden, Germany) or an empty vector control, using 2 μg of cDNA per coverslip and LipofectAMINE Plus (Invitrogen), following the manufacturer's protocol. Cells are cultured for 6–20 h to allow expression of the cDNA encoded proteins.

Indirect Fluorescence Staining of hVps15 and hVps34 in Fixed Cells. The culture medium is removed and the coverslips are washed twice with

PBS before the cells are fixed in 3.7% (w/v) formaldehyde in PBS for 10 min at room temperature. After fixing, the coverslips are washed three times with PBS. (**Note**: in this and subsequent steps, the coverslips must not be allowed to dry out.) The fixed cells are then permeabilized by incubation in 0.1% (v/v) Triton X-100/PBS for 10 min, and washed three times with PBS. Nonspecific binding of antibodies can be minimized by incubating cells in Blocking buffer for 30 min. After blocking, the coverslips are washed three more times with PBS before incubating with epitope tag antibodies.

Optimal antibody concentration should be determined by serial titration; however, in most instances a final antibody concentration of between 1 and 5 μg/ml is sufficient. For incubation with primary and secondary antibodies, coverslips are placed cell-side upward onto a hydrophobic surface such as Parafilm (Pechiney Plastic Packaging, Chicago, IL) placed inside a 15-cm Petri dish. Primary antibodies are diluted into Blocking buffer and 80–100 μl of each diluted primary antibody is overlaid onto the upturned coverslips and incubated for 60 min at room temperature. The coverslips are washed three times with PBS, then incubated with fluorescently labeled secondary antibodies in the dark for 60 min. Alexa-dye-conjugated secondary antibodies (Invitrogen) are diluted 500-fold into PBS and incubated with the coverslips for 60 min at room temperature. The coverslips are washed four times with PBS, mounted onto glass slides in aqueous mounting medium (e.g., 0.33% [w/v] n-propylgallate/PBS/50% glycerol [v/v]), and the edges of the coverslip are then sealed with nail polish.

Comments

1. Overexpression of hVps15 in mammalian cells appears to be toxic, and leads to the precipitation of hVps15 in large perinuclear aggregates. Cells should be analyzed within 6–12 h of transfection with hVps15.

2. If it is essential to stop an incubation/time point rapidly, cells may then be fixed *in situ* by diluting 10-fold a solution of 37% (w/v) formaldehyde directly to the culture media for 15 min. However, this may result in an increase in nonspecific background staining.

3. To prevent coverslips from drying out during incubation with primary and secondary antibodies, a humidified chamber can be improvised by placing a damp tissue into the covered Petri dish.

4. Desorbed, species matched Alexa dye-labeled secondary antibodies have reproducibly given bright staining with low nonspecific cross-reactivity and are more resistant to photobleaching than traditional isothiocyanate derivatives. When examining cells that express variants of green fluorescent protein (GFP), nail polish should be avoided, since significant

photobleaching of the specimen can be observed. As an alternative, a solid mounting medium such as Hydromount (National Diagnostics, Atlanta, GA) containing 2.5% (w/v) 1,4-diazabicyclo[2.2.2]octane (DABCO) as a antibleaching agent can be used. A solid mounting medium avoids the need to permanently seal the coverslip with nail polish.

In Vitro Analysis of the Rab5/hVps15 Binding

A direct interaction between active GTP-bound Rab5 and hVps15 has been established through the use of biochemical and cell biological techniques (Christoforidis and Zerial, 2000; Christoforidis et al., 1999b). As shown in Fig. 2, hVps15 contains an N-terminal Ser/Thr protein kinase domain, a series of central HEAT repeats, and a series of C-terminal WD domains (Murray et al., 2002; Panaretou et al., 1997). We have developed an assay to assess the ability of hVps15, synthesized in an in vitro coupled transcription/translation reaction, to bind to recombinant Rab5 (Murray et al., 2002). This method requires the expression and purification of GST-Rab5 in bacteria, followed by loading of GST-Rab5 with guanosine nucleotides (GDP or GTPγS) to give inactive and active forms of GST-Rab5, respectively. The immobilized GST-Rab5 is then used to precipitate in vitro translated, ^{35}S-labeled, full-length or domain deletion mutants of hVps15. Binding of hVps15 domain mutants is detected by sodium dodecyl sulfate-polyacrylamide gel electrophoresis (SDS-PAGE), separation of the precipitated proteins, and autoradiography (Fig. 3).

Materials

The bacterial expression vector pGEX-2T containing the human Rab5 cDNA was generously provided by Dr. Marino Zerial (Max Plank Institute for Cell Biology, Dresden, Germany) and the purification of GST-Rab5 has been modified from the method of Christoforidis and Zerial (2000). The original hVps15 cDNA clone was a kind gift of Dr. Michael Waterfield (Ludwig Institute for Cancer Research, London, UK).

Buffers

Rab5 Extraction buffer: PBS containing 2 mM ethylenediaminetetraacetic acid (EDTA), 5 mM MgCl$_2$, 5 mM benzamidine, 5 mM 2-mercaptoethanol, 0.35 mg/ml phenylmethylsulfonyl fluoride (PMSF), and 200 μM GDP

Rab5 Wash buffer: PBS containing 5 mM MgCl$_2$, 5 mM 2-mercaptoethanol, and 100 μM GDP

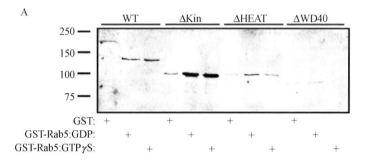

FIG. 2. Domain architecture of hVps15. hVps15 is a myristoylated protein that contains a Ser/Thr protein kinase domain at its N-terminus, followed by multiple HEAT repeats and six WD40 repeats at its C-terminus. The approximate size and position of each domain are indicated.

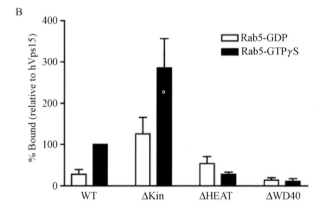

FIG. 3. Domain analysis of hVps15-Rab5 interactions. (A) Radiolabeled *in vitro*-translated wild-type or domain deletion mutants of hVps15 were incubated with recombinant GST-Rab5, or GST control, bound to glutathione-Sepharose beads. GST-Rab5 was preloaded with either GDP or GTPγS before incubation with hVps15. Specifically associated hVps15 was eluted as described, resolved by SDS-PAGE, and visualized by autoradiography. (B) The association of recombinant hVps15 with GST-Rab5 was quantified by densitometry from at least three independent experiments. Reprinted with permission from Murray *et al.* (2002). Role of Rab5 in the recruitment of hVps34/p150 to the early endosome. *Traffic* **3,** 416–427, Blackwell Publishing, Oxford, UK.

Nucleotide Exchange buffer: 20 mM HEPES, 100 mM NaCl, 10 mM
EDTA, 5 mM MgCl$_2$, and 1 mM dithiothreitol (DTT)
Nucleotide Stabilization buffer: 20 mM HEPES, 100 mM NaCl, 5 mM
MgCl$_2$, and 1 mM DTT
hVps15 Binding buffer: 20 mM HEPES, pH 7.5, 100 mM NaCl, 5 mM
MgCl$_2$, 1 mM DTT, and 1 mM either GDP or GTPγS
Mg^{2+}-free Binding buffer: 20 mM HEPES, pH 7.5, 250 mM NaCl, and
1 mM DTT
Elution buffer: 20 mM HEPES, pH 7.5, 1.5 M NaCl, 20 mM EDTA,
and 1 mM DTT.

Method

Purification of GST-Rab5 from Bacteria. A 1-liter flask of Luria–Bertani
(LB) media containing 75 μg/ml ampicillin is inoculated with 10 ml of BL21
pLysS bacteria (Promega, Madison, WI) transformed with pGEX-2T-Rab5
and cultured, with ample aeration, at 37° to a cell density of 0.8 (OD
600 nm). Expression of GST-Rab5 is induced by addition of 1 ml of 1 M
isopropylthiogalactoside (IPTG; 1 mM final concentration). The bacteria
are cultured for an additional 3 h at 37°, then sedimented by centrifugation
at 3000×g for 20 min at 4°.

The bacterial pellet is resuspended in 20 ml of Extraction buffer and the
suspension is sonicated using a probe sonicator with 4 × 20 s bursts at 70%
output. The unwanted insoluble material is removed from the extract by
centrifugation at 28,000×g for 30 min at 4°. The clarified supernatant is
passed through a 0.22-μm filter. Glutathione-Sepharose beads (Pierce Chemicals, Rockford, IL) are prepared by washing three times with Wash
buffer, followed by resuspension in Wash buffer to make a 50% slurry.
The clarified bacterial extract is incubated with 3 ml of the 50% glutathione-Sepharose for 2–4 h at 4° to bind the recombinant GST-Rab5. The
glutathione-Sepharose-bound GST-Rab5 is washed three times with 30 ml
of Wash buffer, each time pelleted by centrifugation at 3000×g for 5 min at
4°. If required the GST-Rab5-coated glutathione-Sepharose can be stored
for a few days at 4° as a 50% slurry in Wash buffer.

*Preparation of Inactive and Active Forms of GST-Rab5 by Nucleotide
Exchange.* Purification of GST-Rab5 by the above method produces GST-
Rab5 predominantly in the inactive, GDP-bound state. This inactive GST-
Rab5:GDP can be converted into the active GST-Rab5:GTP form by a
nucleotide exchange reaction, which we have adapted from the method
described by Christoforidis and Zerial (2000). Although purified GST-
Rab5 is already in the GDP-bound form, preparations of GDP and GTP

Rab5 are processed identically by splitting the preparation into two 1.5-ml aliquots that are then treated as follows.

1. Nucleotide exchange. The immobilized GST-Rab5 beads are washed with 10 ml of Exchange buffer containing either 10 μM GDP or GTPγS. The beads are pelleted by centrifugation at 3000$\times g$ for 2 min, resuspended in 2 ml of Exchange buffer containing 1 mM of either GDP or GTPγS, and incubated for 30 min at room temperature. The beads are recentrifuged at 3000$\times g$ for 2 min and the wash step is repeated twice more.

2. Nucleotide stabilization. GST-Rab5 is stabilized in the GDP- or GTPγS-bound conformation by removing EDTA, which is done by first washing the pelleted beads with 10 ml of Stabilization buffer containing 10 μM GDP or GTPγS, followed by centrifugation at 3000$\times g$ for 2 min. The beads are resuspended in 1.5 ml of Stabilization buffer containing 1 mM GDP or GTPγS, transferred to a 2-ml centrifuge tube and then incubated for 30 min at room temperature. The beads are centrifuged at 3000$\times g$ for 2 min at 4°, and a volume of glycerol equal to the volume of packed beads is then added, and the bead-glycerol slurry is mixed thoroughly by inversion. Nucleotide-loaded GST-Rab5 beads can either be used immediately or stored at −20°.

Comments

1. Before preparation of inactive and active forms of GST-Rab5, 10–20 μl of the 50% slurry of GST-Rab5 coated glutathione-Sepharose beads can be removed and analyzed by SDS-PAGE to determine the purity and yield of GST-Rab5.

2. Incubation of GST-Rab5 with EDTA during the exchange steps will remove the Mg^{2+}-GDP present in the GST-Rab5 purified from bacteria, while the later omission of EDTA from the Stabilization buffer prevents removal of the newly loaded nucleotide.

Coupled In Vitro *Transcription/Translation of hVps15 Mutants*

Using a commercially available rabbit reticulocyte-based coupled *in vitro* transcription/translation (IVT) system (TNT-T7 Quick-Coupled Transcription/Translation System, Catalog no. L1170, Promega, Madison, WI), full-length or domain deletion mutants of [^{35}S]methionine/cysteine-labeled hVps15 can be rapidly synthesized. These *in vitro* translated proteins can then be used in binding assays with the immobilized GST-Rab5 prepared as described previously.

We previously cloned into the mammalian expression vector, pcDNA3.1-V5/His (Invitrogen), full-length hVps15 and hVps15 domain mutants in which the kinase domain (residues 1–300), the HEAT repeats (residues 433–667), and the WD repeats (residues 972–1358) had been individually deleted by polymerase chain reaction (PCR) (Murray *et al.*, 2002). The pcDNA3.1-V5/His vector contains the T7 RNA polymerase promoter upstream of the hVps15 gene, which is required for *in vitro* transcription. The IVT reaction is set up in an RNase/DNase-free microcentrifuge tube with 1 μg of cDNA encoding hVps15 for a 50 μl reaction, according to the manufacturer's protocol. The reactions are incubated for 90 min at 30° and then kept on ice and used immediately. Alternatively, the translation products can be stored at −80° until required.

GST-Rab5/hVps15 In Vitro *Binding Assay*

The GDP- or GTP-loaded GST-Rab5 preparations are used to absorb radiolabeled hVps15 translation products, and GST-Rab5-bound hVps15 is detected by autoradiography. Using this methodology we identified the C-terminal WD repeats as the major Rab5 interacting domain in hVps15 (Murray *et al.*, 2002).

Sixty microliters of either GST-Rab5:GDP or GST-Rab5:GTPγS 50% bead slurry is aliquoted into a 1.5-ml RNase/DNase-free microcentrifuge tube and the beads are pelleted by centrifugation at $3000 \times g$ for 2 min at 4°, followed by resuspension in 300 μl of Binding buffer (see previous description). Ten microliters of an individual IVT reaction is added to the bead suspension and incubated for 2 h at 4° with gentle mixing. The bead complexes are then pelleted by centrifugation at $4000 \times g$ for 2 min at 4°, washed twice with 300 μl of Binding buffer containing 10 μM of either GDP (for GST-Rab5:GDP) or GTPγS (for GST-Rab5:GTPγS), and twice with 300 μl of Binding buffer containing 250 mM NaCl and the relevant nucleotides. Finally, the beads are washed once with 500 μl of Mg^{2+}-free Binding buffer containing nucleotides, and pelleted by centrifugation at $4000 \times g$ for 1 min at 4°.

Proteins that interact specifically with GDP or GTPγS-Rab5 are now eluted by reversing the nucleotide composition of the buffer. Bead complexes are incubated for 10 min at room temperature with 45 μl of Elution buffer containing 1 mM GTPγS to elute proteins specifically bound to GST-Rab5:GDP and 5 mM GDP to elute proteins specifically bound to GST-Rab5:GTPγS. The Sepharose beads are sedimented by centrifugation at $4000 \times g$ for 1 min at 4°, and the eluted proteins can then be transferred to a clean microcentrifuge tube. Fifteen microliters of 4× SDS sample buffer is then added to the eluate, the samples are boiled for

3 min, and eluted proteins are finally separated by SDS-PAGE. The ^{35}S-labeled hVps15 mutants that bind specifically to GST-Rab5:GDP or GST-Rab5:GTPγS are then detected by autoradiography.

References

Auger, K. R., Serunian, L. A., Soltoff, S. P., Libby, P., and Cantley, L. C. (1989). PDGF-dependent tyrosine phosphorylation stimulates production of novel polyphosphoinositides in intact cells. *Cell* **57**, 167–175.

Backer, J. M. (2000). Phosphoinositide 3-kinases and the regulation of vesicular trafficking. *Mol. Cell. Biol. Res. Commun.* **3**, 193–204. [Published erratum appears in *Mol. Cell. Biol. Res. Commun.* **3(5)**, 328.]

Christoforidis, S., and Zerial, M. (2000). Purification and identification of novel Rab effectors using affinity chromatography. *Methods* **20**, 403–410.

Christoforidis, S., McBride, H. M., Burgoyne, R. D., and Zerial, M. (1999a). The Rab5 effector EEA1 is a core component of endosome docking. *Nature* **397**, 621–625.

Christoforidis, S., Miaczynska, M., Ashman, K., Wilm, M., Zhao, L., Yip, S. C., Waterfield, M. D., Backer, J. M., and Zerial, M. (1999b). Phosphatidylinositol-3-OH kinases are Rab5 effectors. *Nat. Cell Biol.* **1**, 249–252.

Corvera, S., D'Arrigo, A., and Stenmark, H. (1999). Phosphoinositides in membrane traffic. *Curr. Opin. Cell Biol.* **11**, 460–465.

Fratti, R. A., Backer, J. M., Gruenberg, J., Corvera, S., and Deretic, V. (2001). Role of phosphatidylinositol 3-kinase and Rab5 effectors in phagosomal biogenesis and mycobacterial phagosome maturation arrest. *J. Cell Biol.* **154**, 631–644.

Fruman, D. A., Meyers, R. E., and Cantley, L. C. (1998). Phosphoinositide kinases. *Annu. Rev. Biochem.* **67**, 481–507.

Futter, C. E., Collinson, L. M., Backer, J. M., and Hopkins, C. R. (2001). Human VPS34 is required for internal vesicle formation within multivesicular endosomes. *J. Cell Biol.* **155**, 1251–1264.

Herman, P. K., and Emr, S. D. (1990). Characterization of VPS34, a gene required for vacuolar protein sorting and vacuole segregation in *Sarccharomyces cerevisiae. Mol. Cell. Biol.* **10**, 6742–6754.

Herman, P. K., Stack, J. H., and Emr, S. D. (1992). An essential role for a protein and lipid kinase complex in secretory protein sorting. *Trends Cell Biol.* **2**, 363–368.

Murray, J. T., Panaretou, C., Stenmark, H., Miaczynska, M., and Backer, J. M. (2002). Role of Rab5 in the recruitment of hVps34/p150 to the early endosome. *Traffic* **3**, 416–427.

Nielsen, E., Severin, F., Backer, J. M., Hyman, A. A., and Zerial, M. (1999). Rab5 regulates motility of early endosomes on microtubules. *Nat. Cell Biol.* **1**, 376–382.

Panaretou, C., Domin, J., Cockcroft, S., and Waterfield, M. D. (1997). Characterization of p150, an adaptor protein for the human phosphatidylinositol (PtdIns) 3-kinase. Substrate presentation by phosphatidylinositol transfer protein to the p150.Ptdins 3-kinase complex. *J. Biol. Chem.* **272**, 2477–2485.

Schu, P. V., Takegawa, K., Fry, M. J., Stack, J. H., Waterfield, M. D., and Emr, S. D. (1993). Phosphatidylinositol 3-kinase encoded by yeast *VPS34* gene essential for protein sorting. *Science* **260**, 88–91.

Siddhanta, U., Mcllroy, J., Shah, A., Zhang, Y. T., and Backer, J. M. (1998). Distinct roles for the p110_ and hVPS34 phosphatidylinositol 3'-kinases in vesicular trafficking, regulation of the actin cytoskeleton, and mitogenesis. *J. Cell Biol.* **143**, 1647–1659.

Tuma, P. L., Nyasae, L. K., Backer, J. M., and Hubbard, A. L. (2001). Vps34p differentially regulates endocytosis from the apical and basolateral domains in polarized hepatic cells. *J. Cell Biol.* **154,** 1197–1208.

Vanhaesebroeck, B., Leevers, S. J., Ahmadi, K., Timms, J., Katso, R., Driscoll, P. C., Woscholski, R., Parker, P. J., and Waterfield, M. D. (2001). Synthesis and function of 3-phosphorylated inositol lipids. *Annu. Rev. Biochem.* **70,** 535–602.

Vieira, O. V., Botelho, R. J., Rameh, L., Brachmann, S. M., Matsuo, T., Davidson, H. W., Schreiber, A., Backer, J. M., Cantley, L. C., and Grinstein, S. (2001). Distinct roles of class I and class III phosphatidylinositol 3-kinases in phagosome formation and maturation. *J. Cell Biol.* **155,** 19–25.

Volinia, S., Dhand, R., Vanhaesebroeck, B., MacDougall, L. K., Stein, R., Zvelebil, M. J., Domin, J., Panaretou, C., and Waterfield, M. D. (1995). A human phosphatidylinositol 3-kinase complex related to the yeast Vps34p-Vps15p protein sorting system. *EMBO J.* **14,** 3339–3348.

Windmiller, D. A., and Backer, J. M. (2003). Distinct phosphoinositide 3-kinases mediate mast cell degranulation in response to G-protein-coupled versus FcepsilonRI receptors. *J. Biol. Chem.* **278,** 11874–11878.

Wishart, M. J., Taylor, G. S., and Dixon, J. E. (2001). Phoxy lipids: Revealing PX domains as phosphoinositide binding modules. *Cell* **105,** 817–820.

Wurmser, A. E., Gary, J. D., and Emr, S. D. (1999). Phosphoinositide 3-kinases and their FYVE domain-containing effectors as regulators of vacuolar/lysosomal membrane trafficking pathways. *J. Biol. Chem.* **274,** 9129–9132.

Further Reading

Li, G., and Stahl, P. D. (1993). Structure-function relationship of the small GTPase rab5. *J. Biol. Chem.* **268,** 24475–24480.

Stenmark, H., Parton, R. G., Steele-Mortimer, O., Lutcke, A., Gruenberg, J., and Zerial, M. (1994). Inhibition of rab5 GTPase activity stimulates membrane fusion in endocytosis. *EMBO J.* **13,** 1287–1296.

Zerial, M., and McBride, H. (2001). Rab proteins as membrane organizers. *Nat. Rev. Mol. Cell. Biol.* **2,** 107–117.

[69] Purification and Functional Properties of Prenylated Rab Acceptor 2

By Pierre-Yves Gougeon and Johnny K. Ngsee

Abstract

PRA2 was found to interact with the ER-localized protein VAMP-associated protein of 33 kDa or VAP-33 by a yeast two-hybrid screen. We describe here the purification of PRA2 and VAP-33 as well as an *in vitro* pull-down procedure to verify the interaction. PRA2 was found to form a large sodium dodecyl sulfate (SDS)–insoluble complex upon heat

METHODS IN ENZYMOLOGY, VOL. 403   0076-6879/05 $35.00
Copyright 2005, Elsevier Inc. All rights reserved.   DOI: 10.1016/S0076-6879(05)03069-7

denaturation, resulting in significant reduction in the Western immunoblot signal. This phenomenon is specific to PRA2 and was not observed with PRA1. We also found that protein interaction with PRA2 is highly sensitive to detergent and describe a covalent cross-linking procedure for mammalian cell extracts to stabilize the PRA2-containing complex prior to membrane solubilization and immunoprecipitation.

Introduction

Rab GTPases are molecular switches that cycle between active GTP-bound states associated with intracellular membranes and inactive GDP-bound states associated with the cytosolic carrier GDP-dissociation inhibitor (GDI). Membrane localization is essential for biological function and requires posttranslational addition of one or two isoprenyl groups to the characteristic Cys-containing motif at the extreme C-terminus. Among the proteins that specifically recognize the prenylated forms of Rab GTPases is a family known as prenylated Rab acceptor or PRA. PRA1 is an ~21-kDa membrane protein broadly expressed in all tissues (Martincic et al., 1997). It binds weakly to GDI and can block the GDI-mediated removal of Rab3A from PC12 membranes (Hutt et al., 2000). It also acts as a GDI displacement factor or GDF for endosomal Rab GTPases (Sivars et al., 2003). Thus, the net outcome is a shift favoring membrane translocation and localization of Rab. Its broad specificity, however, precludes the likelihood that PRA is solely responsible for specific membrane targeting of Rab.

Only three PRA isoforms have been identified in the human genome (Pfeffer and Aivazian, 2004), which is considerably less complex when compared to the over 60 Rab isoforms. Despite this disparity in number, the interaction of PRA with diverse prenylated Rab isoforms suggests that each PRA isoform acts on more than one Rab trafficking pathway. In fact, its ability to interact with a Rab may be constrained by its intracellular localization. PRA1 resides in the Golgi complex/endosome compartment while PRA2 is primarily localized to the endoplasmic reticulum (ER). In both cases, localization is determined by signals inherent within each PRA isoform. A modified diacidic DXE motif at the C-terminal domain acts as an ER exit signal for PRA1, and a modified KKXX motif serves as an ER retention signal for PRA2 (Abdul-Ghani et al., 2001). In addition to Rab GTPases, both isoforms are known to interact with other proteins (Calero and Collins, 2002; Enouf et al., 2003; Evans et al., 2002; Li et al., 2001). Likewise, PRA2 interacts with the neuronal glutamate transporter EAAC1 (Lin et al., 2001) and ARF-like protein ARL6 (Ingley et al., 1999).

Since little is known of the regulation of PRA activity, we have sought to identify proteins that might interact with PRA through a yeast two-hybrid screen. One of the proteins identified in a PRA1 screen of a human brain cDNA library is VAMP-associated protein of 33 kDa (VAP-33). Since VAP-33 is primarily localized to the ER (Skehel *et al.*, 2000), we tested and found that it also interacts with the ER-localized PRA2. Yeast two-hybrid interactions often require independent verification, which typically involves a pull-down and/or coimmunoprecipitation using purified bacterial or mammalian expressed recombinant proteins. We describe here procedures to purify membrane proteins such as PRA2 and VAP-33, which can then be used for an *in vitro* binding assay to verify the interaction. Purification of membrane proteins necessitates the use of detergents, which often interferes with or disrupts protein–protein interaction. In general, a nonionic detergent such as Triton X-100 effectively solubilizes bacterially expressed recombinant proteins. However, *in vitro* protein binding requires a much lower concentration, which favors detergents that can be easily removed or diluted. Since detergent plays such a critical role in hydrophobic or membrane protein interaction, we also provide a covalent cross-linking procedure prior to detergent solubilization and coimmunoprecipitation as a mean to preserve the protein–protein interaction.

Construction of Expression Plasmids

To construct the mammalian expression vector, polymerase chain reaction (PCR) was used to amplify the open reading frame of VAP-33 with the following primer pair: 5'-ATC AGT ATG CGG CCG CTA TGG CGA AGC ACG AGC AG-3' and 5'-AAG AAT TCT ACA AGA TGA ATT TCC C-3'. The PCR product was cloned into the *Not*I and *Eco*RI sites of the expression vector pFLAG-CMV2 (Sigma), in frame with the FLAG tag sequence. pQE10/HA-PRA2, pIRESpuro/HA-PRA2, and pIRESpuro/HA-PRA1 were described previously (Abdul-Ghani *et al.*, 2001; Hutt *et al.*, 2000), while pGEX-KG/VAP-33 was provided by Dr. W. Trimble (University of Toronto).

Purification of Bacterially Expressed Recombinant PRA2 and VAP-33

Solutions

Homogenization buffer (HB): 50 mM sodium phosphate buffer, pH 8.0, 300 mM NaCl

Wash buffer 1: 50 mM sodium phosphate, pH 6.0, 300 mM NaCl, 10% glycerol, 1% Triton X-100

Elution buffer 1: 250 mM imidazole in 50 mM sodium phosphate, pH 6.0, 300 mM NaCl, 10% glycerol, 0.05% Triton X-100

Lysis buffer: 25 mM HEPES-KOH, pH 7.5, 0.3 M sucrose, 2 mM ethylenediaminetetraacetic acid (EDTA), 2 mM dithiothreitol (DTT), 10% glycerol, 1 mM phenylmethylsulfonyl fluoride (PMSF)

Wash buffer 2: 25 mM HEPES-KOH, pH 7.5, 100 mM KCl, 2 mM EDTA, 2 mM DTT, 0.1% Triton X-100, 0.5 mM PMSF

Binding buffer: 25 mM Tris–HCl, pH 7.5, 150 mM KCl, 10% glycerol, and 0.01% Triton X-100

Purification of Recombinant PRA2

The recombinant proteins can be expressed in substantial amounts in *Escherichia coli* and readily purified by either Ni-NTA or glutathione-Sepharose depending on the associated purification tag. For the purification of His$_6$-HA-PRA2, a 1 liter culture of LB + ampicillin (100 μg/ml) grown to mid-log (OD$_{600}$ of 0.5) was induced with 1 mM isopropyl-1-thio-β-D-galactopyranoside (IPTG) for 5–8 h. The cells were collected at 5000×g for 10 min and resuspended in 10 ml of HB. A homogenate was prepared by addition of Triton X-100 to 1% final concentration and sonication (Sonic Dismembrator Model 300, Fisher) to reduce the viscosity. The homogenate was incubated at 4° for 1 h and insoluble material cleared by centrifugation at 10,000×g for 20 min (JA-17 rotor, Beckman). The homogenate was added to 1 ml of equilibrated Ni-NTA resin (Qiagen) per liter of culture. Following a wash with 20–40 volumes of Wash buffer 1, the bound fusion protein was eluted from the resin with Elution buffer 1.

Purification of Recombinant GST-VAP-33

For GST and GST-VAP-33, cultures were grown and induced with-IPTG as described previously. Cells were collected and resuspended in 10 ml of Lysis buffer. Lysozyme (Sigma) was added to 10 mg/ml, and the suspension was incubated at 4° for 30–60 min. Cells were lysed with 1% Triton X-100 and briefly sonicated to reduce viscosity. After a further incubation at 4° for 30–60 min, any insoluble material was cleared by centrifugation at 10,000×g for 30 min. The GST fusion proteins were purified by affinity chromatography on 1 ml of glutathione-Sepharose 4B beads (Amersham Biosciences) per liter of culture. For convenience, the homogenate was incubated with the beads at 4° for 4–16 h before packing into a small disposable column. The beads were washed with 20–40 volumes of Wash buffer 2. The recombinant proteins were left on the beads as a 50% bead slurry in Wash buffer 2 supplemented with 0.01% NaN$_3$ as the subsequent

binding assay is based on preloaded beads. The beads can be stored at 4°
for up to 1 week without substantial loss in binding capacity. All recombinant proteins were quantified by densitometric analysis of varying amounts
of beads or eluted sample by Coomassie blue-stained sodium dodecyl
sulfate (SDS)-polyacrylamide gels using purified bovine serum albumin as
a standard.

Effect of Boiling on Bacterial- and Mammalian-Expressed PRA2

During purification, we noticed that boiling the samples containing
recombinant His_6-HA-PRA2 prior to SDS–polyacrylamide gel electrophoresis (PAGE) greatly decreased the His_6-HA-PRA2 immunoblot signal.
To confirm the effect of heat denaturation, SDS sample buffer was added
to ~20 pmol of purified recombinant His_6-HA-PRA2 containing varying
amounts of Triton X-100. One set was boiled for 10 min while the other
was left at room temperature. The proteins were resolved by SDS-PAGE
and subjected to Western immunoblot with mouse anti-HA antibody
(Roche). Bound antibodies were visualized with Alexa 488 goat anti-mouse
secondary antibody (Invitrogen) on a Typhoon 8600 imager (Amersham
Biosciences). A significant reduction in the PRA2 signal was observed
in the boiled samples (Fig. 1A). Increasing the concentration of nonionic
detergent, Triton X-100, appeared to further reduce the PRA2 signal
in the boiled samples but had no effect on the samples incubated at
room temperature. No detectable amount of PRA2 either as a dimer or
higher oligomer was observed, suggesting that heat denaturation caused
formation of large SDS-insoluble aggregates leading to the loss of signal.

We next examined whether this is a property of bacterially expressed
PRA2 by expressing the protein in mammalian cells. Although transfection
of HA-PRA2 into Chinese hamster ovary (CHO) cells showed a strong ER
staining pattern by indirect immunocytochemistry, extracts from transfected cells showed little, if any, HA-PRA2 signal by Western immunoblot
when boiled prior to SDS-PAGE. To test whether heat denaturation of
PRA2 might be the underlying cause, we prepared extracts of CHO expressing either HA-PRA1 or HA-PRA2. For mammalian expression, 3 ×
10^5 CHO cells were seeded on a 35-mm dish overnight. The cells were
transfected with 1 μg each of pIRESpuro/HA-PRA1 or pIRESpuro/HA-
PRA2 in 3 μl of LipofectAMINE (Invitrogen) and 400 μl Opti-MEM
(Invitrogen). After 15–45 min incubation at 37°, the lipid–DNA mixture
was added to the phosphate-buffered saline (PBS)-washed cells in 800 μl of
Opti-MEM and incubated at 37° for 3–5 h. The medium was replaced with
α-MEM (Invitrogen) supplemented with 5% fetal bovine serum (FBS) and
1% penicillin/streptomycin (PS). After 24 or 48 h, the cells were washed

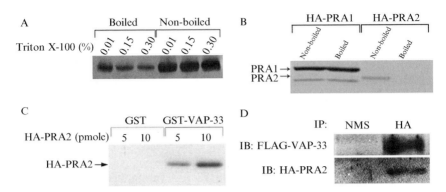

FIG. 1. (A) Effect of heat denaturation on bacterially expressed PRA2: 20 pmol of purified recombinant His_6-HA-PRA2 solubilized in varying amounts of Triton X-100 as indicated was added to SDS sample buffer and was either boiled or left at room temperature for 10 min before SDS–PAGE and Western immunoblot detection with anti-HA antibodies. (B) Effect of heat denaturation on mammalian expressed PRA2: extracts of CHO cells expressing HA-PRA1 or HA-PRA2 were either boiled or left at room temperature for 10 min before SDS-PAGE and Western immunoblot. (C) *In vitro* binding assay of His_6-HA-PRA2 with GST-VAP-33: 5 or 10 pmol of recombinant His_6-HA-PRA2 was incubated with GST or GST-VAP-33 bound beads. The beads were recovered by brief centrifugation and the bound proteins were eluted with SDS sample buffer and subjected to Western immunoblot analysis. (D) Coimmunoprecipitation of FLAG-VAP-33 with HA-PRA2 from transfected CHO cells: extracts of CHO cells coexpressing HA-PRA2 and FLAG-VAP-33 were cross-linked with DSP followed by solubilization with Triton X-100 and immunoprecipitation with control normal mouse serum (NMS) or mouse anti-HA antibodies (HA). The immunoprecipitated complex was processed by Western immunoblot with anti-FLAG and rabbit anti-HA antibodies.

with PBS, removed from the plate with a cell scraper in 1 ml of PBS, and collected by centrifugation at $500 \times g$ for 5 min. The cells were immediately resuspended in 200 μl of SDS sample buffer and sonicated twice for 20 s each to reduce the viscosity. The samples were equally divided into two-aliquot portions where one was boiled for 10 min while the other was left at room temperature. An equal amount (40 μl) of each was subjected to Western immunoblot analysis as described previously. As was observed previously with the bacterially expressed recombinant protein, boiling the cell extracts also decreased the amount of detectable HA-PRA2 (Fig. 1B). Although PRA1 and PRA2 have a similar overall structure, this sensitivity to heat denaturation is specific to HA-PRA2 as HA-PRA1 showed no loss in signal. We also did not observed any significant protein retention in the gel after electrophoretic transfer to nitrocellulose, thereby ruling out the possibility that PRA2 might not transfer well. The most likely explanation is that heat denaturation causes severe aggregation of PRA2 resulting in aberrant migration in SDS-PAGE.

In Vitro Binding Assay of VAP-33 with PRA2

We used the purified bacterially expressed recombinant proteins to confirm the PRA2-VAP-33 interaction observed in the yeast two-hybrid by *in vitro* pull-down assay. A typical pull-down assay contained 5–10 pmol of control glutathione-*S*-transferase (GST) or GST-VAP-33 bound to ~10 μl of glutathione-Sepharose beads. Varying amounts of recombinant His$_6$-HA-PRA2 were added to the beads in a final volume of 250 μl of Binding buffer. A dose–response curve and maximal binding can be determined by adding increasing amounts of His$_6$-HA-PRA2 to the beads. After incubation for 1 h at 4° in a rotator, the beads were recovered by brief centrifugation (15 s pulse in a microfuge) and washed three times with 1 ml of Binding buffer. Proteins bound to the beads were eluted with SDS sample buffer and subjected to quantitative Western immunoblot analysis. We observed binding between His$_6$-HA-PRA2 and GST-VAP-33 beads, but not with an equivalent amount of control GST beads (Fig. 1C). Even though this confirmed the interaction of PRA2 with VAP-33, the possibility remains that the interaction could be nonspecific due to the hydrophobic nature of both proteins. Thus, there is a need to independently confirm the interaction, such as by coimmunoprecipitation.

Coimmunoprecipitation of VAP-33 with PRA2

The interaction between PRA2 and VAP-33 is highly sensitive to detergent and attempts to immunoprecipitate the complex without cross-linking were unsuccessful. Thus, the complex must be stabilized by a reducible cross-linker before detergent solubilization and coimmunoprecipitation. CHO cells at 1×10^6 were seeded onto a 100-mm dish, and cotransfected the next day with 2.5 μg each of pFLAG-CMV2/VAP-33 and pIRESpuro/HA-PRA2 in 7 μl of LipofectAMINE, as described earlier. The transfected cells were maintained in α-MEM supplemented with 5% PBS and 1% PS for 48 h. The cells were then washed two times with prewarmed PBS. After removal of residual PBS, 200 μl of ice-cold 50 mM HEPES-NaOH, pH 7.2, supplemented with 1 mM PMSF was added and the cells were recovered from the plate with a cell scraper. A homogenate was prepared by brief sonication on ice, and cell debris removed by centrifugation at $1000 \times g$ for 5 min at 4°. The resulting supernatant was cross-linked with freshly prepared dithiobis-succinimidyl propionate (DSP from Pierce) diluted from a 25 mM stock dissolved in dimethyl sulfoxide

(DMSO). Final concentrations of 1.25, 2.5, and 5 mM were initially tested to determine the optimal condition, which in this case was 2.5 mM when carried out at room temperature for 20 min. After cross-linking, the DSP was quenched by adding Tris–HCl, pH 7.5, from a 1 M stock to 50 mM final concentration and incubated at room temperature for a further 5 min. The extract was then solubilized by adding an equal volume of 2% NP-40, and 1% deoxycholate, 300 mM NaCl, and 10 mM EDTA. After incubation at 4° for 1 h, insoluble materials were removed by centrifugation at 10,000×g for 5 min, and the supernatants were equally divided into two aliquot portions. Mouse anti-HA antibody (5 μg) or control normal mouse serum was added to the samples and the volume adjusted to 0.5 ml with PBS. After overnight incubation at 4°, the immune complex was recovered with 20 μl of 50% protein G-agarose slurry (Roche). The beads were washed three times with Tris-buffered saline containing 1% Triton X-100 with proteins bound to the beads reduced with DTT and subjected to Western immunoblot analysis as described previously. Rabbit anti-FLAG primary antibody (Sigma) and Alexa 488 goat anti-rabbit secondary antibody (Invitrogen) were used to detect FLAG-VAP-33, while a polyclonal rabbit anti-HA antibody (Zymed) and Alexa 488 goat anti-rabbit secondary antibody were used to detect the HA-PRA2 signal. FLAG-VAP-33 was recovered with HA-PRA2 in the presence of anti-HA antibodies, but not with control mouse serum (Fig. 1D). This demonstrates that PRA2 and VAP-33 interact *in vivo*, most likely on the ER membrane where they predominantly localize. It also shows that chemical cross-linking before detergent solubilization was required to preserve the HA-PRA2-containing complex.

Comments

We have demonstrated that PRA2 can be efficiently purified from bacteria as a His$_6$-HA-tagged recombinant protein and that it can be cross-linked to VAP-33 when coexpressed in mammalian CHO cells. The recovery of the protein, however, is significantly affected by the tendency of PRA2 to aggregate even in the presence of high concentrations of nonionic detergents. Once formed, these aggregates are highly resistant even to strong ionic detergents such as SDS. Heat denaturation seems to exacerbate the aggregation of PRA2 and significantly decrease the signal observed by Western immunoblot. This appears to be an inherent property of PRA2, although we can not rule out the possibility that other cellular proteins may contribute to this aggregate. We also demonstrated that the purified PRA2 and VAP-33 interacted in an *in vitro* pull-down assay. Both PRA isoforms have a tendency to adhere to Sepharose beads, most likely through their extensive hydrophobic domains. This can be easily corrected by incorporating small amounts of a nonionic detergent such as Triton X-100 at 0.05% final concentration. Higher levels of detergent can disrupt PRA-VAP-33 interaction, so it is critical that the interaction be verified independently such as by covalent cross-linking and coimmunoprecipitation. Functionally, it would be interesting to determine whether PRA2 and VAP-33 could affect the rate of membrane translocation of ER-localized Rabs and their effectors. The precise stage in the ER to Golgi transport pathway where this interaction occurs remains to be determined.

References

Abdul-Ghani, M., Gougeon, P.-Y., Prosser, D. C., Da-Silva, L. F., and Ngsee, J. K. (2001). PRA isoforms are targeted to distinct membrane compartments. *J. Biol. Chem.* **276,** 6225–6233.

Calero, M., and Collins, R. N. (2002). *Saccharomyces cerevisiae* Pra1p/Yip3p interacts with Yip1p and Rab proteins. *Biochem. Biophys. Res. Commun.* **290,** 676–681.

Enouf, V., Chwetzoff, S., Trugnan, G., and Cohen, J. (2003). Interactions of rotavirus VP4 spike protein with the endosomal protein Rab5 and the prenylated Rab acceptor PRA1. *J. Virol.* **77,** 7041–7047.

Evans, D. T., Tillman, K. C., and Desrosiers, R. C. (2002). Envelope glycoprotein cytoplasmic domains from diverse lentiviruses interact with the prenylated Rab acceptor. *J. Virol.* **76,** 327–337.

Hutt, D. M., Da-Silva, L. F., Chang, L. H., Prosser, D. C., and Ngsee, J. K. (2000). PRA1 inhibits the extraction of membrane-bound rab GTPase by GDI1. *J. Biol. Chem.* **275,** 18511–18519.

Ingley, E., Williams, J. H., Walker, C. E., Tsai, S., Colley, S., Sayer, M. S., Tilbrook, P. A., Sarna, M., Beaumont, J. G., and Klinken, S. P. (1999). A novel ADP-ribosylation like factor (ARL-6), interacts with the protein-conducting channel SEC61beta subunit. *FEBS Lett.* **459,** 69–74.

Li, L. Y., Shih, H. M., Liu, M. Y., and Chen, J. Y. (2001). The cellular protein PRA1 modulates the anti-apoptotic activity of Epstein-Barr virus BHRF1, a homologue of Bcl-2, through direct interaction. *J. Biol. Chem.* **276,** 27354–27362.

Lin, C.-L. G., Orlov, I., Ruggiero, A. M., Dykes-Hoberg, M., Lee, A., Jackson, M., and Rothstein, J. D. (2001). Modulation of the neuronal glutamate transporter EAAC1 by the interacting protein GTRAP3-18. *Nature* **410,** 84–88.

Martincic, I., Peralta, M. E., and Ngsee, J. K. (1997). Isolation and characterization of a dual prenylated Rab and VAMP2 receptor. *J. Biol. Chem.* **272,** 26991–26998.

Pfeffer, S., and Aivazian, D. (2004). Targeting Rab GTPases to distinct membrane compartments. *Nat. Rev. Mol. Cell Biol.* **5,** 886–896.

Sivars, U., Aivazian, D., and Pfeffer, S. R. (2003). Yip3 catalyses the dissociation of endosomal Rab-GDI complexes. *Nature* **425,** 856–859.

Skehel, P. A., Fabian-Fine, R., and Kandel, E. R. (2000). Mouse VAP33 is associated with the endoplasmic reticulum and microtubules. *Proc. Natl. Acad. Sci. USA* **97,** 1101–1106.

Author Index

A

Aaronson, S. A., 267
Aasland, R., 122
Abdul-Ghani, M., 350, 800, 801
Aberdam, E., 420
Adamian, M., 45
Adams, J., 697, 704
Adesnik, M., 156, 706, 733, 739
Agami, R., 66
Aguilar, R. C., 561
Ahmadi, K., 789
Ahmadian, G., 159
Aida, Y., 768
Aimoto, S., 461, 470, 471
Aiso, S., 263
Aivazian, D., 255, 296, 333, 340, 348, 350, 353, 354, 367, 381, 402, 552, 599, 664, 800
Akanuma, Y., 218
Akhmanova, A., 593, 608
Akiyama, Y., 199
Alam, R., 51
Alarcon, C., 69
Albertson, D. G., 203
Alegret, M., 176, 177
Alessi, D. R., 698
Alexandrov, K., 31, 32, 35, 37, 38, 39, 200, 296, 340, 342
Ali, B. R., 200, 653
Alifano, P., 628, 651, 665, 666, 669, 671, 672, 676, 688, 694, 745, 756
Allaman, M. M., 584
Allan, B. B., 391
Alory, C., 2, 3, 6, 7, 8, 255, 340, 402, 760
Altschul, S. F., 20
Altschuler, Y., 706
Alvarez-Dominguez, C., 296, 300, 301, 369
Alvarez-Mon, M., 176, 177
Alves dos Santos, C. M., 650
An, Y., 255, 340, 402
Anant, J. S., 44
Andermann, F., 262
Anders, R., 778

Anderson, D. H., 541, 542, 543, 544, 546, 547, 548, 550, 553, 558, 559
Anderson, K. V., 760
Anderson, O. R., 72, 77
Anderson, R. G., 744
Andersson, S., 195
Andrews, H. L., 72
Andrews, P. D., 520
Andrulis, E., 333, 338, 350
Ang, A. L., 285
Aniento, F., 374, 375
Antoch, M. P., 2
Antony, C., 137, 608
Aoki, T., 216, 217, 218, 219, 221, 222, 224, 225, 446, 452
Apel, E. D., 44
Applebury, M. L., 44
Araki, K., 218, 285
Araki, S., 254, 401
Araki, Y., 266, 277, 311
Arffman, A., 285
Aridor, M., 382
Ariga, H., 60, 216, 228, 420, 421, 423, 424, 425, 426, 427, 428, 429, 432, 433, 435, 436, 442, 446, 448, 450, 459, 461, 472, 778
Arighi, C. N., 539, 584
Arkinstall, S., 342, 367, 370, 372, 378
Armstrong, J., 285, 541
Aron, L. M., 513
Aronheim, A., 690
Aronson, N. N., Jr., 739
Arpin, M., 183, 724
Artalejo, C. R., 382, 384, 385, 386
Aruga, J., 453, 461, 465, 470, 471, 477
Asakura, T., 254, 401
Asano, T., 218
Asfari, M., 58
Ashman, K., 89, 120, 266, 277, 285, 311, 552, 584, 629, 642, 643, 688, 789, 790, 793
Atlschul, S. F., 21
Auer-Grumbach, M., 665
Auersperg, N., 203, 212
Auger, K. R., 789

Augustine, G. J., 57, 285, 470
Austin, C. D., 524
Authi, K. S., 778
Auvinen, P., 285, 287, 289, 290, 292
Azim, A. C., 262, 266, 272, 310

B

Baass, P. C., 368
Babst, M., 323, 324, 325, 562
Backer, J. M., 552, 629, 631, 642, 643, 644,
    688, 789, 790, 791, 793, 794, 797
Backos, D. S., 154, 155, 156, 158, 159,
    160, 164, 370
Bacq, D., 779
Baehr, W., 44, 45, 48, 49, 51, 52, 53
Bahadoran, P., 420, 421, 432
Bahler, M., 514, 707, 716, 720, 721
Bai, C., 338
Baird, G. S., 121, 156, 171
Baker, A., 768
Baker, J. M., 688
Balch, W. E., 2, 3, 6, 7, 8, 137, 154, 176,
    178, 255, 295, 300, 340, 341, 342, 346,
    347, 382, 383, 384, 385, 391, 402, 760
Balda, M. S., 183, 184, 186, 724
Baldini, G., 59, 60
Baleja, J. D., 492, 519, 520, 707
Balla, T., 247, 248
Ballotti, R., 420, 421, 432
Balsse, R., 13
Bamekow, A., 391, 392
Bananis, E., 93, 98, 101, 102, 105, 106
Bankaitis, V. A., 565
Bannykh, S. I., 382
Banta, L. M., 323, 324, 565, 602
Banting, G., 487
Baralov, S., 2, 3, 7, 8
Barbacid, M., 310
Barbero, P., 357, 358, 360, 362, 696
Barbier, M. A., 285
Barbieri, A. M., 300, 301
Barbieri, M. A., 266, 277, 296, 311, 340,
    369, 370
Barkagianni, E., 420, 421, 432, 436, 446, 470
Barlowe, C., 333, 336, 338
Barnekow, A., 593, 608
Barr, F. A., 391, 393, 398, 400, 514, 593,
    608, 613, 618, 619, 621, 623, 624, 625,
    626, 627, 630

Barral, D. C., 420, 778
Barrat, F. J., 420
Bar-Sagi, D., 230, 241
Bartel, P., 668
Baskerville, C., 642
Baskin, D. G., 69
Bastiaens, P. I., 128
Basu, S. K., 296, 301, 302, 303, 306
Bauer, B., 12
Baumeister, W., 658
Baumert, M., 154
Bayer, M., 391, 392
Beachy, P. A., 759, 760
Beattie, E. C., 159
Beaudet, A. L., 176
Beaumont, J. G., 800
Beavo, J. A., 44, 45, 49
Becker, J., 45
Beckerich, J. M., 13
Bednarski, J. J., 469
Belhumeur, P., 13
Benais-Pont, G., 184
Benedetti, E. L., 183, 186, 193, 724, 725, 726
Ben-Hamida, M., 262, 266, 272, 310
Bennett, D. C., 421
Bensen, E. S., 600
Beranger, F., 195
Beraud-Dufour, S., 347
Berent, E., 203
Berger, E., 137
Berger, K., 72, 665
Bergeron, J. J., 368
Bergmann, D., 510
Berman, D. M., 759
Bernards, R., 66
Berney, C., 128
Beron, W., 296, 300, 301, 665
Berry, G., 126, 127, 128, 132
Berry, T. R., 541, 543, 547, 550, 553,
    558, 559
Bertini, E., 262, 319
Bertolotto, C., 421, 432
Bertram, P. T., 51
Best, C. J., 203
Beuzon, C., 665
Bhakdi, S., 346, 784, 786
Bhartur, S. G., 514, 706, 707, 716, 720, 721
Bialek, W., 20
Bianchi, D., 203, 420, 778
Bielli, A., 102, 108

Bieniasz, P. D., 322
Bifulco, M., 628, 665, 666, 670, 672, 688
Bigay, J., 44
Biggs, S. M., 644
Bilbe, G., 2
Bille, K., 420, 421, 432
Bilodeau, P. S., 584
Bingham, J. B., 93
Bin-Tao, P., 384
Bischoff, J. R., 527
Bittman, R., 169, 170
Bittner, M. A., 458
Bittova, L., 357, 358, 360, 696
Blahos, J. II, 158
Blaine, W. B., 73
Blake, M. S., 301
Blanchard, S., 432, 433, 442, 443, 446
Blanche, S., 779
Blanchette-Mackie, E. J., 176, 177
Block, D., 2, 3, 7, 8
Blocker, A., 93
Bloemendal, H., 193
Blom, N., 198
Blondeau, F., 629
Blott, E. J., 322
Blystad, F. D., 654
Bocher, V., 744, 745, 746, 749
Bock, J. B., 2, 482, 492
Bodin, S., 629
Boeddinghaus, C., 552
Boespflug-Tanguy, O., 262, 319
Boettner, B., 262
Bokoch, G. M., 135, 136
Bole, D. G., 301
Bolshakov, V. Y., 154, 255
Bomsel, M., 374
Bonifacino, J. S., 2, 539, 583, 584
Bonzelius, F., 91, 178
Booth, A., 322
Booth, S., 778
Bordes, N., 598
Bordier, C., 384
Borel, C., 13
Bork, P., 27
Bornens, M., 598
Bortoluzzi, M.-N., 112
Bossi, G., 420
Botelho, R. J., 660, 662, 789
Botstein, D., 16, 20
Bottger, G., 733

Bottinger, E. P., 203
Bourne, J. R., 341, 382
Bowen-Pope, D. F., 541
Bowers, B., 420
Bowers, K., 323
Bowers, M. B., 420, 715
Bowie, D., 160
Bownds, D., 48
Brachmann, S. M., 789
Brady, R. O., 177
Braell, W. A., 300
Brech, A., 552
Bredt, D. S., 154
Breit, S., 394
Bremnes, B., 122, 243
Brenwald, P., 16
Bretz, R., 393
Briggs, B., 665
Briggs, J. A., 487
Brissette, W. H., 199
Brock, S. C., 706
Brodsky, L., 203
Broek, D., 296
Brondyk, W. H., 195, 285, 410
Brown, F. D., 247, 248
Brown, J. C., 167, 169, 170, 177
Brown, M., 670
Brown, M. J., 31, 32
Brown, M. S., 744
Brown, R. H., Jr., 262, 266, 272, 321
Brown, T. C., 154, 155, 156, 158, 159,
    160, 164, 370
Brumell, J., 665
Brummelkamp, T. R., 66
Brunak, S., 198
Brunger, A. T., 200, 324, 469
Bruni, C. B., 350, 354, 628, 651, 665, 666, 669,
    670, 671, 672, 676, 688, 689, 694, 745, 756
Brunsveld, L., 31, 32, 37, 340
Bryant, N. J., 564, 645, 646
Bu, G., 650
Bucci, C., 102, 108, 279, 296, 350, 354,
    482, 483, 486, 514, 515, 519, 628, 651,
    665, 666, 667, 669, 670, 671, 672, 676,
    679, 688, 689, 694, 707, 745, 756
Buchberg, A. M., 420
Buckler, A. J., 310
Buranda, T., 644
Burd, C. G., 311, 333, 564, 565, 568, 578
Burger, P. M., 154

Burgoyne, R. D., 65, 69, 413, 459, 470, 629, 642, 643, 688, 790
Burguete, A. S., 357, 358
Burkhard, P., 483
Burkhardt, J. K., 93
Burnashev, N., 159
Burnette, J. O., 514, 707, 716, 719, 720, 721
Burns, M. E., 57, 285, 470
Burstein, E. S., 195, 410
Burton, J. L., 285
Buscà, R., 420, 421, 432
Busch, G. R., 564
Buschle, M., 768
Bustelo, X. R., 306
Byers, D. M., 744
Byrant, N. J., 565
Byrne, B., 608
Byron, O., 391, 393, 398, 400

C

Cabanie, L., 200
Cadwallader, K., 195
Cagney, G., 333
Calafat, J., 513, 628, 651, 665, 676, 688, 689, 745, 756
Calero, M., 333, 336, 338, 350, 354, 800
Calistri, A., 322, 323
Callaghan, J., 243, 552, 688
Callebaut, I., 45, 779
Calvo, A., 203
Cameron, P., 721
Camonis, J. H., 12
Campbell, C., 85
Campbell, R. E., 121, 156, 171
Cantalupo, G., 102, 108, 482, 483, 486, 514, 515, 519, 628, 651, 665, 666, 669, 671, 672, 676, 688, 694, 707, 745, 756
Cantley, L. C., 540, 541, 553, 619, 632, 789
Caplen, N. J., 69
Capogna, M., 156
Cardelli, J., 230
Carlier, M.-F., 441
Carney, D. S., 561, 562, 564, 568, 570, 579
Caron, E., 665
Carpentier, J.-L., 629, 697, 698
Carroll, K. S., 357, 358, 360, 362
Carroll, R. C., 159
Carson, M., 200
Carstea, E. A., 176, 177

Carter, C., 322
Carter, P., 136
Casanova, J. E., 513, 706
Casari, G., 12, 22
Casaroli-Marano, R. P., 310, 420
Castillo, P. E., 154, 255
Catsicas, S., 61, 62
Cavalli, V., 177, 342, 367, 370, 372, 378, 745, 756
Cavanee, W. K., 203
Celada, A., 74
Cellai, L., 324
Cen, O., 51
Cereijido, M., 184, 186
Cerione, R. A., 43, 44, 267
Cermak, L., 203
Certain, S., 203, 420, 778, 779
Chabre, M., 44
Chaffotte, A., 200
Chakrabarti, R., 540, 541
Chamberlain, C., 135, 136, 541, 542, 543, 544, 547, 548, 550
Chamberlain, M. D., 553, 558, 559
Chan, E. K., 482, 492, 500, 510, 514, 707, 709, 717
Chang, C., 600
Chang, L. H., 350, 800, 801
Changelian, P. S., 199
Chao, J. W., 619
Chapline, C., 387
Chapman, E. R., 469, 470
Chapuis, C., 432, 433, 437, 443
Charbonneau, H., 642
Charollais, A., 59, 60
Chattopadhyay, D., 200
Chavrier, P., 200, 513, 675, 688, 744
Chedeville, G., 779
Chen, C., 386
Chen, C. K., 49, 51, 52, 53, 203
Chen, C. S., 169, 171
Chen, C. Z., 333, 336, 338
Chen, G. L., 198
Chen, J., 72, 77, 469
Chen, J. H., 256
Chen, J. Y., 800
Chen, L., 200, 306
Chen, P. I., 370
Chen, W., 255, 340, 402, 628, 629, 633, 640, 650, 651, 652, 653, 654, 656, 657, 746
Chen, X., 368

Chen, Y. A., 2
Chen, Y. H., 51, 52, 53
Chen, Y. T., 195
Chen, Z., 564, 568, 570, 579
Cheney, C. M., 340
Cheney, R. E., 716
Cheng, H., 510
Cheng, K. W., 203, 212
Chenna, R., 20
Cheviet, S., 57, 65, 69, 459, 470
Chiang, C., 760
Chiariello, M., 350, 354, 628, 665, 666, 670, 672, 688
Chiatiello, M., 628
Chiaverini, C., 421, 432
Chiba, T., 333
Chien, C., 668
Chinetti, G., 744, 745, 746, 749
Ching, K. A., 2, 3, 7, 8
Chong, L., 640
Chou, M. M., 370
Choudhury, A., 167, 168, 169, 170, 171, 172, 174, 175, 176, 177, 179
Chow, V. T., 256
Chowdhry, B. Z., 516
Christensen, H., 441
Christoforidis, S., 89, 108, 120, 122, 124, 368, 370, 398, 552, 584, 600, 601, 629, 631, 642, 643, 688, 789, 790, 793, 795
Chua, J., 296, 661
Chua, N.-H., 441
Chuang, C. C., 676
Chun, J., 519, 520, 522
Chung, H. Y., 322, 323
Chung, S. H., 469, 470
Church-Kopish, J., 51, 52, 53
Chwetzoff, S., 800
Cimbora, D. M., 322, 323
Cimino, D. F., 644
Cinalli, R. M., 745
Citi, S., 183, 724
Clague, M. J., 370, 630, 650
Clark, S., 203
Clarke, M., 72, 77, 608, 618
Clarke, S., 44
Clark-Lewis, I., 30
Cleck, J. N., 600
Clemens, D., 665
Clement, M., 13
Cleveland, D. W., 319

Cockcroft, S., 632, 790, 793
Coen, K., 665
Coers, J., 77
Cohen, J., 800
Cole, N., 262, 266, 272, 310
Colgan, S. P., 724
Collard, J. G., 230
Collawn, J. F., 82
Colley, S., 800
Collins, B. M., 584
Collins, C., 203
Collins, R. N., 333, 336, 338, 350, 354, 800
Collinson, L. M., 420, 421, 778, 789
Colombo, M. I., 195, 628, 665
Comerford, K. M., 724
Comly, M. E., 176
Condeelis, J. S., 203
Conibear, E., 600, 650
Connelly, P. A, 199
Conner, S. D., 367
Connolly, J. L., 200
Conover, D., 333
Constantinescu, A. T., 200
Cook, H. W., 744
Cook, T. A., 44, 49
Cooke, F. T., 697, 704
Cooke, M. P., 2, 3, 7, 8
Cool, R. H., 12
Cooley, L., 697, 704
Cooney, A. M., 177
Cooper, G., 384
Cooper, J. A., 267
Cooper, K. L., 628, 629, 630, 633, 634, 639, 640, 641, 642, 646, 651, 746
Cooper, P. J., 421
Copeland, N. G., 420, 421, 426, 432, 436, 438, 715, 778
Coppola, T., 61, 62, 65, 67, 69, 216, 220, 221, 409, 432, 433, 443, 446, 451, 452, 455, 459, 470
Corbin, J. D., 730
Corcoran, L. M., 299
Corden, J. L., 760
Cordenonsi, M., 183
Cormont, M., 108, 109, 112, 115
Corti, M., 177, 342, 367, 370, 372, 378, 745, 756
Corvera, S., 122, 540, 541, 552, 789, 790
Corwe, J. E., Jr., 706
Cosson, P., 177, 745, 756

Cote, R. H., 44, 45, 48, 49
Cotten, M., 768
Cottrell, J. S., 394
Couedel-Courteille, A., 598
Courtoy, P. J., 203
Couvillon, A. D., 500, 510
Couvreur, M., 203
Coux, O., 660
Cox, A. D., 341
Cox, D., 513
Craig, E. A., 394, 609
Craig, S., 322, 323
Creasy, D. M., 394
Cribier, S., 432, 433, 437, 443
Croizet-Berger, K., 203
Crottet, P., 82, 86, 87, 89, 90, 538
Cuervo, A. M., 375
Cuif, M. H., 598
Cull-Candy, S. G., 160
Cullen, P. J., 65
Cullis, D. N., 492, 519, 520, 707
Czech, M. P., 340, 699

D

Dabbagh, O., 262, 266, 272, 310
Dadak, A. M., 378
Dahlback, H., 195
Dahlmann, B., 654
Dahms, D. M., 696
Dairkee, S. H., 203
Dale, B. M., 513
Daniell, L., 469, 470
Danuser, G., 128
Darchen, F., 57, 59, 60, 408, 432, 433,
    437, 442, 443, 446
Daro, E. A., 82, 102, 733
D'Arrigo, A., 122, 790
Da Silva, J. S., 310
Da-Silva, L. F., 350, 800, 801
Da Silva Xavier, G., 65
D'Atri, F., 183
Daumerie, C., 203
Dauzonne, D., 183, 186, 724, 725, 726
Davidson, H. W., 346, 630, 789
Davidson, T.-L., 263, 272, 274, 311
Davies, B. A., 561, 562, 564, 568, 570, 579
Davies, J. M., 482, 483, 492, 500, 514, 518
Davis, D. L., 195
Davis, G. D., 290

Davis, T., 322, 323
Dawson, P. A., 744
Dawson, P. E., 30
Day, G. J., 296
Debanne, D., 156
De Bernardez, C. E., 37
Deblandre, G. A., 323
De Camilli, P., 59, 154, 255, 285, 469, 470
De Corte, V., 698
Deeb, R., 697, 698, 699, 704
de Felipe, K. S., 72, 77
De Gregorio, L., 665, 688, 689, 694
Deguchi-Tawarada, M., 401, 402, 404
de Gunzberg, J., 195
de Gunzburg, J., 45
de Hoop, M. J., 154
de Hostos, E. L., 93, 95
De Jonghe, P., 665
Delcroix, J. D., 212, 368
de Leeuw, H. J., 137
De Lemos-Chiarandini, C., 156, 706, 733, 739
Delidakis, C., 323
Del Nery, E., 138, 593, 595, 608
Delprato, A., 562, 563
del Real, G., 176, 177
DeLucas, L., 200
Del Villar, K., 256
De Maziere, A. M., 524
Demirov, D. G., 322
Deneka, M., 82, 89, 102, 108, 296, 526, 527,
    531, 532, 533, 536, 552, 733
Deng, H. X., 262, 266, 272, 310
Deng, L., 564, 568, 570, 579
Der, C. J., 310, 341, 382, 676
DeRegis, C. J., 336, 338
de Renzis, S., 89, 108, 130, 156, 526, 733
de Repentigny, L., 13
Deretic, D., 50, 665
Deretic, V., 296, 628, 661, 665, 789
Derewenda, Z., 570
Derre, I., 72
de Saint Basile, G., 203, 420, 432, 433, 437,
    443, 778, 779
Deshusses, J., 58
Desjardins, M., 296
Desnos, C., 432, 433, 437, 442, 443, 446
Desrosiers, R. C., 800
Deterre, P., 44
Devlin, M. J., 44
Devon, R. S., 262, 263, 266, 272, 274, 311, 321

De Vriendt, E., 665
deWit, C., 196
de Wit, H., 82, 108, 526, 536
Dewitte, F., 120
De Zeeuw, C. I., 593, 608
Dhand, R., 789
Dhingra, A., 45, 48, 49
Diaz, E., 357, 358, 359, 360, 697, 704
di Capua, M., 262
Didry, D., 441
Diederichs, S., 358, 360, 361, 362
Diekmann, D., 230, 241
Dietmaier, W., 13
Dietrich, L. E., 552
Dietrich, U., 324
Di Fiore, P. P., 370
DiGiamarrino, E., 200
Di Guglielmo, G. M., 368
Dikic, I., 119, 120
Ding, J., 194, 195, 196, 200
Ding, L., 358, 360, 361, 362
Dirac-Svejstrup, A. B., 296, 348, 349,
    350, 354
Dixon, J. E., 61, 790
Dobrenis, K., 171
Doi, T., 778
Dollet, S., 262
Domin, J., 630, 632, 789, 790, 793
Dominguez, M., 167, 168, 169, 170, 171,
    172, 175, 176, 177
Donaldson, J. G., 243, 244, 247, 248, 249,
    250, 251, 252
Donaldson, K. M., 564, 568
Donevan, S. D., 160
Dong, J., 166, 203, 628, 629, 630, 633, 634,
    639, 640, 641, 642, 646, 650, 651, 652,
    653, 654, 656, 657, 688, 746
Doo, E., 660
Dorn, M. C., 482, 492, 500, 510, 514, 707,
    709, 717, 719
Dortland, B. R., 608
Dotti, C. G., 154, 310
Dove, S. K., 630, 697, 698, 704
Dow, R. L., 199
Dowler, S., 698
Downes, C. P., 698
Downward, D. J., 655
Drechsel, D., 621
Driscoll, P. C., 789
Drummond, F., 102, 108

D'Souza-Schorey, C., 135, 510
Dube, M. P., 262
Dubel, S., 136
Dubensky, T. W., 156
Ducceschi, M. H., 51
Dufourcq-Lagelouse, R., 420
Dugan, J. M., 196
Dulubova, I., 222
Dumas, B., 13
Dumas, J. J., 17, 25, 27, 200
Dumenil, G., 296
Dumont, C., 779
Dunia, I., 183, 186, 193, 724, 725, 726
Dunn, B., 16
Dunn, R., 562
Dunn, W. A., 739
Dunphy, W. G., 300
Dupree, P., 154, 195, 243, 247, 765
Dupuis, S., 203, 420, 778
Durek, T., 31, 32, 37, 39, 340
Dusseljee, S., 628, 651, 665, 676, 688, 689,
    745, 756
Dusterhoft, A., 564
Dwyer, N. K., 177
Dykes-Hoberg, M., 800

E

Echard, A., 137, 593, 608, 618
Eckel, J., 513, 706
Eckert, J., 184
Edinger, A. L., 745
Edwards, J. G., 154
Eggenschwiler, J. T., 760
Ehlers, M. D., 154, 159
Ehnholm, C., 744, 745, 746, 749
Eigenbrot, C., 524
El-Amraoui, A., 62, 65, 67, 69, 432, 433,
    437, 442, 443, 446
Elbashir, S. M., 399, 626
Elferink, L. A., 733
Elia, A. E., 619
Eliason, W. K., 324
Elisee, C., 290
Elkind, N. B., 285
Elledge, S. J., 338
Ellis, S., 514
El Marjou, A., 45
Emans, N., 374
Emmert-Buck, M. R., 203

Emr, S. D., 311, 323, 324, 325, 564, 565, 568, 578, 602, 789, 790
Endo, K., 255, 285
Endo, T., 231, 234, 238, 241
Endow, S. A., 101
Engelke, U., 13
Enouf, V., 800
Epstein, W. W., 31, 32
Erdjument-Bromage, H., 706
Erdman, R. A., 195, 200
Ersoy, F., 203, 420, 778
Escher, G., 59, 60
Eskelinen, E. L., 628, 665, 745
Espinoza, E., 760
Esquerre-Tugaye, M. T., 13
Esteban, J. A., 154, 155, 156, 157, 158, 159, 160, 164, 370
Estepa, E. J., 324
Estepa-Sabal, E. J., 323
Esters, H., 200
Etienne-Manneville, S., 310
Eva, A., 267
Evans, D. T., 800
Evans, G. J., 413, 470
Evans, T., 195, 267
Evans, T. C., Jr., 31
Evans, T. M., 759, 760, 761, 773
Exton, J. H., 500, 510, 514
Eymard-Pierre, E., 262, 319

F

Fabian-Fine, R., 801
Fabiato, A., 786
Fabiato, F., 786
Fabry, S., 13
Fahrenholz, F., 177
Falkow, S., 300
Fan, G. H., 493, 520, 706, 708, 716, 720, 722
Fang, G., 244, 707, 708, 716, 720
Fang, Y., 543, 546, 547
Fanget, I., 432, 433, 437, 443
Farnsworth, C. C., 61, 62
Farquhar, M. G., 382
Fasshauer, D., 324, 552
Faure, J., 177, 745, 756
Fay, F., 540, 541
Feeley, J. C., 73
Feig, L. A., 384, 492, 519, 520, 707, 765
Feinstein, E., 203

Felberbaum-Corti, M., 368
Feldmann, J., 203, 420, 778, 779
Feng, Y., 628, 629, 630, 631, 633, 634, 638, 639, 640, 641, 642, 646, 650, 651, 652, 657, 665, 688, 694, 744, 746
Feramisco, J. R., 230, 241
Ferguson, C., 759, 760, 761, 773
Fernandez, G. E., 340
Fernandez, I., 222
Fernandez-Borja, M., 628, 651, 665, 676, 688, 689, 745, 756
Field, C. M., 115
Fielding, A. B., 500, 510, 514, 520
Fields, S., 108, 333, 668, 680, 751
Figlewicz, D. A., 262, 266, 272, 321
Figueroa, C., 354
Fine, R. E., 85
Finger, F. P., 510
Finlay, B., 665
Finley, R. L., Jr., 697, 698, 699, 704
Fischer, A., 203, 420, 778, 779
Fischer von Mollard, G., 154
Fishel, S., 657
Fishman, D., 203, 212
Fitzgerald, M. L., 194
FitzPatrick, D., 665
Flaus, A. J., 324
Fleischer, S., 301
Fleming, T. P., 183, 184, 724
Fletcher, C. F., 420, 432
Fligger, J., 697, 698, 699, 704
Flores, C., 186
Flores-Maldonado, C., 184
Florio, S. K., 44, 45, 49
Flugge, U. I., 345
Folli, F., 59
Folsch, H., 285
Fontaine, J., 183, 724
Fontaine, J. J., 183, 724
Fontijn, R. D., 608, 618
Forquet, F., 44
Fortner, K. A., 285
Foster, K., 159
Foti, M., 629, 697, 698
Fotsis, T., 394
Fouraux, M. A., 108, 526
Fournier, H., 13
Foutz, T., 644
Francis, S. A., 564, 579
Francis, S. H., 730

Fransen, J. A., 137
Frantz, C., 62, 216, 220, 221, 409, 446,
    451, 452, 455
Fratti, R., 665
Fratti, R. A., 789
Frederick, J. M., 49, 51, 52, 53
Freed, E. O., 322
Freire, E., 516
Freytag, C., 784, 786
Fridmann-Sirkis, Y., 600, 605
Friedberg, A. S., 564, 568, 570, 579
Friedman, H., 74
Fritsch, E. F., 699, 765, 766, 769
Frohlich, V. C., 203
Fruman, D. A., 789
Fry, M. J., 789
Fuerst, T., 388, 636, 656, 667
Fuhrman, S. A., 73
Fujimoto, K., 458
Fujita, M., 510
Fukami, Y., 199
Fukazawa, H., 199
Fukuda, M., 57, 60, 62, 65, 67, 69, 216, 228,
    409, 419, 420, 421, 423, 424, 425, 426, 427,
    428, 429, 432, 433, 434, 435, 436, 437, 438,
    439, 440, 441, 442, 443, 445, 446, 447, 448,
    449, 450, 451, 452, 453, 454, 455, 458, 459,
    460, 461, 462, 463, 464, 465, 466, 467, 469,
    470, 471, 472, 473, 474, 475, 476, 477, 478,
    778, 779, 782
Fukui, K., 254, 401
Fukushige, S., 310
Fuller, S., 640
Funato, K., 296
Fung, B. K. K., 44
Furuhjelm, J., 285, 292, 293
Furuno, K., 779
Furuno, N., 310
Furuse, M., 183, 724
Futter, C. E., 322, 323, 789
Fybin, V., 108

G

Gahwiler, B. H., 156
Gaide-Huguenin, A. C., 48
Gaillardin, C., 13
Galan, J. E., 306
Galjart, N., 593, 608
Galli, T., 45, 183, 513, 608, 724, 725, 726

Gallwitz, D., 13, 333, 338, 350, 564, 600, 604
Galmiche, A., 108, 109, 115
Galperin, E., 124, 128, 129, 130
Gammeltoft, S., 198
Gamou, S., 655
Ganetzky, B., 470
Ganley, I., 357, 358, 360, 361, 362
Gao, M., 250
Garcia, J. A., 600
Garcia-del Portillo, F., 296
Garcia-Ranea, J. A., 12
Gardner, J. P., 199
Garner, C. C., 154
Garrard, S., 570
Garrett, M. D., 184, 340, 724
Garrison, J., 644
Gary, D. S., 204
Gary, J. D., 790
Gascon, G., 262, 266, 272, 310
Gaskin, C., 108
Gattesco, S., 61, 62, 216, 220, 221, 409, 446,
    451, 452, 455
Gatti, E., 285
Gaullier, J. M., 122, 552, 645, 646
Gauthier, N. C., 115
Gautier, N., 112
Gaynor, R. B., 600
Gazit, A., 199
Geissler, H., 348
Gekakis, N., 564, 568
Gelb, M. H., 44, 49
Gelfand, V. I., 421
Gelino, E. A., 470
Geppert, M., 154, 255, 469, 470
Gerard, R. D., 263, 264, 265, 266, 270, 274,
    277, 319
Gerdes, H., 666
Gerez, L., 82, 89, 194, 526
Gerges, N. Z., 154, 155, 156, 158, 160, 164
Gerrard, S. R., 564
Gettemans, J., 698
Geuze, H. J., 82, 108, 323, 526, 775
Ghirlando, R., 564, 568, 579
Ghomashchi, F., 44, 49
Ghosh, P., 696
Gibbs, R. A., 255, 340, 402
Gibson, R. J., 73
Gibson, T. J., 13, 20, 503, 761, 762, 763
Gietz, D., 691
Gigg, R., 698

Gilbert, P. M., 333
Gillespie, P. G., 44
Gillooly, D. J., 645, 646
Gilon, C., 199
Gimpl, G., 177
Giner, A., 368, 688
Ginsberg, L., 665
Giot, L., 333
Girod, A., 593
Gish, W., 20
Gleeson, P. A., 686
Glick, B. S., 2
Glomset, J. A., 61, 62
Goda, Y., 154, 255, 357
Godwin, B., 333
Godzik, A., 45
Goila-Gaur, R., 322
Goishi, K., 695
Goldberg, A. L., 660
Goldenring, J. R., 244, 482, 492, 493, 500, 510, 513, 514, 520, 706, 707, 708, 709, 716, 717, 718, 719, 720, 721, 722
Goldstein, J., 670
Goldstein, J. L., 744
Goldthwaite, L. M., 323
Golemis, E. A., 296
Gomes, A. Q., 420, 421, 778
Gomez-Mouton, C., 176, 177
Gomi, H., 57, 218, 219, 224, 228, 419, 446
Gomim, H., 69
Gonzalez, B., 323
Gonzalez, L., Jr., 514
Gonzalez, S., 186
Gonzalez-Gaitan, M., 369
Goodfellow, A. F., 368
Goodson, H., 177, 745, 756
Goodwin, E. B., 324
Goody, R. S., 31, 32, 35, 37, 38, 39, 200, 340
Gordon, G. W., 126, 127, 128, 132
Gordon, K., 101, 106
Gordon-Walker, A., 48
Gorman, G. W., 73
Gorska, M. M., 51
Gorvel, J.-P., 200, 688, 744
Gotschlich, E. C., 301
Gottlinger, H. G., 322, 323
Goud, B., 57, 82, 89, 108, 136, 137, 138, 141, 149, 152, 200, 408, 482, 483, 486, 513, 514, 515, 519, 593, 594, 595, 598, 608, 618, 653, 675, 707, 733

Gougeon, P.-Y., 350, 800, 801
Gould, G. W., 500, 510, 514, 520
Gould, R., 645, 646
Gould, S. J., 322
Goulding, D., 778
Gournier, H., 527
Govek, E.-E., 310
Grabe, N., 198
Graham, J. R., 256
Gravotta, D., 706
Gray, J. W., 203, 212
Gray, N. W., 263, 264, 265, 266, 270, 274, 277, 319
Gray, P. W., 74
Gray, S. R., 487
Grazio, H. J. III., 44
Green, E. G., 706
Green, J. E., 203
Greenberg, S., 513
Greger, I. H., 158
Griffith, J., 323
Griffiths, G., 93, 296, 305, 374, 640, 679
Griffiths, G. M., 322, 420, 778
Grill, S., 593
Griner, R., 482, 492, 500, 510, 514, 707, 709, 717
Grinstein, S., 660, 662, 665, 671, 672, 745, 756, 789
Griscelli, C., 420
Griswold-Prenner, I., 44
Groisman, E. A., 296
Gros-Louis, F., 262, 263, 272, 274, 311
Grosveld, F., 593, 608
Grove, S. J. A., 698
Gruenberg, J., 119, 120, 167, 177, 200, 295, 299, 305, 322, 342, 367, 368, 370, 372, 374, 378, 526, 640, 665, 696, 745, 756, 766, 789
Gruneberg, U., 608, 618, 619, 626, 627
Grynberg, M., 45
Gu, Q.-M., 699
Guan, K. L., 61
Guarente, L., 668
Guignot, J., 665
Gullick, W. J., 655
Gumusboga, A., 369
Guo, Q., 644
Guo, W., 724
Gupta, R., 198
Gurkan, C., 2, 3, 6, 7, 8, 340

Gutierrez, M. G., 628, 665
Gutkind, J., 666

H

Habermann, B., 368, 688
Habets, G. G. M., 230
Hackstadt, T., 296
Hadano, S., 262, 263, 266, 272, 277, 285, 310, 311, 312, 314, 317, 318, 319, 320, 321
Haefeli, W. E., 176
Hager, J. H., 203
Hahn, K. M., 135, 136
Haire, L. F., 619
Halban, P. A., 58, 416
Hales, C. M., 482, 492, 500, 510, 514, 707, 709, 716, 717, 718, 720, 721
Hall, A., 230, 241, 262, 310, 556
Hall, T. A., 503
Halladay, J., 394, 609
Hama, H., 266, 277, 564, 570, 571, 574
Hammar, E., 183
Hammer, J. A. I., 715
Hammer, J. A. III, 420, 421, 426, 432, 436, 438, 664
Hammer, R. E., 154, 255
Hanada, K., 169
Hanahan, D., 203
Hancock, J. F., 195, 775
Hand, C. K., 262, 266, 272, 321
Hanke, J. H., 199
Hanna, J., 357, 358, 360, 362
Hannah, M. J., 733, 739
Hannon, G. J., 65
Hanson, T. J., 154
Hanzal-Bayer, M., 43, 44, 45
Hao, M., 82
Hara, T., 779
Harbeck, B., 724
Harborth, J., 399, 626
Harbury, P. B., 357, 358
Hardt, W., 306
Hardy, D., 482, 492, 500, 510, 514, 707, 709, 717
Hariono, S., 203
Harris, J., 382
Harris, R., 784, 786
Harrison, R. E., 745, 756
Harrison, R. G., 290, 665, 671, 672
Hart, I. R., 421

Hart, M. J., 267
Hartung, H., 665
Hashim, S., 296, 301, 302, 303, 306
Hashimoto, Y., 263
Haskins, J., 183, 724
Hata, Y., 254, 401
Hatherly, S., 17, 25, 27
Hattori, M., 333
Hattula, K., 285, 292
Haubruck, H., 13
Haugland, R. P., 169, 175
Hauri, H. P., 744
Havas, K. A., 333
Hawkins, W., 721
Hayakawa, M., 2, 3, 7, 8
Hayashi, Y., 157, 159, 160
Hayashida, H., 310
Hayden, M. R., 262, 263, 266, 272, 274, 311, 321
Hayes, B., 285
Haynes, L. P., 65, 69, 413, 459, 470
Hays, L. B., 69
Hays, T. S., 510
He, G. P., 322, 323
He, X., 584, 586, 591
Heidtman, M., 333, 336, 338
Heinemann, I., 31, 32, 35, 37, 39
Heinemann, S., 158
Heinicke, E., 301
Heldin, C. H., 541
Heller-Harrison, R., 17, 25, 27, 122
Hellio, R., 183, 724
Helliwell, S. B., 323
Helm, J. R., 262, 263, 272, 274, 311
Hendershot, L. M., 301
Henderson, M., 608
Hendil, K. B., 654, 657
Hendrick, A. G., 482, 483, 486, 501, 509, 511, 514, 515, 519, 707, 711
Henikoff, S., 136
Henrick, A. G., 482, 707
Henriks, W., 137
Henry, J. P., 432, 433, 437, 442, 443, 446
Hentati, A., 262, 266, 272, 310
Hentati, F., 262, 266, 272, 310
Heo, W. D., 12
Herbert, C., 13
Herman, B., 126, 127, 128, 132, 789
Herman, G. A., 91, 178
Herman, P. K., 789

Hermann, T., 324
Heumann, H., 324
Hewlett, L. J., 322, 323
Hicke, L., 562, 564, 579
Hickenbottom, S. J., 358
Hickson, G. R., 510
Hickson, G. R. X., 514
Hierro, A., 323, 325
Hieter, P., 568
Higashi, T., 778, 779, 782, 786
Higashide, T., 43, 45
Higgins, D. G., 13, 20, 503, 761, 762, 763
Higgs, H. N., 230
Highfield, H., 369
Higuchi, H., 101
Higuchi, R., 198
Hildebrand, A., 32, 37, 39
Hildreth, J. E., 322
Hill, E., 608, 618
Hill, S., 222
Hillebrand, A. M., 35
Hille-Rehfeld, A., 85
Hillig, R. C., 43, 44, 45
Hirano, H., 254, 255, 285, 401
Hirano, M., 262, 266, 272, 310
Hirata, Y., 256
Hirling, H., 61, 62, 216, 220, 221, 409,
    446, 451, 452, 455
Hirosaki, K., 171
Hirosawa, M., 492, 494, 500
Hirst, J., 539
Hisatomi, O., 51
Ho, C. H., 420
Hobdy-Henderson, K. C., 482, 492, 500,
    510, 514, 707, 709, 717
Hodge, T. W., 200
Hodgson, G., 203
Hoepker, K., 619
Hoffman, G. R., 43, 44
Hoflack, B., 120, 628, 634, 639, 657, 665,
    667, 679, 688, 744
Hofman, P., 108, 109, 115
Hofmann, K., 255, 402, 408, 458
Hogenesch, J. B., 2, 3, 6, 7, 8, 340
Holcomb, C., 195
Holden, D., 665
Holicky, E., 171, 177
Hollenberg, S. M., 267
Hollmann, M., 158
Holmes, A. B., 697, 698, 704

Holz, R. W., 63, 154, 258, 285, 405, 458,
    469, 470
Honda, R., 608, 618, 619, 626, 627
Hong, D. H., 45
Hong, W., 382, 608, 665, 676, 679, 683,
    685, 686
Hoogenraad, C. C., 593, 608
Hopkins, C. R., 82, 322, 323, 420, 421,
    778, 789
Horazdovsky, B. F., 263, 264, 265, 266, 270,
    272, 274, 277, 285, 311, 319, 369, 370,
    561, 562, 564, 568, 570, 571, 574, 579
Horgan, C. P., 500, 501, 502, 503, 505, 506,
    507, 509, 510, 511
Horiuchi, H., 89, 120, 266, 277, 279, 285,
    296, 311, 584, 778, 779, 782, 784, 786
Horrevoets, A. J., 608, 618
Horstmann, H., 382, 685
Hortsman, H., 683
Horuichi, H., 89
Horwitz, M. A., 72, 665
Hosaka, M., 216, 217, 218, 219, 221, 222,
    224, 225, 227, 446, 452
Hosier, S., 45, 48, 49
Hosler, B. A., 262, 266, 272, 321
Hotta, I., 706
Houdusse, A., 420
Houlden, H., 665
Houthaeve, T., 394
Howald-Stevenson, I., 323, 565
Howard, J., 95, 97
Howell, K. E., 305, 640
Howell, S. A., 608
Hoyer, D., 2
Hsiao, P.-W., 600
Hsu, K. S., 370
Hsu, V. W., 250
Hu, B., 458, 460, 475
Hu, Y., 679
Hu, Z., 660
Huang, C. C., 370
Huang, J. D., 426, 438
Huang, W., 51, 52, 53
Hubbard, A. L., 739, 789
Huber, H., 13
Huber, L. A., 50, 154, 296, 342, 368, 492
Huet, S., 432, 433, 437, 443
Hughes, D. C., 697, 704
Hughson, F. M., 2
Hull, M., 82, 89, 102, 108, 513, 733

Hume, A. N., 57, 69, 171, 420, 421, 432, 436, 446, 461, 470, 778
Hung, W. Y., 262, 266, 272, 310
Hunt, S. J., 212, 368
Hunziker, Y., 324
Huppi, K., 69
Hurley, J. H., 323, 325, 358, 564, 568, 579, 584
Hust, M., 136
Hutchon, S. P., 657
Hutt, D. M., 350, 800, 801
Huttelmaier, S., 203, 724
Huttner, W. B., 733, 739
Huuskonen, A., 13
Hyatt, S. D., 387
Hyman, A. A., 93, 95, 97, 621, 789

I

Iakovenko, A., 31, 35, 39
Ide, N., 259
Iezzi, M., 59, 60, 62, 65, 67, 69, 432, 433, 443, 458
Igarashi, M., 217, 228, 420, 436
Igarashi, N., 584
Ihrke, G., 487
Iino, M., 311
Ikeda, J.-E., 262, 263, 266, 272, 277, 285, 310, 311, 312, 314, 317, 318, 319, 320, 321
Ikeda, K., 403
Ikeda, W., 259
Ikegami, H., 218
Ikonen, E., 744
Ikonomov, O. C., 629, 697, 698, 699, 702, 704
Illenberger S., 724
Imai, A., 427, 433, 443, 450
Imai, M., 409
Imamura, M., 57
Imanishi, Y., 51
Imashuku, S., 779
Imayoshi, M., 779
Imazumi, K., 254, 401
Immler, D., 513, 706
Imura, H., 416
Inagaki, N., 408, 409, 410, 470
Inana, G., 43, 45
Incardona, J. P., 177
Ingham, P. W., 759
Ingley, E., 800
Inoue, S., 95
Irizarry, R. A., 3

Isberg, R. R., 72, 296, 665
Ishida, H., 416
Ishida, J., 199
Ishii, E., 779
Ishikawa, K., 492, 494, 500
Ishizaki, H., 255, 256, 285
Isomura, M., 254, 401
Israel, D. I., 288
Itin, C., 357, 364
Ito, T., 333
Itoh, H., 778
Itoh, M., 183, 724
Itoh, N., 199
Itoh, T., 420, 421, 425, 426, 427, 442
Ivan, V., 108
Ivins, F. J., 619
Iwanaga, T., 408, 409, 415, 417, 458
Izumi, G., 401, 402, 404
Izumi, T., 57, 69, 216, 217, 218, 219, 220, 221, 222, 223, 224, 225, 226, 227, 228, 408, 419, 420, 436, 445, 446, 451, 452, 455

J

Jabado, N., 420
Jackson, M., 800
Jacobs, A., 665
Jacobsen, A., 13
Jacquet, C., 13
Jagath, 522
Jager, S., 628, 665, 745
Jahn, R., 154, 324, 469, 470
Jain, S., 284
Jaken, S., 387
Jakubowski, J. M., 340
James, G. L., 39
James, P., 394, 609
James, S. R., 698
James, T. L., 420
Janjic, D., 58
Janoueix-Lerosey, I., 149, 593, 594, 595, 598, 608, 618
Jansen, K., 82, 91
Janssen, H., 628, 651, 665, 676, 688, 689, 745, 756
Janssen, L., 628, 651, 665, 676, 688, 689, 745, 756
Janz, R., 154, 255, 470
Jaworski, C. J., 744
Jeanmougin, F., 13, 20

Jefferson, J. R., 177
Jeffries, T. R., 697, 698, 704
Jenkins, N. A., 420, 421, 426, 432, 436, 438, 715, 778
Jiang, X., 665
Jimbow, K., 171
Jimenez-Baranda, S., 176, 177
Jin, F., 296
Jin, H. Y., 171
Joachimiak, A., 584, 586
Joazeiro, C. A., 564, 568
Joberty, G., 61, 62, 183, 470, 724
Jockusch, B. M., 724
Johannes, L., 149, 230, 513, 593, 594, 595, 608
Johannessen, L. E., 654
Johansson, M., 744, 745, 746, 749
Johnson, K. M., 154
Johnson, L. M., 547, 565
Johnson, R., 378
Johnston, C. L., 230, 241
Johnston, K. H., 301
Johnston, M., 333
Joiner, K. A., 73
Jokitalo, E. J., 400
Jollivet, F., 137, 149, 593, 594, 595, 598, 608, 618
Jones, E. A., 564, 568, 579
Jongeneelen, M., 102, 194, 526
Jorcyk, C., 203
Jordan, M., 151
Jordens, I., 628, 651, 665, 676, 688, 689, 745, 756
Jornvall, H., 195
Jovin, T. M., 128
Ju, W., 159
Judson, R. S., 333
Julyan, R., 69
Jung, T. S., 284
Junutula, J. R., 515, 516, 518, 519, 520, 522

K

Kaether, C., 666
Kagan, J. C., 72, 76, 78, 80, 665
Kahn, R. A., 72
Kai, R., 310
Kaibuchi, K., 216, 255, 342, 402, 408, 458, 469, 470
Kajiho, H., 266, 277, 285, 311
Kalbfleisch, T., 333

Kalinin, A., 31, 35
Kamboj, S. K., 160
Kamiya, H., 255
Kandel, E. R., 801
Kandutsch, A. A., 744
Kaneko, K., 256
Kanekura, K., 263
Kang, H. C., 169, 175
Kang, J., 203
Kangawa, K., 311
Kanno, E., 436, 446, 447, 448, 449, 450, 451, 452, 453, 454, 459, 460, 461, 462, 465, 470, 471, 473, 474, 475, 476, 477, 478
Kanzaki, M., 629, 697
Kapeller, R., 540, 541
Kaplan, D. R., 632
Kaplan, J., 301, 322, 323
Karhadkar, S. S., 759
Karin, M., 342, 367, 370, 372, 378
Karylowski, O., 706
Kasai, H., 69, 218, 219, 224
Kasai, K., 57, 69, 218, 219, 224, 228, 419, 446
Kaseda, K., 98
Kashima, Y., 408, 458
Kasper, L. H., 73
Kasuga, M., 218
Katada, T., 266, 277, 285, 311
Katayama, E., 453
Kato, H., 600
Kato, M., 57, 584
Kato, R., 584
Kato, S., 416
Katoh, Y., 584
Katso, R., 789
Katz, L. C., 156
Katzmann, D. J., 323, 325, 564, 568, 570, 579
Kauer, J. A., 154
Kauppi, M., 243
Kawabe, H., 401, 402, 404
Kawaguchi, M., 409, 415, 417
Kawai, R., 510
Kawai, R., 584
Kawasaki, M., 584
Kay, S. A., 2
Kaziro, Y., 199, 778
Kearney, J. F., 301
Kearney, W. R., 584
Kehoe, M., 784, 786
Keinanen, K., 158
Keller, P., 400, 593
Kellogg, D., 621

Kelly, C., 382, 384, 385, 386
Kelly, R. B., 82, 108, 258, 405
Kelso, R., 697, 704
Kendrick-Jones, J., 183
Kent, S. B., 30
Kervina, M., 301
Keslair, F., 108, 115
Kessler, A., 513, 706
Khandani, A., 665
Kharazia, V. N., 274
Khatri, L., 158
Khosravi-Far, R., 341, 382
Kikuchi, A., 254, 401
Kikuno, R., 492, 494, 500
Kikuta, T., 69, 218, 219, 224
Kilimann, M. W., 458, 460, 475
Killisch, I., 195, 243, 247
Kim, J. H., 198, 323, 325
Kim, K., 469
Kim, M. K., 660, 662
Kim, Y. S., 655
Kimm, K., 198
Kimmel, A. R., 358
King, M. C., 697, 704
King, R., 665
King, T. R., 543, 546
Kintner, C., 323
Kirchhausen, T., 2
Kiselev, S. L., 527
Kishida, S., 216, 255, 402, 408, 458,
    469, 470
Kishino, A., 99
Kita, T., 778, 779, 782, 784, 786
Kita, Y., 263
Kitamura, A., 628, 651, 689
Kitamura, H., 409
Kjer-Nielsen, L., 686
Klarlund, J. K., 699
Kleijmeer, M., 323
Klein, T. W., 74
Klein, U., 177
Kleinz, J., 658
Klinken, S. P., 800
Klionsky, D. J., 565, 602
Klueg, K. M., 323
Klumperman, J., 82, 89, 108, 482, 514,
    520, 524, 650, 707
Knapton, H. J., 778
Knecht, E., 375
Knight, J. R., 333

Kobayashi, A., 45
Kobayashi, H., 584
Koenig, M., 198
Koh, I., 198
Kohiki, E., 263, 266, 272, 277, 285, 311,
    312, 314, 317, 319
Kohler, K., 724, 725, 731
Kohn, A. D., 370
Kohrer, K., 323
Koike, K., 779
Koike, T., 20
Koivisto, U. M., 285
Kojima, T., 453, 465, 477
Komazaki, S., 311
Komer, R., 391, 393, 398, 400
Kominami, E., 628, 665, 745
Komuro, R., 470
Kong, C., 370
Kong, X., 158
Kontani, K., 266, 277, 285, 311
Kopajtich, R., 391, 393, 398, 400, 593,
    608, 613, 618, 619, 621, 623, 624,
    625, 626, 627
Kopp, F., 654
Korner, R., 624
Kornfeld, S., 357, 360, 696
Korobko, E. V., 527
Korobko, I. V., 527
Koseki, H., 409, 415, 417
Koshland, D. E., Jr., 676
Koster, M., 391, 392
Kotaka, S., 51
Kotake, K., 408, 409, 410, 470
Kotani, H., 492, 494
Kouri, G., 65
Kowal, A. S., 212, 368
Kowalchyk, J. A., 448, 450, 451, 452
Kowbel, D., 203
Kozasa, T., 778
Kraft, T. W., 51
Krausslich, H. G., 322, 323
Kraynov, V. S., 135, 136
Kreiman, G., 2, 3, 7, 8
Kreis, T. E., 136, 141, 152, 400
Krise, J. P., 357, 358, 360, 361, 362
Kristensen, P., 654
Kronvist, R., 177, 179
Kubota, S., 45
Kuijl, C., 628, 665, 745
Kuismanen, E., 744, 745, 746, 749

Kular, G., 698
Kuma, K., 310
Kumar, N. M., 193
Kumar, R., 482, 492, 500, 510, 513, 514, 706, 707, 709, 716, 717, 720, 721
Kumar, U., 301
Kunita, R., 263, 266, 272, 277, 285, 310, 311, 312, 314, 317, 318, 319, 320
Kuno, T., 510
Kuo, W. L., 203, 212
Kurisu, S., 241
Kuroda, S., 254, 401
Kuroda, T. S., 60, 216, 228, 420, 421, 423, 424, 425, 426, 427, 428, 429, 432, 433, 434, 435, 436, 437, 438, 439, 440, 441, 442, 446, 447, 448, 449, 450, 451, 452, 453, 454, 459, 461, 470, 471, 472, 473, 474, 778
Kurose, T., 416
Kurosu, H., 266, 277, 285, 311
Kushnir, S., 31, 37, 340
Kussel-Andermann, P., 432, 433, 442, 443, 446
Kutateladze, T., 122
Kuznetsov, S. A., 374
Kwiatkowski, T., 262, 266, 272, 321
Kwon, J., 665

L

Lacalle, R. A., 176, 177
Lacapere, J. J., 593, 608, 618
Ladbury, J. E., 516
Laemmli, U. K., 195, 301, 642, 700
Lagace, T. A., 744
LaGrassa, T. J., 552
Lah, J. J., 244, 707, 708, 716, 720
Lahad, J. P., 203, 212
Lai, E. C., 323
Laitinen, S., 744
Lamaze, C., 230, 368
Lambert, N., 779
Lambright, D. G., 17, 25, 27, 122, 200, 562, 563
Lamoreux, L., 200, 653
Lampson, M. A., 706
Lane, W. S., 552
Langford, N. C., 73
Langsley, G., 200, 598
Lanzetti, L., 370, 665

Lapierre, L. A., 482, 492, 493, 500, 510, 513, 514, 520, 706, 707, 708, 709, 716, 717, 719, 720, 721, 722
Laporte, J., 629
Lapp, H., 2, 3, 6, 7, 8, 340
Lapuk, A., 203, 212
Larijani, B., 461
Larsen, A. M., 654
Lattero, D., 350, 354, 628, 665, 666, 670, 672, 688
Lawe, D. C., 17, 25, 27, 122
Lawrence, D. W., 724
Leavitt, B. R., 263, 272, 274
Leavitt, S., 516
Lebrand, C., 177, 745, 756
Le Deist, F., 203, 420, 778, 779
Lee, A., 800
Lee, B., 665
Lee, D. J., 513
Lee, E., 760
Lee, J., 198
Lee, R., 744
Lee, S. H., 159, 274
Lee, S. S., 256
Lee, W. L., 660, 662
Leevers, S. J., 789
Lefrancois, L., 641
Lehavi, D., 203
Lehmann, W. D., 624
Lehto, M., 744, 745, 746, 749
Leijendekker, R., 82, 89, 526
Leinwand, L. A., 324
Leitinger, B., 85
Leitner, Y., 262
Leiva, I., 203
Le Marchand-Brustel, Y., 108, 109, 112, 115
Lemmon, M. A., 697, 704
Lendeckel, W., 399, 626
Leppimäki, P., 177, 179
Lerman-Sagie, T., 262
Le Roy, C., 368
Lesca, G., 262
Leslie, J. D., 301
Lev, D., 262
Lever, D. C., 31, 32
Levey, A. I., 244, 707, 708, 716, 720
Levine, B., 126, 127, 128, 132
Levitt, B. R., 311
Levitzki, A., 199
Lewis, J., 644

Lewis, W. M., 340
Li, A., 744
Li, C., 469, 470
Li, G., 58, 156, 200, 350, 541, 556, 586, 591, 766
Li, G. D., 58
Li, J., 243, 244, 247, 249, 250, 252
Li, L. Y., 800
Li, N., 45, 48, 49
Li, P. M., 199
Li, T., 45
Li, Y., 333
Liang, T. J., 660
Liang, X. H., 126, 127, 128, 132
Liang, Z., 350
Liao, L., 387
Libby, P., 789
Lichtenstein, Y., 82, 108
Lim, S. N., 91, 178
Lim, W. K., 644
Lin, C.-L., 800
Lin, J. W., 159, 350
Lin, L. L., 256
Lin, X. L., 588
Lin, Y. Z., 588
Linari, M., 45
Lindenmaier, H., 176
Lindsay, A. J., 482, 483, 486, 488, 492, 501, 509, 511, 514, 515, 519, 707, 711
Lindsay, M., 645, 646
Lindsay, M. R., 539, 696
Liou, W., 775
Lipman, D. J., 20, 21
Lippe, R., 89, 120, 266, 277, 285, 311, 527, 552, 584
Liscum, L., 176
Litingtung, Y., 760
Liu, J., 200, 203, 212, 586, 591
Liu, K., 541, 556
Liu, M. Y., 800
Liu, X. H., 49, 51
Liu, Y., 642
Ljung, B. M., 203
Lo, D. C., 156
Lockshon, D., 333
Lodge, R., 539, 584
Lombardi, D., 357
Lonart, G., 154
Londos, C., 358
Longva, K. E., 654

Lopez, G. P., 644
Lopez, R., 20
Lord, J. M., 593
Lorenzo, O., 630
Lottridge, J., 323
Loubat, A., 115
Louvard, D., 45, 183, 186, 724, 725, 726, 731
Low, S. H., 91, 178
Lowe, S. L., 679
Lu, H., 72, 77
Lu, K. H., 203, 212
Lu, K. P., 194
Lu, L., 685
Lu, Y., 510
Lucas, P., 176, 177
Luger, K., 324
Lui, M., 706
Luo, H. R., 284
Luo, L., 310
Luo, X., 284
Luo, Z. Q., 72
Lupas, A., 483
Lusis, A. J., 600
Lutcke, A., 119, 120, 167, 195, 243, 247, 765, 766
Luthi, S., 61, 62
Luzio, J. P., 322, 323, 487, 630, 696
Lyles, D. S., 641

M

Ma, J., 751
Macara, I. G., 61, 62, 195, 285, 410, 469, 470
MacDougall, L. K., 789
MacGregor, D. L., 262
Machesky, L. M., 778
Macia, E., 108, 109
Mackel, D. C., 73
Madden, D. R., 158
Madden, T. L., 21
Madshus, I. H., 654
Maduro, M., 43
Maeda, M., 470
Mahon, E. S., 543, 546, 547
Maier, G., 324
Maier, R., 2
Maier, V. H., 514
Mainen, Z. F., 159
Maiorano, M., 350, 354
Malabarba, M. G., 370

Male, P., 82, 89, 513, 733
Malenka, R. C., 154, 159, 255, 470
Malinow, R., 157, 159, 160, 370
Mallard, F., 593, 608
Maltese, W. A., 194, 195, 196, 200
Mammoto, A., 706
Manabe, S., 409
Manafo, D. B., 195
Mancini, M., 50
Mandel, J. L., 629
Manes, S., 176, 177
Maniatis, T., 699, 765, 766, 769
Mann, M., 89, 120, 266, 277, 285, 311, 394, 584
Mansfield, T. A., 333
Manson, F., 45
Mantoux, F., 420
Marcote, M. J., 342, 367, 370, 372, 378
Mari, M., 108, 109, 115
Marie, N., 492
Marinissen, M., 666
Marks, D. L., 167, 168, 169, 170, 171, 172, 174, 175, 176, 177, 179
Marks, M. S., 421
Marsden, J. J., 655
Marsman, M., 628, 651, 665, 676, 688, 689, 745, 756
Martin, O. C., 167, 169, 171, 175, 176, 653
Martin, S. E., 69
Martin, T. F. J., 448, 450, 451, 452
Martincic, I., 350, 594, 800
Martinez, A. C., 176, 177
Martinez, O., 137, 593, 608, 618
Martin-Lluesma, S. M., 136, 141, 152
Martin-Serrano, J., 322
Martinu, L., 370
Marzesco, A. M., 45, 183, 186, 724, 725, 726
Master, S. S., 628
Matanis, T., 593, 608
Matern, H. T., 2, 333, 338, 350, 482, 492
Matesic, L. E., 420, 421, 426, 432, 436, 438
Mather, I., 667
Mather, I. H., 679
Matheson, J., 510, 514
Matozaki, T., 408, 688
Matsubara, K., 285, 310
Matsuda, S., 51
Matsugaki, N., 584
Matsumoto, M., 409, 415, 417
Matsuo, T., 789
Matsuoka, M., 263

Matsuura, Y., 254, 255, 285, 401, 408, 458
Matteoli, M., 59
Matter, K., 183, 184, 186, 724
Mattera, R., 539, 584
Matteson, J., 255, 340, 342, 402
Mattson, M. P., 204
Maury, J., 13
Maxfield, F. R., 82, 243, 247, 295, 482, 552, 733
Mayau, V., 513
Mayer, M. L., 160
Mayer, R. J., 657
Mayer, T. U., 608, 618, 619, 621, 623, 625, 626, 627
Mayorga, L. S., 243, 249, 296
Mayran, N., 177, 745, 756
McAllister, A. K., 156
McBride, H. M., 89, 108, 119, 120, 154, 166, 167, 172, 230, 266, 277, 284, 285, 296, 311, 408, 420, 446, 526, 584, 599, 629, 642, 643, 676, 688, 724, 744, 760, 790
McCaffery, J. M., 323, 382
McCaffrey, M., 89, 122, 124, 200, 584
McCaffrey, M. W., 102, 108, 482, 483, 486, 488, 492, 500, 501, 502, 503, 505, 506, 507, 509, 510, 511, 514, 515, 519, 707, 711
McCallum, A., 183, 724
McCarthy, J., 665, 688, 689, 694, 756
McConlogue, L., 196
McDowell, J. H., 48
McEwen, R. K., 697, 704
McGlinn, E., 759
McGraw, T. E., 82, 108, 243, 247, 482, 552, 706, 733
McIlroy, J., 789
McIntire, W. E., 644
McKiernan, C. J., 195, 285, 410, 469
McKinney, R. A., 156
McLaren, M. J., 45
McMahon, A. P., 759
McMaster, C. R., 744
McNally, E. M., 324
Meda, P., 58, 59, 60, 416
Meehan, E. J., 200
Meerlo, T., 255, 323, 340, 342, 402
Meerschaert, K., 698
Meijer, I. A., 262
Meister, G., 65
Meitinger, T., 45
Melief, C. J., 323

Mellman, I., 73, 82, 89, 102, 108, 230, 285, 513, 733
Menasche, G., 203, 420, 432, 433, 437, 443, 778
Mercer, J. A., 420, 514, 707, 716, 720, 721
Meresse, S., 688, 744
Merithew, E., 17, 25, 27, 562, 563
Mermall, V., 426, 438
Mertins, P., 13
Merzlyak, E., 35
Mesa, R., 243, 249
Meton, I., 108
Meyer, C., 531
Meyer, H. E., 513, 706
Meyer, T., 12
Meyers, J. M., 515, 518, 519, 707
Meyers, R. E., 789
Miaczynska, M., 120, 368, 552, 629, 631, 642, 643, 644, 688, 789, 790, 791, 793, 794, 797
Michell, R. H., 697, 698, 704
Michiels, F., 230
Miettinen, H. M., 73
Mignery, G. A., 154
Mihai, I., 500, 514
Miki, H., 230, 231, 234, 238, 241
Miki, T., 408, 409, 415, 417, 458
Mikoshiba, K., 60, 216, 228, 419, 420, 424, 427, 428, 429, 432, 433, 435, 436, 437, 438, 439, 445, 446, 447, 448, 449, 450, 451, 452, 453, 459, 460, 461, 462, 465, 466, 469, 470, 471, 472, 474, 475, 476, 477, 778
Milano, P. D., 31, 32
Milhavet, O., 204
Miller, B. H., 2
Miller, C. A., 256
Miller, G. J., 584
Miller, J., 680
Miller, W., 20, 21
Mills, G. B., 203, 212
Mills, I. G., 539
Milpetz, F., 27
Minard-Colin, V., 779
Mira, E., 176, 177
Mirey, G., 12
Mische, S., 510
Misra, S., 564, 568, 579, 584
Mitchell, C. A., 514
Mitchison, T., 621
Miura, P., 608

Miyajima, N., 492, 494
Miyamoto, N., 262, 266, 272, 321
Miyata, T., 310
Miyazaki, J., 218
Miyazaki, M., 216, 255, 402, 408, 409, 415, 417, 458, 469, 470
Miyoshi, J., 255, 256, 285
Mizoguchi, A., 255, 256, 285, 403
Mizoguti, H., 403
Mizumura, H., 263, 266, 272, 277, 285, 310, 311, 312, 314, 317, 318, 319, 320
Mizuno, K., 628, 651, 689, 695
Mizuno, S., 199
Mizushima, S., 236, 436, 447, 461, 471
Mizuta, M., 408, 409, 410, 470
Mizutani, S., 57, 69, 218, 219, 224, 228, 419, 446
Mlak, K., 629, 697, 698, 699, 704
Mobley, W. C, 212, 368
Modlin, I. M., 492, 513
Moffatt, B. A., 383
Mohammad, D., 619
Mohler, W. A., 510
Mohrmann, K., 82, 89, 102, 108, 194, 526
Mollat, P., 200
Monahan, C., 77
Monier, S., 138, 149, 594, 595
Monyer, H., 159
Monzo, P., 108, 115
Moor, H. P. H., 195
Moore, D., 203
Moore, I., 27
Moore, K. J., 420
Mooseker, M. S., 426, 438
Mor, O., 203
Mor, Y., 203
Mora, S., 102, 108
Morag, K., 203
Moreira, E., 744
Morgan, A., 413, 470
Morham, S. G., 322, 323
Mori, K., 600
Mori, N., 45
Morimoto, A., 779
Morimoto, B. H., 676
Morimoto, S., 409
Morimoto, T., 706
Morishita, W., 159
Morita, E., 322, 323
Morita, K., 724

Morris, J. A., 176, 177
Morrow, I. C., 645, 646
Moschonas, N. K., 323
Moss, B., 636, 656, 667
Mosteller, R. D., 296
Mou, H., 44
Moutel, S., 136, 138, 141, 152, 595
Moyer, B. D., 391
Mucenski, M. L., 420
Muddle, J., 665
Muir, T. W., 30
Mukherjee, K., 296, 301, 302, 303, 304,
    305, 306, 307, 629
Mukhopadhyay, A., 296, 301, 302, 303,
    304, 305, 306, 307
Mules, E. H., 57, 69, 171, 778
Muller, B., 322, 323
Mullock, B. M., 696
Munafo, D., 665
Munoz, D. G., 301
Munro, S., 686
Murai, J., 266, 277, 311
Murakami, Y., 199
Murotsu, T., 310
Murphy, C., 552
Murray, J. T., 629, 631, 643, 644, 688, 790,
    791, 793, 794, 797
Murray, J. W., 93, 98, 101, 102, 105, 106
Muskavitch, M. A., 323
Mustol, P. A., 311, 564, 565, 568, 578
Muto, E., 98
Myers, E. W., 20

N

Nagahiro, S., 409
Nagai, H., 72
Nagamatsu, S., 69, 218, 219, 220, 222, 224,
    225, 226, 227, 455
Nagano, F., 254, 259, 401, 402, 404
Nagano, T., 285
Nagase, T., 492, 494, 500
Nagashima, K., 217, 228, 420, 436
Nagata, E., 284
Nagata, K., 778
Nagata, S., 236, 436, 447, 461, 471
Nagelkerken, B., 733
Nakagawa, S., 199
Nakajima, Y., 357, 364
Nakamura, H., 340

Nakamura, N., 391
Nakanishi, H., 57, 254, 255, 256, 259, 285,
    401, 402, 404, 469, 695
Nakayama, K., 539, 583, 584
Nancy, V., 45
Narayan, V., 333
Narita, K., 167, 168, 169, 171, 172, 175,
    176, 177
Narumiya, S., 263, 266, 272, 277, 285
Nascimento, A. A., 421
Nashida, T., 427, 433, 443, 450
Nasir, J., 262, 266, 272, 321
Naslavsky, N., 243, 244, 247
Nassar, N., 43, 44
Nath, S., 101, 106
Nativ, O., 203
Naurumiya, S., 311, 312, 314, 317, 319
Navarre, J., 482, 492, 500, 510, 513, 514,
    706, 707, 709, 716, 717, 719, 720, 721
Neef, R., 608, 618, 619, 621, 623, 625,
    626, 627
Neefjes, J., 628, 651, 665, 676, 688, 689,
    745, 756
Neeft, M., 82, 89, 296, 526, 527, 531, 532,
    533, 536, 552, 733
Negalkerken, B., 526, 527
Neubig, R. R., 644
Neufeld, E. B., 176, 177
New, R. R. C., 701
Newey, S. E., 310
Newham, D. M., 290
Newton, C. A., 74
Ng, C. P., 382, 685
Ngsee, J. K., 350, 594, 800, 801
Nguyen, D. G., 322
Nichols, B. J., 686
Nicolas, P., 193
Nicoll, R. A., 154, 255, 470
Nicoziani, P., 665, 688, 689, 694, 756
Niculae, A., 38
Niedzielski, A. S., 158
Nielsen, E., 120, 156, 526, 789
Nigg, E. A., 608, 618, 619, 621, 623, 625,
    626, 627
Niinobe, M., 453, 461, 465, 470, 471, 477
Niles, E. G., 388, 636, 656, 667
Nilsson, J., 541
Nilsson, T., 593, 608
Nio, J., 409, 415, 417
Nishi, M., 311

Nishijima, H., 263, 266, 272, 277, 285, 311, 312, 314, 317, 319
Nishikawa, J., 469, 470
Nishimoto, I., 263
Nishimoto, T., 263, 266, 272, 277, 285, 310, 311, 312, 314, 317, 319
Nishimura, N., 340, 409
Nishioka, H., 57, 778, 779, 782, 784, 786
Nixon, S., 688
Nizak, C., 136, 138, 141, 152, 595
Nogi, T., 584
Nomura, N., 492, 494, 500
Nonet, M. L., 470
Norton, A. W., 45, 48, 49
Novak, E. K., 420, 778
Novak, L., 203
Novick, P., 16, 724
Novick, P. J., 285, 340
Nowak, N., 203
Nozawa, Y., 778
Nuebrand, V. E., 539
Numata, S., 469, 470
Nuoffer, C., 137, 342
Nyasae, L. K., 789

O

Obaishi, H., 254, 255, 285, 401
O'Brien, P. J., 44
O'Connor, B., 358, 360, 361, 362
Odorizzi, G., 526
Ogata, Y., 446, 447, 449, 450, 451, 452, 453, 454, 459, 470, 471, 473, 474, 779
Ogawara, H., 199
Oh, B., 198
Ohara, O., 492, 494, 500
Ohara-Imaizumi, M., 69, 218, 219, 224
Ohga, S., 779
Ohnishi, K., 285
Ohno, K., 177
Ohtsubo, M., 310
Ohtsuka, T., 706
Ohya, T., 57, 259, 584
Oishi, H., 259
Oka, Y., 218
Okada, T., 263, 266, 272, 277, 285, 311, 312, 314, 317, 319
Okada, Y., 262, 266, 272, 321
Okamoto, K., 217, 228, 420, 436
Okamoto, M., 255, 402, 408, 458

Okamoto, Y., 416
Okano, Y., 778
Okayama, H., 386
Oleynikov, Y., 203
Olivo, J. C., 93, 400
Olkkonen, V. M., 119, 167, 172, 195, 243, 247, 541, 675, 677, 744, 745, 746, 749, 760, 765
Olsnes, S., 527
Ong, O. C., 44
Ono, A., 322
Oorschot, V., 82, 89, 526
Opdam, F. J., 137
Orban, P. C., 263, 272, 274, 311
Orenstein, J. M., 322
Orita, S., 470
Orlov, I., 800
Orrell, R., 665
Orsel, J. G., 357
Ortonne, J. P., 420, 421, 432
Osawa, S., 51
Ossendorp, F., 323
Ostermeier, C., 200, 469
Osuga, H., 262, 263, 266, 272, 277, 285, 311, 312, 314, 317, 319, 321
O'Sullivan, T. N., 420, 432, 778
Otomo, A., 262, 263, 266, 272, 277, 285, 310, 311, 312, 314, 317, 318, 319, 320, 321
Ou, S.-H. I., 600
Ou, W. J., 368
Ouachee-Chardin, M., 779
Ouahchi, K., 262, 266, 272, 310
Overduin, M., 122
Overmeyer, J. H., 194, 195, 196, 200
Owen, D., 39
Owen, D. J., 584
Ozaki, H., 786
Ozaki, N., 408, 409, 410, 458, 470
Ozawa, R., 333
Ozawa, S., 255

P

Pabst, M. J., 768
Pagano, A., 82, 86, 87, 89, 90, 538
Pagano, R. E., 167, 168, 169, 170, 171, 172, 174, 175, 176, 177, 179
Page, A., 183, 724
Pakula, T., 13

Pallante, M., 666
Palme, K., 13, 27
Palmer, A. E., 121, 156, 171
Palmer, M., 784, 786
Panaretou, C., 629, 631, 632, 643, 644, 688, 789, 790, 791, 793, 794, 797
Panda, S., 2
Pandjaitan, R., 183, 186, 724, 725, 726
Panhans, B., 324
Pannekoek, H., 608, 618
Pape, T. D., 263, 272, 274, 311
Papermaster, D. S., 50
Pappin, D. J., 394
Parashuraman, S., 304, 305, 307
Parish, E. J., 744
Park, M., 154
Parker, P. J., 697, 698, 704, 789
Parry, D. A., 183
Parsels, L. A., 63, 258, 405
Parton, R. G., 119, 120, 154, 167, 177, 250, 296, 368, 374, 492, 513, 645, 646, 667, 679, 688, 706, 733, 744, 759, 760, 761, 766, 773, 775
Passafaro, M., 159
Passeron, T., 421, 432
Pastor, M C., 541, 543, 547, 550, 553, 558, 559
Pastural, E., 203, 420, 778
Pastuszyn, A., 644
Patel, S., 177
Paterson, H. F., 230, 241
Patki, V., 122, 552
Patterson, M. C., 176, 177
Pavlopoulos, E., 323
Pawlyk, B., 45
Payne, G. S., 600
Payrastre, B., 629
Peak-Chew, S. Y., 600
Pearse, A., 322, 323
Peden, A. A., 2, 482, 492, 515, 516, 518, 519, 520, 522
Pehau-Arnaudet, G., 137
Pelham, H. R., 322, 323, 600, 605
Pelkmans, L., 368
Pena, J. M., 176, 177
Penick, E. C., 154
Pennock, S., 368
Pentchev, P. G., 176, 177
Penttila, M., 13
Pepperkok, R., 400, 593, 608
Peralta, M. E., 350, 594, 800

Peränen, J., 285, 287, 289, 290, 292, 293
Pereira-Leal, J. B., 2, 25, 27, 28, 114, 446, 470
Perelroizen, I., 441
Perez, F., 136, 138, 141, 152, 595
Pericak-Vance, M., 262, 266, 272, 310
Perin, M. S., 154
Perisic, O., 323
Perkins, D. N., 394
Perret-Menoud, V., 61, 62, 216, 220, 221, 409, 446, 451, 452, 455
Persi, N., 203
Pertuiset, B., 13
Pessin, J., 629, 697
Peter, F., 137, 342
Peters, R., 391, 392
Peterson, J. R., 765
Peterson, T. E., 177
Petit, C., 62, 65, 67, 69, 432, 433, 437, 442, 443, 446
Petralia, R. S., 154, 158
Pfeffer, S., 2, 255, 296, 340, 367, 381, 402, 552, 599, 664, 800
Pfeffer, S. R., 30, 154, 166, 167, 174, 255, 296, 333, 340, 348, 349, 350, 353, 354, 357, 358, 359, 360, 361, 362, 364, 402, 408, 676, 688, 696, 697, 704, 724, 744, 800
Phend, K. D., 155
Philip, B., 492, 519, 520, 707
Philippsen, P., 564
Philipson, L. H., 69
Piccini, A., 157, 159
Piech, V., 159
Pilch, P. F., 85
Pilgrim, D., 43
Pind, S. N., 341, 342
Pinkel, D., 203
Piper, R. C., 322, 584, 696, 697, 704
Pisacane, P. I., 524
Pitsouli, C., 323
Pittler, S. J., 44
Plewniak, F., 13, 20
Plutner, H., 255, 340, 341, 342, 346, 382, 402
Pochart, P., 333
Poitout, V., 69
Pollard, T. D., 230
Pollock, N., 93, 95
Pollok, B. A., 199
Poncer, J. C., 157, 159

Ponting, C. P., 27, 564
Ponza, E., 183, 724
Poole, I., 203
Popa, I., 82, 89
Popova, E., 322, 323
Posner, B. I., 368
Possmayer, F., 598
Poulter, C. D., 31, 32
Poupon, V., 696
Powelka, A. M., 250
Powers, S., 195
Prag, G., 323, 325, 564, 568, 579
Prange, R., 13
Prasad, V., 340
Pratt, W. B., 341
Preisinger, C., 391, 393, 398, 400, 593, 608,
    613, 618, 619, 621, 623, 624, 625, 626, 627
Prekeris, R., 482, 483, 492, 500, 510, 513,
    514, 515, 516, 518, 519, 520, 522, 707
Prescianotto-Baschong, C., 82, 86, 87, 89,
    90, 538
Press, B., 628, 629, 631, 633, 634, 638, 639,
    640, 650, 652, 657, 665, 688, 694, 744
Prestwich, G. D., 49, 51, 469, 699
Profiri, E., 195
Prosser, D. C., 350, 800, 801
Prossnitz, E. R., 644
Provance, D. W., Jr., 420, 514, 707, 716,
    720, 721
Pruijn, G. J., 108
Prusti, R. K., 44, 45
Prybylowski, K., 154
Pryor, P. R., 696
Ptashne, M., 751
Puertollano, R., 584
Punn, A., 184
Purcell, M., 72
Puri, V., 167, 168, 169, 170, 171, 172, 175,
    176, 177
Puype, M., 391
Pylypenko, O., 31, 37, 340
Pypaert, M., 72, 76, 78, 80, 285

Q

Qi, H., 370
Qian, Q., 65
Qin, N., 44
Qin, Y., 370
Qureshi-Emili, A., 333

R

Rabinovitch, M., 665
Radhakrishna, H., 251
Rahav, A., 203
Raje, M., 296, 301, 302, 303, 304, 305, 306, 307
Rak, A., 31, 37, 340
Ramalho, J. S., 778
Raman, D., 51
Rameh, L., 789
Ramm, E., 706
Ramm, G., 323
Rancano, C., 348, 357, 364
Ransom, N., 50
Rao, K., 420, 421, 426, 432, 436, 438, 715
Rapak, A., 420, 421, 778
Raposo, G., 184, 432, 433, 437, 443, 779
Rasheed, J. K., 73
Rathman, M., 300
Ray, G. S., 706
Raymackers, J., 108
Raymond, C. K., 323, 565
Reaves, B. J., 487, 630
Recacha, R., 200
Rechsteiner, T. J., 324
Recouvreur, M., 183, 186, 193, 724, 725, 726
Reed, G. L., 194
Reetz, A., 59
Regazzi, R., 57, 58, 59, 60, 61, 62, 65, 67,
    69, 216, 220, 221, 409, 432, 433, 443,
    446, 451, 452, 455, 458, 459, 470
Regev, A., 203
Reggiori, F., 323
Reilly, M., 665
Reinsch, S., 250, 492, 513, 593, 706, 733
Rellos, P., 619
Ren, M., 156, 706, 733, 739
Renault, L., 43, 44, 45
Rescigno, M., 323
Retief, J. D., 15
Reuss, A. E., 420, 432
Rhodes, C. J., 69
Ricard, C. S., 340
Ricciardi-Castagnoli, P., 323
Richmond, A., 493, 520, 706, 708, 716,
    720, 722
Richmond, T. J., 324
Ridgway, N. D., 744
Ridley, A. J., 230, 241
Rieder, S. E., 323

Riederer, M. A., 167, 174, 357
Rietdorf, J., 156, 526
Riggs, B., 510, 514
Riikonen, M., 13
Riley, N. M., 706
Rinchik, E. M., 420
Rinderknect, E., 74
Ritter, B., 539
Rizo, J., 222, 707
Robert, M., 44
Roberti, V., 102, 108, 628, 651, 665, 666, 669, 671, 672, 676, 688, 694, 745, 756
Roberts, L. M., 593
Roberts, R. L., 369
Roberts, T., 384
Roberts, T. M., 632
Robertson, A. D., 584
Robinson, J. S., 565, 602
Robinson, L. J., 299
Robinson, M. S., 230, 513, 539
Robinson, W., 48
Roche, E., 375
Rodgers, K., 586, 591
Rodman, J. S., 230
Rodrigues, O. C., 716
Rodriguez, I. R., 744
Rodriguez-Zapata, M., 176, 177
Roff, C. F., 177
Rogawski, M. A., 160
Rogers, D. A., 263, 272, 274, 311
Roggero, M., 243, 249
Rohrer, J., 360
Roland, J. T., 421
Rosa, J. L., 310
Rose, J. K., 640
Rosenfeld, M., 156, 541
Rosenmund, C., 470
Ross, A. H., 500, 514
Ross, R., 541
Rossi, J. J., 65
Rossman, K. L., 310, 676
Rostkova, E., 31, 35
Roth, D., 724
Roth, R. A., 370
Rothberg, J. M., 333
Rothenstein, D., 203
Rothman, J. E., 178, 300, 699
Rothman, J. H., 565
Rothstein, J. D., 800
Rothwell, W., 510

Rouleau, G. A., 262, 263, 266, 272, 274, 311, 321
Rous, B. A., 487
Rousselet, A., 593, 608, 618
Roux, A., 136, 141, 152
Roversi, P., 43, 44, 45
Row, P. E., 630
Rowe, T., 137, 382, 514
Roy, C. R., 72, 76, 77, 78, 80, 665
Rozelle, A. L., 247, 248
Rozen, A., 203
Rubin, G. M., 323
Rubino, M., 89, 120, 266, 277, 285, 311, 584
Rubsamen-Waigmann, H., 324
Rudensky, A. Y., 323
Rudolph, M. G., 200
Rudolph, R., 37
Ruggiero, A. M., 800
Runge, A., 89, 120
Rusnak, A. S., 323, 325
Russell, D. G., 296
Russell, D. W., 195
Russell, L. B., 420, 426, 438, 778
Russell-Jones, G. J., 301
Rustioni, A., 155
Rutter, G. A., 65
Rybin, V., 89, 120, 122, 124, 266, 277, 285, 311, 370, 527, 584
Rzomp, K., 665

S

Sabatini, D. D., 156, 706, 733, 739
Sachse, M., 650
Sadowski, I., 751
Saegusa, C., 419, 432, 433, 446, 447, 449, 450, 451, 452, 453, 454, 459, 470, 471, 473, 474
Saeki, Y., 218
Safferling, M., 158
Saftig, P., 628, 665, 745
Sagie, T., 262, 266, 272, 321
Sai, J., 493, 708, 716, 720
Saiardi, A., 284
Saidi, L. F., 584
Saint-Pol, A., 608
Saito, K., 266, 277, 285, 311
Sakai, H., 98
Sakai, T., 254
Sakai, Y., 227
Sakaki, Y., 333

Sakata, A., 779
Sakisaka, T., 255, 340, 342, 401, 402, 404
Sakmann, B., 159
Sako, M., 779
Sakoda, T., 255, 402, 408, 458, 469, 470
Sakota, T., 216
Salamero, J., 137, 513
Saloheimo, M., 13
Salomon, C., 243, 249
Salser, S., 621
Saltik, M., 768
Sambrook, J., 699, 765, 766, 769
Sanal, O., 420
Sander, C., 12, 22
Sano, K., 340
Sans, N., 154
Santiago-Walker, A., 370
Santillo, M., 102, 108, 628, 665, 666, 670,
    672, 688
Santorelli, F. M., 262
Saraste, J., 346
Saraya, A., 409, 415, 417
Sargent, D. F., 324
Sarna, M., 800
Sasabe, J., 263
Sasaki, T., 57, 254, 255, 259, 262, 266, 272,
    285, 310, 342, 401, 402, 408, 409, 469,
    470, 584, 628, 651, 688, 689, 695, 706
Sasazuki, T., 779
Satir, P., 101, 102, 105, 106
Sato, M., 285
Sato, T. K., 324
Satoh, A., 254, 255, 285, 401
Satoh, T., 51, 199, 778
Saucier, S. E., 744
Savelkoul, P., 137
Sawa, A., 284
Sawin, K., 621
Sayer, M. S., 800
Sbrissa, D., 629, 697, 698, 699, 702, 704
Scales, S. J., 400
Scarlata, S., 322
Schaffer, A. A., 21
Schaffhausen, B., 632
Schaible, U. E., 296
Schaletzky, J., 593, 608, 613, 630
Schallhorn, A., 151
Scheidig, A. J., 31, 200
Scheidig, A. S., 39
Schell, J., 27

Scheller, R. H., 2, 482, 483, 492, 500, 514,
    515, 516, 518, 519, 520, 522, 524, 707
Scherer, S. W., 262, 266, 272, 321
Schiavo, G., 699
Schiestl, R. H., 691
Schievella, A. R., 256
Schimmöller, F., 697, 704
Schlesinger, S., 156
Schluter, K., 724
Schluter, O. M., 470
Schmedding, E., 665
Schmid, S. L., 367, 368
Schmidt, A., 262, 310, 733, 739
Schmitt, R., 13
Schmitz, F., 154, 255, 402, 408, 458, 470
Schnell, E., 470
Schoch, S., 154
Scholtes, L., 665
Schonn, J. S., 432, 433, 437, 442, 443, 446
Schonteich, E., 515, 516, 518, 519, 520, 522
Schook, A. B., 2
Schreiber, A. D., 660, 662, 665, 789
Schreiber, R. D., 74
Schroder, B., 324
Schroer, T. A., 93, 374, 665, 671, 672, 745, 756
Schu, P., 82, 89
Schu, P. V., 311, 564, 565, 568, 578, 789
Schuebel, K. E., 306
Schultz, J., 27
Schultz, P. G., 2
Schuurhuis, D., 323
Schwab, C., 263, 272, 274, 311
Schwaninger, R., 341
Schwartz, A. L., 295
Schwartz, M. A., 135, 136
Schwarz, E., 37
Schweigerer, L., 394
Schweitzer, E. S., 258, 405
Schweitzer, J. K., 135, 510
Schweizer, A., 360
Scidmore, M., 665
Scita, G., 370
Scott, A., 322, 323
Scott, C., 665
Scriver, C. R., 176
Seabra, M. C., 2, 25, 27, 28, 37, 39, 57, 69,
    114, 171, 200, 296, 420, 421, 432, 436,
    446, 461, 470, 552, 594, 653, 664, 778
Seeburg, P. H., 159
Seemann, J., 400

Seet, L. F., 683
Segal, G., 72, 77
Segall, J. E., 203
Seger, R., 420
Segev, N., 172, 230, 408, 600, 676, 688
Segraves, R., 203
Seino, S., 408, 409, 410, 415, 417, 458, 470
Seino, Y., 416
Sekiguchi, M., 310
Sekiguchi, T., 310
Sekiya, S., 408, 409, 410, 470
Self, A. J., 556
Self, D. W., 2
Sellers, J. R., 420, 421, 426, 432, 436, 438
Senftleben, U., 378
Seni, M. H., 262
Senic-Matuglia, F., 482, 483, 486, 514, 515, 519, 707
Senter, R. A., 63, 258, 405
Senyshyn, J., 154
Seow, K. T., 679
Seperack, P. K., 420
Serunian, L. A., 789
Severin, F. F., 93, 789
Sfeir, A. J., 564, 568, 570, 579
Shah, A. H., 296, 789
Shao, Y., 255, 340, 402
Shapiro, A. D., 167, 174, 296, 348, 349, 354
Sharma, D. K., 167, 168, 169, 170, 171, 172, 174, 175, 176, 177, 179
Shaw, L. M., 250
Sheetz, M. P., 101
Sheff, D. R., 82, 102, 733
Sheffield, P., 570
Shellenberger, K. E., 195, 200
Shen, F., 37
Shen, K., 492
Shen, K. R., 513
Sheng, M., 159, 274
Sheth, B., 183, 724
Shevchenko, A., 368, 394, 688
Shi, M., 644
Shi, S. H., 157, 159, 160
Shia, M., 85
Shiba, T., 584
Shiba, Y., 539, 584
Shibasaki, T., 408, 409, 415, 417, 458
Shibasaki, Y., 218
Shiboleth, Y., 203
Shibuya, M., 199

Shih, H. M., 800
Shih, S. C., 564, 579
Shiku, H., 403
Shimizu, N., 655
Shimomura, H., 427, 433, 443, 450
Shin, H. W., 539, 584
Shin, O. H., 500, 510, 514
Shingai, T., 401, 402, 404
Shinohara, M., 401, 402
Shinozaki, H., 409, 415, 417
Shirakawa, R., 778, 779, 782, 784, 786
Shirakawa, S., 51
Shirataki, H., 216, 255, 402, 408, 458, 469, 470
Shisheva, A., 340, 629, 697, 698, 699, 702, 704
Shoarinejad, F., 296
Shore, D., 183
Short, B., 391, 393, 398, 400, 593, 608, 613, 630
Showguchi-Miyata, J., 262, 263, 266, 272, 277, 285, 310, 311, 312, 314, 317, 318, 319, 320, 321
Shown, E. P., 744
Shpetner, H. S., 552
Shtiegman, K., 562
Shuman, H. A., 72, 77
Shuntoh, H., 510
Shupliakov, O., 119, 120
Sicheritz-Ponten, T., 198
Siddhanta, U., 789
Siddiqi, S. A., 296, 302, 303, 306
Siddique, T., 262, 266, 272, 310
Sidorovitch, V., 32, 39
Siebrasse, J. P., 391, 392
Siegelbaum, S. A., 154, 255
Sikorski, R. S., 568
Simari, R., 177
Simon, G. C., 520
Simon, I., 31, 357, 358, 360, 362
Simonetta, A., 159
Simons, K., 50, 154, 195, 200, 243, 247, 285, 287, 289, 290, 292, 400, 640, 667, 679, 744, 765
Simons, P. C., 644
Simonsen, A., 122, 243, 552
Simpson, J. C., 593
Sincock, P. M., 357, 358, 360, 361, 362
Singaraja, R., 262, 266, 272, 321
Singer, R. H., 203
Singer-Kruger, B., 564
Singh, R. D., 167, 168, 169, 170, 171, 174, 177

Singh, S. B., 296, 628
Siniossoglou, S., 600, 605
Sio, S. O., 510
Sirover, M. A., 382
Sivars, U., 255, 333, 340, 348, 350, 353, 354, 358, 360, 361, 362, 402, 800
Sjaastad, M. D., 300
Skaliter, R., 203
Skaug, J., 262, 266, 272, 321
Skehel, P. A., 801
Sklar, L. A., 644
Skop, A. R., 510
Slabaugh, S., 135, 136
Sleight, R. G., 169
Sliwkowski, M. X., 524
Slot, J. W., 775
Slotte, J. P., 177, 179
Sly, W. S., 176
Smerdon, S. J., 619
Smith, C., 200
Smith, D. B., 299
Smith, J., 285, 721
Smith-McCune, K., 203, 212
Smothers, J. F., 136
Snyder, S. H., 284
Snyder, W. B., 323, 325
Soans, A., 697, 698, 699, 704
Sobek, A., 654
Soden, R., 2, 3, 7, 8
Solari, R., 285
Soldati, T., 296, 348, 354, 357
Solimena, M., 59
Söllner, T. H., 699
Soltoff, S. P., 789
Sommers, S., 69
Somsel Rodman, J., 628, 676
Sondek, J., 310
Song, O., 751
Song, O.-K., 108
Song, W. J., 469
Sonnenberg, A., 250
Sönnichsen, B., 89, 108, 130, 156, 526, 733
Sorkin, A., 124, 128, 129, 130, 540
Soule, G., 194, 195, 196, 200
Spaargaren, M., 527
Sparkes, R. S., 600
Spiess, M., 82, 85, 86, 87, 89, 90, 538
Spijkers, P., 82, 91
Spiro, D. J., 706
Sprong, H., 82, 89

Spudich, J. A., 324
Squicciarini, J., 85
Squire, J. H., 285
Sreerama, N., 587
Srinivasan, M., 333
Sriratana, A., 514
St. Jean, A., 691
Stabila, P. F., 285, 469, 470
Stack, J. H., 789
Staels, B., 744, 745, 746, 749
Stafford, S. J., 51
Stagljar, I., 680
Stahl, P. D., 156, 243, 249, 266, 277, 285, 296, 300, 301, 311, 340, 369, 370, 665, 766
Stam, J. C., 230
Standley, S., 154
Stang, E., 654
Stark, H., 158
Stauber, T., 593, 608
Staubli, W., 393
Staunton, J., 470
Stearns, T., 16
Steele-Mortimer, O., 119, 120, 167, 296, 370, 766
Stefan, C. J., 323
Steffen, P., 621
Stein, A., 203
Stein, M. P., 72, 76, 78, 80, 166, 203, 628, 629, 630, 633, 634, 639, 640, 641, 642, 646, 651, 688, 746
Stein, R., 789
Steinbach, P. A., 121, 156, 171
Stelzer, E. H., 200, 593
Stenius, K., 469, 470
Stenmark, H., 89, 108, 119, 120, 122, 124, 154, 167, 172, 243, 266, 277, 285, 311, 322, 342, 527, 533, 541, 552, 564, 584, 629, 631, 643, 644, 645, 646, 653, 667, 675, 677, 688, 760, 765, 766, 790, 791, 793, 794, 797
Stepanova, T., 593, 608
Stephens, D. J., 487
Sternglanz, R., 333, 338, 350, 668, 680
Stetefeld, J., 483
Stevens, C. F., 154, 255
Stevens, T. H., 323, 564, 565, 600
Stevenson, B. R., 183, 724
Stinchcombe, J. C., 778
Stock, J., 483
Stockert, R. J., 94, 101, 102, 105, 106
Stoorvogel, W., 82, 91, 323, 526

Storrie, B., 593
Strack, B., 322, 323
Straume, M., 2
Strauss, J. F. III, 177
Strelkov, S. V., 483
Strick, D. J., 733
Strobel, M. C., 420, 426, 438
Strom, M., 200, 420, 421, 432, 436, 446, 470, 653
Strous, G. J., 650
Stuchell, M., 322, 323
Studier, F. W., 388, 636, 656, 667
Studier, W. F., 383
Stunnenberg, H., 667, 679
Sturgill-Koszycki, S., 296
Sturley, S. L., 176
Su, A. I., 2, 3, 6, 7, 8, 340
Subramaniam, V. N., 679, 686
Suda, M., 227
Sudar, D., 203
Sudhof, T. C., 154, 222, 255, 340, 402, 408, 458, 469, 470, 707
Sudo, T., 378
Suer, S., 584
Suetsugu, S., 231, 234, 238, 241
Suga, E., 263, 266, 272, 277, 285, 311, 312, 314, 317, 319
Sugawara, H., 20
Sugita, S., 222, 408, 458
Sugiura, R., 510
Sullivan, W., 510, 514
Sumizawa, T., 348, 349, 350, 354
Sun, J., 250, 323, 325
Sun, L., 458
Sun, P., 231, 234, 238, 241
Sun, X., 167, 176, 177
Sunaga, Y., 408, 458
Sundquist, W. I., 322, 323
Supek, F., 564, 568
Superti-Furga, G., 629
Sur, S., 51
Surendhran, K., 584
Suresh, S., 177
Sutanto, M. A., 564, 579
Sutcliffe, J., 608, 618, 619, 621, 623, 625, 626, 627
Suyama, M., 492, 494
Suzuki, K., 176, 317, 318, 319, 320
Suzuki, M., 584
Suzuki-Utsunomiya, K., 310, 311, 317, 318

Svejstrup, A. B., 348
Swank, R. T., 420, 778
Swanson, G. T., 160
Swanson, J. A., 230, 240
Swing, D. A., 420, 432, 778
Symons, M., 310
Sziegoleit, A., 346
Szymkiewicz, I., 119, 120

T

Tabuchi, A., 778, 779, 782, 784, 786
Takada, H., 779
Takahashi, J. S., 2
Takahashi, K., 408, 458
Takahashi, N., 69, 218, 219, 224
Takai, Y., 57, 216, 254, 255, 256, 259, 285, 340, 342, 401, 402, 403, 404, 408, 458, 469, 470, 584, 688, 706
Takaku, F., 218
Takata, K., 216, 217, 218, 219, 221, 222, 224, 225, 446, 452
Takatsu, H., 539, 584
Takegawa, K., 510, 789
Takei, K., 59, 154, 255, 469, 470
Takenawa, T., 230, 231, 234, 238, 241
Takeshima, H., 311
Takeuchi, M., 401, 402, 404
Takeuchi, T., 216, 217, 218, 219, 220, 221, 222, 223, 224, 225, 226, 227, 228, 408, 420, 436, 445, 446, 451, 452, 455
Tall, G. G., 266, 277, 285, 311, 369, 564, 568, 570, 571, 574, 579
Tamary, H., 779
Taminato, T., 416
Tamura, K., 342, 367, 370, 372, 378
Tan, S., 324
Tanaka, A., 492, 494
Tanaka, K., 657
Tanaka, M., 255, 256, 285
Tang, B., 382
Tang, B. L., 608
Tang, J., 584, 586, 588, 591
Tang, W. J., 644
Tanida, I., 628, 665, 745
Tanowitz, M., 733
Tarafder, A. K., 420, 421, 432, 436, 446, 461, 470
Tavitian, A., 108, 183, 724
Taylor, F. R., 744

Taylor, G. A., 628
Taylor, G. S., 642, 790
Taylor, J., 354
Teasdale, R. D., 686
Teis, D., 368
Tempst, P., 706
Ten Broeke, T., 82, 91
Tenza, D., 608
Teo, H., 323
Terew, J. M., 45, 48, 49
Terzyan, S., 200, 584, 586
Tezcan, I., 420
Thoma, N. H., 31, 35, 38, 39
Thompson, C. B., 745
Thompson, J. D., 13, 20, 503, 761, 762, 763
Thompson, N., 285
Thompson, S. M., 156
Thomsen, P., 665, 688, 689, 694, 756
Thornqvist, P., 89, 108, 122, 124, 584
Thuring, J., 697, 704
Thyberg, J., 541
Tichelaar, W., 158
Tilbrook, P. A., 800
Tillman, K. C., 800
Tilney, L. G., 72
Timmerman, V., 665
Timms, J., 789
Tipler, C. P., 657
Tisdale, E. J., 382, 383, 384, 385, 386, 388
Toft, D. O., 341
Togawa, A., 255, 285
Toh, B. H., 552, 688
Toh, D., 200
Tokunaga, F., 51
Tolmachova, T., 778
Tomas, E., 513, 706
Toomre, D., 400
Tooze, S. A., 492
Topp, J. D., 263, 264, 265, 266, 270, 272,
    274, 277, 311, 319, 564, 568, 570, 579
Torii, S., 57, 216, 217, 218, 219, 220, 221,
    222, 223, 224, 225, 226, 227, 228, 419,
    420, 436, 446, 451, 452, 455
Tour, O., 121, 156, 171
Towbin, 48
Tran, I. C., 154, 155, 156, 158, 159, 160,
    164, 370
Tran, V. S., 432, 433, 437, 443
Tranum-Jensen, J., 346, 784, 786
Trepte, H. H., 333, 338, 350

Triezenberg, S., 751
Trimble, W., 665
Troost, J., 176
Trowbridge, I. S., 82, 526
Trugnan, G., 800
Tsai, S., 800
Tsien, R. Y., 121, 156, 171
Tsou, C., 340
Tsuji, K., 416
Tsujita, K., 266, 277, 285, 311
Tsukada, M., 600, 604
Tsukita, S., 183, 707, 724
Tsuura, Y., 416
Tuma, P. L., 789
Turck, C. W., 93, 95
Tuschl, T., 65, 399, 626
Tzonopoulos, T., 470
Tzounopoulos, T., 154, 255
Tzschaschel, B., 593

U

Uchiya, K., 296
Ueda, I., 779
Ueda, T., 403
Ueffing, M., 45
Uehara, Y., 199
Ueno, H., 408, 458
Ueno, T., 628, 665, 745
Uerkvitz, W., 654
Uetz, P., 333
Uhler, M. D., 63, 258, 405
Ulitzur, N., 357
Ullrich, O., 250, 279, 296, 342, 492, 513,
    533, 584, 653, 688, 706, 733
Ungar, D., 2
Ungermann, C., 552
Urbe, S., 250, 492, 513, 650, 706, 733
Uttenweiler-Joseph, S., 120, 368, 688

V

Vaerman, J. P., 492, 707, 716, 718
Vale, R. D., 93, 95
Valencia, A., 12, 22, 653
Valiyaveettil, J. T., 169, 170
Valkonen, M., 13
Valle, D. D., 176
Valletta, J. S., 212, 368
Valony, G., 432

Valtschanoff, J. G., 274
Van Aelst, L., 262, 310, 370
van Dam, E. M., 82, 91
Vandekerckhove, J., 391, 698
van den Hove, M. F., 203
van Der, S. P., 194
Van Der Goot, F. G., 368
van der Heijden, A., 108
van der Kammen, R. A., 230
van der Sluijs, P., 82, 89, 102, 108, 296, 513,
    526, 527, 531, 532, 533, 536, 552, 733
Van der Westhuyzen, D. R., 744
van Deurs, B., 665, 688, 689, 694, 756
Vande Velde, C., 319
van Dijk, S. M., 524
Van Dyke, M., 483
Van Gerwen, V., 665
Vanhaesebroeck, B., 789
Vanier, M. T., 176, 177
van Kerkhof, P., 650
van Marle, J., 608, 618
Van Obberghen, E., 112
van Oort, M., 82, 89
van Raak, M., 194
VanRheenen, S. M., 296
van Suylekom, D., 108
van Venrooij, W. L., 108
van Vliet, C., 686
Vardi, N., 45, 48, 49
Vasara, T., 13
Vater, C. A., 323, 565
Vaughan, H., 492, 513, 721
Vaughan, K. T., 608
Vaughan, P. S., 608
Veitia, R. A., 325
Verbsky, J. W., 340
Verdoorn, T. A., 159
Vergne, I., 296
Verhage, M., 470
Verhoeven, K., 665
Verkhusha, V. V., 128, 130
Vernos, I., 593, 608
Verpoorten, N., 665
Vetter, I. R., 12
Via, L., 665
Vieira, A., 243, 368, 665, 671, 672
Vieira, O. V., 660, 662, 665, 745, 756, 789
Vijayadamodar, G., 333
Vilaró, S., 310
Vilbois, F., 342, 367, 370, 372, 378

Vilmer, E., 779
Virbasius, J., 552
Virta, H., 285, 287, 289, 290, 292
Visser, P., 608
Vitale, G., 89, 108, 122, 124, 526, 533, 584, 688
Vitelli, R., 628, 665, 666, 670, 672, 688
Vogel, J. P., 72
Vogel, M., 340
Vojtek, A. B., 267, 354
Volik, S., 203
Volinia, S., 789
Volpicelli, L. A., 244, 707, 708, 716, 720
von Bonsdorff, C. H., 640
von Schwedler, U. K., 322, 323
von Zastrow, M., 159, 733
Voorberg, J., 137
Vorgias, C., 13

W

Wada, I., 171
Wada, K., 216, 255, 402, 408, 458, 469, 470
Wada, M., 254, 255, 259, 285, 401
Wagner, E., 768
Wagner, K., 665
Wainwright, B. J., 759, 760, 761, 773
Waish, P., 199
Wakatsuki, S., 583, 584
Wakeham, N., 586, 591
Waksman, G., 340
Walch-Solimena, C., 285, 724
Waldmann, H., 31, 32, 35, 37, 39, 340
Walker, C. E., 800
Walker, J. R., 2, 3, 7, 8
Walkley, S., 171
Wallace, D. M., 482, 501, 509, 511, 707, 711
Waller, A., 644
Walsh, M., 500, 501, 502, 503, 505, 506, 507,
    509, 510, 511
Wandinger-Ness, A., 166, 203, 230, 300, 301,
    628, 629, 630, 631, 633, 634, 638, 639, 640,
    641, 642, 646, 650, 651, 652, 653, 654, 656,
    657, 665, 676, 688, 694, 744, 746
Wang, F., 420, 421, 426, 432, 436, 438
Wang, H., 13
Wang, H. E., 322, 323
Wang, J., 216, 218, 408, 445
Wang, P. Y., 744
Wang, T., 665, 676, 679, 683, 685
Wang, W., 203

Wang, X., 458, 460, 475, 513, 706
Wang, Y., 69, 200, 255, 368, 402, 408, 458
Wang, Y. T., 159
Wang, Z., 101, 368
Ward, D. M., 301, 322, 323
Ward, M., 13
Warner, T., 665
Warren, G., 391, 400
Waselle, L., 57, 62, 65, 67, 69, 432, 433, 443
Wasmeier, C., 200, 552, 594, 653, 664
Wasteson, A., 541
Wastney, M., 177
Watanabe, R., 167, 169, 170, 176, 177
Watanabe, S., 199
Watanabe, T., 227
Waterfield, M. D., 552, 629, 632, 642, 643,
    655, 688, 789, 790, 793
Waterman-Storer, C. M., 95
Waters, C. M., 540
Waters, M. G., 382
Watson, P. J., 584
Watts, C., 230, 240, 513
Watzke, A., 31, 32, 37, 340
Waye, M. M., 324
Weber, K., 399, 626
Webster, P., 82, 89, 513, 733
Wei, Q., 420
Weide, T., 391, 392, 593, 608
Weigel, P. H., 93
Weigert, R., 243, 244, 247, 249, 250, 252
Weimbs, T., 91, 178
Weinberg, R. A., 203, 274
Weinberg, R. J., 155
Weiss, E. R., 51
Weiss, J., 176
Welford, A. M., 628, 629, 630, 633, 634, 639,
    640, 641, 642, 646, 650, 651, 652, 653, 654,
    656, 657, 746
Wells, A., 369
Wendland, B., 323, 324, 325, 561, 568
Weng, J., 744
Wennerberg, K., 676
Wenthold, R. J., 154, 158
Wepf, R., 285, 287, 289, 290, 292
Weringer, E. J., 199
Wernick, M., 203
Wessel, D., 345
Wessling-Resnick, M., 706
Westermark, B., 541
Westphal, H., 760

Wheatley, C. L., 167, 168, 169, 170, 171,
    174, 176, 177
Wheatley, C. W., 167, 168, 169, 171, 172,
    175, 176, 177
White, J. G., 203, 400, 510, 593
White, R. L., 203
Whitman, M., 632
Whitney, J. A., 186
Whitt, M. A., 640
Whittaker, G., 665
Wick, P. F., 63, 258, 405
Wicking, C., 759, 760, 761, 773
Wicksteed, B. L., 69
Widen, R., 74
Wilcke, M., 513
Will, E., 600, 604
Wille, H., 91, 178
Willemsen, R., 608
Williams, J. H., 800
Williams, R. L., 323
Williamson, E., 154
Wilm, M., 89, 120, 266, 277, 311, 368, 394,
    552, 584, 629, 642, 643, 688, 789, 790, 793
Wilon, M., 285
Wilson, A. L., 195
Wilson, G. M., 500, 510, 515, 516, 518,
    519, 520
Wilson, I. A., 255, 340, 402
Wilson, S. M., 420, 778
Wiltshire, T., 2, 3, 7, 8
Winand, N. J., 333, 350
Wind, M., 624
Windmiller, D. A., 789
Winistorfer, S. C., 584
Wirtz, K. W., 128
Wishart, M. J., 790
Wittchen, E. S., 183, 724
Wittinghofer, A., 12, 43, 44, 45
Wittmann, J. G., 200
Wolfrum, U., 432, 433, 442, 443, 446
Wolkoff, A. W., 93, 98, 101, 102, 105, 106
Wollheim, C. B., 58, 59, 60, 65, 416, 458
Wong, F., 51
Wong, K. K., 676
Wong, S. H., 679, 683, 686
Wong, S. K., 72
Woodman, P., 89, 120, 266, 277, 285, 311, 584
Woodrum, J. E., 706
Woods, R. A., 691
Woody, R. W., 587

Wordeman, L., 621
Woscholski, R., 789
Wouters, F. S., 128
Wrana, J. L., 368
Wright, A., 45
Wu, C., 212, 368
Wu, F., 600
Wu, M. Y., 370
Wu, W., 510
Wu, X., 420, 715
Wu, X. S., 420, 421, 426, 432, 436, 438, 664
Wu, Z., 3
Wubbolts, R., 628, 651, 665, 676, 688, 689, 745, 756
Wulf, P., 593, 608
Wulffraat, N., 203, 420, 778
Wunderlich, W., 368
Wurm, F. M., 151
Wurmser, A. E., 790
Wyckoff, J. B., 203
Wyles, J. P., 744
Wyszynski, M., 159

X

Xiao, G. H., 296
Xiao, N., 203
Xie, H., 44
Xu, G., 706, 733, 739
Xu, M. Q., 31
Xu, Y., 683, 686
Xue, Y., 198

Y

Yaekura, K., 69
Yaffe, M. B., 619
Yagmour, A., 262, 266, 272, 310
Yalcin, A., 399, 626
Yalman, N., 420
Yamada, K., 203, 212
Yamaguchi, K., 256
Yamaguchi, T., 216, 255, 402, 408, 458, 469, 470
Yamamoto, A., 447, 459, 470, 471, 473, 475, 476, 477, 478, 786
Yamamoto, C., 199
Yamamoto, H., 231, 234, 238, 241
Yamamoto, K., 779
Yamamoto, Y., 74, 409

Yamamura, K., 218
Yamanaka, K., 319
Yamane, H. K., 44
Yamashita, T., 171
Yamato, E., 218
Yamazaki, D., 241
Yan, J., 262, 266, 272, 310
Yanagida, T., 99
Yanagisawa, Y., 262, 263, 266, 272, 277, 285, 310, 311, 312, 314, 317, 318, 319, 320, 321
Yang, C., 200
Yang, M., 333
Yang, X., 333, 338, 350
Yang, Y., 262, 266, 272, 310
Yang, Y.-Z., 263, 272, 274, 311
Yano, H., 408, 458
Yao, X., 198
Yarden, Y., 562
Yasuda, H., 159
Yasuda, M., 263, 266, 272, 277, 285, 311, 312, 314, 317, 319
Yasukawa, M., 779
Yasumi, M., 401, 402, 404
Ye, K., 284
Yeung, A. C., 243, 244, 247, 249, 250, 252
Yeung, B. G., 600
Yeung, R. S., 296
Yi, Z., 216, 217, 218, 219, 220, 221, 222, 223, 224, 225, 228, 420, 436, 446, 451, 452
Yin, H. L., 247, 248, 564, 568
Yip, R., 420, 432, 778
Yip, S. C., 552, 629, 642, 643, 789, 790, 793
Yip, S. S., 688
Yokota, H., 216, 217, 218, 219, 221, 222, 224, 225, 408, 445, 446, 452
Yokota-Hashimoto, H., 217, 219, 222
Yonemura, S., 707
Yoo, J. S., 564
Yoshida, M., 333
Yoshie, S., 427, 433, 443, 450
Yoshino, K., 584
Yoshioka, A., 778, 779, 782, 784, 786
You, J. L., 370
Young, J., 608
Young, J. H., 44, 593
Young, K. E., 760
Yu, H., 93, 284
Yu, J. W., 697, 704
Yu, X., 159, 520
Yue, G., 45

Yue, X., 608
Yurchenko, V., 31

Z

Zacharias, D. A., 121, 156, 171
Zahner, J. E., 340
Zahraoui, A., 45, 108, 183, 186, 724, 725, 726, 731
Zaitzu, M., 779
Zander, H., 598
Zang, T., 322
Zangrilli, D., 267
Zavadil, J., 203
Zeigerer, A., 706
Zeng, J., 156, 706, 733, 739
Zerial, M., 89, 108, 119, 120, 122, 124, 130, 154, 156, 166, 167, 172, 195, 200, 230, 243, 247, 250, 266, 277, 279, 284, 285, 296, 311, 342, 357, 368, 370, 398, 408, 420, 446, 492, 513, 526, 527, 533, 539, 552, 564, 584, 599, 600, 601, 629, 631, 642, 643, 653, 667, 676, 679, 688, 706, 724, 733, 744, 760, 765, 766, 789, 790, 793, 795
Zhai, P., 200, 584, 586, 591
Zhai, Y., 203
Zhang, H., 49, 51, 52, 53, 420, 421, 426, 432, 436, 438, 594
Zhang, J., 2, 3, 7, 8, 21
Zhang, K., 49, 51, 52, 53
Zhang, R., 584, 586
Zhang, T., 679, 686

Zhang, X., 448, 450, 451, 452
Zhang, X. C., 200, 584, 586, 591
Zhang, X. M., 514
Zhang, Y., 420, 778
Zhang, Y. T., 789
Zhang, Z., 21, 350, 660
Zhao, L., 552, 629, 642, 643, 688, 789, 790, 793
Zhao, M., 370
Zhao, S., 216, 217, 218, 219, 220, 221, 222, 223, 224, 225, 446, 451, 452
Zhao, Y., 45
Zhaom, S., 69
Zheng, Y., 267
Zhou, F. F., 198
Zhou, X. E., 200
Zhou, X. Z., 194
Zhu, G., 200, 584, 586, 591
Zhu, J. J., 370
Zhu, W., 333
Zhu, X., 72
Zhu, Z., 200
Ziff, E. B., 158
Zimmerman, C. F., 719
Zimmerman, J., 628, 629, 633, 640
Zimmermann, B., 458, 460, 475
Ziv, N. E., 154
Zorzano, A., 513, 706
Zuk, P. A., 733
Zurawski, T. H., 500, 501, 502, 503, 505, 506, 507, 509, 510, 511
Zvelebil, M. J., 789
Zwickl, P., 658

Subject Index

A

Actin, Slac2-c interactions
  binding domain
    localization on actin filaments in PC12
      cells, 441–442
    mapping, 439–441
  direct interaction assay, 441
  overview, 439
ALS2, see Alsin
ALS2CL
  gel filtration of oligomers, 318–319
  nucleotide exchange assays
    GDP dissociation, 316–317
    GTP binding, 317
    nucleotide loading, 315–316
  purification of recombinant
    protein, 311–314
  Rab5 interaction assay, 318
Alsin
  domains, 262–263, 310–311
  GTPase specificity
    binding assays *in vitro*, 267–268
    overview, 263
    yeast two-hybrid assay, 266–267
  mutation in disease, 262–263, 310
  nucleotide exchange assays
    GDP dissociation, 316–317
    GTP binding, 317
    nucleotide loading, 315–316
    Rab, 268–269
    Rac1, 269–271
  purification of recombinant protein,
    263–265, 311–314
  Rab5 interaction assay, 318
  subcellular distribution in brain
    fractionation, 271–274
    P3 membrane fraction treatment with
      membrane-perturbing agents, 274
α-Amino-3-hydroxy-4-isoxazoleproprionic
    acid receptor
  confocal scanning laser microscopy of
      Rab proteins at dendrites and spines

fusion protein expression, 160–161
large-scale trafficking analysis, 161
local trafficking analysis, 161–164
electrophysiological studies of Rab
    protein function
  electrophysiological tagging, 158–160
  long-term depression, 158
  long-term potentiation, 158
  recombinant Rab protein effects on
      basal synaptic transmission, 157–158
synaptic trafficking and plasticity, 154
AMPA receptor, *see* α-Amino-3-hydroxy-4-
    isoxazoleproprionic acid receptor
Amyotrophic lateral sclerosis, *see* Alsin
AP-1, Rab4–Rabaptin-5α–AP-1 complex
  AP-1 binding to Rabaptin-5α hinge
      fragment, 531–532
  endosomal recycling, 526, 536–539
  immunofluorescence microscopy, 532–533
  pull-down assays, 529–531, 533, 535–536
  yeast two-hybrid analysis, 527, 529
Arf1, *Legionella pneumophila*-containing
    vacuole recruitment, 72
Arfophilin, *see* Rab11-FIPs

B

BicD proteins, Rab6 interactions
  dynein–dynactin complex, 608
  yeast two-hybrid analysis
    competent cell preparation, 611–612
    growth media, 609, 611
    overview, 609
    transformation, 612

C

Cancer, *see* Rab25
CD2AP, Rab4 interactions, *see* Rab4
Cholesterol
  quantification, 177
  Rab effects in fluorescent sphingolipid
      analog trafficking

Cholesterol *(cont.)*
  depletion, 176
  elevation, 176, 177
  Rab4 extraction inhibition, 177–179
  sphingolipid storage diseases, 176
Coimmunoprecipitation
  Rab2/glyceraldehyde-3-phosphate
    dehydrogenase, 385–386
  Rab2/protein kinase C, 385–386
  Rab6 binding partners, 597
  Rab7/oxysterol binding protein related
    protein 1L, 747–749
  Rab7/Rabring7, 692–693
  Rab7/VPS34/p150 phosphatidylinositol
    3-kinase complex, 641–642
  Rab7/XAPC7, 656–657
  Rab27A/granulophilin
    cultured cells, 222–223
    pancreatic islets, 223–224
  Rab27A/Noc2, 475
  Rab27A/Rabphilin, 475
  Slp4-a/Rab, 448–449, 451
  VAP-33/prenylated Rab acceptor
    protein, 805–806
  VPS34/p150 phosphatidylinositol
    3-kinase complex, 641–642
Confocal scanning laser microscopy
  Rab4a interactions with Rabip4 and
    CD2AP, 116–117
  Rab7–oxysterol binding protein related
    protein 1L interactions, 745–747
  Rab7–Rab interacting lysosomal protein
    interaction studies, 669–670
  Rab13 in tight junction structure and
    function, 190–191
  Rabin3 subcellular localization, 293
  Rab proteins at dendrites and spines
    fusion protein expression, 160–161
    large-scale trafficking analysis, 161
    local trafficking analysis, 161–164
CSLM, *see* Confocal scanning laser
    microscopy
CytoTrap system
  advantages in two-hybrid analysis, 694–695
  Rabring7 isolation, 690–691

D

DENN/MADD
  functions, 256

growth hormone release assay and
    calcium-dependent exocytosis
    role, 258–259
  knockout mice, 255–256
  purification from baculovirus–Sf9 system
    anion-exchange chromatography, 257
    lysate preparation, 257
    materials, 256–257
    Rab3 specificity, 254–255
Dense core granule exocytosis
  cell lines for study, 58
  granulophilin in insulin exocytosis,
    see Granulophilin
  Rab GTPase role
    RNA interference studies, 65–69
    subcellular localization, 58–60
    transient transfection studies, 63–65
  Slp4-a role, *see* Slp4-a

E

Eferin, *see* Rab11-FIPs
EGFR, *see* Epidermal growth
    factor receptor
Electron microscopy, *see* Immunoelectron
    microscopy
ELISA, *see* Enzyme-linked immunosorbent
    assay
Enzyme-linked immunosorbent assay
  phosphatidylinositol 3-kinase p85α subunit
    interaction with Rab5, 544–547
  Rab2 interacting domain mapping,
    388–389
  SopE interaction with Rab5, 304–306
  transferrin receptor recycling, 738–740, 742
Epidermal growth factor receptor
  endocytosis
    assay, 654–656
    overview, 367–370
  trafficking assessment through
    Rab11a–myosin Vb recycling
    system, 720
EPL, *see* Expressed protein ligation
ESCRT-II complex
  function, 323–324
  polycistronic expression and purification
    cassettes
    construction, 325–326, 328
    transformation, 328–329
    chromatography, 329

overview, 324–325
pST39 expression vector advantages,
    329–330
Expressed protein ligation, Rab
    applications, 41
    complex formation of Rab with escort
        protein and GDP dissociation
        inhibitor, 33, 37
    crystallization of complexes, 37
    fluorescence measurements
        prenylation assay, 37–39
        tagged protein, 34
    ligation
        prenylated peptides, 33
        unprenylated peptides, 32
    peptide synthesis, 31–32
    prenylation *in vitro*, 33–34, 35, 37
    principles, 30
    protein expression and purification,
        31, 34–35
    vector construction, 31, 34–35

F

Family of Rab11 interacting proteins,
    see Rab11-FIPs
Far Western blot, *see* Western blot
FIPs, *see* Rab11-FIPs
Flow cytometry
    transferrin receptor transport assays,
        522–524
    VPS34/p150 phosphatidylinositol
        3-kinase complex–Rab7 interaction
        studies, 644–645

G

GAP, *see* GTPase-activating protein
GAPDH, *see* Glyceraldehyde-3-phosphate
    dehydrogenase
GDIs, *see* Guanine nucleotide dissociation
    inhibitors
GEFs, *see* Guanine nucleotide exchange
    factors
Gel filtration
    ALS2CL oligomers, 318–319
    Rab6 protein complexes, 597–598
    RIN guanine nucleotide exchange
        factors, 282–283
    Vps9p, 574

Geldanamycin, inhibition of Rab recycling,
    341, 345–346
GGA proteins, *see* Golgi-associated,
    γ-adaptin-related, Arf-binding proteins
Glutathione *S*-transferase pull-down assay
    phosphatidylinositol 3-kinase p85α subunit
        interactions with Rab5, 547–551
    prenyl-binding protein binding partners,
        50–53
    Rab3 protein–protein interactions, 61–62
    Rab3–Noc2 interactions
        bead incubation and washing, 414
        buffers, 413
        constructs, 413
        protein purification, 413–414
    Rab4–Rabaptin-5α–AP-1 complex,
        529–531, 533, 535–536
    Rab4a interactions with Rabip4
        and CD2AP
        expression constructs, 111–112
        fusion protein purification, 112–113
        guanine nucleotide loading, 113–114
        incubation conditions, 113–114
    Rab6 binding partners, 149–150, 594–597
    Rab7–oxysterol binding protein related
        protein 1L interactions, 752–754, 756
    Rab7–Rabring7 interactions, 691–692
    Rab13–protein kinase A interactions,
        729–730
    Rab27 protein–protein interactions, 61–62
    Rab27a/granulophilin interactions,
        220–222
    Rab34–rab interacting lysosomal protein
        interactions, 681, 683
    Rabaptin-5/GGA1 GAT domain
        interactions, 589
    Slac2-c–myosin interactions, 437–438
    Slp4-a binding partners, 449
    SopE interaction with Rab5, 302–304
    VAP-33–prenylated Rab acceptor protein
        interactions, 805
    VPS34/p150 phosphatidylinositol 3-kinase
        complex–Rab7 interaction studies,
        642–645
    VPS34/VPS15 complex–Rab5 interactions,
        793, 795–799
    Ypt6p binding partner identification from
        yeast cytosol
        chromatography, 603
        cytosol preparation, 602–603, 605–606

Glutathione *S*-transferase pull-down assay
  (cont.)
  glutathione *S*-transferase–Ypt6 bead
    preparation and nucleotide
    loading, 601–602
  specificity, 604–605
Glyceraldehyde-3-phosphate dehydrogenase
  Rab2 interactions
    blot overlay assay, 387–388
    coimmunoprecipitation, 385–386
    enzyme-linked immunosorbent assay
      to map Rab2 interacting domains,
      388–389
    mammalian two-hybrid assay, 386
    overview, 382
    quantitative membrane binding assay,
      389–390
    vesicular tubular cluster association, 382
Golgi-associated, γ-adaptin-related,
  Arf- binding proteins
  domains, 584
  isoforms, 583–584
  rabaptin-5 interactions with GGA1
    GAT domain
    competition experiments, 589–590
    constructs, 586
    crystallization of complex, 590–591
    GAT domain fragment preparation,
      585, 587
    materials for study, 585–586
    pull-down assay, 589
    Rabaptin-5 construct preparation,
      587–589
    surface plasmon resonance, 590
Golgin45
  depletion effects on Golgi structure and
    protein transport, 399–400
  function, 391–392
  yeast two-hybrid analysis of Rab2,
    GRASP55, and golgin45 interactions
    competent yeast preparation, 396–397
    growth media, 395–396
    small-scale transformations, 397–398
Granulophilin
  antibody generation, 218–219
  granulophilin-a, *see* Slp4-a
  immunohistochemistry, 219–220
  insulin granule docking assays
    biochemical assay, 227
    morphological assay, 225–227

  overexpression effects on insulin
    secretion, 224
  pancreatic β-cell expression, 216
  Rab27a interactions
    coimmunoprecipitation
      cultured cells, 222–223
      pancreatic islets, 223–224
    evidence, 216–218
    pull-down assay, 220–222
  structure, 216
  Western blot, 220
GRASP55
  function, 391
  purification of complexes from
    Golgi membranes
    affinity column preparation, 393
    affinity purification of proteins, 394
    Golgi membrane isolation from rat
      liver, 392–393
  yeast two-hybrid analysis of Rab2,
    GRASP55, and golgin45
    interactions
    competent yeast preparation, 396–397
    growth media, 395–396
    small-scale transformations, 397–398
GRASP65
  function, 391
  purification of complexes from Golgi
    membranes
    affinity column preparation, 393
    affinity purification of proteins, 394
    Golgi membrane isolation from rat
      liver, 392–393
Growth hormone, release assay
  DENN/MADD effects, 258–259
  rabconnectin-3 effects, 405
GTPase-activating protein
  function, 552–553
  phosphatidylinositol 3-kinase p85α
    subunit interaction with Rab5
    analysis
    findings, 559–560
    GAPette preparation, 555–556
    single turnover assay, 558–559
    steady-state assay, 556–558
Guanine nucleotide dissociation inhibitors
  displacement factor, *see also* Yip3
    assay, 348–349
    discovery, 348–350
    function, 340

heat shock protein-90 role in Rab
   recycling
   geldanamycin inhibition, 341, 345–346
   guanine nucleotide dissociation
      inhibitor extraction assays, 342–344
   overview, 340–341
   Rab1 extraction, 343–344
   Rab3 extraction from synaptic
      membrane, 342
   radicicolol inhibition, 341, 345–346
   recombinant αGDI preparation,
      341–342, 345
mammalian isoforms, 340
Rab5 interaction regulation by p38
   mitogen-activated protein kinase
   capture assay
      capture, 376
      cytosol preparation, 375–376
      endosome fractionation, 374–375
      postnuclear supernatant preparation,
         373–374
      protein expression and purification,
         371–373
      separate analysis of inhibitor
         activation and Rab5 extraction,
         376–377
   endocytosis assays
      bulk marker assay, 378–379
      kinase activation, 378
   overview, 370
   prospects for study, 379
Guanine nucleotide exchange factors,
   see also specific proteins
   GTPase activation, 262
   pathology, 262

H

Heat shock protein-90, role in Rab recycling
   geldanamycin inhibition, 341, 345–346
   guanine nucleotide dissociation inhibitor
      extraction assays, 342–344
   overview, 340–341
   Rab1 extraction, 343–344
   Rab3 extraction from synaptic
      membrane, 342
   radicicolol inhibition, 341, 345–346
   recombinant αGDI preparation,
      341–342, 345
Hsp90, see Heat shock protein-90

I

Immunoblot, see Western blot
Immunoelectron microscopy
   Noc2, 477–478
   Rab proteins at synaptic terminals,
      154–156
   Rab7–XAPC7 interaction, 654
   Rab13 in tight junction structure
      and function
      freeze-fracture, 192
      immunolabeling, 192–193
   Rab23, 775–776
   rabphilin, 477–478
   VPS34/p150 phosphatidylinositol 3-kinase
      complex–Rab7 interaction studies,
      639–640
Immunofluorescence microscopy
   macropinosome formation, 242
   Noc2, 477
   Rab4–Rabaptin-5α–AP-1 complex,
      532–533
   Rab5/VPS34/VPS15 complex, 790–793
   Rab6 antibody probing of conformation,
      147–148, 150–152
   Rab7/XAPC7, 653–654
   Rab11/myosin Vb, 721–722
   Rab11-FIP3, 505–506
   Rab13/protein kinase A, 726–727
   Rab23, 773–775
   Rabphilin, 477
   Rim2, 465–466
   Slac2-a, 423–424
   Slp4-a, 452
   tight junction proteins, 186–187
   VPS34/p150 phosphatidylinositol 3-kinase
      complex, 638–639
Insulin
   granule docking assays
      biochemical assay, 227
      morphological assay, 225–227
   Noc2 knockout mouse and secretion from
      isolated pancreatic islets, 416–417
   Rab27a/granulophilin overexpression
      effects on secretion, 224
Isothermal titration calorimetry,
   Rab11-FIP studies
   protein expression and purification,
      516–517
   Rab11 binding affinities, 515

Isothermal titration calorimetry,
    Rab11-FIP studies *(cont.)*
  running conditions, 517–518
  thermodynamic parameters, 516
ITC, *see* Isothermal titration calorimetry

L

Legionella pneumophila, see Rab1
Lysosome, *see* Rab interacting lysosomal
    protein; Rabring7

M

Macropinosome
  detection with fluorescent dextran, 240
  macropinocytosis, 230
  Rab34 role in formation
    fluorescence microscopy, 241–242
    inducers, 240–241
    overview, 231
Major histocompatibility complex class I
    molecules
  clathrin-independent endocytosis and
      assays, 243–249
  plasma membrane recycling assay,
      250–252
Mammalian two-hybrid system
  Rab3 binding partners, 62–63
  Rab27 binding partners, 62–63
Mannose 6-phosphate receptor
  recycling, 696–697
  TIP47 in transport
    depletion studies, 358, 364–365
    endosome to *trans*-Golgi network
        transport assay, 364
    metabolic labeling and
        immunoprecipitation, 365
    mutant protein-binding studies,
        358–360
    protein–protein interactions, 357–359
    Rab9 effects on binding affinity, 360
MAPK, *see* Mitogen-activated
    protein kinase
Melanophilin, *see* Slac2-a
Membrome datasets, Rab messenger RNA
    expression profiling, 2
Microtubule
  movement assays of Rab4 function
    advantages and limitations, 105–106

antibodies for immunofluorescence
    microscopy, 100
  colocalization parameters, 102,
      104–105
  fluorescent rat liver endosome
      preparation
    fluorescent asialoorosomucoid
        preparation, 93–94
    injection of rats, 94
    modifications for other cell types, 95
    postnuclear supernatant
        preparation, 94
    sucrose density gradient
        centrifugation, 94
  marker segregation, 105
  microscope system, 98
  microtubule-coated microscopy
      chamber preparation
    attachment to chamber, 98
    polymerizing fluorescent microtubule
        preparation, 95, 97–98
  motility measurement, 99–100
  motility parameter quantification
    direction of movement, 101
    fission, 101–102
    frequency of movement, 100–101
    microtubule binding, 100
    run length, 101
    velocity, 101
  overview, 93
Rabkinesin-6/Rab6-KIFL interaction
    assays
  binding and bundling assays, 623
  polymerization, 621, 623
Mitogen-activated protein kinase, *see* p38
    mitogen- activated protein kinase
MKlp2, *see* Rabkinesin-6/Rab6-KIFL
MPR, *see* Mannose 6-phosphate receptor
Multivesicular body, sorting pathway,
    322–323
Munc13-4
  mutation in disease, 779
  permeabilized platelet assay
    overview, 784–785
    platelet isolation, 785
    serotonin
      dense granule labeling, 785–786
      secretion assay, 786–787
  purification of histidine-tagged
      recombinant protein, 781–784

purification using Rab27A affinity
    chromatography
    affinity column preparation, 780–781
    chromatography, 781
    materials, 779
    platelet cytosol preparation, 780
    structure, 779
Myosin
    isoforms, 715–716
    Rab11–myosin Vb interactions
        cotransfection assay, 718–721
        epidermal growth factor receptor
            trafficking assessment through
            Rab11a–myosin Vb recycling
            system, 720
        immunofluorescence microscopy,
            721–722
        yeast two-hybrid analysis
            constructs, 716–717
            reporter assay, 717–718
            screening, 717
    Rab11-FIP2 interactions, 492, 707, 711
    Rab27A linker proteins, see Slac2-a;
        Slac2-c
MyRIP, see Slac2-c

N

Neuropeptide Y, release assay, 453–454,
    466, 478
Noc2
    dense-core vesicle recruitment, 470,
        475–478
    function, 470
    immunoelectron microscopy, 477–478
    immunofluorescence microscopy, 477
    knockout mouse
        insulin secretion assay from isolated
            pancreatic islets, 416–417
        phenotype, 409
    protein–protein interactions, 409
    Rab binding
        assay, 471–472
        competition experiments, 472–474
        Rab27A interactions
            coimmunoprecipitation, 475
            PC12 cell assay, 474–475
            specificity, 469–471, 474
        Rab3 binding assays
            materials, 409–410

pull-down assay
    bead incubation and washing, 414
    buffers, 413
    constructs, 413
    protein purification, 413–414
yeast two-hybrid system
    construct, 410
    β-galactosidase assay, 411–413
    media and solutions, 410–411
    transformation, 411
Rabphilin-3 homology, 409
site-directed mutagenesis, 477
tissue distribution, 409

O

ORP1L, see Oxysterol binding protein
    related protein 1L
Oxysterol binding protein related
    protein 1L
    function, 744
    Rab7 interactions
        binding site, 756
        coimmunoprecipitation, 747–749
        confocal scanning laser microscopy,
            745–747
        pull-down assay, 752–754, 756
        two-hybrid luciferase assay, 749–752

P

p38 mitogen-activated protein kinase,
    Rab5–guanine nucleotide dissociation
    inhibitor interaction regulation
    capture assay
        capture, 376
        cytosol preparation, 375–376
        endosome fractionation, 374–375
        postnuclear supernatant preparation,
            373–374
        protein expression and purification,
            371–373
        separate analysis of inhibitor
            activation and Rab5 extraction,
            376–377
    endocytosis assays
        bulk marker assay, 378–379
        kinase activation, 378
    overview, 370
    prospects for study, 379

p40
    glutathione *S*-transferase fusion protein
        preparation
        affinity chromatography, 700
        bacterial growth, 699–700
        purity analysis, 700–701
        vector construction, 699
    phosphoinositide interactions
        evidence, 697–698
        liposome preparation
            lipid hydration and dispersion, 702
            lipid mixture preparation and
                drying, 701–702
            phospholipid stocks, 701
            purification, 702
        liposome-binding assay, 702–704
    PIKfyve interactions, 697
    Rab9 interactions, 696–697
PCA, *see* Principal components analysis
PCR, *see* Polymerase chain reaction
PDEδ, *see* Prenyl-binding protein
Phage display, selection of Rab6
        conformation-specific antibodies
    biotin coupling of antigen, 139–140
    positive clone isolation, 144–146
    recovery test, 140–141
    screening, 137–138
    selection, 141–144
    target GTPase preparation, 136–137
Phosphatidylinositol 3-kinase
    binding domains, 790
    classes, 789
    function, 553
    p85α subunit interactions with Rab5
        enzyme-linked immunosorbent
            assay-based binding assay,
            544–547
        GTPase-activating protein studies
            findings, 559–560
            GAPette preparation, 555–556
            single turnover assay, 558–559
            steady-state assay, 556–558
        pull-down assay, 547–551
        purification of proteins
            glutathione *S*-transferase and Rab5
                fusion proteins, 541–543,
                553–554
            p85 constructs, 543–544, 554–555
        signal transduction implications,
            540–541

p150/VPS34 complex interactions with
        Rab7, *see* VPS34/p150
    phosphatidylinositol 3-kinase complex
VPS34/VPS15 complex interactions with
        Rab5, *see* VPS34/VPS15 complex
PKA, *see* Protein kinase A
PKC, *see* Protein kinase C
Plk1, *see* Polo-like kinase 1
Polo-like kinase 1
    purification of recombinant protein
        bacteria expression system, 620–621
        baculovirus–Sf9 cell system, 619–620
    Rabkinesin-6/Rab6-KIFL binding
        Far Western blot, 624–625
        overview, 618–619
        phosphorylation assay, 624
Polycistronic expression system,
    see ESCRT-II complex
Polymerase chain reaction
    Rab23 transcript detection with reverse
        transcription–polymerase chain
        reaction
        amplification reactions, 765
        materials, 764
        primer design, 761, 764
    Rab25 quantitative real-time polymerase
        chain reaction in cancer, 205
PRA1, *see* Yip3
PrA2, *see* Prenylated Rab acceptor 2
PrBP, *see* Prenyl-binding protein
Prenyl-binding protein
    binding partners
        competition assay, 53
        pull-down assays, 50–53
        specificity, 44–45
        yeast two-hybrid screening, 45–47
    photoreceptor protein transport, 53
    purification of glutathione *S*-transferase
            fusion proteins
        affinity chromatography, 50
        amphibian protein, 49
        bovine protein, 49
        mouse protein, 49
        thrombin cleavage, 50
    sequence homology
        interspecies, 43
        RhoGDI, 43–44
        unc119, 43–44
    structure, 43
    tissue distribution, 44–45

Western blot, 47–49
Prenylated Rab acceptor 2
  heat denaturation, 803–806
  isoforms, 800
  protein–protein interactions, 800–801
  purification of recombinant protein,
    801–802
  VAP-33 interactions
    coimmunoprecipitation, 805–806
    overview, 801
    pull-down assay, 805
Principal components analysis, Rab
  sequence alignment data, 21–23, 25, 27
Protein kinase A, Rab13 interaction studies
  calcium switch experiments, 725–726
  cell culture, 725
  immunofluorescence microscopy, 726–727
  inhibition of kinase activity, 725
  kinase assay *in vitro*, 730
  kinase inhibitors and activators, 728
  pull-down assay, 729–730
  vasodilator-stimulated phosphoprotein
    phosphorylation and
    immunoprecipitation, 724, 728, 731
  Western blot, 727–728
Protein kinase C
  Rab2 interactions with atypical enzyme
    blot overlay assay, 387–388
    coimmunoprecipitation, 385–386
    enzyme-linked immunosorbent assay
      to map Rab2 interacting
      domains, 388–389
    mammalian two-hybrid assay, 386
    overview, 382
    quantitative membrane binding
      assay, 389–390
    vesicular tubular cluster association, 382
20S Proteasome
  assembly assay, 657–660
  XAPC7 subunit, *see* XAPC7
Pull-down assay, *see* Glutathione
  S-transferase pull-down assay

R

Rab, *see also specific proteins*
  AMPA receptor trafficking, *see* α-Amino-
    3-hydroxy-4-isoxazoleproprionic acid
    receptor
  conformational switching

    overview, 135–136
    Rab6 probing with antibodies, *see* Rab6
  dense core granule release role, *see* Dense
    core granule exocytosis
  dominant negative mutants
    cholesterol effects
      cholesterol quantification, 177
      depletion, 176
      elevation, 176, 177
      Rab4 extraction inhibition, 177–179
      sphingolipid storage diseases, 176
    fluorescent sphingolipid analog
      trafficking studies
      incubation and washes, 169, 171
      overview, 167, 169
      Rab4, 172–174
      Rab9, 174–175
      Rab11, 172–174
    plasmid constructs, 171
    transfection, 171–172
    types, 168
  endocytosis pathways, 243–244
  functional overview, 167, 230, 296, 339–340
  guanine nucleotide cycling, 593–594
  messenger RNA expression profiling
    annotation considerations, 7
    materials, 3
    membrome datasets, 2
    probe sets, 8
    SymAtlas web-application, 3–6
    systems biology approach
      considerations, 8–9
    tissue differences, 7–8
  phosphorylation
    serine/threonine phosphorylation, 194
    tyrosine phosphorylation, *see* Rab24
  phylogenetic analysis
    overview, 2, 20
    principal components analysis of
      alignment data, 21–23, 25, 27
    sequence alignment, 20–21
  prenylation, 30, 33–34, 37–39, 340
  protein semisynthesis, *see* Expressed
    protein ligation
Rab1
  Legionella pneumophila-containing vacuole
    recruitment studies
      Chinese hamster ovary FcγRII cells
        cell fixation and staining, 78
        infection, 77–79

Rab1 *(cont.)*
    Rab1 mutant studies of replication,
        78–80
    transfection, 77, 79
    Dot/Icm transporter, 72
    macrophages
        bone marrow-derived macrophage
            preparation from mouse, 74
        cell fixation and staining, 75–77
        macrophage infection, 75
        materials, 73–74
Rab2
    effector binding assay
        Golgi extract preparation, 399
        immobilized Rab bead preparation,
            398–399
        glyceraldehyde-3-phosphate
            dehydrogenase/atypical protein
            kinase C interactions
            blot overlay assay, 387–388
            coimmunoprecipitation, 385–386
            enzyme-linked immunosorbent assay
                to map Rab2 interacting
                domains, 388–389
            mammalian two-hybrid assay, 386
            overview, 382
            quantitative membrane binding
                assay, 389–390
        GTPase assay, 385
        guanine nucleotide-binding assays,
            384–385
        prenylation *in vitro*, 384
        purification of recombinant protein,
            383–384
        vesicular tubular cluster association, 382
        yeast two-hybrid analysis of Rab2,
            GRASP55, and golgin45 interactions
            competent yeast preparation, 396–397
            growth media, 395–396
            small-scale transformations, 397–398
Rab3
    dense core granule exocytosis role
        RNA interference studies, 65–69
        transient transfection studies, 63–65
    effectors, 401–402, 408
    guanine nucleotide exchange factor,
        see DENN/MADD
    isoforms, 408
    knockout mice, 255
    Noc2 interactions, *see* Noc2

protein–protein interactions
    mammalian two-hybrid system, 62–63
    pull-down assay, 61–62
Rab3A functions, 254–255
rabconnectin-3 regulation,
    see Rabconnectin-3
rabphilin interactions, *see* Rabphilin
Rim2 binding, *see* Rim2
subcellular localization, 58–60
Rab4
    cholesterol effects, 177–179
    endosomal recycling vesicle reconstitution
        cell culture, 85
        cytosol preparation, 85
        immunodepletion studies, 88–89
        inhibitor studies, 90–91
        materials, 84–85
        overview, 82, 84
        permeabilization and cytosol removal,
            85–86
        Rabaptin-5 supplementation studies, 89
        surface biotinylation and stripping, 85
        vesicle formation and detection, 86–88
    fluorescent sphingolipid analog trafficking
        studies with dominant negative
        mutants, 172–174
    microtubule-based movement assays
        advantages and limitations, 105–106
        antibodies for immunofluorescence
            microscopy, 100
        colocalization parameters, 102, 104–105
        fluorescent rat liver endosome
            preparation
            fluorescent asialoorosomucoid
                preparation, 93–94
            injection of rats, 94
            modifications for other cell types, 95
            postnuclear supernatant
                preparation, 94
            sucrose density gradient
                centrifugation, 94
        marker segregation, 105
        microscope system, 98
        microtubule-coated microscopy
            chamber preparation
            attachment to chamber, 98
            polymerizing fluorescent microtubule
                preparation, 95, 97–98
        motility measurement, 99–100
        motility parameter quantification

direction of movement, 101
fission, 101–102
frequency of movement, 100–101
microtubule binding, 100
run length, 101
velocity, 101
overview, 93
Rab4–Rabaptin-5α–AP-1 complex
AP-1 binding to Rabaptin-5α hinge
fragment, 531–532
endosomal recycling, 526, 536–539
immunofluorescence microscopy,
532–533
pull-down assays, 529–531, 533,
535–536
yeast two-hybrid analysis, 527, 529
Rab4a interactions with Rabip4 and
CD2AP
functional overview, 108
glutathione S-transferase pull-down
assay
expression constructs, 111–112
fusion protein purification, 112–113
guanine nucleotide loading,
113–114
incubation conditions, 113–114
subcellular localization
confocal scanning laser microscopy,
116–117
expression constructs, 114
fractionation, 115–116
transfection, 115
yeast two-hybrid system
cotransformation, 109–110
expression vector, 108–109
β-galactosidase assay, 110–111
lysate preparation, 111
yeast strain, 108
Rab5
activation, 369–370
AMPA receptor trafficking,
see α-Amino-3-hydroxy-4-
isoxazoleproprionic acid receptor
cloning and expression in hippocampal
neurons, 156–157
effectors, 552–553, 562, 688
fluorescence resonance energy transfer
microscopy assays
calculations, 124–127
cell culture, 123

EEA.1-based sensor, 122–123, 128–130
expression constructs, 120–123
GTP-bound Rab5 visualization, 124
imaging, 123–124
Rabaptin5-based sensor, 122, 125–126
three-protein complex detection in
endosomes, 130–133
transfection, 123
functional overview, 119–120, 552
guanine nucleotide dissociation inhibitor
interaction regulation by p38
mitogen-activated protein kinase
capture assay
capture, 376
cytosol preparation, 375–376
endosome fractionation, 374–375
postnuclear supernatant preparation,
373–374
protein expression and purification,
371–373
separate analysis of inhibitor
activation and Rab5 extraction,
376–377
endocytosis assays
bulk marker assay, 378–379
kinase activation, 378
overview, 370
prospects for study, 379
guanine nucleotide exchange factors,
see Alsin; RIN guanine nucleotide
exchange factors; SopE; Vps9p
immunogold electron microscopy at
synaptic terminals, 154–156
phosphatidylinositol 3-kinase p85α subunit
interactions with Rab5
enzyme-linked immunosorbent assay-
based binding assay, 544–547
GTPase-activating protein studies
findings, 559–560
GAPette preparation, 555–556
single turnover assay, 558–559
steady-state assay, 556–558
pull-down assay, 547–551
purification of proteins
glutathione S-transferase and Rab5
fusion proteins, 541–543, 553–554
p85 constructs, 543–544, 554–555
signal transduction implications,
540–541
protein–protein interactions, 120

Rab5 *(cont.)*
    purification of recombinant proteins,
        265–266, 314–315
    regulation of activity, 277
    signaling modulation, 369
    SopE recruitment on *Salmonella*-
        containing phagosomes, *see* SopE
    VPS34/VPS15 complex interactions
        immunofluorescence microscopy,
            790–793
        pull-down assays, 793, 795–799
Rab6
    antibody probing of conformation
        antibody libraries, 136
        immunofluorescence microscopy,
            147–148, 150–152
        immunoprecipitation, 148–149
        phage display selection of
            conformation-specific antibodies
            biotin coupling of antigen, 139–140
            positive clone isolation, 144–146
            recovery test, 140–141
            screening, 137–138
            selection, 141–144
            target GTPase preparation, 136–137
        recombinant scFv production and
            purification, 146–147, 150
    BicD protein interactions
        dynein–dynactin complex, 608
        yeast two-hybrid analysis
            competent cell preparation,
                611–612
            growth media, 609, 611
            overview, 609
            transformation, 612
    function, 593, 600, 608
    MKlp2, *see* Rabkinesin-6/Rab6-KIFL
    protein interaction studies
        cell extract assays, 613, 616
        coimmunoprecipitation, 597
        gel filtration, 597–598
        materials, 594
        pull-down assay, 594–597
        purified protein assays, 616–617
        Rab6 expression and purification,
            613–615
    pull-down assay of binding partners,
        149–150
    TMF/ARA160 interactions, 600, 606
    yeast homolog, *see* Ypt6p

Rab7
    effectors, 628–629, 650–651
    function, 628–629, 650, 665, 688, 744–745
    mutation in disease, 665
    oxysterol binding protein related protein
        1L interactions
        binding site, 756
        coimmunoprecipitation, 747–749
        confocal scanning laser microscopy,
            745–747
        pull-down assay, 752–754, 756
        two-hybrid luciferase assay, 749–752
    Rab interacting lysosomal protein
        interactions, *see* Rab interacting
        lysosomal protein
    Rabring7 interactions, *see* Rabring7
    VPS34/p150 phosphatidylinositol
        3-kinase complex interaction studies
        antibody generation
            Rab7, 633–634
            VPS34, 634–635
        coimmunoprecipitation, 641–642
        constructs, 629–632
        immunoelectron microscopy, 639–640
        immunofluorescence microscopy,
            638–639
        morphological evaluation of late
            endocytic transport, 640–641
        phosphatidylinositol 3-kinase activity
            assay, 632–633
        phosphatidylinositol 3-phosphate
            formation assay, 645–646
        pull-down assays, 642–645
        transfection, 636, 638
    XAPC7 interaction
        coimmunoprecipitation, 656–657
        epidermal growth factor receptor
            endocytosis assay, 654–656
        functions, 651
        immunoelectron microscopy, 654
        immunofluorescence microscopy, 653–654
        phagocytosis assay, 660–662
        proteasome assembly assay, 657–660
        XAPC7 antibody generation, 652–653
        XAPC7 construct preparation, 652
Rab8
    actin reorganization role, 285
    AMPA receptor trafficking, *see* α-Amino-
        3-hydroxy-4-isoxazoleproprionic acid
        receptor

cloning and expression in hippocampal
neurons, 156–157
guanine nucleotide exchange factor,
see Rabin3
immunogold electron microscopy at
synaptic terminals, 154–156
Rab9
fluorescent sphingolipid analog trafficking
studies with dominant negative
mutants, 174–175
mannose 6-phosphate receptor transport
role, *see* TIP47
p40 effector, *see* p40
Rab11
AMPA receptor trafficking, *see* α-Amino-
3-hydroxy-4-isoxazoleproprionic acid
receptor
endocytic recycling compartment, 733
family of proteins and functions, 513
fluorescent sphingolipid analog trafficking
studies with dominant negative
mutants, 172–174
functions, 492
isoforms, 492
myosin Vb interactions
cotransfection assay, 718–721
epidermal growth factor receptor
trafficking assessment through
Rab11a–myosin Vb recycling
system, 720
immunofluorescence microscopy,
721–722
yeast two-hybrid analysis
constructs, 716–717
screening, 717
reporter assay, 717–718
overexpression effects on endosome
morphology, 249–250
subcellular localization, 492
Rab11-FIPs, *see also* Rab-coupling protein
classes, 514
domain structures, 514
flow cytometry-based transport assays,
522–524
isothermal titration calorimetry
protein expression and purification,
516–517
Rab11 binding affinities, 515
running conditions, 517–518
thermodynamic parameters, 516

Rab-binding domains
identification and characterization,
518–519
sequence homology, 515–516
Rab11-FIP2
chemokine receptor complexes,
492–493
discovery, 492, 707
endocytic recycling assay, 495–496
homodimerization studies, 711–713
myosin interactions, 492, 707, 711
purification of recombinant protein,
494–495
Rab11 interaction analysis
antibody generation, 708
Far Western blot, 709–711
recombinant protein production,
708–709
structure, 492
transferrin internalization and
recycling assay, 497–498
Rab11-FIP3
antibody production
characterization, 505
epitope prediction, 503–504
immunization of rabbits, 505
binding proteins, 500
discovery, 500
exogenous expression studies
mutant proteins, 510–511
plasmid construction, 507, 509
wild-type protein, 509–510
purification of recombinant protein
glutathione *S*-transferase fusion
protein, 502–503
histidine-tagged protein, 501–502
plasmid construction, 501
subcellular localization
cell culture and transfection, 505
drug treatment studies, 507
immunofluorescence microscopy,
505–506
mitotic synchronization
experiments, 506
RNA interference studies of membrane
trafficking, 520–522
Rab13
function, 724–725
gene cloning and site-directed
mutagenesis, 184–185

Rab13 *(cont.)*
  protein kinase A interaction studies
    calcium switch experiments, 725–726
    cell culture, 725
    immunofluorescence microscopy,
      726–727
    inhibition of kinase activity, 725
    kinase assay *in vitro*, 730
    kinase inhibitors and activators, 728
    pull-down assay, 729–730
    vasodilator-stimulated phosphoprotein
      phosphorylation and
      immunoprecipitation, 724, 728, 731
    Western blot, 727–728
  recombinant protein purification from
    Escherichia coli, 729
  tight junction structure and function
    regulation
    assembly, 183
    electron microscopy
      freeze-fracture, 192
      immunolabeling, 192–193
    epithelial cell culture, 184–186
    fence function analysis
      confocal scanning laser microscopy,
        190–191
      fluorescent sphingomyelin
        synthesis, 190
    immunofluorescence microscopy of
      tight junction proteins, 186–187
    monolayer integrity assessment,
      188–189
    overview, 183–184
    paracellular gate analysis
      flux assay, 191–192
      transepithelial electrical resistance
        measurement, 191
Rab15 effector protein, transferrin
  receptor recycling assays
  endocytic recycling assay with cell
    surface biotinylation
    cell culture, 736
    incubation and analysis, 736–737
    materials, 734–735
    optimization, 737–738
    overview, 733–734
  enzyme-linked immunosorbent assay
    of recycling, 738–740, 742
Rab22a, clathrin-independent
  endocytosis role

assay, 245–247
  distribution in Arf6-Q67L-expressing
    cells, 248
  imaging, 249
  mechanisms, 247
  overexpression and depletion effects
    endosome morphology, 249–250
    plasma membrane recycling of cargo,
      250–252
  overview, 244–245
Rab23
  immunoelectron microscopy, 775–776
  immunofluorescence microscopy, 773–775
  site-directed mutagenesis, 767–768
  Sonic Hedgehog regulation, 760–761
  transcript detection with reverse
    transcription–polymerase chain
    reaction
    amplification reactions, 765
    materials, 764
    primer design, 761, 764
  transfection
    cloning, 765–767
    endotoxin removal from plasmids,
      768–769
    lipofection, 769–770
    stable cell line preparation, 770–771
  transferrin internalization assay, 771–772
Rab24
  function, 195
  subcellular localization, 195
  tyrosine phosphorylation analysis
    expression, 195
    functional effects, 200
    immunoprecipitation, 197–198
    inhibitors, 198–199
    mutagenesis of phosphorylation
      site, 198
    subcellular distribution effects, 199–200
    Western blot, 195–197
Rab25
  cancer
    candidate gene identification, 203–204
    overexpression, 203, 209
    quantitative real-time polymerase chain
      reaction, 205, 210
  overexpression effects
    apoptosis assay, 207–208
    colony formation assay, 207, 210
    expression constructs, 205–206

transfection and clonal selection, 206–207

tumorigenicity assay, 208

reverse-phase protein lysate array analysis of signaling pathways, 209, 212, 214

RNA interference studies, 208, 212

Rab27

dense core granule exocytosis role

RNA interference studies, 65–69

transient transfection studies, 63–65

effectors, 778

functions, 778–779

Munc13-4 association, see Munc13-4

protein–protein interactions

mammalian two-hybrid system, 62–63

pull-down assay, 61–62

subcellular localization, 58–60

Rab27A

antibody generation, 219

granulophilin interactions

coimmunoprecipitation

cultured cells, 222–223

pancreatic islets, 223–224

evidence, 216–218

pull-down assay, 220–222

immunohistochemistry, 219–220

myosin linker proteins, see Slac2-a; Slac2-c

Noc2 interactions, see Noc2

overexpression effects on insulin secretion, 224

rabphilin interactions, see Rabphilin

Slp4-a interactions, see Slp4-a

Western blot, 220

Rab34

function, 676

GTP binding/GTPase assays

overview, 234–235

in vitro, 235–236

in vivo, 236, 238

macropinosome formation role

fluorescence microscopy, 241–242

inducers, 240–241

overview, 231

Rab interacting lysosomal protein interactions, see Rab interacting lysosomal protein

recombinant protein expression and purification, 231, 233

site-directed mutagenesis, 231–232

subcellular localization

epitope tagging, 238–240

green fluorescent protein tagging, 238–240

macropinosome detection with fluorescent dextran, 240

Rabaptin-5

functions, 584

GGA1 GAT domain interaction studies

competition experiments, 589–590

constructs, 586

crystallization of complex, 590–591

GAT domain fragment preparation, 585, 587

materials, 585–586

pull-down assay, 589

Rabaptin-5 construct preparation, 587–589

surface plasmon resonance, 590

Rab interactions, see Rab4; Rab5

Rab4–Rabaptin-5α–AP-1 complex

AP-1 binding to Rabaptin-5α hinge fragment, 531–532

endosomal recycling, 526, 536–539

immunofluorescence microscopy, 532–533

pull-down assays, 529–531, 533, 535–536

yeast two-hybrid analysis, 527, 529

Rabconnectin-3

calcium-dependent exocytosis and growth hormone release assay, 405

purification of rat brain protein

immunoprecipitation, 404

materials, 402–403

storage, 406

sucrose density gradient centrifugation, 404–405

synaptic vesicle preparation, 403–404

Rab3 association, 402

subunits, 402

Rab-coupling protein, see also Rab11-FIPs

discovery, 482

far-Western analysis of binding partners, 488

function, 482–483

Rab binding specificity, 482

site-directed mutagenesis, 483–486

structure, 482, 492

subcellular localization, 488–490

Rab-coupling protein *(cont.)*
  yeast two-hybrid analysis
    plasmids, 486
    reporter assay, 487
    transformation, 486
    yeast strain, 486
Rabin3
  expression quantification, 292
  guanine nucleotide exchange assay
    constructs, 290
    protein expression and purification, 290–291
    radioassay of GDP/GTP exchange, 291–292
  Rab8 binding
    assay, 289
    specificity, 285
  subcellular localization
    cell culture and transfection, 293
    constructs, 292
    microscopy, 293
  yeast two-hybrid screen, 287–289
Rab interacting lysosomal protein
  Rab specificity, 665
  Rab7 interaction studies
    confocal scanning laser microscopy, 669–670
    low-density lipoprotein degradation assay, 670–671
    lysosomal transport, 672–673
    organelle migration promotion, 671–672
    overlay assay, 669
    yeast two-hybrid system, 667–669
  Rab34 interaction studies of lysosome morphogenesis
    direct binding assay, 683
    expression constructs and mutagenesis, 676–677
    GTP overlay assay, 679–680
    lysosomal positioning assay, 683, 685–686
    pull-down assay, 681, 683
    Rab34 antibody generation, 677, 679
    yeast two-hybrid analysis, 680–681
  recombinant protein expression in mammalian cells
    cell culture, 666
    transient transfection, 666–667
    vectors, 665–666

Rabip4, Rab4 interactions, *see* Rab4
Rabkinesin-6/Rab6-KIFL
  cell cycle regulation, 618
  cytokinesis role, 625–627
  function, 618
  microtubule interaction assays
    binding and bundling assays, 623
    polymerization, 621, 623
  polo-like kinase 1 binding
    Far Western blot, 624–625
    overview, 618–619
    phosphorylation assay, 624
  purification of recombinant protein
    bacteria expression system, 620–621
    baculovirus–Sf9 cell system, 619–620
  RNA interference studies, 625–627
Rabphilin
  dense-core vesicle recruitment, 470, 475–478
  function, 470
  immunoelectron microscopy, 477–478
  immunofluorescence microscopy, 477
  Rab binding
    assay, 471–472
    competition experiments, 472–474
    Rab27A interactions
      coimmunoprecipitation, 475
      PC12 cell assay, 474–475
    specificity, 469–471, 474
  site-directed mutagenesis, 477
Rabring7
  function, 689
  isolation with CytoTrap system, 690–691
  lysosome subcellular distribution and acidity effects, 694
  Rab7 interaction studies
    coimmunoprecipitation, 692–693
    epidermal growth factor degradation assay, 694
    pull-down assay, 691–692
    recruitment assay, 693
Rac1
  guanine nucleotide exchange factor, *see* Alsin
  immunofluorescence microscopy of macropinosome formation, 242
  purification of recombinant protein, 265–266
Radicicolol, inhibition of Rab recycling, 341, 345–346

Ras, immunofluorescence microscopy of macropinosome formation, 242
RCP, *see* Rab-coupling protein
REP15, *see* Rab15 effector protein
Reverse-phase protein lysate array, Rab25 signaling pathway analysis, 209, 212, 214
Rhodopsin kinase, prenyl-binding protein transport, 53
RILP, *see* Rab interacting lysosomal protein
Rim1
 splice variants, 458–460
 structure, 458
 subcellular localization, 458
Rim2
 immunofluorescence microscopy, 465–466
 neuropeptide Y release assay, 466
 Rab3A binding
  binding domain, 459–460
  competition experiments, 462–463
  cotransfection assay, 461–462
  dense core vesicle localization, 465–466
  effector function, 458–459
  isoform differences, 463–465
 site-directed mutagenesis, 465
 splice variants, 458–460
 structure, 458
 subcellular localization, 458
RIN guanine nucleotide exchange factors
 assays
  GDP dissociation, 279–280
  guanine nucleotide binding, 280–281
  materials, 279
 gel filtration analysis, 282–283
 Rab5b binding assay
  expression and purification from mammalian cells, 281–282
  nucleotide binding to Rab5b, 281
  pull down, 282
 recombinant protein purification from baculovirus–Sf9 cell system
  baculovirus construction and selection, 277–278
  expression, 278
  FLAG-tagged protein immunoaffinity chromatography, 278–279
 structure, 277
RNA interference
 Rab effector studies of dense core granule exocytosis, 65–69

Rab11-FIP studies of membrane trafficking, 520–522
Rab22a and endosome morphology, 249–250
Rab25 studies, 208, 212
Rabkinesin-6/Rab6-KIFL studies, 625–627
Slac2-a studies of melanosome transport, 427–428
RPPA, *see* Reverse-phase protein lysate array

S

Salmonella-containing phagosomes, *see* SopE
Sec4p
 function, 11
 functional gene substitution, 13
 Ypt1p comparison for domain function specificity
  algorithm
   features, 12–13
   implementation, 16
   output, 16
  alignment, 11, 13
  caveats, 16–18
  functional residue identification, 15
Serotonin
 dense granule labeling, 785–786
 secretion assay for Munc13-4 function studies, 786–787
Sgm1, Ypt6p binding, 600
Site-directed mutagenesis
 Noc2, 477
 Rab13, 184–185
 Rab23, 767–768
 Rab34, 231–232
 Rab-coupling protein, 483–486
 Rabphilin, 477
 Rim2, 465
 Slac2-a, 425–426
 Slp4-a, 453
Slac2-a
 gene, 420
 Rab27A–myosin linker protein
  evidence, 420–421
  melan-a cell assays
   immunofluorescence microscopy, 423–424

Slac2-a *(cont.)*
    melanocyte culture and transfection,
        421–423
    melanosome distribution assay, 423
    RNA interference studies of
        melanosome transport,
        427–428
    PEST-like sequence deletion effects,
        426–427
    protein–protein interaction assays
        domain constructs, 424
        site-directed mutagenesis studies,
            425–426
        weak interaction with myosin
            globular tail, 426
    Slac2-c homology, 432
    structure, 419–420, 424
Slac2-c
    actin interactions with carboxy-terminus
        binding domain
            localization on actin filaments in
                PC12 cells, 441–442
            mapping, 439–441
        direct interaction assay, 441
        overview, 439
    function, 432–433, 443
    myosin interaction assays
        constructs, 437
        cotransfection assay, 438
        pull-down assay, 437–438
        Rab27A–Slac2-c–myosin complex
            formation from purified components,
            438–439
    Slac2-a homology, 432
    SLP homology domain interactions with
        Rab27
        assay, 435–436
        domain expression and purification,
            433, 435
        overview, 433
        Rab27A-GTP studies, 436
    structure, 432
    tissue distribution, 432–433
Slp4-a
    dense core vesicle studies of Rab27A
        interactions
        antibody production, 450–451
        coimmunoprecipitation, 451
        exocytosis inhibition, 452–455
        immunofluorescence microscopy, 452

PC12 cell studies, 451–452
        subcellular fractionation, 452
    discovery, 445
    Rab binding partner characterization
        coimmunoprecipitation, 448–449
        cotransfection assay, 450
        pull-down assay, 449
        transfection, 447–448
        site-directed mutagenesis, 453
        structure, 446
Sonic Hedgehog
    embryogenesis signaling, 759–760
    Rab23 regulation, 760–761
SopE
    guanine nucleotide exchange assay, 306
    Rab5 recruitment on *Salmonella*-
        containing phagosomes
        bacteria growth and harvesting,
            299–300
        binding analysis with enzyme-linked
            immunosorbent assay, 304–306
        materials for study, 297–299
        phagosome preparation, 300–301
        pull-down assay, 302–304
        Salmonella viability dependence,
            301–302
        survival mechanism, 296, 307
        Western blot of Rab5, 301
SPR, *see* Surface plasmon resonance
Surface plasmon resonance, Rabaptin-5/
    GGA1 GAT domain interactions, 590
SymAtlas, Rab messenger RNA expression
    profiling, 3–6
Synaptic plasticity, *see* α-Amino-3-hydroxy-
    4-isox acid receptor
Syntaxin-1a
    immunohistochemistry, 219–220
    Western blot, 220

T

Thin-layer chromatography
    GTPase-activating protein assays, 557–558
    Rab34 GTP binding/GTPase assays
        overview, 234–235
        in vitro, 235–236
        in vivo, 236, 238
Tight junction, *see* Rab13
TIP47
    mannose 6-phosphate receptor transport

depletion studies, 358, 364–365
endosome to *trans*-Golgi network
   transport assay, 364
metabolic labeling and
   immunoprecipitation, 365
mutant protein-binding studies,
   358–360
protein–protein interactions, 357–359
Rab9 effects on binding affinity, 360
purification
   bovine kidney protein, 360–362
   recombinant protein, 362–364
TLC, *see* Thin-layer chromatography
TMF/ARA160, Rab6 binding, 600, 606
Transferrin
clathrin-independent endocytosis and
   assay, 243–247
internalization and recycling assays,
   497–498, 771–772
plasma membrane recycling assay,
   250–252
Transferrin receptor
endocytic recycling assay with cell surface
   biotinylation
   cell culture, 736
   incubation and analysis, 736–737
   materials, 734–735
   optimization, 737–738
   overview, 733–734
enzyme-linked immunosorbent assay of
   recycling, 738–740, 742
flow cytometry-based transport assays,
   522–524
Rab11-FIP in trafficking, 519
recycling pathways, 733
Transforming growth factor-β receptor,
   endosome targeting, 368

U

Ubiquitination
endocytic pathway proteins, 562
Vps9p, *see* Vps9p

V

VAP-33
prenylated Rab acceptor protein
   interactions
   coimmunoprecipitation, 805–806

overview, 801
pull-down assay, 805
purification of recombinant protein,
   801–803
Vasodilator-stimulated phosphoprotein,
   protein kinase A phosphorylation
   and immunoprecipitation, 724,
   728, 731
VASP, *see* Vasodilator-stimulated
   phosphoprotein
Vps proteins, *see* ESCRT-II complex;
   specific proteins
VPS34/p150 phosphatidylinositol 3-kinase
   complex
functions, 629, 646
Rab7 interaction studies
   antibody generation
      Rab7, 633–634
      VPS34, 634–635
   coimmunoprecipitation, 641–642
   constructs, 629–632
   immunoelectron microscopy, 639–640
   immunofluorescence microscopy,
      638–639
   morphological evaluation of late
      endocytic transport, 640–641
   phosphatidylinositol 3-kinase activity
      assay, 632–633
   phosphatidylinositol 3-phosphate
      formation assay, 645–646
   pull-down assays, 642–645
   transfection, 636, 638
VPS34/VPS15 complex
function, 789–790
Rab5 interactions
   immunofluorescence microscopy,
      790–793
   pull-down assays, 793, 795–799
Vps9p
carboxypeptidase Y sorting assay of
   Rab5 activation, 565–567
deletion mutant studies, 565
purification of recombinant proteins
   anion-exchange chromatography, 573
   bacteria culture and induction, 572
   gel filtration, 574
   materials, 570–572
   nickel affinity chromatography,
      572–573
   overview, 570

Vps9p *(cont.)*
  Rab5
    activation, 562
    nucleotide release assay
      fluorescence resonance energy
        transfer, 576–577
      mantGDP loading, 575–576
      Rab5 preparation, 574–575
    signal transduction pathway
      integration, 562
    structure, 564
    ubiquitin binding, 562, 581
    ubiquitination assays
      in vivo, 577–579
      in vitro, 579–581
    ubiquitination mutant studies of Ste3p
      trafficking, 568–569

W

WAVE2, immunofluorescence microscopy
  of macropinosome formation, 242
Western blot
  Far Western blot
    Rab11/Rab11-FIP2 interactions,
      709–711
    Rab-coupling protein binding
      partners, 488
    Rabkinesin-6/Rab6-KIFL binding,
      624–625
  prenyl-binding protein, 47–49
  Rab5 on *Salmonella*-containing
    phagosomes, 301
  Rab27a, 220
  syntaxin-1a, 220
  tyrosine phosphorylation in Rab
    proteins, 195–197
  Yip1p, 335

X

XAPC7, Rab7 interaction
  coimmunoprecipitation, 656–657
  epidermal growth factor receptor
    endocytosis assay, 654–656
  functions, 651
  immunoelectron microscopy, 654
  immunofluorescence microscopy, 653–654
  phagocytosis assay, 660–662

proteasome assembly assay, 657–660
XAPC7 antibody generation, 652–653
XAPC7 construct preparation, 652

Y

Yeast two-hybrid system
  Alsin binding partners, 266–267
  prenyl-binding protein binding partner
    screening, 45–47
  Rab2, GRASP55, and golgin45
    interactions
    competent yeast preparation, 396–397
    growth media, 395–396
    small-scale transformations, 397–398
  Rab4–Rabaptin-5α–AP-1 complex,
    527, 529
  Rab4a interactions with Rabip4 and
    CD2AP
    cotransformation, 109–110
    expression vector, 108–109
    β-galactosidase assay, 110–111
    lysate preparation, 111
    yeast strain, 108
  Rab6–BicD protein interactions
    competent cell preparation, 611–612
    growth media, 609, 611
    overview, 609
    transformation, 612
  Rab7–Rab interacting lysosomal protein
    interaction studies, 667–669
  Rab11–myosin Vb interactions
    constructs, 716–717
    reporter assay, 717–718
    screening, 717
  Rab34–rab interacting lysosomal protein
    interactions, 680–681
  Rab-coupling protein
    plasmids, 486
    reporter assay, 487
    transformation, 486
    yeast strain, 486
  Rabin3 binding partners, 287–289
  Yip1p binding partners, 338
Yip1p
  functions, 333
  fusion constructs, 334–335
  genetic analysis of mutant alleles,
    335–338
  Western blot, 335

yeast two-hybrid analysis, 338
Ypt interactions, 333
Yip3
  guanine nucleotide dissociation inhibitor
    displacement factor activity
      assay, 353–355
    overview, 350–351
  purification of recombinant protein,
    351–353
  Rab interactions, 350–351
Ypt1p
  function, 11
  functional gene substitution, 13
  Sec4p comparison for domain function
    specificity
      algorithm
        features, 12–13

        implementation, 16
        output, 16
      alignment, 11, 13
      caveats, 16–18
      functional residue identification, 15
Ypt6p
  binding partner identification from
    yeast cytosol
    chromatography, 603
    cytosol preparation, 602–603,
      605–606
    glutathione *S*-transferase–Ypt6 bead
      preparation and nucleotide
      loading, 601–602
    specificity, 604–605
  function, 600
  Sgm1 binding region, 600

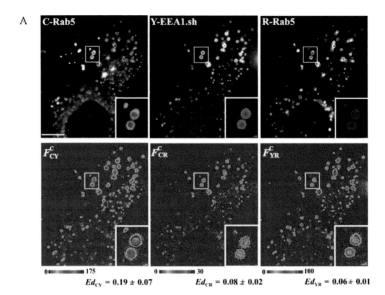

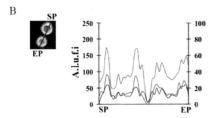

$Ed_{CY} = 0.19 \pm 0.07$ $Ed_{CR} = 0.08 \pm 0.02$ $Ed_{YR} = 0.06 \pm 0.01$

GALPERIN AND SORKIN, CHAPTER 11, FIG. 4. 3-FRET microscopy analysis of Rab5 microdomains in single endosomes of living cells. (A) mRFP-Rab5, YFP-EEA1.sh, and CFP-Rab5 were coexpressed in Cos-1 cells, the cells were treated with nocodazole for 15 min, and six images were acquired as described in the text. $FRET^C$ images are presented in a pseudocolor mode. Insets show an enlargement of the outlined regions of the images. Mean *Ed* values measured for individual endosomes of the presented cell are shown below the corresponding image inset. Bar = 10 μm. (B) Three $FRET^C$ images in RGB color format obtained in 3-FRET experiments ($FRET^C_{CY}$ is green, $FRET^C_{YR}$ is red, and $FRET^C_{CR}$ is blue) were merged. "White" designates the overlap of red, blue, and green. The arbitrary fluorescence intensities of $FRET^C$ signals across two endosomes were plotted. $FRET^C_{CR}$ is plotted on the right axes. SP is the starting point and EP is the end point. A.l.u.f.i. is arbitrary linear units of fluorescence intensity.

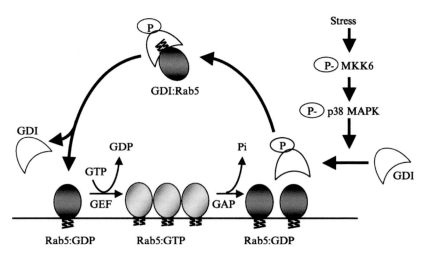

FELBERBAUM-CORTI *ET AL.*, CHAPTER 32, FIG. 1. Endocytosis is modulated by the p38 MAPK pathway. Activation of p38 by stress (phosphorylated p38, P-p38) leads to phosphorylation and activation of GDI, thus stimulating the formation of the cytosolic GDI:Rab5 complex. Regulation of GDI activity may provide a mechanism by which Rab5-GDP is delivered and reactivated specifically and rapidly to active regions of the membrane. The observed increase in internalization rates upon p38 MAPK activation may also reflect a direct role of Rab5 in the formation of clathrin-coated vesicles.

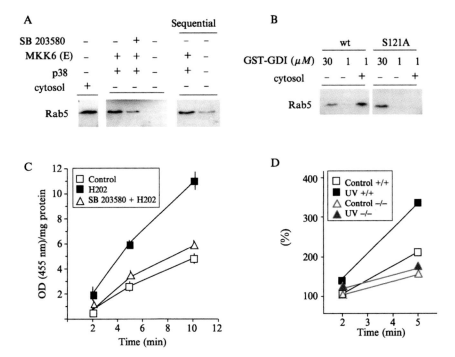

FELBERBAUM-CORTI *ET AL.*, CHAPTER 32, FIG. 3. (A) Rab5 capture assay with pure, recombinant proteins. The assay is supplemented with 150 nM recombinant GST-p38, 50 nM GST-MKK6(E) in the presence or absence of 10 μM SB203580. As indicated, the assay is also carried out sequentially: GST-GDI is first activated with 150 nM recombinant His-p38 and 50 nM GST-MKK6(E), and then the phosphorylated GDI is mixed with endosomes, so that it can capture membrane-associated Rab5. (B) Rab5 capture assay with GDI phosphorylation mutant. Wild-type GST-GDI or mutant are first incubated with (+) or without (−) cytosol, and then the assay is carried out sequentially. (C) Endocytosis *in vivo*. BHK cells are mock treated (open squares), treated with 50 μM H_2O_2 for 10 min (closed squares), or preincubated for 30 min with 10 μM SB203580 prior to adding 50 μM H_2O_2 for 10 min (open triangles). Then, cells are incubated for the indicated time periods with 2 mg/ml HRP as a marker of fluid-phase uptake. (D) Endocytosis in p38α($^{-/-}$) cells *in vivo*. p38 MAPKα (+/+) (square) or (−/−) (triangle) MEFs are treated with (close symbols) or without (open symbols) UV light and then incubated with 2 mg/ml HRP. Reprinted from Cavalli *et al.* (2001). Copyright (2001), with permission from Elsevier.

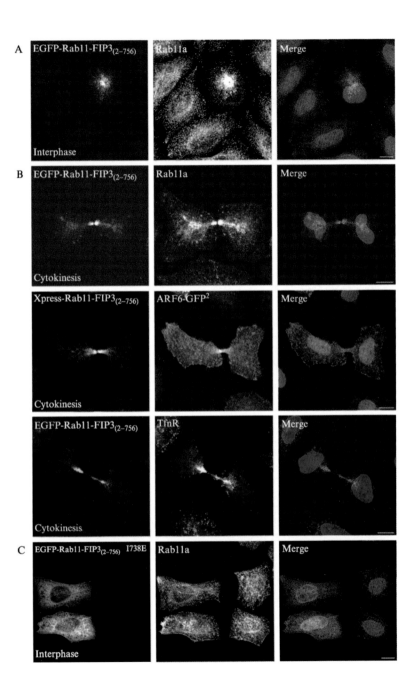

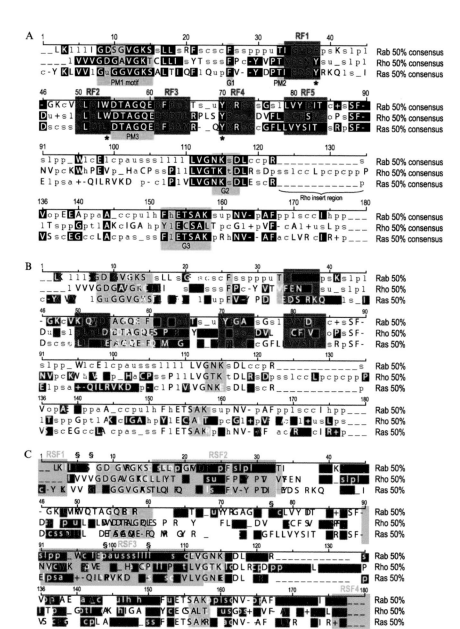

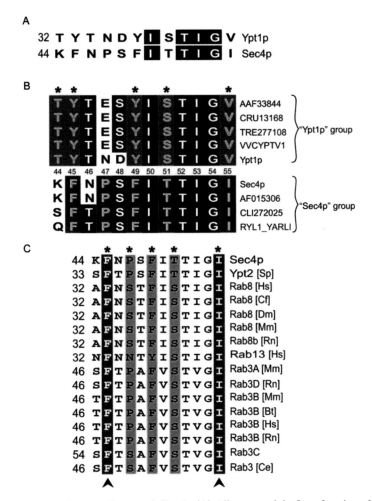

NUSSBAUM AND COLLINS, CHAPTER 2, FIG. 3. (A) Alignment of the Loop2 region of Sec4p and Ypt1p. The identity of 7 amino acids differs out of a total of 12 possible positions. (B) Alignment of the Loop2 region of Sec4p and Ypt1p together with other members of the Sec4p and Ypt1p groups. Amino acids considered potentially significant for Sec4p function are shaded blue and those significant for Ypt1p function are shaded red and are indicated with an asterisk; the algorithm will subsequently rank these positions according to conservation and divergence distances. (C) Qualitative overview of position ranking. The Loop2 region of Sec4p is aligned together with Sec4p-related sequences. The position of potentially significant residues is indicated with shading; of these positions, F45 and I55 (shaded blue and indicated with arrows) are strictly conserved and P47, F49, and T51 are less conserved. The algorithm will rank these residues and others over the entire Sec4p sequence according to conservation, chemical identity, and divergence from the Ypt1p sequence.

COLLINS, CHAPTER 3, FIG. 3. Comparison of the core domain of Ras superfamily sequences between Rab, Ras, and Rho families. The core domain is aligned showing in uppercase bold letters, those residues conserved at the 50% consensus level (i.e., 50% or greater sequences) show this residue at the position indicated. Bold is also used for positions conserved for positive (+, H, K, R) or negative charge (−, D, E). In lowercase letters is shown the consensus sequence at nonconserved positions designated according to the amino acid class abbreviation; o (alcohol, S,T), l (aliphatic (I, L,V), a (aromatic, F, H,W,Y), c (charged, D, E,H,K,R), h (hydrophobic, A,C,F,G,H,I,K,L,M,R,T,V,W,Y), p (polar, C,D,E,H,K,N,Q,R,S, T), s (small, A,C,D,G,N,P,S,T,V), u (tiny, A,G,S), and t (turn-like, A,C,D,E,G,H,K,N,Q,R,S, T). The consensus sequence data were obtained from the SMART database (Schultz *et al.*, 1998) (http://smart.embl-heidelberg.de/) and are derived from 339 Ras domains, 460 Rho domains, and 1120 Rab domains. The location of the Rho insert region is marked; this insert is not contained in the Ras or Rab families. For greater clarification, the G protein-conserved sequence elements are shown highlighted in yellow. Numbering is arbitrary and intended as a descriptive guide. Highlighted in red are Rab family (RF) regions that have been proposed to uniquely distinguish the Rab subfamily of the Ras superfamily (Pereira-Leal and Seabra, 2000). Indicated below each row with an asterisk is a triad of conserved hydrophobic residues that provides structural plasticity in stabilization of the activated conformation of Rab3A and Rab5C (Merithew *et al.*, 2001). (A) Conserved positions among Rab/Rho/Ras subfamilies. In this representation, all residues that are conserved at the 50% consensus level within a subfamily and shared with at least one other subfamily member are shaded in black. An asterisk marks the positions of a triad of hydrophobic residues (position 38, 54, 70) that stabilizes the active conformation. (B) Signature motifs of Rab/Rho/Ras subfamilies. In this representation, all residues that are both conserved at the 50% consensus level within one of the subfamilies and unique within that subfamily are shaded in blue. (C) Subclass discriminant residues of Rab/Rho/Ras subfamilies. The alignment of core domains highlights nonconserved residue positions with green shading. Highlighted in pink are Rab subfamily (RSF) regions that have been proposed to identify subclasses of Rab proteins (Moore *et al.*, 1995; Pereira-Leal and Seabra, 2000). Indicated with a § above each row are the positions of amino acids that form part of the hydrophobic core between switch regions of the Rab3A and Rab5C GTPases (Merithew *et al.*, 2001) and that are predicted to be key specificity determinants as their packing in turn dictates the particular conformation of the invariant hydrophobic triad (see A).

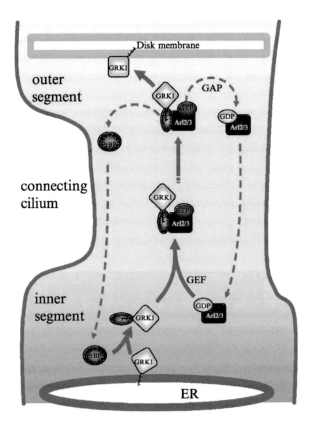

Zhang et al., Chapter 5, Fig. 6. Hypothetical model of PrBP(PDEδ) transport of rhodopsin kinase (GRK1) from the inner segment to the outer segment of rod photoreceptors. PrBP(PDEδ) solubilizes GRK1 from the endoplasmic reticulum (ER) membrane by binding the C-terminal farnesyl group, and the resulting soluble protein complex interacts with a trafficking protein such as Arl2/3-GDP. An unidentified guanine nucleotide exchange factor (GEF) exchanges GDP with GTP to activate Arl2/3, and the complex is then transported through the connecting cilium. When a GTPase activating protein (GAP) accelerates GTP hydrolysis on Arl2/3, the complex falls apart and GRK1 associates with the disk membrane in the outer segment. Arl2/3-GDP and PrBP(PDEδ) return to the inner segment to participate in another round of protein transport.

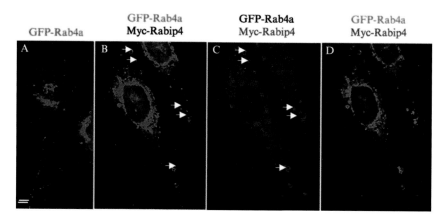

| GFP-Rab4a | GFP-Rab4a
Myc-Rabip4 | **GFP-Rab4a**
Myc-Rabip4 | GFP-Rab4a
Myc-Rabip4 |

MONZO *ET AL.*, CHAPTER 10, FIG. 2. Evidence for the interaction of Rab4 with Rabip4: the morphology of the Rab4a-positive endosome is changed by Rabip4 overexpression. CHO cells are transiently transfected with pEGFP-Rab4a alone (A), or together with pcDNA3-myc-Rabip4 (B–D). GFP-Rab4a labels small vesicular structures at the periphery of the cell and in a perinuclear region (A). When myc-Rabip4 is coexpressed, GFP-Rab4a is detected in enlarged vesicles (arrows) as well as in the perinuclear region (B). The same enlarged vesicles are also labeled by myc-Rabip4 (C, arrows), and the colocalization is indicated by the yellow color obtained when the green and the red images are merged (D). Cells with equal intensity in the GFP channel, i.e., that overexpressed similar amounts of Rab4a, are analyzed. Similar observations are made when myc-Rabip4′ is expressed instead of myc-Rabip4. Bar = 1 μm.

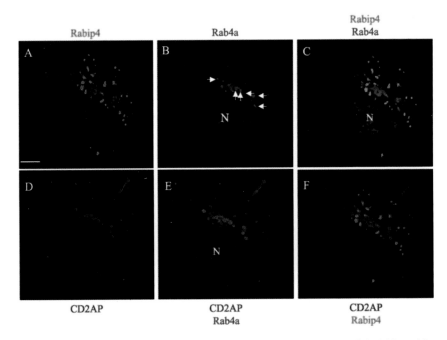

MONZO *ET AL.*, CHAPTER 10, FIG. 3. Rab4a is shared between Rabip4 and CD2AP-positive structures. CHO cells are transiently cotransfected with pEGFP-Rabip4, pcDNA3-myc-Rab4a, and pDsRed-CD2AP. GFP-Rabip4 appears in green (shown in A), pDsRed-CD2AP appears in red (shown in D), and myc-Rab4a, revealed with anti-myc antibody followed by Cy5-coupled anti-mouse antibody, appears in blue (shown in B). The PhotoShop software (Adobe Systems) is used to merge the images corresponding to GFP-Rabip4 and Rab4a (C), DsRed-CD2AP and Rab4 (E), or GFP-Rabip4 and DsRed-CD2AP (F). Colocalization of Rabip4 and Rab4a in (C) results in a light blue color. Arrows in (B) point to structures containing Rab4a but not Rabip4. These structures contain in fact CD2AP as indicated by the purple color appearing in (E). CD2AP and Rabip4 do not mainly colocalize, since the merged image of CD2AP and Rabip4 (F) does not result in a yellow color. N is for nucleus with nonspecific labeling due to the Cy5-coupled anti-mouse antibodies used in this experiment. Bar = 1 μm.

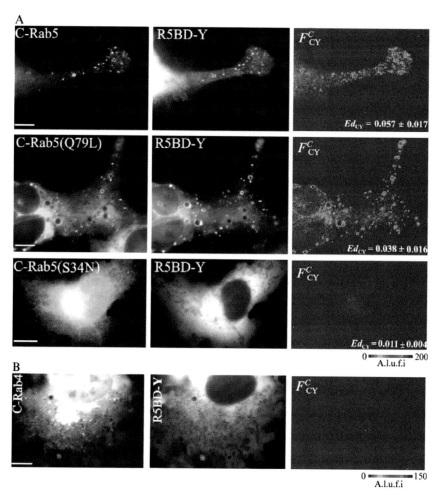

GALPERIN AND SORKIN, CHAPTER 11, FIG. 2. Detection of GTP-bound Rab5 in living cells. CFP-Rab5, CFP-Rab5(Q79L), and CFP-Rab5(S34N) were coexpressed with R5BD-YFP (R5BD-YFP in Cos1 cells) or in Cos-1 cells (A). CFP-Rab4 was coexpressed with R5BD-YFP in Cos-1 cells (B). YFP, CFP, and FRET images were acquired from living cells at room temperature. FRET^C images were calculated as described and presented in a pseudocolor mode. Mean Ed values measured for individual endosomes of the presented cell are shown next to the corresponding image. Intensity bars are presented in arbitrary linear units of fluorescence intensity. Bar $= 10\,\mu$m. A.l.u.f.i. is arbitrary linear units of fluorescence intensity.

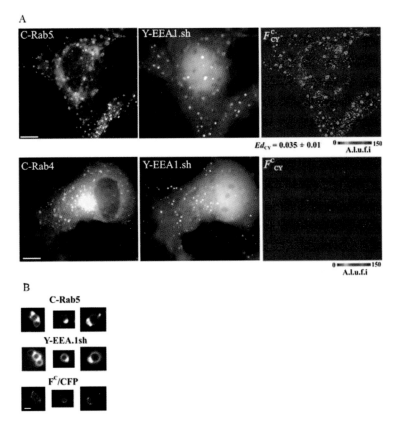

GALPERIN AND SORKIN, CHAPTER 11, FIG. 3. Specificity of YFP-EEA.1sh sensors for Rab5. (A) CFP-Rab5 was coexpressed with YFP-EEA.1sh in Cos-1 cells. CFP-Rab4 was coexpressed with YFP-EEA.1sh in Cos-1 cells. YFP, CFP, and FRET images were acquired from living cells at room temperature. FRET^C images were calculated as described and presented in a pseudocolor mode. Mean Ed values measured for individual endosomes of the presented cell are shown next to the corresponding image. Intensity bars are presented in arbitrary linear units of fluorescence intensity. Bar = 10 μm. (B) Gallery of high magnification images shows individual endosomes or tethered endosomes in cells coexpressing CFP-Rab5 and YFP-EEA.1sh. FRET^C images are presented as pseudocolor intensity-modulated images (FRET^C/ CFP). Bar = 2 μm. A.l.u.f.i. is arbitrary linear units of fluorescence intensity.

HORGAN *ET AL.*, CHAPTER 44, FIG. 3. Localization of exogenously expressed EGFP-fused Rab11-FIP3 and Rab11-FIP3 mutants. (A and B, upper and lower panels) HeLa cells were transfected with pEGFP-C1/Rab11-FIP3$_{(2-756)}$ (green in merge). Sixteen to eighteen hours posttransfection the cells were processed for immunofluorescence as described in Methods and immunostained with antibodies to Rab11a or TfnR (red in merge). (B, middle panel) HeLa cells were double transfected with pcDNA3.1HisB/Rab11-FIP3$_{(2-756)}$ (red in merge) and pARF6/GFP^2-N3 (green in merge). Sixteen hours posttransfection the cells were immunostained with anti-Xpress antibody (red in merge). (C) HeLa cells were transfected with pEGFP-C1/Rab11-FIP3$_{(2-756)}$ I738E (green in merge). Sixteen hours posttransfection the cells were immunostained with an antibody to Rab11a (red in merge). Hoechst dye (blue) was used to visualize the nuclei. Bar = 10 μm.

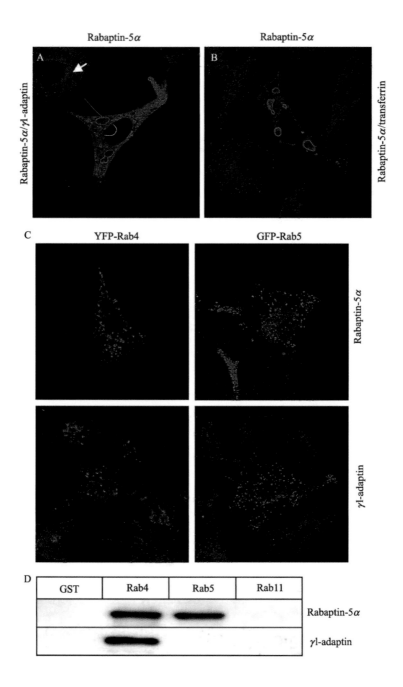

Popa *et al.*, Chapter 46, Fig. 2. Rab4-dependent localization of the rabaptin-5α/AP-1 complex to endosomes. (A) HeLa cells were transfected with pcDNA3.1His-rabaptin-5α. The cells are labeled for rabaptin-5α (green) and endogenous γ1-adaptin (red). Note the colocalization of rabaptin-5α and γ1-adaptin in transfected cells and the distinct localization of γ1-adaptin in nontransfected cells (arrow). (B) HeLa cells were incubated with Alexa594-Tf for 60 min at 37° and subsequently labeled with anti-rabaptin-5α antibody and Alexa488-IgG, showing that the rabaptin-5α/AP-1 complex localizes to early endosomes. (C) Localization of γ1-adaptin to endosomes depends on Rab4. HeLa cells expressing His$_6$-rabaptin-5α in combination with YFP-Rab4 or GFP-Rab5 and labeled with anti-γ1-adaptin antibody and anti-rabaptin-5α antibody, detected with Cy3-labeled secondary antibodies. Note that GFP-Rab5 colocalizes with rabaptin-5α, but not with γ1-adaptin. (D) Rab4 recruits rabaptin-5α/AP-1 complex. GST-Rab4, GST-Rab5, and GST-Rab11 were isolated on GSH beads, loaded with GTPγS and incubated with pig brain cytosol. Bound fractions were immunoblotted with antibodies against rabaptin-5α and γ1-adaptin. Note that GST-Rab4 and GST-Rab5 both retrieve rabaptin-5α, but that only GST-Rab4 bound material includes γ1-adaptin (Adapted with permission from Denaka *et al.*, 2003d).

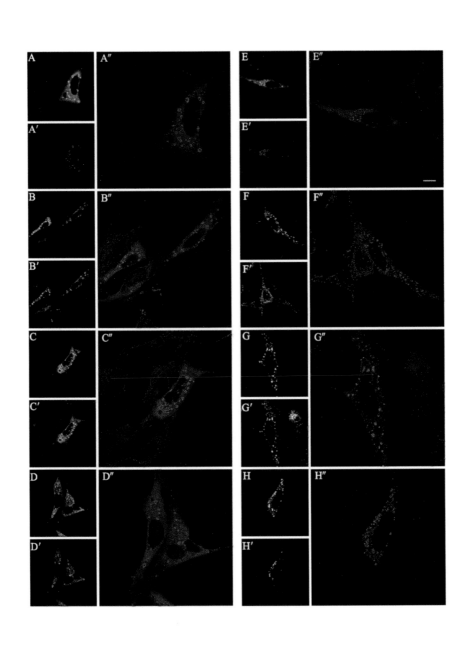

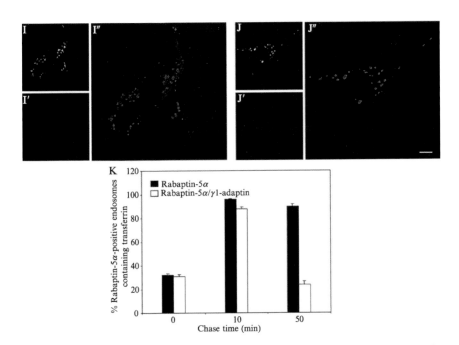

POPA *ET AL.*, CHAPTER 46, FIG. 3. Transfected rabaptin-5α delays transferrin recycling. HeLa cells were transfected with pcDNA3.1 His-rabaptin-5α (A–E) alone or in combination with pcDNA3-γ1-adaptin (F–J). The cells were incubated with 15 μg/ml Alexa488-Tf (A–E) or Alexa594-Tf (F–J) at 16° for 30 min, subsequently chased at 37° for 0 min (A′, F′), 10 min (B′,G′), 20 min (C′, H′), 40 min (D′, I′), and 50 min (E′, G′) and labeled with anti-Xpress followed by Cy3-labeled anti-mouse IgG (A–E) or Alexa488-labeled anti-mouse IgG (F–J). Merged images are shown in A″–J″. Bar = 10 μm. (K) Quantitation of the fraction of rabaptin-5α-positive endosomes containing Alexa-Tf at 0, 10, and 50 min of chase. Error bars denote the standard deviation (*n* = 10). Note that entry of Alexa-Tf into early endosomes is the same in single rabaptin-5α transfectants as in the double transfectants, showing that Rab5-dependent internalization is not affected. Recycling of Alexa488-Tf in the pcDNA3.1 His-rabaptin-5α transfectants, however, is strongly diminished compared to pcDNA3.1 His-rabaptin-5α/ pcDNA3-γ1-adaptin transfected (or control) cells. Others reported a similar negative regulatory function of rabaptin-5 in an *in vitro* endosome recycling assay (Pagano *et al.*, 2004) (Adapted with permission from Denaka *et al.*, 2003a).

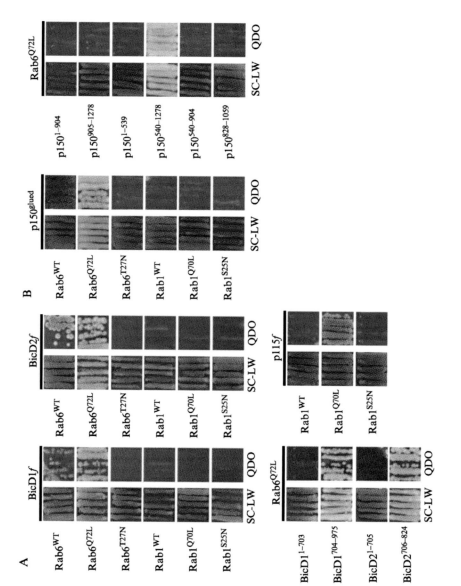

A

BicD1f

	SC-LW	QDO
Rab6^WT		
Rab6^Q72L		
Rab6^T27N		
Rab1^WT		
Rab1^Q70L		
Rab1^S25N		

BicD2f

	SC-LW	QDO
Rab6^WT		
Rab6^Q72L		
Rab6^T27N		
Rab1^WT		
Rab1^Q70L		
Rab1^S25N		

p115f

	SC-LW	QDO
Rab1^WT		
Rab1^Q70L		
Rab1^S25N		

Rab6^Q72L

	SC-LW	QDO
BicD1^1–703		
BicD1^704–975		
BicD2^1–705		
BicD2^706–824		

B

p150^glued

	SC-LW	QDO
Rab6^WT		
Rab6^Q72L		
Rab6^T27N		
Rab1^WT		
Rab1^Q70L		
Rab1^S25N		

Rab6^Q72L

	SC-LW	QDO
p150^1–904		
p150^905–1278		
p150^1–539		
p150^540–1278		
p150^540–904		
p150^828–1059		

Fuchs *et al.*, Chapter 53, Fig. 1. BicD1, BicD2, and p150 interact with Rab6 in Y2H. (A) Full-length BicD1 and BicD2 show interaction with Rab6 but not with Rab1. The interaction with wild-type Rab6 (Rab6^{WT}) is weaker than with the GTP-locked Rab6 mutant (Rab6^{Q72L}); note that Rab6Q72L colonies are less pink on QDO. The inactive GDP-locked Rab6 (Rab6^{T27N}) does not bind any of the effector proteins. As a positive control for Rab1, the interaction with the *cis*-Golgi tethering protein p115 and known Rab1 effector is shown. Note that the GTP-locked Rab1^{Q72L} but not the wild-type Rab1 shows an interaction in the yeast two-hybrid assay. To map the binding region for Rab6 various constructs of BicD1 and BicD2 were tested against Rab6^{Q72L}. BicD1 and BicD2 bind Rab6 via the C-terminal quarter of the protein (aa 704–975 in BicD1, aa 706–824 in BicD2). (B) Full-length p150^{glued} shows a strong interaction with activated Rab6^{Q72L} and a weaker interaction with wild type Rab6, but no interaction with Rab1. The minimal binding region in p150 covers the C-terminal half of the protein (aa 540–1278).

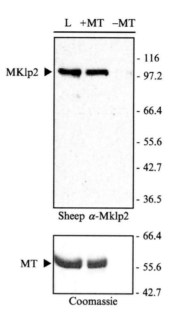

NEEF *ET AL.*, CHAPTER 54, FIG. 1. Binding of MKlp2 to microtubules. MKlp2 was incubated in the presence (+MT) or absence of microtubules (−MT) for 20 min at room temperature. The microtubules and bound proteins were pelleted by centrifugation, then the washed pellets and a sample equivalent to the input material (L) were analyzed by SDS–PAGE and Coomassie brilliant blue staining to check the recovery of microtubules and Western blotting with antibodies to detect MKlp2.

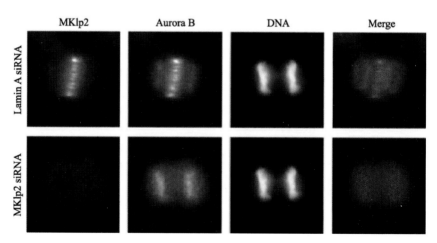

NEEF *ET AL.*, CHAPTER 54, FIG. 5. Aurora B is unable to relocate from kinetochores to the central spindle in MKlp2-depleted cells. HeLa cells were transfected with siRNA duplexes for Lamin A or MKlp2 for 24 h, fixed with 3% (w/v) paraformaldehyde, and stained with a sheep antibody to MKlp2 (Hill *et al.*, 2000), mouse monoclonal antibodies to Aurora B (AIM-1, Becton Dickinson), and DAPI for DNA. In the merged image MKlp2 is green, Aurora B is red, and DNA is blue.

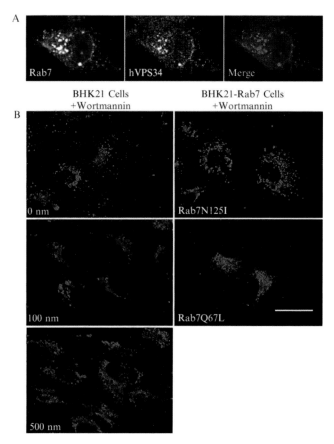

STEIN *ET AL.*, CHAPTER 55, FIG. 6. Colocalization of hVPS34 and Rab7 by immunofluorescence microscopy. (A) BHK21 cells were infected/transfected as detailed in the text to overexpress hVPS34 and Rab7. Samples were processed for immunofluorescence staining. Overexpressed Rab7 was visualized with our chicken anti-Rab7 (1:800) and overexpressed hVPS34 was detected with our rabbit anti-hVPS34 NW259 (1:250). FITC-conjugated donkey anti-chicken and Texas Red-conjugated donkey anti-rabbit were used as secondary detection antibodies. Samples were imaged on a Zeiss LSM510. Merge shows the overlap between the two channels. (B) Parental BHK21 cells were left untreated or treated with 100 or 500 nm wortmannin for 30 min. In comparison, stable cell lines expressing dominant-negative (Rab7N125I) or dominant-active (Rab7Q67L) Rab7 were treated with 100 nM wortmannin for 30 min. Cells were fixed and stained with rabbit pAb against mannose 6-phosphate receptor (red) (kind gift of A. R. Robbins, NIDDK, Bethesda, MD) and mouse mAb 4A1 against lgp120 (green) (kind gift of J. Gruenberg, University of Geneva, Switzerland) and imaged on a Zeiss LSM410. Note the increased dilation of endosomes in wortmannin-treated cells expressing the dominant-negative Rab7N125I protein (comparable to control cells treated with a 5-fold higher concentration of wortmannin) versus the normal appearance of endosomes (comparable to untreated control cells) in cells expressing the dominant-positive Rab7Q67L protein.

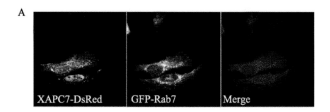

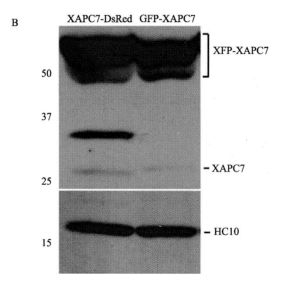

MUKHERJEE *ET AL.*, CHAPTER 56, FIG. 3. Proteasome incorporation of XFP-variants of XAPC7. (A) Human cervical carcinoma (HeLa) cells were cotransfected with plasmids encoding XAPC7-DsRed and GFP-Rab7, treated with 0.05% saponin, fixed with 3% paraformaldehyde in PBS as detailed in the text, and imaged using a Zeiss LSM510 fitted with a 63× objective. (B) BHK21 cells were transfected with plasmids encoding GFP-XAPC7 or XAPC7-DsRed and cells were grown in complete media for 18 h. 20S proteasome preparations were isolated by immunoprecipitation with a rabbit polyclonal antibody directed against the 20S complex, resolved by SDS-PAGE on 12% gels, and immunoblotted with a mouse monoclonal antibody against XAPC7 (upper panel) and a mouse monoclonal antibody against the $\beta3$(HC10) subunit of the 20S proteasome (lower panel). The bands corresponding to the chimeric and endogenous XAPC7 are indicated. An additional band is observed in the XAPC7-DsRed-transfected sample that is thought to be a degradation fragment. Prestained kaleidoscope molecular weight standards were from Bio-Rad (Hercules, CA) and migration is as indicated.

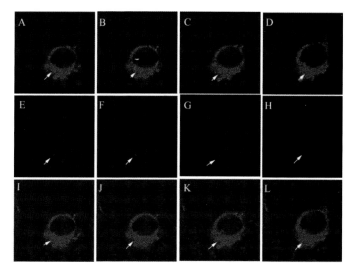

MUKHERJEE *ET AL.*, CHAPTER 56, FIG. 4. XAPC7 is involved in phagocytosis. Raw 264.7 mouse macrophages were cotransfected with expression plasmids encoding GFP-Rab7 and XAPC7-DsRed. Live cells were imaged at 30-s intervals for 10 min with a digital camera. In the sequential frames shown in (A)–(D) Rab7 (green) and (E)–(H) XAPC7 (red) are seen to incorporate into latex bead phagosomes (arrows). The time frame for incorporation is 7–10 min. In (I)–(L) a merged image shows the DIC image with the fluorescence images for both proteins superimposed.

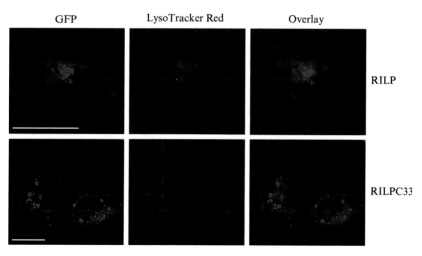

GFP	LysoTracker Red	Overlay	

COLUCCI *ET AL.*, CHAPTER 57, FIG. 1. Confocal fluorescence analysis of HeLa cells expressing GFP-RILP or GFP-RILPC33. Cells were transfected with the appropriate constructs as indicated. Two hours before fixation cells were incubated with LysoTracker Red. Bars = 10 μm.

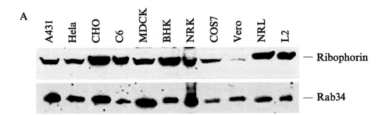

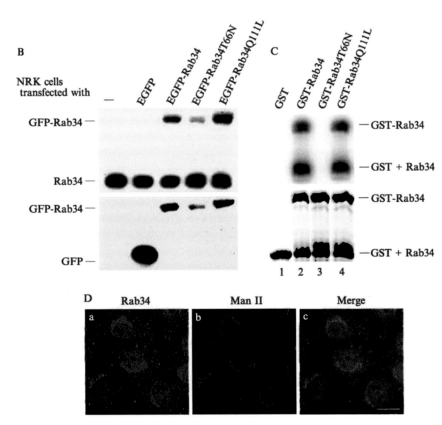

WANG AND HONG, CHAPTER 58, FIG. 1. Biochemical and cellular characterizations of Rab34. (A) Antibodies raised against GST-Rab34(1–58) recognize a 29-kDa protein in different cells. About 30 μg of total cell lysate derived from the indicated cell lines is resolved by SDS-PAGE and transferred to a filter. The filter is probed with anti-Rab34 antibodies or antibodies against ribophorin (kindly provided by D. I. Meyer, University of California, Los Angeles). (B) Characterization of EGFP-Rab34 fusion proteins. Control NRK cells and cells transfected with constructs for expressing EGFP, EGFP-Rab34, GDP-restricted EGFP-Rab34T66N, and GTP-restricted EGFP-Rab34Q111L as indicated are analyzed by Western blot using anti-Rab34 antibodies (upper panel) or anti-EGFP antibodies (lower panel). In addition to endogenous 29-kDa Rab34 polypeptide, anti-Rab34 also detects the fusion proteins of about 59 kDa. (C) Wild-type and Rab34Q111L but not Rab34T66N bind GTP as revealed by a GTP overlay assay. The indicated proteins resolved by SDS–PAGE are either stained with Coomassie blue (lower panel) or transferred to a filter and then incubated with [^{32}P]GTP (upper panel). In addition to the GST-Rab34, autocleavage of fusion proteins into GST and Rab34 is observed. (D) Rab34 is present in the Golgi apparatus and cytosol. NRK cells are fixed, permeabilized, and incubated with anti-Rab34 antibodies (a) and monoclonal antibody against Golgi mannosidase II (Man II) (b). The merged image is also shown (c). As shown, Rab34 colocalizes with the Golgi apparatus marked by Man II. Bar = 10 μm. This figure is adopted from Wang and Hong (2002) with copyright permission of *Mol. Biol. Cell* of the American Society of Cell Biology.

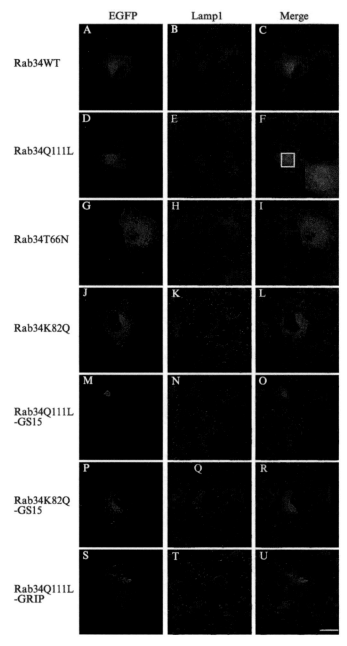

WANG AND HONG, CHAPTER 58, FIG. 3. Rab34 regulates lysosomal distribution. NRK cells are transfected to express EGFP-Rab34wt (A–C), EGFP-Rab34Q111L (D–F), EGFP-Rab34T66N (G–I), EGFP-Rab34K82Q (J–L), EGFP-Rab34Q111L-GS15 (M–O), EGFP-Rab34K82Q-GS15 (P–R), or EGFP-Rab34Q111L-GRIP (S–U). Transfected cells are fixed, permeabilized, and then labeled with monoclonal antibody against lysosomal marker Lamp1. As shown, expression of a wild-type GTP-restricted but not GDP-restricted form of Rab34 specifically shifts lysosomes from the periphery to the peri-Golgi region. This effect is dependent on K82 and can occur when Rab34 is tethered to the Golgi apparatus by fusion to GS15 or to the *trans*-Golgi network by fusion with the GRIP domain of Golgin-97. Bar = 10 μm. This figure is adopted from Wang and Hong (2002) with copyright permission of *Mol. Biol. Cell* of the American Society of Cell Biology.

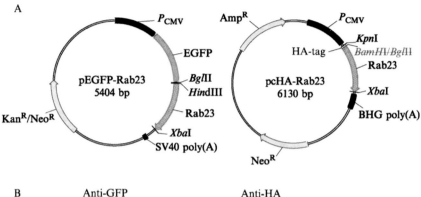

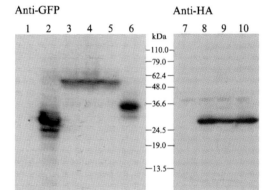

Evans *et al.*, Chapter 66, Fig. 4. Cloning of Rab23 for mammalian expression. (A) Schematic representation of Rab23 expression constructs, pEGFP-Rab23 and pcHA-Rab23, as used in this study. Kan[R], kanamycin resistance; Neo[R], neomycin resistance; Amp[R], ampicillin resistance; P_{CMV}, cytomegalovirus promoter; EGFP, enhanced green fluorescent protein; SV40 poly(A), simian virus 40 polyadenylation signal; BHG poly(A), bovine growth hormone polyadenylation signal. (B) Protein expression from Rab23 constructs analyzed by Western blotting according to standard techniques (Sambrook *et al.*, 1989). BHK-21 cells were transfected with the following: 1 and 7, untransfected control; 2, pEGFP-C1; 3, wild-type pEGFP-Rab23; 4, pEGFP-Rab23S23N; 5, pEGFP-Rab23Q68L; 6, pEGFP-Rab23R80X; 8, wild-type pcHA-Rab23; 9, pcHA-Rab23S23N; 10, pcHA-Rab23Q68L. Lanes 1–6 were probed with rabbit anti-GFP antibody and lanes 7–10 were probed rabbit anti-HA antibody followed by anti-rabbit HRP antibody. The sizes of detected bands are as predicted: GFP, ~28 kDa; Rab23-GFP fusion, ~54 kDa; Rab23R80X-GFP, ~36 kDa; Rab23-HA fusion proteins, ~27 kDa.

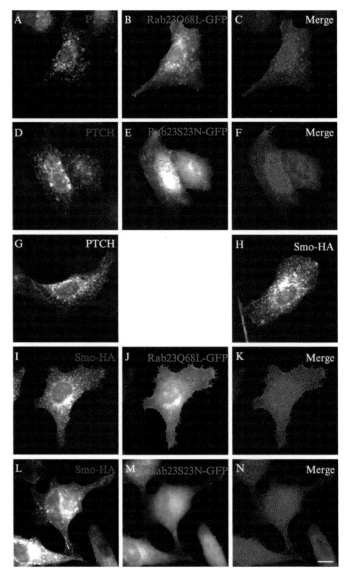

EVANS *ET AL.*, CHAPTER 66, FIG. 5. Patched or smoothened distribution is unaltered in the presence of mutant Rab23. Human patched, PTCH, was transfected with Rab23Q68L-GFP (A–C), Rab23S23N-GFP (E–F), or alone (G). Rat smoothened, Smo-HA, was transfected with Rab23Q68L-GFP (I–K), Rab23S23N-GFP (L–N), or alone (H). Cells were labeled for PTCH with chicken anti-PTCH followed by anti-chicken Alexa fluor 594 (A, D, G, C red, F red) antibodies or labeled for Smo-HA with rabbit anti-HA followed by anti-rabbit Alexa fluor 594 antibodies (H, I, L, K red, N red). Rab23Q68L-GFP fluorescence (B, C green, J, K green) and Rab23S23N-GFP (E, F green, M, N green) fluorescence is shown. The distribution of PTCH or Smo-HA remains unaltered in the presence of these Rab23 mutants compared with cells transfected with PTCH or Smo-HA alone. Bar = 10 μm. Reproduced with permission from Evans *et al.* (2003).